Meteorology Today

AN INTRODUCTION TO WEATHER, CLIMATE, AND THE ENVIRONMENT

FIRST CANADIAN EDITION

Meteorology Today

AN INTRODUCTION TO WEATHER, CLIMATE, AND THE ENVIRONMENT

C. Donald Ahrens
Emeritus, Modesto Junior College

Peter L. Jackson
University of Northern British Columbia

Christine E. J. Jackson
University of Northern British Columbia

NELSON / EDUCATION

NELSON / EDUCATION

Meteorology Today: An Introduction to Weather, Climate, and The Environment, First Canadian Edition

by C. Donald Ahrens, Peter L. Jackson, Christine E. J. Jackson

Vice President, Editorial Director:
Evelyn Veitch

**Editor-in-Chief,
Higher Education:**
Anne Williams

Publisher:
Paul Fam

Executive Editor:
Jackie Wood

Senior Marketing Manager:
Sean Chamberland

Developmental Editor:
Suzanne Simpson Millar

Photo and Permissions Researcher:
Melody Tolson

Content Production Manager:
Christine Gilbert

Production Service:
MPS Limited, a Macmillan Company

Copy Editor:
Holly Dickinson

Proofreader:
Dianne Fowlie

Indexer:
Edwin Durbin

**Manufacturing Manager—Higher
Education:**
Joanne McNeil

Design Director:
Ken Phipps

Managing Designer:
Franca Amore

Interior Design:
Cari Sherriff/Peter Papayanakis

Interior icons:
Focus on an Environmental Issue
(Iceberg): NOAA; Focus on an
Observation (Red sun setting
and silhouettes): NOAA; Focus
on a Special Topic (Sun on
horizon): Shutterstock; Focus on
an Advanced Topic (Antarctic
sunset): NOAA

Cover Design:
Peter Papayanakis

Cover Image:
Philip and Karen Smith/Getty
Images

Compositor:
MPS Limited, a Macmillan Company

Printer:
Transcontinental

**Library and Archives Canada
Cataloguing in Publication**

Ahrens, C. Donald
    Meteorology today : an
introduction to weather, climate,
and the environment / C. Donald
Ahrens, Peter L. Jackson, Christine
E. J. Jackson. — 1st Canadian ed.

Includes bibliographical references
and index.
ISBN: 978-0-17-650039-9

    1. Meteorology—Textbooks.
2. Meteorology—Canada—
Textbooks. I. Jackson, Peter
L. (Peter Lawrence), 1961–
II. Jackson, Christine E. J., 1961–
III. Title.

QC861.3.A47 2011
551.5   C2010-906419-4

ISBN-13: 978-0-17-650039-9
ISBN-10: 0-17-650039-1

MIX
Paper from
responsible sources
FSC® C011825
FSC
www.fsc.org

BRIEF CONTENTS

CONTENTS

© Frans Lanting/www.lanting.com

CHAPTER 3

Seasonal and Daily Temperatures   65

CHAPTER 4

Atmospheric Humidity   99

© C. Donald Ahrens

© C. Donald. Ahrens

© C. Donald Ahrens

NASA

CHAPTER 14

Thunderstorms and Tornadoes 415

CHAPTER 15

Hurricanes 457

© C. Donald Ahrens

CHAPTER 16

Earth's Changing Climate 487

CHAPTER 17

Global Climate 525

CHAPTER 18

Air Pollution   557

© J. L. Medeiros

CHAPTER 19

Light, Colour, and Atmospheric Optics   593

About This Book

Meteorology Today: An Introduction to Weather, Climate, and the Environment, First Canadian Edition, was written specifically for Canadian university and college students studying the atmospheric environment. It is based on the U.S. ninth edition by C. Donald Ahrens: a classic textbook that is well known for its clear writing, nontechnical explanations, and excellent graphic illustrations of the key concepts in meteorology. In producing the First Canadian Edition, we have benefited immensely from this solid foundation and have maintained the organization, philosophy, and style that Professor Ahrens honed and refined over nearly three decades. At the same time, we have made this edition more relevant to Canadian students by ensuring that Canadian content, practices, conventions, and examples are used throughout.

This book is a comprehensive survey of the atmosphere that emphasizes understanding and application of meteorological principles. The reader is encouraged to directly apply the book's content to the world by watching the weather. To assist with this, a detachable colour cloud chart is included at the back of this book. Full-colour photographs and figures are used throughout the book to clearly show concepts and engage the reader.

The book is organized into 19 chapters covering the same topics as the U.S. ninth edition. This organization was maintained to facilitate its use by instructors who are familiar with the earlier U.S. editions. The chapters are largely self-contained, allowing instructors the flexibility of using them in almost any desired order to meet the needs of a specific course. The coverage is comprehensive, with enough material for two semesters. For a one-semester course, instructors can choose which chapters or parts of chapters to cover.

After an introductory chapter on the atmosphere's origin, composition, and structure, the next two chapters discuss concepts of energy and temperature. After this, four chapters discuss humidity, condensation and clouds, stability, and precipitation. Winds are covered in the next three chapters, from the factors causing winds in Chapter 8 to local and global wind systems. A chapter covering fronts and air masses is followed by chapters dealing with mid-latitude cyclones and weather forecasting. Chapter 14 discusses thunderstorms and tornadoes and is followed by a chapter on hurricanes. Next are chapters on climate change, global climate, and air pollution. The last chapter covers atmospheric optical phenomena.

Each chapter has features to engage students and enable comprehension of the chapter content while facilitating an understanding of how specific chapter content links to other chapters and to the whole Earth.

About the Authors

C. Donald Ahrens

C. Donald Ahrens is Emeritus Professor at Modesto Junior College and the author of several best-selling textbooks. The Textbook and Academic Authors Association awarded Professor Ahrens its 2009 McGuffey Longevity Award in the physical science category for his market-leading text *Meteorology Today*, 9e. Dr. Ahrens has influenced not only professionals in the field of atmospheric science, but has brought better understanding of the science to hundreds of thousands of non-atmospheric science majors who used his books to expand their knowledge of weather and climate. In 2007, the National Weather Association awarded Professor Ahrens a lifetime achievement award for these accomplishments.

Peter L. Jackson

Peter L. Jackson is a professor of atmospheric science in the Environmental Science and Engineering Program and the Natural Resources and Environmental Studies Institute at the University of Northern British Columbia (UNBC). Prior to joining UNBC he was a faculty member at the University of Western Ontario, and before that a meteorologist with the Atmospheric Environment Service (now Meteorological Service) of Environment Canada. He is a meso- and synoptic scale meteorologist whose research interests focus on the interaction between the atmosphere and Earth's surface in mountains and along coastlines and environmental applications like air pollution and the movement of biota. At UNBC he teaches undergraduate and graduate courses on weather and climate, storms, biometeorology and air pollution, as well as research methods.

Christine E. J. Jackson

Christine E. J. Jackson is a senior laboratory instructor in the Geography Program at the University of Northern British Columbia (UNBC) with degrees in Physical Geography and Education. She develops, coordinates, and teaches experiential labs in courses on weather and climate, geomorphology, and soils. Prior to joining UNBC she taught climatology and geomorphology at the College of New Caledonia, preceded by several years working in museum education, science and technology community development, as an educator in the school system, and as a lab demonstrator and coordinator at the University of British Columbia. She is interested in improving understanding of the natural environment through applied learning and gives workshops to train educators to do this in the school system.

FEATURES AND PEDAGOGY

EARTH SYSTEMS GUIDE   New to the First Canadian Edition, an *Earth systems guide* has been included on each chapter's opening page and discussed in the introduction.

The purpose of this feature is to help readers understand the "big picture" and how each chapter's content is related to other chapters and to Earth as a whole.

The Earth systems guide pictorially represents Earth and its systems: atmosphere, hydrosphere (including the cryosphere), lithosphere, and biosphere (including the anthrosphere).

For a particular chapter, the pertinent Earth system components are highlighted to illustrate the linkages between systems, and a graphic is added as a "visual table of contents" to key concepts in that chapter. The relationships between the chapter's content and Earth systems are expanded on in the introduction of each chapter.

FOCUS ON ...

Each chapter contains at least two Focus sections that either expand on specific chapter content or provide more in-depth coverage on one aspect of the material.

There are four types of Focus boxes:

- Special Topics
- Environmental Issues
- Observations
- Advanced Topics

We invited 16 renowned Canadian scientists and experts to contribute Focus sections on topics of particular relevance to Canadians, such as drought on the Canadian Prairies, East Coast storms, the pineapple express, changing sea ice in the arctic and its impacts, and smog in Southern Ontario and British Columbia's Lower Fraser Valley. We have also added several other Focus sections and revised many others to make them more relevant to Canadians. Quantitative discussions of important equations such as the geostrophic wind equation and the hydrostatic equation, as well as concepts such as the tephigram (new to the First Canadian Edition), are found in Focus sections on advanced topics.

WEATHER WATCH These mini-boxes throughout each chapter provide anecdotes or unexpected weather facts related to the chapter topic and are included to stimulate reader interest in the content.

ADDITIONAL LEARNING AIDS

- Opening-page introductory pieces are designed to draw the reader into the main text. Many of these are new in the First Canadian Edition.
- Key terms are in **bold**. These are repeated in a list at the end of the chapter and are defined in the Glossary at the end of the book.
- Important phrases are *italicized*.
- One or two brief reviews of the main points are provided at the end of major sections within each chapter.
- Summaries at the end of the chapter review the main ideas.
- Problem material checks how well students assimilate the material.
- Tables and Figures, when referenced in the text, include a small icon, so students can easily look at the figure, then find their way back to the reference in the text.

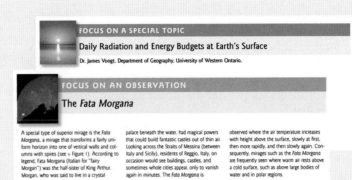

FOCUS ON A SPECIAL TOPIC
Daily Radiation and Energy Budgets at Earth's Surface
Dr. James Voogt, Department of Geography, University of Western Ontario.

FOCUS ON AN OBSERVATION
The *Fata Morgana*

A special type of superior mirage is the *Fata Morgana*, a mirage that transforms a fairly uniform horizon into one of vertical walls and columns with spires (see ● Figure 1). According to legend, Fata Morgana (Italian for "fairy Morgan") was the half-sister of King Arthur. Morgan, who was said to live in a crystal palace beneath the water, had magical powers that could build fantastic castles out of thin air. Looking across the Straits of Messina (between Italy and Sicily), residents of Reggio, Italy, on occasion would see buildings, castles, and sometimes whole cities appear, only to vanish again in minutes. The *Fata Morgana* is observed where the air temperature increases with height above the surface, slowly at first, then more rapidly, and then slowly again. Consequently, mirages such as the *Fata Morgana* are frequently seen where warm air rests above a cold surface, such as above large bodies of water and in polar regions.

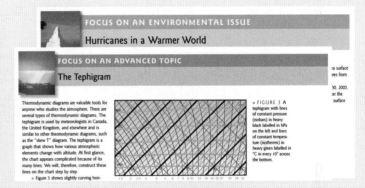

FOCUS ON AN ENVIRONMENTAL ISSUE
Hurricanes in a Warmer World

FOCUS ON AN ADVANCED TOPIC
The Tephigram

Thermodynamic diagrams are valuable tools for anyone who studies the atmosphere. There are several types of thermodynamic diagrams. The tephigram is used by meteorologists in Canada, the United Kingdom, and elsewhere and is similar to other thermodynamic diagrams, such as the "skew T" diagram. The tephigram is a graph that shows how various atmospheric elements change with altitude. At first glance, the chart appears complicated because of its many lines. We will, therefore, construct these lines on the chart step by step.

● Figure 3 shows slightly curving hori-

● FIGURE 3 A tephigram with lines of constant pressure (isobars) in heavy black labelled in hPa on the left and lines of constant temperature (isotherms) in heavy green labelled in °C in every 10° across the bottom.

WEATHER WATCH

If you take a walk on a bitter cold, yet clear, winter morning, when the air is calm and a strong subsidence inversion exists, the air aloft—thousands of metres above you—may be more than 17°C warmer than the air at the surface. You can even notice this when skiing in mountainous British Columbia and Alberta—under cold, clear conditions, it is often colder in the valley bottoms than it is at the mountain tops.

water is everywhere. It is the only substance that exists as a gas, a liquid, and a solid at temperatures and pressures normally found near Earth's surface (see ● Figure 1.3). Water, through the **hydrologic cycle**, transforms and circulates between the *atmosphere* and *hydrosphere* and is like the lifeblood linking all of Earth's systems.

▼ Table 1.1 shows the relative distribution, sources, sinks, and residence times of the various permanent and variable gases present in a volume of air near Earth's surface. Permanent gases are also called constant gases because their concentrations are nearly constant throughout the atmosphere and have not changed much over recent Earth history,

- Appendices provide additional information and a reference for students throughout their course.

TEXT RESOURCES

END OF CHAPTER MATERIAL

- Summaries at the end of each chapter review the main ideas.
- Three kinds of questions are provided to review and test students' knowledge and as supplements for lab activities:
  - *Questions for Review* act to check how well students have assimilated chapter content.
  - *Questions for Thought* require students to synthesize concepts for deeper understanding.
  - *Problems and Exercises* provide a technical challenge for students through activities or calculations based on chapter content.

END OF TEXT MATERIAL   For easy reference, there are 10 appendices with important tables, charts, maps, definitions, symbols, equations, and constants at the back of the book.

A new feature of the First Canadian Edition is the inclusion of a tephigram, the chart used in Canada to assess the vertical structure of the atmosphere:

There is also a tear-out laminated colour cloud chart intended for readers to use outside when learning cloud identification.

The glossary defines all of key terms used throughout the book.

On the back inside cover of the book is a geophysical map of North America. This quick reference will help readers locate important physiographic features such as water bodies, mountain ranges, political boundaries, and features of interest referred to in the text:

Digital Material

In addition to the textbook, a number of electronic resources have been adapted and provided as aids for in learning about and understanding the atmosphere. These resources, unless otherwise noted, are available on the text's website www.ahrensmeteorology1ce.nelson.com

Nelson Education Teaching Advantage

ABOUT NETA The **Nelson Education Teaching Advantage (NETA)** program delivers research-based instructor resources that promote student engagement and higher-order thinking to enable the success of Canadian students and educators.

Instructors today face many challenges. Resources are limited, time is scarce, and a new kind of student has emerged: one who is juggling school with work, has gaps in his or her basic knowledge, and is immersed in technology in a way that has led to a completely new style of learning. In response, Nelson Education has gathered a group of dedicated instructors to advise us on the creation of richer and more flexible ancillaries that respond to the needs of today's teaching environments.

The members of our editorial advisory board have experience across a variety of disciplines and are recognized for their commitment to teaching. They include:

Norman Althouse, Haskayne School of Business, University of Calgary

Brenda Chant-Smith, Department of Psychology, Trent University

Scott Follows, Manning School of Business Administration, Acadia University

Jon Houseman, Department of Biology, University of Ottawa

Glen Loppnow, Department of Chemistry, University of Alberta

Tanya Noel, Department of Biology, York University

Gary Poole, Director, Centre for Teaching and Academic Growth and School of Population and Public Health, University of British Columbia

Dan Pratt, Department of Educational Studies, University of British Columbia

Mercedes Rowinsky-Geurts, Department of Languages and Literatures, Wilfrid Laurier University

David DiBattista, Department of Psychology, Brock University

Roger Fisher, PhD

In consultation with the editorial advisory board, Nelson Education has completely rethought the structure, approaches, and formats of our key textbook ancillaries. We've also increased our investment in editorial support for our ancillary authors. The result is the Nelson Education Teaching Advantage and its key components: *NETA Engagement* and *NETA Assessment*. Each component includes one or more ancillaries prepared according to our best practices, and a document explaining the theory behind the practices.

NETA Engagement presents materials that help instructors deliver engaging content and activities to their classes. Instead of Instructor's Manuals that regurgitate chapter outlines and key terms from the text, NETA Enriched Instructor's Manuals (EIMs) provide genuine assistance to teachers. The EIMs answer questions like *What should students learn?*, *Why should students care?*, and *What are some common student misconceptions and stumbling blocks?* EIMs not only identify the topics that cause students the most difficulty, but also describe techniques and resources to help students master these concepts. Dr. Roger Fisher's *Instructor's Guide to Classroom Engagement (IGCE)* accompanies every Enriched Instructor's Manual.

NETA Assessment relates to testing materials: not just Nelson's Test Banks and Computerized Test Banks, but also in-text self-tests, Study Guides and web quizzes, and homework programs like CNOW. Under *NETA Assessment*, Nelson's authors create multiple-choice questions that reflect research-based best practices for constructing effective questions and testing not just recall but also higher-order thinking. Our guidelines were developed by David DiBattista, a 3M National Teaching Fellow whose recent research as a professor of psychology at Brock University has focused on multiple-choice testing. All Test Bank authors receive training at workshops conducted by Prof. DiBattista, as do the copyeditors assigned to each Test Bank. A copy of *Multiple Choice Tests: Getting Beyond Remembering*, Prof. DiBattista's guide to writing effective tests, is included with every Nelson Test Bank/Computerized Test Bank package.

SUPPLEMENTAL RESOURCES

Instructor Resources

INSTRUCTOR'S RESOURCE CD (IRCD)   The *Instructor's Resource CD* (ISBN 0-17-661012-X), gives instructors the ultimate tool for customizing lectures and presentations by providing key ancillaries in one place. Included on the IRCD are the Enriched Instructor's Manual; Test Bank and Computerized Test Bank; PowerPoint Presentations; and Image Library.

NETA ENRICHED INSTRUCTOR'S MANUAL   The *Enriched Instructor's Manual* for use with *Meteorology Today* has been prepared according to the NETA Engagement best practices, providing genuine support for your in-class instruction and strategies for your students. Adapted by Mark Moscicki from the University of Windsor, the *EIM* includes:

- Introduction: What is this chapter teaching?
- Earth Systems: How does this chapter connect to the world of practice, and the Earth's interconnected systems?
- Student Motivation: Why should students care?
- Barriers to Learning: What are some common student misconceptions and stumbling blocks?
- Engagement Strategies: Teaching and learning activities
- Student Projects (also useful for assessment)
- Assessment Tools
- Reflections on Teaching: How can I assess my own "performance"?
- Additional resources (such as online videos, Active Figures, interesting websites)

NETA TEST BANK AND EXAMVIEW® COMPUTERIZED TEST BANK   Prepared using the NETA Assessment framework, the tests are provided in rich text format for easy editing and printing with all common word-processing formats. All Test Bank questions are also provided in the ExamView® computerized version. This easy-to-use software is compatible with Microsoft Windows and Mac. Create tests by selecting questions from the question bank, modifying these questions as desired, and adding new questions you write yourself. You can administer quizzes online and export tests to WebCT, Blackboard, and other formats. The Test Bank was adapted for *Meteorology Today* by Luc Bernier of McMaster University.

MICROSOFT® POWERPOINT®   Key concepts from *Meteorology Today* are presented in PowerPoint format, with generous use of figures, photographs, and short tables from the text. The PowerPoint presentation for *Meteorology Today* was created by Henry Leighton of McGill University.

IMAGE LIBRARY   Customize your own PowerPoint presentations using figures, tables, illustrations, and photographs from the book. These items are provided in jpeg format for your use. (Note: a small number of items may not be available due to copyright restrictions.)

Additional Resources

ACTIVE FIGURES   Scattered throughout the text, select figures are labeled *Active Figures*. *Active Figures* are computer animations or visualizations of the figures in this text. By expanding and further illustrating important but difficult concepts presented in the text, these *Active Figures* aid student comprehension.

When a student comes across an *Active Figure* in the text, they can go to the *Meteorology Today* companion website for its animated version. Aside from just a visual, many of these are also accompanied by descriptions and questions to test whether the concepts have been learned. Instructors can use them in class to help in their own lectures as well (they are listed in the Enriched Instructor's Manual).

Examples of *Active Figures* include such topics as the Coriolis force, Bergeron process, hail formation, the life cycle of a thunderstorm, and development of a cumulus cloud.

COMPANION WEBSITE   Additional resources can be found at www.ahrensmeteorology1ce.nelson.com. This companion website includes a password-protected Instructor's Resource section which includes the ancillaries listed above, as well as a Student Resources section with additional self-testing review quizzes and *Active Figures*.

GLOBAL GEOSCIENCE WATCH PRINTED ACCESS CARD [ISBN 978-1-111-42905-3]   Updated several times a day, the Global Geoscience Watch is a focused portal into GREENR—our Global Reference on the Environment, Energy, and Natural Resources—an ideal one-stop site for classroom discussion and research projects for all things geosciences.

What's New in the First Canadian Edition

In addition to the major features described above, each chapter underwent a thorough revision to streamline and simplify the descriptions, to ensure that SI units and other units common in Canada are used throughout, and to include content that reflects meteorology as it is practised in Canada. All figures were critically examined and 333 new or revised figures or photos were produced for this edition. In all the chapters and in the end of chapter materials, Canadian examples were used wherever appropriate. Many of these changes resulted from reviewer feedback. The main changes are listed in the following summaries.

CHAPTER 1—EARTH AND ITS ATMOSPHERE

- New section, "Earth as a System," to discuss Earth's systems and introduce the Earth systems guide
- Table on atmospheric constituents to add sources, sinks, and residence times for each, as well as a discussion of bio-geochemical cycles, revised
- Focus section "The Radiosonde" updated
- Focus section "What Is a Meteorologist?" revised to include Canadian context
- Section "Storms of All Sizes" revised for relevance to Canada
- New local examples included in "Weather and Climate in Our Lives"
- Discussion of Earth's early atmosphere expanded

CHAPTER 2—ENERGY: WARMING AND COOLING EARTH AND THE ATMOSPHERE

- Definitions of energy, heat, and the forms of energy expanded
- Order of the radiation material covered in this chapter changed to discuss incoming radiation and factors that reduce it prior to absorption and the greenhouse effect
- Discussion of different radiation scattering processes (Rayleigh, Mie, and geometric) expanded
- Canadian UV index and more Canadian examples incorporated
- New section on surface radiation and energy budgets added, linking cause (energy) and effect (temperature) together more closely
- New Focus section, *Daily Radiation and Energy Budgets at Earth's Surface*, written by Dr. James Voogt, provides Canadian examples of radiation and energy flux measurements

CHAPTER 3—SEASONAL AND DAILY TEMPERATURES

- Energy budget concepts introduced in Chapter 2 used to explain temperature variations throughout this chapter
- Physical controls of temperature section added which integrates with energy balance concepts.
- Temperature measurement section expanded and updated
- Temperature advection concept introduced

CHAPTER 4—ATMOSPHERIC HUMIDITY

- Discussion of the hydrologic cycle expanded
- Discussion on controls of evaporation added
- Coverage of humidity measurement to include more details on modern electronic methods and data loggers
- New Focus section on psychrometry
- Discussion of humidex used in Canada added

CHAPTER 5—CONDENSATION: DEW, FOG, AND CLOUDS

- New Focus section, *Why Are Canada's Coastal Regions So Foggy?* by Dr. Ismail Gultepe
- Cloud classification discussion expanded
- New discussion of satellite images that combine visible and infrared bands
- Details and figure on water collection from fog added
- New information and imagery from CloudSat

CHAPTER 6—STABILITY AND CLOUD DEVELOPMENT

- Focus section on adiabatic charts revised to discuss and demonstrate use of the tephigram chart used in Canada
- Examples and figures modified where appropriate to indicate variable saturated adiabatic lapse rates (curved rather than straight lines)

CHAPTER 7—PRECIPITATION

- Terminology revised for Canadian standards (e.g. ice pellets versus sleet)
- Discussion of solute versus curvature effects on droplet growth enhanced
- New discussion of transport and deposition of pollutants to Arctic by snow
- Discussion of precipitation measurement in Canada enhanced
- New Focus section on the 1998 ice storm
- New Focus section, *Measuring Snow Depth*, by Dr. Stephen Déry

CHAPTER 8—AIR PRESSURE AND WINDS

- Discussion of gas law and pressure revised to emphasize SI units
- Discussion of the hypsometric equation added to the Focus section on the hydrostatic law
- Discussion of forces and naming of sections revised to explicitly list the forces

CHAPTER 9—WINDS AT DIFFERENT SCALES—SMALL AND LOCAL

- Discussion of turbulence and fluxes expanded
- Discussion of wind and the landscape (Aeolian processes) expanded
- Thermal circulation discussion enhanced and discussion of urban–rural breezes added; differentiation between large- (monsoon) and small-scale thermal circulations clarified

- New examples of local winds from Canada (chinook and Squamish)
- New Focus section, *Measuring Wind and Transport in the Boundary Layer*, by Dr. Ian Strachan
- Focus section on wind power enhanced by adding Canadian details as well as a discussion of meteorological factors to consider in locating wind farms

CHAPTER 10—WIND: GLOBAL SYSTEMS

- Discussion of atmosphere–ocean interactions enhanced
- Concept of ocean general circulation models introduced
- New Weather Watch on ENSO conditions and the 2010 Olympics

CHAPTER 11—AIR MASSES AND FRONTS

- New chapter opening vignette about lake-effect snow in Ontario
- Air masses affecting Canada and analyzed by Canadian meteorologists used throughout
- Canadian three-front/four air mass model introduced
- Discussion of arctic front and continental arctic air enhanced
- Canadian TROWAL instead of surface occlusion discussed and used throughout
- Focus section *Arctic Outbreaks* revised by Cliff Raphael to reflect impacts on Canada
- Discussion of lake-effect and ocean-effect snows modified to include other locations, as well as the Great Lakes, where these occur

CHAPTER 12—MIDDLE-LATITUDE CYCLONES

- Canadian frontal model for surface analyses used
- Middle-latitude cyclones developing along fronts other than the polar front discussed
- Many Canadian examples added, such as a discussion of Hurricane Hazel, the Saguenay flood, etc.
- New Focus section, *East Coast Storms,* by Dr. Chris Fogarty
- New Focus section, *Pineapple Express,* by Ruping Mo added
- New case study of Toronto's "snowstorm of the century", (January 1999)
- New section on vorticity advection
- New section on tropical–extratropical linkages

CHAPTER 13—WEATHER FORECASTING

- New chapter opener about weather proverbs
- Chapter substantially revised to reflect Meteorological Service of Canada (MSC) practices
- New descriptions and graphics of the MSC forecast office system, models, products, technology, etc.
- Numerical weather prediction discussion enhanced and parameterization issues discussed
- New forecasting rules of thumb (e.g., snow level versus thickness over mountains)
- New section on the SCRIBE forecast production system used by the MSC
- U.S. examples and major case studies replaced by Canadian ones throughout
- Focus section *TV Weathercasters—How Do They Do It?* revised by Claire Martin

CHAPTER 14—THUNDERSTORMS AND TORNADOES

- New chapter opener describing the 1985 Barrie tornado
- Information on flooding and flash flooding in Canada added
- Weather Watch on the sinking of the *SV Concordia* by a microburst added
- New Focus section, *Canada's Tornado Alley,* by Patrick King
- New Focus section, *The Edmonton Tornado,* by Dr. Gerhard Reuter

CHAPTER 15—HURRICANES

- Information on aspects of tropical weather other than hurricanes expanded
- New information and examples on the impacts of hurricanes in Canada added, including discussion of extratropical transitions
- New Focus section, *Hurricane Juan Strikes Halifax,* by Dr. Yongsheng Chen

CHAPTER 16—EARTH'S CHANGING CLIMATE

- New opener leading in with a report on the Intergovernmental Panel on Climate Change (IPCC) Fourth Assessment Report
- "Reconstructing Past Climates" section reorganized and expanded. Proxy methods are listed and enhanced
- "Prehistoric Climates" section extensively revised
- Discussion of climate feedbacks expanded
- Section on plate tectonics and climate change revised
- Discussion on radiative forcing of climate and causes for climate change enhanced
- Focus section on nuclear winter updated
- New Focus section, *Climate Models,* by Dr. René Laprise
- New Focus section, *Changing Sea Ice in the Arctic and Its Impact,* by Dr. William Gough
- Current and future climate change information updated to be consistent with the IPCC Fourth Assessment and data through 2009–10
- New Weather Watches on climate change, mountain pine beetle populations, and Inuit traditional climate knowledge

CHAPTER 17—GLOBAL CLIMATE

- New Focus section, *Drought on the Canadian Prairies,* by Dr. Ronald Stewart
- Statistics and climate station/region examples updated to include Canadian data
- Updated Köppen climate classification map and description provided
- Canadian and British Columbian systems of climate classification discussed

CHAPTER 18—AIR POLLUTION

- History of air pollution in Canada added
- Distinction between hazardous air pollutants and criteria air contaminants made and expanded on
- Pollutant emissions and ambient levels for Canada included
- Aerosol (particulate matter) discussion enhanced, and concept of aerosol optical thickness for visibility reduction introduced

- Canada-wide standards for air quality introduced and used
- Canadian Air Quality Health Index described
- New Focus section, *Long-Range Transport of Dust and Air Pollution to Western Canada,* by Dr. Ian McKendry
- New Focus section, *Smog in Southern Ontario and British Columbia's Lower Fraser Valley,* by Dr. Douw Steyn
- Discussion of meteorological factors affecting air quality enhanced

CHAPTER 19—LIGHT, COLOUR, AND ATMOSPHERIC OPTICS

- Section on scattering revised to clarify Mie and Rayleigh scattering and better differentiate them from nonselective scattering

- Some sections reorganized and renamed
- Discussion of polar day/polar night and twilight expanded

APPENDICES

- SI units emphasized throughout
- Additional equations, constants, and tables
- Complete listing of weather map symbols
- Listing of humidity equations and inclusion of detailed saturation vapour pressure curve
- New full-colour tephigram added

Acknowledgements

Writing and adapting a textbook is a team effort. That is a good thing because the task was much larger than we thought at the outset, and if there wasn't a team behind us, this book would never have been possible. We have been privileged to work with and learn from the excellent team of professionals that Nelson Education put together for this project. This began with Paul Fam, who started us on this adventure by asking us to undertake the project in the first place. Jackie Wood supervised the whole process. Suzanne Simpson held our hands, encouraging and guiding us every step of the way as we developed the manuscript. Holly Dickinson's sharp copy editing makes us look more literate than we actually are. Melody Tolson researched many of the new pictures and sorted out copyright permissions. Christine Gilbert led the design team and managed production. Dave McKay designed and drew the Earth systems guide for each chapter. Michael Borop of World Sites Atlas revised and created many of the maps throughout the text. Deborah Crowle also contributed to the map art. Gunjan Chandola of MPS Limited supervised the team that masterfully produced the book.

We would like to thank Bruce Ainslie for creating the North American humidity and degree-day figures, Maarten Ambaum for providing computer code to plot tephigram charts, and Ian Okabe for providing pictures of Meteorological Service of Canada weather forecasting technology and advising us on current MSC practices.

We received invaluable advice, suggestions, and feedback from expert reviewers who teach introductory meteorology courses at Canadian universities. Many of the modifications that we made were guided by this input, which was provided on the U.S. eighth and ninth editions. Other very helpful feedback was provided after we completed first drafts of the manuscript for this edition. We would therefore like to thank the following reviewers:

Luc Bernier	McMaster University
Norm Catto	Memorial University of Newfoundland
Adam Cornwell	Lakehead University
William A. Gough	University of Toronto, Scarborough
Henry Leighton	McGill University
Glen Lesins	Dalhousie University
John Maclachlan	McMaster University
R.A. McGinn	Brandon University
Richard L. Raddatz	University of Winnipeg
Ian Saunders	University of British Columbia
Ronald Stewart	University of Manitoba
Ian B. Strachan	McGill University
Gerard Szejwach	McGill University
Mary J. Thornbush	Lakehead University
Stanton Tuller	University of Victoria
James Voogt	University of Western Ontario

An important source of new Canadian content in this book was provided by 16 renowned researchers and experts who wrote or revised Focus sections on topics of particular relevance to Canadians. This new material has added considerable value to the book. We would therefore like to thank the following contributors (in the order they appear in the book):

Dr. James Voogt, University of Western Ontario (Chapter 2: Focus on a Special Topic: *Daily Radiation and Energy Budgets at Earth's Surface*)

Dr. Ismail Gultepe, Environment Canada (Chapter 5: Focus on an Observation: *Why Are Canada's Coastal Regions so Foggy?*)

Dr. Stephen Déry, University of Northern British Columbia (Chapter 7: Focus on an Observation: *Measuring Snow Depth*)

Dr. Ian Strachan, McGill University (Chapter 9: Focus on an Observation: *Measuring Wind and Transport in the Planetary Boundary Layer*)

Cliff Raphael, College of New Caledonia (Chapter 11: Focus on a Special Topic: *Arctic Outbreaks*)

Dr. Chris Fogarty (Chapter 12: Focus on a Special Topic: *East Coast Storms*)

Ruping Mo, Meteorological Service of Canada (Chapter 12: Focus on a Special Topic: *Pineapple Express*)

Claire Martin, CBC NEWS: Weather Centre (Chapter 13: Focus on an Observation: *TV Weathercasters—How Do They Do It?*)

Patrick King, Environment Canada (retired) (Chapter 14: Focus on a Special Topic: *Canada's Tornado Alley*)

Dr. Gerhard Reuter, University of Alberta (Chapter 14: Focus on a Special Topic: *The Edmonton Tornado*)

Dr. Yongsheng Chen, York University (Chapter 15: Focus on a Special Topic: *Hurricane Juan Strikes Halifax*)

Dr. William Gough, University of Toronto Scarborough (Chapter 16: Focus on a Special Topic: *Changing Sea Ice in the Arctic and Its Impact*)

Dr. René Laprise, Université du Québec à Montréal (Chapter 16: Focus on an Advanced Topic: *Climate Models*)

Dr. Ronald Stewart, University of Manitoba (Chapter 17: Focus on a Special Topic: *Drought on the Canadian Prairies*)

Dr. Ian McKendry, University of British Columbia (Chapter 18: Focus on an Environmental Issue: *Long-Range Transport of Dust and Air Pollution to Western Canada*)

Dr. Douw Steyn, University of British Columbia (Chapter 18: Focus on an Environmental Issue: *Smog in Southern Ontario and British Columbia's Lower Fraser Valley*)

We are deeply indebted to C. Donald Ahrens, the author of the original book, for providing the excellent foundation for this Canadian edition. We hope that our adaptation has continued the student-centred tradition of excellence that is typified by Professor Ahrens's series of books. Lastly, we'd like to thank our family and friends for their patience and support as we were preoccupied with this project.

To the Student

The atmosphere is all around us and plays a major role in our ever-changing natural world. Weather and climate affect almost everything we do; they affect what we wear, when we wear it, and even where we live. Causing everything from extreme winds to drought to floods, atmospheric phenomena such as mid-latitude cyclones, hurricanes, tornadoes, and thunderstorms underlie many of the natural hazards and disasters we face. The atmosphere is the most dynamic Earth system; a cumulus cloud can develop into a towering thunderstorm and spawn a tornado in just an hour or two. The atmosphere is also implicated in most long-term global environmental issues, from air pollution to ozone depletion to climate change. These are the defining issues of this century. For these reasons, interest in meteorology (the study of the atmosphere) continues to grow.

As you will discover, meteorology is an exciting, rapidly developing, and highly relevant subject. Modern technology, from satellites to radar to supercomputers, has revolutionized the way we examine and know the atmosphere. But many mysteries and interesting problems remain for the next generation of meteorologists. This book is designed to guide your personal understanding and appreciation of Earth's dynamic atmosphere. But don't just read this book—go outside, look at the sky, and question what you see there. If you can find answers to your questions in this book, then it will have been successful.

Peter L. Jackson and Chris E. J. Jackson

Earth's atmosphere: the view from space.
Photos.com

Earth and Its Atmosphere

To fly in space is to see the reality of Earth, alone. To touch the earth after is to see beauty for the first time.

Roberta Bondar, scientist, neurologist, physician, Canada's first female astronaut aboard the Space Shuttle Discovery Mission, January 22–30, 1992.

The Ukrainian Weekly, Nov. 2, 2003, Volume 71, Number 44, pg. 13.

CONTENTS

Our **atmosphere** is a delicate, life-giving blanket of air that surrounds Earth. In one way or another, it influences everything we see and hear—it is intimately connected to our lives. Air is with us from birth. We cannot detach ourselves from its presence. At Earth's surface, we can travel for many thousands of kilometres (km) in any horizontal direction, but should we move a mere 8 km above the surface, we would suffocate. We may be able to survive without food for a few weeks or without water for a few days, but without air, we would not survive more than a few minutes. Just as fish are confined to water, we are confined to an ocean of air.

Earth without its atmosphere would not have lakes or oceans. There would be no sounds, no clouds, no colourful sunsets. The beautiful pageantry of the sky would be absent. It would be unimaginably cold at night and unbearably hot during the day. Everything on Earth would be at the mercy of an intense sun beating down on a parched planet.

Living on Earth's surface, we have adapted so completely to our airy environment that we sometimes forget how truly remarkable air is. Even though it is tasteless, odourless, and invisible, it protects us from the sun's scorching rays and provides us with a mixture of gases that allow life to flourish. Because we usually cannot see, smell, or taste air, it may seem surprising that between your eyes and the pages of this book there are trillions of air molecules. Some of them may have been in a cloud yesterday or over another continent last week. Some may have been part of a life-giving breath for something that lived hundreds, thousands, or even millions of years ago. Air truly connects everything on Earth.

In this chapter, we will examine a number of important concepts and ideas about Earth's atmosphere, many of which will be expanded on in subsequent chapters. However, we will start by discussing Earth as a set of interconnected systems; the atmosphere is one.

Earth as a System

Earth is made up of several interlinked systems, one of which is the atmosphere. A system is a set of interacting interrelated elements forming a complex whole.

Each system can be clearly defined. Systems interact with each other, and their parts interact within the system. We will define four major Earth systems, some with subsystems (refer to the illustration on the chapter opening page):

1. The *atmosphere* includes the gaseous part of Earth from its surface to the exosphere, where the atmosphere gradually merges with space.
2. The **lithosphere** (sometimes called the **geosphere**) encompasses the solid Earth. It includes all the rock and geologic material making up the planet. It includes the soil, which is sometimes treated as a separate system called the **pedosphere**.
3. The **hydrosphere** includes Earth's watery parts, both fresh, salt, and frozen water (i.e., snow and ice). The frozen part is sometimes treated separately as the **cryosphere**.

4. The **biosphere** encompasses all life on Earth—plants, animals, and humans. We sometimes separate ourselves into a human system called the **anthrosphere**, which encompasses our human presence in the world. It includes our economy, culture, technology, communications, structures, and any activities associated with these.

To fully and correctly understand phenomena in nature, we must *holistically* consider the interactions within the system, as well as with other environmental systems. For example, understanding Earth's changing climate involves understanding the climate within the *atmospheric system* and its interaction with the *hydrosphere* through the ocean because ocean conditions, especially sea surface temperature, have a major impact on weather and climate. Over longer periods of time, climate and the atmosphere's composition are governed by the gas exchanges of plants in the *biosphere* and the weathering of rocks in the *lithosphere*. Even considering a single system requires knowledge of the various disciplines that study the processes occurring in that system. Understanding the atmosphere as a physical system requires interdisciplinary knowledge.

Chapters in this book describe parts of the atmospheric system. These are connected to other systems. We have created a "systems icon" on the opening page of each chapter as a guide to illustrate the linkages between the systems. This will assist you in understanding what systems are involved and how they interrelate as you read each chapter.

Overview of Earth's Atmosphere

The universe contains billions of galaxies, and each galaxy is made up of billions of stars. Stars are hot, glowing balls of gas that generate energy by converting hydrogen into helium near their centres. Our sun is an average-sized star situated near the edge of the Milky Way galaxy. Revolving around the sun are eight planets, including Earth (see • Figure 1.1),* and the other material (e.g, comets, asteroids, meteors, dwarf planets) that comprise our solar system.

Warmth for our solar system is provided primarily by the sun's energy. At an average distance of nearly 150 million kilometres from the sun, Earth intercepts only a very small fraction of the sun's total energy output. A portion of this solar **radiation**[†] is converted into other forms of energy, warming Earth and atmosphere, evaporating water, and driving the atmosphere into the patterns of everyday wind and weather we experience. Radiation allows Earth to maintain a global average surface temperature of about 15°C. This seems comfortable, but because it is a global average temperature, it is composed of widely ranging temperatures from all parts of the world. Thermometer readings can drop below

---

*Pluto was previously classified as a true ninth planet but recently was reclassified as a planetary object called a *dwarf planet*.

[†]Radiation or **radiant energy** is energy transferred in the form of waves that have electrical and magnetic properties. Light that we see, as well as ultraviolet (UV) light, is radiation. Chapter 2 contains more on this important topic.

● FIGURE 1.1 Relative sizes and positions for planets in our solar system. Pluto is included as an object called a dwarf planet. (Planet positions are not to scale.)

−85°C during a frigid Antarctic night and climb above 50°C during the day in hot subtropical deserts.

Although our atmosphere extends upward for many hundreds of kilometres, almost 99 percent of it lies within 30 km of Earth's surface (see ● Figure 1.2). In fact, if Earth were to

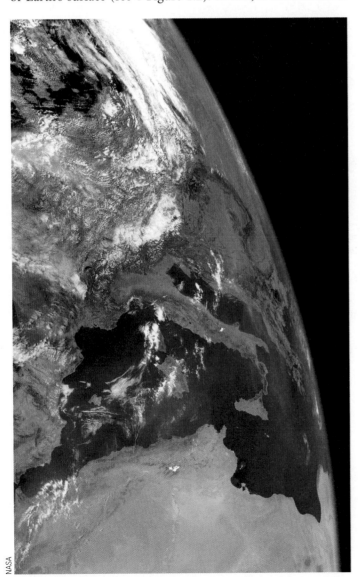

● FIGURE 1.2 Earth's atmosphere as viewed from space. The atmosphere is the thin blue region along the edge of Earth.

shrink to the size of a basketball, its inhabitable atmosphere would be thinner than a piece of paper. This thin blanket of air, composed primarily of nitrogen and oxygen, constantly shields Earth's surface and its inhabitants from the sun's dangerous ultraviolet radiant energy and from the onslaught of material from interplanetary space. Nestled in this thin atmosphere are clouds of liquid water and ice crystals that are part of the global water cycle. There is no definite upper limit to the atmosphere; it just becomes thinner and thinner, eventually merging with the empty space that surrounds all the objects in our solar system.

Composition of the Atmosphere

Earth's atmosphere is a thin, gaseous envelope composed mostly of **nitrogen (N_2)** (about 78%) and **oxygen (O_2)** (about 21%), with small to trace amounts of other gases, primarily **water vapour**, **argon**, and **carbon dioxide (CO_2)**. Many of the gases in the atmosphere have cycles of production (sources) and removal (sinks), so the composition of air for several gases is a dynamic process. The study of the cycling of molecules and nutrients on Earth is called *biogeochemistry* because most of the cycles involve interaction between the *biosphere*, with the other Earth systems.

▼ Table 1.1 shows the relative distribution, sources, sinks, and residence times of the various permanent and variable gases present in a volume of air near Earth's surface. Permanent gases are also called constant gases because their concentrations are nearly constant throughout the atmosphere and have not changed much over recent Earth history, whereas variable gases exist in small and variable amounts. As many of these gases occupy only a small fraction of a percent in a volume of air near the surface, they are referred to collectively as **trace gases**. (For a closer look at the composition of air at Earth's surface, read Focus on a Special Topic: A Breath of Fresh Air on p. 8.)

The relative amounts of nitrogen and oxygen are fairly constant in the atmosphere up to an elevation of about 80 km. At Earth's surface, there is a balance between destruction (output) and production (input) of these two gases. For example, nitrogen is removed from the atmosphere primarily by biological processes that involve soil bacteria and by tiny ocean-dwelling plankton that convert it into nutrients that help fortify the ocean's food chain. Nitrogen is returned to the

▼ Table 1.1   Composition of the Atmosphere near Earth's Surface

PERMANENT GASES

Gas	Symbol	Percent Dry Air (by volume)	Parts per Million* ppm (by volume)	Atmospheric Sources	Atmospheric Sinks (removal mechanism)	Atmospheric Residence Time[†] (in years)
Nitrogen	N_2	78.084	780,840	decaying plants and animals combustion	nitrogen-fixing bacteria in soil and oceans lightning	14,000,000
Oxygen	O_2	20.946	209,460	photosynthesis water and nitrous oxide breakdown by ultraviolet radiation in the stratosphere	plant and animal respiration decaying plants and animals chemical rock weathering growth of shellfish	4,500
Argon	Ar	0.93	9,300	radioactive decay of potassium	no sinks	forever, gradually accumulating
Neon	Ne	0.0018	18	radioactive decay of Earth materials	no sinks	forever, gradually accumulating
Helium	He	0.0005	5	radioactive decay of uranium and thorium	drifts into space	2,000,000
Hydrogen	H_2	0.00006	0.6	oxidation of methane automobile exhaust volcanoes	drifts into space	6.5
Xenon	Xe	0.000009	0.09	radioactive decay of Earth materials	no sinks	forever, gradually accumulating

VARIABLE GASES

Gas and Particles	Symbol	Percent Dry Air (by volume)	Parts per Million* ppm (by volume)	Atmospheric Sources	Atmospheric Sinks (removal mechanism)	Atmospheric Residence Time[†] (in years)
Water vapour	H_2O	0 to 4	0 to 40,000	evaporation transpiration	precipitation	0.026 or 9.5 days
Carbon dioxide	CO_2	0.0389	389	respiration combustion, (especially fossil fuels) industrial activity volcanoes oceans	absorbed by oceans photosynthesis burying organic material (landfills)	5 to 200 plus, depending on source
Methane	CH_4	0.00018	1.8	wetlands growing rice agriculture ruminant digestion (cattle, sheep, bison, deer, etc.) landfill decay biomass burning sewage treatment termites ocean bacteria	atmospheric oxidation (breaks down when it reacts with OH (hydroxyl) radicals) uptake in soils	8.4

VARIABLE GASES-CONT'D

Gas and Particles	Symbol	Percent Dry Air (by volume)	Parts per Million* ppm (by volume)	Atmospheric Sources	Atmospheric Sinks (removal mechanism)	Atmospheric Residence Time[†] (in years)
Nitrous oxide	N_2O	0.0000314	0.314	nitrogen breakdown by bacteria in soils and oceans agricultural soils and manure fossil fuel combustion sewage	destruction through reaction with ultraviolet radiation and oxygen in the stratosphere uptake in soils	120
Ozone	O_3	0.000004	0.04[‡]	oxygen breakdown by ultraviolet radiation in the stratosphere photochemical smog	recombines to form oxygen (O_2) in the stratosphere reacts with vegetation in the troposphere	0.25 or 91 days
Particles (dust, soot, etc.)		0.000001	0.01–0.15	volcanoes dust from soil fires sea spray combustion (fossil fuels, biomass)	removed by rain and settling by gravity	0 to 0.04 (minutes to 14 days, depending on size in the troposphere and longer in the stratosphere)
Chlorofluorocarbons	CFCs	0.00000002	0.0002	production by humans for refrigerants, propellants, and solvents	destroyed by ultraviolet radiation in the stratosphere	55 (CFC11) 140 (CFC12)

*Parts per million (ppm) measure very small amounts as 1 part in 1 million parts $\left(\frac{1}{1,000,000}\right)$. For example, 389 CO_2 parts per million (by volume) means that there are 389 CO_2 molecules in every 1,000,000 air molecules.

[†]Residence time indicates the time that the substance remains in the atmosphere.

[‡]In the stratosphere (altitudes between 11 and 50 km), values are about 5 to 12 parts per million (ppm).

atmosphere mainly through decaying plant and animal matter. This conversion and use of nitrogen by the *biosphere* is critical to its productivity because nitrogen, in forms other than N_2, is an important macronutrient. Oxygen, on the other hand, is removed from the atmosphere when organic matter decays; when oxygen combines with other substances to produce oxides; and during breathing as lungs take in oxygen and release carbon dioxide (CO_2). Oxygen is added to the atmosphere during photosynthesis as plants combine carbon dioxide and water to produce sugar and oxygen in the presence of sunlight.

Water vapour (H_2O) is an invisible gas whose concentration varies greatly from place to place and from time to time. Close to the surface in warm, steamy, tropical locations, water vapour may account for up to 4 percent of Earth's atmospheric gases, whereas in frigid polar areas, its concentration may dwindle to a fraction of a percent (see Table 1.1). Water vapour molecules are invisible. They become visible only when they transform into larger liquid or solid particles, such as cloud droplets and ice crystals, which may eventually grow large enough in size to fall from the sky as rain or snow. The process of water vapour changing into liquid water is called **condensation**, whereas the conversion of liquid water to water vapour

is called **evaporation**. Falling rain, snow, or some combination of these is called **precipitation**. In the lower atmosphere, water is everywhere. It is the only substance that exists as a gas, a liquid, and a solid at temperatures and pressures normally found near Earth's surface (see ● Figure 1.3). Water, through the **hydrologic cycle**, transforms and circulates between the *atmosphere* and *hydrosphere* and is like the lifeblood linking all of Earth's systems.

Water vapour is an extremely important atmospheric gas. As it changes from gas to liquid to ice, it releases large amounts of energy, called **latent heat**; the reverse transformations, from ice to liquid to gas, require the addition of energy, which is then stored as latent heat. *Latent heat is an important source of atmospheric energy, especially for storms, such as thunderstorms and hurricanes.* Moreover, *water vapour is a potent greenhouse gas because it strongly absorbs a portion of Earth's outgoing radiant energy.* Thus, water vapour plays a significant role in Earth's heat–energy balance.

Carbon dioxide (CO_2) gas is a small (about 0.0389 percent) but important naturally occurring component of Earth's air. Carbon dioxide enters the atmosphere mainly through decaying vegetation, but it also comes from volcanic eruptions, exhaling breaths of animals, burning fossil fuels

FOCUS ON A SPECIAL TOPIC

A Breath of Fresh Air

If we could examine a breath of air, we would see that air, like everything else, is composed of atoms. Although we cannot see atoms individually, they are composed of electrons whirling about an extremely dense centre. The centre, or nucleus, contains the atom's protons and neutrons. Almost all of the atom's mass is concentrated here, in a trillionth of the atom's entire volume. In the nucleus, the proton carries a positive charge, whereas the neutron is electrically neutral. Each circling electron carries a negative charge. As long as the total number of protons in the nucleus equals the number of orbiting electrons, the atom is balanced and electrically neutral (see ● Figure I).

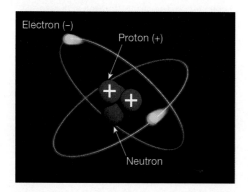

● FIGURE I An atom has protons and neutrons at its centre (called a nucleus) with electrons orbiting this centre. Molecules are combinations of two or more atoms. The air we breathe is mainly molecular nitrogen (N_2) and molecular oxygen (O_2).

Most air particles are molecules, which are combinations of two or more atoms (e.g., nitrogen, N_2, and oxygen, O_2). Most molecules are electrically neutral, but a few are electrically charged as they have lost or gained some of their electrons. Charged atoms and molecules are called ions, and these can react with other atoms or molecules.

An average breath of fresh air contains a tremendous number of molecules. With every deep breath, trillions of molecules from the atmosphere enter your body. Some of these inhaled gases become a part of you, whereas others are exhaled.

The volume of an average-sized breath of air is about a litre. Near sea level, there are roughly 10 thousand million, million, million (10^{22})* air molecules in a litre. So,

$$1 \text{ breath of air} \approx 10^{22} \text{ molecules}$$

We can appreciate the size of this number when we compare it to the number of stars in the universe. Astronomers have estimated that there are about 100 billion (10^{11}) stars in an average-sized galaxy and that there may be as many as 10^{11} galaxies in the universe. To determine the total number of stars in the universe, we multiply the number of stars in a

*The notation 10^{22} means the number one followed by 22 zeroes. For further explanation of this system of notation (called scientific notation), see Appendix A.

galaxy by the total number of galaxies and obtain

$$10^{11} \times 10^{11} = 10^{22} \text{ stars in the universe}$$

Therefore, each breath of air contains about as many molecules as there are stars in the known universe.

In Earth's entire atmosphere, there are nearly 10^{44} molecules. To imagine this, remember that 10^{44} is 10^{22} squared and there are 10^{22} molecules in a single breath. Consequently, there are about 10^{22} breaths of air in the entire atmosphere or

$$10^{22} \times 10^{22} = 10^{44} \text{ molecules in the atmosphere}$$

In other words, there are as many molecules in a single breath as there are breaths in the atmosphere.

Each time we breathe, the molecules we exhale enter the turbulent atmosphere. If we wait a long time, those molecules will eventually become thoroughly mixed with all the other air molecules. If none of the molecules are consumed in other processes, eventually, there would be a molecule from that single breath in every breath that is out there. So, considering the many breaths people exhale during their lifetimes, it is possible that our lungs contain molecules that were once in the lungs of people who lived hundreds or even thousands of years ago. In a very real way, we all share the same atmosphere.

(such as coal, oil, and natural gas), and deforestation. Carbon dioxide is removed from the atmosphere during **photosynthesis** as plants consume CO_2 and transform it into carbon stored in their roots, branches, and leaves. Oceans act as huge reservoirs for CO_2 as phytoplankton (tiny, drifting water plants) fix* CO_2 into their organic tissues. Carbon dioxide that dissolves directly into surface water mixes downward and circulates through greater depths. Estimates are that the oceans hold more than 50 times the total atmospheric CO_2

*Carbon *fixation* is a process that converts CO_2 gas into solid carbon, usually by photosynthesis.

content. ● Figure 1.4 illustrates important ways carbon dioxide enters and leaves the atmosphere.

● Figure 1.5 reveals that the concentration of atmospheric CO_2 has risen more than 23 percent since 1958, when it was first measured at the Mauna Loa Observatory in Hawaii. This increase means that more CO_2 is entering the atmosphere than is being removed. The increase appears to be mainly due to fossil fuel burning; however, deforestation also plays a role as trees that are cut, burned, or left to rot release CO_2 directly into the air, which also may result in soil CO_2 being released. Deforestation is thought to account for about 20 percent of the observed increase. CO_2 measurements for earlier time

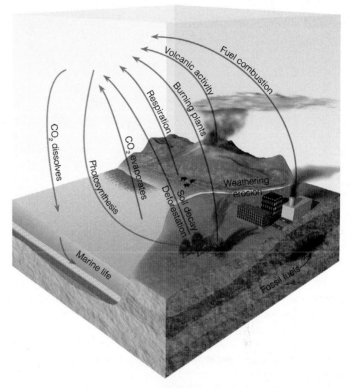

● **FIGURE 1.3** Earth's atmosphere is a rich mixture of many gases, with clouds of condensed water vapour and ice crystals. Water in gas, liquid, and solid forms exists here. The ocean's surface evaporates, forming invisible water vapour. Rising air currents condense water vapour into many billions of tiny liquid droplets that appear as puffy cumulus clouds. When rising air in the cloud extends to greater and colder heights, some of the liquid droplets freeze into minute ice crystals.

● **FIGURE 1.4** The main components of the atmospheric carbon dioxide cycle. The grey lines show processes that put carbon dioxide into the atmosphere; the red lines show processes that remove carbon dioxide from the atmosphere.

periods come from ice cores (see ● Figure 1.6). In Greenland and Antarctica, tiny bubbles of air trapped within the ice sheets reveal that for several thousand years before the industrial revolution, CO_2 levels were relatively stable at about 280 parts per million (ppm), although over longer time periods, CO_2 levels have fluctuated considerably. Since the early 1800s, CO_2 levels have increased more than 38 percent. With CO_2 levels presently increasing by about 0.4 percent annually (1.9 ppm/year), scientists now estimate that the concentration of CO_2 will likely rise from its current value of about 389 ppm in 2010 to a value near 500 ppm toward the end of this century.

Carbon dioxide is another important greenhouse gas. Like water vapour, it traps a portion of Earth's outgoing radiant energy. Consequently, as the atmospheric concentration of CO_2 increases, so should the average global surface air temperature. In fact, in the past century, Earth's average surface temperature has warmed by approximately 0.74°C. Mathematical climate models that predict future atmospheric conditions estimate that if levels of CO_2 (and other greenhouse gases) continue at their present rates, Earth's air temperature near the surface could warm by an additional 3°C by the end of this century. As we shall see in Chapter 16, the

negative consequences of global warming, such as rising sea levels and the rapid melting of polar ice, will be felt worldwide.

Carbon dioxide and water vapour are not the only greenhouse gases. Recently, others have been gaining notoriety, primarily because they are becoming more concentrated and are more effective greenhouse gases than CO_2. Such gases include methane (CH_4), nitrous oxide (N_2O), and chlorofluorocarbons (CFCs).

Levels of **methane (CH_4)**, for example, have been rising over the past century, increasing recently by about one-half of 1 percent per year. Most methane appears to derive from the breakdown of plant material by certain bacteria in rice paddies, wet oxygen-poor soil, the biological activity of termites, and biochemical reactions in the stomachs of cows. Why methane is increasing so rapidly is currently under study. Levels of **nitrous oxide (N_2O)**, commonly known as laughing gas, have been rising annually at the rate of about one-quarter of a percent. Nitrous oxide forms in the soil through a chemical process involving bacteria and certain microbes. Ultraviolet light from the sun destroys it.

Chlorofluorocarbons (CFCs) represent a group of greenhouse gases that, up until recently, had been increasing in concentration. At one time, they were the most widely used propellants in spray cans and were also used as refrigerants, as propellants for blowing plastic foam insulation, and as solvents for cleaning electronic microcircuits. Although their average

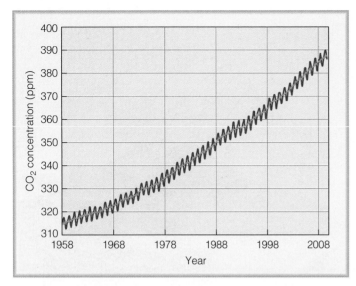

NOAA. Dr. Pieter Tans, NOAA/ESRL (www.esrl.noaa.gov/gmd/ccgg/trends/)

● FIGURE 1.5 Measurements of carbon dioxide (CO_2) in parts per million (ppm) at Mauna Loa Observatory, Hawaii. Higher readings occur in winter, when plants die and release CO_2 to the atmosphere. Lower readings occur in summer, when more abundant vegetation absorbs CO_2 from the atmosphere. The solid line is the average yearly value. Notice that the concentration of CO_2 has increased by more than 23 percent since 1958.

concentration in a volume of air is quite small (see Table 1.1, p. 7), they have important effects on our atmosphere. Not only are they efficient greenhouse gases, they also play a part in destroying ozone, a protective gas in the upper atmosphere (or stratosphere, a region in the atmosphere located between about 11 and 50 km above Earth's surface). As a result of the recognition of their effect on the stratospheric ozone layer in the 1980s, they have been phased out and replaced with less damaging *hydrochlorofluorocarbons* (HCFCs).

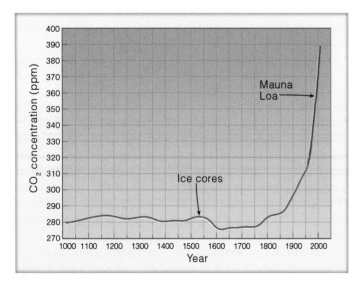

● FIGURE 1.6 Carbon dioxide (CO_2) values in parts per million during the past 1000 years from ice cores in Antarctica (blue line) and from Mauna Loa Observatory in Hawaii (red line).

Data courtesy Carbon Dioxide Information Analysis Center, Oak Ridge National Laboratory.

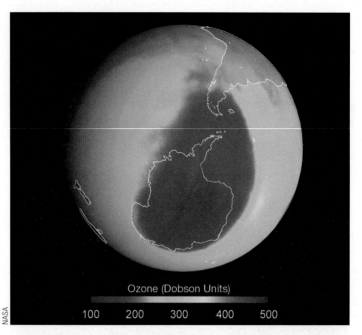

● FIGURE 1.7 The darkest colour represents the area of lowest ozone concentration, or ozone hole, over the Southern Hemisphere on September 22, 2004. Notice that the hole is larger than the continent of Antarctica. A Dobson unit (DU) is the physical thickness of the diffuse and deep ozone layer if it were concentrated as pure ozone and then brought to Earth's surface, where 500 DU equals 5 mm.

At the surface, **ozone (O_3)** is the primary ingredient in **photochemical smog,*** which irritates the eyes and throat and damages vegetation. But the majority of atmospheric ozone (about 97 percent) is found in the upper atmosphere or stratosphere, where it forms naturally as oxygen atoms combine with oxygen molecules. Here the concentration of ozone averages less than 0.002 percent by volume. This small quantity is important, however, because it shields plants, animals, and humans from the sun's harmful ultraviolet rays. It is ironic that ozone, which damages plant life in a polluted environment, provides a natural protective shield in the upper atmosphere so that plants on the surface may survive.

When CFCs enter the stratosphere, ultraviolet rays break them apart, and the CFCs release ozone-destroying chlorine. Because of this effect, ozone concentration in the stratosphere has been decreasing over parts of the Northern and Southern hemispheres. The reduction in stratospheric ozone levels over springtime Antarctica has plummeted at such an alarming rate that during September and October, there is an **ozone hole** over the region. ● Figure 1.7 illustrates the extent of the ozone hole above Antarctica during September 2004. A similar situation can occur over the Arctic during the Northern Hemisphere's spring; however, it is normally much less intense than in the Antarctic because the extreme cold

---

*Originally, the word *smog* meant the combining of smoke and fog. Today, however, the word usually refers to the type of pollution that forms in large cities, such as Los Angeles, California, as well as Toronto, Ontario, and Vancouver, British Columbia. Because this type of smog forms when chemical reactions take place in the presence of sunlight, it is termed *photochemical smog*.

© David Weintraub/Photo Researchers

● FIGURE 1.8 Erupting volcanoes can send tons of particles into the atmosphere, along with vast amounts of water vapour, carbon dioxide, and sulphur dioxide.

stratospheric temperatures that enhance the ozone destruction do not normally occur in the Arctic.

Impurities from both natural and human sources are also present in the atmosphere: wind picks up dust and soil from Earth's surface and carries it aloft; small saltwater drops from ocean waves are swept into the air (on evaporating, these drops leave microscopic salt particles suspended in the atmosphere); smoke from forest fires is often carried high above Earth; and volcanoes spew many tonnes of fine ash particles and gases into the air (see ● Figure 1.8). Many kinds of human activity, especially combustion in industrial or other settings, can directly release particles or gases that subsequently condense to form particles. Some of these pollutants can be carried by winds for a long distance: for example, some organic pollutants and brominated flame retardants that are produced in the middle latitudes can be transported to the Arctic, where they accumulate, causing health and environmental problems. Collectively, these tiny solid or liquid suspended particles of various composition are called particulates or **aerosols**.

Some impurities found in the atmosphere are natural and can be quite beneficial. Small, floating particles, for instance, act as surfaces on which water vapour condenses to form clouds. However, most human-made impurities (and some natural ones) are a nuisance, as well as a health hazard. These we call **pollutants**. For example, automobile engines emit copious amounts of **nitrogen dioxide (NO_2)**, **carbon monoxide (CO)**, and **hydrocarbons**. In sunlight, nitrogen dioxide reacts with hydrocarbons and other gases to produce ozone. Carbon monoxide is a major pollutant of city air. Colourless and odourless, this poisonous gas forms during the incomplete combustion of carbon-containing fuel. Hence, over 75 percent of carbon monoxide in urban areas comes from road vehicles.

The burning of sulphur-containing fuels (such as coal and oil) releases the colourless gas **sulphur dioxide (SO_2)** into the air. When the atmosphere is sufficiently moist, the SO_2 may transform into tiny dilute drops of sulphuric acid. Rain containing sulphuric acid corrodes metals and painted surfaces and turns freshwater lakes acidic. **Acid rain** is a major environmental problem, especially downwind from major industrial areas. In addition, high concentrations of SO_2 produce serious respiratory problems in humans, such as bronchitis and emphysema, and have an adverse effect on plant life.

These gas exchanges between the *atmosphere* and the *biosphere*, *hydrosphere*, and *lithosphere*, which lead to the current composition of air, illustrate some of the interconnections between systems that characterize Earth. These interconnections also account for the development of the atmosphere's composition over the course of Earth's evolution.

The Early Atmosphere   The atmosphere that originally surrounded Earth was probably much different from the air we breathe today. Earth's first atmosphere (some 4.6 billion years ago) most likely consisted of hydrogen and helium, the two most abundant gases found in the universe, as well as hydrogen compounds, such as methane (CH_4) and ammonia (NH_3). Most scientists feel that this early atmosphere escaped into space from Earth's hot surface.

A second, more dense atmosphere gradually enveloped the Earth as gases from molten rock within Earth's hot interior escaped through volcanoes and steam vents. We assume that volcanoes spewed out the same gases then as they do today: mostly water vapour (about 80 percent), carbon dioxide (about 10 percent), and sulphur dioxide or hydrogen sulphide, with up to a few percent nitrogen. These gases (mostly water vapour and carbon dioxide) probably created Earth's second atmosphere.

As millions of years passed, the constant outpouring of gases from the hot interior, known as **outgassing**, provided a rich supply of water vapour, which formed into clouds. Some of Earth's water may have originated from numerous collisions with small meteors and disintegrating comets when Earth was very young. Rain fell on the Earth for many thousands of years, forming the rivers, lakes, and oceans of the world. During this time, large amounts of CO_2 were dissolved in the oceans. Through chemical and biological processes, much of the CO_2 became locked up in carbonate sedimentary rocks, such as limestone. With much of the water vapour already condensed and the concentration of CO_2 dwindling, the atmosphere gradually became rich in N_2, which is usually not chemically active.

It appears that O_2, the second most abundant gas in today's atmosphere, probably began an extremely slow increase in concentration as energetic rays from the sun split water vapour (H_2O) into hydrogen and oxygen during a process called **photodissociation**. The hydrogen (H_2), being lighter, probably rose and escaped into space, whereas the oxygen remained in the atmosphere. Similarly, photodissociation of CO_2 produced oxygen in the early atmosphere by splitting into CO and O, which then reacted with OH to produce O_2. The concentration of O_2 in the early atmosphere was kept in check, however, by the production of H_2 in volcanoes, which reacts with O_2 to remove it.

About 2 to 3 billion years ago, the slow increase in oxygen may have been enough for primitive plants to evolve. Or the plants may have evolved in an almost oxygen-free (anaerobic) environment. At any rate, plant growth greatly enriched our atmosphere with oxygen. The reason for this enrichment is that, during the process of photosynthesis, plants, in the presence of sunlight, combine carbon dioxide and water to produce oxygen. Of course, as plants respire and decay, they take up oxygen and release carbon dioxide, reversing this process. How, then, do plants result in increased atmospheric oxygen? Some plants eventually become embedded in sediments and join the *lithosphere*, becoming fossil fuels and organic sedimentary rocks such as limestone. In this case, they effectively remove CO_2 and enhance O_2 in the atmosphere. Hence, after plants and the *biosphere* evolved, the atmospheric oxygen content increased more rapidly, probably reaching its present composition about several hundred million years ago.

BRIEF REVIEW

Before going on to the next several sections, here is a review of some of the important concepts presented so far:

- Earth's atmosphere is a mixture of many gases. In a volume of dry air near the surface, nitrogen (N_2) occupies about 78 percent and oxygen (O_2) about 21 percent.

- Water vapour varies spatially and temporally. It normally occupies less than 4 percent in a volume of air near the surface and

can condense into liquid cloud droplets or transform into delicate ice crystals. Water is the only substance in our atmosphere that is found naturally as a gas (water vapour), as a liquid (water), and as a solid (ice).

- Both water vapour and carbon dioxide (CO_2) are important greenhouse gases. Some trace gases are also effective greenhouse gases.

- Ozone (O_3) in the stratosphere protects life from harmful ultraviolet (UV) radiation. At the surface, ozone is a harmful main ingredient of photochemical smog.

- The majority of water on our planet is believed to have come from Earth's hot interior through outgassing.

Vertical Structure of the Atmosphere

A vertical profile in the atmosphere identifies how properties change with altitude. The atmosphere can be viewed as a series of layers as one moves from space to Earth's surface. Each layer can be defined in a number of ways: by the manner in which air temperature varies through it, by the gases that comprise it, or even by its electrical properties. Before we can examine these various atmospheric layers, we need to understand the vertical profile of two important variables: air pressure and air density.

A BRIEF LOOK AT AIR PRESSURE AND AIR DENSITY Earlier in this chapter, we learned that our atmosphere is more crowded close to Earth's surface. This occurs because air molecules (as well as everything else) are held near Earth by *gravity*. This strong, invisible force pulls everything toward Earth's centre. In the atmosphere, it squeezes or compresses air molecules closer together, which causes their number in a given volume to increase. The more air there is above any level in the atmosphere, the more weight, the greater the squeezing or compression effect, and the greater the number of air molecules in a given volume.

Consequently, gravity has an effect on the weight of objects, including air. In fact, *weight* is the force acting on an object due to gravity. Weight is defined as the mass of an object multiplied by the acceleration of gravity or

$$\text{weight} = \text{mass} \times \text{gravity}$$

An object's *mass* is the amount of matter in the object. The mass of air in a sealed container is the same everywhere in the universe. However, if you were to instantly travel to the moon, where the acceleration of gravity is much less than it is on Earth, the mass of air in that container would be the same, but its weight would decrease.

The **density** of any substance, including air, is determined by the mass of atoms and molecules that make up the substance and the amount of space between them. In other words, density tells us how much matter exists in a given space or volume. We can express density in a variety of ways. The molecular density of air is the number of molecules in a given volume. Most commonly, density is given as the mass of air in a given volume or

$$\text{density} = \frac{\text{mass}}{\text{volume}}$$

In the SI system of units (see Appendix A), mass is given in kilograms (kg) and volume is given in cubic metres (m³). Near sea level, air density is about 1.2 kilograms per cubic metre (1.2 kg m⁻³).*

There are appreciably more molecules within the same-sized volume of air near Earth's surface than there are at higher levels of the atmosphere. Consequently, air density is greatest at the surface and decreases as we move to higher altitudes. Notice in ● Figure 1.9 that because air near the surface is compressed, air density normally decreases very rapidly at first and then more slowly as we move farther away from the surface. This is an example of an **exponential rate of change**. The term *exponential change* describes the situation when the rate at which a property changes is proportional to the current size of the property. In the case of air density, it decreases rapidly near the surface, where it is large, and then decreases less rapidly in the upper atmosphere, where it is smaller.

Air molecules are in constant motion. On a mild spring day near the surface, an air molecule will collide about 10 billion times each second with other air molecules. It will also bump against objects around it—houses, trees, flowers, the ground, and even people. Each time an air molecule bounces against a person, it gives a tiny push. This small push or force divided by the area on which it pushes is called **pressure** and can be written as

$$\text{pressure} = \frac{\text{force}}{\text{area}}$$

In the atmosphere, the pressure resulting from multiple molecular "pushes" is surprisingly large. If we could weigh a column of air that has a cross section of one square metre and extends from sea level to the top of the atmosphere, its mass would be over 10,000 kg or 10 metric tonnes[†] (see Figure 1.9). Under normal conditions, this results in atmospheric pressures near sea level that are close to 101,325 new-

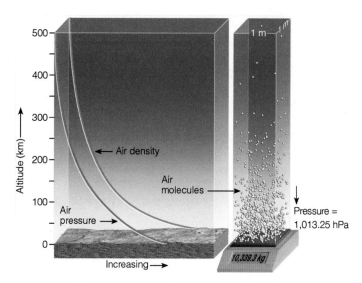

● **FIGURE 1.9** Both air pressure and air density decrease exponentially with increasing altitude. The average mass per square metre of all the air molecules above Earth's surface is 10,339.3 kg, which produces an average pressure of 1,013.25 hPa.

tons per square metre.[††] So if more molecules are packed into the air column, the air becomes more dense, it weighs more, and the surface pressure goes up. On the other hand, when fewer molecules are in the column, the air weighs less, and the surface pressure goes down. In summary, the surface air pressure can be changed by changing the mass of air above the surface.

Billions of air molecules push constantly on the human body. This force is exerted equally in all directions and is what we call pressure. We are not crushed by it because billions of molecules inside our body push outward just as hard. Even though we do not actually feel this constant bombardment of air, we can detect quick changes in it. For example, if we climb rapidly in elevation, our ears may "pop." This happens because the air collisions outside our eardrums lessen. The popping occurs as the air collisions between the inside and the outside of our ears equalize. A drop in the number of collisions informs us that the pressure has decreased. The force exerted by air molecules is less as there are fewer air molecules the higher you are above Earth's surface. A similar type of ear popping occurs as we drop in elevation.

---

*The notation "m⁻³" means $\frac{1}{m^3}$, so kg m⁻³ means $\frac{kg}{m^3}$ or kilograms per cubic metre.

[†]A common misunderstanding occurs because the everyday usage of *kilograms* confuses the terms *weight* and *mass* and treats them as though they are the same. They are not. Here the one square metre column extending from sea level to the top of the atmosphere has a mass of 10,339.3 kg of air. When working with the SI unit system, *weight* is the mass of air (10,339.3 kg) multiplied by gravity (9.8 m s⁻²), so weight is actually measured in units of force (kg m s⁻²) called newtons (N). Thus, the air's weight = mass × g = 10,339.3 kg × 9.8 m s⁻² = 101,325 kg m s⁻² = 101,325 N.

[††]To calculate the pressure for the one square metre column of air extending from sea level to the top of the atmosphere referred to here, we must first compute the *weight* of the air column. The air's weight = mass × g = 10,339.3 kg × 9.8 m s⁻² = 101,325 kg m s⁻² = 101,325 N. Since this weight is distributed over one square metre, the pressure is 101,325 N m⁻², and a newton per square metre (N m⁻²) is a pressure term also known as a pascal (abbreviated as Pa). Hectopascals (1 hPa = 100 Pa) are commonly used units of pressure, as are kilopascals (1 kPa = 1000 Pa).

WEATHER WATCH

Located in the U.S. Rocky Mountains, Denver, Colorado, has an elevation of 1609 m and a Major League Baseball franchise. The air density in this "mile-high" city is normally about 15 percent less than the air density at sea level. Less air density causes less drag force on a baseball as it moves through the air. A baseball hit in Denver will travel farther than one hit in a city closer to sea level, such as Toronto. Consequently, a "hit" that is a home run in Denver could be an "out" at the SkyDome in Toronto because of air density.

The Atmospheres of Other Planets

Earth is unique. Not only does it lie at just the right distance from the sun so that life as we know it flourishes, it also provides its inhabitants with an atmosphere rich in nitrogen and oxygen—two gases that are not abundant in the atmospheres of Venus or Mars, our closest planetary neighbours.

The Venusian atmosphere is 95 percent carbon dioxide with minor amounts of water vapour and nitrogen. An opaque acid-cloud deck encircles the planet, hiding its surface. Measurements reveal a turbulent atmosphere with twisting eddies and fierce winds in excess of 200 km hr^{-1}. This thick, dense atmosphere produces a surface air pressure of about 90,000 hPa, which is 90 times greater than that on Earth. On Earth, one would have to descend to a depth of about 900 m in the ocean to experience a similar pressure. Moreover, this thick atmosphere of CO_2 produces a strong greenhouse effect, with a scorching hot surface temperature of 480°C.

The atmosphere of Mars, like that of Venus, is mostly carbon dioxide with small amounts of other gases. Unlike Venus, the Martian atmosphere is very thin and heat rapidly escapes from the surface. Surface temperatures on Mars are much lower, averaging around −60°C. The combination of evidence from the Martian surface gathered by NASA's *Phoenix*

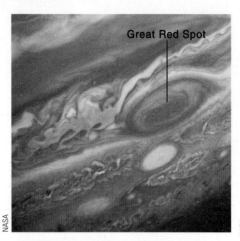

● FIGURE 2 An image of Jupiter extending from the equator to the southern polar latitudes. The spots, including the Great Red Spot, are spinning eddies similar to the storms that exist in Earth's atmosphere.

Mars Lander and the planet's thin, cold atmosphere, with virtually no cloud cover, currently has scientists believing that there is no liquid water on the Martian surface, although ice was found just under the surface. This thin atmosphere produces an average surface air pressure of about 7 hPa, which is less than one-hundredth of that experienced at the surface of Earth. On Earth, similar pressures are observed at altitudes of nearly 35 km. Occasionally, huge dust storms develop near the Martian surface.

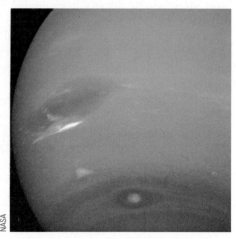

● FIGURE 3 Neptune's Great Dark Spot. White wispy clouds below this spot are similar to the high wispy cirrus clouds we have on Earth. However, on Neptune, they are probably composed of methane ice crystals.

Such storms may be accompanied by winds of several hundreds of kilometres per hour. These winds carry fine dust around the entire planet. The dust gradually settles out, coating the landscape with a thin, reddish veneer.

The atmosphere of the largest planet, Jupiter, is much different from that of Venus and Mars. Jupiter's atmosphere is mainly hydrogen (H_2) and helium (He), with minor amounts of methane (CH_4) and ammonia (NH_3). A prominent feature on Jupiter is the Great Red Spot, a

---

Air molecules not only take up space, freely darting, twisting, spinning, and colliding with everything around them, but as we have seen, these same molecules also have weight. In fact, air is surprisingly heavy. The weight of all the air surrounding Earth is a staggering 5136 trillion metric tonnes, or about 5.136×10^{18} kg. This weight of air molecules acts as a downward force on the Earth. The amount of force exerted over an area of surface is called *atmospheric pressure* or, simply, **air pressure.*** The pressure at any level in the atmosphere may be measured in terms of the total mass of air per unit area above any point. As we climb in elevation, fewer air molecules are above us; hence, *atmospheric pressure*

always decreases with increasing height. Like air density, air pressure decreases rapidly at first and then more slowly at higher levels, as illustrated in Figure 1.9.

● Figure 1.10 also illustrates how rapidly air pressure decreases with height. Near sea level, atmospheric pressure is usually close to 1000 hPa. Normally, just above sea level, atmospheric pressure decreases by about 10 hPa for every 100-metre (m) increase in elevation. At higher levels, air pressure decreases much more slowly with height. Much like air density, air pressure shows an exponential decrease with height. With a sea-level pressure near 1000 hPa, we can see in Figure 1.10 that at an altitude of only 5.5 km, the air pressure is about 500 hPa, or half of the sea-level pressure. This situation means that if you were at a mere 5.5 km above Earth's surface, you would be above one-half of all the molecules in the atmosphere.

---

*Because air pressure is measured with an instrument called a *barometer*, atmospheric pressure is often referred to as *barometric pressure*.

huge atmospheric storm that measures about three times larger than Earth. This storm spins counterclockwise in Jupiter's southern hemisphere (see ● Figure 2). Large white ovals near the Great Red Spot are similar smaller storm systems. Unlike Earth's weather machine, which is driven by the sun, Jupiter's massive swirling clouds appear to be driven by a collapsing core

of hot hydrogen. Energy from this lower region rises toward the surface; then it, along with Jupiter's rapid rotation, stirs the cloud layer into more or less horizontal bands of various colours.

Swirling storms exist on other planets, such as Saturn and Neptune. In fact, the large dark oval on Neptune (see ● Figure 3) appears to be a storm similar to Jupiter's Great Red

Spot. The white wispy clouds in the photograph are probably composed of methane ice crystals. Studying the atmospheric behaviour of other planets may give us added insight into the workings of our own atmosphere. Additional information about size, surface temperature, and atmospheric composition of our solar system's planets is given in ▼ Table 1.

▼ Table 1   Our Solar System: Surface Temperatures and Atmospheric Components

	DIAMETER	AVERAGE DISTANCE FROM SUN	AVERAGE SURFACE TEMPERATURE	MAIN ATMOSPHERIC COMPONENTS
	Kilometres	Millions of Kilometres	°C	
Sun	1392×10^3		5505	—
Mercury	4880	58	260*	—
Venus	12,112	108	480	CO_2
Earth	12,742	150	15	N_2, O_2
Mars	6800	228	−60	CO_2
Jupiter	143,000	778	−110	H_2, He
Saturn	121,000	1427	−190	H_2, He
Uranus	51800	2869	−215	H_2, CH_4
Neptune	49000	4498	−225	N_2, CH_4
Pluto	3100	5900	−235	CH_4

*This value is for the side of Mercury that receives sunlight.

At the elevation of the highest mountain peak on Earth, Mount Everest (8.850 km), the air pressure would be about 300 hPa. This summit is above nearly 70 percent of all the air molecules in the atmosphere. At an altitude approaching 50 km, the air pressure is about 1 hPa, which means that 99.9 percent of all the air molecules are below this level. Yet the atmosphere extends upwards for many hundreds of kilometres, gradually becoming thinner and thinner until it ultimately merges with outer space. (Up to now, we have concentrated on Earth's atmosphere. For a brief look at the atmospheres of the other planets, read Focus on a Special Topic: The Atmospheres of Other Planets on p. 14.)

LAYERS OF THE ATMOSPHERE   We have seen that both air pressure and density decrease *exponentially* with height above

Earth's surface. *Air temperature,** however, has a more complicated vertical profile.

Look closely at ● Figure 1.11 and notice that air temperature normally decreases from Earth's surface up to an altitude of about 11 km. This decrease in air temperature with increasing height is primarily due to sunlight warming Earth's surface, which then warms the air above the surface (see Chapter 2 for more details). The rate at which the air temperature decreases with height is called the temperature **lapse rate**. The *average* or *standard lapse rate* in the lower atmosphere is about 6.5°C for every 1000 m rise in elevation.

Air temperature is a quantity measured by a thermometer that represents the degree of hotness or coldness of the air. It is also a measure of the *kinetic energy* of the air molecules, which is proportional to their speed squared, as we will see in Chapter 2.

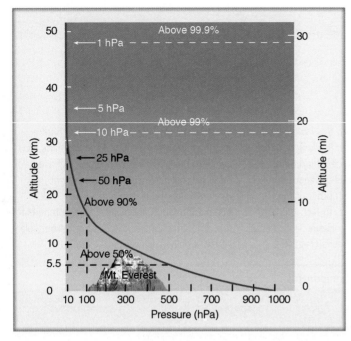

● FIGURE 1.10 Atmospheric pressure decreases rapidly with height. Climbing to an altitude of only 5.5 km, where the pressure is 500 hPa, would put you above one-half of the atmosphere's molecules. Climbers to the peak of Mount Everest (8850 m above sea level) know this all too well as breathing becomes so difficult above 8000 metres above sea level that the last section of the climb is known as the "death zone."

Remember that these are average rates, and on any given day, real temperature lapse rates can differ from the average. Lapse rates fluctuate day to day and season to season. Occasionally, air temperatures actually *increase* with height (so there is a negative lapse rate), creating a **temperature inversion**.

The atmosphere from the surface up to about 11 km contains all of the weather we are familiar with on Earth. This region is kept well stirred by rising and descending air currents. Here it is common for air molecules to circulate through a depth of more than 10 km in just a few days. This region of circulating air extending upward from Earth's surface to where the air stops becoming colder with height is called the **troposphere**—from the Greek *tropein*, meaning to turn or change.

Notice in Figure 1.11 that just above 11 km the air temperature normally stops decreasing with height. Here the lapse rate is zero. This region, where, on average, the air temperature remains constant with height, is referred to as an *isothermal* (equal temperature) zone.* The bottom of this zone marks the top of the troposphere and the beginning of another atmospheric layer, the **stratosphere**. The boundary separating the troposphere from the stratosphere is called the **tropopause**. The height of the tropopause varies. It is normally found at higher elevations over equatorial regions and decreases in elevation as we travel poleward. Generally all over the world, the tropopause occurs at higher altitudes in summer and at lower ones in winter. In some regions, the tropopause

---

*In many instances, the isothermal layer is not present, and an inversion occurs where the air temperature begins to increase with increasing height.

WEATHER WATCH

If you are flying in a jet aircraft at 9000 m above Earth's surface, the air temperature outside your window would typically be about −43°C. As air temperatures normally decrease with increasing height, the air temperature outside your window may be more than 60°C colder than air at the ground directly below where you are flying.

"breaks" and is difficult to locate, and here scientists have observed tropospheric air mixing with stratospheric air and vice versa. These breaks also mark the position of *jet streams*—high-altitude winds that meander in a narrow channel, like a river, often at speeds exceeding 100 knots (185 km h⁻¹).*

From Figure 1.11, we can see that, in the stratosphere, the air temperature begins to increase with height, producing a temperature inversion. The inversion region, along with the lower isothermal layer, tends to keep the vertical air currents of the troposphere from spreading into the stratosphere. The inversion also tends to reduce the amount of vertical motion in the stratosphere, making the stratosphere a stratified, stable layer.

Even though air temperature increases with height, the air at an altitude of 30 km is extremely cold, averaging less than −46°C. Above polar latitudes, at this altitude, air temperatures can change dramatically from one week to the next. A *sudden warming* can raise the temperature in one week by more than 50°C. (Such a rapid warming, although not well understood, is probably due to sinking air associated with circulation changes that occur in late winter or early spring, as well as with the poleward displacement of strong jet stream winds in the lower stratosphere.)

How do we measure the atmosphere's temperature profile? **Radiosondes**, or weather balloons, are instruments that measure the air's vertical temperature profile up to elevations exceeding 30 km. See Focus on an Observation: The Radiosonde on p. 18 for more information about them.

The reason for the inversion in the stratosphere is that ozone gas, which is concentrated in the upper atmosphere, plays a major role in heating the air at this altitude. Recall that ozone is important because of its protective capacity to absorb energetic ultraviolet solar energy. Some of this absorbed energy warms the stratosphere, which explains why there is an inversion. If ozone were not present, the air would probably continue to become colder with height, as it does in the troposphere.

Figure 1.11 represents the average temperature profile for Earth's middle latitudes. Notice that the level of maximum ozone concentration is observed near 25 km, yet the stratospheric air temperature reaches a maximum near 50 km. This occurs because ozone absorbs only certain wavelengths of

---

*A knot is a nautical mile per hour. A nautical mile was originally defined as a minute of latitude (1/60th of a degree of latitude). Although not an SI unit, the knot is a common measure for wind speed used in aviation, boating, and meteorology. One knot is equal to 1.852 kilometres per hour (km hr⁻¹) or 0.51 metres per second (m s⁻¹).

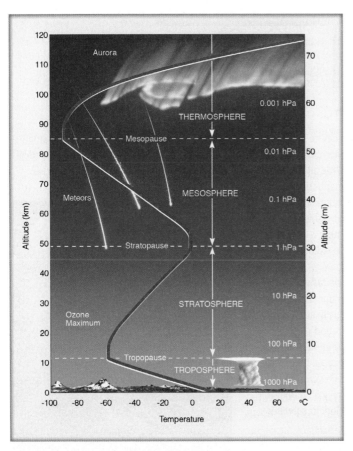

● FIGURE 1.11 Thermal layers of the atmosphere as defined by the average air temperature profile (heavy line) above Earth's surface.

ultraviolet radiation from the sun. Even though there are fewer ozone molecules, much of the energy responsible for heating the stratosphere is absorbed in the upper part of this layer and removed before it reaches the lower layers containing the ozone maximum. So even though there is more ozone lower down, there is not much radiation left at the right wavelengths to be absorbed, and the heating is less. Also, the low air density and the layer's stratification make the transfer of energy from the upper to the lower stratosphere quite slow.

Above the stratosphere is the **mesosphere** or middle sphere. The boundary near 50 km separates these layers and is called the *stratopause*. At this level, the stratosphere reaches its highest temperature. Here air is extremely thin and air pressure is quite low, averaging about 1 hPa, which means that only one-thousandth of all the atmosphere's molecules occurs above this level and 99.9 percent of the atmosphere's mass is located below it.

The percentage of nitrogen and oxygen in the mesosphere is about the same as it is at sea level, but the air's low density makes it impossible to get enough oxygen, and we would suffocate in a matter of minutes without the proper breathing equipment. Being adapted to living nearer sea level, our brains soon become oxygen starved—a condition known as hypoxia. High-elevation mountaineers and pilots who fly above 3 km for too long without an oxygen-breathing apparatus experience this. The first symptoms of hypoxia usually involve no pain, just a feeling of exhaustion. Soon visual impairment sets in, and routine tasks become difficult. Some people drift into an incoherent state, neither realizing nor caring what is happening to them. If this oxygen deficiency persists, a person will lapse into unconsciousness and death.

Suffocating is not the only problem that would be experienced in the mesosphere. Exposure to ultraviolet solar energy would cause severe burns on exposed body parts. Also, given the low air pressure, the blood in one's veins would begin to boil at normal body temperatures.

The air temperature in the mesosphere decreases with height, partly because there is little ozone in the air to absorb solar radiation. Consequently, the molecules near the top of the mesosphere absorb less energy than those near the bottom of the layer. This results in decreasing temperature with height. So we find air in the mesosphere becoming colder with height up to an elevation near 85 km. At this altitude, the temperature of the atmosphere reaches its lowest average value of $-90°C$.

The "hot layer" above the mesosphere is the **thermosphere**. The boundary that separates the lower, colder mesosphere from the higher temperature thermosphere is the *mesopause*. In the thermosphere, oxygen molecules absorb energetic solar rays, increasing the **kinetic energy** of the molecules and therefore the temperature.* Because there are very few atoms and molecules in the thermosphere, the absorption of a small amount of solar energy can cause a large increase in kinetic energy and therefore air temperature. Furthermore, because the amount of solar energy affecting this region depends strongly on solar activity, temperatures in the thermosphere vary from day to day (see ● Figure 1.12). The low density of the thermosphere also means that an air molecule will move an average distance of over one kilometre before colliding with another molecule. A similar air molecule at Earth's surface will move an average distance of less than one-millionth of a centimetre before it collides with another molecule. Because of this, it would not feel "warm" in the thermosphere, even though the temperature might be quite high. This is because our perception of air's "warmth" has to do with both the average speed of molecules colliding with our body (i.e., the temperature) and the number of molecules colliding with our body. In the thermosphere, molecules are zipping around quickly, but there are too few of them to impart much heat, so it would feel extremely cold. Moreover, it is in the thermosphere where charged particles from the sun interact with air molecules to produce dazzling aurora or northern lights displays. We will investigate aurora in Chapter 2.

Air density in the upper thermosphere is so low that air temperatures in this region are not measured directly. They are determined by observing the orbital change of satellites caused by the drag of the atmosphere. Even though the air is extremely thin, enough air molecules strike a satellite to slow it down, making it drop into a slightly lower orbit. This is the reason the

---

*As will be discussed in Chapter 2, air temperature is actually a measure of the kinetic energy of the molecules in air. In other words, it represents how fast the molecules are moving about.

FOCUS ON AN OBSERVATION

The Radiosonde

Up to an altitude of about 30 km, the vertical distribution of temperature, pressure, and humidity can be obtained using a measuring device called a *radiosonde*.* The radiosonde is a small, lightweight box equipped with electronic weather sensors, a battery, an antenna, and a radio transmitter. It is attached to a tightly tied helium or hydrogen gas-filled balloon by a cord (see ● Figure 4). Some radiosondes also have a parachute. As the balloon rises, the attached radiosonde measures air temperature with a small electrical thermometer, called a thermistor, located outside the box. The radiosonde measures humidity electrically by sending an electric current across a carbon-coated plate. Air pressure is obtained by a small barometer located inside the box. Every second, all of this information is transmitted to the surface by radio, where it is processed and stored every two seconds. Some units use special equipment such as a Global Positioning System (GPS) or weather radar to track the radiosonde's position as it moves through the sky. These types of radiosondes also provide a vertical profile of winds.

*A radiosonde that is dropped by parachute from an aircraft is called a *dropsonde*.

When winds are added, the device is called a rawinsonde, although the term *radiosonde* is often used generically to include all such instruments. When plotted on a graph, the vertical distribution of temperature, humidity, and wind is called a sounding. Eventually, the balloon bursts—usually somewhere near 33 km altitude for helium balloons—and the radiosonde returns to Earth.

Selected weather stations tasked with releasing radiosondes, called upper-air stations, release them twice a day, usually at the time that corresponds to midnight and noon in Greenwich, England.* Releasing radiosondes is an expensive operation because the instruments are never retrieved, and even when spent ones are found, they are usually so damaged that they are not reusable. Modern satellites

*Since weather is global, it is important to standardize weather observations around a standard time. Coordinated Universal Time (UTC) is used. UTC is very similar to Greenwich Mean Time (GMT), corresponding to the local solar time at Greenwich, near London, England. Sometimes UTC is abbreviated as Z; so UTC time is also called "Zulu" time. Appendix F on p. A-18 gives conversions between North American local standard times and UTC.

complement radiosondes by using instruments that measure radiant energy to provide vertical temperature profiles in inaccessible regions.

● FIGURE 4 A radiosonde and balloon. Canada's upper-air radiosonde network consists of 31 stations where weather balloons are launched twice daily.

Solar Maximum Mission spacecraft fell to Earth in December 1989, and the Russian space station *Mir* did the same in March 2001. The amount of drag is related to the density of the air, and the density is related to the temperature. So, by determining air density, scientists are able to construct a vertical profile of air temperature through the entire atmosphere.

At the top of the thermosphere, about 500 km above Earth's surface, molecules can move distances of 10 km before they collide with other molecules. Here many of the lighter, faster moving molecules traveling in the right direction actually escape Earth's gravitational pull. The region where atoms and molecules can shoot off into space is referred to as the **exosphere**, and it represents the upper limit of our atmosphere.

So far, we have examined the atmospheric layers based on the vertical profile of temperature. However, the atmosphere may also be divided into layers based on chemical composition. The composition of the atmosphere begins to slowly change in the lower part of the thermosphere. Below the thermosphere, the composition of air remains

fairly uniform at 78 percent nitrogen, 21 percent oxygen due to turbulent mixing. When classifying layers chemically, this lower, well-mixed region is known as the **homosphere** (see Figure 1.12). In the thermosphere, collisions between atoms and molecules are infrequent, and the air is unable to keep itself stirred. As a result, diffusion takes over as heavier atoms and molecules (such as oxygen and nitrogen) tend to settle to the bottom of the layer, whereas lighter gases (such as hydrogen and helium) float to the top. The region from approximately the base of the thermosphere to the top of the atmosphere is also called the **heterosphere**.

THE IONOSPHERE The **ionosphere** is not really a layer but rather an electrified region within the upper atmosphere where fairly large concentrations of ions and free electrons exist. Ions are atoms and molecules that have lost or gained one or more electrons. Atoms lose electrons and become positively charged when they cannot absorb all of the energy transferred to them by a colliding energetic particle or the sun's energy.

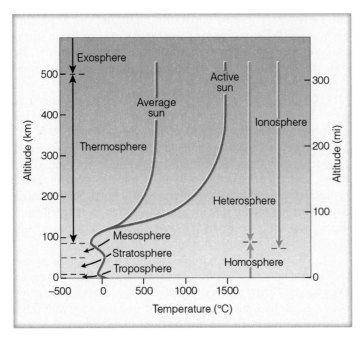

• FIGURE 1.12 The various layers and regions of the atmosphere: layers based on temperature are represented by the red line (an active sun is associated with larger numbers of solar eruptions.), layers based on chemical composition are shown by the green line, and regions where electrical properties occur are represented by the dark blue line.

The ionosphere usually starts about 60 km above Earth's surface and extends to the outer limits of the atmosphere. As illustrated in Figure 1.12, the bulk of the ionosphere is in the thermosphere.

The ionosphere plays a major role in AM radio communications, as shown in • Figure 1.13. The lower part of the ionosphere, called the *D* region, reflects standard AM radio waves back to Earth, but at the same time, it seriously weakens them through absorption. At night, the *D* region gradually disappears and AM radio waves are able to penetrate higher

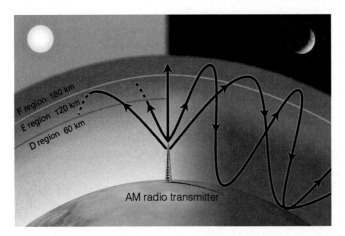

• FIGURE 1.13 At night, the higher region of the ionosphere, the *F* region, strongly reflects AM radio waves, allowing them to be sent over great distances. During the day, the lower *D* region strongly absorbs and weakens AM radio waves, preventing them from being picked up by distant receivers.

into the ionosphere, into the *E* and *F* regions, where the waves are reflected back to Earth. Consequently, at night, there is little absorption of radio waves in the higher reaches of the ionosphere and AM radio waves bounce repeatedly from the ionosphere to the Earth's surface and back to the ionosphere again. In this way, standard AM radio waves are able to travel for many hundreds of kilometres at night.

Around sunrise and sunset, AM radio stations usually make "necessary technical adjustments" to compensate for the changing electrical characteristics of the *D* region. Because they can broadcast over a greater distance at night, most AM stations reduce their output near sunset. This reduction prevents two stations—both transmitting at the same frequency but hundreds of kilometres apart—from interfering with each other's radio programs. At sunrise, as the *D* region intensifies, the power supplied to AM radio transmitters is normally increased. FM stations do not need to make these adjustments because FM radio waves are shorter than AM waves and are able to penetrate through the ionosphere without being reflected.

BRIEF REVIEW

We have examined our atmosphere from a vertical perspective. The main points are as follows:

- Atmospheric pressure at any level represents the total mass of air above that level, and atmospheric pressure always decreases with increasing height above the surface.
- The rate at which the air temperature decreases with height is called the *lapse rate*. A measured increase in air temperature with height is called an *inversion*.
- The atmosphere may be divided into layers according to its vertical profile of temperature and its gaseous composition. The atmosphere can also be divided into regions based on its electrical properties.
- The atmospheric layer with the highest temperature is the thermosphere; the layer with the coldest temperature is the mesosphere. Ozone gas is found in the stratosphere.
- We live at the bottom of the troposphere, which is an atmospheric layer where the air temperature normally decreases with height. The troposphere is a region that contains all of the weather we are familiar with.
- The ionosphere is an electrified region of the upper atmosphere that normally extends from about 60 km to the top of the atmosphere.

Weather and Climate

We will now turn our attention to weather events that take place in the lower atmosphere. The remainder of this chapter serves as a broad overview of material in later chapters. Many of the concepts and ideas you will encounter here are designed to familiarize you with items you might read about in newspapers or magazines or see on television.

Meteorology is the study of the atmosphere and its phenomena. When we talk about the **weather**, we are talking about the condition of the atmosphere at any particular time and place. Weather is always changing. It is composed of the following:

1. *air temperature*—the degree of hotness or coldness of the air, which corresponds to the kinetic energy of air molecules
2. *air pressure*—the force of the air above an area
3. *humidity*—a measure of the amount of water vapour in the air
4. *clouds*—visible masses of tiny water droplets and/or ice crystals that are above Earth's surface
5. *precipitation*—any form of water, either liquid (rain) or solid (snow), that falls from clouds and reaches the ground
6. *wind*—the horizontal movement of air
7. *visibility*—the greatest distance one can see to identify prominent objects

If we measure and observe these *weather elements* over a specified interval of time, say, for many years, we would obtain the "average weather" of a particular area. In addition, if we keep track of the variability in each weather element, we can define the **climate** of that area. Climate, therefore, represents the accumulation of daily and seasonal weather events and their variability, including extreme weather events such as heat waves in summer and cold spells in winter over a long period of time.

If we were able to watch Earth for thousands to millions of years, even the climate would change. We might see huge glaciers, rivers of ice moving down stream-cut valleys, and continental sheets of moving snow and ice spreading over large portions of North America. Over a time span of about two million years, we might see multiple glaciations where the ice advances and retreats several times. Of course, for this to happen, the average temperature of North America would have to decrease and then rise in a cyclic manner.

If we could photograph Earth once every thousand years, for many hundreds of millions of years we could watch Earth in time-lapse photography. This would show that climate and Earth itself are changing: mountains would form and rise up, only to be torn down by erosion; isolated puffs of smoke and steam would appear as volcanoes spew hot gases and fine dust into the atmosphere; and Earth's entire surface would undergo a gradual transformation as certain ocean basins widen and others shrink.*

In summary, Earth is composed of a number of dynamic systems, the *atmosphere, hydrosphere, lithosphere,*

---

*The movement of Earth's continents and ocean floor is explained by the theory of **plate tectonics**. In this theory, Earth's surface is composed of about eight major plates that move in relation to each other. Plate tectonics explains how the *lithosphere* evolves, how volcanoes and earthquakes occur, how and where mountains, build, and accounts for the changing distribution of land and ocean surfaces over geologic time. These changes have greatly affected Earth's climate.

and *biosphere.* These are constantly changing and impacting each other. Whereas major transformations of Earth's *lithosphere* are completed only after long spans of time, the state of the atmosphere can change in a matter of minutes.

METEOROLOGY: A BRIEF HISTORY   The term **meteorology** goes back to the Greek philosopher Aristotle, who wrote a book on natural philosophy entitled *Meteorologica* in about 340 B.C. This work represented the sum of knowledge on weather and climate at that time, as well as material on astronomy, geography, and chemistry. Some of the topics covered included clouds, rain, snow, wind, hail, thunder, and hurricanes. In those days, anything seen in the air and all substances that fell from the sky were called meteors. The term *meteorology* comes from the Greek word *meteoros*, meaning "high in the air."

In *Meteorologica*, Aristotle attempted to explain atmospheric phenomena in a philosophical and speculative manner. Even though many of his speculations were found to be erroneous, Aristotle's ideas were accepted without reservation for almost 2000 years. In fact, the birth of meteorology as a genuine natural science did not take place until the invention of weather instruments (the thermometer at the end of the 16th century, the barometer for measuring air pressure in 1643, and the hygrometer for measuring humidity in the late 1700s).

As more and better instruments were developed in the 1800s, the science of meteorology progressed. The invention of the telegraph in 1843 allowed for the transmission of routine weather observations. The understanding of the concepts of wind flow and storm movement became clearer, and in 1869, crude weather maps with lines of equal pressure (isobars) were drawn. Around 1920, the concepts of air masses and weather fronts were formulated in Norway. By the 1940s, daily upper-air balloon observations of temperature, humidity, and pressure gave a three-dimensional view of the atmosphere, and high-flying military aircraft discovered the existence of jet streams.

Meteorology took another step forward in the 1950s, when high-speed computers were developed to solve mathematical equations. At the same time, a group of scientists in Princeton, New Jersey, developed numerical means for predicting the weather. Prior to this advance, the mathematical equations that represent the atmosphere had to be simplified to by solved "by hand," with less accurate results. Today, computers plot the observations, draw the lines on the map, and forecast the state of the atmosphere at some desired time in the future.

After World War II, surplus military radars became available, and many were transformed into precipitation-measuring tools. In the mid-1990s, these were replaced by the more sophisticated *Doppler radars*, which have the ability to use radio waves to image storms, their winds, and precipitation (see ● Figure 1.14).

In 1960, the first weather satellite, *TIROS I*, was launched, ushering in space-age meteorology. Subsequent satellites provided a wide range of useful information, ranging from time-lapse images of clouds and storms to images that depict swirling ribbons of water vapour flowing around the globe.

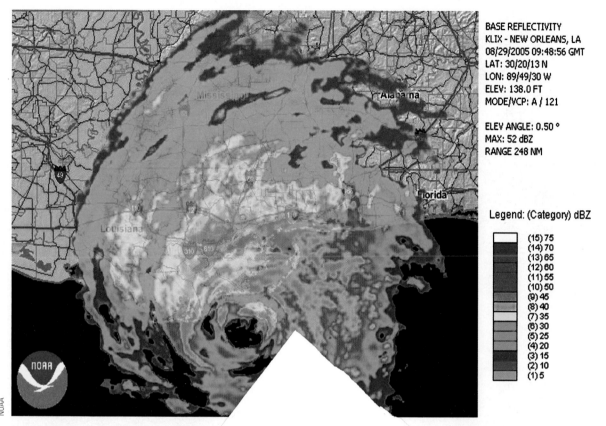

BASE REFLECTIVITY
KLIX - NEW ORLEANS, LA
08/29/2005 09:48:56 GMT
LAT: 30/20/13 N
LON: 89/49/30 W
ELEV: 138.0 FT
MODE/VCP: A / 121

ELEV ANGLE: 0.50 °
MAX: 52 dBZ
RANGE 248 NM

Legend: (Category) dBZ

(15) 75
(14) 70
(13) 65
(12) 60
(11) 55
(10) 50
(9) 45
(8) 40
(7) 35
(6) 30
(5) 25
(4) 20
(3) 15
(2) 10
(1) 5

NOAA

● FIGURE 1.14 Radar image showing t'... ...orms embedded in Hurricane Katrina as it makes landfall near New Orleans on Augu... ...e precipitation that is just offshore is the hurricane eye.

Throughout the 1990s and ... cated satellites have supplied ... range of imagery and measur... tracking and forecasts.

A SATELLITE'S VIEW OF THE WE... weather can be seen from a weathe... satellite image centred over the North... in the infrared band such that cold clo... the image. The image was obtained by c... two *geostationary satellites* situated ab... ...ve Earth—one located over the western par... ...nd the other over the eastern part. At this e... ...llite travels at the same rate as Earth spins, whicl... ...main positioned above the same spot on Earth so ... ...itinu-ously monitor what is taking place beneath it.

The solid white grid lines running from n... ...uth on the satellite image are called *meridians*, or lines of longitude. Since the zero meridian (or prime meridian) runs through Greenwich, England, the *longitude* of any place on Earth is simply how far east or west, in degrees, it is from the prime meridian. Most of North America lies between 52°W and 130°W longitude.

The solid white lines that parallel the equator are called *parallels of latitude.* The latitude of any place is how far north or south, in degrees, it is from the equator. The latitude of the ...ereas the latitude of the North Pole is 90°N ...e South Pole is 90°S. The latitudes between ...°N are commonly referred to as the **middle lati-** ...nid-latitudes.

...ns of All Sizes   Infrared satellite images such as ...ure 1.15 provide a snapshot related to temperatures in the ...tmosphere and on Earth. The clouds appear white because they are colder than the ground below them, and cold objects are assigned a white colour in this type of image. Organized cloud masses are storms. Superimposed on the satellite image are areas of low pressure called "lows" that correspond to storm centres (indicated by large red "L" symbols) and their adjoining weather fronts in red and blue. These **middle-latitude cyclonic storms** have winds spinning about their centre. In the case of the system over the Great Lakes, there are two centres (Ls): one over western Lake Superior and the other over Lake Huron. Another middle-latitude cyclone is over British Columbia, and a third one is crossing the Maritime provinces. Fronts are discussed in Chapter 11 and middle-latitude cyclones in Chapter 12.

A smaller but more vigorous storm, called a **hurricane**, occurs over tropical oceans (see Figure 1.14). These storms have diameters of a few hundred kilometres and often have a zone of clear skies at their centre, called the *eye*. Near the surface, in the eye, winds are light, skies are generally clear,

● FIGURE 1.15 This satellite image (taken in the infrared) shows a variety of cloud patterns and storms in Earth's atmosphere on September 28, 2009, at 12:00 UTC. Clouds are colder than the ground and are coloured white in the image. The coloured base map and weather fronts are added after the image is taken.

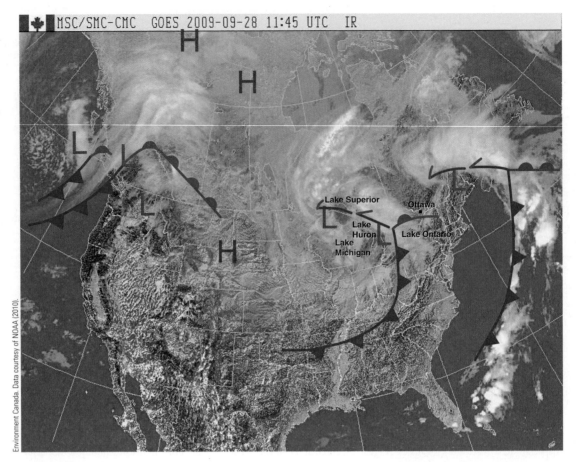

MSC/SMC-CMC GOES 2009-09-28 11:45 UTC IR

Lake Superior
Ottawa
Lake Huron
Lake Ontario
Lake Michigan

Environment Canada. Data courtesy of NOAA (2010).

and the atmospheric pressure is lowest. Around the eye, however, is an extensive region where heavy rain and high surface winds are reaching peak gusts of 100 knots (185 km h^{-1}). To be categorized as a hurricane, surface winds must exceed 64 knots* or 119 km h^{-1}. Hurricanes are discussed further in Chapter 15.

Trailing from the Maritimes' storm system in Figure 1.15 is a line of smaller storms off Florida's Atlantic coast. This line of cloud represents clusters of towering *cumulus* clouds that have developed into **thunderstorms**, one of the most common types of storms. ● Figure 1.16 is an example of the tall, churning cumulonimbus clouds that are accompanied by lightning, thunder, strong gusty winds, and heavy rain. Thunderstorms, at their most intense, can spawn the most violent disturbance in the atmosphere, a **tornado**.

A tornado or twister is an intense, rotating column of air that extends downward from the base of a severe thunderstorm. Tornadoes can appear as ropes or large cylinders. The majority are less than a kilometre wide, and many are smaller than a football field. Tornado winds can exceed 200 knots (370 km h^{-1}), but most peak at less than 125 knots (232 km h^{-1}). Some rapidly rotating clouds, called funnel clouds, that appear to hang from the base of a parent cloud never reach the ground to form a tornado.

*1 knot = 1.852 km hr^{-1} or 0.51 m s^{-1}.

A Look at a Weather Map   We can obtain a better picture of the middle-latitude storm by examining a simplified surface weather map (see ● Figure 1.17) for the same time as the Figure 1.15 satellite image. In Figure 1.17, the red letter "L"s on the map indicate regions of low atmospheric pressure, called **lows**, or **cyclones**, which mark the centre of the mid-latitude storm. The blue letter "H"s on the map represent

© C. Donald Ahrens

● FIGURE 1.16 Thunderstorms developing and advancing along an approaching cold front.

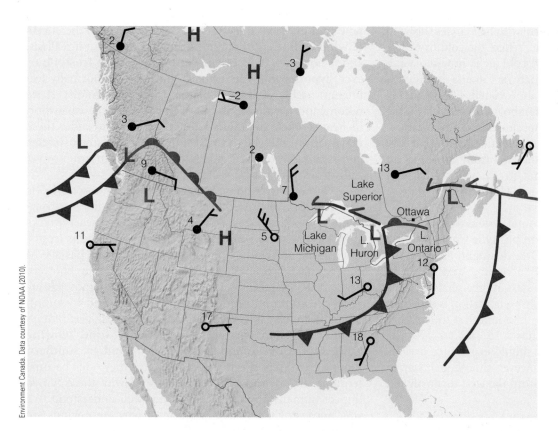

Environment Canada. Data courtesy of NOAA (2010).

● FIGURE 1.17 Simplified surface weather map taken at the same time as the satellite image shown in Figure 1.15. The numbers on the map represent air temperatures in °C.

regions of high pressure, called **highs**, or **anticyclones**. The circle symbols on the map represent either individual weather stations or cities where observations are taken. **Wind** is the horizontal movement of air and has both direction and speed. **Wind direction** is defined as the direction that the wind is blowing from.* On weather maps, wind direction is shown by the shaft lines that point toward the weather station symbol. Wind speed is indicated by the size and number of barbs on the wind-direction shafts. Each barb represents 10 knots (18.5 km h^{-1} or 5.1 m s^{-1}).

Notice how the wind blows around the highs and the lows. The horizontal pressure differences create a force that starts the air moving from higher pressure toward lower pressure. The winds are slowed down near Earth's surface because of its roughness, or *friction*. Because of Earth's rotation, the winds are deflected from their path toward the right in the Northern Hemisphere.† This deflection, combined with friction due to Earth's surface, causes the winds to blow *clockwise* and *outward* from the centre of highs and *counterclockwise* and *inward* toward the centre of lows.

Also notice by comparing Figures 1.15 and 1.17 that in the regions of high pressure, skies are generally clear. As the surface air flows outward away from the centre of a high, air sinking from above must replace the laterally spreading

surface air. Since sinking air does not usually produce clouds, we find generally clear skies and fair weather associated with the regions of high atmospheric pressure.

The swirling air around areas of high and low pressure is the major weather producer for the middle latitudes. Look at the middle-latitude storm and the surface temperatures in Figure 1.17 and notice that to the southeast of the storm affecting the Great Lakes, southerly winds from the Gulf of Mexico are bringing warm, humid air northward over much of the southeastern portion of the continent. On the storm's western side, cool, dry, northerly breezes combine with sinking air to create generally clear weather over the Prairies and U.S. Rocky Mountains. The boundary that separates the warm and cool air appears as heavy, coloured lines on the map—**fronts**, across which there is a sharp change in temperature, humidity, and wind direction.

Where the cool air from Canada replaces the warmer air from the Gulf of Mexico, a **cold front** is drawn in blue, with arrowheads showing the front's general direction of movement. Where the warm Gulf air is replacing cooler air to the

---

*For example, if you are facing north and the wind is blowing in your face, the wind is called a "north wind."

†This deflecting force, known as the *Coriolis force*, is discussed more completely in Chapter 8, as are the winds.

north, a **warm front** is drawn in red, with half circles showing its general direction of movement. Where the cold front has caught up to the warm front and cold air is now replacing cool air, an **occluded front**, which in Canada is called a **TRO**ugh of **W**arm **A**ir **AL**oft (abbreviated TROWAL), is drawn with blue and red hook symbols. Along each of the fronts, warm air is rising, producing clouds and precipitation. Notice in the satellite image (see Figure 1.15) that the TROWAL (occluded front) and the cold front appear as an elongated, curling cloud band that stretches from the low-pressure areas over lakes Superior and Huron into the state of Pennsylvania south of Lake Erie.

In Figure 1.17, observe the frontal system over southern Ontario. As the westerly winds aloft push the front eastward, a person on the outskirts of Ottawa might observe the approaching front as gradually increasing and lowering clouds, eventually followed by the start of precipitation. On a Doppler radar image, the advancing precipitation is shown in ● Figure 1.18, which is taken some 15 hours after the weather map and satellite image in Figures 1.15 and 1.17. As the system passes through Ottawa, it should experience periods of rain or showers. All of this, however, should give way to clearing skies and surface winds from the west or northwest after the front has moved on.

WEATHER AND CLIMATE IN OUR LIVES   Weather and climate play a major role in our lives. Weather, for example, often dictates the type of clothing we wear, whereas climate influences the type of clothing we buy. Climate determines when to plant crops as well as what types of crops can be planted. Weather determines if these same crops will grow to maturity.

Even when we are properly dressed for the weather, wind, humidity, and precipitation change our perception of how cold or warm it feels. On a cold, windy day, the effects of **wind chill** tell us that it feels much colder than it really is, and if not properly dressed, we run the risk of **frostbite** or even **hypothermia**. On a hot, humid day, we normally feel uncomfortably warm and blame it on the humidity. If we become too warm, our bodies overheat, and heat exhaustion or **heat stroke** may result. Those most likely to suffer these maladies are the elderly with impaired circulatory systems and infants, whose heat regulatory mechanisms are not yet fully developed.

Weather affects how we feel in other ways, too. Arthritic pain is most likely to occur when rising humidity is accompanied by falling pressures. The incidence of heart attacks shows a statistical peak after the passage of warm fronts, when rain and wind are common, and after the passage of cold fronts, when an abrupt change takes place as showery precipitation is accompanied by cold, gusty winds. Headaches are common on days when we are forced to squint, often due to hazy skies or a thin, bright, overcast layer of high clouds. For some people, a warm, dry wind blowing downslope (for example, a **chinook wind** in southern Alberta) adversely affects their behaviour (they often become irritable and depressed). Just how and why these winds impact humans physiologically is not well understood. We will, however, take up the question of why these winds are warm and dry in Chapter 9.

When the weather turns colder or warmer than normal, it impacts directly on the lives and pocketbooks of many people. For example, the exceptionally warm winter of 1997–98 over North America saved over $6.7 billion in heating costs in the United States, whereas Canadian homes saved an average of $200 each. The exceptional warmth (2 to 8°C above normal) in the winter of 1997–98 was due to the effects of a particularly strong El Niño that year. El Niño is a

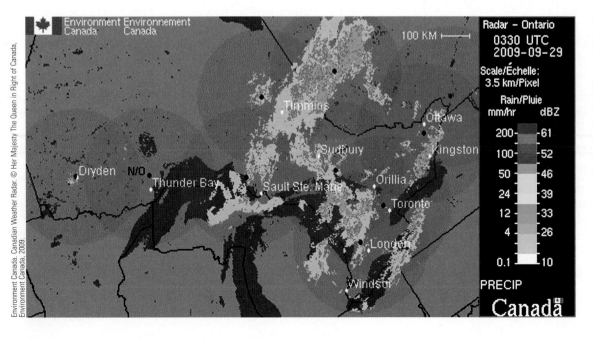

● FIGURE 1.18 Radar image showing the light rain associated with passing of a low-pressure system and TROWAL over Ontario on September 29, 2009, at 11:30 p.m. EDT. The image is a composite from 10 radars. The colours denote the intensity of rainfall.

phenomenon discussed in Chapter 10 that affects global weather patterns and arises when warmer than normal sea surface temperatures form in the eastern tropical Pacific Ocean. On the other side of the coin, the colder than normal winter of 2000–01 over much of North America sent heating costs soaring as the demand for heating fuel escalated.

Major cold spells accompanied by heavy snow and ice can play havoc by snarling commuter traffic, curtailing airport services, closing schools, and downing power lines (see Figure 1.19). For example, a huge ice storm during January 1998, affecting eastern Ontario through southern Quebec and Nova Scotia, as well as the northern New England states, left millions of people without power for as long as a month and caused over $3 billion in damage in Canada, making it Canada's costliest natural disaster. When frigid air settles into the U.S. Deep South, many millions of dollars worth of temperature-sensitive fruits and vegetables may be ruined, the eventual consequence being higher produce prices in the supermarket. Winter weather does not have to be cold to cause damage. When frontal systems stretch from the subtropics and cross the North American West Coast, they can bring moist, warm winds and copious amounts of rain with

● FIGURE 1.19 The ice storm of 1998 that affected southern Quebec, eastern Ontario, New Brunswick, and Nova Scotia was Canada's costliest natural disaster, at a cost of over $3 billion. About 1.5 million customers were without electricity for up to 30 days when the weight of the ice crumpled transmission towers.

snow over the higher mountains. These "pineapple express" storms can result in severe flooding and loss of electricity due to downed power lines.

Prolonged dry spells, especially when accompanied by high temperatures, can lead to a shortage of food and, in some places, widespread starvation. Parts of Africa, for example, have periodically suffered through major droughts and famine. During the years from 1999 through 2004, the Canadian Prairies experienced the worst prolonged drought in over 100 years, which had devastating impacts on agriculture. In 2002 alone, this cost the Canadian economy $3.6 billion and some 41,000 jobs.

When the climate turns hot and dry, animals suffer too. In 1986, over 500,000 chickens perished in the U.S. state of Georgia during a two-day period at the peak of a summer heat wave. Severe drought also has an effect on water reserves, often forcing communities to ration water and restrict its use. During periods of extended drought, vegetation often becomes tinder-dry, and sparked by lightning or a careless human, such a dried-up region can quickly become a raging inferno. During 2002, in the midst of the prairie drought, Alberta experienced a fivefold increase in the number of wildfires.

Every summer, scorching *heat waves* take many lives. During the past 20 years, an annual average of more than 300 deaths in North America is attributed to excessive heat exposure. Europe suffered through a devastating heat wave during the summer of 2003, when it is estimated that 70,000 people died, including 14,802 during one two-week period in France alone. The high death tolls mainly affected the elderly, many of whom were in understaffed care facilities in the cities during the traditional August holiday period. This left them without support resources that might have helped them cope with these unusual conditions. Daily maximum temperatures across the region exceeded 35 to 40°C in many parts of Europe, and several all-time maximum temperature records were set.

Every year, the violent side of weather influences the lives of millions. It is amazing how many people whose family roots are in the U.S. Midwest know the story of someone who was severely injured or killed by a tornado. Tornadoes have not only taken many lives, but, annually, they also cause damage to buildings and property totalling in the hundreds of millions of dollars as a single large tornado can level an entire section of a town (see ● Figure 1.20). Although not as frequent as in the U.S. Midwest, tornadoes in Canada can also have significant impacts: in the 1980s, tornadoes in Barrie, Ontario, and Edmonton, Alberta, killed 35 people.

Although the gentle rains of a typical summer thunderstorm are welcome over much of North America, the heavy downpours, high winds, and hail of **severe thunderstorms** are not. Cloudbursts from slowly moving, intense thunderstorms can provide too much rain too quickly, creating **flash floods** as small streams become raging rivers composed of mud and sand entangled with uprooted plants and trees (see • Figure 1.21). On average, flooding and flash flooding cause more property damage in Canada and more deaths in the United States than any other natural disaster. Strong downdrafts originating inside an intense thunderstorm (a **downburst**) create turbulent winds that are capable of destroying crops and inflicting damage on surface structures. Several airline crashes have been attributed to the turbulent **wind shear** zone within the downburst. Annually, hail damages crops worth millions of dollars, and lightning takes the lives of about seven people each year in Canada and 62 in the United States. Forty-five percent of the 8000 average annual wildfires in Canada are started by lightning, which causes 81 percent of the total area burned. Each year, wildfires destroy between 0.7 and 7.6 million hectares and directly cost $500 million to $1 billion to control, for a total annual average cost of about $14 billion (see • Figure 1.22).

Even the quiet side of weather has its influence. When winds die down and humid air becomes more tranquil, fog may form. Dense fog can restrict visibility over the water, affecting shipping, and at airports, causing flight delays and cancellations. Every winter, deadly, fog-related automobile accidents occur along our busy highways. But fog has a positive side, too, especially during a dry spell, as fog moisture collects on tree branches and drips to the ground, where it provides water for the root system.

Weather and climate have become so much a part of our lives that the first thing many of us do in the morning is listen to the local weather forecast. For this reason, many radio and television newscasts have their own "weather-

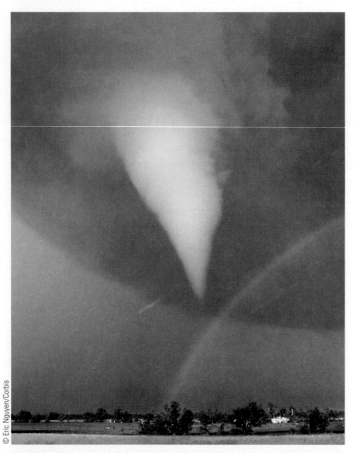

• FIGURE 1.20 A tornado and a rainbow form over south-central Kansas during June 2004. White streaks in the sky are descending hailstones.

• FIGURE 1.21 Flooding in Winnipeg during April and May 1997. The Red River reached its highest levels since 1826, causing widespread flooding affecting southern Manitoba as well as North Dakota and Minnesota.

FOCUS ON A SPECIAL TOPIC

What Is a Meteorologist?

Most people associate the term "meteorologist" with the weatherperson they see on television or hear on the radio. Many television and radio weathercasters are, in fact, professional meteorologists, but some are not. A professional meteorologist is usually considered to be a person who has completed the requirements for a university degree in meteorology or atmospheric science. This individual has strong, fundamental knowledge concerning how the atmosphere behaves, along with a substantial background of coursework in mathematics, physics, chemistry, and the environmental sciences.

A meteorologist uses scientific principles to explain and to forecast atmospheric phenomena. About two-thirds of the approximately 1350 meteorologists and atmospheric scientists in Canada (about half of the 9000 in the United States) work doing weather forecasting for the Meteorological Service of Canada (the National Weather Service in the United States). The rest work for the military or television or radio stations, work in research, teach atmospheric science courses in colleges and universities, or do meteorological consulting work.

Scientists who do atmospheric research may be investigating how the climate is changing, how snowflakes form, or how pollution impacts the environment. Aided by supercomputers, some research meteorologists simulate the atmosphere using computer models, to see how it behaves (see ● Figure 5). Researchers often work closely with scientists from other fields, such as biologists, environmental scientists, chemists, physicists, oceanographers, and mathematicians, as well as

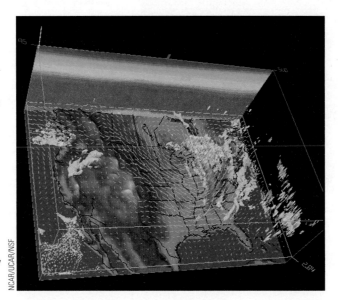

NCAR/UCAR/NSF

● FIGURE 5 A model that simulates a three-dimensional view of the atmosphere. This computer model predicts how winds and clouds over North America will change with time.

planners and social scientists, to determine how the atmosphere interacts with the entire ecosystem. Scientists doing work in physical meteorology may study how radiant energy warms the atmosphere; those at work in the field of dynamic meteorology might be using the mathematical equations that describe airflow to learn more about jet streams. Scientists working in operational meteorology might be preparing a weather forecast by analyzing upper-air information over North America. A climatologist, or climate scientist, might be studying the interaction of the atmosphere and ocean to see what influence such an interchange might have on Earth many years from now. Consulting meteorologists might be conducting air pollution dispersion modelling to understand the impact of an industrial facility on air quality.

Meteorologists also provide a variety of services not only to the general public in the form of weather forecasts but also to city planners, contractors, farmers, and large corporations. Meteorologists working for private weather firms create the forecasts and graphics that are found in newspapers, on television, and on the Internet. Overall, there are many exciting jobs that fall under the heading of "meteorologist"—too many to mention here. However, for more information on this topic, try some of the following websites:

- http://www.cmos.ca/
- http://www.msc-smc.ec.gc.ca/jobs_emplois/ Careers/Meteorologist_e.cfm
- http://www.ametsoc.org/ and click on "Students"

person" to present weather information and give daily forecasts. More and more of these people are professionally trained in meteorology, and many stations require that the weathercaster obtain endorsement from the Canadian Meteorological and Oceanographic Society (CMOS) (or a seal of approval from the American Meteorological Society). To make their weather presentation as up-to-the-minute as possible, most stations in Canada take advantage of the

information provided by the Meteorological Service of Canada (MSC), such as computerized weather forecasts, time-lapse satellite images, and colour Doppler radar displays. (At this point, it is interesting to note that many viewers believe that the weatherperson they see on TV is a meteorologist and that all meteorologists forecast the weather. If you are interested in learning what a meteorologist or atmospheric scientist is and what he or she might do

● **FIGURE 1.22** Estimates are that lightning strikes Earth about 100 times every second. About 25 million lightning strikes hit the United States each year, and there are about 2.7 million in Canada. Consequently, lightning is a very common, and sometimes deadly, weather phenomenon.

WEATHER WATCH

During late August and September 2005, Hurricane Katrina slammed into Mississippi and Louisiana (see Figure 1.14). In the city of New Orleans, several levees (that protected the city from flooding) broke, and flood waters over six metres deep inundated parts of the city, killing over 1200 people.

for a living [other than forecast the weather], read Focus on a Special Topic: What Is a Meteorologist? on p. 27.)

For many years, a staff of trained professionals at the "Weather Network"/"Météo Média" in Canada and "The Weather Channel" in the United States have provided weather information 24 hours a day on cable television. As well, the Meteorological Service of Canada operates *Weatheradio Canada*, a national network broadcasting weather and environmental information 24 hours a day on the VHF band directly from Environment Canada's Storm Prediction Centres. A special warning tone can be issued when a weather warning is made, which will activate a weather radio's internal alert system to turn on the radio. A similar system called *NOAA Weather Radio* exists in the United States.

SUMMARY

This chapter provided an overview of Earth's atmosphere. Most of the topics touched on, such as the various storms and weather systems, will be examined in much more depth in subsequent chapters. We saw that the atmosphere is one of a set of interconnected systems that make up Earth: atmosphere, biosphere/anthrosphere, hydrosphere/cryosphere, and lithosphere. Each of these systems is linked to each of the others in various ways, and these linkages will be highlighted at the start of each chapter.

Our atmosphere is one rich in nitrogen and oxygen as well as smaller amounts of other gases, such as water vapour, carbon dioxide, and other greenhouse gases whose increasing levels are resulting in global warming. We examined Earth's early atmosphere and found it to be much different from the air we breathe today.

We investigated the various layers of the atmosphere: the troposphere (the lowest layer), where almost all weather events occur, and the stratosphere, where ozone protects us from a portion of the sun's harmful rays. In the stratosphere, ozone has decreased in concentration over polar regions during spring, especially in the Southern hemisphere. Above the stratosphere lies the mesosphere, where the air temperature drops dramatically with height. Above the mesosphere lies the thermosphere, where temperatures are highest. At the top of the thermosphere is the exosphere, where collisions between gas molecules and atoms are so infrequent that fast-moving lighter molecules can actually escape Earth's gravitational pull and shoot off into space. The ionosphere represents that portion of the upper atmosphere where large numbers of ions and free electrons exist.

We looked briefly at the weather map and a satellite image and observed that dispersed throughout the atmosphere are storms and clouds of all sizes and shapes. The movement, intensification, and weakening of these systems, as well as the dynamic nature of air itself, produce a variety of weather events that we described in terms of weather elements. The sum total of weather and its extremes over a long period of time is what we call climate. Although sudden changes in weather may occur in a moment, climatic change takes place gradually over many years. The study of the atmosphere and all of its related phenomena is called *meteorology*, a term whose origin dates back to the days of Aristotle. Finally, we discussed some of the many ways weather and climate influence our lives.

KEY TERMS

The following terms are listed (with page numbers) in the order they appear in the text. Define each. Doing so will aid you in reviewing the material covered in this chapter.

atmosphere, 4	geosphere, 4
lithosphere, 4	pedosphere, 4

QUESTIONS FOR REVIEW

1. What is the primary source of energy for Earth's atmosphere?

2. List the four most abundant gases in today's atmosphere.

3. Of the four most abundant gases in our atmosphere, which one shows the greatest variation at Earth's surface?

4. What are some of the important roles that water plays in our atmosphere?

5. Briefly explain the production and natural destruction of carbon dioxide near Earth's surface. Give two reasons for the increase of carbon dioxide over the past 100 years.

6. List the two most abundant greenhouse gases in Earth's atmosphere. What makes them greenhouse gases?

7. Explain how the atmosphere "protects" inhabitants at Earth's surface.

8. What are some of the aerosols in our atmosphere?

9. How has the composition of Earth's atmosphere changed over time? Briefly outline the evolution of Earth's atmosphere.

10. (a) Explain the concept of air pressure in terms of mass of air above some level.

    (b) Why does air pressure always decrease with increasing height above the surface?

11. What is standard atmospheric pressure at sea level in (a) millimetres of mercury, (b) hectopascals, and (c) kilopascals?

12. What is the average or standard temperature lapse rate in the troposphere?

13. Briefly describe how the air temperature changes from Earth's surface to the lower thermosphere.

14. On the basis of temperature, list the layers of the atmosphere from the lowest to the highest layer.

15. What atmospheric layer contains all of our weather?

16. (a) In what atmospheric layer do we find the lowest average air temperature?

    (b) The highest average temperature?

    (c) The highest concentration of ozone?

17. Above what region(s) of the world would you find an ozone hole?

18. How does the ionosphere affect AM radio transmission during the day versus during the night?

19. Even though the actual concentration of oxygen is close to 21 percent (by volume) in the upper stratosphere, explain why, without proper breathing apparatus, you would not be able to survive there.

20. Define meteorology and discuss the origin of this word.

21. When someone says that "the wind direction today is south," does this mean that the wind is blowing toward the south or from the south?

22. Describe some of the features observed on a surface weather map.

23. Explain how wind blows around low-pressure areas in the Northern Hemisphere.

24. How are fronts defined?

25. Rank the following storms in size from largest to smallest: hurricane, tornado, middle-latitude cyclonic storm, thunderstorm.

26. Weather in the middle latitudes tends to move in what general direction?

27. How does weather differ from climate?

28. Describe some of the ways weather and climate influence the lives of people.

QUESTIONS FOR THOUGHT

1. Which of the following statements relate more to weather and which relate more to climate?

    (a) The summers here are warm and humid.

    (b) Cumulus clouds presently cover the entire sky.

    (c) Our lowest temperature last winter was −29°C.

    (d) The air temperature outside is 22°C.

    (e) December is our foggiest month.

    (f) The highest temperature ever recorded in Midale, Saskatchewan, was 45.0°C on July 5, 1937.

    (g) Snow is falling at the rate of 5 cm per hour.

    (h) The average temperature for the month of January in Edmonton, Alberta, is −13.5°C.

2. A standard pressure of 1013.25 hectopascals is also known as one atmosphere (1 ATM).

    (a) Look at Figure 1.10 and determine at approximately what levels you would record a pressure of 0.5 ATM and 0.1 ATM.

    (b) The surface air pressure on Mars is about 0.007 ATM. If you were standing on Mars, the surface air pressure would be equivalent to a pressure observed at approximately what elevation in Earth's atmosphere?

3. If you were suddenly placed at an altitude of 100 km above Earth, would you expect your stomach to expand or contract? Explain.

PROBLEMS AND EXERCISES

1. Keep track of the weather. On an outline map of North America, mark the daily position of fronts and pressure systems for a period of several weeks or more. (This information can be obtained from newspapers, the television news, or the Internet.) Plot the general upper-level flow pattern on the map. Observe how the surface systems move. Relate this information to the material on wind, fronts, and cyclones covered in later chapters.

2. Compose a one-week journal, including daily newspaper weather maps and weather forecasts from the newspaper or from the Internet. Provide a commentary for each day regarding the coincidence of actual and predicted weather.

3. Formulate a short-term climatology for your city for one month by recording maximum and minimum temperatures and precipitation amounts every day. You can get this information from television, newspapers, the Internet, or your own measurements. Compare this data to the actual climatology for that month. How can you explain any large differences between the two?

The aurora borealis, which forms as energetic particles from the sun interact with Earth's atmosphere, is seen here over Edmonton, Alberta.

© Carson Ganci/Design Pics/Corbis

Energy: Warming and Cooling Earth and the Atmosphere

2

At high latitudes after darkness has fallen, a faint, white glow may appear in the sky. Lasting from a few minutes to a few hours, the light may move across the sky as a yellow-green arc much wider than a rainbow, or it may faintly decorate the sky with flickering draperies of blue, green, and purple light that constantly change in form and location, as if blown by a gentle breeze.

For centuries, curiosity and superstition have surrounded these eerie lights. Inuit legend says they are the lights from demons' lanterns as they search the heavens for lost souls. Nordic sagas called them a reflection of fire that surrounds the seas of the north. Even today there are those who proclaim that the lights are reflected sunlight from polar ice fields. Actually, this light show in the Northern Hemisphere is the aurora borealis—the northern lights—which is caused by invisible energetic particles bombarding our upper atmosphere. Anyone who witnesses this, one of nature's spectacular colour displays, will never forget it.

CONTENTS

Energy is everywhere. It is the basis for life. It comes in various forms: it can warm a house, melt ice, and drive the atmosphere, producing our everyday weather events. When the sun's energy interacts with our upper atmosphere, we see energy at work in yet another form, a shimmering display of light from the sky—the aurora. Energy, as it flows within and between systems in its various forms, makes Earth such a dynamic place. In fact, the essential role of the atmosphere is to move energy from areas of surplus to areas of deficit. In Earth's lower latitudes (close to the equator), the intense sunlight creates an energy surplus and warm temperatures, whereas at the frigid poles, there is much less energy. All the storms, winds, jet streams, and other weather phenomena that are discussed in this book fundamentally exist to move this excess energy from the low latitudes to try to balance the energy deficit at the high latitudes. Energy enters our atmosphere from the sun as radiation and warms Earth's surface and soil (affecting the *lithosphere*), which, in turn, heats the lower atmosphere both directly and by evaporating water. The water vapour eventually condenses into clouds, releasing its heat, and falls to the surface again as precipitation, thus playing an essential role in the *hydrosphere*. Radiation from the sun is essential to photosynthesis that allows plants to grow, and it is the presence of sufficient energy, in the form of heat, as measured by temperature that makes Earth's *biosphere* viable for the life it supports.

What, precisely, is this common, yet mysterious, quantity we call "energy"? What is its primary source? How does it warm Earth and provide the driving force for each of its systems, including our atmosphere? And in what form does it reach our atmosphere to produce a dazzling display like the aurora? To answer these questions, we must first begin with the concept of energy itself. Then we will examine energy in its various forms and how energy is transferred from one form to another. We will look more closely at the sun's energy and its influence on our atmosphere. Finally, we will examine how energy acts globally and locally to affect conditions on Earth.

Energy, Temperature, and Heat

Energy is a property of matter that can make things happen. (Matter is anything that has mass and occupies space.) What does "make things happen" mean? It means to make things move or change condition in some way. In a physical system, by definition, *energy* is the ability or capacity to do work on some form of matter. Work is done when matter is pushed, pulled, or lifted over some distance. When we lift a brick, for example, we exert a force against the pull of gravity—we "do work" on the brick. The higher we lift the brick, the more work we do. So, by doing work on something, we give it "energy," which it can, in turn, use to do work on other things. The brick that we lifted, for instance, can now do work on your toe—by falling on it. Work or energy in a physical system like the brick example is equal to the force applied to an object times the distance the force is applied. In the SI system, force is measured in newtons (N), and distance in metres (m), so that newtons times metres = joules (J) is used to represent work or energy.

Before we discuss the various forms of energy and how energy is converted and transferred, let's examine temperature scales because some forms of energy involve temperature.

TEMPERATURE SCALES Suppose we take a small volume of air (like the one shown in ● Figure 2.1a) and allow it to cool. As the air slowly cools, its atoms and molecules would move slower and slower until the air reaches a temperature of −273.15°C, which is the lowest temperature possible. At this temperature, called **absolute zero**, the atoms and molecules would possess a minimum amount of energy and, theoretically, no thermal motion. At absolute zero, we can begin a temperature scale called the *absolute scale*, or **Kelvin scale** after Lord Kelvin (1824–1907), a famous British scientist who first introduced it. Since the Kelvin scale begins at absolute zero, it contains no negative numbers, so zero kelvin really does mean no temperature (i.e., no thermal energy). It is therefore quite convenient for scientific calculations.

● FIGURE 2.1 Air temperature is a measure of the average speed of the molecules. In the cold volume of air, the molecules move more slowly and crowd closer together. In the warm volume, they move faster and farther apart.

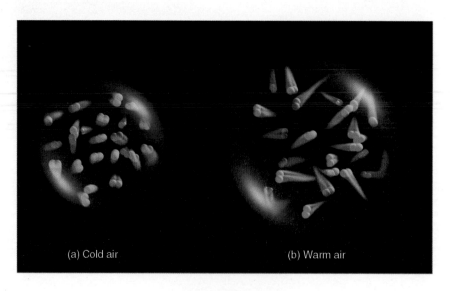

(a) Cold air

(b) Warm air

Two other temperature scales commonly used today are the Celsius (formerly centigrade) and Fahrenheit scales. The **Celsius scale**, denoted as °C (degrees Celsius), was introduced in the late 1700s: 0°C on this scale is the temperature at which pure water freezes, and 100°C is the temperature at which pure water boils at sea level. The **Fahrenheit scale**, denoted as °F (degrees Fahrenheit), was developed in the early 1700s: 32°F is the temperature at which water freezes, and 212°F is the temperature at which water boils. Since there are 180 equal divisions in the Fahrenheit scale between freezing and boiling, each Celsius degree is 180/100 or 1.8 times larger than a Fahrenheit degree.

On the Kelvin scale, the increments are called *kelvins* (abbreviated K; note that there is no "degree" kelvin). Each increment on the Kelvin scale is exactly the same size as a degree Celsius, and a temperature of 0 K is equal to −273.15°C. Converting from °C to K can be made by simply adding 273.15 to the Celsius temperature as

$$K = °C + 273.15$$

● Figure 2.2 compares the Kelvin, Celsius, and Fahrenheit scales. A more complete table of conversions, together with formulae, is given in Appendix A.

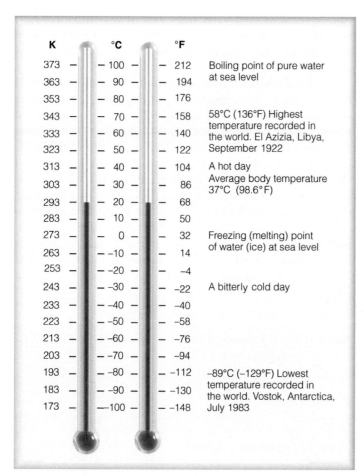

● FIGURE 2.2 Comparison of the Kelvin, Celsius, and Fahrenheit scales.

In most of the world, temperature readings are taken in degrees Celsius. In the United States, however, temperatures at the surface are typically read in degrees Fahrenheit, so on some U.S. surface weather maps and on U.S. weather forecasts and reports, temperatures are given in Fahrenheit. This book will mainly give temperatures in degrees Celsius, except in some calculations when kelvins are appropriate.

FORMS OF ENERGY   The total amount of energy stored in any object determines how much work that object is capable of doing. A lake behind a dam contains energy by virtue of its position. This is called *gravitational potential energy* or simply **potential energy** because it represents the potential to do work—a great deal of destructive work if the dam were to break. The potential energy (*PE*) of any object is given as

$$PE = mgh$$

where *m* is the object's mass, *g* is the acceleration of gravity, and *h* is the object's height above a specified level; often sea level is used.

A volume of air aloft has more potential energy than the same-sized volume of air just above the surface. This is because the air aloft has the potential to sink and warm through a greater depth of atmosphere. A substance also possesses potential energy if it can do work when a chemical change takes place. Thus, coal, natural gas, and food all contain chemical potential energy.

Any moving substance possesses energy of motion, or **kinetic energy**. The kinetic energy (*KE*) of an object is equal to half its mass multiplied by its velocity squared; thus,

$$KE = ½ \, mv^2$$

Consequently, the faster something moves, the greater its kinetic energy; hence, a strong wind possesses more kinetic energy than a light breeze. Since kinetic energy also depends on the object's mass, a volume of liquid water and an equal volume of air may be moving at the same speed, but because the water has greater mass, it has more kinetic energy.

The atoms and molecules that comprise all matter have kinetic energy due to their motion. This form of kinetic energy is often referred to as **sensible heat** (the heat we can feel, "sense," and measure with a thermometer). The sensible heat contained within matter is equal to its mass times its **specific heat*** times its temperature:

$$H = mCT$$

where *m* is the mass, *C* is the specific heat, and *T* is the temperature of the matter in kelvins.

Probably the most important form of energy in terms of weather and climate is the energy we receive from the sun—**radiant energy**. Radiation is unique because it is both a *form of energy* and a *mode of energy transfer*. In other words, energy

---

*Specific heat is a property of matter that specifies how much energy is required to raise the temperature of 1 kg of matter by 1°C (or 1 kelvin).

can exist as radiation, which also moves it around. Radiant energy will be focused on in the last half of this chapter, but for now let's just look at the amount of energy held as radiation by a **photon**. The radiant energy of a photon is a constant times the speed of light (*c*) divided by the wavelength* of the radiation (λ):

$$RE = \frac{hc}{\lambda}$$

where *h* is called the **Planck constant** ($= 6.62606896 \times 10^{-34}$ J s).

Energy, therefore, takes on many forms and can change from one form into another. But the total amount of energy in the universe remains constant. *Energy cannot be created, nor can it be destroyed.* It merely changes from one form to another in many ordinary physical or chemical processes. In other words, the energy lost during one process must equal the energy gained during another. This is what we mean when we say that energy is conserved. This statement is known as the *law of conservation of energy* and is also called the *first law of thermodynamics.*

We know that air is a mixture of countless billions of atoms and molecules. If they could be seen, they would appear to be moving about in all directions, freely darting, twisting, spinning, and colliding with one another like an angry swarm of bees. Close to Earth's surface, each individual molecule will travel only about a thousand times its diameter before colliding with another molecule. Moreover, we would see that all the atoms and molecules are not moving at the same speed as some are moving faster than others. The temperature of the air (or any substance) is a measure of the average kinetic energy of its molecules. Simply stated, for a given substance, **temperature** *is a measure of the average speed of the atoms and molecules,* where higher temperatures correspond to faster average speeds.

Suppose we examine a volume of surface air enclosed by a large, flexible balloon, as shown in Figure 2.1a. If we warm the air inside, the molecules would move faster, but they also would move slightly farther apart—the air becomes less dense, as illustrated in Figure 2.1b. Conversely, if we cool the air back to its original temperature, the molecules would slow down and crowd closer together, and the air would become more dense. This molecular behaviour is why, in many places throughout the book, we refer to surface air as either *warm, less dense air* or as *cold, more dense air.*

Imagine that you are sipping a hot cup of tea on a small raft in the middle of a lake. The tea has a much higher temperature than the lake, yet the lake contains more sensible heat because it is composed of many more molecules. If the cup of tea is allowed to float on top of the water, the tea would cool rapidly. The energy that would be transferred from the hot tea to the cool water (because of their temperature difference) is called *heat*.

In essence, **heat** is energy in the process of being transferred from one object to another because of the temperature difference between them. How is this energy transfer process accomplished? In the atmosphere, heat is moved by conduction, convection, and radiation. We will examine these mechanisms of energy transfer after we look at the important concepts of specific heat and latent heat.

SPECIFIC HEAT   A watched pot never boils, or so it seems. The reason for this is that water requires a relatively large amount of sensible heat to bring about a small temperature change. The *heat capacity* of a substance is the ratio of the amount of sensible heat absorbed by that substance to its corresponding temperature rise. The heat capacity of a substance per unit mass is called **specific heat**; in other words, it is the amount of heat needed to raise the temperature of one kilogram (kg) of a substance one kelvin (or 1°C since the increment is the same).

If we heat 1 kg of liquid water on a stove, it would take about 4186 J of energy to raise its temperature by 1 K. So water has a specific heat of 4186 J kg^{-1} K^{-1}. If, however, we put the same amount (i.e., same mass) of compact dry soil on the flame, we would see that it would take about one-fifth the heat (about 800 J) to raise its temperature by 1 K. The specific heat of water is therefore five times greater than that of dry soil. In other words, water must absorb five times as much heat as the same quantity of soil to raise its temperature by the same amount. The specific heat of various substances is given in ▼ Table 2.1.

Not only does water heat slowly, it cools slowly as well. It has a much higher capacity for storing energy than other common substances, such as soil and air. A given volume of water can lose or gain a large amount of energy while undergoing only a small temperature change. Because of this attribute, water has a strong modifying effect on weather and climate. Near large bodies of water, for example, winters usually remain warmer and summers cooler than nearby inland regions—a fact well known to people who live adjacent to oceans or large lakes. There are other reasons besides specific heat for moderate temperatures over and near large water bodies. These are discussed in Chapter 3, The Geographic Controls of Temperature, on p. 83.

▼ Table 2.1   Specific Heat of Various Substances

SUBSTANCE	SPECIFIC HEAT (J kg^{-1} K^{-1})
Water (pure)	4186
Wet mud	2512
Ice (0°C)	2093
Sandy clay	1381
Dry air (sea level)	1005
Quartz sand	795
Granite	794

---

*Wavelength is a fundamental property of radiation that exists as a wave. It is the distance from the crest of one wave to the crest of the next one, much like for waves on a water surface.

LATENT HEAT—THE HIDDEN WARMTH   We know from Chapter 1 that water vapour is an invisible gas that becomes visible when it changes into larger liquid or solid (ice) particles. This process of transformation is known as a *change of state* or, simply, a *phase change*. The energy required to change a substance, such as water, from one phase to another is called **latent heat**. But why is this heat referred to as "latent"? To answer this question, we will begin with something familiar to most of us—the cooling produced by evaporating water.

Suppose we microscopically examine a small drop of pure liquid water. At the drop's surface, molecules are constantly escaping (evaporating). Because the more energetic, faster moving molecules escape most easily, the average motion of all the molecules left behind decreases as each additional molecule evaporates. Since temperature is a measure of average molecular motion, the slower motion suggests a lower water temperature. *Evaporation is, therefore, a cooling process*. Stated another way, evaporation is a cooling process because the energy needed to evaporate the water— that is, to change its phase from a liquid to a gas—may come from the water or other sources, including the air.

In the everyday world, we experience evaporational cooling as we step out of a shower or swimming pool into a dry area. Because some of the energy used to evaporate the water comes from our skin, we may experience a rapid drop in skin temperature, even to the point where goose bumps form. In fact, on a hot, dry, windy day in Regina, Saskatchewan, cooling may be so rapid that we begin to shiver even though the air temperature is hovering around 37°C, which is the same as the normal body internal temperature.

The energy lost by liquid water during evaporation can be thought of as carried away by, and "locked up" within, the water vapour molecule. The energy is thus in a "stored" or "hidden" condition and is, therefore, called *latent heat*. It is latent (hidden) in that the temperature of the substance changing from liquid to vapour is still the same. However, the energy will reappear as sensible heat when the vapour condenses back into liquid water. Therefore, condensation (the opposite of evaporation) *is a warming process* in which latent heat is transformed into sensible heat.

The energy released when water vapour condenses to form liquid droplets is called *latent heat of condensation*. Conversely, the energy used to change liquid into vapour at the same temperature is called *latent heat of evaporation* (vaporization). Nearly 2,500,000 J are required to evaporate a single kilogram of water at room temperature. With hundreds of grams of water evaporating from the body, it is no wonder that after a shower we feel cold before drying off.

In a similar way, latent heat is responsible for keeping a cold drink with ice colder than one without ice. As ice melts, its temperature does not change. The reason for this fact is that the heat added to the ice breaks down only the rigid crystal pattern, changing the ice to a liquid without changing its temperature. The energy used in this process is called *latent heat of fusion* (melting). Roughly 335,000 J are required to melt a kilogram of ice. Consequently, heat added to a cold drink with ice primarily melts the ice, whereas heat added to a cold drink without ice warms the beverage. If a kilogram of water at 0°C changes back into ice at 0°C, this same amount of heat (335,000 J) would be released as sensible heat to the environment. Therefore, when ice melts, heat is taken in; when water freezes, heat is liberated.

The energy required to change ice into vapour (a process called *sublimation*) is referred to as *latent heat of sublimation*. For a kilogram of ice to transform completely into vapour at 0°C requires nearly 2,850,000 J—335,000 J for the latent heat of fusion plus 2,500,000 J for the latent heat of evaporation. If this same vapour transformed back into ice (a process called *deposition*), approximately 2,850,000 J would be released.

● Figure 2.3 summarizes the concepts examined so far. When the change of state is from left to right, energy is

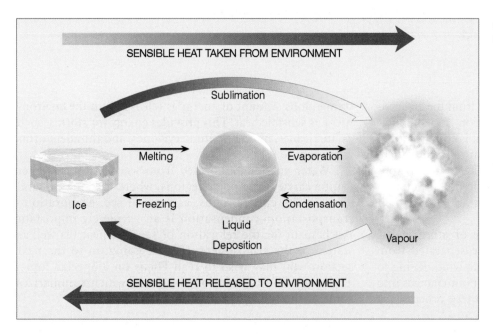

● FIGURE 2.3 Energy absorbed and released by phase changes of water.

FOCUS ON A SPECIAL TOPIC

The Fate of a Sunbeam

Energy is neither created nor destroyed, but it can change its form. Consider sunlight in the form of radiant energy striking a large lake (see ● Figure 1). Part of the incoming energy heats the water, causing greater molecular motion and, hence, an increase in the water's kinetic energy and therefore its sensible heat as the water temperature increases. This greater kinetic energy allows more water molecules to evaporate from the surface. As each molecule escapes, work is done to break it away from the remaining water molecules. This energy becomes the latent heat that is carried with the water vapour.

Above the lake, a large bubble* of warm, moist air rises and expands. For this expansion to take place, the gas molecules inside the bubble must use some of their kinetic energy to do work against the bubble's sides. This results in a slower molecular speed and a lower temperature as sensible heat is used to expand the bubble. Well above the surface, the water vapour in the rising, cooling bubble of moist air condenses into clouds. The condensation of water vapour releases latent heat into the atmosphere, where it is converted to sensible heat, warming the air. The tiny suspended cloud droplets possess potential energy, which

● FIGURE 1 Solar energy striking a large body of water goes through many transformations.

becomes kinetic energy when these droplets grow into raindrops that fall earthward.

When the drops reach the surface, their kinetic energy erodes the land. As rain-swollen streams flow into a lake behind a dam, there is a buildup of potential energy, which can be transformed into kinetic energy as water is harnessed to flow down a chute. If the moving water drives a generator, kinetic energy is

converted into electrical energy, which is sent to cities. There it heats, cools, and lights the buildings in which people work and live. Meanwhile, some of the water in the lake behind the dam evaporates and is free to repeat the cycle. Hence, the energy from the sunlight on a lake can undergo many transformations and help provide the moving force for many natural and human-made processes.

*A bubble of rising (or sinking) air about the size of a large balloon is often called a **parcel of air**.

---

absorbed by the substance and taken away from the environment. The processes of melting, evaporation, and sublimation all cool the environment: sensible heat is converted into latent heat. When the change of state is from right to left, energy is given up by the substance and added to the environment: latent heat is converted into sensible heat. The processes of freezing, condensation, and deposition all warm their surroundings.

Latent heat is an important source of atmospheric energy. Once vapour molecules become separated from Earth's surface, they are swept away by the wind. Rising to high altitudes where the air is cold, the vapour changes into liquid and ice cloud particles. During these processes, a

tremendous amount of energy is released into the environment as sensible heat. This provides energy for storms, such as hurricanes, middle-latitude cyclones, and thunderstorms (see ● Figure 2.4).

Water vapour evaporated from warm, tropical water can be carried into polar regions, where it condenses and gives up its energy. Thus, as we will see, evaporation–transportation–condensation is an extremely important mechanism for the relocation of sensible heat (as well as water) in the atmosphere. (Before going on to the next section, you may wish to read Focus on a Special Topic: The Fate of a Sunbeam on this page, which summarizes some of the concepts considered thus far.)

● FIGURE 2.4 Every time a cloud forms, it warms the atmosphere. Inside this developing thunderstorm a vast amount of stored energy (latent heat) is given up to the air as sensible heat and increased air temperature, when invisible water vapour becomes countless billions of water droplets and ice crystals. In fact, for the duration of this storm alone, more energy is released inside this cloud than is unleashed by a small nuclear bomb.

Heat Transfer in the Atmosphere

CONDUCTION   The transfer of heat from molecule to molecule within a substance is called **conduction**. Imagine holding one end of a metal straight pin between your fingers and placing a flaming candle under the other end (see ● Figure 2.5). Because of the energy they absorb from the flame, the molecules in the pin vibrate faster. The faster vibrating molecules cause adjoining molecules to vibrate

faster. These, in turn, pass vibrational energy on to their neighbouring molecules, and so on, until the molecules at the finger-held end of the pin begin to vibrate rapidly. These fast-moving molecules eventually cause the molecules of your finger to vibrate more quickly. Heat is now being transferred from the pin to your finger, and both the pin and your finger feel hot. If enough heat is transferred, you will drop the pin. The transmission of heat from one end of the pin to the other, and from the pin to your finger, occurs by conduction. Heat transferred in this fashion always flows from *warmer to colder* regions. Generally, the greater the temperature difference, the more rapid the heat transfer.

When materials can easily pass energy from one molecule to another, they are considered to be good conductors of heat. How well they conduct heat depends on how their molecules are structurally bonded together. ▼ Table 2.2 shows that solids, such as metals, are good heat conductors. It is often difficult, therefore, to judge the temperature of metal objects. For example, if you grab a metal pipe at room temperature, it will seem to be much colder than it actually is because the metal conducts heat away from the hand quite rapidly. Conversely, *air is an extremely poor conductor of heat*, which is why most insulating materials have a large number of air spaces trapped within them. Air is such a poor heat conductor that in calm weather, the hot ground warms only a shallow layer of air a few millimetres thick by conduction. Yet air can carry this energy rapidly from one region to another. How, then, does this phenomenon happen?

▼ Table 2.2   Thermal Conductivity* of Various Substances

SUBSTANCE	THERMAL CONDUCTIVITY WATTS† PER METRE PER °C (W m⁻¹ °C⁻¹)
Still air	0.023 (at 20°C)
Wood	0.08
Dry soil	0.25
Water	0.60 (at 20°C)
Snow	0.63
Wet soil	2.1
Ice	2.1
Sandstone	2.6
Granite	2.7
Iron	80
Copper	401
Silver	427

*Thermal conductivity describes a substance's ability to conduct heat as a consequence of molecular motion.

†A watt (W) is a unit of power where one watt equals one joule (J) per second (J s⁻¹). One joule equals 0.24 calories.

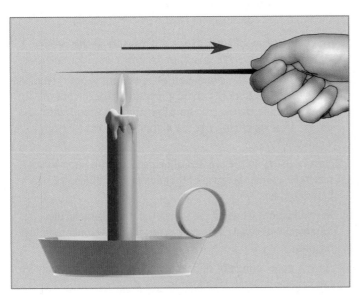

● FIGURE 2.5 The transfer of heat from the hot end of the metal pin to the cool end by molecular contact is called conduction.

CONVECTION   The transfer of heat by the mass movement of a fluid (such as water and air) is called **convection**. This type of heat transfer takes place in liquids and gases because they can move freely, and it is possible to set up currents within them.

Convection happens naturally in the atmosphere. On a warm, sunny day, certain areas of Earth's surface absorb more heat from the sun than others; as a result, the air near Earth's surface is heated somewhat unevenly. Air molecules adjacent to these hot surfaces bounce against them, thereby gaining some extra energy by conduction. The heated air expands and becomes less dense than the surrounding cooler air. The expanded warm air is buoyed upward and rises. In this manner, large bubbles of warm air rise and transfer sensible heat upward. Cooler, heavier air flows toward the surface to replace the rising air. This cooler air becomes heated in turn and rises, and the cycle is repeated. In meteorology, this vertical exchange of heat is called convection, and the rising air bubbles are known as **thermals** (see ● Figure 2.6).

The rising air expands and gradually spreads outward. It then slowly begins to sink. Near the surface, it moves back into the heated region, replacing the rising air. In this way, a convective circulation, or thermal "cell," is produced in the atmosphere. In a convective circulation, the warm, rising air cools. In our atmosphere, *any air that rises will expand and cool*, and *any air that sinks is compressed and warms*.

This type of convection is called **free convection** and is discussed in some detail in Chapter 6. Another type of convection can occur on a smaller scale when wind blows over a rough surface, creating twisting and swirling air currents called **turbulence**. When turbulent air flows over a hot surface, the swirling eddies transport warm air upward and cold air downward. This type of convection is called **forced convection** and is covered in Chapter 9. In the atmosphere, above the first few millimetres from Earth's surface where conduction occurs, both sensible and latent heat are transferred entirely by convection—both free and forced.

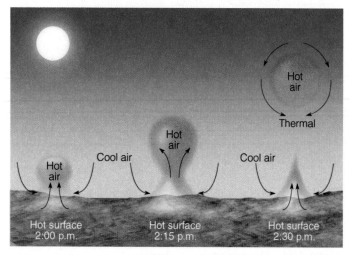

● FIGURE 2.6   The development of a thermal. A thermal is a rising bubble of air that carries sensible heat upward by convection.

Although the entire process of heated air rising, spreading out, sinking, and finally flowing back toward its original location is known as convective circulation, meteorologists usually restrict the term *convection* to the process of the rising and sinking parts of the circulation.

The horizontally moving part of the circulation (called *wind*) carries properties of the air in that particular area with it. The transfer of these properties by horizontally moving air is called **advection**. For example, wind blowing across a body of water will "pick up" water vapour from the evaporating surface and transport it elsewhere in the atmosphere. If the air cools, the water vapour may condense into cloud droplets and release latent heat, converting it into sensible heat. In this way, energy is advected (carried) by the water vapour as it is swept along with the wind. Earlier, we saw that this is an important way to redistribute energy in the atmosphere.

BRIEF REVIEW

Before moving on to the next section, here is a summary of some of the important concepts and facts we have covered:

● The temperature of a substance is a measure of the average kinetic energy (average speed) of its atoms and molecules.

● Evaporation (the transformation of liquid into vapour) requires energy and therefore is a cooling process that can cool the air, whereas condensation (the transformation of vapour into liquid) gives up energy and is a warming process that can warm the air.

● Heat is energy in the process of being transferred from one object to another because of the temperature difference between them.

● In conduction, which is the transfer of heat by molecule-to-molecule contact, heat always flows from warmer to colder regions.

● Air is a poor conductor of heat, so conduction occurs only in the first few millimetres above Earth's surface.

● Convection is an important mechanism of heat transfer as it represents the vertical movement of warmer air upward and cooler air downward.

Radiation

There is yet another mechanism for the transfer of energy—radiation, or **radiant energy**, which is what we receive from the sun. In this method, energy may be transferred from one object to another without the space between them necessarily being heated.

On a summer day, you may have noticed how warm your face feels as you stand facing the sun. Sunlight travels through the surrounding air with little effect on the air itself. Your face, however, absorbs this energy and converts it to sensible heat. Thus, sunlight warms your face without actually warming the air. The energy transferred from the sun to your face is called radiant energy, or **radiation**. It travels in the form of waves that release energy when they are absorbed by an object. Because these waves have magnetic and electrical properties, we call them **electromagnetic waves**. Electromagnetic waves do not need molecules to propagate them. In a vacuum, they travel at a constant speed of nearly 300,000 km per second—the speed of light.

● Figure 2.7 shows some of the different wavelengths of radiation. Notice that the **wavelength** (which is usually expressed by the Greek letter lambda, λ) is the distance measured along a wave from one crest to another. Also notice that some of the waves have exceedingly short lengths. For example, radiation that we can see (visible light) has an average wavelength of less than one-millionth of a metre—a distance nearly one-hundredth the diameter of a human hair. Units of **micrometres** (μm), which equal one-millionth of a metre, are used to represent these wavelengths; thus,

$$1 \ \mu m = 0.000001 \ m = 10^{-6} \ m$$

In Figure 2.7, we can see that the average wavelength of visible light is about 0.0000005 m or 0.5 μm. To help visualize this quantity, the average height of a letter on this page is about 2000 μm (2 mm), whereas the thickness of this page is about 100 μm.

We can also see in Figure 2.7 that the longer waves carry less energy than do the shorter waves. When comparing the energy carried by various waves, it is useful to give electromagnetic radiation characteristics of particles to explain some of the waves' behaviour. We can actually think of radiation as streams of particles or photons that are discrete packets of energy.*

An ultraviolet photon carries more energy than a photon of visible light. In fact, certain ultraviolet photons have enough energy to produce sunburns and penetrate skin tissue, sometimes causing skin cancer. As we discussed in Chapter 1, it is ozone in the stratosphere that protects us from the vast majority of these harmful rays. ● Figure 2.8 illustrates the concept of radiation along with the other forms of heat transfer—conduction and convection.

RADIATION AND TEMPERATURE *All things (whose temperature is above absolute zero), no matter how big or small, emit radiation.* This book, your body, flowers, trees, air, Earth, and the stars are all radiating a wide range of electromagnetic waves. The energy originates from rapidly vibrating electrons, billions of which exist in every object.

The wavelengths that each object emits depend on the object's temperature. The higher the temperature, the faster

---

*Packets of photons make up waves, and groups of waves make up a beam of radiation.

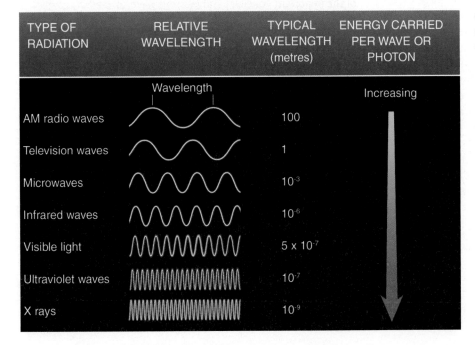

TYPE OF RADIATION	RELATIVE WAVELENGTH	TYPICAL WAVELENGTH (metres)	ENERGY CARRIED PER WAVE OR PHOTON
	Wavelength		Increasing
AM radio waves		100	
Television waves		1	
Microwaves		10^{-3}	
Infrared waves		10^{-6}	
Visible light		5×10^{-7}	
Ultraviolet waves		10^{-7}	
X rays		10^{-9}	

● FIGURE 2.7 Radiation characterized according to wavelength. As the wavelength decreases, the energy carried per wave increases.

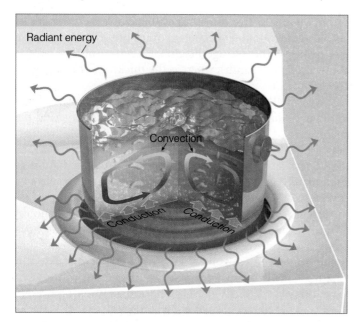

● FIGURE 2.8 The hot burner warms the bottom of the pot by conduction. The warm pot, in turn, warms the water in contact with it. The warm water rises, setting up convection currents. The pot, water, burner, and everything else constantly emit radiant energy (orange arrows) in all directions.

the electrons vibrate, and the shorter are the wavelengths of the emitted radiation. This can be visualized by attaching one end of a rope to a post and holding the other end. If the rope is shaken rapidly (high temperature), numerous short waves travel along the rope; if the rope is shaken slowly (lower temperature), longer waves appear on the rope. Although objects at a temperature of about 500°C radiate waves with many lengths, some of them are short enough to be detected by the human eye as a red glow. Objects cooler than this radiate at wavelengths that are too long for us to see. The page of this book, for example, is radiating electromagnetic waves. But because its temperature is only about 20°C, the waves emitted are much too long to stimulate vision. We are able to see the page, however, because light waves in the visible wavelength region from other sources (such as light bulbs or the sun) are being *reflected* off the paper. If this book were carried into a completely dark room, it would continue to radiate, but the pages would appear black because there are no visible light waves in the room to reflect off the pages.

Objects that have a high temperature emit radiant energy at a greater rate or intensity than objects at a lower temperature. Thus, *as the temperature of an object increases, more total radiation is emitted each second.* This can be expressed mathematically as

$$E = \sigma T^4 \text{ (Stefan–Boltzmann law)}$$

where E is the rate of radiation emitted by each square metre of surface area of the object, σ (the Greek letter sigma) is the Stefan-Boltzmann constant,* and T is the object's surface temperature in kelvins. This relationship, called the

Stefan–Boltzmann law after Josef Stefan (1835–93) and Ludwig Boltzmann (1844–1906), who derived it, states that all objects with temperatures above absolute zero (0 K or −273.15°C) emit radiation at a rate proportional to the fourth power of their absolute temperature. Consequently, a small increase in temperature results in a large increase in the amount of radiation emitted.

RADIATION OF THE SUN AND EARTH Most of the sun's energy is emitted from its surface, where the temperature is about 5778 K. Earth, on the other hand, has an average surface temperature of 288 K (15°C). The sun, therefore, radiates a great deal more energy than does Earth (see ● Figure 2.9). At what wavelengths do the sun and Earth radiate most of their energy? Fortunately, the sun and Earth both have characteristics (discussed in a later section) that enable us to use the following relationship called **Wien's law** (or *Wien's displacement law*) after the German physicist Wilhelm Wien (1864–1928), who discovered it:

$$\lambda_{max} = \frac{\text{constant}}{T} \text{ (Wien's law)}$$

where λ_{max} is the wavelength in micrometres at which maximum radiation emission occurs, T is the object's temperature in kelvins, and the constant is 2897 μm K.

For the sun, with a surface temperature of about 5778 K, the equation becomes

$$\lambda_{max} = \frac{2897 \ \mu m \ K}{5778 \ K} = 0.5013 \ \mu m$$

*The Stefan–Boltzmann constant σ in SI units is $5.67 \times 10^{-8} \text{ W m}^{-2} \text{ K}^{-4}$. A watt (W) is a unit of power, where one watt equals one joule (J) per second (J s⁻¹). More information on units and conversions is given in Appendix A.

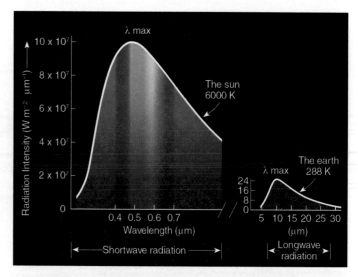

● FIGURE 2.9 The hotter sun not only radiates more energy than the cooler Earth (the area under the curve) but also radiates the majority of its energy at much shorter wavelengths. (The area under the curves is equal to the total energy emitted, and the scales for the two curves differ by a factor of 100,000.)

Thus, the sun emits a maximum amount of radiation at wavelengths near 0.5 μm; note the peak in the left curve in Figure 2.9.

The cooler Earth, with an average surface temperature of 288 K, emits maximum radiation near wavelengths of 10 μm since

$$\lambda_{max} = \frac{2897\ \mu m\ K}{288\ K} = 10.1\ \mu m$$

Thus, Earth emits most of its radiation at longer wavelengths between about 5 and 25 μm, whereas the sun emits the majority of its radiation at wavelengths less than 2 μm. For this reason, Earth's radiation (*terrestrial radiation*) is often called **longwave radiation**, whereas the sun's energy (*solar radiation*) is referred to as **shortwave radiation**.

Wien's law demonstrates that as the temperature of an object increases, the wavelength at which maximum emission occurs is shifted toward shorter values. For example, if the sun's surface temperature were to double to 12,000 K, its wavelength of maximum emission would be halved to about 0.25 μm. If, on the other hand, the sun's surface cooled to 3000 K, it would emit its maximum amount of radiation near 1.0 μm.

Even though the sun radiates at a maximum rate at a particular wavelength, it nonetheless emits some radiation at almost all other wavelengths. If we look at the amount of radiation given off by the sun at each wavelength, we obtain the sun's *electromagnetic spectrum*. A portion of this spectrum is shown in ● Figure 2.10.

Since our eyes are sensitive to radiation between 0.4 and 0.7 μm, these waves reach the eye and stimulate the sensation

WEATHER WATCH

The large ears of a jackrabbit are efficient emitters of infrared energy. Its ears help the rabbit survive the heat of a summer's day by radiating a great deal of infrared energy to the cooler sky above. Similarly, the large ears of the African elephant greatly increase its radiating surface area and promote cooling of its large mass.

of colour. This portion of the spectrum is referred to as the **visible region**, and the radiant energy that reaches our eye is called *visible light*. The sun emits nearly 44 percent of its radiation in this zone, with the peak of energy output found at the wavelength corresponding to the colour blue-green. The colour violet is the shortest wavelength of visible light. Wavelengths shorter than violet (between 0.01 and 0.4 μm) are **ultraviolet (UV)**. X-rays and gamma rays have exceedingly short wavelengths less than 0.01 μm. The sun emits only about 7 percent of its total energy at ultraviolet or shorter wavelengths. (Additional information on radiation intensity and its effect on humans is given in Focus on an Environmental Issue: Wave Energy, Sun Burning, and Ultraviolet Rays, on p. 44.)

The longest wavelengths of visible light correspond to the colour red. Wavelengths longer than red (0.7 μm) are **infrared (IR)**. These waves cannot be seen by humans. Nearly 37 percent of the sun's energy is radiated between 0.7 and 1.5 μm, with an additional 12 percent radiated at wavelengths longer than 1.5 μm. To learn more about the sun, read Focus on a Special Topic: Characteristics of the Sun on p. 46.

Whereas the hot sun emits only a part of its energy in the infrared portion of the spectrum, the relatively cool Earth emits almost all of its energy at infrared wavelengths. Although we cannot see infrared radiation, there are instruments called *infrared sensors* that can. Weather satellites that orbit the globe use these sensors to observe radiation emitted by Earth's surface, the clouds, and the atmosphere. Since objects of different temperatures radiate their maximum energy at different wavelengths, infrared photographs can distinguish among objects of different temperatures. Clouds always radiate infrared energy; thus, cloud images using infrared sensors can be taken during both day and night.

In summary, both the sun and Earth emit radiation. The *hot sun* (5778 K) radiates nearly 88 percent of its energy at wavelengths less than 1.5 μm, with maximum emission in the *visible region* near 0.5 μm. The *cooler earth* (288 K) radiates

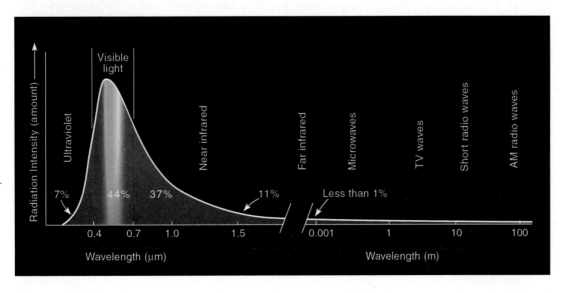

● FIGURE 2.10 The sun's electromagnetic spectrum and some of the descriptive names of each region. The numbers underneath the curve approximate the percentage of energy the sun radiates in various regions.

Wave Energy, Sun Burning, and Ultraviolet Rays

Standing close to a fire makes us feel warmer than we do when we stand at a distance from it. Does this mean that as we move away from a hot object, the waves carry less energy and are, therefore, weaker? Not really. The intensity of radiation decreases as we move away from a hot object because radiant energy spreads outward in all directions. ● Figure 2 illustrates that as the distance from a radiating object increases, a given amount of energy is distributed over a larger area, so the energy received over a given area and over a given time decreases. In fact, at twice the distance from the source, the radiation is spread over four times the area.

Another interesting fact about radiation that we learned earlier in this chapter is that shorter waves carry much more energy than do longer waves. Hence, a photon of ultraviolet light carries more energy than a photon of visible light. In fact, ultraviolet (UV) wavelengths in the range between 0.20 and 0.29 μm (known as ultraviolet C or *UVC radiation*) are harmful to living things as certain waves can cause chromosome mutations, kill single-celled organisms, and damage the cornea of the eye. Fortunately, virtually all the ultraviolet radiation at wavelengths in the UVC range is absorbed by oxygen and ozone in the stratosphere.

Ultraviolet wavelengths between about 0.29 and 0.32 μm (known as *UVB radiation*) are mostly removed by ozone in the stratosphere but can reach Earth's surface in small amounts, especially in locations and at times of the year when stratospheric ozone depletion occurs. UVB levels are normally highest in the tropics year-round, but high levels can also occur over middle and high latitudes during the

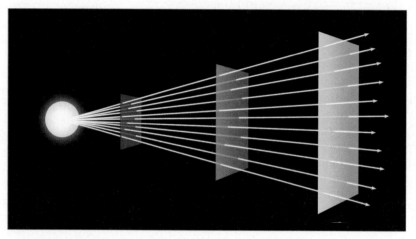

● FIGURE 2 The intensity, or amount, of radiant energy transported by electromagnetic waves decreases as we move away from a radiating object because the same amount of energy is spread over a larger area.

spring and summer months, when total solar radiation increases. Stratospheric ozone depletion, especially during the austral spring (September and October) over Antarctica, and to a lesser extent during spring in the Northern Hemisphere, increases UVB levels (see Chapter 18 starting at p. 560). Photons in the UVB wavelength range have enough energy to produce sunburns and penetrate skin tissues, sometimes causing skin cancer. About 90 percent of all skin cancers are linked to sun exposure and UVB radiation. Oddly enough, these same wavelengths activate provitamin D in the skin and convert it into vitamin D, which is essential to health.

Longer ultraviolet waves with lengths of about 0.32 to 0.40 μm (called *UVA radiation*) are less energetic but can still tan the skin.

Although UVB is mainly responsible for burning the skin, UVA can cause skin redness. It can also interfere with the skin's immune system and cause long-term skin damage that shows up years later as accelerated aging and skin wrinkling. Moreover, recent studies indicate that longer UVA exposures needed to create a tan pose about the same cancer risk as a UVB tanning dose.

Upon striking the human body, ultraviolet radiation is absorbed beneath the outer layer of skin. To protect the skin from these harmful rays, the body's defence mechanism kicks in. Certain cells (when exposed to ultraviolet radiation) produce a dark pigment (*melanin*) that begins to absorb some of the ultraviolet radiation. (It is the production of melanin that produces a tan.) Consequently, a body that

nearly all its energy between 5 and 25 μm, with a peak intensity in the *infrared region* near 10 μm (see Figure 2.9). The sun's surface is nearly 20 times hotter than Earth's surface. From the Stefan–Boltzmann relationship, this fact means that a unit area on the sun emits nearly 160,000 (20^4) times more energy during a given time period than the same-sized area on Earth. And since the sun has such a huge surface area

from which to radiate, the total energy emitted by the sun each minute amounts to a staggering 23 billion, billion, billion joules (i.e., 3.865×10^{26} Watts)! To learn how incoming solar energy, in the form of particles, produces a dazzling light show known as the aurora, you may wish to read Focus on a Special Topic: Solar Particles and the Aurora, on p. 58 near the end of this chapter.

produces little melanin—one with pale skin—has little natural protection from UVB.

Additional protection can come from a sunscreen. Unlike the old lotions that simply moisturized the skin before it baked in the sun, sunscreens today block ultraviolet rays from ever reaching the skin. Some contain chemicals (such as zinc oxide) that reflect ultraviolet radiation. (These are the white pastes once seen on the noses of lifeguards.) Others consist of a mixture of chemicals (such as benzophenone and para-aminobenzoic acid, PABA) that actually absorb ultraviolet radiation, usually UVB, although new products with UVA-absorbing qualities are now on the market. The sun protection factor (SPF) number on every container of sunscreen dictates how effective the product is in protecting from UVB—the higher the number, the better the protection.

Protecting oneself from excessive exposure to the sun's energetic ultraviolet rays is certainly wise. Estimates are that in a single year, over 3,900 Canadians (30,000 Americans) will be diagnosed with malignant melanoma, the most deadly form of skin cancer, and over 800 Canadians will die from it. And if the protective ozone shield should diminish even more over certain areas of the world, there is an ever-increasing risk of problems associated with UVB. Using a good sunscreen and proper clothing can certainly help. The best way to protect yourself from too much sun, however, is to limit your time in direct sunlight, especially between the hours of 11 a.m. and 3 p.m., when the sun is highest in the sky and its rays are the most direct.

Environment Canada's UV Index

UV Index	Description	Sun Protection Actions
0–2	Low	• Minimal sun protection required for normal activity • Wear sunglasses on bright days. If outside for more than one hour, cover up and use sunscreen • Reflection off snow can nearly double UV strength. Wear sunglasses and apply sunscreen
3–5	Moderate	• Take precautions—cover up, wear a hat, sunglasses and sunscreen—especially if you will be outside for 30 minutes or more • Look for shade near midday when the sun is strongest
6–7	High	• Protection required—UV damages the skin and can cause sunburn • Reduce time in the sun between 11 a.m. and 4 p.m. and take full precautions—seek shade, cover up, wear a hat, sunglasses, and sunscreen
8–10	Very high	• Extra precautions required—unprotected skin will be damaged and can burn quickly • Avoid the sun between 11 a.m. and 4 p.m. and take full precautions—seek shade, cover up, wear a hat, sunglasses, and sunscreen
11+	Extreme	• Values of 11 or more are very rare in Canada. However, the UV Index can reach 14 or more in the tropics and southern U.S. • Take full precautions. Unprotected skin will be damaged and can burn in minutes. Avoid the sun between 11 a.m. and 4 p.m., cover up, wear a hat, sunglasses, and sunscreen • White sand and other bright surfaces reflect UV and increase UV exposure

● FIGURE 3 The UV index.

Environment Canada. UV Index. Found at: http://www.ec.gc.ca/Publications/default.asp?lang=En&xml=9C0A3543-2DB7-4673-905D-0541E9622C68. © Her Majesty The Queen in Right of Canada, Environment Canada, 2010.

Presently, Environment Canada's Meteorological Service of Canada makes a daily prediction of UV radiation levels for 48 locations across Canada, as well as producing a forecast UV map for North America. The forecast, known as the UV index, gives the UV level at its peak, around noon standard time or 1 p.m. daylight savings time. The index corresponds to five exposure categories set by Environment Canada. An index value of between 0 and 2 is considered "low," with minimal sun protection required, whereas a value of 8 to 10 is considered "very high," and values greater than 11 are deemed "extreme" (see ● Figure 3), with full sun protection and avoidance between 11 a.m. and 4 p.m. recommended. Depending on skin type, a UV index of 10 means that in direct sunlight (without sunscreen protection), a person's skin will likely begin to burn in about 6 to 30 minutes.

Incoming Solar Energy

As the sun's radiant energy travels through space, essentially nothing interferes with it until it reaches the atmosphere. At the top of the atmosphere, solar energy received on a surface perpendicular to the sun's rays appears to remain fairly constant at 1367 W m^{-2}—a value called the **solar constant**.*

*By definition, the solar constant (which, in actuality, is *not* "constant") is the rate at which radiant energy from the sun is received on a surface at the outer edge of the atmosphere perpendicular to the sun's rays when Earth is at an average distance from the sun. Since Earth's orbit around the sun is shaped like an *ellipse*, not a circle, it is closest to the sun in January and furthest in July. Therefore, each year, the solar "constant" increases in January and decreases in July. Also, satellite measurements from the *Earth Radiation Budget Satellite* suggest that the sun's radiant output varies slightly, so the annual average solar constant varies between 1365 and 1372 W m^{-2}.

FOCUS ON A SPECIAL TOPIC

Characteristics of the Sun

The sun is our nearest star. It is some 150 million kilometres from Earth. The next star, Alpha Centauri, is more than 250,000 times farther away. Even though Earth receives only about one two-billionths of the sun's total energy output, this energy allows life to flourish. Sunlight determines the rate of photosynthesis in plants and strongly regulates the amount of evaporation from the oceans. It warms this planet and drives the atmosphere into the dynamic patterns we experience as everyday wind and weather. Without the sun's radiant energy, Earth would gradually cool, in time becoming encased in a layer of ice! Evidence of life on the cold, dark, and barren surface would be found only in fossils. Fortunately, the sun has been shining for billions of years and is likely to shine for at least several billion more.

The sun is a giant celestial furnace. Its core is extremely hot, with a temperature estimated to be near 15 million degrees Celsius. In the core, hydrogen nuclei (protons) collide at such fantastically high speeds that they fuse together to form helium nuclei. This thermonuclear process generates an enormous amount of energy, which gradually works its way to the sun's outer luminous surface—the photosphere ("sphere of light"). Temperatures here are much cooler than in the interior, generally near 5778°C. We have noted already that a body with this surface temperature emits radiation at a maximum rate in the visible region of the spectrum. The sun is, therefore, a shining example of such an object.

Dark blemishes on the photosphere, called sunspots, are huge, cooler regions that typically average more than five times the diameter of Earth. Although sunspots are not well understood, they are known to be regions of strong magnetic fields. They are cyclic, with the maximum number of spots occurring approximately every 11 years.

Above the photosphere are the chromosphere and the corona (see ● Figure 4). The chromosphere ("colour sphere") acts as a boundary between the relatively cool (5778°C) photosphere and the much hotter (2,000,000°C) corona, the outermost envelope of the solar atmosphere. During a solar eclipse, the corona is visible. It appears as a pale, milky cloud encircling the sun. Although much hotter than the photosphere, the corona radiates much less energy because its density is extremely low. This very thin solar atmosphere extends into space for many millions of kilometres.*

Violent solar activity occasionally occurs in the regions of sunspots. The most dramatic of these events are *prominences* and *flares*. Prominences are huge, cloudlike jets of gas that often shoot up into the corona in the form of an arch. Solar flares are tremendous, but brief, eruptions. They emit large quantities of high-energy ultraviolet radiation, as well as energized charged particles, mainly protons and electrons,

---

*During a solar eclipse or at any other time, you should not look at the sun's corona either with sunglasses or through exposed negatives. Take this warning seriously. Viewing just a small area of the sun directly permits large amounts of ultraviolet radiation to enter the eye, causing serious and permanent damage to the retina. View the sun by projecting its image onto a sheet of paper, using a telescope or pinhole camera.

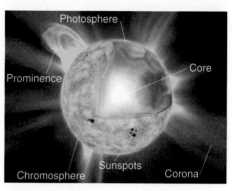

● FIGURE 4 Various regions of the sun.

which stream outward away from the sun at extremely high speeds.

An intense solar flare can disturb Earth's magnetic field, producing a so-called magnetic storm. Because these storms can intensify the electrical properties of the upper atmosphere, they are often responsible for interruptions in radio and satellite communications. They can also induce ground currents that can actually melt the copper windings of transformers that are part of electric power distribution systems. One such storm knocked out electricity throughout Quebec during March 1989, causing revenue losses of hundreds of millions of dollars. And in May 1998, after a period of intense solar activity, a communications satellite failed, causing 45 million pagers to suddenly go dead.

More recently, a sudden burst of radio waves from an energetic flare overwhelmed dozens of radio receivers linked to the Global Positioning System (GPS) satellites, causing a widespread loss of GPS signals in New Mexico and Colorado.

When solar radiation enters the atmosphere, a number of interactions can take place. It can be *absorbed* by particles or gases such as ozone and converted into sensible heat, warming the air. When sunlight strikes small objects, such as air molecules and dust particles, as well as clouds, the radiation can be deflected, or *scattered* in all directions—forward, sideways, and backward (see ● Figure 2.11). The distribution of light in this manner is called **scattering**. Forward scattered light is also called *diffuse light*. When more of the light striking an object bounces backward than forward, the radiation is *reflected*. Finally, solar radiation that passes through the air unimpeded is called *transmitted* radiation. ● Figure 2.12 shows the global distribution of transmitted solar radiation reaching Earth's surface. Notice that the annual solar radiation amount generally decreases with latitude, except that the cloudy equatorial region has less solar radiation than the subtropics.

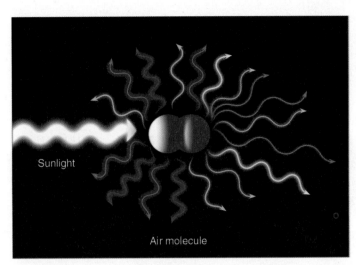

● FIGURE 2.11 The scattering of light by air molecules. Air molecules tend to selectively scatter the shorter (violet, green, and pink) wavelengths of visible wavelength light more effectively than the longer (orange, yellow, and red) wavelengths. This type of scattering is called *Rayleigh scattering.*

SCATTERED AND REFLECTED LIGHT Because air molecules are much smaller than the wavelengths of visible light, they are more effective scatterers of the shorter (blue) wavelengths than the longer (red) wavelengths. Hence, when we look away from the direct beam of sunlight, blue light strikes our eyes from all directions, turning the daytime sky blue. This preferential scattering of the shortest wavelengths by air molecules creating a blue sky is called **Rayleigh scattering**. (More information on the effect of scattered light and what we see is given in Focus on an Observation: Blue Skies, Red Suns, and White Clouds on p. 48.)

Sunlight can be **reflected** from objects. Generally, reflection differs from scattering in that during the process of reflection, more light is sent *backward.* **Albedo** is the percentage of shortwave radiation returning from a given surface compared to the amount of radiation initially striking that surface. Albedo, then, represents the *reflectivity* of the surface to shortwave radiation. In ▼ Table 2.3, notice that thick clouds have a higher albedo than thin clouds. On average, the albedo of clouds is near 60 percent. When solar energy strikes a surface freshly covered with snow, up to 95 percent of the sunlight may be reflected. Most of this energy is in the visible and ultraviolet wavelengths. Consequently, reflected radiation, coupled with direct sunlight, can produce severe sunburns on the exposed skin of unwary snow skiers, and unprotected eyes can suffer the agony of snow blindness.

Water surfaces, on the other hand, reflect only a small amount of solar energy. For an entire day, a smooth water surface will have an average albedo of about 10 percent. Water has the highest albedo (and can therefore reflect sunlight best) when the sun is low on the horizon and the water is a little choppy. This may explain why people who wear brimmed hats while fishing from a boat in choppy water on a sunny day can still get sunburned during midmorning or midafternoon. Averaged for an entire year, Earth and its atmosphere (including its clouds) will redirect about

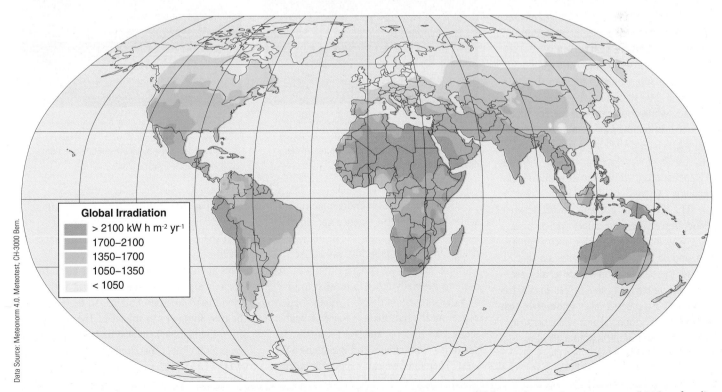

Data Source: Meteonorm 4.0. Meteotest. CH-3000 Bern.

Global Irradiation
- \> 2100 kW h m⁻² yr⁻¹
- 1700–2100
- 1350–1700
- 1050–1350
- < 1050

● FIGURE 2.12 Global distribution of total annual incoming solar radiation. Units are annual kilowatt hours per square metre (kW h m⁻² yr⁻¹).

FOCUS ON AN OBSERVATION

Blue Skies, Red Suns, and White Clouds

We know that the sky is blue because air molecules selectively scatter the shorter wavelengths of visible light—green, violet, and blue waves—more effectively than the longer wavelengths of red, orange, and yellow (see Figure 2.11)—a process called Rayleigh scattering. When these shorter waves reach our eyes, the brain processes them as the colour "blue." Therefore, on a clear day when we look up, blue light strikes our eyes from all directions, making the sky appear blue.

At noon, the sun is perceived as white because all the waves of visible sunlight strike our eyes (see ● Figure 5). At sunrise and sunset, the white light from the sun must travel a longer path through a thick portion of the atmosphere. Scattering of light by air molecules (and particles) removes the shorter waves (blue light) from the beam, leaving the longer waves of red, orange, and yellow to pass through. This situation often creates the image of a ruddy sun at sunrise and sunset. The scattering of the remaining red and orange light can be further enhanced by small particles of smoke or air pollution in a process called **Mie scattering**. An observer at sunrise or sunset in Figure 5 might see a sun similar to the one shown in ● Figure 6. Mie scattering also accounts for the hazy-white appearance of the sky that can occur in polluted air at midday: the aerosols (particles), if they are a range of sizes, will scatter all wavelengths, making the clear sky appear more white than blue and reduce visibility.

The sky is blue, but why are clouds white? Cloud droplets are much larger than air molecules as well as other particles, such as pollution, and do not selectively scatter sunlight. Instead, these larger droplets scatter all wavelengths of visible light more or less equally (see ● Figure 7) in a process called **geometric scattering**. Hence, clouds appear white because millions of cloud droplets scatter all wavelengths of visible light about equally in all directions.

● **FIGURE 5** At noon, the sun usually appears a bright white. At sunrise and at sunset, sunlight must pass through a thick portion of the atmosphere. Much of the blue light is scattered out of the beam, causing the sun to appear more red.

© C. Donald Ahrens

● **FIGURE 6** A red sunset produced by the process of Rayleigh scattering in the upper atmosphere removing the blue and green light, leaving red and oranges at sunset.

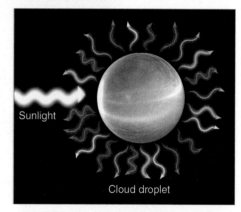

● **FIGURE 7** Mie scattering by cloud droplets scatters all wavelengths of visible white light about equally. This type of scattering by millions of tiny cloud droplets makes clouds appear white.

Together, Rayleigh and geometric scattering provide the science behind the weather lore rhyme "Red sky at night, sailor's delight; red sky at morning, sailor take warning." When we consider that weather systems generally move from west to east and the sun sets in the west, then a red sunset means that there is Rayleigh scattering from clear air and not geometric scattering from clouds and moisture to the west. Therefore, we can expect fair weather tomorrow. The "red sky at morning" part is not quite as straightforward; it implies that there are clear skies to the east. Since weather systems are often spaced a few days apart, clear skies to the east mean that there is a great distance to the east with no clouds and weather systems, so there may be one imminent to the west. The red light from the sunrise may also be lighting incoming clouds that are to the observer's west.

▼ Table 2.3   Typical Albedo of Various Surfaces

SURFACE	ALBEDO (PERCENT)
Fresh snow	75 to 95
Ice	30 to 40
Sand	15 to 45
Grassy field	10 to 30
Dry, ploughed field	5 to 20
Water	10*
Forest	3 to 10
Clouds (thick)	60 to 90
Clouds (thin)	30 to 50
Earth and atmosphere	30
Moon	7
Venus	78
Mars	17
*Daily average.	

30 percent of the sun's incoming radiation back to space, which gives Earth and its atmosphere a combined albedo of 30 percent (look ahead to ● Figure 2.17).

Radiation Absorption, Emission, and Equilibrium

If Earth and all things on it are continually radiating energy, why doesn't everything get progressively colder? The answer is that all objects not only radiate energy, they absorb it as well. There is also the possibility of heat being transferred to or from an object by conduction or convection—this will change its temperature as well. In the absence of convective or conductive heat transfers, if an object radiates more energy than it absorbs, it gets colder; if it absorbs more energy than it emits, it gets warmer. On a sunny day, Earth's surface warms by absorbing more energy from the sun and the atmosphere than it radiates, whereas at night, Earth cools by radiating more energy than it absorbs from its surroundings. When an object emits and absorbs energy at equal rates, its temperature remains constant.

The rate at which something radiates and absorbs energy depends strongly on its surface characteristics, such as colour, texture, and moisture, as well as temperature. For example, a black object in direct sunlight is a good absorber of visible radiation. It converts radiant energy from the sun into sensible heat, and its temperature increases. You need only walk barefoot on a black asphalt road on a summer afternoon to experience this. At night, the blacktop road will cool quickly by emitting infrared radiation, and by early morning, it may be cooler than surrounding surfaces.

© C. Donald Ahrens

● FIGURE 2.13   The melting of snow outward from the trees causes small depressions to form. The melting is caused mainly by the snow's absorption of the infrared energy being emitted from the warmer tree and its branches. The trees are warmer because they are better absorbers of sunlight than is the snow.

Any object that is a perfect absorber (i.e., absorbs all the radiation that strikes it) and a perfect emitter (i.e., emits the maximum radiation possible at its given temperature) is called a **blackbody**. Blackbodies do not have to be coloured black; they simply must absorb and emit all possible radiation at the wavelengths that correspond to their temperature. Since Earth's surface and the sun absorb and radiate with nearly 100 percent efficiency for their respective temperatures, they both behave as blackbodies. This is the reason we were able to use Wien's law and the Stefan–Boltzmann law to determine the characteristics of radiation emitted from the sun and Earth. It is important to note, however, that the term *blackbody* applies only to particular wavelengths of radiation. Therefore, since Earth's surface behaves as a blackbody for emission of longwave radiation (it is nearly a perfect emitter of longwave), this means that it is also nearly a perfect absorber of longwave. However, being a blackbody for longwave radiation does not say anything about how well it absorbs shortwave radiation. We know that due to its albedo, Earth's surface does not absorb all the shortwave radiation it receives but reflects a portion of it.

Objects that do not emit and absorb the maximum possible radiation for their temperature are called *greybodies* because they emit and absorb only a fraction of the radiation of a blackbody. The fraction of radiation that a greybody emits or absorbs compared to a blackbody is called its **emissivity**. By definition, a blackbody has an emissivity of 1.0, and a greybody has a value less than 1.0. The Stefan–Boltzmann law we saw previously can be modified to account for the reduced emission from a greybody by multiplying it by the object's emissivity:

$$E = \epsilon \sigma T^4 \text{ (Stefan–Boltzmann law for a greybody)}$$

where ϵ (the Greek letter epsilon) is the object's emissivity.

When we look at Earth from space, we see that half of it is in sunlight and the other half is in darkness. The outpouring of solar energy constantly bathes Earth with radiation, whereas Earth and its atmosphere, in turn, constantly emit infrared radiation out to space. Since Earth is surrounded by the near vacuum of space, radiation is the only effective way that energy can enter or leave the system. If Earth's average temperature is not changing much, this means that solar energy absorbed by Earth and its atmosphere must equal infrared energy leaving Earth and its atmosphere. In other words, on average, a state of *radiative equilibrium* is achieved. The infrared radiation leaving Earth required to balance the solar radiation received by Earth can be estimated using the Stefan–Boltzmann law ($E = \sigma T_{re}^4$). The average temperature at which this balance occurs (T_{re}) is called the **radiative equilibrium temperature**. At this temperature, Earth is emitting infrared radiation at the same rate it is absorbing solar radiation, so its average temperature does not change. Because Earth is about 150 million kilometres from the sun and absorbs about 70 percent of the solar radiation it receives, Earth's *radiative equilibrium temperature* is about 255 K ($-18°C$). But this temperature is *much* lower than Earth's observed average surface temperature of 288 K ($15°C$). Why is there such a large difference?

The answer lies in the fact that *Earth's atmosphere absorbs and emits infrared radiation*. Unlike Earth's surface, the atmosphere does not behave like a blackbody as it absorbs some wavelengths of radiation and is transparent to others. Therefore, only a portion of the infrared radiation leaving Earth's surface reaches space—much of it is absorbed by the atmosphere. Objects that selectively absorb and emit radiation, such as gases in our atmosphere, are known as **selective absorbers**. Let's examine this concept more closely.

SELECTIVE ABSORBERS AND THE ATMOSPHERIC GREENHOUSE EFFECT

Most substances in our environment are selective absorbers; that is, they absorb only certain wavelengths of radiation. Glass is a good example of a selective absorber in that it absorbs some of the infrared and ultraviolet radiation it receives but not the visible radiation that is transmitted through the glass. As a result, it is difficult to get a sunburn through the windshield of your car, although you can see through it.

Objects that selectively absorb radiation also selectively emit radiation at the same wavelength. This phenomenon is called **Kirchhoff's law**. This law states that *good absorbers are good emitters at a particular wavelength, and poor absorbers are poor emitters at the same wavelength.*

• Figure 2.14 shows some of the most important selectively absorbing gases in our atmosphere. The shaded area represents the absorption characteristics of each gas at various wavelengths. Notice that both water vapour (H_2O) and carbon dioxide (CO_2) are strong absorbers of infrared radiation and poor absorbers of visible solar radiation. Other important selective absorbers include nitrous oxide (N_2O), methane (CH_4), and ozone (O_3), which is most abundant in the strato-

*Strictly speaking, this law applies only to gases.

Snow is a good absorber as well as a good emitter of infrared energy (white snow actually behaves as a blackbody in the infrared wavelengths). The bark of a tree absorbs sunlight and emits infrared energy, which the snow around it absorbs. During the absorption process, the infrared radiation is converted into sensible and latent heat, and the snow melts outward away from the tree trunk, producing a small depression that encircles the tree, like the ones shown in Figure 2.13.

sphere. As these gases absorb infrared radiation emitted from Earth's surface, they gain sensible heat (kinetic energy of gas molecules increase). The gas molecules share this energy by colliding with neighbouring air molecules, such as oxygen and nitrogen (both of which are poor absorbers of infrared energy). These collisions increase the average kinetic energy of the air, which results in an increase in air temperature. Thus, most of the infrared energy emitted from Earth's surface keeps the lower atmosphere warm, increasing its sensible heat.

Besides being selective absorbers, water vapour and CO_2 selectively emit radiation at infrared wavelengths.* This radiation travels away from these gases in all directions. A portion of this energy is radiated toward Earth's surface and absorbed, thus heating the ground. Earth's surface, in turn, constantly radiates infrared energy upward, where it is absorbed and warms the lower atmosphere. In this way, water vapour and CO_2 absorb and radiate infrared energy and act as an insulating layer around Earth, keeping part of Earth's infrared radiation from escaping rapidly into space. Consequently, Earth's surface and the lower atmosphere are much warmer than they would be if these selectively absorbing gases were not present. In fact, as we saw earlier, Earth's mean radiative equilibrium temperature without CO_2 and water vapour would be around $-18°C$, or about $33°C$ lower than at present.

The absorption characteristics of water vapour, CO_2, and other gases such as methane and nitrous oxide (depicted in Figure 2.14) were, at one time, thought to be similar to the glass of a greenhouse. In a greenhouse, the glass allows visible radiation to come in but inhibits to some degree the passage of outgoing infrared radiation. For this reason, the absorption of infrared radiation from Earth by water vapour and CO_2 is popularly called the **greenhouse effect**. However, studies have shown that the warm air inside a greenhouse is caused more by the air's inability to circulate and mix with the cooler outside air rather than by the entrapment of infrared energy. Because of these findings, some scientists suggest that the greenhouse effect should be called the *atmosphere effect*. To accommodate everyone, we will usually use the term *atmospheric greenhouse effect* when describing the role that water vapour, CO_2, and other

*Nitrous oxide, methane, and ozone also emit infrared radiation, but their concentration in the atmosphere is much smaller than water vapour and carbon dioxide (see Table 1.1, p. 6).

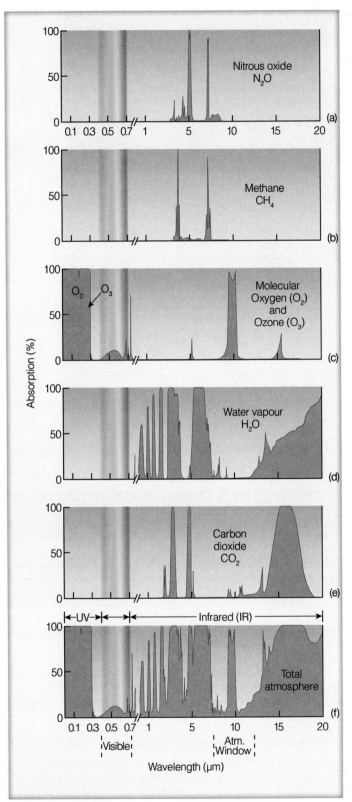

greenhouse gases* play in keeping Earth's mean surface temperature higher than it otherwise would be.

Look again at Figure 2.14 and observe that, in the bottom diagram, there is a region between about 8 and 11 µm where neither water vapour nor CO2 readily absorbs infrared radiation. Because these wavelengths of emitted energy pass upward through the atmosphere and out into space, the wavelength range (between 8 and 11 µm) is known as the **atmospheric window**.

Clouds can enhance the atmospheric greenhouse effect. Tiny liquid cloud droplets are selective absorbers in that they are good absorbers of infrared radiation but poor absorbers of visible solar radiation. Clouds even absorb the wavelengths between 8 and 11 µm, which are otherwise "passed up" by water vapour and CO_2. Thus, they have the effect of enhancing the atmospheric greenhouse effect by closing the atmospheric window.

Clouds—especially low, thick ones—are excellent emitters of infrared radiation. Their tops radiate infrared energy upward, and their bases radiate energy back to Earth's surface, where it is absorbed. This process keeps calm, cloudy nights warmer than calm, clear ones. If the clouds remain into the next day, they prevent much of the sunlight from reaching the ground by reflecting it back to space. Since the ground does not heat up as much as it would in full sunshine, cloudy, calm days are normally cooler than clear, calm days. Hence, the presence of clouds tends to keep nighttime temperatures higher and daytime temperatures lower.

In summary, the atmospheric greenhouse effect occurs because water vapour, CO_2, and other greenhouse gases are selective absorbers. They allow most of the sun's visible radiation to reach the surface, but they absorb a good portion of Earth's outgoing infrared radiation, preventing it from escaping into space (see ● Figure 2.15). It is the atmospheric greenhouse effect, then, that keeps the temperature of our planet at a level where life can survive. The greenhouse effect is not just a "good thing"; it is essential to life on earth.

ENHANCEMENT OF THE GREENHOUSE EFFECT In spite of the inaccuracies that have plagued temperature measurements in the past, studies suggest that, during the past century, Earth's

ACTIVE FIGURE 2.14 Absorption of radiation by gases in the atmosphere. The shaded area represents the percentage of radiation absorbed by each gas. The strongest absorbers of infrared radiation are water vapour and carbon dioxide. The bottom figure represents the percentage of radiation absorbed by all the atmospheric gases. Visit the textbook's website to view this and other Active Figures at www.ahrensmeteorology1ce.nelson.com.

---

*The term "greenhouse gases" derives from the standard use of "greenhouse effect." Greenhouse gases include, among others, water vapour, carbon dioxide, methane, nitrous oxide, and ozone.

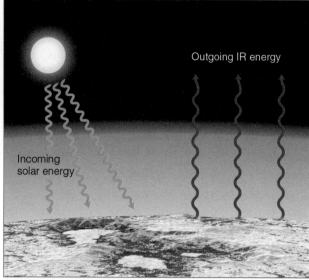

(a) Without greenhouse gases

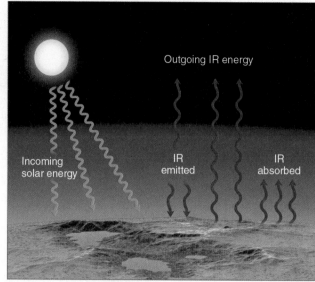

(b) With greenhouse gases

● FIGURE 2.15 (a) Near the surface in an atmosphere with little or no greenhouse gases, Earth's surface would constantly emit infrared (IR) radiation upward, both during the day and at night. Averaged over a year, incoming energy from the sun would equal outgoing energy from the surface, but the surface would receive virtually no IR radiation from its lower atmosphere (no atmospheric greenhouse effect). Earth's surface air temperature would be quite low, and most water found on the planet would be in the form of ice. (b) In an atmosphere with greenhouse gases, Earth's surface not only receives energy from the sun but also IR energy from the atmosphere. Annual average incoming energy still equals outgoing energy, but the added IR energy from the greenhouse gases raises Earth's average surface temperature to a more habitable level.

surface air temperature has undergone a warming of about 0.74°C. In recent years, this *global warming* trend has not only continued but has also increased. In fact, climate models that mathematically simulate the physical processes of the atmosphere, oceans, and ice predict that if such a warming should continue unabated, global annual average temperatures would increase an additional 1.8 to 4°C by the end of the 21st century depending on CO_2 emissions. We would be irrevocably committed to the effects of climate change, such as a continuing rise in sea level and a shift in global precipitation patterns.

The main cause of this global warming is the greenhouse gas CO_2, whose concentration has been increasing primarily due to the burning of fossil fuels and to deforestation. (Look back at Figure 1.5 and Figure 1.6 on p. 10.) However, increasing concentrations of other greenhouse gases, such as methane (CH_4), nitrous oxide (N_2O), and chlorofluorocarbons (CFCs), have collectively been shown to have an effect almost equal to that of CO_2. Look at Figure 2.14 and notice that both CH_4 and N_2O absorb strongly at infrared wavelengths. Moreover, a particular CFC (CFC-12) absorbs in the region of the atmospheric window between 8 and 11 μm. Thus, in terms of its absorption impact on infrared radiation, the addition of a single CFC-12 molecule to the atmosphere is the equivalent of adding 10,000 molecules of CO_2. Overall, water vapour accounts for about 60 percent of the atmospheric greenhouse effect, CO_2 accounts for about 26 percent, and the remaining greenhouse gases contribute about 14 percent. The increasing concentrations of greenhouse gases affect

Earth's climate system by a series of interactions that are discussed more fully in Chapter 16.

WEATHER WATCH

Talk about an enhanced greenhouse effect! The atmosphere of Venus, which is mostly carbon dioxide, is considerably more dense than that of Earth. Consequently, the greenhouse effect on Venus is exceptionally strong, producing a surface air temperature of about 500°C.

BRIEF REVIEW

In the last several sections, we have explored examples of some of the ways radiation is absorbed and emitted by various objects. Before reading the next several sections, let's review a few important facts and principles:

- *All* objects with a temperature above absolute zero emit radiation.
- The higher an object's temperature, the greater the amount of radiation emitted per unit surface area and the shorter the wavelength of maximum emission.
- Earth absorbs solar radiation only during the daylight hours; however, it emits infrared radiation continuously, both during the day and at night.
- Earth's surface behaves as a blackbody for longwave radiation, making it a much better absorber and emitter of longwave radiation than the atmosphere.

● Water vapour and carbon dioxide are important atmospheric greenhouse gases that selectively absorb and emit infrared radiation, thereby keeping Earth's average surface temperature warmer than it otherwise would be.

● Cloudy, calm nights are often warmer than clear, calm nights because clouds strongly emit infrared radiation back to Earth's surface.

● It is *not* the greenhouse effect itself that is of concern but the *enhancement* of it due to increasing levels of greenhouse gases.

● As greenhouse gases continue to increase in concentration, the average surface air temperature is projected to rise substantially by the end of this century.

With these concepts in mind, we will first examine how the air near the ground warms; then we will consider how Earth and its atmosphere maintain a yearly energy balance and examine how energy flows affect temperature and moisture near Earth's surface on a daily basis.

Annual and Daily Energy Balances

WARMING THE AIR FROM BELOW   If you look back at Figure 2.14 (p. 51), you will notice that the atmosphere does not readily absorb radiation with wavelengths between 0.3 and 1.0 μm, the region where the sun emits most of its energy. Consequently, on a clear day, solar energy passes through the lower atmosphere with little effect on the air. Ultimately, it reaches the surface, warming it (see ● Figure 2.16). Air molecules in contact with the heated surface bounce against it, gain energy by *conduction*, and then shoot upward like freshly popped kernels of corn, carrying their energy with them. Because the air near the ground is very dense, these molecules travel only a short distance (about 10^{-7} m) before they collide with other molecules. During the collision, these more rapidly moving molecules share their energy with less energetic molecules, raising the average temperature of the air. But air is such a poor heat conductor that this process is important only within a few millimetres of the ground.

As the surface air warms, it actually becomes less dense than the air directly above it. The warmer air rises and the cooler air sinks, setting up thermals, or *free convection cells*, that transfer heat upward and distribute it through a deeper layer of air. The rising air expands and cools, and if sufficiently moist, the water vapour condenses into cloud droplets, releasing latent heat that warms the air. Meanwhile, Earth's surface constantly emits infrared energy. Some of this energy is absorbed by greenhouse gases (such as water vapour and carbon dioxide) that emit infrared energy upward and downward, back to the surface. Since the concentration of water vapour decreases rapidly above Earth, most of the

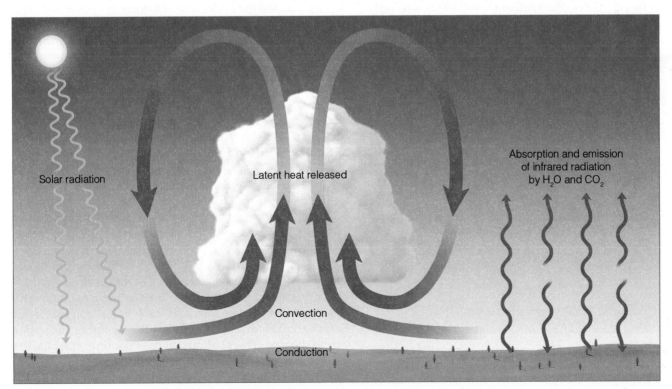

● FIGURE 2.16 Air in the lower atmosphere is heated from the ground upward. Sunlight warms the ground, and the air above is warmed by conduction, convection, and infrared radiation. Further warming occurs during condensation as latent heat is given up to the air inside the cloud.

absorption occurs in a layer near the surface. Hence, the lower atmosphere is mainly heated from the ground upward.

EARTH'S ANNUAL ENERGY BALANCE Although the average temperature at any one place may vary considerably from year to year, Earth's overall average equilibrium temperature changes only slightly from one year to the next. This fact indicates that each year, Earth and its atmosphere combined must send off into space just as much energy as they receive from the sun. The same type of energy balance must exist between Earth's surface and the atmosphere. That is, each year, Earth's surface must return to the atmosphere the same amount of energy that it absorbs. If this did not occur, Earth's average surface temperature would change. How do Earth and its atmosphere maintain this yearly energy balance?

Suppose 100 units of solar energy reach the top of Earth's atmosphere. We can see in ● Figure 2.17 that, on average, clouds, Earth, and the atmosphere reflect and scatter 30 units back to space (the symbol $K\uparrow$ represents the reflected solar radiation from the surface) and that the atmosphere and clouds together absorb 19 units, which leaves 51 units of direct and diffuse solar radiation to be absorbed at Earth's surface (we will use the symbol K^* to represent this net flux of solar radiation absorbed by the surface, whereas the symbol $K\downarrow$ represents the solar radiation that strikes the surface before a portion of it is reflected, so $K^* = K\downarrow - K\uparrow$). ● Figure 2.18 shows approximately what happens to the solar radiation that is absorbed by the surface and the atmosphere. Of 51 units absorbed by the surface, a large amount (23 units) is used to evaporate water (the symbol Q_E will be used to represent this latent heat flux), and about 7 units are lost to the

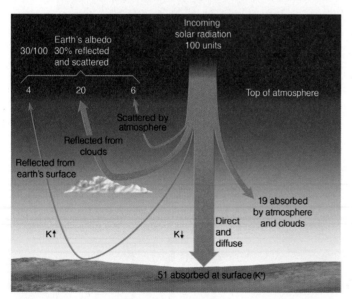

● **FIGURE 2.17** On average, of all the solar energy that reaches Earth's atmosphere annually, about 30 percent (30/100) is reflected and scattered back to space, giving Earth and its atmosphere an albedo of 30 percent. Of the remaining solar energy, about 19 percent is absorbed by the atmosphere and clouds, and 51 percent is absorbed at the surface.

atmosphere through conduction and convection as sensible heat (Q_H represents the sensible heat flux), which leaves a net amount of 21 units to be radiated away as infrared energy (L^* is used to represent the net longwave flux of radiation). Look closely at Figure 2.18 and notice that Earth's surface actually radiates upward a whopping 117 units ($L\uparrow$ represents the outgoing longwave flux). It does so because although it receives solar radiation only during the day, it constantly emits infrared energy both during the day and at night. Additionally, the atmosphere above allows only a small fraction of this energy (6 units) to pass through into space. The majority of it (111 units) is absorbed mainly by the greenhouse gases water vapour and CO_2 and by clouds. Much of this energy (96 units—given the symbol $L\downarrow$) is radiated back to earth, producing the atmospheric greenhouse effect. Hence, Earth's surface receives nearly twice as much longwave infrared energy from its atmosphere as it does shortwave radiation from the sun. In all these exchanges, notice that the energy lost at Earth's surface (147 units) is exactly balanced by the energy gained there (147 units).

A similar balance exists between Earth's surface and its atmosphere. Again in Figure 2.18, observe that the energy gained by the atmosphere (160 units) balances the energy lost. Moreover, averaged for an entire year, the solar energy received at Earth's surface (51 units) and that absorbed by Earth's atmosphere (19 units) balances the infrared energy lost to space by Earth's surface (6 units) and its atmosphere (64 units). This illustrates that the Earth and atmosphere together are in a state of radiative equilibrium.

We can see the effect that conduction, convection, and latent heat play in the warming of the atmosphere by looking at the net radiation of the surface and atmosphere. Earth's surface receives 147 units of radiant energy from the sun and its own atmosphere, whereas it radiates away 117 units, producing a *surplus* of 30 units. The atmosphere, on the other hand, receives 130 units (19 units from the sun and 111 from Earth), whereas it loses 160 units, producing a *deficit* of 30 units. The balance (30 units) is the warming of the atmosphere produced by the heat transfer processes of conduction and convection of sensible heat ($Q_H = 7$ units) and latent heat ($Q_E = 23$ units) between the surface and the atmosphere. These convective transfers of sensible (Q_H) and latent heat (Q_E) between the surface and the atmosphere largely characterize the climate at a location because they cause warming and cooling of the air (Q_H), as well as changes in its humidity (Q_E).

So Earth and the atmosphere absorb energy from the sun, as well as from each other. In all of the energy exchanges, a delicate balance is maintained. Essentially, there is no yearly gain or loss of total energy, and the average temperature of Earth and the atmosphere remains fairly constant from one year to the next. This equilibrium does not imply that Earth's average temperature does not change but that the changes are small from year to year (usually less than one-tenth of a degree Celsius) and become significant only when measured over many years.

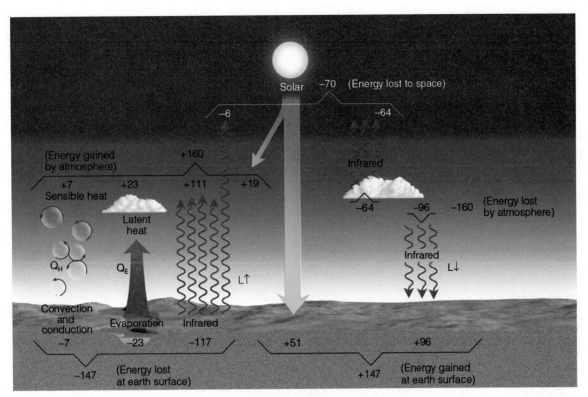

● FIGURE 2.18
Earth–atmosphere
energy balance. Num-
bers represent annual
average approximations
for the whole Earth
based on surface obser-
vations and satellite
data. Although the
actual value of each pro-
cess may vary by several
percent, it is the relative
size of the numbers that
is important.

Even though Earth and the atmosphere together main-tain an annual energy balance, such a balance is not main-tained at each latitude. High latitudes tend to lose more radiant energy to space each year than they receive from the sun, whereas low latitudes tend to gain more radiant energy during the course of a year than they lose. From ● Figure 2.19, we can see that only at middle latitudes near 38° does the amount of radiant energy received each year bal-ance the amount lost. From this situation, we might conclude that polar regions are growing colder each year, whereas trop-ical regions are becoming warmer. But this does not happen. To compensate for these gains and losses of energy, winds in the atmosphere and currents in the oceans circulate warm air and water toward the poles and cold air and water toward the equator. Thus, the transfer of sensible heat by atmospheric and oceanic circulations (a process called *advection*) prevents low latitudes from steadily becoming warmer and high lati-tudes from steadily growing colder. These circulations are key roles of the atmosphere and ocean from an energy perspec-tive: virtually all weather systems and phenomena exist to create a net poleward transport of heat to move the energy surplus at low latitudes in attempts to balance the energy deficit at high latitudes. This process will be treated more completely in Chapter 10.

DAILY ENERGY BALANCE AT EARTH'S SURFACE   Whereas global annual radiation balances give the "big picture" of Earth's climate, daily radiation and energy budgets determine the

climate at each location. For an example of this in Montreal, Quebec, see Focus on a Special Topic: Daily Radiation and Energy Budgets at Earth's Surface. These budgets of radiation and energy at Earth's surface largely determine how air tem-perature and humidity change over the course of a day, and their variation over a year results in the annual climate of a location.

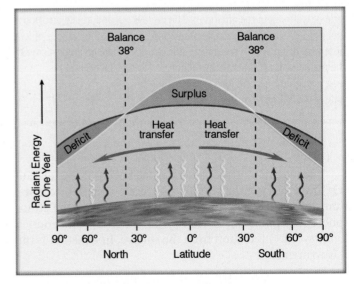

● FIGURE 2.19 The average annual incoming solar radiation (yellow lines) absorbed by Earth and the atmosphere along with the average annual infrared radiation (red lines) emitted by Earth and the atmosphere to space.

FOCUS ON A SPECIAL TOPIC

Daily Radiation and Energy Budgets at Earth's Surface

Dr. James Voogt, Department of Geography, University of Western Ontario.

The radiation and energy budgets describe the energy status for a surface and help determine the climate over a surface. • Figure 8 shows the radiation balance for an urban residential area of Montreal on two clear days, one in winter and the other in summer. In winter, incoming shortwave radiation (K↓) is much less than in summer because the sun is lower in the sky. The daily pattern of K↓ is very symmetrical under these clear sky conditions. Snow cover on the roofs and yards in the neighbourhood in winter increases the surface albedo (45%) compared to that in the summer (12%). The winter albedo of urban areas remains lower than that of natural surfaces due to snow removal and the presence of snow-free vertical surfaces in the city that absorb K↓. Incoming longwave radiation (L↓) is less on the winter day because of the colder air. It increases slightly in the late afternoon on both days as air temperature near Earth's surface rises. The more abrupt increase of L↓ in the late evening of both days is because clouds were present at these times, providing a stronger source of longwave radiation. Outgoing longwave radiation (L↑) exceeds L↓ because Earth's surface is warmer than the atmosphere and has a daily cycle of warming in the afternoon. The cycle is stronger in the summer because of the larger input and absorption of shortwave radiation. The net radiation (Q*) is positive during most daylight hours and follows the daily cycle of K↓. At night, in the early morning,

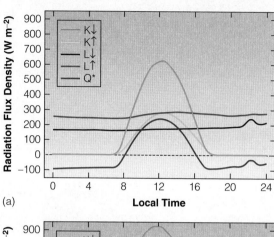

(a)

(b)

• FIGURE 8 Radiation budget for an urban residential area of Montreal. Values observed on (a) February 16, 2008, and (b) July 2, 2008. Measurements made by a CNR1 radiometer mounted 25 m above the surface as part of the Environmental Prediction in Canadian Cities Network.

Data provided by the Environmental Prediction in Canadian Cities (EPiCC) network funded by the Canadian Foundation for Climate and Atmospheric Sciences (CFCAS). Thanks to Onil Bergeron and Ian Strachan of McGill University.

and in late afternoon, Q* is negative because of the net loss of longwave radiation (L*). The energy budget over this urban neighbourhood (• Figure 9) is

$$Q^* + Q_F = Q_H + Q_E + \Delta Q_S$$

It describes both how surplus radiation (positive Q*) at the surface is dissipated and how a radiation deficit (negative Q*) is supplied through heat transfer with the atmosphere by convection, either as sensible heat (Q_H) or latent heat (Q_E), and by the net heat stored via

The radiation budget describes the net input of radiation (Q*) to a surface: the sum of shortwave radiation received from the sun (K↓) and longwave radiation from the atmosphere (L↓) minus the reflected shortwave (K↑) and emitted longwave (L↑) radiation from the surface. In equation form, it is written as

$$Q^* = K\downarrow - K\uparrow + L\downarrow - L\uparrow \quad \text{or}$$

$$Q^* = K^* + L^*$$

where K* is the net shortwave radiation (K* = K↓ – K↑) and L* is the net longwave radiation (L* = L↓ – L↑). The surface is

the interface that separates atmosphere from ground. It does not have mass and cannot store energy. Positive values of Q* indicate that radiation is resulting in a surplus of energy at the surface, which must be transferred away from the surface down into the ground and/or up into the atmosphere. Negative values of Q* indicate a deficit of energy at the surface, which is compensated by a flow of energy from some combination of the ground and atmosphere toward the surface.

The energy budget distributes the net radiation (Q*). It states that Q* is the sum of conductive and convective heating between the surface and the air called the sensible heat flux (Q_H), latent heat transfer between surface and air called the

● FIGURE 9 Energy budget terms from the same Montreal observing tower used for the radiation budget. (a) February 16, 2008; (b) July 2, 2008. The net radiation (Q^*), sensible heat (Q_H), and latent heat (Q_E) are directly measured; the anthropogenic heat term (Q_F) is estimated from energy consumption data, and the residual term, ΔQ_S, represents the net storage and release of heat by the volume of urban surfaces below the tower measurement height.

Data provided by the Environmental Prediction in Canadian Cities (EPiCC) network funded by the Canadian Foundation for Climate and Atmospheric Sciences (CFCAS). Thanks to Onil Bergeron and Ian Strachan of McGill University.

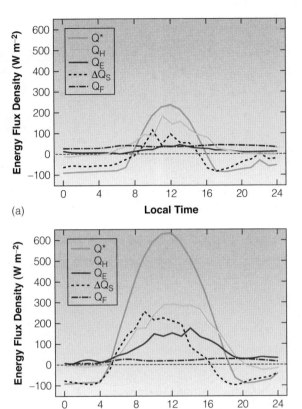

(a)

(b)

conduction (ΔQ_S) in the volume of surface and subsurface materials (note that ΔQ_S is equivalent to Q_G). Positive values of Q_H, Q_E, and ΔQ_S represent an energy flow away from the surface, whereas negative values represent an energy flow toward the surface. The Q^* is augmented in urban areas by an input of heat from human activities such as space heating, electricity use, and transportation, especially where energy use is high (Q_F).

In the winter, Q_H is the largest component of the energy budget during the day, whereas at night, the loss of longwave radiation is mostly made up with heat released from storage in the ground (ΔQ_S). The latent heat flux (Q_E) is small because plants are dormant, temperatures are low and frozen, and snow-covered ground does not have much water available to evaporate. In summer, Q_E is much larger but remains smaller than Q_H through the day. The ratio between sensible and latent heat (called the **Bowen ratio**, $\beta = Q_H/Q_E$) is determined by the amount of vegetation and whether or not water for evaporation is available from natural or human (e.g., irrigation) sources. In urban areas, the Bowen ratio may range from 0.5 to 4 depending on the amount of urbanization and moisture availability. A large amount of heat goes into storage (positive ΔQ_S) in the morning, but this pattern reverses by early afternoon, and by the late afternoon, release of heat from storage helps sustain positive fluxes of Q_H and Q_E from the surface as Q^* decreases. The summer night-time situation suggests that heat released from storage (negative ΔQ_S) and a heat flux from the atmosphere to the surface (negative Q_H) are helping to sustain a small positive evapotranspiration (Q_E) from the surface. Urban areas often have small but positive nighttime convective heat fluxes due to the relatively warm urban surface below and the input of heat and moisture from human energy use (Q_F), whereas more open rural areas often have fluxes of heat and moisture toward the surface at night.

latent heat flux (Q_E), and conduction of sensible heat into or out of the ground (Q_G). In equation form, the energy budget is

$$Q^* = Q_H + Q_E + Q_G$$

The radiation and energy budgets describe the processes that determine the climate near Earth's surface for situations where advection* is not important. They tell us how much energy and moisture are exchanged between Earth and atmosphere due to the radiation that drive these sensible and latent heat fluxes. ▼ Table 2.4 summarizes the variables in these two important relationships.

To understand how the radiation and energy budgets work, imagine what happens at Earth's surface on a typical sunny day. Just before sunrise, there is no solar radiation ($K^* = 0$), so the radiation budget states that net radiation (Q^*) is equal to the net longwave radiation ($Q^* = L^* = L\downarrow - L\uparrow$). The surface is losing longwave radiation depending on its temperature ($L\uparrow = \sigma T_{sfc}^4$ using the Stefan–Boltzmann law) and is getting longwave radiation from the sky, depending on the temperature of the gases and clouds in the atmosphere. The net longwave radiation is normally negative ($L^* = (L\downarrow - L\uparrow) < 0$),

*Advection is the horizontal transport of heat and moisture by wind and weather systems.

Solar Particles and the Aurora

From the sun and its tenuous atmosphere comes a continuous discharge of particles. This discharge happens because at extremely high temperatures, gases become stripped of electrons by violent collisions and acquire enough speed to escape the sun's gravitational pull. As these charged particles (ions and electrons) travel through space, they are known as *plasma*, or **solar wind**. When the solar wind moves close enough to Earth, it interacts with Earth's magnetic field.

The magnetic field that surrounds Earth is much like the field around an ordinary bar magnet (see ● Figure 10). Both have north and south magnetic poles, and both have invisible lines of force (field lines) that link the poles. On Earth, these field lines form closed loops as they enter near the magnetic North Pole and leave near the magnetic South Pole. Most scientists believe that an electric current coupled with fluid motions deep in Earth's hot molten core is responsible for its magnetic field. This field protects Earth, to some degree, from the onslaught of the solar wind.

Observe in ● Figure 11 that when the solar wind encounters Earth's magnetic field, it severely deforms it into a teardrop-shaped cavity known as the *magnetosphere*. On the side facing the sun, the pressure of the solar wind compresses the field lines. On the opposite side, the *magnetosphere* stretches out into a long tail—the *magnetotail*—which reaches far beyond the moon's orbit. In a way, the magnetosphere acts as an obstacle to the solar wind by causing some of its particles to flow around Earth. Inside Earth's magnetosphere are ionized gases. Some of these gases are solar wind particles, whereas others are ions from Earth's upper atmosphere that have moved upward along electric field lines into the magnetosphere.

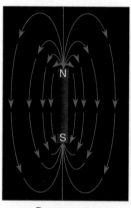 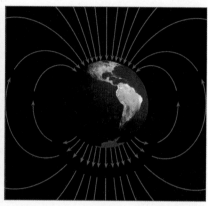

Bar magnet   Earth

● FIGURE 10 A magnetic field surrounds Earth just as it does a bar magnet.

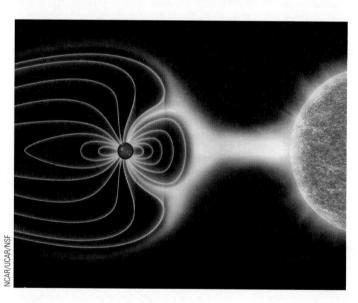

NCAR/UCAR/NSF

● FIGURE 11 The stream of charged particles from the sun—called the *solar wind*—distorts Earth's magnetic field into a teardrop shape known as the *magnetosphere*.

Normally, the solar wind approaches Earth at an average speed of 400 km s^{-1}. However, during periods of high solar activity (many sunspots and flares), the solar wind is denser, travels much faster, and carries more energy.

When these energized solar particles reach Earth, they cause a variety of effects, such as changing the shape of the magnetosphere and disturbing the magnetosphere to produce auroral displays. The disturbance involves

representing a net loss of longwave radiation because the sky is normally colder than Earth's surface, so the net radiation is also less than zero ($Q^* = L^* < 0$). This means that Earth's surface is in a deficit of energy from radiation. The deficit in radiant energy is balanced by transfers of sensible heat from the atmosphere (Q_H) and ground (Q_G) and latent energy (Q_E) from the atmosphere according to the energy budget ($Q^* = Q_H + Q_E + Q_G$). This means that some combination of the air cooling ($Q_H < 0$), or ground cooling ($Q_G < 0$) or condensation ($Q_E < 0$) is providing the energy to balance the net loss of radiation ($Q^* < 0$) (see ● Figure 2.20a). When the sun rises, the net longwave radiation stays about the same, but the amount of incoming solar radiation ($K\downarrow$) increases. Due to the surface albedo, a percentage of this radiation is reflected back to the atmosphere ($K\uparrow$), but much of it remains to be absorbed by the surface, resulting in a net surplus of high-energy particles within the magnetosphere being ejected into Earth's upper atmosphere, where they excite atoms and molecules. The excited atmospheric gases emit visible radiation, which causes the sky to glow like a neon light. Let's examine this process more closely.

A high-energy particle from the magnetosphere will, upon colliding with an air molecule (or atom), transfer some of its energy to the molecule. The molecule then becomes excited, and its electrons jump into a higher energy level as they orbit its centre. When molecules de-excite, they release the energy originally received from the energetic particle, either all at once (one big jump) or in steps (several smaller jumps). This emitted energy is given up as radiation. If its wavelength is in the visible range, we see it as visible light. In the Northern Hemisphere, we call this light show the **aurora borealis**, or *northern lights*; its counterpart in the Southern Hemisphere is the **aurora australis**, or *southern lights*.

Since each atmospheric gas has its own set of energy levels, each gas has its own characteristic colour. For example, the de-excitation of atomic oxygen can emit green or red light. Molecular nitrogen gives off red and violet light. The shades of these colours can be spectacular as they brighten and fade, sometimes in the form of waving draperies, sometimes as unmoving, yet flickering, arcs and soft coronas. On a clear night, the aurora is an eerie yet beautiful spectacle. (See the chapter opening photograph on p. 32.)

The aurora is most frequently seen in polar latitudes. Energetic particles trapped in the magnetosphere move along Earth's magnetic field lines. Because these lines emerge from Earth near the magnetic poles, it is here that the particles interact with atmospheric gases to produce an

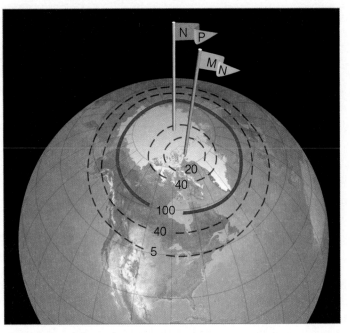

● FIGURE 12 The aurora belt (solid red line) represents the region where you would most likely observe the aurora on a clear night. (The numbers represent the average number of nights per year on which you might see an aurora if the sky were clear.) The flag MN denotes the magnetic North Pole, where Earth's magnetic field lines emerge from Earth. The flag NP denotes the geographic North Pole, about which Earth rotates.

aurora. Notice in ● Figure 12 that the zone of most frequent auroral sightings (aurora belt) is not at the magnetic pole (marked by the flag MN) but equatorward of it, where the field lines emerge from Earth's surface. At lower latitudes, where the field lines are oriented almost parallel to Earth's surface, the chances of seeing an aurora diminish rapidly. In Canada, the aurora belt passes close to Dawson City, Yukon; Yellowknife, Northwest Territories; Churchill, Manitoba; Kuujjuaq, Quebec; and Nain, Labrador.

How high above Earth is the aurora? The exact height appears to vary, but it is almost always observed within the thermosphere. The base of an aurora is rarely lower than 80 km, and it averages about 105 km. Since the light of an aurora gradually fades, it is difficult to define

an exact upper limit. Most auroras, however, are observed below 200 km.

But there is other light coming from the atmosphere—a faint glow at night much weaker than the aurora. This feeble luminescence, called **airglow**, is detected at all latitudes and shows no correlation with solar wind activity. Apparently, this light comes from ionized oxygen and nitrogen and other gases that have been excited by solar radiation.

In summary, energy for the aurora comes from the solar wind, which disturbs Earth's magnetosphere. This disturbance causes energetic particles to enter the upper atmosphere, where they collide with atoms and molecules. The atmospheric gases become excited and emit energy in the form of visible light.

solar radiation ($K^* = K{\downarrow} - K{\uparrow} > 0$). The net radiation is now made up of both the net longwave radiation and the net solar radiation ($Q^* = K^* + L^*$). Shortly after sunrise, usually K^* becomes a large enough positive number to exceed the negative L^*, so the net radiation, Q^*, is positive. Once this happens, there is now a surplus of energy from radiation at the surface ($Q^* > 0$). This excess energy is transferred away from the surface as some combination of heating the ground

($Q_G > 0$), warming the air ($Q_H > 0$), or evaporating water into the air ($Q_E > 0$) according to the energy budget ($Q^* = Q_H + Q_E + Q_G$) (see Figure 2.20b).

On a cloudy day, a similar situation would arise, except the reduced incoming solar radiation ($K{\downarrow}$) due to the clouds would decrease the net radiation (Q^*) available to drive the energy budget during the day. This means that the latent (Q_E) and sensible (Q_H) heat fluxes would be less, so there would be

▼ Table 2.4   Radiation and Energy Budgets at Earth's Surface

RADIATION BUDGET: $Q^* = K\downarrow - K\uparrow + L\downarrow - L\uparrow \ (= K^* + L^*)$								
Net Radiation		Incoming Shortwave (solar radiation)		Outgoing Shortwave (reflected by surface)		Incoming Longwave (emitted by atmosphere)		Outgoing Longwave (emitted by surface)
Q^*	=	$K\downarrow$	−	$K\uparrow$	+	$L\downarrow$	−	$L\uparrow$

ENERGY BUDGET: $Q^* = Q_H + Q_E + Q_G$						
Net Radiation (from the radiation budget)		Sensible Heat Flux to/from Atmosphere		Latent Heat Flux to/from Atmosphere		Ground Heat Flux to/from Ground
Q^*	=	Q_H	+	Q_E	+	Q_G
Meaning of positive or negative values						
Positive: surface has a surplus of radiant energy that can provide heat to the air and ground or evaporate water		Positive: sensible heat flux away from the surface into the air—air warms		Positive: latent heat flux away from the surface into the air—evaporation from the surface		Positive: heat conducts away from the surface into the ground—ground warms
Negative: surface has a deficit of radiant energy that must be made up by heat flow from the air or ground or by condensation		Negative: sensible heat flux toward the surface—air cools		Negative: latent heat flux toward the surface—condensation on the surface		Negative: heat conducts from the ground toward the surface—ground cools

less evaporation and warming of the air during the day. On a cloudy night, the incoming longwave ($L\downarrow$) increases because the cloud bases are very effective emitters of infrared radiation. This reduces the net loss of longwave radiation overnight, so Q^* is not as large a negative number. In this case, the air does not have to cool as much (less of a negative Q_H) to supply energy to balance the negative Q^*. Consequently, the daily range in temperature and evaporation is reduced on cloudy days.

In summary, the daily air temperature and humidity changes are caused by the fluxes of sensible heat (Q_H) and latent heat (Q_E), which, in turn, are driven by net radiation (Q^*). Many of these ideas will be looked at further in Chapter 3.

What happens with net radiation and energy over the course of a year is similar to what happens over a day. During winter months, incoming solar radiation during the day ($K\downarrow$) is less than in summer because the sun is lower in the sky. Also, if snow is on the ground, the higher albedo increases $K\uparrow$, so K^* is much less in winter. This means that net radiation (Q^*) by day is lower, or even remains negative, so the surface is more often in a radiation deficit ($Q^* < 0$). In this situation, the deficit in radiation has to be balanced more often by heat flowing to the surface from the ground ($Q_G < 0$) and from the air ($Q_H < 0$, $Q_E < 0$), so the ground and air are cooled by Earth's surface. This explains why it is cold in winter and warm in summer: it is due to the energy budget that drives temperatures, responding to annual changes in solar and net radiation. If you wish to learn what can happen to solar radiation as it interacts with Earth's magnetic field, read Focus on a Special Topic: Solar Particles and the Aurora on p. 58.

● FIGURE 2.20 Energy budgets at Earth's surface for (a) Q^* deficit (the normal situation at night), illustrating flows of energy from the ground and atmosphere that balance the deficit, and (b) Q^* surplus (the normal situation during the day), illustrating how the excess energy from radiation is used.

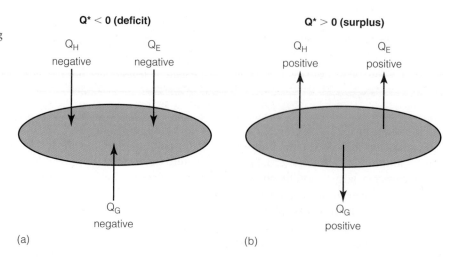

SUMMARY

In this chapter, we have discussed how energy and its movement are the driving force behind weather, affecting all of Earth's systems. We have seen how the concepts of heat and temperature differ and how heat is transferred in our environment. We learned that latent heat is an important source of atmospheric energy. We also learned that conduction, the transfer of sensible heat by molecular collisions, is most effective in solids. Because air is a poor heat conductor, conduction in the atmosphere is important only in the shallow layer of air in contact with Earth's surface. A more important process of atmospheric sensible and latent heat transfer is convection, which involves the mass movement of air (or any fluid) with its energy from one region to another. Perhaps the most important heat transfer process is radiation—the transfer of energy by means of electromagnetic waves.

The hot sun emits most of its radiation as shortwave radiation. A portion of this energy heats Earth's surface, and the surface, in turn, warms the air above. The cool Earth emits most of its radiation as longwave (infrared) radiation. Selective absorbers in the atmosphere, such as water vapour and carbon dioxide, absorb some of Earth's infrared radiation and radiate a portion of it back to the surface, where it warms the surface, producing the atmospheric greenhouse effect. Because clouds are both good absorbers and good emitters of infrared radiation, they keep calm, cloudy nights warmer than calm, clear nights. The average equilibrium temperature of Earth and the atmosphere remains fairly constant from one year to the next because the amount of energy they absorb each year is equal to the amount of energy they lose. Radiation drives the energy budget, which, in turn, determines how the air and ground warm and cool, as well as how the air's humidity changes.

KEY TERMS

The following terms are listed (with page numbers) in the order they appear in the text. Define each. Doing so will aid you in reviewing the material covered in this chapter.

energy, 34	heat, 36
absolute zero, 34	specific heat, 36
Kelvin scale, 34	latent heat, 37
Celsius scale, 35	parcel of air, 38
Fahrenheit scale, 35	conduction, 39
potential energy, 35	convection, 40
kinetic energy, 35	thermals, 40
sensible heat, 35	free convection, 40
specific heat, 35	turbulence, 40
radiant energy, 35	forced convection, 40
photon, 36	advection, 40
Planck constant, 36	radiant energy, 41
temperature, 36	radiation, 41
electromagnetic waves, 41	Mie scattering, 48
wavelength, 41	geometric scattering, 48
micrometre, 41	blackbody, 49
Stefan–Boltzmann law, 42	emissivity, 49
Wien's law, 42	radiative equilibrium temperature, 50
longwave radiation, 43	selective absorbers, 50
shortwave radiation, 43	Kirchhoff's law, 50
visible region, 43	greenhouse effect, 50
ultraviolet (UV) radiation, 43	atmospheric window, 51
infrared (IR) radiation, 43	Bowen ratio, 57
solar constant, 45	solar wind, 58
scattering, 46	aurora borealis, 59
Rayleigh scattering, 47	aurora australis, 59
reflected (light), 47	airglow, 59
albedo, 47	

QUESTIONS FOR REVIEW

1. How does the average speed of air molecules relate to the air temperature?

2. Distinguish between temperature and heat.

3. (a) How does the Kelvin temperature scale differ from the Celsius scale?

   (b) Why is the Kelvin scale often used in scientific calculations?

   (c) Based on your experience, would a temperature of 250 K be considered warm or cold? Explain.

4. Explain how in winter heat is transferred by
   (a) conduction
   (b) convection
   (c) radiation

5. How is latent heat an important source of atmospheric energy?

6. In the atmosphere, how does advection differ from convection?

7. How does the temperature of an object influence the radiation that it emits?

8. How does the amount of radiation emitted by Earth differ from that emitted by the sun?

9. How do the wavelengths of most of the radiation emitted by the sun differ from those emitted by the surface of Earth?

10. Which photon carries the most energy—infrared, visible, or ultraviolet?

11. When a body reaches a radiative equilibrium temperature, what is taking place?

12. If Earth's surface continually radiates energy, why doesn't it become colder and colder?

13. Why are carbon dioxide and water vapour called selective absorbers?

14. Explain how Earth's atmospheric greenhouse effect works.

15. What gases appear to be responsible for the enhancement of Earth's greenhouse effect?

16. Why do most climate models predict that Earth's average surface temperature will increase by an additional 1.8 to 4°C by the end of this century?

17. What processes contribute to Earth's albedo being 30 percent?

18. Explain how the atmosphere near Earth's surface is warmed and cooled from below.

19. Explain what causes the air to warm in the day and cool at night.

20. If a blackbody is a theoretical object, why can both the sun and Earth be treated as blackbodies?

21. What is the solar wind?

22. Explain how the aurora is produced.

QUESTIONS FOR THOUGHT

1. Explain why the bridge in the diagram (● Figure 2.21) is the first to become icy.

● FIGURE 2.21

2. Explain why the first snowfall of the winter usually "sticks" better to tree branches than to bare ground.

3. At night, why do materials that are poor heat conductors cool to temperatures less than the surrounding air?

4. Explain how, in winter, ice can form on puddles (in shaded areas) when the temperature above and below the puddle is slightly above freezing.

5. In northern latitudes, the oceans are warmer in summer than they are in winter. In which season do the oceans lose heat most rapidly to the atmosphere by conduction and convection? Explain.

6. How is heat transferred away from the surface of the moon? (Hint: The moon has no atmosphere.)

7. Why is ultraviolet radiation more successful in dislodging electrons from air atoms and molecules than is visible radiation?

8. Why must you stand closer to a small fire to experience the same warmth you get when standing farther away from a large fire?

9. If water vapour were no longer present in the atmosphere, how would Earth's energy budget be affected?

10. Which will show the greatest increase in temperature when illuminated with direct sunlight: a ploughed field or a blanket of snow? Explain.

11. Why does the surface temperature often increase on a clear, calm night as a low cloud moves overhead?

12. Which would have the greatest effect on Earth's greenhouse effect: removing all of the CO_2 from the atmosphere or removing all of the water vapour? Explain why you chose your answer.

13. Explain why an increase in cloud cover surrounding Earth would increase Earth's albedo yet not necessarily lead to a lower earth surface temperature.

14. Could a liquid thermometer register a temperature of −273°C when the air temperature is actually 1000°C? Where would this happen in the atmosphere, and why?

15. Why is it that auroral displays can be forecast several days in advance?

PROBLEMS AND EXERCISES

1. Suppose that 0.50 kg of water vapour condenses to make a cloud about the size of an average room. If we assume that the latent heat of condensation is 2500 kJ kg^{-1}, how much heat would be released to the air? If the total mass of air before condensation is 100 kg, how much warmer would the air be after condensation? Assume that the air is not undergoing any pressure changes. (Hint: Use the specific heat of air in Table 2.1, p. 36.)

2. Suppose planet A is exactly twice the size (in surface area) of planet B. If both planets have the same surface temperature (1500 K), which planet would be emitting the most radiation? Determine the wavelength of maximum energy emission of both planets using Wien's law.

3. Suppose, in question 2, that the temperature of planet B doubles.
   (a) What would be its wavelength of maximum energy emission?
   (b) In what region of the electromagnetic spectrum would this wavelength be found?
   (c) If the temperature of planet A remained the same, determine which planet (A or B) would now be emitting the most radiation (use the Stefan–Boltzmann relationship). Explain your answer.

4. Suppose your surface body temperature averages 30°C. How much radiant energy in W m^{-2} would be emitted from your body?

A warm fall day near Fort St. John, British Columbia. Here air temperatures may climb well above freezing during the day and drop to well below freezing at night.
Daryl Benson/Getty Images

Seasonal and Daily Temperatures

3

The Sun doesn't rise or fall: it doesn't move, it just sits there, and we rotate in front of it. Dawn means that we are rotating around into sight of it, while dusk means we have turned another 180° and are being carried into the shadow zone. The Sun never "goes away from the sky." It's still there sharing the same sky with us; it's simply that there is a chunk of opaque earth between us and the Sun which prevents our seeing it. Everyone knows that, but I really see it now. No longer do I drive down a highway and wish the blinding Sun would set; instead I wish we could speed up our rotation a bit and swing around into the shadows more quickly.

Michael Collins, Astronaut, Gemini 10, 1963, and Apollo 11, 1969; pilot for the first moon walk.

Michael Collins, *Carrying the Fire.* Cooper Square Press.

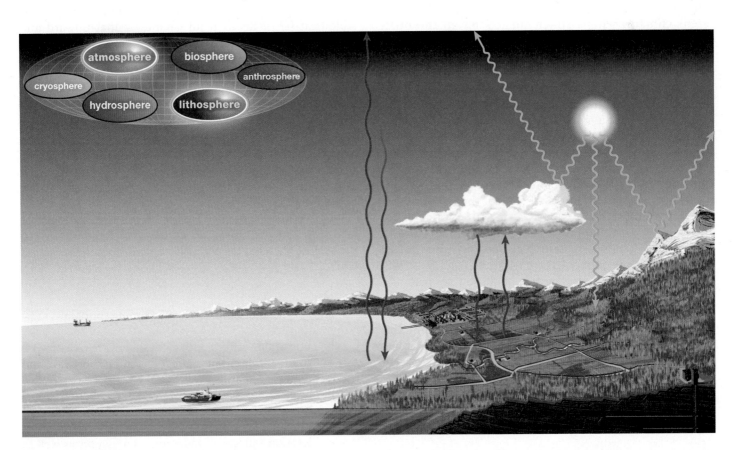

CONTENTS

Temperatures drop when air loses energy and rise when air gains it. This energy is exchanged in the form of sensible heat. This chapter uses the energy concepts developed in Chapter 2 to describe how temperature varies seasonally and daily. Reading these two chapters as a unit will enhance your understanding of the linkages between energy and temperature.

Daily and seasonal temperature variations are defining features of climate for an area and are key components of the *atmospheric* system. Temperature patterns affect the *hydrosphere*, changing the availability and form of precipitation and impacting the amount of moisture that is available for other systems. Snow has a very different effect than rain. When temperatures fall below freezing, water becomes part of the *cryosphere*, storing moisture for later use. Temperature controls whether water storage is seasonal or lasts for ice ages. Heat with a lack of moisture creates deserts, whereas heat with moisture forms tropical rain forests. This interaction of the atmosphere and the hydrosphere creates two very different climates with distinct impacts on all the other Earth systems. Temperature interacting with the hydrosphere controls the type and rate of chemical and biological interaction with the *lithosphere* and through the disintegration of Earth's surface forms rocks, soils, and the diversity of landscapes comprising Earth's environments. These have a major impact on the *biosphere*, influencing the types of vegetation and animal life that inhabit an area. Within the biosphere, temperature plays a key role in where and how humans live. We add heat to make cold environments comfortable and remove heat when it is too hot. Human attempts to control the environment create the *anthrosphere*. We need to better understand and apply the principles of energy and temperature to create more livable, sustainable, and energy-efficient human living environments.

As you sit quietly reading this book, you are part of a moving experience. Earth is speeding around the sun at thousands of kilometres per hour, while at the same time, it is spinning on its axis. We normally do not think of it in this way, but this is what causes our seasons and the daily rising of the sun, moon, and stars in the east and setting in the west. After discussing the controls of temperature, we will examine how Earth's motions and the sun's energy produce seasonal temperature variations and later examine temperature variations on a daily basis.

The Physical Controls of Temperature

What makes air temperature near Earth's surface rise and fall? To answer this question, we need to go back to some of the concepts of energy discussed in Chapter 2. Air temperature was defined as a measure of the random motion of air molecules—as the kinetic energy of air at a microscopic level. Temperature rises when energy enters air, increasing the speed of molecules within the air, and falls when energy is removed. This energy is called **sensible heat**. So, to understand why, when, and how air temperature changes, we must look at the cause of that change: sensible heat changes. If more sensible heat enters a layer of air than leaves it, the air temperature rises.

The sensible heat in air near Earth's surface can change in two main ways. The first is when the wind horizontally blows warmer or colder air to the location—this is called **advection** and results when weather systems such as storms and fronts affect an area. Advection is also important when wind blows from one surface type to another, such as between ocean and continent. We will be discussing these sorts of phenomena in Chapters 9 to 15.

The second factor affecting air's sensible heat content, and the most important for the discussions here, is when there are upward and downward **convective** fluxes of sensible heat (Q_H). Recall from Chapter 2 that upward and downward sensible heat flows are determined by two relationships: the radiation and energy balances at Earth's surface. The radiation balance considers shortwave and longwave radiation entering and leaving the surface and determines the **net radiation** (Q^*) available for the energy balance. The energy balance, in turn, distributes Q^* between heating of the ground (ground heat flux, Q_G), evaporation of water (latent heat flux, Q_E), and heating of the air (sensible heat flux, Q_H). The amount of Q^* going into the ground heat flux and latent heat flux affects the energy left over for the sensible heat flux.

Let us then list the key factors that control the sensible heat flux and therefore also control the temperature. In the rest of the chapter, we will explain how these factors work.

1. Net radiation (Q^*)
   a. Latitude and season (affect incoming shortwave radiation)
   b. Time of day (affects incoming shortwave radiation)
   c. Clouds and humidity (affect incoming shortwave and longwave radiation)
   d. Local topography and surface effects such as slope, aspect, surface type, vegetation, and soil (affect all radiation components)
   e. Elevation (affects all radiation components as well as temperature directly because rising air cools as it expands)
2. Latent heat flux (Q_E) and ground heat flux (Q_G)
   These are important factors because if the available net radiation is used for these processes, less will be available for the sensible heat flux to heat the air.
   a. Humidity, surface type, soil moisture, vegetation, and wind (affect the latent heat flux)
   b. Soil type and moisture (affect the ground heat flux)

3. Sensible heat flux (Q_H)
   a. Difference in temperature between Earth's surface and air near the surface (affects the direction and rate of the sensible heat flux from the surface)
   b. Wind (affects the rate at which sensible heat is mixed from the near-surface layer into the deeper atmosphere by forced convection)
   c. Atmospheric stability (affects the rate at which sensible heat is mixed from the near-surface layer into the deeper atmosphere by free convection)
4. Advection due to weather systems and air moving between oceans and continents (Phenomena causing advection of various types are discussed in Chapters 9 to 15.)

Why Earth Has Seasons

Earth revolves around the sun, completing its elliptical path in slightly longer than 365 days. As it revolves around the sun, it spins, completing one spin every 24 hours. The average distance from Earth to the sun is 150 million kilometres, but because Earth's orbit is an ellipse instead of a circle, the actual distance varies during the year. As shown in • Figure 3.1, Earth travels closer to the sun in January (147 million kilometres) than it does in July (152 million kilometres).* Consequently, we might think that our warmest weather should occur in January and our coldest weather in July, but in the Northern Hemisphere, the opposite occurs. We experience colder weather in January, when we are closer to the sun, and warmer weather in July, when we are farther away. So if nearness to the sun does not control the seasons, what does?

Seasons are regulated by the amount of solar energy received at Earth's surface. The angle that sunlight strikes the surface and how long the sun shines on any latitude are the key factors. Let's look more closely at these.

Solar energy that strikes Earth's surface perpendicularly is much more intense than solar energy that strikes it at an angle. Think of shining a flashlight straight at a wall—you get a bright, small, circular spot of light (see • Figure 3.2). Now tilt the flashlight so it shines at an angle and notice how the spot of light diffuses and spreads over a larger area. The same phenomenon occurs with sunlight. Solar radiation striking Earth at an angle spreads out and heats a larger region less intensely than sunlight shining directly on Earth. Everything else being equal, an area experiencing more direct solar rays will receive more energy in the form of heat than the same-sized area being struck by sunlight at an angle.†

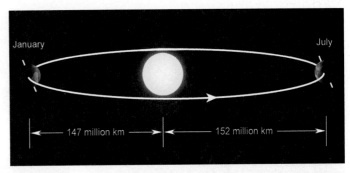

• **FIGURE 3.1** The elliptical path of Earth about the sun brings Earth slightly closer to the sun in January than in July. Note that the path is highly exaggerated in this image.

In addition, more angled solar rays must penetrate more atmosphere, and the more atmosphere they penetrate, the more they can be scattered and absorbed before reaching Earth's surface. So when the sun is high in the sky, it can heat a surface to a much higher temperature than when it is low on the horizon.*

The second important factor determining how warm Earth's surface becomes is the length of time the sun shines each day. Longer daylight hours mean that more energy is available from sunlight. For any given location, more solar energy reaches Earth's surface and is absorbed as surface heating on a clear, long day than on a clear, shorter day.

We know that summer days have more daylight hours than winter days. Also, the noontime summer sun is higher in the sky than is the noontime winter sun. Both of these events occur because our spinning planet is tilted on its axis as it revolves around the sun. As • Figure 3.3 illustrates, the angle of tilt is 23½° from a plane that is perpendicular to Earth's orbit (represented by the dashed line in this figure). Earth's axis points in the same direction in space all year long, making the Northern Hemisphere tilt toward the sun in summer (June) and away from the sun in winter (December). In our continuing discussion, we will use latitude and longitude lines to describe positions on Earth. To see how this coordinate system works, refer to • Figure 3.4.

SEASONS IN THE NORTHERN HEMISPHERE   In the warm summer season, the northern half of the world is directed toward the sun (see Figure 3.3, June 21). Peak solar exposure occurs on the **summer solstice**. At noon on this day, solar rays beat down on the Northern Hemisphere more directly than during any other time of year. The sun is at its highest position directly above the Tropic of Cancer at 23½° north (N) latitude. If you are standing at this latitude on June 21,† the

---

*Around January 3, the Earth is closest to the sun. This point is called *perihelion* (from the Greek *peri*, meaning "near," and *helios*, meaning "sun"). Earth is farthest from the sun around July 4. This point is called *aphelion* (from the Greek *ap*, meaning "away from").

†This phenomenon is summarized by the **cosine law of illumination**, which states that the radiation intensity (R) received by an inclined surface is equal to the maximum radiation intensity (R_0, the radiation striking the surface directly) times the cosine of the zenith angle (Z, the angle between the angled radiation beam and a line that is perpendicular to the surface). Mathematically, this is $R = R_0 \cos(Z)$.

As discussed in Chapter 2, air temperature increases when the net radiation (Q) is positive, and this excess net radiation drives a sensible heat flux (Q_H) that warms the air. The annual cycle of Q* is determined by annual variation in incoming solar radiation (K↓), which depends mostly on latitude and season.

†The solstice and equinox calendar dates can vary by a day from year to year because the exact timing of these events occurs roughly six hours later each year. Things get caught up with a leap year every fourth year, but occasionally the date changes into the next day first.

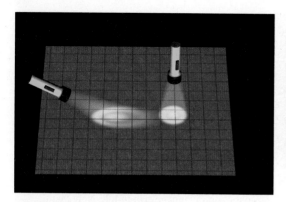 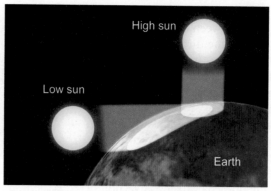

● FIGURE 3.2 Sunlight that strikes a surface at an angle is spread over a larger area than sunlight that strikes the surface directly. Oblique sun rays are less intense and deliver less energy to a surface than direct sun rays. This phenomenon is described by the cosine law of illumination.*

---

*This phenomenon is summarized by the **cosine law of illumination**, which states that the radiation intensity (R) received by an inclined surface is equal to the maximum radiation intensity (R_0, the radiation striking the surface directly) times the cosine of the zenith angle (Z, the angle between the angled radiation beam and a line that is perpendicular to the surface). Mathematically, this is $R = R_0 \text{cosine} (Z)$.

sun at noon would be directly overhead. This day is the summer solstice, the astronomical first day of summer in the Northern Hemisphere.

Notice in Figure 3.3 that as Earth spins on its axis each day, half the globe is receiving sunlight, whereas the other half is in darkness. If Earth's axis were not tilted, the sun at noon would always be directly overhead at the Equator, and every latitude except the North and South poles would have 12 hours of daylight and 12 hours of darkness every day of the year. However, Earth is tilted. So, day length varies throughout the year as more or less of the globe at different latitudes is exposed to the sun. For example, the Northern Hemisphere faces toward the sun on June 21, and each latitude in the Northern Hemisphere will have more than

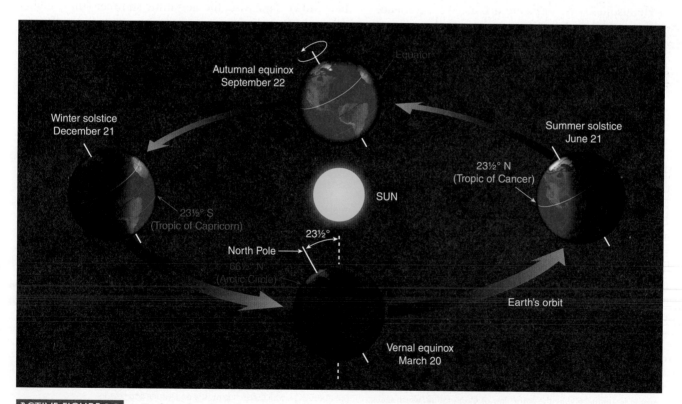

ACTIVE FIGURE 3.3   As Earth revolves about the sun, it is tilted on its axis by an angle of 23½°. If we could see our solar system from space, we would see that this axis always points at the same area in space. Thus, in June, when the Northern Hemisphere is tipped toward the sun, more direct sunlight and long hours of daylight cause warmer weather than in December, when the Northern Hemisphere is tipped away from the sun. (The diagram is not to scale.) Visit the textbook's website to view this and other Active Figures at www.ahrensmeteorology1ce.nelson.com

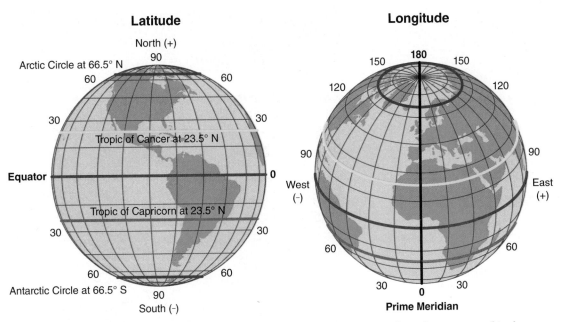

• FIGURE 3.4 Longitude and latitude lines are used to define locations on Earth. Both are measured in degrees, with latitudes counted from zero to 90° in both the Northern and Southern hemispheres and longitudes measured from 0 to 180° in the west and east directions. Key latitudes are identified.

12 hours of daylight—the farther north, the longer the daylight. When we reach the Arctic Circle at 66½°N, daylight lasts 24 hours (notice in Figure 3.3 on June 21 how the region above 66½°N never gets into the "shadow" zone as Earth spins). At the North Pole, the sun actually rises above the horizon on March 20 and does not set again until six months later, on September 22. No wonder this region is called the "Land of the Midnight Sun"! (See • Figure 3.5.)

Do longer days near polar latitudes mean that the highest daytime summer temperatures are experienced there? Not normally. Toronto, Ontario, located at 43° 40′ N, experiences much hotter summer weather than Qausuittuq (Resolute), Nunavut, at 74° 43′ N. Yet the days in Qausuittuq are much longer, so why isn't it warmer? To figure this out, we must examine the *incoming solar radiation* (called **insolation**) accumulated over the whole day on June 21. • Figure 3.6 shows two curves: the upper curve represents the amount of insolation at the top of Earth's atmosphere on June 21, whereas the bottom curve represents the amount of radiation that eventually reaches

• FIGURE 3.5 Land of the Midnight Sun. Time-lapse photography of the sun taken before, during, and after midnight in northern Alaska during July.

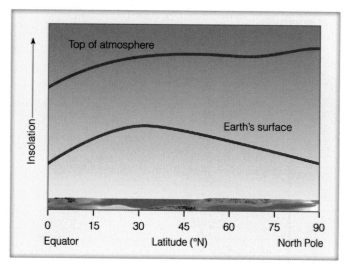

• FIGURE 3.6 The relative amount of radiant energy received at the top of Earth's atmosphere and at Earth's surface on June 21—the Northern Hemisphere's summer solstice.

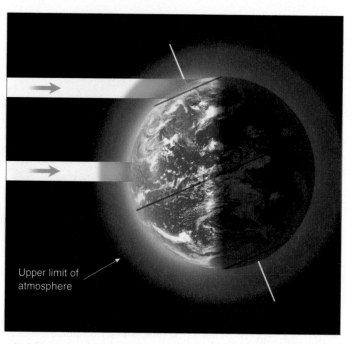

• FIGURE 3.7 During summer in the Northern Hemisphere, sunlight that reaches Earth's surface in far northern latitudes has passed through a thicker layer of absorbing, scattering, and reflecting atmosphere than sunlight that reaches Earth's surface farther south. Sunlight is lost both through the thickness of the pure atmosphere and by impurities in the atmosphere. As the sun's rays become more angled, these effects become more pronounced.

Earth's surface on the same day. The upper curve increases from the equator to the North Pole, indicating that more solar radiation reaches the top of Earth's atmosphere above the poles than above the equator. Because the sun shines for 24 hours a day, this occurs even though the sun shines at a relatively low angle at polar latitudes. The lower curve shows that the amount of solar radiation eventually reaching Earth's surface on June 21 is maximum near 30°N and decreases as we move poleward.

The two curves are different because once sunlight enters the atmosphere, fine dust and air molecules scatter it, clouds reflect it, and some of it is absorbed by atmospheric gases. What remains reaches the surface. Generally, the thicker the atmosphere that sunlight must penetrate, the greater the chances it will be scattered, reflected, or absorbed by the atmosphere. During the summer in far northern latitudes, the sun is never very high above the horizon, so its radiant energy must pass through a thick portion of atmosphere before it reaches Earth's surface (see • Figure 3.7). This causes a reduction in the amount of energy that reaches the surface. Also, increased cloud cover during arctic summers causes much of the sunlight to be reflected before it reaches the ground.

Solar energy that eventually reaches Earth's surface near the poles does not heat the surface as effectively as it would at lower latitudes. A greater portion of the sun's energy is reflected by ice and snow, and some of it is used in the phase change from solid to liquid when frozen soil melts. So even though northern towns, such as Qausuittuq, experience 24 hours of continuous sunlight on June 21, they are not warmer than cities farther south. Overall, they receive less radiation at the surface, and what radiation they do receive does not produce sensible heat as effectively.

In our discussion of Figure 3.6, we saw that on June 21, solar energy incident on Earth's surface is maximum near latitude 30°N. On this day, the sun is shining directly above

latitude 23½°N. Why, then, isn't the most sunlight received here? A quick look at a world map shows that the major deserts of the world are centred near 30°N. Cloudless skies and drier air predominate near this latitude. At latitude 23½°N, the climate is more moist and cloudy, causing more sunlight to be scattered and reflected before reaching the surface. In addition, day length is longer at 30°N than at 23½°N on June 21. For these reasons, more radiation falls on 30°N latitude than at the Tropic of Cancer (23½°N).

Each day past June 21, the noon sun is slightly lower in the sky. Summer days in the Northern Hemisphere begin to shorten. June eventually gives way to September, and fall begins.

Revisit Figure 3.3 (p. 68). Notice that by September 22, Earth will have moved so that the sun is directly above the equator. Except at the poles, the days and nights throughout the world are of equal length. This day is called the **autumnal** (or fall) **equinox**, and it marks the astronomical beginning of

WEATHER WATCH

Contrary to popular belief, it is not the first frost that causes the leaves of deciduous trees to change colour. The yellow and orange colours, which are actually in the leaves, begin to show through several weeks before the first frost as shorter days and cooler nights cause a decrease in the production of the green pigment chlorophyll.

Photos.com

• FIGURE 3.8 The pageantry of fall colours in Ontario. The weather most suitable for an impressive display of fall colours is warm, sunny days followed by clear, cool nights, with temperatures dropping below 7°C but remaining above freezing.

fall in the Northern Hemisphere. At the North Pole, the sun appears at the horizon for the entire 24-hour day. A few days afterward, the sun disappears from view and will not rise again for six months.

Throughout the northern half of the world, there are fewer hours of daylight, and the sun at noon is slightly lower in the sky on each successive day. Less direct sunlight and shorter hours of daylight spell cooler weather for the Northern Hemisphere. Reduced radiation, lower air temperatures, and cooling breezes stimulate the beautiful pageantry of fall colours (see • Figure 3.8).

On December 21, three months after the fall equinox, the Northern Hemisphere is tilted as far away from the sun as it gets all year (see Figure 3.3, p. 68). Nights are long and days are short. ▼ Table 3.1 shows how daylight decreases from 12 hours at the equator to nothing (0 hours) at latitudes above 66½°N. This shortest day of the year, called the (Northern Hemisphere) **winter solstice**, is the astronomical beginning of winter in the north. On this day, the sun shines directly above the Tropic of Capricorn at latitude 23½°S. In the northern half of the world, the sun is at its lowest position in the sky at noon. Its rays pass through a thick section of atmosphere and spread over a large area on the surface, so the sunlight is weak.

With so little incident sunlight, Earth's land surface cools quickly. A blanket of clean snow covering the ground reduces incoming solar energy by reflecting much of the sunlight that reaches the surface. The surface is also continually radiating away infrared energy during the long nights. In northern Canada and Alaska, the arctic air rapidly becomes extremely cold. Periodically, this cold arctic air moves into southern Canada and the United States, producing a rapid drop in temperature called a *cold wave*, which occasionally reaches far into the southern United States during the winter. Sometimes these cold spells arrive well before the "official" first day of winter (solstice), bringing with them heavy snow and blustery winds. (See Focus on a Special Topic: Is December 21 Really the First Day of Winter? on p. 72.)

▼ Table 3.1   Length of Time from Sunrise to Sunset for Various Latitudes on Different Dates in the Northern Hemisphere

Latitude	March 20	June 21	September 22	December 21
0°	12 hr	12.0 hr	12 hr	12.0 hr
10°	12 hr	12.6 hr	12 hr	11.4 hr
20°	12 hr	13.2 hr	12 hr	10.8 hr
30°	12 hr	13.9 hr	12 hr	10.1 hr
40°	12 hr	14.9 hr	12 hr	9.1 hr
50°	12 hr	16.3 hr	12 hr	7.7 hr
60°	12 hr	18.4 hr	12 hr	5.6 hr
70°	12 hr	2 months	12 hr	0 hr
80°	12 hr	4 months	12 hr	0 hr
90°	12 hr*	6 months	12 hr*	0 hr

*The sun rises on March 20 and sets on September 20.

FOCUS ON A SPECIAL TOPIC

Is December 21 Really the First Day of Winter?

In the Northern Hemisphere, on December 21 or 22 depending on the year, after a month or so of cold weather and a snowstorm or two (see ● Figure 1), some media maven has the audacity to proclaim, "Today is the first official day of winter." If the last several weeks were not winter, then what season was it?

The "official" beginning of any season is simply the day on which the Sun passes over a particular latitude and has nothing to do with how cold or warm the weather is. December 21 marks the astronomical first day of winter in the Northern Hemisphere (NH), just as June 21 marks the astronomical first day of summer (NH). Since Earth is tilted on its axis by 23½°, the sun is directly above 23½° south latitude on December 21 and directly above 23½° north latitude on June 21. The astronomical first day of spring (NH) occurs around March 20 as the sun crosses the equator moving northward, and likewise, the astronomical first day of autumn (NH) occurs around September 22 as the sun crosses the equator moving southward.

In the middle latitudes, summer is defined as the warmest season and winter as the coldest season. Typically, we think the year is divided into four seasons, with each season consisting of three months, and then the meteorological definition of summer over much of the Northern Hemisphere would be the three warmest months of June, July, and August. Winter would be the three coldest months of December, January, and February. Autumn would be September, October, and November—the transition between summer and winter. And spring would be March, April, and May—the transition between winter and summer.

More realistically, if seasons could have unequal lengths, and winter is defined meteorologically as the period when average daily temperatures are below freezing, whereas summer is defined as the frost-free period, then seasons would more closely match our weather experiences, and many parts of Canada would have winters lasting longer than three months. In Winnipeg, for example, winter would extend from late October through early April and summer from late May through mid-September, whereas spring and fall would be curtailed to less than two months each in length between these dates.

So the next time you hear someone remark on December 21 that "winter officially begins today," remember that this is the astronomical definition of the first day of winter. According to the meteorological definition, winter has been around for several weeks or more.

THE CANADIAN PRESS/Paul Chiasson

● FIGURE 1 A woman makes her way through a park during a snowstorm expected to dump some 25 cm of snow in Montreal on Wednesday, December 9, 2009. Since the snowstorm occurred before the winter solstice, is this a late fall storm or an early winter storm?

Each winter day after December 21, the sun climbs a bit higher in the midday sky and the periods of daylight grow longer until March 20, when days and nights are of equal length again, and we have another equinox. This is called the **vernal** (or spring) **equinox**. At this equinox, the sun at noon is shining directly on the equator, while at the North Pole, the sun rises above the horizon after hiding for six months. Three months after the vernal equinox, it is June again and the cycle repeats itself.

Up to now, we have seen that the seasons are controlled by solar energy striking our tilted planet as it makes its annual voyage around the sun. Earth's tilt causes seasonal variation in both the length of daylight and the intensity of sunlight that reaches the surface. These facts are summarized in ● Figure 3.9, which shows how the sun would appear in the sky at different times of the year to an observer at various latitudes. The sun's path is shown by the yellow orbits in each image. Notice that in the polar latitudes (see Figures 3.9, a and b), the sun never climbs very high in the sky, even in summer, compared to the height of the sun in middle, low latitudes and the equator (see Figures 3.9, c, d, and e). Figure 3.9f shows the situation in the Southern Hemisphere.

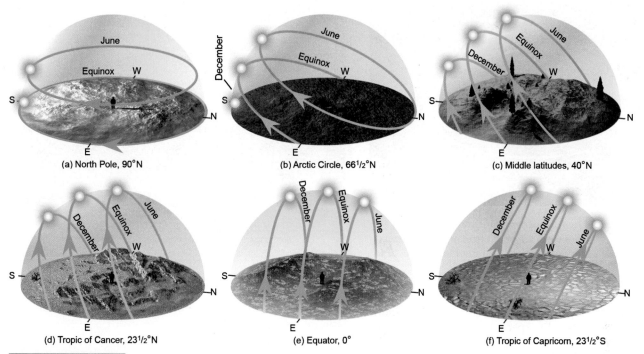

(a) North Pole, 90°N

(b) Arctic Circle, 66½°N

(c) Middle latitudes, 40°N

(d) Tropic of Cancer, 23½°N

(e) Equator, 0°

(f) Tropic of Capricorn, 23½°S

ACTIVE FIGURE 3.9 The sun's path across the sky as observed from different latitudes on the summer solstice (June 21), the winter solstice (December 21), and the spring and fall equinoxes (March 20 and September 22). Visit the textbook's website to view this and other Active Figures at www.ahrensmeteorology1ce.nelson.com

WEATHER WATCH

Does darkness (constant night) really occur at the Arctic Circle (66½°N) on the winter solstice? The answer is no. Due to the bending and scattering of sunlight by the atmosphere, the sky is not totally dark at the Arctic Circle on December 21. In fact, on this date, total darkness only happens north of about 82° latitude. Even at the North Pole, total darkness does not occur from September 22 through March 20 but rather from about November 5 through February 5.

It is interesting to note that although sunlight is most intense in the Northern Hemisphere on June 21, the warmest weather in middle latitudes normally occurs several weeks later, in July or August. This *lag in seasonal temperature* arises because sensible heat is gained as more energy comes into the system than leaves it. More specifically, incoming energy from the sun is greatest in June and exceeds the outgoing energy from Earth for a period of several weeks after. When incoming solar energy and outgoing Earth energy are in balance, the highest average temperature is attained. When outgoing energy exceeds incoming energy, the temperature drops. Because outgoing Earth energy exceeds incoming solar energy well past the winter solstice (December 21), we normally find our coldest weather

occurring in January or February. As we will see later in this chapter, there is a similar lag in daily temperature between the time of most intense sunlight and the time of highest air temperature for the day.

SEASONS IN THE SOUTHERN HEMISPHERE Earth's tilt affects the Southern Hemisphere in the opposite way to what happens in the Northern Hemisphere. The June solstice marks the astronomical beginning of winter in the Southern Hemisphere and the December solstice marks the beginning of summer.

We know Earth comes nearer to the sun in January than in July. Even though this difference in distance amounts to only about 3 percent, the energy that strikes the top of Earth's atmosphere is almost 7 percent greater on January 3 than on July 4. These facts should indicate that summer is warmer in the Southern Hemisphere than it is in the Northern Hemisphere, but this is not the case. A close examination of the Southern Hemisphere reveals that nearly 81 percent of the surface is water compared to 61 percent in the Northern Hemisphere. The added solar energy due to the closeness of the sun is absorbed by large bodies of water, becoming well mixed and circulated within them. This process keeps the average summer (January) temperatures in the Southern Hemisphere cooler than average summer (July) temperatures in the Northern Hemisphere. Because of water's large heat

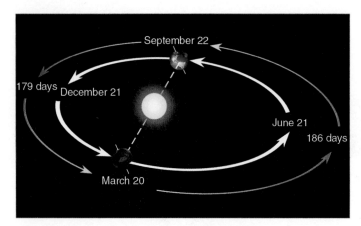

● FIGURE 3.10 Because Earth travels more slowly when it is farther from the sun, it takes Earth a little more than 7 days longer to travel from March 20 to September 22 than from September 22 to March 20.

● FIGURE 3.11 In areas where small temperature changes can cause major changes in soil moisture, sparse vegetation on the south-facing slopes will often contrast with lush vegetation on the north-facing slopes.

capacity, it also tends to keep winters in the Southern Hemisphere warmer than we might expect. A comparison of global January and July temperatures is given in Figures 3.19 and 3.20 on pp. 84 and 85.

Another difference between seasons in the Northern and Southern hemispheres concerns their length. Because Earth travels in an ellipse around the sun, the total number of days from the spring (March 20) to the fall (September 22) equinox is about 7 days longer than from the fall to spring equinox (see ● Figure 3.10). This means that spring and summer in the Northern Hemisphere are about a week longer than fall and winter. Also, spring and summer in the Southern Hemisphere are shorter than they are in the Northern Hemisphere. The longer spring and summer in the Northern Hemisphere offset the extra insolation received by being closer to the sun in summer in the Southern Hemisphere.

Up to now, we have considered the seasons on a global scale. We will now shift to more local considerations.

Local Seasonal Variations

Revisit Figure 3.9c (p. 73) and observe that in the middle latitudes of the Northern Hemisphere, objects facing south will receive more sunlight during a year than those facing north. This fact becomes strikingly apparent in hilly or mountainous country.

Hills or mountains that face south receive more sunshine and, hence, become warmer than the partially shielded north-facing hills. Higher temperatures usually mean greater rates of evaporation and slightly drier soil conditions. Thus, south-facing slopes are usually warmer and drier when compared to north-facing slopes at the same elevation. In many dry areas, only sparse vegetation grows on south-facing slopes, whereas dense vegetation grows on the cool, moist hills that face north (see ● Figure 3.11).

In northern latitudes, south-facing slopes usually have a longer growing season and often more developed soils. Because air temperatures normally decrease with increasing height, trees found on the cooler north-facing side of mountains are often those that usually grow at higher elevations, whereas the warmer, south-facing side of the mountain often supports trees usually found at lower elevations.

In the mountains, snow usually lingers on the ground for a longer time on north-facing slopes than on the warmer, south-facing slopes. Similarly, there are usually more and larger glaciers on north-facing slopes than on south-facing slopes. For this reason, some ski runs are built facing north wherever possible. Also, homes and cabins built on the north side of a hill usually have a steep pitched roof as well as a reinforced deck to withstand the added weight of snow from successive winter storms.

The seasonal change in the sun's position during the year can have an effect on the vegetation around the home. In winter, a large two-story home can shade its own north side, keeping it much cooler than its south side. Trees that require warm, sunny weather should be planted on the south side, where sunlight reflected from the house can even add to the warmth.

WEATHER WATCH

Seasonal changes can affect how we feel. For example, some people face each winter with a sense of foreboding, especially at high latitudes, where days are short and nights are long and cold. If the depression is lasting and disabling, the problem is called *seasonal affective disorder* (SAD). People with SAD tend to sleep longer, overeat, and feel tired and drowsy during the day. The treatment usually involves extra doses of bright light.

FOCUS ON AN ENVIRONMENTAL ISSUE

Solar Heating and the Noon Sun

The amount of solar energy that falls on a typical Canadian home over two summer days is enough energy to heat the inside for a year. Thus, many people are turning to the sun as a clean, safe, and virtually inexhaustible source of energy. If solar collectors are used to heat a home, to heat hot water, or to generate electricity, they should be placed on south-facing roofs to take maximum advantage of the energy provided. The roof itself should be constructed as nearly perpendicular to winter sun rays as possible. To determine the proper roof angle at any latitude, we need to know how high the sun will be above the southern horizon at noon.

The angle of the sun at noon can be calculated in the following manner:

1. Determine the number of degrees between your latitude and the latitude where the sun is currently directly overhead.
2. Subtract the number you calculated in step 1 from 90°. This will give you the sun's elevation above the southern horizon at noon at your latitude.

For example, suppose you live in Calgary, Alberta (latitude 51°N), and the date is December 21. The difference between your latitude and where the sun is currently overhead is

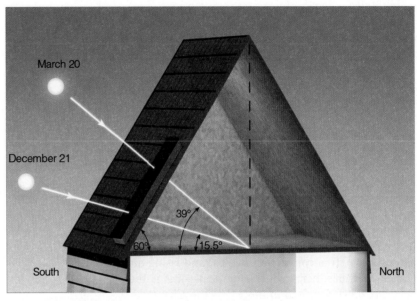

● FIGURE 2 The roof of a solar-heated home constructed in Calgary, Alberta, at an angle of 60° absorbs the sun's energy in midwinter at nearly right angles.

74.5° (51°N to 23½°S), so the sun is 15.5° (90° to 74.5°) above the southern horizon at noon. On March 20 in Calgary, the angle of the sun is 39° (90° to 51°). To determine a reasonable roof angle, we must consider the average altitude of the midwinter sun (about 27° for Calgary), building costs, and snow loads.

● Figure 2 illustrates that a roof constructed in Calgary, Alberta, at an angle of about 60° will be nearly perpendicular to much of the winter sun's energy. Solar panel installers use a rule of thumb to calculate the best roof pitch: latitude plus 10 to 15 degrees for applications where the major load is expected to be in winter.

The design of a home can be important in reducing heating and cooling costs. Large windows should face south, allowing sunshine to penetrate into the home in winter. When solar heating a home, proper roof angle is important in capturing much of the winter sun's energy. (The information needed to determine the angle at which sunlight will strike a roof is given in Focus on an Environmental Issue: Solar Heating and the Noon Sun above.) To block out excess sunlight during the summer, a small eave or overhang should be built. A kitchen with windows facing east will let in enough warm morning sunlight to help heat this area. Because the west side warms rapidly in the afternoon, rooms with small windows (such as garages) should be placed here to act as a thermal buffer. Deciduous trees planted on the west or south side of a home provide shade in the summer. In winter, they drop their leaves, allowing the winter sunshine to warm the house. If you like the bedroom slightly cooler than the rest of the home, face it

toward the north. Let nature help with the heating and air conditioning. Proper house design, orientation, and landscaping can help cut the demand for electricity, as well as for natural gas and fossil fuels, which are being depleted and produce the greenhouse gas carbon dioxide.

Daily Temperature Variations

Each day the air goes through a daily cycle of warming and cooling, somewhat similar to the seasonal cycle. The air warms during the morning hours as the sun gradually rises higher in the sky, spreading a blanket of heat energy over the ground. The sun reaches its highest point around local solar noon, after which it begins its slow journey toward the western horizon. It is around noon when Earth's surface receives the most intense solar rays. However, noon is usually not the warmest part of the day. Rather, the air continues to be heated, often reaching a maximum temperature later in the afternoon. To find out why this lag in temperature occurs, we need to consider how sensible heat flows into and out of a shallow layer of air in contact with the ground.

DAYTIME WARMING As the sun rises in the morning, sunlight warms the ground, and the ground warms the air in contact with it by conduction. However, air is such a poor heat conductor that this process takes place within only a few millimetres of the ground. As the sun rises higher in the sky, the air in contact with the ground becomes even warmer, and a thermal boundary separating the hot surface air from the slightly cooler air above exists. Given their random motion, some air molecules will cross this boundary: the "hot" molecules below bring greater kinetic energy to the cooler air; the "cool" molecules above bring a deficit of energy to the hot, surface air. However, on a windless day, this form of heat exchange is slow, and a substantial temperature difference usually exists just above the ground (see ● Figure 3.12). This explains why joggers on a clear, windless, summer afternoon may experience air temperatures of over 50°C at their feet and only 32°C at their waist.

Near the surface, **free convection** begins, and rising air bubbles, called thermals, help redistribute heat—this is the sensible heat flux (Q_H) discussed in more detail in Chapter 2. In calm weather, these thermals are small and do not effectively mix the air near the surface. Thus, large vertical temperature gradients are able to exist. On windy days, however, turbulent

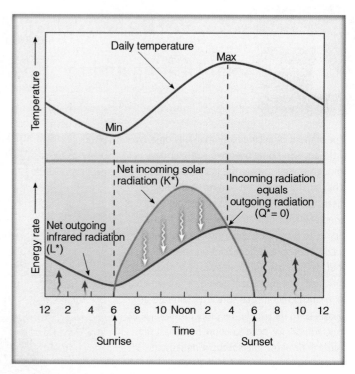

● FIGURE 3.13 The daily variation in air temperature is controlled by the net radiation (Q^*), which is the balance between incoming and outgoing solar and infrared radiation. Where incoming radiation exceeds outgoing radiation (orange shade), Q^* is positive, creating a sensible heat flux into the air (Q_H is positive), and the air temperature rises. Where outgoing energy exceeds incoming energy (blue shade), Q^* is negative, making Q_H negative, so the air loses sensible heat to the surface and the air temperature falls.

eddies are able to mix hot surface air with the cooler air above. This form of mechanical stirring, sometimes called **forced convection**, helps the thermals transfer heat away from the surface more efficiently. Therefore, on sunny, windy days, the molecules near the surface are more quickly carried away than on sunny, calm days. Figure 3.12 shows the typical vertical profiles of air temperature on windy days and on calm days in summer. If the atmosphere is calm and stable,* then any thermals generated are particularly weak, so the sensible heat flux is confined to a shallow depth of the atmosphere, and near-surface temperatures are able to rise rapidly.

We can now see why the warmest part of the day is usually in the afternoon. Around noon, the sun's rays are the most intense. However, even though incoming solar radiation decreases in intensity after noon, it still exceeds outgoing radiation from the surface for a time. In other words, the net radiation (Q^*) is still positive. This situation yields an energy surplus for two to four hours after noon and causes a lag between the time of maximum solar heating and the time of maximum air temperature above the surface (see ● Figure 3.13).

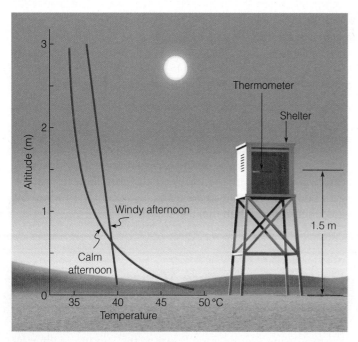

● FIGURE 3.12 On a sunny, calm day, the air near the surface can be substantially warmer than the air a metre or so above the surface. Notice that the near-surface air temperature is warmer on a calm day than on a windy day as convection on the windy day is more efficient at mixing the warm near-surface air with cooler air above.

---

*Stability depends on the atmosphere's vertical temperature profile or **lapse rate**, which was introduced in Chapter 1 and will be discussed in detail in Chapter 6.

The exact time of the highest temperature reading varies somewhat. Where the summer sky remains cloud-free all afternoon, the maximum temperature may occur sometime between 3:00 and 5:00 p.m. Where there is afternoon cloudiness or haze, the temperature maximum usually occurs an hour or two earlier. If clouds persist throughout the day, the overall daytime temperatures are usually lower.

Adjacent to large bodies of water, cool air moving inland as part of a **sea breeze** (see Chapter 9) may modify the rhythm of temperature change such that the warmest part of the day occurs at noon or before. In winter, storms circulating warm air northward can even cause the highest temperature to occur at night.

Just how warm the air becomes depends on such factors as the type of soil, its moisture content, and vegetation cover. If you recall the energy budget concepts discussed in Chapter 2, the net radiation (Q^*) gets distributed as the sum of conduction of heat into or out of the ground (Q_G—causing the ground to warm up or cool down), a sensible heat flux into air (Q_H—causing air to warm up or cool down), and a latent heat flux into air (Q_E—associated with evaporation of water into air or condensation of water out of air). Mathematically, this is expressed as

$$Q^* = Q_H + Q_E + Q_G$$

When the soil is a poor heat conductor (as loosely packed dry sand is), daytime energy from the net radiation does not readily transfer into the ground, so Q_G is small. This fact allows the surface to reach a higher temperature, increasing the temperature gradient in the air near the surface, resulting in a larger sensible heat flux (Q_H), which provides more energy to warm the air above. On the other hand, if the soil is moist or covered with vegetation, much of the available energy from Q^* evaporates water, so Q_E is large, leaving less energy for Q_H to heat the air. As you might expect, the highest summer temperatures usually occur over desert regions, where clear skies coupled with low humidity and meagre vegetation permit the surface and the air above to warm up rapidly.

Where the air is humid, haze and cloudiness lower the maximum temperature by preventing some of the sun's rays from reaching the ground. In the desert-like Osoyoos, located near the U.S. border in the British Columbia interior, the average daily maximum temperature in July is 29.2°C. In contrast, Cochrane, Ontario, at the same latitude, has an average July maximum of 24°C. An even larger difference exists between humid Atlanta, Georgia, where the average maximum temperature for July is 30.5°C, and Phoenix, Arizona. Phoenix, in the desert southwest at the same latitude as Atlanta, experiences an average July maximum of 40.5°C. (Additional information on high daytime temperatures is given in Focus on a Special Topic: Record High Temperatures on p. 78.)

NIGHTTIME COOLING Lower sun angles mean that solar energy is spread over a larger area. This reduces the heat available to warm the surface. Observe in Figure 3.13 that sometime in late afternoon or early evening, Earth's surface and the air above begin to lose more energy than they receive. They start to cool by radiating infrared energy, a process called **radiational** or **radiative cooling**. This process happens day and night, but during daytime, the incoming solar radiation usually offsets the infrared radiative cooling. At night, when solar radiation is not present, the net radiation (Q^*) is simplified to the difference between the incoming infrared radiation emitted from the atmosphere ($L\downarrow$) and the outgoing infrared radiation emitted by the surface ($L\uparrow$). Mathematically, this is written as $Q^* = L^* = L\downarrow - L\uparrow$. Recall from Chapter 2 that the amount of radiation emitted by an object depends on its temperature to the fourth power. Since Earth's surface is generally warmer than the atmosphere, $L\uparrow$ is larger than $L\downarrow$, so, usually, Q^* is negative at night. The ground, being a much better radiator than air, is able to cool more quickly. Consequently, shortly after sunset, Earth's surface becomes cooler than the air directly above it. The air transfers some energy to the surface by both convection and conduction of sensible heat. Quantitatively, this is expressed as Q_H is negative as the air loses heat and cools, providing energy to offset the negative Q^* at the surface.

As the night progresses, the ground and the air in contact with it continue to cool more rapidly than does the air a few metres above the ground. There is some downward transfer of warmer upper air, but the process is slow due to weak convection, which often occurs with stable nighttime air, and the air's poor thermal conductivity. Therefore, by late night or early morning, the coldest air is found next to the surface, with slightly warmer air above it (see ● Figure 3.14).

This measured increase in air temperature just above the ground is known as a **radiation inversion** because it forms mainly through radiative cooling of the surface. Because radiation inversions occur on most clear, calm nights, they are also known as **nocturnal inversions**.

Radiation Inversions Strong radiation inversions occur when the air near the ground is much colder than the air higher up. Ideal conditions for a strong inversion exist when the air is calm, the night is long, there are no clouds, and the air is fairly dry. Let's examine these ingredients one by one.

A windless night is essential for a strong radiation inversion because winds tend to mix the colder air at the surface

FOCUS ON A SPECIAL TOPIC

Record High Temperatures

Most people are aware of the extreme heat that exists during the summer in the desert southwest of the United States. But how hot does it get there? On July 10, 1913, Greenland Ranch in Death Valley, California (● Figure 3), reported the highest temperature ever observed in North America: 57°C. This is only 1°C below the world record high temperature of 58°C measured in El Azizia, Libya, in 1922. Death Valley air temperatures are persistently hot throughout the summer, with the average maximum for July being 47°C. During the summer of 1917, there was an incredible period of 43 consecutive days when the maximum temperature reached 48°C or higher. In 1974, Death Valley recorded a high temperature of at least 38°C on 134 days.

In more humid climates, the maximum temperature rarely climbs above 41°C. However, during the record heat wave of 1936 that affected all of North America, the air temperature reached 49.4°C near Alton, Kansas, whereas in St. Albans, Manitoba, the temperature peaked at 44.4°C. In Ontario, a provincial high temperature record of 42.2°C was recorded at both Atikokan and Fort Frances. The 1936 heat wave killed 780 in Canada, mostly in Ontario.

These readings, however, do not hold a candle to the hottest place in the world. That distinction probably belongs to Dallol, Ethiopia. Dallol is located south of the Red Sea near latitude 12°N, in the hot, dry Danakil Depression. A

● **FIGURE 3** The hottest place in North America, Death Valley, California, where the air temperature reached 57°C, just 1°C below the world record high temperature of 58°C measured in El Azizia, Libya.

prospecting company kept weather records at Dallol from 1960 to 1966. During this time, the average daily maximum temperature exceeded 38°C every month of the year, except during December and January, when the average maximum lowered to 36.7°C and 36.1°C, respectively. On many days, the air temperature exceeded 49°C. The average annual temperature for the six years at Dallol was 34°C. In comparison, the average annual temperature in Yuma is 23°C and at Death Valley is 24°C. The highest temperature reading on Earth (under standard conditions) occurred northeast of Dallol at El Azizia, Libya (32°N), when, on

September 13, 1922, the temperature reached a scorching 58°C. ▼ Table 1 gives record high temperatures throughout the world.

Why don't daytime high temperatures climb even higher? Luckily, there are some limits. The first is the available net radiation that provides energy for the sensible heat flux. Net radiation ultimately depends on the amount of solar radiation, and this is limited by the solar constant, latitude, and date. The second limit is one set by the atmosphere. When surface air temperature rises, free convection increases and thermals mix this heat through a depth of the atmosphere. This mixing limits the temperature rise near the surface.

▼ Table 1 Some Record High Temperatures Throughout the World

LOCATION (LATITUDE)	RECORD HIGH TEMPERATURE (°C)	RECORD FOR	DATE
El Azizia, Libya (32°N)	58	The world	September 13, 1922
Death Valley, California, U.S.A. (36°N)	57	Western Hemisphere	July 10, 1913
Tirat Tsvi, Israel (32°N)	54	Middle East	June 21, 1942
Cloncurry, Queensland (21°S)	53	Australia	January 16, 1889
Seville, Spain (37°N)	50	Europe	August 4, 1881
Rivadavia, Argentina (35°S)	49	South America	December 11, 1905
Midale and Yellow Grass, Saskatchewan (49°N)	45	Canada	July 5, 1937
Esparanza, Antarctica (63°S)	14	Antarctica	October 20, 1956

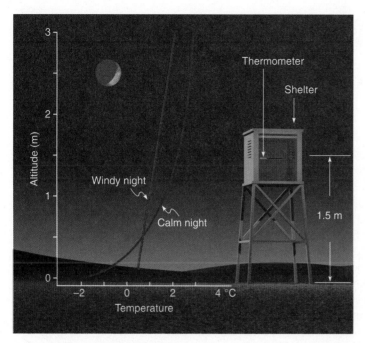

● **FIGURE 3.14** On a clear, calm night, the air near the surface can be much colder than the air above. The increase in air temperature with increasing height above the surface is called a radiation temperature inversion. Notice that the radiation inversion develops better on the calm night.

with the warmer air above, creating a layer with no thermal variation. In the absence of wind, the cooler, more dense surface air does not readily mix with the warmer, less dense air above it, and an inversion is more strongly developed, as illustrated in Figure 3.14.

A long night also contributes to a strong inversion. Generally, the longer the night, the longer the time of radiative cooling and the better the chances that air near the ground will be much colder than the air above. Consequently, winter nights provide the best conditions for strong radiation inversions.

Finally, radiation inversions are more likely with a clear sky and dry air. Under these conditions, the ground is able to radiate its infrared energy to outer space and thereby cool rapidly. This happens because the clear, dry atmosphere is a poor radiator, so incoming longwave radiation (L↓) is small and does little to offset the longwave radiation emitted by the surface (L↑). In this situation, the net longwave, $L^* = L\downarrow - L\uparrow$, is a large negative number, and this radiative energy loss rapidly cools the near-surface air. However, clouds and water vapour are more effective radiators than the clear atmosphere, so the incoming longwave radiation (L↓) is larger, making the net infrared radiation ($L^* = L\downarrow - L\uparrow$) less of an energy loss, and the surface cannot cool as quickly. Also, on humid nights, condensation in the form of fog or dew will release latent heat (Q_E), which warms the air (Q_E is negative as the air loses latent heat). This tends to offset the radiative cooling, limiting a radiation inversion that might be developing. In summary, radiation inversions may occur

on any night, but during long winter nights, when the air is still, cloud-free, and relatively dry, inversions can become strong and deep.

On winter nights in middle latitudes, it is common to experience below-freezing temperatures near the ground and air temperatures that are 5°C warmer at your waist. In middle latitudes, the top of the inversion, the region where the air temperature stops increasing with height, is usually less than 100 m above the ground. In dry, polar regions, where winter nights are measured in months, the top of the inversion is often 1000 m above the surface, but it can extend to as high as 3000 m.

So how cold air becomes at night depends primarily on the length of the night, the moisture content of the air, cloudiness, and the wind. Even though wind may initially bring cold air into a region, the coldest nights usually occur when the air is clear and relatively calm. But there are other factors that determine how cold the night air becomes. For example, a surface that is wet or covered with vegetation can add water vapour to the air, retarding nighttime cooling. If the air becomes saturated when cooled, then condensation will occur as dew forms on the surface. This will release latent heat, warming the air and limiting the nighttime cooling. Likewise, if the soil is a good heat conductor, the ground can lose heat to the surface (Q_G is negative) during the night, adding warmth to the surface and the air, restricting cooling. On the other hand, snow covering the ground acts as an insulating blanket and prevents heat stored in the soil from reaching the air. Snow, a good emitter of infrared energy, radiates energy away rapidly at night, which helps keep the air temperature above a snow surface quite low. Additional information on this topic is given in Focus on a Special Topic: Record Low Temperatures on p. 80.

Revisit Figure 3.13 (p. 76). Notice that the lowest temperature on any given day is usually observed around sunrise. However, the cooling of the ground and surface air may even continue beyond sunrise for a half-hour or so as outgoing energy can exceed incoming energy. This happens because light from the early-morning sun passes through a thick section of atmosphere and strikes the ground at a low angle. Consequently, the sun's energy does not effectively heat the surface. Also, surface heating may be reduced further when the ground is moist and the available energy is used for evaporation. (Any duck hunter who has waited for his prey lying flat in a marsh will recognize this experience as the sudden cooling that occurs as evaporation chills the air just after sunrise.) Hence, the lowest temperature may occur shortly after the sun has risen.

Cold, heavy surface air slowly drains downhill during the night and eventually settles in low-lying basins and valleys. Consequently, valley bottoms are colder than the surrounding hillsides (see ● Figure 3.15). In middle latitudes, these warmer hillsides, called **thermal belts**, are less likely to experience freezing temperatures than the valley below. This encourages farmers to plant trees on hillsides that are unable to survive the valley's low temperature. On the valley floor, the cold, dense air is unable to rise. Smoke and other pollutants trapped in this heavy air restrict visibility. Therefore,

FOCUS ON A SPECIAL TOPIC

Record Low Temperatures

The coldest areas in North America are found in the Yukon, Northwest Territories, and Nunavut. Qausuittuq (Resolute), Nunavut (latitude 75°N), has an average temperature of −32°C for the month of January. The coldest month on record in Canada is at Eureka, Nunavut, with a mean minimum temperature of −50.1°C in February 1979.

The lowest temperatures and coldest winters in the Northern Hemisphere are found in the interior of Siberia and Greenland. For example, the average January temperature in Yakutsk, Siberia (latitude 62°N), is −43°C. There the mean temperature for the entire year is a bitterly cold −11°C. At Eismitte, Greenland, the average temperature for the coldest month, February, is −47°C, with the mean annual temperature being a frigid −30°C. Even though these temperatures are extremely low, they do not come close to the coldest area of the world: the Antarctic.

At the geographic South Pole, over 2800 m above sea level, where the Amundsen–Scott scientific station has been keeping records for more than 40 years, the average temperature for their peak winter month of July is −59°C and the mean annual temperature is −49°C.

The lowest temperature ever recorded there was −83°C. It occurred under clear skies with a light wind on the morning of June 23, 1983. Cold as it was, it is not the world's lowest recorded temperature. That distinction belongs to the Russian station at Vostok, Antarctica, located at latitude 78°S, where air temperatures plummeted to −89.2°C on July 21, 1983. At these temperatures, a drop of saliva falling from the lips of a person taking an observation would have frozen solid before reaching the ground. See ▼ Table 2 for record low temperatures throughout the world.

▼ Table 2   Some Record Low Temperatures Throughout the World

LOCATION (LATITUDE, ELEVATION)	RECORD LOW TEMPERATURE (°C)	RECORD FOR	DATE
Vostok, Antarctica (78°S, 3420 m)	−89	The world	July 21, 1983
Verkhoyansk, Russia (67°N, 101 m)	−68	Northern Hemisphere	February 7, 1892
Northice, Greenland (72°N, 2343 m)	−66	Greenland	January 9, 1954
Snag, Yukon (62°N, 646 m)	−63	North America	February 3, 1947
Prospect Creek, Alaska (66°N, 309 m)	−62	U.S.A.	January 23, 1971
Sarmiento, Argentina (34°S, 268 m)	−33	South America	June 1, 1907
Ifrane, Morocco (33°N, 1635 m)	−24	Africa	February 11, 1935
Charlotte Pass, Australia (36°S, 1755 m)	−22	Australia	July 22, 1949

valley bottoms are not only colder but are also more frequently polluted than nearby hillsides. Even when the land is only gently sloped, cold air settles into lower lying areas, such as river basins and floodplains. Because the flat floodplains are agriculturally rich areas, cold air drainage often forces farmers to seek protection for their crops, as described in the next section.

So far, we have looked at how and why the air temperature near the ground changes during the course of a 24-hour day. We saw that during the day, the air near Earth's surface can become quite warm, whereas at night, it can cool off dramatically. The temperature responds to sensible heating: when sensible heat flows into the air, the temperature rises, and when sensible heat leaves the air, the temperature falls. ● Figure 3.16 summarizes these observations by illustrating

how the average air temperature above the ground can change over a span of 24 hours. Notice in the figure that although the air a few metres above the surface both cools and warms, it does so at a slower rate than air at the surface. Also observe that the warmest part of the day a few metres above the surface occurs at 3 p.m. (local solar time*), whereas the surface reaches its maximum temperature at solar noon, when the sun's energy is the most intense.

The discussions of daily temperature variations so far are typical scenarios that occur over a day when no major weather systems are affecting the area. When storms are nearby, they

*Solar time is defined so that at solar noon, the sun is directly in the south (Northern Hemisphere) at a given location.

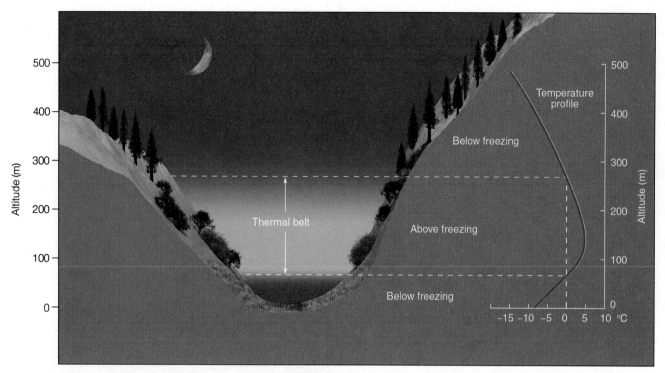

• **FIGURE 3.15** On cold, clear nights, the settling of cold air into valleys makes them colder than surrounding hillsides. The region along the side of the hill where the air temperature is above freezing is known as a *thermal belt*.

can cause temperature to change in different ways. This is caused by temperature *advection*—the horizontal movement of heat and temperature by wind that occurs with many storms. Advection is especially important when warm or cold **fronts** are expected as part of the storm. Fronts are zones that separate warm air masses from cold air masses, so when fronts move through, the air temperature can be expected to change. Fronts and air masses are discussed in more detail in Chapter 11.

Protecting Crops from the Cold   The concepts we have been discussing have practical applications that have been used by farmers and gardeners for some time. On unexpectedly cold nights, early frost damage can be reduced for many crops, small plants, or shrubs by covering them with straw, cloth, plastic sheeting, or any material that covers the plants. This prevents ground heat from being radiated away to the colder surroundings. Even though covering the plant does not change its outgoing longwave emission ($L\uparrow$), the cloth or plastic cover has the effect of increasing the incoming longwave radiation ($L\downarrow$), so the net longwave radiation ($L^* = L\downarrow - L\uparrow$) is less of a loss, and the plant cools less. This occurs because the cover is a more effective radiator than the open sky and its temperature is warmer than the effective temperature of the open sky, so the infrared radiation emitted by the cover back to the plant is higher that it would be if the cover were not there. If you are a household gardener concerned about outside flowers and

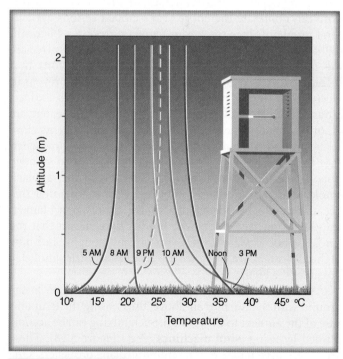

ACTIVE FIGURE 3.16   An idealized distribution of air temperature above the ground during a 24-hour day. The temperature curves represent the variations in average air temperature above a grassy surface for a mid-latitude city during the summer under clear, calm conditions. Visit the textbook's website to view this and other Active Figures at www.ahrensmeteorology1ce.nelson.com

© C. Donald Ahrens

● FIGURE 3.17 Orchard heaters circulate the air by setting up convection currents and warming the trees.

© C. Donald Ahrens

● FIGURE 3.18 Wind machines mix cooler surface air with warmer air above.

plants during cold weather, simply wrap them in plastic or cover each with a paper cup.

Fruit trees are particularly vulnerable to cold weather during their spring blossoming period. Frost damage at this time can obliterate the entire year's crop, a serious agricultural problem. Since the lowest temperatures on a clear, still night occur near the surface, the lower branches of a tree are the most susceptible to this damage, and solutions that increase air temperature close to the ground can prevent it. **Orchard heaters** are one way to warm the trees and air around them. As shown in ● Figure 3.17, orchard heaters are oil or gas-fired burners that set up convection currents close to the ground and emit infrared radiation as they glow when they burn. Early forms of these heaters were called smudge pots because they produced large amounts of dense, black smoke that caused severe pollution. People tolerated this because they believed that the smoke acted like a blanket, trapping some of Earth's heat. Studies have shown that the smoke is not as significant as previously thought, and now orchard heaters are designed to produce as little smoke as possible (see Figure 3.17).

Another way to protect trees is to mix the cold air at the ground with the warmer air above, thus raising the temperature of the air next to the ground. Such mixing can be accomplished by using **wind machines** (see ● Figure 3.18). These are power-driven fans that resemble airplane propellers. One significant benefit of wind machines is that they can be thermostatically controlled to turn on and off at prescribed temperatures. Farmers without their own wind machines sometimes hire helicopters to perform the same operation; although effective in mixing the air, they are expensive.

Where sufficient water is available, trees can be protected by irrigation. On potentially cold nights, farmers can flood the orchard. Because water has a high heat capacity, it cools more slowly than dry soil. Consequently, the surface does not become as cold as it would if it were dry. Furthermore, wet soil has a higher thermal conductivity than dry soil. Hence, in wet soil, heat is conducted upward from subsurface soil more rapidly (Q_G is negative as the ground loses heat to the surface), which helps keep the surface warmer.

Another problem that can affect our food supply occurs when, unexpectedly, freezing air moves into a growing region where the plants are not able to tolerate these conditions. This is known as a **freeze** and occurs when the surface air temperature remains below freezing long enough to damage certain agricultural crops. The terms *frost* and *freeze* are often used interchangeably by various segments of society. But to growers of perennial crops, such as apples and citrus fruits, who have to protect them against damaging low temperatures, it makes no difference if visible "frost" is present or not. The concern is whether or not the plant tissue has been exposed to temperatures equal to or below 0°C. The actual freezing point of the plant, however, can vary because perennial plants develop hardiness in the fall that usually lasts through the winter and then wears off gradually in the spring. A single freeze in California or Florida can cause several million dollars in damage to citrus crops. For example, several freezes during the spring of 2001 caused millions of dollars in damage to California's north coast vineyards, which resulted in higher wine prices.

Protecting an orchard from this type of damaging cold air is difficult. Wind machines will not help because they would only mix cold air at the surface with the colder air above. Orchard heaters and irrigation are of little value as they would protect only the branches just above the ground. However, one form of protection does work: an orchard's sprinkling system may be turned on so that it emits a fine spray of water. In the cold air, the water freezes around the branches and buds, coating them with a thin veneer of ice. As long as the spraying continues, the latent heat that is given off as the water changes into ice keeps the ice temperature at 0°C. The ice acts as a protective coating against the subfreezing air by keeping the buds or fruit at a temperature higher than their damaging point. Care must be taken because too much ice can cause the branches to break. The fruit may be saved from the cold air, whereas the tree itself may be damaged by too much protection. Sprinklers work well when the air is fairly humid. They do not work well when the air is dry as a good deal of the water may be lost through evaporation.

BRIEF REVIEW

Up to this point, we have examined temperature variations on a seasonal and daily basis. Here is a review of some of the important concepts and facts we have covered:

- Temperature changes in response to sensible heating or cooling of the air.
- The seasons are caused by Earth being tilted on its axis as it revolves around the sun. The tilt causes annual variations in the amount of sunlight that strikes the surface as well as variations in the length of time the sun shines at each latitude.
- During the day, Earth's surface and air above will continue to warm as long as incoming energy (mainly sunlight) exceeds outgoing energy from the surface.
- At night, Earth's surface cools, mainly by giving up more infrared radiation than it receives, a process called radiational or radiative cooling.
- The coldest nights normally occur in winter when the nights are long, the air is calm and fairly dry (i.e., has a low water vapour content), and the sky is cloud-free.
- The highest temperatures during the day and the lowest temperatures at night are normally observed at Earth's surface due to radiation absorption and losses.
- Radiation inversions exist usually at night, when the air near the ground is colder than the air above it.
- Near storms and fronts, the typical daily temperature patterns are masked by temperature changes resulting from the movement of warmer or colder air into the area due to the storm (i.e., warm or cold air advection).

The Geographic Controls of Temperature

The geographic controls of temperature are the main factors that cause variations in temperature from one place to another. Earlier, we listed the physical controls of temperature and saw that the greatest factor in determining temperature is the amount of solar radiation that reaches Earth's surface. This, of course, is determined by the length of daylight hours and the intensity of incoming solar radiation. Both of these factors are a function of latitude; hence, latitude is considered an important geographic control of temperature. The main geographic controls are

1. latitude
2. land and water distribution
3. ocean currents
4. elevation

We can obtain a better picture of these controls by examining ● Figure 3.19 and ● Figure 3.20, which show the average monthly temperatures throughout the world for January and July. The colour-filled lines on the map are isotherms, lines connecting places that have the same temperature. Figures 3.19 and 3.20 show the importance of latitude on temperature. Notice that the average temperatures decrease as you move poleward in both January and July. Also, in the Northern Hemisphere, the January isotherms are closer together than they are in July. This happens because the difference in solar radiation between the equator and poles is larger in winter than it is in summer. For example, if you travel from New Orleans to Toronto in January, you are more likely to experience a greater temperature change than if you make the same trip in July. Also notice that in both figures, the isotherms do not run horizontally; in many places, they bend, especially where they approach an ocean–continent boundary and over higher terrain, such as the Rocky Mountains, the Andes, and the Himalayas.

On the January map, temperatures are much lower in the middle of continents than they are for the same latitude near the oceans. On the July map, the reverse is true. These temperature variations can be attributed to the unequal heating and cooling properties of land and water. Solar energy reaching land is absorbed in a thin layer of soil, but when reaching water, it penetrates deeply. Because water is able to circulate, it distributes heat from solar energy through a much deeper layer. Also, some of the energy striking the water evaporates rather than heats it. Another important reason for the temperature contrasts is that water has a high specific heat. As we saw in Chapter 2, it takes a great deal more heat to raise the temperature of 1 kg of water 1°C than it does to raise the temperature of 1 kg of soil or rock by 1°C. Water not only heats more slowly than land, it cools more slowly as well, so the oceans act as huge heat reservoirs. Thus, mid-ocean surface temperatures change relatively little from

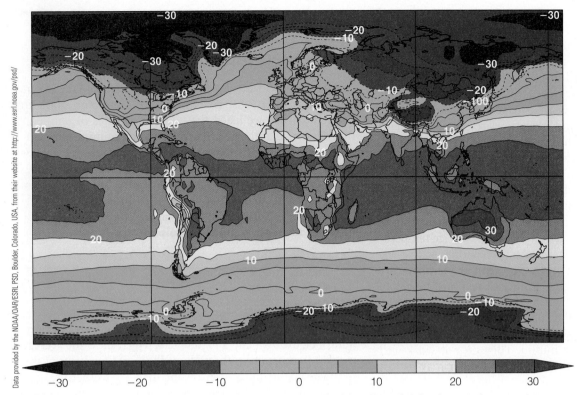

● FIGURE 3.19 Average January surface air temperatures in degrees Celcius (°C) for the period 1980 to 2009. Dashed contours represent temperatures below zero.

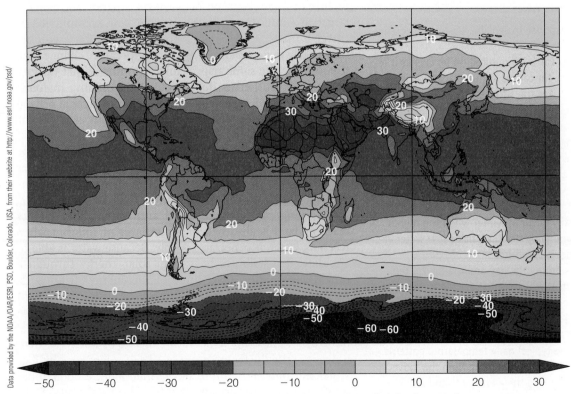

● FIGURE 3.20 Average July surface air temperatures in degrees Celcius (°C) for the period 1980 to 2009. Dashed contours represent temperatures below zero.

summer to winter compared to the much larger annual temperature changes over the middle of continents.

Notice that the isotherms indicate cooler temperatures over higher terrain. This is because temperature generally cools with elevation, so near-surface air in the mountains is cooler than near-surface air in the surrounding lowlands and plains.

Along the margins of continents, ocean currents often influence air temperatures over land. For example, along eastern coasts, warm ocean currents transport warm water poleward, whereas along western coasts, they transport cold water toward the equator. As we will see in Chapter 10, some coastal areas also experience upwelling, a process that brings cold water from ocean depths to the surface.

Even large lakes can modify the temperature around them. In summer, Ontario's Great Lakes remain cooler than the land. As a result, refreshing breezes blow inland, bringing relief from the sometimes sweltering heat. As winter approaches, the water cools more slowly than the land. The first blast of cold air from the Arctic is modified as it crosses the lakes, so the first freeze is delayed downwind of the Great Lakes. A similar phenomenon occurs over other large lakes, such as Lake Winnipeg, Manitoba. The late frost in these areas effectively extends the growing season near these large lakes. In spring, the reverse happens—the cold water takes longer to warm up after winter than the adjacent land, so areas near the lakes have delayed warming in spring and summer. This can also benefit agriculture as plants delay their initial spring growth and are therefore less likely to suffer damage from any late-spring frosts. A similar situation delays the onset of spring in the Maritime provinces, where cold ocean temperatures offshore delay warming over land and the onset of spring.

Air Temperature Data

The careful recording and application of temperature data are tremendously important to us all. Without this accurate information, farmers, power company engineers, weather analysts, and many others would find their work much more difficult. In these next sections, we will study the ways temperature data are organized and used. We will also examine the significance of daily, monthly, and yearly temperature ranges and averages in terms of practical application to everyday living.

DAILY, MONTHLY, AND YEARLY TEMPERATURES The greatest variation in daily temperature occurs right at Earth's surface. The difference between the daily maximum and minimum temperature, called the **diurnal** or **daily range of temperature**, is greatest next to the ground and becomes progressively smaller as we move away from the surface. This is experienced by the residents living in the high-rise apartment shown in • Figure 3.21. The diurnal temperature variation is also much larger on clear days than on cloudy ones.

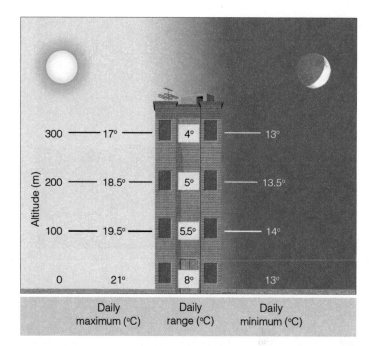

• FIGURE 3.21 The daily range of temperature decreases as we climb away from Earth's surface. Hence, there is less day-to-night variation in air temperature near the top of a high-rise apartment complex than at the ground level.

The largest diurnal temperature variations occur over high-elevation deserts, where the air is often cloud-free, and there is less water vapour. With these elements missing from the atmosphere, the emission of incoming longwave (L↓) is less, so at night, the surface cools rapidly. By day, clear summer skies allow the sun's energy to reach the surface. The dry surface allows all the energy to be used in warming the ground, which, in turn, warms the air above it. In these environments, daytime temperatures sometimes exceed 45°C, whereas at night, the surface cools rapidly by radiating infrared energy to space, and minimum temperatures can dip below freezing, giving large daily temperature ranges. In Tucson, Arizona, diurnal temperature variations as high as 57°C have been reported. At Puntzi Mountain on the Chilcotin plateau of British Columbia's southwest interior, diurnal temperature ranges of more than 32°C have been observed.

Clouds can have a large effect on the daily temperature range. As we saw in Chapter 2, clouds, especially low, thick ones, are good reflectors of incoming solar radiation, so they prevent much of the sun's energy from reaching the surface. This effect tends to lower daytime temperatures (see • Figure 3.22a). If the clouds persist into the night, they tend to keep nighttime temperatures higher as they are excellent absorbers and emitters of infrared radiation. Clouds actually emit a great deal of infrared energy back to the surface, increasing the L↓, which makes the net longwave radiation, L*, less of an energy loss, so the surface cools less. Clouds, therefore, have the effect of lowering the daily range of temperature. In clear weather

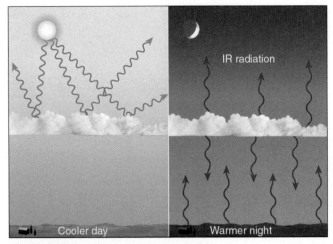

(a) Small daily temperature range

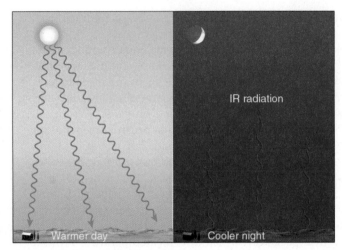

(b) Large daily temperature range

● FIGURE 3.22 (a) Clouds tend to keep daytime temperatures lower and nighttime temperatures higher, producing a small daily temperature range. (b) In the absence of clouds, days tend to be warmer and nights cooler, producing a larger daily temperature range.

(see ● Figure 3.22b), daytime air temperatures tend to be higher as the sun's rays impinge directly on the surface, whereas nighttime temperatures are usually lower due to rapid radiational cooling. Therefore, clear days and clear nights combine to promote a large diurnal temperature ranges.

Humidity can also have an effect on diurnal temperature ranges. In humid regions, the diurnal temperature range is usually small as haze and clouds lower the maximum temperature by preventing some of the sun's energy from reaching the surface. At night, the moist air keeps the minimum temperature high by increasing L↓, the amount of infrared radiation reaching the surface. This reduces the net longwave radiation loss, L*, so there is less cooling. An example of a humid city with a small diurnal temperature range is tropical Iquitos, Peru, at 3.75°S latitude. Here in July, the average daily maximum temperature is 30.9°C and the average daily minimum is 20.8°C, making the diurnal range only 10.1°C.

Cities near large bodies of water typically have smaller diurnal temperature ranges than cities farther inland. This phenomenon is caused in part by the additional water vapour in the air and by the fact that water warms and cools much more slowly than land. For example, in July, Tofino, located on British Columbia's West Coast, has an average daily maximum temperature of 18.5°C and an average daily minimum of 10.2°C, for a diurnal range of 8.3°C. Moreover, rural locations often have larger diurnal temperature ranges than urban areas. The reason for this fact is that nighttime temperatures in cities tend to be warmer than those in outlying rural areas. This nighttime city warmth, called the **urban heat island**, is due to industrial and urban development (for more on this, see page 585 in Chapter 18).

The average of the highest and lowest temperatures for a 24-hour period is known as the **mean** or **average daily temperature**. Most media list the mean daily temperature along with the highest and lowest temperatures for the preceding day when they report weather. Mean daily temperatures for a particular date that are averaged over a 30-year period are called the average or "normal" temperatures for that date. The average temperature for each month is determined by averaging the daily mean temperatures for that month. The concept of "*normal*" temperature is discussed in more detail in the Focus on a Special Topic: When It Comes to Temperature, What's Normal? section on page 87.

At any location, the difference between the average temperature of the warmest and coldest months is called the **annual range of temperature**. Usually, the largest annual ranges occur over land, whereas the smallest occur over water. For an example of this, see ● Figure 3.23, illustrating temperatures at Cape Scott on the northern tip of Vancouver Island, compared to Kamloops at the same latitude in the British Columbia southern interior. The annual temperature range at Cape Scott is 8.9°C, whereas in Kamloops, it is 25.2°C. Winnipeg, Manitoba, is located slightly further south but has an annual temperature range of 37.3°C.

Near the equator, annual temperature ranges are small, usually less than 3°C, because daylight length has little variation and the sun is always high in the sky at noon. Quito, Ecuador, on the equator at an elevation of 2850 m, experiences an annual temperature range of less than 1°C. In the middle and high latitudes, large seasonal variations in the amount of sunlight reaching the surface produce large temperature contrasts between winter and summer. Consequently, the annual temperature ranges are large, especially in the middle of a continent. Yakutsk, in northeastern Siberia near the Arctic Circle, has an extremely large annual temperature range of 62°C.

The average temperature of any location for the entire year is known as the **mean annual temperature**. (The mean annual temperature may be obtained by taking the sum of the 12 monthly means and dividing that total by 12, or by obtaining the sum of the daily means and dividing that total by 365.) When two cities have the same mean annual temperature, it might first seem that their temperatures throughout the year are quite similar. However, often this is not the case.

FOCUS ON A SPECIAL TOPIC

When It Comes to Temperature, What's Normal?

When a meteorologist reports that "the normal low for today is −15°C," does this mean that the day's low temperature is usually −15°C? Or does it mean that we should expect a low temperature near −15°C? Actually, it means neither.

Remember that the word "normal," or "norm," refers to weather data averaged over a period of 30 years. Graphically, it might look like ● Figure 4, which shows the low temperatures measured on January 15 over 30 years in Fort St. James (in the central interior of British Columbia). The average, hence the "normal" low temperature for this period, is −15°C and is represented by the dashed line on the graph. But looking more closely at these data shows that only two days during this 30-year period (highlighted with red dots) had a low temperature of −15°C. The most common low temperature was −19°C, occurring on three days (blue dots).

So if −15°C is the normal low temperature, what temperature should we expect on this date? To answer this, we need to represent the range of values that could be expected. Here is where a statistic that measures the spread of the data, called the standard deviation, can help. Two standard deviations represent 95 percent of the spread in most data sets. So in our case, 95 percent of the time the low temperature for this day will be between −33°C and +3°C or two times the standard deviation of 9°C on either side of −15°C.

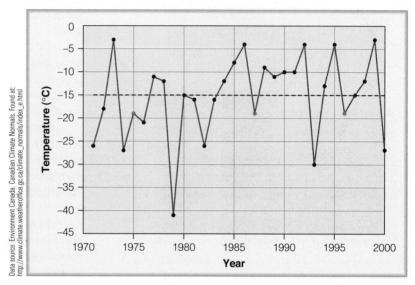

Data source: Environment Canada. Canadian Climate Normals. Found at: http://www.climate.weatheroffice.gc.ca/climate_normals/index_e.html

● FIGURE 4 The daily low temperature measured (for 30 years) on January 15 in Fort St. James in the central interior of British Columba (latitude 54.4° N). The dashed line represents the normal daily low temperature for the 30-year period.

For example, Cape Scott, British Columbia, and Kamloops, British Columbia, are at the same latitude (about 50.7°N). Both have similar hours of daylight during the year. Both have the same mean annual temperature, of 8.9°C. But here the similarities end. The temperature differences between the two cities are apparent to anyone travelling to Cape Scott during the summer with a suitcase full of clothes suitable for summer weather in Kamloops. Figure 3.23 summarizes the average temperatures for Cape Scott and Kamloops. Notice that the coldest month for both cities is January. Even though, on average, January temperatures in Kamloops are 9.1°C colder than those in Cape Scott, residents' experience of their weather is quite different. People in Kamloops awaken to average minimum temperatures of −7.6°C; this is almost the lowest temperature ever recorded at Cape Scott in January (−8.3°C). Plants that thrive at Cape Scott would find it difficult to survive the winter in Kamloops. So even though Cape Scott and Kamloops have the same mean annual temperature, the pattern and range of their temperatures differ greatly.

THE USE OF TEMPERATURE DATA Engineers estimate societal energy needs by using daily temperature data to develop the concept of a **heating degree-day**. A heating degree-day is based on the assumption that people will use their furnaces when the mean daily temperature drops below 18°C. It is determined by subtracting the mean temperature for the day from 18°C. For example, if the mean temperature for a particular day is 17°C, there would be one heating degree-day on this day (18°C − 17°C = 1°C). Lower average daily temperatures will result in more heating degree-days and greater predicted

WEATHER WATCH

One of the greatest temperature ranges ever recorded in the Northern Hemisphere occurred at Browning, Montana, on January 23, 1916, when the air temperature plummeted from 7°C to −49°C—a 56°C temperature range in less than 24 hours.

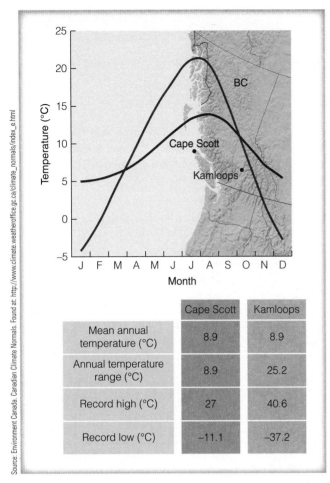

	Cape Scott	Kamloops
Mean annual temperature (°C)	8.9	8.9
Annual temperature range (°C)	8.9	25.2
Record high (°C)	27	40.6
Record low (°C)	−11.1	−37.2

Source: Environment Canada. Canadian Climate Normals. Found at: http://www.climate.weatheroffice.gc.ca/climate_normals/index_e.html

● FIGURE 3.23 Temperature data for Cape Scott and Kamloops, British Columbia. Two cities at the same latitude with the same mean annual temperature, but Cape Scott is a coastal location, whereas Kamloops is an interior location.

Figures 3.24 and 3.25 can indicate the energy requirements for an area over the year.

Farmers use an index called **growing degree-days** to guide their planting schedule and determine when a crop should be ready for harvesting. There are a variety of methods of computing growing degree-days. The most common uses the mean daily temperature because air temperature is a key factor in the physiological development of plants. Normally, a growing degree-day is defined as a day when the mean daily temperature is one degree above the *base or zero temperature* (the minimum temperature required for growth) of a particular crop. ▼ Table 3.2 provides base temperatures for particular crops. For sweet corn, the base temperature is 10°C. If the mean daily temperature on a summer day in southern Ontario, was 23°C, sweet corn would accumulate 13 (23 − 10) growing degree-days on that day. From Table 3.2, sweet corn could be harvested after 1220 growing degree-days. So if sweet corn is planted in early April and averages 11 growing degree-days every day after it was planted, the corn could be harvested about 110 days later, or around the middle of July.

In humid southern Ontario, summer nighttime temperatures are also high, and this increases the daily mean temperature (remember that daily mean temperatures are averaged over 24 hours). Consequently, growing degree-days accumulate much faster here, and corn matures in fewer days. In drier areas, summer nighttime temperatures are lower, daily mean temperatures are lower, and each day accumulates fewer growing degree-days. So although moisture is not taken into account, air temperature is affected through humid air's greater absorption and downward emission of infrared energy, which increases the daily mean air temperature. The benefits to crops of warmer temperatures are not limitless, and for every crop, temperatures above a certain point are stressful. For corn, temperatures above 30°C slow growth, so any daily temperature over 30°C is reduced to 30°C when computing the mean daily air temperature.

fuel consumption. When the number of heating degree-days for a whole year is calculated, the heating fuel requirements for any location can be estimated. ● Figure 3.24 shows the yearly average number of heating degree-days across North America.

As the mean daily temperature rises above 18°C, many people cool their indoor environments. So a similar index, called the **cooling degree-day**, is used to estimate energy needs for air conditioning. Cooling degree-days are calculated by subtracting 18°C from the mean daily temperature. For example, a day with a mean temperature of 21°C would correspond to three cooling degree-days (21 − 18). High cooling degree-day values indicate warm weather and higher power consumption for air cooling (see ● Figure 3.25). Knowledge of the number of cooling degree-days in an area allows builders to match the size and type of equipment that should be installed to provide adequate air conditioning. Also, by forecasting cooling degree-days, power companies can anticipate energy demands during peak energy-use periods. Maps of heating and cooling degree-days such as

Air Temperature and Human Comfort

The same air temperature can feel differently on different occasions. For example, a temperature of 20°C on a clear, windless March afternoon can feel balmy after a long, hard winter. Yet this same temperature can feel uncomfortably cool on a summer afternoon when accompanied by a stiff breeze. The reason for these perceptions is related to how we exchange heat energy with our environment.

By converting food into heat, our body's *metabolism* stabilizes its temperature. To maintain a constant temperature, the heat we produce and the heat our bodies absorb must equal the heat we lose to our surroundings. Consequently, there is a constant exchange of heat between the body, especially at the surface of the skin, and the environment. We lose heat by emitting infrared energy, but we absorb it as well. We also absorb solar radiation, so it feels warmer in the sun than in the shade even when the air temperature is the same.

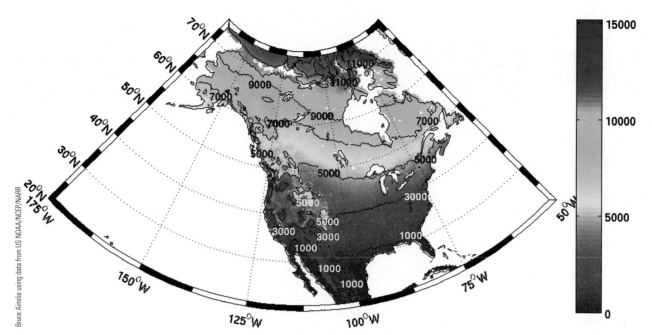

Bruce Ainslie using data from US NOAA/NCEP/NARR

● FIGURE 3.24 North American average heating demand as represented by annual heating degree-days. The contour colours indicate the relative amount of energy needed for space heating over a year. Higher numbers occur in areas with colder temperatures (represented in bluer colours), while lower values have warmer temperatures and redder colours.

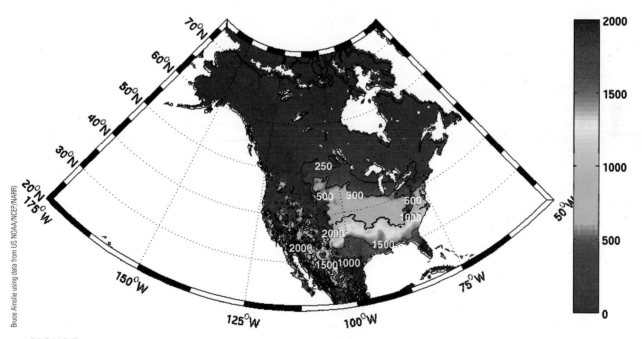

Bruce Ainslie using data from US NOAA/NCEP/NARR

● FIGURE 3.25 North American average cooling demand as represented by annual cooling degree-days. The contour colours indicate the relative amount of energy needed for space cooling over a year. Higher numbers occur in areas with warmer temperatures (represented in redder colours), while lower values have cooler temperatures and bluer colours.

We also lose and gain heat by conduction and convection. For example, on a cold day, a thin layer of warm, insulating air molecules* forms close to the skin, protecting us

from the surrounding cooler air and the rapid transfer of heat by convection. This is why when the air is calm, the temperature we perceive, called the **sensible temperature**, is often higher than a thermometer would indicate. (Could the opposite effect occur where the air temperature is very high and a person might feel exceptionally cold? If you are unsure, read Focus on an Observation: A Thousand Degrees and Freezing to Death, p. 91.)

*This thin layer, called the **laminar boundary layer**, is only a few millimetres thick. There is no convection, so conduction is the only way that sensible heat transfers through this layer. Conduction in air is not very efficient, so this allows our skin temperature to be much greater than the air temperature without having large heat losses from our body.

Table 3.2 Estimated Growing Degree-Days for Certain Naturally Grown Agricultural Crops to Reach Maturity

CROP (VARIETY, LOCATION)	BASE TEMPERATURE (°C)	GROWING DEGREE-DAYS TO MATURITY
Beans (snap/South Carolina)	10	670–720
Corn (sweet/Indiana)	10	1220–1560
Cotton (Delta Smooth Leaf/Arkansas)	15.6	1050–1390
Peas (early/Indiana)	4.4	610–670
Rice (Vegold/Arkansas)	15.6	940–1170
Wheat (Indiana)	4.4	1170–1340

Once the wind starts to blow, the insulating layer of warm air is swept away, and the constant bombardment of cold air rapidly removes heat from the skin by convection. When all other factors are the same, the faster the wind blows, the greater the heat loss, and the colder we feel. How cold the wind makes us feel is expressed as a **wind-chill index (WCI)**, better known as wind chill.

The modern wind-chill index (see Table 3.3) was formulated in 2001 and is used by Environment Canada and the U.S. National Weather Service to indicate how the cooling effect of wind makes us feel. The new index recognizes that the wind speed we feel occurs at about 1.5 m above the ground (instead of at 10 m above the ground where "official" wind readings are taken). In addition, it translates the ability of the air to take heat away from a person's exposed skin into a wind-chill equivalent temperature.* For example, Table 3.3 indicates that an air temperature of −20°C with a wind speed of 20 km h^{-1} produces a wind-chill equivalent temperature of −30.5°C. This means that a person's exposed skin would lose as much heat in air with a temperature of −20°C and a wind speed of 20 km h^{-1} as it would in calm air with a temperature of −30.5°C. Of course, how cold we feel actually depends on a number of factors in addition to wind chill, including the fit and type of clothing we wear, the amount of sunshine striking the body, and the actual amount of exposed skin. High winds, in below-freezing air, can remove heat from exposed skin so quickly that the skin may actually freeze. Called **frostbite**, it occurs on extremities first because they are furthest away from the source of body heat.

In cold weather, wet skin can make us feel colder. A cold rainy, drizzly, or even foggy day often feels colder than a "dry" one because water evaporates and cools the skin. This removes heat from the body in the forms of latent heat and sensible heat, whereas dry skin feels only sensible heat loss. In cold, wet, and windy weather, a person may lose body heat faster than the body can produce it, causing a drop in core body temperature and resulting in a condition known as **hypothermia**. This can occur in relatively mild weather with air temperatures as high as 10°C. In cases of severe hypothermia, rapid, progressive mental and physical collapse accompanies the lowered human body temperature. The first symptom of hypothermia is exhaustion. If exposure continues, judgment

*The wind-chill equivalent temperature formula is wind chill (°C) = $13.12 + 0.6215T - 11.37\,(V^{0.16}) + 0.3965T\,(V^{0.16})$, where T is the air temperature in °C and V is the wind speed in km h^{-1}.

Table 3.3 Wind-Chill Equivalent Temperature (°C)*

WIND SPEED (km h^{-1})	AIR TEMPERATURE (°C)											
Calm	10	5	0	−5	−10	−15	−20	−25	−30	−35	−40	−45	−50
10	8.6	2.7	−3.3	−9.3	−15.3	−21.1	−27.2	−33.2	−39.2	−45.1	−51.1	−57.1	−63.0
15	7.9	1.7	−4.4	−10.6	−16.7	−22.9	−29.1	−35.2	−41.4	−47.6	−51.6	−59.9	−66.1
20	7.4	1.1	−5.2	−11.6	−17.9	−24.2	−30.5	−36.8	−43.1	−49.4	−55.7	−62.0	−68.3
25	6.9	0.5	−5.9	−12.3	−18.8	−25.2	−31.6	−38.0	−44.5	−50.9	−57.3	−63.7	−70.2
30	6.6	0.1	−6.5	−13.0	−19.5	−26.0	−32.6	−39.1	−45.6	−52.1	−58.7	−65.2	−71.7
35	6.3	−0.4	−7.0	−13.6	−20.2	−26.8	−33.4	−40.0	−46.6	−53.2	−59.8	−66.4	−73.1
40	6.0	−0.7	−7.4	−14.1	−20.8	−27.4	−34.1	−40.8	−47.5	−54.2	−60.9	−67.6	−74.2
45	5.7	−1.0	−7.8	−14.5	−21.3	−28.0	−34.8	−41.5	−48.3	−55.1	−61.8	−68.6	−75.3
50	5.5	−1.3	−8.1	−15.0	−21.8	−28.6	−35.4	−42.2	−49.0	−55.8	−62.7	−69.5	−76.3
55	5.3	−1.6	−8.5	−15.3	−22.2	−29.1	−36.0	−42.8	−49.7	−56.6	−63.4	−70.3	−77.2
60	5.1	−1.8	−8.8	−15.7	−22.6	−29.5	−36.5	−3.4	−50.3	−57.2	−64.2	−71.1	−78.0

*Shaded areas represent conditions where frostbite occurs in 30 minutes or less

FOCUS ON AN OBSERVATION

A Thousand Degrees and Freezing to Death

Is there somewhere in our atmosphere where the air temperature can be exceedingly high (say above 1000°C), yet a person might feel extremely cold? There is such a region, but it is not at Earth's surface.

You may recall from Chapter 1 (see Figure 1.10, p. 16) that in the upper part of our atmosphere—the middle and upper thermosphere—air temperatures can exceed 1000°C. However, a thermometer shielded from the sun in this region would indicate an extremely low temperature. This apparent discrepancy lies in the meaning of air temperature and how we measure it.

In Chapter 2, we learned that air temperature is directly related to the average speed at which the air molecules are moving—faster speeds correspond to higher temperatures. In the middle and upper thermosphere, at altitudes approaching 300 km, air molecules are zipping

about at speeds corresponding to extremely high temperatures. However, to transfer enough energy to heat something up by conduction, an extremely large number of molecules must collide with the object. In the "thin" air of the upper atmosphere, air molecules are moving extraordinarily fast, but there are simply not enough of them bouncing against the thermometer bulb for it to register a high temperature. In fact, when properly shielded from the sun, the thermometer bulb loses far more energy than it receives and would read a temperature near absolute zero. This explains why an astronaut, when walking in space, will not only survive temperatures exceeding 1000°C but also feels a profound coldness when shielded from the sun's radiant energy (see ● Figure 5). Consequently, at these high altitudes, the traditional meaning of air temperature, how "hot" or "cold" something feels, is no longer applicable.

● FIGURE 5 How can an astronaut survive when the "air" temperature is 1000°C?

and reasoning begin to disappear. Prolonged exposure, especially at temperatures near or below freezing, produces stupor and collapse. Death occurs when the internal body temperature drops to 26°C. Most cases of hypothermia occur when the air temperature is between 0°C and 10°C. This may be because many people apparently do not realize that wet clothing in windy weather greatly enhances the loss of body heat, even when the temperature is well above freezing.

In cold weather, heat is more easily dissipated through the skin. To counteract this rapid heat loss, the peripheral blood vessels of the body constrict, cutting off the flow of blood to the outer layers of the skin. In hot weather, the blood vessels enlarge, allowing a greater loss of heat energy to the surroundings. In addition, we perspire. If our sweat can evaporate, it cools our skin as large amounts of energy are removed through the latent heat of vaporization (about 2500 J of heat for each gram of sweat that evaporates). When the air contains a great deal of water vapour and is close to being saturated, perspiration does not readily evaporate from our skin because the evaporation is "cancelled out" by the condensation from the air that is also occurring on our skin. Less evaporative cooling causes most people to feel hotter than the temperature alone would indicate, and a number of people start to complain about the "heat and humidity." (A closer look at how we feel in hot, humid weather will be given in Chapter 4 after we have examined the concepts of relative humidity and wet-bulb temperature.)

Measuring Air Temperature

Thermometers were developed to measure air temperature. There are various ways of doing this, ranging from the familiar liquid-in-glass types to ones that use the electrical properties of materials to measure temperature electronically.

Liquid-in-glass thermometers are often used for measuring surface air temperature because they are easy to read, are inexpensive to construct, and can be very accurate. These thermometers have a glass bulb attached to a sealed, graduated tube of varying lengths depending on the temperature range and precision required. A very small cylindrical opening or bore extends from the thermometer's bulb to just before the end of the tube. A liquid in the bulb (usually mercury or coloured alcohol) is free to move from the bulb up through the bore and into the tube. When the air temperature increases, the liquid in the bulb expands and rises up the tube. When the air temperature decreases, the liquid contracts and moves down the tube. Hence, the length of the liquid in the tube represents the air temperature. Because the bore is very narrow, a small temperature change will show up as a relatively large change in the length of liquid in the thermometer tube (see ● Figure 3.26). The narrower the thermometer bore, the more precise the instrument will be, whereas the more even the bore diameter is throughout its entire length, the more accurate the thermometer will be.

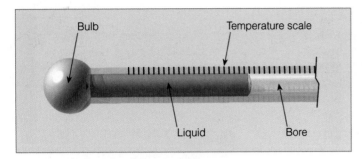

● FIGURE 3.26 A liquid-in-glass thermometer.

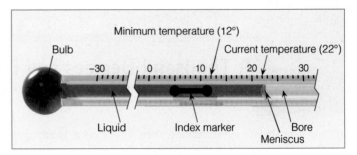

● FIGURE 3.28 Cross section of a minimum thermometer in °C. The meniscus shows the current air temperature, whereas the end of the index marker that is furthest from the bulb indicates the minimum temperature.

Maximum thermometers and minimum thermometers are liquid-in-glass thermometers that are used for determining daily maximum and minimum temperatures. A **maximum thermometer** looks like any other liquid-in-glass thermometer with one exception: it has a small constriction in the bore just above the bulb (see ● Figure 3.27). As the air temperature increases, the mercury expands and is able to move freely past the constriction up the tube, until the maximum temperature occurs. However, as the air temperature begins to drop, the small constriction near the bulb is just the right size to prevent the mercury from flowing past it into the bulb as the mercury contracts. The constriction creates a small gap in the mercury column, and the mercury left in the thermometer tube indicates the maximum temperature for the day. The mercury will stay at this position until either the air warms to a higher reading or the thermometer is reset. It is reset by swinging it back and forth in an arc until the motion has pushed the mercury through the constriction and the mercury column reads the current air temperature.*

A **minimum thermometer** measures the lowest temperature reached during a given period. Most minimum thermometers use alcohol as their liquid because it freezes at a temperature of −130°C compared to −39°C for mercury. The minimum thermometer is similar to other liquid-in-glass thermometers except that it contains a small barbell-shaped index marker in the bore (see ● Figure 3.28). This marker is free to slide back and forth within the liquid, but it cannot move out of the liquid because the surface tension at

---

*Liquid-in-glass thermometers that measure body temperature are maximum thermometers.

the end of the liquid column, the meniscus, holds it in the liquid. As the air temperature drops, the contracting liquid moves back into the bulb and brings the index marker down the bore with it (i.e., the meniscus pulls the marker down the bore). When the air temperature stops decreasing, the liquid and the index marker stop moving down the bore. As the air warms, the alcohol expands and moves freely up the tube, around and past the stationary index marker. The minimum temperature is read by observing the upper end of the marker. It is reset simply by tipping it upside down, allowing the index marker to slide up to the meniscus at the top of the alcohol column. The thermometer is then remounted horizontally so that the marker will move toward the bulb as the air temperature decreases.

Another common type of thermometer that reads both maximum and minimum temperature at the same time is called a "max-min thermometer" or "Six's thermometer." It is constructed as a "U"-shaped tube with a bulb at one end and temperature scales on both sides of the U. As shown in ● Figure 3.29, on the maximum side, the temperature scale increases from bottom to top, whereas on the minimum side, the scale decreases from bottom to top. Alcohol fills the bulb on the minimum temperature side, mercury fills the majority of the U, and there is a vacuum on the maximum temperature side. As the temperature increases, the alcohol expands and pushes the mercury around the U. The position of the mercury on each side of the U marks the current air temperature. Pins, whose edges are barbed, are placed on top of the mercury on each side of the U. These are pushed by the mercury as it expands, but the barbs are large enough to catch on the sides of the tube when the mercury falls, so the pins get left behind, marking the highest and lowest temperatures. The bottom of the pins indicate the maximum and minimum temperatures that were experienced since the instrument was last reset. The pins are reset by using a magnet to overcome the tension provided by the pin barbs and drag the pin back in contact with the mercury. Now the instrument is ready to use again.

Increasingly, temperature measurements are made with **electrical thermometers**. These instruments are highly accurate, respond quickly to fluctuating temperature, and are

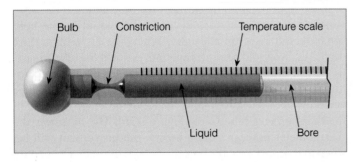

● FIGURE 3.27 Cross section through a maximum thermometer.

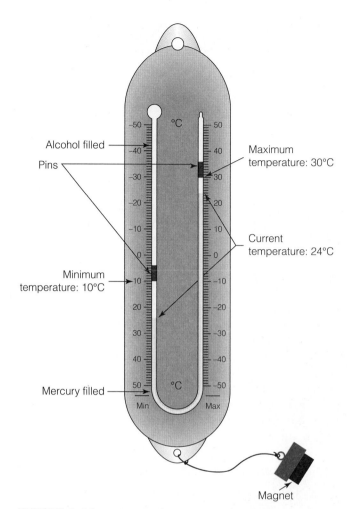

● **FIGURE 3.30** Automated weather stations monitor weather electronically. They record data from wind sensors, thermometers, and humidity and pressure sensors. The max-min temperature shelter is the middle box.

● **FIGURE 3.29** A max-min thermometer in °C. The meniscus shows the current air temperature. The bottom of the pins indicate the maximum temperature on one side and the minimum temperature on the other.

convenient because the measurement can be processed and stored digitally for automated measurements. At this point, it should be noted that the replacement of liquid-in-glass thermometers with electrical thermometers has raised concern among climatologists. Since they respond more quickly, they may reach a brief extreme reading, which could have been missed by the slower responding liquid-in-glass thermometer. In addition, many temperature readings, which were taken at airport weather offices, are now taken at automatic weather station locations that sit near or between runways at the airport. This change in instrumentation and relocation of the measurement site can sometimes introduce a small, but significant, temperature change at the reporting station.

Electrical thermometers are connected to an electrical meter that converts the electrical signal into a temperature measurement that can either be displayed on a gauge or stored digitally. Commonly, they are connected to a **data logger**, which is a special-purpose programmable microcomputer that can measure electrical signals, convert them into usable data with recognizable units, process these data (e.g., average them, find the maximum or minimum, identify when these occur, etc.), and store the data electronically. Data log-

gers can also often communicate with other computers over telephone, Internet, radio, cellular phone, or satellite links to transmit the data and are integral to modern environmental data collection systems.

One type of electrical thermometer is the *electrical resistance thermometer*, which measures the resistance of a wire, usually platinum or nickel, whose resistance increases as the temperature increases. Electrical resistance thermometers are the type of thermometers used in the measurement of air temperature at automated surface weather stations that exist at many airports and other weather stations around the world (see ● Figure 3.30). Hence, many of the liquid-in-glass thermometers have been replaced with electrical thermometers.

Thermistors are another type of electrical thermometer. They are made of ceramic material whose resistance increases as the temperature decreases. A thermistor is the temperature-measuring device in a radiosonde, the instrument that is carried by a balloon from the surface to altitudes of nearly 30 km to measure air temperature and other atmospheric properties.

Another electrical thermometer is the *thermocouple*. This device operates on the principle that the temperature difference between the junction of two dissimilar metals creates a weak electrical current. When one end of the junction is maintained at a temperature different from that of the other end, an electrical current will flow in the circuit. This current is proportional to the temperature difference between the junctions.

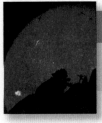

FOCUS ON AN OBSERVATION

Should Thermometers Be Read in the Shade?

When we measure air temperature with a liquid-in-glass thermometer, an incredible number of air molecules bombard the bulb, transferring energy either to or away from it. When the air is warmer than the thermometer, the liquid gains energy, expands, and rises up the tube; the opposite happens when the air is colder than the thermometer. The liquid stops rising (or falling) when equilibrium between incoming and outgoing energy is established. At this point, we can read the temperature by observing the height of the liquid in the tube.

It is impossible to measure air temperature accurately in direct sunlight because the thermometer absorbs radiant energy from the sun in addition to energy it absorbs from the air molecules. The thermometer gains energy at a much faster rate than it can radiate it away, and the liquid keeps expanding and rising until there is equilibrium between incoming and outgoing energy. Because of the direct absorption of solar energy, the level of the liquid in the thermometer indicates a temperature much *higher* than the actual air temperature, so the statement "Today, the air temperature measured 40° in the sun" has no meaning. Hence, a thermometer must be kept in a shady place to measure the temperature of the air accurately (see ● Figure 6).

© Ross DePaola

● FIGURE 6 Louvered instrument shelters such as this Stevenson screen provide airflow but shield thermometers from the sun for accurate temperature measurements.

Surface and air temperatures may also be obtained with instruments called *infrared sensors*, or **radiometers**. Radiometers do not measure temperature directly; rather, they measure emitted radiation (usually infrared). By measuring the intensity of radiant energy within different wavelength intervals, radiometers in orbiting satellites are now able to provide surface temperature readings and air temperatures at selected levels in the atmosphere.

A **bimetallic thermometer** consists of two different pieces of metal (usually brass and iron) welded together to form a single strip. As the temperature changes, the brass expands more than the iron, causing the strip to bend. The small amount of bending is amplified through a system of levers to a pointer on a calibrated scale. The bimetallic thermometer is usually the temperature-sensing part of the **thermograph**, an older style of instrument that measures temperature and records it on a paper chart (see ● Figure 3.31).

Chances are you may have heard someone exclaim something such as "Today, the thermometer measured 30° in the shade!" Does this mean that the air temperature is sometimes measured in the sun? If you are unsure of the answer, read Focus on an Observation: Should Thermometers Be Read in the Shade? before reading the next section on instrument shelters.

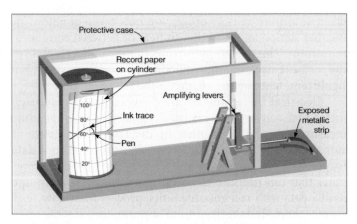

● FIGURE 3.31 The thermograph with a bimetallic thermometer.

Thermometers and other instruments are usually housed in an **instrument shelter** such as a Stevenson screen. The shelter completely encloses the instruments, protecting them from rain, snow, and the sun's direct rays. It is painted white to reflect sunlight, faces north to avoid direct exposure to sunlight when opened, and has louvered sides so that air is free to flow through it. This construction helps keep the air inside the shelter at the same temperature as the air outside and is essential to accurately measure air temperature. As new meteorological monitoring equipment is developed, new shelters that accommodate the needs of this equipment also evolve, such as the ones shown in Figure 3.30. These newer systems often have fans that draw fresh air past the thermometer to ensure that the actual air temperature is measured and not the temperature inside the shelter.

The thermometers inside a standard shelter are mounted about 1.5 to 2 m above the ground. As we saw in an earlier section, on a clear, calm night, the air at ground level may be much colder than the air at the level of the shelter. As a result, on clear winter mornings, it is possible to see ice or frost on the ground even though the minimum thermometer in the shelter did not reach the freezing point. Because air temperatures vary considerably above different types of surfaces, where possible, shelters are placed over grass to ensure that the air temperature is measured at the same elevation over the same type of surface. This allows the measurements across an observing network to be compared to each other and to be compared globally. This is why national agencies such as the Meteorological Service of Canada, as well as the World Meteorological Organization internationally, set observing standards for proper measurements.

Unfortunately, however, some shelters, especially "nonofficial" ones, are placed on asphalt, others sit on concrete, and others are located on the tops of tall buildings, making it difficult to compare air temperature measurements from different locations. In fact, if either the maximum or minimum air temperature in your area seems suspiciously different from those of nearby towns, find out where the instrument shelter is situated.

SUMMARY

Earth has seasons because it is tilted on its axis as it revolves around the sun. The tilt of Earth causes a seasonal variation in both the length of daylight and the intensity of sunlight that reaches the surface. When the Northern Hemisphere is tilted toward the sun, the Southern Hemisphere is tilted away from the sun. Longer hours of daylight and more intense sunlight produce summer in the Northern Hemisphere, whereas in the Southern Hemisphere, shorter daylight hours and less intense sunlight produce winter. On a smaller scale, Earth's inclination influences the amount of solar energy received on the north and south sides of a hill, as well as around a home.

The daily variation in air temperature near Earth's surface is controlled mainly by the input of energy from the sun and the output of energy from the surface. On a clear, calm day, the surface air warms as long as heat input, mainly sunlight and infrared radiation from the sky, exceeds heat output, mainly the convection of sensible and latent heat and radiated infrared energy from the near surface air. The surface air cools at night as long as heat output exceeds input. Because the ground at night cools more quickly than the air above, the coldest air is normally found at the surface. This is where a radiation inversion often forms. When the air temperature in agricultural areas unexpectedly drops to dangerously low readings, fruit trees and grape vineyards can be protected from the cold by a variety of means, from mixing the air to spraying the trees and vines with water.

The greatest daily variation in air temperature occurs at Earth's surface. Both the diurnal and annual ranges of temperature are greater in dry climates than in humid ones. Even though two cities may have similar average annual temperatures, the range and extreme of their temperatures can differ greatly. Temperature information impacts our lives in many ways, from influencing decisions on what clothes to take on a trip to providing critical information for energy-use predictions and agricultural planning. We reviewed some of the many types of temperature sensors that are in use. Those designed to measure air temperatures near the surface must be housed in instrument shelters to protect them from direct sunlight and precipitation.

KEY TERMS

The following terms are listed (with page number) in the order they appear in the text. Define each. Doing so will aid you in reviewing the material covered in this chapter.

sensible heat, 66	sea breeze, 77
advection, 66	radiational (radiative)
convective, 66	cooling, 77
net radiation, 66	radiation inversion, 77
cosine law of illumination, 67	nocturnal inversions, 77
	thermal belts, 79
summer solstice, 67	fronts, 81
insolation, 69	orchard heaters, 82
autumnal equinox, 70	wind machines, 82
Indian summer, 71	freeze, 82
winter solstice, 71	daily (diurnal) range of temperature, 85
vernal equinox, 72	
free convection, 76	urban heat island, 86
forced convection, 76	mean (average) daily temperature, 86
lapse rate, 76	

QUESTIONS FOR REVIEW

1. In the Northern Hemisphere, why are summers warmer than winters, even though Earth is actually closer to the sun in January?

2. What are the main factors that determine seasonal temperature variations?

3. During the Northern Hemisphere's summer, the daylight hours in northern latitudes are longer than in middle latitudes. Explain why northern latitudes are not warmer.

4. If it is winter and January in Toronto, Canada, what is the season in Sydney, Australia?

5. Explain why Southern Hemisphere summers are not warmer than Northern Hemisphere summers.

6. Explain why the vegetation on the north-facing side of a hill is frequently different from the vegetation on the south-facing side of the same hill.

7. Look at Figures 3.12 and 3.14, which show vertical profiles of air temperature during different times of the day. Explain why the temperature curves are different.

8. What are some of the factors that determine the daily fluctuation of air temperature just above the ground?

9. Explain how incoming energy and outgoing energy regulate the daily variation in air temperature.

10. On a calm, sunny day, why is the air next to the ground normally much warmer than the air just above?

11. Explain why the warmest time of the day is usually in the afternoon, even though the sun's rays are most direct at noon.

12. Explain how radiational cooling at night produces a radiation temperature inversion.

13. What weather conditions are best suited for the formation of a cold night and a strong radiation inversion?

14. Explain why thermal belts are found along hillsides at night.

15. List some of the measures farmers use to protect their crops against the cold. Explain the physical principle behind each method.

16. Why are the lower tree branches most susceptible to damage from low temperatures?

17. Describe each of the controls of temperature.

18. Look at Figure 3.19 (temperature map for January) and explain why the isotherms dip southward (equatorward) over the Northern Hemisphere continents.

19. Explain why the daily range of temperature is normally greater (a) in dry regions than in humid regions and (b) on clear days than on cloudy days.

20. Why is the largest annual range of temperatures normally observed over continents away from large bodies of water?

21. Two cities have the same mean annual temperature. Explain why this fact does not mean that their temperatures throughout the year are similar.

22. During a cold, calm, sunny day, why do we usually feel warmer than a thermometer indicates?

23. What atmospheric conditions can bring on hypothermia?

24. During the winter, white frost can form on the ground when the minimum thermometer indicates a low temperature above freezing. Explain.

25. Why do daily temperature ranges decrease as you increase in altitude?

26. Why do the first freeze in autumn and the last freeze in spring occur in low-lying areas?

27. Someone says, "The air temperature today measured 40°C in the sun." Why does this statement have no meaning?

28. Briefly describe how the following thermometers measure air temperature:
    (a) liquid-in-glass
    (b) bimetallic
    (c) electrical
    (d) radiometer

QUESTIONS FOR THOUGHT

1. Explain (with the aid of a diagram) why the morning sun shines brightly through a south-facing bedroom window in December but not in June.

2. Consider these two scenarios: (a) The tilt of Earth decreased to 10°. (b) The tilt of Earth increased to 40°. How would this change the summer and winter temperatures in your area? Explain, using a diagram.

3. At the top of Earth's atmosphere during the early summer (Northern Hemisphere), above what latitude would you expect to receive the most solar radiation in one day? During the same time of year, where would you expect to receive the most solar radiation at the surface? Explain why the two locations are different. (If you are having difficulty with this question, refer to Figure 3.6, p. 70.)

4. If a construction company were to build a solar-heated home in middle latitudes in the Southern Hemisphere, in which direction should the solar panels on the roof be directed for maximum daytime heating?

5. Aside from the aesthetic appeal (or lack of such), explain why painting the outside north-facing wall of a middle latitude house one colour and the south-facing wall another colour is not a bad idea.

6. How would the lag in daily temperature experienced over land compare to the daily temperature lag over water?

7. Where would you expect to experience the smallest variation in temperature from year to year and from month to month? Why?

8. On a warm summer day, one city experienced a daily range of 22°C, whereas another had a daily range of 10°C. One of these cities is located in Nova Scotia and the other in Alberta. Which location most likely had the highest daily range, and which one had the smallest? Explain.

9. Minimum thermometers are usually read during the morning yet are reset in the afternoon. Explain why.

10. If clouds arrive at 2 a.m. in the middle of a calm, clear night, it is quite common to see temperatures rise after 2 a.m. How does this happen?

11. In the Northern Hemisphere, south-facing mountain slopes normally have a greater diurnal range in temperature than north-facing slopes. Why?

12. If the poles have 24 hours of sunlight during the summer, why is the average summer temperature still below −18°C?

PROBLEMS AND EXERCISES

1. Draw a graph similar to Figure 3.6 (p. 70). Include in it the amount of solar radiation reaching Earth's surface in the Northern Hemisphere on the equinox.

2. Each day past the winter solstice, the noon sun is a little higher above the southern horizon. (a) Determine how much change takes place each day at your latitude. (b) Does the same amount of change take place at each latitude in the Northern Hemisphere? Explain.

3. On approximately what dates will the sun be overhead at noon at latitudes (a) 10°N and (b) 15°S?

4. Design a solar-heated home that sits on the north side of an east–west running street. If the home is located at 40°N, draw a proper roof angle for maximum solar heating. Design windows, doors, overhangs, and rooms with the intent of reducing heating and cooling costs. Place trees around the home that will block out excess summer sunlight and yet let winter sunlight inside. Choose a paint colour for the house that will add to the home's energy efficiency.

5. Suppose peas are planted in Ontario on May 1. If the peas need 660 growing degree-days before they can be picked, and if the average maximum temperature for May and June is 27°C and the average minimum temperature is 15°C, on about what date will the peas be ready to pick? (Assume a base temperature of 13°C.)

6. What is the wind-chill equivalent temperature when the air temperature is −10°C and the wind speed is 50 km h^{-1} (Use Table 3.3, p. 90.)

As the sun disappears behind an approaching deck of clouds, the air above the snow-covered landscape slowly cools. As the air temperature drops, the relative humidity increases, and the air gradually approaches saturation, creating this misty sky.
© Brad Perks

Atmospheric Humidity

Sometimes it rains and still fails to moisten the desert—the falling water evaporates halfway down between cloud and earth. Then you see curtains of blue rain dangling out of reach in the sky while the living things wither below for want of water. Torture by tantalizing, hope without fulfillment. And the clouds disperse and dissipate into nothingness. . . . The sun climbed noon-high, the heat grew thick and heavy on our brains, the dust clouded our eyes and mixed with our sweat. My canteen is nearly empty and I'm afraid to drink what little water is left—there may never be any more. I'd like to cave in for a while, crawl under yonder cottonwood and die peacefully in the shade, drinking dust.

Edward Abbey, *Desert Solitaire—A Season in the Wilderness.* 1968. Random House.

CONTENTS

Humidity is the amount of water vapour that is in the air. Although it is usually less than a few percent of all the molecules in our atmosphere (see Chapter 1 for more details), this invisible gas is extremely important. It effectively absorbs longwave radiation, adding several degrees of warming to Earth's average surface temperature. It transports heat as it changes phase between solid, liquid, and gas states. It affects weather when it transforms into the liquid water droplets or ice crystals that form clouds, and where these particles can grow large enough, it falls to the ground as precipitation.

Humidity is a general term that can describe the amount of water vapour in the air in multiple ways. For example, a moist day suggests that there is high humidity. However, there is usually more water vapour in the hot, "dry" air of the Sahara Desert than there is in the cold, "damp" air of the Maritime provinces. This raises the interesting question: Does the desert air have a higher humidity? As we will see, the answer is yes and no, depending on the type of humidity being discussed. Consequently, when representing humidity, we have to be more specific.

To better understand the concept of humidity, we will begin this chapter by examining the circulation of water in the atmosphere, discussing the phases of water, and defining the concept of saturation. Then we will look at different ways to express humidity. At the end of the chapter, we will investigate various ways to measure humidity.

Circulation of Water in the Atmosphere

Water is the glue that binds the atmosphere to all the other Earth systems. Most of the water in the atmosphere is composed of water vapour. Water transforms between vapour, liquid, and solid forms and interacts with the hydrosphere, cryosphere, lithosphere, biosphere, and anthrosphere. In liquid or solid forms, it falls to the surface, nourishing the other systems, and, in turn, it accepts vapour from those systems.

The atmosphere is constantly circulating and is connected to other Earth systems through the water or **hydrologic cycle**. Since oceans occupy over 70 percent of Earth's surface, we can visualize this circulation as beginning over the ocean, where the sun's energy transforms enormous quantities of liquid water into water vapour in a process called **evaporation**. Winds transport this moist air to other regions, where the water vapour changes back into liquid, forming clouds in a process called **condensation**. Where the atmosphere is cold enough, clouds can also be made of ice particles, which form when water vapour transforms into liquid and then ice, or transforms directly into ice. (We will discuss these details under the topic of precipitation in

Chapter 7 and focus here on the key humidity process—transformations of water between vapour and liquid.) Under certain conditions, liquid or ice cloud particles may grow large enough to fall to the surface as **precipitation**—rain, snow, or hail. If precipitation falls into an ocean, the water is ready to begin its cycle again. If precipitation falls on a continent, a more complex journey begins, but, eventually, this water also returns to the ocean. The cycle of moving and transforming water molecules from liquid or solid, to vapour, and back to liquid again is called the hydrologic or water cycle.

● Figure 4.1a illustrates the variety of complex journeys water molecules can take as part of the hydrologic cycle. For example, before falling rain ever reaches the ground, a portion of it can evaporate back into the atmosphere (only rain that reaches the ground counts as precipitation). Also, some precipitation is intercepted by vegetation, where it evaporates or drips to the ground long after a storm has ended. Snowfall can become trapped in the *cryosphere* by glaciers for thousands of years, remain in a snowpack for a season, or melt instantly. Once on the surface, water may nourish animals or plants in the *biosphere*, feed the ever-thirsty human *anthrosphere*, or enter the *lithosphere* by soaking into the ground—percolating downward through small openings in soil or rock and forming groundwater. What does not soak in sustains the *hydrosphere*, collecting as puddles of standing water, forming lakes, or running into streams and rivers, which find their way back to the ocean. Eventually, even slower moving groundwater can surface and evaporate or be carried to the sea.

Over land, water vapour is added to the atmosphere through evaporation from soil, lakes, and streams. Plants also release moisture through a process called **transpiration**, where liquid water, absorbed by a plant's root system, moves upward through the plant and emerges as water vapour through numerous small openings on the underside of its leaves. The term **evapotranspiration** incorporates both evaporation and transpiration and represents the total flow of water vapour into the atmosphere, allowing us to simplify these complex processes.

If we consider how much water is involved with the hydrologic cycle and how it is distributed, we need to recognize that water in its various forms is both "stored" and "in circulation." In Figure 4.1b, the average annual percentage of water stored by Earth is shown in purple bubbles, whereas the average annual percentage of water exchanged is indicated in yellow boxes with orange arrows showing the direction of movement. The similarly coloured tables report the actual amounts of water involved. Notice how Earth's stored water is distributed; almost all of it (~97 percent) is in the oceans, whereas fresh water (~3 percent), in all its forms, is very limited. (Available drinking water is a very much smaller component of fresh water. These numbers emphasize the value of our freshwater resources.)

The amount of water stored in the atmosphere is exceedingly small—only 0.0009 percent of the average annual global

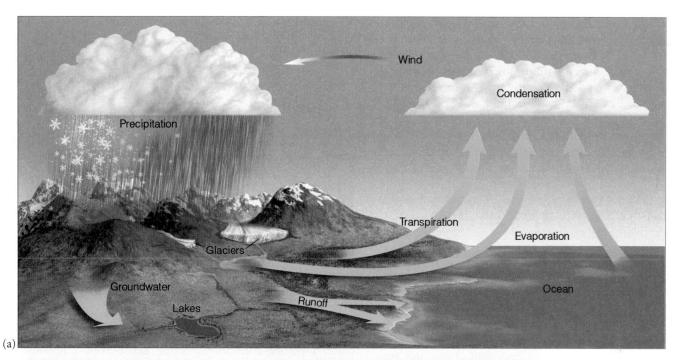

(a)

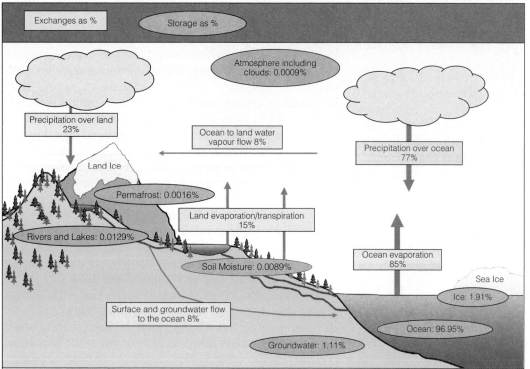

Average Annual Global Water Exchanges	× 1000 km³ y⁻¹	%
Ocean precipitation	373	77
Ocean evaporation	413	85
Ocean to land water vapour transport	40	8
Land precipitation	113	23
Land evaporation/transpiration	73	15
Surface & groundwater flow to ocean	40	8

Average Annual Global Water Storage	× 1000 km³	%
Ocean	1,335,040	96.95
Ice (all types)	26,350	1.91
Groundwater	15,300	1.11
Rivers/lakes	178	0.0129
Soil moisture	122	0.0089
Permafrost	22	0.0016
Atmosphere	12.7	0.0009
Total	1,377,025	100.00

(b)

● **FIGURE 4.1**

(a) The hydrologic cycle. Note the many and complex ways that water interacts with all Earth systems as it transforms and moves between solid, liquid, and vapour forms. (b) Quantifying average annual global water storage and exchange in the hydrologic cycle. The percentage of global water stored by various components of the water cycle is shown in purple bubbles, whereas the percentage of global water exchanged on an annual basis is given in yellow boxes. Orange arrows represent the relative sizes of the exchange. The accompanying tables also report the amounts of water stored or exchanged.

water storage. If all the water vapour in the atmosphere suddenly condensed and fell as rain, it would cover the entire globe with only 2.5 cm of water, or just a little over a week's supply of the world's precipitation. However, water exchanges are not distributed equally around Earth. To generalize, the amounts of precipitation, evaporation, and transpiration differ most significantly between ocean and land areas. More water evaporates from the oceans than precipitates over them. This creates an excess of water vapour over the ocean, which is transported to the land by wind. Conversely, more precipitation occurs over the land than evaporates from it. This leads to surface water runoff and groundwater flow that moves from the land back to the ocean. These exchanges balance the hydrologic cycle and make it exceedingly efficient in circulating water through the atmosphere.

The Many Phases of Water

Water is everywhere in the lower atmosphere. Chemically, it is a molecule composed of two hydrogen atoms and one oxygen atom, written as H_2O. If we could magnify an individual water molecule a billion times, it would look like a basketball-sized "head" with "ears" (see ● Figure 4.2) that resemble those of the cartoon character "Mickey Mouse." The "head" is the oxygen atom, which forms the bulk of the water molecule. Its "mouth" region has excess negative charge, whereas its "ears" have excess positive charge, caused by partially exposed protons on the hydrogen atoms. These charges control the water molecule's behaviour as it changes from a gas, to a liquid, to a solid—processes known as *phase changes*.

The molecular behaviour of water varies according to its phase or state. As a gas, water vapour molecules move about quite freely, mixing well with neighbouring atoms and molecules (see ● Figure 4.3). As we learned in Chapter 2 when discussing kinetic energy, the higher the temperature of a gas,

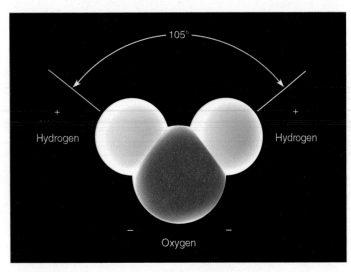

● FIGURE 4.2 The water molecule. Notice the relative positions of the atoms and their associated charges.

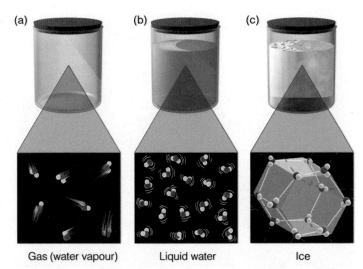

Gas (water vapour)   Liquid water   Ice

● FIGURE 4.3 Water can exist in three states in Earth's atmosphere: as a gas, a liquid, and a solid.

the faster its molecules move. As a liquid, water molecules are closer together, constantly jostling and bumping into one another. If we cool liquid water, these molecules move slower and slower until they freeze and arrange themselves into solid ice. Ice has an orderly pattern, with each molecule more or less locked into a rigid position; molecules can vibrate but are not able to move about freely. The shape and charges on water molecules cause them to arrange themselves into six-sided hexagonal ice crystals (see Figure 4.3). If warmth is applied to an ice crystal, its molecules vibrate faster. With enough warmth, some of the molecules vibrate out of their rigid crystal pattern into a less orderly state—the ice melts and becomes liquid.

If we observe an ice crystal in freezing air, we will occasionally see an ice molecule gain enough energy to break away from its neighbours and join the other gas molecules in the surrounding air. It changes from solid ice directly into water vapour without passing through the liquid state. This ice-to-vapour phase change is called **sublimation**. If a water vapour molecule should attach itself to an ice crystal, the vapour-to-ice phase change is called **deposition**.

So water vapour is an invisible atmospheric gas that changes phase to become visible water (or ice) when millions of molecules join together to form tiny cloud droplets (or ice crystals). In this process, water changes only its disguise, not its identity.

Evaporation, Condensation, and Saturation

Suppose we are able to observe individual liquid water molecules in a beaker. What we would see are water molecules jiggling, bouncing, and moving about. We would also see that the molecules are not all moving at the same speed. Some are moving much faster than others. At the surface, molecules travelling with enough speed in the right direction occasionally

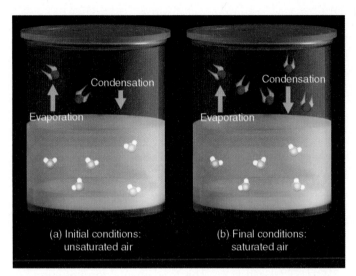

● FIGURE 4.4 Evaporation, condensation, and saturation over time in a sealed beaker. The arrows indicate the amount of evaporation and condensation. (a) Initially, the beaker contains unsaturated air over liquid water. (b) Later, when equilibrium is reached, the air is saturated. (For clarity, only the water molecule component of the air is illustrated.)

break through the liquid's surface and enter the air above. These molecules are *evaporating, changing from a liquid state into a vapour state.* Whereas some water molecules are leaving the liquid, others are returning to it by *condensing, changing from a vapour to a liquid.*

If we want to see what happens to these water vapour molecules as unsaturated (almost dry) air becomes saturated, we need a sealed beaker with "dry" air sitting above liquid water as illustrated in ● Figure 4.4. The water molecules at the surface are evaporating (transforming from liquid to vapour) and condensing (changing from vapour to liquid). The length of the arrows in Figure 4.4 indicates the amount of evaporation or condensation that is occurring at different times. The amount of evaporation in the beaker depends on the temperature of the water. Warm water evaporates more readily than cool water as molecules speed up when they are heated, and more of them have enough speed to break through the liquid's surface into the air above. *So the warmer the water, the greater the evaporation.* However, the amount of condensation depends on the number of water vapour molecules in the air above the water. More vapour molecules can have more encounters with the liquid water's surface, which results in more condensation. Since the air in the beaker in Figure 4.4a is almost dry, evaporation greatly exceeds condensation. We would say that there is *net evaporation* in the beaker and the air above the liquid is unsaturated with water vapour. As evaporation continues, we reach the point where the number of water molecules evaporating from the liquid is balanced by those returning through condensation (see Figure 4.4b). For every molecule that evaporates, one condenses. Now the air's water vapour content is no longer changing, and a state of **equilibrium** is reached; the air above the liquid is **saturated** with water vapour.

If we cooled the beaker in Figure 4.4b, the evaporation rate would decrease, so condensation is larger than evaporation. This removes water vapour from the air until the condensation rate reaches a new equilibrium with the reduced evaporation rate. *So cooling the air leads to net condensation.* Even though condensation is more likely to occur when air cools, it is important to understand that no matter how cold air becomes, there are always a few molecules with sufficient speed (or energy) to remain as water vapour.

If we remove the beaker's cover and blow across the top of the water, some of the vapour molecules are blown away, creating a difference between the number of vapour molecules in the air and the total number required for **saturation**. Now the air is unsaturated again, and the situation resembles the initial conditions seen in Figure 4.4a, where there was more evaporation than condensation. This shows how wind enhances evaporation.

So what happens in the real atmosphere (instead of a beaker)? Here the air molecules are mixed with microscopic bits of dust, smoke, salt, and other small particles called **condensation nuclei** (so called because water vapour condenses on them). When the air is warm, fast-moving vapour molecules strike the nuclei with such impact that they simply bounce away (see ● Figure 4.5a). However, if the air is chilled (Figure 4.5b), vapour molecules move more slowly and are more likely to stick and condense on the nuclei. These start forming water droplets in the atmosphere, which act in the same way as the surface of the water in the beaker. When the air around the droplets becomes saturated, condensation on the droplets balances evaporation from the droplets. If the temperature drops, the evaporation rate from the droplets decreases, there is net condensation, and the droplets grow. When many billions of water vapour molecules condense onto these nuclei, tiny liquid cloud droplets form. So, *through condensation, water droplets form and grow as the air cools.*

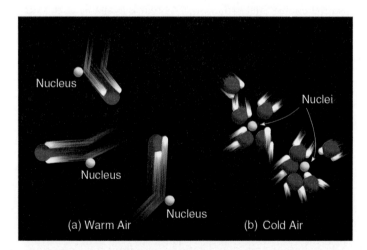

● FIGURE 4.5 Condensation at a molecular scale. Condensation is more likely to occur as the air cools. (a) In warm air, fast-moving water vapour molecules tend to bounce away after colliding with nuclei. (b) In cool air, slow-moving vapour molecules are more likely to join together on nuclei. Condensing many billions of water molecules produces tiny liquid water droplets.

Conversely, if the temperature rises, the evaporation rate from the droplets increases, there is net evaporation, and the droplets shrink or even disappear. *So, through evaporation, warming causes droplets to disappear as they turn into vapour.*

The statements that "warm air can hold more water vapour molecules before becoming saturated than can cold air" or "warm air has a greater capacity for water vapour than does cold air" are commonly misinterpreted. It is important to realize that although these statements are correct, using the words "hold" and "capacity" can be misleading because air's ability to contain water is dependent only on whether or not it is saturated. (It is not related to the amount of room available in the air for water vapour in comparison to other constituents—this is the misconception that occurs.) If the air is unsaturated, then more water can still evaporate into the air. If the air is saturated, then condensation will remove the excess vapour. So saturation is effectively an upper limit to adding water vapour to the atmosphere. Since warmer air has a higher saturation level, it can have more water vapour in it.

What factors control the rate of evaporation from Earth's surface into the atmosphere? We have already seen that net evaporation happens when evaporation rates are greater than condensation rates. The largest net evaporation from a surface will occur when the condensation rate is low—when the air is unsaturated or "dry." Hence, atmospheric humidity levels control evaporation. We have also seen that temperature plays a role as higher temperatures have more energy and increased evaporation rates. Moreover, we saw that blowing across an open beaker's water surface enhanced evaporation by carrying away water vapour molecules and exposing the water surface to drier air. Consequently, wind is another important control on evaporation, with stronger winds more effectively carrying away water vapour from the surface and thereby increasing evaporation. In discussing energy (Chapter 2), we saw that there is a link between the amount of water that evaporates and the energy (net radiation, Q^*) that is available at the surface for processes such as evaporation, warming the air, and warming the ground. Therefore, net radiation is another control on evaporation as evaporation rates can increase when the net radiation is larger.* So far, we have assumed that water for evaporation is readily available, as was the case in the beaker, and would be over a lake or the ocean. But not all land is moist enough to evaporate water. Surface materials and soil types affect how water in the landscape is absorbed and retained. Vegetation types, or the lack of vegetation, control how plants use (transpire) available water. It is possible to have all the meteorological conditions for large evaporation rates but insufficient water for any evaporation to occur. The atmosphere cannot "squeeze water from a stone"! Therefore, the amount of soil moisture and the type of surface material—rock, sediment, soil, or vegetation—also impact evaporation.

---

*Greater net radiation may not always translate into more evaporation as site-specific conditions can cause available net radiation (Q^*) energy to be used to heat the air or heat the ground rather than for evaporation. See Chapter 2 for a more detailed discussion.

To summarize, evaporation from Earth's surface is driven by meteorological controls: atmospheric humidity, temperature, wind, and net radiation. It is also mediated by water availability and land surface characteristics such as land surface types, soil moisture conditions, and vegetation types. These land surface constraints illustrate the linkage between atmospheric moisture, the *biosphere,* and the *lithosphere.*

Humidity

We are now ready to look more closely at the concept of humidity and the number of different ways the amount of water vapour in the air (see ● Figure 4.6) can be specified.

ABSOLUTE HUMIDITY   This fundamental form of humidity represents the water vapour density (mass of water vapour per volume of air). Although absolute humidity is easy to visualize, it is not commonly used in atmospheric studies. Let's see why. Suppose we enclose a volume of air in an imaginary thin, elastic container—a parcel—about the size of the large balloon shown in Figure 4.6. With a chemical drying agent, we extract the water vapour from the air, weigh it, and obtain its mass. If we compare the vapour's mass to the volume of air in the parcel, we have determined the **absolute humidity** of the air (normally expressed as grams of water vapour in a cubic metre of air):

$$\text{Absolute humidity} = \frac{\text{mass of water vapour}}{\text{volume of air}}$$

As absolute humidity depends on volume, changing the volume of air also changes the absolute humidity, even though the air's vapour content has remained the same

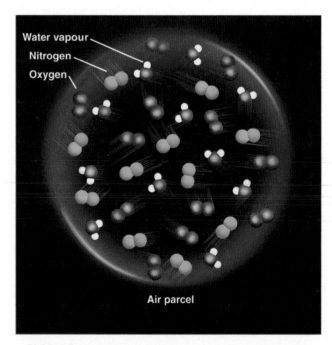

● **FIGURE 4.6** The water vapour content (humidity) inside this air parcel can be expressed in a number of ways.

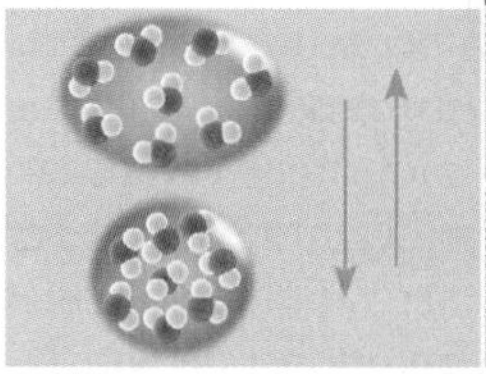

Parcel Size	Mass of H₂O Vapour	Absolute Humidity
2 m³	10 g	5 g m⁻³
1 m³	10 g	10 g m⁻³

● FIGURE 4.7 Absolute humidity changes. If the amount of water vapour in a parcel of air remains the same, an increase in the parcel's volume decreases absolute humidity, whereas a decrease in volume increases absolute humidity. So air volume changes that occur when air pressure or temperature changes affect absolute humidity—even if the vapour content has not changed.

(see ● Figure 4.7). (Remember from Chapter 2 that a rising or descending parcel of air changes its volume because of changes in the surrounding air pressure.) Consequently, it is difficult to differentiate a change in absolute humidity that results from moisture loss or gain from a change due to varying air volume (i.e., changing pressure or changing temperature at constant pressure). This is why absolute humidity is not commonly used in atmospheric studies.

SPECIFIC HUMIDITY AND MIXING RATIO   Humidity can be expressed in ways that are not influenced by air volume changes through using the related concepts of specific humidity or mixing ratio. When the mass of water vapour in the air parcel in Figure 4.6 is compared to the mass of <u>all</u> the air in the parcel (including water vapour), the result is called **specific humidity**:

$$\text{Specific humidity} = \frac{\text{mass of water vapour}}{\text{total mass of air}}$$

Similarly, another convenient way to express humidity is to compare the mass of water vapour in the parcel of air to the mass of remaining dry air. Humidity expressed in this manner is called the **mixing ratio**:

$$\text{Mixing ratio} = \frac{\text{mass of water vapour}}{\text{mass of dry air}}$$

Both specific humidity and mixing ratio are usually expressed as grams of water vapour per kilogram of air (g kg⁻¹).*

---

*These can also be expressed as a dimensionless quantity by using kilograms of water vapour per kilogram of air (kg kg⁻¹). Dimensionless quantities are often used in models of the atmosphere.

The specific humidity and mixing ratio of a parcel of air remain constant as long as water vapour is not added or removed from the parcel. Here the total number of molecules (or mass of the parcel of air) remains the same even when the parcel expands or contracts (see ● Figure 4.8). Since changes in parcel size do not affect the specific humidity or mixing ratio, these measures are used extensively to represent humidity in the atmosphere.

● Figure 4.9 shows how specific humidity varies with latitude (position north or south on Earth). The average specific humidity is highest in the warm, muggy tropics; as we move away from the tropics, it decreases, reaching its lowest average value in the polar latitudes. Although the world's major deserts are located near latitude 30° in both the Northern and Southern hemispheres, Figure 4.9 shows that, on average, the air contains nearly twice the water vapour content at 30° as it does at 50° latitude. Hence, the air of a desert is certainly not "dry," nor is the water vapour content extremely low. Since the hot, desert air of the Sahara often contains more water vapour than the cold, polar air of the Arctic, we can say that, on average, Saharan air has a higher specific humidity. Later we will discuss what we mean when considering desert air to be "dry."

VAPOUR PRESSURE   Air's moisture content may also be described by measuring the pressure exerted by water vapour in the air. Suppose the air parcel in Figure 4.6 is near sea level. The total pressure inside the parcel is due to the collision of all the molecules against the inside surface of the parcel. In other words, the total pressure inside the parcel is equal to the sum of the pressures of all the individual gases. This is known as **Dalton's law of partial pressure**. If we are at sea level, a reasonable pressure inside the parcel is one thousand hectopascals* (expressed as 1000 hPa), and, typically, the gases inside the parcel consist of approximately 78 percent nitrogen, 21 percent oxygen, and 1 percent water vapour. Hence, the partial pressure exerted by nitrogen would be 780 hPa; by oxygen, 210 hPa; and by water vapour, 10 hPa. Since the number of water vapour molecules in any volume of air is small when compared to the total number of air molecules in that volume, the partial pressure due to water vapour is normally a small fraction of the air pressure. The partial pressure of water vapour is called the **actual vapour pressure** (denoted by the symbol "e"), and it indicates the air's water vapour content.

---

*A hectopascal (hPa) is a unit of pressure equal to 100 pascals (Pa). A Pascal measures pressure as a force per area. See Chapter 8 for more information about pressure.

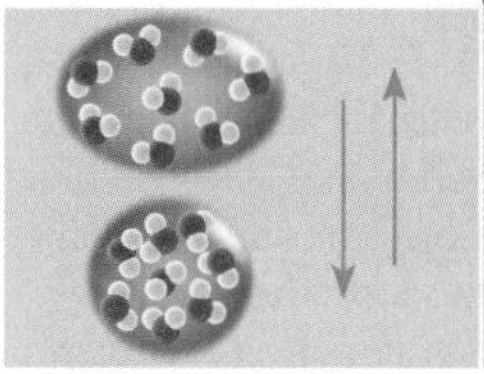

Mass of Parcel	Mass of H₂O Vapour	Specific Humidity	Mass of Dry Air	Mixing Ratio
1 kg	1 g (0.001 kg)	1 g kg⁻¹	0.999 kg	1.001 g kg⁻¹
1 kg	1 g (0.001 kg)	1 g kg⁻¹	0.999 kg	1.001 g kg⁻¹

● FIGURE 4.8 Specific humidity and mixing ratio do not change with air volume changes. (Note: Mass of dry air = mass of the parcel − mass of the water vapour or 1 kg − 0.001 kg = 0.999 kg.)

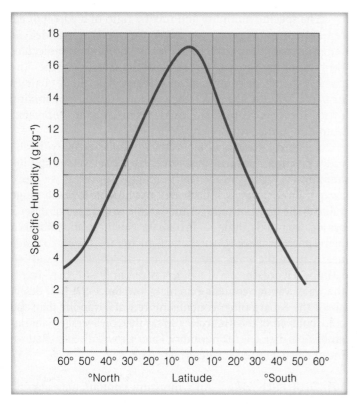

● FIGURE 4.9 The average specific humidity for each latitude. The highest average values are observed in the tropics, whereas the lowest values are in the polar regions.

The more air molecules there are in a parcel, the greater the air pressure. When you blow up a balloon, you increase its pressure by putting more air into the balloon. Similarly, an increase in the number of water vapour molecules will increase the actual vapour pressure. Hence, the actual vapour pressure is a fairly good measure of the amount of water vapour in the air. *High actual vapour pressure indicates a large number of water vapour molecules, whereas low actual vapour pressure indicates a smaller numbers of vapour molecules.* It is important to realize that vapour pressure also changes with atmospheric pressure and therefore with altitude. If the air pressure goes up, the vapour pressure will increase by the same fraction. Similarly, if air pressure falls, vapour pressure will decrease by the same amount. So vapour pressure depends on atmospheric pressure and the amount of water vapour in the air.

This is contrasted with **saturation vapour pressure** (denoted by the symbol "e^*"), which describes how much water vapour will make the air saturated at a given temperature. Put another way, *saturation vapour pressure is the pressure that water vapour molecules exert if the air at a specified temperature is saturated.* It represents the maximum amount of moisture content that could be absorbed by an air parcel at a given temperature. We can better understand the concept of saturation vapour pressure by revisiting our example of molecules evaporating from a

beaker of water. Look back at Figure 4.4b (p. 103) and recall that when the air is saturated, the number of molecules escaping from the water's surface equals the number returning. Since the number of "fast-moving" molecules increases as the temperature increases, the number of water molecules escaping per second also increases. To maintain equilibrium, this situation causes an increase in the number of water vapour molecules in the air above the liquid until the condensation rate equals the evaporation rate. Consequently, at higher air temperatures, it takes more water vapour to saturate the air, and more vapour molecules exert a greater pressure. So *saturation vapour pressure depends only on air temperature.* (Read Focus on a Special Topic: Vapour Pressure and Boiling—The Higher You Go, the Longer Cooking Takes to see how pressure affects the boiling point of water.) For example, the graph in ● Figure 4.10 indicates that at an air temperature of 10°C, the saturation vapour pressure is about 12 hPa, but at 30°C, it is about 42 hPa.

The insert in Figure 4.10 shows that when both liquid water and ice exist at the same temperature below freezing,* *the saturation vapour pressure associated with the water is*

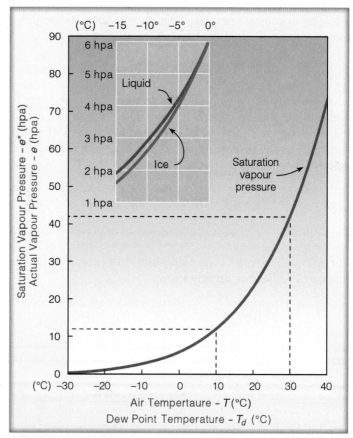

ACTIVE FIGURE 4.10 Saturation vapour pressure increases with increasing air temperature as shown by the graph when using the purple axis labels. At a temperature of 10°C, the saturation vapour pressure is about 12 hPa, whereas at 30°C, it is about 42 hPa. The inset graph illustrates that the saturation vapour pressure over water is greater than the saturation vapour pressure over ice. The second set of axis labels (in blue) indicate that relationship between actual vapour pressure and dew point temperature follows the same plot. Visit the textbook's website to view this and other Active Figures at www.ahrensmeteorology1ce.nelson.com

---

*Water can exist in all three phases at the same time within clouds when the air temperature is between 0 and –40°C. This will be explained in more detail in Chapter 7.

FOCUS ON A SPECIAL TOPIC

Vapour Pressure and Boiling—The Higher You Go, the Longer Cooking Takes

If you camp in the mountains, you may have noticed that the higher you camp, the longer it takes vegetables to cook in boiling water. To understand this observation, we need to examine the relationship between vapour pressure and boiling. As water boils, bubbles of vapour rise to the top of the liquid and escape. For this to occur, the saturation vapour pressure exerted by the bubbles must equal the pressure of the atmosphere; otherwise, the bubbles would collapse. Boiling, therefore, occurs when the saturation vapour pressure of the escaping bubbles is equal to the total atmospheric pressure.

Because the saturation vapour pressure is directly related to the temperature of the liquid, higher water temperatures produce higher vapour pressures. Hence, any change in atmospheric pressure will change the temperature at which water boils: An increase in air pressure raises the boiling point, whereas a decrease in air pressure lowers it. Notice in • Figure 1 that to make pure water boil at sea level, the water must be heated to a temperature of 100°C. At Banff, Alberta, which is situated about 1383 m above sea level, the air pressure is near 850 hPa, and water boils at 95°C. At Potossi, Bolivia, one of the highest cities in the world at 4090 m above sea level, the air pressure is near 630 hPa. Potossi would have a boiling point of about 86°C.

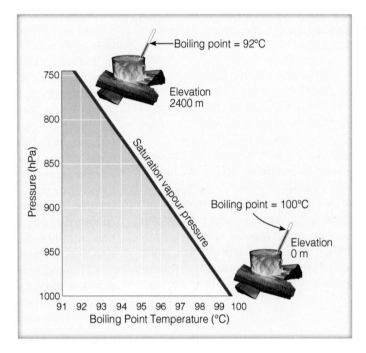

• FIGURE 1 The lower the air pressure, the lower the saturation vapour pressure and, hence, the lower the boiling point temperature.

Once water starts to boil, its temperature remains constant, even if you continue to heat it. This happens because energy supplied to the water is used to convert liquid water into the gas, water vapour, rather than to increase its temperature. To be thoroughly cooked, food must boil for a longer time at higher elevations because the boiling water is cooler than it is at lower levels. In Halifax, which is near sea level, it takes about five minutes to hard boil an egg. An egg boiled for five minutes in Banff is still runny. Now we can see why vegetables take longer to cook in the mountains.

---

greater than the saturation vapour pressure associated with the ice. In other words, at any temperature below freezing, it takes more vapour molecules to saturate air directly above water than it does to saturate air directly above ice. This situation occurs because it is harder for molecules to escape an ice surface than a water surface. Since fewer molecules escape the ice surface, fewer vapour molecules are required to maintain equilibrium. Similarly, salts in solution bind water molecules together and reduce the number that can escape. Consequently, saltwater also has a lower saturation vapour pressure. As we will see in Chapter 7, these concepts are important because they play a role in the formation of rain.

For now we will forgo discussing the second set of axes labels on Figure 4.10 as their connection to this graph will be clearer after we have discussed the concept of dew point. Also, examples of how to read and use all parts of this graph are given in Focus on a Special Topic: Computing Relative Humidity and Dew Point, which comes later in this chapter.

So far, we've described a number of different ways of representing humidity. If we want to report the moisture content of the air around us, we have these options:

- specific humidity, which measures the mass of water vapour in the total mass of air
- mixing ratio, which measures the mass of water vapour in the mass of remaining dry air
- actual vapour pressure, which is the partial pressure exerted by the water vapour molecules in the air

Although absolute humidity indicates humidity in the form of water vapour density, it is not commonly used in atmospheric science as it also responds to volume changes. This makes it difficult to discern when a change in humidity is caused by changing moisture, changing air pressure, or changing temperature.

Each of these measures has its uses, but the concepts of vapour pressure and saturation vapour pressure (or mixing ratio and saturation mixing ratio) are critical to understanding the sections that follow.

RELATIVE HUMIDITY   Although relative humidity is the most common way of describing atmospheric moisture, unfortunately, it is the most misunderstood. The concept of relative humidity differs from the types of humidity we have already described as it does not indicate the actual amount of water vapour in the air. Instead, it tells us how close the air is to being saturated. The **relative humidity** *(RH) is the ratio of the amount of water vapour actually in the air to the maximum amount of water vapour that would be in the air at saturation.* In other words, RH is the ratio of the air's water vapour content to its capacity to contain water vapour. As vapour pressure and mixing ratio are both ways of representing water vapour content, and saturation vapour pressure and saturation mixing ratio* are both ways of representing the air's capacity to absorb water vapour, they can be used to determine RH as follows:

$$RH = \frac{\text{actual vapour pressure}}{\text{saturation vapour pressure}} \times 100\%$$

$$RH = \frac{\text{actual mixing ratio}}{\text{saturation mixing ratio}} \times 100\%$$

Relative humidity is always given as a percent (% RH). Air with a 50 percent relative humidity contains one-half the amount of water vapour required for saturation. Air with a 100 percent relative humidity is at saturation, and any additional moisture will condense. Air with a relative humidity greater than 100 percent is said to be **supersaturated.** Supersaturation can occur in air surrounding tiny cloud droplets. It is an important concept for understanding precipitation processes, which are discussed in Chapter 7.

Since relative humidity is used so much in the everyday world, let's examine it more closely and discuss its uses. A change in relative humidity occurs when the water vapour content of the air changes or the temperature of the air changes. Sometimes both occur.

● Figure 4.11a shows that an increase in the air's water vapour content, with no change in air temperature, increases the air's relative humidity. As more and more water vapour molecules are added to the air, it gradually approaches saturation, and the relative humidity increases. (Conversely, removing water vapour from the air dries the air, it moves further from saturation, and its relative humidity decreases.) Figure 4.11b

*Saturation mixing ratio is the mixing ratio of saturated air at a particular temperature and pressure.

shows that an increase in air temperature with no change in the water vapour content decreases the air's relative humidity. This occurs because the water vapour capacity (the saturation vapour pressure or saturation mixing ratio) increases in warmer air. More vapour molecules can be held in warmer air. So with no change in water vapour content, when the saturation vapour pressure increases, the air moves farther away from saturation and causes the relative humidity to decrease.

As we can see, relative humidity does not respond only to changes in water vapour content. This makes it an inappropriate indicator of the actual water vapour in the air. But by indicating how far the air is from saturation at any temperature, it is an excellent way to represent how we sense moisture and how muggy or dry the air seems.

In many places, the air's actual vapour content varies only slightly during an entire day, so it is the changing air temperature that primarily regulates the daily variation in relative humidity (see ● Figure 4.12). As air cools during the night, the relative humidity increases. Normally, the highest relative humidity occurs in the early morning, during the coolest part of the day. As air warms during the day, the relative humidity decreases, with the lowest values usually occurring during the warmest part of the afternoon.

These changes in relative humidity are important in determining the amount of evaporation from vegetation and wet surfaces. If you water your lawn on a hot afternoon, when the relative humidity is low, much of the water will evaporate instead of soaking into the ground. Watering the same lawn in the evening or during the early morning, when the relative humidity is higher, will reduce evaporation and increase the effectiveness of the watering.

RELATIVE HUMIDITY AND DEW POINT   Suppose it is early morning and the outside air is saturated. The air temperature is 10°C, and the relative humidity is 100 percent. We know from the previous section that relative humidity is the ratio of water vapour content to water vapour capacity. Looking back at Figure 4.10, we see that air with a temperature of 10°C has a saturation vapour pressure of 12 hPa. Since the air is saturated and the relative humidity is 100 percent, the actual vapour pressure *must* be the same as the saturation vapour pressure. This is shown as

$$RH = \frac{\text{actual vapour pressure}}{\text{saturation vapour pressure}} = \frac{12\,\text{hPa}}{12\,\text{hPa}} \times 100\% = 100\%$$

Suppose during the day the air warms to 30°C, with no change in water vapour content (or air pressure). Because there is no change in water vapour content, the actual vapour pressure of 12 hPa must be the same as it was in the early morning when the air was saturated. But the saturation vapour pressure has increased because the air temperature has increased. From Figure 4.10, air with a temperature of 30°C has a saturation vapour pressure of 42 hPa. The relative humidity of this unsaturated, warmer air is now much lower:

$$RH = \frac{12\,\text{hPa}}{42\,\text{hPa}} \times 100\% = 29\%$$

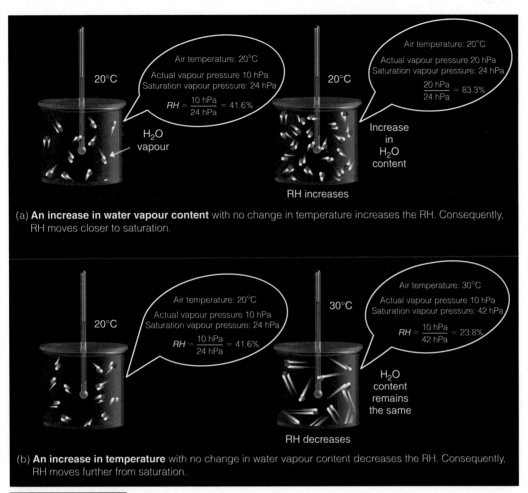

(a) **An increase in water vapour content** with no change in temperature increases the RH. Consequently, RH moves closer to saturation.

(b) **An increase in temperature** with no change in water vapour content decreases the RH. Consequently, RH moves further from saturation.

ACTIVE FIGURE 4.11 Changes in relative humidity can be caused by either changing the water vapour content as shown in (a), changing the temperature as shown in (b), or both. This makes relative humidity an often misunderstood humidity measurement. Visit the textbook's website to view this and other Active Figures at www.ahrensmeteorology1ce.nelson.com

To what temperature must this air (with a temperature of 30°C) be cooled so that it is once again saturated? The answer, of course, is 10°C. For this amount of water vapour in the air, 10°C is called the **dew point temperature**, or simply the **dew**

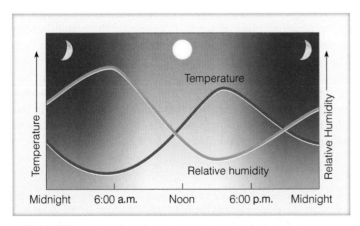

● FIGURE 4.12 When the air is cool (morning), the relative humidity is high. When the air is warm (afternoon), the relative humidity is low. These conditions exist in clear weather conditions, when the air is calm or has a constant wind speed.

point. It represents *the temperature to which air must be cooled for saturation to occur (with no change in air pressure or moisture content)*. The dew point is determined when air is saturated over a flat surface of water. When the dew point is determined with respect to a flat surface of ice, it is called the **frost point**.

Since dew point changes with atmospheric pressure as well as the actual vapour content, and atmospheric pressure varies only slightly at Earth's surface, *the dew point is a good indicator of the air's actual water vapour content. High dew points indicate high water vapour content; low dew points indicate low water vapour content*. The addition of water vapour to the air increases the dew point; removing water vapour lowers it. Consequently, dew point and vapour pressure are equivalent measures of humidity. It is important to realize that Figure 4.10 shows two relationships: the relationship between saturation vapour pressure and temperature and the relationship between actual vapour pressure and dew point. This occurs because they are, in fact, the same relationship.

The dew point temperature is an important measurement used in predicting the formation of dew, frost, and fog. As shown in ● Figure 4.13, dew point relates to radiative processes and can be used to predict minimum temperatures (more

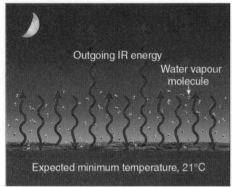
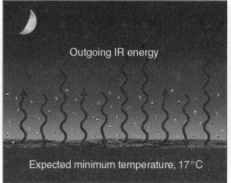
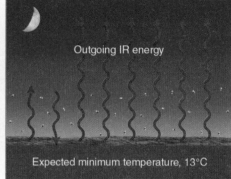

Dew point temperature, 20°C   Dew point temperature, 16°C   Dew point temperature, 10°C

● FIGURE 4.13 On a calm clear night, the lower the dew point temperature, the lower the expected minimum temperature. With the same initial evening air temperature (27°C) and with no change in weather conditions during the night, as the dew point drops, the expected minimum temperature drops. This situation occurs because a lower dew point means that there is less water vapour in the air to radiate infrared energy back to the surface. Consequently, more infrared energy is lost from the surface, producing more rapid radiational cooling at the surface. (Dots in each diagram represent the amount of water vapour in the air. Red wavy arrows represent infrared (IR) radiation. Chapters 2 and 3 contain more information about radiative and temperature processes.)

detailed discussions occur in Chapter 2 and Chapter 3). Dew point can also be used to estimate the height of cumulus clouds, as illustrated in the Focus on an Observation: Determining Convective Cloud Bases in Chapter 6 (p. 181).

On a larger scale, dew point temperatures exhibit continental humidity patterns. ● Figure 4.14 shows the average vapour pressure values and dew point temperatures across North America for January and July. In January, notice that the dew point temperatures are lowest in the Arctic and higher along coastal zones (Figure 4.14a). The highest values are along the southern coasts of the United States and Central America, with the highest Canadian values occurring along the temperate West Coast. At this time of year, the water vapour in the air (shown by the average vapour pressure values; see Figure 4.14b) is generally low everywhere in Canada. Again, the highest Canadian values are along the temperate West Coast. During the winter, cold, dry air dominates and winds from northern Canada flow into North America's interior, keeping this area dry, whereas warm, moist winds push air from the Gulf of Mexico and the Pacific onto the continent, creating higher dew points and water vapour values in these regions.

Figure 4.14c shows the average North American dew point temperatures for July. Notice that at this time of year, moisture is much greater everywhere. Now humidity from the Gulf of Mexico clearly dominates the central and eastern half of the continent. Large areas of the southeastern United States experience average dew point temperatures over 20°C. The highest Canadian dew points extend from the southern Prairies to the Maritimes. The Canadian West Coast is higher than the rest of western Canada due to the influence of warmer Pacific Ocean waters. The lowest dew points are found in the Arctic due to colder air temperatures and in the west, where mountain chains effectively shield these areas from significant amounts of moisture moving in from the Pacific and the Gulf of Mexico. Moisture patterns in July are

indicated by the vapour pressures in Figure 4.14d; they closely follow the dew point temperature pattern.

The difference between air temperature and dew point can indicate whether the relative humidity is low or high. When the air temperature and dew point are far apart, the relative humidity is low; when they are close to the same value, the relative humidity is high. When the air temperature and dew point are equal, the air is saturated and the relative humidity is 100 percent. Even though the relative humidity may be 100 percent, under certain conditions, air may still be considered to be "dry" as there is little moisture in the air. ● Figure 4.15 demonstrates this. The polar air in Figure 4.15a has the same air temperature and dew point temperature; consequently, the air is saturated and the relative humidity is 100 percent. The desert air in Figure 4.15b has a large separation between air temperature and dew point and, consequently, a much lower relative humidity of 21 percent.* But since dew point is a measure of the amount of water vapour in the air, the desert air with a higher dew point must contain more water vapour. (The same pattern would result if we had compared the actual vapour pressure, the specific humidity, or the mixing ratio.) So even though the polar air has a higher relative humidity, the desert air contains more moisture. In cold polar air, the dew point and air temperature are normally close together, but the low dew point temperature means that there is little water vapour in this air. This is why polar air is often described as "dry" when its relative humidity is high (even close to 100 percent).

---

*Relative humidity is computed by determining the saturation vapour pressure and actual vapour pressure from Figure 4.10 (p. 106). Desert air with an air temperature of 35°C has a saturation vapour pressure of about 56 hPa. A dew point temperature of 10°C indicates an actual vapour pressure of about 12 hPa. These values produce a relative humidity of 21 percent:

$$\left(\frac{12}{56}\right) \times 100\% = 21\%$$

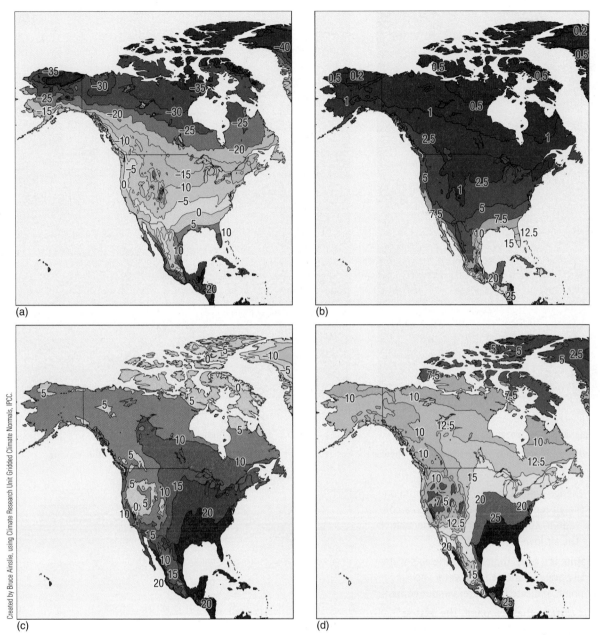

● FIGURE 4.14 Average surface dew point and vapour pressure values for North America. (a) Average January dew points (°C). (b) Average January vapour pressures (hPa). (c) Average July dew points (°C). (d) Average July vapour pressures (hPa).

BRIEF REVIEW

Up to this point, we have looked at a number of different ways to describe humidity. We have also discussed the differences between relative humidity and dew point and indicated how these are used. Before going on, here is a review of some of the key ideas so far:

● Humidity describes the amount of moisture in the air.

● The hydrologic cycle describes the movement of water through Earth's systems and its processes of transformation between liquid, solid, and vapour forms.

● Saturation is an equilibrium between evaporation and condensation that represents the upper limit of water vapour that can be in air at a given temperature.

● Specific humidity, mixing ratio, and actual vapour pressure represent humidity in different ways. In meteorology, these are the most commonly used ways of indicating amounts of moisture in the air.

● Relative humidity tells us how close the air is to being saturated; it does <u>not</u> tell us how much water vapour is in the air.

(a) POLAR AIR: Air temperature –2°C
                Dew point –2°C
                Relative humidity 100 percent

(b) DESERT AIR: Air temperature 35°C
                 Dew point 10°C
                 Relative humidity 21 percent

● FIGURE 4.15 The polar air has a higher relative humidity, whereas the desert air, with the higher dew point, contains more water vapour.

● Relative humidity can change when the air's water vapour content changes or when the air temperature changes. With a constant amount of water vapour, cooling the air raises the relative humidity and warming the air lowers it.

● The dew point temperature is a good indicator of the air's water vapour content: high dew points indicate high water vapour content, and low dew points indicate low water vapour content.

● If the air temperature is close to the dew point, the relative humidity is high. Conversely, if the air temperature and dew point are far apart, the relative humidity is low.

COMPARING HUMIDITIES   On a global scale, ● Figure 4.16 shows how the average relative humidity varies from the equator to the poles. High relative humidities are normally found in the tropics and near the poles, where there is little separation between air temperature and dew point. The average relative humidity is lowest near 30° latitude, where we find the deserts of the world girdling the globe. Contrast this with Figure 4.9 (p. 106), which shows how specific humidity, a measure of the actual water vapour in the air, decreases continually with latitude.

Of course, not all locations near 30°N are deserts. Take, for example, humid New Orleans, Louisiana. During July, the air in New Orleans has an average dew point temperature of 22°C and contains a great deal of water vapour—nearly 50 percent more than does the air along the southern California coast.

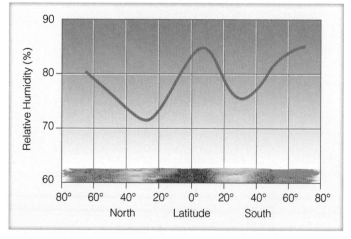

● FIGURE 4.16 The average relative humidity at each latitude for latitudes north and south of the equator.

Since both locations are adjacent to large bodies of water, let's find out why New Orleans is more humid.

● Figure 4.17 shows a typical summertime situation. Air from the northern Pacific Ocean blows onto the West Coast from California to British Columbia. The cooler water temperatures along this coast are caused by cool ocean currents and upwelling, where deeper, colder sea water is pushed to the surface. (Ocean upwelling and its climatic effects are

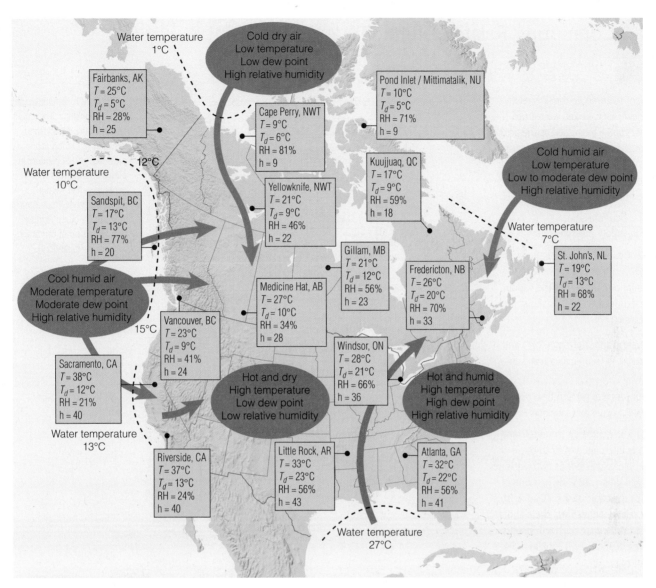

● FIGURE 4.17 Typical summertime afternoon temperature and humidity values over North America. South-central and eastern Canada are affected by hot, humid air originating in the Gulf of Mexico, whereas other parts of Canada are affected by cooler, less humid air from the Pacific or cold air from the Arctic oceans. For each city, T represents the air temperature, T_d the dew point, RH the relative humidity, and h the humidex value. (All data represent typical July midafternoon conditions.)

discussed more completely in Chapter 10.) Notice that temperatures in the Pacific Ocean are much cooler than those in the Gulf of Mexico. As water affects the air above it, each area has air temperatures near those of the water below it and both areas are nearly saturated with water vapour. Consequently, dew point temperatures over the cooler Pacific are much lower than dew point temperatures over the warmer Gulf. So, as seen in the dew point values for cities near these areas, air from the Gulf of Mexico contains a great deal more water vapour than air from the Pacific.

Air from the Gulf of Mexico moves into the southeastern United States and then on to eastern Canada. As it moves inland, away from its source of moisture, air temperatures increase and relative humidities drop, but the amount of water vapour in the air (the dew point temperature) hardly changes. This creates the humid, muggy conditions with high relative humidities that we see over the eastern United States and Canada. These are most extreme over cities such as New Orleans, which are just inland from the Gulf of Mexico.

On the western coast of North America, as air moves inland, a similar pattern occurs, but the Pacific Ocean's lower dew point temperatures combined with high inland air temperatures produce low relative humidities. These form the deserts we see around 30° latitude through the US Southwest and northern Mexico. In Canada's semiarid B.C. interior and southern Prairies, we see a similar effect, but northern air temperatures are reduced, and airflow from colder Arctic areas has a cooling influence.

Computing Relative Humidity and Dew Point

Suppose we want to compute the air's relative humidity and dew point. Earlier, we learned that relative humidity can be expressed as the actual vapour pressure divided by the saturation vapour pressure times 100 percent. Using the symbols previously described in this chapter, the expression for relative humidity (RH) becomes

$$RH = \frac{\text{actual vapour pressure}}{\text{saturation vapour pressure}} = \frac{e}{e^*} \times 100\%$$

• Figure 2, Figure 4.10, and Figure D.1 (in Appendix D—a more precise version of the same information) show that the vapour pressure curve defines two relationships: the relationship between

• saturation water vapour pressure (e^*) and air temperature (T) (purple axis labels in Figure 2) and

• actual vapour pressure (e) and dew point temperature (T_d) (blue axis labels in Figure 2)

Since it can be confusing to understand that this graph is interpreted in both these ways, some people visualize it as two separate graphs, with each graph showing just one of the relationships indicated above. Reading this graph for both relationships gives the e and e^* values we need to calculate relative humidity using the above equation. Let's look at a practical example of this.

Suppose the air temperature in a room is 28°C. Because the saturation vapour pressure (e^*) depends on the air temperature, we can read Figure 2 and obtain e^* by determining the intersection of 28°C with the saturation vapour pressure curve. We get a e^* value of about 38 hPa (values in purple; if we could read the graph precisely, the true value is just under 38—actually 37.8 hPa).

Also suppose that from other measurements, we know that the dew point temperature is 20°C. Using Figure 2 again, but this time to obtain the actual vapour pressure, we determine that e is 23 hPa (values in blue). Essentially, we used Figure 2 to obtain the saturation vapour pressure (e^*) when the air temperature (T) was known; following the same process, we determined the actual vapour pressure (e) by knowing the dew point temperature (T_d). With this information, we can calculate relative humidity as follows:

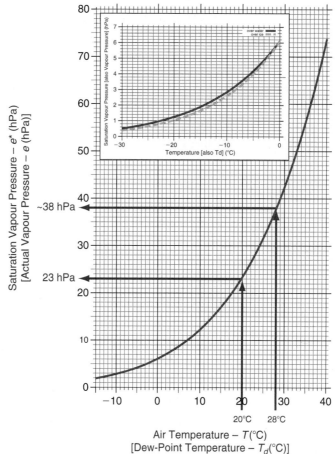

$$RH = \frac{e}{e^*} \times 100\%$$

$$RH = \frac{23 \text{ hPa}}{38 \text{ hPa}} \times 100\%$$

$$RH = 61\%$$

These values could also have been determined from a table showing values based on both relationships (see Table D.3 in Appendix D) and following the same calculation. Before leaving this example, understand that by following this process in reverse, we could also use Figure 2 to determine the air temperature and dew point temperature if we know the saturation vapour pressure and actual vapour pressure.

Now let's try using these relationships another way. If we know the air temperature is 26°C and the relative humidity is 54 percent, what is the dew point temperature of the air? From Figure 2, an air temperature of 26°C produces a saturation vapour pressure (e^*) of 34

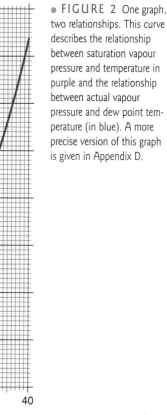

• FIGURE 2 One graph, two relationships. This curve describes the relationship between saturation vapour pressure and temperature in purple and the relationship between actual vapour pressure and dew point temperature (in blue). A more precise version of this graph is given in Appendix D.

hPa. Here is how to obtain the actual vapour pressure (e) from these numbers:

$$RH = \frac{e}{e^*} \times 100\%$$

$$54\% = \frac{e}{34 \text{ hPa}} \times 100\%$$

$$34 \text{ hPa} \times 0.54 = e$$

$$e = 18 \text{ hPa}$$

Finally, using the vapour pressure curve in Figure 2 again, we determine that an actual vapour pressure of 18 hPa corresponds to a dew point temperature of 16°C.

The key to these calculations is understanding that two relationships are shown by the same vapour pressure curve:

• saturation vapour pressure (e^*) is related to air temperature (T)

• actual vapour pressure (e) is related to the dew point temperature (T_d)

FOCUS ON A SPECIAL TOPIC

Is Humid Air "Heavier" than Dry Air?

Does a volume of humid air weigh more than a similar-sized volume of dry air? The answer is no! At the same temperature and pressure, humid air weighs less than dry air. (Keep in mind that we are referring strictly to water vapour—a gas—and not suspended liquid droplets.) To understand why, we must first see what determines the weight of atoms and molecules.

Almost all of the weight of an atom is concentrated in its nucleus, where the protons and neutrons are found. Neutrons weigh nearly the same as protons. To get some idea of how heavy an atom is, we simply add up the number of protons and neutrons in the nucleus. (Electrons are so light that we can ignore them in comparing weights.) The larger this total, the heavier the atom. Now we can compare one atom's weight with another's. For example, hydrogen, the lightest known atom, has only one proton in its centre (it has no neutrons). Thus, it has an atomic weight of 1. Nitrogen, with seven protons and seven neutrons in its nucleus, has an atomic weight of 14. Oxygen, with eight protons and eight neutrons, weighs in at 16.

A molecule's weight is the sum of the atomic weight of its atoms. For example, molecular oxygen, with two oxygen atoms (O_2), has a molecular weight of 32. The most abundant atmospheric gas, molecular nitrogen (N_2), has a molecular weight of 28.

When we determine the weight of air, we are dealing with the weight of a mixture. As you might expect, a mixture's weight is a little more complex. We cannot just add the weights of all its atoms and molecules because the mixture might contain more of one kind than another. Air, for example, has far more nitrogen (78 percent) than oxygen (21 percent). We allow for this by multiplying the molecule's weight by its share in the mixture, as shown in ▼ Table 1. Since dry air is essentially composed of N_2 and O_2 (99 percent), we ignore the other parts of air for this rough average.

The symbol ≈ means "is approximately equal to." Therefore, dry air has a molecular weight of about 29. How does this compare to humid air?

Water vapour is composed of two atoms of hydrogen and one atom of oxygen (H_2O). It is an invisible gas, just as oxygen and nitrogen are invisible. Its molecular weight is 18, composed of two atoms of hydrogen (each with atomic weight of 1) and one atom of oxygen (atomic weight 16). Obviously, air, at nearly 29, weighs appreciably more than water vapour.

If we replace dry air molecules one for one with water vapour molecules, the total number of molecules remains the same, but the total weight of the air decreases. Since density is mass per volume, *humid air at the surface is less dense or lighter than dry air.*

This fact can have an important influence on our weather. The lighter the air becomes, the more likely it is to rise. All other factors being equal, humid (less dense) air will rise more readily than dry (more dense) air, and this causes warm, moist surface air to rise, cool, and condense, creating clouds and precipitation.

Of lesser importance to weather but of greater importance to sports is the fact that a baseball will "carry" farther in moist, less dense air. Consequently, without the influence of wind, a ball will travel slightly farther on a humid day than it will on a dry day. So when the sports announcer proclaims, "The air today is heavy because of the high humidity," implying that this makes the air denser, remember that this statement is wrong and the ball will actually travel farther on a humid day due to the less dense air than it will on a dry day.

▼ Table 1   Molecular Weight of Dry Air

GAS	WEIGHT		NUMBER OF ATOMS		MOLECULAR WEIGHT		PERCENT BY VOLUME	GRAMS PER MOLE OF AIR
Oxygen	16	×	2	=	32	×	21%	≈ 7
Nitrogen	14	×	2	=	28	×	78%	≈ 22
							Molecular weight of dry air ≈ 29	

Air entering Canada from the Arctic Ocean and the Labrador Sea is cold, with a low dew point temperature. As it moves southward over the land, the air gradually becomes warmer and more humid. This air is warmed by the land, and evaporation from it and its many lakes moistens the air as it moves further from the cold ocean.

RELATIVE HUMIDITY IN THE HOME   Question: How does the relative humidity of the winter air in your home compare to that in the Sahara Desert? Surprisingly, during cold winters, some homes actually have a lower relative humidity than the desert, and, usually, the inhabitants are unaware of it. Remember that cold polar air contains only a little water vapour. Even when saturated, air with a temperature and dew point of −15°C has an actual vapour pressure of only 1.9 hPa. When this air is brought indoors and heated to 20°C, its saturation vapour pressure increases to 23.4 hPa—about 12 times what it was outside. Notice in ● Figure 4.18 that the relative humidity of the heated air inside the house drops to 8 percent.* This relative humidity is lower than what you would normally experience in a desert during the hottest day!

Very low relative humidities in a house can have an adverse effect on things living inside. For example, house plants have a difficult time surviving because the moisture from their leaves and the soil evaporates rapidly. Hence, house plants usually need watering more frequently in winter

---

*($RH = (1.9\ hPa)/(23.4\ hPa) \times 100\% = 8\%$

INSIDE AIR

$T = 20°C$

$T_d = -15°C$

RH = 8%

OUTSIDE AIR

$T = -15°C$

$T_d = -15°C$

RH = 100%

● FIGURE 4.18 The relative humidity drops to 8 percent when outside air, with air temperature and a dew point of –15°C, is brought indoors without adding any water vapour to it and heated to a temperature of 20°C. This places stress on the plants, animals, and humans living inside. (T represents temperature; T_d, dew point; and RH, relative humidity.)

than in summer. People also suffer when the relative humidity is quite low. The rapid evaporation of moisture from exposed flesh causes skin to crack, dry, flake, or itch. These low humidities also irritate the mucous membranes in the nose and throat, producing an "itchy" throat and dry nasal passages that permit inhaled bacteria to incubate, causing persistent infections. The remedy for most of these problems is simply to increase the relative humidity. But how?

The relative humidity in a home can be increased just by heating water and allowing it to evaporate into the air. The added water vapour raises the relative humidity to a more comfortable level. In modern homes, a humidifier associated with the furnace adds moisture to the air and circulates it throughout the home through the furnace's forced-air heating system. In this way, all rooms get their fair share of moisture—not just the room where the vapour is added. Surprisingly, this can amount to about four litres of water per room each day.

Homes can experience the opposite problem during the summertime, when high temperatures and humidities are common. To lower the air's moisture content, as well as its temperature, many homes are air conditioned. Outside air cools as it passes through a system of cold coils located in the air conditioning unit. The cooling increases the air's relative humidity, the air reaches saturation, and the water vapour condenses into liquid water, which is drained away. The cooler, dehumidified air is now circulated through the home.

In hot regions, where the relative humidity is low, evaporative cooling systems, also known as "swamp coolers," are used to cool the air. These systems operate by having a fan blow hot, dry outside air across pads that are saturated with water. Evaporation cools the air, which is circulated through the home, bringing some relief from the hot weather. Swamp coolers do not work well in hot, muggy weather because high relative humidity greatly reduces the rate of evaporation. Besides, swamp coolers add water vapour to the air—something that is not needed when the air is already uncomfortably humid. That is why swamp coolers can be found in Saskatchewan or Arizona but not in southern Ontario or Florida homes.

RELATIVE HUMIDITY AND HUMAN DISCOMFORT　On a hot, muggy day when the relative humidity is high, it is common to hear someone complain, "It's not the heat; it's the humidity." Even though most people do not accurately sense humidity, this statement actually has validity. In warm weather, our bodies cool mainly by evaporation, or, as we know it, perspiring. In Chapter 2, we learned that evaporation is a cooling process, so when the air temperature is high and the relative humidity is low, perspiration on the skin evaporates quickly, making us feel that the air temperature is lower than it really is. However, when both the air temperature and the relative humidity are high, the air is nearly saturated, so body moisture does not evaporate very well—it collects on our skin as beads of perspiration. Less evaporation means less cooling, so we usually feel warmer than we would in a similar air temperature with a lower relative humidity.

We can measure how much cooling can occur by knowing the **wet-bulb temperature**—*the temperature caused by the cooling effect of evaporating water into the air.* We will discuss wet-bulb measurement more fully later in this chapter, but here we will focus on how this measurement helps us anticipate heat-related health problems. On a hot, dry day, the wet-bulb temperature is low and rapid evaporation causes cooling on our skin's surface. Conversely, on a hot humid day, the wet-bulb temperature is closer to the air temperature, there is less evaporation so less cooling occurs, and our skin temperature rises. If the wet-bulb temperature exceeds our skin's temperature, no net evaporation can occur, and our body temperature can rise quite rapidly. Fortunately, the wet-bulb temperature is considerably below our skin temperature most of the time.

When weather is hot and muggy, a number of heat-related health problems can occur. For example, in hot weather, as our body temperature rises, the hypothalamus

WEATHER WATCH

A man actually survived 45 minutes in a sauna with an extremely low relative humidity (less than 1%) and an air temperature approaching 125°C. Even though a roast would cook at this incredibly high temperature, the man survived to tell about it because his rapidly evaporating perspiration formed a layer of cool air around his body, protecting him from the extreme heat. If somehow this protective layer had been blown away, the man's skin would have burned in a matter of seconds.

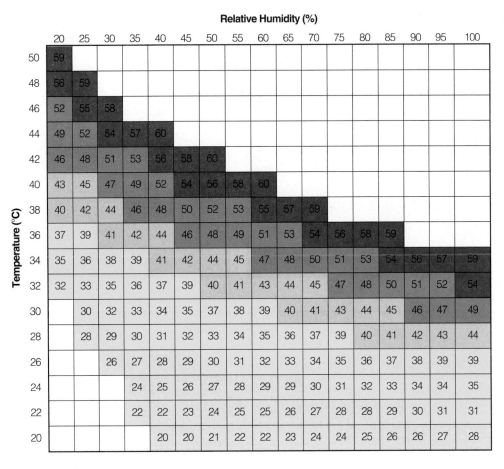

Relative Humidity (%)

● FIGURE 4.19 Air temperature (°C) and humidity are combined to determine an apparent temperature or humidex. An air temperature of 30°C with a relative humidity of 60% produces a humidex value of 39, which would feel like a temperature of 39°C under low humidity conditions.

(a gland in the brain that regulates human body temperature) activates our heat-regulating mechanism, and over 10 million sweat glands pour as much as two litres of liquid per hour over our bodies. As this perspiration evaporates, rapid loss of water and salt can result in a chemical imbalance that may lead to painful *heat cramps*. Excessive water loss through perspiring, coupled with increasing body temperature, may result in *heat exhaustion*—fatigue, headache, nausea, and even fainting. If our body temperature rises above about 41°C, **heat stroke** can occur, resulting in complete failure of our circulatory functions. If body temperature continues to rise, death may result. In fact, each year across North America, hundreds of people die from heat-related maladies. Even strong, healthy individuals can succumb to heat stroke. During a Minnesota heat wave, the NFL's all-pro offensive lineman Korey Stringer died after exhibiting classic but unrecognized—even by team doctors—symptoms of heat stroke. He collapsed after an intense practice on July 31, 2001, and died 15 hours later. Before he fainted, temperatures on the playing field were in the mid-30s (°C) and the relative humidity was above 55 percent.

To highlight the seriousness of weather-related health hazards, summertime weather forecasts now usually also report an index that represents the perceived temperature incorporating the impact of humidity. In Canada, this is the unitless **humidex**; the United States uses a similar concept,

▽ Table 4.1   The Humidex and Degree of Discomfort

HUMIDEX	DEGREE OF DISCOMFORT
20–29	No discomfort
30–39	Some discomfort
40–45	Great discomfort; avoid exertion
46 and over	Dangerous; probable heat stroke
54 and over	Heat stroke imminent

the heat index. The humidex combines air temperature with dew point temperature and determines the apparent temperature we would experience based on the current humidity.* For example, in ● Figure 4.19, an air temperature of 40°C and a relative humidity of 50 percent produce a humidex of 56, which means that based on the current humidity, the air feels as if it is 56°C even though it is only 40°C outside. As we can see in ▽ Table 4.1, heat stroke is imminent when the humidex reaches this level. However, heat stroke–related deaths can occur when the humidex is considerably lower than 56. In the previously discussed case of Korey Stringer, the conditions on

---

*These indexes (humidex, heat index) do not consider other factors that affect how hot it feels. Factors such as radiation and a lack of wind also increase our perception of the air's temperature.

FOCUS ON AN ADVANCED TOPIC

Psychrometry—Two Thermometers, a Wet Sock, and 200 Years of Humidity Measurements

Psychrometry is the simplest accurate method for determining humidity. Based on the principle that evaporation uses energy, a wet thermometer in unsaturated air will be cooler than an ordinary thermometer. Measuring humidity using this concept was developed by several scientists in the 1750s, but the term "psychrometry" was first applied by the German inventor Ernst Ferdinand August in 1818.

A psychrometer basically consists of two identical liquid-in-glass thermometers mounted side by side (see ● Figure 3). One thermometer is covered by a wick or sock dipped in clean, deionized or distilled water (called a wet-bulb thermometer), and the other is unaltered (called a dry-bulb thermometer). Water evaporates from the wet-bulb thermometer and, after a few minutes, cools to the lowest temperature that can be reached by evaporating water into the air,* resulting in a **wet-bulb temperature** reading. Drier air experiences a greater amount of evaporative cooling. The dry-bulb thermometer measures the current air temperature, called the **dry-bulb temperature**. The temperature difference between the dry-bulb and wet-bulb temperatures is known as the **wet-bulb depression**. A large depression indicates that a lot of water vapour has evaporated from the

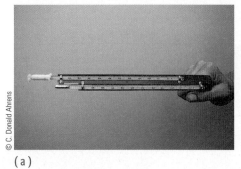

(a)   (b)

● FIGURE 3   (a) The sling psychrometer. (b) The Assman-type psychrometer.

wet bulb and that the relative humidity is low. A small depression indicates little evaporation from the wet bulb and that the relative humidity is high and the air is close to saturation. When there is no depression, the relative humidity is 100 percent, the air is saturated, and the dry-bulb, wet-bulb, and dew point temperatures are all the same (since dew point is the temperature at saturation).

There are three main types of psychrometers: unventilated and two variations of ventilated psychrometers. Unventilated psychrometers rely on ambient wind conditions.* Ventilated psychrometers maximize

evaporation from the wet thermometer by either swinging the instrument (these are called sling psychrometers; see Figure 3a) or by using a mechanical (wind-up) or electric fan to draw air past the thermometers (these are called aspirated psychrometers or Assman-type psychrometers; see Figure 3b).

From wet-bulb and dry-bulb temperatures, all the other humidity values, such as actual water vapour pressure, mixing ratio, absolute humidity, and relative humidity, can be determined. Let's see how this works.

Once wet-bulb and dry-bulb temperature readings are made, you can use either tables (see Appendix D, Tables D.1 and D.2) or calculations as described here to find all the other humidity measures. For example, actual vapour pressure (*e*) can be calculated using the *psychrometric equation* as follows:

$$e = e^*_{(Tw)} - \gamma(T - T_w)$$

---

*The wet-bulb temperature always lies between the dew point and the air temperature, unless the air is saturated when all these temperatures are the same. Notice that the wet-bulb temperature and the dew point temperature are different entities. The wet-bulb temperature is attained by *evaporating water* into the air, whereas the dew point temperature is reached by *cooling* the air.

*Although unventilated psychrometers are commonly used, they generally suffer from underventilation as they rely on ambient wind conditions to remove the saturated air from the wet thermometer so that evaporation can occur. This leads to unknown ventilation rates that often have less evaporation than needed for accurate measurements.

---

the day he died translated to a humidex of 47—just into the category identified as "dangerous—probable heat stroke."

An example of heat-related impacts comes from the Chicago heat wave of July 1995. Tragically, about 600 people died during the event, which lasted several days. On July 13, the afternoon air temperature reached 40°C. With a dew point temperature of 24°C and a relative humidity near 40 percent, the apparent temperature (humidex) soared to 52 (see Table 4.1). A disproportionate number of poor elderly people died, often

because they were unable to cool their rooms or were too concerned about possible crimes to open windows. The combination of high temperature and humidity can also affect parts of eastern and central Canada, especially in summer, when air from the Gulf of Mexico moves over the region. Humidex values higher than 35 occur an average of 10.5 days each July in Windsor, Ontario, and 6.5 days each July in Toronto.

During hot, humid weather, some people remark about how "heavy" or dense the air feels. Is hot, humid air really more

The terms in the psychrometric equation are as follows:

- e is the actual vapour pressure of the air

- $e^*_{(Tw)}$ is the **saturation vapour pressure** at the wet-bulb temperature. To find this value, take the wet-bulb temperature (T_w), and using Figure D.1 in Appendix D, determine the saturation vapour pressure from the graph. For a complete explanation of how to use Figure D.1, see Focus on a Special Topic: Computing Relative Humidity and Dew Point (p. 115).

- T_w is the wet-bulb temperature

- T is the dry-bulb temperature, which is the same as the air temperature

- γ is the psychrometric coefficient (γ is the Greek letter gamma). Its value changes depending on whether the water is frozen (ice) or unfrozen (water). For water at sea level, it is approximately equal* to 0.66 hPa °C^{-1}; for ice, it is 0.582 hPa °C^{-1}.

Let's try an example. Suppose we use a psychrometer and measure a dry-bulb temperature (T) of 20°C and a wet-bulb temperature (T_w) of 10°C. We find a wet-bulb saturation vapour pressure $e^*_{(Tw)}$ of 12 hPa. Let's put these values into the psychrometric equation:

---

*As we saw earlier in this chapter, saturation vapour pressure behaves differently for frozen than unfrozen conditions, leading to two different psychrometric coefficients: one over water, the other over ice. Both coefficients also have a slight variation with total atmospheric pressure.

$e = 12$ hPa $- (0.66$ hPa °C$^{-1})(20 - 10)$°C

$e = 5.4$ hPa $= 540$ Pa

Now that we have determined the actual vapour pressure (e), we can calculate the relative humidity and find the dew point, as done in Focus on a Special Topic: Computing Relative Humidity and Dew Point (p. 114). We can also use the vapour pressure to calculate other measures of humidity. To determine the mixing ratio (r) (remember from discussions earlier in this chapter that r has the units of g kg^{-1}),

$$r = \frac{0.622 \times e}{P - e} \times 1000 \text{ g kg}^{-1}$$

where

- e is always the actual vapour pressure.

- P is the atmospheric pressure. Note that the units of P and e have to be the same for the pressure values to cancel each other; either both are in hPa or both are in Pa.

- 0.622 is a unitless number that represents a ratio of the gas constant for water vapour to the gas constant for dry air. It arises from the ideal gas law and is used to scale pressure values to mass values.

So, to complete our calculation and determine the mixing ratio (r) for the vapour pressure (e) of 540 Pa, determined above, we also need to know the atmospheric pressure (P). This depends on elevation and the weather, but at sea level, it is approximately 1000 hPa = 100,000 Pa, so we will use this value:

$$r = \frac{0.622 \times 540 \text{ Pa}}{(100000 - 540)\text{Pa}} \times 1000 \text{ g kg}^{-1}$$
$$= 3.55 \text{ g kg}^{-1}$$

Finally, let's revisit what happens when the wet-bulb temperature goes well below zero and the psychrometer wick freezes solid. In this case, the temperature is called the **frost-bulb temperature**, and there is sublimation from a frozen wick instead of evaporation from a liquid-covered wick. Because frozen water is held more tightly than liquid water, the sublimation rate and amount of cooling are less for the frozen wick. (This is the reason why in Figure D.1 [see Appendix D] the saturation vapour pressure over ice is lower than it is over water.) In calculations, we must use the saturation vapour pressure over ice (instead of over water), and the psychrometric coefficient is replaced by 0.582 hPa °C^{-1}. Temperatures around 0°C are most problematic as both water and ice can exist on the wick. This creates several complications. Thick layers of ice can insulate the thermometer and cause it to read higher than the true frost-bulb temperature (so, ideally, we want a very thin layer of ice on the wick). Also, right at 0°C, water changes from liquid to ice on the wick, and we can see the temperature reading on the thermometer rise due to heat released during freezing (latent heat of fusion). So when making wet-bulb measurements around zero, the observer must carefully note whether the wick is frozen or liquid before making calculations.

---

dense than hot, dry air? To find out, read Focus on a Special Topic: Is Humid Air "Heavier" Than Dry Air? (p. 115).

MEASURING HUMIDITY Instruments that measure humidity are commonly called **hygrometers**. One type—called a **psychrometer**—is based on the principle that evaporation cools the air so that in unsaturated air, a thermometer that is kept wet (called a wet-bulb thermometer) will be cooler than an ordinary thermometer (called a dry-bulb ther-

mometer). The section Focus on an Advanced Topic: Psychrometry—Two Thermometers, a Wet Sock, and 200 Years of Humidity Measurements (above) provides a more detailed explanation of how a psychrometer works and how wet-bulb and **dry-bulb temperature** readings can be used to find other measures of humidity, such as vapour pressure, absolute humidity, mixing ratio, and relative humidity. Another type of hygrometer, called a **hair hygrometer**, is based on the principle that the length of human or horse hair

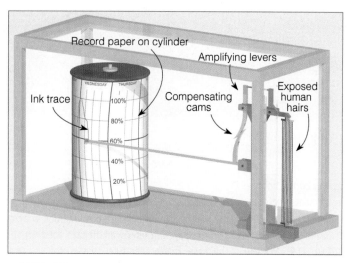

● FIGURE 4.20 The hair hygrometer measures relative humidity by amplifying and measuring changes in the length of human (or horse) hair.

increases by 2.5 percent as relative humidity changes from 0 to 100 percent. In this mechanical instrument, clean hair strands are mounted as shown in ● Figure 4.20 so that their length according to changes in humidity can be observed on a dial or tracked on a clock-driven rotating chart that produces a continuous record of relative humidity. Hair hygrometers are not as accurate as psychrometers and require frequent calibration. Generally, these devices are being replaced by electronic sensors, but they are still used in settings where it is difficult to obtain reliable electricity.

An *electrical hygrometer* consists of a flat plate coated with a film of carbon that changes its electrical resistance (how electricity flows through) as the relative humidity changes and moisture is absorbed or lost by the carbon. This device is commonly used in radiosondes—balloon-mounted atmospheric monitoring equipment, which is discussed in more detail in Chapter 1.

A *thin-film capacitance hygrometer* has a sensor with two gold electrodes that are separated by a very thin layer of material made from a dielectric polymer (a poor electrical conductor) that can absorb and release moisture. When an electrical voltage is applied to the electrodes, the sensor acts as a capacitor by storing an electrical charge. The sensor's ability to hold an electrical charge, its *capacitance*, changes with changes in relative humidity because the polymer absorbs and releases moisture. The hygrometer converts the capacitance into a relative humidity reading, which is usually stored electronically in an attached data logger.* Sensors of

*A data logger is a special-purpose, programmable microcomputer that can measure electrical signals, convert them into usable data with recognizable units, process these data (i.e., average them, find the maximum or minimum, identify when these occur, etc.), and store the data in a digital format. They can also often transmit their stored data by communicating over telephone, Internet, radio, cellular phone, or satellite communication systems.

this type are common in modern automated weather stations.

The *infrared hygrometer* quantifies atmospheric humidity by measuring the amount of infrared energy absorbed by water vapour in a sample of air. It operates by emitting a known amount of infrared radiation and then measures the radiation's intensity after it has passed through a sample of air. The amount of radiation absorbed by water vapour in the air is related to the current humidity.

The *dew point hygrometer* measures the dew point temperature by cooling the surface of a mirror until condensation (dew) forms and recording the temperature at this point. This sensor is the type that measures dew point temperature in the hundreds of government-run automated weather stations that exist throughout Canada and the United States.

Finally, a commonly used instrument, the *dew cell*, is a sophisticated hygrometer that uses the salt lithium chloride to measure the air's dew point temperature. A dew cell consists of a thin layer of this salt, in solution, on a fibreglass tape within a probe. Basically, by changing the evaporative equilibrium of this solution, the instrument determines the air's current humidity. Lithium chloride is hygroscopic—it holds on to water. This means that at normal air temperatures, the solution will collect water vapour (something like the desiccant crystals we use in unheated air spaces to keep the area dry). Being hygroscopic, the solution has a very low evaporation rate that is not in equilibrium with its condensation rate; the condensation rate is dependent only on the air's humidity. The instrument works because we heat the solution and increase its evaporation rate until it matches the condensation rate of the air. Now the solution is in equilibrium. By knowing the temperature where equilibrium is reached, we can determine the dew point temperature of the air.

In summary, there are a variety of ways to detect atmospheric humidity. New equipment and techniques will evolve with developments in current technology.

SUMMARY

This chapter examined the concept of atmospheric humidity. The chapter began by looking at the hydrologic cycle and the circulation of water in our atmosphere. It then looked at the different phases of water, showing how evaporation, condensation, and saturation occur at the molecular level. The next several sections looked at the many ways of describing the amount of water vapour in the air. Here we learned that there are many ways of describing humidity. The absolute humidity represents the density of water vapour in a given volume of air. Specific humidity measures the mass of water vapour in a fixed mass of air, whereas the mixing ratio expresses humidity as the mass of water vapour in the fixed mass of remaining dry air. The actual vapour pressure indicates the air's total water vapour content by expressing the amount of water vapour in terms of the amount of pressure that the water vapour molecules exert. The saturation vapour pressure describes how much water vapour the air could hold at any given temperature in terms of how much pressure the water vapour molecules would exert if the air were saturated at that temperature. A good indicator of the air's actual water vapour content is the dew point—the temperature to which air would have to be cooled (at constant pressure) for saturation to occur.

Relative humidity is a measure of how close the air is to being saturated. Air with a high relative humidity does not necessarily contain a great deal of water vapour; it is simply close to being saturated. With a constant water vapour content, cooling the air causes the relative humidity to increase, whereas warming the air causes the relative humidity to decrease. When the air temperature and dew point are close together, the relative humidity is high, and when they are far apart, the relative humidity is low. High relative humidity in hot weather makes us feel hotter than it really is by retarding the evaporation of perspiration. The humidex is a measure of how hot it feels to an average person for various combinations of air temperature and humidity. Although relative humidity can be confusing (because it can change with either air temperature or moisture content), it is nevertheless the most widely used way of describing the air's moisture content.

The chapter concludes by examining the various instruments that measure humidity.

KEY TERMS

The following terms are listed (with page number) in the order they appear in the text. Define each. Doing so will aid you in reviewing the material covered in this chapter.

humidity, 100
hydrologic cycle, 100
evaporation, 100
condensation, 100
precipitation, 100

transpiration, 100
evapotranspiration, 100
sublimation, 102
deposition, 102
equilibrium, 103

saturated (air), 103
saturation, 103
condensation nuclei, 103
absolute humidity, 104
specific humidity, 105
mixing ratio, 105
Dalton's law of partial
   pressure, 105
actual vapour pressure, 105
saturation vapour
   pressure, 106
relative humidity, 108
supersaturation, 108

dew point temperature
   (dew point), 109
frost point, 109
wet-bulb temperature, 116
heat stroke, 117
humidex, 117
dry-bulb temperature, 118
wet-bulb depression, 118
saturation vapour pressure, 119
frost-bulb temperature, 119
hygrometer, 119
psychrometer, 119
hair hygrometer, 119

QUESTIONS FOR REVIEW

1. Briefly explain the movement of water in the hydrologic cycle.

2. Basically, how do the three states of water differ?

3. What are the primary factors that influence evaporation?

4. Explain why condensation occurs primarily when the air is cooled.

5. How are evaporation and condensation related to saturated air above a flat water surface?

6. How does condensation differ from precipitation?

7. Why are specific humidity and mixing ratio more commonly used in representing atmospheric moisture than absolute humidity? What is the only way to change the specific humidity or mixing ratio of an air parcel?

8. In a volume of air, how does the actual vapour pressure differ from the saturation vapour pressure? When are they the same?

9. What does saturation vapour pressure primarily depend on?

10. Explain why it takes longer to cook vegetables in the mountains than at sea level.

11. (a) What does the relative humidity represent?
    (b) When the relative humidity is given, why is it also important to know the air temperature?
    (c) Explain two ways the relative humidity may be changed.

12. Explain why, during a summer day, the relative humidity will change as shown in Figure 4.12 (p. 109).

13. Why do hot and humid summer days usually feel hotter than hot and dry summer days?

14. Why is the wet-bulb temperature a good measure of how cool human skin can become?

15. Explain why the air on a hot, humid day is less dense than on a hot, dry day.

16. (a) What is the dew point temperature?
    (b) How is the difference between dew point and air temperature related to the relative humidity?

17. Why is cold polar air described as "dry" when the relative humidity of that air is very high?

18. How can a region have a high specific humidity and a low relative humidity? Give an example.

19. Why is the air from the Gulf of Mexico so much more humid than air from the Pacific Ocean at the same latitude?

20. How are the dew point temperature and wet-bulb temperature different? Can they ever read the same? Explain.

21. When outside air is brought indoors on a cold winter day, the relative humidity of the heated air inside often drops below 25 percent. Explain why this situation occurs.

22. Describe how a sling psychrometer works. What does it measure? Does it give you dew point and relative humidity? Explain.

23. Why are human hairs often used in a hair hygrometer?

QUESTIONS FOR THOUGHT

1. Would you expect water in a glass to evaporate more quickly on a windy, warm, dry summer day or on a calm, cold, dry winter day? Explain.

2. How can frozen clothes "dry" outside in subfreezing weather? What exactly is taking place?

3. Explain how and why each of the following will change as a parcel of air with an unchanging amount of water vapour rises, expands, and cools: (a) absolute humidity; (b) relative humidity; (c) actual vapour pressure; and (d) saturation vapour pressure.

4. Where in Canada would you go to experience the least variation in dew point (actual moisture content) from January to July?

5. After completing a gruelling semester of meteorological coursework, you call your travel agent to arrange a much-needed summer vacation. When your agent suggests a trip to the desert, you decline because of a concern that the dry air will make your skin feel uncomfortable. The travel agent assures you that almost daily "desert relative humidities are above 90 percent." Could the agent be correct? Explain.

6. On a clear, calm morning, water condenses on the ground in a thick layer of dew. As the water slowly evaporates into the air, you measure a slow increase in dew point. Explain why.

7. Two cities have exactly the same amount of water vapour in the air. The 6:00 a.m. relative humidity in one city is 93 percent, whereas the 3:00 p.m. relative humidity in the other city is 28 percent. Explain how this can come about.

8. Suppose the dew point of cold outside air is the same as the dew point of warm air indoors. If the door is opened, and cold air replaces some of the warm inside air, would the new relative humidity indoors be (a) lower than before, (b) higher than before, or (c) the same as before? Explain your answer.

9. On a warm, muggy day, the air is described as "close." What are several plausible explanations for this expression?

10. Outside, on a very warm day, you swing a sling psychrometer for about a minute and read a dry-bulb temperature of 38°C and a wet-bulb temperature of 24°C. After swinging the instrument again, the dry bulb is still 38°C, but the wet bulb is now 26°C. Explain how this could happen.

11. Why are evaporative coolers used in Arizona, Nevada, and southern Alberta but not in Florida, Georgia, or southern Ontario?

12. Devise a way of determining elevation above sea level if all you have is a thermometer and a pot of water.

PROBLEMS AND EXERCISES

1. On a bitterly cold, snowy morning, the air temperature and dew point of the outside air are both −7°C. If this air is brought indoors and warmed to 21°C, with no change in vapour content, what is the relative humidity of the air inside the home? (Hint: Use Figure D.1 on p. A-14 or Table D.3 on p. A-12.)

2. (a) With the aid of Figure 4.14c (p. 111), determine the average July dew points in Vancouver, British Columbia; New Orleans, Louisiana; and Toronto, Ontario.

   (b) If the high temperature on a particular summer day in all three cities is 30°C, then calculate the afternoon relative humidity at each of the three cities. (Hint: Figure D.1 on p. A-14 or Table D.3 on p. A-12 will be helpful.)

3. Suppose with the aid of a sling psychrometer that you obtain an air temperature of 30°C and a wet-bulb temperature of 25°C. What is (a) the wet-bulb depression, (b) the dew point, and (c) the relative humidity of the air? (Use tables in Appendix D at the back of the book or the equations in Focus on an Advanced Topic: Psychrometry—Two Thermometers, a Wet Sock, and 200 Years of Humidity Measurements on p. 118.)

4. If the air temperature is 35°C and the dew point is 21°C, determine the relative humidity using Figure D.1 on p. A-14 or Table D.3 on p. A-12.

5. Suppose the average vapour pressure in Banff, Alberta, is about 8 hPa.

   (a) Use Figure D.1 (p. A-14) to determine the average dew point of this air.

   (b) Banff is about 1383 m above sea level, where the normal pressure is about 15 percent less than at sea level. If the air over Banff were brought down to sea level, without any change in vapour content, what would be the new vapour pressure of the air?

6. At Yellowstone National Park in the United States, there are numerous ponds of boiling water. If Yellowstone is about 2200 m above sea level (where the air pressure is

normally about 775 hPa), what is the normal boiling point of water in Yellowstone? (Hint: See Figure 1 on p. 108.)

7. Three cities have the following temperatures (T) and dew points (T_d) during a July afternoon:

Atlanta, Georgia, $T = 32°C$; $T_d = 24°C$

London, Ontario, $T = 27°C$; $T_d = 21°C$

Prince Rupert, British Columbia, $T = 15°C$; $T_d = 13°C$

(a) Which city appears to have the highest relative humidity?

(b) Which city appears to have the lowest relative humidity?

(c) Which city has the most water vapour in the air?

(d) Which city has the least water vapour in the air?

(e) Use Figure D.1 or Table D.3 to calculate the relative humidity for each city.

(f) Using both the relative humidity calculated in (e) and the air temperature, determine the humidex for each city using Figure 4.19 (p. 117).

A lighthouse keeps a constant vigil as clouds and fog approach the coast and the Swallowtail Lighthouse at dawn on Grand Manan Island, New Brunswick.
Radius Images/Jupiter

Condensation: Dew, Fog, and Clouds 5

The weather is an ever-playing drama before which we are a captive audience. With the lower atmosphere as the stage, air and water as the principal characters, and clouds for costumes, the weather's acts are presented continuously somewhere about the globe. The script is written by the Sun; the production is directed by the earth's rotation; and, just as no theatre scene is staged exactly the same way twice, each weather episode is played a little differently; each is marked with a bit of individuality.

Dr. Clyde Orr was an atmospheric scientist who wrote one of the early publicly accessible books about meteorology. It was printed in three editions between 1959 and 1961.

Clyde Orr, Jr., *Between Earth and Space.* Macmillan. 1959.

CONTENTS

Have you walked barefoot across a lawn on a summer morning and felt the wet grass under your feet? Did you ever wonder how those glistening droplets of dew could form on a clear summer night? Or why they formed on the grass but not on the bushes several metres above the ground? In this chapter, we will investigate the formation of dew and frost, examine different types of fog, and conclude with the identification and observation of clouds.

Water is the only molecule on Earth that commonly exists in all three phases. This fact has many implications for all the systems that interact on Earth. First, let us consider the atmospheric system. After radiation, latent energy that is taken up and given off as water changes between the phases of solid, liquid, and gas is the most important way of transferring energy between Earth's surface and the atmosphere. Condensation is the process of water changing from an invisible gas to the visible droplets that make up dew, clouds, and fog. This process releases latent heat that is converted into sensible heat and is the potent energy source fueling thunderstorms, tornadoes, and hurricanes.

The movement of water through Earth's systems and its three phases is an important link between the *atmosphere* and the *hydrosphere* that has implications for the *cryosphere, biosphere, anthrosphere,* and *lithosphere.* As water evaporates from Earth's surface and condenses into clouds, it cycles through the **hydrologic cycle.** Clouds have a profound impact on climate. They reduce incoming solar radiation (K↓), causing a decrease in the daytime air temperature. But as we saw in Chapters 2 and 3, they also increase incoming infrared radiation (L↓), causing an increase in air temperature at night. Additionally, clouds are a necessary prerequisite for precipitation, which controls the availability of fresh water on Earth. Precipitation is a significant determinant of climate and consequently impacts the *biosphere* through its effect on vegetation, animal, and human distributions.

© Jennifer McNab. Used with permission.

● FIGURE 5.1 Dew forms on clear, calm nights when objects on the surface cool below their dew point temperature, become saturated, and condensation occurs.

The Formation of Dew and Frost

On clear, calm nights, objects near Earth's surface cool rapidly by emitting infrared radiation. The ground and objects on it often become much colder than the surrounding air. Air that contacts these cold surfaces cools by conduction. Eventually, this air can cool to its dew point temperature, the temperature where saturation occurs.

For dew point temperatures above freezing, as surfaces such as twigs, leaves, and blades of grass cool below their dew point temperature, saturation occurs and water vapour begins to condense on them, forming tiny visible specks of water called **dew** (see ● Figure 5.1). If the air temperature subsequently drops to freezing or below, dew will freeze, becoming tiny beads of ice called **frozen dew.** If the dew point temperature is below freezing (called the *frost point*), then the water vapour will form directly into ice crystals and create **frost.**

Since the atmosphere cools from the surface, the coolest air on clear, calm nights is usually at ground level, and dew is more likely to form on blades of grass than on objects several metres above the surface. This thin coating of dew dampens bare feet, but, more importantly, it provides a valuable source of moisture for many plants, especially during periods of low rainfall. For locations in the mid-latitudes, dew accounts for a blanket of water between 12 and 50 mm thick each year. To put this in perspective, this is as much as one-quarter of the annual rainfall received by dry, subtropical deserts such as the Australian outback.

Dew is more likely to form on clear, calm nights than on cloudy or windy nights. Clear nights allow objects near the ground to cool rapidly by emitting infrared radiation, and calm winds mean that there is less air mixing, so the coldest air will be located at ground level. These atmospheric conditions are usually associated with large fair-weather, high-pressure systems. Conversely, cloudy, windy weather inhibits the rapid cooling near the ground that causes dew and often signifies an approaching rainstorm. These observations inspired the following folk rhyme:

> When the dew is on the grass,
> rain will never come to pass.
> When grass is dry at morning light,
> look for rain before the night!

Frost is visible on cold, clear, calm mornings when the dew point temperature is at or below freezing. When air temperature cools to its frost point, water vapour changes directly to ice without being a liquid. This process is called *deposition.**

---

*When ice changes back into vapour without melting, the process is called *sublimation.*

It results in delicate, white ice crystals properly known as *hoarfrost* or *white frost*, although until these crystals become large or spectacular, we usually just refer to them as frost. Frost has a treelike branching pattern that easily distinguishes it from frozen dew, which can form spherical beads of ice when there is only a little dew, or a more transparent layer of thin ice called a glaze, when heavy dew forms before it freezes. Black ice, a major winter driving hazard, can also form in these ways. So, in summary, frost results from water vapour changing to ice directly by deposition, and frozen dew results from water vapour condensing to liquid water first and then freezing.

On cold winter mornings, deposition often forms frost on vehicle windshields. Frost can be distinguished from frozen dew or glaze because it is much easier to scrape off. If the dew point temperature is above 0°C, dew will form on the windshield and then freeze into a hard glaze once the windshield temperature dips below freezing. Frost also forms on the inside of poorly insulated windows. The cold glass chills the adjacent indoor air below freezing, and water vapour, due to increased indoor humidity associated with human activities, forms a light, feathery deposit of frost (see • Figure 5.2). Beautiful hoarfrost displays as shown in • Figure 5.3 form in still air when trees and other objects are surrounded by a fog made up of liquid water droplets that are supercooled below 0°C. When hoarfrost forms on the surface of snow, it is known as *surface hoar*, and it is of great interest when trying to identify avalanche risks. Surface hoar crystals range in size from less than a millimetre to more than a centimetre. An unusually large surface hoar crystal is shown in • Figure 5.4. When surface hoar forms on steeper slopes and is buried by a subsequent snowfall, it can result in a weak layer in the snow pack that may cause a future avalanche.

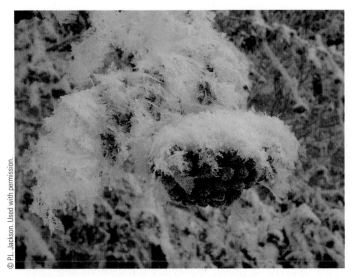

• FIGURE 5.3 These delicate ice crystals called hoarfrost result from *deposition*, the process where water vapour transforms into ice without being a liquid. Hoarfrost can form overnight when the radiative cooling of air at and near the surface goes below the frost point.

In low humidity, the air temperature can drop below freezing without ever reaching the frost point; it is too dry for frost to form. In this situation, the temperature may be cold enough to freeze the liquids in plants and kill them. This is known as "**dry freeze**" (as it occurs in low humidity) or "**black frost**" as the plant dies and turns black without any obvious signs of frost. (See Chapter 3, p. 82, for methods that may protect against frost damage.)

So dew, frozen dew, and frost form in the rather shallow layer of air near the ground on clear, calm nights. But what happens to air as a deeper layer adjacent to the ground is cooled?

• FIGURE 5.2 Frost forms delicate ice-crystal patterns on a window during a cold winter morning.

• FIGURE 5.4 Surface hoar crystals form by the *deposition* of water vapour onto the surface of a snow pack. Once buried, they can create weak layers in the snow pack, which are often responsible for avalanches. This is an unusually large surface hoar crystal.

Condensation Nuclei

As we have seen in Chapter 4, when a deeper layer of air adjacent to the ground is cooled to the dew point, the air becomes saturated and the relative humidity reaches 100 percent. Continued cooling condenses some of the vapour into tiny cloud droplets that form on airborne particles. These particles must be present to produce cloud droplets.

Although air may look clean, it never really is. On an ordinary day, a volume of air about the size of your index finger contains between 1000 and 150,000 particles. Many of these serve as surfaces on which water vapour can condense and are called **condensation nuclei**. Without them, relative humidities of several hundred percent would be required before condensation could begin.

Some condensation nuclei are quite small and have a radius less than 0.1 μm; these are referred to as *small nuclei* or *Aitken nuclei*, after the British physicist who discovered that water vapour condenses on nuclei. Particles ranging in size from 0.1 to 1 μm are called *large nuclei*. Others, called *giant nuclei*, are much larger and have radii exceeding 1 μm (see ▼ Table 5.1). The condensation nuclei most favourable for producing clouds (called *cloud condensation nuclei*) have radii of 0.1 μm or more. Usually, between 100 and 1000 nuclei of this size exist in a cubic centimetre of air. These particles enter the atmosphere in a variety of ways: dust, volcanoes, air pollution, forest fires, salt from ocean spray, and even sulphate particles emitted by phytoplankton in the oceans. In fact, studies show that sulphates provide the major source of cloud condensation nuclei over oceans. Because most particles are released into the atmosphere near the surface, the largest concentrations of nuclei are observed in the lower atmosphere.

Condensation nuclei are extremely light. As many have a mass less than one-trillionth of a gram, they can remain suspended in the air for many days. They are most abundant over industrial cities, where highly polluted air may contain nearly 1 million particles per cubic centimetre. They decrease in cleaner air and over the oceans, where concentrations may dwindle to only a few nuclei per cubic centimetre.

Some particles such as ocean salt and common salt are **hygroscopic** ("water seeking"), and water vapour condenses on these surfaces when the relative humidity is considerably lower than 100 percent. For example, in humid weather, without the addition of additives that prevent "caking," it becomes difficult to pour salt from a shaker because water vapour condenses onto the salt crystals, sticking them together. This is why you sometimes see uncooked rice in restaurant salt shakers. Salty potato chips left outside in an uncovered bowl can turn soggy. Other hygroscopic nuclei include sulphuric and nitric acid particles. But not all particles are good condensation nuclei. Some are **hydrophobic** ("water repelling"), such as oils, gasoline, paraffin waxes, and synthetic hydrophobics such as polytetrafluoroethylene (PTFE), or Teflon (a material used in rain-repellent fabric). These resist condensation even when the relative humidity is above 100 percent (see ● Figure 5.5). In summary, condensation may begin on some particles when the relative humidity is well below 100 percent and on others only when the relative humidity is much higher than 100 percent. At

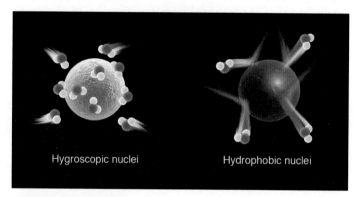

Hygroscopic nuclei   Hydrophobic nuclei

ACTIVE FIGURE 5.5 Hygroscopic nuclei are "water seeking," and water vapour rapidly condenses on their surfaces. Hydrophobic nuclei are "water repelling" and resist condensation. Visit the textbook's website to view this and other Active Figures at www.ahrensmeteorology1ce.nelson.com

▼ Table 5.1   Characteristic Sizes and Concentration of Condensation Nuclei and Cloud Droplets

TYPE OF PARTICLE	APPROXIMATE RADIUS IN MICROMETRES* (μm)	NO. OF PARTICLES PER CUBIC CENTIMETRE (cm⁻³)	
		Range	Typical
Small (Aitken) condensation nuclei	< 0.1	1000 to 10,000	1000
Large condensation nuclei	0.1 to 1.0	1 to 1000	100
Giant condensation nuclei	> 1.0	< 1 to 10	1
Fog and cloud droplets	> 10	10 to 1000	300

*1 micrometre (μm) = 10^{-6} metres = $\frac{1}{1,000,000}$ m or one one-millionth of a metre.

any given time, there are usually many hygroscopic nuclei present, so haze, fog, and clouds will form at relative humidities near or below 100 percent.

Up to this point, we have seen that condensation in air occurs on hygroscopic nuclei, which form droplets. If there are enough droplets, they can form fog and clouds. However, droplets with a diameter of about 20 μm are generally too small to fall as precipitation. For comparison, a typical raindrop diameter is 2000 μm. So condensing air forms fog and clouds, but not all clouds or droplets form precipitation. In Chapter 7, we will discuss both the microscale processes that allow tiny cloud droplets to grow into larger cloud droplets and the processes that encourage cloud droplets to grow large enough to fall as precipitation.

Haze

Haze is defined as reduced visibility due to the scattering of light by a layer of particles suspended in the air. Suppose you visit an area where dust, salt, or other air pollution particles form a layer of haze, and you noticed that distant objects were more visible in the afternoon than in the morning, even when the concentration of particles in the air had not changed. Why? During the warm afternoon, the relative humidity of the air drops, often below the point where water vapour begins to condense, even on active hygroscopic nuclei. Therefore, the floating particles remain small, usually no larger than about one-tenth of a micrometre (0.1 μm). These tiny *dry haze* particles selectively scatter some rays of sunlight while allowing others to penetrate the air. The scattering effect of dry haze produces different optical effects—a bluish colour when viewed against a dark background or a yellowish tint when viewed against a light-coloured background.

As the air cools during the night, the relative humidity increases. When the relative humidity reaches about 75 percent, condensation may begin on the most active hygroscopic nuclei, producing a wet haze. As water collects on the nuclei, the particles increase in size and become large enough to scatter light much more efficiently. In fact, as the relative humidity increases from about 60 percent to 80 percent, the scattering effect increases by a factor of nearly three. Since relative humidities are normally high during cool mornings, much of the light from distant objects is scattered by these wet haze particles before reaching our eyes; hence, it is difficult to see these distant objects.

Not only does wet haze restrict visibility more than dry haze, it also appears dull grey or white (see ● Figure 5.6). Near seashores and in clean air over the open ocean, large salt particles suspended in air with a high relative humidity often produce a thin white veil across the horizon.

Fog

By now, it should be apparent that condensation is a continuous process beginning when water vapour condenses onto hygroscopic nuclei at relative humidities as low as 75 percent. As the air's relative humidity increases, the visibility decreases, and the landscape becomes masked with a greyish tint. As the relative humidity approaches 100 percent, the haze particles grow larger, and condensation begins on the less active nuclei. Now a large fraction of the available nuclei have water condensing onto them, causing the droplets to grow even bigger until, eventually, they become visible to the naked eye. The increasing size and concentration of droplets further restrict visibility, and the air is wet with countless millions of tiny, floating water droplets. Now the wet haze becomes a cloud

● FIGURE 5.6 The high relative humidity of the cold air above the lake is causing a layer of haze to form on a still winter morning.

© C. Donald Ahrens

resting near the ground, which we call **fog**. The official international definition for fog occurs when the visibility is less than 1 km. The Meteorological Service of Canada* reports fog when visibility is less than half a mile (0.8 km); at visibilities less than a quarter of a mile, fog is considered dense. When the visibility is between half a mile and six miles (between 0.8 and 10 km), it is called mist. Fog is rarely observed when the difference between the air temperature and the dew point temperature is greater than 2°C. This and other processes that promote and limit the growth of water droplets to form fog and clouds are described in more detail in Chapter 7 on p. 193.

Given the same water content, fog that forms in dirty city air is often thicker than fog that forms over the ocean. Normally, the smaller number of condensation nuclei over the middle of the ocean produce fewer, but larger, fog droplets. City air, with its abundant nuclei, produces many tiny fog droplets. These greatly increase the thickness or opaqueness of the fog and reduce visibility. A dramatic example of a thick fog forming in air with abundant nuclei occurred in London, England, during the early 1950s. The fog became so thick and the air so laden with smoke particles that sunlight could not penetrate the smoggy air, requiring that street lights be left on at midday. Additionally, fog that forms in polluted air can turn acidic as the tiny liquid droplets combine with gaseous impurities, such as oxides of sulphur and nitrogen. **Acid fog** poses a threat to human health, especially to people with pre-existing respiratory problems. We will examine the health problems associated with acid fog and other pollution more closely in Chapter 18.

As tiny fog droplets grow larger, they become heavier and tend to fall toward Earth. A fog droplet with a diameter of 25 μm settles toward the ground at about 5 cm each second. At this rate, most of the droplets in a fog layer 180 m thick would reach the ground in less than one hour. Therefore, two questions arise: How does fog form? How is fog maintained once it does form?

Fog, like any cloud, forms in one of two ways:

- by cooling—air cools below its dew point and condensation occurs
- by adding moisture—water vapour is added to the air by evaporation, and the moist air mixes upward, cools, and condenses

Once fog forms, it is maintained by new fog droplets, which constantly form on available nuclei. In other words, the air must maintain its degree of saturation either by continual cooling or by evaporation and mixing the vapour into the air to maintain the humidity levels. Let's examine both processes.

RADIATION FOG   How can the air cool so that a cloud will form near the surface? Radiation and conduction are the primary means for cooling nighttime air near the ground.

*Note that the definition of fog and mist, even in Canada, still uses imperial measures—statute miles over land and nautical miles over water rather than kilometres. This is because the imperial measures are still in common use in aviation and marine operations, where imperial units are still standards.

Fog produced by radiative cooling of Earth's surface is called **radiation fog**, or *ground fog*. It forms best on clear nights when a shallow layer of moist air near the ground is overlain by drier air. Since the moist layer is shallow, it does not absorb much of Earth's outgoing infrared radiation, and the ground cools rapidly. So does the air directly above it, and this forms a surface inversion with cooler air at the surface and warmer air above. The moist lower layer quickly becomes saturated, and fog forms. The longer the night is, the longer the time for cooling and the greater the likelihood of fog. Hence, radiation fogs are most common over land in late fall and winter.

Another factor promoting the formation of radiation fog is a light breeze of less than 2.5 m s^{-1} (9 km h^{-1}). Although radiation fog may form in calm air, slight air movement brings more moist air in direct contact with the cold ground. Now the transfer of heat occurs more rapidly as convection and conduction are involved. A strong breeze tends to prevent radiation fog from forming by mixing the air near the surface with drier air above. The ingredients of clear skies and light winds are usually associated with large high-pressure areas (called anticyclones). Consequently, during the winter, when an anticyclone becomes stagnant over an area, radiation fog may form on many consecutive days.

Because cold, heavy air drains downhill and collects in valley bottoms, we normally see radiation fog forming in low-lying areas. Hence, radiation fog is frequently called *valley fog*. The cold air and high moisture content in river valleys make them susceptible to radiation fog. Since radiation fog normally forms in lowlands, hills may be clear all day long, whereas adjacent valleys are fogged in (see ● Figure 5.7 and ● Figure 5.8).

Radiation fog layers grow upward from the ground as the night progresses and are usually deepest around sunrise. However, fog may occasionally form after sunrise, especially when evaporation and mixing take place near the surface. This usually occurs at the end of a clear, calm night as radiative cooling brings the air temperature close to the dew point in a rather shallow layer above the ground. Here the air becomes saturated, forming a thick blanket of dew on the grass. At daybreak, the sun's rays evaporate the dew, adding water vapour to the air. A light breeze then stirs the moist air with drier air above it, causing saturation, which forms fog in a shallow layer near the ground.

Often a shallow fog layer will dissipate or "*burn off*" by the afternoon. Of course, the fog does not "burn"; rather, sunlight penetrates the fog and warms the ground, which warms the air in contact with the ground. The warm air rises and mixes with the foggy air above, which increases the temperature of the foggy air. In this slightly warmer air, some of the fog droplets evaporate, allowing more sunlight to reach the ground; this produces more heating, and soon the fog completely evaporates and disappears.

Satellite images show that blankets of radiation fog tend to evaporate ("burn off") first around their periphery, where

● FIGURE 5.7 Radiation/valley fog near Revelstoke, British Columbia.

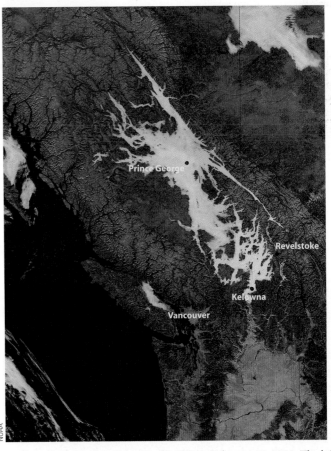

● FIGURE 5.8 Satellite imagery showing dense radiation (valley) fog in British Columbia, taken at 11:35 p.m. PST on February 21, 2010. The left picture is a normal visible image, whereas the right image is a *false colour* visible image in which snow has been coloured red, so the valley fog stands out. The smooth, evenly textured, white areas are radiation fog that frequently forms and persists in British Columbia's valleys during the long nights of fall and winter. (Images are from the US NASA *Modis Terra* satellite.)

FOCUS ON AN OBSERVATION

Why Are Canada's Coastal Regions so Foggy?

Dr. Ismail Gultepe

Dr. Ismail Gultepe, Cloud Physics and Severe Weather Research Section, Science and Technology Branch, Meteorological Research Division, Environment Canada

The Canadian marine environments, covering the areas adjacent to the Atlantic, Pacific, and Arctic oceans, as well as lakes, tend to be the foggiest areas in Canada. However, large river basins and valleys can also have dense fog conditions. Fog formation in each area can be explained by examining the large-scale weather patterns, atmospheric conditions such as temperature, humidity, and wind, near Earth's surface, as well as conditions over the water body, especially its surface temperature. All the coastal areas are foggy, with the largest probability of fog formation found along the coasts of the Canadian Maritime provinces, with a maximum of about 150 fog days per year (see ● Figure 5.15).

Canada's Maritime provinces are prone to advection fog in the warm seasons that forms due to the onshore, upslope flow of warm air over the cold ocean surface. Cold sea surface temperatures (2 to 10°C) around Nova Scotia and Newfoundland are caused by the cold, southward-flowing Labrador Current and are responsible for the fog formation. ● Figure 1 shows the probability of fog at Halifax airport by time of year and time of day, indicating the

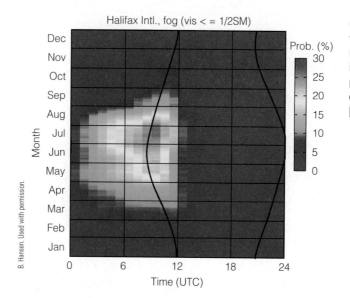

Halifax Intl., fog (vis < = 1/2SM)

B. Hansen. Used with permission.

● FIGURE 1 Probability of visibility less than 0.5 statute miles (800 m) in fog over Halifax Stanfield International Airport by time of day and month of the year (local standard time [EST] is UTC minus 5 hours, so 12:00 UTC is 7 a.m. EST).

predominance of fog overnight between spring and fall. The fog formation over land occurs mainly at night, when its formation is enhanced by radiative cooling; during daytime, the summer sun warms the surface and air, usually evaporating the fog. Over the oceans, the fog banks can persist day and night.

● Figure 2 shows a warm season advection fog formation in Lunenburg, Nova Scotia, that occurred during the FRAM (Fog Remote Sensing and Modeling) project on June 16, 2006.

Conversely, during winter, northerly airflow from the continent and frozen ocean brings cold air (T < −10°C) over the relatively warm ocean, leading to steam fog formation, which is sometimes called *sea smoke* as evaporation from the warm water condenses, creating a thin layer of fog above the ocean surface.

Along the coasts of British Columbia, fog formation is usually related to advection of warm and moist air from the Pacific Ocean

the fog is usually thinnest. Sunlight rapidly warms this region, causing the fog to dissipate as the warmer air mixes in toward the denser foggy area.

If the fog is thick, with little sunlight penetrating it, and there is little mixing along the outside edges, the fog may not dissipate. This is often the case in British Columbia's deeper, less windy valleys during winter. A fog layer, which may be up to several hundred metres thick, settles in the valley, whereas a strong inversion normally keeps the warmest air above the top of the fog. Because of the low winter sun angles, much of the available sunlight reflects off the top of the fog. Only a small amount of light penetrates and warms the ground. This gradually warms the air from below, and the fog dissipates

upward, forming a shallow, fog-free layer at the surface that is less than 150 m thick. This creates the illusion that the fog is lifting as it no longer touches the ground and is called *high-inversion fog, high fog,* or simply *stratus* (low cloud). But because of the strong inversion, the fog layer does not dissipate, and as soon as the sun sets, radiative cooling reduces the air temperature, and fog once again forms on the ground. This daily lifting and lowering of the fog without the sun ever breaking through it may last for many days or even weeks during winter (see Figure 5.8). Residents who are tired of continuous cloudy gloom can go up a nearby mountain that rises above the fog and stratus layer to find bright sunshine, as shown in Figure 5.7.

over the coastal waters that are cooled by the upwelling of cold water from below, toward the mountainous coastline. The fog formation is enhanced by the abundance of cloud condensation nuclei over the ocean, the availability of water vapour, and upward air motion as it flows up the steep coastal mountains. When moist air with cloud condensation nuclei from the ocean moves over the high topography, the upward motion generates cooling, resulting in condensation in the form of fog or low clouds (see Figure 2).

Along Canada's Arctic coastlines, fog formation is related to colder air temperatures, the presence of ice condensation nuclei, and the availability of moisture from breaks in the ice-covered ocean surfaces called *leads* and *polynyas*. Cold Arctic air moving over the open water results in enhanced evaporation. The cold temperatures result in rapid condensation of vapour and freezing of droplets and the formation of **ice fog**. Figure 3 shows the hoarfrost that formed on an ice particle sensor during an ice fog event at Barrow, Alaska. The ice fog resulting from radiative cooling usually occurs early in the morning following clear skies overnight with temperatures less than −15°C. During this ice fog episode, visibility went below 200 m, and it was very difficult for aircraft to land at Barrow Airport, causing flight cancellations.

● **FIGURE 2** Advection fog (left) at Lunenburg, Nova Scotia, on June 16, 2006, during FRAM fog project and mountain fog (right) over Whistler, British Columbia.

● **FIGURE 3** Ice fog occurred at the U.S. Department of Energy's Atmospheric Radiation Measurement (ARM) North Slope of Alaska (NSA) site at Barrow, Alaska, on April 10, 2008. The sensors shown were used in collecting fog-related parameters. Hoarfrost that formed on a York University ice particle counter (at left) during the ice fog episode resulted in malfunctioning of the instrument. Mountain fog (right) over Whistler, British Columbia, on January 17, 2010, during the SNOW-V10 project that was part of Environment Canada's preparations for weather forecasting at the 2010 Olympics.

ADVECTION FOG Cooling surface air to saturation may also be accomplished by warm, moist air moving over a cold surface. The surface must be sufficiently cooler than the air above so that the transfer of heat from the air to the surface will cool the air to its dew point and produce fog. Fog that forms in this manner is called **advection fog.**

Advection fog is often observed along Canada's coastal areas. The section "Focus on an Observation: Why Are Canada's Coastal Regions So Foggy?" helps us understand why this is so. Advection fog also occurs along North America's Pacific Coast during summer, when surface water near the coast is much colder than the surface water farther offshore. Warm, moist air from the Pacific Ocean is carried by westerly winds

(advected) over the cold coastal waters. Chilled from below, the air temperature drops to the dew point, and fog is produced. It is also more common to see advection fog forming at headlands that protrude seaward than in the mouths of bays. If you are curious as to why, read Focus on an Observation: Why Are Headlands Usually Foggier than Beaches? on p. 135.

As winds carry the fog inland over the warmer land, the fog near the ground dissipates, leaving a sheet of low-lying grey clouds that block out the sun. Farther inland, the air is sufficiently warm, and even these low clouds evaporate and disappear. Since the fog is more likely to burn off during the warmer part of the day, a typical coastal summertime weather forecast would read, "Morning fog and low clouds along the

● FIGURE 5.9 Many fog droplets gather to form the tiny drops that drip from the needles of this tree and provide a valuable source of moisture during the otherwise dry summer along the coastal temperate rain forests of North America.

coast, extending inland locally. Sunny in the afternoon with fog reforming overnight."

Advection fogs are valuable sources of moisture. Along the Pacific Coast, they contribute important summertime moisture to the temperate rain forests in Canada and the United States. When fog comes into an area, it collects on surfaces. In these coastal forests, the needles and branches of trees amass fog moisture until it drips to the ground as *fog drip*, where it can be used by the trees (see ● Figure 5.9). Studies have shown that 34 percent of the summer moisture used by the massive redwood trees of northern California's coastal

redwood forests comes from fog drip. Without this moisture, these trees would have trouble surviving the dry Californian summers. In British Columbia's wet, temperate rain forests, fog also provides important moisture during the shorter, relatively dry summers, when summer moisture stress would otherwise limit the luxuriant growth in this ecosystem. But because annual precipitation is greater and the temperatures are cooler, trees there are not as reliant on this moisture as they are further south.

Fog can also be exploited as a source of drinking and irrigation water in otherwise arid regions. One such region is the fishing village of Chungungo, Chile, a dry region that experiences advection fog and cloud from the Pacific Ocean. In the 1980s, the Meteorological Service of Canada started a fog collection experiment in Chungungo, initially to understand the chemical composition and acidity of the fog. They set up nylon nets on poles on a hillside along the coast, as shown in ● Figure 5.10. The fog condensed on the nets and ran down into troughs, where the fog water was collected. They found that the system produced an average of 15,000 litres of water a day, with some days as much as 100,000 litres, and decided to build a pipeline to the town to supply them with drinking water. News of this experiment revived interest in fog water sources in many dry regions around the world, and now you may find fog collection systems in areas such as Morocco, Ethiopia, Ecuador, Peru, Guatemala, Israel, South Africa, and Nepal—anywhere there are water shortages and significant fog of any type.

Advection fogs also prevail where two ocean currents with different temperatures flow next to one another. Such is the case in the Atlantic Ocean's Grand Banks, off the coast of Newfoundland, where the cold, southward-flowing Labrador Current lies almost parallel to the warm, northward-flowing

● FIGURE 5.10 Fog collection provides otherwise unavailable drinking and irrigation water in many arid coastal and mountain areas.

FOCUS ON AN OBSERVATION

Why Are Headlands Usually Foggier Than Beaches?

If you drive along a highway that parallels an irregular coastline, you may have observed that advection fog is more likely to form on headlands that protrude seaward than it is on beaches that are nestled in the mouths of bays. Why?

As air moves onshore, it hits the coastline almost at right angles (see ● Figure 4). This causes the air to flow together or converge in the vicinity of coastal headlands. This converging of air onto the headlands causes the surface air to rise and cool just a little (a process called small-scale convergence). If the rising air is close to being saturated, it will cool to its dew point, and fog will form.

Meanwhile, near the beach area, the surface air spreads apart or diverges as it crosses the coastline. This creates sinking and slightly warmer air (a process called small-scale divergence). Because the sinking air is warmed, the separation between air temperature and dew point is increased, and fog is less likely to form. Hence, the headlands can be shrouded in fog while the beaches are basking in sunshine.

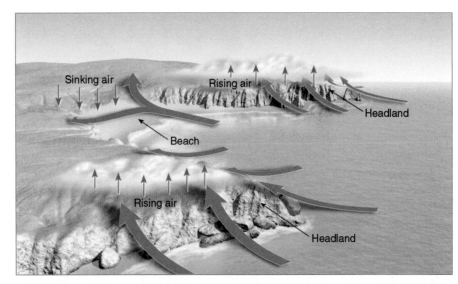

● FIGURE 4 Along an irregular coastline, advection fog is more likely to form at the headlands, where moist surface air converges and rises, than at the beach, where air diverges and sinks.

Gulf Stream. This situation allows warm air from the south to flow over the cold water and produce fog. This happens so frequently that during summer, fog occurs approximately two out of every three days, making this area one of the foggiest places on Earth. St. John's, Newfoundland and Labrador, is the foggiest city in Canada, with an average of 121 fog days each year (see ● Figure 5.11). For more information, read Focus on an Observation: Why Are Canada's Coastal Regions So Foggy? on p. 132.

Advection fog also forms over land. In winter, warm, moist air from the Gulf of Mexico moves northward over progressively colder and slightly elevated land. As the air cools to its dew point temperature, fog forms in the southern or central United States. Because the cold ground is often the result of radiative cooling, fog that forms in this manner is sometimes called **advection–radiation fog**. During this same time of year, air moving from the warm Gulf Stream encounters the colder land of the British Isles and produces the thick fogs of England. Similarly, fog forms as marine air moves over an ice or snow surface. In extremely cold arctic air, ice

© Elena Elisseeva/Alamy

● FIGURE 5.11 Advection fog forms as the wind moves moist air over a cooler surface. Here a fog bank that formed over the cold ocean is rolling into the harbour in St. John's, Newfoundland. As the fog moves inland, the air warms, and the fog may lift above the surface. Eventually, the air becomes warm enough to totally evaporate the fog.

• FIGURE 5.12 (a) Radiation fog tends to form on clear, relatively calm nights when cool, moist surface air is overlain by drier air and rapid radiative cooling occurs. (b) Advection fog forms when the wind moves moist air over a cold surface and the moist air cools to its dew point.

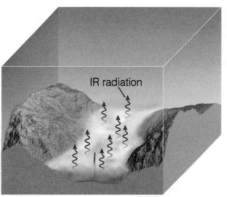

(a) Radiation fog

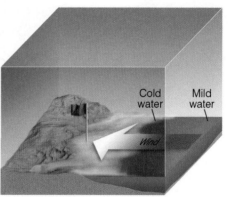

(b) Advection fog

crystals form instead of water droplets, producing an *ice fog*. • Figure 5.12 summarizes the ideas behind the formation of both advection and radiation fog.

UPSLOPE FOG   Fog that forms as moist air flows up along an elevated plain, hill, or mountain is called **upslope fog**. Typically, in North America, upslope fog forms during the winter and spring on the eastern side of the Rockies, where the eastward-sloping plains rise above the land farther east. Occasionally, cold air moves from the lower eastern plains westward. The air gradually rises, expands, and becomes cooler, and if it is sufficiently moist, a fog forms (see • Figure 5.13). Upslope fogs that form over an extensive area may last for many days.

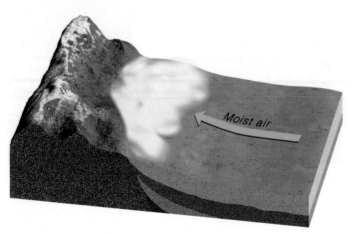

• FIGURE 5.13 Upslope fog forms as moist air slowly rises, cools, and condenses over elevated terrain.

MIXING FOG   Up to now, we have seen how cooling air produces fog. But fog may also form by mixing two unsaturated air masses that are at different temperatures. Fog that forms this way is called **mixing fog** because it is the combination of the two that creates the right conditions for fog to form. When the resulting mixed air has a high enough moisture content and a low enough temperature to be saturated, then condensation can occur, forming fog. On a cold day, you may have unknowingly produced mixing fog. When moist air from your mouth or nose meets the cold air and mixes with it, the air becomes saturated, and a tiny cloud forms with each exhaled breath. For a better understanding of how mixing can produce fog, read Focus on a Special Topic: Fog That Forms by Mixing on p. 138.

EVAPORATION FOG   When warm water is in contact with colder air, evaporation from the water enriches the air with water vapour and increases the humidity of the air. The warm water also heats the air just above its surface, which mixes with the colder air, and cools, creating a layer of saturated air that condenses and forms fog. This is called **evaporation fog**.

Steam fog is one type of evaporation fog and is associated with warm bodies of water such as swimming pools, lakes, puddles, or hot, wet pavement. As a type of evaporation fog, it forms because of the evaporation and subsequent mixing when cold air moves over warm water. This type of fog forms above a heated outside hot tub or swimming pool in winter. From a distance, the rising condensing vapour appears as "steam." It is common to see steam fog forming over lakes on autumn mornings as cold air settles over water that is still warm from the long summer. On occasion, over the Great Lakes, columns of condensed vapour rise from the fog layer, forming whirling *steam devils*, which appear similar to the dust devils on land. If you travel to natural hot springs, such as the Liard Hot Springs of northern British Columbia (see • Figure 5.14), the Takhini in the Yukon, numerous ones along the Canadian Rocky Mountains, or Yellowstone National Park in the United States, you will see steam fog forming above thermal ponds. In high-temperature hot springs, steam fog is visible all year long. Over the ocean in polar regions, steam fog is referred to as *arctic sea smoke*.

FOCUS ON AN OBSERVATION

Why Are Headlands Usually Foggier Than Beaches?

If you drive along a highway that parallels an irregular coastline, you may have observed that advection fog is more likely to form on headlands that protrude seaward than it is on beaches that are nestled in the mouths of bays. Why?

As air moves onshore, it hits the coastline almost at right angles (see ● Figure 4). This causes the air to flow together or converge in the vicinity of coastal headlands. This converging of air onto the headlands causes the surface air to rise and cool just a little (a process called small-scale convergence). If the rising air is close to being saturated, it will cool to its dew point, and fog will form.

Meanwhile, near the beach area, the surface air spreads apart or diverges as it crosses the coastline. This creates sinking and slightly warmer air (a process called small-scale divergence). Because the sinking air is warmed, the separation between air temperature and dew point is increased, and fog is less likely to form. Hence, the headlands can be shrouded in fog while the beaches are basking in sunshine.

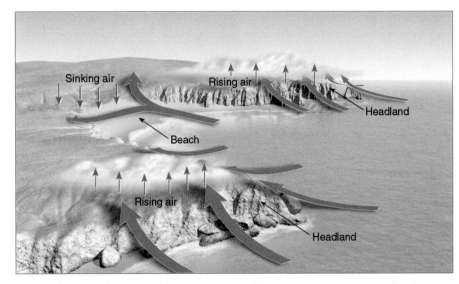

● FIGURE 4 Along an irregular coastline, advection fog is more likely to form at the headlands, where moist surface air converges and rises, than at the beach, where air diverges and sinks.

Gulf Stream. This situation allows warm air from the south to flow over the cold water and produce fog. This happens so frequently that during summer, fog occurs approximately two out of every three days, making this area one of the foggiest places on Earth. St. John's, Newfoundland and Labrador, is the foggiest city in Canada, with an average of 121 fog days each year (see ● Figure 5.11). For more information, read Focus on an Observation: Why Are Canada's Coastal Regions So Foggy? on p. 132.

Advection fog also forms over land. In winter, warm, moist air from the Gulf of Mexico moves northward over progressively colder and slightly elevated land. As the air cools to its dew point temperature, fog forms in the southern or central United States. Because the cold ground is often the result of radiative cooling, fog that forms in this manner is sometimes called **advection–radiation fog**. During this same time of year, air moving from the warm Gulf Stream encounters the colder land of the British Isles and produces the thick fogs of England. Similarly, fog forms as marine air moves over an ice or snow surface. In extremely cold arctic air, ice

● FIGURE 5.11 Advection fog forms as the wind moves moist air over a cooler surface. Here a fog bank that formed over the cold ocean is rolling into the harbour in St. John's, Newfoundland. As the fog moves inland, the air warms, and the fog may lift above the surface. Eventually, the air becomes warm enough to totally evaporate the fog.

● FIGURE 5.12 (a) Radiation fog tends to form on clear, relatively calm nights when cool, moist surface air is overlain by drier air and rapid radiative cooling occurs. (b) Advection fog forms when the wind moves moist air over a cold surface and the moist air cools to its dew point.

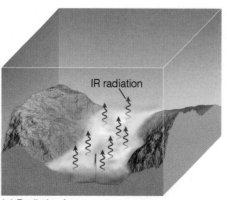

(a) Radiation fog

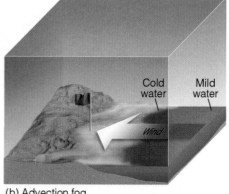

(b) Advection fog

crystals form instead of water droplets, producing an *ice fog*. ● Figure 5.12 summarizes the ideas behind the formation of both advection and radiation fog.

UPSLOPE FOG   Fog that forms as moist air flows up along an elevated plain, hill, or mountain is called **upslope fog**. Typically, in North America, upslope fog forms during the winter and spring on the eastern side of the Rockies, where the eastward-sloping plains rise above the land farther east. Occasionally, cold air moves from the lower eastern plains westward. The air gradually rises, expands, and becomes cooler, and if it is sufficiently moist, a fog forms (see ● Figure 5.13). Upslope fogs that form over an extensive area may last for many days.

WEATHER WATCH

Ever hear of caribou fog? It is not the fog that forms in the Cariboo region of British Columbia's central interior but a mixing fog that forms around herds of caribou. In very cold weather, just a little water vapour added to the air will saturate it. Consequently, the perspiration and breath from large herds of caribou add enough water vapour to the air to create a blanket of fog that hovers around the herd.

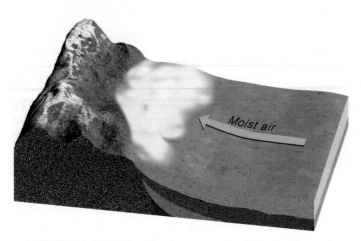

● FIGURE 5.13 Upslope fog forms as moist air slowly rises, cools, and condenses over elevated terrain.

MIXING FOG   Up to now, we have seen how cooling air produces fog. But fog may also form by mixing two unsaturated air masses that are at different temperatures. Fog that forms this way is called **mixing fog** because it is the combination of the two that creates the right conditions for fog to form. When the resulting mixed air has a high enough moisture content and a low enough temperature to be saturated, then condensation can occur, forming fog. On a cold day, you may have unknowingly produced mixing fog. When moist air from your mouth or nose meets the cold air and mixes with it, the air becomes saturated, and a tiny cloud forms with each exhaled breath. For a better understanding of how mixing can produce fog, read Focus on a Special Topic: Fog That Forms by Mixing on p. 138.

EVAPORATION FOG   When warm water is in contact with colder air, evaporation from the water enriches the air with water vapour and increases the humidity of the air. The warm water also heats the air just above its surface, which mixes with the colder air, and cools, creating a layer of saturated air that condenses and forms fog. This is called **evaporation fog**.

Steam fog is one type of evaporation fog and is associated with warm bodies of water such as swimming pools, lakes, puddles, or hot, wet pavement. As a type of evaporation fog, it forms because of the evaporation and subsequent mixing when cold air moves over warm water. This type of fog forms above a heated outside hot tub or swimming pool in winter. From a distance, the rising condensing vapour appears as "steam." It is common to see steam fog forming over lakes on autumn mornings as cold air settles over water that is still warm from the long summer. On occasion, over the Great Lakes, columns of condensed vapour rise from the fog layer, forming whirling *steam devils*, which appear similar to the dust devils on land. If you travel to natural hot springs, such as the Liard Hot Springs of northern British Columbia (see ● Figure 5.14), the Takhini in the Yukon, numerous ones along the Canadian Rocky Mountains, or Yellowstone National Park in the United States, you will see steam fog forming above thermal ponds. In high-temperature hot springs, steam fog is visible all year long. Over the ocean in polar regions, steam fog is referred to as *arctic sea smoke*.

● FIGURE 5.14 Warm air rising above thermal pools at Liard Hot Springs on the Alaska Highway in northern British Columbia condenses into a type of steam fog.

Steam fog may even form above a wet surface on a sunny day. This type of fog is commonly observed on roads after a summer thunderstorm as rain quickly evaporates from the hot asphalt. Fog that forms in this manner is short-lived and disappears as the road surface dries.

Another form of evaporation fog occurs when warm rain falling through a layer of cold, moist air produces fog. Here the warm water takes the form of a raindrop. Remember from Chapter 4 that the saturation vapour pressure depends on temperature: higher temperatures correspond to higher saturation vapour pressures. When a warm raindrop falls into a cold layer of air, the saturation vapour pressure over the raindrop is greater than that of the air. This vapour-pressure difference causes water to evaporate from the raindrop into the air. This process may saturate the air, and if mixing occurs, fog forms. Fog of this type is often associated with warm air riding up and over a mass of colder surface air. The fog usually develops in the shallow layer of cold air just ahead of an approaching warm front or behind a cold front, which is why this type of evaporation fog is also known as *precipitation fog*, or **frontal fog**. Snow covering the ground is an especially favourable condition for the formation of frontal fog. The melting snow extracts heat from the environment, thereby cooling the already rain-saturated air.

FOGGY WEATHER   The foggiest regions in Canada are shown in ● Figure 5.15. Notice that dense fog is more prevalent in coastal margins (especially those regions lapped by cold ocean currents) than in the centre of the continent.

Unfortunately, fog also has many negative aspects. Along a gently sloping highway, the elevated sections may have excellent visibility, whereas in lower regions only a few kilometres away, fog may cause poor visibility. Driving from

● FIGURE 5.15 Average annual number of days with fog (visibility less than 0.5 statute miles or 800 m) through Canada averaged over the period 1951 to 1980 (after Phillips, 1990).

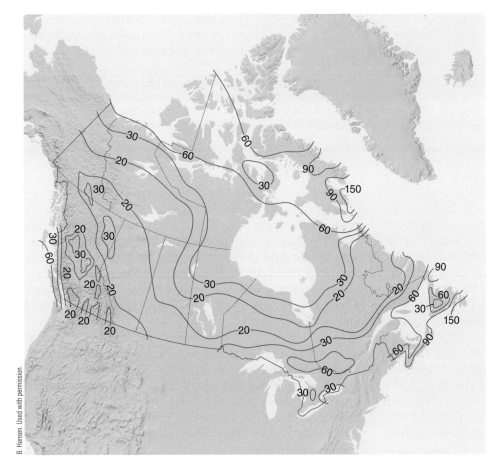

Fog That Forms by Mixing

How can unsaturated bodies of air mix together to produce fog (or a cloud)? To answer this question, let's first examine two separate unsaturated air parcels. Later, we will look at the parcels mixed together. The two parcels are essentially the same size and have a mass of 1 kg each. Yet each has a different temperature and a different relative humidity. (We will assume that the parcels are near sea level, where the atmospheric pressure is close to 1000 hPa.)

The situation is shown in ● Figure 5. Parcel A has an air temperature (T) of 20°C and a dew point temperature (T_d) of 15°C. Remember from Chapter 4 that the air's relative humidity (RH) can be expressed as

$$RH = \frac{\text{actual mixing ratio } (r)}{\text{saturation mixing ratio } (r_s)} \times 100\%$$

where the actual mixing ratio (r) is the mixing ratio of ambient air. It is determined by measuring the mass of water vapour (g) per kilogram (kg) of dry air. It is expressed as grams per kilogram (g kg^{-1}) or

$$\text{actual mixing ratio } (r) = \frac{\text{mass of water vapour (g)}}{\text{kilogram of dry air (kg)}}$$

The saturation mixing ratio (r_s) is also expressed in g kg^{-1}.

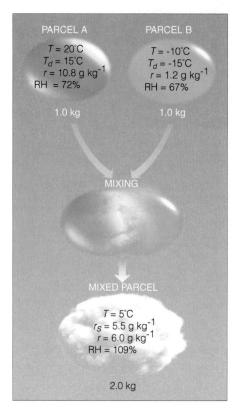

PARCEL A
$T = 20°C$
$T_d = 15°C$
$r = 10.8$ g kg^{-1}
RH = 72%
1.0 kg

PARCEL B
$T = -10°C$
$T_d = -15°C$
$r = 1.2$ g kg^{-1}
RH = 67%
1.0 kg

MIXING

MIXED PARCEL
$T = 5°C$
$r_s = 5.5$ g kg^{-1}
$r = 6.0$ g kg^{-1}
RH = 109%
2.0 kg

To obtain the *saturation mixing ratio*, we look at ▼ Table 1 and read the value that corresponds to the parcel's *air temperature*. (For a

● FIGURE 5 Mixing two unsaturated air parcels can produce fog. Notice that the mixed parcel has an actual mixing ratio (r) that is greater than the saturation mixing ratio (r_s). This produces an RH of 109% and results in fog. (As the mixed parcel cools below its dew point temperature, water vapour will condense onto condensation nuclei, producing the liquid droplets that create fog.)

temperature of 20°C, the saturation mixing ratio is 15.0 g kg^{-1}.) Likewise, the *actual mixing ratio* is obtained by reading the value in Table 1 that corresponds to the parcel's *dew point temperature*. (For a dew point temperature of 15°C, the actual mixing ratio is 10.8 g kg^{-1}.) Consequently, the relative humidity of the air in parcel A is 72% as shown:

$$RH = \frac{r}{r_s} = \left(\frac{10.8}{15.0}\right) \times 100\% = 72\%$$

Air parcel B is considerably colder than parcel A. It has a temperature of $-10°C$ and is considerably drier, with a dew point temperature of $-15°C$. These temperatures yield a relative humidity of 67 percent, as follows:

$$RH = \frac{r}{r_s} = \left(\frac{1.2}{1.8}\right) \times 100\% = 67\%$$

Suppose we now thoroughly mix these two parcels of air. After mixing, the combined

---

the clear area into fog on a major highway can be extremely dangerous. In fact, every winter, many people are involved in fog-related automobile accidents. These usually occur when a car enters the fog and, because of the reduced visibility, the driver puts on the brakes to slow down. The car behind then slams into the slowed vehicle, causing a chain-reaction accident with many cars involved. One of Canada's worst recorded highway accidents occurred this way on Highway 401 east of Windsor, Ontario, on September 3, 1999 (Labour Day weekend). Thick fog instantly reduced visibility to zero on a section of the highway, causing an 82-vehicle pileup, with many cars fused together in the intense heat, killing eight and injuring 33 more.

High-beam headlights reduce the already limited visibility experienced while driving at night in thick fog. The light scattered back to the driver's eyes from the fog droplets

makes it difficult to see very far down the road. However, even in thick fog, there is usually a drier and therefore clearer region extending about 35 cm above the road surface. People who drive a great deal in foggy weather take advantage of this by installing extra head lamps—called *fog lamps*—just above the front bumper. These lights are directed downward into the clear space, where they provide improved visibility.

Fog-related problems are not confined to land. Even with sophisticated electronic equipment, dense fog can still hamper navigation. Interestingly, Canada's largest sea disaster is associated with fog. It happened on May 29, 1914, when the *Empress of Ireland* collided with a Norwegian coal freighter in foggy conditions on the St. Lawrence River near Rimouski, Quebec. The ship sank in 14 minutes, killing 1,012 people.

▼ Table 1   Saturation Mixing Ratios of Water Vapour for Various Air Temperatures

AIR TEMPERATURE (°C)	SATURATION MIXING RATIO (g kg⁻¹)
20	15.0
15	10.8
10	7.8
5	5.5
0	3.8
−5	2.6
−10	1.8
−15	1.2
−20	0.8

*Air pressure is 1000 hPa.

parcel's temperature will be the average of parcels A and B, or about 5°C. The water vapour content of the mixed parcel (i.e., actual mixing ratio) will be the average of the mixing ratios of parcels A and B, or

$$\frac{10.8 \text{ g kg}^{-1} + 1.2 \text{ g kg}^{-1}}{2} = \frac{12 \text{ g kg}^{-1}}{2}$$
$$= 6.0 \text{ g kg}^{-1}$$

Look at Table 1 and observe that the saturation mixing ratio for a saturated parcel at 5°C is only 5.5 g kg⁻¹. This means that the water vapour content of the mixed parcel, at 6.0 g kg⁻¹, is above that required for saturation and that the parcel is *supersaturated*, with a relative humidity of 109 percent. Of course, such a high relative humidity is almost impossible to obtain as water vapour would certainly condense on condensation nuclei, producing liquid water droplets as the two parcels mix together and the relative humidity approaches 100 percent. Hence, mixing two initially unsaturated masses of air can produce fog or a cloud.

Another way to look at this mixing process is to place the two unsaturated air parcels into ● Figure 6, which is a graphic representation of Table 1. The solid blue line in Figure 6 represents the saturation mixing ratio. If the air temperature and actual mixing ratio of an air parcel lie on the blue line, then the air parcel is saturated with a relative humidity of 100 percent. If an air parcel is located to the right of the blue line, the air parcel is unsaturated. If an air parcel lies to the left of the line, the parcel is supersaturated, and condensation will occur.

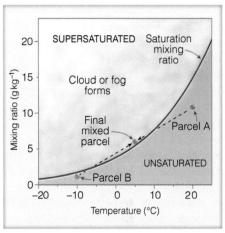

● FIGURE 6 The blue line is the saturation mixing ratio. The mixing of two unsaturated air parcels (A and B) can produce a saturated air parcel and fog.

Notice that when parcel A and parcel B are plotted in Figure 6, both unsaturated air parcels fall to the right of the blue line. However, when parcel A and parcel B are mixed, the mixed air parcel (with an air temperature of 5°C and an actual mixing ratio of 6.0 g kg⁻¹) falls to the left of the blue line, indicating that water vapour inside the mixed parcel will condense either into fog or a cloud. Fog that forms in this manner is called mixing fog.

In the skies, a series of errors on a fog-covered runway in the Canary Islands caused two 747 jet airliners to collide, taking the lives of over 570 people in March 1977. Airports suspend flight operations when fog causes visibility to drop below a prescribed minimum, which depends on the airport and the type of equipment the aircraft has. The resulting delays and cancellations become costly to the airline industry and irritate passengers. With fog-caused problems such as these, it is no wonder that scientists have been seeking ways to disperse, or at least "thin," fog. (For more information on fog-thinning techniques, read Focus on an Environmental Issue: Fog Dispersal on p. 140.)

Up to this point, we have looked at the different forms of condensation that occur on or near Earth's surface. In particular, we learned that fog is simply many billions of tiny liquid droplets (or ice crystals) that form near the ground. In the following sections, we will see how these same particles, forming well above the ground, are classified and identified as clouds.

BRIEF REVIEW

However, before going on to the section on clouds, here is a brief review of some of the facts and concepts we have covered so far:

- Dew, frozen dew, frost, hoarfrost, and surface hoar form on clear nights when the temperature of objects on the surface cools below the air's dew point temperature.

- Frost, hoarfrost, and surface hoar form in saturated air when the air temperature is at or below freezing. Under these conditions, water vapour can change directly to ice, in a process called deposition.

FOCUS ON AN ENVIRONMENTAL ISSUE

Fog Dispersal

In any airport fog-clearing operation, the problem is to improve visibility so that aircraft can take off and land. Experts have tried various methods, which can be grouped into four categories: (1) increase the size of the fog droplets so that they become heavy and settle to the ground as a light drizzle; (2) seed cold fog with dry ice (solid carbon dioxide) so that fog droplets are converted into ice crystals; (3) heat the air so that the fog evaporates; and (4) mix the cooler, saturated air near the surface with the warmer, unsaturated air above.

To date, only one of these methods has been reasonably successful—the seeding of cold fog. *Cold fog* forms when the air temperature is below freezing, and most of the fog droplets remain as liquid water. (Liquid fog in below-freezing air is also called *supercooled fog*.) The fog can be cleared by injecting several hundred pounds of dry ice into it. As the tiny pieces of cold (−78°C) dry ice descend, they freeze some of the supercooled fog droplets in their path, producing ice crystals. As we will

see in Chapter 7, these crystals then grow larger at the expense of the remaining liquid fog droplets. Hence, the fog droplets evaporate and the larger ice crystals fall to the ground as precipitation, which leaves a "hole" in the fog for aircraft takeoffs and landings.

Unfortunately, in many parts of the world, the fogs that close airports are *warm fogs* that form when the air temperature is above freezing. Since dry ice seeding does not work in warm fog, other techniques must be tried.

One method involves injecting hygroscopic particles into the fog. Large salt particles and other chemicals absorb the tiny fog droplets and form into larger drops. More large drops and fewer small drops improve the visibility; plus, the larger drops are more likely to fall as a light drizzle. Since the chemicals are expensive and the fog clears for only a short time, this method of fog dispersal is not economically feasible.

Another technique for fog dispersal is to warm the air enough so that the fog droplets

evaporate and visibility improves. Tested at Los Angeles International Airport in the early 1950s, this technique was abandoned because it was smoky, expensive, and not very effective. In fact, the burning of hundreds of dollars worth of fuel cleared the runway only for a short time. And the smoke particles, released during the burning of the fuel, provided abundant nuclei for the fog to recondense upon.

A final method of warm fog dispersal uses helicopters to mix the air. The chopper flies across the fog layer, and the turbulent downwash created by the rotor blades brings drier air above the fog into contact with the moist fog layer (see ● Figure 7). The aim, of course, is to evaporate the fog. Experiments show that this method works well as long as the fog is a shallow radiation fog, with relatively low liquid water content. But many fogs are thick, have a high liquid water content, and form by other means. An inexpensive and practical method of dispersing warm fog has yet to be discovered.

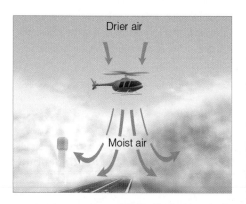

© V.G. Plank/Air Force Geophysics

● FIGURE 7 Helicopters hovering above an area of shallow fog (left) can produce a clear area (right) by mixing the drier air into the foggy air below.

- Condensation nuclei act as surfaces on which water vapour condenses. Those nuclei that have an affinity for water vapour are called hygroscopic.

- Fog is a cloud resting on the ground. It can be composed of water droplets, ice crystals, or a combination of both.

- Radiation fog, advection fog, and upslope fog all form as the air cools. The cooling for radiation fog is mainly radiative cooling at

Earth's surface; for advection fog, the cooling is mainly warmer air moving over a colder surface; for upslope fog, the cooling occurs as moist air gradually rises and expands along sloping terrain.

- Mixing fog forms when moist air of different temperatures mixes and results in saturation.

- Evaporation fog, such as steam fog and frontal fog, is a type of mixing fog that forms as water evaporates and mixes with drier air.

Clouds

Clouds are aesthetically appealing and add excitement to the atmosphere. Without them, there would be no rain, snow, thunder, lightning, rainbows, or halos. How monotonous it would be if we only had clear blue sky. A *cloud* is a visible aggregate of tiny water droplets or ice crystals suspended in the air. Some are found only at high elevations, whereas others nearly touch the ground. Clouds can be thick or thin, big or little—they exist in a seemingly endless variety of forms. To impose order on this variety, we divide clouds into 10 basic types. With a careful and practised eye, observers can become proficient in correctly identifying them.

CLASSIFICATION OF CLOUDS   Although ancient astronomers named the major stellar constellations about 2000 years ago, clouds were not formally identified and classified until the early 19th century. The French naturalist Lamarck (1744–1829) proposed the first system for classifying clouds in 1802; however, his work did not receive wide acclaim. One year later, Howard, an English naturalist, developed a cloud classification system that found general acceptance. In essence, Howard's innovative system employed Latin words to describe clouds as they appear to a ground observer. He named a sheet-like cloud *stratus* (Latin for "layer"—clouds like this are now grouped as **stratiform**); a puffy cloud *cumulus* ("heap"—clouds like this are now grouped as **cumuliform**); a wispy cloud *cirrus* ("curl of hair"); and a rain cloud *nimbus* ("violent rain"). In Howard's system, these were the four basic cloud forms. Other clouds could be described by combining the basic types. For example, nimbostratus is a rain cloud that shows layering, whereas cumulonimbus is a rain cloud with pronounced vertical development.

In 1887, Abercromby and Hildebrandsson expanded Howard's original system and published a classification system that is basically still in use today. This classification groups clouds by their altitude, the details of their appearance, and how they are formed. ▼ Table 5.2 summarizes the system that uses altitude to define **cloud levels** or **étages**. These identify

▼ Table 5.2   Cloud Classification*

LEVEL (ÉTAGES)	GENERA	SPECIES	VARIETIES	SUPPLEMENTARY FEATURES AND ACCESSORY CLOUDS
(Names Given to the Altitude of a Cloud's Base)	(The 10 Basic Cloud Types)	(A Subdivision of Genera Based on Shape and a Cloud's Internal Structure)	(A Subdivision of Genera Indicating Cloud Arrangement and Transparency)	(Terms Used to Describe Cloud Add-Ons)
High clouds	cirrus (Ci)	fibratus	intortus	**incus**
	cirrostratus (Cs)	uncinus	vertebratus	**mamma (also mammatus)**
	cirrocumulus (Cc)	spissatus	**undulatus**	
Middle clouds	altostratus (As)	**castellanus**	radiatus	virga
	altocumulus (Ac)	floccus	lacunosus	praecipitatio
Low clouds	stratus (St)	stratiformis	duplicatus	arcus
	stratocumulus (Sc)	nebulosus	**translucidus**	tuba
	nimbostratus (Ns)	**lenticularis**	perlucidus	**pileus**
Clouds with vertical development†	cumulus (Cu)	**fractus**	opacus	velum
	cumulonimbus (Cb)	**humilis**		pannus
		mediocris		
		congestus		
		calvus		
		capillatus		

*Genera define 10 main cloud types. The other subdivisions (species, varieties, and supplementary features and accessory clouds) further distinguish cloud forms and are used less often. The most commonly used of the other subdivision is indicated in **bold** text and described in Table 5.4 on page 143.

†Clouds with vertical development are sometimes grouped with low clouds.

▼ Table 5.3 Approximate Height of Cloud Bases Above the Surface for Various Locations*

CLOUD GROUP	TROPICAL REGION	MIDDLE LATITUDE REGION	POLAR REGION
High			
Ci, Cs, Cc	6000 to 18,000 m	5000 to 13,000 m	3000 to 8000 m
Middle			
As, Ac	2000 to 8000 m	2000 to 7000 m	2000 to 4000 m
Low			
St, Sc, Ns	0 to 2000 m	0 to 2000 m	0 to 2000 m
Vertical Development			
Cu, Cb	0 to 2000 m	0 to 2000 m	0 to 2000 m

*Note that altitudes are lower in winter than in summer and lower at high latitudes. Also, in the mountains, cloud bases can be closer to the surface than indicated.

the height of the cloud's base above the surface. Officially, there are three cloud levels: *high clouds, middle clouds,* and *low clouds*. Often a fourth cloud level, *clouds with vertical development*, is used as it distinguishes clouds that grow vertically in the atmosphere. The system also defines two main cloud types based on their appearance and formation process: stratiform and cumuliform. Within the four levels and two types are 10 principal cloud forms, defined by their shape and appearance, called **genera**. The genera are further subdivided into 14 **species** and nine **varieties**. Species differentiate clouds by deviations in their shape and the cloud's internal structure. Varieties distinguish clouds by their arrangements and levels of transparency. In addition, there are nine **supplementary features and accessory clouds** that are associated minor cloud forms and identify any remaining clouds that are not already categorized.

Clouds are always identified at least by their genera as this classifies the level, type, and formation process. High clouds all start with the prefix "cirro" or "cirrus," middle clouds use the prefix "alto," and low clouds have no prefix. Stratiform clouds are indicated by a root word, "strato" or "stratus," whereas cumuliform clouds employ the root word "cumulo" or "cumulus." Clouds that are always associated with precipitation use the word "nimbo" or "nimbus." So names for the 10 cloud genera are created by combining these prefixes and root words, as shown in Table 5.2.

The approximate base height of each cloud level is given in ▼ Table 5.3. Note that the altitude separating the high and middle cloud groups overlaps and varies with latitude on the Earth. For example, high clouds are composed almost entirely of ice crystals. In tropical regions, only altitudes above 6000 m are cold enough to freeze all liquid water. In polar regions, altitudes as low as 3000 m are cold enough. Hence, although cirrus clouds are observed at 3600 m over the Yukon, they are not found at that elevation in the Caribbean.

In practice, cloud classification usually stops at the genera level and does not often include the species, variety, and supplementary features and accessory clouds subdivisions. This is because these features are often subtle, are difficult to identify, or rarely occur. Some of the more common descriptions in these lesser used categories are given in ▼ Table 5.4.

CLOUD IDENTIFICATION

High Clouds High clouds in middle and low latitudes generally form above 6000 m. Because the air at these elevations is quite cold and "dry," high clouds are composed almost exclusively of ice crystals and are also rather thin.* High clouds usually appear white, except near sunrise and sunset, when the unscattered (red, orange, and yellow) components of sunlight are reflected from the underside of the clouds. Precipitation does not develop in high clouds.

The most common high clouds are the **cirrus** (Ci), which are thin, wispy clouds blown by high winds into long streamers called "mares' tails." Notice in ● Figure 5.16 that they can look like a white, feathery patch with a faint wisp of a tail at one end. Cirrus clouds usually move across the sky from west to east, indicating the prevailing winds at their elevation. They are generally associated with fair, pleasant weather, but some types can be early indicators of approaching storm systems.

Cirrocumulus (Cc) clouds, seen less frequently than cirrus, appear as small, rounded, white puffs that may occur individually or in long rows (see ● Figure 5.17). When in rows, cirrocumulus clouds have a rippling appearance that distinguishes them from silky-looking cirrus and sheetlike cirrostratus types. Cirrocumulus seldom covers more than a small portion of the sky. Their dappled appearance reflects the red or yellow light of sunsets and makes this one of the most beautiful of all clouds. Small ripples in cirrocumulus clouds strongly resemble fish scales; hence, the expression "mackerel sky" is commonly used to describe a sky full of cirrocumulus.

Thin, sheetlike, high clouds that often cover the entire sky are **cirrostratus** (Cs) (see ● Figure 5.18). These are so thin that the sun and moon can be clearly seen through them. The ice

---

*Small quantities of liquid water in cirrus clouds at temperatures as low as −36°C were discovered during research conducted above Boulder, Colorado.

▼ Table 5.4   The Most Common Species, Varieties, and Supplementary Features and Accessory Clouds Terms

SUBDIVISION	TERM	LATIN ROOT AND MEANING	DESCRIPTION
Species	lenticularis	(*lens, lenticula*, lentil)	Clouds with the shape of a lens or an almond, often elongated and usually with well-defined outlines. This term modifies these basic cloud types to cirrocumulus, altocumulus, and stratocumulus and indicates wave clouds that form downwind of mountains.
	fractus	(*frangere*, to break or fracture)	Clouds that have a ragged or torn appearance; applies only to stratus and cumulus
	humilis	(*humilis*, of small size)	Cumulus clouds with generally flattened bases and small amounts of vertical development
	congestus	(*congerere*, to bring together; to pile up)	Cumulus clouds of great vertical extent that from a distance may resemble a head of cauliflower. Sometimes these are called towering cumulus.
	calvus	(*calvus*, bald)	Cumulonimbus in which at least some of the upper part is beginning to lose its cumuliform outline
	capillatus	(*capillus*, hair; having hair)	Cumulonimbus characterized by the presence in the upper part of cirriform clouds with fibrous or striated structure
	castellanus	(*castellum*, a castle)	Clouds that show vertical development and produce towerlike extensions, often in the shape of small castles. This term modifies altocumulus and cumulus clouds.
Variety	undulatus	(*unda*, wave; having waves)	Clouds in patches, sheets, or layers showing undulations
	translucidus	(*translucere*, to shine through; transparent)	Clouds that cover a large part of the sky and are sufficiently translucent to reveal the position of the sun or moon
Supplementary Features and Accessory Clouds	incus	(*incus*, anvil)	The anvil-shaped, smooth cirriform mass of cloud in the upper part of a cumulonimbus
	mamma (mammatus)	(*mamma*, mammary)	Baglike clouds that hang like a cow's udder on the underside of a cloud; may occur with cirrus, altocumulus, altostratus, stratocumulus, and cumulonimbus. These usually appear during their decaying phase.
	pileus	(*pileus*, cap)	A cloud in the form of a cap or hood above or attached to the upper part of a cumuliform cloud. These occur most often during its developing stage.

crystals in these high clouds bend the sun or moon light passing through them and often produce a halo—a ring of light that encircles the sun or moon. In fact, the veil of cirrostratus may be so thin that a halo is the only clue to its presence. Thick cirrostratus clouds give the sky a glary white appearance and frequently form ahead of an advancing storm; hence, they can be used to predict rain or snow within 12 to 24 hours, especially if they are followed by middle-type clouds.

Middle Clouds
With bases forming between 2000 and 7000 m in the middle latitudes, these clouds are composed of water droplets and some ice crystals, when temperatures become low enough. Precipitation can form in middle clouds if they become thick enough.

Altocumulus (Ac) clouds are middle clouds composed mostly of water droplets and are rarely more than 1 km thick. They appear as grey, puffy masses (see ● Figure 5.19) that sometimes seem rolled out in parallel waves or bands. Usually, one part of the cloud is darker than another, which helps separate it from higher cirrocumulus clouds. Also, the individual puffs of an altocumulus cloud appear larger than those of a cirrocumulus. A layer of altocumulus may sometimes be confused with altostratus. They are called altocumulus if rounded masses or rolls are present.

© C. Donald Ahrens

● FIGURE 5.16 Cirrus clouds. Notice the silky, "mare's tail" appearance.

© C. Donald Ahrens

● FIGURE 5.18 Cirrostratus clouds. Notice the faint halo encircling the sun. If you look closely, you will see the sun as the bright white area in the centre of the circle.

Altocumulus clouds that look like "little castles" because of their small, turretlike structures are called **altocumulus castellanus** (Acc) (see Figure 6.28 on p. 186). Altocumulus castellanus in the sky indicate the presence of rising air or instability in the atmosphere at the cloud level. The appearance of these clouds on warm, humid summer mornings often forecasts thunderstorms by late afternoon.

Altostratus (As) is a grey or blue-grey cloud composed of ice crystals and water droplets. Altostratus often cover the entire sky over an area that extends many hundreds of square kilometres. In thinner sections of the cloud, the sun (or moon) may be *dimly visible* as a round disk, looking as if it were shining through frosted glass. This appearance is sometimes referred to as a "watery sun" (see ● Figure 5.20). Thick cirrostratus clouds are occasionally confused with thin altostratus. Look for grey colour, a low cloud height, and dimness of the

sun as good clues when identifying altostratus. The fact that halos occur only in cirrus clouds also helps distinguish them. Another way to separate these two clouds is to look at the ground for shadows. If there are none, it is a good bet that the clouds are altostratus because cirrostratus are usually transparent enough to produce shadows. Altostratus often forms ahead of storms that have widespread and relatively continuous precipitation. If precipitation forms in altostratus, the clouds become thicker, and their base altitude usually lowers. When precipitation falls from these clouds and reaches the ground, they are reclassified as nimbostratus. The precipitation is steady and not showery, as it would be from cumuliform clouds.

Low Clouds   The bases of low clouds lie below 2000 m and are almost always composed of water droplets. However,

© C. Donald Ahrens

● FIGURE 5.17 Cirrocumulus clouds cover the sky, giving rise to the name "mackerel sky."

© C. Donald Ahrens

● FIGURE 5.19 Altocumulus clouds. Notice the puffiness and dark-to-light contrasting patterns that distinguish these clouds.

● FIGURE 5.20 Altostratus clouds. A dimly visible "watery sun" appearing through a deck of light grey clouds is usually a good indicator of altostratus.

● FIGURE 5.21 Nimbostratus clouds are dark, sheetlike, and associated with rain or snow. (The ragged-appearing clouds beneath these nimbostratus are stratus fractus, or scud, an amorphous cloud that is commonly associated with frontal precipitation.)

in cold weather, they may contain ice particles. Low clouds that do not extend into middle cloud altitudes (St, Sc, Cu) generally do not produce precipitation. These are clouds with little vertical development; consequently, they have little ability to produce large enough precipitation drops or crystals to create rain. Sometimes they can produce smaller droplets we call drizzle, and in winter, snow flurries can come from low clouds.

Nimbostratus (Ns) is a dark grey, "wet-looking" cloud layer associated with more or less continuously falling rain or snow (see ● Figure 5.21). The intensity of precipitation is usually light to moderate. (It is not the short-lived, heavy, showery precipitation associated with thunderstorms, unless well-developed cumulus clouds are embedded within the nimbostratus cloud.) Precipitation makes the base of nimbostratus clouds difficult to see, but they are thick and their tops extend over 3 km high. Nimbostratus is easily confused with altostratus. Thin nimbostratus is darker grey than thick altostratus, and the sun or moon is not visible through a layer of nimbostratus. Visibility below a nimbostratus cloud deck is usually quite poor as rain will evaporate and mix with the air in this region, often creating misty conditions. If this air becomes saturated, fog or a lower layer of clouds may form beneath the original cloud base. Since these lower clouds drift rapidly with the wind, they usually form irregular shreds that have a ragged appearance called **stratus fractus**, or *scud*.

Stratocumulus (Sc) are low, lumpy clouds that appear in rows, patches, or rounded masses with blue sky visible between individual clouds (see ● Figure 5.22). Occasionally, the sun will shine through breaks in these cloud, producing bands of light (called **crepuscular rays**) that appear to reach down to the ground. Stratocumulus range in colour from light to dark grey. As can be seen by comparing Figure 5.19 with Figure 5.22, they differ from altocumulus as they have a

lower base and larger individual clouds. To see this, hold your hand at arm's length and point toward one of these clouds. Altocumulus will be about the size of your thumbnail, whereas stratocumulus are about the size of your fist. Although precipitation rarely falls from stratocumulus, light rain showers or winter snow flurries may occur if the cloud develops vertically into a much thicker cloud and the temperature at their top is colder than about −5°C. When this occurs, generally these are changing into cumulus types of clouds, which produce precipitation.

Stratus (St) is a uniform greyish cloud that often covers the entire sky. It resembles a fog that does not reach the ground (see ● Figure 5.23). Actually, when a thick fog "lifts," the resulting cloud is a deck of low stratus. Normally, no

● FIGURE 5.22 Stratocumulus clouds. Notice that the rounded masses of stratocumulus are larger than those of altocumulus.

● FIGURE 5.23 A layer of low-lying stratus clouds hides these mountains in Iceland.

precipitation falls from the stratus, but sometimes it is accompanied by a light mist or drizzle. A thick layer of stratus might be confused with nimbostratus, but the distinction between them can be made by observing the low base of the stratus cloud and remembering that light-to-moderate precipitation occurs with nimbostratus. Moreover, stratus clouds often have a more uniform base than does nimbostratus. Stratus clouds cannot be confused with a layer of altostratus by remembering that stratus are lower and darker grey and that the "watery" sun cannot be seen through stratus.

Clouds with Vertical Development   Officially, clouds in this category are included with the low cloud level (étage) of the classification system as their bases are in the low cloud altitudes. But identifying them as a separate group helps those learning to identify clouds recognize and distinguish

the importance of vertical motion in the atmospheric processes that produce these clouds. This group includes all low cumuliform types of clouds: cumulus and cumulonimbus clouds.

Familiar to almost everyone, puffy **cumulus** (Cu) clouds take on a variety of shapes, but most often they look like cotton balls with clearly defined outlines and flat bases (see ● Figure 5.24). Their bases appear white to light grey, and on humid days, they may be only 1000 m above the ground and about a kilometre wide. The top of the cloud is usually not very high and forms rounded towers denoting the limit of rising air within the cloud. Cumulus can be distinguished from stratocumulus by the larger amount of blue sky between each cloud that makes them appear as individual clouds. (Stratocumulus clouds usually occur in groups or patches.) Also, cumulus have puffy dome- or tower-shaped tops that

● FIGURE 5.24 Cumulus clouds. Small cumulus clouds such as these are sometimes called fair-weather cumulus, or cumulus humilis.

© C. Donald Ahrens

are evidence of turbulence in the cloud. This is opposed to the generally flat tops of stratocumulus clouds. Cumulus clouds with only slight vertical development are called **cumulus humilis** or "fair-weather cumulus" and do not have precipitation. Ragged-edge cumulus clouds that are smaller than cumulus humilis and scattered across the sky are called **cumulus fractus**.

Harmless-looking cumulus often develop on warm summer mornings and, by afternoon, become much larger and more vertically developed, extending into the middle cloud altitudes. When the growing cumulus resembles a head of cauliflower, it becomes a **cumulus con-gestus**, or **towering cumulus** (Tcu). Usually, this is a single large cloud, but, occasionally, several grow into each other, forming a line of towering clouds, as shown in ● Figure 5.25. Precipitation that falls from a cumulus congestus is always showery, with frequent changes in intensity. If a cumulus congestus cloud continues to grow vertically, it develops into a **cumulonimbus** (Cb) cloud or *thundercloud* (see ● Figure 5.26). Whereas its dark base may be no more than 600 m above Earth's surface, its top often extends to the tropopause, over 12,000 m higher. A cumulonimbus can occur as an isolated cloud or as part of a line or "wall" of clouds.

© T. Ansel Toney

● FIGURE 5.26 A cumulonimbus cloud. These may produce thunderstorms. Strong upper-level winds blowing from right to left produce the well-defined anvil shape at the top of this cloud. Sunlight scattered by falling ice crystals produces the bright white area beneath the anvil. Notice the heavy rain shower falling from the base of the cloud.

The tremendous amounts of energy released by the condensation of water vapour within a cumulonimbus cloud result in the development of violent updrafts and downdrafts, which may exceed 35 m s⁻¹ (126 km h⁻¹). The lower, warmer part of the cloud is usually composed of only water droplets. Higher up, water droplets and ice crystals both abound. Toward the cold top, there are only ice crystals. Swift winds at these higher altitudes can reshape the top of the cloud into a huge, flattened anvil* shape (called **cumulonimbus incus**). These great thunderheads may contain all forms of precipitation—large raindrops, snowflakes, snow pellets, and some-

*An anvil is a heavy block of iron or steel with a smooth, flat top on which blacksmiths shape metals by hammering.

times hailstones—which can fall to Earth in the form of heavy showers. Lightning, thunder, and even tornadoes are associated with the cumulonimbus. More information on the violent nature of thunderstorms and tornadoes is given in Chapter 14.

Cumulus congestus and cumulonimbus frequently look alike, making it difficult to distinguish them. However, looking at the top of the cloud helps. If the sprouting upper part of the cloud is sharply defined and has a turbulent, cauliflower-shaped top, it is usually a cumulus congestus. If the top of the cloud becomes flattened, loses its sharpness, and has the appearance of being fibrelike or wispy, it is usually a cumulonimbus. Compare Figure 5.25 with Figure 5.26. The weather associated with these clouds also differs: lightning, thunder, and large hail typically occur with cumulonimbus, whereas less violent weather is associated with cumulus congestus.

So far, we have focused on the 10 primary cloud forms, summarized pictorially in ● Figure 5.27. This figure, along with the cloud photographs, descriptions, and the cloud chart at the back of the book, should help you identify the more common cloud forms. Try to estimate the cloud type and their height, but do not worry if you find heights difficult to estimate. This requires much practice and noticing that clouds on the horizon often appear at different heights than they do overhead. You can use local objects (hills, mountains, tall buildings) of known heights as references on which to base your height estimates.

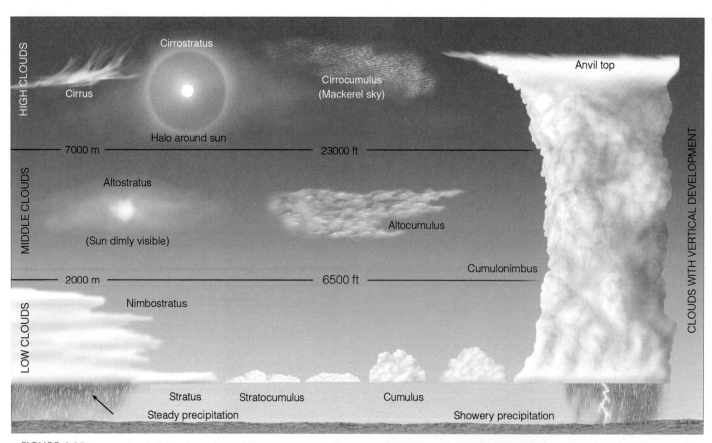

● FIGURE 5.27 A generalized illustration of basic cloud types (genera) based on height above Earth's surface and the extent of vertical development.

● FIGURE 5.28 Lenticular clouds over Jasper National Park, Alberta.

To better describe a cloud's shape and form, a number of descriptive words may be used in conjunction with the cloud name or genera. We mentioned a few in the previous section; for example, a stratus cloud with a ragged appearance is a stratus fractus, and a cumulus cloud with marked vertical growth is a cumulus congestus. Table 5.4 on page 143 lists some of the terms that are used when describing more uncommon clouds.

SOME UNUSUAL CLOUDS Although the 10 basic cloud genera are the most frequently seen types of clouds, there are some unusual clouds that deserve mentioning. For example, moist air crossing a mountain barrier often forms into waves. The clouds that form in the wave crest usually have a lens shape and are called **lenticular clouds** (see ● Figure 5.28). Frequently, they form one above the other like a stack of pancakes and at a distance may resemble a hovering spacecraft. It is no wonder that a large number of UFO sightings take place when lenticular clouds are present. Another cloud formed when moving over topography is a **banner cloud**, which forms over and extends downwind of an isolated mountain peak, as shown in ● Figure 5.29.

Similar to lenticular clouds are *cap clouds*, or **pileus**, which resemble a smooth cap made of cirrus cloud capping the top of a growing cumulus cloud (see ● Figure 5.30). Pileus clouds form when moist winds are deflected up and over the top of a cumulus congestus or cumulonimbus cloud. If the air flowing over the top of the cloud condenses, a pileus often forms.

Most clouds form in rising air, but the mamma or mammatus* form in sinking air. **Mamma clouds** derive their name from their appearance as baglike sacs that hang beneath the cloud and resemble a cow's udder (see ● Figure 5.31). Although mamma most frequently form on the underside of cumulonimbus, they may develop beneath cirrocumulus, altostratus, altocumulus, and stratocumulus. For mamma to form, the sinking air must be cooler than the air around it and have a high liquid water or ice content. As

---

*Mammatus is now considered an older form of the term *mamma*.

● FIGURE 5.29 The cloud forming over and downwind of Mount Assiniboine, Alberta, is called a banner cloud.

● FIGURE 5.30 Pileus clouds forming above a developing cumulus cloud.

● FIGURE 5.31 Mamma clouds (also known as mammatus) form beneath a thunderstorm.

● FIGURE 5.32 A contrail forming behind a jet aircraft.

saturated air sinks, it warms, but the warming is reduced because heat is taken from the air to evaporate the liquid or melt ice particles. If the sinking air remains saturated and cooler than the air around it, the sinking air can extend below the cloud base, appearing as rounded masses we call mamma clouds.

Jet aircraft flying at high altitudes often produce a cirrus-like trail of condensed vapour called a *condensation trail* or **contrail** (see ● Figure 5.32). Contrails evaporate rapidly when the relative humidity of the surrounding air is low. If the relative humidity is high, however, contrails may persist for many hours. They can form from water vapour that is added to the air from the jet's engine exhaust and condense when there is sufficient mixing of the hot exhaust gases with the cold air to produce saturation. The release of particles in the exhaust may even provide nuclei on which ice crystals can form. Contrails may also form by a cooling process. The reduced pressure produced by air flowing over the wing causes the air to cool. This cooling may supersaturate the air, producing an *aerodynamic contrail*. This type of contrail usually disappears quickly in the turbulent wake of an aircraft.

Aside from the cumulonimbus cloud that sometimes penetrates into the stratosphere, all of the clouds described so far are observed in the troposphere. Occasionally, clouds may be seen above the troposphere. For example, soft, pearly-looking clouds called **nacreous clouds**, or *mother-of-pearl clouds*, form in the stratosphere at altitudes above 30 km (see ● Figure 5.33). They are best viewed in polar latitudes during the winter months when the sun is just below the horizon, and it illuminates them because of their high

● FIGURE 5.33 The clouds in this photograph are nacreous clouds. They form in the stratosphere and are most easily seen at high latitudes.

© Pekka Parviainen

● FIGURE 5.34 The wavy clouds in this photograph are noctilucent clouds. They occur at altitudes between 75 and 90 km above Earth's surface. They are usually seen at twilight, during the summer, at latitudes that are closer to the poles than 50°.

altitude. Their exact composition is not known, although they appear to be composed of water in either solid or liquid (supercooled) form.

Wavy, bluish-white clouds, so thin that stars shine brightly through them, may sometimes be seen in the upper mesosphere at altitudes above 75 km. They are best seen at twilight, during the summer, at latitudes poleward of 50°. This happens because the clouds are at such a high altitude that they are still in sunshine, and to a ground observer, they appear bright against a dark background. For this reason, they are called **noctilucent clouds**, meaning "luminous night clouds" (see ● Figure 5.34). Studies reveal that these clouds are composed of tiny ice crystals. The water that makes this ice may originate in meteoroids that disintegrate when entering the upper atmosphere or from the chemical breakdown of methane gas at high levels in the atmosphere.

CLOUD OBSERVATIONS

Determining Sky Conditions

Often a daily weather forecast will include a phrase such as "cloudy with sunny periods, clearing this evening." To the average person, this means that the cloudiness will diminish, but to the meteorologist, the terms *cloudy with sunny periods* and *clearing* have a more specific meaning. In meteorology, descriptions of sky conditions are defined by the fraction of sky covered by clouds. Internationally, this is the number of eighths (called *oktas*) of the sky obscured by cloud; however, it is also very common to report cloud cover as the number of tenths. In Canada, cloud cover is reported in both tenths and eighths depending on the purpose. For hourly observations from airports, cloud amounts are reported in tenths. But on surface weather maps, cloud amounts are reported in eighths, conforming to international standards for weather maps, which compile **synoptic** information every six hours.

Meteorologists also use words such as *clear, few, scattered, broken,* and *overcast* to indicate increasing amounts of cloud cover. ▼ Table 5.5 presents a summary of sky cover terms and their meanings. Notice that there is not always direct correspondence between cloud amounts given in eighths, tenths, and percents. Agencies do the best they can to make these values compatible but have decided not to subdivide their observation units, and this causes small discrepancies between systems.

Observing sky conditions can be tricky. Observers have to look up and focus on the sky instead of just looking ahead as we normally do. With this point of view, objects on the horizon occupy less of the sky than those directly above us. This is how cloud amounts are reported. Fisheye lens cameras can do this for us and can be used to precisely determine sky amounts. As ● Figure 5.35 shows, objects on the horizon account for a smaller amount of the sky than we normally perceive. This can sometimes fool even trained observers when looking at clouds that are far away. A *broken* cloud deck near the horizon usually appears as *overcast* because the open spaces between the clouds are less visible at a distance. Therefore, cloudiness is usually overestimated when clouds are near the horizon. Viewed from afar, clouds not normally associated with precipitation may appear darker and thicker than they actually are. The reason for this observation is that light from a distant cloud travels through more atmosphere and is more attenuated than the light from the same type of cloud closer to the observer. (Information on measuring the height of cloud bases is given in Focus on an Observation: Measuring Cloud Ceilings on p. 153.)

▼ Table 5.5　Description of Sky Conditions*

OBSERVATION					
Official Term (Official Abbreviation)	Eighths (8ths, Oktas)	Tenths (10ths)	AWS† Percent (%)	Environment Canada Public Forecast Wording	Human Meaning
Clear (CLR)	0	0	0 to 5	Sunny (0–2/10)	No clouds
				Clear (0–1/10)	
Few (FEW)	≤ 2/8	≤ 3/10	> 5 to ≤ 25	Sunny with cloudy periods (3–4/10)	Few clouds visible
				A few clouds (2–3/10)	
Scattered (SCT)	3/8 to 4/8	4/10 to 5/10	> 25 to ≤ 50	A mix of sun and clouds (5–6/10)	Partly cloudy
				Cloudy periods (4–6/10)	Mostly sunny/clear
Broken (BKN)	5/8 to 7/8	6/10 to 9/10	> 50 to ≤ 87	Cloudy with sunny periods (7–8/10)	Mostly cloudy
				Cloudy (7–10/10)	Partly sunny/clear
Overcast (OVC)	8/8	10/10	> 87 to 100	Overcast (10/10)	Sky is covered by clouds
Sky obscured (X)	—	—	—	—	Sky is hidden by surface-based phenomena, such as fog, blowing snow, smoke, and so forth, rather than by cloud cover

*Symbol definitions: > means greater than; < means less than; ≥ means greater than or equal to; ≤ means less than or equal to.

†Automated weather station.

Image Source/Getty Images

● FIGURE 5.35 A fisheye lens photograph of the sky showing how it looks to an observer who is viewing cloud amounts.

● Figure 5.36a and b shows the average percentage cloud cover over North America in January and July. In January (see Figure 5.36a), a band of cloudiness in the northern mid-latitudes is associated with cyclonic storms that characterize winter weather in these regions. Notice that the mountainous areas over much of British Columbia and areas near major moisture sources such as the Maritimes, Great Lakes, and, to a lesser extent, the Gulf of Mexico also tend to be cloudier. In July (see Figure 5.36b), as winter storms weaken and their tracks shift northward, the cloudiness also moves northward. Average summer cloud cover increases in northern Canada and Alaska and diminishes elsewhere in North America. Increased clouds over Mexico in July result from the summer monsoon circulation affecting this region (monsoon weather is discussed in more detail in Chapter 9).

Up to this point, we have seen how clouds look from the ground. We will now look at clouds from a different vantage point—the satellite view.

FOCUS ON AN OBSERVATION

Measuring Cloud Ceilings

In addition to knowing about sky conditions, it is usually important to have a good estimate of the height of cloud bases. Aircraft cannot operate safely without accurate cloud height information, particularly at lower elevations.

The term *ceiling* is defined as the height of the lowest layer of clouds above the surface that are either broken or overcast but not thin. Direct information on cloud height can be obtained from pilots who report the altitude at which they encounter the ceiling. Less directly, ceiling balloons can measure the height of clouds. A small balloon filled with a known amount of hydrogen or helium rises at a fairly constant and known rate. The ceiling is determined by measuring the time required for the balloon to enter the lowest cloud layer.* Ceiling balloon observations can be made at night simply by attaching a small battery-operated light to the balloon.

For many years, the *rotating-beam ceilometer* provided information on cloud

---

*For example, if the balloon rises 125 m each minute, and it takes three minutes to enter a broken layer of stratocumulus, the ceiling would be 375 m above ground.

ceiling, especially at airports. This instrument consists of a ground-based projector that rotates vertically from horizon to horizon. As it rotates, it sends out a powerful light beam that moves along the base of the cloud. A light-sensitive detector, some known distance from the projector, points upward and picks up the light reflected from the cloud base. By knowing the projector angle and its distance from the detector, the cloud height is determined mathematically.

Most of the rotating-beam ceilometers have been phased out and replaced with *laser-beam ceilometers*. The laser ceilometer is a fixed-beam type whose transmitter and receiver point straight up at the cloud base (see • Figure 8). Short, intense pulses of infrared radiation from the transmitter strike the cloud base, and a portion of this radiation is reflected back to the receiver. The time interval between pulse transmission and its return from the cloud determines the cloud-base height.

Typical automated weather stations at airports use a laser-beam ceilometer to measure cloud height. The ceilometer measures the cloud height and then infers the amount of

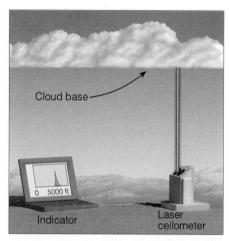

• FIGURE 8 The laser-beam ceilometer sends pulses of infrared radiation up to the cloud. Part of this beam is reflected back to the ceilometer. The interval of time between pulse transmission and return is a measure of cloud height, as displayed on the indicator screen.

cloud cover by averaging the amount of clouds that have passed over the sensor during a period of 30 minutes. The laser ceilometer is unable to measure clouds that are not above the sensor. To help remedy this situation, a second laser ceilometer may be located nearby.

WEATHER WATCH

Many of the cumulonimbus clouds that form in the middle latitudes have tops that do not exceed 9 km. However, over the tropical Pacific Ocean, between northern Australia and Indonesia, the tops of cumulonimbus clouds can exceed 18 km, making them some of the tallest clouds in the world.

Satellite Observations
Weather satellites are cloud-observing platforms in Earth's orbit. They provide extremely valuable cloud images of the whole Earth, including where there are no ground-based observations. Because water covers over 70 percent of Earth's surface, there are vast regions where

few (if any) surface cloud observations are made. Before weather satellites were available, tropical storms, such as hurricanes and typhoons, often went undetected until they moved dangerously near inhabited areas. Residents of the regions affected had little advance warning. Today, satellites spot these storms while they are still far out in the ocean and track them accurately.

Two primary types of weather satellites are used for viewing clouds. The first are called **geostationary satellites** (or *geosynchronous satellites*) because they orbit the equator at the same rate Earth spins and remain nearly 36,000 km above a fixed spot on Earth's surface (see • Figure 5.37). These types of satellites allow continuous monitoring of a specific region.

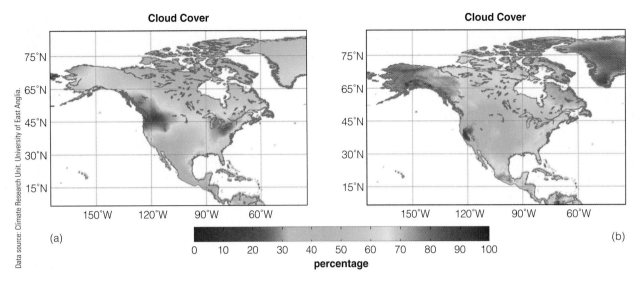

Cloud Cover

Data source: Climate Research Unit. University of East Anglia.

(a)   (b)

percentage

● FIGURE 5.36 Percent average cloud cover based on surface observations from 1961 to 1990: (a) January; (b) July.

Geostationary satellites are also important because they use a "real-time" data system, meaning that the satellites transmit images to the receiving system on the ground as soon as the image is made. Successive cloud images from these satellites can be put into a time-lapse movie sequence (an animation) to show the cloud movement, dissipation, or development associated with weather fronts and storms. This information greatly helps forecasting the progress of large weather systems. Wind directions and speeds at various levels may also be approximated by monitoring cloud movement with geostationary satellites.

Polar-orbiting satellites, which have orbit paths that approximately parallel Earth's lines of longitude, complement geostationary satellites. These satellites pass over the north and south polar regions on each revolution. As Earth rotates to the east beneath the satellite, each pass of the satellite monitors an area to the west of the previous pass (see ● Figure 5.38), and, eventually, the satellite covers the entire Earth.

Polar-orbiting satellites have the advantage of scanning clouds directly beneath them. Thus, they provide sharp images in polar regions, where images from a geostationary satellite are distorted because of the low angle at which the satellite "sees" this region. Polar orbiters also circle Earth at a much lower altitude (about 850 km) than geostationary satellites and therefore provide more detailed images about objects, such as violent storms and cloud systems.

Continuously improved detection devices make weather observation by satellites more versatile than ever. Early satellites, such as *TIROS I*, launched on April 1, 1960, used

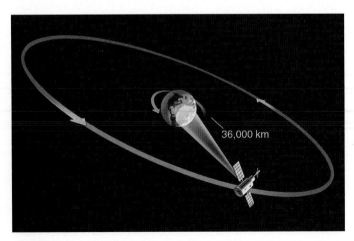

● FIGURE 5.37 The geostationary satellite moves through space at the same rate that Earth rotates, so it remains above a fixed spot on the equator and monitors one area constantly.

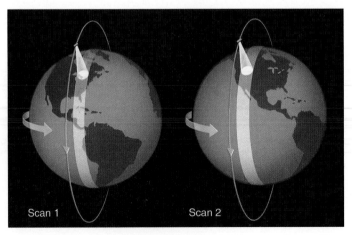

● FIGURE 5.38 Polar-orbiting satellites scan from north to south, and on each successive orbit, the satellite scans an area farther to the west.

television cameras to photograph clouds. Contemporary satellites use radiometers, which can observe clouds both day and night by detecting different wavelengths of radiation that emanate from the top of the clouds. Additionally, satellites have the capacity to obtain vertical profiles of atmospheric temperature and moisture by detecting emitted radiation from atmospheric gases, such as carbon dioxide and water vapour. In modern satellites, advanced radiometers (called *imagers*) provide satellite images with much better resolution than did previous imagers. Moreover, another type of radiometer (called a *sounder*) gives a more accurate profile of temperature and moisture at different levels in the atmosphere than did earlier instruments. In the latest Geostationary Operational Environment Satellite (GOES) series, the imager and sounder are able to operate independently of each other.

Information on cloud thickness and height can be deduced from satellite images. Visible satellite images show sunlight that is reflected from a cloud's upper surface. Because thick clouds have a higher albedo (reflectivity) than thin clouds, they appear brighter on a visible satellite image. However, these images cannot identify cloud height as thick clouds can occur at all altitudes. To recognize cloud heights, infrared images are used as they measure an object's radiation and report it as a temperature. The various temperatures on an infrared image are assigned a colour (or a shade of grey), and these colours and their corresponding temperatures are usually printed on the image, so their meaning is obvious. Where colour codes are not given, standard infrared colours that span the continuum of white to grey to black tones are used. Here the coldest temperatures are assigned the brightest, whitest colours and the warmest temperatures are the darkest colours. Because the tops of low clouds are warmer than those of high clouds, infrared images can distinguish cloud heights (see ● Figure 5.39).

● Figure 5.40 shows satellite images of the same storm system in the eastern Pacific, taken at about the same time by a geostationary satellite using different wavelengths of radiation. Figure 5.40a is a visible satellite image. Notice that all of the clouds in this image appear white and that the clouds over Oregon and northern California appear relatively thin compared to the thicker, bright clouds over the ocean to the west. Figure 5.40b is an infrared image. Here the clouds appear to have many shades of grey, indicating that they have different temperatures. Furthermore, the thin clouds over Oregon and California must be cold because they also appear bright and are therefore high. The elongated band of clouds off the coast marks the position of an approaching frontal system. Here the clouds appear white and bright in both Figure 5.40a and Figure 5.40b, indicating a zone of thick, heavy clouds. Behind the front, the lumpy clouds are probably cumulus because they appear grey in the infrared image, indicating that their tops are lower and relatively warm. Figure 5.40c is based on the same infrared image as Figure 5.40b except it has been *enhanced*

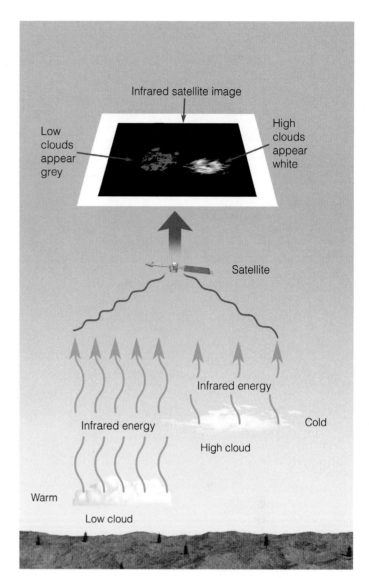

● FIGURE 5.39 Generally, the lower the cloud, the warmer its top. Warm objects emit more infrared energy than do cold objects. Thus, an infrared satellite image can distinguish warm, low clouds from cold, high clouds. Often the warm (low) clouds are given grey shades, whereas the cold, high clouds are given white shades in the images, although, increasingly, other colours are being used.

to emphasize particular cloud temperatures. Often the enhancement is a distinct change in colour that makes it easier to identify significant cloud features. For example, blue, red, or orange is assigned to clouds with the coldest (highest) tops. Hence, the red areas embedded along the front in Figure 5.40c represent the region where the coldest and, therefore, highest and thickest clouds are found. Generally, this is where the stormiest weather is occurring. Also notice the line of red blotches near the southern tip of the frontal system. These are thunderstorms that have developed over warm tropical waters. They show up clearly as thick white clouds in both the visible (see Figure 5.40a) and

(a)

(b)

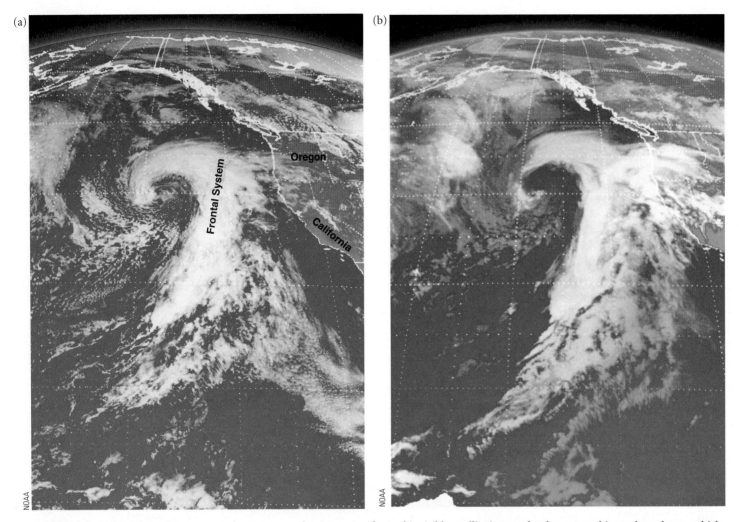

Oregon

California

Frontal System

NOAA

NOAA

● FIGURE 5.40 (a) Satellite imagery of a storm over the eastern Pacific. In this visible satellite image, clouds appear whitest where they are thick. Notice that they are thinner over Oregon and northern California. (b) Satellite imagery of a storm over the eastern Pacific. In this infrared image, high clouds are colder and whiter, whereas low clouds are warmer and appear in various shades of grey.

infrared (see Figure 5.40b) images. By examining the movement of these clouds on successive satellite images, forecasters can predict the arrival of clouds and storms, as well as the passage of weather fronts.

Meteorologists can combine infrared and visible images taken at the same time into a single image, as shown in ● Figure 5.41, to emphasize different layers of clouds. This is called "visible + infrared" imagery. It is created by overlaying an infrared imaged that is coloured in shades of blue with a visible image that is coloured in shades of yellow. The combination of yellow + blue makes a white colour in the resulting image. In this way, the cold (high) clouds that are thick will appear white, whereas low clouds (and snow-covered ground) will appear yellow. If the cold clouds are thin, they appear blue on the image because they do not reflect light strongly. Of course, the visible + infrared image is available only during daylight hours.

In regions where there are no clouds, it is difficult to observe the movement of the air or its moisture content. To help with this situation, geostationary satellites can make images by detecting infrared radiation at wavelengths of water vapour emission and absorption. These water vapour images represent the temperature of atmospheric water vapour that is emitting radiation to the satellite. The images give a picture of water vapour content in the middle and upper troposphere (see ● Figure 5.42). In time-lapse animations, the patterns of moisture clearly show moist regions and dry regions, as well as middle tropospheric swirling wind patterns and jet streams.

The *TRMM* (*Tropical Rainfall Measuring Mission*) satellite provides information on clouds and precipitation from about 35°N to 35°S. A joint venture of NASA of the United States and the National Space Agency of Japan, this polar-orbiting satellite orbits Earth 16 times each day at an altitude of about 400 km. From this vantage point, the satellite, when looking straight down, can pick out individual cloud features as small as 2.4 km in diameter. Some of the instruments onboard the *TRMM* satellite include a visible

(c)

Environment Canada. Found at: http://www.weatheroffice.gc.ca/data/satellite/goes_nam_vvi_100.jpg. © Her Majesty The Queen in Right of Canada, Environment Canada, 2010.

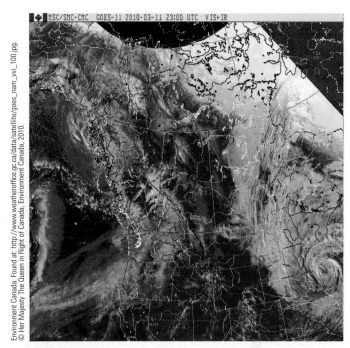

● **FIGURE 5.41** This is a "visible + infrared" satellite image that combines a visible image (in shades of yellow) with an infrared image (in shades of blue). White areas indicate cold, thick clouds; yellow areas indicate low clouds (or snow-covered ground); and blue areas indicate thin, high clouds.

● **FIGURE 5.40** (*Continued*) (c) Satellite imagery of a storm over the eastern Pacific. In this enhanced infrared image, the coldest, highest cloud temperatures have been emphasized using red to display significant weather features.

and infrared scanner, a microwave imager, and precipitation radar. These instruments help provide three-dimensional images of clouds and storms, along with the intensity and distribution of precipitation (see ● Figure 5.43). Additional onboard instruments send back information

● **FIGURE 5.42** Infrared water vapour image. The darker areas represent dry air aloft; the brighter the grey is, the more moist the air in the middle or upper troposphere. Bright white areas are dense cirrus clouds with the coldest tops.

FOCUS ON A SPECIAL TOPIC

Satellites Do More Than Observe Clouds

The use of satellites to monitor weather is not restricted to observing clouds. For example, satellites measure radiation from Earth's surface and atmosphere, giving us information about the Earth–atmosphere energy balance, discussed in Chapter 2. The infrared radiation measurements from different levels in the atmosphere are transformed into vertical profiles of temperature and moisture, which are fed into **numerical weather prediction** models.

Radiation intensities from the ocean surface are translated into sea surface temperature readings (see ● Figure 9). This information is valuable to the fishing industry, as well as to meteorologists and oceanographers. In fact, the *Tropical Rainfall Measuring Mission* (*TRMM*) satellite obtains sea surface temperatures with a microwave scanner, even through clouds and atmospheric particles.

Satellites also monitor the amount of snow cover in winter, the extent of ice fields in the Arctic and Antarctic, the movement of large icebergs that drift into shipping lanes, and the height of the ocean's surface. One polar-orbiting satellite carries equipment that can detect faint distress signals anywhere on the globe and relay them to rescue forces on the ground.

Infrared sensors on polar-orbiting satellites are able to assess conditions of crops, areas of deforestation, and regions of extensive drought. Satellites are also able to detect volcanic eruptions and follow the movement of ash clouds. During the winter, *GOES* satellites are able to monitor the southward progress of freezing air in Florida and Texas, allowing forecasters to warn growers of impending low temperatures so that they can take necessary measures to protect sensitive crops.

The *Global Positioning System* (*GPS*) consists of 24 polar-orbiting satellites that transmit radio signals to ground receivers, which then use the signals for navigation and relative positioning on Earth. The signal the

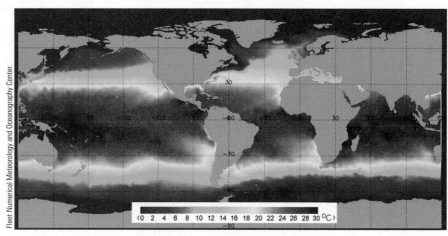

● FIGURE 9 Sea surface temperatures for February 19, 2004. Temperatures are derived mainly from satellites, but temperature information also comes from buoys and ships.

satellites send to Earth is slowed by the amount of water vapour in the air. Because of this effect, the GPS can estimate the atmosphere's precipitable water vapour (the total atmospheric water vapour contained in a vertical column of air).

Geostationary satellites, such as *GOES*, are equipped with systems that receive environmental information from remote data collection platforms on the surface. These platforms include instrumented buoys, river gauges, automated weather stations, seismic and tsunami ("tidal" wave) stations, and ships. This information is transmitted to the satellite, which relays it to a central receiving station.

Normally, a network of five geostationary satellites positioned over the equator gives nearly complete global coverage from about latitudes 60°N to 60°S. Along with monitoring clouds and the atmosphere, the latest GOES series provides forecasters and researchers with data from Doppler radars and networks of automated weather stations. Geostationary satellites detect pollution and haze and provide accurate cloud-height measurements during the day. They even have the

capacity to monitor the seasonal and daily trend in atmospheric ozone.

Satellites specifically designed to monitor the natural resources of Earth such as *LandSat* circle Earth 14 times a day in a near-polar circular orbit. Images taken in several wavelength bands provide valuable information about this planet's geology, hydrology, oceanography, and ecology. *LandSat* also collects data transmitted from remote ground stations in North America. These stations monitor a variety of environmental data, with water quality, rainfall amount, and snow depth of particular interest to meteorologists and hydrologists.

Satellite information is not confined to the lower atmosphere. There are satellites that monitor the concentrations of ozone, air temperature, and winds in the upper atmosphere. And both geostationary and polar-orbiting satellites carry instruments that monitor solar activity. Even with all this information available, there are more sophisticated satellites on the drawing board that will provide more and improved data in the future.

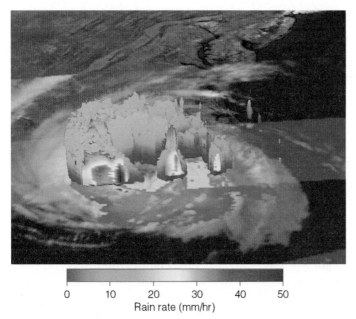

Rain rate (mm/hr)

● FIGURE 5.43 A three-dimensional TRMM satellite image of Hurricane Ophelia along the North Carolina coast on September 14, 2005. The light green areas in the cutaway view represent the region of lightest rainfall, whereas dark red and orange indicate regions of heavy rainfall.

concerning Earth's energy budget and lightning discharges in storms.

CloudSat is satellite-based radar, developed by NASA in partnership with the Canadian Space Agency, that uses millimetre wavelengths of microwave (radar) radiation to detect very fine structures in the makeup of clouds. This polar-orbiting satellite radar can detect cloud top and base altitude, cloud thickness, optical properties of the cloud, and the abundance of liquid and ice particles, as well as where precipitation is located in a cloud. It does this by scanning through clouds to produce a cross section of these cloud properties. ● Figure 5.44b shows a *CloudSat* profile through a storm on June 15, 2010, that affected France and dumped 350 mm of rain that caused flash flooding over the next 24 hours. The *CloudSat* profile shows the clouds stretched from south to north (B to A) as the warm, moist, southerly winds from the Mediterranean provided moisture to the system.

At this point, it should be apparent that today's satellites do a great deal more than simply observe clouds. More information on satellites and the information they provide is given in Focus on a Special Topic: Satellites Do More Than Observe Clouds on p. 158.

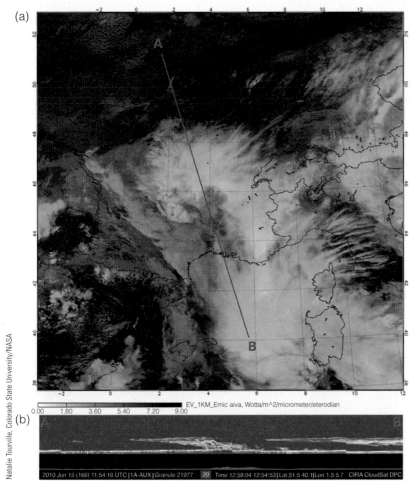

● FIGURE 5.44 (a) Infrared satellite image and (b) *CloudSat* radar reflectivity profile for a developing middle-latitude cyclone over France on June 15, 2010, at 12:55 UTC. Image (a) shows the location of the profile (blue line from A to B) through the cloud that produces the profile shown in (b). The top part of (b) shows radar reflectivity between 0 and 30 km elevation. It indicates the amount of ice and liquid water in the cloud. The bottom part of (b) shows the topography, with blue for water and brown for the land surface.

SUMMARY

In this chapter, we examined the different forms of condensation. We saw that dew forms when air temperature cools to the dew point in a shallow layer of air near the surface. If the dew should freeze, it produces tiny beads of ice called frozen dew. Frost forms when the air cools to a dew point that is at freezing or below.

As the air cools in a deeper layer near the surface, the relative humidity increases and water vapour begins to condense on hygroscopic condensation nuclei, forming wet haze. As the relative humidity approaches 100 percent, condensation occurs on most nuclei, and the air becomes filled with tiny liquid droplets (or ice crystals) called fog.

Fog forms in two primary ways: cooling of air and evaporating and mixing water vapour into the air. Radiation fog, advection fog, and upslope fog form by the cooling of air, whereas steam fog and frontal fog are two forms of evaporation fog. Although fog has some beneficial effects—providing water for thirsty redwoods in California—in many places, it is a nuisance because it disrupts air traffic and is the primary cause of a number of automobile accidents.

Condensation above Earth's surface produces clouds. Clouds are classified into 10 genera according to the height of their bases (high middle and low) and physical appearance based on their process of formation (stratiform and cumuliform). Since each cloud has physical characteristics that distinguish it from all the others, careful cloud observations normally lead to correct identification.

Satellites enable scientists to obtain a bird's-eye view of clouds on a global scale. Polar-orbiting satellites obtain data covering Earth from pole to pole, whereas geostationary satellites located above the equator continuously monitor a portion of Earth. Both types of satellites use radiometers (imagers) that detect emitted infrared radiation. As a consequence, clouds can be observed both day and night.

Visible satellite images, which show sunlight reflected from a cloud's upper surface, can distinguish thick clouds from thin clouds. Infrared images show an image of the cloud's radiating top and can distinguish low clouds from high clouds. To increase the contrast between cloud features, infrared photographs are enhanced.

Satellites do a great deal more than simply photograph clouds. They provide us with a wealth of physical information about Earth and the atmosphere.

KEY TERMS

The following terms are listed (with page number) in the order they appear in the text. Define each. Doing so will aid you in reviewing the material covered in this chapter.

hydrologic cycle, 126
dew, 126
frozen dew, 126
frost, 126
dry freeze (black frost), 127
condensation nuclei, 128
hygroscopic nuclei, 128
hydrophobic nuclei, 128
haze, 129
fog, 130
acid fog, 130
radiation fog, 130
advection fog, 133
ice fog, 133
advection–radiation fog, 135
upslope fog, 136
mixing fog, 136
evaporation fog, 136
steam fog, 136
frontal fog, 137
stratiform, 141
cumuliform, 141
cloud levels (étages), 141
genera, 142
species, 142
varieties, 142
supplementary features and accessory clouds, 142
cirrus, 142
cirrocumulus, 142
cirrostratus, 142
altocumulus, 143
altocumulus castellanus, 144
altostratus, 144
nimbostratus, 145
stratus fractus, 145
stratocumulus, 145
crepuscular rays, 145
stratus, 145
cumulus, 146
cumulus humilis, 147
cumulus fractus, 147
cumulus congestus (towering cumulus), 147
cumulonimbus, 147
cumulonimbus incus, 148
lenticular clouds, 149
banner cloud, 149
pileus, 149
mamma clouds, 149
contrail, 150
nacreous clouds, 150
noctilucent clouds, 151
synoptic, 151
geostationary satellites, 153
polar-orbiting satellites, 154
numerical weather prediction, 158

QUESTIONS FOR REVIEW

1. Explain how dew, frozen dew, and visible frost each form.
2. Distinguish among dry haze, wet haze, and fog.
3. Why is fog that forms in industrial areas normally thick?
4. How can fog form when the air's relative humidity is less than 100 percent?
5. Name and describe four types of fog. What conditions are necessary for the formation of radiation fog?
6. Why do ground fogs usually "burn off" by early afternoon?
7. List as many positive consequences of fog as you can.
8. List and describe three methods of fog dispersal.
9. How does radiation fog normally form?
10. What atmospheric conditions are necessary for the development of advection fog?
11. How does evaporation fog form?
12. Clouds are most generally classified by height above Earth's surface. List the major height categories and the cloud types associated with each.
13. List at least two distinguishable characteristics of each of the 10 basic clouds.
14. Why are high clouds normally thin? Why are they composed almost entirely of ice crystals?

15. How can you distinguish altostratus from cirrostratus?
16. Which clouds are associated with each of the following characteristics?
    (a) lightning
    (b) heavy rain showers
    (c) mackerel sky
    (d) mares' tails
    (e) halos
    (f) light continuous rain or snow
    (g) hailstones
    (h) anvil top
17. Why does a broken layer of clouds near the horizon often appear as overcast?
18. How do geostationary satellites differ from polar-orbiting satellites?
19. Explain why visible and infrared images can be used to distinguish: (a) high clouds from low clouds; (b) thick clouds from thin clouds.
20. Why are infrared images enhanced?
21. Name two clouds that form above the troposphere.
22. List and explain the various types of environmental information obtained from satellites.

QUESTIONS FOR THOUGHT

1. Explain the reasoning behind the wintertime expression, "Clear moon, frost soon."
2. Explain why icebergs are frequently surrounded by fog.
3. During a summer visit to New Orleans, you stay in an air-conditioned motel. One afternoon, you put on your sunglasses and step outside, and within no time, your glasses are "fogged up." Explain what has apparently caused this.
4. While driving from cold air (well below freezing) into much warmer air (well above freezing), frost forms on the windshield of the car. Does the frost form on the inside or outside of the windshield? How can the frost form when the air is so warm?
5. Why are really clean atmospheres and really dirty atmospheres undesirable?
6. Why do relative humidities seldom reach 100 percent in polluted air?
7. Why are advection fogs rare over tropical water?

8. A January snowfall covers southwest Ontario with 15 cm of snow. The following day, a south wind brings heavy fog to this region. Explain what has apparently happened.
9. If all fog droplets gradually settle earthward, explain how fog can last (without disappearing) for many days at a time.
10. Explain why steam fog is more likely to form during the autumn and advection fog in early spring near the shore of large lakes such as one of the Great Lakes.
11. The air temperature during the night cools to the dew point in a deep layer, producing fog. Before the fog formed, the air temperature cooled each hour about 2°C. After the fog formed, the air temperature cooled by only 0.5°C each hour. Give two reasons why the air cooled more slowly after the fog formed.
12. On a winter night, the air temperature cooled to the dew point and fog formed. Before the formation of fog, the dew point remained almost constant. After the fog formed, the dew point began to decrease. Explain why.
13. Why can you see your breath on a cold morning? Does the air temperature have to be below freezing for this to occur?
14. Explain why altocumulus clouds might be observed at 6400 m above the surface in Mexico City, Mexico, but never at that altitude above Inuvik, Northwest Territories.
15. The sky is overcast, and it is raining. Explain how you could tell if the cloud above you is a nimbostratus or a cumulonimbus.
16. Suppose it is raining lightly from a deck of nimbostratus clouds. Beneath the clouds are small, ragged, puffy clouds that are moving rapidly with the wind. What would you call these clouds? How did they probably form?
17. You are sitting inside your house on a sunny afternoon. The shades are drawn, and you look at the window and notice the sun disappear for about 10 seconds. The alternate light and dark period lasts for nearly 30 minutes. Are the clouds passing in front of the sun cirrocumulus, altocumulus, stratocumulus, or cumulus? Give a reasonable explanation for your answer.

PROBLEMS AND EXERCISES

1. The data in ▼ Table 5.6 below represent the dew point temperature and expected minimum temperature near the ground for various clear winter mornings. Assume

▼ Table 5.6

	MORNING 1	MORNING 2	MORNING 3	MORNING 4	MORNING 5
Dew point temperature	2°C	−7°C	1°C	−4°C	3°C
Expected minimum temperature	4°C	−3°C	0°C	−4.5°C	2°C

that the dew point remains constant throughout the night. Answer the following questions about the data.

(a) On which morning would there be the greatest likelihood of observing visible frost? Explain why.

(b) On which morning would frozen dew most likely form? Explain why.

(c) On which morning would there be black frost with no sign of visible frost, dew, or frozen dew? Explain.

(d) On which morning would you probably only observe dew on the ground? Explain why.

2. If a ceiling balloon rises at 120 m each minute, what is the ceiling of an overcast deck of stratus clouds 1500 m thick if the balloon disappears into the clouds in five minutes?

3. Compare the visible satellite image (see Figure 5.40a) with the infrared image (see Figure 5.40b). With the aid of the infrared image, label on the visible image the regions of middle, high, and low clouds. On the enhanced infrared image (see Figure 5.40c), label where the highest and thickest clouds appear to be located.

Viewed from an airplane, these convective clouds exhibit all stages of development from cumulus, to towering cumulus, to cumulonimbus.

Stability and Cloud Development

6

In July and August on the high desert the thunderstorms come. Mornings begin clear and dazzling bright, the sky as blue as the Virgin's cloak, unflawed by a trace of cloud in all that emptiness. . . . By noon, however, clouds begin to form over the mountains, coming it seems out of nowhere, out of nothing, a special creation.

The clouds multiply and merge, cumulonimbus piling up like whipped cream, like mashed potatoes, like sea foam, building on one another into a second mountain range greater in magnitude than the terrestrial range below.

The massive forms jostle and grate, ions collide, and the sound of thunder is heard over the sun-drenched land. More clouds emerge from empty sky, anvil-headed giants with glints of lightning in their depths. An armada assembles and advances, floating on a plane of air that makes it appear, from below, as a fleet of ships must look to the fish in the sea. *Edward Abbey*

Edward Abbey was a controversial essayist and environmental advocate.

Edward Abbey, *Desert Solitaire—A Season in the Wilderness.* Random House. 1968.

CONTENTS

Clouds, spectacular features in the sky, add beauty and colour to the natural landscape. Yet clouds are important for nonaesthetic reasons, too. As they form, vast quantities of heat are released into the atmosphere, providing essential energy that fuels storms. Clouds help regulate Earth's climate by reflecting and scattering solar radiation and by absorbing Earth's infrared energy. And, of course, without clouds, there would be no precipitation, the source of water in the *hydrosphere* that is distributed to the *biosphere* and *lithosphere* by the *atmosphere*. But clouds are also significant because they visually indicate the physical processes taking place in the atmosphere; to a trained observer, they are signposts in the sky. This chapter will examine the atmospheric processes these signposts point to, the first of which is atmospheric stability.

Atmospheric Stability

Most clouds form as air rises and cools. Why does air rise on some occasions but not on others? And why do the size and shape of clouds vary so much when the air does rise? Let's see how knowing about the air's stability will help us answer these questions.

When we speak of atmospheric stability, we are referring to a condition of equilibrium. For example, rock A resting in the depression in ● Figure 6.1 is in stable equilibrium. If the rock is pushed up along either side of the hill and then let go, it will quickly return to its original position. On the other hand, rock B, resting on the top of the hill, is in a state of unstable equilibrium as a slight push will set it moving away from its original position. Applying these concepts to the atmosphere, we can see that air is in stable equilibrium when, after being lifted or lowered, it tends to return to its original position—it resists upward and downward air motions. Air that is in unstable equilibrium will, when given a little push, move farther away from its original position—it favours vertical air currents.

To explore the behaviour of rising and sinking air, we must first put some air in an imaginary thin elastic wrap, like

a balloon. This small volume of air is referred to as a **parcel of air**. Although the air parcel can expand and contract freely, it does not break apart but remains as a single unit. At the same time, neither external air nor heat can mix with the air inside the parcel. The space occupied by the air molecules within the parcel defines the air density, which is the mass of air molecules divided by the parcel's volume. The average speed of the molecules is directly related to the air temperature, and the molecules colliding against the parcel walls determine the air pressure inside.

At Earth's surface, the parcel has the same temperature and pressure as the air surrounding it. Suppose we lift the air parcel up into the atmosphere. We know from Chapter 1 that air pressure decreases with height. Consequently, the air pressure surrounding the parcel lowers. The lower pressure outside allows the air molecules inside to push the parcel walls outward, expanding the parcel. Because there is no other energy source, the air molecules inside must use some of their own energy to expand the parcel. This shows up as slower average molecular speeds, which result in a lower parcel temperature. If the parcel is lowered to the surface, it returns to a region where the surrounding air pressure is higher. The higher pressure squeezes (compresses) the parcel back into its original (smaller) volume. This squeezing increases the average speed of the air molecules, and the parcel temperature rises. Hence, *a rising parcel of air expands and cools, whereas a sinking parcel is compressed and warms.*

If a parcel of air expands and cools, or compresses and warms, with no interchange of heat with its surroundings, this situation is called an **adiabatic process**. As long as the air in the parcel is unsaturated (the relative humidity is less than 100 percent), the rate of adiabatic cooling or warming remains constant. This rate of heating or cooling is about 10°C for every 1000 m* of change in elevation and applies only to unsaturated air. For this reason, it is called the **dry adiabatic lapse rate** (see ● Figure 6.2). It is called a *lapse rate* because temperature decreases with increased elevation.

As the rising air cools, its relative humidity increases as the air temperature approaches the dew point temperature. If the rising air cools to its dew point temperature, the relative humidity becomes 100 percent. Further lifting results in condensation, a cloud forms, and latent heat is released inside the rising air parcel. Because the heat added during condensation offsets some of the cooling due to expansion, the air no longer cools at the dry adiabatic lapse rate but at a lesser rate called the saturation adiabatic lapse rate or **saturated adiabatic lapse rate**. If a saturated parcel containing water droplets were to sink, it would compress and warm at the saturated adiabatic lapse rate because evaporation of the liquid droplets would offset the rate of compressional warming. Hence, the rate at

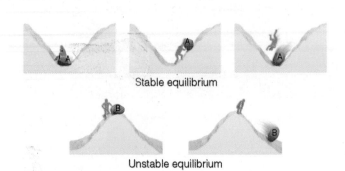

Stable equilibrium

Unstable equilibrium

● FIGURE 6.1 When rock A is disturbed, it will return to its original position; rock B, however, will accelerate away from its original position.

---

*The exact dry adiabatic lapse rate is 9.8°C of cooling for every 1000 m rise in elevation.

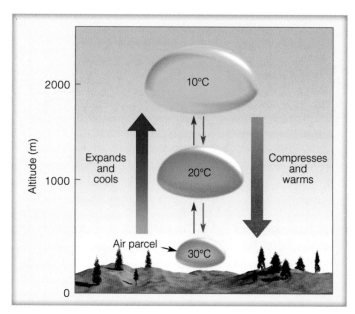

● **FIGURE 6.2** The dry adiabatic lapse rate. As long as the air parcel remains unsaturated, it expands and cools by 10°C per kilometre; the sinking parcel compresses and warms by 10°C per kilometre.

▼ Table 6.1 Saturated Adiabatic Lapse Rate for Different Temperatures and Pressures (°C km^{-1})

PRESSURE (hPa)	TEMPERATURE (°C)				
	−40	−20	0	20	40
1000	9.5	8.6	6.4	4.3	3.0
800	9.4	8.3	6.0	3.9	
600	9.3	7.9	5.4		
400	9.1	7.3			
200	8.6				

which rising or sinking saturated air changes temperature—the saturated adiabatic lapse rate—is less than the dry adiabatic lapse rate.[*]

Unlike the dry adiabatic lapse rate, the saturated adiabatic lapse rate is not constant but varies greatly with temperature. This is because warm saturated air has a much higher water vapour content than cold air—refer back to Figure 4.10 on p. 106, which shows how the saturation vapour pressure increases dramatically with increased temperature. As a result of this, a warm saturated air parcel, when forced to cool, condenses more liquid water than cold saturated air does when it is forced to cool the same amount. The added condensation in warm saturated air liberates more latent heat. Consequently, the saturated adiabatic lapse rate is much less than the dry adiabatic lapse rate when the rising air is warm; however, the two rates are nearly the same when the rising air is very cold (see ▼ Table 6.1). Although the saturated adiabatic lapse rate does vary, to make the numbers easy to deal with, we generally use a representative average cooling rate of 6°C per 1000 m.

---

[*]Consider an air parcel initially at rest. Suppose the air parcel rises and cools, and a cloud forms. Further suppose that no precipitation (rain or snow) falls from the cloud (leaves the parcel). If the parcel should descend to its original level, the latent heat released inside the parcel during condensation will be the same amount that is absorbed as the cloud evaporates. This process is called a *reversible adiabatic process*. If, on the other hand, rain or snow falls from the cloud during uplift and leaves the parcel, the sinking parcel will not have the liquid to evaporate back into the air, so there will not be the same amount of latent heat taken up by evaporation as was released during condensation. Therefore, the sinking air parcel will warm at the dry adiabatic lapse rate instead of the saturated adiabatic lapse rate. This process is known as an *irreversible saturated adiabatic process* and is a closer approximation to what usually happens in nature.

Determining Stability

We determine the stability of the air by comparing the temperature of a rising parcel to that of its surroundings. If the rising air is colder than its environment, it will be more dense[*] (heavier) and tend to sink back to its original level. In this case, the air is **stable** because it resists upward movement. If the rising air is warmer and, therefore, less dense (lighter) than the surrounding air, it will be buoyant and continue to rise until it reaches the same temperature as its environment. This is an example of **unstable air**. To determine the air's stability, we need to measure the temperature of both the rising air and its environment at various levels above Earth's surface. The type of stability we are discussing here is more precisely called *static stability* because we are not considering the effects of wind in creating turbulence and mixing; however, we will refer to it simply as stability.

A STABLE ATMOSPHERE Suppose we release a balloon-borne instrument—a radiosonde (see Chapter 1, Figure 2, p. 18)—and it sends back temperature data, as shown in ● Figure 6.3. (Such a vertical profile of temperature is called a *sounding*.) We measure the air temperature in the vertical and find that it decreases by 4°C for every 1000 m. The rate at which the air temperature decreases with elevation is called the *lapse rate*. Because this is the rate at which the air temperature surrounding us will be changing if we were able to fly upward into the atmosphere, we will refer to it as the **environmental lapse rate**. Now suppose in Figure 6.3a that a parcel of unsaturated air with a temperature of 30°C is lifted from the surface. As it rises, it cools at the dry adiabatic lapse rate (10°C per 1000 m), and the temperature inside the parcel at 1000 m would be 20°C, or 6°C lower than the air surrounding it. Look at Figure 6.3a closely and notice that as the air parcel rises higher, the temperature difference between it and the surrounding air becomes even greater. Even if the parcel is initially saturated (see Figure 6.3b),

---

[*]When, at the same level in the atmosphere, we compare parcels of air that are equal in size but vary in temperature, we find that cold air parcels are more dense than warm air parcels; that is, in the cold parcel, there are more molecules that are crowded closer together.

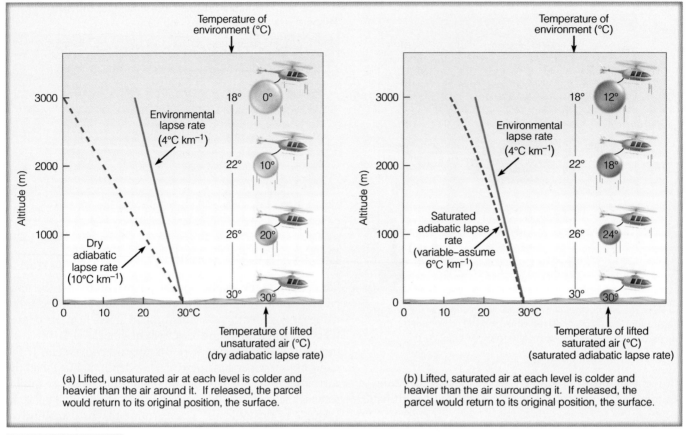

ACTIVE FIGURE 6.3 An absolutely stable atmosphere occurs when the environmental lapse rate is less than the saturated adiabatic lapse rate. In a stable atmosphere, a rising air parcel is colder and denser than the air surrounding it, and if given the chance, it will return to its original position. Visit the textbook's website to view this and other Active Figures at www.ahrensmeteorology1ce.nelson.com

it will cool at the saturated lapse rate 6°C per 1000 m—and will be colder than its environment at all levels. In both cases, the rising air is colder and heavier than the air surrounding it. In this example, the atmosphere is **absolutely stable**. *The atmosphere is always absolutely stable when the environmental lapse rate is less than the saturated adiabatic lapse rate.*

Since air in an absolutely stable atmosphere strongly resists upward vertical motion, it will tend to spread out horizontally *if forced to rise*. If clouds form in this rising air, they, too, will spread horizontally in relatively thin layers and usually have flat tops and bases. We might expect to see **stratiform** clouds—such as cirrostratus, altostratus, nimbostratus, or stratus—forming in stable air.

What conditions are necessary to bring about a stable atmosphere? As we have just seen, the atmosphere is stable when the environmental lapse rate is small; that is, when the difference in temperature between the surface air and the air aloft is relatively small. Consequently, the atmosphere tends to become more stable—that is, it stabilizes—as the air aloft warms or the surface air cools. If the air aloft is being replaced by warmer air brought in by the wind (warm advection), and the surface air is not changing appreciably, the environmental lapse rate decreases and the atmosphere becomes more stable. Similarly, the environmental lapse rate decreases and the

atmosphere becomes more stable when the lower layer cools (see ● Figure 6.4). The *cooling* of the *surface* air may be due to

1. nighttime radiative cooling of the surface
2. an influx of cold surface air brought in by the wind (cold advection)
3. air moving over a cold surface

Consequently, on any given day, the atmosphere is usually most stable in the early morning around sunrise, when the lowest surface air temperature is recorded. If the surface air becomes saturated in a stable atmosphere, a persistent layer of haze or fog may form (see ● Figure 6.5).

Another way the atmosphere becomes more stable is when an entire layer of air sinks. For example, if a layer of unsaturated air over 1000 m thick and covering a large area subsides, the entire layer will warm by adiabatic compression. As the layer subsides, it becomes compressed by the weight of the atmosphere and shrinks vertically. The upper part of the layer sinks farther and, hence, warms more than the bottom part. This phenomenon is illustrated in ● Figure 6.6. After subsiding, the top of the layer is actually warmer than the bottom, and an inversion* is formed. Inversions that form as air slowly sinks

---

*An inversion represents an atmospheric condition where the air becomes warmer with height. See Chapter 3 for more about inversions.

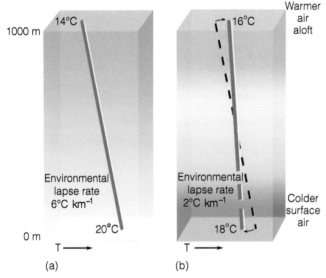

● FIGURE 6.4 The initial environmental lapse rate in diagram (a) will become more stable (stabilize) as the air aloft warms and the surface air cools, as illustrated in diagram (b).

over a large area are called **subsidence inversions**. They sometimes occur at the surface, but more frequently, they are observed aloft and are often associated with large high-pressure areas because of the sinking air motions associated with these systems. During the winter, a large high-pressure area called the "Canadian High" often dominates weather over the central part of North America, resulting in frequent subsidence inversions.

An inversion represents an atmosphere that is absolutely stable. Why? Within the inversion, warm air overlies cold air, and if air rises into the inversion, it is becoming colder, whereas the air around it is getting warmer. Obviously, the colder air would tend to sink. Inversions, therefore, are very stable air layers that act as lids on vertical air motion. When an inversion exists near the ground, stratus clouds, fog, haze, and pollutants

are all kept close to the surface. In fact, as we will see in Chapter 18, most air pollution episodes occur with subsidence inversions. (For additional information on subsidence inversions, read Focus on a Special Topic: Subsidence Inversions—Put a Lid on It on p. 171.)

Another common form of inversion is a nocturnal **radiation inversion** that is discussed in Chapter 3 on p. 77. These shallow, surface-based inversion layers (usually less than 100 m thick) form overnight due to radiative cooling of the surface, especially when winds are calm and skies are clear.

Before we turn our attention to unstable air, let's first examine a condition known as **neutral stability**. If the environmental lapse rate is exactly equal to the dry adiabatic lapse rate, rising or sinking unsaturated air will cool or warm at the same rate as the air around it. At each level, it would have the same temperature and density as the surrounding air. Because this air tends neither to continue rising nor sinking, the atmosphere is said to be neutrally stable. For saturated air, *neutral stability* exists when the environmental lapse rate is equal to the saturated adiabatic lapse rate.

AN UNSTABLE ATMOSPHERE Suppose a radiosonde sends back the temperatures above Earth as plotted in ● Figure 6.7a. Once again, we determine the atmosphere's stability by comparing the environmental lapse rate to the saturation and

● FIGURE 6.5 Cold surface air, on this morning, produces a stable atmosphere that inhibits vertical air motions and allows the fog and haze to linger close to the ground.

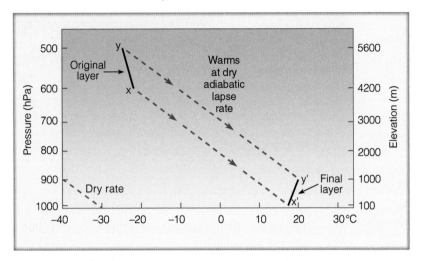

● FIGURE 6.6 The layer x–y is initially 1400 m thick. If the entire layer slowly subsides, it shrinks in the more dense air near the surface. As a result of the shrinking, the top of the layer lowers more and hence warms more than the bottom, and the entire layer (x′ –y′) becomes more stable, and in this example forms an inversion called a subsidence inversion.

dry adiabatic lapse rates. In this case, the environmental lapse rate is 11°C per 1000 m. A rising parcel of unsaturated surface air will cool at the dry adiabatic lapse rate. Because the dry adiabatic lapse rate is less than the environmental lapse rate, the parcel will be warmer than the surrounding air and so it will be buoyant, like a hot-air balloon, and continue to rise, constantly moving upward, away from its original position. The atmosphere is unstable. Of course, a parcel of saturated air cooling at the lower saturated adiabatic lapse rate will be even warmer than the air around it (see Figure 6.7b). In both cases, the air parcels, once they start upward, will continue to rise on their own because the rising air parcels are warmer and

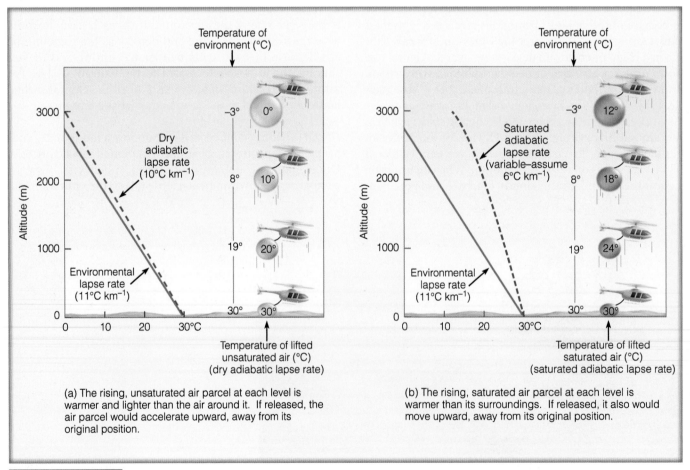

(a) The rising, unsaturated air parcel at each level is warmer and lighter than the air around it. If released, the air parcel would accelerate upward, away from its original position.

(b) The rising, saturated air parcel at each level is warmer than its surroundings. If released, it also would move upward, away from its original position.

ACTIVE FIGURE 6.7 An absolutely unstable atmosphere occurs when the environmental lapse rate is greater than the dry adiabatic lapse rate. In an unstable atmosphere, a rising air parcel will continue to rise because it is warmer and less dense than the air surrounding it. Visit the textbook's website to view this and other Active Figures at www.ahrensmeteorology1ce.nelson.com

FOCUS ON A SPECIAL TOPIC

Subsidence Inversions—Put a Lid on It

● Figure 1 shows a typical summertime vertical profile of air temperature and dew point along North America's West Coast. Notice that the air temperature decreases from the surface up to an altitude of about 300 m. Notice also that where the air temperature reaches the dew point, a cloud forms.

Above about 300 m, the air temperature increases rapidly up to an altitude near 900 m. This region of increasing air temperature with increasing height marks the region of the subsidence inversion. Within the inversion, air from aloft warms by compression. The sinking air at the top of the inversion is not only warm (about 24°C) but also dry, with a low relative humidity, as indicated by the large spread between air temperature and dew point. The subsiding air, which does not reach the surface, is associated with a large high-pressure area, called the Pacific high, located over the eastern Pacific Ocean in summer.

Immediately below the base of the inversion lies cool, moist air. The cool air is unable to penetrate the inversion because a lifted parcel of cool, marine air within the inversion would be much colder and heavier than the air surrounding it. Since the colder air parcel would fall back to its original position, the atmosphere is absolutely stable within the inversion. The subsidence inversion, therefore, acts as a lid on the air below, preventing the air from mixing vertically into the inversion. And so the marine air with its pollution

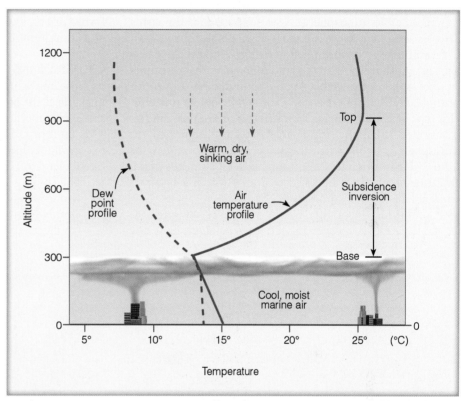

● FIGURE 1 A summertime strong subsidence inversion along the West Coast of North America. The base of the stable inversion acts as a cap or lid on the cool, marine air below. An air parcel rising into the inversion layer would sink back to its original level because the rising air parcel would be colder and more dense than the air surrounding it.

and clouds is confined to a relatively shallow region near Earth's surface. It is this trapping of air near the surface, associated with a strong subsidence inversion, that worsens air pollution problems in West Coast cities such as Vancouver, British Columbia, and Los Angeles, California.

---

less dense than the air around them. The atmosphere in this example is said to be **absolutely unstable**.* *Absolute instability occurs when the environmental lapse rate is greater than the dry adiabatic lapse rate.*

It should be noted, however, that deep layers in the atmosphere are seldom, if ever, absolutely unstable. This is because if such layers existed, convection and mixing would tend to move warm air up and cold air down. In this process, an

unstable profile is destroyed and becomes neutral over time. Therefore, absolute instability is usually limited to a very shallow layer near the ground on hot, sunny days because continual heating of the surface by solar radiation maintains the hot air temperature over the surface even though convection and thermals are mixing it upward. Here the environmental lapse rate can exceed the dry adiabatic lapse rate, and the lapse rate is called *superadiabatic*.

So far, we have seen that the atmosphere is absolutely stable when the environmental lapse rate is less than the saturated adiabatic lapse rate and absolutely unstable when the environmental lapse rate is greater than the dry adiabatic lapse rate. However, a typical type of atmospheric instability

---

*When an air parcel is warmer (less dense) than the air surrounding it, there is an upward-directed force (called a *buoyant force*) acting on it. The warmer the air parcel compared to its surroundings, the greater the buoyant force and the more rapidly the air rises.

exists when the lapse rate lies between the saturation and dry adiabatic lapse rates.

A CONDITIONALLY UNSTABLE ATMOSPHERE

The environmental lapse rate in • Figure 6.8 is 7°C per 1000 m. When a parcel of unsaturated air rises, it cools dry adiabatically and is colder at each level than the air around it (see Figure 6.8a). It will, therefore, tend to sink back to its original level because it is in a stable atmosphere. Now suppose that the rising parcel is saturated. As we can see in Figure 6.8b, the rising air is warmer than its environment at each level. Once the parcel is given a push upward, it will tend to move in that direction; the atmosphere is unstable for the saturated parcel. In this example, the atmosphere is said to be **conditionally unstable**. This type of stability depends on whether or not the rising air is saturated. When the rising parcel of air is unsaturated, the atmosphere is stable; when the parcel of air is saturated, the atmosphere is unstable. Conditional instability means that if unsaturated air could be lifted to a level where it becomes saturated, instability would result.

Conditional instability occurs whenever the environmental lapse rate is between the saturated adiabatic lapse rate and the dry adiabatic lapse rate. Recall from Chapter 1 that the average lapse rate in the troposphere is about 6.5°C per 1000 m. Since this value lies between the dry adiabatic lapse rate and the average saturated adiabatic lapse rate, the atmosphere is ordinarily in a state of conditional instability. (• Figure 6.9 summarizes the concept of unstable, conditionally unstable, and stable atmospheres.) When clouds form in a conditionally unstable atmosphere, they tend to be cumuliform, with types such as cirrocumulus, altocumulus, cumulus, and cumulonimbus.

CAUSES OF INSTABILITY

What causes the atmosphere to become more unstable? The atmosphere becomes more unstable as the environmental lapse rate steepens—that is, as the air temperature drops rapidly with increasing height. This circumstance may be brought on by either air aloft becoming colder or the surface air becoming warmer (see • Figure 6.10).

The cooling of the air aloft may be due to
1. winds bringing in colder air (cold advection)
2. clouds (or the air) emitting infrared radiation to space (radiational cooling)

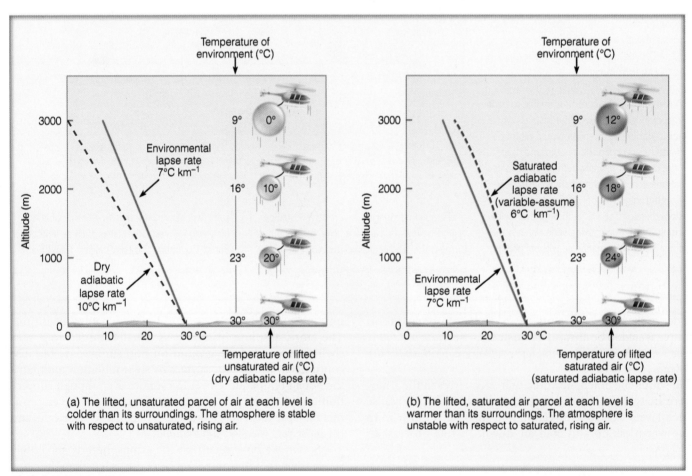

(a) The lifted, unsaturated parcel of air at each level is colder than its surroundings. The atmosphere is stable with respect to unsaturated, rising air.

(b) The lifted, saturated air parcel at each level is warmer than its surroundings. The atmosphere is unstable with respect to saturated, rising air.

ACTIVE FIGURE 6.8 Conditionally unstable atmosphere. The atmosphere is stable if the rising air is *unsaturated* (a) but *unstable* if the rising air is *saturated* (b). A conditionally unstable atmosphere occurs when the environmental lapse rate is between the saturated adiabatic lapse rate and the dry adiabatic lapse rate. Visit the textbook's website to view this and other Active Figures at www.ahrensmeteorology1ce.nelson.com

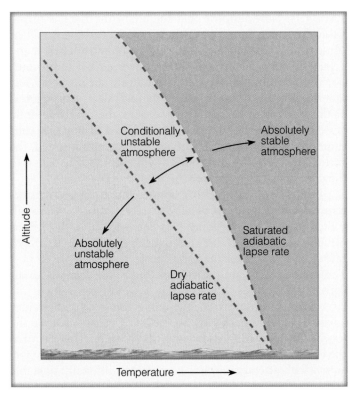

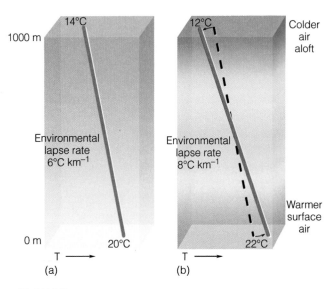

● FIGURE 6.9 When the environmental lapse rate is greater than the dry adiabatic lapse rate, the atmosphere is absolutely unstable. When the environmental lapse rate is less than the saturated adiabatic lapse rate, the atmosphere is absolutely stable. And when the environmental lapse rate lies between the dry adiabatic lapse rate and the saturated adiabatic lapse rate (shaded green area), the atmosphere is conditionally unstable.

● FIGURE 6.10 The initial environmental lapse rate in diagram (a) will become more unstable (i.e., destabilize) as the air aloft cools and the surface air warms, as illustrated in diagram (b).

The warming of the surface air may be due to
1. daytime solar heating of the surface
2. an influx of warm air brought in by the wind (warm advection)
3. air moving over a warm surface

The combination of cold air aloft and warm surface air can produce a steep lapse rate and atmospheric instability (see ● Figure 6.11).

At this point, we can see that the stability of the atmosphere changes during the course of a day. In clear, calm weather

● FIGURE 6.11 The warmth from this forest fire in Jasper National Park, Alberta, heats the air, causing instability near the surface. Warm, less dense air (and smoke) bubbles upward, expanding and cooling as it rises. Eventually, the rising air cools to its dew point, condensation begins, and a cumulus cloud forms. This type of cloud, initiated by a fire, is called a *pyrocumulus* cloud.

Todd Gipstein/Getty Images

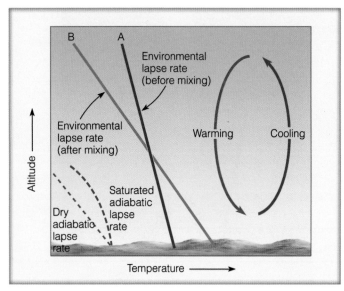

● FIGURE 6.12 Mixing a stable layer tends to increase the lapse rate. Rising, cooling air lowers the temperature toward the top of the layer, whereas sinking, warming air increases the temperature near the bottom. Over time, the environmental lapse rate becomes similar to the dry adiabatic lapse rate (B).

around sunrise, surface air is normally colder than the air above it, a radiation inversion exists, and the atmosphere is quite stable, as indicated by smoke or haze lingering close to the ground. As the day progresses, sunlight warms the surface and the surface warms the air above. As the air temperature near the ground increases, the lower atmosphere gradually becomes more unstable—that is, it *destabilizes*—with maximum instability usually occurring during the hottest part of the day.

Up until now, we have seen that a layer of air may become more unstable by either cooling the air aloft or warming the air at the surface. A layer of air may also become less stable by either mixing or lifting. Let's look at mixing first. In ● Figure 6.12, the environmental lapse rate before mixing is less than the saturated adiabatic lapse rate, and the layer is

stable (A). Now suppose that the air in the layer is mixed either by convection or by wind-induced turbulent eddies. Air is cooled adiabatically as it is brought up from below and heated adiabatically as it is mixed downward. The up–and-down motion in the layer redistributes the air in such a way that the temperature at the top of the layer decreases, whereas at the base, it increases. This steepens the environmental lapse rate and makes the layer less stable. If this mixing continues for some time, and the air remains unsaturated, the vertical temperature distribution will eventually be equal to the dry adiabatic lapse rate (B).

Just as lowering an entire layer of unsaturated air makes it more stable, the lifting of a layer makes it more unstable. In ● Figure 6.13, the air lying between 1000 and 900 hPa is initially absolutely stable because the environmental lapse rate of layer x–y is less than the saturated adiabatic lapse rate. The layer is lifted, and as it rises, the rapid decrease in air density aloft causes the layer to stretch out vertically. If the layer remains unsaturated, the entire layer cools at the dry adiabatic lapse rate. Due to the stretching effect, however, the top of the layer cools more than the bottom. This steepens the environmental lapse rate. Note that the absolutely stable layer x′–y′, after rising, has become conditionally unstable between 500 and 600 hPa (layer x′–y′).

A very stable air layer may be converted into an absolutely unstable layer when the lower portion of a layer is moist and the upper portion is quite dry. In ● Figure 6.14, the inversion layer between 900 and 850 hPa is absolutely stable. Suppose that the bottom of the layer is saturated, whereas the air at the top is unsaturated. If the layer is forced to rise, even a little, the upper portion of the layer cools at the dry adiabatic lapse rate and grows cold quite rapidly, whereas the air near the bottom cools more slowly at the saturated adiabatic lapse rate. It does not take much lifting before the upper part of the layer is much colder than the bottom part; the environmental lapse rate steepens, and the entire layer becomes absolutely unstable (layer a′–b′). The instability, brought about by the lifting of a stable layer whose surface is humid and whose top

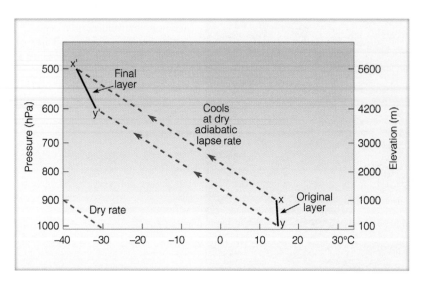

● FIGURE 6.13 The lifting of an entire layer of unsaturated air tends to increase the instability of the layer. The initial stable layer (x − y) after lifting is now a conditionally unstable layer (x′ − y′).

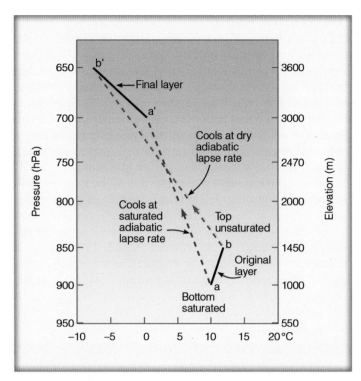

● FIGURE 6.14 Convective instability (also called *potential instability*). The layer a–b is initially absolutely stable. The lower part of the layer is saturated, and the upper part is "dry." After lifting, the entire layer (a′−b′) becomes absolutely unstable.

is "dry," is called *convective instability* or *potential instability*. Convective instability is associated with the development of severe storms, such as thunderstorms and tornadoes, which are investigated more thoroughly in Chapter 14.

BRIEF REVIEW

Up to now, we have looked briefly at stability as it relates to cloud development. The next section describes how atmospheric stability influences the physical mechanisms responsible for the development of individual cloud types. However, before going on, here is a brief review of some of the facts and concepts concerning stability:

● The air temperature in a rising parcel of unsaturated air decreases at the dry adiabatic lapse rate, whereas the air temperature in a rising parcel of saturated air decreases at the saturated adiabatic lapse rate.

● The dry adiabatic lapse rate and the saturated adiabatic lapse rate are different because latent heat is released in a rising parcel of saturated air, and this latent heat warms the air, partly offsetting the adiabatic cooling due to expansion as a parcel moves upward to lower pressure.

● In a stable atmosphere, a lifted parcel of air will be colder (heavier) than the air surrounding it. Because of this fact, the lifted parcel will tend to sink back to its original position.

● In an unstable atmosphere, a lifted parcel of air will be warmer (lighter) than the air surrounding it and thus will continue to rise upward, away from its original position.

● The atmosphere becomes more stable (stabilizes) as the surface air cools, the air aloft warms, or a layer of air sinks (subsides) over a vast area.

● The atmosphere becomes more unstable (destabilizes) as the surface air warms, the air aloft cools, or a layer of air is either mixed or lifted.

● A conditionally unstable atmosphere exists when the environmental lapse rate is between the saturated adiabatic lapse rate and the dry adiabatic lapse rate.

● The atmosphere is normally most stable in the early morning and most unstable in the afternoon.

● Stratiform clouds tend to form in a stable atmosphere, whereas cumuliform clouds tend to form in a conditionally unstable atmosphere.

Cloud Development

We know that clouds require rising air currents to exist. Why is this? There are two main reasons:

1. Rising air results in adiabatic cooling, which brings the air closer to saturation so that condensation can occur.
2. Clouds are made up of water droplets and ice crystals. Gravity "pulls" these particles toward Earth's surface. Each of these particles will therefore fall downward until the force of gravity pulling them down is balanced by the upward drag on the particle due to air friction as it falls. When these two forces are in balance, the particle has reached its downward **terminal velocity**.* So upward air currents are needed to counteract the terminal velocity if the particles in the cloud are to remain in the atmosphere and not fall below the cloud, where they can evaporate.

---

*The terminal velocity of cloud droplets is small and depends on their size. A typical cloud droplet falls at a rate of about 1 cm s⁻¹. Look ahead to Table 7.1 on p. 194 in Chapter 7 for terminal velocities of different-sized particles.

What is it, then, that causes the air to rise so that clouds are able to form? Basically, the following mechanisms shown in ● Figure 6.15, are responsible for the development of the majority of clouds we observe:

1. surface heating and free convection
2. uplift along topography
3. ascent due to horizontal convergence of surface air (covered in Chapter 12)
4. uplift along weather fronts (covered in Chapter 11)

The first mechanism that can cause the air to rise is convection. Although we briefly looked at convection in Chapter 2 when we examined rising thermals and how they transfer heat upward into the atmosphere, we will now look at convection from a slightly different perspective—how rising thermals are able to form into cumulus clouds. After that, we will see how topography influences clouds.

CONVECTION AND CLOUDS Some areas of Earth's surface are better absorbers of sunlight than others and, therefore, heat up more quickly. The air in contact with these "hot spots" becomes warmer than its surroundings. A hot "bubble" of air—a *thermal*—breaks away from the warm surface and rises, expanding and cooling as it ascends. As the thermal rises, it mixes with the cooler, drier air around it and gradually loses its identity. Its upward movement now slows. Frequently, before it is completely diluted, subsequent rising thermals penetrate it and help the air rise a little higher. If the rising air cools to its saturation point, the moisture will condense, and the thermal becomes visible to us as a cumulus cloud.

Observe in ● Figure 6.16 that the air motions are downward on the outside of the cumulus cloud. The downward motions are caused in part by evaporation around the outer edge of the cloud, which cools the air, making it heavy.

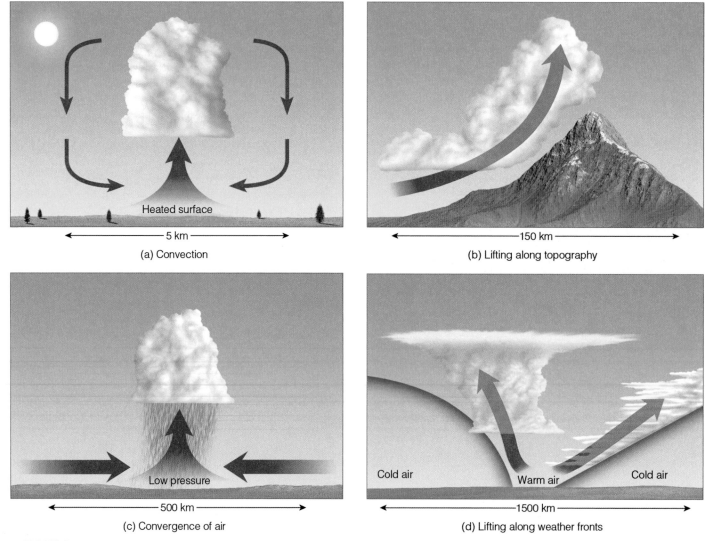

(a) Convection — 5 km

(b) Lifting along topography — 150 km

(c) Convergence of air — 500 km

(d) Lifting along weather fronts — 1500 km

● FIGURE 6.15 The primary ways clouds form: (a) surface heating and convection; (b) forced lifting along topographic barriers; (c) convergence of surface air; and (d) forced lifting along weather fronts.

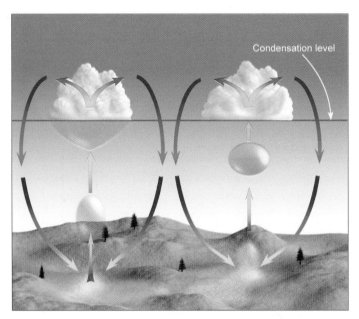

• FIGURE 6.16 Cumulus clouds form as hot, invisible air bubbles detach themselves from the surface and then rise and cool to the condensation level. Below and within the cumulus clouds, the air is rising. Around the cloud, the air is sinking.

• FIGURE 6.17 Cumulus clouds building on a warm summer afternoon. Each cloud represents a region where thermals are rising from the surface. The clear areas between the clouds are regions where the air is sinking.

Another reason for the downward motion is the completion of the convection current started by the thermal. Cool air slowly descends to replace the rising warm air. Therefore, we have rising air in the cloud and sinking air around it. Since subsiding air greatly inhibits the growth of thermals beneath it, small cumulus clouds usually have a great deal of blue sky between them (see • Figure 6.17).

As the cumulus clouds grow, they shade the ground from the sun. This, of course, cuts off surface heating and upward convection. Without the continual supply of rising air, the cloud begins to erode as its droplets evaporate. Unlike the sharp outline of a growing cumulus, the cloud now has indistinct edges, with cloud fragments extending from its sides. As the cloud dissipates (or moves along with the wind), surface heating begins again and regenerates another thermal, which becomes a new cumulus. This is why you often see cumulus clouds form, gradually disappear, and then reform in the same spot.

Suppose that it is a warm, humid summer afternoon and the sky is full of cumulus clouds. The cloud bases are all at nearly the same level above the ground and the cloud tops extend only about a thousand metres higher. The development of these clouds depends primarily on the air's stability and moisture content. To illustrate how these factors influence the formation of a convective cloud, we will examine the temperature and moisture characteristics within a rising bubble of air. Since the actual air motions that go into forming a cloud are rather complex, we will simplify matters by making these assumptions:

1. No mixing takes place between the rising air and its surroundings.
2. Only a single thermal produces the cumulus cloud.

3. The cloud forms when the relative humidity is 100 percent.
4. The rising air in the cloud remains saturated.

The environmental lapse rate on this particular day is plotted in • Figure 6.18 and is represented as a dark grey line on the far left of the illustration. The changing environmental air temperature indicates changes in the atmosphere's stability. The environmental lapse rate in layer A is greater than the dry adiabatic lapse rate, so the layer is absolutely unstable. The air layers above it—layer B and layer C—are both absolutely stable because the environmental lapse rate in each layer is less than the saturated adiabatic lapse rate. However, the overall environmental lapse rate from the surface up to the base of the inversion (2000 m) is 7.5°C per 1000 m, which indicates a conditionally unstable atmosphere.

Now suppose that a warm bubble of air with an air temperature of 35°C and a dew point temperature of 27°C breaks away from the surface and begins to rise (which is illustrated in the middle of Figure 6.18). Notice that a short distance above the ground, the air inside the bubble is warmer than the air around it, so it is buoyant and rises freely. This level in the atmosphere where the rising air becomes warmer than the surrounding air is called the *level of free convection*. The rising bubble will continue to rise as long as it is warmer than the air surrounding it.

The rising air cools at the dry adiabatic lapse rate and the dew point falls, but not as rapidly.* The rate at which the dew

---

*The decrease in dew point temperature is caused by the decrease in air pressure within the rising air. As the air pressure drops, so does the actual vapour pressure. Since the dew point is directly related to the actual vapour pressure, the dew point temperature decreases in the rising air.

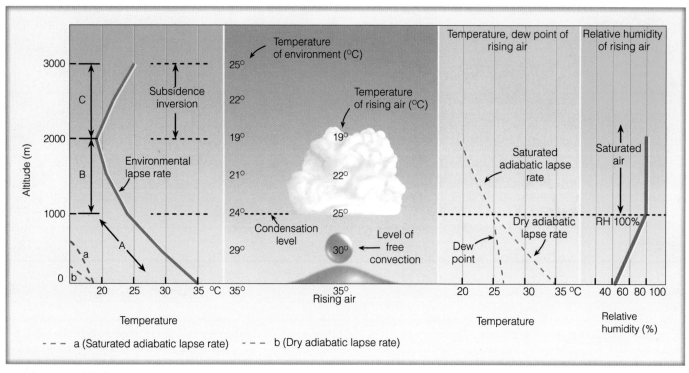

ACTIVE FIGURE 6.18 The development of a cumulus cloud. Visit the textbook's website to view this and other Active Figures at www.ahrensmeteorology1ce.nelson.com

point drops varies with the moisture content of the rising air, but an approximation of 2°C per 1000 m is commonly used. So as unsaturated rising air cools, the air temperature and dew point approach each other at the rate of 8°C per 1000 m. This process causes an increase in the air's relative humidity (illustrated in the far right-hand side of Figure 6.18 by the dark green line).

At an elevation of 1000 m, the air has cooled to the dew point, the relative humidity is 100 percent, condensation begins, and a cloud forms. The elevation where the cloud forms is called the **condensation level**. Above the condensation level, the rising air is saturated and cools at the saturated adiabatic lapse rate. Condensation continues to occur, and because water vapour is transforming into liquid cloud droplets, the dew point within the cloud now drops more rapidly with increasing height than before. The air remains saturated as both the air temperature and dew point decrease at the saturated adiabatic lapse rate (illustrated in the area of Figure 6.18 shaded tan).

Notice that inside the cloud, the rising air remains warmer than the environment and continues its spontaneous rise upward through layer B. The top of the bulging cloud at 2000 m represents the top of the rising air, which has now cooled to a temperature equal to its surroundings. The air would have a difficult time rising much above this level because of the stable subsidence inversion directly above it. The subsidence inversion, associated with the

downward air motions of a high-pressure system, prevents the clouds from building very high above their bases. Hence, an afternoon sky full of flat-base cumuli with little vertical growth indicates fair weather. (Recall from Chapter 5 that the proper name of these fair-weather cumulus clouds is *cumulus humilis.*)

As we can see, the stability of the air above the condensation level plays a major role in determining the vertical growth of a cumulus cloud. Notice in • Figure 6.19 that when a deep, stable layer begins a short distance above the cloud base, only cumulus humilis are able to form. If a deep, conditionally unstable layer exists above the cloud base, cumulus congestus are likely to grow, with billowing, cauliflower-like tops. When the conditionally unstable layer is extremely deep—usually greater than 4 km—the cumulus congestus may even develop into a cumulonimbus.

Seldom do cumulonimbus clouds extend very far above the tropopause. The stratosphere is quite stable, so once a cloud penetrates the tropopause, it usually stops growing vertically and spreads horizontally. The low temperature at this altitude produces ice crystals in the upper section of the cloud. In the middle latitudes, high winds near the tropopause blow the ice crystals laterally, producing the flat, anvil-shaped top so characteristic of cumulonimbus clouds (see • Figure 6.20).

The vertical development of a convective cloud also depends on the mixing that takes place around its periphery.

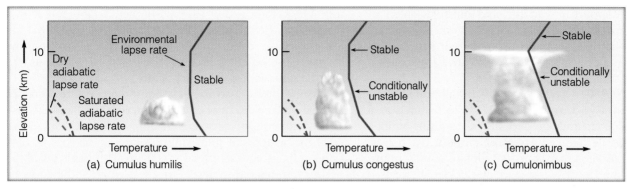

● FIGURE 6.19 The air's stability greatly influences the growth of cumulus clouds.

The rising, churning cloud mixes cooler air into it. Such mixing is called **entrainment**. If the environment around the cloud is very dry, the cloud droplets quickly evaporate. The effect of entrainment, then, is to increase the rate at which the rising air cools by the injection of cooler air into the cloud and the subsequent evaporation of the cloud droplets. If the rate of cooling approaches the dry adiabatic lapse rate, the air stops rising and the cloud no longer builds, even though the lapse rate may indicate a conditionally unstable atmosphere.

Up to now, we have looked at convection over land. Convection and the development of cumulus clouds also occur over large bodies of water. As cool air flows over a body of relatively warm water, the lowest layer of the atmosphere becomes warm and moist. This induces instability—convection begins and cumulus clouds form. If the air moves over progressively warmer water, as is sometimes the case over the open ocean, more active convection occurs and a cumulus cloud can build into cumulus congestus and finally into cumulonimbus. This sequence of cloud development is observed from satellites as cold, northerly winds move southward over the northern portions of the Atlantic and Pacific oceans (see ● Figure 6.21).

Once a convective cloud forms, stability, humidity, and entrainment all play a part in its vertical development. The level at which the cloud initially forms, however, is determined

● FIGURE 6.20 Cumulus clouds developing into thunderstorms in a conditionally unstable atmosphere over the Great Plains. Notice that, in the distance, the cumulonimbus with the anvil top has reached the stable part of the atmosphere.

● FIGURE 6.21 Satellite views of stratocumulus clouds forming in rows over the Atlantic Ocean as cold, dry arctic air sweeps over Canada and then out over warmer water offshore from Nova Scotia. Notice that the clouds are absent over the landmass and directly along the coast but form and gradually thicken as the surface air warms and destabilizes farther offshore.

primarily by the surface temperature and moisture content of the original thermals. (Focus on an Observation: Determining Convective Cloud Bases on p. 181 uses this information and a simple formula to determine the bases of convective clouds.)

TOPOGRAPHY AND CLOUDS   Horizontally moving air obviously cannot go through a large obstacle, such as a mountain, so the air must go over it. Forced lifting along a topographic barrier is called **orographic uplift**. Often large masses of air rise when they approach a long chain of mountains such as the Coast Mountains or Rockies of western North America. This lifting produces cooling, and if the air is humid, clouds form. Clouds produced in this manner are called *orographic clouds*. The type of cloud that forms will depend on the air's stability and moisture content. On the leeward (downwind) side of the mountain, as the air moves downhill, it warms. This sinking air is now drier because much of its moisture was removed by precipitation on the windward (upwind) side. This region on the leeward side, where precipitation is noticeably less, is called a **rain shadow**. Rain shadows are common features downwind of mountains and in Canada are especially notable east of the south Coast Mountains in British Columbia and east of the Rocky Mountains in southern Alberta.

An example of orographic uplift and cloud development is given in ● Figure 6.22. Before rising up and over the barrier, the air at the base of the mountain (0 m) on the windward side has an air temperature of 20°C and a dew point temperature of 12°C. Notice that the atmosphere is conditionally

unstable, as indicated by the environmental lapse rate of 8°C per 1000 m. (Remember from our earlier discussion that the atmosphere is conditionally unstable when the environmental lapse rate falls between the dry adiabatic lapse rate and the saturated adiabatic lapse rate.)

As the unsaturated air rises, the air temperature decreases at the dry adiabatic lapse rate (10°C per 1000 m) and the dew point temperature decreases at approximately 1.85°C per 1000 m. Notice that the rising, cooling air reaches its dew point and becomes saturated at 1000 m. This level (called the **lifting condensation level**, or **LCL**) marks the base of the cloud that has formed as air is lifted (in this case, by the mountain). As the rising saturated air condenses into many billions of liquid cloud droplets, and as latent heat is liberated by the condensing vapour, both the air temperature and dew point temperature decrease at the saturated adiabatic lapse rate—which, for this example, has an average value of 5.5°C per kilometre.

At the top of the mountain, the air temperature and dew point are both −1°C. Note in Figure 6.22 that this temperature (−1°C) is higher than that of the surrounding air (−4°C). Consequently, the rising air at this level is not only warmer but also unstable with respect to its surroundings. Therefore, the rising air should continue to rise and build into a much larger cumuliform cloud.

Suppose, however, that the air at the top of the mountain (temperature and dew point of −1°C) is forced to descend to the base of the mountain (0 m) on the leeward side. If we assume that the cloud remains on the windward side and

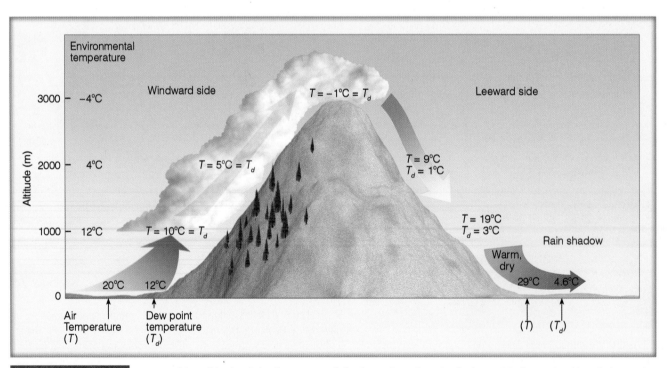

ACTIVE FIGURE 6.22   Orographic uplift, cloud development, and the formation of a rain shadow. Visit the textbook's website to view this and other Active Figures at www.ahrensmeteorology1ce.nelson.com

FOCUS ON AN OBSERVATION

Determining Convective Cloud Bases

The bases of cumulus clouds that form by convection on warm, sunny afternoons can be estimated quite easily when the surface air temperature and dew point are known. If it is not too windy, we can assume that entrainment of air will not change the characteristics of a rising thermal. Since the rising air cools at the dry adiabatic lapse rate of about 10°C per 1000 m, and the dew point drops at about 2°C per 1000 m,* the air temperature and dew point approach each other at the rate of 8°C for every 1000 m of rise. Rising surface air with an air temperature and dew point spread of 8°C would produce saturation and a cloud at an elevation of 1000 m. Put another way, a 1°C difference between the surface air temperature and the dew point produces a cloud base at 125 m. Therefore, by finding the difference between surface air temperature (T) and dew point (T_d) and multiplying this value by 125, we can estimate the base of the convective cloud forming overhead as $H = 125\,(T - T_d)$,[†] where H is the height of the base of the cumulus cloud in metres above the surface, with both T and T_d measured in degrees Celsius.

To illustrate the use of the formula, let's determine the base of the cumulus cloud in Figure 6.18. Recall that the surface air temperature and dew point were 35°C and 27°C, respectively. The difference, $T - T_d$, is 8°C. This value multiplied by 125 gives us a cumulus cloud with a base at 1000 m above the ground. This agrees

---

*More exactly, the dry adiabatic lapse rate is 9.8°C km⁻¹ and the dew point lapse rate is closer to 1.85°C km⁻¹. However, the difference between the lapse rates is still near 8°C km⁻¹.

[†]The formula works best when the air is well mixed from the surface up to the cloud base, such as in the afternoon on a sunny day. The formula does not work well in mountains, at night, or in the early morning.

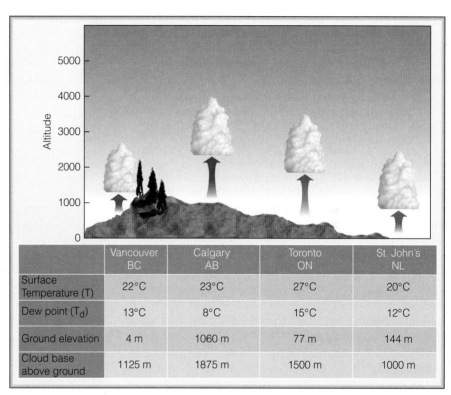

	Vancouver BC	Calgary AB	Toronto ON	St. John's NL
Surface Temperature (T)	22°C	23°C	27°C	20°C
Dew point (T$_d$)	13°C	8°C	15°C	12°C
Ground elevation	4 m	1060 m	77 m	144 m
Cloud base above ground	1125 m	1875 m	1500 m	1000 m

● FIGURE 2 During the summer, cumulus cloud bases typically increase in elevation above the ground as one moves westward from the Maritime provinces to the Prairies.

with the condensation level we originally calculated.

Along the Maritime provinces in summer, when the air is warm and muggy, the separation between air temperature and dew point may be 8°C or less. The bases of afternoon cumulus clouds over cities, such as St. John's and Halifax, are typically about 1000 m above the surface (see ● Figure 2). In Quebec and Ontario, the moisture content is even higher; however, so are the air temperatures, so the cumulus cloud bases will also be higher above ground. In the Prairie provinces, the air becomes much drier and the spread between the surface air temperature and the dew point even greater, and we find the highest cloud

bases in Canada. For example, in Calgary, the average daily high temperature in July is around 23°C, with an average dew point of only 8°C, resulting in cloud bases that are 1875 m above ground on average. On the West Coast, cooler air temperatures and more moisture result in low cloud bases, similar to the situation on the East Coast. Other places in North America, such as the Central Valley of California, where the summer afternoon spread between air temperature and dew point usually exceeds 22°C, the air must rise to almost 2700 m before a cloud forms. Due to sinking air aloft, thermals in this area are unable to rise to that elevation, and afternoon cumulus clouds are seldom observed forming overhead there.

does not extend beyond the mountain top, the temperature of the sinking air will increase at the dry adiabatic lapse rate (10°C per 1000 m) all the way down to the base of the mountain. (The dew point temperature increases at a much lower rate of about 1.85°C per 1000 m.)

We can see in Figure 6.22 that on the leeward side, after descending 3000 m, the air temperature is 29°C and the dew point temperature is 4.6°C. The air is now 9°C warmer than it was before being lifted over the barrier. The higher air temperature on the leeward side is the result of latent heat being converted into sensible heat during condensation on the windward side. (In fact, the rising air at the top of the mountain is considerably warmer than it would have been had condensation not occurred.) The lower dew point temperature and, hence, drier air on the leeward side are the result of water vapour condensing and then remaining as liquid cloud droplets and precipitation on the windward side.

This example illustrates the key features of the chinook wind—a warm, dry wind periodically affecting the east slopes of the Rockies from southern Alberta through to northern New Mexico (see Chapter 9, p. 275). Of course, in the actual chinook wind, some of the cloud forming on the upwind side of the mountain may spill over onto the downwind side. Also, the air on the downwind side does not descend back to sea level but to the terrain elevation near the mountains over the Prairies, which in southern Alberta is about 1000 m above sea level. And when the chinook wind starts, it often displaces cold air, so the legendary warming is due as much to the removal of the cold air as to the warm, dry, descending air. (A graphic representation of the preceding example is given in Focus on an Advanced Topic: The Tephigram, on p. 184.)

Although clouds are more prevalent on the windward side of mountains, they may, under certain atmospheric conditions, form on the leeward side as well. For example, stable air flowing over a mountain often moves in a series of waves that may extend for several hundred kilometres on the leeward side (see ● Figure 6.23). These waves resemble the waves that form in a river downstream from a large boulder. Recall from Chapter 5 that *wave clouds* often have a characteristic lens shape and are commonly called *lenticular clouds*.

The formation of lenticular clouds is shown in ● Figure 6.24. As moist air rises on the upwind side of the wave, it cools and condenses, producing a cloud. On the downwind side of the wave, the air sinks and warms; the cloud evaporates. Viewed from the ground, the clouds appear motionless as the air rushes through them; hence, they are often referred to as *standing wave clouds*. Since they most frequently form at altitudes where middle clouds form, they are called *altocumulus standing lenticularis*.

When the air between the cloud-forming layers is too dry to produce clouds, lenticular clouds will form one above the other. Actually, when a strong wind blows almost perpendicular to a high mountain range, mountain waves may extend into the stratosphere, producing a spectacular

NASA

● FIGURE 6.23 Satellite views of wave clouds over Ireland forming many kilometres downwind of the mountains in Scotland to the northeast.

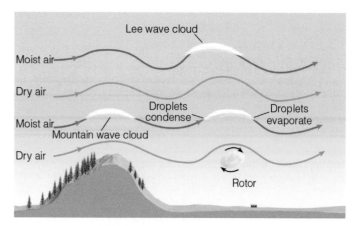

● FIGURE 6.24 The formation of lenticular clouds. Clouds that form in the wave directly over the mountains are called *mountain wave clouds*, whereas those that form downwind of the mountain are called *lee wave clouds*.

● FIGURE 6.25 Lenticular clouds forming one on top of the other over the Sierra Nevada.

display, sometimes resembling a fleet of hovering spacecraft (see ● Figure 6.25).

Notice in Figure 6.24 that beneath the lenticular cloud downwind of the mountain range, a large, swirling eddy forms. The rising part of the eddy may cool enough to produce **rotor clouds**. The air in the rotor is extremely turbulent and presents a major hazard to aircraft in the vicinity. Dangerous flying conditions also exist near the leeside of the mountain, where strong, downwind air motions are present. (These types of winds will be treated in more detail in Chapter 9.)

Now, having examined the concept of stability and the formation of clouds, we are ready to see what role stability might play in changing a cloud from one type into another.

CHANGING CLOUD FORMS   Under certain conditions, a layer of altostratus may change into altocumulus. This happens if the top of the original cloud deck cools while the bottom warms. Because clouds are such good absorbers and emitters of infrared radiation, the top of the cloud will often cool as it radiates infrared energy to space more rapidly than it absorbs solar energy. Meanwhile, the bottom of the cloud will warm as it absorbs infrared energy from below more quickly than it radiates this energy away. This process makes the cloud layer conditionally unstable to the point that small convection cells begin within the cloud itself. The up-and-down motions in a layered cloud produce globular elements that give the cloud a lumpy appearance. The cloud forms in the rising part of a cell, and clear spaces appear where descending currents occur.*

When the wind is fairly uniform throughout a stratocumulus cloud layer with sufficient moisture that is unstable

---

*An example of an altocumulus cloud with a lumpy appearance is given in Figure 5.19 on p. 144.

● FIGURE 6.26 Satellite views of cloud streets, rows of stratocumulus clouds forming over the warm Georgia, U.S.A., landscape.

below and capped by an inversion, the horizontal axes of the convection cells may align close to the average direction of the wind. The new cloud elements may then become arranged in rows and are given the name **cloud streets** (see ● Figure 6.26). Sometimes when the wind speed and direction change strongly with height, reaching a critical value, and an inversion caps the cloud-forming layers, wavelike clouds called **billow clouds** may form along the top of the cloud layer (see ● Figure 6.27).

Occasionally, altocumulus show vertical development and produce towerlike extensions. The clouds often resemble

● FIGURE 6.27 Billow clouds forming in a region where wind speed changes rapidly with altitude. This is called a region of strong vertical *wind shear*.

FOCUS ON AN ADVANCED TOPIC

The Tephigram

Thermodynamic diagrams are valuable tools for anyone who studies the atmosphere. There are several types of thermodynamic diagrams. The tephigram is used by meteorologists in Canada, the United Kingdom, and elsewhere and is similar to other thermodynamic diagrams, such as the "skew T" diagram. The tephigram is a graph that shows how various atmospheric elements change with altitude. At first glance, the chart appears complicated because of its many lines. We will, therefore, construct these lines on the chart step by step.

    Figure 3 shows slightly curving horizontal lines of pressure, called *isobars*, decreasing with altitude with major intervals highlighted in black, and labelled in hPa on the left side. The lines of temperature in degrees Celsius, called *isotherms*, sloping and increasing toward the right are coloured in green. The height values on the right are approximate elevations above sea level every hundred hPa that have been computed assuming that the air temperature decreases at a standard rate of 6.5°C per kilometre.

    In Figure 4, the slanted red lines are called *dry adiabats*. They show how the air temperature would change inside a rising or descending *unsaturated* air parcel. Suppose, for example, that an unsaturated air parcel at the surface (pressure 1000 hPa) with a temperature of 10°C rises and cools at the dry adiabatic lapse rate (10°C per kilometre). What would be the parcel temperature at a pressure of 900 hPa? To find out, simply follow the dry adiabat from the surface temperature of 10°C up to where it crosses the 900 hPa line. Answer: about 1.5°C. If the same parcel returns to the surface, follow the dry adiabat back to the surface and read the temperature, 10°C.

    The dry adiabats are also called lines of **potential temperature**, which can be expressed either in kelvins or degrees Celsius. The potential temperature is the temperature an air parcel would have if it were moved dry adiabatically to a pressure of 1000 hPa. Moving parcels to the same level allows them to be observed under identical conditions. Thus, it can be determined which parcels are potentially warmer than others. Therefore, the potential

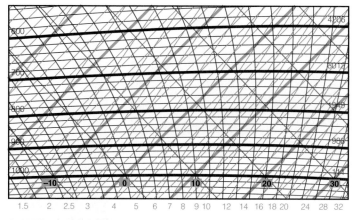
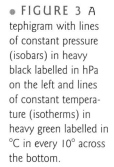

● **FIGURE 3** A tephigram with lines of constant pressure (isobars) in heavy black labelled in hPa on the left and lines of constant temperature (isotherms) in heavy green labelled in °C in every 10° across the bottom.

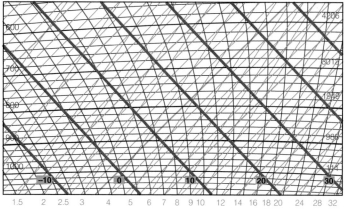

● **FIGURE 4** A tephigram with the dry adiabats (lines of constant potential temperature) in heavy red.

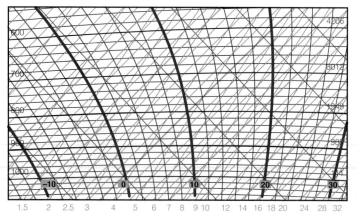

● **FIGURE 5** A tephigram with the saturated adiabats in heavy blue.

temperature of a dry adiabat can be read as the temperature value where it crosses 1000 hPa.

    The curving blue lines in ● Figure 5 are called *saturated adiabats*. They show how the air temperature would change inside a rising or descending parcel of *saturated* air. In other words, they represent the saturated adiabatic

lapse rate for a rising or sinking saturated air parcel, such as in a cloud.

    The sloping orange lines in ● Figure 6 are lines of constant *saturation mixing ratio*. At any given temperature and pressure, they show how much water vapour the air could hold if it were saturated—the *saturation mixing ratio* (r_s)

in grams of water vapour per kilogram of dry air (g kg^{-1}). At a given dew point temperature, they show how much water vapour the air is actually holding—the *actual mixing ratio* (r) in g kg^{-1}. Hence, given the air temperature and dew point temperature at some level, we can compute the relative humidity of the air.*

For example, suppose at the surface (pressure 1000 hPa) that the air temperature and dew point temperature are 30°C and 14°C, respectively. In Figure 6, observe that at 30°C, the saturation mixing ratio (r_s) is 28 g kg^{-1}, and with a dew point temperature of 14°C, the actual mixing ratio (r) is 10 g kg^{-1}. This produces a relative humidity of 10/28 × 100 percent, or 36 percent.

The mixing ratio lines also show how the dew point temperature changes in a rising or sinking unsaturated air parcel. If an unsaturated air parcel with a dew point of 14°C rises from the surface (pressure 1000 hPa) up to where the pressure is 700 hPa (approximately 3 km), notice in Figure 6 that the dew point temperature inside the parcel would have dropped to a temperature near 8.5°C.

Let's use the tephigram to see what happens to air that rises up and over a mountain range.

Suppose we use the example given in Figure 6.22 on p. 180. Air at a pressure of 1013 hPa (elevation at sea level), with a temperature of 20°C (T_1) and a dew point temperature of 12°C (T_{d1}), first ascends and then descends a 3000 m high mountain range. Look at ● Figure 7 closely and observe that the surface air with a temperature of 20°C indicates a saturation mixing ratio of about 15 g kg^{-1}, and at 12°C, the dew point temperature indicates an actual mixing ratio of about 8.8 g kg^{-1}. Hence, the relative humidity of the air before rising over the mountain is 8.8/15, or 59 percent.

Now, as the unsaturated air rises (as indicated by arrows in Figure 7), the air temperature follows a dry adiabat (red line), and the dew point temperature follows a saturation mixing ratio line (orange line). Carefully follow the

*The relative humidity (RH) of the air can be expressed as RH = r/r_s × 100%.

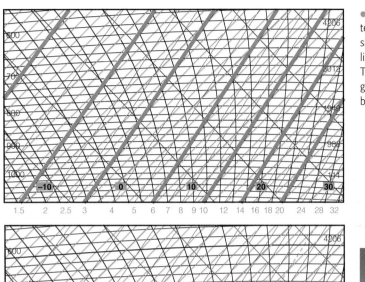

● FIGURE 6 A tephigram with the saturation mixing ratio lines in heavy orange. They are labeled in g kg^{-1} across the bottom.

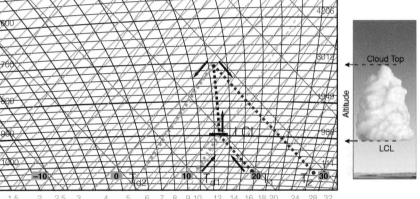

● FIGURE 7 The tephigram. The arrows illustrate the example given in the text. The cloud on the right side represents the base and height of the cloud given in the example.

saturation mixing ratio line in Figure 7 from 12°C up to where it intersects the dry adiabat that slopes upward from 20°C. Notice that the intersection occurs at a pressure slightly less than 900 hPa (an elevation near 1000 m). This, of course, marks the base of the cloud—the *lifting condensation level* (*LCL*)—where the relative humidity is 100 percent and condensation begins. Above this level, the rising air is saturated. Consequently, the air temperature and dew point temperature together follow a saturated adiabat (solid blue curved lines on the tephigram, dashed for the parcel) to the top of the mountain.

Notice in Figure 7 that at the top of the mountain (at 3 km or about 700 hPa), both the air temperature and dew point are −1°C. If we

assume that all the condensed water in the cloud stays on the windward side, then from 3 km (700 hPa), the descending air follows a dry adiabat all the way back to 1013 hPa. Notice that, after descending, the air has a temperature of 29°C (T_2). From the mountaintop, the dew point temperature follows a saturation mixing ratio line and reaches the surface (1013 hPa) with a temperature of 4.6°C (T_{d2}). Observe in Figure 7 that with an air temperature of 29°C, the saturation mixing ratio is about 25 g kg^{-1}, and with a dew point of 4.6°C, the actual mixing ratio is about 5.2 g kg^{-1} (the same value the parcel had at 700 hPa). Thus, the relative humidity of the air after descending is 5.2/25, or 21 percent. A larger blank tephigram is provided in Appendix J.

© C. Donald Ahrens

● FIGURE 6.28 An example of altocumulus castellanus.

floating castles, and for this reason, they are called *altocumulus castellanus* (see ● Figure 6.28). They form when rising currents within the cloud extend into conditionally unstable air above the cloud. The buoyancy for the rising air comes from the latent heat released during condensation within the cloud. This process can occur in cirrocumulus clouds, producing *cirrocumulus castellanus*. When altocumulus castellanus

appear, they indicate that the midlevel of the troposphere is becoming more unstable (destabilizing). This destabilization is often the precursor to shower activity. So a morning sky full of altocumulus castellanus will likely become afternoon showers and even thunderstorms.

Occasionally, the stirring of a moist layer of stable air will produce a deck of stratocumulus clouds. In ● Figure 6.29, the air is stable and close to saturation. Suppose that a strong wind mixes the layer from the surface up to an elevation of 600 m. The turbulent mixing will make the temperature profile tend toward neutral as rising air cools at the dry adiabatic lapse rate and sinking air warms at the dry adiabatic lapse rate. As the lapse rate steepens, the upper part of the layer cools and the lower part warms. At the same time, mixing will make the moisture distribution in the layer more uniform. The warmer temperature and decreased moisture content cause the lower part of the layer to dry out. On the other hand, the decrease in temperature and increase in moisture content saturate the top of the mixed layer, producing a layer of stratocumulus clouds with a lapse rate approaching the saturated adiabatic lapse rate. Figure 6.29 indicates that the air above the region of mixing is still stable and inhibits further mixing. In some cases, an inversion may actually form above the clouds. However, if the surface warms substantially, rising thermals may penetrate the stable region and the stratocumulus clouds may change into more widely separated clouds, such as cumulus or cumulus congestus. A stratocumulus layer changing to a sky dotted with growing cumulus clouds often occurs as surface heating increases on a warm, humid summer day.

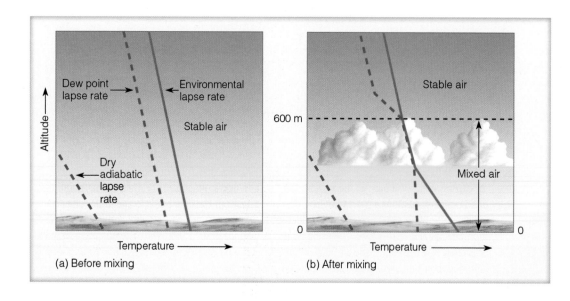

● FIGURE 6.29 The mixing of a moist layer of air near the surface can produce a deck of stratocumulus clouds.

SUMMARY

In this chapter, we tied together the concepts of stability and the formation of clouds. We learned that rising unsaturated air cools at the dry adiabatic lapse rate, and due to the release of latent heat, rising saturated air cools at the saturated adiabatic lapse rate. In a stable atmosphere, a lifted parcel of air will be colder (heavier) than the air surrounding it at each new level and will sink back to its original position. Because stable air tends to resist upward vertical motions, clouds forming in a stable atmosphere often spread horizontally and form stratiform clouds, such as cirrostratus and altostratus. A stable atmosphere may be caused by either cooling the surface air, warming the air aloft, or the sinking (subsidence) of an entire layer of air, in which case, a very stable subsidence inversion usually forms.

In an unstable atmosphere, a lifted parcel of air will be warmer (lighter) than the air surrounding it at each new level and will continue to rise upward away from its original position. In a conditionally unstable atmosphere, an unsaturated parcel of air can be lifted to a level where condensation begins, latent heat is released, and instability results as the temperature inside the rising parcel becomes warmer than the air surrounding it. In a conditionally unstable atmosphere, rising air tends to form clouds that develop vertically, such as cumulus congestus and cumulonimbus. Instability may be caused by warming the surface air, cooling the air aloft, or the lifting or mixing an entire layer of air.

On warm, humid days, the instability generated by surface heating can produce cumulus clouds at a height determined by the temperature and moisture content of the surface air. Instability may cause changes in existing clouds as convection changes an altostratus into an altocumulus. Also, mixing can change a clear day into a cloudy one.

KEY TERMS

The following terms are listed (with page numbers) in the order they appear in the text. Define each. Doing so will aid you in reviewing the material covered in this chapter.

parcel of air, 166
adiabatic process, 166
dry adiabatic lapse rate, 166
saturated adiabatic lapse rate, 166
stable air, 167
unstable air, 167
environmental lapse rate, 167
absolutely stable atmosphere, 168
stratiform, 168
subsidence inversion, 169
radiation inversion, 169
neutral stability, 169

absolutely unstable atmosphere, 171
conditionally unstable atmosphere, 172
terminal velocity, 175
condensation level, 178
entrainment, 179
orographic uplift, 180
rain shadow, 180
lifting condensation level (LCL), 180
rotor clouds, 183
cloud streets, 183
billow clouds, 183
potential temperature, 184

QUESTIONS FOR REVIEW

1. What is an adiabatic process?
2. Why are saturated and dry adiabatic lapse rates of cooling different?
3. Under what conditions would the saturated adiabatic lapse rate of cooling be almost equal to the dry adiabatic lapse rate?
4. Explain the difference between environmental lapse rate and dry adiabatic lapse rate.
5. How would one normally obtain the environmental lapse rate?
6. What is a stable atmosphere, and how can it form?
7. Describe the general characteristics of clouds associated with stable and unstable atmospheres.
8. List and explain several processes by which a stable atmosphere can be made unstable.
9. If the atmosphere is conditionally unstable, what condition is necessary to bring on instability?
10. Explain why cumulus clouds are conspicuously absent over a cool water surface.
11. Why are cumulus clouds more frequently observed during the afternoon than at night?
12. Explain why an inversion represents an absolutely stable atmosphere.
13. How and why does lifting or lowering a layer of air change its stability?
14. List and explain several processes by which an unstable atmosphere can be made stable.
15. Why do cumulonimbus clouds often have flat tops?
16. Why are there usually large spaces of blue sky between cumulus clouds?
17. List four primary ways clouds form and describe the formation of one cloud type by each method.
18. (a) Why are lenticular clouds also called standing wave clouds? (b) On which side of a mountain (windward or leeward) would lenticular clouds most likely form?
19. Explain why rain shadows form on the leeward side of mountains.
20. How can a layer of altostratus change into one of altocumulus?
21. Describe the conditions necessary to produce stratocumulus clouds by mixing.
22. Briefly describe how each of the following clouds forms:
    (a) lenticular
    (b) rotor
    (c) billow
    (d) castellanus

QUESTIONS FOR THOUGHT

1. How is it possible for a layer of air to be convectively unstable and absolutely stable at the same time?

2. Are the bases of convective clouds generally higher during the day or the night? Explain.

3. Where would be the safest place to build an airport in a mountainous region? Why?

4. Use Figure 4.14c on p. 111 (Chapter 4) to help you explain why the bases of cumulus clouds, which form from rising thermals during the summer, increase in height above the surface as you move due west of a line that runs north–south through central Manitoba.

5. To minimize air pollution, what would be the best time of day for a farmer to burn agricultural debris?

6. Suppose that surface air on the windward side of a mountain rises and descends on the leeward side. Recall from Chapter 4 that the dew point temperature is a measure of the amount of water vapour in the air. Explain, then, why the relative humidity of the descending air drops as the dew point temperature of the descending air increases.

7. Usually, when a cumulonimbus cloud begins to dissipate, the bottom half of the cloud dissipates first. Give an explanation as to why this situation might happen.

PROBLEMS AND EXERCISES

1. Under which set of conditions would a cumulus cloud base be observed at the highest level above the surface? Surface air temperatures and dew points are as follows: (a) 35°C, 14°C; (b) 30°C, 19°C; (c) 34°C, 9°C; (d) 29°C, 7°C; (e) 32°C, 6°C.

2. If the height of the base of a cumulus cloud is 1000 m above the surface, and the dew point at Earth's surface beneath the cloud is 20°C, determine the air temperature at Earth's surface beneath the cloud.

3. The condensation level over Montreal, Quebec, on a warm muggy afternoon is 800 m. If the dew point temperature of the rising air at this level is 21°C, what is the approximate dew point temperature and air temperature at the surface? Determine the surface relative humidity. (Hint: See Focus on a Special Topic: Computing Relative Humidity and Dew Point in Chapter 4 on p. 114.)

4. Suppose that the air pressure outside a conventional jet airliner flying at an altitude of 10 km is 250 hPa. Further suppose that the air inside the aircraft is pressurized to 1000 hPa. If the outside air temperature is −50°C, what would be the temperature of this air if brought inside the aircraft and compressed at the dry adiabatic lapse rate to a pressure of 1000 hPa? (Assume that a pressure of 1000 hPa is equivalent to an altitude of 0 m.)

5. In ● Figure 6.30, a radiosonde is released and sends back temperature data as shown in the diagram. (This is the environment temperature.)

(a) Calculate the environmental lapse rate from the surface up to 3000 m.

(b) What type of atmospheric stability does the sounding indicate?
Suppose that the wind is blowing from the west and a parcel of surface air with a temperature of 10°C and a dew point of 2°C begins to rise upward along the western (windward) side of the mountain.

(c) What is the relative humidity of the air parcel at 0 m (pressure 1000 hPa) before rising? (Hint: See Focus on a Special Topic: Computing Relative Humidity and Dew Point in Chapter 4 on p. 114 or use the tephigram in Appendix J.)

(d) As the air parcel rises, at approximately what elevation would condensation begin and a cloud start to form?

(e) What is the air temperature and dew point of the rising air at the base of the cloud?

(f) What is the air temperature and dew point of the rising air inside the cloud at an elevation of 3000 m? (To simplify, assume a constant saturated adiabatic lapse rate of 6°C per 1000 m.)

(g) At an altitude of 3000 m, how does the air temperature inside the cloud compare with the air temperature outside the cloud, as measured by the radiosonde? What type of atmospheric stability (stable or unstable) does this suggest? Explain.

(h) At an elevation of 3000 m, would you expect the cloud to continue to develop vertically? Explain.

6. Answer the same questions in problem 5 except, this time, use the tephigram provided in Appendix J.

(i) What would be the name of the cloud that is forming? Suppose that a parcel of air inside the cloud descends from the top of the mountain at 3000 m (pressure

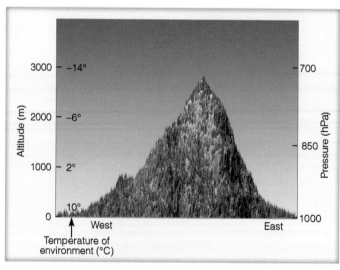

● FIGURE 6.30 The information in this illustration is to be used in answering questions 5 and 6 in the Problems and Exercises section.

700 hPa) down the eastern (leeward) side of the mountain to sea level at an elevation of 0 m (pressure 1000 hPa).

(j) If the descending air warms at the dry adiabatic lapse rate from the top of the mountain all the way down to 0 m, what is the sinking air's temperature and dew point when it reaches 0 m?

(k) What would be the relative humidity of the sinking air at 0 m? (Hint: See Focus on a Special Topic: Computing Relative Humidity and Dew Point in Chapter 4 on p. 114 or use the tephigram.)

(l) What accounts for the sinking air being warmer at the base of the mountain on the eastern side?

(m) Explain why the sinking air is drier (its dew point is lower) on the eastern side at 0 m.

Heavy snow covers mountains in the Selkirk range, near Revelstoke, British Columbia.
Ian Jackson

Precipitation

Two days later the weather was bad, wonderfully bad, I should say, for I loved to see the snow puff up and rise and wander high in the sky in ceaselessly changing shapes that might even be half alive, for I believed I could hear them shouting with happiness at being finally set free by the storm.

Gabrielle Roy, from "The Road Past Altamont"
Gabrielle Roy was an influential French Canadian author and three-times winner of the Governor General's Award for fiction.

Gabrielle Roy, *The Road Past Altamont*. McClelland & Stewart Ltd. 2010.

If Earth and its interconnected systems are viewed as a living organism,* then water as it circulates through the *atmosphere, hydrosphere, cryosphere, lithosphere,* and *biosphere* could be thought of as its lifeblood. **Precipitation,** as any form of water (liquid or solid) that falls from a cloud and reaches the ground, is the movement of this lifeblood from the atmosphere to the other systems. All the weather systems that produce precipitation serve to distribute this lifeblood. Precipitation, of course, is a defining feature of Earth's climate, providing input to the hydrologic cycle and the hydrosphere, and is the source of moisture in the biosphere over all parts of the lithosphere. Precipitation that falls as snow either melts in spring, recharging groundwater and providing runoff to rivers and lakes, or is stored in mountain glaciers and ice sheets, becoming part of the cyrosphere. Precipitation that falls as rain provides moisture to the lithosphere, soils, and biosphere that is needed for life.

This chapter raises a number of interesting questions regarding precipitation. Why, for example, does the heaviest form of precipitation—hail—often fall during the warmest time of the year? Why does it sometimes rain on one side of the street but not on the other? What are ice pellets, and how do they differ from hail? First, we will examine the processes that produce rain and snow; then we will look closely at the other forms of precipitation. Our discussion will conclude with a section on how precipitation is measured.

Precipitation Processes

Although cloudy weather is necessary for rain or snow, the presence of clouds does not necessarily mean that it *will* rain or snow. In fact, clouds can form, linger for many days, and never produce precipitation. For example, Eureka, in the Redwood Forest region of northern California, experiences August daytime skies that are overcast more than 50 percent of the time, yet the average precipitation for August is merely 2.5 mm. (See Chapter 5, p. 134, for more about moisture in redwood forests.) We know that clouds form by condensation, yet condensation alone is not sufficient to produce rain. Why not? To answer this question, we need to closely examine the tiny world of cloud droplets.

HOW DO CLOUD DROPLETS GROW LARGER? An ordinary cloud droplet is extremely small, with an average diameter of 20 μm[†] or 0.02 mm. Notice in ● Figure 7.1 that a typical cloud droplet is 100 times smaller in diameter than a typical raindrop and 100 times larger than a typical condensation nucleus. This observed difference in size leads to the distinction between a *droplet* and a *drop.* Droplets are small and make up clouds but are unable to fall as precipitation, whereas

---

*The theory regarding Earth and its systems as a self-regulating system that behaves like an organism is called the *Gaia Theory* and was proposed by James Lovelock in the 1970s.

[†]Remember that one micrometre (μm) equals one-millionth of a metre.

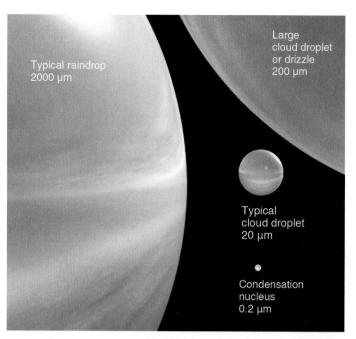

● **FIGURE 7.1** Relative diameters of raindrops, drizzle, cloud droplets, and condensation nuclei in micrometres (μm).

drops are much larger and are able to fall as precipitation. So, for precipitation, size matters. Small droplets fall very slowly and cannot overcome weak updrafts, or evaporation in drier air below clouds, to reach Earth. However, precipitation-sized drops fall quickly enough to overcome updrafts in clouds, as well as evaporation below clouds, to survive their journey from cloud to Earth. Let us first look at the processes that allow droplets to grow from tiny condensation nuclei to the size of cloud droplets. Later we will see how cloud droplets, in turn, are able to grow large enough to fall as raindrops.

If a cloud droplet is in equilibrium with its surroundings, the size of the droplet does not change because the water molecules condensing onto the droplet will be exactly balanced by those evaporating from it. If, however, it is not in equilibrium, the droplet size will either increase or decrease, depending on whether condensation or evaporation predominates.

Consider a cloud droplet in equilibrium with its environment. The total number of vapour molecules around the droplet remains fairly constant and defines the droplet's *saturation vapour pressure.* Since the droplet is in equilibrium, the saturation vapour pressure is also called the **equilibrium vapour pressure.** ● Figure 7.2 shows a cloud droplet and a flat water surface, both of which are in equilibrium. Because more vapour molecules surround the droplet, it has a greater equilibrium vapour pressure. The reason for this fact is that water molecules are less strongly attached to a curved (convex) water surface; hence, they evaporate more readily.

To keep the droplet in equilibrium, more vapour molecules are needed around it to replace those molecules that are constantly evaporating from its surface. Smaller cloud droplets exhibit a greater curvature, which causes a more rapid rate of evaporation. As a result of this process (called the

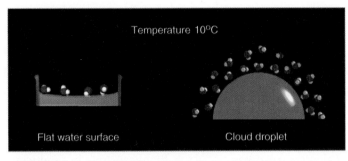

● FIGURE 7.2 At equilibrium, the vapour pressure over a curved droplet of water is greater than that over a flat surface.

curvature effect), smaller droplets require an even greater vapour pressure to keep them from evaporating away. Therefore, *when air is saturated with respect to a flat surface, it is unsaturated with respect to a curved droplet of pure water*, and the droplet evaporates. So, to keep tiny cloud droplets in equilibrium with the surrounding air, the air must be supersaturated; that is, the relative humidity must be greater than 100 percent. The smaller the droplet is, the greater its curvature, and the higher the supersaturation needed to keep the droplet in equilibrium.

● Figure 7.3 shows the curvature effect for pure water. The dark blue line represents the relative humidity needed to keep a droplet with a given diameter in equilibrium with its environment. Note that when the droplet's size is less than 2 μm, the relative humidity (measured with respect to a flat surface) must be above 100.1 percent for the droplet to survive. As droplets become larger, the effect of curvature lessens; for a droplet whose diameter is greater than 20 μm, the curvature effect is so small that the droplet behaves as if its surface were flat.

Just as relative humidities less than that required for equilibrium permit a water droplet to evaporate and shrink,

those greater than the equilibrium value allow the droplet to grow by condensation. From Figure 7.3, we can see that a droplet whose diameter is 1 μm will grow larger as the relative humidity approaches 101 percent. But relative humidities, even in clouds, rarely become greater than 101 percent. How, then, do tiny cloud droplets of less than 1 μm grow to the size of an average cloud droplet?

Recall from Chapter 5 (fog formation discussion) that condensation begins on tiny particles called *cloud condensation nuclei*. Because many of these nuclei are *hygroscopic* (i.e., they have an affinity for water vapour), condensation may begin on such particles when the relative humidity is well below 100 percent. When condensation begins on hygroscopic salt particles, for example, they dissolve, forming a solution. Since the salt ions in solution bind closely with water molecules, it is more difficult for the water molecules to evaporate. This condition reduces the equilibrium vapour pressure, a phenomenon known as the **solute effect**. Due to the solute effect, once an impurity (such as a salt particle) replaces a water molecule in the droplet, the equilibrium vapour pressure surrounding the droplet is lowered. As a result of the solute effect, a droplet containing salt can be in equilibrium with its environment when the atmospheric relative humidity is much lower than 100 percent. Should the relative humidity of the air increase, water vapour molecules would attach themselves to the droplet at a faster rate than they would leave, and the droplet would grow larger in size.

As the droplet grows, however, the hygroscopic solution becomes more dilute, and the solute effect diminishes. So the curvature effect and the solute effect are in competition: the curvature effect acts to inhibit the growth of small droplets, and the solute effect acts to enhance their growth. The combination of these two effects results in the growth of an individual droplet. Initially, for the droplet to grow, the solute effect must dominate; as the droplet grows and the solute becomes more diluted, the curvature effect becomes important, so there is generally a critical level of supersaturation within a cloud that must exist for droplets to grow (see ● Figure 7.4).

Imagine that we place cloud condensation nuclei of varying sizes into moist but unsaturated air. As the air cools, the relative humidity increases. When the relative humidity reaches a value near 78 percent, condensation occurs on the majority of nuclei. As the air cools further, the relative humidity increases, with the droplets containing the most salt reaching the largest sizes. And since the smaller nuclei are more affected by the curvature effect, only the larger nuclei are able to become cloud droplets.

Over landmasses where large concentrations of nuclei exist, there may be many hundreds of droplets per cubic centimetre, all competing for the available supply of water vapour. Over the oceans, where the concentration of nuclei is less, there are normally fewer (typically less than 100 per cubic centimetre) but larger cloud droplets. So, in a given volume, we tend to find more cloud droplets in clouds that form over land and fewer, but larger, cloud droplets in clouds that form over the ocean.

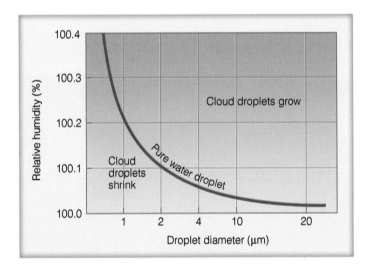

● FIGURE 7.3 The curved line represents the relative humidity needed to keep a droplet in equilibrium with its environment. For a given droplet size, the droplet will evaporate and shrink when the relative humidity is less than that given by the curve. The droplet will grow by condensation when the relative humidity is greater than the value on the curve.

We now have a cloud composed of many small droplets—too small to fall as rain. These minute droplets require only slight upward air currents to keep them suspended. Those droplets that do fall descend slowly and evaporate in the drier air beneath the cloud. It is evident, then, that *most clouds cannot produce precipitation.* The condensation process by itself is entirely too slow to produce rain. Even under ideal conditions, it would take several days for this process alone to create a raindrop. However, observations show that clouds can develop and begin to produce rain in less than an hour. Since it takes about 1 million average-sized (20 μm) cloud droplets

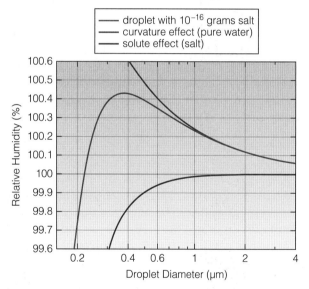

● **FIGURE 7.4** The red line shows the equilibrium relative humidity for a droplet that forms on a tiny, hygroscopic, salt, cloud condensation nucleus relative to the droplet's diameter. (The actual amount of salt in the condensation nucleus is 10^{-16} g.) This line represents the combined impact on the droplet of the curvature and solute effects. When the humidity is higher than the red line, the droplet will grow; when it is lower, it will diminish. The peak in this curve (at 100.43 percent relative humidity) represents the minimum environmental humidity needed for the droplet to grow beyond 0.39 μm in diameter. If we separate the impacts of the combined effects, the blue line illustrates how the solute effect allows droplets to grow even when the relative humidity is less than 100 percent. The purple line illustrates the impact of the curvature effect, which is the only effect when a droplet has no condensation nucleus (i.e., in pure water). In this case, high supersaturation levels—humidity higher than the green line—are needed for droplets to grow.

to make an average-sized (2000 μm) raindrop, there must be some other process by which cloud droplets grow large and heavy enough to fall as precipitation. Even though all of the intricacies of how rain is produced are not yet fully understood, two important processes stand out: (1) the collision–coalescence process and (2) the ice-crystal (Bergeron) process.

COLLISION AND COALESCENCE PROCESS In clouds with tops warmer than approximately –15°C, the **collision–coalescence process** can play a significant role in producing precipitation. To produce the many collisions necessary to form a raindrop, some cloud droplets must be larger than others. Larger drops may form on large condensation nuclei, such as salt particles, or they may form through random collisions of droplets. Studies suggest that turbulent mixing between the cloud and its drier environment may play a role in producing larger droplets.

As cloud droplets fall, air retards the falling drops. The amount of air resistance depends on the size of the drop and on its rate of fall: the greater its speed, the more air molecules the drop encounters each second. The speed of the falling drop increases until the air resistance equals the pull of gravity. At this point, the drop continues to fall, but at a constant speed, which is called its **terminal velocity**. Because larger drops have a smaller surface area to weight ratio, they must fall faster before reaching their terminal velocity. Thus, *larger drops fall faster than smaller drops* (see ▼ Table 7.1). Note in Table 7.1 that in calm air, a typical raindrop falls over 600 times faster than a typical cloud droplet! This also means that there must be upward motion in the cloud at least equal to the terminal velocity for a given-sized droplet to form.

Large droplets overtake and collide with smaller drops in their path. This merging of cloud droplets by collision is called **coalescence**. Laboratory studies show that collision does not always guarantee coalescence; sometimes the droplets actually

▼ **Table 7.1** Terminal Velocity of Different-Sized Particles Involved in Condensation and Precipitation Processes

DIAMETER		TERMINAL VELOCITY	TYPE OF PARTICLE
(μm)	(mm)	(m s^{-1})	
0.2	0.0002	0.0000001	Condensation nuclei
20	0.02	0.01	Typical cloud droplet
100	0.1	0.27	Large cloud droplet
200	0.2	0.70	Large cloud droplet or drizzle
1000	1	4.0	Small raindrop
2000	2	6.5	Typical raindrop
5000	5	9.0	Large raindrop

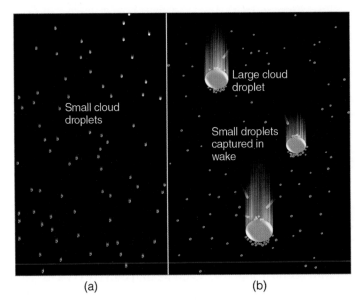

● FIGURE 7.5 Collision and coalescence. (a) In a warm cloud composed only of small cloud droplets of uniform size, the droplets are less likely to collide as they all fall very slowly at about the same speed. Those droplets that do collide frequently do not coalesce because of the strong surface tension that holds each tiny droplet together. (b) In a cloud composed of different-sized droplets, larger droplets fall faster than smaller droplets. Although some tiny droplets are swept aside, some collect on the larger droplet's forward edge, whereas others (captured in the wake of the larger droplet) coalesce on the droplet's backside.

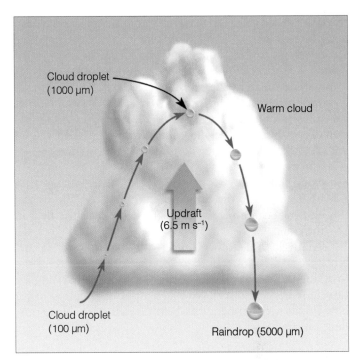

● FIGURE 7.6 A cloud droplet rising and then falling through a warm cumulus cloud can grow by collision and coalescence and emerge from the cloud as a large raindrop.

bounce apart during collision. For example, the forces that hold a tiny droplet together (surface tension) are so strong that if the droplet were to collide with another tiny droplet, chances are they would not stick together (coalesce) (see ● Figure 7.5). Coalescence appears to be enhanced if colliding droplets have opposite (and, hence, attractive) electrical charges.* An important factor influencing cloud droplet growth by the collision process is the amount of time the droplet spends in the cloud. A very large cloud droplet of 200 μm falling in still air takes about 12 minutes to travel through a cloud 500 m thick and over an hour if the cloud thickness is 2500 m. Rising air currents in a forming cloud slow the rate at which droplets fall toward the ground. Consequently, a thick cloud with strong updrafts maximizes the time cloud droplets spend in the cloud and, hence, the size to which they can grow.

A warm stratus cloud is typically less than 500 m thick and has slow upward air movement (generally less than 0.1 m s^{-1}). Under these conditions, a large droplet would be in the cloud for a relatively short time and grow by coalescence to only about 200 μm. If the air beneath the cloud is moist, the droplets may reach the ground as *drizzle*, the lightest form of rain.

If, however, the stratus cloud base is fairly high above the ground, the drops will evaporate before reaching the surface, even when the relative humidity is 90 percent.

Clouds that have above-freezing temperatures at all levels are called *warm clouds*. In such clouds, precipitation forms by the collision and coalescence process. For example, in tropical regions, where warm cumulus clouds build to great heights, convective updrafts of at least 1 m s^{-1} (and some exceeding many tens of metres per second) occur. Look at the warm cumulus cloud in ● Figure 7.6. Suppose a cloud droplet of 100 μm is caught in an updraft whose velocity is 6.5 m s^{-1}. As the droplet rises, it collides with and captures smaller drops in its path and grows until it reaches a size of about 1000 μm. At this point, the updraft in the cloud is just able to balance the pull of gravity on the drop. Here the drop remains suspended until it grows just a little bigger. Once the fall velocity of the drop is greater than the updraft velocity in the cloud, the drop slowly descends. As the drop falls, larger cloud droplets are captured by the falling drop, which then grows larger. By the time this drop reaches the bottom of the cloud, it will be a large raindrop with a diameter of over 5000 μm (5 mm). Because raindrops of this size fall faster and reach the ground first, they typically occur at the beginning of a rain shower originating in these warm, convective cumulus clouds.

Raindrops that reach Earth's surface are seldom larger than about 5 mm. The collisions between raindrops (whether glancing or head-on) tend to break them up into many smaller drops. Additionally, a large drop colliding with another large drop may result in oscillations within the combined drop. As

---

*It was once thought that atmospheric electricity played a significant role in the production of rain. Today, many scientists feel that the difference in electrical charge that exists between cloud droplets results from the bouncing collisions between them. It is felt that the weak separation of charge and the weak electrical fields in developing, relatively warm clouds are not significant in initiating precipitation. However, studies show that coalescence is often enhanced in thunderstorms, where strongly charged droplets exist in a strong electrical field.

the resultant drop grows, these oscillations may tear the drop apart into many fragments, all smaller than the original drop.

So far, we have examined the way cloud droplets in warm clouds (i.e., those clouds with temperatures above freezing) grow large enough by the collision–coalescence process to fall as raindrops. Rain that falls from warm clouds is sometimes called warm rain. The most important factor in the production of raindrops is the cloud's *liquid water content*. In a cloud with sufficient water, other significant factors are the

1. range of droplet sizes
2. cloud thickness
3. updrafts within the cloud
4. electrical charge of the droplets and the electric field in the cloud

Relatively thin stratus clouds with slow, upward air currents are, at best, only able to produce drizzle, whereas the towering cumulus clouds associated with rapidly rising air can cause heavy showers. Now let's turn our attention to see how clouds with temperatures below freezing are able to produce precipitation.

ICE-CRYSTAL PROCESS The **ice-crystal** (or **Bergeron**)* **process** of rain formation is extremely important in middle and high latitudes, where clouds extend upward into regions where the air temperature is well below freezing. Such clouds are called *cold clouds*. ● Figure 7.7 illustrates a typical cold cloud that has formed over the Prairies, where the "cold" part is well above the 0°C isotherm.

---

*The ice-crystal process is also known as the *Bergeron process* after the Swedish meteorologist Tor Bergeron, who proposed that essentially all raindrops begin as ice crystals.

Suppose we take an imaginary balloon flight up through the cumulonimbus cloud in Figure 7.7. Entering the cloud, we observe cloud droplets growing larger by processes described in the previous section. As expected, only water droplets exist here as the base of the cloud is warmer than 0°C. Surprisingly, in the below-freezing air just above the 0°C isotherm, almost all of the cloud droplets are still composed of liquid water. Water droplets existing at temperatures below freezing are referred to as **supercooled**. Even at higher levels, where the air temperature is −10°C, there is only one ice crystal for every million liquid droplets. Near 5500 m, where the temperature becomes −20°C, ice crystals become more numerous but are still outnumbered by water droplets.* The distribution of ice crystals, however, is not uniform as the downdrafts contain more ice than the updrafts.

Not until we reach an elevation of 7600 m, where temperatures drop below −40°C, do we find only ice crystals. The region of a cloud where only ice particles exist is called *glaciated*. Why are there so few ice crystals in the middle of the cloud, even though temperatures there are well below freezing? Laboratory studies reveal that the smaller the amount of pure water is, the lower the temperature at which water freezes. Since cloud droplets are extremely small, it takes very low temperatures to turn them into ice. (More on this topic is given in Focus on a Special Topic: The Freezing of Tiny Cloud Droplets on p. 197.)

---

*In continental clouds, such as the one in Figure 7.7, where there are many small cloud droplets less than 20 μm in diameter, the onset of ice-crystal formation begins at temperatures between −9°C and −15°C. In clouds where larger but fewer cloud droplets are present, ice crystals begin to form at temperatures between −4°C and −8°C. In some of these clouds, glaciation can occur at −8°C, which may be only 2500 m above the surface.

● FIGURE 7.7 The distribution of ice and water in a cumulonimbus cloud.

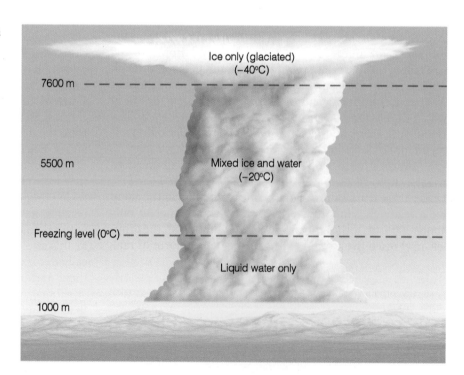

FOCUS ON A SPECIAL TOPIC

The Freezing of Tiny Cloud Droplets

Over large bodies of fresh water, ice ordinarily forms when the air temperature drops slightly below 0°C. Yet a cloud droplet of pure water about 25 μm in diameter will not freeze spontaneously until the air temperature drops to about −40°C or below.

The freezing of pure water (without the benefit of some nucleus) is called spontaneous or homogeneous freezing. For this type of freezing to occur, enough molecules within the water droplet must join together in a rigid pattern to form a tiny ice structure, or ice embryo. When the ice embryo grows to a critical size, it acts as a nucleus. Other molecules in the droplet then attach themselves to the nucleus of ice, and the water droplet freezes.

Tiny ice embryos form in water at temperatures just below freezing, but at these temperatures, thermal agitations are large enough to weaken their structure. The ice embryos simply form and then break apart. At lower temperatures, thermal motion is reduced, making it

easier for bigger ice embryos to form. Hence, freezing is more likely.

The chances of an ice embryo growing large enough to freeze water before the embryo is broken up by thermal agitation increase with larger volumes of water. Consequently, only larger cloud droplets will freeze by homogeneous freezing at air temperatures higher than

● FIGURE 1 This cirrus cloud is probably composed entirely of ice crystals because any liquid water droplet, no matter how small, must freeze spontaneously at the very low temperature (below −40°C) found at this altitude, 9 km.

−40°C. In air colder than −40°C, however, it is almost certain that an ice embryo will grow to critical size in even the smallest cloud droplet. Thus, any cloud that forms in extremely cold air (below −40°C), such as cirrus clouds (see ● Figure 1), will almost certainly be composed of ice because any cloud droplets that form will freeze spontaneously.

Just as liquid cloud droplets form on condensation nuclei, ice crystals may form in subfreezing air on particles called **ice nuclei**. The number of ice-forming nuclei available in the atmosphere is small, especially at temperatures above −10°C. However, as the temperature decreases, more particles become active and promote freezing. Although some uncertainty exists regarding the principal source of ice nuclei, it is known that clay minerals, such as kaolinite, become effective nuclei at temperatures near −15°C. Some types of bacteria in decaying plant leaf material and ice crystals themselves are also excellent ice nuclei. Moreover, particles serve as excellent ice-forming nuclei if their geometry resembles that of an ice crystal. However, it is difficult to find substances in nature that have a lattice structure similar to ice because there are so many possible lattice structures. In the atmosphere, it is easy to find hygroscopic ("water seeking") particles. Consequently, ice-forming nuclei are rare compared to cloud condensation nuclei.

In a cold cloud, several types of ice-forming nuclei may be present. For example, certain ice nuclei allow water vapour to deposit as ice directly onto their surfaces in cold saturated air. These are called *deposition nuclei* because, in this situation, water vapour changes directly into ice without going through the liquid phase. Ice nuclei that promote the

freezing of supercooled liquid droplets are called *freezing nuclei*. Some freezing nuclei cause freezing after they are immersed in a liquid drop; some promote condensation and then freezing; yet others cause supercooled droplets to freeze if they collide with them. This last process is called **contact freezing**, and the particles involved are called *contact nuclei*. Studies suggest that contact nuclei can be just about any substance and that contact freezing may be the dominant force in the production of ice crystals in some clouds.

We can now understand why there are so few ice crystals in the cold mixed region of some clouds. Cloud droplets may freeze spontaneously, but only at the very low temperatures usually found at high altitudes. Ice nuclei may initiate the growth of ice crystals, but they do not abound in nature. Because there are many more cloud condensation nuclei than ice nuclei, we are left with a cold cloud that contains many more liquid droplets than ice particles, even at temperatures as low as −10°C. Neither the tiny liquid nor solid particles are large enough to fall as precipitation. How, then, does the ice-crystal (Bergeron) process produce rain and snow?

In the subfreezing air of a cloud, many supercooled liquid droplets will surround each ice crystal. Suppose that the ice crystal and liquid droplet in ● Figure 7.8 are part of a

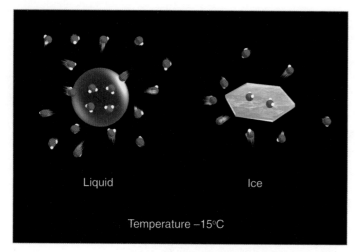

● FIGURE 7.8 In a saturated environment, the water droplet and the ice crystal are in equilibrium as the number of molecules leaving the surface of each droplet and ice crystal equals the number returning. The greater number of vapour molecules above the liquid indicates, however, that the saturation vapour pressure over water is greater than it is over ice.

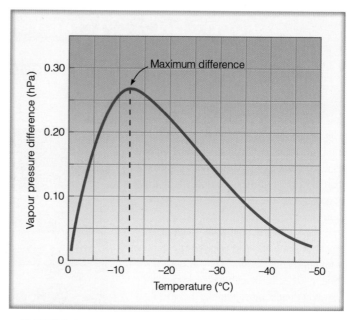

● FIGURE 7.9 The difference in saturation vapour pressure between supercooled water and ice at different temperatures.

cold (−15°C), supercooled, saturated cloud. Since the air is saturated, both the liquid droplet and the ice crystal are in equilibrium, meaning that the number of molecules leaving the surface of both the droplet and the ice crystal must equal the number of molecules returning. Observe, however, that there are more vapour molecules above the liquid. The reason for this fact is that molecules escape the surface of water much more easily than they escape the surface of ice. Consequently, more molecules escape the water surface at a given temperature, requiring more in the vapour phase to maintain saturation. This situation reflects the important fact discussed briefly in Chapter 4: at the same subfreezing temperature, *the saturation vapour pressure just above a water surface is greater than the saturation vapour pressure above an ice surface.* This difference in saturation vapour pressure between water and ice is illustrated in ● Figure 7.9.

This difference in vapour pressure causes water vapour molecules to move (diffuse) from the water droplet toward the ice crystal. The removal of vapour molecules reduces the vapour pressure above the water droplet. Since the droplet is now out of equilibrium with its surroundings, it evaporates to replenish the diminished supply of water vapour above it. This process provides a continuous source of moisture for the ice crystal, which absorbs the water vapour by deposition and grows rapidly (see ● Figure 7.10). Hence, during the *ice-crystal (Bergeron) process, ice crystals grow larger at the expense of the surrounding water droplets.*

The constant supply of moisture to the ice crystal allows it to enlarge rapidly. At some point, the ice crystal becomes heavy enough to overcome updrafts in the cloud and begins to fall. But a single falling ice crystal does not comprise a snowstorm; consequently, other ice crystals must quickly form.

In some clouds, especially those with relatively warm tops, ice crystals might collide with supercooled droplets. On contact, the liquid droplets freeze into ice and stick together.

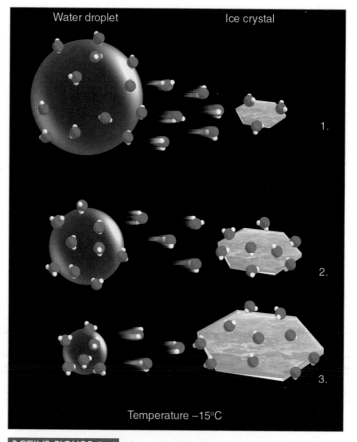

ACTIVE FIGURE 7.10 The ice-crystal (Bergeron) process. (1) The greater number of water vapour molecules around the liquid droplet causes water molecules to diffuse from the liquid droplet toward the ice crystal. (2) The ice crystal absorbs the water vapour by deposition and grows larger, whereas (3) the water droplet grows smaller. Visit the textbook's website to view this and other Active Figures at www.ahrensmeteorology1ce.nelson.com

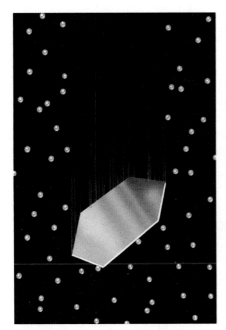

(a) Falling ice crystals may freeze supercooled droplets on contact (accretion), producing larger ice particles.

(b) Falling ice particles may collide and fracture into many tiny (secondary) ice particles.

(c) Falling ice crystals may collide and stick to other ice crystals (aggregation), producing snowflakes.

● FIGURE 7.11   Ice particles in clouds.

This process of ice crystals growing larger as they collide with supercooled cloud droplets is called **accretion**. The icy matter that forms is called **snow pellets** (or **graupel**). As the snow pellets fall, they may fracture or splinter into tiny ice particles when they collide with cloud droplets. These splinters may grow to become new snow pellets, which, in turn, may produce more splinters.

In colder clouds, the delicate ice crystals may collide with other crystals and fracture into smaller ice particles, or tiny seeds, which freeze hundreds of supercooled droplets on contact. In both cases, a chain reaction may develop, producing many ice crystals. As the ice crystals fall, they may collide and stick to one another. The process of ice crystals colliding and then sticking together is called **aggregation**.* The end product of this clumping together of ice crystals is a **snowflake** (see ● Figure 7.11). If the snowflake melts before reaching the ground, it continues its fall as a raindrop. Therefore, much of the rain falling in middle and high latitudes—even in summer—begins as snow.

For ice crystals to grow large enough to produce precipitation, there must be many, many times more water droplets than ice crystals. Generally, the ratio of ice crystals to water droplets must be on the order of 1:100,000 to 1:1,000,000. When there are too few ice crystals in the cloud, each crystal grows large and falls out of the cloud, leaving the majority of cloud behind (unaffected). Since there are very few ice crystals, there is very little precipitation. If, on the other hand,

there are too many ice crystals (such as an equal number of crystals and droplets), then each ice crystal receives the mass of one droplet. This would create a cloud of many tiny ice crystals, each too small to fall to the ground, and no precipitation. Now if the ratio of crystals to droplets is on the order of 1:100,000, then each ice crystal would receive the mass of 100,000 droplets. Most of the cloud would convert to precipitation as the majority of ice crystals would grow large enough to fall to the ground as precipitation. To learn more about how these ideas can be used to enhance precipitation, read Focus on a Special Topic: Cloud Seeding and Precipitation on p. 200., and Focus on an Environmental Issue: Does Cloud Seeding Enhance Precipitation? on p. 202.

The first person to formally propose the theory of ice-crystal growth due to differences in the vapour pressure between ice and supercooled water was Alfred Wegener (1880–1930), a German climatologist who also proposed the geologic theory of continental drift. In the early 1930s, important additions to this theory were made by the Swedish meteorologist Tor Bergeron. Several years later, the German meteorologist Walter Findeisen made additional contributions to Bergeron's theory; hence, the ice-crystal theory of rain formation has come to be known as the *Wegener–Bergeron–Findeisen process* or, simply, the *Bergeron* process.

PRECIPITATION IN CLOUDS   In cold, strongly convective clouds, precipitation may begin only minutes after the cloud forms and may be initiated by either the collision–coalescence or the ice-crystal (Bergeron) process. Once either process begins, most precipitation growth is by accretion. Although

---

*Significant aggregation seems possible only when the air is relatively warm, usually warmer than –10°C.

Cloud Seeding and Precipitation

The primary goal in many cloud-seeding experiments is to inject (or seed) a cloud with small particles that will act as nuclei so that the cloud particles will grow large enough to fall to the surface as precipitation. The first ingredient in any seeding project is, of course, the presence of clouds as seeding does not generate clouds. However, not just any cloud will do. For optimum results, the cloud must be cold; that is, at least a portion of it (preferably the upper part) must be supercooled because **cloud seeding** uses the ice-crystal (Bergeron) process to cause the cloud particles to grow.

The idea in cloud seeding is to first find clouds that have too low a ratio of ice crystals to droplets and then to add enough artificial ice nuclei so that the ratio of crystals to droplets is about 1:100,000. However, it should be noted that the natural ratio of ice nuclei to cloud condensation nuclei in a typical cold cloud is about 1:100,000, just about optimal for producing precipitation.

Some of the first experiments in cloud seeding were conducted by Vincent Schaefer and Irving Langmuir during the late 1940s. To seed a cloud, they dropped crushed pellets of

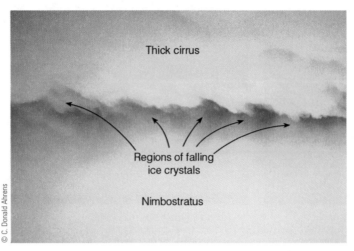

● FIGURE 2 Ice crystals falling from a dense cirriform cloud into a lower nimbostratus cloud. This photograph was taken at an altitude near 6 km above western Pennsylvania, U.S.A. At the surface, moderate rain was falling over the region.

dry ice (solid carbon dioxide) from a plane. Because dry ice has a temperature of $-78°C$, it acts as a cooling agent. As the extremely cold, dry ice pellets fall through the cloud, they quickly cool the air around them. This cooling causes the air around the pellet to become supersaturated. In this supersaturated air, water vapour forms directly into many tiny cloud droplets. In the very cold air created by the falling pellets (below $-40°C$), the tiny droplets instantly freeze into tiny ice crystals. The newly formed ice crystals then grow larger by deposition as the water vapour molecules attach themselves to the ice crystals at the expense of the nearby liquid droplets. On reaching a sufficiently large size, they fall as precipitation.

---

precipitation is commonly absent in warm-layered clouds, such as stratus, it is often associated with such cold-layered clouds as nimbostratus and altostratus. This precipitation is thought to form principally by the ice-crystal (Bergeron) process because the liquid-water content of these clouds is generally lower than that in convective clouds, thus making the collision–coalescence process much less effective. Nimbostratus clouds are normally thick enough to extend to levels where air temperatures are quite low, and they usually last just long enough for the ice-crystal (Bergeron) process to initiate precipitation. ● Figure 7.12 illustrates how ice crystals produce precipitation in clouds of both low and high liquid-water content.

BRIEF REVIEW

In the last few sections, we encountered a number of important concepts and ideas about how cloud droplets can grow large enough to fall as precipitation. Before examining the various types

of precipitation, here is a summary of some of the important ideas presented so far:

● Cloud droplets are very small, much too small to fall as rain.

● The smaller the cloud droplet is, the greater its curvature, and the more likely it will evaporate.

● Cloud droplets form on cloud condensation nuclei. Hygroscopic nuclei, such as salt, allow condensation to begin when the relative humidity is less than 100 percent.

● Cloud droplets, in above-freezing air, can grow larger as faster falling, bigger droplets collide and coalesce with smaller droplets in their path.

● In the ice-crystal (Bergeron) process of rain formation, both ice crystals and liquid cloud droplets must coexist at below-freezing temperatures. The difference in saturation vapour pressure between liquid and ice causes water vapour to diffuse from the liquid droplets (which shrink) toward the ice crystals (which grow).

In 1947, Bernard Vonnegut demonstrated that silver iodide (AgI) could be used as a cloud-seeding agent. Because silver iodide has a crystalline structure similar to an ice crystal, it acts as an effective ice nucleus at temperatures of −4°C and lower. Silver iodide causes ice crystals to form in two primary ways:

1. Ice crystals form when silver iodide crystals come in contact with supercooled liquid droplets.
2. Ice crystals grow in size as water vapour deposits onto the silver iodide crystal.

Silver iodide is much easier to handle than dry ice because it can be supplied to the cloud from burners located either on the ground or on the wing of a small aircraft. Although other substances, such as lead iodide and cupric sulphide, are also effective ice nuclei, silver iodide still remains the most commonly used substance in cloud-seeding projects. (Additional information on this controversial topic, the effectiveness of cloud seeding, is given in Focus on an Environmental Issue: Does Cloud Seeding Enhance Precipitation? on p. 202.)

Under certain conditions, clouds may be seeded naturally. For example, when

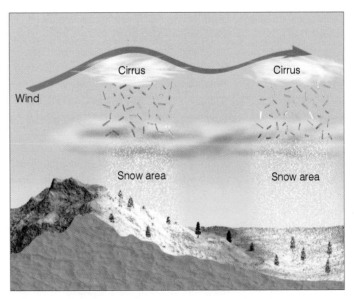

● FIGURE 3 Natural seeding by cirrus clouds may form bands of precipitation downwind of a mountain chain.

cirriform clouds lie directly above a lower cloud deck, ice crystals may descend from the higher cloud and seed the cloud below (see ● Figure 2). As the ice crystals mix into the lower cloud, supercooled droplets are converted to ice crystals, and the precipitation process is enhanced. Sometimes the ice crystals in the lower cloud may settle out, leaving a clear area or "hole" in the cloud. When the cirrus clouds form waves downwind from a mountain chain, bands of precipitation often form (see ● Figure 3).

● Most of the rain that falls over middle latitudes results from melted snow that formed from the ice-crystal (Bergeron) process.
● Cloud seeding with silver iodide can be effective in coaxing precipitation from a cloud only if the cloud is supercooled and the proper ratio of cloud droplets to ice crystals exists.

Precipitation Types

Up to now, we have seen how cloud droplets are able to grow large enough to fall to the ground as rain or snow. While falling, raindrops and snowflakes may be altered by atmospheric conditions encountered beneath the cloud and transformed into other forms of precipitation that can profoundly influence our environment. ● Figure 7.13 shows the average amount of precipitation of all types that falls over Canada.

Notice the general decrease in precipitation in the higher latitudes as air's ability to hold moisture is less due to low temperatures. Also notice the enhanced precipitation in western Canada, especially along the windward (west facing) slopes of the Coastal and Vancouver Island Mountains and the Rocky Mountains. Enhanced precipitation can also be found near moisture sources such as the Great Lakes region, the Maritime provinces, and Newfoundland.

RAIN   Most people consider **rain** to be any falling drop of liquid water. To the meteorologist, to be considered rain, that falling drop must have a diameter equal to or greater than 0.5 mm. Fine, uniform drops of water whose diameters are smaller than 0.5 mm are called **drizzle**. Most drizzle falls from stratus clouds; however, small raindrops may fall through air that is unsaturated, partially evaporate, and reach the ground as drizzle. Occasionally, the rain falling from a cloud never reaches the surface because the low humidity causes rapid evaporation. As the drops become

FOCUS ON AN ENVIRONMENTAL ISSUE

Does Cloud Seeding Enhance Precipitation?

Just how effective is artificial seeding with silver iodide in increasing precipitation? This is a much-debated question among meteorologists. First, it is difficult to evaluate the results of a cloud-seeding experiment. When a seeded cloud produces precipitation, the question always remains as to how much precipitation would have fallen had the cloud not been seeded.

Other factors must be considered when evaluating cloud-seeding experiments: the type of cloud, its temperature, its moisture content, droplet size distribution, and updraft velocities in the cloud.

Although some experiments suggest that cloud seeding does not increase precipitation, others seem to indicate that seeding under the right conditions may enhance precipitation between 5 and 20 percent. And so the controversy continues.

Some cumulus clouds show an "explosive" growth after being seeded. The latent heat given off when the droplets freeze functions to warm the cloud, causing it to become more buoyant. It grows rapidly and becomes a longer lasting cloud, which may produce more precipitation.

The business of cloud seeding can be a bit tricky because overseeding can produce too many ice crystals. When this phenomenon occurs, the cloud becomes glaciated (all liquid droplets become ice), and the ice particles, being very small, do not fall as precipitation. Since few liquid droplets exist, the ice crystals cannot grow by the ice-crystal (Bergeron) process; rather, they evaporate, leaving a clear area in a thin, stratified cloud. Because dry ice can produce the most ice crystals in a supercooled cloud, it is the substance most suitable for deliberate overseeding. Hence, it is the substance most commonly used to dissipate cold fog at airports (see Focus on an Environmental Issue: Fog Dispersal in Chapter 5, p. 140).

Warm clouds with temperatures above freezing have also been seeded in an attempt to produce rain. Tiny water drops and particles of hygroscopic salt are injected into the base (or top) of the cloud. These particles (called seed drops), when carried into the cloud by updrafts, create large cloud droplets, which grow even larger by the collision–coalescence process. Apparently, the seed drop size plays a major role in determining the effectiveness of seeding with hygroscopic particles. To date, however, the results obtained using this method are inconclusive.

Cloud seeding may be inadvertent. Some industries emit large concentrations of condensation nuclei and ice nuclei into the air. Studies have shown that these particles are at least partly responsible for increasing precipitation in, and downwind of, cities. On the other hand, studies have also indicated that the burning of certain types of agricultural waste may produce smoke containing many condensation nuclei. These produce clouds that yield less precipitation because they contain numerous, but very small, droplets.

In summary, cloud seeding in certain instances may lead to more precipitation; in others, to less precipitation; and in still others, to no change in precipitation amounts. Many of the questions about cloud seeding have yet to be resolved.

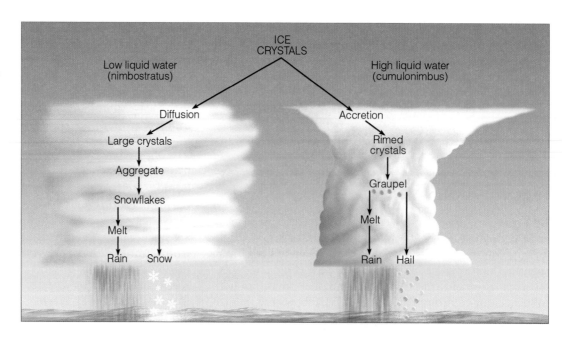

● FIGURE 7.12 How ice crystals grow and produce precipitation in clouds with a low liquid-water content and a high liquid-water content.

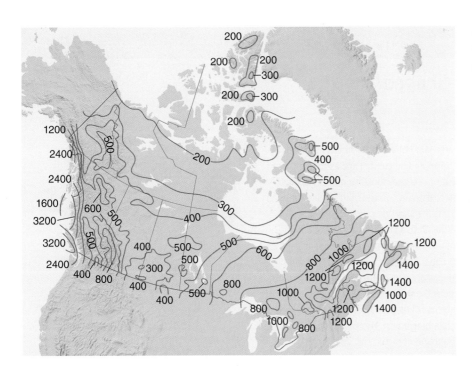

● FIGURE 7.13 Average annual precipitation over Canada. (Modified after D. Phillips, 1990.)

The Climates of Canada. D. Phillips. Environment Canada, 1990. Pg. 31. © Her Majesty The Queen in Right of Canada, Environment Canada, 2010.

smaller, they fall more slowly and appear to hang in the air. These evaporating streaks of precipitation are called **virga** (see ● Figure 7.14).

Raindrops may also fall from a cloud and not reach the ground if they encounter rapidly rising air. Large raindrops have a terminal velocity of about 9 m s^{-1}, and if they encounter rising air whose speed is greater than 9 m s^{-1}, they will not reach the surface. If the updraft weakens or changes direction and becomes a downdraft, the suspended drops will fall to the ground as a sudden rain **shower**. Showers falling from convective or cumuliform clouds are usually brief and sporadic and occur as the cloud moves overhead and then drifts by. If a shower is excessively heavy, it is popularly known as a cloudburst. Beneath cumulonimbus clouds, which normally contain large convection currents of rising and descending air, it is entirely possible that one side of a street may be dry (updraft side) while a heavy shower is occurring across the street (downdraft side). Continuous rain, on the other hand, usually falls from layered or stratiform clouds such as nimbostratus. These clouds cover larger areas and have smaller vertical air currents.

Raindrops that reach Earth's surface are seldom larger than about 6 mm because the collisions (whether glancing or head-on) between raindrops tend to break them up into many smaller drops. Additionally, when raindrops grow too large, they become unstable and break apart. What is the shape of the falling raindrop? Is it tear-shaped, or is it round? If you are unsure of the answer, read Focus on a Special Topic: Are Raindrops Tear-Shaped? on p. 204.

After a rainstorm, visibility usually improves primarily because precipitation removes or scavenges many of the suspended particles. When rain combines with gaseous pollutants, such as oxides of sulphur and nitrogen, it becomes acidic. Acid rain has an adverse effect on plants and water resources and is a major problem in many industrialized regions of the world.

It is important to know the interval of time over which rain falls. Did it fall over several days, gradually soaking into the soil? Or did it come all at once in a cloudburst, rapidly eroding the land, clogging city gutters, and causing floods

Gail Ross

● FIGURE 7.14 The streaks of falling precipitation that evaporate before reaching the grounds are called *virga*.

WEATHER WATCH

Does rain have an odour? Often before it rains, the air has a distinctive (somewhat earthy) smell to it. This odour may originate from soil bacteria that produce aromatic gases. As rain falls onto the soil, it pushes these gases into the air, where winds carry them out ahead of the advancing rain shower.

FOCUS ON A SPECIAL TOPIC

Are Raindrops Tear-Shaped?

As rain falls, the drops take on a characteristic shape. Choose the shape in ● Figure 4 that you feel most accurately describes that of a falling raindrop. Did you pick number 1? The tear-shaped drop has been depicted by artists for many years. Unfortunately, *raindrops are not tear-shaped*. Actually, the shape depends on the drop size. Raindrops less than 2 mm in diameter are nearly spherical and look like raindrop number 2. The attraction among the molecules of the liquid (surface tension) tends to squeeze the drop into a shape that has the smallest surface area for its total volume—a sphere.

Large raindrops, with diameters exceeding 2 mm, take on a different shape as they fall. Believe it or not, they look like number 3, slightly elongated, flattened on the bottom, and rounded on top. As the larger drop falls, the air pressure against the drop is greatest on the bottom and least on the sides. The pressure of the air on the bottom flattens the drop, whereas the lower pressure on its sides allows it to expand a little. This mushroom shape has been described as everything from a falling parachute to a loaf of bread or even a hamburger bun. You may call it what you wish, but remember that it is not tear-shaped.

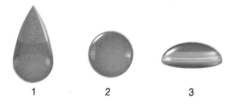

● FIGURE 4 Which of these drops represents the real shape of a falling raindrop?

---

along creeks and rivers unable to handle the sudden increased flow? The intensity of rain is the amount that falls in a given period; the intensity of rain is always based on the accumulation during a certain interval of time (see ▼ Table 7.2).

SNOW We know that much of the precipitation reaching the ground actually begins as **snow**. In summer, the freezing level is usually above 3600 m, and the snowflakes falling from a cloud melt before reaching the ground. However, in winter, freezing levels are much lower, and falling snowflakes often reach the surface. Snowflakes generally last just 300 m below the freezing level before completely melting. Occasionally, you can spot the melting level when you look in the direction of the sun if it is near the horizon. Because snow scatters incoming sunlight better than rain, the darker region beneath the cloud contains falling snow, whereas the lighter region is falling rain. The melting zone, then, is the transition between the light and dark areas (see ● Figure 7.15).

The sky will look different, however, if you are looking directly up at the precipitation. Because snowflakes are so effective at scattering light, they redirect the light beneath the cloud in all directions—some of it eventually reaching your eyes, making the region beneath the cloud appear a lighter shade of grey. Falling raindrops, on the other hand, scatter very little light toward you, and the underside of the cloud appears dark. This change in shading enables some observers to predict with uncanny accuracy whether falling precipitation will be in the form of rain or snow.

When ice crystals and snowflakes fall from high cirrus clouds, they are called **fallstreaks**. Fallstreaks behave in much the same way as virga. As the ice particles fall into drier air, they usually sublimate (i.e., change from ice into vapour). Because winds at higher levels move cloud and ice particles horizontally more quickly than do slower winds at lower levels, fallstreaks appear as dangling white streamers (see ● Figure 7.16). Moreover, fallstreaks descending into lower, supercooled clouds may actually seed them.

▼ Table 7.2   Rainfall Intensity

RAINFALL DESCRIPTION	RAINFALL RATE (mm h⁻¹)
Light	≤ 2.5
Moderate	2.6 to 7.5
Heavy	≥ 7.6

● FIGURE 7.15 Snow scatters sunlight more effectively than rain. Consequently, when you look toward the sun, the region of falling precipitation looks darker above the melting level than below it.

FOCUS ON A SPECIAL TOPIC

Snowing When the Air Temperature Is Well Above Freezing

At the beginning of this chapter, we learned that it is never too cold to snow. So when is it too warm to snow? A person who has never been in a snowstorm might answer, "When the air temperature rises above freezing." However, snow falls when air temperatures are considerably above freezing (see ● Figure 5). Why doesn't the falling snow melt in this air? Actually, it does melt, at least to some degree. Let's examine this in more detail.

For falling snowflakes to survive in air with temperatures much above freezing, the air must be unsaturated (relative humidity is less than 100 percent), and the wet-bulb temperature must be at freezing or below. You may recall from our discussion on humidity in Chapter 4 that the wet-bulb temperature is the lowest temperature that can be attained by evaporating water into the air. Consequently, it is a measure of the amount of cooling that can occur in the atmosphere as water evaporates into the air. When rain falls into a layer of dry air with a low wet-bulb temperature, rapid evaporation and cooling occur, which is why the air temperature often decreases when it begins to rain. During the winter, as raindrops evaporate in this dry air, rapid cooling may actually change a rainy day into a snowy one. This same type of cooling allows snowflakes to survive in above-freezing temperatures.

Suppose it is winter and the sky is overcast. At the surface, the air temperature is 2°C, the dew point is −6°C, and the wet-bulb temperature is 0°C.* The air temperature drops

---

*The wet-bulb temperature is always higher than the dew point, except when the air is saturated. At that point, the air temperature, wet-bulb temperature, and dew point are all the same.

● FIGURE 5 It is snowing at 4.4°C during the middle of July near the summit of Beartooth Mountain, Montana.

© C. Donald Ahrens

sharply with height, from the surface up to the cloud deck. Soon flakes of snow begin to fall from the clouds into the unsaturated layer below. In the above-freezing temperatures, the snowflakes begin to partially melt. However, the air is dry, so the water quickly evaporates, cooling the air. In addition, evaporation cools the falling snowflake to the wet-bulb temperature, which retards the flake's rate of melting. As snow continues to fall, evaporative cooling causes the air temperature to continue to drop. The addition of water vapour to the air increases the dew point, whereas the wet-bulb temperature remains essentially unchanged. Eventually, the entire layer of air cools to the wet-bulb temperature and becomes saturated at 0°C. As long as the wind does not bring in warmer air, the precipitation remains as snow.

We can see that when snow falls into warmer air (say at 8°C), the air must be extremely dry to have a wet-bulb temperature at freezing or below. In fact, with an air temperature of 8°C and a wet-bulb temperature of 0°C, the dew point would be −23°C and the relative humidity 11 percent. Conditions such as these are extremely unlikely at the surface before the onset of precipitation. Actually, the highest air temperature possible with a below-freezing wet-bulb temperature is about 10°C. Hence, snowflakes will melt rapidly in air with a temperature above this value. However, it is still possible to see flakes of snow at temperatures greater than 10°C, especially if the snowflakes are swept rapidly earthward by the cold, relatively dry downdraft of a thunderstorm.

Snowflakes and Snowfall Snowflakes that fall through moist air that is slightly above freezing slowly melt as they descend. A thin film of water forms on the edge of the flakes, which acts like glue when other snowflakes come in contact with them. In this way, several flakes join to produce giant snowflakes often measuring several centimetres or more in diameter. These large, soggy snowflakes are associated with moist air and temperatures near freezing. However, when snowflakes fall through extremely cold air with a low moisture content, small, powdery flakes of "dry" snow accumulate on the ground. (To understand how snowflakes can survive in air that is above freezing, read Focus on a Special Topic: Snowing When the Air Temperature Is Well Above Freezing on p. 205.)

● FIGURE 7.16 The dangling white streamers of ice crystals beneath these cirrus clouds are known as fallstreaks. The bending of the streaks is due to the wind speed changing with height.

© C. Donald Ahrens

If you catch falling snowflakes on a dark, cold object* and examine them closely, you will see that the most common snowflake form is a fernlike, branching star shape called a *dendrite*. Since there are many types of ice crystals (see ● Figure 7.17), why is the dendrite crystal the most common shape for snowflakes? The type of crystal that forms, as well as its growth rate, depends on the air temperature and relative humidity. ▼ Table 7.3 summarizes the crystal forms (properly called crystal habits) that develop when supercooled water and ice coexist in saturated air. Note that dendrites are common at temperatures between −12°C and −16°C. The maximum growth rate of ice crystals depends on the difference between the saturation vapour pressure of water and that of ice. This difference reaches a maximum at about −14°C, in the temperature range where dendrite crystals are most likely to grow. (Look back at Figure 7.9, p. 198.) Therefore, this type of crystal grows more rapidly than the other crystal forms. As ice crystals fall through a cloud, they are constantly exposed to changing temperature and moisture conditions.

*Try this yourself. During freezing conditions, put a dark frying pan in a shady location outside to cool and then use it to catch falling snowflakes so that you can examine them more closely.

Since many ice crystals can join together (aggregate) to form a much larger snowflake, snow crystals may assume many complex patterns (see ● Figure 7.18 on p. 207).

Snow falling from developing cumulus clouds is often in the form of **flurries** or **snow showers**. These are usually light showers that fall intermittently for short durations and produce only light accumulations. A more intense snow shower is called a **snow squall**. These brief but heavy snowfalls are comparable to summer rain showers. Like snow flurries, they usually fall from cumuliform clouds and are associated with cold fronts that form in the lee of large bodies of water such as the Great Lakes. These *lake-effect* snow squalls occur when cold, dry air crosses the lake, picking up moisture and heat from the warm water, which creates convection. This forms convective clouds, snow showers, or snow squalls in snow belts downwind of the lakes. (For a more complete discussion

▼ Table 7.3 Ice Crystal Habits That Form at Various Temperatures*

AIR TEMPERATURE (°C)	CRYSTAL HABIT
0 to −4	thin plates
−4 to −6	needles
−6 to −10	columns
−10 to −12	plates
−12 to −16	dendrites, plates
−16 to −22	plates
−22 to −40	hollow columns

*Note that at each temperature, the type of crystal that forms (e.g., hollow columns versus solid columns) will depend on the difference in saturation vapour pressure between ice and supercooled water.

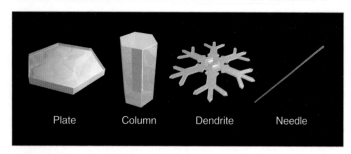

Plate Column Dendrite Needle

● FIGURE 7.17 Common ice crystal forms (habits).

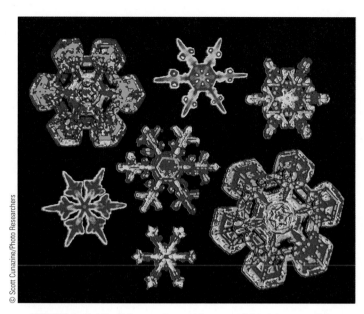

● FIGURE 7.18 Colour-enhanced images of dendrite snowflakes.

of this, look ahead to Focus on a Special Topic: Lake-Effect Snows on p. 328 in Chapter 11.) More continuous snowfall, often steadily, for several hours, accompanies nimbostratus and altostratus clouds. The measure of *snowfall intensity* (light, moderate, or heavy) is based on an observation of the reduction in horizontal visibility (see ▼ Table 7.4).

When a strong wind is blowing at the surface, snow can be picked up and deposited into huge drifts. Drifting snow is usually accompanied by *blowing snow*—that is, snow lifted from the surface by the wind and blown about in such quantities that horizontal visibility is greatly restricted. The combination of drifting and blowing snow, accompanied by wind chill, is called a **blizzard**. Environment Canada defines a blizzard as weather lasting for four hours or more that is characterized by winds of at least 40 km h⁻¹, and visibility of less than 1 km in snow and/or blowing snow.

A Blanket of Snow A mantle of snow covering the landscape is much more than a beautiful setting—it is a valuable resource provided by nature. A blanket of snow stores fresh

▼ Table 7.4 Snowfall Intensity

SNOWFALL DESCRIPTION	VISIBILITY
Light	Greater than ½ mile* (> 800 m)
Moderate	Greater than ¼ mile, less than or equal to ½ mile (> 400 m ≤ 800 m)
Heavy	Less than or equal to ¼ mile (≤ 400 m)

*In Canada and the United States, government weather services determine visibility (the greatest horizontal distance you can see) in statute miles.

water for later use and is a good insulator as it stops heat from being conducted out of the ground. In fact, the more air spaces there are between the individual snowflake crystals, the better insulators they become. A light, fluffy covering of snow protects sensitive plants and their root systems from damaging low temperatures. In this way, snow can prevent the ground from freezing downward to great depths. In cold climates that receive little snow, it is often difficult to grow certain crops because the frozen soil makes spring cultivation almost impossible. Similarly, if you become lost in a cold and windy snowstorm, build a snow cave and climb inside. Not only will this protect you from the wind, but the insulating effect of the snow will also protect you from the cold air. The snow cave also acts as a barrier between your body and the cold sky, increasing the incoming longwave radiation to your body and thereby reducing its loss of infrared radiation.

Frozen ground also prevents early spring rains from percolating downward into the soil, leading to rapid water runoff and flooding. If subsequent rains do not fall, the soil could even become moisture deficient. In areas with an insulating snowpack, melting snow provides soil moisture to the ground as it percolates through the snowpack to the unfrozen soil below.

Many think that the accumulation of snow just provides opportunities for winter recreation, but, more importantly, melting snow in spring and the continued melting of mountain snow in summer supply fresh water to streams, reservoirs, and groundwater. This water is critical for ecosystems, agriculture, communities, industry, and our drinking water. For example, snowmelt flowing to streams cools water temperatures, which is important for many fish species. Fish breathe oxygen that is dissolved in water, and dissolved oxygen decreases at higher water temperatures. Species such as salmon have trouble surviving when stream water temperatures become too high.

Winter snows are not without hardships and potential hazards. In spring, rapid melting of the snowpack may flood low-lying areas. Too much snow on the side of a steep hill or mountain may become an avalanche. The added weight of snow on the roof of a building may cause it to collapse, leading to costly repairs and even loss of life, as occurred in March 2008 in Montreal and several smaller southern Quebec communities. Annually, winter snows disrupt transportation, causing road and air travel advisories, delays, and expensive plowing, sanding, and/or salting operations in an effort to

WEATHER WATCH

According to urban legend, Arctic aboriginal peoples, such as the Inuit, have many names for snow (ranging from 40 to 400!). This is a myth. The Inuit language is polysynthetic, which means that expressions that in English or French would be several words can be combined into one "word" in Inuit. In Inuit, there are really only a few base words for snow: one is *aput*, which means snow as a material, and another is *qanniq*, which normally is like the verb "to snow." Starting with these base words, Inuit speakers can add descriptive modifiers to describe snow well.

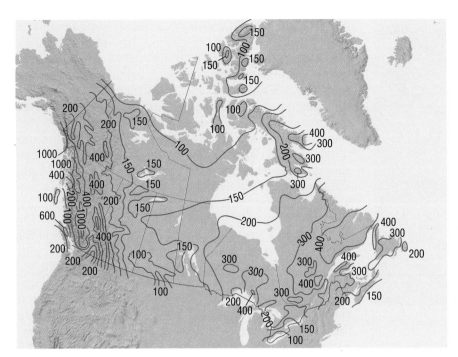

● FIGURE 7.19 Average annual snowfall over Canada.

The Climates of Canada. D. Phillips. Environment Canada, 1990. Pg. 31 © Her Majesty The Queen in Right of Canada, Environment Canada, 2010.

reduce winter travel hazards. A January snowfall of several centimetres in Vancouver, British Columbia, can bring traffic to a standstill, whereas a snowfall of several centimetres in most of the rest of Canada would go practically unnoticed. ● Figure 7.19 gives the annual average snowfall across Canada. Notice the enhanced annual snowfall near moisture sources such as oceans and the Great Lakes, as well as over higher terrain. In coastal British Columbia, warm winter temperatures explain the low snowfall amounts, despite the high annual precipitation in this region (refer back to Figure 7.13). (A blanket of snow also has an effect on the way sounds are transmitted. More on this subject is given in Focus on a Special Topic: Sounds and Snowfalls on p. 209.)

Snow, and the pollutants it carries, can have important implications for arctic ecosystems. Current studies indicate that mercury, a dangerous neurotoxin, is transported in the atmosphere from southern latitudes and deposited in the Arctic by rain and snow. In wet environments, the mercury changes into a more active form, called methylmercury. When the arctic sun begins to rise in the spring, mercury that was deposited on the snow over the winter is converted into methylmercury and released to the arctic ecosystem in a pulse as the snow and ice melt. Once in the environment, methylmercury bioaccumulates by entering arctic food chains through the smallest organisms, phytoplankton, which are ingested by fish. The fish are eaten by mammals such as seals, which are then eaten by those at the top of the food web, polar bears and humans, who accumulate these toxins at high enough levels to result in health impacts. In this way, melting snow appears to be involved in ecosystem-wide health issues.

ICE PELLETS AND FREEZING RAIN Consider the falling snowflake in ● Figure 7.20. As it falls into warmer air, it begins

to melt. When it falls through the deep, subfreezing surface layer of air, the partially melted snowflake or cold raindrop turns back into ice, not as a snowflake but as an **ice pellet.*** Generally, these are transparent (or translucent), with diameters of 5 mm or less. They bounce when striking the ground and produce a tapping sound when they hit a window or piece of metal.

The cold surface layer beneath a cloud may be too shallow to freeze falling raindrops. Consequently, they reach the surface as supercooled liquid drops. On striking a cold object,

---

*In the United States, ice pellets are referred to as **sleet**. This causes confusion because in Commonwealth countries, such as Canada, the word *sleet* refers to mixed rain and snow.

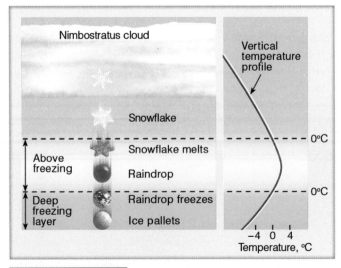

ACTIVE FIGURE 7.20 Ice pellets form when a partially melted snowflake or a cold raindrop freezes into a pellet of ice before reaching the ground. Visit the textbook's website to view this and other Active Figures at www.ahrensmeteorology1ce.nelson.com

FOCUS ON A SPECIAL TOPIC

Sounds and Snowfalls

A blanket of snow is not only beautiful, it can also affect what we hear. You may have noticed that after a snowfall, it seems quieter than usual. Freshly fallen snow can absorb sound—just like acoustic tiles. As the snow gets deeper, this absorption increases. Anyone who has walked through the woods on a snowy evening knows the quiet created by a thick blanket of snow (see ● Figure 6). As snow becomes older and more densely packed, its ability to absorb sound is reduced. That is why sounds you could not hear right after a snowstorm become more audible several days later.

New snow covering a pavement will sometimes squeak as you walk in it. The sound produced is related to the snow's temperature. When the air and snow are only slightly below freezing, the pressure from the heel of a boot partially melts the snow. The snow can then

flow under the weight of the boot, and no sound is made. However, on cold days when the snow temperature drops below −10°C, the heel of the boot will not melt the snow, and the ice crystals are crushed. The crunching of the crystals produces the creaking sound.

● FIGURE 6 This freshly fallen blanket of snow absorbs sound waves so effectively that even the water flowing in the tiny stream is difficult to hear.

© C. Donald Ahrens

---

the drops spread out and almost immediately freeze, forming a thin veneer of ice. This form of precipitation is called **freezing rain**, **glaze**, or **verglas**. If the drops are small (less than 0.5 mm in diameter), the precipitation is called **freezing drizzle**. When small, supercooled cloud or fog droplets strike an object whose temperature is below freezing, and the tiny droplets freeze rapidly, they can form an accumulation of white or milky granular ice called **rime*** (see ● Figure 7.21).

Freezing rain can create a beautiful winter wonderland by coating everything with silvery, glistening ice. At the same time, highways turn into skating rinks for automobiles, and the destructive weight of the ice—which can be many tonnes on a single tree—breaks tree branches, power lines, and telephone cables. A case in point is the huge ice storm of January 4 to 10, 1998, which left millions of people without power in southern Quebec, eastern Ontario, and the northern New England states, causing 35 deaths (28 in Canada) and over 900 injuries. It was the costliest natural disaster in Canadian history, at over $3 billion in damages. (For more information on this event, see Focus on a Special Topic: The Ice Storm of 1998 on p. 210.) The areas in Canada most frequently hit by these storms are in

southern Ontario and parts of the Maritime provinces through Newfoundland and Labrador (see ● Figure 7.22). (For additional information on freezing rain and its effect on aircraft, read Focus on an Observation: Aircraft Icing on p. 213.)

In summary, ● Figure 7.23 shows various winter temperature profiles and the type of precipitation associated with each. In Figure 7.23a, the air temperature is below freezing at

© Guillem Lopez/Alamy

● FIGURE 7.21 An accumulation of rime forms as supercooled fog droplets freeze on contact in the below-freezing air.

---

*When the ice covering a road surface appears black, it is called *black ice*, which commonly forms when light rain, drizzle, or supercooled fog droplets come in contact with surfaces (especially those of bridges and overpasses) that have cooled to a temperature below freezing. Regardless of its formation process, when a sheet of ice covers pavement and appears relatively dark, it is referred to as black ice.

FOCUS ON A SPECIAL TOPIC

The Ice Storm of 1998

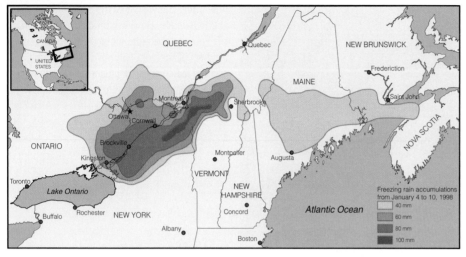

● **FIGURE 7** Freezing rain accumulations in millimetres between January 4 and 10, 1998. (Data from Environment Canada).

Source: Environment Canada. Ice Storm 1998 Precipitation. Found at: http://www.msc-smc.ec.gc.ca/media/icestorm98/maps/national_web_zr.gif; Wikipedia. Ice Storm map. Found at: http://upload.wikimedia.org/wikipedia/commons/f/ff/1998_Ice_Storm_map.png

The ice storm of January 4 to 10, 1998, left millions of people without power in eastern Ontario, southern Quebec, and the northern New England states. It forced about 100,000 Canadians to leave their homes, caused 35 deaths and over 900 injuries, and was the costliest natural disaster in Canadian history, with over $3 billion in damages. During the six-day event, the water equivalent of the freezing rain and ice pellets exceeded 100 mm in many places—more than normally falls in two years (see ● Figure 7). The storm brought down millions of trees, 130 transmission towers (see Chapter 1, Figure 1.19, on p. 25), 30,000 utility poles, and 120,000 km of power and telephone lines (see ● Figure 8).

The freezing rain was caused by two meteorological features that remained nearly stationary for the six-day period: a front that was situated just south of the Canada–U.S. border and a cold arctic air mass associated with an area of high pressure over northern Quebec (see ● Figure 9). Warm air from south of the front rode up the front, creating a warm, above-freezing layer that was elevated above the surface in a zone parallel to the front and just north of the surface front location. Meanwhile, cold air from the anticyclone over northern Quebec maintained a shallow, below-freezing layer of air near the surface below the warm air layer. This combination of circumstances created a vertical profile similar to that shown in Figure 7.23c and set the stage for freezing rain. The unique part of this freezing rain event was that these meteorological features persisted over such a long time in about the same location. Normally, the front moves through, so freezing rain that often occurs ahead of a front is

all levels, and snowflakes reach the surface. In Figure 7.23b, a zone of above-freezing air causes snowflakes to partially melt; then, in the deep, subfreezing air at the surface, the liquid freezes into ice pellets. In the shallow, subfreezing surface air in Figure 7.23c, the melted snowflakes, now supercooled liquid drops, freeze on contact, producing freezing rain. In Figure 7.23d, the air temperature is above freezing in a sufficiently deep layer so that precipitation reaches the surface as rain. (Weather symbols for these and other forms of precipitation are presented in Appendix B.)

SNOW GRAINS AND SNOW PELLETS   **Snow grains** are small, fairly flat or elongated, opaque grains of ice with diameters that are generally less than 1 mm. They fall in small quantities from stratus clouds, but never in a shower. They are the solid equivalent of drizzle. On striking a hard surface, they neither bounce nor shatter. Alternatively, snow pellets, or graupel, are white, opaque grains of ice, with diameters less than 5 mm, that form near 0°C by accretion of supercooled

droplets onto a falling ice crystal. They are similar to hail, except smaller, and are also sometimes confused with snow grains. The distinction between snow pellets and snow grains is easily made by remembering that unlike snow grains, snow pellets are brittle and crunchy and bounce (or break apart) on hitting a hard surface. They usually fall as showers, especially from towering cumulus (cumulus congestus) clouds.

To understand how snow pellets form, consider the cumulus congestus (towering cumulus) cloud with a high liquid-water content in ● Figure 7.24. The freezing level is near the surface, and because the atmosphere is conditionally unstable, the air temperature drops quickly with height. An ice crystal falling into the cold (−23°C) middle region of the cloud would be surrounded by many supercooled cloud droplets and ice crystals. In the very cold air, the crystals tend to rebound after colliding rather than stick to one another. However, when the ice crystals collide with the supercooled water droplets, they immediately freeze the droplets, producing a spherical accumulation of icy matter (rime) containing many tiny air spaces.

• **FIGURE 8** Freezing rain covers Syracuse, New York, during January 1998, causing tree limbs to break and power lines to sag.

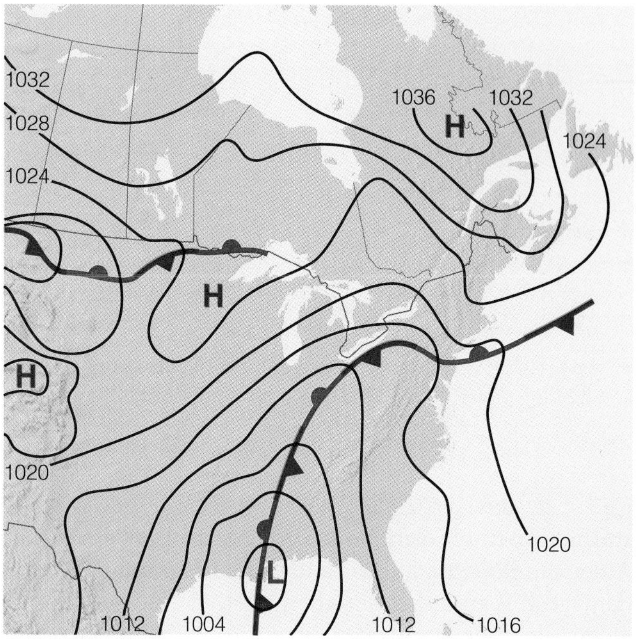

• **FIGURE 9** Surface weather map for January 7, 1998. The front just south of the Great Lakes was near this position for the period between January 4 and 10, as was the area of high pressure over northern Quebec.

replaced by rain behind a front. The particular circumstances resulting in this storm have been linked by some researchers to the strong El Niño that occurred that winter, although it is difficult to say for certain whether the El Niño is to blame. (For more information on El Niño, a tropical ocean and climate phenomenon that can affect worldwide weather, read ahead in Chapter 10, p. 310.)

These small air bubbles have two effects on the growing ice particle: (1) they keep its density low, and (2) they scatter light, making the particle opaque. By the time the ice particle reaches the lower half of the cloud, it has grown in size and its original shape is gone. When the ice particle accumulates a heavy coating of rime, it is called graupel. Since the freezing level is at a low elevation, the graupel reaches the surface as a light, round clump of snowlike ice called a snow pellet (see • Figure 7.25).

On the surface, the accumulation of snow pellets sometimes gives the appearance of tapioca pudding; hence, it can be referred to as *tapioca snow*. In a thunderstorm, graupel that reaches the ground is sometimes called *soft hail* (although real hail is over 5 mm in size, which is larger than graupel). During summer, the graupel may melt and reach the surface as a large raindrop. In vigorously convective clouds, however, the graupel may develop into full-fledged hailstones.

HAIL **Hailstones** are pieces of ice, either transparent or partially opaque, measuring 5 mm in diameter or larger (the size

of small peas, to golf balls, or larger). Some are round; others take on irregular shapes (see • Figure 7.26). Currently, the largest authenticated hailstone fell on Aurora, Nebraska, during June 2003 and had a diameter of 17.8 cm and a circumference of 47.6 cm (see • Figure 7.27). But as it broke on landing, its weight was never officially determined, although it was estimated to weigh about 800 g. The heaviest recorded hailstone, weighing 1.02 kg, fell in the Gopalganj district of Bangladesh in April 1986 in conjunction with a hailstorm that killed 92 people. As Canada generally has less intense thunderstorms, the record hailstone is smaller. It fell on Cedoux, Saskatchewan, in August 1973, weighed 290 g, and measured about 11.4 cm in diameter during a storm that caused about $10 million in damage.

Needless to say, large hailstones are quite destructive. They harm wildlife, break windows, dent cars, batter roofs, kill livestock, and cause extensive damage to crops. A single hailstorm can destroy crops in a matter of minutes, which is why they are known as "the white plague." In North America, hail damage amounts to hundreds of millions of dollars annually.

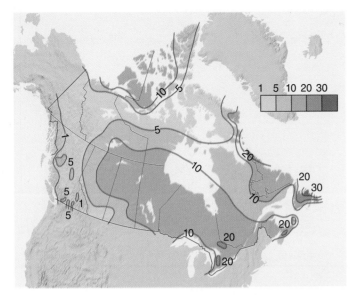

● FIGURE 7.22 Average annual number of days with freezing rain and freezing drizzle over Canada. (Modified after D. Phillips, 1990.)

The Climates of Canada. D. Phillips. Environment Canada, 1990. Pg. 61 © Her Majesty The Queen in Right of Canada, Environment Canada, 2010.

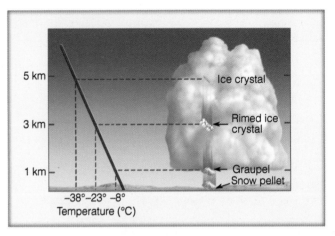

● FIGURE 7.24 The formation of snow pellets. In the cold air of a convective cloud, with a high liquid-water content, ice particles collide with supercooled cloud droplets, freezing them into clumps of icy matter called graupel. On reaching the relatively cold surface, the graupel is classified as snow pellets.

Currently, the costliest hailstorm on record, the Tristate hailstorm, moved for eight hours along a 586 km path across the American Midwest in 2001 and caused $1.9 billion in insured damage. It is among the costliest weather events in the United States. In Canada, a hailstorm on August 2, 2009, touched Calgary but devastated Carstairs, Alberta, and the surrounding agricultural areas, causing an estimated $500 million in damage. Although hailstones are potentially lethal, fatalities are rare in North America due to sparse populations in hail-affected areas. Worldwide death tolls caused by hail are greater and vary widely depending on the region's weather and people's ability to find shelter in a storm.

Hail is produced in a cumulonimbus cloud, usually associated with an intense thunderstorm. Inside the cloud, just about any particle—graupel, large frozen raindrops, even insects—acts as a hailstone *embryo* that grows by accumu-

lating supercooled liquid droplets in a process called *accretion*. It takes a million cloud droplets to form a single raindrop, but 10 billion cloud droplets are needed to form a golf ball–sized hailstone. For hail to grow this large, it must remain in a cloud for 5 to 10 minutes where violent, upsurging air currents carry small ice particles high above the freezing level. There hailstones grow by colliding with supercooled liquid cloud droplets. Violent rotating updrafts in severe thunderstorms are even capable of sweeping the growing ice particles latterly through the cloud. In fact, it appears that the best trajectory for hailstone growth is one that is nearly horizontal through a storm (see ● Figure 7.28).

As growing ice particles pass through regions of varying liquid water content, a coating of ice forms around them, causing them to grow larger and larger. In a strong updraft, the larger hailstones ascend very slowly and may appear to "float" in the updraft, where they continue to grow rapidly by colliding with numerous supercooled liquid droplets. When

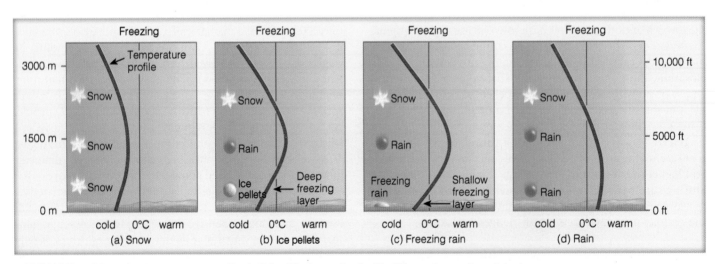

● FIGURE 7.23 Vertical temperature profiles (solid red line) associated with different forms of precipitation.

FOCUS ON AN OBSERVATION

Aircraft Icing

Consider an aircraft flying through an area of freezing rain or through a region of large, supercooled droplets in a cumuliform cloud. As the large, supercooled drops strike the leading edge of the wing, they break apart and form a film of water, which quickly freezes into a solid sheet of ice. This smooth, transparent ice—called clear ice—is similar to the freezing rain or glaze that coats trees during ice storms. Clear ice can build up quickly; it is heavy and difficult to remove, even with modern deicers.

When an aircraft flies through a cloud composed of tiny, supercooled liquid droplets, rime ice may form. Rime ice forms when some of the cloud droplets strike the wing and freeze before they have time to spread, thus leaving a rough and brittle coating of ice on the wing. Because the small, frozen droplets trap air between them, rime ice usually appears white (see Figure 7.21 on page 213). Even though rime ice redistributes the flow of air over the

● FIGURE 10 An aircraft undergoing deicing during inclement winter weather.

wing more than clear ice does, it is lighter in weight and is more easily removed with deicers.

Because the raindrops and cloud droplets in most clouds vary in size, a mixture of clear and rime ice usually forms on aircraft. Also, because concentrations of liquid water tend to be greatest in warm air, icing is usually heaviest and most severe when the air temperature is between 0°C and −10°C.

A major hazard to aviation, icing reduces aircraft efficiency by increasing weight. Icing has other adverse effects, depending on where it forms. On a wing or fuselage, ice can disrupt the airflow and decrease the plane's flying capability. When ice forms in the air intake of the engine, it robs the engine of air, causing a reduction in power. Icing may also affect the operation of brakes, landing gear, and instruments. Because of the hazards of ice on an aircraft, its wings are usually sprayed with a type of antifreeze before taking off during cold, inclement weather (see ● Figure 10).

---

winds aloft carry the large hailstones away from the updraft or when the hailstones reach appreciable size, they become too heavy to be supported by the rising air and fall.

In the warmer air below the cloud, hailstones begin to melt. Small hail often melts completely before reaching the ground, but in violent thunderstorms, hailstones often grow large enough to remain frozen when they reach the surface. Because instability provides the conditions for hail formation, we find the largest form of frozen precipitation occurring during the warmest time of the year when, generally, the atmosphere is most unstable and in areas most affected by convective storms.

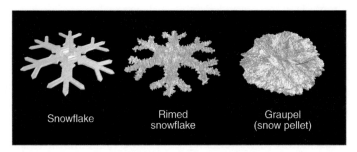

● FIGURE 7.25 A snowflake becoming a rimed snowflake and then finally graupel (a snow pellet).

● Figure 7.29 shows a cut section of a very large hailstone. Notice that it has distinct concentric layers of milky white and clear ice. We know that a hailstone grows by accumulating supercooled water droplets. If the growing hailstone enters a region inside the storm where the liquid water content is relatively low (called the *dry growth regime*), supercooled droplets will freeze immediately on the hail, producing a coating of white or opaque rime ice containing many air bubbles. As supercooled water droplets freeze onto a hailstone's surface, the liquid-to-ice transformation releases latent heat, which keeps the hailstone's surface temperature (which is below freezing) warmer than that of its environment. As long as the hailstone's surface temperature remains below freezing, liquid supercooled droplets freeze on contact, producing a coating of rime.

Should, however, the hailstone get swept into a region of the storm where the liquid-water contact is higher (called the *wet growth regime*), supercooled water droplets will collect so rapidly on the hail that, due to the release of latent heat, the stone's surface temperature will remain at 0°C. Now the supercooled droplets no longer freeze on impact; instead, they spread a coating of water around the hailstone, filling in the porous regions. As the water coating the hailstone slowly freezes, air bubbles are able to escape, leaving a layer of clear ice around the stone. Therefore, as a hailstone passes through

• FIGURE 7.26 The accumulation of small hail after a thunderstorm. Hail forms as supercooled cloud droplets collect on ice particles called graupel inside cumulonimbus clouds.

a thunderstorm of changing liquid water contents (the dry and wet growth regimes), alternating layers of opaque and clear ice form, as illustrated in Figure 7.29.

As a thunderstorm moves along, it may deposit its hail in a long narrow band, often several kilometres wide and about 10 km long, known as a **hailstreak**. If the storm should remain almost stationary for a period of time, substantial accumulation of hail is possible. Because hailstones are so damaging, various methods have been tried to prevent them from forming in thunderstorms. One method employs the seeding of clouds with large quantities of silver iodide. These nuclei freeze supercooled water droplets and convert them into ice

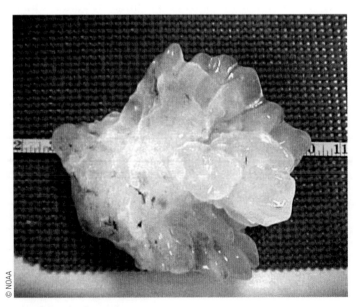

• FIGURE 7.27 This giant hailstone—the largest ever reported in the United States with a diameter of 17.8 cm—fell on Aurora, Nebraska, in June 2003.

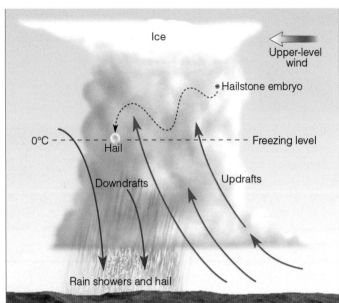

ACTIVE FIGURE 7.28 Hailstones begin as embryos that remain suspended in the cloud by violent updrafts. When the updrafts are tilted, the ice particles are swept horizontally through the cloud, producing the optimal trajectory for hailstone growth. Along their path, the ice particles collide with supercooled liquid droplets, which freeze on contact. The ice particles eventually grow large enough and heavy enough to fall toward the ground as hailstones. Visit the textbook's website to view this and other Active Figures at www.ahrensmeteorology1ce.nelson.com

WEATHER WATCH

February 11, 1999, was a snowy day in Tahtsa Lake, located in the northern Coast Mountains of British Columbia. On this day, the area received 145 cm of snow—the greatest 24-hour snowfall ever observed in Canada.

crystals. The ice crystals grow larger as they come in contact with additional supercooled cloud droplets. In time, the ice crystals grow large enough to be called graupel, which then becomes a hailstone embryo. Large numbers of embryos are produced by seeding in the hope that competition for the remaining supercooled droplets may be so great that none of the embryos will be able to grow into large and destructive hailstones. Russian scientists claim great success in suppressing hail using ice nuclei, such as silver iodide and lead iodide. Hail suppression experiments in the United States and Canada's Alberta Hail Project have been largely inconclusive.

Up to this point, we have examined the various types of precipitation. ▼ Table 7.5 summarizes the different types (from drizzle to hail).

Measuring Precipitation

Precipitation measurements seem to be the simplest of all weather measurements; all you need is a container to collect rain and a ruler to measure its depth. But what does this

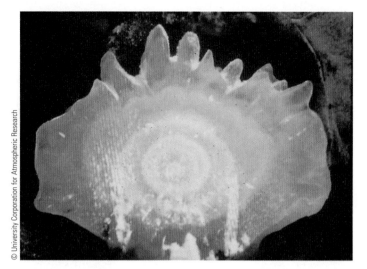
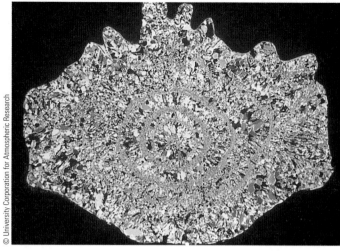

● FIGURE 7.29 To reveal a hailstone's layered structure, it is cut in cross section and then photographed under regular light (left) and polarized light (right).

▼ Table 7.5 Summary of Precipitation Types

PRECIPITATION TYPE (CODE USED IN WEATHER REPORTS)	WEATHER SYMBOL	DESCRIPTION
Drizzle (DZ)	〞 (light)	Tiny water drops with diameters less than 0.5 mm that fall slowly, usually from a stratus cloud
Rain (RA)	•• (light)	Falling liquid drops that have diameters greater than 0.5 mm
Snow (SN)	✱✱ (light)	White (or translucent) ice crystals in complex hexagonal (six-sided) shapes, most of which are branched. At temperatures higher than about −5°C, they often join together to form snowflakes.
Ice pellets (PL)	△	These take two forms: (a) Frozen raindrops that form as cold raindrops (or partially melted snowflakes) refreeze while falling through a relatively deep subfreezing layer (b) Pellets of snow covered by a thin layer of ice that has formed from freezing, either of water from partial melting of the pellets or of droplets intercepted by the pellets
Freezing rain (FZRA)	⌢ (light)	Supercooled raindrops that fall through a relatively shallow subfreezing layer and freeze on contact with cold objects at the surface
Freezing drizzle (FZDZ)	⌢ (light)	Supercooled drizzle droplets that fall through a relatively shallow subfreezing layer and freeze on contact with cold objects at the surface
Snow grains (SG), also called granular snow	⟁	White or opaque particles of ice less than 1 mm in diameter that usually fall from stratus clouds or fog in small amounts and are the solid equivalent of drizzle. The grains are often flat or elongated and do not bounce or shatter when they hit ground.
Snow pellets (SHGS)	⬦ (light showers)	Brittle, soft white (or opaque), usually round particles of ice with diameters less than 5 mm (graupel) that generally fall as showers from cumuliform clouds; they are softer and larger than snow grains
Hail (SHGR)	⬦ (moderate or heavy showers)	Transparent or partially opaque ice particles in the shape of balls or irregular lumps that range in size from that of a pea to that of a softball; *small hail* (graupel) has a diameter less than 5 mm and is abbreviated SHGS like snow pellets; hail almost always is produced in a thunderstorm
Ice crystals (IC), also called diamond dust	↔	A fall of nonbranched needle, plate, or column ice crystals, often so tiny that they seem suspended in air. They may fall from a cloud or a cloudless sky and are made visible when they glitter in sunshine, when they can produce a halo. Ice crystals occur only at very low temperatures under stable conditions and are frequent in polar regions.

mean? Is this the precipitation for the region? Or the area just around the measurement? How do you interpret this information? For example, a typical rain gauge (such as the Type B gauge described below) has a collector that is 100 cm^2 in area, but the measurements from this gauge may be supposed to represent an area that is 10,000 km^2 or more in size. In this case, the rain gauge precipitation sample is representing an area that is a trillion (10^{12}) times larger. Since we know that precipitation can be quite variable over an area, there are always questions about how well the precipitation measured at one spot represents the precipitation over a larger area. Some of the remote sensing methods, such as radar and satellite, which are discussed later in this chapter, allow for better estimation of precipitation across an area.

Another factor that will affect precipitation measurements is the details of the site where the precipitation is being measured. Precipitation, especially snow, can be blown about by wind currents, and this can affect how much precipitation reaches the collector. For this reason, to make measurements that will better represent precipitation across an area, it is important to locate precipitation collectors close to the ground and in clear areas away from obstructions that affect local wind patterns.

INSTRUMENTS   Any instrument that can collect and measure rainfall is called a rain gauge. Although there are many shapes and types of these, most consist of a circular funnel-shaped collector attached to a long measuring tube.* Note how the cross-sectional area of the collector is many times that of the measuring tube. Hence, rain falling into the collector is amplified in the tube, permitting measurements of greater precision. Because of this amplification, rainfall can be measured when amounts as small as 0.2 mm are collected. Any rainfall less than 0.2 mm is called a **trace (of precipitation)**.

In Canada, official precipitation measurements are made using Type B rain gauges (see ● Figure 7.30). Here the measuring tube is graduated with markings indicating the water's depth scaled to the area of the collecting funnel so that precipitation is measured as a depth collected over the area of the rain gauge opening (rather than as a volume). The Canadian Type B gauge measuring tube can hold 25 mm of rain in the measuring tube. Amounts above this overflow into an outer cylinder that can hold up to 250 mm of rain. Here the excess rainfall is stored and protected from appreciable evaporation. When the gauge is emptied, the overflow is carefully poured into the tube, measured, and added to the 25 mm that was collected before overflowing. Type B rain gauges can be used when precipitation is frozen (e.g., snow or hail), but to obtain the reading, a measured amount of warm water is added to melt the collected precipitation. The resulting total amount is measured, and the known amount of added warm

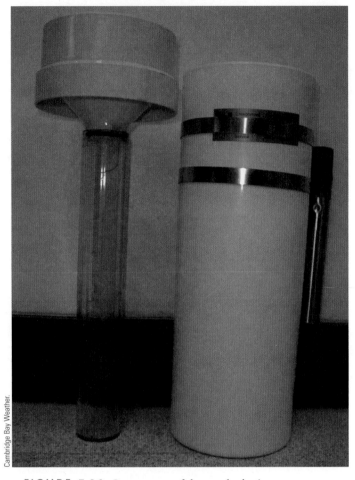

Cambridge Bay Weather.

● FIGURE 7.30 Components of the standard rain gauge.

water is subtracted to arrive at the actual precipitation amount.

For all rain gauges, evaporation of the collected precipitation is a problem that must be eliminated. This is usually done through making frequent enough measurements, instrument design, or changing the surface tension of the collected precipitation with additives so that evaporation is reduced.

Another instrument that measures rainfall is the **tipping bucket rain gauge**. In ● Figure 7.31, notice that this gauge has a receiving funnel leading to two small collecting buckets. Each bucket can align beneath the funnel and collect 0.2 mm of water. When a bucket is full, the water's weight tips that bucket; it empties and is counted, and the other bucket moves under the funnel and continues to catch water. This continues until there is not enough water to fill one of the buckets. Each time a bucket tips, an electric contact is made, causing the water in the bucket to be counted on a *data logger** or, for older instruments, on an automatic recording chart. Adding up the total number of buckets that filled results in the rainfall

---

*Outside North America, basic rain gauges are often a simple water storage container with a funnel top. Measurements are made by pouring the water into a graduated cylinder and measuring it as a volume, which is subsequently converted to a depth measurement.

---

*Data loggers are electronic devices based on microcomputers that can collect, process, and store measurements from electronic sensors, including various types of rain gauges. They are the "brain" of modern automated weather stations and can often transmit data to other computers using Internet, cellular phone, telephone, radio, and satellite communications.

• FIGURE 7.31 The tipping bucket rain gauge. Each time the bucket fills with 0.2 mm of rain, it tips, sending an electric signal to the remote recorder.

rate or amount of rainfall for a certain time period. A problem with tipping bucket rain gauges, especially during heavy downpours, is that during each "tip," they lose a small amount of rainfall and undermeasure rainfall amounts.

Remote recording of precipitation can also be made with a **weighing-type rain gauge**, where precipitation is caught in a cylinder and accumulates in a bucket. The bucket sits on a sensitive weighing platform. Special gears translate the accumulated weight of rain or snow into millimetres of precipitation. Precipitation totals are recorded on a data logger or automated chart paper recorder (a pen on a clock-driven drum). As for other electronic measurements, a data logger with communications equipment can transmit data collected from rain gauges in remote areas to satellites or land-based stations, thus providing precipitation totals from previously inaccessible regions.

Snow depth can be determined by measuring its depth at three or more representative areas. The amount of snowfall is defined as the average of these measurements. Since snow often blows around and accumulates into drifts, finding a representative area can be a problem, especially in confined areas or under tree canopies. Determining the actual depth of snow can include considerable educated guesswork. Snow depth may also be measured by removing the collector and inner cylinder of a standard rain gauge and allowing snow to accumulate in the outer tube. Turbulent air around the edge of the tube often blows flakes away from the gauge. This makes the amount of snow collected less than the actual snowfall. To remedy this, slatted windshields are placed around the cylinder to block the wind and ensure a more correct catch. (For more information on measuring snow, see Focus on an Observation: Measuring Snow Depth on p. 218.)

The depth of water that would result from the melting of a snow sample is called the **water equivalent**, or **snow water equivalent** (SWE). In a typical fresh snowpack, about 10 cm of snow will melt down to about 1 cm of water, giving a snow to water equivalent ratio of 10:1. This ratio, however, will vary greatly, depending on the density (i.e., texture and packing) of the snow. Very wet snow falling in air near freezing may have a water equivalent of 6:1. On the other hand, in dry, powdery snow, the ratio may be as high as 30:1. Toward the end of the winter, large, compacted drifts representing the accumulation of many storms may have a water equivalent of less than 2:1.

Determining the snow water equivalent is a fairly straightforward process: the snow accumulated in a rain gauge is melted and its depth (adjusted for the collecting area of the rain gauge) is measured. Another method uses a long, hollow tube pushed into the snow to a desired depth. This snow sample is then melted and poured into a rain gauge for measuring its depth. Knowing the water equivalent of snow can provide valuable information about spring runoff and the potential for flooding.

Precipitation is a highly variable weather element. Often rain can drench one section of a town while leaving another completely dry. Given this variability, it should be apparent that multiple rain gauges are necessary to accurately characterize precipitation for any particular region.

RADAR AND PRECIPITATION **Radar** (radio detection and ranging) has become an essential tool of the atmospheric scientist as it gathers information about storms and precipitation in previously inaccessible regions. Atmospheric scientists use radar to examine the inside of a cloud much like physicians use X-rays to examine the inside of a human body. Essentially, the radar unit consists of a transmitter that sends out short, powerful microwave pulses. When this energy encounters a foreign object—called a *target*—a fraction of the energy is scattered back toward the transmitter and is detected by a receiver (see • Figure 7.32). The returning signal

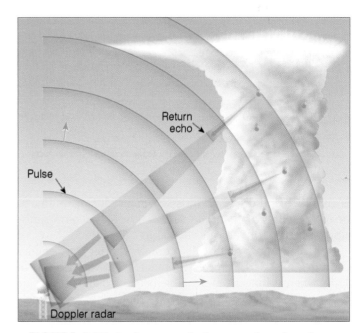

• FIGURE 7.32 A microwave pulse is sent out from the radar transmitter. The pulse strikes precipitation-sized particles, and a fraction of its energy is reflected back to the radar unit, where it is detected and displayed, as shown in Figure 7.33.

FOCUS ON AN OBSERVATION

Measuring Snow Depth

Dr. Stephen J. Déry, Environmental Science and Engineering Program, University of Northern British Columbia

Snow forms a ubiquitous feature of the winter landscape across most of Canada and indeed much of the world poleward of the middle latitudes. Snow has distinct physical properties that influence local weather conditions and even global climate. For instance, it is a cold (snow is always at or below the freezing point) and highly reflective surface* that generally cools ambient air temperatures. It also forms a seasonal reservoir for fresh water that is quickly released during spring, leading to snowmelt pulses (known as freshets) in streams and rivers. The accumulation of snow typically begins in the fall, when air temperatures decrease sufficiently for precipitation to fall in the solid phase. Snowfall leads to increases in snow depth, whereas melting, settling, and compaction of the snow, as well as **sublimation** and **evaporation** from wet snow packs, reduce its depth. Winds that cause the snow to drift in open environments and avalanches in alpine terrain may also affect snow depths.

Given its prevalence and its highly dynamic nature, snow is routinely measured by meteorologists. In its most basic form, snow depth is typically measured on a daily basis with a ruler or a graduated stick. The simplicity of this approach has been exploited for many decades, providing some of the longest environmental records in places characterized by snow, such as

*The **albedo**, or reflectivity of snow to solar radiation, ranges from 40 percent for old snow to as high as 95 percent for fresh snow.

Matthew J. Beedle

● FIGURE 11 Photograph of a meteorological tower deployed at Castle Creek Glacier in the Cariboo Mountains of British Columbia. The acoustic snow depth sensor is the device mounted at the end of the second (from the top) cross-arm.

Canada. More recently, sonic technology has allowed frequent automatic snow depth measurements (as often as once a minute). These snow depth sensors work by transmitting sound waves toward the surface, which are then reflected back to the device (see ● Figure 11). The time elapsed between transmission and reception of the sound wave depends on the air temperature and the distance between the sensor and the snow surface and is used to determine snow depth. The information is then recorded by a data logger, which can remain unattended for months at a time. This method is

therefore particularly useful in gauging snow depths in remote regions such as mountainous terrain or the high Arctic.

However, both of these approaches provide information only at one point and may not necessarily be representative of conditions across a region. Thus, single-point assessments may be augmented with comprehensive snow surveys during which many more snow depth measurements (perhaps 100 over an area of 1 km^2) using snow probes are taken. Although labour intensive, this field work provides essential information on the spatial variability of

is amplified and displayed on a screen, producing an image, or echo, from the target. The elapsed time between transmission and reception indicates the target's distance.

Smaller targets require detection by shorter wavelengths. Cloud droplets are detected by radar using wavelengths of 1 cm, whereas longer wavelengths (between 3 and 10 cm) are only weakly scattered by tiny cloud droplets but are strongly scattered by larger precipitation particles. The brightness of the echo is directly related to the amount (intensity) of rain falling in the cloud. So the radar screen shows not only where precipitation is occurring but also how *intense* it is. The radar

image is usually displayed using various colours to denote the intensity of precipitation, typically with grey or light blue representing the lightest precipitation and reds and oranges representing the heaviest precipitation.

During the 1990s, **Doppler radar** replaced conventional radar that was put into service shortly after World War II. Like conventional radar, Doppler radar detects areas of precipitation and measures rainfall intensity (see ● Figure 7.33a). Using a computer program, the rainfall intensity over an area for a given time can be computed and displayed as an estimate of total rainfall. But because Doppler radar uses the snow depths, whereas the point data indicate how the depth changes over time.

Recent developments in remote sensing technologies now allow the detection of snow depth information from satellites, as shown in ● Figure 12. Remote detection devices record a variety of radiation, including gamma waves, microwaves, and light, both emitted and reflected by Earth's surface. Complex algorithms then interpret the spectrum of this radiation, which depends in part on surface type and cover. Through this approach, global maps of snow depth averaged over areas of about 100 km² are derived on a daily basis. Despite its unique advantages, remote sensing sometimes has difficulty establishing snow depth over rough terrain or when snowpacks are deep (greater than one metre).

Maps of snow depth in many areas, such as Canada, are also currently generated by models that incorporate a variety of information, including point observations at meteorological stations and remote sensing data. These analyses provide day-to-day variations in snow depths that are especially useful for water resource managers, avalanche forecasters, and ski operators. Recently, there has been renewed interest in tracking snow accumulation as it is a sensitive indicator of climate change. Routine measurements of snow depth through a variety of techniques are likely to continue as we seek to better understand the impacts of climate change on the Canadian environment.

NASA/GSFC, MODIS Rapid Response

● FIGURE 12 Image from the *Terra* satellite showing snow cover over eastern North America on March 30, 2008.

principle called the *Doppler shift*,* it can also measure the speed that falling precipitation is moving horizontally toward or away from the radar antenna (see Figure 7.33b). Since falling precipitation moves with the wind, Doppler radar allows meteorologists to peer into a storm and observe its wind. We will investigate these ideas further in Chapter 14 when we consider the formation of severe thunderstorms and tornadoes.

In some instances, radar displays indicate precipitation when there is none reaching the surface. This situation happens because the radar signal travels at an angle just above the horizon (0.5° in Figures 7.33a and 7.33b), whereas Earth's surface curves away from it. Hence, the radar signal travels upward in altitude as it travels away from the radar antenna,

---

*The Doppler shift, also called the Doppler effect, is the change in the frequency of waves that occurs when either or both the emitter and the receiver are moving toward or away from each other. For example, a similar phenomenon occurs when a high-speed train whistles as it approaches. The higher pitched (higher frequency) whistle heard as the train approaches will shift to a lower pitch (lower frequency) after the train passes. The change in pitch is related to the speed of the train. For radar, the change in microwave frequency is used to calculate the speed of the precipitation in the direction of the radar beam.

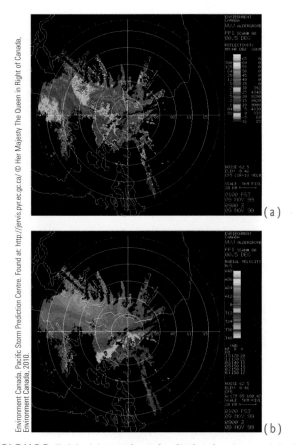

● **FIGURE 7.33** (a) Doppler radar display showing precipitation intensity over the Lower Mainland of British Columbia, near Vancouver, for November 9, 1999. The lightest precipitation is shown as green and purple; heavier rainfall is indicated by blue and red colours. The numbers under the letters DBZ represent the logarithmic scale for measuring the radar echo caused by precipitation particles. The corresponding rainfall rate in millimetres per hour is listed beside each colour. (b) Doppler radar display showing wind speed in the direction of the radar beam. Winds blowing toward the radar are coloured in blues and purples, and winds blowing away from the radar are in reds and oranges, with speeds in metres per second shown in the colour bar to the right of the image. The change in colour from shades of blue to shades of red indicates zero wind speed in the radar beam direction. Therefore, the wind is blowing at right angles to this line: from the south–southeast toward the north–northwest.

and the return echo is not necessarily from precipitation reaching the ground but of raindrops in a cloud. So if radar indicates that it is raining in your area, and outside you observe that it is not, remember that it is raining, but the raindrops are probably evaporating before reaching the ground.

The next improvement for Doppler radar is *polarimetric radar*. This type of Doppler radar transmits both a vertical and a horizontal pulse that will make it easier to determine whether falling precipitation is in the form of rain or snow.

MEASURING PRECIPITATION FROM SPACE
To resolve the problems with a lack of surface precipitation measurements over most of Earth's oceans (over 71% of the globe) and spotty or inconsistent measurements over large parts of con-

tinents, space-based measurements are being developed. Starting in the 1980s, the concept of measuring precipitation from satellite-based observing platforms has led to large international collaborative research projects, dominantly supported by the United States, Japan, and other international agencies. Beginning with the *TRMM* (Tropical Rainfall Measuring Mission) and moving in 2013 to the GPM (Global Precipitation Measurement Mission), the technology is increasingly successful at remotely measuring precipitation. As it circles Earth at an altitude of about 400 km, the *TRMM* satellite is able to measure rainfall intensity in previously inaccessible regions of the tropics and subtropics. The onboard precipitation radar is capable of detecting rainfall rates down to about 0.7 mm per hour while at the same time providing vertical profiles of rain and snow intensity from the surface up to about 20 km. The microwave imager complements the precipitation radar by measuring emitted microwave energy from Earth, the atmosphere, clouds, and precipitation, which is translated into rainfall rates over a broad area. The visible and infrared scanner onboard the satellite measures visible and infrared energy from Earth, the atmosphere, and clouds. This information is used to delineate the location of clouds and the temperature of cloud tops, which complements the rainfall rate information provided by the radar and microwave sensors. The satellite also carries instruments to measure Earth's radiant energy and to detect lightning. A *TRMM* satellite image of Hurricane Humberto and its pattern of precipitation is provided in ● Figure 7.34.

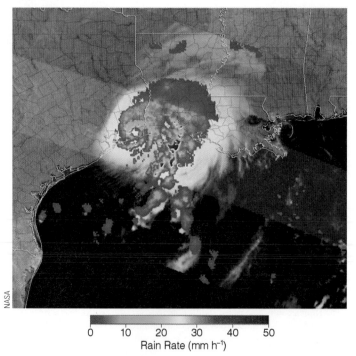

● **FIGURE 7.34** A satellite and radar image of Hurricane Humberto obtained by the *TRMM* satellite on September 13, 2007. Precipitation rates (lowest in blue, highest in dark red) were obtained by the satellite's precipitation radar and microwave imager. The rainfall estimates are overlain on the infrared image of the storm.

Launched in April 2006, the satellite *CloudSat*, developed by NASA in partnership with the Canadian Space Agency, circles Earth in an orbit about 700 km above the surface. Onboard *CloudSat*, a very sensitive radar (called the cloud profiling radar) is able to peer into a cloud and provide a vertical view of its tiny cloud droplets and ice particles, as well as precipitation. An example *CloudSat* scan through a storm affecting France is shown in Chapter 5, Figure 5.44. It is hoped that such vertical profiling of liquid water and ice will provide scientists with a better understanding of precipitation processes that go on inside the cloud and the role that clouds play in Earth's global climate system.

SUMMARY

In this chapter, we have seen that cloud droplets are too small and light to reach the ground as rain. Cloud droplets do grow larger by condensation, but this process by itself is much too slow to produce substantial precipitation. Because larger cloud droplets fall faster and farther than smaller ones, they grow larger as they fall by coalescing with drops in their path. If the air temperature in a cloud drops below freezing, then ice crystals play an important role in producing precipitation. Some ice crystals may form directly on ice nuclei, or they may result when an ice nucleus makes contact with and freezes a supercooled water droplet. Because of differences in vapour pressures between water and ice, an ice crystal surrounded by water droplets grows larger at the expense of the droplets. As the ice crystal begins to fall, it grows even larger by colliding with supercooled liquid droplets, which freeze on contact. In an attempt to coax more precipitation from them, clouds can be seeded with silver iodide.

Precipitation can reach the surface in a variety of forms. In winter, raindrops may freeze on impact, producing freezing rain, which can disrupt electrical service by downing power lines. Raindrops may freeze into tiny pellets of ice above the ground and reach the surface as ice pellets. Depending on conditions, snow may fall as pellets, grains, or flakes, all of which can influence how far we see and hear. Strong updrafts in a cumulonimbus cloud may keep ice particles suspended above the freezing level, where they acquire a further coating of ice and form destructive hailstones. Although the rain gauge is still the most commonly used method of measuring precipitation, Doppler radar has become an important instrument for determining precipitation intensity and estimating rainfall amount. In tropical regions, rainfall estimates can be obtained from radar and microwave scanners onboard satellites.

KEY TERMS

The following terms are listed in the order they appear in the text. Define each. Doing so will aid you in reviewing the material covered in this chapter.

precipitation, 192
equilibrium vapour pressure, 192
curvature effect, 193
solute effect, 193
collision–coalescence process, 194
terminal velocity, 194
coalescence, 194
ice-crystal (Bergeron) process, 196
supercooled (water droplets), 196
ice nuclei, 197
contact freezing, 197
accretion, 199
snow pellets (graupel), 199
aggregation, 199
snowflake, 199
cloud seeding, 200
rain, 201
drizzle, 201
virga, 203
shower (rain), 203
snow, 204
fallstreaks, 204
flurries (snow showers), 206
snow squall, 206
blizzard, 207
ice pellet, 208
sleet, 208
freezing rain (glaze, verglas), 209
freezing drizzle, 209
rime, 209
snow grains, 210
hailstones, 211
hailstreak, 214
trace (of precipitation), 216
tipping bucket rain gauge, 216
weighing-type rain gauge, 217
water equivalent (snow water equivalent), 217
radar, 217
sublimation, 218
evaporation, 218
albedo, 218
Doppler radar, 218

QUESTIONS FOR REVIEW

1. What is the primary difference between a cloud droplet and a raindrop?

2. Why do typical cloud droplets seldom reach the ground as rain?

3. Describe how the process of collision and coalescence produces precipitation.

4. Would the collision and coalescence process work better at producing rain in (a) a warm, thick nimbostratus cloud or (b) a warm, towering cumulus congestus cloud? Explain.

5. List and describe three ways in which ice crystals can form in a cloud.

6. When the temperature in a cloud is $-30°C$, are larger cloud droplets more likely to freeze than smaller cloud droplets? Explain.

7. In a cloud where the air temperature is $-10°C$, why are there many more cloud droplets than ice crystals?

8. How does the ice-crystal (Bergeron) process produce precipitation? What is the main premise describing this process?

9. Why do heavy showers usually fall from cumuliform clouds? Why does steady precipitation normally fall from stratiform clouds?

10. Why are large snowflakes usually observed when the air temperature near the ground is just below freezing?

11. In a cloud composed of water droplets and ice crystals, is the saturation vapour pressure greater over the droplets or over the ice?

12. Why is it foolish to seed a clear sky with silver iodide?

13. When seeding a cloud to promote rainfall, is it possible to overseed the cloud so that it prevents rainfall? Explain.

14. Explain how clouds can be seeded naturally.

15. What atmospheric conditions are necessary for snow to fall when the air temperature is considerably above freezing?

16. List the advantages and disadvantages of heavy snowfall.

17. How do the atmospheric conditions that produce ice pellets differ from those that produce hail?

18. What is the difference between freezing rain and ice pellets?

19. Describe how hail might form in a cumulonimbus cloud.

20. Why is hail more common in summer than in winter?

21. List the common precipitation gauges that measure rain and snow.

22. (a) What is Doppler radar?
    (b) How does radar measure the intensity of precipitation?

QUESTIONS FOR THOUGHT

1. Ice crystals that form by accretion are fairly large. Explain why they fall slowly.

2. Why is a warm, tropical cumulus cloud more likely to produce precipitation than a cold stratus cloud?

3. Explain why very small cloud droplets of pure water evaporate even when the relative humidity is 100 percent.

4. Suppose that a thick nimbostratus cloud contains ice crystals and cloud droplets all about the same size. Which precipitation process will be most important in producing rain from this cloud? Why?

5. Clouds that form over water are usually more efficient in producing precipitation than clouds that form over land. Why?

6. During a recent snowstorm, Calgary, Alberta, received 7 cm of snow. Sixty kilometres east of Calgary, a city received no measurable snowfall, whereas 150 km east of Calgary, another city received 10 cm of snow. Since Calgary is located to the east of the Rockies, and the upper-level winds were westerly during the snowstorm, give an explanation as to what could account for this snowfall pattern.

7. Raindrops rarely grow larger than 5 mm. Two reasons were given on p. 195. Can you think of a third? (Hint: See the Focus section on p. 204 and look at the shape of a large drop.)

8. Lead iodide is an effective ice-forming nucleus. Why do you think it has not been used for that purpose?

9. When cirrus clouds are above a deck of altocumulus clouds, occasionally, a clear area, or "hole," will appear in the altocumulus cloud layer. What do you suppose could cause this to happen?

10. It is −12°C in Montreal, Quebec, and freezing rain is falling. Can you explain why? Draw a vertical profile of the air temperature (a sounding) that illustrates why freezing rain is occurring at the surface.

11. When falling snowflakes become mixed with ice pellets, why is this condition often followed by the snowflakes changing into rain?

12. A major snowstorm occurred in a city in New Brunswick. Three volunteer weather observers measured the snowfall. Observer #1 measured the depth of newly fallen snow every hour. At the end of the storm, Observer #1 added up the measurements and came up with a total of 30 cm of new snow. Observer #2 measured the depth of new snow twice, once in the middle of the storm and once at the end, and came up with a total snowfall of 25 cm. Observer #3 measured the new snowfall only once, after the storm had stopped, and reported 21 cm. Which of the three observers do you feel has the correct snowfall total? List at least five possible reasons why the snowfall totals were different.

PROBLEMS AND EXERCISES

1. In the daily newspaper, a city is reported as receiving 1.32 cm of precipitation over a 24-hour period. If all the precipitation fell as snow, and if we assume a normal snow to water equivalent ratio of 10:1, how much snow did this city receive?

2. How many times faster does a large raindrop (diameter 5000 μm) fall than a cloud droplet (diameter 20 μm) if both are falling at their terminal velocity in still air?

3. (a) How many minutes would it take drizzle with a diameter of 200 μm to reach the surface if it falls at its terminal velocity from the base of a cloud 1000 m above the ground? (Assume that the air is saturated beneath the cloud, the drizzle does not evaporate, and the air is still.)

   (b) Suppose the drizzle in problem 3a evaporates on its way to the ground. If the drop size is 200 μm for the first 450 m of descent, 100 μm for the next 450 m, and 20 μm for the final 100 m, how long will it take the drizzle to reach the ground if it falls in still air?

4. Suppose that a large raindrop (diameter 5000 μm) falls at its terminal velocity from the base of a cloud 1500 m above the ground.

   (a) If we assume that the raindrop does not evaporate, how long would it take the drop to reach the surface?

   (b) What would be the shape of the falling raindrop just before it reaches the ground?

   (c) What type of cloud would you expect this raindrop to fall from? Explain.

5. In ● Figure 7.35, a drawing of a large hailstone, explain how the areas of clear ice and rime ice could have formed.

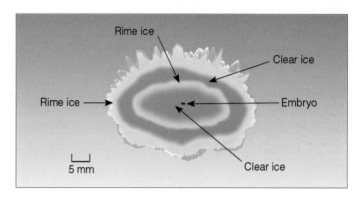

● FIGURE 7.35 Cross section of a large hailstone.

Strong swirling winds help develop these clouds over the Labrador Sea, off the coast of Newfoundland.
© John Eastcott and Yva Momatiuk/National Geographic Image Collection

NEL

Air Pressure and Winds

8

Who has seen the wind?
Who has seen the wind?
Neither I nor you:
But when the leaves hang trembling
The wind is passing thro'

Who has seen the wind?
Neither you nor I:
But when the trees bow down their heads
The wind is passing by.

Christina Rossetti

CONTENTS

Just as ocean currents are the movement of water, winds are the movement of air. They transport heat from hot regions to cold regions. Moisture and pollution are also carried with the wind. Even though the air is invisible, it has mass, and winds redistribute the air's mass to equalize pressure. Meteorologists use the term **advection** to describe the horizontal movement of air. Winds operate within the *atmospheric* system, but by moving air with its properties of heat, mass, moisture, and particles (dust, smog) from one region to another, the impacts of wind can be seen in other Earth systems. Storms, such as middle-latitude cyclones,* tornadoes, and hurricanes, are the mechanisms the atmosphere uses to accomplish this.

Why do winds blow? From Chapter 1, we know that winds result from horizontal differences in pressure. This phenomenon happens when we open a vacuum-packed can—air rushes from the higher pressure region outside the can toward the region of lower pressure inside. In the atmosphere, winds blow in an attempt to equalize imbalances in air pressure. Does this mean that the wind always blows directly from high to low pressure? Not really, because the movement of air is controlled by other forces as well as pressure.

In this chapter, we will first consider how and why atmospheric pressure varies. Then we will look at the forces that influence atmospheric motion. Through studying these forces and examining surface and upper atmosphere weather maps, we will be able to see how the wind should blow in a particular region.

Atmospheric Pressure

In Chapter 1, we learned several important concepts about atmospheric pressure. On p. 12, we learned that **air pressure** results from the weight of air above a given level. If we think of a column of air extending from Earth's surface to the top of the atmosphere, air pressure is the force of gravity acting on the mass of air above any level (i.e., the air's weight) divided by the area of the column.† As we climb in elevation above Earth's surface, there are fewer air molecules above us; hence, atmospheric pressure always decreases with increasing height. Another concept we learned was that most of our atmosphere is crowded close to Earth's surface, which causes air pressure to decrease with height, rapidly at first and then more slowly at higher altitudes.

So one way to change air pressure is to simply move up or down in the atmosphere. But what causes the air pressure to change at a fixed altitude (i.e., horizontally)? And why does the air pressure change at the surface?

---

*Middle-latitude cyclones are also known as mid-latitude cyclones.

†Pressure (*P*) is defined as a force (*F*) acting on area (*A*), where the force (*F*) results from the weight of the object causing the force (remember that weight is the object's mass [*m*] times the acceleration of gravity [*g*]). This is written mathematically as

$$P = \frac{F}{A} = \frac{(m)(g)}{A}.$$

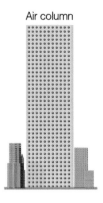

Air column

● FIGURE 8.1 A model of the atmosphere where air density remains constant with height. The air pressure at the surface is related to the number of molecules above. When more air molecules are stuffed into the column, its mass increases and the surface air pressure rises. When air is removed from the column, the surface pressure falls.

HORIZONTAL PRESSURE VARIATIONS—A TALE OF TWO CITIES Constructing *models* allows us to eliminate some of the atmosphere's complexities and answer these questions. ● Figure 8.1 shows a simple atmospheric model—a column of air, extending well up into the atmosphere. In this column, dots represent air molecules, and our model assumes that

1. the air molecules are not crowded close to the surface, and, unlike the real atmosphere, the air density remains constant from the surface up to the top of the column
2. the width of the column does not change with the height
3. air is unable to freely move into or out of the column

Suppose that we somehow force more air into the column in Figure 8.1. What would happen? The added air would increase the column's mass, resulting in increased surface air pressure. Likewise, if air were removed from the column, the surface air pressure would decrease. Consequently, to change the surface air pressure, we need to change the mass of air in the column above the surface. But how can this be accomplished?

Now let's look at ● Figure 8.2, where the same model assumptions apply. Suppose that both columns in Figure 8.2a are located at the same elevation, both have the same air temperature, and both have the same surface air pressure. This condition, of course, means that there must be the same number of molecules (same mass of air) in each column above both cities. Further suppose that the surface air pressure for both cities remains the same, whereas the air above City 1 cools and the air above City 2 warms (see Figure 8.2b).

As the air in column 1 cools, the molecules move more slowly and crowd closer together—the air becomes denser. In the warm air above City 2, the molecules move faster and spread farther apart—the air becomes less dense. Since the width of the columns does not change and air cannot flow in or out of the columns, the total number of molecules above each city remains the same, and the surface pressure does not change. Therefore, in the denser cold air above City 1, the column shrinks (coloured blue in Figure 8.2), whereas the column rises in the less dense, warm air above City 2 (coloured red). From this situation, we can conclude that it *takes a shorter column of cold, more dense air to exert the same surface pressure as a taller column of warm, less dense air.* This concept has a great deal of meteorological significance.

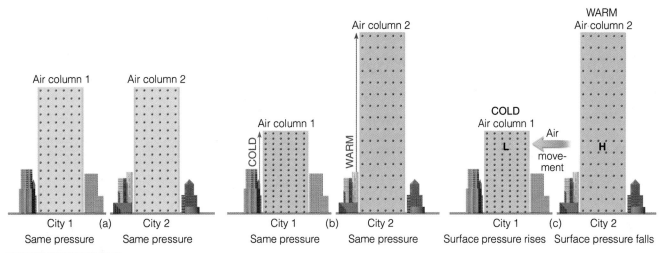

ACTIVE FIGURE 8.2 (a) Two air columns, each with identical mass, have the same surface air pressure. (b) A shorter column of cold air exerts the same pressure as a taller column of warm air. So, as column 1 cools, it shrinks, and as column 2 warms, it expands. (c) At the same level in the atmosphere, there is more air above the H in the warm column than above the L in the cold column. Warm air above the surface is associated with high pressure and cold air aloft with low pressure. The pressure differences aloft create a force that causes the air to move from a region of higher pressure toward a region of lower pressure. The removal of air from column 2 causes its surface pressure to drop, whereas the addition of air into column 1 causes its surface pressure to rise. (Note that the difference in height between the two columns is greatly exaggerated.) Visit the textbook's website to view this and other Active Figures at www.ahrensmeteorology1ce.nelson.com

Atmospheric pressure decreases more rapidly with height in the cold column of air. As you move up this column (cold air above City 1 in Figure 8.2b), notice how quickly you pass through the more densely packed molecules. In the warmer, less dense air, the pressure does not decrease as rapidly with height simply because you climb above fewer molecules in the same vertical distance.

In Figure 8.2c, move up the warm (red) column until you come to the letter H. Now move up the cold (blue) column the same distance until you reach the letter L. Notice that there are more molecules *above* the letter *H* in the warm column than *above* the letter *L* in the cold column. The number of molecules above any level is a measure of the atmospheric pressure, and this leads to an important concept: *warm air aloft is normally associated with high atmospheric pressure, and cold air aloft is associated with low atmospheric pressure.*

In Figure 8.2c, the difference in temperature creates a horizontal difference in pressure. This pressure difference establishes a force (called the *pressure gradient force*) that causes air to move from higher pressure toward lower pressure. Consequently, if we allow the air above the surface to move horizontally, the air will move from column 2 toward column 1. As the air aloft leaves column 2, the mass of the air in the column decreases, and so does the surface air pressure. Meanwhile, the accumulation of air in column 1 causes the surface air pressure to increase.

Higher air pressure at the surface in column 1 and lower air pressure at the surface in column 2 cause the surface air to move from City 1 toward City 2 (see ● Figure 8.3). As the surface air moves away from City 1, the air aloft slowly sinks to replace this outwardly spreading surface air. As the surface air flows into City 2, it slowly rises to replace the depleted air aloft.

In this manner, a complete circulation of air is established due to the heating and cooling of air columns. In summary, we can see how heating and cooling columns of air can establish horizontal variations in air pressure that cause the wind to blow.

Air temperature, air pressure, and air density are all interrelated; if one of these variables changes, the other two usually also change. The relationship among these three

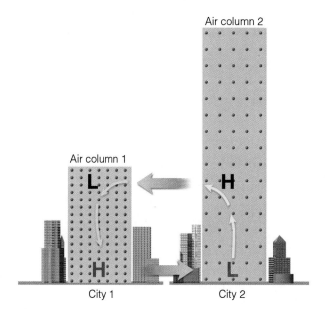

● **FIGURE 8.3** The heating and cooling of air columns cause horizontal pressure variations aloft and at the surface. These pressure variations force the air to move from areas of higher pressure toward areas of lower pressure. In conjunction with these horizontal air motions, the air slowly sinks above the surface high and rises above the surface low.

FOCUS ON A SPECIAL TOPIC

The Atmosphere Obeys the Gas Law

The relationship among pressure, temperature, and density of air can be expressed by

Pressure = Density × temperature

This simple relationship, often referred to as the *ideal gas law* or *equation of state for the atmosphere*, tells us that the pressure of a gas is equal to its temperature times its density times a constant. For air and ignoring the constant, this equation becomes*

$$P \propto \rho \times T$$

where P is air pressure, T is air temperature, ρ (this is the Greek letter rho—pronounced "row") represents air density, and $\propto$ is a mathematical symbol

---

*From chemistry, this gas law may be more familiar in the form

$$PV = nRT$$

where P = pressure of the gas, V = volume, n = number of moles of molecules, R = universal gas constant = 8.314 J K^{-1} mol^{-1} (remember: 1 mole is 6.022×10^{23} molecules), and T = temperature in kelvins. In this form of the equation, ρ (from $P \propto \rho \times T$) is separated into its components which are V, n, and the molecular weight of the gas, in this case dry air (M_d), as follows:

$$\rho = \frac{nM_d}{V}$$

meaning "is proportional to." This relationship shows that a change in one variable causes a corresponding change in the other two variables. Thus, it will be easier to understand the behaviour of air if we keep one variable from changing and observe the behaviour of the other two.

First let's hold the temperature constant. This relationship becomes

$$P \propto \rho \times (\text{constant } T)$$

Now the air pressure is proportional to its density—as long as its temperature does not change. Under these conditions, as the pressure increases, the density increases, and as the pressure decreases, the density decreases. In other words, *at the same temperature, air at a higher pressure is denser than air at a lower pressure.* As shown in ● Figure 1, if we apply this concept to the atmosphere, air above a region of surface high pressure is denser than air above a region of surface low pressure. Or stated more simply, air in an anticyclone is denser than in a mid-latitude cyclone. Surface air pressure increases when the wind causes more air molecules to move into an air column than are able to leave it. This is called *net convergence.*

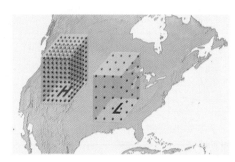

● FIGURE 1 The air above a region of surface high pressure is denser than air above a region of surface low pressure when the two are at the same temperature. The dots in each column represent air molecules.

Surface air pressure decreases when the wind causes more air molecules to move out of an air column than are able to enter it. This is called *net divergence.* The concepts of convergence and divergence will be discussed later in this chapter.

Earlier, we considered how pressure and density are related when the temperature remains constant. What happens to the gas law when the air pressure remains constant? Now the law becomes

$$(\text{constant } P) \propto \rho \times T$$

---

variables is expressed by the gas law, which is described in Focus on a Special Topic: The Atmosphere Obeys the Gas Law, above.

DAILY PRESSURE VARIATIONS From what we have learned so far, we might expect to see the surface pressure dropping as the air temperature rises, and vice versa. Over large continental areas, especially the southwestern United States in summer, hot surface air is accompanied by surface low pressure. Likewise, in winter, bitterly cold arctic air is often accompanied by surface high pressure over the Arctic. Yet, on a daily basis, any cyclic change in surface pressure associated with daily temperature changes is largely concealed by the pressure changes created by the warming of the upper atmosphere.

In the tropics, for example, pressure rises and falls in a regular pattern twice a day (see ● Figure 8.4). Maximum pressures occur around 10:00 a.m. and 10:00 p.m., and minimum pressures occur near 4:00 a.m. and 4:00 p.m. The

largest pressure difference, about 2.5 hPa, occurs near the equator. A smaller difference also shows up in higher latitudes. This **diurnal** (daily) pressure fluctuation appears to be due primarily to the absorption of solar energy by ozone in the upper atmosphere and by water vapour in the lower atmosphere. The warming and cooling of the air create density oscillations known as *thermal* (or *atmospheric*) *tides* that show up as small pressure changes near Earth's surface.

In middle latitudes, surface pressure changes are primarily the result of large high- and low-pressure areas that move toward or away from a region. Generally, when an area of high pressure approaches, surface pressure rises. When it moves away, pressure usually falls. Likewise, an approaching low causes the air pressure to fall, and one moving away causes surface pressure to rise.

PRESSURE MEASUREMENTS Meteorological instruments that detect and measure pressure are describing the force of

This relationship tells us that when air pressure is not changing, air becomes less dense as the temperature rises and denser as the temperature drops. Therefore, at a given atmospheric pressure (i.e., height in the atmosphere), cold air is denser than warm air.

The relationships described by the gas law allow us to determine information about the atmosphere when we know some values and not others. For example, in the mid-latitudes at an altitude of about 5600 m above sea level, the air pressure is approximately 500 hectopascals (hPa), and the average air density is 0.690 kg m^{-3}. By using the gas law ($P = \rho \times T \times$ constant), we can calculate the average air temperature. First, rearrange the gas law equation so that you can solve for temperature. Then substitute the known values (in SI units) into the rearranged equation. In SI units,

- P is the pressure is in pascals (Pa), so 500 hPa = 50,000 Pa = 50,000 kg m^{-1}s^{-2} (in its most basic SI units)
- T is the temperature in kelvins
- ρ = the density of air is in kilograms per cubic metre, ρ = 0.690 kg m^{-3}

- constant = R, called the gas constant. For the atmosphere, we use R_d for dry air: R_d = 287 J kg^{-1} K^{-1} = 287 m^2 s^{-2} K^{-1} (in its most basic SI units)

Following these steps, we have

$$P = \rho \times T \times R_d$$

rearranged as

$$T = \frac{P}{\rho \times R_d}$$

now substitute the known values into the equation:

$$T = \frac{50{,}000 \text{ Pa}}{0.690 \frac{\text{kg}}{\text{m}^3} \times 287 \frac{\text{J}}{\text{kg K}}}$$

$$T = \frac{50{,}000 \frac{\text{kg}}{\text{m s}^2}}{0.690 \frac{\text{kg}}{\text{m}^3} \times 287 \frac{\text{m}^2}{\text{s}^2 \text{K}}} = \frac{252.5 \text{ kg m}^3 \text{ s}^2 \text{ K}}{\text{m s}^2 \text{ kg m}^2}$$

$$= 252.5 \text{ K}$$

Finally, convert kelvins to degrees Celsius by subtracting 273.15 K:

$$T = 252.5 \text{ K} - 273.15 \text{ K} = -20.65\,^\circ\text{C}$$

$$T = -20.65\,^\circ\text{C}$$

If we know the value of temperature and density, we can use the gas law to obtain the air pressure. In Chapter 1, we saw that the average global temperature near sea level is 15°C, which is the same as 288.15 K. If the average air density at sea level is 1.225 kg m^{-3}, what would be the standard (average) sea-level pressure? Using the gas law, we obtain

$$P = \rho \times T \times R_d$$

$$P = 1.225 \frac{\text{kg}}{\text{m}^3} \times 288.15 \text{ K} \times 287 \frac{\text{m}^2}{\text{s}^2 \text{K}}$$

$$P = 101306 \text{ kg m}^{-1}\text{s}^{-1}$$

$$= 101306 \text{ Pa}$$

$$\cong 1013 \text{ hPa}$$

Since the air pressure is related to both temperature and density, a small change in either or both of these variables can bring about a change in pressure.

WEATHER WATCH

What is sea-level pressure? Although 1013.25 hPa is the **standard atmospheric pressure** at sea level, it is not Earth's average sea-level pressure, which is 1011.0 hPa. These values vary because "standard pressure" is an accepted constant used to simplify and standardize sea-level pressure under representative atmospheric conditions, whereas actual pressure observations determine the average pressure, which can vary over time. Because much of Earth's surface is above sea level, Earth's annual average *surface pressure* is estimated to be 984.43 hPa.

air over a given area and are called **barometers**. The term *barometer* literally means an instrument that measures bars*— one of several units for pressure measurement. For meteorological purposes, the bar is a large unit, so air pressure is measured in **millibars** (mb); translating this into the metric

system, these are **hectopascals** (hPa). Conveniently, 1 mb = 1hPa, which are both one one-thousandth of a bar. So

$$1 \text{ bar} = 1000 \text{ mb} = 1000 \text{ hPa}$$

The unit of pressure designated by the International System (SI) of measurement is the *pascal*, named in honour of Blaise Pascal (1632–62), whose experiments on atmospheric pressure greatly increased our knowledge of the atmosphere. A pascal (Pa) is the force of one newton acting on a surface area of one square metre. One hundred pascals equals one hectopascal. In Canada, hectopascals and kilopascals (kPa, which are 1000 Pa) are the

---

*A bar is a force of 100,000 newtons acting on a surface area of 1 square metre. A *newton* (N) is the amount of force required to move an object with a mass of 1 kg so that it increases its speed every second at a rate of 1 metre per second. In SI units, 1 bar equals 100,000 N m^{-2} = 100,000 Pascals (Pa).

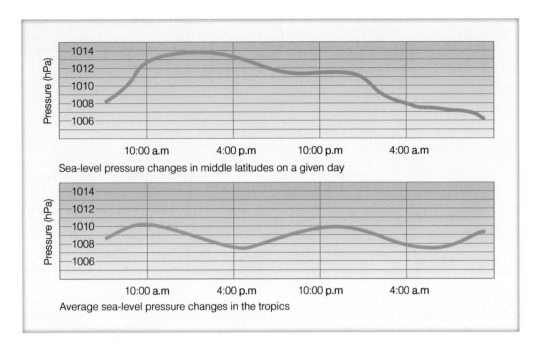

• FIGURE 8.4 Diurnal surface pressure changes in the middle latitudes and in the tropics.

official measurement units for weather-related pressures. In the United States, millibars are generally used. A common pressure unit used in aviation and engineering is *inches of mercury* (in Hg); its metric equivalent is *millimetres of mercury* (mm Hg). Additional pressure units and conversions between different measurement systems are given in Appendix A. • Figure 8.5 compares pressure readings in different unit systems and indicates

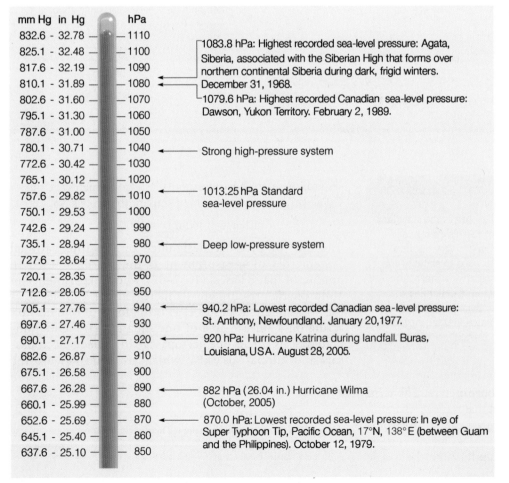

• FIGURE 8.5 Atmospheric pressure in millimetres of mercury (mm Hg), inches of mercury (in Hg), and hecto-pascals (hPa).

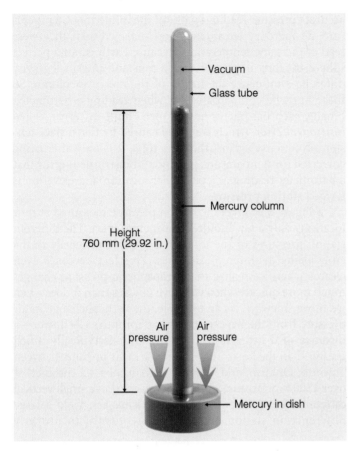

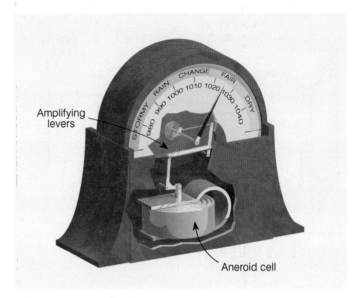

• FIGURE 8.7 An aneroid barometer.

• FIGURE 8.6 The mercury barometer. The height of the mercury column is a measure of atmospheric pressure.

significant pressure points or events. At sea level, *standard atmospheric pressure** is

1013.25 hPa = 29.92 in Hg = 760 mm Hg = 1013.25 mb

Because we measure atmospheric pressure with an instrument called a *barometer*, atmospheric pressure is also referred to as *barometric pressure*. Evangelista Torricelli, a student of Galileo, invented the **mercury barometer** in 1643. His barometer, similar to those in use today, consisted of a long glass tube open at one end and closed at the other (see • Figure 8.6). Filling the tube with mercury and covering the open end, Torricelli inverted the tube and immersed the open end into a dish of mercury. He removed the cover, and the mercury settled in the tube at nearly 760 mm above the mercury level in the dish. Torricelli correctly concluded that the column of mercury in the tube was balancing the weight of the air above the dish; hence, its height was a measure of atmospheric pressure. Mercury-in-glass or Fortin-type barometers, similar to this one, are still used today as a reliable way of determining pressure and calibrating other pressure sensors.

Why is mercury used in a barometer rather than water? The primary reason is convenience (also, water can evaporate from the tube). Mercury seldom rises to a height above 800 mm. A water barometer, however, presents a problem. Because water is 13.6 times less dense than mercury, an atmospheric pressure of 760 mm of mercury would be equivalent to 10.34 m of water. A water barometer resting on the ground near sea level would have to be read from a ladder over 10 m tall.

The most common type of home-use barometer is an **aneroid barometer**. It has no mercury and contains a small, flexible metal box called an *aneroid cell*. Before the cell is tightly sealed, air is partially removed, so small changes in external air pressure can cause the cell to expand or contract. The size of the cell is calibrated to represent different pressures; any change in cell size is amplified by levers and transmitted to an indicating arm, which points to the current atmospheric pressure (see • Figure 8.7). Notice the descriptive weather-related words that are generally printed above specific pressure values. Generally, higher readings are associated with clear weather and lower readings with inclement weather. This occurs because surface high-pressure areas are associated with sinking air and normally fair weather, whereas surface low-pressure areas are associated with rising air and usually cloudy, wet weather. A steady rise in atmospheric pressure (a rising barometer) usually indicates clearing or fair weather, whereas a steady drop in atmospheric pressure (a falling barometer) often signals an approaching storm. **Altimeters** and **barographs** are two specialized types of aneroid barometers. Altimeters measure pressure but are calibrated to indicate altitude. Barographs are aneroid cells with a pen attached to an indicating arm so that the pen continuously records pressure on a chart paper. The time of a pressure reading can be interpreted from the chart paper as it is mounted on a drum that rotates according to an internal mechanical clock (see • Figure 8.8).

*Standard atmospheric pressure at sea level is the pressure exerted by a column of mercury 760 mm (29.92 in.) high, with a mercury density of 1.35951×10^4 kg m^{-3} and subject to an acceleration of gravity of 9.80665 m s^{-2}.

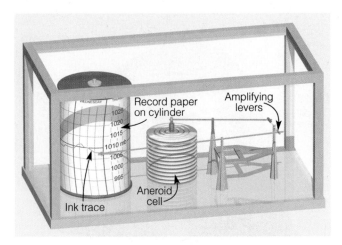

● FIGURE 8.8 A recording barograph.

Modern pressure instruments use a variety of electronic pressure sensors that are connected to an electronic data logger that processes and stores the data.

PRESSURE READINGS The seemingly simple task of reading the height of the mercury column to obtain the air pressure is actually not all that simple. Being a fluid, mercury is sensitive to changes in temperature; it will expand when heated and contract when cooled. Consequently, to obtain

accurate pressure readings without the influence of temperature, all mercury barometers are corrected as if they were read at the same temperature. Because Earth is not a perfect sphere, the force of gravity is not a constant, and small gravity differences influence the height of the mercury column. So these must be compensated for when reading a barometer. Finally, each barometer has its own "built-in" error, called *instrument error*, which is mostly caused by the surface tension of mercury against the glass tube. It is only after being corrected for temperature, gravity, and instrument error that a barometer reading at a particular location and elevation is termed **station pressure**.

● Figure 8.9a gives the station pressure measured at four locations only a few hundred kilometres apart. The different station pressures of the four cities are due primarily to the cities being at different altitudes. This fact becomes even clearer when we realize that atmospheric pressure changes much more quickly when we move upward than it does when we move sideways. As an example, the vertical change in air pressure from the base to the top of Toronto's CN Tower—a distance of a little more than 0.5 km—is typically much greater than the horizontal difference in air pressure between Toronto, Ontario, and Winnipeg, Manitoba—a distance of over 1500 km. Because of this, we can see that a small vertical difference between two observation sites can yield a large difference in station pressure. Consequently, to properly

● FIGURE 8.9 The top diagram (a) shows four locations (A, B, C, and D) at varying elevations above sea level, all with different station pressures. The middle diagram (b) represents sea-level pressures of the four locations plotted on a sea-level chart. The bottom diagram (c) shows sea-level pressure readings of the four locations plus other sea-level pressure readings, with isobars drawn on the chart (grey lines) at intervals of 4 hPa.

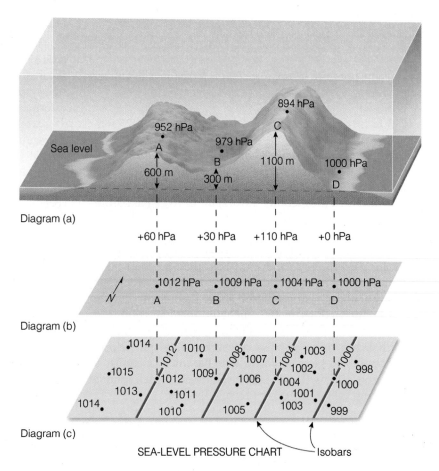

monitor horizontal changes in pressure, barometer readings must be corrected for altitude.

Altitude corrections are made so that a barometer reading taken at one elevation can be compared to a barometer reading taken at another. Station pressure observations are normally adjusted to the altitude of *mean sea level* (the altitude representing the average surface of the ocean). This adjusted reading is called **sea-level pressure**. The size of the correction depends primarily on how high the station is above sea level.

Near Earth's surface, atmospheric pressure decreases on average by about 10 hPa for every 100 m increase in elevation.* Notice in Figure 8.9a that location A has a station pressure of 952 hPa. Notice also that location A is 600 m above sea level. Adding 10 hPa per 100 m to its station pressure yields a sea-level pressure of 1012 hPa (see Figure 8.9b). After all the station pressures are adjusted to sea level (see Figure 8.9c), we are able to see the horizontal variations in sea-level pressure—something we were not able to see from the station pressures alone in Figure 8.9a.

When more pressure data are added (see Figure 8.9c), the chart can be analyzed and the pressure pattern visualized. **Isobars** (lines connecting points of equal pressure) are normally drawn at intervals of 4 hPa, with 1000 hPa being the base value. Note that the isobars do not pass through each

---

*This decrease in atmospheric pressure with height (10 hPa per 100 m) is an approximate value near sea level. Because atmospheric pressure decreases more rapidly with height in cold (denser) air than it does in warm (less dense) air, the vertical rate of pressure change is typically greater than 10 hPa per 100 m in cold air and less than that in warm air.

point but, rather, between many of them, with the exact values being interpolated from the data shown on the chart. For example, follow the 1008 hPa line from the top of the chart southward and observe that there is no plotted pressure of 1008 hPa. The 1008 hPa isobar comes closer to the station reporting a sea-level pressure of 1007 hPa than it does to the station with a pressure of 1010 hPa. The bottom chart containing the isobars (see Figure 8.9c) is called a *sea-level pressure chart* or simply a **surface map**. When weather data are added to the map, it becomes a *surface weather map*.

Surface and Upper-Level Charts

The isobars on the surface map in ● Figure 8.10a are drawn precisely; each observation is taken into account when determining the isobar's location. Notice that many of the lines are irregular, especially in mountainous regions over the Rockies. The reason for the wiggle is due mostly to local small-scale variations in pressure and to errors introduced by correcting observations that were taken at high-altitude stations. An extreme case of this type of error occurs at Leadville, Colorado (elevation 3096 m), the highest city in the United States. Here the station pressure is typically near 700 hPa. This means that nearly 300 hPa must be added to obtain a sea-level pressure reading! A mere 1 percent error in estimating the exact correction would result in a 3 hPa error in sea-level pressure. For this reason, isobars are smoothed through readings from high-altitude stations and from

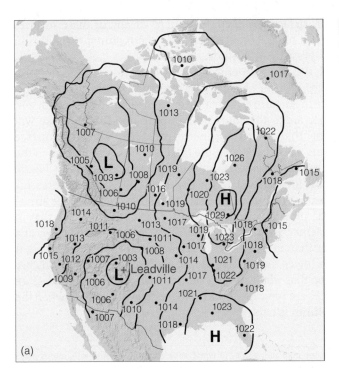

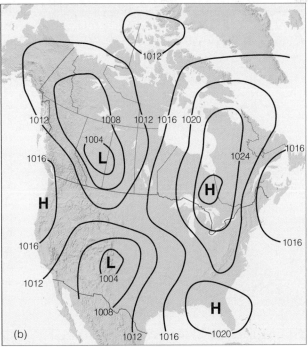

● FIGURE 8.10   (a) Sea-level isobars drawn so that each observation is taken into account. Not all observations are plotted.
(b) Sea-level isobars after smoothing.

stations that might have small observational errors. Figure 8.10b shows how the isobars appear on the surface map after they are smoothed.

The sea-level pressure chart described so far is called a *constant height chart* because it represents the atmospheric pressure at a constant altitude; in this case, the height is sea level. The same type of chart could be drawn to show the horizontal variations in pressure at any level in the atmosphere, for example, at 3000 m (see ● Figure 8.11).

Another type of chart commonly used in studying the weather is the *constant pressure chart*, or **isobaric chart**. Instead of showing pressure variations at a constant altitude, these charts are constructed to show height variations along an equal pressure (*isobaric*) surface. Constant pressure charts are convenient to use because the height variables they show are easier to deal with in meteorological equations than the variables of pressure. Since isobaric charts are in common use, let's examine them in detail.

Imagine that the dots inside the air column in ● Figure 8.12 represent evenly distributed air molecules extending from the surface up to the tropopause. If we climb halfway up the air column and stop at 5600 m, and then draw a sheetlike surface at this level, we will have made a constant height surface. Under standard conditions, the pressure at 5600 m measures 500 hPa. Since the air molecules are evenly distributed, there are an equal number of molecules everywhere above the tan-shaded surface in the diagram. This condition means that the

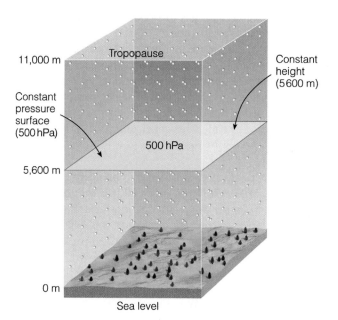

● FIGURE 8.12 When there are no horizontal variations in pressure, both constant pressure and constant height surfaces are just horizontal surfaces. Consequently, a measured pressure of 500 hPa corresponds to 5600 m above sea level everywhere. (Dots in the diagram represent air molecules.)

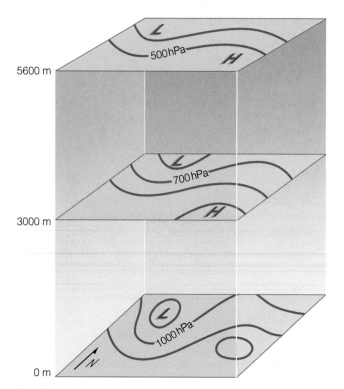

● FIGURE 8.11 Each map shows isobars on a constant height chart. The isobars represent variations in horizontal pressure at that altitude. An average isobar at sea level would be about 1000 hPa; at 3000 m, about 700 hPa; and at 5600 m, about 500 hPa.

level of constant height also represents a level of constant pressure. At every point on this constant pressure or *isobaric surface*, the height is 5600 m above sea level and the pressure is 500 hPa. We could cut any number of horizontal slices through the air column, each one at a different altitude, and each slice would represent both an isobaric and constant height surface. But a contour map* of any of these surfaces would be blank because there are no horizontal variations in either pressure or altitude.

If the air temperature in any portion of the column changes, the air density and pressure would also change. Notice in ● Figure 8.13 that we have colder air to the north and warmer air to the south. To simplify this situation, we will assume that the atmospheric pressure at Earth's surface remains constant. Hence, the total number of molecules in the column above each region must remain constant. In Figure 8.13, the area shaded grey at the top of the column represents a constant pressure surface (isobaric surface), where the atmospheric pressure along this surface is 500 hPa. Notice that in the warmer, less dense air, the 500 hPa pressure surface is at a higher than average level, whereas in the colder, denser air, it is at a lower than average level. From these observations, we can see that when the air column is warm, constant pressure surfaces are typically found at higher elevations than when the air column is cold.

---

*A contour map shows lines of equal value. Usually, we think of contour maps showing lines of equal elevation, but the term is broader. In the case discussed here, the lines of equal value would be lines of equal pressure or lines of equal height.

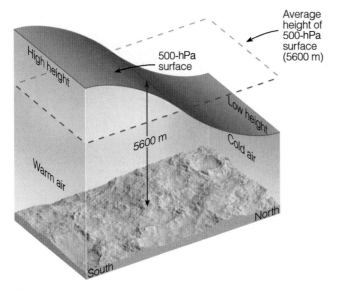

• FIGURE 8.13 The area shaded grey in the above diagram represents a surface of constant pressure, or isobaric surface. Because of the changes in air density, the isobaric surface rises in warm, less dense air and drops in cold, denser air. Where the horizontal temperature changes most quickly, the isobaric surface changes elevation most rapidly.

Look again at Figure 8.13 and observe that in the warm air at an altitude of 5600 m, the atmospheric pressure must be greater than 500 hPa, whereas in the cold air, at the same altitude (5600 m), the atmospheric pressure must be less than 500 hPa. Therefore, we can conclude that *high heights on an isobaric chart correspond to higher-than-normal pressures at any given altitude, and low heights on an isobaric chart correspond to lower-than-normal pressures.*

The variations in height of the isobaric surface in Figure 8.13 are shown in • Figure 8.14. Where the constant altitude lines intersect the 500 hPa pressure surface, **contour lines** (lines connecting points of equal elevation) are drawn on the 500 hPa map. Each contour line tells us the altitude above sea level where the pressure reading is 500 hPa. In the warmer air to the south, the elevations are high, whereas in the cold air to the north, the elevations are low. The contour lines are more crowded in the middle of this chart, where the pressure surface changes rapidly due to the changing air temperatures. Where there is little horizontal temperature change, there are also few contour lines. Although contour lines are lines of equal height, keep in mind that they illustrate pressure because contour lines of low height represent a region of lower pressure and contour lines of high height represent a region of higher pressure.

A column of cold air is normally associated with low heights or low pressures and one of warm air with high heights or high pressures. In the Northern Hemisphere, on upper-air charts, contour lines and isobars usually decrease in value from south to north. This occurs because the air is typically warmer to the south and colder to the north. However, the contour lines are not straight; they bend and turn, indicating **ridges**, which are *elongated highs* where the air is warm, and depressions or **troughs**, which are *elongated lows* where the air is cold. In • Figure 8.15, we can see how wavy contours on the map relate to changes in the altitude of the isobaric surface.

Although we have examined only the 500 hPa chart, other isobaric charts are also commonly used. ▼ Table 8.1 lists these charts and their approximate heights above sea level.

Upper-level charts are a valuable tool as they show wind-flow patterns, the movement of weather systems, and the behaviour of surface pressure areas that are extremely important for weather forecasting. To a small aircraft pilot, a constant pressure chart can help determine whether the plane is flying higher or lower than its altimeter* indicates.

---

*An altimeter is an instrument used to measure the altitude of an object above a fixed level, usually using pressure.

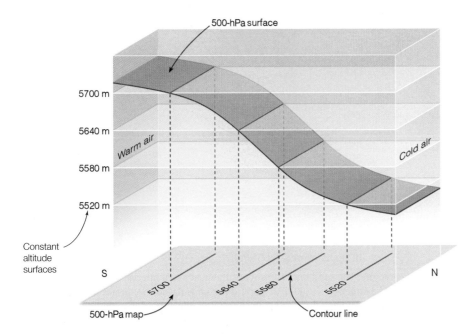

• FIGURE 8.14 Changes in the elevation of a constant pressure or isobaric surface show up as contour lines on an isobaric map. In this case, the surface and the map are at 500 hPa. The contour lines are closer where the surface pressure changes most rapidly.

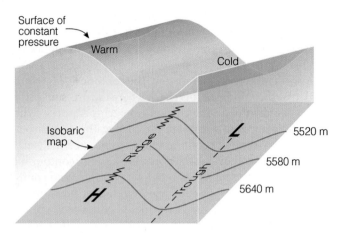

• FIGURE 8.15 The wavelike patterns of an isobaric surface reflect the changes in air temperature. A region of warm air appears on an isobaric map as a ridge of higher heights, whereas colder air is a trough of lower heights.

(Read more about this in Focus on an Observation: Flying on a Constant Pressure Surface—High to Low, Look Out Below on p. 237.)

• Figure 8.16a is a simplified surface map showing areas of high and low pressure and arrows indicating *wind direction*, reported as the direction the wind is blowing from. The large

▼ Table 8.1 Common Isobaric Charts and Their Approximate Elevations Above Sea Level

ISOBARIC SURFACE CHARTS (hPa)	APPROXIMATE ELEVATIONS (m)
1000	120
850	1,457
700	3,012
500	5,574
250	10,363
100	16,200

Hs on the map indicate high-pressure centres, also called **anticyclones**. The large Ls represent low-pressure centres, also known as *depressions* or **middle-latitude cyclones** because these storms form in the middle latitudes, outside the tropics. The solid dark lines are isobars in units of hectopascals. Notice that the surface winds tend to blow across the isobars toward regions of lower pressure. In fact, as we briefly observed in Chapter 1, in the Northern Hemisphere, winds blow counterclockwise and inward toward the centre of lows and clockwise and outward from the centre of highs.

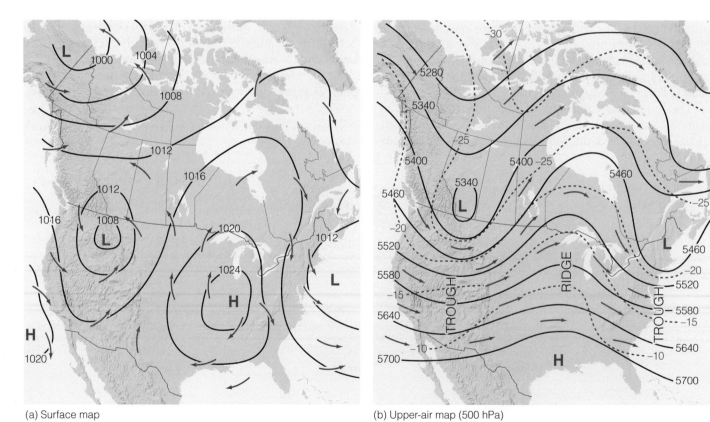

(a) Surface map

(b) Upper-air map (500 hPa)

• FIGURE 8.16 (a) Surface map showing areas of high and low pressure. The solid lines are isobars drawn at 4 hPa intervals. The arrows represent wind direction. Notice that the wind blows across the isobars. (b) The upper-level, 500 hPa, map for the same day as the surface map. Solid lines on the map are contour lines in metres above sea level. Dashed red lines are isotherms in °C. Arrows show wind direction. Notice that on this upper-air map, the wind blows parallel to the contour lines.

FOCUS ON AN OBSERVATION

Flying on a Constant Pressure Surface—High to Low, Look Out Below

Aircraft that use pressure altimeters typically fly along a constant pressure surface rather than fly along a constant altitude surface. This is because an altimeter is simply an aneroid barometer calibrated to convert atmospheric pressure to an approximate elevation.* The elevation indicated by an altimeter assumes a standard atmosphere,[†] where the air temperature decreases at the rate of 6.5° per kilometre. Since the air temperature seldom, if ever, decreases at exactly this rate, altimeters generally do not indicate their true elevation.

● Figure 2 shows a standard air[†] column bounded on each side by air with a different temperature and density. On the left side, the air is warm; on the right, it is cold. The orange line represents the 700 hPa constant pressure surface as seen from the side. In standard air, the 700 hPa surface is located at 3000 m above sea level.

In warm air, the 700 hPa surface rises; in cold air, it descends. An aircraft flying along the 700 hPa surface would be at an altitude less than 3000 m in cold air, equal to 3000 m in standard air, and greater than 3000 m in warmer air. With no corrections for temperature, the altimeter would indicate the same altitude at all three positions because the air pressure does not change. But we can see that an aircraft flying into warm air will increase in altitude and fly higher than its altimeter indicates. Unfortunately, the converse is also true.

Flying from standard air into cold air represents a potentially dangerous situation. As an aircraft flies into cold air, it flies along a pressure surface that is dropping in elevation. If no correction for temperature is made, the altimeter shows no change in elevation even though the aircraft is losing altitude. This can be a serious problem, especially for planes flying above mountainous terrain in poor

---
*To see how this calibration works, see Focus on an Advanced Topic: The Hydrostatic and Hypsometric Equations on p. 250.
[†]The terms "standard atmosphere" or "standard air" describe the International Standard Atmosphere (ISA), an atmospheric model of how the pressure, temperature, density, and viscosity of Earth's atmosphere change over a wide range of altitudes. See Appendix H.

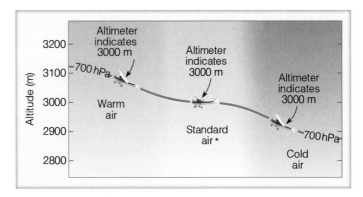

● FIGURE 2 An aircraft flying along a surface of constant pressure (orange line) will change altitude as the air temperature changes, but the plane's pressure altimeter will continue to read the same elevation if no temperature correction is made.

visibility, where high winds and turbulence can drastically reduce air pressure. To ensure adequate clearance under these conditions, pilots fly their aircraft higher than they normally would, record the air's temperature, and compute a more realistic altitude by resetting their altimeters to reflect these conditions.

Even without sharp temperature changes, pressure surfaces may dip suddenly (see ● Figure 3). An aircraft flying into an area of decreasing pressure will lose altitude unless corrections are made. For example, suppose that a pilot has set the altimeter for sea-level pressure above station A. At this location, the plane is flying along an isobaric surface at a true altitude of 150 m. As the plane flies toward station B, the pressure surface and the plane dip, but the altimeter continues to read 150 m, which is too

high. To correct for these pressure changes, a pilot can obtain a current altimeter setting from ground control. Now the altimeter reading will more closely match the aircraft's actual altitude.

Because of inaccuracies inherent in a pressure altimeter, high performance and commercial aircraft are equipped with radio altimeters. This device is like a small radar unit that measures the altitude of the aircraft by sending out radio waves, which bounce off the terrain. The time it takes these waves to reach Earth's surface and return is a measure of the aircraft's altitude. If used in conjunction with a pressure altimeter, a pilot can determine the variations in a constant pressure surface simply by flying along that surface and observing how the true elevation measured by the radio altimeter changes.

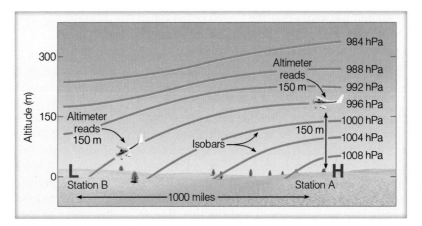

● FIGURE 3 Pressure surfaces can also dip toward Earth without changes in temperature and cause an aircraft flying along a constant pressure surface to lose or gain altitude, depending on the plane's direction.

Figure 8.16b shows an upper-air chart, a 500 hPa isobaric map, for the same day as the surface map in Figure 8.16a. The solid grey lines on the map are contour lines given in metres above sea level. The difference in elevation between each contour line, called a *contour interval*, is 60 m. Superimposed on this map are dashed red lines, which represent lines of equal temperature called isotherms. Observe how the contour lines tend to parallel the isotherms. As we would expect, the contour lines tend to decrease in value from south to north.

The arrows on the 500 hPa map show the wind direction. Notice that unlike the surface winds that cross the isobars in Figure 8.16a, winds on the 500 hPa chart tend to flow *parallel* to the contour lines in a wavy, west-to-east direction. Why does the wind tend to cross the isobars on a surface map yet blow parallel to the contour lines or isobars on an upper-air chart? To answer this question, we will now examine the forces that affect winds.

Newton's Laws of Motion and Forces

Our understanding of why the wind blows stretches back several centuries, with many scientists contributing to our knowledge. When we think of understanding the movement of air, one great scholar stands out: Isaac Newton (1642–1727), who formulated several fundamental laws of motion.

Newton's first law of motion states that an object at rest will remain at rest, and an object in motion will remain in motion (and travel at a constant velocity along a straight line) as long as no force is exerted on the object. For example, a baseball in a pitcher's hand will remain there until a force—a push— acts on the ball. Once the ball is pushed—thrown—it would move forever except that the force of air friction slows it down, the force of gravity pulls it toward the ground, and the catcher's mitt exerts an equal but opposite force to stop it. So we can see that to start air moving, to speed it up, to slow it down, or even to change its direction requires an external force. This brings us to Newton's second law.

Newton's second law states that the force exerted on an object equals its mass times the acceleration produced or force equals mass times acceleration. In symbolic form, this law is written as

$$F = ma$$

This relationship states that when the mass of an object is constant, the force acting on it is directly related to the acceleration that is produced. A force in its simplest form is a push or a pull. *Acceleration is the speeding up, slowing down, or change of direction of an object.* More precisely, acceleration is the change in velocity over a period of time, and velocity indicates both the speed and the direction of an object in motion.

Because more than one force may act on an object, Newton's second law always refers to the *net force* or result of all the forces that act on an object. An object will always accelerate in the direction of the net force. Therefore, to determine which direction the wind will blow, we must identify and examine all the forces that affect air movement:

Horizontal forces

- **horizontal pressure gradient force**—causes horizontal motion

- **Coriolis force**—occurs after motion starts and causes a change in horizontal wind direction

- **centripetal force**—occurs after motion starts and appears as an imbalance in other forces, causing a change in horizontal wind direction

- **friction**—occurs after motion starts and causes a decrease in horizontal wind speed near Earth's surface

Vertical forces

- **vertical pressure gradient force**—causes upward or downward vertical motion

- **gravity**—causes downward vertical motion

We will focus on the horizontal forces because they influence the horizontal winds. The vertical forces are discussed in more detail in Focus on an Advanced Topic: The Hydrostatic and Hypsometric Equations near the end of this chapter. We will first study the forces that influence flow of air higher in the atmosphere (where there is no friction). Then we will see which forces modify winds near the ground.

Forces That Influence the Horizontal Winds

We already know that horizontal differences in atmospheric pressure cause air to move and, hence, the wind to blow. Since air is an invisible gas, it may be easier to see how pressure differences cause motion if we examine a visible fluid, such as water.

In ● Figure 8.17, the two large tanks are connected by a pipe. Tank A is two-thirds full, and tank B is only one-half full. Since the water pressure at the bottom of each tank is proportional to the weight of water above, the pressure at the bottom of tank A is greater than the pressure at the bottom of tank B. Moreover, since fluid pressure is exerted equally in all directions, there is a greater pressure in the pipe directed from tank A toward tank B than from B toward A.

Since pressure is force per area, there must also be a net force directed from tank A toward tank B. This force causes the water to flow from left to right, from higher pressure toward lower pressure. The greater the pressure difference is, the stronger the force, and the faster the water moves. In a similar way, horizontal differences in atmospheric pressure cause air to move.

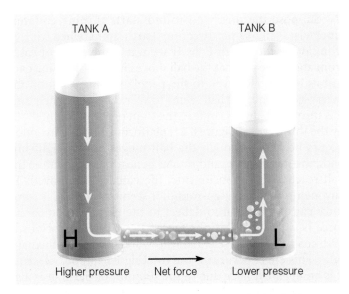

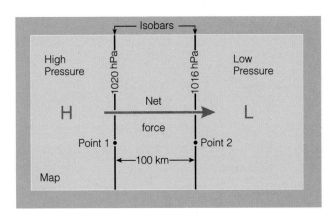

● FIGURE 8.17 The higher water level creates higher fluid pressure at the bottom of tank A and a net force directed toward the lower fluid pressure at the bottom of tank B. This net force causes water to move from higher pressure toward lower pressure.

● FIGURE 8.18 The pressure gradient between point 1 and point 2 is 4 hPa per 100 km. The red arrow is the *pressure gradient force*.

PRESSURE GRADIENT FORCE ● Figure 8.18 shows a region of higher pressure on the map's left side and lower pressure on the right. The isobars show how the horizontal pressure is changing. If we compute the amount of pressure change that occurs over a given distance, we have the **pressure gradient**:

$$\text{Presssure gradient} = \frac{\text{change in pressure}}{\text{change in distance}}$$

If we let the symbol delta (Δ) mean "a change in," we can simplify the expression and write the pressure gradient as

$$PG = \frac{\Delta P}{\Delta x}$$

where ΔP is the pressure difference between two places that are some distance (Δx) apart. For example, in Figure 8.18, the pressure gradient between points 1 and 2 is

$$PG = \frac{1020 \text{ hPa} - 1016 \text{ hPa}}{100 \text{ km}} = \frac{4 \text{ hPa}}{100 \text{ km}} \text{ or hPa per 100 km}$$

Now suppose that the pressure in Figure 8.18 were to change and the isobars become closer together. This produces a rapid change in pressure over a relatively short distance: a strong (or steep) pressure gradient. However, if the isobars spread farther apart, then the pressure difference would be small over a relatively large distance, producing a weak (or gentle) pressure gradient.

Notice in Figure 8.18 that when differences in horizontal air pressure exist, there is a net force acting on the air. All forces have both magnitude (size) and direction. This force, called the **pressure gradient force** (*PGF*), *is directed from higher to lower pressure, at right angles to the isobars* (represented by the red arrow). The *PGF* always acts from high pressure toward low pressure. The magnitude of this force is

directly related to the pressure gradient so that steep pressure gradients correspond to strong pressure gradient forces and vice versa. To calculate the magnitude of the *PGF*, divide the pressure gradient by the air density (ρ).* This gives you the pressure gradient force per kilogram of air (often phrased as the pressure gradient force per unit mass). Symbolically, this is written as

$$PGF = \frac{1}{\rho} PG = \frac{1 \Delta P}{\rho \Delta x}$$

where ρ is the air density and $\frac{\Delta P}{\Delta x}$ is the pressure gradient (*PG*) represented by its components: ΔP is the pressure difference, and Δx is the distance over which the pressure difference occurs. ● Figure 8.19 shows the relationship between pressure gradient and pressure gradient force.

The *pressure gradient force is the force that causes the wind to blow.* Closely spaced isobars on a weather map indicate steep pressure gradients, strong forces, and high winds. On the other hand, widely spaced isobars indicate gentle pressure gradients, weak forces, and light winds. An example of a steep pressure gradient and strong winds is given in ● Figure 8.20. Notice that the tightly packed isobars along the green line are producing a steep pressure gradient of 32 hPa per 500 km and strong surface winds of 74 km h^{-1} (40 knots[†]).

If the pressure gradient force were the only force acting on air, we would always find winds blowing directly from higher toward lower pressure. However, the moment air starts to move, it is deflected in its path by the *Coriolis force*.

CORIOLIS FORCE The **Coriolis force** describes an *apparent force* that is due to the rotation of Earth. It is seen (thus called

---

*Recall that air density is the number of kilograms in 1 cubic metre of air. The Greek letter rho (ρ) represents air density.

[†]The knot is a non-SI unit for speed equal to one nautical mile per hour, or exactly 1.852 km h^{-1}. Worldwide, the knot is widely used in meteorology, as well as maritime and air navigation. The term originates from a method for determining sailing ship speed by counting how many evenly spaced knots on a rope are released during a specific time when the rope with a wooden board on the end is placed in the water.

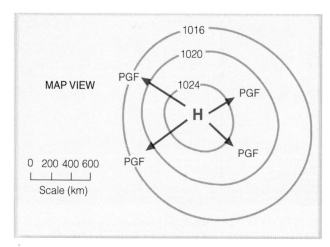

● FIGURE 8.19 The closer the spacing of the isobars is, the greater the pressure gradient. The greater the pressure gradient is, the stronger the pressure gradient force (PGF). The stronger the PGF is, the greater the wind speed. Here the red arrows represent the relative magnitude of the force, which is always directed from higher toward lower pressure.

apparent) only when moving objects are viewed by an observer who is rotating with Earth. It does not exist when moving objects are viewed from outside Earth's rotation. To understand how it works, consider two people playing catch as they sit opposite one another on the platform of a merry-go-round (see ● Figure 8.21, platform A). If the merry-go-round is not moving, each time the ball is thrown, it moves in a straight line to the other person.

Suppose the merry-go-round starts turning counter-clockwise—the same direction Earth spins when viewed from above the North Pole. If we watch the game of catch from above, we see that the ball moves in a straight-line path just as before. However, to the people playing catch on the merry-go-round, the ball seems to veer to its right each time it is thrown. It always lands to the right of the point intended by the thrower (see Figure 8.21, platform B). This perception occurs because although the ball moves in a straight-line path, the merry-go-round rotates beneath it. By the time the ball reaches the opposite side, the catcher has moved. To anyone on the merry-go-round, it seems as if there is some force causing the ball to deflect to the right. This apparent force is called the Coriolis force after Gaspard Coriolis, a 19th century French scientist who worked it out mathematically. This effect occurs on the rotating Earth, too. (Because it is an *apparent force* due to the rotation of Earth, it is also called the *Coriolis effect*.) All free-moving objects, such as ocean currents, aircraft, artillery projectiles, and air molecules, seem to deflect from a straight-line path because Earth rotates under them.

The Coriolis force causes the wind to deflect to the right of its intended path in the Northern Hemisphere and to the left of its intended path in the Southern Hemisphere. To illustrate this, consider a satellite in circular orbit around the Poles. If Earth were not rotating, an observer on Earth would see it move from north to south, parallel to Earth's meridian lines. However, Earth does rotate, carrying us and meridians eastward with it. Because of this rotation in the

● FIGURE 8.20 A surface weather map showing the relationship between sea-level pressure and surface wind for a major winter storm that passed through the U.S. Midwest states of Iowa, Wisconsin, and Minnesota before crossing southern Ontario (see the following Weather Watch box on page 242 to better understand the impacts). The map is from 6 a.m. Central Standard Time, Tuesday, November 10, 1998. The dark grey lines are isobars in hectopascals. In this image, a deep low with a central pressure of 972 hPa is moving over northwestern Iowa. The distance along the green line from X to X′ is 500 km; the difference in pressure between X and X′ is 32 hPa. This produces a pressure gradient of 32 hPa per 500 km. The tightly packed isobars along the green line are associated with strong northwesterly winds of 74 km h^{-1} (20.6 m s^{-1}, or 40 knots) and higher wind gusts. On the map, surface winds are represented by the wind-direction shaft lines with barb and flag symbols shown in the accompanying table. Wind directions are shown by the shaft orientation, with the wind blowing from the tail to the tip. Wind speeds are indicated by barbs and flags. So, for example, a wind indicated by the symbol ⌐ would be a wind from the northwest at 18.5 km h^{-1} or 10 knots. The barbed blue line is a cold front, the semicircular dotted red line is a warm front, and the alternating blue and red "hook" shapes indicate an occluded front or TROWAL (TROugh of Warm-air ALoft).

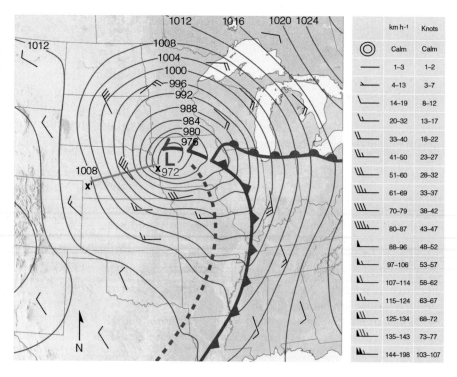

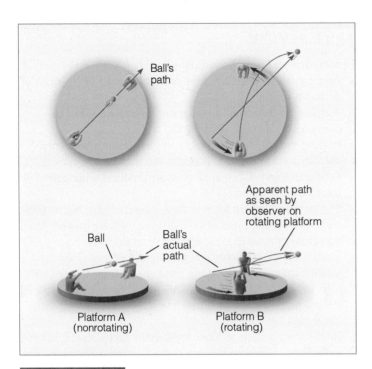

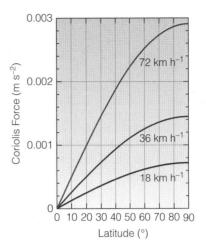

ACTIVE FIGURE 8.21 On nonrotating platform A, the thrown ball moves in a straight line. On platform B, which rotates counterclockwise, the ball continues to move in a straight line. However, platform B is rotating while the ball is in flight; thus, to anyone on platform B, the ball appears to deflect to the right of its intended path. Visit the textbook's website to view this and other Active Figures at www.ahrensmeteorology1ce.nelson.com

Northern Hemisphere, we see the satellite moving southwest instead of south; it seems to veer off its path and move toward its right. In the Southern Hemisphere, Earth's direction of rotation is clockwise when viewed from above the

South Pole. Consequently, a satellite moving northward from the South Pole would appear to move northwest and would veer to the left of its path.

The magnitude of the Coriolis force varies with the speed and latitude of the moving object. ● Figure 8.22 shows the variation for various wind speeds at different latitudes. In each case, as the wind speed increases, the Coriolis force increases; hence, *the stronger the wind speed is, the greater the deflection.* Also, note that the Coriolis force increases for all wind speeds from a value of *zero at the equator to a maximum at the poles.* We can see this latitude effect better by examining ● Figure 8.23.

Imagine in Figure 8.23 that there are three aircraft, each at a different latitude and each flying along a straight-line path, with no external forces acting on them. The destination of each aircraft is due east and is marked on the diagram (see Figure 8.23a). Each plane travels in a straight path relative to

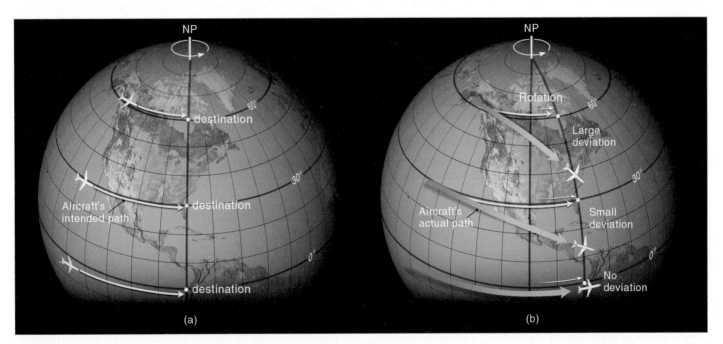

● FIGURE 8.23 Except at the equator, a free-moving object heading either east or west (or any other direction) will appear from Earth to deviate from its path as Earth rotates beneath it. The deviation caused by the Coriolis force is greatest at the poles and decreases to zero at the equator.

The deep, low-pressure storm of November 10, 1998, illustrated in Figure 8.20 was quite an event. As it passed through the region, the low intensified to a record low pressure for Minnesota of 963 hPa. Its tightly packed isobars and strong pressure gradient produced extremely high winds that gusted over 167 km h^{-1} (90 knots), blowing roofs off homes by the time it reached Wisconsin. The extreme winds caused blizzards over North and South Dakota, closed many interstate highways, shut down airports, and overturned trucks.

an observer positioned at a fixed spot in space. Earth rotates beneath the moving planes, causing the destination points at latitudes 30° and 60° to change direction slightly (to the observer in space) (see Figure 8.23b). To an observer standing on Earth, however, it is the plane that appears to deviate. The amount of deviation is greatest toward the pole and nonexistent at the equator. Therefore, the Coriolis force has a far greater effect on the plane at high latitudes (large deviation) than on the plane at low latitudes (small deviation). On the equator, it has no effect at all. The Coriolis force has the same impact on winds.

In summary, to an observer on Earth, objects moving in *any direction*, north, south, east, or west, are deflected to the *right* of their intended path in the Northern Hemisphere and to the *left* of their intended path in the Southern Hemisphere. The *Coriolis force acts at right angles to the wind, influencing only wind direction and never wind speed.* The amount of deflection by the Coriolis force* depends on the

● Earth's rotation rate (Ω)

● latitude (ϕ)

● object's speed (V)

The Coriolis force behaves as a real force, constantly tending to "pull" the wind to its right in the Northern Hemisphere and to its left in the Southern Hemisphere. Moreover, this effect is present in all motions relative to Earth's surface. However, in most of our everyday experiences, the Coriolis force is so small (compared to other forces involved in those experiences) that it is negligible and, contrary to popular belief, does not cause water to turn clockwise or counterclockwise when draining from a sink. The Coriolis force is also minimal on small-scale winds, such as those that blow inland along coasts in summer. Here the Coriolis force might be strong because of high winds, but the force cannot produce much deflection over the relatively short distances. Only where winds blow over vast regions is the effect significant.

---

*The Coriolis force (*CF*) is related to Ω, ϕ, V through the equation

$$CF = m(2\Omega \, sin\phi)V$$

where *m* is the object's mass, Ω is Earth's rotation rate, *V* is the speed of the object, and ϕ is the latitude. Note that ($2\Omega \, sin\phi$) is called the *Coriolis parameter* (*f*) so that the above equation can be written as: $CF = mfV$

BRIEF REVIEW

In summary, we know the following:
● Atmospheric air pressure is the pressure exerted by the mass of air above a region.
● A change in surface air pressure can be brought about by changing the mass of air above the surface.
● Heating and cooling columns of air can establish horizontal variations in atmospheric pressure above and at the surface.
● A difference in horizontal air pressure produces a horizontal pressure gradient force.
● The pressure gradient force is always directed from higher pressure toward lower pressure, and it is the pressure gradient force that causes the air to move and the wind to blow.
● Steep pressure gradients, seen as tightly packed isobars on a weather map, indicate strong pressure gradient forces and high winds; gentle pressure gradients, seen as widely spaced isobars on weather maps, indicate weak pressure gradient forces and light winds.
● Once the wind starts to blow, the Coriolis force causes it to bend to the right of its intended path in the Northern Hemisphere and to the left of its intended path in the Southern Hemisphere.

With this information in mind, we will first examine how the pressure gradient force and the Coriolis force produce straight-line winds aloft. We will then see what influence the centripetal force has on winds that blow along a curved path.

STRAIGHT-LINE FLOW ALOFT—GEOSTROPHIC WINDS
Earlier in this chapter, we saw that the winds aloft on an upper-level chart blow more or less parallel to the isobars or contour lines. We can see why this happens by examining
● Figure 8.24, which shows the evolution of the balance of forces associated with increasing wind in the Northern Hemisphere, above Earth's frictional influence.* The figure shows a map of the horizontal pressure variations at an altitude of about 1 km above Earth's surface. The evenly spaced isobars indicate a constant pressure gradient force directed from south toward north, as indicated by the red arrow at the left. Why, then, does the map show a wind blowing from the west? We can answer this question by placing a parcel of air at position 1 in the diagram and watching its behaviour.

At position 1, the *PGF* acts immediately on the air parcel, accelerating it northward toward lower pressure. However,

If it were not for the friction between your tires and the road surface, the Coriolis force would "pull" your vehicle to the right by about 500 m for every 160 km you travel as you drive your car along a highway at the speed limit.

---

*The friction layer (the layer where the wind is influenced by frictional interaction with objects on Earth's surface) usually extends from the surface up to about 1000 m above the ground.

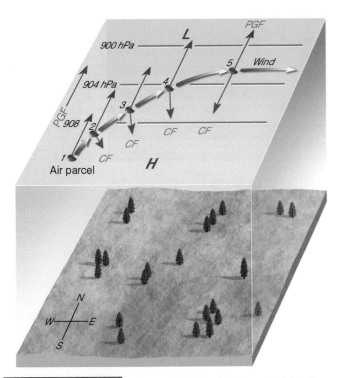

ACTIVE FIGURE 8.24 Above the level of friction, air initially at rest will accelerate until it flows parallel to the isobars at a steady speed with the pressure gradient force (PGF) balanced by the Coriolis force (*CF*). Wind blowing under these conditions is called geostrophic. Visit the textbook's website to view this and other Active Figures at www.ahrensmeteorology1ce.nelson.com

the instant the air begins to move, the Coriolis force deflects the air toward its right, curving its path. As the parcel of air increases in speed (positions 2, 3, and 4), the magnitude of the Coriolis force increases (as shown by the longer arrows), bending the wind more and more to its right. Eventually, the wind speed increases to the point where the Coriolis force just balances the *PGF*. At this point (position 5), the wind no longer accelerates because the net force is zero. Here the wind flows in a straight path, parallel to the isobars at a constant speed.* This flow of air is called a **geostrophic** (*geo* = earth; *strophic* = turning) **wind**. Notice that the geostrophic wind blows in the Northern Hemisphere with lower pressure to its left and higher pressure to its right.

When the flow of air is purely geostrophic, the isobars or contours are straight and evenly spaced, and the wind speed is constant. In the atmosphere, isobars are rarely straight or evenly spaced, and the wind normally changes speed as it flows along. So the geostrophic wind is usually only an approximation of the real wind. However, the approximation is generally close enough to help us more clearly understand the behaviour of the winds aloft.

As we would expect from our previous discussion of winds, the speed of the geostrophic wind is directly related to the pressure gradient. In ● Figure 8.25, we can see that a geostrophic wind flowing parallel to the isobars is similar to water in a stream flowing parallel to its banks. At position 1, the wind is blowing at a low speed; at position 2, the pressure gradient increases and the wind speed picks up. Notice also that at position 2, where the wind speed is greater, the Coriolis force is greater and balances the stronger pressure gradient force. A more mathematical approach to the concept of geostrophic wind is given in Focus on an Advanced Topic: A Mathematical Look at the Geostrophic Wind on p. 244.

In ● Figure 8.26, we can see that the geostrophic wind direction can be determined by studying the orientation of the isobars; its speed can be estimated from the spacing of the isobars. On an isobaric chart, the geostrophic wind direction and speed are related in a similar way to the contour lines. Therefore, if we know the isobar or contour patterns on an upper-level chart, we also know the direction and relative speed of the geostrophic wind, even for regions where no direct wind measurements have been made. Similarly, if we know the geostrophic wind direction and speed, we can estimate the orientation and spacing of the isobars, even if we do not have a current weather map. (It is also possible to estimate the wind flow and pressure patterns aloft by watching the movement of clouds. Focus on an Observation: Estimating Wind Direction and Pressure Patterns Aloft by Watching Clouds on p. 245 illustrates this further.)

---

*At first, it may seem odd that the wind blows at a constant speed with no net force acting on it. But when we remember that the net force is necessary only to accelerate (F = ma) the wind, it makes more sense. For example, it takes a considerable net force to push a car and get it rolling from rest. But once the car is moving, it only takes a force large enough to counterbalance friction to keep it going. There is no net force acting on the car, yet it rolls along at a constant speed.

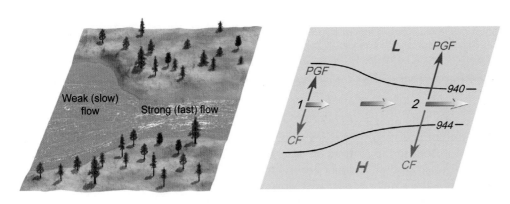

● FIGURE 8.25 The isobars and contours on an upper-level chart are like the banks along a flowing stream. When they are widely spaced, the flow is weak; when they are narrowly spaced, the flow is stronger. The increase in winds on the chart results in a stronger Coriolis force (*CF*), which balances a larger pressure gradient force (*PGF*).

FOCUS ON AN ADVANCED TOPIC

A Mathematical Look at the Geostrophic Wind

We know from an earlier discussion that the geostrophic wind gives us a good approximation of the real wind above the level of friction, about 1000 m above Earth's surface. Above the friction layer, the winds tend to blow parallel to the isobars, or contours. We know that for any given latitude, the speed of the geostrophic wind is proportional to the pressure gradient. This may be represented as

$$V_g \, \alpha \, \frac{\Delta P}{\Delta x}$$

where V_g is the geostrophic wind and ΔP is the pressure difference between two places some horizontal distance (Δx) apart. From this, we can see that the greater the pressure gradient is, the stronger the geostrophic wind.

In a previous section, we saw that the geostrophic wind represents a balance of forces between the Coriolis force and the pressure gradient force. Here it should be noted that the Coriolis force (per unit mass) can be expressed as

$$CF = (2\Omega \, \sin\phi) \, V$$

where CF is the Coriolis force, Ω is Earth's angular spin (a constant), V is the speed of the wind, and ϕ is the latitude. (sin ϕ is a trigonometric function that takes into account the variation of the Coriolis force with latitude. At the equator ($0°$), sin ϕ is 0. At $30°$ north latitude, sin ϕ is 0.5. At the North Pole ($90°$), sin ϕ is 1. For latitudes south of the equator, sin ϕ has the same numeric values, but they are negative.)

This balance between the Coriolis force and the pressure gradient force (per unit mass) can be written as

$$CF = PGF$$
$$2(\sin\phi) \, V_g = \frac{1}{\rho} \frac{\Delta P}{\Delta x}$$

Solving for V_g, the geostrophic wind, the equation becomes

$$V_g = \frac{1}{(2\Omega \, \sin\phi) \, \rho} \frac{\Delta P}{\Delta x} \qquad (1)$$

Customarily, the rotational (2Ω) and latitudinal (sin ϕ) factors are combined into a single value f, called the *Coriolis parameter*. Thus, we have the geostrophic wind equation written as

$$V_g = \frac{1}{f\rho} \frac{\Delta P}{\Delta x} \qquad (2)$$

Suppose that we compute the geostrophic wind for the example given in ● Figure 4. Here the wind is blowing parallel to the isobars in the Northern Hemisphere at latitude $50°$. The spacing between the isobars is 200 km, and the pressure difference is 4 hPa. The altitude is 5600 m above sea level, where the air temperature is $-25°C$ and the air density is 0.70 kg m^{-3}. First, we list our data and put them in the proper units as

$$\Delta P = 4 \, hPa = 400 \, Pa = 400 \, N \, m^{-2}$$
$$= 400 \, kg \, m^{-1}s^{-2}$$
$$\Delta x = 200 \, km = 2 \times 10^5 \, m$$
$$\sin \phi = \sin(50) = 0.77$$
$$\rho = 0.70 \, kg \, m^{-3}$$
$$2\Omega = 1.46 \times 10^{-4} \, radian \, s^{-1}*$$

---

*The rate of Earth's rotation (Ω) is $360°$ in one day, actually a sidereal day consisting of 23 hours, 56 minutes, and 4 seconds, or 86,164 seconds. (The sidereal day is not quite 24 hours as it takes into account Earth rotating around the sun, as well as on its own axis.) This gives a rate of rotation of 4.18×10^{-3} degrees per second. Most often, Ω is given in radians, where 2π radians equals $360°$ ($\pi = 3.14$). Therefore, the rate of Earth's rotation can be expressed as 2π radians divided 86,164 seconds $\frac{2\pi}{(86,164 \, s)}$ or 7.29×10^{-5} radian s^{-1}, and the constant 2Ω becomes 1.46×10^{-4} s^{-1}. Here the word "radian" is dropped from the units because unlike measurements of angles in degrees, the units in radians "cancel out," so radians have no units.

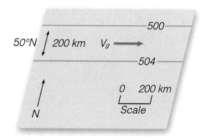

● FIGURE 4 A portion of an upper-air chart for part of the Northern Hemisphere at an altitude of 5600 m above sea level. The lines on the chart are isobars, where 500 equals 500 hPa. The air temperature is $-25°C$, and the air density is 0.70 kg m^{-3}.

When we use equation (1) to compute the geostrophic wind, we obtain

$$V_g = \frac{1}{(2\Omega \, \sin\phi) \, \rho} \frac{\Delta P}{\Delta x}$$

$$V_g = \frac{1}{(1.46 \times 10^{-4}) \times 0.77 \times 0.70} \times \frac{400}{2 \times 10^5}$$

units of V_g: $\frac{1}{s^{-1} \, kg \, m^{-3}} \times \frac{kg \, m^{-1}s^{-2}}{m} = m \, s^{-1}$

$$V_g = 25.4 \, m \, s^{-1} = 91.5 \, km \, h^{-1} = 49.4 \, knots$$

Since this calculation can be tedious, meteorologists have devised ways of estimating the geostrophic wind more quickly. Here is a quick rule of thumb to estimate the geostrophic wind that is reasonably accurate in mid-latitudes: *The geostrophic wind in knots is approximately five times the number of hectopascals when measured perpendicular to the isobars across a distance of four degrees of longitude at a latitude of interest.* For example, suppose that on a weather map, four degrees of longitude measures 1.5 cm and the pressure changes by 6 hPa over this 1.5 cm distance. The geostrophic wind speed would then be $6 \times 5 = 30$ knots.

We know that the winds aloft do not always blow in a straight line; frequently, they curve and bend into meandering loops as they tend to follow the patterns of the isobars. In the Northern Hemisphere, winds blow counterclockwise around lows and clockwise around highs. The next section explains why.

CENTRIPETAL FORCE AND GRADIENT WINDS—CURVED WINDS AROUND LOWS AND HIGHS ALOFT Because lows are also known as cyclones, the counterclockwise flow of air around them is often called *cyclonic flow*. Likewise, the clockwise flow of air around a high, or anticyclone, is called

FOCUS ON AN OBSERVATION

Estimating Wind Direction and Pressure Patterns Aloft by Watching Clouds

Both the upper-air wind direction and orientation of the isobars can be estimated by observing middle- and high-level clouds from Earth's surface. Suppose we are in the Northern Hemisphere watching clouds directly above us move from southwest to northeast at an elevation of about 3000 m (see ● Figure 5a). This indicates that the geostrophic wind at this level is southwesterly. Looking downwind, the geostrophic wind blows parallel to the isobars with lower pressure on the left and higher pressure on the right. Thus, if we *stand with our backs to the direction from which the clouds are moving, lower pressure aloft will always be to our left and higher pressure to our right.* From this observation, we can draw a rough upper-level chart (see Figure 5b), which shows isobars and wind direction for an elevation of approximately 3000 m.

The isobars aloft will not continue in a southwest–northeast direction indefinitely;

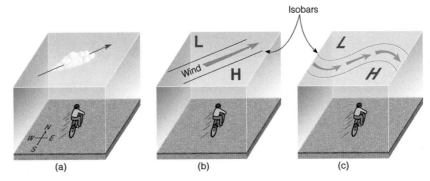

● **FIGURE 5** This drawing of a simplified upper-level chart is based on cloud observations. Upper-level clouds moving from the southwest (a) indicate isobars and winds aloft (b). When extended horizontally, the upper-level chart appears as in (c), where a trough of low pressure is to the west and a ridge of high pressure is to the east.

rather, they will often bend into wavy patterns. We may carry our observation one step further, then, by assuming a bending of the lines (see Figure 5c). Thus, with a southwesterly wind aloft, a trough of low pressure will be found to

our west and a ridge of high pressure to our east. What would the pressure pattern be if the winds aloft were blowing from the northwest? Answer: A trough would be to the east and a ridge to the west.

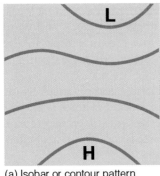
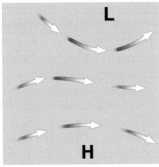

(a) Isobar or contour pattern   (b) Wind pattern

● FIGURE 8.26 By observing the orientation and spacing of the isobars or contours in diagram (a), the geostrophic wind direction and speed can be determined in diagram (b).

anticyclonic flow. Look at the wind flow around the Northern Hemisphere upper-level low in ● Figure 8.27. At first, it appears as though the wind is defying the Coriolis force by bending to the left as it moves counterclockwise around the system. Let's see why the wind blows in this manner.

Suppose that we consider a parcel of air initially at rest at position 1 in Figure 8.27a. The pressure gradient force accelerates the air inward toward the centre of the low, and the Coriolis force deflects the moving air to its right, until

the air is moving parallel to the isobars at position 2. If the wind were geostrophic, at position 3, the air would move northward parallel to straight-line isobars at a constant speed. The wind is blowing at a constant speed but parallel to *curved isobars*. A wind that blows at a constant speed parallel to curved isobars above the level of frictional influence is termed a **gradient wind**.

Earlier in this chapter, we learned that an object accelerates when there is a change in its speed or direction (or both). Therefore, the gradient wind blowing *around* the low-pressure centre is constantly accelerating because it is constantly changing direction. This acceleration, called the **centripetal acceleration**, is directed at right angles to the wind, inward toward the low centre.

Remember from Newton's second law that if an object is accelerating, there must be a net force acting on it. In this case, the *net force* acting on the wind must be directed toward the centre of the low so that the air will keep moving in a circular path. This inward-directed force is called the **centripetal force** (*centri* = centre; *petal* = to push toward). The magnitude of the centripetal force is related to the wind velocity (V) and the radius of the wind's curved path (r) by the formula

$$\text{Centripetal force} = \frac{V^2}{r}$$

● FIGURE 8.27 Winds and related forces around areas of low and high pressure above the friction level in the Northern Hemisphere. Notice that the pressure gradient force (PGF) is in red, whereas the Coriolis force (*CF*) is in blue.

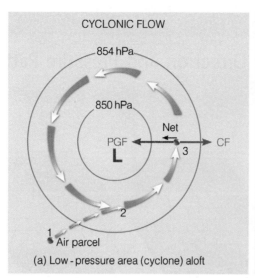

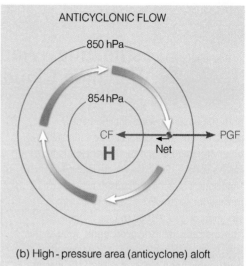

Where wind speeds are light and there is little curvature (a large radius), the centripetal force is weak and, compared to other forces, may be considered insignificant. However, where the wind is strong and blows in a tight curve (a small radius as in the case of tornadoes and tropical hurricanes), the centripetal force is large and becomes quite important.

The centripetal force results from an imbalance between the Coriolis force and the pressure gradient force.* Again, look closely at position 3 in Figure 8.27a and observe that the inward-directed pressure gradient force (PGF) is greater than the outward-directed Coriolis force (*CF*). The difference between these two forces—the net force—is the inward-directed centripetal force. In Figure 8.27b, the wind blows clockwise around the centre of the high. The spacing of the isobars tells us that the magnitude of the PGF is the same as in Figure 8.27a. However, to keep the wind blowing in a circle, the inward-directed Coriolis force must now be greater in magnitude than the outward-directed pressure gradient force, so that the centripetal force again, the net force, is directed inward.

The greater Coriolis force around the high results in an interesting observation. Because the Coriolis force (at any

given latitude) can increase only when the wind speed increases, we can see that for the same pressure gradient (the same spacing of the isobars), the winds around a high-pressure area (or a ridge) must be greater than the winds around a low-pressure area (or a trough). Normally, however, the winds blow much faster around an area of low pressure (a cyclonic storm) than they do around an area of high pressure because the isobars around the low are usually spaced much closer together, resulting in a much stronger pressure gradient.

In the Southern Hemisphere, the pressure gradient force starts the air moving, and the Coriolis force deflects the moving air to the *left*, thereby causing the wind to blow *clockwise around lows* and *counterclockwise around highs*. ● Figure 8.28 shows a satellite image of clouds and wind flow (dark arrows) around a low-pressure area in the Northern Hemisphere (see Figure 8.28a) and in the Southern Hemisphere (see Figure 8.28b).

Near the equator, where the Coriolis force is small, winds may blow around intense tropical storms, with the centripetal force being almost as large as the pressure gradient force. In this type of flow, the Coriolis force is considered negligible, and the wind is called *cyclostrophic*.

So far, we have seen how winds blow in theory, but how do they appear on an actual map?

WINDS ON UPPER-LEVEL CHARTS
On the upper-level 500 hPa map shown in ● Figure 8.29, notice that the winds tend to parallel the contour lines in a wavy, west-to-east direction, as we would expect. Also notice that the contour lines tend to decrease in elevation from south to north. This situation occurs because the air at this level is warmer to the south and colder to the north. On the map, where horizontal temperature contrasts are large, there is also a large height gradient—the contour lines are close together, and the winds are strong. Where the horizontal temperature contrasts are small, there is a small height gradient—the contour lines are spaced farther apart, and the winds are weaker. In general, on

---

*In some cases, it is more convenient to express the centripetal force (and the centripetal acceleration) as the *centrifugal force*, an apparent force that is equal in magnitude to the centripetal force but directed outward from the centre of rotation. The gradient wind is then described as a balance of forces between the centrifugal force,

$$\frac{V^2}{r}$$

the pressure gradient force,

$$\frac{1}{\rho}\frac{\Delta P}{\Delta x}$$

and the Coriolis force,

$$(2\Omega\sin\phi)V.$$

Under these conditions, the *gradient wind equation* for a unit mass of air is expressed as

$$\frac{V^2}{r} + \frac{1}{\rho}\frac{\Delta P}{\Delta x} + 2\Omega V\sin\phi = 0$$

(a) Northern Hemisphere

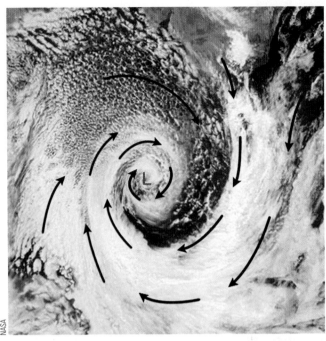

(b) Southern Hemisphere

● FIGURE 8.28 Clouds and related wind-flow patterns (black arrows) around low-pressure areas. (a) In the Northern Hemisphere, winds blow counterclockwise around an area of low pressure. (b) In the Southern Hemisphere, winds blow clockwise around an area of low pressure.

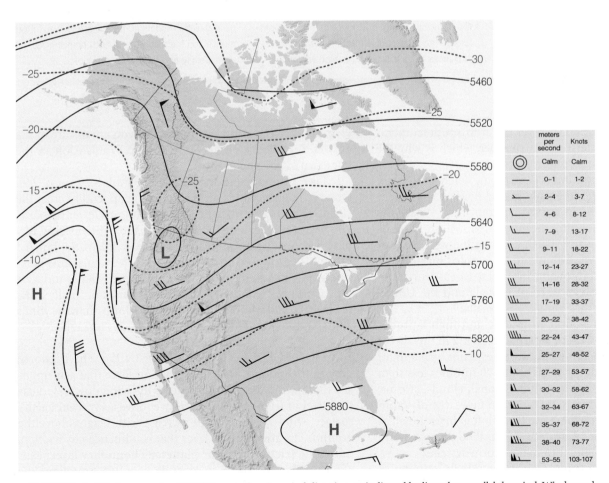

	meters per second	Knots
◎	Calm	Calm
	0–1	1–2
	2–4	3–7
	4–6	8–12
	7–9	13–17
	9–11	18–22
	12–14	23–27
	14–16	28–32
	17–19	33–37
	20–22	38–42
	22–24	43–47
	25–27	48–52
	27–29	53–57
	30–32	58–62
	32–34	63–67
	35–37	68–72
	38–40	73–77
	53–55	103–107

● FIGURE 8.29 An upper-level 500 hPa map showing wind direction, as indicated by lines that parallel the wind. Wind speeds are indicated by barbs and flags. (See the blue insert.) Solid grey lines are contours in metres above sea level. Dashed red lines are isotherms in °C.

FOCUS ON AN OBSERVATION

Winds Aloft in the Southern Hemisphere

In the Southern Hemisphere, just as in the Northern Hemisphere, winds aloft blow because of horizontal differences in pressure. The pressure differences are caused by variations in temperature. Recall from an earlier discussion of pressure that warm air aloft is associated with high pressure and cold air aloft with low pressure. Look at ● Figure 6. It shows an upper-level chart that extends from the Northern Hemisphere into the Southern Hemisphere. Over the equator, where the air is warmer, the pressure aloft is higher. North and south of the equator, where the air is colder, the pressure aloft is lower.

To begin with, let's assume that there is no wind on the chart. In the Northern Hemisphere, the pressure gradient force directed northward starts the air moving toward lower pressure. Once the air is set in motion, the Coriolis force bends it to the right until it is a west wind, blowing parallel to the isobars. In the Southern Hemisphere, the pressure gradient force directed southward starts

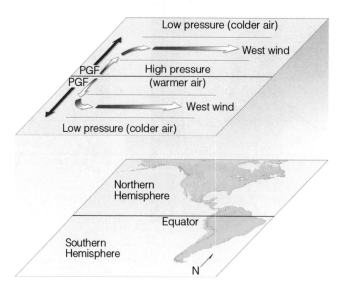

● FIGURE 6 Upper-level chart that extends over the Northern and Southern hemispheres. Grey lines on the chart are isobars.

the air moving south. But notice that the Coriolis force in the Southern Hemisphere bends the moving air to its left, until the wind is blowing

parallel to the isobars *from the west*. Hence, in the middle and high latitudes of both hemispheres, we generally find westerly winds aloft.

---

maps such as this, we find stronger north-to-south temperature contrasts in winter than in summer, which is why the winds aloft are usually stronger in winter.

In Figure 8.29, the wind is geostrophic where it blows in a straight path parallel to evenly spaced lines; it is gradient where it blows parallel to curved contour lines. Where the wind flows in large, looping meanders, following a more or less north–south trajectory (such as along the west coast of North America), the wind-flow pattern is called **meridional**. Where the winds are blowing in a west-to-east direction (such as over the eastern third of Canada and the United States), the flow is termed **zonal**.

Because the winds aloft in middle and high latitudes generally blow from west to east, planes flying in this direction have a beneficial tail wind, which explains why a flight from Tokyo to Vancouver takes about 70 minutes less than the return flight. If the flow aloft is zonal, clouds, storms, and surface anticyclones tend to move more rapidly from west to east. However, where the flow aloft is meridional, as we will see in Chapter 12, surface storms tend to move more slowly, often intensifying into major storm systems.

We know that the winds aloft in the middle latitudes of the Northern Hemisphere tend to blow in a west-to-east pattern. Does this mean that the winds aloft in the Southern

Hemisphere blow from east to west? If you are unsure of the answer, read Focus on an Observation: Winds Aloft in the Southern Hemisphere, above.

Take a minute and look back at Figure 8.20 on p. 240. Observe that the winds on this surface map tend to cross the isobars, blowing from higher pressure toward lower pressure. Observe also that along the green line, the tightly packed isobars are producing a steady surface wind of 74 km h^{-1} (40 knots). However, on an upper-level chart, this same pressure gradient with the same air temperature would produce a much stronger wind. Why do surface winds normally cross the isobars, and why do they blow more slowly than the winds aloft? The answer to both of these questions is *friction*.

FRICTION FORCE AND SURFACE WINDS   The frictional drag of the ground results in a friction force that acts opposite to the wind direction and slows the wind down. Because the effect of friction decreases as we move away from Earth's surface, wind speeds tend to increase with height above the ground. The atmospheric layer that is influenced by friction, called the **friction layer** or **planetary boundary layer**, usually extends upward to an altitude near 1000 m above the surface, but this altitude may vary due to strong winds, irregular terrain, and atmospheric stability. (We will examine

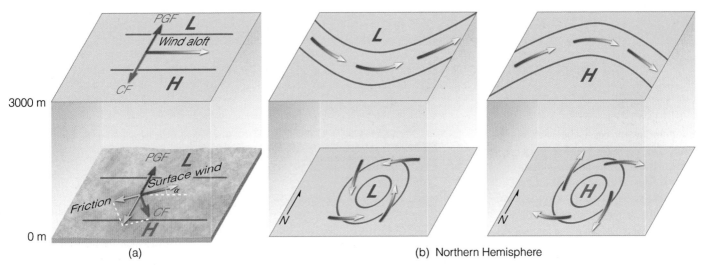

                                    (a)   (b) Northern Hemisphere

● FIGURE 8.30 (a) The effect of surface friction is to slow down the wind so that, near the ground, the wind crosses the isobars and blows toward lower pressure. (b) This phenomenon at the surface produces an inflow of air around a low and an outflow of air around a high. Aloft, away from the influence of friction, the winds blow parallel to the lines, usually in a wavy, west-to-east pattern.

the planetary boundary layer winds more thoroughly in Chapter 9.)

In ● Figure 8.30a, the wind aloft is blowing at a level above the frictional influence of the ground. At this level, the wind is approximately geostrophic and blows parallel to the isobars, with the pressure gradient force (*PGF*) on its left balanced by the Coriolis force (*CF*) on its right. At Earth's surface, the friction force must be considered. The same pressure gradient will not produce the same wind speed, and the wind will not blow in the same direction.

Near the surface, *the friction force reduces the wind speed, which, in turn, reduces the Coriolis force.* Consequently, the weaker Coriolis force no longer balances the pressure gradient force, and the wind blows across the isobars toward lower pressure. The angle (α in Figure 8.30a) at which the wind crosses the isobars varies but averages about 30°.* As we can see in Figure 8.30a, at the surface, *there is a three-way force balance with the pressure gradient force balanced by the*

sum of the friction force and the Coriolis force. Therefore, in the Northern Hemisphere, we find surface winds blowing counterclockwise and *into* a low; they flow clockwise and *out of* a high, as seen in Figure 8.30b. In the Southern Hemisphere, winds blow clockwise and inward around surface lows and counterclockwise and outward around surface highs, as seen in ● Figure 8.31. ● Figure 8.32 on page 252 illustrates a surface weather map and the general wind-flow pattern on a particular day in South America.

---

*The angle at which the wind crosses the isobars depends on the size of the friction force, which to a large degree depends on the roughness of the terrain. Everything else being equal, the rougher the surface is, the larger the friction force and the angle. Over hilly land, the angle might average between 35° and 40°, whereas over an open body of relatively smooth water, it may average between 10° and 15°. Taking into account all types of surfaces, the average is near 30°. This angle also depends on the wind speed. Typically, the angle is smallest for high winds and largest for gentle breezes. As we move upward through the friction layer, the wind becomes more and more parallel to the isobars.

● FIGURE 8.31 Winds around an area of (a) low pressure and (b) high pressure in the Southern Hemisphere.

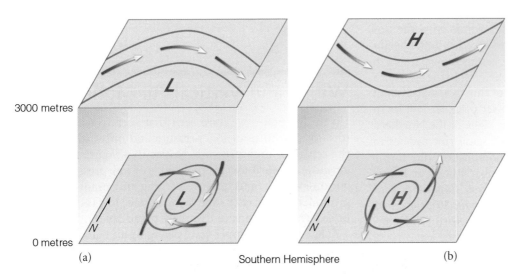

         (a)   Southern Hemisphere   (b)

FOCUS ON AN ADVANCED TOPIC

The Hydrostatic and Hypsometric Equations

Air is in hydrostatic equilibrium when the upward-directed pressure gradient force is exactly balanced by the downward force of gravity. ● Figure 7 shows air in hydrostatic equilibrium. Since there is no net vertical force acting on the air, there is no net vertical acceleration, and the sum of the forces is equal to zero, all of which is represented by

$$PGF_{vertical} + g = 0$$

$$\frac{1}{\rho}\frac{\Delta P}{\Delta z} = g = 0$$

where $PGF_{vertical}$ is the vertical pressure gradient force, ρ is the air density, ΔP is the decrease in pressure along a small change in height Δz, and g is the force of gravity. This expression is usually given as

$$\frac{\Delta P}{\Delta z} = -\rho g$$

This equation is called the *hydrostatic equation*. The hydrostatic equation tells us that the rate at which the pressure decreases with height is equal to the air density times the acceleration of gravity.* The minus sign indicates that as the air pressure decreases,

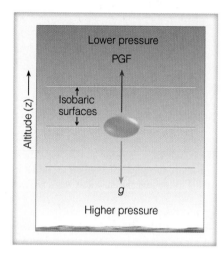

● **FIGURE 7** When the vertical pressure gradient force (PGF) is in balance with the force of gravity (g), the air is in hydrostatic equilibrium.

---

*Where (ρ g) is actually the force of gravity per unit volume. This is because the force of gravity is equal to the mass of air times the acceleration of gravity (m g), and density is mass per volume so that $\rho g = \frac{mg}{V} = \frac{Force\ of\ gravity}{Volume}$.

the height increases. When the hydrostatic equation is given as

$$\Delta P = -\rho g\ \Delta z$$

it tells us something important about the atmosphere that we learned earlier: the air pressure decreases more rapidly with height in cold (denser) air than it does in warm (less dense) air. In addition, we can use the hydrostatic equation to determine how rapidly the air pressure decreases with increasing height above the surface. For example, suppose at the surface a 1000 m thick layer of air (under standard conditions) has an average density of 1.1 kg m^{-3} and an acceleration of gravity of 9.8 m s^{-2}. Therefore, we have

$$\rho = 1.1\ \text{kg m}^{-3}$$
$$g = 9.8\ \text{m s}^{-2}$$
$$\Delta z = 1000\ \text{m}$$

This value is the height difference from the surface (0 m) to an altitude of 1000 m. Using the hydrostatic equation to compute ΔP, the difference in pressure in a 1000 m thick layer of air, we obtain

---

We know that because of friction, surface winds move more slowly than do the winds aloft with the same pressure gradient. Surface winds also blow across the isobars toward lower pressure. The angle at which the winds cross the isobars depends on surface friction, wind speed, and the height above the surface. Aloft, however, the winds blow parallel to contour lines, with lower pressure to their left in the Northern Hemisphere. Consequently, because of this fact, if you stand with the wind aloft to your back, lower pressure will be to your left and higher pressure to your right in the Northern Hemisphere (see ● Figure 8.33a). The same rule applies to the surface wind but with a slight modification due to the fact that here the wind crosses the isobars. Look at Figure 8.33b and notice that at the surface, *if you stand with your back to the wind and then turn clockwise about 30°, the centre of lowest pressure will be to your left** in the Northern Hemi-

---

*In the Southern Hemisphere, stand with your back to the wind and then turn counterclockwise about 30°—the centre of lowest pressure will then be to your right.

sphere. This relationship between wind and pressure is often called **Buys-Ballot's law**, after the Dutch meteorologist Christoph Buys-Ballot (1817–1890), who formulated it.

Winds and Vertical Air Motions

Up to this point, we have seen that surface winds blow in toward the centre of low pressure and outward away from the centre of high pressure. Notice in ● Figure 8.34 that as air moves inward toward the centre of low pressure, it must go somewhere. Since this converging air cannot go into the ground, it slowly rises. Above the surface low (at about 6 km or so), the air begins to diverge (spread apart).

As long as the upper-level diverging air balances the converging surface air, the central pressure in the surface low does not change. However, the surface pressure *will change* if

$$\Delta P = -\rho g \, \Delta z$$

$$\Delta P = -1.1 \frac{kg}{m^3} \times 9.8 \frac{m}{s^2} \times 1000 \, m$$

$$\Delta P = -10{,}780 \, kg \, m \, s^{-2} \, m^{-2} = -10{,}780 \, Pa$$

Since 1 hPa = 100 Pa:

$$\Delta P = -108 \, hPa$$

Hence, air pressure decreases by about 108 hPa in a standard 1000 m layer of air near the surface. This closely approximates the pressure change of 10 hPa per 100 m we used as an estimate in converting station pressure to sea-level pressure earlier in this chapter.

The hydrostatic equation can be combined with the gas law to allow us to use pressure and temperature measurements to determine altitude. This is the basis for altimetry using barometers—the way most airplanes find their altitude. To see how this works, start with the hydrostatic equation

$$\Delta P = -\rho g \, \Delta z$$

Divide both sides by $-\rho g$ to find an equation for Δz:

$$\Delta z = -\frac{\Delta P}{\rho g}$$

Remember from earlier in this chapter that the gas law ($P = \rho R_d T$) relates density to temperature and pressure and can be used to find an expression for density by dividing both sides by $R_d T$:

$$\rho = \frac{P}{R_d T}$$

Combining the two equations gives

$$\Delta z = -\frac{\Delta P}{P} \frac{R_d T}{g}$$

This equation is called the *hypsometric equation.** So if we know the pressure difference between two elevations (ΔP) and the average temperature in the layer between the two elevations (T), we can find the difference in elevation (Δz), given that P represents the average pressure and both R_d and g are constants.

---

*A slightly more exact version of this equation is found using natural logarithms and written as

$$\Delta z = \frac{R_d T}{g} \ln \frac{P_1}{P_2}$$

where P_1 is the pressure at the bottom of the layer and P_2 is the pressure at the top of the layer.

Let's try an example. Suppose that the pressure at sea level is 1000 hPa and the pressure on a hillside is 940 hPa. The average temperature in the layer is 20°C. Let's find the altitude above sea level on the hillside.

- ΔP is $940 - 1000 = -60$ hPa or -6000 Pa.
- P is the average pressure of 970 hPa or 97,000 Pa.
- T is the temperature in kelvins, which equals $20 + 273.15 = 293.15$ K.
- g is the acceleration of gravity $= 9.8 \, m \, s^{-2}$.
- R_d is the gas constant for air, which is 287 J kg^{-1} K^{-1}.

Putting it all together,

$$\Delta z = -\frac{\Delta P}{P} \frac{R_d T}{g}$$

$$\Delta z = -\frac{-6000 \, Pa}{97000 \, Pa} \times \frac{287 \frac{J}{kg \, K} \times 293.15 \, K}{9.8 \frac{m}{s^2}}$$

$$\Delta z = 531 \frac{J \, s^2}{kg \, m} = \frac{531 \frac{kg \, m^2}{s^2} s^2}{kg \, m} = 531 \, m$$

So the elevation on the hillside is 531 m above sea level.

---

upper-level divergence and surface convergence are not in balance. For example, as we saw earlier in this chapter, when we examined the air pressure above two cities, the surface pressure will change if the mass of air above the surface changes. Consequently, if upper-level divergence exceeds surface convergence (i.e., more air is removed at the top than is taken in at the surface), the air pressure at the centre of the surface low will decrease. This situation increases the pressure gradient (and, hence, the pressure gradient force), resulting in the isobars around the low becoming more tightly packed and increasing the surface winds.

Surface winds move outward (diverge), away from the centre of a high-pressure area. To replace this laterally spreading air, the air aloft converges and slowly descends, as shown in Figure 8.34. Again, as long as upper-level converging air balances surface diverging air, the air pressure in the centre of the high will not change. (Convergence and divergence of air are so important to the development or weakening of surface pressure systems that we will examine this topic again when we

look more closely at the vertical structure of pressure systems in Chapter 12.)

The rate at which air rises above a low or descends above a high is small compared to the horizontal winds that spiral about these systems. Generally, the vertical motions are usually only several centimetres per second, or about 1.5 km per day.

Earlier in this chapter, we learned that air moves in response to pressure differences. Because air pressure decreases rapidly with increasing height above the surface, there is always a strong pressure gradient force directed upward, much stronger than in the horizontal. Why, then, doesn't the air rush off into space? Air does not rush off into space because the upward-directed pressure gradient force is nearly always exactly balanced by the downward force of gravity. When these two forces are in exact balance, the air is said to be in **hydrostatic equilibrium**. When air is in hydrostatic equilibrium, there is no net vertical force acting on it, so there is no net vertical acceleration.

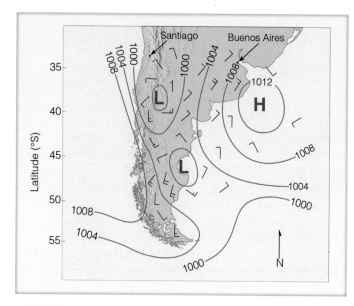

● FIGURE 8.32 Surface weather map showing isobars and winds on a day in December in South America.

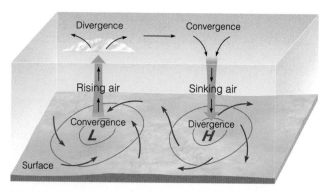

● FIGURE 8.34 Winds and air motions associated with surface highs and lows in the Northern Hemisphere.

Most of the time, the atmosphere approximates hydrostatic balance, even when air slowly rises or descends at a constant speed. However, this balance does not exist in violent thunderstorms and tornadoes, where the air shows appreciable vertical acceleration. But these occur over relatively small vertical distances, considering the total vertical extent of the atmosphere. (A more mathematical look at *hydrostatic equilibrium*, expressed by the hydrostatic equation, is given in Focus on an Advanced Topic: The Hydrostatic and Hypsometric Equations on p. 250.)

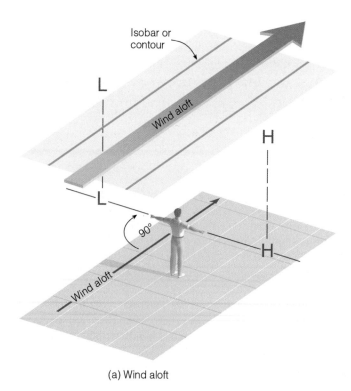

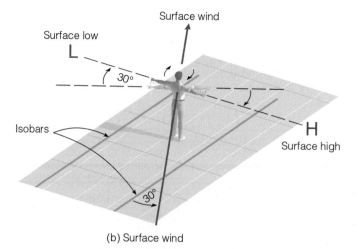

(a) Wind aloft

(b) Surface wind

Northern Hemisphere

● FIGURE 8.33 (a) In the Northern Hemisphere, if you stand with the wind aloft at your back, lower pressure aloft will be to your left and higher pressure to your right. (b) At the surface, the centre of lowest pressure will be to your left if, with your back to the surface wind, you turn clockwise about 30°.

SUMMARY

This chapter gave a broad view of how and why the wind blows. We examined constant pressure charts and found that low heights correspond to low pressure and high heights to high pressure. In regions where the air aloft is cold, the air pressure is normally lower than average; where the air aloft is warm, the air pressure is normally higher than average. Where horizontal variations in temperature exist, there is a corresponding horizontal change in pressure. The difference in pressure establishes a force, the pressure gradient force, which starts the air moving from higher toward lower pressure.

Once the air is set in motion, the Coriolis force bends the moving air to the right of its intended path in the Northern Hemisphere and to the left in the Southern Hemisphere. Above the level of surface friction, the wind is bent enough so that it blows nearly parallel to the isobars, or contours. Where the wind blows in a straight-line path, and a balance exists between the pressure gradient force and the Coriolis force, the wind is termed geostrophic. Where the wind blows parallel to curved isobars (or contours), the centripetal acceleration becomes important, and the wind is called a gradient wind. When the wind-flow pattern aloft is west to east, the flow is called *zonal*; where the wind flow aloft is more north–south, the flow is called *meridional*.

The interaction of the forces causes the wind in the Northern Hemisphere to blow clockwise around regions of high pressure and counterclockwise around areas of low pressure. In the Southern Hemisphere, the wind blows counterclockwise around highs and clockwise around lows. The effect of surface friction is to slow down the wind. In this case, the pressure gradient force is now balanced by the sum of the Coriolis force and the friction force. This causes the surface air to blow across the isobars from higher pressure toward lower pressure. Consequently, in both hemispheres, surface winds blow outward, away from the centre of a high, and inward, toward the centre of a low.

When the upward-directed pressure gradient force is in balance with the downward force of gravity, the air is in hydrostatic equilibrium. Since there is no net vertical force acting on the air, it does not rush off into space.

KEY TERMS

The following terms are listed (with page numbers) in the order they appear in the text. Define each. Doing so will aid you in reviewing the material covered in this chapter.

QUESTIONS FOR REVIEW

1. Why does air pressure decrease with height more rapidly in cold air than in warm air?

2. What can cause the air pressure to change at the bottom of a column of air?

3. What is considered standard sea-level atmospheric pressure in hectopascals? In inches of mercury? In kilopascals?

4. How does an aneroid barometer differ from a mercury barometer?

5. How does sea-level pressure differ from station pressure? Can the two ever be the same? Explain.

6. On an upper-level chart, is cold air aloft generally associated with low or high pressure? What about warm air aloft?

7. What do Newton's first and second laws of motion tell us?

8. Explain why, in the Northern Hemisphere, the average height of contour lines on an upper-level isobaric chart tends to decrease northward.

9. What is the force that initially sets the air in motion?

10. What does the Coriolis force do to moving air (a) in the Northern Hemisphere? (b) in the Southern Hemisphere?

11. Explain how each of the following influences the Coriolis force: (a) rotation of Earth; (b) wind speed; (c) latitude.

12. How does a steep (or strong) pressure gradient appear on a weather map?

13. Explain why on a map, closely spaced isobars (or contours) indicate strong winds, and widely spaced isobars (or contours) indicate weak winds.

14. What is a geostrophic wind? Why would you not expect to observe a geostrophic wind at the equator?

15. Why do upper-level winds in the middle latitudes of both hemispheres generally blow from the west?

16. Describe how the wind blows around highs and lows aloft and near the surface (a) in the Northern Hemisphere and (b) in the Southern Hemisphere.

17. What are the forces that affect the horizontal movement of air?

18. What factors influence the angle at which surface winds cross the isobars?

19. Describe the type of vertical air motions associated with surface high- and low-pressure areas.

20. Since there is always an upward-directed pressure gradient force, why doesn't the air rush off into space?

21. How does Buys-Ballot's law help locate regions of high and low pressure aloft and at the surface?

22. Explain the effect surface friction has on wind speed and direction.

23. Explain how on a 500 hPa chart you would be able to distinguish a trough from a ridge?

QUESTIONS FOR THOUGHT

1. Explain why, on a sunny day, an aneroid barometer would indicate "stormy" weather when carried to the top of a hill or mountain.

2. The gas law states that pressure is proportional to temperature times density. Use the gas law to explain why a balloon will deflate when placed inside a refrigerator. Use the gas law to explain why the same balloon will inflate when removed from the refrigerator and placed in a warm room.

3. In ● Figure 8.35, suppose that the air column above city Q is completely saturated with water vapour, and the air column above city T is completely dry. If the temperature of the air in both columns is the same, which column will have the highest atmospheric pressure at the surface? Explain. (Hint: Refer back to Focus on a Special Topic: Is Humid Air "Heavier" Than Dry Air? on p. 115 in Chapter 4.)

4. Could station pressure ever exceed sea-level pressure? Explain.

5. Suppose that you are in the Northern Hemisphere watching altocumulus clouds 4000 m above you drift from the northeast. Draw the orientation of the isobars above you. Locate and mark regions of lowest and highest

pressure on this map. Finish the map by drawing isobars and the upper-level wind-flow pattern hundreds of kilometres in all directions from your position. Would this type of flow be zonal or meridional? Explain.

6. Pilots often use the expression "high to low, look out below." In terms of upper-level temperature and pressure, explain what this can mean.

7. Suppose an aircraft using a pressure altimeter flies along a constant pressure surface from standard temperature into warmer-than-standard air without any corrections. Explain why the altimeter would indicate an altitude lower than the aircraft's true altitude.

8. If Earth were not rotating, how would the wind blow with respect to centres of high and low pressure?

9. Why are surface winds that blow over the ocean closer to being geostrophic than those that blow over the land?

10. If the wind aloft is blowing parallel to curved isobars, with the horizontal pressure gradient force being of greater magnitude than the Coriolis force, would the wind flow be cyclonic or anticyclonic? In this example, what would be the relative magnitude of the centripetal acceleration, and how would it be directed?

11. With your present outside surface wind, use Buys-Ballot's law to determine where regions of surface high- and low-pressure areas are located. If clouds are moving overhead, use the relationship to locate regions of higher and lower pressure aloft.

12. If you live in the Northern Hemisphere and a region of surface low pressure is directly west of you, what would probably be the surface wind direction at your home? If an upper-level low is also directly west of your location, describe the probable wind direction aloft and the direction in which middle-type clouds would move. How would the wind direction and speed change from the surface to where the middle clouds are located?

13. In the Northern Hemisphere, you observe surface winds shift from N to NE, to E, and then to SE. From this observation, you determine that a west-to-east moving high-pressure area (anticyclone) has passed north of your location. Describe with the aid of a diagram how you were able to come to this conclusion.

14. The Coriolis force causes winds to deflect to the right of their intended path in the Northern Hemisphere, yet around a surface low-pressure area, winds blow counterclockwise, appearing to bend to their left. Explain why.

15. Why is it that on the equator, winds may blow either counterclockwise or clockwise with respect to an area of low pressure?

16. Use the gas law in Focus on a Special Topic: The Atmosphere Obeys the Gas Law on p. 228 to explain why a car with tightly closed windows will occasionally have a window "blow out" or crack when exposed to the sun on a hot day.

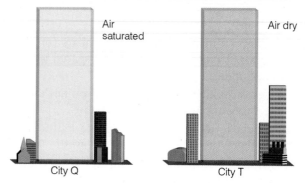

Air saturated   Air dry

City Q   City T

●FIGURE 8.35

17. Consider wind blowing over a land surface that crosses a coastline and then blows over a lake. How will the wind speed and direction change as it moves from the land surface to the lake surface?

18. As a cruise ship crosses the equator, the entertainment director exclaims that water in a tub will drain in the opposite direction now that the ship is in the Southern Hemisphere. Give two reasons to the entertainment director why this assertion is not so.

PROBLEMS AND EXERCISES

1. A station 300 m above sea level reports a station pressure of 994 hPa. What would be the sea-level pressure for this station, assuming standard atmospheric conditions? If the observation were taken on a hot summer afternoon, would the sea-level pressure be greater or less than that obtained during standard conditions? Explain.

2. ● Figure 8.36 is a sea-level pressure chart (Northern Hemisphere) with isobars drawn for every 4 hPa. Answer the following questions, which refer to this map.

   (a) What is the lowest possible pressure in whole hectopascals that there can be in the centre of the closed low? What is the highest pressure possible?

   (b) Place a dashed line through the ridge and a dotted line through the trough.

   (c) What would be the wind direction at point A and at point B?

   (d) Where would the stronger wind be blowing, at point A or B? Explain.

   (e) Compute the pressure gradient between points 1 and 2 and between points 3 and 4. How do the computed pressure gradients relate to the pressure gradient force?

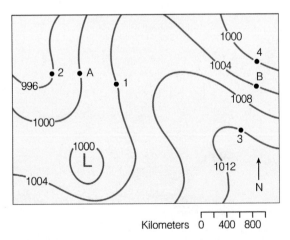

● FIGURE 8.36 Sea-level pressure chart.

(f) If point A and point B are located at 30°N, and if the air density is 1.2 kg m^{-3}, use the geostrophic wind equation in Focus on an Advanced Topic: A Mathematical Look at the Geostrophic Wind on p. 244 to compute the geostrophic wind at point A and point B. (Hint: Be sure to convert km to m and hPa to Pa, where 1 hPa = 100 Pa = 100 kg m^{-1} s^{-2}.)

(g) Would the actual winds at point A and point B be greater than, less than, or equal to the wind speeds computed in problem (f)? Explain.

3. (a) Suppose the atmospheric pressure at the bottom of a deep air column 5.6 km thick is 1000 hPa. If the average air density of the column is 0.91 kg m^{-3} and the acceleration of gravity is 9.8 m s^{-2}, use the hydrostatic equation on p. 251 to determine the atmospheric pressure at the top of the column. (Hint: Be sure to convert km to m and hPa to Pa, where 1 hPa = 100 Pa = 100 kg m^{-1} s^{-2}.)

   (b) If the air in the column of problem (a) becomes much colder than average, would the atmospheric pressure at the top of the new column be greater than, less than, or equal to the pressure computed in problem (a)? Explain.

   (c) Determine the atmospheric pressure at the top of the air column in problem (a) if the air in the column is quite cold and has an average density of 0.97 kg m^{-3}.

4. Use the gas law in the Focus on a Special Topic: The Atmosphere Obeys the Gas Law on p. 228 to calculate the air pressure in hectopascals when the air temperature is −23°C and the air density is 0.700 kg m^{-3}. (Hint: Be sure to use the temperature in kelvins.) At approximately what elevation would you expect to observe this pressure?

5. Suppose air in a closed container has a pressure of 1000 hPa and a temperature of 20°C.

   (a) Use the gas law to determine the air density in the container.

   (b) If the density in the container remains constant, but the pressure doubles, what would be the new temperature?

6. A large balloon is filled with air so that the air pressure inside just equals the air pressure outside. The volume of the filled balloon is 3 m^3, the mass of air inside is 3.6 kg, and the temperature inside is 20°C. What is the air pressure? (Hint: Density = mass/volume.)

7. If the clouds overhead are moving from north to south, would the upper-level centre of low pressure be to the east or west of you? Draw a simplified map to explain.

Chinook arch cloud and wind farm near Pincher Creek, Alberta.
Rogier Gruys. BluePeak Travel Photography.

Winds at Different Scales—Small and Local

9

On December 30, 1997, a United Airlines' Boeing 747 carrying 374 passengers was en route to Hawaii from Japan. Dinner had just been served, and the aircraft had reached a cruising altitude of 10 km. Suddenly, without warning, this routine flight turned tragic when the aircraft entered a region of severe turbulence and a vibration ran through the aircraft. The plane nosed upward and then plunged about 300 m before stabilizing. Screaming, terrified passengers not fastened to their seats were flung against the walls of the aircraft and then dropped. Bags, serving trays, and luggage that slipped out from under the seats were flying about inside the plane. Within seconds, the entire ordeal was over. At least 110 people were injured, 12 seriously. Tragically, there was one fatality. A 32-year-old woman, who had been hurled against the ceiling of the plane, died of severe head injuries. What sort of atmospheric phenomenon could cause such turbulence?

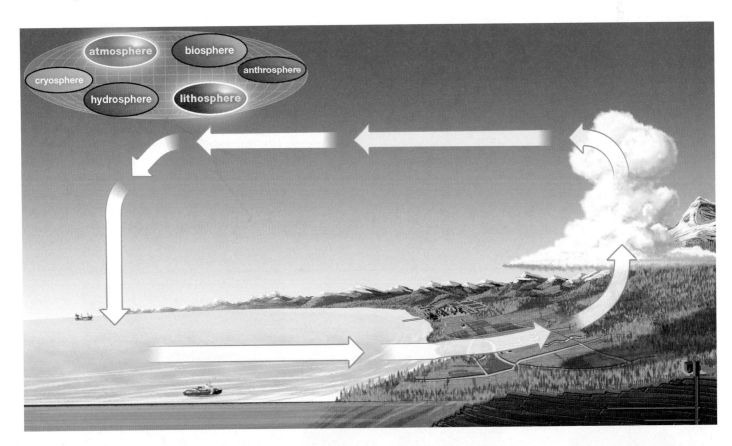

CONTENTS

The aircraft in the opening vignette encountered a turbulent eddy—an "air pocket"—in perfectly clear weather. Luckily, such violent impacts are rare. But they can occur, especially in the air above the highest clouds, in the vicinity of jet streams. How and why these events occur will be discussed in more detail later in this chapter in the section Focus on an Observation: Eddies, Clear Air Turbulence, and "Air Pockets."

The air in motion—what we commonly call wind—is invisible, yet we see evidence of it nearly everywhere. It sculpts rocks, moves leaves, blows smoke, and lifts water vapour upward, where it can condense into clouds. Wind is with us wherever we go. On a hot day, it cools us; on a cold day, it makes us shiver. A breeze can sharpen our appetite when it blows bakery aromas in our direction. The wind is a powerful workhorse of weather. It moves storms and large fair-weather systems around the globe. It transports heat, moisture, dust, insects, bacteria, and pollen from one area to another. The movement of snow, silt, and sand by the wind shapes Earth's landscapes. At small scales, turbulent eddies* transport heat, moisture, and other properties up and down between the surface and the atmosphere. Eddies move water vapour upward from the surface, and winds redistribute it until the water eventually re-enters the *hydrosphere* as precipitation. Similarly, heat is lofted upward by eddies, where winds transport it across the Earth. In these ways, wind links the *atmosphere* to all of Earth's systems: *hydrosphere, cryosphere, lithosphere,* and *biosphere.*

Circulations of all sizes exist within the atmosphere. Little whirls form inside bigger whirls, which encompass even larger whirls—one huge mass of turbulent, twisting eddies.

---

*Eddies are spinning regions of air that can interact with each other. Turbulent flow (turbulence) represents any irregular or disturbed flow in the atmosphere that produces gusts and eddies.

For clarity, meteorologists arrange circulations according to their size. This hierarchy of motion from tiny gusts to giant storms is called the **scales of motion**. In this chapter, we will examine a variety of eddy circulations. First, we will see how eddies form. Next, we will see how eddies and other small-scale winds interact with our environment. Then we will examine slightly larger systems, called **local winds**, as well as larger seasonal circulations, describing how they form and the type of weather they generally bring. Finally, we will look at how wind is measured.

Scales of Motion

Consider smoke rising from a chimney into the otherwise clean air in an industrial section on the outskirts of a large city (see ● Figure 9.1a). Within the smoke, small chaotic motions— tiny **eddies**—cause it to tumble and turn. These eddies constitute the smallest scale of motion—the **microscale**. Here eddies with diameters of a few metres or less disperse smoke, sway branches, swirl dust and paper, and transport quantities vertically between the surface and the atmosphere. Eddies form by convection or by wind blowing past obstructions. They are usually short-lived, lasting from seconds to a few minutes at best. The condensation, deposition, evaporation, and sublimation of water vapour to and from Earth's surface are transported in the atmosphere by turbulent eddies. Other gases, such as carbon dioxide, that are used and released by the biosphere are also transported in this way.

In Figure 9.1b, observe that as the smoke rises, it drifts even higher and is carried many kilometres downwind toward the centre of town. This circulation of city air is one example of the next larger scale—the **mesoscale** or middle scale. Typical mesoscale winds range from a few kilometres to a few

—2 m—

(a) Microscale

—20 km—

(b) Mesoscale

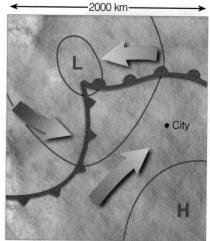

—2000 km—

(c) Synoptic scale

● FIGURE 9.1 Visualizing scales of atmospheric motion. Tiny microscale motions constitute a part of larger mesoscale motions, which, in turn, are part of much larger synoptic scale motions. Notice that as the scale becomes larger, the motions observed at a smaller scale are no longer visible.

hundred kilometres in size. Generally, they last longer than microscale motions, lasting minutes, hours, or as long as a day. Mesoscale circulations include local winds such as those that form along shorelines, mountains, and urban environments, as well as thunderstorms, tornadoes, and small tropical storms.

When we look for the smoke stack on the surface weather map in Figure 9.1c, we see neither the smoke stack nor the circulation of city air. Only the circulations around the high- and low-pressure areas over the region are visible. We are now looking at the **synoptic scale**, or weather map scale. Circulations of this magnitude dominate regions of hundreds to even thousands of square kilometres. Although the life span of these features' varies, they typically last for days and sometimes weeks. Circulations on this scale are responsible for the weather and precipitation patterns we observe on a daily basis. They transport or **advect** heat and moisture horizontally from warm, humid tropical regions to cold, dry polar regions.

The largest wind patterns are seen at the **planetary or global scale**. Here we have wind patterns ranging over the entire Earth. Synoptic and global scales can be combined and referred to as the **macroscale**. • Figure 9.2 summarizes the various scales of motion, their typical size, and their average life span. The next sections will concentrate on microscale winds and the effect they have on our environment.

Microscale Winds Interacting with the Environment

FRICTION AND TURBULENCE IN THE BOUNDARY LAYER   We are all familiar with friction. If we rub our hand over the top of a table, friction tends to slow its movement because of irregularities in the table's surface. On a microscopic level, friction arises as atoms and molecules of the two surfaces seem to adhere and then snap apart as our hand slides over the table. Friction is not restricted to solid objects; it occurs in moving fluids, too. Consider the example of standing in a steadily flowing stream. By looking downstream, you will see whirling, turbulent eddies forming in the water flowing past your legs. These eddies take energy from the main stream flow, use it in the eddy, and, in the process, slow the flow in the main stream. This is fluid friction, which is called **viscosity**. There are two types of viscosity, or fluid friction. Let's examine them in more detail.

When a fluid—such as air—slows down because of the random motion of gas molecules, this is referred to as **molecular viscosity**, our first type of viscosity. Consider a mass of air gliding horizontally and smoothly over a stationary mass of air (i.e., the air is exhibiting **laminar flow**). Even though the molecules in the stationary air are not moving horizontally,

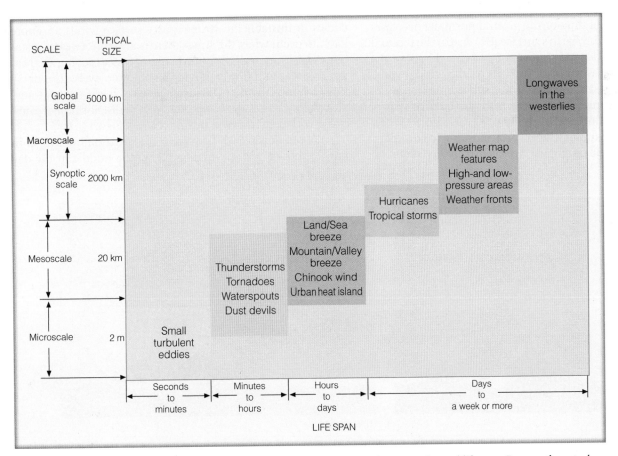

• **FIGURE 9.2** Scales of atmospheric motion indicating the phenomenon's average size and life span. Because the actual size of certain features can vary, some features fall into more than one category.

they are still darting about and colliding with each other. At the boundary separating the air layers, there is a constant exchange of molecules between the stationary air and flowing air. The overall effect of this molecular exchange is to slow down the moving air. If molecular viscosity were the only type of friction acting on moving air, the effect of friction would disappear in a thin layer just above the surface. But there is another type of fluid frictional effect that is more important in reducing wind speeds.

When laminar flow gives way to irregular turbulent motion, the internal friction produced by turbulent whirling eddies is called **eddy viscosity**, our second type of viscosity. Its effect is similar to that of molecular viscosity, but it occurs throughout a much larger portion of the moving air. Near the surface, it is related to wind speed and the roughness of the ground. As wind blows over a landscape dotted with trees and buildings, it breaks into a series of irregular, twisting eddies that can influence the airflow for hundreds of metres above the surface. Within each eddy, wind speed and direction fluctuate rapidly, producing the irregular air motion known as *wind gusts*. Eddies occur at different scales and can be thought of as being created in two different ways. Let's examine this more closely.

Eddy motions created by obstructions to the wind are commonly referred to as **mechanical turbulence** or **forced convection**. Mechanical turbulence creates a drag on the airflow that is far greater than that caused by molecular viscosity. The frictional drag of the ground normally decreases as we move away from Earth's surface. Because of the reduced friction, wind speeds tend to increase with height above the ground. In fact, at a height of only 10 m, the wind is often moving twice as fast as it does at the surface. As we saw in Chapter 8, the atmospheric layer near the surface that is influenced by friction and turbulence is called the *friction layer*, **planetary boundary layer**, or **atmospheric boundary layer**. The top of the boundary layer is usually near 1000 m, but this height varies considerably because strong winds, rough terrain, or instability can extend the region of frictional influence.

Surface heating and instability result in convection, and this causes turbulence to extend to greater altitudes. As Earth's surface heats, thermals rise and convection cells form. The resulting vertical motion creates **thermal turbulence** or **free convection**, which increases with the intensity of surface heating and the degree of atmospheric instability. During the early morning, when the air is most stable, thermal turbulence is normally at a minimum. As surface heating increases, instability is induced and thermal turbulence becomes more intense. If this heating produces convective clouds that rise to great heights, there may be turbulence extending from Earth's surface to the base of the stratosphere.

Although we have treated thermal and mechanical turbulence separately, they occur together in the atmosphere—each magnifying the influence of the other. Let's consider a simple example, the eddy forming behind the barn in ● Figure 9.3. In stable air with weak winds, the eddy is nonexistent or small. As wind speed and surface heating increase, instability develops, and the eddy becomes larger and extends through a greater depth. The rising side of the eddy carries slow-moving surface air upward, causing a frictional drag on the faster flow of air aloft. Some of the faster moving air is brought down with the descending part of the eddy, producing a momentary gust of wind. Because of the increased depth of circulating eddies in unstable air, strong, gusty surface winds are more likely to occur when the atmosphere is unstable. Greater instability also leads to a greater exchange of faster moving air from upper levels with slower moving air at lower levels. In general, this exchange increases the average wind speed near the surface and decreases it aloft, producing the distribution of wind speed with height shown by line (b) in ● Figure 9.4.

We can now see why surface winds are usually stronger in the afternoon. Vertical mixing during the middle of the day

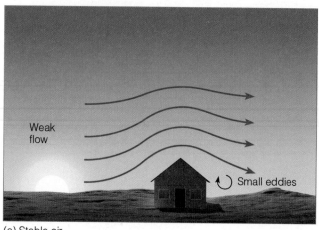

(a) Stable air

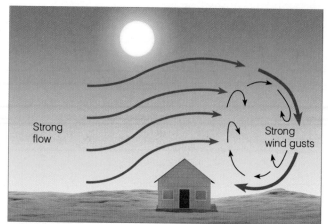

(b) Unstable air

● FIGURE 9.3 Winds flowing past an obstacle. (a) In stable air, light winds produce small eddies and little vertical mixing. (b) Greater winds in unstable air create deep, vertically mixing eddies that produce strong, gusty surface winds.

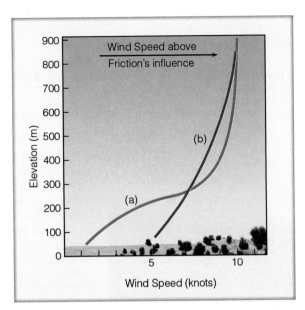

• FIGURE 9.4 Depth of the surface friction layer under different conditions: When the air is stable and the terrain fairly smooth, as in (a), vertical mixing is minimal, and the effect of friction extends a short distance above the surface to about 400 m in this example. When the air is unstable and the terrain rough, as in (b), vertical mixing is maximized, and the effect of friction extends to a greater height—about 800 m in this example. In (b), within the region of frictional influence, vertical mixing increases the wind speed near the ground and decreases it aloft.

links surface air with the faster moving air aloft. The result is that the surface air is pulled along more quickly. At night, when convection is reduced, the interchange between the air at the surface and the air aloft is at a minimum. Hence, the wind near the ground is less affected by the faster wind flow above, so it blows more slowly.

In summary, the friction of air flow is fluid friction and is called viscosity. It is a result of an exchange of air molecules moving at different speeds. The exchange brought about by random molecular motions (molecular viscosity) is quite small in comparison to the exchange brought about by turbulent motions (eddy viscosity). Therefore, the frictional effect of the surface on moving air depends largely on mechanical and thermal turbulent mixing in the planetary boundary layer. The depth of mixing, and hence frictional influence in the boundary layer, depends primarily on these three factors:

1. surface heating—producing a steep lapse rate* and creating instability and strong thermal turbulence
2. strong wind speeds—producing strong mechanical turbulence
3. rough or hilly landscape—producing strong mechanical turbulence

When these three factors occur simultaneously, the frictional effect of the ground is transferred upward to considerable heights, and the wind at the surface is typically strong and gusty.

---

*A lapse rate is a rate of temperature decrease (cooling) with height.

Turbulent air within eddies causes eddy viscosity and the vertical transport of quantities such as heat and moisture. It is responsible for the movement of these between Earth's surface and the atmosphere. To visualize how eddies can transport moisture, imagine that the atmosphere has humid air near the surface and dry air higher up. Turbulent eddy motions that move air upward will transport moist air upward. Eddy motions that move air downward will transport dry air downward. Consequently, downward motions remove drier air from the overlying layer, which is the same as increasing the moisture of the overlying layer. The net effect of both upward and downward turbulent motions is to transport moisture upward. (To learn more about how eddies transport quantities in the atmosphere and how this is measured, see Focus on an Observation: Measuring Wind and Transport in the Planetary Boundary Layer on p. 262.)

EDDIES—BIG AND SMALL When the wind encounters a solid object, a whirl of air—an eddy—forms on the object's downwind (called leeward) side, as shown in • Figure 9.5. The size and shape of the eddy often depend on the size and shape

• FIGURE 9.5 A satellite image of horizontal eddies forming on the leeward (downwind) side of the Cape Verde Islands (in the Atlantic Ocean off Africa's west coast) during April 2004. As the air moves past the islands, it breaks into a variety of swirling eddies, which are indicated by the cloud pattern.

Measuring Wind and Transport in the Planetary Boundary Layer

Dr. Ian Strachan, Department of Natural Resource Sciences, McGill University, Ste. Anne de Bellevue, Quebec

What governs the transfer of heat, carbon dioxide, and water vapour from the surface to the atmosphere? How do different surfaces influence these exchanges? What is the role of wind at very small scales in these exchanges? These are the questions posed by researchers in micrometeorology. These scientists are interested in the exchanges between the surface and the atmosphere that occur at time scales from fractions of a second to multiyear cycles and range in size from metres to tens of kilometres. Micrometeorologists use sophisticated instrumentation to measure winds at the smallest scales. They study the movement of turbulent eddies within the atmospheric layer closest to the surface—the atmospheric boundary layer.

In a turbulent layer of air, heat, momentum, and gases such as water vapour and CO_2 are transported by the seemingly chaotic movement of turbulent "eddies." These are generated mechanically by forced convection as air moves over a surface or convectively by free convection as temperature increases at the surface. This transport by eddies is called a turbulent *flux*.* It works in the following way: if a

---

*Turbulent fluxes are usually represented as *flux densities*. A flux density is a flow (an amount per time) of some property, divided by the area the property is flowing through. For example, a flux of heat is measured in watts in the SI sysem (where a watt = a joule per second, and a joule represents a unit for energy). This is normally represented as a flux density of heat (watts per square metre—W m^{-2} in the SI unit system).

turbulent eddy causes a brief updraft (*a positive vertical wind speed fluctuation*), and the air in that updraft is warmer than average (*a positive temperature fluctuation*), then the eddy is transporting warm air upward. Likewise, if the eddy causes a downdraft (*a negative vertical wind speed fluctuation*), and the air in the downdraft is colder than average (*a negative temperature fluctuation*), then the eddy is transporting cold air downward. In both cases, there is an upward flux of sensible heat that will warm the air. Since in either case the fluctuations covary similarly, the product of two negative fluctuations or two positive fluctuations results in a positive flux. Over a period of time, a consistent turbulent flux results when fluctuations in the vertical wind and fluctuations in the transported property are synchronized, or co-vary. This phenom-

enon is called **eddy covariance** and is the basis for measuring turbulent transport near Earth's surface.

To measure turbulence, instruments are deployed on towers that extend over the surface and provide continuous observations of atmospheric properties (see ● Figure 1). The turbulent wind moves past a point at a variety of frequencies from very fast-moving gusts that make leaves flutter to more sustained motions that can last several minutes. Observations must capture all types of turbulence, so instruments must be able to measure as quickly as 10 to 20 hertz—in other words, 10 to 20 measurements each second!

Instrumentation has developed rapidly over the past several decades, and the majority

I. Strachan

● FIGURE 1 A micrometeorological flux tower extends upward from a black spruce forest canopy in the James Bay region of Quebec. Such towers are used to support flux measurement systems as well as equipment that continuously monitors the surface radiation budget, wind speed, temperature, and humidity.

---

of the obstacle and on the speed of the wind. Light winds produce small, stationary eddies. Wind moving past trees, shrubs, and even your body produces small eddies. (You may have

experienced dropping a piece of paper on a windy day only to have it carried away by a swirling eddy as you bend down to pick it up.) Air flowing over a building produces larger eddies that will at most be about the size of the building. Strong winds blowing past an uncovered sports stadium can produce eddies that may rotate and create surface winds on the playing field that move in the opposite direction to the wind flow above the stadium. Wind blowing over a fairly smooth surface produces few eddies, but when the surface is rough, many eddies form.

The eddies that form downwind from obstacles can produce a variety of interesting effects. For instance, at much

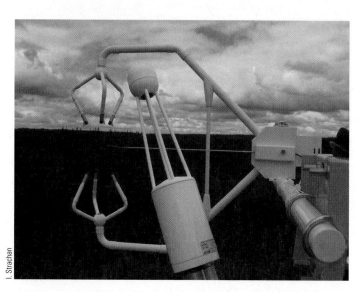

I. Strachan

● FIGURE 2 An eddy covariance system for measuring fluxes of sensible heat, latent heat, and CO_2 consists of an infrared gas analyzer (foreground) and a sonic anemometer (background). A fast-response thermocouple for measuring temperature extends along the sonic anemometer.

of micrometeorologists now use the eddy covariance method to measure surface fluxes. A three-dimensional sonic anemometer (see ● Figure 2) is used to measure the three directional components of the wind speed: two horizontal and one vertical. When the vertical component of the wind speed is combined with high-frequency measurements of trace gases or temperature, we can observe the exchanges of these quantities between the surface and the atmosphere (e.g., CO_2 fluxes or sensible heat fluxes). Expressed generally, the vertical flux (F_s) of a quantity (water vapour, CO_2, or sensible heat) is found using eddy covariance methods by taking the average product of the instantaneous departure from the mean vertical wind speed (w')

and the instantaneous departure from the mean quantity (s') multiplied by a constant (C). Mathematically, this is

$$F_s = C\,\overline{w's'}$$

The overbar in ($\overline{w's'}$) indicates that the quantities under it are averaged for a time period, which is typically 30 minutes. In this way, all turbulent fluctuations are measured, but longer cycles that are attributable to day-to-night or synoptic weather patterns are excluded because they take longer than 30 minutes.

In the case of water vapour, the flux (F_{H_2O}) becomes

$$F_{H_2O} = L_v\,\overline{w'\rho_v'}$$

where L_v is a constant, the latent heat of vapourization of water, w' is the instantaneous departure from the mean vertical wind speed, and ρ_v' is the instantaneous departure from the mean water vapour density.

If, instead of water vapour, we want to determine the sensible heat flux, temperature, measured by a fast-response thermocouple, is used as the s' value. Many micrometeorological researchers are also interested in carbon exchange due to the importance of CO_2 as a **greenhouse gas**, so it is typical to measure both water vapour and CO_2 concentrations using a fast-response infrared gas analyzer (IRGA), as shown in Figure 2.

There are now many tower flux stations in use around the world as part of Fluxnet, a global network to measure the exchanges of CO_2, water vapour, and heat between the surface and the atmosphere. In Canada, micrometeorologists are actively conducting research in settings such as temperate, coastal and boreal forests; wetlands; subarctic and arctic environments; agroecosystems; and urban centres. Micrometeorologists work closely with scientists who develop models to further our knowledge of how Earth's diverse surfaces influence the atmosphere. Understanding the weather begins with an understanding of the surface and how its temperature and humidity profiles are influenced by the local surface conditions—this is micrometeorology.

larger scales, wind moving over a mountain range, at speeds greater than 70 km h^{-1} in a stable atmosphere, often produces *mountain waves*. Mountain waves can take the form of **lee waves** or **downslope windstorms**.

Lee waves oscillate up and down every 5 to 10 km after the air has past the mountain, as shown in ● Figure 9.6. If the air is sufficiently humid, *lee wave clouds* form on the wave crests, resulting in cloud bands that parallel the mountains (also see Figure 6.23 and 6.24 in Chapter 6 on p. 182). In Figure 9.6, we can see that eddies form beneath each wave crest. These are called *roll eddies*, or **rotors**, and they have

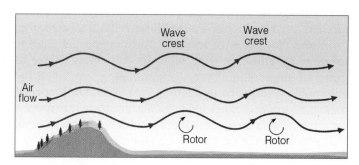

● FIGURE 9.6 Under stable conditions, air flowing past a mountain range can create eddies many kilometres downwind from the mountain.

violent vertical motions that produce extreme turbulence and hazardous flying conditions.

Downslope windstorms are a particularly extreme variation of a mountain wave. These form in very windy high-elevation areas and are experienced in winter along the eastern slopes of the Rocky Mountains as well as downwind of other large mountains. They occur when strong winds blow over a large mountain range with a stable layer near the mountain top. The result is a single, large mountain wave with extreme winds and turbulence that can down airplanes, rip roofs from buildings, and create hazards for the residents of affected areas. When this happens, wind in the lowest kilometre of the air resembles the flow of water over a river rock that speeds up as it moves over the rock and flows back on itself in a large rotor at the base of the rock. In a downslope windstorm, winds accelerate dramatically, to speeds as high as 185 km h^{-1} (100 knots), down the leeward (downwind) slope. One of the best known examples of a downslope windstorm is called the *Boulder windstorm*, named for Boulder, Colorado, which is situated on the lee side of the U.S. Rocky Mountains, where this phenomenon is observed and studied.

The largest atmospheric eddies form as the flow of air becomes organized horizontally into huge spiraling whirls, which can have diameters that are greater than 1000 km. These are the cyclones and anticyclones of middle latitudes. As these migrating systems make weather in the mid-latitudes so changeable, we will examine their formation and movement in Chapters 11 and 12.

Turbulent eddies form aloft as well as near the surface. Turbulence aloft can occur suddenly and unexpectedly, especially where the wind abruptly changes its speed, direction, or both. This is called **wind shear**. Shearing creates forces that produce eddies along a mixing zone. When these eddies form in clear air—a sky condition with no visible sign of clouds or weather—it is called **clear air turbulence** (or **CAT**). Additional information on this topic is given in Focus on an Observation: Eddies, Clear Air Turbulence, and "Air Pockets" on p. 265.

THE FORCE OF THE WIND   The force of the wind on an object is proportional to the wind speed squared ($F \propto V^2$). This means that a small increase in wind speed can greatly increase the force of the wind on an object, so strong winds may blow down trees, overturn mobile homes, and even move railroad cars. For example, in February 1965, the wind presented people in North Dakota with a "ghost train" as it pushed five railroad cars about 125 km, from Portal to Minot, without a locomotive.

Wind blowing with sufficient force to move a train is uncommon. However, wind blowing with enough force to move an automobile is very common, especially when the automobile is exposed to a strong crosswind. On a normal road, the force of a crosswind is usually insufficient to move a car sideways because of the reduced wind flow near the ground. However, when the car crosses a high bridge, where the frictional influence of the ground is reduced, the increased wind speed can be felt by the driver. Near the top of a high bridge, where the wind flow is typically strongest, complicated eddies

● **FIGURE 9.7** Strong winds flowing past an obstruction, such as these hills, can produce a reverse flow of air that strikes an object from the side opposite the general wind direction.

pound against the car's side as the air moves past obstructions, such as guard railings and posts. In a strong wind, these eddies may even break into extremely turbulent whirls that buffet the car, causing driving difficulty as it moves from side to side. If there is a wall on the bridge, the wind may swirl around and strike the car from the side opposite the wind direction.

A similar effect occurs where the wind moves over low hills paralleling a highway (see ● Figure 9.7). When the vehicle moves by the obstruction (the hill in Figure 9.7), a wind gust from the opposite direction can suddenly and without warning push the car to the opposite side. This wind hazard is especially problematic for trucks, campers, and trailers, which have large side-areas. Highway signs warning of "strong crosswinds in the area" are often posted.

As stable air flows over a ridge, it increases in speed. Thus, winds blowing over mountains tend to be stronger than winds blowing at the same level on either side. In fact, the second greatest surface wind speed ever recorded occurred in this type of situation. At the summit of Mount Washington, New Hampshire, at an elevation 1909 m, the wind gusted to 372 km h^{-1} (201 knots) on April 12, 1934. A similar increase in wind speed occurs where air accelerates as it funnels through a narrow constriction, such as a low pass or saddle in a mountain crest.

Up to now, we have seen that when the wind meets a barrier, it exerts a force on it. If the barrier does not move, the wind moves around, up, and over it. If the barrier is long and low like a water wave, the slight updrafts created on the windward side can support the wings of birds, allowing them to skim the water in search of food without having to flap their wings. Elongated hills and cliffs that face into the wind increase wind speed at the hilltops and create upward air motions (updrafts) that can support a hang glider for a long time. Wind speeds greater than about 28 km h^{-1} (15 knots) blowing over a smooth, moderately sloping ridge may provide excellent ridge soaring for sailplanes and gliders.

FOCUS ON AN OBSERVATION

Eddies, Clear Air Turbulence, and "Air Pockets"

To better understand how eddies form along a zone of wind shear, imagine that, high in the atmosphere, there is a stable layer of air with vertical wind speed shear (wind speed changing with height) as depicted in ● Figure 3a. The top half of the layer slowly slides over the bottom half, and the relative speed of both halves is low. As long as the wind shear between the top and bottom of the layer is small, few, if any, eddies form. However, if the shear and corresponding relative speed of these layers increase (see Figures 3b and 3c), wavelike undulations may form. When the shearing exceeds a certain value (see Figure 3d), the waves break into large swirls, with significant vertical movement, as shown in ● Figure 4. As seen in the opening vignette on p. 257, eddies such as these often form in the upper troposphere near jet streams,

where large wind speed shears exist. They also occur in conjunction with mountain waves, which may extend upward into the stratosphere. When these huge eddies develop in clear air—a sky condition with no visible sign of clouds or weather—they are referred to as clear air turbulence, or CAT. Clear air turbulence has occasionally caused structural damage to aircraft by breaking off vertical stabilizers and tail structures.

The eddies that form in clear air may have diameters ranging from a couple of metres to several hundred metres. An unsuspecting aircraft entering such a region may be in for more than just a bumpy ride. If the aircraft flies into a zone of descending air, it may drop suddenly, producing the sensation that there is no air supporting the wings. Consequently, these regions have come to be known as "air pockets" or "air holes."

Air pockets have caused commercial aircraft to drop hundreds of metres, injuring passengers and flight attendants who were not strapped into their seats. This chapter's opening vignette on p. 257 provides a very dramatic example. Often the consequences are less severe, although terrifying for those involved. On April 25, 2010, on a United Arab Emirates flight between Dubai and India, a passenger jet carrying 364 people experienced severe turbulence associated with an air pocket before landing. The actual drop caused by the air pocket was only 60 m, but it still resulted in 18 injuries (cuts, bruises, and a dislocated shoulder), with many noninjured passengers in shock. This is one of the important reasons we are told to keep our seat belts on while flying.

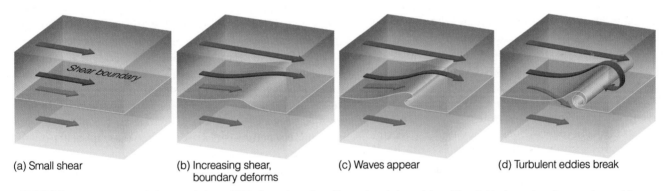

(a) Small shear   (b) Increasing shear, boundary deforms   (c) Waves appear   (d) Turbulent eddies break

● FIGURE 3 The formation of clear air turbulence (CAT) along a boundary of increasing wind speed shear. The wind in the top layer increases in speed from (a) through (d) as it flows over the bottom layer.

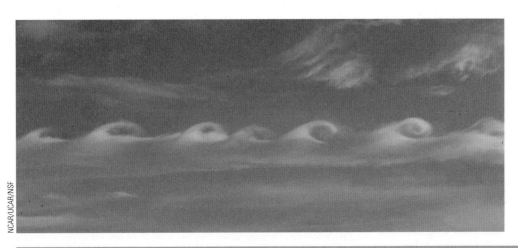

● FIGURE 4 Following the process shown in Figure 3, turbulent eddies forming in a wind shear zone produce these waves, called *Kelvin Helmholtz waves*. The visible clouds that form are called *billow clouds*.

IMPACTS OF MICROSCALE WINDS

Wind and the Landscape Where the wind blows over exposed sediments and soil, it can take an active role in shaping the landscape. Called aeolian processes, wind's ability to entrain, transport, and deposit material illustrates a key link between the *atmosphere* and the *lithosphere*. Wind has been and is an important mover of earth. *Loess*, typically silt sized, is the term applied to sediments that are transported by wind. As wind speed determines the size of particles that are transported, the presence of larger sediments (sands) and smaller ones (clays, dust) indicates fluctuating wind flow. Loess is deposited when the wind slows and particles are dropped, forming a blanketlike cover that fills hollows, buries existing terrain, and forms plains. As ice ages ended, melting ice sheets, higher winds, massive unvegetated surfaces, and fluctuating glacial rivers exposed huge amounts of sediment that became entrained by the wind and deposited downwind. During the last North American ice age, much material was blown from Canada to the United States, where these loess deposits now form some of the best agricultural soils. Non–ice sheet–related loess is still being blown and deposited downwind from unvegetated areas that range in size from small, recently ice-free mountainous areas, to regional volcanic ash deposits, to huge areas such as the large deserts of northern China, Africa, and Australia.

The impact of wind is still particularly noticeable in deserts, where, in addition to the movement and the expansion of the desert itself, wind creates sand dunes and pavement features. As tiny, loose particles of sand, silt, and dust are lifted and carried away by the wind, the ground level lowers and leaves the surface covered with gravel and pebbles, which are too large to be transported. Such a landscape is termed **desert pavement**. The blowing sand will eventually come to rest behind an obstacle (debris, rock, or a clump of vegetation). As sand grains accumulate, they also act as obstacles. If the wind speed is strong enough and blows in the same direction for long enough, sand piles grow high enough to become a *sand dune*. On the dune's surface, the sand rolls, slides, and gradually creeps along, producing wavelike patterns called *sand ripples*. Ripples and dunes form perpendicular to the wind direction, with a gentle slope on the upwind side and a steeper slope on the downwind side. If the wind direction changes frequently, the shape becomes more symmetrical.

Wind and Snow Features created as the wind moves sand can also be seen where the wind moves snow. Strong winds blowing over an open snow-covered landscape gather small accumulations of snow and form *snow drifts* and even *snow dunes*. Wind blowing over a snow-covered landscape may also create wavelike patterns several centimetres high and oriented at right angles to the wind. These *snow ripples* are similar to sand ripples. Snow drifts are a common prairie winter phenomenon that can accumulate significant amounts of snow in areas that are far removed from where the snow originally fell. After the snow stops falling, strong winds can whip it into the

● FIGURE 9.8 Snow drifts accumulating behind snow fences.

air, leaving fields snow-free, removing insulation, and exposing the soil to cold dry winds that rob it of any remaining moisture. As occurs with sand dunes, the snow settles out of the air when the wind slows down as it encounters obstacles such as built-up areas. In this way, snowfall accumulations in town can be much greater than those in the surrounding countryside.

To help remedy this situation, *snow fences* are constructed in open spaces (see ● Figure 9.8). Behind a snow fence, the wind speed is reduced because the air is broken into small eddies, which allow the snow to settle to the ground. The added snow cover acts like an insulating blanket that protects the open ground from the bitterly cold air, which often follows a major snowstorm. In regions of low rainfall, moisture from snowmelt that becomes stored in groundwater can be critical during long, dry summers. Snow fences are also built to protect roadways by placing them in such a way that snow accumulates behind the fence rather than in drifts on the road, where it poses a hazard to drivers.

If the snow on the ground is moist and sticky, some of it may be picked up by the wind and sent rolling, collecting snow as it rolls along. These **snow rollers** range from egg- to barrel-sized cylinders. The tracks they make in the snow are typically less than 1 cm deep and several metres long. Snow rollers are rare, but when they occur, they create striking winter scenes, such as that seen in ● Figure 9.9.

Wind and Vegetation Strong winds can have an effect on vegetation, too. Armed with sand, winds can sandblast objects, eroding landscapes and damaging or destroying tender new vegetation. In agricultural areas, this decreases crop productivity. Most plants increase their rate of transpiration as wind speed increases,* leading to rapid water loss, especially in warmer areas with low humidities. This can actually dry out plants, which stunts their growth. In some windy, dry regions, mature trees that should be many metres

---

*This effect actually drops above a certain wind speed and varies greatly among plant species.

© University Corporation for Atmospheric Research

● FIGURE 9.9 Snow rollers—natural cylindrical rolls of snow— grow larger as the wind blows them along.

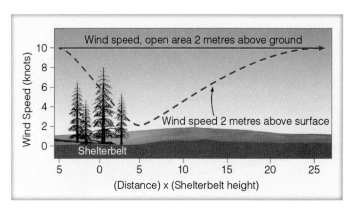

● FIGURE 9.10 A properly designed shelterbelt can reduce the air flow downwind for a distance of 25 times the height of the belt. The minimum wind flow behind the belt is typically measured downwind at a distance of about four times the belt's height.

tall grow only to the height of a small shrub. Wind-dried vegetation can cause high fire danger. Additionally, fires in windy areas are stronger and more volatile as winds help fires along, directing their movement, adding oxygen for combustion, and carrying burning embers elsewhere to start new fires.

Vegetation screens the surface from the direct force of the wind, anchors the soil, and greatly reduces wind-caused soil erosion. Soil moisture also helps resist wind erosion by cementing earth particles together, increasing their cohesiveness. Land where the vegetation cover has been removed followed by several years of drought is ripe for wind erosion. This happened in drier parts of the Canadian Prairies and the U.S. Great Plains in the mid-1930s when agricultural land became desiccated and known as the "dust bowl." During this time, winds carried millions of tons of dust into the air, creating vast dust storms that buried whole farmhouses, reduced millions of acres of soil to unproductive wasteland, and financially ruined thousands of families.

To protect crops and soil, **windbreaks (shelterbelts)** are planted. Shelterbelts usually consist of rows of mixed conifer and deciduous trees or shrubs planted perpendicular to the prevailing wind. They greatly reduce the wind speed behind them (see ● Figure 9.10). As air filters through the belt, the flow is broken into small eddies, which have little mixing effect on the air near the surface. This slows near-surface winds and reduces evapotranspiration from the soil and vegetation downwind of the shelterbelt. The use of properly designed shelterbelts has benefited agriculture through stabilizing the soil, increasing snow accumulation by acting as a snow fence, and improving wheat yields. However, if trees are planted too close together, unwanted effects may result: the air moving past the belt becomes broken into larger, more turbulent eddies, which remove soil. Furthermore, in high winds, strong downdrafts may damage the crops. Despite their advantages, many of the shelterbelts planted during the 1930s drought years have been removed. Since these years, nondrought conditions and irrigation have made them seem economically unfeasible as they occupy valuable crop land or interfere with the large centre-pivot sprinkler systems used today. One wonders how the absence of shelterbelts would impact the region if a 1930s-type drought struck again.

Wind and Water Waves formed by wind blowing over the water's surface are known as **wind waves**. Just as air blowing over the top of a water-filled pan creates tiny ripples, so waves are created as the frictional drag of the wind transfers energy to the water. In general, the greater the wind speed is, the greater the amount of energy added, and the higher the waves. The amount of energy transferred to the water (and thus the height to which a wave can build) depends on four factors:
1. wind speed
2. length of time that the wind blows over the water
3. *fetch*—the distance the wind blows over water
4. depth of water

A sustained 93 km h^{-1} (50 knot) wind blowing steadily for nearly three days over a minimum distance of 2600 km can generate waves with an average height of 15 m. Thus, a stationary storm system centred somewhere over the open sea is capable of creating large waves with heights reaching over 31 m.

Microscale winds actually help waves grow taller. For example, the wind blowing over the small wave depicted in ● Figure 9.11 and the wave are moving in the same direction. Notice that the wave crest deflects the wind upward, producing an undulation in the airflow just above the water. This looping air motion establishes a small eddy of air between the two crests. The upward-and-downward motion of the eddy reinforces the upward-and-downward motion of the water. Consequently, the eddy helps the wave build in height.

Travelling in the open ocean, waves represent a form of energy. As they move into a region of weaker winds, they gradually change: Their crests become lower and more rounded, forming what are commonly called swells. When waves reach a shoreline, they grow in height as the water depth decreases and they transfer their energy to the coast. Sometimes the effects of this are catastrophic. High, storm-induced

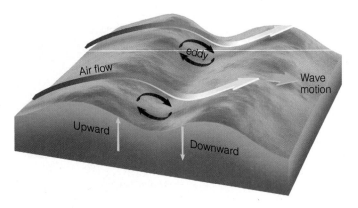

● FIGURE 9.11 Wind blowing over a wave creates a small eddy of air that helps reinforce the up-and-down motion of the water.

waves can hurl thousands of tonnes of water against the shore, causing erosion and damaging shoreline structures such as piers, homes, and roads. If this happens during an unusually high tide, impacts are seen even further inland. Usually, the storms that created these waves are thousands of kilometres away, and they never reach the affected shore.

In summary, we have discussed how the wind blowing over a surface can produce a variety of features, from snow rollers to large ocean waves. To learn how the force of the wind can influence someone riding a bicycle, read Focus on a Special Topic: Pedalling into the Wind on p. 269.

BRIEF REVIEW

Up to this point, we have been examining microscale winds and how they affect our environment. Before we turn our attention to winds on a larger scale, here is a brief review of some of the main points presented so far:

● Viscosity is the friction of fluid flow. The small-scale fluid friction that is due to the random motion of the molecules is called molecular viscosity. The larger scale internal friction produced by turbulent flow is called eddy viscosity.

● Mechanical turbulence (forced convection) is created by twisting eddies that form as the wind blows past obstructions. Thermal turbulence (free convection) results as rising and sinking air forms when Earth's surface is heated unevenly by the sun.

● Turbulence in air that has a vertical gradient such as a temperature or moisture gradient results in transport of the quantity in the atmosphere. For example, if the air is hot near the surface and cooler aloft, turbulent eddies will move temperature and consequently heat upward.

● The planetary boundary layer (or friction layer) is approximately the first 1000 m above the surface.

● Wind shear is a sudden change in wind speed or wind direction (or both).

● The wind can shape a landscape, influence crop production, transport material from one area to another, and generate waves.

Local Wind Systems

Local winds are mesoscale phenomena. There are two general categories of local wind systems:
1. thermal circulations driven by heating and cooling caused by the rising and setting sun
2. topographically forced flows that occur when wind is forced to flow over and through mountainous regions

In the next sections, we will begin examining local winds by looking at how thermal circulations form. Then we will discuss types of local thermal circulation. Finally, we will look at topographically forced flows and their different forms.

THERMAL CIRCULATIONS   Circulations caused by horizontal changes in air temperature, in which warmer air rises and colder air sinks, are termed **thermal circulations**. To see how and why these occur, let us start by considering the vertical distribution of pressure shown in ● Figure 9.12a. The isobaric surfaces all lie parallel to Earth's surface; thus,

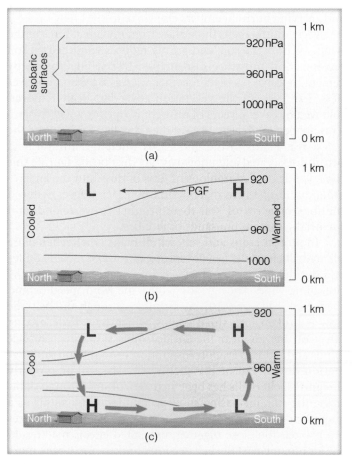

● FIGURE 9.12 A thermal circulation produced by the heating and cooling of the atmosphere near the ground. The **H** and **L** symbols refer to high and low atmospheric pressure. The lines represent surfaces of constant pressure (isobaric surfaces) where the numbers are in hectopascals (hPa). For more information on isobaric surfaces, see Chapter 8, p. 234.

FOCUS ON A SPECIAL TOPIC

Pedalling into the Wind

Anyone who rides a bicycle knows that it is much easier to pedal with the wind than against it. The reason is obvious: as the wind blows against an object, it exerts a force on it. The amount of force exerted by the wind over an area increases as the square of the wind velocity. This relationship is shown by

$$F \propto V^2$$

where: F is the wind force, V is the wind velocity, and $\propto$ means "proportional to." From this we can see that if the wind velocity doubles, the force goes up by a factor of 2^2, or 4, which means that pedalling into a 70 km h^{-1} wind requires four times as much effort as pedalling into a 35 km h^{-1} wind (see ● Figure 5).

Wind striking an object exerts a pressure on it. The amount of pressure depends on the object's shape and size, as well as on the pressure drop on its downwind side. We can approximate the wind pressure on an object with a simple formula:

$$\frac{F}{A} = P = \frac{1}{2}\rho V^2 C_d$$

where F is the wind force, A is the object's surface area, P is the wind pressure, ρ is the air density, and V is the wind velocity. C_d is called the drag coefficient—a factor depending on an object's shape. For the case of a typical bicycle rider, the drag coefficient, C_d, is approximately 0.9. With the equation in SI units, and assuming an air density, ρ, of 1.2 kg m^{-3}, it simplifies to

$$\frac{F}{A} = P = 0.54V^2$$

We can look at a practical example of this expression if we consider a bicycle rider going 18 km h^{-1} (5 m s^{-1})* into a head wind of

● FIGURE 5 Pedalling into a 30 km h^{-1} wind requires nine times as much effort as pedalling into a 10 km h^{-1} wind.

54 km h^{-1} (15 m s^{-1}). This creates a total velocity of wind against the rider (i.e., wind speed plus bicycle speed) of 72 km h^{-1} (20 m s^{-1}). Using this value, the pressure of the wind is

$$P = 0.54 \ V^2$$

$$P = 0.54 \ \frac{\text{kg}}{\text{m}^3}\left(20^2 \frac{\text{m}^2}{\text{s}^2}\right)$$

$$P = 216 \ \text{kg m}^{-1} \text{ s}^{-2} \ (= 216 \text{ Pa})$$

If the rider has a surface body area of 0.5 m^2, the total force exerted by the wind becomes

$$F = P \times A$$

---

*Metres per second (ms^{-1}) are also given as they are more easily used in the calculation because they allow the units of the other quantities to cancel without converting them.

$$F = 216 \text{ N m}^{-2} \times 0.5 \text{ m}^2$$

$$F = 108 \text{ N}$$

This force is the same as the downward force from gravity on an 11 kg object—enough to make pedalling into the wind extremely difficult. To remedy this adverse effect, cyclists—especially racers—bend forward as low as possible to expose a minimum surface area to the wind.

Runners also experience wind impacts. At competitions, wind affects records set during track events to the extent that when runners race with a tail wind of more than 2 m s^{-1}, their results are asterisked with the qualifier "wind-aided."

---

there is no horizontal variation in pressure (or temperature), and there is no pressure gradient and no wind. Suppose in Figure 9.12b that the atmosphere is cooled to the north and warmed to the south. In the cold, dense air above the surface, the isobars bunch closer together, whereas in the warm, less dense air, they spread farther apart. This dipping of the

isobars produces a horizontal pressure gradient force aloft that causes the air to move from higher pressure (warm air) toward lower pressure (cold air).

At the surface, the air pressure changes as the air aloft begins to move. As the air aloft moves from south to north, air leaves the southern area and "piles up" above the northern area. This redistribution of air reduces the surface air pressure to the south and raises it to the north. Consequently, a pressure gradient force is established at Earth's surface from north to south; hence, surface winds begin to blow from north to south.

We now have a distribution of pressure and temperature and a circulation of air, as shown in Figure 9.12c. As the cool surface air flows southward, it warms and becomes less dense. In the region of surface low pressure, the warm air slowly rises, expands, cools, and flows out the top at an elevation of about 1 km above the surface. At this level, the air flows horizontally northward toward lower pressure, where it completes the circulation by slowly sinking and flowing out the bottom of the surface high. This circulation, brought on by horizontal changes in air temperature, in which warmer air rises and colder air sinks, is a thermal circulation.

The regions of surface high and low atmospheric pressure created as the atmosphere either cools or warms are called **thermal** *cold-core* **highs** or **thermal** *warm-core* **lows**. In general, they are shallow systems, usually extending no more than a few kilometres above the ground. These systems weaken with height. For example, at the surface, atmospheric pressure is lowest in the centre of the warm thermal low in Figure 9.13. In the warm air above the low, the isobars spread apart, and at some intermediate level, the thermal low disappears and actually changes into a high. A similar phenomenon happens above the cold thermal high. The surface pressure is greatest in its centre, but because the isobars aloft are crowded together due to the cold dense air, the surface thermal high becomes a low a kilometre or so above the ground. We can summarize the typical characteristics of thermal pressure systems as being shallow, weakening with height, and dominantly being maintained by local surface heating and cooling.

Thermal circulations are mesoscale wind patterns that take many forms and can range in horizontal size from hundreds of metres to hundreds of kilometres. They can form wherever contrasting temperature surfaces lie next to each other. An example of this is the urban–rural breeze that can form due to the *urban heat island* effect in which urban areas are normally a few degrees warmer than the surrounding countryside. In this case, a country breeze can develop with surface air from the rural area blowing toward the urban area. Local scale examples, where moisture differences create the circulation, occur when irrigated land lies next to dry land on a hot summer day or forested land lies next to an area that has been clear-cut. In this case, air over the irrigated (or forested) surface remains relatively cool due to evaporation when compared to air over the dry surface. Consequently, a low-level breeze from the cool surface toward the warm surface develops.

Now let's look at two of the most important mesoscale thermal circulations, the sea and land breeze, as well as mountain and valley breezes.

Sea and Land Breezes   A **sea breeze** is a type of thermal circulation. If a similar circulation occurred near a large freshwater body, it would be called a **lake breeze**. The uneven heating rates of land and water (described in Chapter 3) cause these mesoscale circulations. During the day, the land heats more quickly than the adjacent water, and the intensive heating of the air above the land produces a shallow thermal low. The air over the water remains cooler than the air over the land; hence, a shallow thermal high exists over the water. The overall effect of this pressure distribution is a **sea breeze** that blows at the surface from the sea toward the land (see Figure 9.14a). Since the strongest gradients of temperature and pressure occur near the land–water boundary, the strongest winds typically occur right near the beach and diminish inland. Since the greatest contrast in temperature between land and water usually occurs in the afternoon, sea breezes are strongest at this time.

At night, the land cools more quickly than the water. The air above the land becomes cooler than the air over the water, producing a distribution of pressure such as the one shown in Figure 9.14b. With higher surface pressure now over the land, the surface wind reverses itself and becomes a **land breeze**—a breeze that flows from the land toward the water. Temperature contrasts between land and water are generally much smaller at night. Hence, land breezes are usually weaker than their daytime counterpart—the sea breeze. In regions where there are greater nighttime temperature contrasts, stronger land breezes occur over the water near the shoreline. They are not usually noticed by those on shore but are frequently observed by ships in coastal waters.

Sea breezes are best developed where large temperature differences exist between land and water. Such conditions prevail year-round in many tropical regions, but in the mid-latitudes, sea breezes are invariably spring and summer phenomena. During the summer, a sea breeze usually sets in about midmorning, after the land has been warmed. By early afternoon, the breeze has increased in strength and depth. By late afternoon, the cool ocean air may reach a depth of more

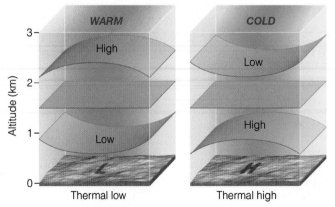

● FIGURE 9.13 The vertical distribution of pressure with thermal (cold) highs and thermal (warm) lows.

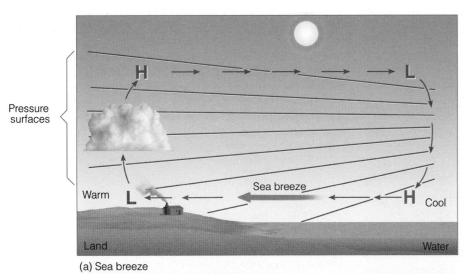

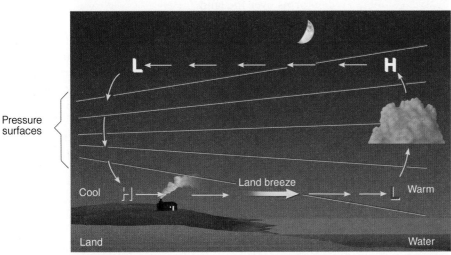

(a) Sea breeze

(b) Land breeze

● FIGURE 9.14 Development of a sea breeze and land breeze. (a) At the surface, a sea breeze blows from the water onto the land, whereas (b) the land breeze blows from the land toward the water. Notice that the pressure at the surface changes more rapidly with the sea breeze. This situation indicates a stronger pressure gradient force and higher winds with a sea breeze.

than 300 m and extend inland for more than 20 km in middle latitudes. These depths are deeper and distances are further in the tropics.

The leading edge of the sea breeze is called the **sea breeze front**. As the front moves inland, a rapid drop in temperature usually occurs just behind it. In some locations, this temperature change may be 5°C or more during the first hours—a refreshing experience on a hot, sultry day. In regions where the water temperature is warm, the cooling effect of the sea breeze is hardly evident. Since cities near the ocean usually experience the sea breeze by noon, their highest temperature usually occurs much earlier than in inland cities. The passage of the sea breeze front is marked by a wind shift, usually to an onshore direction. In the cool ocean air, the relative humidity rises as the temperature drops. If the relative humidity increases to above 70 percent, water vapour begins to condense on particles of sea salt or industrial smoke, producing haze. When the ocean air is highly concentrated with pollutants, the sea breeze front may meet relatively clear air and

thus appear as a *smoke front*, or a *smog front*. If the ocean air becomes saturated, a mass of low clouds and fog will mark the leading edge of the marine air.

When there is a sharp contrast in air temperature across the frontal boundary, the warmer, lighter air will converge and rise. If this rising air is sufficiently moist, a line of cumulus clouds will form along the sea breeze front. If the air is also conditionally unstable, the initial uplift provided by the sea breeze front can trigger deeper convection, leading to thunderstorms. Examples of this are seen on a hot, humid day, near Southwestern Ontario's Great Lakes, where one can drive toward the lake, encounter heavy showers several kilometres from the shore, and arrive at the beach to find it sunny with a steady onshore breeze.

Converging lake breezes occur in Southwestern Ontario stretching on either side of a line from Sarnia (at the bottom of Lake Huron, near the Michigan–Ontario border) through London to Toronto. In this area during summer, lake breezes from Lake Huron, Lake Erie, and Lake Ontario converge,

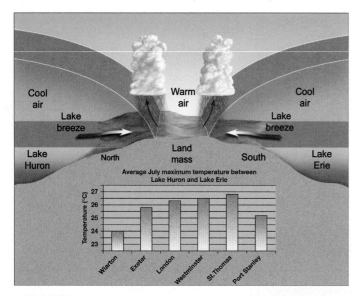

● FIGURE 9.15 The convergence of two lake breezes and their influence on the maximum temperature during July in Southwestern Ontario.

Source: Environment Canada. Canadian Climate Normals of Averages 1971-2000. Found at: http://www.climate.weatheroffice.gc.ca/climate_normals/index_e.html

resulting in frequent thunderstorms, to the extent that the area is known as Canada's Tornado Alley. In the meanwhile, the lakeshore area remains sunny, pleasantly cool, and dry. (To learn more about this, read Focus on a Special Topic: Canada's Tornado Alley on p. 446 in Chapter 14.) As can be seen in ● Figure 9.15, the lake breezes also have a significant effect on the daily high temperature in the region.

As shown in ● Figure 9.16, converging, moist, large-scale sea breezes in Florida help make it the "thunderstorm capital" of North America. On the Atlantic side of the

peninsula, a sea breeze blows in from the east; on the Gulf of Mexico shore, a sea breeze moves in from the west. Their convergence, coupled with daytime convection, produces cloudy conditions and showery weather with frequent thunderstorms over the land.

Local Winds and Water Frequently, local winds will change speed and direction as they cross a large body of water. ● Figure 9.17 shows the wind speed and direction as air flows over a large lake. At position A, on the upwind side, the wind is blowing at 10 knots (18.5 km h^{-1}) from the northwest; at position B, the wind speed is 15 knots (28 km h^{-1}) and has shifted toward a more northerly direction; at position C, the wind is again blowing at 10 knots (18.5 km h^{-1}) from the northwest. Why does the wind blow faster and from a slightly different direction in the centre of the lake? As the air moves from the rough land over the relatively smooth lake, friction with the surface lessens, and the wind speed increases. The increase in wind speed, however, increases the Coriolis force, which turns the wind flow to the right, as shown by the wind report at position B. When the air reaches

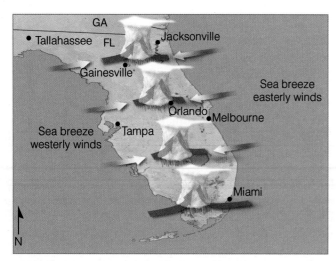

● FIGURE 9.16 Typically, during the summer over Florida, converging sea breezes in the afternoon produce uplift that enhances thunderstorm development and rainfall over the land. However, when westerly surface winds dominate and a ridge of high pressure forms over the area, thunderstorm activity diminishes, and dry conditions prevail.

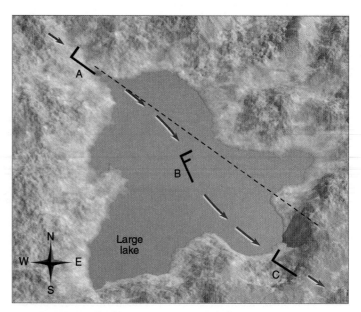

● FIGURE 9.17 Wind can change in both speed and direction when crossing a large lake.

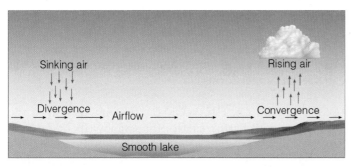

● FIGURE 9.18 Sinking air develops where surface winds move offshore, speed up, and diverge. Rising air develops as surface winds move onshore, slow down, and converge.

the opposite side of the lake, it again encounters rough land, and its speed slows. This process reduces the Coriolis force, and the wind responds by shifting to a more westerly direction, as shown by the report at position C.

Changes in wind speed along the shore of a large lake can inhibit cloud formation on one side and enhance it on the other. Suppose that warm, moist air flows over a lake, as illustrated in ● Figure 9.18. Observe that clouds are forming on the downwind side but not on the upwind side. The lake is slightly cooler than the air. Consequently, by the time the air reaches the downwind side of the lake, it will be cooler, denser, and less likely to rise. Why, then, are clouds forming on this side of the lake? As air moves from the land over the water, it travels from a region of greater friction into a region of less friction, so it increases in speed, which causes the surface air to diverge—to spread apart. Such spreading of air forces air from above to slowly sink, which, of course, inhibits the formation of clouds. Hence, there are no clouds on the upwind side of the lake. Out over the lake, the separation between air temperature and dew point lessens. As this nearly saturated air moves onshore, friction with the

rougher ground slows it down, causing it to "bunch up" or converge (which forces the air upward). This slight upward motion, coupled with surface heating, is often sufficient to initiate the formation of clouds along the downwind side of the lake.

Strong winds blowing over an open body of water, such as a lake, can cause the water to slosh back and forth rhythmically. This sloshing causes the water level to periodically rise and fall, much like water does at both ends of a bathtub when the water is disturbed. Water waves like these, which oscillate back and forth, are called **seiches** (pronounced "sayshes"). In addition to strong winds, seiches may also be generated by sudden changes in atmospheric pressure or by earthquakes.* Around the Great Lakes, the term *seiche* applies to any sudden rise in water level whether or not it oscillates. In November 2003, strong westerly winds, gusting to more than 93 km h⁻¹ (50 knots), created a seiche on Lake Erie that caused a 4 m difference in lake level between its western shore and its eastern shore.

Mountain and Valley Breezes Mountain and valley breezes develop along mountain slopes. Observe in ● Figure 9.19 that during the day, sunlight warms the valley walls, which, in turn, warm the air column above the slope over a depth of a few hundred metres. In a manner similar to the formation of the sea breeze, the heated air column over the slope expands and becomes less dense than the air of the same altitude above the valley. This results in decreased air pressure near the surface on the slope and an increase in air pressure aloft, causing a pressure gradient that pushes air toward the slope near the surface. Consequently, the heated

---

*Earthquakes and other disturbances on a lake floor can cause the water to slosh back and forth, producing a seiche. Earthquakes on the ocean basin floor can cause a tsunami, a Japanese word meaning "harbour waves," because these waves build in height as they enter a bay or harbour.

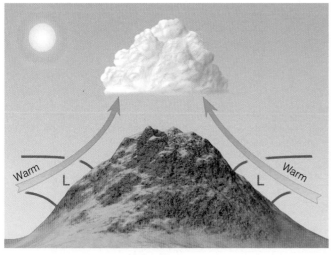

Valley breeze

Mountain breeze

● FIGURE 9.19 Valley breezes blow uphill during the day; mountain breezes blow downhill at night. (The Ls and Hs represent low and high pressure. The purple lines represent surfaces of constant pressure.)

air rises as a gentle upslope wind known as an **anabatic wind** or **valley breeze**. At night, the flow reverses. The mountain slopes cool quickly, chilling the air in contact with them. The cooler, more dense air has increased pressure near the slope and glides downslope into the valley, providing a **katabatic** downslope **wind** known as a **mountain breeze**. (Because gravity is the force that directs these winds downhill, they are also referred to as *gravity winds*, or *nocturnal drainage winds*.) This daily cycle of wind flow is best developed in clear summer weather when prevailing winds are light.

In many areas, the upslope winds begin early in the morning, reach a peak speed of about 10 km h^{-1} by midday, and reverse direction by late evening. The downslope mountain breeze increases in intensity, reaching its peak in the early-morning hours, usually just before sunrise. In the Northern Hemisphere, valley breezes are particularly well developed on south-facing slopes, where sunlight is most intense. On partially shaded north-facing slopes, the upslope breeze may be weak or absent. Since upslope winds begin soon after the sun's rays strike a hill, valley breezes typically begin first on the hill's east-facing side. In the late afternoon, this side of the mountain goes into shade first, triggering the onset of downslope winds at an earlier time than experienced on west-facing slopes. Hence, it is possible for campfire smoke to drift downslope on one side of a mountain and upslope on the other side.

When the upslope winds are well developed and have sufficient moisture, they reveal themselves by building cumulus clouds above mountain summits (see ● Figure 9.20). Since valley breezes usually reach their maximum strength in the early afternoon, cloudiness, showers, and even thunderstorms are common over mountains during the warmest part of the day—a fact well known to climbers and hikers, who avoid being caught high in the mountains at midday to avoid dangerous thunderstorms.

LOCAL WINDS OF THE WORLD   The interaction of the atmosphere with Earth's surface topography (its *lithosphere*) results in a rich assortment of winds in specific geographic areas that are often given local names. In the following sections, we will examine some of the more prominent local winds.

Katabatic Winds, Fall Winds, and Gap Winds   Although any wind flowing down a slope is technically a katabatic wind, the term is usually reserved for downslope flows such as the *mountain breeze*, which are generated on the slope itself by surface cooling creating dense air that flows downslope due to gravity. In some circumstances, katabatic winds can be semipermanent features and very strong if the slope is large and cold, as is the case over Antarctica and the Greenland ice sheet, where katabatic winds of over 185 km h^{-1} (100 knots) have been observed on the coast. The fact that Antarctica is the windiest continent on Earth is due to the katabatic winds that persist there. For example, Cape Denison on the Antarctic coast of Adélie Land has an average wind speed of over 74 km h^{-1} (40 knots) due to the cold katabatic winds that collect over the continent and are channeled to the coast in this region. On a smaller scale, *glacier winds* are shallow katabatic flows that pass over alpine glaciers and are

● FIGURE 9.20 As mountain slopes warm during the day, air rises and often condenses into cumuliform clouds, such as these.

particularly prominent during summer months. They occur as the surface of a melting glacier is kept at 0°C, cooling the overlying warmer air to produce a downslope wind both day and night.

In some circumstances, the cold downslope winds are part of a larger scale phenomenon that extends beyond an individual slope. If the cold air accumulates over an elevated plateau or spills through a mountain pass before descending the slope, the winds may be much stronger than a mountain breeze and are called **fall winds**. These winds are distinguished from katabatic winds since they typically accelerate as they approach the slope or pass before the terrain starts going downhill. The ideal setting for a fall wind is an elevated plateau surrounded by mountains, with an opening that slopes rapidly downhill. When winter snows accumulate on the plateau, the overlying air grows extremely cold and a shallow dome of high pressure forms near the surface (see ● Figure 9.21). Along the edge of the plateau, the horizontal pressure gradient force is usually strong enough to cause the cold air to flow across the isobars through gaps and saddles in the hills. Along the slopes of the plateau, the wind continues downhill as a gentle or moderate cold breeze. If the horizontal pressure gradient increases substantially, such as when a storm approaches, or if the wind is confined to a narrow canyon or channel, the flow of air can increase, often destructively, as cold air rushes downslope like water flowing over a fall.

Fall winds are observed in various regions of the world. For example, along the northern Adriatic coast of Bosnia and Croatia, a polar invasion of cold air from Russia descends the slopes from a high plateau and reaches the lowlands as the *bora*—a cold, gusty, northeasterly wind with speeds sometimes in excess of 185 km h^{-1} (100 knots). It is now believed that the bora and other strong fall winds are similar to the

mountain waves and downslope windstorms discussed previously. A similar, but often less violent, cold wind known as the *mistral* descends the western mountains into the Rhône Valley of France and then out over the Mediterranean Sea. It frequently causes frost damage to exposed vineyards and makes people bundle up in the otherwise mild climate along the Riviera.

In western North America, during winter, when cold air accumulates over interior plateaus, it can flow toward the coast through passes, fjords, and valleys that dissect the coastal mountains. This connects the cold interior plateaus of British Columbia and Yukon Territory (Canada) and Washington and Oregon (U.S.A.) to the warm coast and results in cold, strong, offshore winds that blow from the land onto the water. Two types of winds behave this way and are distinguished as follows: fall winds occur if the winds are strongest on the slope, and **gap winds** result if the winds accelerate in flat valleys or through gaps in the mountain barrier. Pressure gradients cause gap winds to accelerate along the channel or valley. High pressure in the cold interior air and lower pressure in the warmer coastal air move the air from high pressure to low pressure, from the interior to the coast. Examples of these are the *Stikine* and *Taku* winds that run through valleys of the same names on the Alaskan panhandle, as well as *outflow* or *Squamish* winds that form in the fjords along the British Columbia coast, such as Howe Sound in the south and Portland Inlet further north. Gap winds also occur through the Strait of Juan de Fuca separating Vancouver Island from the Olympic Peninsula.* Similarly, the *Columbia Gorge wind* (called the *coho*) forms in the Columbia River Gorge between the states of Washington and Oregon.

All of these gap winds occur when wintertime cold arctic air and its associated thermal high-pressure area track southward from Alaska and the Yukon Territory to lie over the plateaus east of the coastal mountains. The pressure gradient that forms across the coastal mountains between the cold, high-pressure area inland and the warmer, lower pressure area on the coast causes the winds to blow over the mountains, through passes and valleys, onto the coast. When the cold continental air reaches the coast, temperatures plummet well below freezing and snow may occur. In some regions, such as the east coasts of British Columbia's Vancouver Island and Haida Gwaii, the cold outflowing air picks up moisture as it crosses the ocean and dumps heavy snow amounts as the moisture-laden cold air is forced to rise again over the land once it reaches the island.

Chinook (Foehn) Winds

The **chinook wind** or **foehn** is distinguished from bora-type fall winds by being a warm, dry, downslope wind. The chinook descends the eastern slope

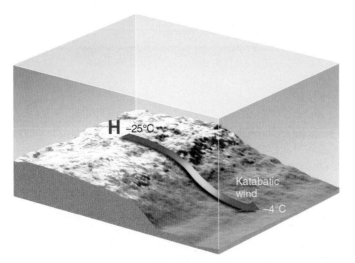

● FIGURE 9.21 Strong katabatic winds can form where cold winds rush downhill from an elevated plateau covered with snow.

*Information on geographic features and their location in North America is provided in the back of the book.

FOCUS ON A SPECIAL TOPIC

Snow Eaters and Rapid Temperature Changes

Chinooks are thirsty winds. As they move over a heavy snow cover, they can melt and evaporate 30 cm of snow in less than a day. This situation has led to some tall tales about these so-called "snow eaters." Canadian folklore has it that a sled-driving traveller once tried to outrun a chinook. During the entire ordeal, his front runners were in snow while his back runners were on bare soil.

Actually, the chinook is important economically. It not only brings relief from the winter cold, but it also uncovers prairie grass, so livestock can graze on the open range. Also, these warm winds have reduced snow plowing costs for Calgarians and kept railroad tracks clear of snow, so trains can keep running. On the other hand, the drying effect of a chinook can create an extreme fire hazard. And when a chinook follows spring planting, the seeds may die in the parched soil. Along with the dry air comes a buildup of static electricity, making a simple handshake a shocking experience. These warm, dry winds have sometimes adversely affected human behaviour. During periods of chinook winds, some people feel irritable and depressed and others become ill and have migraine headaches. The exact reason for this phenomenon is not clearly understood.

In Canada, chinooks are most prevalent in south-central Alberta, around the Lethbridge

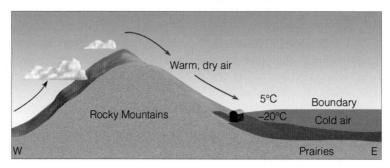

● FIGURE 6 Cities near the warm air–cold air boundary can experience sharp temperature changes if cold air should rock up and down like water in a bowl.

area and extending up to Calgary. Chinooks occur an average of 50 to 52 days each winter in the Crowsnest Pass to the Lethbridge area and 48 to 50 days in Calgary and diminish to the north and east.

The temperature change at the onset of a chinook can be dramatic. On January 11, 1983, the temperature in Calgary rose 26.4°C (from −16.8°C to 10.6°C) in 5 hours. The temperature can also go up and down rapidly in a chinook. In fact, on January 30, 2004, due to a chinook wind, the air temperature in Pincher Creek, Alberta, went from −19.0°C at 1 p.m. to 0.5°C at 2 p.m., where it hovered until crashing back to −20.1°C suddenly at 5 p.m. as the wind shifted from southeasterly (cold) to southwesterly (warm for the chinook) and

back to easterly again. How such rapid changes in temperature can occur is illustrated in ● Figure 6. Notice that a shallow layer of extremely cold air is resting against the Rocky Mountains. The cold air behaves just as any fluid, and in some cases, atmospheric conditions may cause the air to move up and down much like water does when a bowl is rocked back and forth. This rocking motion can cause extreme temperature variations for cities located at the base of the hills along the periphery of the cold air–warm air boundary as they are alternately in and then out of the cold air. Such a situation is thought to be responsible for the extremely rapid two-minute temperature change of 27°C recorded at Spearfish, South Dakota, during the morning of January 22, 1943.

of the Rocky Mountains in a rather narrow region (only several hundred kilometres wide) that extends from Canada southward into northeastern New Mexico. Similar winds occur along the leeward slopes of mountains in other regions of the world. In the European Alps, such a wind is called a foehn; in Argentina, it is called a *zonda*. When these winds move through an area, the temperature rises sharply, sometimes 20°C or more in one hour, and a corresponding sharp drop in the relative humidity occurs, occasionally to relative humidities of less than 5 percent. (More information on temperature changes associated with chinooks is given in Focus on a Special Topic: Snow Eaters and Rapid Temperature Changes above.)

Chinooks occur when strong westerly winds aloft flow over a north–south-trending mountain range, such as the Rockies. Such conditions can produce a trough of low pressure on the mountain's eastern side, a trough that tends to force the air downslope. As the air descends, it is compressed and warms at the dry adiabatic rate (9.8°C km^{-1}). So the main source of warmth for a chinook is *compressional heating* as potentially warmer (and drier) air is brought down. Usually, the warm descending air is displacing and replacing much colder air, which contributes to the sudden warming observed along the eastern slopes.

When clouds and precipitation occur on the mountain's windward side, they can enhance the chinook.

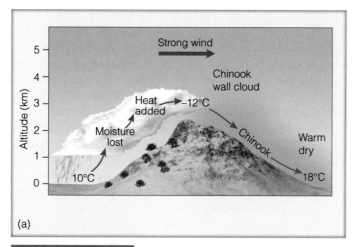

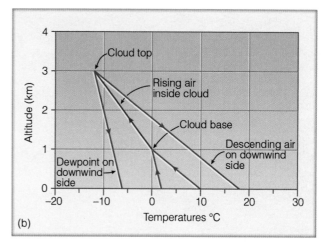

ACTIVE FIGURE 9.22 (a) A chinook wind can be enhanced when clouds form on the mountain's windward side. Heat added and moisture lost on the upwind side produce warmer and drier air on the downwind sides. (b) A graphic representation of the rising and sinking air as it moves over the mountain. Visit the textbook's website to view this and other Active Figures at www.ahrensmeteorology1ce.nelson.com

For example, as the cloud forms on the upwind side of the mountain in ● Figure 9.22a, the release of latent heat inside the cloud supplements the compressional heating on the downwind side. This phenomenon makes the descending air at the base of the mountain on the downwind side warmer than it was before it started its upward journey on the windward side. The air is also drier because much of its moisture was removed as precipitation on the windward side (see Figure 9.22b).

Along the east slopes of the Rockies, two distinctive cloud types are often associated with chinook winds. A bank of clouds forming over the windward slope of the mountains is a telltale sign of an impending chinook. This *chinook wall cloud* (which looks like a wall of clouds) usually remains stationary as air rises, condenses, and then evaporates as the air rapidly descends the leeward slopes, often causing strong winds in foothill communities. A *chinook arch* is a mountain wave cloud that occurs along the east slopes of the Rockies and often precedes chinook winds reaching the surface in Southern Alberta and Montana. It consists of middle and high clouds that form as air rises in a mountain wave over the lee (eastern) slope of the Rockies. The chinook arch cloud extends from south to north, paralleling the mountain chain. To an observer on the Prairies looking west, the sharp western edge of the wave cloud has an arch shape as it extends from the northern to the southern horizons. There is often clear sky between the cloud and the mountain peaks that further distinguishes it. ● Figure 9.23 shows how a chinook arch cloud appears as one looks west toward the Rockies from Calgary, Alberta.

The **Santa Ana wind** is a type of warm, dry chinook or foehn wind that blows from the east or northeast into southern California. As with all chinooks, the air descends and warms but in this case from an elevated desert plateau, which results in very dry air. The wind often blows with exceptional speed—occasionally over 167 km h^{-1}. These conditions set the stage for serious bush fires—especially in autumn, when the vegetation is already parched.

Similar downslope hot, dry winds occur in other locations, such as the *California norther* in northern California; the *puelche*, an east wind that descends the Andes along the west coast of South America; and the *berg* wind that flows from the South African interior plateau onto the coast.

Desert Winds
Winds of all sizes develop over deserts. Huge *dust storms* form in dry regions, where strong winds are able to lift and fill the air with particles of fine dust. An exceptional dust storm the size of Spain formed over the African Sahara during February 2001 and swept westward off the

● FIGURE 9.23 A chinook arch cloud forming over the Alberta Rockies (viewed from Calgary).

NASA

● FIGURE 9.24 A large dust storm over the African Sahara Desert during February 2001 sweeps westward off the coast and then northward into a mid-latitude cyclonic storm west of Spain, as indicated by the red arrow.

African coast and then northeastward (see ● Figure 9.24). In desert areas where loose sand is more prevalent, *sandstorms* develop as high winds enhanced by surface heating rapidly carry sand particles close to the ground. A spectacular example of a storm composed of dust or sand is the **haboob** (from Arabic *hebbe*: blown). The haboob forms as cold downdrafts along the leading edge of a thunderstorm lift dust or

sand into a huge, tumbling dark cloud that may extend horizontally for over 150 km and rise vertically to the base of the thunderstorm (see ● Figure 9.25). Spinning whirlwinds of dust frequently form along the turbulent cold air boundary. Haboobs are most common in the African Sudan (about 24 occur each year) and in the U.S. Southwest desert, especially southern Arizona.

On a smaller scale, spinning vortices commonly seen on hot days in dry areas are called **dust devils** or **whirlwinds**. Generally, dust devils form on clear, hot days over a dry surface, where most of the sunlight goes into heating the surface rather than evaporating water from vegetation. The atmosphere directly above the hot surface becomes unstable, convection sets in, and the heated air rises. Wind, often deflected by small topographic barriers, flows into this region, rotating the rising air (see ● Figure 9.26). Depending on the nature of the topographic feature, the spin of a dust devil around its central eye may be cyclonic or anticyclonic, and both directions occur with about equal frequency.

With diameters of only a few metres and heights of less than a hundred metres, most dust devils are small and last only a short time (see ● Figure 9.27). There are, however, some larger dust devils capable of considerable damage; winds exceeding 139 km h^{-1} (75 knots) may overturn mobile homes and tear the roofs off buildings. Also keep in mind that dust devils *are not* tornadoes. The circulation of many tornadoes (as we will see in Chapter 14) descends downward from the base of a thunderstorm, whereas the circulation of a dust devil begins at the surface, normally in sunny weather, although some form beneath convective-type clouds.

There are other desert winds that should be mentioned. Winds originating over the Sahara Desert are given local

● FIGURE 9.25 A haboob approaching Phoenix, Arizona. The dust cloud is rising to a height of about 450 m above the valley floor.

NOAA

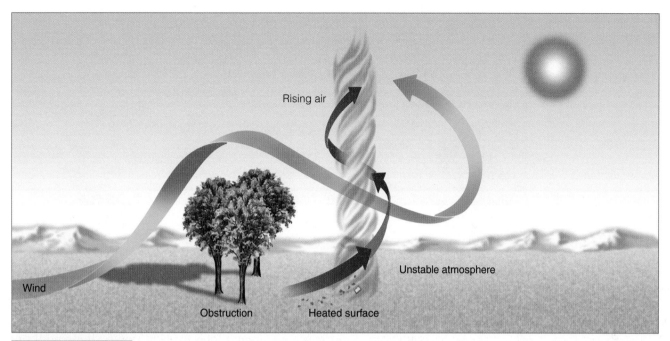

ACTIVE FIGURE 9.26 The formation of a dust devil. On a hot, dry day, the atmosphere next to the ground becomes unstable. As the heated air rises, wind blowing past an obstruction twists the rising air, forming a rotating air column or dust devil. Air from the sides rushes into the rising column, lifting sand, dust, leaves, or any other loose material from the surface. Visit the textbook's website to view this and other Active Figures at www.ahrensmeteorology1ce.nelson.com

● FIGURE 9.27 A dust devil forming on a clear, hot summer day just south of Phoenix, Arizona.

names as they move into different regions. For example, the normal flow of surface air over North Africa is from the north; however, when a storm system is located west of Africa or southern Spain (see position 1, ● Figure 9.28), a hot, dry, and dusty easterly or southeasterly wind—the *leste*—blows over Morocco and out into the Atlantic. If the wind crosses the Mediterranean, it becomes the leveche when it enters southern Spain.

When a low-pressure centre is located at position 2, a warm, dry, dust-laden south or southeast wind originates over the Sahara Desert and blows across North Africa. This wind is known as the *sirocco*. As it moves over the Mediterranean, it picks up moisture and arrives in Sicily and southern Italy as a warm but more humid wind. A storm located still farther to the east (position 3) can cause a dry, hot southerly wind—the *khamsin* or *sharav*—to blow over Egypt, the Red Sea, and Saudi Arabia. These exceedingly hot winds can raise the air temperature to 50°C while lowering the relative humidity to less than 10 percent. Because storm systems are not common over the Mediterranean in summer, scorching breezes such as these occur in spring or fall.

Other Local Winds   Up to now, we have examined a number of wind systems recognized more than just locally. There are many other locally named winds in many parts of the world. ▼ Table 9.1 lists winds of local significance observed in other regions of the world that have not yet been discussed.

● FIGURE 9.28 Local winds that originate over North Africa.

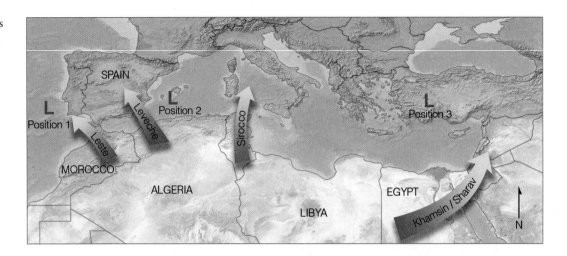

Large-Scale Thermal Circulations

Sometimes thermal circulations can occur at larger scales than are typical of most sea or valley breezes. One example is the seasonal monsoon circulation, which will be discussed in the next section. Another example that is similar to a monsoon occurs in western North America during summer.

Intense heating of the continent, combined with cool coastal waters, results in the formation of a semipermanent large-scale thermal low over the western edge of the continent. The thermal low intensifies during the day and weakens at night as the sun warms and cools the land surface. It extends from California northward, often into British Columbia, and results in a sea-level pressure gradient force directed from the cool ocean toward the hot interior of the continent.

▼ Table 9.1   Some Local Winds of the World

NAME	DESCRIPTION
Cold Winds	
Athos	A strong northeasterly fall wind that descends from Mount Athos over the Aegean Sea
Buran	A strong, cold wind that blows over Russia and central Asia
Purga	A buran accompanied by strong winds and blowing snow
Pampero	A cold wind blowing from the south over Argentina and Uruguay and into the Amazon Basin
Burga	A cold northeasterly wind in Alaska usually accompanied by snow; similar to the buran and purga of Russia
Bise	Generally a cold north or northeast wind that blows over southern France; often brings damaging spring frosts
Papagayo	A cold northeasterly wind along the Pacific coast of Nicaragua and Guatemala; occurs when a cold air mass overrides the mountains of Central America
Tehuantepecer	A strong wind from the north or northwest funneled through the gap between the Mexican and Guatemalan mountains and out into the Gulf of Tehuantepec
Texas Norther	Cold northerly winds behind an intense middle-latitude cyclone crossing the U.S. Great Plains, which may penetrate into Central America, where the wind is called a *norte*
Vardar	A cold fall northwesterly fall wind down the Vardar valley in Greece to the Gulf of Salonica
Mild Winds	
Levanter	A mild, humid, and often rainy east or northeast wind that blows across southern Spain
Harmattan	A dry, dusty, but mild wind from the northeast or east that originates over the cool Sahara in winter and blows over the west coast of Africa; brings relief from the hot, humid weather along the coastal region
Hot Winds	
Simoom	A strong, dry, and dusty desert wind that blows over the African and Arabian deserts; name means "poison wind" because it is often accompanied by temperatures in excess of 52°C, which may cause heat stroke

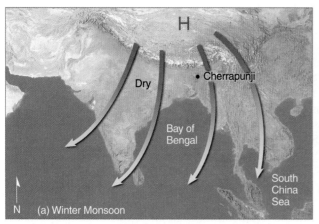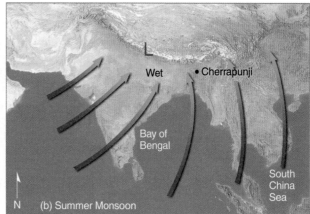

● FIGURE 9.29 Changing annual wind-flow patterns associated with the winter and summer Asian monsoon.

This pressure gradient force pushes cool marine air inland, especially through the sea-level channels that dissect the complex coastline, resulting in a variety of onshore flows that are stronger and larger in scale than normal sea breezes. These winds occur in many places up and down the west coast, including Alberni Inlet on Vancouver Island; Howe Sound, near the town of Squamish between Whistler and Vancouver, British Columbia; the Strait of Juan de Fuca between Vancouver Island and the Olympic Peninsula of Washington State; and the Columbia River gorge between the states of Washington and Oregon. They are the reason places such as Squamish and the Columbia gorge are such famous windsurfing destinations: the onshore winds are strong and steady and occur on a daily basis.

When cool, dense, stable marine air encounters an obstacle, such as a row of hills, the heavy air tends to flow around them rather than over them. When the opposing sea breezes meet on the opposite side of the obstruction, they form what is called a *sea breeze convergence zone*. An example of this at a larger scale than a normal sea breeze is the "Puget Sound Convergence Zone," where onshore winds converge south of Vancouver, British Columbia, in an area east of the high Olympic Peninsula mountains of northwest Washington State.

SEASONALLY CHANGING WINDS—THE MONSOON The word *monsoon* derives from the Arabic *mausim*, which means seasons. A **monsoon wind system** is one that *changes direction seasonally*, blowing from one direction in summer and from the opposite direction in winter. This seasonal reversal of winds is especially well developed in eastern and southern Asia.

WEATHER WATCH

Cherrapunji, India, received 26.5 m of rain between August 1860 and July 1861, most of which fell over the summer monsoon period (August to October 1860 and April to July 1861). In fact, during July 1861, Cherrapunji recorded a whopping 9.3 m of rain.

In some ways, the monsoon is similar to a large-scale sea breeze. During the winter, the air over the continent becomes much colder than the air over the ocean. A large, shallow, high-pressure area develops over continental Siberia, producing a *clockwise* circulation of air that flows out over the Indian Ocean and South China Sea (see ● Figure 9.29a). Subsiding air of the anticyclone and the downslope movement of northeasterly winds from the inland plateau provide eastern and southern Asia with generally fair weather. Hence, the winter monsoon, which lasts from about December through February, means clear skies (*dry season*), with surface winds that blow from land to sea.

In summer, the wind flow pattern reverses itself as air over the continents becomes much warmer than air above the water. A shallow thermal low develops over the continental interior. The heated air within the low rises, and the surrounding air responds by flowing *counterclockwise* into the low centre. This condition results in moisture-bearing winds sweeping into the continent from the ocean. The humid air converges with a drier westerly flow, causing it to rise; further lifting is provided by hills and mountains. Lifting cools the air to its saturation point, resulting in heavy showers and thunderstorms. Thus, the *summer monsoon* of southeastern Asia, which lasts from about June through September, means wet, rainy weather (*wet season*) with winds that blow from sea to land (see Figure 9.29b). Although the majority of rain falls during the wet season, it does not rain all the time. In fact, rainy periods of between 15 to 40 days are often followed by several weeks of hot, sunny weather.

Many factors help create the monsoon wind system. The latent heat given off during condensation aids in the warming of the air over the continent and strengthens the summer monsoon circulation. Rainfall is enhanced by weak, westward–moving, low-pressure areas called *monsoon depressions*. The formation of these depressions is aided by the upper-level jet stream. Where winds in the jet diverge, surface pressures drop, the monsoon depressions intensify, and surface winds increase. The greater inflow of moist air supplies larger quantities of latent heat, which, in turn, intensifies the summer monsoon circulation.

The strength of the Indian monsoon appears to be related to the reversal of surface air pressure that occurs at irregular intervals about every two to seven years at opposite ends of the tropical South Pacific Ocean. As we will see in Chapter 10, this reversal of pressure (which is known as the Southern Oscillation) is linked to an ocean warming phenomenon known as El Niño. During a major El Niño event, surface water near the equator becomes much warmer over the central and eastern Pacific. Over the region of warm water, we find rising air, huge convective clouds, and heavy rain. Meanwhile, to the west of the warm water (over the region influenced by the summer monsoon), sinking air inhibits cloud formation and convection. Hence, during El Niño years, monsoon rainfall is likely to be deficient.

Summer monsoon rains over southern Asia can reach record amounts. Located about 300 km inland on the southern slopes of the Khasi Hills in northeastern India, Cherrapunji receives an average of 1080 cm of rainfall each year, most of it during the summer monsoon between April and October (see ● Figure 9.30). The summer monsoon rains are essential to the agriculture of that part of the world. With a population of nearly 1.2 billion people, India depends heavily on the summer rains so that food crops will grow. The

people also depend on the rains for drinking water. Unfortunately, the monsoon can be unreliable in both duration and intensity. Since the monsoon is vital to the survival of so many people, it is no wonder that meteorologists have investigated it extensively. They have tried to develop methods of accurately forecasting the intensity and duration of the monsoon. With the aid of current research projects and the latest climate models, there is hope that monsoon forecasts will begin to improve in accuracy.

Monsoon wind systems exist in other regions of the world, such as Australia, Africa, and North and South America, where large seasonal contrasts in temperature develop between oceans and continents. (Usually, however, these systems are not as pronounced as in southeast Asia.) For example, a monsoonlike circulation exists in the southwestern United States, especially in Arizona, New Mexico, Nevada, and the southern part of California, where spring and early summer are normally dry as warm westerly winds sweep over the region. By mid-July, however, the thermal low discussed previously forms over California, and humid southerly or southeasterly winds with afternoon showers and thunderstorms are more common (see ● Figure 9.31).

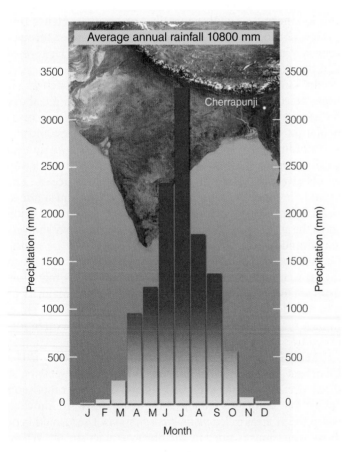

● FIGURE 9.30 Average annual precipitation for Cherrapunji, India. Note the abundant rainfall during the summer monsoon (April through October) with the lack of rainfall during the winter monsoon (November through March).

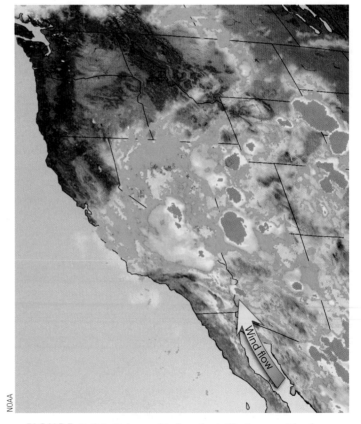

● FIGURE 9.31 Enhanced infrared satellite image with a heavy arrow showing strong monsoonal circulation. Moist, southerly winds are causing showers and thunderstorms (yellow and red areas) to form over the southwestern section of the United States during July 2001.

Determining Wind Direction and Speed

Wind—the horizontal movement of air—is characterized by its direction, speed, and gustiness. Wind speed is the rate at which air moves by a stationary observer. This movement can be expressed as the distance in nautical miles traveled in one hour (knots), as the distance traveled in kilometres over an hour (km h^{-1}), or as the number of metres traveled in one second (m s^{-1}), which are the units used in the SI system.

Since air is invisible, we see things being moved by it. Therefore, we can determine wind direction by watching the movement of objects as air passes them. For example, the rustling of small leaves, smoke drifting near the ground, and flags waving on a pole all indicate wind direction. In a light breeze, a tried and true method of determining wind direction is to raise a wet finger into the air. The dampness quickly evaporates on the wind-facing side, cooling the skin. Traffic sounds carried from nearby railroads or airports can be used to help figure out the direction of the wind. Even your nose can alert you to the wind direction as the smell of fried chicken or broiled hamburgers drifts with the wind from a local restaurant.

We already know that wind direction is given as the direction *from which* it is blowing—a north wind blows from the north toward the south. However, near large bodies of water and in hilly regions, wind direction may be expressed differently. For example, wind blowing from the water onto the land is referred to as an **onshore wind**, whereas wind blowing from land to water is called an **offshore wind**. Consequently, a sea breeze is an onshore wind and a land breeze an offshore wind. Air moving uphill is an *upslope wind*; air moving downhill is a *downslope wind*. Hence, valley breezes are upslope winds, and mountain breezes are downslope winds. The wind direction may also be given as degrees about a 360° circle. These directions are expressed by the numbers shown in ● Figure 9.32. For example, a wind direction of 360° is a north wind, an east wind is 90°, a south wind is 180°, and calm is expressed as zero. It is also common practice to express the wind direction in terms of compass points, such as N, NW, NE, and so on. (Helpful hints for estimating wind speeds from surface observations may be found in the *Beaufort Wind Scale*, located in Appendix C on p. A-8.)

THE INFLUENCE OF PREVAILING WINDS

At many locations, the wind blows more frequently from one direction than from any other. The **prevailing wind** is the name given to the wind direction most often observed during a given time period. Prevailing winds can greatly affect the climate of a region. For example, where the prevailing winds are upslope, the rising, cooling air makes clouds, fog, and precipitation more likely than where the winds are downslope. Prevailing onshore winds in summer carry moisture, cool air,

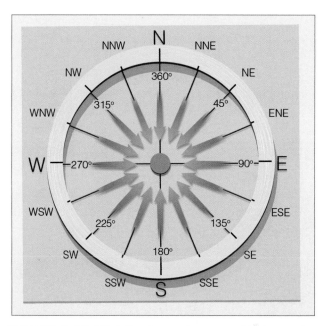

● FIGURE 9.32 Wind direction can be expressed in degrees about a circle or as compass points.

and fog into coastal regions, whereas prevailing offshore breezes carry warmer and drier air into the same locations.

The prevailing wind should be considered when planning a city or building a house. Sources of air pollution such as sewage plants, landfills, and industrial centres should be located downwind of populated areas. Airport runways should be aligned to the prevailing wind direction for safe takeoff and landing. Homes should be built with windows facing the prevailing summer winds to provide summer ventilation but away from prevailing winter winds to reduce heat loss.

Many local ground and landscape features show the effect of a prevailing wind. For example, in windy areas and at high elevations, strong prevailing winds can bend and twist tree branches toward the downwind side, producing *wind-sculptured "flag trees"* (see ● Figure 9.33). Wind blowing over surfaces of snow and sand produces ripples and dunes with a more gentle slope facing into the prevailing wind. Look at ● Figure 9.34 and see if you can determine the prevailing wind when this cinder cone in Iceland erupted.

The prevailing wind can be represented by a **wind rose**, which indicates the percentage of time the wind blows from different directions. Extensions from the centre of a circle point to the wind direction, and the length of each extension indicates the percentage of time the wind blew from that direction.

● Figure 9.35 shows a wind rose for a city averaged over a period of 10 years during the month of January. Observe that the longest extension points toward the northwest and that the wind blew from this direction 25 percent of the time. This is the prevailing wind for this time period. Of course, a wind rose can be made for any time of the day, and it can represent the wind direction for any month or season of the year. It is also possible to use different colours or widths on

© C. Donald Ahrens

● FIGURE 9.33 In the high country, trees standing unprotected from the wind are often sculpted into "flag" trees, such as these trees in Wyoming.

the wind rose direction arms to indicate different wind speeds so that the frequency by wind direction and speed can be seen simultaneously. The prevailing wind in a town does not always represent the prevailing wind of an entire region. In mountainous regions, the wind is usually guided by topography and is often deflected by obstructions that cause its direction to change abruptly over a small area.

© C. Donald Ahrens

● FIGURE 9.34 Was the wind blowing from the right or from the left when this cinder cone in Iceland erupted? (*The answer is given in the footnote below.)

---

*During eruption, the prevailing wind was from left to right. We can tell this by the volcano's shape. Particles ejected from the volcano were blown by the wind to the right, where they accumulated, producing a more gentle slope.

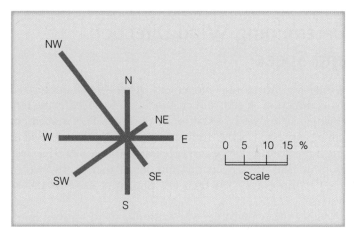

● FIGURE 9.35 This wind rose represents the percentage of time the wind blew from different directions at a given site during the month of January for the past 10 years. The prevailing wind is from the NW and the least frequent wind direction is NE.

In an attempt to harness some of the prevailing wind's energy and turn it into electricity, many countries are building wind power facilities called wind farms. More information on this topic is given in Focus on a Special Topic: Wind Power on p. 285.

WIND MEASUREMENTS   A very old, yet reliable, weather instrument for determining wind direction is the **wind vane**. Most wind vanes consist of a long arrow with a tail, which is allowed to move freely about a vertical post (see ● Figure 9.36). The arrow always points into the wind and, hence, always gives the wind direction. Wind vanes can be made of almost

© C. Donald Ahrens

● FIGURE 9.36 A wind vane and a cup anemometer. These instruments are part of an automated weather station. (For a complete picture of the system, see Figure 3.30, p. 93.)

FOCUS ON A SPECIAL TOPIC

Wind Power

For many decades, thousands of small windmills—their arms spinning in a stiff breeze—have pumped water, sawed wood, and even supplemented the electrical needs of small farms. It was not until the energy crisis of the early 1970s, however, that we seriously considered using wind-driven turbines to run generators that produce electricity.

Wind power seems an attractive way of producing energy—it is nonpolluting and, unlike solar power, is not restricted to daytime use. It does, however, pose some problems because the power produced can be intermittent and the cost of a single wind turbine can exceed $1 million. Other problems include impacts on wildlife, aesthetics, and community concerns about health impacts.

Wildlife issues are almost solely related to impacts on birds and bats. Each year, spinning turbines kill enough birds that many companies hire avian specialists to study bird behaviour at their sites. Some actually shut down their turbines during nesting season. Research into ameliorating effects on bats is ongoing, but indications are that lower pressure areas near the turbines can damage bat lungs. Public perceptions of turbine aesthetics are highly divergent; in some areas, they even attract tourists. Reports of widespread adverse human health impacts (due to low-frequency noise, blade glint, and electromagnetic interference) are not currently substantiated in peer-reviewed literature, although governments worldwide are starting to create legislation governing the location and operating practices for wind farms.

● FIGURE 7 The Melancthon wind farm near Shelburne, Ontario.

If the wind turbine is to produce electricity, there must be winds that are neither too weak nor too strong. A slight breeze will not turn the blades, and a powerful wind gust could severely damage the machine. Thus, regions with the greatest potential for wind-generated power have moderate, steady winds. Wind turbines actually produce energy in winds as low as 9 km h^{-1} and as high as 80 km h^{-1}. They generally become economical at mean annual wind speeds of 18 to 21.6 km h^{-1} depending on the cost of conventional electricity in an area and the cost of the specific turbine being installed. The wind power available is related to the third power of the wind speed by this equation:

$$P = 0.5 \times \rho \times A \times V^3$$

where P is the wind power in watts, ρ is the air density, A is the area swept by the turbine rotors, and V is the wind speed. Depending on the turbine design and efficiency, the actual power generated is typically about 25 percent of the wind power available.

Before a wind turbine is built, an assessment of the wind resource available is usually made by meteorologists who specialize in the emerging field of "wind prospecting." Many of the factors discussed in this chapter that affect winds on a local scale are considered in choosing sites that will have enough wind. An assessment of the wind power potential at a location usually means monitoring the wind and temperature on a tall tower (often 30 m or higher). (The temperature is used to calculate air density.) A complete assessment of wind power potential for a site means a detailed analysis of the distribution of winds by speed and by direction, in combination with consideration of the characteristics of the turbine, such as the minimum and maximum operating wind speeds.

As fossil fuels diminish and concerns over the impact of CO_2 emitted from burning fossil fuels on climate increase, the wind can help provide a pollution-free alternative form of energy. For example, in 2010, more than 2185 wind machines on 106 wind farms had the capacity to generate more than 3,432 megawatts of electricity in Canada, enough to power 1 million homes (see ● Figure 7). Of this total, 950 megawatts were installed in 2009 alone, making it the most rapidly growing energy sector. In 2010, there were 66 additional wind farms in the planning stages in Canada slated to begin production by the middle of the decade. Present estimates are that wind power may be able to furnish up to a few percent of the nation's total energy needs during the first half of this century.

---

any material. At airports, a cone-shaped bag opened at both ends so that it extends horizontally as the wind blows through it is mounted on a pole near the runway. This form of wind vane, called a wind sock, enables pilots to tell the surface wind direction when landing.

The instrument that measures wind speed is called an **anemometer**. The oldest type of anemometer is the *pressure plate anemometer* developed by Robert Hooke in 1667. It consists of a rectangular metal plate, which is free to swing in

the vertical. As the speed of the wind increases, the force of the wind on the plate pushes it outward at a greater angle. The wind speed is read from a scale mounted adjacent to the arm of the swinging plate. Most anemometers today consist of three (or more) hemispherical cups (cup anemometer) mounted on a vertical shaft, as shown in Figure 9.36. The difference in wind pressure from one side of a cup to the other causes the cups to spin about the shaft. The rate at which they rotate is directly proportional to the speed of the wind. The

© Peter Jackson

● FIGURE 9.37 Close-up of a propeller-type anemometer—an R.M. Young Wind Monitor®—which measures both wind speed and direction.

spinning of the cups is usually translated into wind speed through a system of gears or electronically by opening and closing a switch with each rotation or by generating an electric current and may be read from a dial or transmitted to a data logger for recording.

Propeller-type anemometers indicate both wind speed and direction. They consist of a bladed propeller that rotates at a rate proportional to the wind speed. The streamlined shape and a vertical fin keep the blades facing into the wind (see ● Figure 9.37). When attached to a recorder or data logger, a continuous record of both wind speed and direction is obtained.

Sonic anemometers are starting to replace mechanical cup and propeller anemometers in some automated weather stations. Since they work without any moving parts, they can be superior to mechanical anemometers in harsh environments. These devices work by transmitting sound pulses back and forth across a gap of 10 to 20 cm. Based on the difference in time it takes sound to travel in one direction versus the other, the wind speed in the direction of the gap is measured by the instrument. With two gaps oriented horizontally in different directions, the horizontal wind speed and direction can be found. By adding a gap in the vertical direction, the full three-dimensional wind can be observed. To learn more about how sonic anemometers are used to measure turbulence, read Focus on an Observation: Measuring Wind and Transport in the Planetary Boundary Layer on p. 262.

The wind-measuring instruments described thus far are "ground based" and only give wind speed or direction at a particular fixed location. But the wind is influenced by local conditions, such as buildings and trees. Also, wind speed normally increases rapidly with height above the ground. Thus, wind instruments should be exposed to freely flowing air at a height of at least the standard height of 10 m above the surface and well above the roofs of buildings. In practice, unfortu-

nately, anemometers are placed at various levels; the result, then, is often erratic wind observations.

A simple way to obtain wind data above the surface is with a **pilot balloon**. A small balloon filled with helium is released from the surface. The balloon rises at a known rate but drifts freely with the wind. It is manually tracked with a small telescope called a *theodolite*. Every minute (or half-minute), the balloon's vertical angle (height) and horizontal angle (direction) are measured. The data from the observations are used to calculate the wind speed and direction at specific intervals—usually every 300 m—above the surface.

The pilot balloon principle can be used to obtain wind information during a radiosonde observation. During this type of observation, a balloon rises from the surface carrying a *radiosonde* (an instrument package designed to measure the vertical profile of temperature, pressure, and humidity; see Chapter 1, p. 16). Equipment located on the ground constantly tracks the balloon. From this information, a computer determines the vertical profile of wind from the surface up to where the balloon normally pops, typically in the stratosphere near 30 km altitude. The observation of winds using a radiosonde balloon is called a **rawinsonde observation**. Of course, both pilot balloons and radiosondes make an observation at only one point in time. To obtain another profile, a new balloon must be launched and its data recorded and processed.

Above about 30 km, rockets and radar provide information about the wind flow. One type of rocket ejects an instrument attached to a parachute that drifts with the wind as it slowly falls to Earth. While descending, the instrument is tracked by a ground-based radar unit that determines wind information for that region of the atmosphere. Other rockets eject metal strips at some desired level. Again, radar tracks these drifting pieces of chaff, which provide valuable wind speed and direction data for elevations outside the normal radiosonde range.

Doppler radar has been employed to obtain a vertical profile of wind speed and direction up to an altitude of 16 km or so above the ground. Such a profile is called a *wind sounding*; the radar is called a **wind profiler** (or simply a *profiler*). Doppler radar, like conventional radar, emits pulses of microwave radiation that are returned (backscattered) from a target, in this case, the irregularities in moisture and temperature created by turbulent, twisting eddies that move with the wind. Doppler radar works on the principle that as these eddies move toward or away from the receiving antenna, the returning radar pulse will change in frequency. The Doppler radar wind profilers are so sensitive that they can translate the backscattered energy from these eddies into a vertical picture of wind speed and direction at various levels in a column of air 16 km thick (see ● Figure 9.38). Presently, there is a network of wind profilers scattered across the central United States.

A device similar to radar called **lidar** (LIght Detection And Ranging) uses infrared or visible light in the form of a laser beam to determine wind information. Basically, it sends

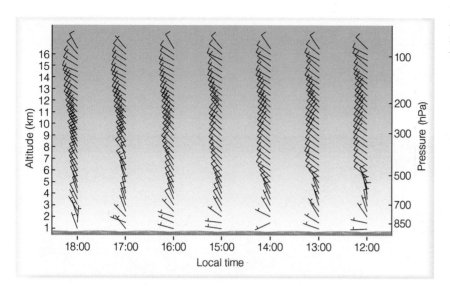

● FIGURE 9.38 A profile of wind direction and speed above Hillsboro, Kansas, on June 28, 2006, from a wind profiler.

out a narrow beam of light that is reflected from particles, such as smoke or dust—it measures wind velocity by measuring the movement of these particles.

Similar in principle to lidar and radar, a Doppler **sodar** (SOund Detection And Ranging) emits audible sound tones and then listens for their echo (i.e., backscatter) as they reflect off turbulence and temperature structures in the lowest kilometre of the atmosphere (see ● Figure 9.39). By detecting the change in tone (i.e., frequency) of the echoed sound, a Doppler sodar can calculate a vertical profile of the horizontal wind, in much the same way as a Doppler radar wind profiler. ● Figure 9.40 shows Doppler sodar wind profiles that were measured during the passage of a strong cold front.

In remote regions of the world where upper-air observations are lacking, wind speed and direction can be obtained from satellites. Geostationary satellites can track the movement of clouds at 30-minute intervals. The direction of cloud movement indicates wind direction, and the horizontal distance the cloud moves during a given time period indicates the wind speed. Satellites can also measure surface winds above the ocean by observing the roughness of the sea. (More information on this topic is given in Focus on an Observation: Observing Winds from Space on p. 288.)

● FIGURE 9.39 A Scintec® phased-array Doppler sodar system. Each of the 64 white "eggs" houses a speaker that both emits sound and listens for echoes. By emitting the sound at slightly staggered time intervals from one row to the next, the sound can be steered in four different directions, as well as straight up. Using the change in tone of the echoed sound, a profile of the horizontal wind can be constructed.

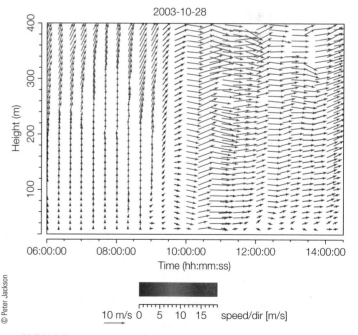

● FIGURE 9.40 A profile of wind direction and speed on October 28, 2003, at Prince George, British Columbia, from a Doppler sodar. The strong shift in wind direction from southerly to westerly and dramatic increase in speed starting at 9:30 marks the passage of a cold front.

FOCUS ON AN OBSERVATION

Observing Winds from Space

Oceans cover more than 70 percent of Earth's surface. For many years, our only observations of surface winds over the open seas came from a few ships and buoys. Today, however, NASA's *QuickScat* satellite, equipped with a sophisticated instrument, called a scatterometer, is able to provide a clear picture of wind speed and wind direction over the open ocean. This scatterometer, named SeaWinds, is a type of radar capable of obtaining wind information during all types of weather.

The scatterometer sends out a microwave pulse of energy that travels through the clouds, down to the sea surface. A portion of this energy is scattered (bounced) back to the satellite. The amount of energy returning to the scatterometer, called the echo, depends on the roughness of the sea—rougher seas have a stronger echo because they scatter back more incoming energy. Since the sea's roughness depends on the strength of the wind blowing over it, the echo's intensity can be translated into an image such as ● Figure 8 showing surface wind speed and direction.

Surface wind information of this nature can be extremely valuable to the shipping

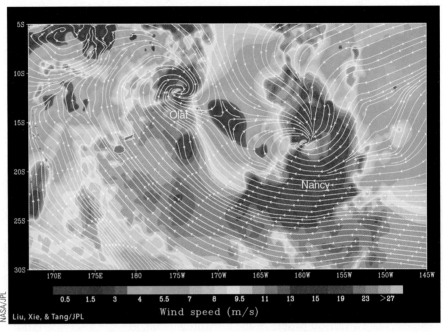

● FIGURE 8 A *QuickScat* satellite image of wind direction and wind speed associated with tropical cyclones Olaf and Nancy over the South Pacific Ocean on February 15, 2005. Wind direction is shown with arrows. Wind speed is indicated by colours, where purple represents the lightest winds and light pink the strongest winds.

industry, as well as to coastal communities. Storms over the open ocean can be carefully monitored to see how their winds are changing. Incorporating sea surface wind information into computer forecast models will improve weather forecasts.

SUMMARY

In this chapter, we concentrated on microscale and mesoscale winds. At the beginning of the chapter, we considered both how our environment influences the wind and how the wind influences our environment. We saw that the friction of airflow—viscosity—can be brought about by the random motion of air molecules (molecular viscosity) or by turbulent whirling eddies of air (eddy viscosity). The depth of the atmospheric layer near the surface that is influenced by surface friction (the boundary layer) depends on atmospheric stability, the wind speed, and the roughness of the terrain. Although it varies, the top of the boundary layer is typically near 1000 m. The turbulent eddies are responsible for convective exchanges of all quantities (heat, water vapour, momentum) within the planetary boundary layer.

Winds blowing past obstructions can produce a number of effects, from gusty winds at a sports stadium to howling winds on a blustery night. Aloft, winds blowing over a mountain range may generate hazardous rotors downwind of the range. And the eddies that form in a region of strong wind shear, especially in the vicinity of a jet stream, can produce extreme turbulence, even in clear air.

Wind blowing over Earth's surface can create a variety of features. In deserts, we see sand dunes and desert pavement. Over a snow surface, the wind produces snow ripples and snow rollers. Where high winds blow over a ridge, trees may be sculpted into "flag" trees. In unprotected areas, shelterbelts are planted to protect crops and soil from damaging winds and from drying out.

We also examined winds on larger scales. Land and sea breezes are true mesoscale winds that blow in response to local pressure differences created by the uneven heating and cooling rates of land and water. When winds move across a large body of water, they often change in speed and direction.

Local winds that blow uphill during the day are called valley breezes, and those that blow downhill at night are called mountain breezes.

Downslope winds that occur due to cooling of air over a slope are called katabatic winds. Fall winds, in which the cold air originates over a region beyond the slope, may be much stronger than typical katabatic winds. When the winds accelerate through a gap in the mountains or along a level valley, they are called gap winds.

A warm, dry wind that descends the eastern side of the Rocky Mountains is the chinook. The same type of wind in the Alps is called a foehn. The Santa Ana wind is a warm, dry, usually strong downslope wind that blows into southern California from the east or northeast.

Local intense heating of the surface can produce small rotating winds, such as the dust devil, whereas downdrafts in a thunderstorm are responsible for the desert haboob. Some winds, such as the sirocco, are dust bearing. Finally, the wind's name may express the direction from which it blows (the Texas norther), or it may represent the region that it blows from (the Santa Ana).

At larger scales, where the winds change direction seasonally, they are termed monsoon winds. Monsoon winds exist in many parts of the world, including North America, Asia, Australia, and Africa.

The prevailing wind is the wind direction most frequently observed during a given time. Onshore winds blow from water to land; offshore winds blow from land to water. A wind rose gives the percentage of time the wind blows from different directions.

Wind direction and speed at the surface are measured by wind vanes and anemometers, whereas above the surface, they can be obtained with pilot balloons, radiosondes, rockets, satellites, Doppler radar, lidar, and Doppler sodar.

KEY TERMS

The following terms are listed (with page numbers) in the order they appear in the text. Define each. Doing so will aid you in reviewing the material covered in this chapter.

QUESTIONS FOR REVIEW

1. Describe the various scales of motion and give an example of each.

2. How does Earth's surface influence the flow of air above it?

3. What causes wind gusts?

4. How does mechanical turbulence differ from thermal turbulence?

5. Why are winds near the surface typically stronger and more gusty in the afternoon?

6. Describe several ways in which an eddy might form.

7. A friend has just returned from a trans-Atlantic jet aircraft flight and reported that the plane dropped about 100 m when it entered an "air pocket." Explain to your friend what apparently happened to cause this drop.

8. Explain why the car in the diagram in ● Figure 9.41 may experience a west wind as it travels past the wall.

● FIGURE 9.41

9. What is wind shear, and how does it relate to clear air turbulence?

10. Explain how shelterbelts protect sensitive crops from damage.

11. With the same wind speed, explain why a camper is more easily moved by the wind than a car.

12. What are the necessary conditions for the development of large wind waves?

13. How can a coastal area have heavy waves on a clear, non-stormy day?

14. If you are standing directly south of a smoke stack and the wind from the stack is blowing over your head, what would be the wind direction?

15. An upper wind direction is reported as 315°. From what compass direction is the wind blowing?

16. List as many ways as you can of determining the wind direction and the wind speed.

17. Name and describe three instruments used to measure wind speed and direction.

18. Using a diagram, explain how a thermal circulation develops.

19. Why do winds usually change direction and speed when moving over a large body of water?

20. Discuss the factors that contribute to the formation of the summer monsoon and the winter monsoon in India.

21. You are fly fishing in a mountain stream during the early morning; would you expect the wind to be blowing upstream or downstream? Explain.

22. Which wind will most likely produce clouds: a valley breeze or a mountain breeze? Why?

23. Explain why chinook winds are warm and dry.

24. Name some of the benefits of a chinook wind.

25. What atmospheric conditions contribute to the development of a strong Santa Ana condition? Why is a Santa Ana wind warm?

26. How do katabatic winds form? What are some examples of locations that experience strong katabatic winds?

27. How do fall winds form?

28. Where might you expect to find gap winds?

29. Describe how dust devils usually form.

30. In what part of the world would you expect to encounter each of the following winds, and what type of weather would each wind bring?

    (a) foehn

    (b) California norther

    (c) Santa Ana

    (d) zonda

    (e) chinook

    (f) Squamish winds

    (g) sirocco

    (h) mistral

QUESTIONS FOR THOUGHT

1. A pilot enters the weather service office and wants to know what time of the day she can expect to encounter the least turbulent winds at 760 m above southern Manitoba. If you were the weather forecaster, what would you tell her?

2. Why is it dangerous during hang-gliding to enter the downwind side of the hill when the wind speed is strong?

3. Why is the difference in surface wind speed between morning and afternoon typically greater on a clear, sunny day than on a cloudy, overcast day?

4. Might it be possible to have a city/suburb breeze? If so, would you expect it to be more prominent during the day or night? Describe how it would form. Use a diagram to help you.

5. Average annual wind speed information in km h^{-1} is given in the table below for two cities located on the Prairies. If a wind turbine has a "cut-in" speed (the minimum speed at which it begins producing power) of 12 km h^{-1}, which city would probably be the best site for a wind turbine? Why?

	TIME								
	Mid-night	3	6	9	Noon	3	6	9	Average Annual Wind Speed (km h^{-1})
City A	24	14	16	26	30	36	28	26	25
City B	16	12	12	26	40	44	30	20	25

6. Which of the sites in ● Figure 9.42 would probably be the best place to construct a wind turbine? A, B, or C? Which would be the worst? Explain.

7. Explain why cities near large bodies of cold water in summer experience well-developed sea breezes but only poorly developed land breezes.

8. Why do clouds tend to form over land with a sea breeze and over water with a land breeze?

Prevailing wind

● FIGURE 9.42

9. If campfire smoke is blowing uphill along the east-facing side of the hill and downhill along the west-facing side of the same hill, are the fires cooking breakfast or dinner? From the drift of the smoke, how were you able to tell?

10. Why don't chinook winds form on the east side of the Appalachians?

11. Show, with the aid of a diagram, what atmospheric and topographic conditions are necessary for an area in the Northern Hemisphere to experience hot summer breezes from the north.

PROBLEMS AND EXERCISES

1. A model city is to be constructed in the middle of an uninhabited region. The wind rose seen here (see ● Figure 9.43) shows the annual frequency of wind directions for this region. With the aid of the wind rose, on a square piece of paper determine where the following should be located:
   (a) industry
   (b) parks
   (c) schools
   (d) shopping centres
   (e) sewage disposal plants
   (f) housing development
   (g) an airport with two runways

2. What would be the total force exerted on a camper 5 m long and 2.5 m high if a wind of 70 km h^{-1} blows perpendicular to one of its sides?

3. On the map of North America (see ● Figure 9.44), label where each of the following winds might be observed and then show with arrows the general direction of air flow that occurs with each of the winds.
   (a) Santa Ana wind
   (b) chinook wind
   (c) Tehuantepecer

● FIGURE 9.44 Base map of North America

   (d) Squamish wind
   (e) Texas norther
   (f) sea breeze in Nova Scotia
   (g) lake breeze near Lake Ontario

4. On the same map of North America (see Figure 9.44), show where the centres of atmospheric pressure should be located to produce the following winds. Place a large L on the map for the centre of low pressure and a large H for the centre of high pressure. (Be sure to place the letter representing the wind next to the L or the H.)
   (a) high-pressure area for a Santa Ana wind
   (b) low-pressure area for a chinook wind
   (c) high-pressure area for a Tehuantepecer
   (d) high-pressure area for a Squamish wind
   (e) high- and low-pressure areas for a Texas norther
   (f) high-pressure area for a sea breeze along the Nova Scotia shore
   (g) low-pressure area for a lake breeze in Toronto, Ontario

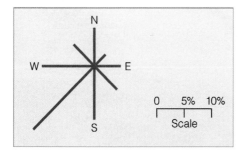

● FIGURE 9.43

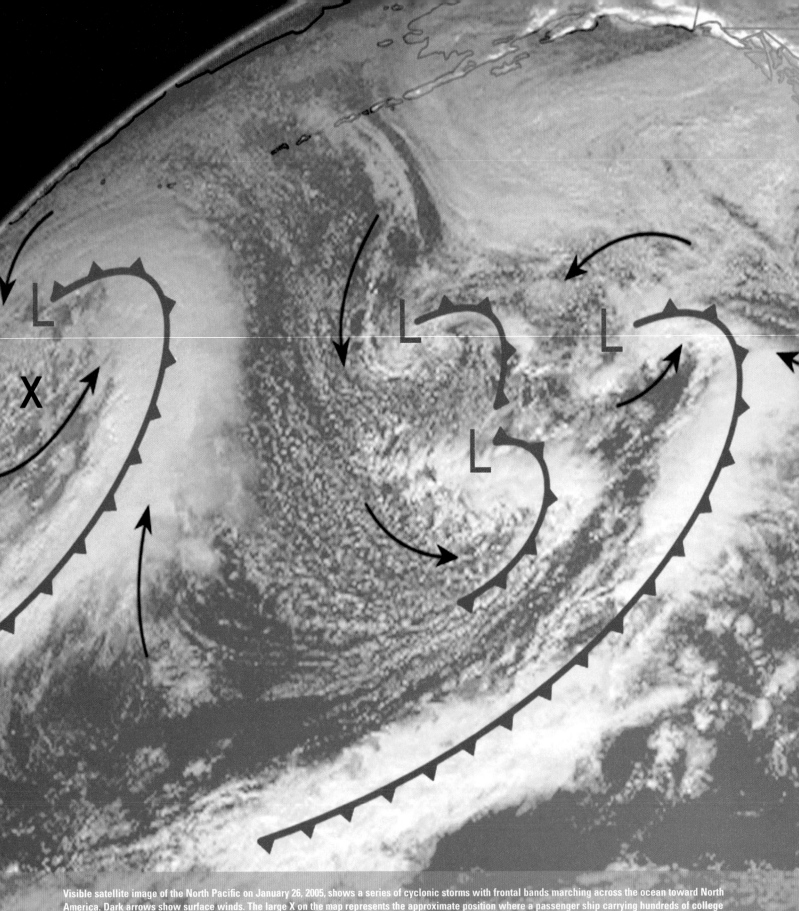

Visible satellite image of the North Pacific on January 26, 2005, shows a series of cyclonic storms with frontal bands marching across the ocean toward North America. Dark arrows show surface winds. The large X on the map represents the approximate position where a passenger ship carrying hundreds of college students encountered high winds, huge waves, and rough seas.

NOAA

Wind: Global Systems

10

Most of us know someone who at one time was "sick of school." But what about someone who was "sick in school"? That's what happened to many of the nearly 700 college students who took off from Vancouver, British Columbia, on January 18, 2005, for a 100-day "semester at sea."

After days of riding the huge waves of the North Pacific, the weather turned ugly as a storm greeted the ship (see the X on the chapter opening satellite image). Early in the morning on January 26, hurricane-force winds and huge waves—one estimated at 15 m high—smashed the glass on the bridge and shorted out the ship's electrical and navigational systems. With three of the four engines disabled, the 180 m vessel swayed from side to side, knocking students from their bunks onto the floor, where they dodged flying books, television sets, coffee pots, and furniture.

The captain ordered everyone to put on their life jackets and get into the ship's narrow hallways. As the students huddled together, the ship continued to roll from side to side. Students found themselves tumbling over one another as they slid across the floor. The storm gradually subsided, and the ship was soon under control. The vessel, with its cargo of anxious students, then limped into Honolulu, Hawaii, for repairs. Fortunately, except for a few bumps and bruises and many nauseated passengers, no one was seriously injured, and many of the students continued their "semester at sea" in the air, flying from one destination to another.

What these students learned firsthand on this adventure was how the interaction between the atmosphere and ocean can have a rather exciting, if not violent, outcome.

CONTENTS

In Chapter 9, we learned that local winds vary considerably from day to day and from season to season. As you may suspect, these local winds are part of a much larger circulation—the little whirls within larger whirls that we spoke of. Imagine that the rotating high- and low-pressure areas we see on a satellite image are spinning eddies in a huge river, and the flow of air around the globe is like a giant meandering river. When winds throughout the world are averaged over a long period of time, the local wind patterns vanish, and what we see is a picture of the winds on a global scale—what is commonly called the **general circulation of the atmosphere**. Just as the eddies in a river are carried along by the overall flow of water, so the highs and lows in the atmosphere are swept along by the general circulation. Global winds, just like the winds discussed in Chapters 8 and 9, are an important part of the Earth system, operating in the *atmospheric system* as mechanisms for moving atmospheric properties (such as moisture and heat) on a global scale. This has a major effect on the other systems as it transports heat and moisture from the warm, humid tropics to the cold, dry polar regions. This redistribution of heat and moisture is essential for Earth to sustain its *biosphere*. Global wind patterns can also transport material injected (like volcanic ash or pollution) or lofted (like dust eroded by wind from dry desert areas) into the atmosphere great distances. We will examine this large-scale circulation of air and its effects and features in this chapter.

General Circulation of the Atmosphere

Before we study the general circulation, we must remember that it represents only the average airflow around the world. Actual winds at any one place and at any given time may vary considerably from this average. Nevertheless, the average can answer why and how the winds blow around the world the way they do. For example, why are prevailing surface winds northeasterly in Honolulu, Hawaii, and southwesterly in St. John's, Newfoundland? The average can also give a picture of the driving mechanism behind these winds, as well as a model of how heat, moisture, and momentum are transported from equatorial regions poleward, keeping the climate in middle latitudes tolerable.

The underlying cause of the general circulation is the unequal heating of Earth's surface. We learned in Chapter 2 that, averaged over the entire globe, incoming solar radiation absorbed by the Earth system is roughly equal to outgoing infrared radiation emitted to space. However, we also know that this energy balance is not maintained for each latitude because the tropics experience a net gain in energy, whereas polar regions suffer a net loss. To balance these inequities, the atmosphere transports warm air poleward and cool air equatorward. Although seemingly simple, the actual airflow is complex; certainly, not everything is known about it. To better understand it, we will first look at some models that simplify some of the complexities of the general circulation.

SINGLE-CELL MODEL   The first model is the single-cell model, in which we assume that

1. Earth's surface is uniformly covered with water (so differential heating between land and water does not come into play)
2. the sun is always directly over the equator (so the winds will not shift seasonally)
3. Earth does not rotate (so the only force we need to deal with is the pressure gradient force)

With these assumptions, the general circulation of the atmosphere on the side of Earth facing the sun would look much like the representation in ● Figure 10.1a, a huge, thermally driven convection cell in each hemisphere. (Figure 10.1b indicates the names of the different regions of the world and their approximate latitudes.)

The circulation of air described in Figure 10.1a is the **Hadley cell** (named after the 18th century English meteorologist George Hadley, who first proposed the idea). It is referred to as a *thermally direct cell* because it is driven by energy from the sun as warm air rises and cold air sinks. Excessive heating of the equatorial area produces a broad region of surface low pressure, whereas at the poles, excessive cooling creates a region of surface high pressure. This horizontal pressure gradient causes cold surface polar air to flow toward the equator, whereas at higher levels, tropical air flows toward the poles. The entire circulation consists of a closed loop with rising air near the equator and sinking air over the poles. In this manner, excess energy from the tropics is transported as sensible and latent heat to the regions of energy deficit at the poles.*

This simple cellular circulation does not actually exist on Earth. For one thing, Earth rotates, so the Coriolis force would deflect the southward-moving surface air in the Northern Hemisphere to the right, producing easterly surface winds at practically all latitudes. These winds would be moving in the opposite direction to Earth's rotation, and friction with the surface would slow down Earth's spin, making our days longer. We know that this does not happen and that prevailing winds in middle latitudes actually blow from the west. Therefore, observations alone tell us that this is not the proper model for a rotating globe. But this model does show us how a nonrotating planet would balance an excess of energy at the equator and a deficit at the poles. How, then, does wind blow on a rotating planet? To answer, we will keep our model simple by retaining our first two assumptions—that is, that Earth is covered with water and that the sun is always directly above the equator.

THREE-CELL MODEL   If we allow Earth to spin, the simple convection system shown in Figure 10.1 breaks into a series of cells, as shown in ● Figure 10.2. Although this model is

---

*Additional information on thermal circulations is found on p. 268 in Chapter 9.

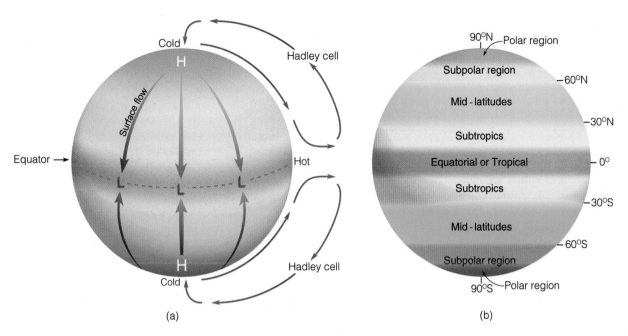

(a)

(b)

● FIGURE 10.1 Diagram (a) shows the general circulation of air on a nonrotating Earth uniformly covered with water and with the sun directly above the equator (single-cell model). Note that the vertical air motions are highly exaggerated. Diagram (b) indicates the names of different regions of the world and their approximate latitudes.

considerably more complex than the single-cell model, there are some similarities. The tropical regions still receive an excess of heat, whereas the poles have a deficit. In each hemisphere, three cells instead of one have the task of energy redistribution. A surface high-pressure area is located at the poles, and a broad trough of surface low pressure still exists at the equator. From the equator to latitude 30°, the circulation is the *Hadley cell*. Let's look at this model more closely by examining what happens to the air above the equator. Refer to Figure 10.2 as you read the following section.

● FIGURE 10.2 Three-cell model: the idealized wind and surface-pressure distribution over a uniformly water-covered rotating earth.

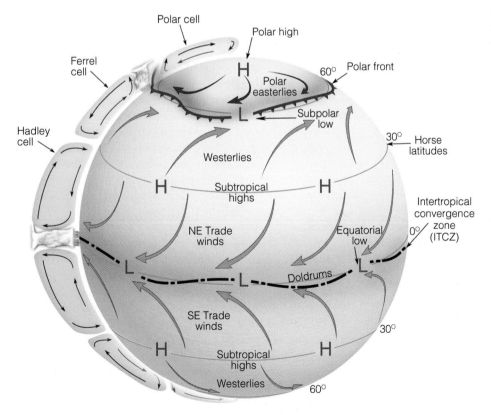

Christopher Columbus was a lucky man. The year he set sail for the New World, the trade winds had edged unusually far north, and a steady northeast wind glided his ships along. Only for about 10 days did he encounter the light and variable wind more typical of this notorious region (30°N)—the horse latitudes (see Figure 10.2 for their location).

Over equatorial waters, the air is warm, horizontal pressure gradients are weak, and winds are light. This region is referred to as the **doldrums**. The monotony of the weather in this area created the expression "down in the doldrums." Here warm air rises, often condensing into huge cumulus clouds and thunderstorms called *convective "hot" towers* because enormous amounts of latent heat are released as water vapour condenses into clouds. This heat warms the air, makes it even more buoyant, and provides energy to drive the Hadley cell. The rising air reaches the tropopause, which acts like a barrier, causing the air to move laterally toward the poles. The Coriolis force deflects this air poleward flow toward the right in the Northern Hemisphere (to the left in the Southern Hemisphere), providing westerly winds aloft in both hemispheres. We will see later that these westerly winds reach maximum velocity and produce jet streams near 30° and 60° latitudes.

As air moves poleward from the tropics, it constantly cools by giving up infrared (longwave) radiation. At the same time, the air also begins to converge as Earth's circumference decreases, especially as it approaches the middle latitudes.* Convergence (piling up) of air aloft increases the mass of air above the surface, which, in turn, increases air pressure at the surface. Hence, near 30° latitude, the convergence of air aloft produces belts of high pressure called **subtropical highs** (or anticyclones). As the converging, relatively dry air above the highs slowly descends, it warms by compression. This subsiding air produces generally clear skies and warm surface temperatures; it is here that we find the world's major deserts (e.g., Sahara). Over the ocean, the weak pressure gradients in the centre of the high produce only weak winds. According to legend, sailing ships traveling to the New World were frequently becalmed in this region, and as food and supplies dwindled, horses were either thrown overboard or eaten. Consequently, this region is also known as the **horse latitudes.**

From the horse latitudes (near 30°), some of the surface air moves back toward the equator. However, it does not flow straight back because the Coriolis force deflects the air, causing it to blow from the northeast in the Northern Hemisphere (from the southeast in the Southern Hemisphere). Sailing ships used these steady winds as an ocean highway to the New World; hence, these winds are called the **trade winds**. Near the

equator, the *northeast trades* converge with the *southeast trades* along a boundary called the **intertropical convergence zone (ITCZ)**, as shown in Figure 10.2. In this region of surface convergence, air rises and continues its cellular journey.

Meanwhile, from 30° to 60° latitude, not all of the surface air moves toward the equator. Some air moves toward the poles and deflects toward the east, resulting in a more or less westerly airflow in both hemispheres. These are called the *prevailing westerlies* or, simply, **westerlies**. Consequently, between northern Mexico and northern Canada, it is much more common to experience winds blowing from the west than from the east. However, this westerly flow is not constant as migrating areas of high and low pressure break up the surface flow pattern. In the Southern Hemisphere, where the surface is mostly water between 30° and 60° latitude, winds blow more steadily from the west.

As mild air carried by the westerlies travels poleward, it encounters cold air moving down from the poles. These two air masses do not readily mix and are separated by a boundary called the **polar front** (shown at about 60° in Figure 10.2, although its position is variable), which lies through a zone of low pressure—the **subpolar low**. Here surface air converges and rises, developing clouds and storms. Some of the rising air returns at high levels to the horse latitudes (around 30°), where it sinks back to the surface in the vicinity of the subtropical high and then flows along the surface toward the polar front. In this three-cell model, the middle cell, called the *Ferrel cell* after the American meteorologist William Ferrel, is a *thermally indirect cell*, in which cool air rises and warm air sinks. The subpolar low of the Ferrel cell represents the middle-latitude cyclones that characterize weather in this region and are discussed fully in Chapter 12.

At both poles (90°), cold air moving equatorward is deflected by the Coriolis force and flows toward the polar front. These are the regions of the **polar easterlies** (as seen in Figure 10.2 for the Northern Hemisphere). Viking explorers likely exploited the polar easterlies to reach Iceland, Greenland, and Newfoundland from Scandinavia. In winter, the polar front's cold air can move into middle and subtropical latitudes, producing a cold polar outbreak. Along the front, a portion of the rising air moves poleward, and the Coriolis force deflects the air into a westerly wind at high levels. Air aloft eventually reaches the poles, slowly sinks to the surface, and flows back toward the polar front, completing the weak *polar cell*.

We can summarize all of this by referring back to Figure 10.2 and noting that there are two major areas of high pressure and two major areas of low pressure at the surface. Areas of high pressure exist near latitude 30° and the poles (90°); areas of low pressure exist over the equator (0°) and near 60° latitude in the vicinity of the polar front. By knowing the way the winds blow around these systems, we have a generalized picture of surface winds throughout the world. The trade winds extend from the subtropical high to the equator, the westerlies from the subtropical high to the polar front, and the polar easterlies from the poles to the polar front.

---

*Use a globe or Google Earth® to see why the air converges in the mid-latitudes. Identify the equator (widest part of the globe that runs through South America, Africa, and the islands of Southeast Asia). Identify longitude lines (they bisect the globe through the poles). Trace two longitude lines that are close to each other from the equator to the pole. Notice how these lines bunch together in the middle latitudes (between northern Mexico and Canada's northern territories).

But how does the three-cell model compare to actual observations of winds and pressure? There are differences, which are especially obvious with the middle, Ferrel cell. Observations tell us that upper-level winds in the middle latitudes generally blow from the west—the westerly jet stream. But the Ferrel cell suggests an east wind aloft as air flows toward the equator. This model does, however, agree closely with the winds and pressure distribution at the *surface*, so we will examine this next.

AVERAGE SURFACE WINDS AND PRESSURE: THE REAL WORLD   The real world, with its continents, oceans, mountains, and ice fields, has an average distribution of sea-level pressure and winds as shown in ● Figure 10.3a for January and Figure 10.3b for July. Look closely at both maps and observe that there are regions where pressure systems appear to persist throughout the year. These systems are referred to as **semipermanent highs and lows** because they move only slightly during the course of a year.

In Figure 10.3a, we can see that there are four semipermanent pressure systems in the Northern Hemisphere during January. In the eastern Atlantic, between latitudes 25° and 35°N is the *Bermuda–Azores high*, often called the **Bermuda high**, and in the Pacific Ocean its counterpart, the **Pacific high**. These are the subtropical anticyclones that develop in response to the convergence of air aloft, where the Hadley cell and Ferrel cell meet near an upper-level jet stream. Since surface winds blow clockwise around these systems, we find the trade winds to the south and the prevailing westerlies to the north.

Where we would expect to observe the polar front (between latitudes 40° and 65°), there are two semipermanent subpolar lows. In the North Atlantic, there is the *Greenland–Icelandic* low, or simply **Icelandic low**, which covers Iceland and southern Greenland, whereas the **Aleutian low** sits over the Gulf of Alaska and Bering Sea near the Aleutian Islands in the North Pacific. These zones of cyclonic activity actually represent regions where numerous storms, having travelled eastward, tend to converge, especially in winter.[*] In the Southern Hemisphere, the subpolar low forms a continuous trough that completely encircles the globe.

On the January map (see Figure 10.3a), there are other pressure systems, which are seasonal and not semipermanent in nature. Over Asia, for example, there is a huge (but shallow) thermal anticyclone called the **Siberian high**, which forms because of the intense cooling of the land. South of this system, the winter monsoon shows up clearly as air flows away from the high across Asia and out over the ocean. A similar (but less intense) anticyclone (called the **Canadian high**) is evident over North America. The Canadian high is normally located east of the Rocky Mountains over Canada but can move about. It is associated with a shallow dome of very cold, arctic air, and its intensity and movement affect winter weather over much of North America.

As summer approaches, the land warms and the cold, shallow highs disappear. In some regions, areas of surface low pressure replace areas of high pressure. The lows that form over the warm land are **thermal lows**. On the July map (see Figure 10.3b), warm thermal lows are found over the desert southwest of the United States and over the plateau of Iran. Notice that these systems are located at the same latitudes as the subtropical highs. We can understand why they form when we realize that during the summer, the subtropical high-pressure belt girdles the world *aloft* near 30° latitude.[*] Within this system, the air sinks and warms, producing clear skies (which allow intense surface heating by the sun). This air near the ground warms rapidly, rises only slightly, and then flows laterally several hundred metres above the surface. The outflow lowers the surface pressure, and as we saw in Chapter 9, a shallow thermal low forms. The thermal low over India, also called the *monsoon low*, develops when the continent of Asia warms. As the low intensifies, warm, moist air from the ocean is drawn into it, producing the wet summer monsoon so characteristic of India and Southeast Asia. Where these surface winds converge with the general westerly flow, rather weak monsoon depressions form. These enhance the position of the monsoon low on the July map.

When we compare the January and July maps, we can see several changes in the semipermanent pressure systems. The strong subpolar lows so well developed in January over the Northern Hemisphere are hardly discernible on the July map. The subtropical highs, however, remain dominant in both seasons. Because the sun is overhead in the Northern Hemisphere in July and overhead in the Southern Hemisphere in January, the zone of maximum surface heating shifts seasonally. In response to this shift, the major pressure systems, wind belts, and ITCZ (the heavy red line in Figure 10.3) *shift toward the north in July and toward the south in January*.[†] ● Figure 10.4 illustrates a winter weather map where the main features of the general circulation have been displaced southward.

THE GENERAL CIRCULATION AND PRECIPITATION PATTERNS   The position of the major features of the general circulation and their latitudinal displacement (which annually averages about 10° to 15° of latitude)

---

[*]For a better picture of these cyclonic storms in the Gulf of Alaska during the winter, see the opening satellite image on p. 292.

[*]This belt of high pressure aloft shows up well in Figure 10.8b on p. 301, the average 500 hPa map for July.

[†]An easy way to remember the seasonal shift of pressure systems is to think of birds—in the Northern Hemisphere, they migrate south in the winter and north in the summer.

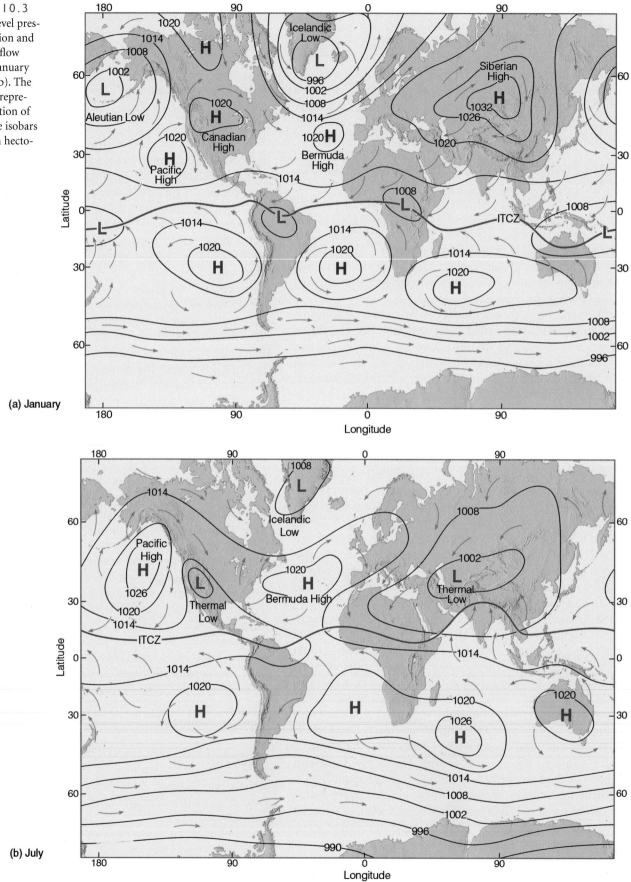

● FIGURE 10.3
Average sea-level pressure distribution and surface wind-flow patterns for January (a) and July (b). The solid red line represents the position of the ITCZ. The isobars are labelled in hectopascals (hPa).

(a) January

(b) July

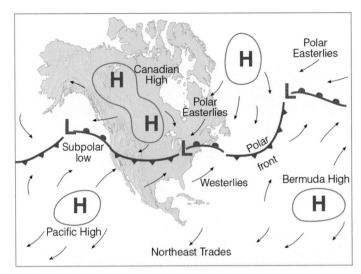

● FIGURE 10.4 A winter weather map depicting the main features of the general circulation over North America. Notice that the Canadian high, polar front, and subpolar lows have all moved southward into the United States and that the prevailing westerlies exist south of the polar front. The arrows on the map illustrate wind direction.

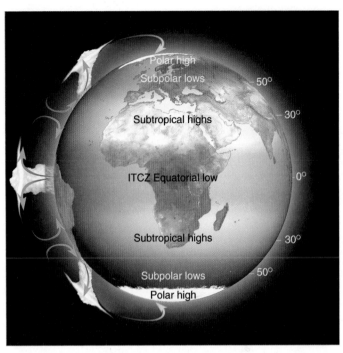

● FIGURE 10.5 Rising and sinking air associated with the major pressure systems of Earth's general circulation. Where the air rises, precipitation tends to be abundant (blue shade); where the air sinks, drier regions prevail (tan shade). Note that the sinking air of the subtropical highs produces the major desert regions of the world.

strongly influence the precipitation of many areas. For example, on the global scale, we would expect abundant rainfall where the air rises and very little where the air sinks. This is because upward motion both causes adiabatic cooling that brings air to saturation and is required to counteract the fall velocity of cloud droplets and ice particles. Consequently, areas of high rainfall exist in the tropics, where humid air rises in conjunction with the ITCZ, and between 40° and 55° latitude, where middle-latitude storms and the polar front force air upward. Areas of low rainfall are found near 30° latitude in the vicinity of the subtropical highs and in polar regions where the air is cold and dry (see ● Figure 10.5).

Poleward of the equator, between the doldrums and the horse latitudes, the area is influenced by both the ITCZ and the subtropical high. In summer (high sun period), the

subtropical high moves poleward and the ITCZ invades this area, bringing with it ample rainfall. In winter (low sun period), the subtropical high moves equatorward, bringing with it clear, dry weather.

During the summer, the Pacific high drifts northward to lie as an elongated ridge extending northward from off the California coast (see Figure 10.3b and ● Figure 10.6). Sinking air on its eastern side produces a strong, upper-level subsidence inversion, which tends to keep summer weather along the West Coast of North America relatively dry. The rainy season typically occurs in winter, when the high moves south

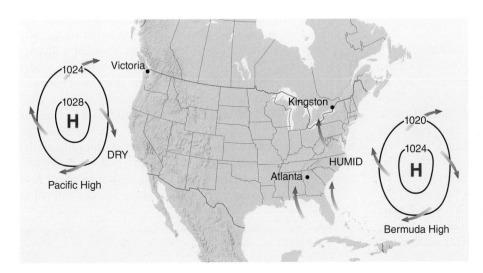

● FIGURE 10.6 During the summer, the Pacific high moves northward. Sinking air along its eastern margin (over California) produces a strong subsidence inversion, which causes relatively dry weather to prevail. Along the western margin of the Bermuda high, southerly winds bring in humid air, which rises, condenses, and produces abundant rainfall.

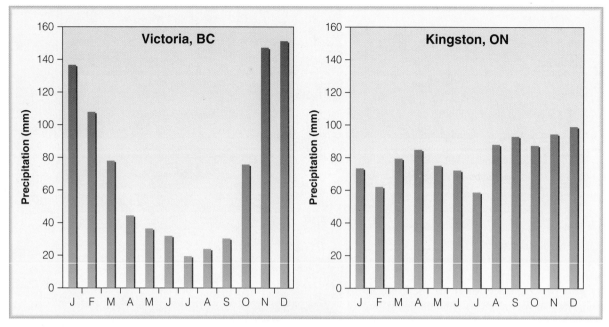

● FIGURE 10.7 Average annual precipitation for Victoria, British Columbia, and Kingston, Ontario.

Source: Environment Canada. Climate Normals. Found at: http://www.climate.weatheroffice.gc.ca/climate_normals/index_e.html. Accessed June 2010. Graphed by Peter Jackson.

and the polar front and storms are able to penetrate the region. Observe in Figure 10.6 that along the East Coast, the clockwise circulation of winds around the Bermuda high brings warm tropical air northward into the United States and southern Canada from the Gulf of Mexico and the Atlantic Ocean. Because subsiding air is not as well developed on this side of the high, the humid air can rise and condense into towering cumulus clouds and thunderstorms. The warm, humid air of maritime tropical origin (to learn more about air masses, look ahead to Chapter 12) is the reason summers are often uncomfortably hot and humid in eastern Canada and the United States. So, in part, it is the air motions associated with the subtropical highs that keep summer weather dry in southwest British Columbia and moist in Ontario and eastern Canada. (Compare the rainfall patterns for Victoria, British Columbia, and Kingston, Ontario, in ● Figure 10.7.)

AVERAGE WIND FLOW AND PRESSURE PATTERNS ALOFT

● Figures 10.8a and 10.8b are average global 500 hPa charts for the months of January and July, respectively. Look at both charts carefully and observe that some of the surface features of the general circulation are reflected on these upper-air charts. On the January map, for example, both the Icelandic low and the Aleutian low are located to the west of their surface counterparts. On the July map, the subtropical high-pressure areas of the Northern Hemisphere appear as belts of high height (high pressure) that tend to circle the globe south of 30°N. In both hemispheres, the air is warmer over low latitudes and colder over high latitudes. This horizontal temperature gradient establishes

a horizontal pressure (contour) gradient that causes the winds to blow from the west, especially in middle and high latitudes.* Notice that the temperature gradients and the contour gradients in the Northern Hemisphere are steeper in January than in July. Consequently, the winds aloft are stronger in winter than in summer. The westerly winds, however, do not extend all the way to the equator as easterly winds appear on the equatorward side of the upper-level subtropical highs.

In middle and high latitudes, the westerly winds continue to increase in speed above the 500 hPa level. We already know that the wind speed increases up through the friction layer, but why should it continue to increase at higher levels? You may remember from Chapter 8 that the geostrophic wind at any latitude is directly related to the pressure gradient and inversely related to the air density. Therefore, a greater pressure gradient will result in stronger winds, and so will a decrease in air density. Owing to the fact that air density decreases with height, the same pressure gradient will produce stronger winds at higher levels. In addition, the north-to-south temperature gradient causes the horizontal pressure (contour) gradient to increase with height up to the tropopause. As a result, the winds increase in speed up to the tropopause. Above the tropopause, the temperature gradients reverse. This changes the pressure gradients and reduces the strength of the westerly winds. Where strong winds tend to concentrate into narrow bands at the tropopause, we find "rivers" of fast-flowing air—jet streams. (In the following

---

*Remember that at this level (about 5600 m above sea level), the winds are approximately geostrophic and tend to blow more or less parallel to the contour lines.

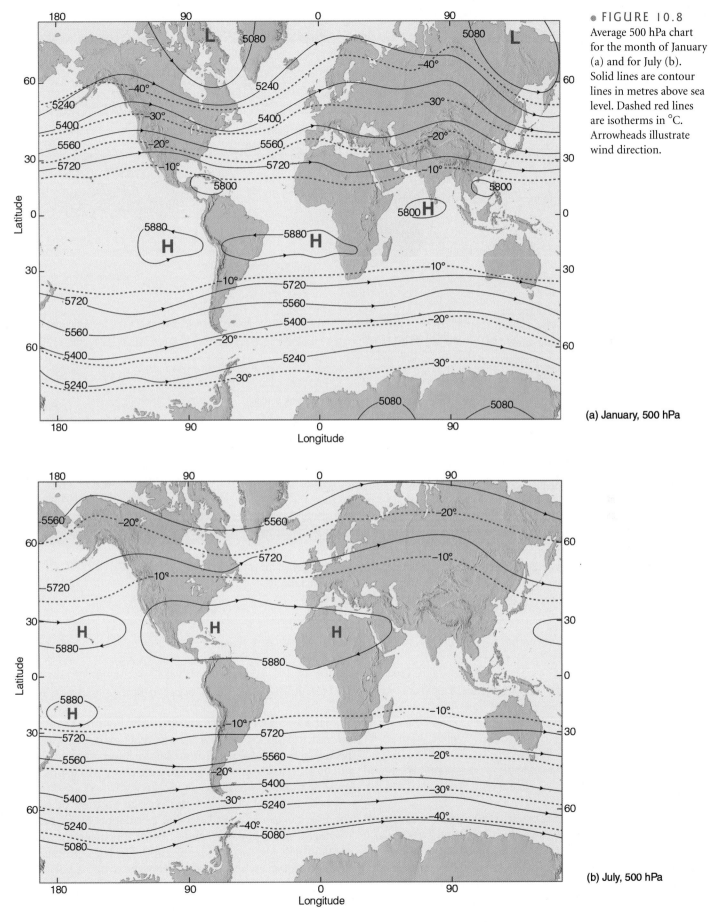

● FIGURE 10.8
Average 500 hPa chart
for the month of January
(a) and for July (b).
Solid lines are contour
lines in metres above sea
level. Dashed red lines
are isotherms in °C.
Arrowheads illustrate
wind direction.

(a) January, 500 hPa

(b) July, 500 hPa

FOCUS ON AN OBSERVATION

The "Dishpan" Experiment

We know that the primary cause of the atmosphere's general circulation is the unequal heating that occurs between tropical and polar regions. A laboratory demonstration that tries to replicate this situation is the *"dishpan"* experiment.

The dishpan experiment consists of a flat, circular pan filled with water several centimetres deep. The pan is positioned on a rotating table (see ● Figure 1a). Around the edge of the pan, a heating coil supplies heat to the pan's "equator." In the centre of the pan, a cooling cylinder represents the "pole," and ice water is continually supplied here. When the pan is rotated counterclockwise, the temperature difference between "equator" and "pole" produces a thermally driven circulation that transports heat poleward.

Aluminum powder (or dye) is added to the water so that the motions of the fluid can be seen. If the pan rotates at a speed that corresponds to the rotation of Earth, the flow develops into a series of waves and rotating eddies similar to those shown in Figure 1b. The atmospheric counterpart of these eddies is the cyclones and anticyclones of the middle latitudes. At Earth's surface, they occur as winds circulating around centres of low and high pressure. Aloft, the waves appear as a series of troughs and ridges that encircle the

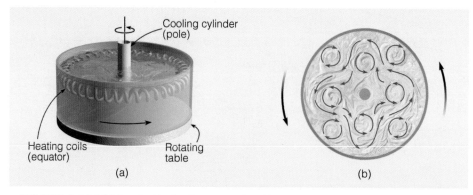

● FIGURE 1 (a) A "dishpan" with a hot "equator" and a cold "pole" rotating at a speed corresponding to that of Earth (b) produces troughs, ridges, and eddies, which appear (when viewed from above) very similar to the patterns we see on an upper-level chart.

● FIGURE 2 The circulation of the air aloft is in the form of waves—troughs and ridges—that encircle the globe.

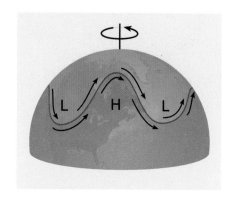

globe and slowly migrate from west to east (see ● Figure 2). The waves represent a fundamental feature of the atmosphere's circulation. Later in this chapter, we will see how they transfer momentum, allowing the atmosphere to maintain its circulation. In Chapter 12, we will see that these waves are instrumental in the development of surface mid-latitude cyclonic storms.

---

section, you will read about a wavy jet stream. The Focus section above describes an experiment that illustrates how these waves form.)

Jet Streams

Atmospheric **jet streams** are swiftly flowing air currents thousands of kilometres long, a few hundred kilometres wide, and only a few kilometres thick. Wind speeds in the central core of a jet stream often exceed 185 km h^{-1} (100 knots) and occasionally exceed 370 km h^{-1} (200 knots). Jet streams are usually found at the tropopause at elevations between 10 and

15 km, although they may occur at both higher and lower altitudes.

Jet streams were first encountered by high-flying military aircraft during World War II, but their existence was suspected before the war. Ground-based observations of fast-moving cirrus clouds had revealed that westerly winds aloft must be moving rapidly.

● Figure 10.9 illustrates the average position of the jet streams, tropopause, and general circulation of air for the Northern Hemisphere in winter. From this diagram, we can see that there are two jet streams, both located at tropopause discontinuities, where mixing between tropospheric and stratospheric air takes place. The jet stream situated near 30° latitude at about 13 km above the subtropical high is the

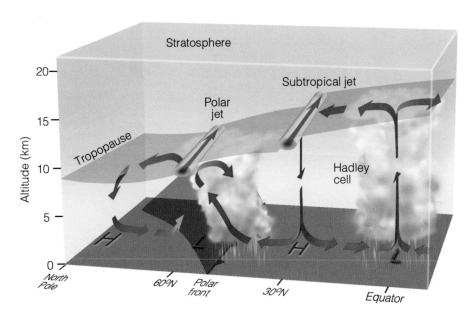

● FIGURE 10.9 Average position of the polar jet stream and the subtropical jet stream with respect to a model of the general circulation in winter. Both jet streams are flowing from west to east.

subtropical jet stream.* The jet stream situated at about 10 km near the polar front is known as the **polar front jet stream** or, simply, the *polar jet stream*. Since both are found at the tropopause, they are referred to as **tropopause jets**.

In Figure 10.9, the wind in the centre of the jet stream would be flowing as a westerly wind away from the viewer. This direction, of course, is only an average as jet streams often flow in a wavy, west-to-east pattern. When the polar jet stream flows in broad loops that sweep north and south, it may even merge with the subtropical jet. Occasionally, the polar jet splits into two jet streams. The jet stream to the north is often called the northern branch of the polar jet, whereas the one to the south is called the southern branch. ● Figure 10.10 illustrates how the polar jet stream and the subtropical jet stream might appear as they sweep around Earth in winter.

We can better see the looping pattern of the jet by studying ● Figure 10.11a, which shows the position of the polar jet stream and the subtropical jet stream at the 250 hPa level (near 10.4 km) on April 25, 2010. The fastest flowing air, or jet core, is represented by the heavy dark arrows. The map shows a strong polar jet stream approaching the British Columbia coast, with a weaker subtropical jet approaching Mexico and crossing the Gulf of Mexico states. Notice that the polar jet has a number of loops, with one over the West Coast of North America and another crossing New Brunswick. Observe in the satellite image (see Figure 10.11b) that the polar jet stream (blue arrows) is directing cold, polar air into the middle of North America, whereas the subtropical jet stream (orange arrow) is sweeping subtropical moisture, in the form of a dense cloud cover, over the southeastern states.

The looping (meridional) pattern of the polar jet stream has an important function. In the Northern Hemisphere, where the air flows southward, swiftly moving air directs cold air equatorward; where the air flows northward, warm air is

carried toward the poles. Jet streams, therefore, play a major role in the global transfer of heat. Moreover, since jet streams tend to meander around the world, we can easily understand how pollutants or volcanic ash injected into the atmosphere in one part of the globe could eventually settle to the ground many thousands of kilometres downwind. And as we will see

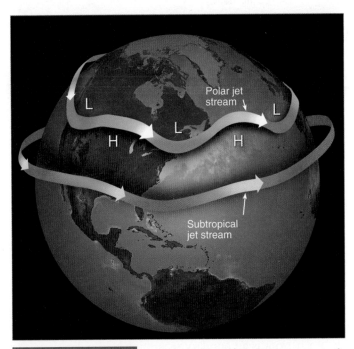

ACTIVE FIGURE 10.10 A jet stream is a swiftly flowing current of air that moves in a wavy, west-to-east direction. The figure shows the position of the polar jet stream and subtropical jet stream in winter. Although jet streams are shown as one continuous river of air, in reality, they are discontinuous, speeding up and slowing down along their path with their position varying from one day to the next. Visit the textbook's website to view this and other Active Figures at www.ahrensmeteorology1ce.nelson.com

---

*The subtropical jet stream is normally found between 20° and 30° latitude.

(a)

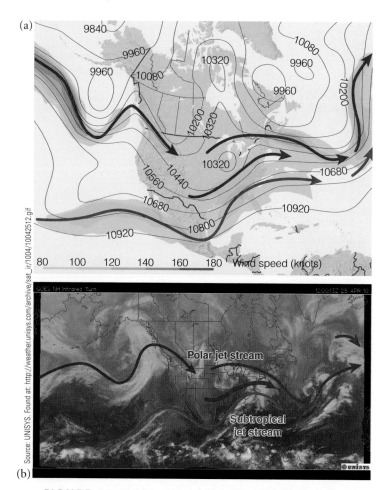

(b)

● **FIGURE 10.11** (a) Position of the polar jet stream (blue arrows) and the subtropical jet stream (orange arrows) at the 250 hPa level (about 10.4 km above sea level) on April 25, 2010, at 12:00 UTC. Solid lines are 250 hPa height contours and colour filled contours are lines of equal wind speed (isotachs) in knots. (b) Satellite image showing clouds and positions of the jet streams for the same day.

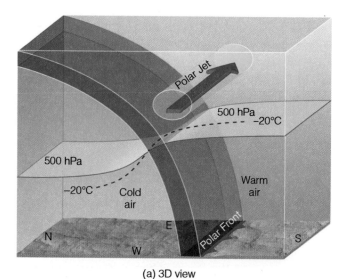

(a) 3D view

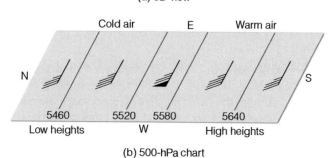

(b) 500-hPa chart

● **FIGURE 10.12** Model of the polar front. Diagram (a) shows a vertical 3-D view of the polar front in association with a sharply dipping 500 hPa pressure surface, an isotherm (dashed line), and the position of the polar front jet stream in winter. The diagram is highly exaggerated in the vertical. Diagram (b) represents a 500 hPa chart that cuts through the polar front as illustrated by the dipping 500 hPa surface in (a). Sharp temperature contrasts along the front produce tightly packed contour lines and strong winds (contour lines are in metres above sea level).

in Chapter 12, the looping nature of the polar jet stream has an important role in the development of middle-latitude cyclonic storms.

The ultimate cause of jet streams is the energy imbalance that exists between high and low latitudes. How, then, do jet streams actually form?

THE FORMATION OF THE POLAR FRONT JET AND THE SUBTROPICAL JET Horizontal variations in temperature and pressure offer clues to the existence of the polar jet stream. ● Figure 10.12 is a three-dimensional model that shows a side view of the atmosphere in the region of the polar front. Since the polar front is a boundary separating cold polar air to the north from warm subtropical air to the south, the greatest contrast in air temperature occurs along the frontal zone. We can see this contrast as the −20°C isotherm dips, sharply crossing the front. This rapid change in temperature produces a rapid change in pressure (as shown by the sharp bending of the constant pressure [isobaric] 500 hPa surface as it passes through the front). In Figure 10.12b, we can see that *the sudden*

change in pressure along the front sets up a steep pressure (height contour) gradient that intensifies the wind speed and causes the jet stream. Observe in Figure 10.12b that the wind is blowing along the front (from the west), parallel to the contour lines, with cold air on its left side.* The north–south temperature contrast along the polar front is strongest in winter and weakest in summer. This situation explains why the polar-front jet shows seasonal variations. In winter, the winds blow stronger and the jet moves farther south as the leading edge of the cold air extends into subtropical regions. In summer, the jet is weaker and is usually found over more northern latitudes.

The subtropical jet stream, which is usually strongest slightly above the 200 hPa level (above 12 km), tends to form along the poleward side of the Hadley cell, as shown in Figure 10.9. Here warm air carried poleward by the Hadley

---

*Recall from Chapter 8 that any horizontal change in temperature causes the isobaric surfaces to dip or slant. The greater the temperature difference is, the greater the slanting, and the stronger the winds. The changing wind speed with height due to horizontal temperature variations is referred to as the *thermal wind.* The thermal wind always blows with cold air on its left side in the Northern Hemisphere and on its right side in the Southern Hemisphere.

cell produces sharp temperature contrasts along a boundary sometimes called the *subtropical front*. In the vicinity of the subtropical front (which does not have a frontal structure extending to the surface), sharp contrasts in temperature produce sharp contrasts in pressure and strong winds.

When we examine jet streams carefully, we see that another mechanism (other than a steep temperature gradient) causes a strong, westerly flow aloft. The cause appears to be the same as that which makes an ice skater spin faster when the arms are pulled in close to the body—the *conservation of angular momentum*.

At the equator, Earth rotates toward the east at a speed close to 1700 km h^{-1} (900 knots). On a windless day, the air above moves eastward at the same speed. If, somehow, Earth should suddenly stop rotating, the air above would continue to move eastward until friction with the surface brought it to a halt; the air keeps moving because it has momentum.

Straight-line momentum—called *linear momentum*—is the product of the mass of the object times its velocity. An increase in either the mass or the velocity (or both) produces an increase in momentum. Air on a spinning planet moves about an axis in a circular path and has angular momentum. Along with the mass and the speed, angular momentum depends on the distance (r) between the mass of air and the axis about which it rotates. *Angular momentum* is defined as the product of the mass (m) times the velocity (V) times the radial distance (r). Thus,

$$\text{Angular momentum} = mVr$$

As long as there are no external twisting forces (torques) acting on the rotating system, the angular momentum of the system does not change. We say that angular momentum is conserved; that is, the product of the quantity mVr at one time will equal the numerical quantity mVr at some later time. Hence, a decrease in radius must produce an increase in speed and vice versa. An ice skater, for instance, with arms fully extended rotates quite slowly. As the arms are drawn in close to the body, the radius of the circular path (r) decreases, which causes an increase in rotational velocity (V), and the skater spins faster. As arms become fully extended again, the skater's speed decreases. The conservation of angular momentum, when applied to moving air, will help us understand the formation of fast-flowing air aloft.

Consider heated air parcels rising from the surface near the equator on a calm day. As the parcels approach the tropopause, they spread laterally and begin to move poleward in the

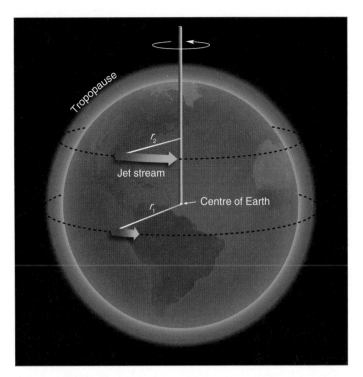

● **FIGURE 10.13** Air flowing poleward at the tropopause moves closer to the rotational axis of Earth (r_2 is less than r_1). This decrease in radius is compensated for by an increase in velocity and the formation of a jet stream.

Hadley cell. If we follow the air moving northward (see ● Figure 10.13), we see that because of the curvature of Earth, air constantly moves closer to its axis of rotation (r decreases). Because angular momentum is conserved (and since the mass of air is unchanged), the decrease in radius must be compensated for by an increase in speed. The air must, therefore, move faster to the east than a point on Earth's surface does. To an observer, this is a west wind. Hence, the conservation of angular momentum of northward-flowing air leads to the generation of strong westerly winds and the formation of a jet stream.

OTHER JET STREAMS   There is another jet stream that forms in summer near the tropopause above Southeast Asia, India, and Africa. Here the altitude of the summer tropopause and the jet stream is near 15 km. Because the jet forms on the equatorward side of the upper-level subtropical high, its winds are from the east; hence, it is known as the **tropical easterly jet stream**. Although the exact causes of this jet have yet to be completely resolved, its formation appears to be at least partly related to the warming of the air over large, elevated land masses, such as the Tibetan Plateau. During the summer, the air above this region (even at high elevations) is warmer than the air above the ocean to the south. This contrast in temperature produces a north-to-south pressure gradient and strong easterly winds that usually reach a maximum speed near 15°N latitude.

Not all jet streams form at the tropopause. For example, there is a jet stream that forms near the top of the stratosphere over polar latitudes. Because little, if any, sunlight reaches the polar region during the winter, air in the upper stratosphere is able to cool to low temperatures. By comparison, in equatorial regions, sunlight prevails all year long, allowing stratospheric ozone to absorb solar energy and warm the air. The horizontal temperature gradients between the cold poles and the warm tropics create steep horizontal pressure gradients, and a strong westerly jet forms in polar regions at elevations near 50 km. Because this wind maximum occurs in the stratosphere during the dark polar winter, it is known as the *stratospheric polar night jet stream.*

In summer, the polar regions experience more hours of sunlight than do tropical areas. Stratospheric temperatures over the poles increase more than those at the same altitude above the equator. This causes the horizontal temperature gradient to reverse itself. The jet stream disappears, and in its place are weaker winds from the east.

Jet streams also form in the upper mesosphere and in the thermosphere. Not much is known about the winds at these high levels, but they are probably related to the onslaught of charged particles that constantly bombard this region of the atmosphere.

Jet streams form near Earth's surface as well. One such jet develops over the central plains of the United States, where it occasionally attains speeds of 111 km h^{-1} (60 knots) several hundred metres above the surface. This wind speed maximum, which usually flows from the south or southwest, is known as a **low-level jet**. It typically forms at night above a temperature inversion, so it is sometimes called a *nocturnal jet stream.* The stable air reduces the interaction between the air within the inversion and the air directly above it. Consequently, the air in the vicinity of the jet is able to flow faster because it is not being slowed by the lighter winds below. Also, the north–south trending Rocky Mountains tend to funnel the air northward. Another important element contributing to the formation of the low-level jet is the downward sloping of the land from the Rockies to the Mississippi Valley, which causes nighttime air above regions to the west to be cooler than air at the same elevation to the east. This horizontal contrast in temperature causes pressure surfaces to dip toward the west. The dipping of pressure surfaces produces strong pressure gradient forces directed from east to west, which cause strong, southerly winds. During the summer, these strong, southerly winds carry moist air from the Gulf of Mexico into the Central Plains. This moisture, coupled with converging, rising air of the low-level jet, enhances thunderstorm formation. Therefore, on warm, moist, summer nights, when the low-level jet is present, it is common to have nighttime thunderstorms over the plains.

BRIEF REVIEW

Before going on to the next section, which describes some of the many interactions between the atmosphere and the ocean, here is a review of some of the important concepts presented so far:

- The two major semipermanent subtropical highs that influence the weather of North America are the Pacific high situated off the West Coast and the Bermuda high situated off the southeast coast.
- The polar front is a zone of low pressure where cyclonic storms often form. It separates the mild westerlies of the middle latitudes from the cold, polar easterlies of the high latitudes.
- In equatorial regions, the intertropical convergence zone (ITCZ) is a boundary where air rises in response to the convergence of the northeast trades and the southeast trades.
- In the Northern Hemisphere, the major global pressure systems and wind belts shift northward in summer and southward in winter.
- The northward movement of the Pacific high in summer tends to keep summer weather along the West Coast of North America relatively dry.
- Jet streams exist where strong winds become concentrated in narrow bands. The polar front jet stream is associated with the polar front. The polar jet meanders in a wavy, west-to-east pattern, becoming strongest in winter, when the contrast in temperature along the front is greatest.
- The subtropical jet stream is found on the poleward side of the Hadley cell, between 20° and 30° latitude.
- The conservation of angular momentum plays a role in producing strong, westerly winds aloft. As air aloft moves from lower latitudes toward higher latitudes, its axis of rotation decreases, which results in an increase in its speed.

Atmosphere–Ocean Interactions

The *atmosphere* and the ocean part of the *hydrosphere* are both dynamic, fluid systems that play essential roles in redistributing energy on Earth. Both transfer heat from the hot tropics to the cold polar regions: The atmosphere uses winds and weather. Oceans use currents. In addition to transferring heat and other quantities, the oceans are an enormous storehouse, holding about 1000 times more heat than the atmosphere. These two systems do not operate independently—they are intimately linked by vertical exchanges of sensible and latent heat, as well as exchanges of momentum and gases such as CO_2. (Chapters 2 and 3 discuss atmospheric latent heat [Q_E] and sensible heat [Q_H] exchanges.)

Oceans dominate Earth's surface, occupying over 70 percent of its area. Evaporation over the oceans takes up latent heat and provides the atmosphere with surplus water that falls over land, as indicated in the hydrologic cycle (see Chapter 4). The latent heat held by water vapour turns to sensible heat when condensation forms clouds, fuelling the atmosphere's weather and distributing precipitation over Earth. Areas of warm ocean surface waters provide sensible heat as well as moisture to the atmosphere, which results in convection and storm development that alters the wind and weather. The warm waters are, in turn, slowly cooled by the atmosphere unless the heat is replenished by solar radiation or warm currents. Conversely, cold surface waters chill the

Wind: Global Systems **307**

overlying air, which releases its sensible heat to slowly warm the oceans. Air blowing over the oceans loses momentum at the sea surface due to friction. This momentum from the wind is turned into waves and also creates ocean currents. So the atmosphere causes changes to heat and motion in the oceans, and, in turn, the oceans cause changes in heat, wind, and weather patterns in the atmosphere.

However, there is a difference in the time it takes for the two systems to respond when they are forced to change. The atmosphere can respond quickly: winds of 10 or 100 km h^{-1} can develop in minutes, a thunderstorm can form in minutes to hours, and a middle-latitude cyclone goes through its life cycle in a week. The oceans, on the other hand, generally respond much more gradually because they move slowly and have a high heat capacity. Although winds can form waves in a few hours, it takes days to weeks of persistent winds to start surface currents and decades before the surface currents adjust across a whole ocean basin. The slow, deep ocean currents can take centuries to adjust to new conditions in the atmosphere. Ocean current speeds are typically a few centimetres per second but can reach one or two metres per second in rapid surface currents such as the Gulf Stream.

The time difference in how the two systems respond to change accounts for the way oceans influence weather and especially how they influence climate over longer time periods. The atmosphere over time alters the ocean's temperature and currents. The new ocean pattern persists into the future, where it affects weather and climate, possibly for months or years, until the atmosphere once again changes the ocean. The complexity and global scale of the interactions between the ocean and atmosphere make scientific understanding difficult and far from complete. We have observed patterns and can recognize their implications, but current research is focused on understanding what causes them and why. The rest of this chapter summarizes what we know, starting with currents, and then discusses some of the most important climate oscillations that result from these atmosphere–ocean interactions.

GLOBAL WIND PATTERNS AND SURFACE OCEAN CURRENTS Wind blowing over the oceans causes surface water to travel with it. The moving water gradually piles up, creating pressure differences within the water itself. This extends the motion several hundreds of metres down into the water. This is how general wind flow around the globe starts the major surface ocean currents moving. The relationship between the general circulation and ocean currents can be seen by comparing Figure 10.3 (p. 298) and ● Figure 10.14.

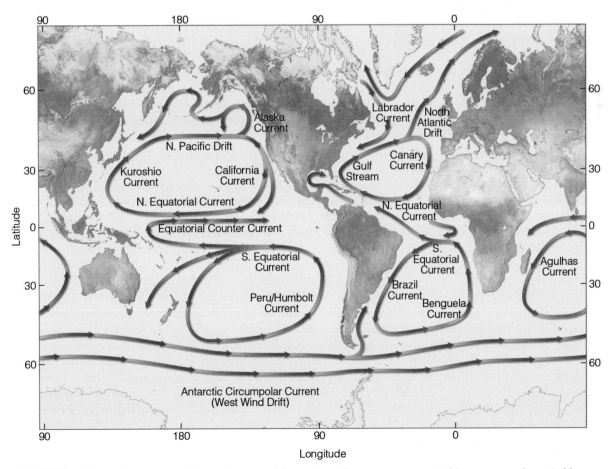

● FIGURE 10.14 The average position and extent of the major surface ocean currents. Cold currents are shown in blue; warm currents are shown in red.

NEL

Because of the larger frictional drag in water, ocean currents move more slowly than the prevailing wind. Typically, these currents range in speed from several kilometres per day to several kilometres per hour. As we can see in Figure 10.14, major ocean currents do not follow the wind pattern exactly; rather, they spiral in semiclosed circular whirls called gyres. In the North Atlantic and North Pacific, prevailing winds blow clockwise and outward from the subtropical highs. As the water moves beneath the wind, the Coriolis force deflects the water to the right in the Northern Hemisphere (to the left in the Southern Hemisphere). This deflection causes the surface water to move at an angle between 20° and 45° to the direction of the wind. Hence, surface water tends to move in a circular pattern as winds blow outward, away from the centre of the subtropical highs.

Important interactions between the atmosphere and the ocean can be seen by examining the huge gyre in the North Atlantic. Flowing northward along the East Coast of the United States is a current called the *Gulf Stream*. Like a vast warm river within the ocean, the Gulf Stream carries tremendous quantities of warm tropical water into higher latitudes. To the north, as part of a smaller subpolar gyre, the *Labrador Current* moves cold water southward along the Atlantic coast of North America as far south as Massachusetts in summer and North Carolina in winter. Near the Grand Banks of Newfoundland,* these two opposing currents flow side by side, creating a sharp temperature gradient. When warm Gulf Stream air blows over the cold Labrador Current water, the stage is set for the formation of the fog so common to this region.

Meanwhile, steered by the prevailing westerlies, the Gulf Stream swings away from the North American coast and moves eastward toward Europe. Gradually, it widens and slows as it merges into the broader *North Atlantic Drift*. As this current approaches Europe, it divides into two currents. A portion flows northward along the coasts of Great Britain and Norway, bringing warm water, which helps keep winter temperatures much warmer than one would expect this far north. The other part flows southward as the *Canary Current*, which transports cool northern water toward the equator. Eventually, the Atlantic gyre is completed as the Canary Current merges with the westward-moving *North Equatorial Current*, which derives its energy from the northeast trades.

The ocean circulation in the North Pacific is similar to that in the North Atlantic. On the western side of the ocean is the Gulf Stream's counterpart, the warm, northward-flowing *Kuroshio Current*, which gradually merges into the slower moving *North Pacific Drift*. As this current approaches the North American coast, it also splits. One portion flows northward as the smaller, relatively warmer, shallow *Alaska Current* and contributes to the temperate climate along the British Columbian and Alaskan coast. The other flows southward along the coastline of the western United States as the *California Current*, where it is

colder than the surrounding waters. The California current joins with the North Equatorial Current that flows westward and joins the Kuroshio Current to complete this gyre.

In the Southern Hemisphere, surface ocean circulations are much the same except that the gyres move counterclockwise in response to the winds around the subtropical highs. Notice that the ocean currents at higher latitudes tend to move in a more west-to-east pattern than do the currents of the Northern Hemisphere. This zonal pattern limits, to some extent, the poleward transfer of warm tropical water. Hence, there is a much smaller temperature difference between the ocean's surface and the atmosphere than exists over Northern Hemisphere oceans. This situation tends to limit the development of vigorous convective activity over the oceans of the Southern Hemisphere. In the Indian Ocean, monsoon circulations tend to complicate the general pattern of ocean currents.

To sum up, on the eastern edge of continents, there usually is a warm current that carries huge quantities of warm water from the equator toward the pole; whereas on the western side of continents, a cool current typically flows from the pole toward the equator.

Up to now, we have seen that atmospheric circulations and ocean circulations are closely linked; wind blowing over the oceans produces surface ocean currents. The currents, along with the wind, transfer surplus heat energy from tropical areas to polar regions, where there is a deficit. This helps equalize the latitudinal energy imbalance, with about 40 percent of the total heat transport in the Northern Hemisphere coming from surface ocean currents. The environmental implications of this heat transfer are tremendous. If the energy imbalance were to go unchecked, yearly temperature differences between low and high latitudes would increase greatly, and, as we will see in Chapter 16, the climate would change.

Satellite pictures reveal that distinct temperature and water density gradients exist along the boundaries of surface ocean currents. For example, where the warm Gulf Stream meets the cold Labrador Current to the north, sharp temperature contrasts are often present. The boundary separating the two masses of water with contrasting temperatures and densities is called an **oceanic front**. Along this frontal boundary, a portion of the meandering Gulf Stream occasionally breaks away and develops into a closed circulation of either cold or warm water—a whirling eddy (see ● Figure 10.15). Because these eddies transport heat and momentum from one region to another, they may have a far-reaching effect on the climate and a more immediate impact on coastal waters.

To understand the impact of ocean eddies, as well as other aspects of ocean circulation on climate, scientists collect and analyze ocean data and also use computer models of the oceans. These models solve the mathematical equations relating ocean currents with temperature, sea level, salinity, and exchanges with the atmosphere. The models are called *Ocean General Circulation Models* (OGCMs) and are the cousins of the models used to simulate climate in the atmosphere, called *Atmospheric General Circulation Models*

---

*To locate the places, see the map provided on this text's back cover: Massachusetts is abbreviated as MA and North Carolina as NC.

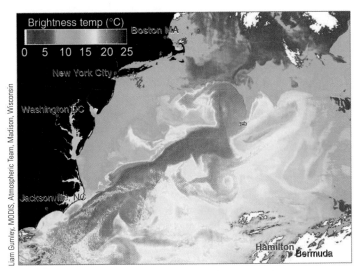

● FIGURE 10.15 The Gulf Stream (dark red band) and its eddies are revealed in this satellite mosaic of sea surface temperatures of the western North Atlantic during May 2001. Bright red shows the warmest water, followed by orange and yellow. Green, blue, and purple represent the coldest water.

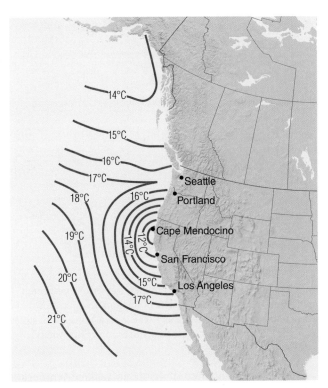

● FIGURE 10.16 Average sea surface temperatures (°C) along the West Coast of North America during August.

(AGCMs—these will be discussed in more detail in Chapter 16). More recent advances in modeling to better understand the implications of climate change have coupled atmosphere with ocean models—these are called *Atmosphere–Ocean General Circulation Models* (AOGCMs).

UPWELLING   Earlier, we saw that the cool California Current flows roughly parallel to the West Coast of North America. From this observation, we might conclude that summer surface water temperatures would be cool along the coast of British Columbia and gradually warm as we move south. A quick glance at the August water temperatures along the U.S. West Coast quickly alters that notion (see ● Figure 10.16). The coldest water is observed along the northern California coast near Cape Mendocino. Why there? To answer this, we need to examine how the wind influences the movement of surface water.

As the wind blows over an open stretch of ocean, the surface water beneath it is set in motion. The Coriolis force bends the moving water to the right in the Northern Hemisphere. Thus, if we look at a shallow surface layer of water, it moves at an average angle of about 45° to the direction of the wind (as in ● Figure 10.17). If we imagine the top layer of ocean water to be broken into a series of layers, then each layer will exert a frictional drag on the layer below. Each successive layer will not only move a little slower than the one above, but because of the Coriolis effect, each layer will also rotate slightly to the right of the layer above. (The rotation of each layer is to the right in the Northern Hemisphere and to the left in the Southern Hemisphere.) Consequently, descending from the surface, we would find water slowing and turning to the right until at some depth (usually about 100 m), the water actually moves in a direction opposite to

the flow of water at the surface. This turning of water with depth is known as the **Ekman spiral**.* The Ekman spiral in Figure 10.17 shows us that the average movement of water from the surface down to a depth of about 100 m is at right angles (90°) to the surface wind direction. The Ekman spiral helps explain why surface water is cold during the summer along the West Coast of North America.

The summertime position of the Pacific high and the coastal mountains cause winds to blow parallel to the coastline from southwestern British Columbia to California (see ● Figure 10.18). The net transport of surface water (called the **Ekman transport**) is at right angles to the wind, in this case, out to sea. As surface water drifts away from the coast, cold, nutrient-rich water from below rises to replace it. The rising of cold water is known as **upwelling**. Upwelling is strongest and surface water is coolest in the area near Cape Mendocino because here the winds are strong, steady, and parallel the coast best. Swimming is only for the hardiest souls as the average summer surface water temperature along the northern California coast is nearly 10°C colder than the average coastal water temperature at the same latitude along the Atlantic Coast. However, ecologically, upwelling produces good fishing as higher concentrations of nutrients are brought to the surface. Upwelling generally occurs in the coastal waters on the west coasts of continents.

*The Ekman spiral is also present in the atmospheric boundary layer, from the surface up to the top of the friction layer, which is usually about 1000 m above the surface.

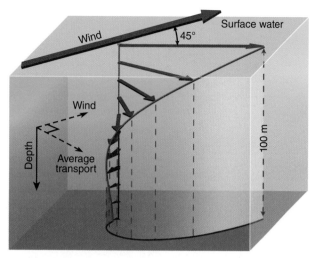

● FIGURE 10.17 The Ekman spiral. Winds move the water, and the Coriolis force deflects the water to the right (Northern Hemisphere). Below the surface, each successive layer of water moves more slowly and is deflected to the right of the layer above. The average transport of surface water in the Ekman layer is at right angles to the prevailing winds.

Another consequence of cool sea surface temperature caused by upwelling along the west coasts of continents is the formation of extensive areas of low clouds—stratus and stratocumulus—that persist in these regions. These form as the cold sea surface cools the atmosphere above it, often to its dew point temperature, resulting in condensation and cloud formation. These low clouds, besides spoiling a sunny day at the beach, also play an important role in Earth's climate by reflecting solar radiation out to space.

Between the ocean surface and the atmosphere, there is an exchange of heat, moisture, and momentum that depends, in part, on temperature differences between water and air. In winter, when air–water temperature contrasts are greatest, there is a substantial transfer of sensible and latent heat from the ocean surface into the atmosphere. This energy helps maintain the global airflow. Because of the difference in heat capacity between water and air, even a relatively small change in surface ocean temperatures will modify atmospheric circulations. Such a change could have far-reaching effects on global weather patterns. One ocean–atmosphere phenomenon linked to worldwide weather events is a warming of the eastern tropical Pacific Ocean known as *El Niño*.

EL NIÑO AND THE SOUTHERN OSCILLATION Along the west coast of South America, where the cool *Peru Current* sweeps northward, southerly winds promote upwelling of cold, nutrient-rich water that gives rise to large fish populations, especially anchovies. The abundance of fish supports a large population of sea birds whose droppings (called guano) produce huge, phosphate-rich deposits, which support the fertilizer industry. Near the end of the calendar year, a warm current of nutrient-poor tropical water often moves southward, replacing the cold, nutrient-rich surface water. Because this condition frequently occurs around Christmas, local residents call it *El Niño* (Spanish for boy child), referring to the Christ child.

In most years, the **El Niño** warming lasts for a few weeks to a month, after which weather patterns usually return to normal and fishing improves. These are known as normal years. However, when El Niño conditions last for many months, and more extensive ocean warming occurs, the economic results are catastrophic. Extremely warm episodes occur at irregular intervals of two to seven years and can cover an area of the tropical Pacific Ocean about the size of Canada.

During a major El Niño event, large numbers of fish and marine plants may die. Dead fish and birds may litter the water and beaches of Peru; their decomposing carcasses deplete the water's oxygen supply, which leads to the bacterial production of huge amounts of smelly hydrogen sulphide. The El Niño of 1972–73 reduced the annual Peruvian anchovy catch from 10.3 million tonnes in 1971 to 4.6 million tonnes in 1972. Since much of the harvest of this fish is converted into fishmeal and exported for use in feeding livestock and poultry, the world's fishmeal production in 1972 was greatly reduced. Countries such as the United States that rely on fishmeal for animal feed had to use soybeans as an alternative. This raised poultry prices in the United States by more than 40 percent.

Why does the ocean become so warm over the eastern tropical Pacific? In a normal year, the trade winds are persistent and blow westward from a region of higher pressure over the eastern Pacific toward a region of lower pressure centred near Indonesia (see ● Figure 10.19a). The "trades" create upwelling that brings cold water to the surface. As this

● FIGURE 10.18 As winds blow parallel to the West Coast of North America, surface water is transported to the right (out to sea). Cold water moves up from below (upwells) to replace the surface water.

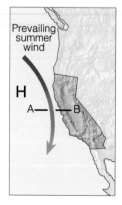

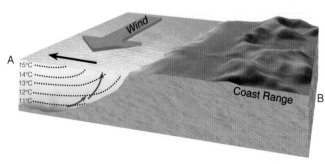

water moves westward, it is heated by sunlight and the atmosphere. Consequently, in the Pacific Ocean, surface water along the equator is usually cool in the east and warm in the west. In addition, the trade winds drag surface water, and it raises the sea level in the western Pacific and lowers it in the eastern Pacific. This produces a thick layer of warm water over the tropical western Pacific Ocean and a weak ocean current (called the *countercurrent*—note that this is not shown in Figure 10.19a) that flows slowly eastward toward South America. When the trades are exceptionally strong, water along the equator in the eastern Pacific becomes cooler than normal, and this event is called **La Niña**.

When an El Niño occurs, the surface atmospheric pressure patterns break down, and air pressure rises over the western Pacific and falls over the eastern Pacific. These pressure changes weaken the trade winds. During strong pressure reversals, winds that normally flow toward the west are replaced by

winds that flow eastward, and strengthen the countercurrent (see Figure 10.19b). Surface water warms over a broad area of the tropical Pacific and heads eastward toward South America in a surge known as a Kelvin wave. This is an enormous wave, perhaps 15 cm high but extending for hundreds of kilometres north and south of the equator (see ● Figure 10.20). Toward the end of the El Niño warming period, which may last between one and two years, atmospheric pressure over the eastern Pacific returns to its normal condition and begins to rise, whereas over the western Pacific, it falls. This seesaw pattern of reversing surface air pressure at opposite ends of the Pacific Ocean is called the **Southern Oscillation**. Because the atmospheric pressure reversals (Southern Oscillation) and ocean warming (El Niño) are more or less simultaneous, scientists call this phenomenon the **El Niño–Southern Oscillation**, or **ENSO**, for short. To summarize, the term *ENSO* refers to three states in the tropical Pacific Ocean and atmosphere: the normal state (see Figure 10.19a); a warm phase, El Niño (see Figure 10.19b); and a cool phase, La Niña (a more intense version of the normal state). Although ENSO patterns are recognizable, each has its own personality, differing in both strength and behaviour.

During especially strong El Niño events (such as in 1982–83 and 1997–98), the easterly trades may actually

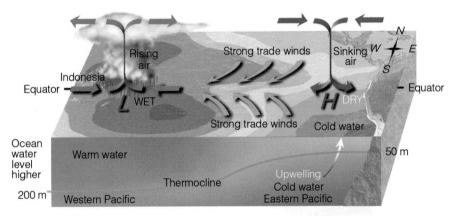

(a) Non-El Niño conditions

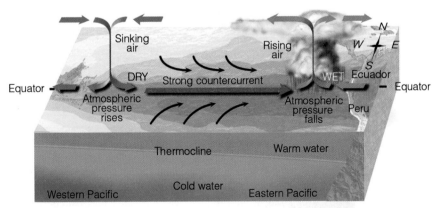

(b) El Niño Conditions

● FIGURE 10.19 The components of El Niño–Southern Oscillation (ENSO). In diagram (a), under normal conditions, higher pressure over the southeastern Pacific and lower pressure near Indonesia produce easterly trade winds along the equator. These winds promote upwelling and cooler ocean water in the eastern Pacific, whereas warmer water prevails in the western Pacific. The trades are part of a circulation (called the *Walker circulation*) that typically has rising air and heavy rain over the western Pacific and sinking air and generally dry weather over the eastern Pacific. When the trade winds are exceptionally strong, water along the equator in the eastern Pacific becomes cooler than normal, and this event is called La Niña. During El Niño conditions, diagram (b), atmospheric pressure decreases over the eastern Pacific and rises over the western Pacific. This change in pressure causes the trades to weaken or reverse direction. This situation enhances the countercurrent that carries warm water from the west over a vast region of the eastern tropical Pacific. The **thermocline**, a boundary that separates the warm water of the upper ocean from the cold water below, changes position as ocean conditions change from normal to El Niño conditions.

● FIGURE 10.20 These three images depict the evolution of a warm water Kelvin wave moving eastward in the equatorial Pacific Ocean during the March and April 1997 El Niño event. The white areas near the equator represent ocean levels about 20 cm higher than average, whereas the red areas represent ocean levels about 10 cm higher than average. Notice how the wave (high region) moves eastward across the tropical Pacific Ocean. These data were collected by the altimeter on board the joint U.S./French *TOPEX/Poseidon* satellite.

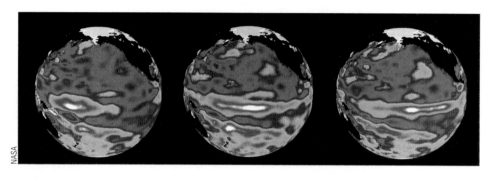

become westerly winds, as shown in Figure 10.19b. As these winds push eastward, they drag surface water with them. This dragging raises sea level in the eastern Pacific and lowers sea level in the western Pacific. The eastward-moving water gradually warms under the tropical sun, becoming as much as 6°C warmer than normal in the eastern equatorial Pacific. Gradually, a thick layer of warm water pushes into coastal areas of Ecuador and Peru, choking off the upwelling that supplies cold, nutrient-rich water to South America's coastal region. The unusually warm water may extend from South America's coastal region for many thousands of kilometres westward along the equator (see ● Figure 10.21a).

During El Niño events, the warm water that is usually held in a relatively deep layer in the western Pacific is spread across a greater surface area of the ocean. Such a large area of abnormally warm water can have an effect on global temperature, wind, and precipitation patterns as warm tropical water fuels the atmosphere with additional warmth and moisture, which the atmosphere turns into additional storminess and rainfall. The warmer sea surface temperatures over a greater area provide more heat to the atmosphere and account for warmer than normal global temperatures in El Niño years. For example, the global temperature was estimated to be 0.17°C higher in 1997–98 due to the strong El Niño that year. The added warmth from the oceans and the release of latent heat during condensation influence the westerly winds aloft so that certain regions of the world experience too much rainfall, whereas others have too little. Over

the warm tropical central Pacific, the frequency of typhoons usually increases. However, over the tropical Atlantic, between Africa and Central America, the winds aloft tend to disrupt the organization of thunderstorms that are necessary for hurricane development. Hence, there are fewer hurricanes in this region during strong El Niño events. And as we saw in

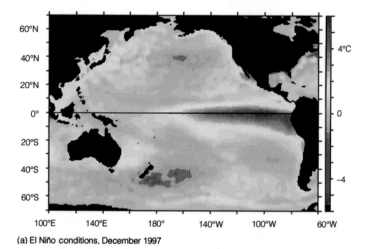

(a) El Niño conditions, December 1997

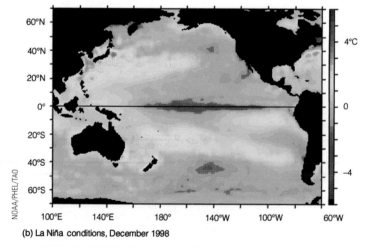

(b) La Niña conditions, December 1998

● FIGURE 10.21 Warm and cold phases of ENSO. (a) Average sea surface temperature departures from normal as measured by satellite. During El Niño conditions, upwelling is greatly diminished and warmer than normal water (deep red colour) extends from the coast of South America westward, across the Pacific. (b) During La Niña conditions, strong trade winds promote upwelling, and cooler than normal water (dark blue colour) extends over the eastern and central Pacific.

WEATHER WATCH

Unfortunately, an El Niño was developing during the Vancouver, British Columbia, Winter Olympic Games in February 2010. January 2010 was Vancouver's warmest since record keeping started in 1937. The result was a very mild winter with many rain-on-snow events, a lot of fog and visibility issues, and a lack of snow at some venues that created havoc for Games organizers. On Cypress Mountain, just north of downtown Vancouver, organizers were forced to cancel thousands of tickets as spectator stadium areas washed out and became hazardous. To keep snowboarding and freestyle skiing events running, snow was trucked in from the British Columbia interior several hours away and flown in from higher elevations on surrounding mountains.

Chapter 9, during El Niño events, there is a tendency for monsoon conditions over India to weaken.

Although the mechanisms for surface ocean temperature changes to influence global wind patterns are not fully understood, the byproducts are obvious. For example, during exceptionally warm El Niños, drought is normally felt in Indonesia, southern Africa, and Australia, whereas heavy rains and flooding often occur in Ecuador and Peru. In the Northern Hemisphere, a strong, subtropical, westerly jet stream normally directs storms into California and heavy rain into the states of Texas through Florida. During El Niño winters, most of Canada is warmer than normal, with peak warming in Manitoba and western Ontario. Southern Canada is also drier than normal, and there is below-average snowfall over British Columbia. Worldwide, the total damage due to flooding, winds, and drought may exceed many billions of dollars.

Following an ENSO event, the trade winds usually return to normal. However, if the trades are exceptionally strong, unusually cold surface water moves over the central and eastern Pacific, and the warm water and rainy weather are confined mainly to the western tropical Pacific (see Figure 10.21b). This cold-water episode, which is the opposite of El Niño conditions, has been termed La Niña (the girl child).

● Figure 10.22 shows the warm and cold phases of ENSO (warm El Niño years in red and cold La Niña years in blue). Notice that the two strongest El Niños were 1982–83 and 1997–98. The 1997–98 event had remarkable impacts worldwide. In 1998, the global average temperature was the warmest on record until 2005. In South America, rains were 10 times normal, causing flooding, erosion, mudslides, and deaths. California and the southern United States had near-record rainfall. In Canada, the winter was warmer than normal

nearly everywhere, setting records in many locations, with temperatures as much as 7.5°C warmer than normal in southern Manitoba. Worldwide, 24,000 people died due to droughts, flooding, famine, and disease associated with this severe El Niño. The weaker El Niño events also have an effect on the Northern Hemisphere's weather patterns. For example, during the El Niño of 1986–87, the subtropical jet stream (being fuelled by the warm tropical waters and huge thunderstorms) curved its way over the southeastern United States, where it brought abundant rainfall to a region that, during the previous summer, had suffered through a devastating drought. During the El Niño of 1991–92, the subtropical jet stream once again swung over North America. Water evaporating from the warm tropical oceans fueled huge thunderstorms. The subtropical jet stream initially swept this moisture into Texas, where it caused extensive flooding.

● Figure 10.23a illustrates typical winter weather patterns over North America during El Niño conditions (ENSO warm phase). Notice that a persistent area of low pressure (L) forms over the North Pacific, and to the south of the low, the jet stream steers wet weather and storms into California, the southern United States, and northern Mexico. The polar jet stream that forms over eastern Canada is weaker than normal, allowing warmer weather to prevail over a large part of North America.

Figure 10.23b shows typical winter weather patterns with a La Niña (ENSO cold phase). Notice that a persistent high-pressure area (called a *blocking high*) forms south of Alaska, forcing the polar jet stream into Alaska and then southward, bringing colder than normal temperatures into Canada and the western United States. The southern branch of the polar jet, which forms south of the high, directs moist air from the ocean into the Pacific Northwest, producing a wet winter

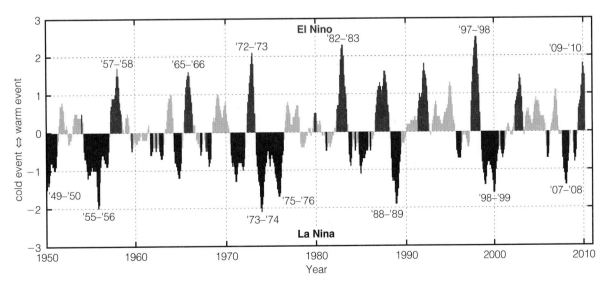

● FIGURE 10.22 The Ocean Niño Index (ONI) showing ENSO patterns from 1950 to 2010. The ONI represents a running three-month mean of sea surface temperature variations (from normal) over the tropical Pacific Ocean from latitude 5°N to 5°S and from longitude 120°W to 170°W. Warm El Niño episodes are in red; cold La Niña episodes are in blue. Warm and cold phases occur when the deviation from the normal is 0.5 or greater. An index value between 0.5 and 0.9 is considered weak, an index value between 1.0 and 1.4 is considered moderate, and an index value of 1.5 or greater is considered strong.

Source: NOAA

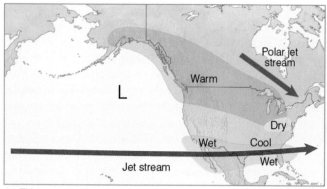
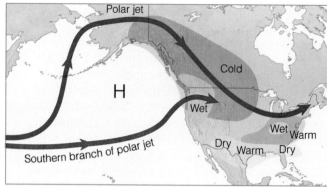

(a) El Niño winter conditions

(b) La Niña winter conditions

● FIGURE 10.23 Warm and cold phases of ENSO. Typical winter weather patterns across North America during an El Niño warm event (a) and during a La Niña cold event (b). During El Niño conditions, a persistent trough of low pressure forms over the North Pacific, and to the south of the low, the jet stream (from off the Pacific) steers wet weather and storms into California and the southern part of the United States. During La Niña conditions, a persistent high-pressure area forms south of Alaska, forcing the polar jet stream and accompanying cold air over much of western North America. The southern branch of the polar jet stream directs moist air from the ocean into the Pacific Northwest, producing a wet winter for that region.

there. Winters are cooler than normal across western Canada but tend to be warmer and drier than normal in the southern part of the United States.

As we have seen, ENSO is part of a large-scale ocean–atmosphere interaction that occurs over several years and enormous distances. Certain regions in the world experience significant climatic responses to this weather phenomenon. Using data from previous ENSO warm and cold phases, scientists have obtained a global picture of where climatic abnormalities are most likely (see ● Figure 10.24). Such ocean–atmosphere interactions where warm or cold surface ocean temperatures can influence precipitation patterns in distant parts of the world are called **teleconnections**.

Some scientists feel that the trigger necessary to start an El Niño event lies within the changing seasons, especially the transition periods of spring and fall. Others feel that the winter monsoon plays a major role in triggering a major El Niño event. As explained in Chapter 9 (p. 282), it appears that an El Niño and the monsoon system are intricately linked, so a change in one brings about a change in the other.

Is there a similar pattern in the Atlantic that compares to the Southern Oscillation in the Pacific? Typically, in the eastern Atlantic (off the coast of Africa), the water is cool and the weather is drier than in the tropical western Atlantic. Periodically, the cool water along the African coast is replaced by warm water and heavy rainfall. This Atlantic warming, however, occurs more sporadically and is not as strong as the warming in the tropical Pacific.

Presently, scientists using general circulation models are trying to simulate atmospheric and oceanic conditions, so ENSO can be anticipated. At this point, several models have been formulated that show promise in predicting the onset and evolution of an ENSO event a season or more in advance.

Up to this point, we have looked at El Niño and the Southern Oscillation and how the reversal of surface ocean temperatures and atmospheric pressure combine to influence

regional and global weather and climate patterns. There are other atmosphere–ocean interactions that can have an effect on large-scale weather patterns. Some of these are described in the following sections.

PACIFIC DECADAL OSCILLATION Over the Pacific Ocean, changes in surface water temperatures appear to influence winter weather along the West Coast of North America. In the mid-1990s, scientists at the University of Washington, while researching connections between Alaskan salmon production and Pacific climate, identified a long-term Pacific Ocean temperature fluctuation, which they called the **Pacific Decadal Oscillation (PDO)** because the ocean surface temperature patterns reverse every 20 to 30 years. The Pacific Decadal Oscillation is like ENSO in that it has a warm phase and a cool phase, but its temperature behaviour is much different from that of ENSO.

During the warm (or positive) phase, unusually warm surface water exists along the West Coast of North America, whereas over the central North Pacific, cooler than normal surface water prevails (see ● Figure 10.25a). At the same time, the Aleutian low in the Gulf of Alaska strengthens, which causes more Pacific storms to move into Alaska and California, missing British Columbia, which becomes warmer and drier than normal. Elsewhere, winters tend to be drier over the Great Lakes and cooler and wetter in the southern United States. During a warm phase, salmon populations increase in Alaska and diminish along the Pacific Northwest coast. The latest warm phase began in 1977 and continued until mid-1998, after which it oscillated between cool and warm phases through the 2000s (see ● Figure 10.26).

Cool (or negative) PDO phases have cooler-than-average surface water along the West Coast of North America and an area of warmer-than-normal surface water extending from Japan into the central North Pacific (see Figure 10.25b). Winters in the cool phase tend to be cooler and wetter than

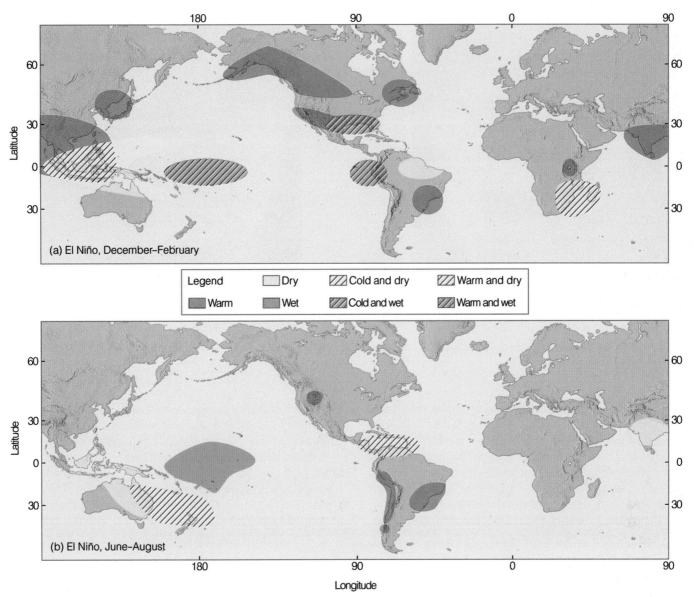

• FIGURE 10.24 Global climatic abnormalities associated with El Niño–Southern Oscillation (ENSO) conditions (a) during December through February and (b) during June through August. A strong ENSO event may trigger a response in nearly all indicated areas, whereas a weak event will likely play a role in only some areas.

Source: NOAA Climate Prediction Center.

average over northwestern North America, wetter over the Great Lakes, and warmer and drier in the southern United States. Salmon fishing diminishes in Alaska and picks up along the Pacific Northwest coast.

The climate patterns described so far only represent average conditions as individual years within either phase may vary considerably. In fact, the Pacific Ocean temperature pattern in a particular phase may even reverse for a few years, as it did from 1958 to 1960 and from mid-1998 through the 2000s (see Figure 10.26). These small reversals can make it difficult to decipher exactly when the ocean temperature changes from one phase to another. Hopefully, as our understanding of the interactions between the ocean and atmosphere improves, climate forecasts will also improve.

NORTH ATLANTIC OSCILLATION Over the Atlantic, there is a reversal of pressure called the **North Atlantic Oscillation (NAO)** that has an effect on the weather in Europe and along the East Coast of North America. For example, in winter, if the atmospheric pressure in the vicinity of the Icelandic low drops, and the pressure in the region of the Bermuda–Azores high rises, there is a corresponding large difference in atmospheric pressure between these two regions that strengthens the westerlies. The strong westerlies, in turn, direct strong storms on a more northerly track into northern Europe, where winters tend to be wet and mild. During this positive phase of the NAO, winters in the eastern United States tend to be wet and relatively mild, whereas northern Canada and Greenland are usually cold and dry (see • Figure 10.27a).

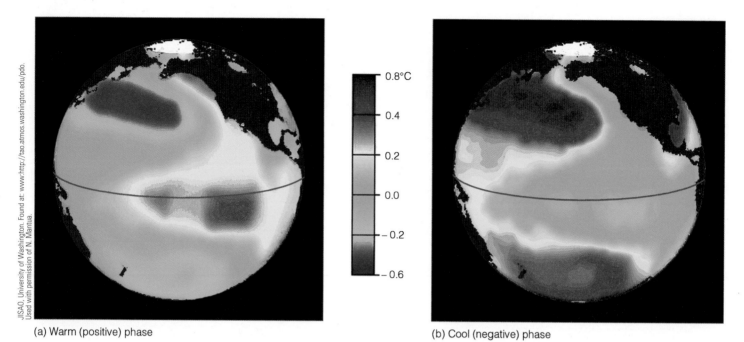

JISAO, University of Washington. Found at: www.http://tao.atmos.washington.edu/pdo.
Used with permission of N. Mantua.

(a) Warm (positive) phase

(b) Cool (negative) phase

● FIGURE 10.25 Typical winter sea surface temperature departure from normal in degrees Celsius during the Pacific Decadal Oscillation's warm phase (a) and cool phase (b).

The *negative phase* of the NAO occurs when the atmospheric pressure in the vicinity of the Icelandic low rises, whereas the pressure drops in the region of the Bermuda high (see Figure 10.27b). This pressure change results in a reduced pressure gradient and weaker westerlies that steer fewer and weaker winter storms across the Atlantic in a more westerly path. These storms bring wet weather to southern Europe and to the region around the Mediterranean Sea. Meanwhile, winters in northern Europe are usually cold and dry, as are the winters along the East Coast of North America. Greenland and northern Canada usually experience mild winters.

Although the NAO varies from year to year (and sometimes from month to month), it may exhibit a tendency to remain in one phase for several years. The winter NAO was in a positive phase from 1980 until 2008; however, recently, it has entered a negative phase.

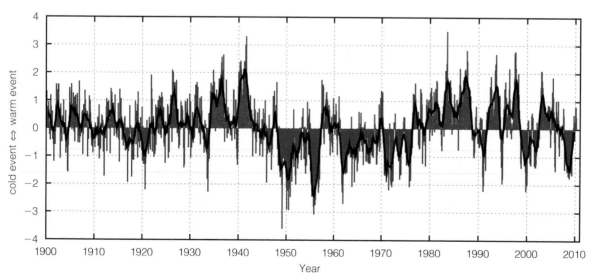

● FIGURE 10.26 The Pacific Decadal Oscillation (PDO) Index is based on spatially averaged sea surface temperatures (SST) of the Pacific Ocean north of 20° N. Positive values (in red) indicate the warm phase, whereas negative values (in blue) represent the cool phase. The heavy black line is a 12-month average of the index.

Source: JISAO, University of Washington. Found at: http://jisao.washington.edu/pdo/PDO.latest. Courtesy of N. Mantua.

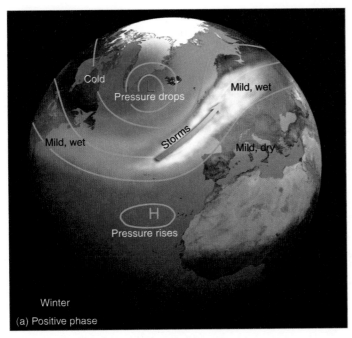

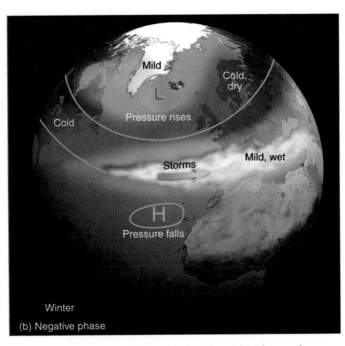

● FIGURE 10.27 Change in surface atmospheric pressure and typical winter weather patterns associated with the (a) positive phase and (b) negative phase of the North Atlantic Oscillation.

ARCTIC OSCILLATION Closely related to the North Atlantic Oscillation is the **Arctic Oscillation (AO)**, where changes in atmospheric pressure between the Arctic and regions to the south cause changes in upper-level westerly winds. During the

positive warm phase of the AO (see ● Figure 10.28a), strong pressure differences produce strong, westerly winds aloft that prevent cold arctic air from invading southern Canada and the United States, so winters in this region tend to be warmer than

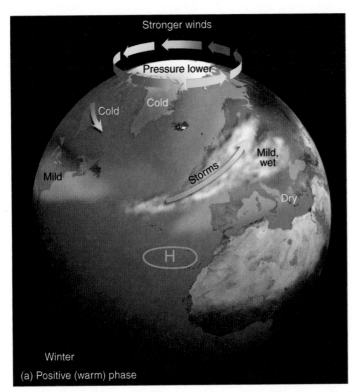

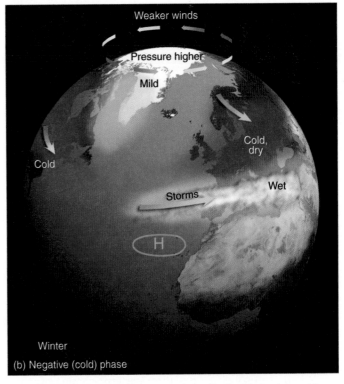

● FIGURE 10.28 Change in surface atmospheric pressure in polar regions and typical winter weather patterns associated with the (a) positive (warm) phase and the (b) negative (cold) phase of the Arctic Oscillation.

normal. With cold arctic air to the north, winters over Canada's north, as well as Newfoundland and Greenland, tend to be very cold. Meanwhile, strong winds over the Atlantic direct storms into northern Europe, bringing wet, mild weather.

During the *negative cold phase* of the *AO* (see Figure 10.28b), small pressure differences between the arctic and regions to the south produce weaker westerly winds aloft. Cold arctic air is now able to penetrate farther south, producing colder than normal winters over southern Canada and much of the United States. Cold air also invades northern Europe and Asia, whereas Newfoundland and Greenland experience warmer than normal winters.

So when Greenland has mild winters, northern Europe has cold winters and vice versa. This seesaw in winter temperatures between Greenland and northern Europe has been known for many years. What was not known until recently is that during the warm Arctic Oscillation phase, relatively warm, salty water from the Atlantic is able to move into the Arctic Ocean, where it melts sea ice, causing it to thin by more than 40 cm. During the cold phase, surface winds tend to keep warmer Atlantic water to the south, which promotes thicker sea ice. Although the Arctic Oscillation switches from one phase to another on an irregular basis, one phase may persist for several years, bringing with it a succession of either cold or mild winters.

SUMMARY

In this chapter, we described the large-scale patterns of wind and pressure that persist around the world. We found that at the surface in both hemispheres, the trade winds blow equatorward from the semipermanent high-pressure areas centred near 30° latitude. Near the equator, the trade winds converge along a boundary known as the intertropical convergence zone (ITCZ). On the poleward side of the subtropical highs are the prevailing westerly winds. The westerlies meet cold polar easterly winds along a boundary called the polar front, a zone of low pressure where middle-latitude cyclonic storms often form. The annual shifting of the major pressure areas and wind belts—northward in July and southward in January—strongly influences the annual precipitation of many regions.

Warm air aloft (high pressure) over low latitudes and cold air aloft (low pressure) over high latitudes produce westerly winds aloft in both hemispheres, especially at middle and high latitudes. Near the equator, easterly winds exist. The jet streams are located where strong winds concentrate into narrow bands. The polar jet stream forms in response to temperature contrasts along the polar front, whereas the subtropical jet stream forms at higher elevations above the subtropics, along an upper-level boundary called the subtropical front.

Near the surface, we examined the interaction between the atmosphere and oceans. We found the interaction to be an ongoing process where everything, in one way or another, seems to influence everything else. On a large scale, winds blowing over the surface of the water drive the major ocean currents. The oceans, in turn, release energy to the atmosphere, which helps maintain the general circulation of winds. Where winds and the Ekman spiral move surface water away from a coastline, cold, nutrient-rich water upwells to replace it, creating good fishing and cooler surface water.

When atmospheric circulation patterns change over the tropical Pacific, and the trade winds weaken or reverse direction, warm tropical water is able to flow eastward toward South America, where it chokes off upwelling and produces disastrous economic conditions. When the warm water extends over a vast area of the tropical Pacific, the warming is called a major El Niño event, and the associated reversal of pressure over the Pacific Ocean is called the Southern Oscillation (these are components of the warm phase of ENSO). The large-scale interaction between the atmosphere and the ocean during ENSO affects global atmospheric circulation patterns. The sweeping winds aloft provide too much rain in some areas and not enough in others. La Niña is the name given to the cold phase of ENSO, a situation where the surface water of the central and eastern tropical Pacific turns cooler than normal.

Over the north-central Pacific and along the West Coast of North America, there is a reversal of surface water temperature, called the Pacific Decadal Oscillation, that occurs every 20 to 30 years. Over the Atlantic Ocean, there is a reversal of pressure called the North Atlantic Oscillation that influences weather in various regions of the world. Surface atmospheric pressure changes over the Arctic also seem to influence weather patterns over regions of the Northern Hemisphere. Studies now in progress are designed to determine how the interaction between the atmosphere and the ocean can influence climate patterns in various regions of the world.

KEY TERMS

The following terms are listed (with page numbers) in the order they appear in the text. Define each. Doing so will aid you in reviewing the material covered in this chapter.

general circulation of the atmosphere, 294
Hadley cell, 294
doldrums, 296
subtropical highs, 296
horse latitudes, 296

trade winds, 296
intertropical convergence zone (ITCZ), 296
westerlies, 296
polar front, 296
subpolar low, 296

QUESTIONS FOR REVIEW

1. Draw a large circle. Now place the major surface pressure and wind belts of the world at their appropriate latitudes.

2. Explain how and why the average surface pressure features shift from summer to winter.

3. Why is it impossible on Earth for a Hadley cell to extend from the pole to the equator?

4. Along a meridian line running from the equator to the poles, how does the general circulation help explain zones of abundant and sparse precipitation?

5. Most of the Canada is in what wind belt?

6. Explain why the winds in the middle and upper troposphere tend to blow from west to east in both the Northern and Southern hemispheres.

7. How does the polar front influence the development of the polar front jet stream?

8. Describe how the conservation of angular momentum plays a role in the formation of a jet stream.

9. Why is the polar front jet stream stronger in winter than in summer?

10. Explain the relationship between the general circulation of air and the circulation of ocean currents.

11. List at least four important interactions that exist between the ocean and the atmosphere.

12. Describe how the Ekman spiral forms.

13. What conditions are necessary for upwelling to occur along the West Coast of North America? The East Coast of North America?

14. (a) What is a major El Niño event?
    (b) What happens to the surface pressure at opposite ends of the Pacific Ocean during the Southern Oscillation?
    (c) Describe how the Southern Oscillation influences a major El Niño event.

15. What are the conditions over the tropical eastern and central Pacific Ocean during the phenomenon known as La Niña?

16. What type of weather (cold/warm, wet/dry) would you expect over North America during a strong El Niño? During a strong La Niña?

17. Describe the ocean surface temperatures associated with the Pacific Decadal Oscillation. What climate patterns (cool/warm, wet/dry) tend to exist during the warm phase and the cool phase?

18. How does the positive phase of the North Atlantic Oscillation differ from the negative phase?

19. During the negative cold phase of the Arctic Oscillation, when Greenland is experiencing mild winters, what type of winters (cold or mild) is northern Europe usually experiencing?

QUESTIONS FOR THOUGHT

1. What effect would continents have on the circulation of air in the single-cell model?

2. How would the general circulation of air appear in summer and winter if Earth were tilted on its axis at an angle of 45° instead of 23.5°?

3. Summer weather in the southwestern section of the United States is influenced by a subtropical high-pressure cell, yet Figure 10.3b (p. 298) shows an area of low pressure at the surface. Explain.

4. Explain why icebergs tend to move at right angles to the direction of the wind.

5. Over the open ocean in the vicinity of the Pacific high, observations have indicated that ozone concentrations hundreds of metres above the surface are greater than expected. Give a possible explanation for this.

6. Give *two* reasons why pilots would prefer to fly in the core of a jet stream rather than just above or below it.

7. Why do the major ocean currents in the North Indian Ocean reverse direction between summer and winter?

8. Explain why the surface water temperature along the northern California coast is warmer in winter than it is in summer.

9. You are given an upper-level map that shows the position of two jet streams. If one is the polar front jet and the other the subtropical jet, how would you be able to tell which is which?

10. The Coriolis force deflects moving water to the right of its intended path in the Northern Hemisphere and to the left of its intended path in the Southern Hemisphere. Why, then, does upwelling tend to occur along the western margin of continents in both hemispheres?

PROBLEMS AND EXERCISES

1. Locate the following cities on a world map. Then, based on the general circulation of surface winds, predict the prevailing wind for each one during July and January.

   (a) Nashville, Tennessee

   (b) London, Ontario

   (c) Melbourne, Australia

   (d) London, England

   (e) Paris, France

   (f) Reykjavik, Iceland

   (g) Grande Prairie, Alberta

   (h) Vancouver, British Columbia

2. In the next column is a list of average weather conditions that prevail during the month of July at San Francisco, California, and Atlantic City, New Jersey. Both cities lie adjacent to an ocean at nearly the same latitude; however, the average weather conditions vary greatly. In terms of the average surface winds and pressure systems (see Figure 10.3b, p. 298) and the interaction between the atmosphere and ocean, explain what accounts for the variation of each weather element between the two cities.

Atlantic City, New Jersey
(latitude 39°N)
Average weather, July

Temperature maximum	29°C
minimum	19°C
Dew point	18°C
Precipitation	94 mm
Prevailing wind	S
Weather: Clear to partly cloudy with the possibility of afternoon showers	
Water temperature	21°C

San Francisco, California
(latitude 37°N)
Average weather, July

Temperature maximum	18°C
minimum	12°C
Dew point	12°C
Precipitation	0.3 mm
Prevailing wind	NW
Weather: Fog and low clouds during the night and morning with afternoon clearing	
Water temperature	12°C

Clouds developing along an approaching cold front.
© C. Donald Ahrens

Air Masses and Fronts

On December 10, 2009 as residents of Ontario and Quebec shovelled out from the season's first general snowfall, cold westerly winds set up across the relatively warm open waters of Georgian Bay and Lake Huron. An intense snow squall locked into 30-km-wide bands stretching across Muskoka and Haliburton. The monster squall blew relentlessly for two days dumping between 50 and 125 cm of snow in locales to the lee of the lakes. Yet, just 20 km on either side of the streamers, snow flurries fell with little accumulation. Bursts of heavy snowfall and driving winds created zero visibility and forced officials in Hunstville and Bracebridge to call a snow emergency—the first one that long-time residents could remember in 40 years.

Environment Canada. Canada's Top Ten Weather Stories for 2009. Found at: http://www.ec.gc.ca/meteo-weather/default.asp?lang=En&n=6B9C6658-1#r3

CONTENTS

The opening paragraph describes a lake-effect snow squall. These are common downwind of large lakes such as the Great Lakes and Lake Winnipeg in early winter. Large bodies of water change temperature slowly; consequently, lakes such as these remain unfrozen into midwinter, providing moisture in the area. These storms often occur following a cold front, when cold, dry winter air crosses warm, unfrozen water, resulting in narrow bands of intense snowfall downwind from the lake. Later in winter, these types of storms are diminished as lakes freeze over and no longer provide a moisture or heat source.

Snow squalls are a hazard that can have significant impacts on the *anthrosphere*—large snow drifts and reduced visibility in snow may strand drivers and cause automobile accidents in snowbelt regions as motorists suddenly enter the intense but localized storms. These systems show how a storm can occur within an air mass rather than being associated with a front. As we shall see, most "weather" in the middle latitudes is associated with fronts. The movement of air masses and fronts results in weather changes and is how the atmosphere transports heat and moisture from one region to another. Therefore, air masses and fronts underlie Earth's cycles of water and energy, which directly affect the *hydrosphere*, *biosphere*, and *cryosphere*, as well as the *anthrosphere*, through the natural hazards associated with them.

In this chapter, in addition to special cases such as lake-effect snow squalls, we will examine the more typical weather associated with cold fronts and warm fronts. We will address questions such as Why are cold fronts usually associated with showery weather? How can warm fronts during the winter cause freezing rain and ice pellets to form over a vast area? And how can one read the story of an approaching warm front by observing its clouds? But, first, so that we may better understand fronts, we will examine air masses. We will look at where and how they form and the type of weather usually associated with them.

Air Masses

An **air mass** is an extremely large body of air whose properties of temperature and humidity are fairly similar in any horizontal direction at any given altitude. Air masses may cover many thousands of square kilometres. In ● Figure 11.1, a large winter air mass associated with a high-pressure area covers the eastern half of North America. Note that although the surface air temperature and dew point vary across the map, everywhere near the area of high pressure the air is cold and dry. This cold, shallow anticyclone will drift eastward, carrying with it the temperature and moisture characteristic of the region where the air mass formed; hence, in a day or two, cold air will be located over the central Atlantic Ocean. Part of weather forecasting is, then, a matter of determining air mass characteristics, predicting how and why they change, and in what direction the systems will move.

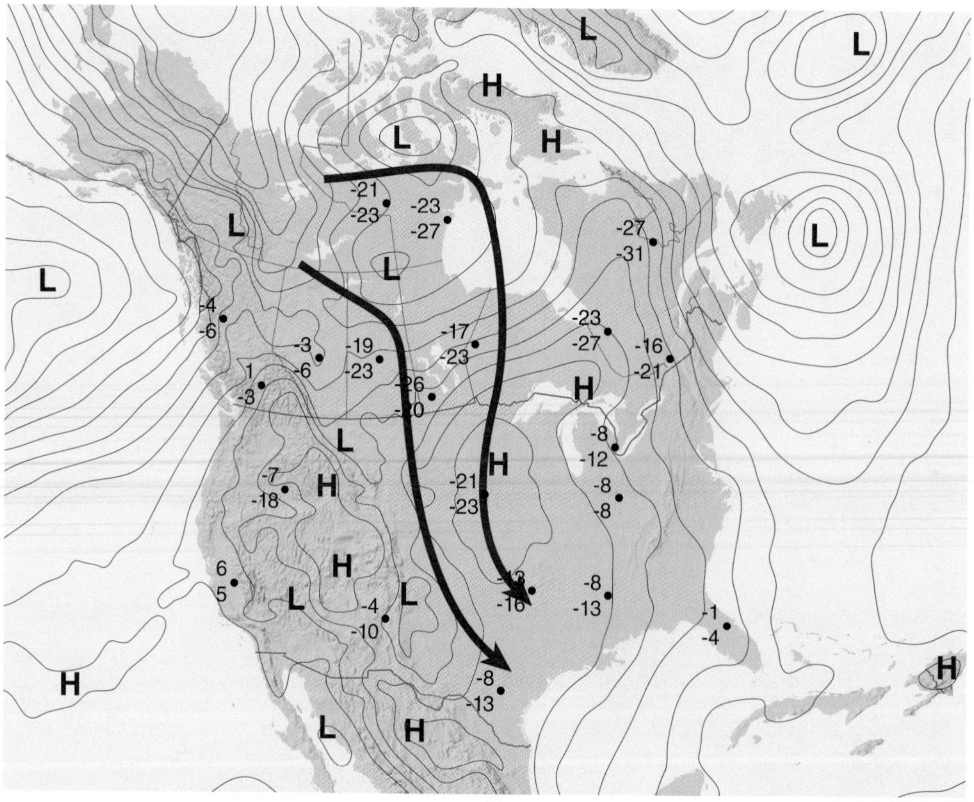

● FIGURE 11.1 On January 9, 2010, at 7:00 a.m. EST, a large, cold winter air mass is dominating the weather over the eastern half of North America. At almost all cities within the area of high pressure, the air is cold and dry. The upper number is air temperature (°C); the bottom number is dew point (°C). Heavy arrows indicate the upper-level airflow.

Source: Environment Canada. Found at http://www.weatheroffice.gc.ca/analysis/index_e.html

SOURCE REGIONS   Regions where air masses originate are known as **source regions**. For a huge mass of air to develop uniform characteristics, its source region should be generally flat and of uniform composition with light surface winds. As air resides over a surface, it acquires properties of the surface by exchanging radiation with the surface, as well as heat and moisture by convection. The longer the air remains stagnant over its source region, or the longer the path over which the air moves, the more likely it is that it will acquire properties of the surface below. Consequently, ideal source regions are usually those areas dominated by surface high pressure. They include the ice- and snow-covered arctic plains in winter and subtropical oceans in summer. The middle latitudes, where surface temperatures and moisture characteristics vary considerably, are not good source regions. Instead, this region is a transition zone where air masses with different physical properties move in, clash, and produce an exciting array of weather activity.

CLASSIFICATION   Air masses are usually classified according to their temperature and humidity, both of which usually remain fairly uniform in any horizontal direction.* There are cold and warm air masses and humid and dry air masses. Air masses are grouped into six general categories according to their source region. Air masses that originate in the high arctic latitudes are designated by the capital letter "A" (for arctic); those that form in middle latitudes are designated by the capital letter "P" (for *polar*); those that form in warm tropical regions are designated by the capital letter "T" (for *tropical*). If the source region is land, the air mass will be dry and the lowercase letter "c" (for *continental*) precedes the A, P, or T. If the air mass originates over water, it will be moist—at least in the lower layers—and the lowercase letter "m" (for *maritime*) precedes the A, P, or T. We can now see that arctic air originating over land will be classified cA on a surface weather map, whereas tropical air originating over water will be marked as mT. In winter, northern Canada and the Arctic Ocean consist

*In classifying air masses, it is common to use the *potential temperature* of the air. The potential temperature is the temperature that unsaturated (dry) air would have if it moved from its original level to a pressure of 1000 hPa at the dry adiabatic rate (10°C per 1000 m).

of one large frozen surface providing a source region for continental arctic (cA) air. During summer, the Arctic Ocean largely melts and warms, so the region is now a source for maritime arctic (mA) air. Canada is largely affected by four air masses, cA, mA, mP, and mT, during the winter and mA, mP, and mT during the summer. Continental tropical air can form over northern Mexico during summer; however, the source region is small, and this air seldom reaches Canada. Continental polar air can form over North America during winter when continental arctic air moves over a snow-free surface, but this category is rarely used by Canadian meteorologists.*
▼ Table 11.1 lists the six basic air masses, and ● Figure 11.2 shows the air masses and their source regions.

After the air mass spends some time over its source region, it usually begins to move in response to the winds aloft. As it moves away from its source region, it encounters surfaces that may be warmer or colder than itself. When the air mass is colder than the underlying surface, it is warmed from below, which produces instability at low levels. In this case, increased convection and turbulent mixing near the surface usually produce good visibility, cumuliform clouds, and showers of rain or snow. On the other hand, when the air mass is warmer than the surface below, the lower layers are chilled by contact with the cold Earth. Warm air above cooler air produces stable air with little vertical mixing. This situation causes the accumulation of dust, smoke, and pollutants, which restricts surface visibilities. In moist air, stratiform clouds accompanied by drizzle or fog may form.

AIR MASSES OF NORTH AMERICA   The principal air masses (with their source regions) that affect North America that are analyzed by Canadian meteorologists are shown in Figure 11.2. We are now in a position to study the formation and modification of each of these air masses and the variety of weather that accompanies them.

*Note that U.S. and international usage identify continental arctic (cA) air that has moved over continents away from the high arctic as continental polar (cP), but this air is normally considered to be modified cA by Canadian meteorologists unless it is further modified by residing for a period of time over a snow-free surface, when it might be called cP. Also, the designation of maritime arctic (mA) is not used in the United States.

▼ Table 11.1   Air Masses and their Characteristics

SOURCE REGION (symbol) & *characteristics*	ARCTIC (A)	POLAR (P)	TROPICAL (T)
CONTINENTAL (c) *land*	†**Continental Arctic (cA)**	**Continental Polar (cP)**	**Continental Tropical (cT)**
	extremely cold, dry stable; ice- and snow-covered surfaces	*cold, dry, stable*	*hot, dry, stable air aloft; unstable surface air*
MARITIME (m) *water*	†**Maritime Arctic (mA)**	†**Maritime Polar (mP)**	†**Maritime Tropical (mT)**
	cold, moist, unstable	*cool, moist, unstable*	*warm, moist; usually unstable*

†These air masses commonly affect Canada.

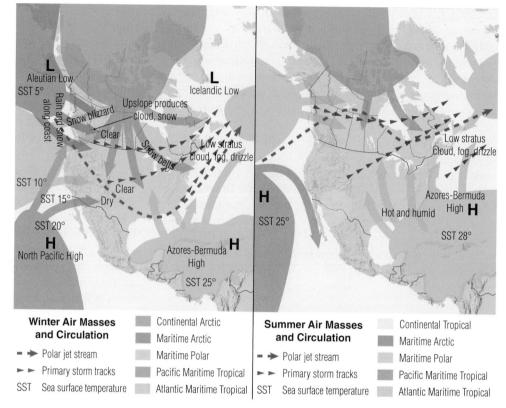

● FIGURE 11.2 Air mass source regions and their paths during winter and summer.

Source: Environment Canada. *Climates of Canada*. Dave Phillips. Pg. 12. 1990

Winter Air Masses and Circulation

- ➤ Polar jet stream
- ► ► Primary storm tracks
- SST Sea surface temperature

Continental Arctic
Maritime Arctic
Maritime Polar
Pacific Maritime Tropical
Atlantic Maritime Tropical

Summer Air Masses and Circulation

- ➤ Polar jet stream
- ► ► Primary storm tracks
- SST Sea surface temperature

Continental Tropical
Maritime Arctic
Maritime Polar
Pacific Maritime Tropical
Atlantic Maritime Tropical

cA (Continental Arctic) and cP (Continental Polar) Air Masses

The bitterly cold weather that affects Canada and the United States in winter is associated with **continental arctic** air masses. These air masses originate over the ice- and snow-covered regions of the arctic, northern Canada and Alaska, and the frozen Arctic Ocean, where long, clear nights allow for strong radiational cooling of the surface. Air in contact with the surface becomes very cold and stable. Since little moisture is added to the air, it is also quite dry, and dew point temperatures are often less than −30°C. Eventually, a portion of this cold air breaks away and, under the influence of the air flow aloft, moves southward as an enormous shallow high-pressure area, as illustrated in ● Figure 11.3.

As the cold air moves into the interior plains, there are no topographic barriers to restrain it, so it continues southward, bringing with it cold wave warnings and frigid temperatures. When cA air encounters a snow-free surface, it may eventually be modified into cP (**continental polar**) air according to the Canadian system of air mass identification. However, in the middle latitudes, the air seldom resides over a snow-free surface long enough to form a distinct air mass, so the cP air mass is seldom analyzed by Canadian meteorologists. In the United States, meteorologists commonly use the term "continental polar" (cP), but these air masses would usually be identified as modified cA air by Canadian meteorologists. Similarly, summer air that is labeled cP in the United States would be considered modified mA or mP air according to the Canada system of air mass identification.

The infamous *Texas norther* is associated with cA air. As the air mass moves over warmer land to the south, the air temperature moderates slightly as it is heated from below. However, even during the afternoon, when the surface air is most unstable, cumulus clouds are rare because of the extreme dryness of the air mass. At night, when the winds die down, rapid radiational surface cooling and clear skies combine to produce low minimum temperatures. If the cold air moves as far south as central or southern Florida, the winter vegetable crop may be severely damaged. When the cold, dry air mass moves over

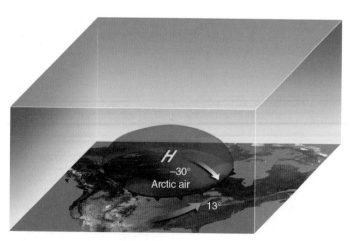

● FIGURE 11.3 A shallow but large dome of extremely cold air—a continental arctic air mass—moves slowly southeastward across the upper plains. The leading edge of the air mass is marked by a cold front. (Numbers represent air temperature, °C.)

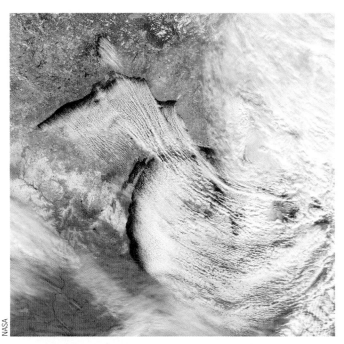

NASA

● FIGURE 11.4 Visible satellite image of lake-effect clouds and snow that forms when cA air moved across the warm Great Lakes on December 5, 2000. The convective cloud streamers are aligned with the northwesterly surface winds. For more information on this phenomenon, read Focus on a Special Topic: Lake-Effect Snows, on p. 328.

WEATHER WATCH

Heavy lake-effect snows can bury a city in a matter of days. For example, in less than four days, Buffalo, New York, received nearly 2 m of snow during December 2001. And Montague, New York (which lies on the eastern side of Lake Ontario), received over 2 m of snow in less than 48 hours during January 1997.

a relatively warm body of water, such as the Great Lakes, heavy snow showers—called **lake-effect snows** (or **snow squalls**)—often form on the downwind shores (see ● Figure 11.4). (More information on lake-effect snows is provided in Focus on a Special Topic: Lake-Effect Snows on p. 328.)

In winter, the cold but generally fair weather accompanying cA air is due to the stable nature of the atmosphere aloft. Sinking air develops above the large dome of high pressure. The subsiding air warms by compression and creates warmer air, which lies above colder surface air. Therefore, a strong upper-level subsidence inversion often forms. Should the anticyclone stagnate over a region for several days, the visibility gradually drops as pollutants become trapped in the cold air near the ground. Usually, however, winds aloft move the cold air mass either eastward or southeastward.

The Rockies, Coastal Mountains, and, in the United States, the Sierra Nevada and Cascades normally protect the Pacific coast of North America from the onslaught of cA air, but, occasionally, very cold air masses do invade these regions. When the upper-level winds blow from the north or northeast

on a trajectory beginning over northern Canada or Alaska, cold cA air can slip over the mountains and extend its icy fingers all the way to the Pacific Ocean. As the air moves off the high plateau, over the mountains, through the passes and fjords and the lower valleys, compressional heating of the sinking air causes its temperature to rise, so by the time it reaches the lowlands, it is warmer than it was originally. However, in no way would this air be considered warm. When the cold cA air emerges on the west coasts of British Columbia and Washington State, it may result in snow—a relatively infrequent occurrence—as well as uncharacteristically cold temperatures. In some cases, the subfreezing temperatures extend southward into the coastal areas of southern California.

The arrows in Figure 11.1 show the upper-air wind pattern that led to the cold outbreak of arctic air on January 9, 2010, over eastern North America. Upper-level winds typically blow from west to east, but in arctic outbreak cases, the flow has a strong north–south (meridional) trajectory. Notice how the upper-level wind directs the paths of the air masses.

When an air mass moves over a large body of water, its original properties may change considerably. For instance, cold, dry cA air moving over the Atlantic Ocean warms rapidly and gains moisture. The air quickly assumes the qualities

Cold, Dry continental arctic (cA) air

NASA

● FIGURE 11.5 Visible satellite image on January 17, 2007, at 10:00 a.m. EST showing the modification of cold continental arctic air as it moves from the Maritime provinces and New England states over the warmer Atlantic Ocean.

FOCUS ON A SPECIAL TOPIC

Lake-Effect Snows

During the winter, when the weather over most of Canada is dominated by clear and cold arctic air, people living on the eastern shores of the Great Lakes brace themselves for heavy snow showers, called **snow squalls**. Snowstorms that form on the downwind side of one of these lakes are known as **lake-effect snows**. Since the lakes are responsible for enhancing the amount of snow that falls on its downwind side, these snowstorms are also called *lake-enhanced snows*, especially when the snow accompanies a cold front or middle-latitude cyclone. These storms are highly localized, extending from just a few kilometres to more than 100 km inland. The snow usually falls as a heavy shower or squall in a concentrated zone. The snowfall is so centralized that one part of a city may accumulate many centimetres of snow, whereas in another part, the ground is bare.

Lake-effect snows near the Great Lakes are most numerous from November to January. During these months, cold air moves over the lakes when they are relatively warm and not quite frozen. The contrast in temperature between water and air can be as much as 25°C. Studies show that the greater the contrast in temperature, the greater the potential for snow showers. In ● Figure 1, we can see that as the cold air moves over the warmer water, the air mass is quickly warmed from below, making it more buoyant and less stable. Rapidly, the air sweeps up moisture, soon becoming saturated.

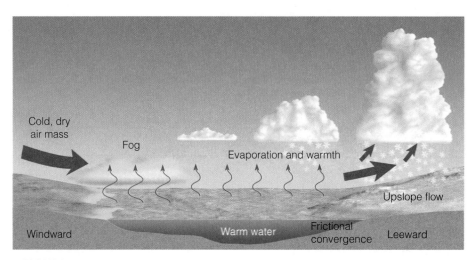

● FIGURE 1 The formation of lake-effect snows. Cold, dry air crossing the lake gains moisture and warmth from the water. The more buoyant air now rises, forming clouds that deposit large quantities of snow on the lake's leeward (downwind) shores.

Out over the water, the vapour condenses into steam fog. As the air continues to warm, it rises and forms billowing cumuliform clouds, which continue to grow as the air becomes more unstable. Eventually, these clouds produce heavy showers of snow, which make the lake seem like a snow factory. Once the air and clouds reach the downwind side of the lake, additional lifting is provided by low hills and the convergence of air due to friction as it slows down over the rougher terrain. In late winter, the frequency and intensity of lake-effect snows often taper off as the temperature

contrast between water and air diminishes and larger portions of the lakes freeze.

Generally, the longer the stretch of water over which the air mass travels (the longer the **fetch**), the greater the amount of warmth and moisture derived from the lake, and the greater the potential for heavy snow showers. In fact, studies show that for significant snowfall to occur, the air must move across 80 km of open water. Consequently, forecasting lake-effect snowfalls depends to a large degree on determining the trajectory of the air as it flows over the lake. Regions that

---

of a maritime air mass. Notice in ● Figure 11.5 that rows of cumulus clouds (**cloud streets**) are forming over the Bay of Fundy and the Atlantic Ocean parallel to northwesterly surface winds as cA air is being warmed by the water beneath it, causing the air mass to destabilize. As a result, on this day, snow showers were reported over Prince Edward Island, Nova Scotia, and Newfoundland. As the cA air encounters the much warmer Gulf Stream water, it warms rapidly and becomes conditionally unstable. Vertical mixing brings down faster flowing cold air from aloft. This mixing creates strong, gusty surface winds and choppy seas, which can be hazardous to shipping. Hence, a once cold, dry, and stable air mass can

be modified to such an extent that its original characteristics are no longer discernible. When this happens, the air mass is given a new designation.

In summary, arctic air masses are responsible for the bitter cold winter weather that can cover wide sections of North America. When the air mass originates over the Canadian Northwest Territories, frigid air can bring record-breaking low temperatures. Such was the case on Christmas Eve, 1983, when arctic air covered most of North America. (A detailed look at this air mass and its accompanying record-setting low temperatures is given in Focus on a Special Topic: Arctic Outbreaks on p. 330.)

• FIGURE 2 White shaded areas show regions near the Great Lakes that experience heavy lake-effect snows.

experience heavy lake-effect snowfalls are shown in • Figure 2.*

As the cold air moves farther away from the lake, the heavy snow showers usually taper off; however, the western slope of the Appalachian Mountains in the U.S. states of Pennsylvania, New York, and Vermont produces further lifting, enhancing the possibility of more and heavier showers.

Lake-effect snows are not confined to the Great Lakes. In fact, any large unfrozen lake (such as lakes Great Bear, Slave, and Salt, as well as lakes Athabasca, Winnipeg, Winnipe-

___
*Buffalo, New York, is a city that experiences heavy lake-effect snows. The U.S. National Weather Service Buffalo Forecast Office has a lake-effect website with information on lake-effect snowstorms measured in feet, as well as interesting weather stories. (Search on the keywords "lake effect buffalo" to find the URL.)

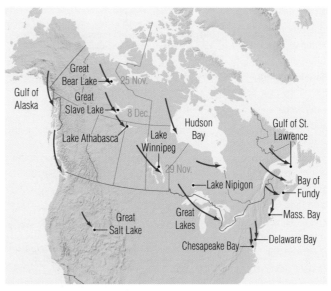

• FIGURE 3 Areas of North America prone to lake- effect and ocean-effect snow. Arrows indicate typical paths of cold air, Dates indicate the average date of freezing for lakes that normally freeze over.
Source: MetEd/UCAR. Found at: http://www.meted.ucar.edu/norlat/snow/lake_effect/print_whole.html

gosis, and Nipigon) can enhance snowfall when cold, relatively dry air sweeps over them (see • Figure 3). The lake enhancement is reduced considerably once the lakes freeze over, so the more northerly lakes in Canada experience this effect mainly in the fall (average freeze dates for some lakes are shown in Figure 3). Moreover, a type of lake-effect snow occurs when cold air moves over a relatively warm ocean and then lifts slightly as it

moves over a landmass. Such **ocean-effect snows** in winter are common in places such as Juneau, Alaska, on the northern and eastern coasts of Vancouver Island, along the shores of Hudson Bay and James Bay, on the west coast of Nova Scotia along the Bay of Fundy and over the Cape Breton Highlands, on Prince Edward Island, along the west and southeast coasts of Newfoundland, and over Cape Cod, Massachusetts.

mA (Maritime Arctic) Air Masses

In winter, **maritime arctic** air originates in the Chukchi and Bering Seas, north of the Aleutian Islands in the North Pacific, and can affect the northwest coast of North America. In summer, after the winter snow and ice melts, mA air forms over the Arctic Ocean and archipelago, affecting the northern interior of the continent. The mA air that moves into Canada and the United States in summer has properties much different from those of its winter cA counterpart. The source region remains the same but is now characterized by long summer days that melt snow and warm the land. The air is only moderately cool, and surface evaporation adds water vapour to the air. A summertime mA

air mass usually brings relief from the oppressive heat in central and eastern North America as cooler air lowers the air temperature to more comfortable levels. Daytime heating warms the lower layers, producing surface instability. The water vapour in the rising air may condense and create a sky dotted with fair-weather cumulus clouds (cumulus humilis). Typical profiles of temperatures for a summer mA and a winter cA air mass are given in • Figure 11.6. Notice that the strong inversion so prevalent in winter is absent in summer. When mA air moves far from its source region over the North American continent, it may be sufficiently modified to be named a new air mass—continental polar. Canadian meteorologists use

FOCUS ON A SPECIAL TOPIC

Arctic Outbreaks

Cliff Raphael, College of New Caledonia, Prince George, British Columbia

Arctic outbreaks are regular features of Canadian winters. A winter arctic outbreak event can be defined as that period of time when the maximum and minimum temperatures at a location are both below normal for three or more days.

Source regions for the air masses associated with arctic outbreaks include the far reaches of the Yukon, Alaska, and the Northwest Territories, where radiative cooling of the surface during the long arctic night creates the cold continental arctic air mass. Sweeping down across the Prairies, Ontario, and Quebec, the cold air may extend as far south as the Gulf of Mexico. Under the right conditions, the arctic air may also reach coastal British Columbia. These events have adverse effects on the weather, economic activities, transportation, etc., of the locations they impact. The occurrence of an arctic outbreak is linked to the development of a large upper-level ridge of high pressure. The location and orientation of the upper level ridge determine the area of impact of the cold air and its associated area of high pressure, or anticyclone, at the surface.

Two major arctic outbreaks during the winter of 1983–84 provide good examples of the impact of these events on winter temperatures. Dubbed the "Siberian Express" by the media in the United States, these events were responsible for one of the coldest winters on record for many locations across North America. The upper-level pattern and surface configuration for

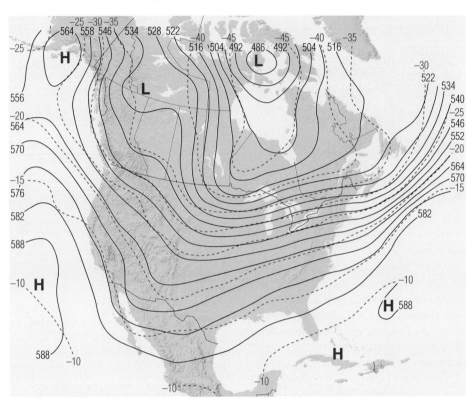

● FIGURE 4 A 500 hPa weather map for December 18, 1983, at 12:00 UTC (07:00 EST). Note the large upper-level high and its ridge of high pressure located near Alaska. Solid contours are the height of the 500 hPa surface in tens of metres (decametres). Dashed contours are the Celcius temperatures at 500 hPa.
Source: NOAA

the strong arctic outbreak event of December 1983 are shown in ● Figure 4 and ● Figure 5. There is a well-developed upper-level ridge off the coast of British Columbia and Alaska (see

Figure 4) with an associated surface high extending from the northern reaches of Canada to the southern parts of the United States (see Figure 5). During the week of December 18

---

the mA classification, whereas their U.S. counterparts use the term cP to label this air.

mP (Maritime Polar) Air Masses

During the winter, cA air originating over Asia and frozen polar regions is carried eastward and southward over the Pacific Ocean by the circulation around the Aleutian low. The ocean water modifies these cold air masses by adding warmth and moisture to them. Since this air travels across many hundreds or even thousands of kilometres of water, it gradually changes into *maritime polar air*.

By the time this **maritime polar** air mass reaches the Pacific Coast, it is cool, moist, and conditionally unstable. The ocean's effect is to keep air near the surface warmer than the air aloft. Temperatures between 5 and 15°C are common near the surface, whereas air at an altitude of about a kilometre or so above the surface may be at the freezing point. Within this colder air, characteristics of the original cold, dry air mass may still prevail. As the air moves inland, coastal mountains force it to rise, and much of its water vapour condenses into rain-producing clouds. In the colder air aloft, the rain changes to snow, with heavy amounts accumulating in

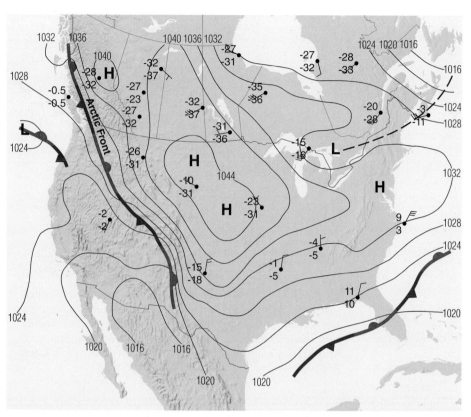

● **FIGURE 5** Surface weather map for December 18, 1983, at 12:00 UTC (07:00 EST). Note the large area of high pressure covering the central portion of North America. Contours are sea level pressures in hPa.
Source: NOAA

to 25, the cold air's icy grip stretched across Canada from coast to coast. Maximum and minimum temperatures were below normal by as much as 27°C. On December 20, the minimum temperature of −18.9°C recorded at Halifax,

Nova Scotia, was just 5° warmer than the record (−23.3°C) set in 1980. Vancouver's minimum temperature of −12.7°C on December 23 was also just 5° warmer than the record (−17.8°C) set in 1968. The severest conditions were experi-

enced in southern Saskatchewan and southern Manitoba. Record-breaking low temperatures (−39.5°C) in Estevan, Saskatchewan, and (−40.4°C) Brandon, Manitoba, were experienced on December 23. Accompanying wind-chill temperatures reached a bitter −57°C. On December 23 at 9:00 a.m., the temperature at Yellowknife, Northwest Territories (−28°C), was more than 12° warmer than the temperature (−40.3°C) at Brandon, Manitoba. Intense frost penetration into the ground resulted in numerous broken water mains in Winnipeg, Manitoba. By December 26, conditions began to ameliorate as a more westerly flow was established in the upper levels.

Just a few weeks later, in mid-January 1984, a second pulse of cold air pushed southward. From January 16 to 26, bitter cold gripped northern locations such as Whitehorse, Yukon, with temperatures reaching as low as −42.3°C. Below-normal temperatures were experienced at locations across southern Canada but not to the same extent as in December. Of note were the temperatures experienced in Fredericton, New Brunswick, where the minimum temperature of −31.8°C on January 22 was only 4°C warmer than the record of −35.6°C set in 1971.

The winter of 1983–84 was one of the coldest on record for several Canadian locations. The severity of the winter was associated with the occurrence of these two major Arctic outbreaks, regular features of Canadian winter weather.

mountain regions. Over the relatively warm open ocean, the cool, moist air mass produces cumulus clouds.

When the maritime polar air moves inland, it loses much of its moisture as it crosses a series of mountain ranges. Beyond these mountains, it travels over a cold, elevated plateau that chills the surface air and modifies it slowly. If the air resides for a sufficiently long time over a snow-free surface, it may transform into dry, stable continental polar air. East of the Rockies, this air mass is referred to as *Pacific air* (see ● Figure 11.7). Here it often brings fair weather and temperatures that are cool but not nearly as cold as the continental

arctic air that invades this region from the north. In fact, just east of the Rockies, when Pacific air from the west replaces retreating cold air from the north, chinook winds often develop. Furthermore, when the modified maritime polar air replaces moist tropical air, storms can form along the boundary separating the two air masses.

Along the East Coast, mP air originates in the North Atlantic. (Look back at Figure 11.2 on p. 326.) Steered by northeasterly winds, mP air swings southwestward toward the Maritime provinces and states. Because the water of the North Atlantic is very cold and the air mass travels only a

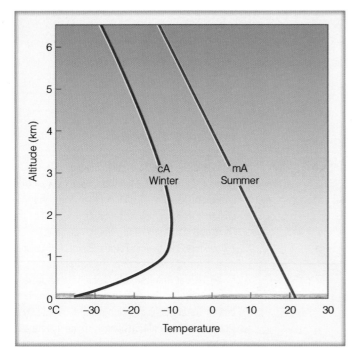

● FIGURE 11.6 Typical vertical temperature profile over land for a summer mA and a winter cA air mass.

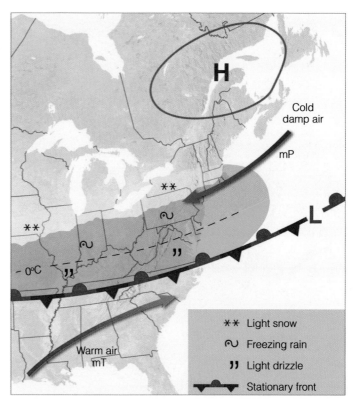

● FIGURE 11.8 A winter and early spring surface weather pattern that usually prevails during the invasion of cold, moist mP air into the mid-Atlantic and New England states. (The green-shaded area represents light rain and drizzle; the pink-shaded region represents freezing rain and ice pellets; and the white-shaded area is experiencing snow.)

short distance, wintertime Atlantic mP air masses are usually much colder than their Pacific counterparts. Because the prevailing winds aloft are westerly, Atlantic mP air masses are also much less common.

● Figure 11.8 illustrates a typical late winter or early spring surface weather pattern that carries mP air from the Atlantic into the Maritime provinces and New England states. A slow-moving, cold anticyclone over Quebec drifting to the east causes a northeasterly flow of mP air to the south. The boundary separating this invading colder air from warmer air even farther south is marked by a stationary front. North of this front, northeasterly winds provide generally undesirable weather, consisting of damp air and low, thick clouds from which light precipitation falls in the form of rain, drizzle, or snow. When upper atmospheric conditions are right, storms may develop along the stationary front, move eastward, and intensify over the ocean near Cape Hatteras. Such storms, called Hatteras lows, sometimes swing northeastward along

the coast, where they become *northeasters* (commonly called *nor'easters*), bringing with them strong northeasterly winds, heavy rain or snow, and coastal flooding. (Such developing storms will be treated in detail in Chapter 12.)

mT (Maritime Tropical) Air Masses The wintertime source region for Pacific **maritime tropical** air masses is the subtropical east Pacific Ocean. (Look back at Figure 11.2 on p. 326.) Air from this region must travel over 1600 km of water before it reaches the West Coast of North America. Consequently, these air masses are very warm and moist by the time they arrive along the West Coast. In winter, the warm

● FIGURE 11.7 After crossing several mountain ranges, cool, moist mP air from off the Pacific Ocean descends the eastern side of the Rockies as modified, relatively dry Pacific air.

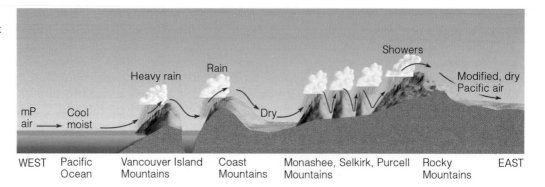

Hawaii

● FIGURE 11.9 An infrared satellite image that shows maritime tropical air (heavy yellow arrow) moving into northern California on January 1, 1997. The warm, humid airflow (sometimes called "the pineapple express") produced heavy rain and extensive flooding in northern and central California after hammering southwest British Columbia with snow at the end of December 1996.

air produces heavy precipitation usually in the form of rain, even sometimes at high elevations. Melting snow and rain quickly fill rivers, which overflow into the low-lying valleys. The rapid snowmelt leaves local ski slopes barren, and the heavy rain can cause disastrous mud slides in the steep canyons.

● Figure 11.9 shows maritime tropical air (usually referred to as subtropical air) streaming into northern California on January 1, 1997. The humid, subtropical air, which originated near the Hawaiian Islands, is called a *pineapple express* by many forecasters. After battering southern British Columbia and the U.S. Pacific Northwest with heavy snow and rain just after Christmas 1996, the pineapple express roared into northern and central California, causing catastrophic floods that sent over 100,000 people fleeing from their homes, mudslides that closed roads, property damage (including crop losses) that amounted to more than $1.5 billion, and eight fatalities.

The warm, humid subtropical air that influences much of the weather east of the Rockies originates over the Gulf of Mexico and Caribbean Sea. In winter, cold arctic and polar air tends to dominate the continental weather scene, so maritime tropical air is usually confined to the Gulf and extreme southern U.S. states. Occasionally, a slow-moving storm system over the Central Plains of North America draws warm, humid air northward. Gentle south or southwesterly winds carry this air into the central and eastern parts of the United States in advance of the storm. Since the land is still extremely cold, air near the surface is chilled to its dew point. Fog and low clouds form in the early morning, dissipate by midday, and reform in the evening. This mild winter weather lasts, at best, only a few days. Soon cold arctic air will move down from the north behind the eastward-moving storm system. Along the boundary between the two air masses, the warm, humid air is lifted above the more dense cold, arctic air—a

situation that often leads to heavy and widespread precipitation and storminess.

When a large mid-latitude cyclonic storm system stalls over the Central Plains of North America, a constant supply of warm, humid air from the Gulf of Mexico can bring record-breaking maximum temperatures to the eastern half of the continent. Sometimes the air temperatures are higher in the mid-Atlantic states than they are in the deep southern United States as compressional heating warms the air even more as it moves downslope after crossing the Appalachian Mountains.

● Figure 11.10 shows a surface weather map and the associated upper airflow (heavy arrow) that brought unseasonably warm maritime tropical air into the central and eastern states during April 1976. A large surface high-pressure area centred off the southeast coast of North America coupled with a strong southwesterly flow aloft carried warm moist air into the eastern half of the continent, causing a record-breaking April heat wave. The flow aloft prevented the surface low and the cold arctic (cA) air behind it from making much eastward progress, so the warm spell lasted for five days. Note that on the west side of the surface low, the winds aloft funneled cold air from the north into the western half of the continent, creating unseasonably cold weather from California to the Rockies. Hence, while people in the U.S. Southwest were huddled around heaters, others several thousand kilometres away in southeast Canada and the U.S. Northeast were turning on air conditioners. We can see that it is the upper-level meridional flow, directing cold arctic air southward and warm subtropical air northward, that makes these contrasts in temperature possible.

In summer, the circulation of air around the Bermuda High (which sits off the southeast coast of North America—see Figure 10.3b on p. 298) pumps warm, humid (mT) air northward from off the Gulf of Mexico and from off the

● FIGURE 11.10 Weather conditions during an unseasonably hot spell in eastern North America that occurred between April 15 and April 20, 1976. The surface low-pressure area and fronts are shown for April 17. Numbers to the east of the surface low (in red) are maximum temperatures recorded during the hot spell, whereas those to the west of the low (in blue) are minimum temperatures reached during the same time period. The heavy arrow is the average upper-level flow during the period. The purple L and H show average positions of the upper-level trough and ridge.

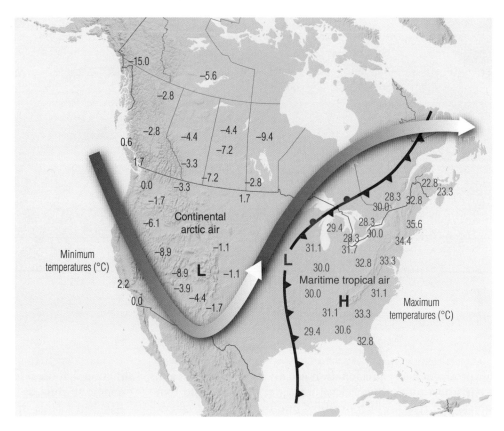

Atlantic Ocean into the eastern half of the United States and southeastern Canada. As this humid air moves inland, it warms even more, rises, and frequently condenses into cumuliform clouds, which produce afternoon showers and thunderstorms. You can almost count on thunderstorms developing along the Gulf of Mexico coast every summer afternoon. As evening approaches, thunderstorm activity typically dies off. Nighttime cooling lowers the temperature of this hot, muggy air only slightly. Should the air become saturated, fog or low clouds usually form, and these normally dissipate by late morning as surface heating warms the air again.

During the summer, humid subtropical air originating over the southeastern Pacific and Gulf of California normally remains south of California. Occasionally, an upper-level southerly flow will spread this humid air northward into the southwestern United States, most often Arizona, Nevada, and the southern part of California, in a type of monsoon circulation. In many places, the moist, conditionally unstable air aloft shows up only as middle and high cloudiness, especially altocumulus and cirrocumulus castellanus. However, where the moist flow meets a mountain barrier, it usually rises and condenses into towering, shower-producing clouds. (For an exceptionally strong flow of subtropical air into this region, look at Figure 9.31 on p. 282.)

cT (Continental Tropical) Air Masses

The only real source region for hot, dry **continental tropical** air masses in North America is found during the summer in northern Mexico and the adjacent arid southwestern United States. Here the air mass is hot, dry, and conditionally unstable at low levels, with frequent dust devils forming during the day. Because of the low relative humidity (typically less than 10 percent during the afternoon), air must rise over 3000 m before condensation begins. Furthermore, an upper-level ridge usually produces weak subsidence over the region, tending to make the air aloft rather stable and the surface air even warmer. Consequently, skies are generally clear, the weather is hot, and rainfall is practically nonexistent where continental tropical air masses prevail. If this air mass moves outside its source region and into the U.S. Great Plains and stagnates over that region for any length of time, a severe drought may result.

So far, we have examined the various air masses that enter North America annually. The characteristics of each depend on the air mass source region and the type of surface over which the air mass moves. The winds aloft determine

---

WEATHER WATCH

A continental tropical air mass, stretching from southern California to the heart of Texas, brought record warmth to the desert southwest during the last week of June 1990. The temperature, which on June 26 soared to a sweltering peak of 50°C in Phoenix, Arizona, caused officials to suspend aircraft takeoffs at Sky Harbor Airport. The extreme heat had lowered air density to the point where it reduced aircraft lift.

the trajectories of these air masses. Occasionally, an air mass will control the weather in a region for some time. These persistent weather conditions are sometimes referred to as *air mass weather.*

Air mass weather is especially common in the southeastern United States during summer as, day after day, humid subtropical air from the Gulf brings sultry conditions and afternoon thunderstorms. It is also common in the West Coast of Canada and the U.S. Pacific Northwest in winter when conditionally unstable, cool moist air accompanied by widely scattered showers dominates the weather for several days or more. The real weather action, however, usually occurs not within air masses but at their margins, where air masses with sharply contrasting properties meet—in the zone marked by weather fronts.*

BRIEF REVIEW

Before we examine fronts, here is a review of some of the important facts about air masses:

- An air mass is a large body of air whose properties of temperature and humidity are fairly similar in any horizontal direction.
- Source regions for air masses tend to be generally flat, of uniform composition, and in an area of light winds, dominated by surface high pressure.
- Continental air masses form over land. Maritime air masses form over water. Polar air masses originate in cold, middle-latitude regions, and extremely cold arctic air masses form over arctic regions. Tropical air masses originate in warm, tropical regions.
- Continental arctic (cA) air masses are extremely cold and dry; maritime arctic (mA) air masses are cold and moist; continental tropical (cT) air masses are hot and dry but seldom reach Canada; maritime tropical (mT) air masses are warm and moist; maritime polar (mP) air masses are cool and moist.

Fronts

Although we briefly looked at fronts in Chapter 1, we are now in a position to study them in depth, which will aid us in forecasting the weather. We will now learn about the general nature of fronts—how they move and what weather patterns are associated with them.

A **front** is the transition zone between two air masses of different densities. Since density differences are most often caused by temperature differences, fronts usually separate air masses with contrasting temperatures. Often they separate air masses with different humidities as well. Remember that air masses have both horizontal and vertical extent; conse-

quently, the upward extension of a front is referred to as a *frontal surface*, or a frontal zone.

- Figure 11.11a illustrates the vertical extent of two frontal zones—the *polar front* and the *arctic front* that separate three air masses, as can frequently be analyzed on a weather map. Figure 11.11b shows the frontal and air mass model used by Canadian meteorologist for winter weather situations in which four air masses (cA, mA, mP, mT) are divided by three fronts (continental arctic, maritime arctic, and polar). In practice, however, it is often not possible to be able to identify all four distinct air masses separated by three fronts at one time over every part of a weather map. More commonly, in a winter weather situation, the West Coast of North America might be affected by two fronts separating mA, mP, and mT air, whereas the interior of the continent is affected by two fronts separating cA, mP, and mT air. The polar front and maritime arctic front boundaries, which extend upward to over 5 km, separate warm, humid air to the

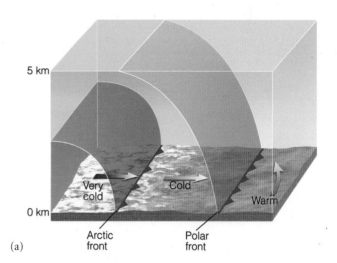

(a)

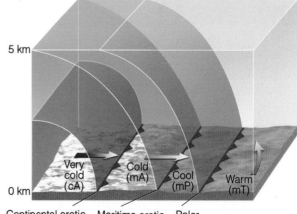

(b)

- FIGURE 11.11 (a) The polar front represents a cold frontal boundary that separates colder air from warmer air at the surface and aloft. The more shallow arctic front separates cold air from extremely cold air. (b) Canadian meteorologists often use a "three-front" model in which four air masses are identified (cA, mA, mP, and mT) that are divided by three fronts.

---

*The word *front* is used to denote the clashing or meeting of two air masses, probably because it resembles the fighting in western Europe during World War I, when the term originated.

south from cold arctic air to the north. The continental arctic front, which separates extremely cold continental arctic air from maritime arctic or maritime polar air, is much more shallow than the polar front and extends upward only to an altitude of about 1 or 2 km. In the next several sections, as we examine fronts on a flat surface weather map, keep in mind that all fronts have vertical as well as horizontal extent.

● Figure 11.12 shows a surface weather map illustrating five different fronts. Notice that the fronts are associated with lower pressure and that they separate differing air masses. As we move from west to east across the map, the fronts appear in the following order: a stationary front between points A and B; a cold front between points B and C; a warm front

between points C and D; and an occluded front and a TROWAL* (TROugh of Warm air ALoft) between points C and L. Let's examine the properties of each of these fronts.

STATIONARY FRONTS
A **stationary front** has essentially no movement.[†] On a coloured weather map, it is drawn as an alternating red and blue line. Red semicircles face toward colder air on the red line and blue triangles point toward

---

*Normally, either the TROWAL or occluded front (but not both) is analyzed on a surface weather map. Canadian meteorologists use the TROWAL.

[†]They are usually called *quasistationary fronts* because they can show some movement.

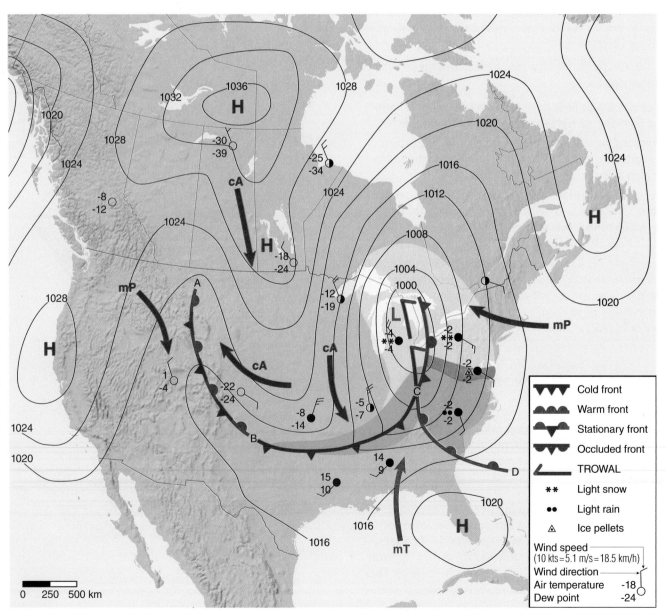

● FIGURE 11.12 A surface weather map showing surface pressure systems, air masses, fronts, and isobars (in hectopascals) as solid grey lines. Large arrows in colour show airflow. (The green-shaded area represents rain; the pink-shaded area represents freezing rain and ice pellets; and the white-shaded area represents snow.)

warmer air on the blue line. The stationary front between points A and B in Figure 11.12 marks the boundary where cold, dense continental arctic (cA) air from Canada butts up against the north–south trending Rocky Mountains. Unable to cross the barrier, the cold air shows little or no westward movement. The stationary front is drawn along a line separating the cA air from the milder, more humid maritime polar (mP) air to the west. Notice that the surface winds tend to blow parallel to the front but in opposite directions on either side of it. Upper-level winds often blow parallel to a stationary front. Since the front is moved by the colder, more dense air, the fact that the wind direction in the colder air is parallel to the front is the reason the front is stationary—it is unable to "push" or "pull" the front along.

The weather along the front is clear to partly cloudy, with much colder air lying on its eastern side. Because both air masses are relatively dry, there is no precipitation. This is not, however, always the case. When warm, moist air rides up and over the cold air, widespread cloudiness with light precipitation can cover a vast area. These are the conditions that prevail north of the east–west running stationary front depicted in Figure 11.8 (p. 332).

If the cooler air to the east begins to move away from the front, it will allow the warmer air to the west to move and replace the colder air. If this happens, the front in Figure 11.12 will no longer remain stationary; it will become a warm front. If, on the other hand, the colder air advances toward the front, causing it to slide up over the mountain, and replaces the warmer air, the front will become a cold front. If either a cold front or a warm front stops moving, it becomes a stationary front.

COLD FRONTS   The **cold front** between points B and C on the surface weather map (in Figure 11.12) represents a zone where cold, dry, stable arctic air is replacing warm, moist, conditionally unstable subtropical air. The front is drawn as a solid blue line, with the triangles along the front showing its direction of movement. How did the meteorologist know to draw the front at that location? A closer look at the front will give us the answer.

The weather in the immediate vicinity of this cold front in the southern United States is shown in • Figure 11.13. The data plotted on the map represent the current weather at selected cities. The station model used to represent the data at each reporting station is a simplified one that shows temperature, dew point, present weather, cloud cover, sea-level pressure, wind direction, and speed. The little line to the right of each station shows the *pressure tendency*—the pressure change, whether rising (/) or falling (\)—during the last three hours. With all of this information, together with a history of the front location and its progression, the front can be properly located.* (Appendix B explains the weather symbols and the station model more completely.)

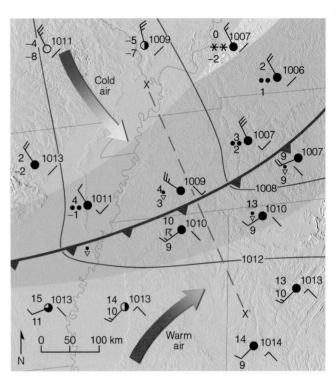

• FIGURE 11.13 A closer look at the surface weather associated with the cold front situated in the southern United States in Figure 11.12. (Grey lines are isobars. The green-shaded area represents rain; the white-shaded area represents snow.)

The following criteria are used to locate a front on a surface weather map:

1. sharp temperature changes over a relatively short distance
2. changes in the air's moisture content (as shown by marked changes in the dew point)
3. shifts in wind direction
4. pressure and pressure changes
5. clouds and precipitation patterns

In Figure 11.13, we can see a large contrast in air temperature and dew point on either side of the front. There is also a wind shift from southwesterly ahead of the front to northwesterly behind it. Notice that each isobar kinks as it crosses the front, forming an elongated area of low pressure—a trough—which accounts for the wind shift. Since surface winds normally blow with low pressure on the left but across the isobars toward lower pressure, we find winds with a southerly component ahead of the front and winds with a northerly component behind it.

Since the cold front is normally in a trough of low pressure, sharp changes in pressure can be significant in locating the front's position. One important fact to remember is that the lowest pressure usually occurs just as the front passes a station. Notice that as you move toward the front, the pressure drops, and as you move away from it, the pressure rises. This is clearly shown by the pressure tendencies for each station on the map, keeping in mind that this front is moving from the northwest toward the southeast. Just

---

*Locating any front on a weather map is not always a clear-cut process. Even meteorologists can disagree on an exact position.

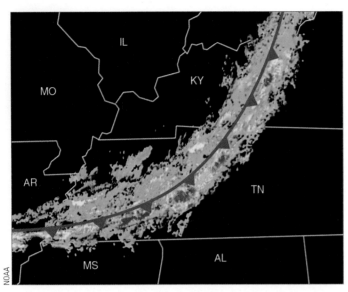

● FIGURE 11.14 A Doppler radar image of radar reflectivity showing precipitation patterns along a cold front similar to the cold front in Figure 11.13. Green represents light-to-moderate precipitation; yellow represents heavier precipitation; and red represents the most likely areas for thunderstorms. (The cold front is superimposed on the radar image.)

before the front passes, the pressure tendency shows that the atmospheric pressure is falling (\), whereas just behind the front, the pressure is now beginning to rise (✓), and farther behind the front, the pressure is rising steadily (/) (Refer to Appendix B for a full description of the weather map plotting symbols.)

The precipitation pattern along the cold front in Figure 11.13 is consistent with the Doppler radar image of precipitation intensity shown in ● Figure 11.14. The region in colour extending from northeast to southwest represents precipitation along a cold front. Notice that light-to-moderate rain (colour green) occurs over a wide area along the front, whereas the heavier precipitation (colour yellow) tends to occur in a narrow band along the front itself. Thunderstorms

(colour red) do not occur everywhere but only in certain areas along the front.

The cloud and precipitation patterns in Figure 11.13 are shown in a side view of the front along the line X–X′ as illustrated in ● Figure 11.15. We can see from Figure 11.15 that at the front, the cold, dense air wedges under the warm air, forcing the warm air upward, much like a snow shovel forces snow upward as the shovel glides through the snow. As the moist, conditionally unstable air rises, it condenses into a series of cumuliform clouds. Strong, upper-level westerly winds blow the delicate ice crystals (which form near the top of the cumulonimbus) into cirrostratus (Cs) and cirrus (Ci). These clouds usually appear far in advance of the approaching front. At the front itself, a relatively narrow band of thunderstorms (Cb) produces heavy showers with gusty winds. Behind the front, the air cools quickly. (Notice how the freezing level dips as it crosses the front.) The winds shift from southwesterly to northwesterly, pressure rises, and precipitation ends. As the air dries out, the skies clear, except for a few lingering fair-weather cumulus clouds.

Observe that the leading edge of the front is steep. The steepness is due to friction, which slows the airflow near the ground. The air aloft pushes forward, blunting the frontal surface. If we could walk from where the front touches the surface back into the cold air, a distance of 50 km, the front would be about 1 km above us. Thus, the slope of the front—the ratio of vertical rise to horizontal distance—is 1:50. This is typical for a fast-moving cold front—those that move about 25 knots (46 km h⁻¹). In a slower-moving cold front—one that moves about 15 knots—the slope is much more gentle.

With slow-moving cold fronts, clouds and precipitation usually cover a broad area behind the front. When the ascending warm air is stable, stratiform clouds, such as nimbostratus, become the predominant cloud type, and even fog may develop in the rainy area. Occasionally, out ahead of a fast-moving front, a line of active showers and thunderstorms, called a **squall line**, develops parallel to and often ahead of the advancing front, producing heavy precipitation and strong, gusty winds.

ACTIVE FIGURE 11.15 A vertical view of the weather across the cold front in Figure 11.13 along the line X–X′. Visit the textbook's web site to view this and other Active Figures at www.ahrensmeteorology1ce.nelson.com

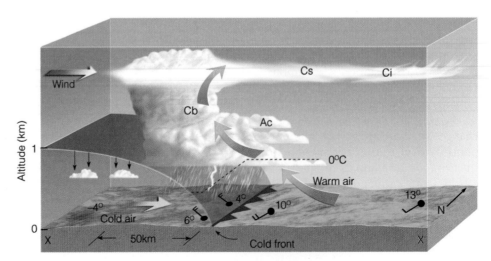

As the temperature contrast across a front lessens, the front will often weaken and dissipate. Such a condition is known as **frontolysis**. On the other hand, an increase in the temperature contrast across a front can cause it to strengthen and regenerate into a more vigorous frontal system, a condition called **frontogenesis**.

An example of a regenerated front is shown in the infrared satellite images in ● Figure 11.16. The cold front in Figure 11.16a is weak, as indicated by the low clouds (grey tones) along the front. As the front moves offshore, over the warm Gulf Stream (see Figure 11.16b), it intensifies into a more vigorous frontal system as surface air becomes conditionally unstable and convective activity develops. Notice that the area of cloudiness is more extensive and thunderstorms are now forming along the frontal zone.

So far, we have considered the general weather patterns of "typical" cold fronts. There are, of course, many exceptions. In fact, no two fronts are exactly alike. In some, the cold air is very shallow; in others, it is much deeper. If the rising warm air is dry and stable, scattered clouds are all that form, and there is no precipitation. In extremely dry weather, a marked change in the dew point, accompanied by a slight wind shift, may be the only clue to a passing cold front.

During the winter, a series of cold arctic outbreaks may travel southward across North America so quickly that warm air is unable to develop ahead of the front. In this case, frigid arctic air associated with a continental arctic front usually replaces cold polar (or maritime arctic) air, and a drop in temperature is the only indication that a front has moved through an area. Along the West Coast of North America, the Pacific Ocean modifies the air so much that cold fronts originating over the ocean, such as those described in the previous section, are seldom seen. In fact, as a cold front moves inland from the Pacific Ocean, the surface temperature contrast

across the front may be quite small. Topographic features usually distort the wind pattern so much that locating the position of the front and the time of its passage is exceedingly difficult. In this case, the pressure tendency, in combination with other evidence, such as satellite imagery, is the most reliable indication of a frontal passage.

In some instances along the West Coast, an approaching cold front (or upper-level trough) will cause cool marine air at the surface to *surge* into coastal and inland valleys. The cool air (which is often accompanied by a wind shift) may produce a sharp drop in air temperature. This may give the impression that a rather strong cold front has moved through when, in reality, the front may be many kilometres offshore.

Cold fronts usually move toward the south, southeast, or east. But sometimes they will move southwestward. In the Maritime provinces and New England states, this movement occurs when northeasterly surface winds, blowing clockwise around an anticyclone centred to the north over Canada, push a cold front southwestward, often as far south as Boston. Because the cold front moves in from the east or northeast, it is known as a **"back door" cold front**. As the front passes, westerly surface winds usually shift to easterly or northeasterly and temperatures drop as mP air flows in off the Atlantic Ocean.

An example of a back door cold front is shown in ● Figure 11.17. This is a springtime situation where, behind

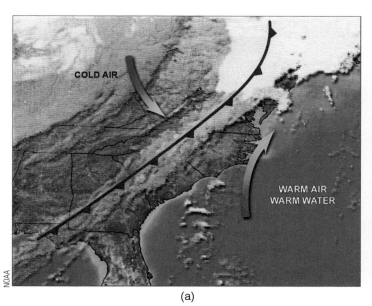

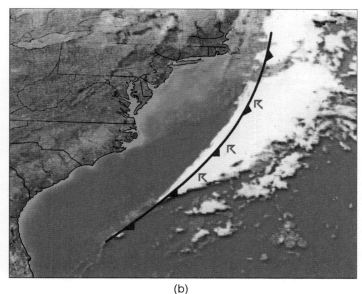

(a)　　　　　　　　　　　　　　　　(b)

● FIGURE 11.16　The infrared satellite image (a) shows a weakening cold front over land on Tuesday morning, November 21, intensifying into (b) a vigorous front over warm Gulf Stream water on Wednesday morning, November 22.

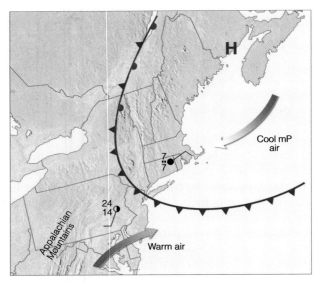

● FIGURE 11.17 A "back door" cold front moving into the New England states during the spring. Notice that behind the front, the weather is cold and damp with drizzle, whereas to the south, ahead of the front, the weather is partly cloudy and warm.

the front, the weather is cold and damp with drizzle as northeasterly winds sweep into the region from off the chilly Atlantic. To the south of the front, the weather is much warmer. Should the front move through this area, the more summerlike weather would change, in a matter of hours, to more winterlike. The cold, dense air behind the front is rather shallow. Consequently, the Appalachian Mountains act as a dam to the front's forward progress, halting its westward movement. This situation, where the cold, damp air is confined to the eastern side of the mountains, is called **cold air damming**. The stalled cold front now becomes a stationary front. The cool air behind the front may linger for some time as warmer, less dense air to the south rides up and over it.

Forecasting how far south the back door cold front will move and when the entrenched cold air will leave can be a bit tricky.

Even though cold-front weather patterns have many exceptions, learning these patterns can be to your advantage if you live in an area that experiences well-defined cold fronts. Knowing them improves your own ability to make short-range weather forecasts. For your reference, ▼ Table 11.2 summarizes idealized cold-front weather in the Northern Hemisphere.

WARM FRONTS   In Figure 11.12 (p. 336), a **warm front** is drawn along the solid red line running from points C to D. Here the leading edge of advancing warm, moist subtropical (mT) air from the Gulf of Mexico replaces the retreating cold maritime polar air from the North Atlantic. The direction of frontal movement is given by the half-circles, which point into the cold air; this front is heading toward the northeast. As the cold air recedes, the warm front slowly advances. The average speed of a warm front is about 10 knots (18.5 km h^{-1}), or about half that of an average cold front. During the day, as mixing occurs on both sides of the front, its movement may be much faster. Warm fronts often move in a series of rapid jumps, which show up on successive weather maps. At night, however, radiational cooling creates cool, dense surface air behind the front. This inhibits both lifting and the front's forward progress. When the forward surface edge of the warm front passes a station, the wind shifts, the temperature rises, and the overall weather conditions improve. To see why, we will examine the weather commonly associated with the warm front at both the surface and aloft.

● Figure 11.18 is a surface weather map showing the position of a warm front and its associated weather. ● Figure 11.19 is a vertical view of the warm front in Figure 11.18. Look at these two figures and observe that the warmer, less dense air rides up and over the colder, more dense surface air. This rising of warm air over cold, called **overrunning**, produces clouds and precipitation well in advance of the front's surface position. The warm front that separates the two air masses has an average slope of

▼ Table 11.2   Typical Weather Conditions Associated with a Cold Front in the Northern Hemisphere

WEATHER ELEMENT	BEFORE PASSING	WHILE PASSING	AFTER PASSING
Winds	south or southwest	gusty, shifting	west or northwest
Temperature	warm	sudden drop	steadily dropping
Pressure	falling steadily	minimum, then sharp rise	rising steadily
Clouds	increasing Ci, Cs, then either Tcu* or Cb*	Tcu or Cb	often Cu, Sc* when ground is warm
Precipitation	short period of showers	heavy showers of rain or snow, sometimes with hail, thunder, and lightning	decreasing intensity of showers, then clearing
Visibility	fair to poor in haze	poor, followed by improving	good, except in showers
Dew point	high; remains steady	sharp drop	lowering

*Tcu stands for towering cumulus, such as cumulus congestus, whereas Cb stands for cumulonimbus. Sc stands for stratocumulus.

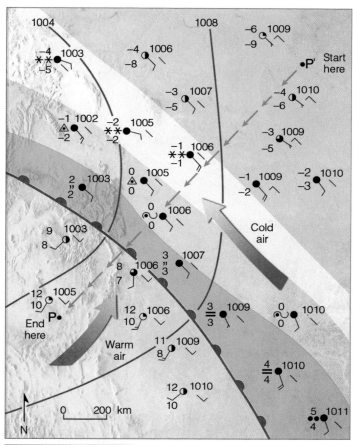

ACTIVE FIGURE 11.18 Surface weather associated with a typical warm front. A vertical view along the dashed line P–P′ is shown in Figure 11.19. (The green-shaded area represents rain; the pink-shaded area represents freezing rain and ice pellets; and the white-shaded area represents snow.) Visit the textbook's web site to view this and other Active Figures at www.ahrensmeteorology1ce.nelson.com

about 1:300*—a much more gentle or inclined shape than that of a typical cold front. Warm air overriding the cold air creates a stable atmosphere (see the vertical temperature profile in Figure 11.19b). Notice that a temperature inversion—called a **frontal inversion**—exists in the region of the upper-level front at the boundary where the warm air overrides the cold air. Another fact to notice in Figure 11.19b is that the wind veers (shifts clockwise) with altitude, so the southeasterly (SE) surface winds become southwesterly (SW) and westerly (W) aloft.

Suppose that we are standing at the position marked P′ in Figures 11.18 and 11.19. Note that we are over 1200 km ahead of where the warm front is touching the surface. Here the surface winds are light and variable. The air is cold, and about the only indication of an approaching warm front is the high cirrus clouds overhead. We know that the front is moving slowly toward us and that within a day or so it will pass our area. Suppose that instead of waiting for the front to pass us, we drive toward it, observing the weather as we go.

Heading toward the warm front, we notice that the cirrus (Ci) clouds gradually thicken into a thin, white veil of cirrostratus (Cs) whose ice crystals cast a halo around the sun.* Almost imperceptibly, the clouds thicken and lower, becoming altocumulus (Ac) and altostratus (As), through which the sun shows only as a faint spot against an overcast grey sky. Snowflakes begin to fall, and we are still over 600 km from the surface front. The snow increases, and the clouds thicken into a sheetlike covering of nimbostratus (Ns). The winds become brisk and out of the southeast, while the atmospheric pressure slowly falls. Within 400 km of the front, the cold surface air mass is now quite shallow. The surface air temperature moderates, and as we approach the front, the light snow changes first into ice pellets (called sleet in the United States). It then becomes freezing rain and finally rain and drizzle as the air temperature climbs above freezing. Overall, the precipitation remains light or moderate but covers a broad area. Moving still closer to the front, warm, moist air mixes with cold, moist air, producing ragged, wind-blown stratus (St) and fog. (Thus, flying in the vicinity of a warm front is quite hazardous.)

Finally, after a trip of over 1200 km, we reach the warm front's surface boundary. As we cross the front, the weather changes are noticeable but much less pronounced than those experienced with the cold front; they show up more as a gradual transition rather than a sharp change. On the warm side of the front, the air temperature and dew point rise, the wind shifts from southeast to south or southwest, and the air pressure stops falling. The light rain ends, and, except for a few stratocumulus (Sc), the fog and low clouds vanish.

This scenario of an approaching warm front represents idealized or average warm-front weather in winter. In some instances, the weather can differ from this dramatically. For example, if the overrunning warm air is relatively dry and stable, only high and middle clouds will form, and no precipitation will occur. On the other hand, if the warm air is relatively moist and conditionally unstable (as is often the case during the summer), heavy showers can develop as thunderstorms become embedded in the cloud mass. In the southern U.S. Great Plains, warm, humid air may be separated from warm, dry air along a boundary called a **dryline**. More on drylines is given in Focus on a Special Topic: Drylines on p. 344.

Along the West Coast of North America, the Pacific Ocean significantly modifies the surface air so that warm fronts are difficult to locate on a surface weather map. Also, not all warm fronts move northward or northeastward. On rare occasions, a front will move into the eastern seaboard from the Atlantic Ocean as the front spins all the way around a deep storm positioned off the coast. Cold northeasterly winds ahead of the front usually become warm northeasterly winds behind it. Even with these exceptions, knowing the normal sequence of warm-front weather will be useful, especially if you live where warm fronts become well developed. You can look for certain cloud and weather patterns and make reasonably accurate short-range forecasts of

---

*The slope of 1:300 is a much more gentle slope than that of most warm fronts. Typically, the slope of a warm front is on the order of 1:150 to 1:200.

*If the warm air is relatively unstable, ripples or waves of cirrocumulus clouds will appear as a "mackerel sky."

● FIGURE 11.19 Vertical view of clouds, precipitation, and winds across the warm front in Figure 11.18 along the line P–P′.

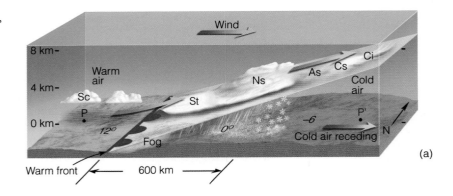

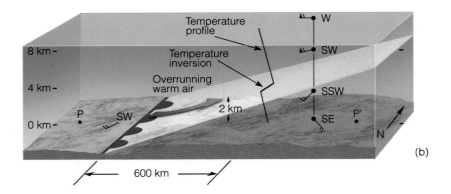

your own. ▼ Table 11.3 summarizes typical warm-front weather. (Before going on to the next section, you may wish to read Focus on a Special Topic: The Wavy Warm Front on p. 345, which gives additional information about warm fronts.)

OCCLUDED FRONTS   If a cold front catches up to and overtakes a warm front, the frontal boundary created between the two air masses is called an **occluded front** or, simply, an **occlusion** (meaning "closed off"). On the surface weather map, it can be represented as a purple line with alternating cold-front triangles and warm-front half-circles; both symbols point in the direction toward which the front is moving. Look back at Figure 11.12 (p. 336) and notice that the air behind the occluded front is colder than the air ahead of it. This is known as a *cold-type occluded front*, or **cold occlusion**. Let's see how this front develops.

The development of a cold occlusion is shown in ● Figure 11.20. Along line A–A′, the cold front is rapidly approaching the slower-moving warm front. Along line B–B′, the cold front overtakes the warm front and, as we can see in the vertical view across C–C′, undercuts and lifts off the ground both the warm front and the warm air mass. As a cold-occluded front approaches, the weather sequence is similar to that of a warm front, with high clouds lowering and thickening into middle and low clouds, with precipitation forming well in advance of the surface front. Since the front represents a trough of low pressure, southeasterly winds and falling atmospheric pressure occur ahead of it. The frontal passage, however, brings weather similar to that of a cold front: heavy, often showery precipitation with winds shifting to west or northwest. After a period of wet weather, the sky begins to clear, atmospheric pressure rises, and

▼ Table 11.3   Typical Weather Conditions Associated with a Warm Front in the Northern Hemisphere

WEATHER ELEMENT	BEFORE PASSING	WHILE PASSING	AFTER PASSING
Winds	south or southeast	variable	south or southwest
Temperature	cool to cold, slow warming	steady rise	warmer, then steady
Pressure	usually falling	levelling off	slight rise, followed by fall
Clouds	in this order: Ci, Cs, As, Ns, St, and fog; occasionally Cb in summer	stratus type	clearing with scattered Sc, especially in summer; occasionally Cb in summer
Precipitation	light-to-moderate rain, snow, ice pellets, or drizzle; showers in summer	drizzle or none	usually none; sometimes light rain or showers
Visibility	poor	poor, but improving	fair in haze
Dew point	steady rise	steady	rise, then steady

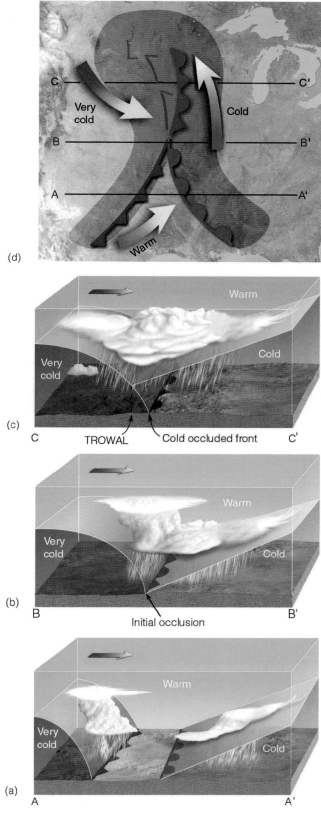

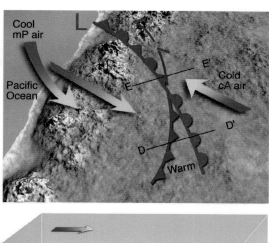

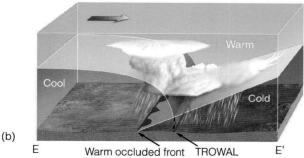

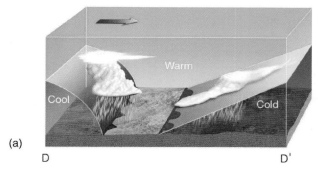

● FIGURE 11.20 The formation of a cold-occluded front. The faster moving cold front (a) catches up to the slower moving warm front (b) and forces it to rise off the ground (c). The green-shaded area in (d) represents precipitation.

● FIGURE 11.21 The formation of a warm-type occluded front. The faster moving cold front in (a) overtakes the slower moving warm front in (b). The lighter air behind the cold front rises up and over the denser air ahead of the warm front. Diagram (c) shows a surface map of the situation.

the air turns colder. The most violent weather usually occurs where the cold front is just overtaking the warm front, at the point of occlusion, where the greatest contrast in temperature occurs. Cold occlusions are the most prevalent type of front that moves onto the west coast and into the interior North America. Occluded fronts frequently form over the North Pacific and North Atlantic, as well as in the vicinity of the Great Lakes. In Canada, the surface occlusion is seldom analyzed on a weather map. Instead, the location of the elevated front, where the cold air lifts the warmer air above the surface, is indicated as a **TROWAL** (standing for TROough of Warm air ALoft). The TROWAL position is the surface projection of where the cold and warm fronts intersect, and a "hook" symbol is used to indicate it.

Continental arctic air over eastern British Columbia and the states of Washington and Oregon may be much colder than milder maritime polar air moving inland from the Pacific Ocean. ● Figure 11.21 illustrates this situation. Observe

FOCUS ON A SPECIAL TOPIC

Drylines

Drylines are not warm fronts or cold fronts but represent a narrow boundary where there is a steep horizontal change in moisture, so drylines separate moist air from dry air. Because dew point temperatures may drop along this boundary by as much as 9°C per kilometre, drylines have been referred to as dew point fronts.* Although drylines can occur in the United States as far north as the Dakotas and as far east as the Texas–Louisiana border, they are most frequently observed in the western half of Texas, Oklahoma, and Kansas, especially during spring and early summer. In these locations, drylines tend to move eastward during the day and then westward toward evening. Drylines are observed in other regions of the world, too. They occur, for example, in central West Africa and in India before the onset of the summer monsoon.

• Figure 6 shows a dryline moving across Texas during May 2001. Notice that the dryline is represented as a line with brown half-circles. Notice also that to the west of the dryline, warm, dry continental tropical air is moving in from the southwest. Consequently, on this side, the weather is usually hot and dry, with gusty southwesterly winds. To the east of the dryline, warm, very humid maritime tropical air is sweeping northward from the Gulf of

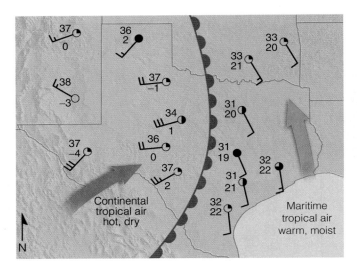

• FIGURE 6 A dryline represents a narrow boundary where there is a steep horizontal change in moisture as indicated by a rapid change in dew point temperature. Here a dryline (drawn in brown) moving across Texas and Oklahoma separates warm, moist air from warm, dry air during an afternoon in May.

Mexico. Here we typically find air temperatures to be slightly lower and the humidity (as indicated by the higher dew points) considerably higher than on the western side. The semicircles of the dryline point toward this humid air.

Even though the dryline represents a moisture boundary, its actual position on a weather map is plotted according to a shift in surface winds. When insects and insect-eating birds congregate along the dryline, Doppler radar may be able to locate it. On the radar screen, the echo from insects and birds shows up as a thin line, called a *fine line*.

Sometimes drylines are associated with middle-latitude cyclones; sometimes they are not. Cumulus clouds and thunderstorms

often form along or to the east of the dryline. This cloud development is caused in part by daytime convection and a sloping terrain. The Central Plains area of North America is higher to the west and lower to the east. Convection over the elevated western Plains carries dry air high above the surface. Westerly winds sweep this dry air eastward over the lower plains, where it overrides the slightly cooler but more humid air at the surface. This situation sets up a potentially unstable atmosphere that finds warm, dry air above warm, moist air. In regions where the air rises, cumulus clouds and organized bands of thunderstorms can form. We will examine in more detail the development of these storms in Chapter 14.

*Recall from Chapter 4 that the dew point temperature is a measure of the amount of water vapour in the air.

---

that the air ahead of the warm front is colder than the air behind the cold front. Consequently, when the cold front catches up to and overtakes the warm front, the milder, lighter air behind the cold front is unable to lift the colder, heavier air off the ground. As a result, the cold front rides "piggyback" along the sloping warm front. This produces a *warm-type occluded front*, or a **warm occlusion**. The surface weather associated with a warm occlusion is similar to that of a warm front.*

Contrast Figure 11.20 and Figure 11.21. Note that the primary difference between the warm- and cold-type occluded front is the location of the TROWAL or upper-level front. In a warm occlusion, the TROWAL precedes the surface occluded

*Due to the relatively mild winter air that moves into Europe from the North Atlantic, many of the occlusions that move into this region in winter are of the warm occlusion variety.

FOCUS ON A SPECIAL TOPIC

The Wavy Warm Front

Up to this point, we have examined idealized warm fronts on a surface weather map—like the one shown in Figure 11.12 (p. 336). Some warm fronts do look like this example; others, however, have an entirely different appearance. For instance, look at the warm front in ● Figure 7. Notice that it has a wavelike shape as it approaches North Carolina from three different directions. So what causes the warm front to bend in this manner?

Look carefully at Figure 7 and notice that at the surface, cold air is flowing southwestward around a high-pressure area centred over southern Canada. As cold, dense surface air pushes south into the southern states, it flows up against the Appalachian Mountains, which impede its westward progress. Since the shallow layer of cold air is unable to ride up and over the mountains, it becomes wedged along the mountains' eastern foothills. Recall from an earlier discussion that this trapping of cold air is called *cold air damming*.

Since the cold air is more dense, the warm air pushing northward from the Gulf of Mexico rides up and over the cold, dry surface air. Clouds and precipitation often form in this rising warm air. When rain falls into the shallow, cold air, it may evaporate, chilling the air even more. Sometimes the rain freezes before reaching the ground, producing ice pellets; other times, the rain freezes on impact, producing freezing rain. If the frozen precipitation falls for many hours, severe ice storms may result, with heavy accumulations of ice causing treacherous driving conditions and downed power lines.

The shallow layer of cold air usually becomes entrenched in low-lying areas and therefore retreats northward very slowly. As the cold air slowly recedes northward, warmer air pushes in from different directions, and the leading edge of the warm air (the warm front) no longer has a nice curved shape but begins to take on a move wavy shape, such as the warm front in Figure 7.

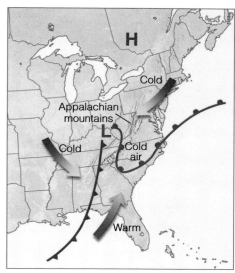

● FIGURE 7 Surface weather map for 11:00 p.m. (EST), February 13, 2007.

front, whereas in a cold occlusion, the TROWAL follows the surface occluded front.

In the world of weather fronts, occluded fronts are the mavericks. In our discussion, we treated occluded fronts as forming when a cold front overtakes a warm front. Some may form in this manner, but others apparently form as new fronts, which develop when a surface middle-latitude cyclonic storm intensifies in a region of cold air after its trailing cold and warm fronts have broken away and moved eastward. The new occluded front shows up on a surface chart as a trough of low pressure separating two cold air masses. Because of this, locating and defining occluded fronts at the surface are often difficult for the meteorologist. Similarly, you, too, may find it hard to recognize an occlusion. Satellite images can help identify the location of fronts, as well as whether or not a front is occluded. ● Figure 11.22 shows an occluded system with the fronts superimposed. A *dry slot* or notch in the cloud pattern often indicates the presence of an occluded front. The general weather conditions associated with occluded fronts are described in ▼ Table 11.4.

The frontal systems described in this chapter are actually part of a much larger storm system—the middle-latitude cyclone. Figure 11.22 shows the cold front, warm front, and occluded front in association with a mid-latitude cyclonic storm. Notice that, as we would expect, clouds and precipitation form in a rather narrow band along the cold front and in a much wider band with the warm front and TROWAL (occluded front). In Chapter 12, we will look more closely at middle-latitude cyclonic storms, examining where, why, and how they form. Before we move on, however, we need to look at fronts that form in the upper troposphere—which may, or may not, show up at the surface.

UPPER-AIR FRONTS   An **upper-air front** (which is also known as *upper front* or *upper-tropospheric front*) is a front that is present aloft. It may or may not extend down to the surface. ● Figure 11.23 shows a north-to-south side view of an idealized upper-air front. Notice that the front forms when the tropopause—the boundary separating the troposphere from the stratosphere—dips downward and folds under the polar jet stream. In the fold, the isotherms are tightly packed, marking the position of the upper front. Although the upper front may not connect with a surface front, the position of the surface front is shown in the diagram.

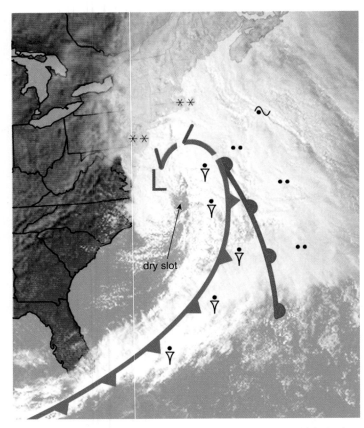

● FIGURE 11.22 A visible satellite image showing a mid-latitude cyclonic storm with its weather fronts over the Atlantic Ocean during March 2005. Superimposed on the photograph are the positions of the surface cold front, warm front, and TROWAL. Precipitation symbols indicate where precipitation is reaching the surface. See appendix XX for a key to precipitation symbols.

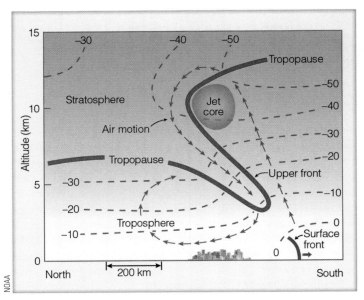

● FIGURE 11.23 An idealized vertical view of an upper-air front showing tropopause (heavy red line), isotherms in °C (dashed grey lines), and vertical air motions. The polar jet stream core (maximum winds) is flowing into the page (from west to east).

The small arrows in Figure 11.23 show air motion associated with the upper front. On the north side of the front (and the north side of the jet stream), the air is slowly sinking. Here, in the folded troposphere, ozone-rich air from the stratosphere descends into the troposphere. To the south of the front (and south of the jet stream), the air slowly rises. These rising and descending air motions can aid in the development of middle-latitude cyclonic storms described in the next chapter.

▼ Table 11.4 Typical Weather Most Often Associated with TROWALs/Occluded Fronts in North America

WEATHER ELEMENT	BEFORE PASSING	WHILE PASSING	AFTER PASSING
Winds	east, southeast, or south	variable	west or northwest
Temperature (a) Cold type occluded (b) Warm type occluded	cold or cool cold	dropping rising	colder milder
Pressure	usually falling	low point	usually rising
Clouds	in this order: Ci, Cs, As, Ns	Ns, sometimes Tcu and Cb	Ns, As, or scattered Cu
Precipitation	light, moderate, or heavy precipitation	light, moderate, or heavy continuous precipitation or showers	light-to-moderate precipitation followed by general clearing
Visibility	poor in precipitation	poor in precipitation	improving
Dew point	steady	usually slight drop, especially if cold occluded	slight drop, although may rise a bit if warm occluded

SUMMARY

In this chapter, we considered the different types of air masses and the various kinds of weather each brings to a particular region. Continental arctic air masses are responsible for the extremely cold (arctic) outbreaks of winter, whereas maritime arctic air masses bring cold, moist weather in winter to northwestern North America and cool weather in summer to the continent. U.S. meteorologists use the label *continental polar* to describe this air mass giving cool, pleasant weather in summer. Maritime polar air, having travelled over an ocean for a considerable distance, brings cool, moist weather to an area. The hot, dry summer weather affecting the U.S. Southwest and Mexico is associated with continental tropical air masses. The warm, humid conditions mainly affecting the eastern half of the continent are due to maritime tropical air masses. Where air masses with sharply contrasting properties meet, we find weather fronts.

A front is a boundary between two air masses of different densities. Stationary fronts have essentially no movement, with cold air on one side and warm air on the other. Winds tend to blow parallel to the front but in opposite directions on either side of it. Along the leading edge of a cold front, where colder air replaces warmer air, showers are prevalent, especially if the warmer air is moist and conditionally unstable. Along a warm front, warmer air rides up and over colder surface air, producing widespread cloudiness and light-to-moderate precipitation that can cover thousands of square kilometres. When the rising air is conditionally unstable (as it often is in summer), showers and thunderstorms may form ahead of the advancing warm front. Cold fronts typically move faster and are more steeply sloped than warm fronts. Occluded fronts, which are often difficult to locate and define on a surface weather map, may have characteristics of both cold and warm fronts. Fronts that form in the upper troposphere, in the vicinity of the polar-front jet stream, are called upper-air fronts.

KEY TERMS

The following terms are listed (with page numbers) in the order they appear in the text. Define each. Doing so will aid you in reviewing the material covered in this chapter.

air mass, 324
source regions (for air masses), 325
continental arctic (air mass), 326
continental polar (air mass), 326
lake-effect snows (snow squalls), 327
fetch, 328
cloud streets, 328
ocean-effect snows, 329
maritime arctic (air mass) 329
maritime polar (air mass), 330
maritime tropical (air mass), 332
continental tropical (air mass), 334
front, 335
stationary front, 336
cold front, 337
squall line, 338
frontolysis, 339
frontogenesis, 339
"back door" cold front, 339
cold air damming, 340
warm front, 340
overrunning, 340
frontal inversion, 341
dryline, 341
occluded front (occlusion), 342
cold occlusion, 342
TROWAL, 343
warm occlusion, 344
upper-air front, 345

QUESTIONS FOR REVIEW

1. If an area is described as a "good air mass source region," what information can you give about it?

2. It is summer. What type of afternoon weather would you expect from an air mass designated as mT? Explain.

3. Why is continental arctic air not welcome in the Prairies in winter, yet its summer counterpart, maritime arctic air, is very welcome in summer?

4. Explain why central North America is not a good air mass source region.

5. Why do air temperatures tend to be a little higher on the eastern side of the Appalachian Mountains than on the western side, even though the same winter cA air mass dominates both areas?

6. Explain how the airflow aloft regulates the movement of air masses.

7. List the temperature and moisture characteristics of each of the major air mass types.

8. What are lake-effect snows, and how do they form? On which side of a lake do they typically occur?

9. Why are maritime polar air masses along the East Coast of Canada usually colder than those along Canada's West Coast? Why are they also less prevalent?

10. The boundaries between neighbouring air masses tend to be more distinct during the winter than during the summer. Explain why.

11. What type of air mass would be responsible for the weather conditions listed as follows?
    (a) heavy snow showers and low temperatures at London, Ontario
    (b) hot, muggy summer weather in Montreal
    (c) daily afternoon thunderstorms along the Gulf Coast
    (d) heavy snow showers along the western slope of the Rockies
    (e) refreshing, cool, dry breezes after a long summer hot spell on the Prairies
    (f) heavy summer rain showers in southern Arizona
    (g) drought with high temperatures over the U.S. Southwest

(h) persistent cold, damp weather with drizzle along the East Coast

(i) summer afternoon thunderstorms forming across the central interior plateau of British Columbia

(j) record low winter temperatures in south Saskatchewan

12. On a surface weather map, what do you know about a region where the word frontogenesis is marked?

13. Explain why barometric pressure usually falls with the approach of a cold front or occluded front.

14. How does the weather usually change along a dryline?

15. Based on the following weather forecasts, what type of front will most likely pass the area?

(a) Light rain and cold today, with temperatures just above freezing. Southeasterly winds shifting to westerly tonight. Turning colder with rain becoming heavy and possibly changing to snow.

(b) Cool today with rain becoming heavy at times by this afternoon. Warmer tomorrow. Winds southeasterly becoming westerly by tomorrow morning.

(c) Increasing cloudiness and warm today, with the possibility of showers by evening. Turning much colder tonight. Winds southwesterly, becoming gusty and shifting to northwesterly by tonight.

(d) Increasing high cloudiness and cold this morning. Clouds increasing and lowering this afternoon, with a chance of snow or rain tonight. Precipitation ending tomorrow morning. Turning much warmer. Winds light easterly today, becoming southeasterly tonight and southwesterly tomorrow.

16. Sketch side views of a typical cold front, warm front, and cold-occluded front. Include in each diagram cloud types and patterns, areas of precipitation, surface winds, and relative temperature on each side of the front.

17. During the spring, on a warm, sunny day in Halifax, Nova Scotia, the wind shifts from southwesterly to northeasterly and the weather turns cold, damp, and overcast. What type of front moved through the Halifax area? From what direction did the front apparently approach Halifax?

18. How does the tropopause indicate where an upper-level front is located?

QUESTIONS FOR THOUGHT

1. Suppose an mP air mass moving eastward from the Pacific Ocean travels across Canada. Describe all of the modifications that could take place as this air mass moves eastward in winter and in summer.

2. Explain how an anticyclone during autumn can bring record-breaking low temperatures and continental arctic air to the southeastern United States and, only a day or so later, very high temperatures and maritime tropical air to the same region.

3. In Figure 11.6 (p. 332), there is a temperature inversion. How does this inversion differ from the frontal inversion illustrated in Figure 11.19b (p. 342)?

4. For Hamilton, Ontario, to experience heavy lake-effect snows, from what direction would the wind have to be blowing?

5. When a very cold air mass covers half of North America, a very warm air mass often covers the other half. Explain how this happens.

6. Explain why freezing rain more commonly occurs with warm fronts than with cold fronts.

7. In winter, cold-front weather is typically more violent than warm-front weather. Why? Explain why this is not necessarily true in summer.

8. When a cold front passes a station in the Northern Hemisphere, the wind shifts in a clockwise manner. How would the winds shift during the passage of a cold front in the Southern Hemisphere?

9. You are in southern Quebec and observe the wind shift from easterly to southerly. This shift in wind is accompanied by a sudden rise in both the air temperature and dew point temperature. What type of front passed?

10. If Lake Erie freezes over in January, is it still possible to have lake-effect snows off Lake Erie in February? Why or why not?

PROBLEMS AND EXERCISES

1. Make a sketch of North America and show the upper-air wind-flow pattern that would produce the following:

(a) very cold cA air moving into coastal British Columbia in winter

(b) cold cA air over the Prairies in winter

(c) warm mT air over Ontario in winter

(d) cold, moist mA air over Manitoba during the summer

2. You are presently taking a weather observation. The sky is full of wispy cirrus clouds estimated to be about 6 km overhead. If a warm front is approaching from the south, about how far away is it (assuming a slope of 1:200)? If it is moving toward you at an average warm-front speed of about 10 knots, how long will it take before it passes your area?

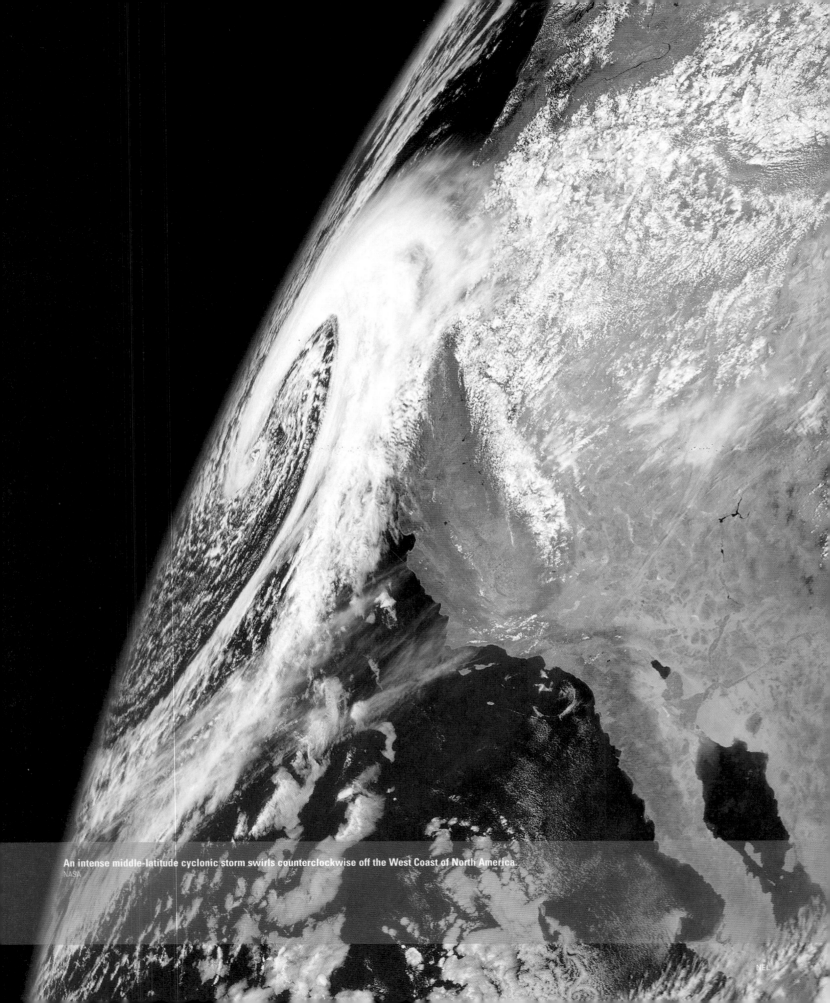

An intense middle-latitude cyclonic storm swirls counterclockwise off the West Coast of North America.

NASA

NEL

Middle-Latitude Cyclones

12

*"The lake, it is said, never gives up her dead,
when the skies of November turn gloomy."[1]*

The sinking of the *Edmund Fitzgerald* during a fateful Lake Superior storm on November 10, 1975, was memorialized in song by Canadian folk-music legend Gordon Lightfoot. At 220 m long, the "Mighty Fitz" was one of the largest ships ever to have sailed the Great Lakes and, like the *Titanic*, was thought to be unsinkable.

*"That good ship and true was a bone to be chewed
when the 'Gales of November' came early."[1]*

The storm tracked across the middle of Lake Superior from the southwest toward the northeast. Winds ahead of the storm were from the northeast, whereas behind the storm, winds shifted abruptly to come from the northwest. The north-westerly winds blowing along the length of Lake Superior created the large waves in eastern Lake Superior where the *Edmund Fitzgerald* eventually sank.

Ernest McSorley, captain of the *Edmund Fitzgerald*, radioed that his ship was taking on water and had developed a list, or tilt to one side. It was "one of the worst seas I've ever been in," McSorley said.

Although the precise reasons for the ship's demise remain unclear, there is no doubt that the sustained winds of 50 knots (83 km h^{-1}) with gusts to over 75 knots (139 km h^{-1}), creating waves as high as 4.9 m, contributed to the sudden sinking and loss of all 29 crew members.

*"And all that remains is the faces and the names
of the wives and the sons and the daughters."[1]*

CONTENTS

The *Edmund Fitzgerald* storm in our opening was a **middle-latitude cyclone**—a type of storm system that characterizes middle latitude weather, especially during winter. What precisely are these storm systems? What atmospheric conditions are needed for such a devastating storm to develop? This chapter seeks to answer these questions.

Early weather forecasters were aware that precipitation generally accompanied falling barometers and areas of low pressure. However, it was not until the early part of the 20th century that scientists began to piece together the information that yielded the ideas of modern meteorology and storm development.

Working largely from surface observations, a group of scientists in Bergen, Norway, developed a model explaining the life cycle of an *extratropical* or *middle-latitude cyclonic storm*—that is, a storm that forms at middle and high latitudes outside the tropics. This extraordinary group of meteorologists included Vilhelm Bjerknes, his son Jakob, Halvor Solberg, and Tor Bergeron. They published their *Norwegian cyclone model* shortly after World War I. It was widely acclaimed and became known as the "**polar-front theory** of a developing wave cyclone" or, simply, the polar-front theory. What these meteorologists gave to the world was a working model of how a mid-latitude cyclone progresses through the stages of birth, growth, and decay. An important part of the model involved the development of weather along the polar front. As new information became available, the original work was modified so that, today, it serves as a convenient way to describe the structure and weather associated with a migratory middle-latitude* cyclonic storm system. The storm experienced by the *Edmund Fitzgerald* is an example of such a storm system.

Middle-latitude cyclones characterize and largely determine the weather in the middle latitudes, especially in fall, winter, and spring. They bring clouds and precipitation, cause the winds to strengthen and change direction, and are associated with fronts that bring about rapid changes in temperature. They may result in extreme precipitation, causing flooding and storm-to-hurricane-force winds. These affect the *anthrosphere* by creating hazards to aviation, the general public, and marine operations—as was the case with the *Edmund Fitzgerald*. Mid-latitude cyclones are the main mechanism for the *atmosphere* to transfer heat and moisture from equatorial areas poleward through and to the middle latitudes. Mid-latitude cyclones therefore largely account for the patterns of precipitation and temperature in these regions, which directly affect the climate of an area and link the *hydrosphere* with the *biosphere*. Since mid-latitude cyclones are dominant in the cool seasons, they are the primary weather system providing the *cryosphere* with snow. This allows water to be stored and used for biological uses and evaporation during warmer times of the year. Melting snowfall and glaciers at higher elevations also provide rivers and lakes with water during spring and summer.

---

*The terms "middle-latitude cyclone" and "mid-latitude cyclone" will be used interchangeably.

In the following sections, we will first examine, from a surface perspective, how a middle-latitude cyclone develops along a front. Then we will examine how the winds aloft influence the developing surface storm. Later, we will obtain a three-dimensional view of a middle-latitude cyclone by observing how ribbons of air glide through the storm system.

Polar-Front Theory

The development of a middle-latitude cyclone, according to the Norwegian model, begins along the polar front. Remember from the discussion of the general circulation in Chapter 10 and fronts in Chapter 11 that the polar front is a semicontinuous global zone separating cold polar air from warm subtropical air. Although the polar front is normally the main frontal zone, middle-latitude cyclones can also form along colder fronts, such as the arctic front. Because middle-latitude cyclonic storms form and move along these fronts in a wavelike manner, the developing storms are called **wave cyclones**. The stages of a developing wave cyclone are illustrated in the sequence of surface weather maps shown in ● Figure 12.1.

Figure 12.1a shows a segment of the polar front as a stationary front. It represents a trough of lower pressure with higher pressure on both sides. Cold air to the north and warm air to the south flow parallel to the front, but in opposite directions. This type of flow sets up a cyclonic wind shear. You can conceptualize the shear more clearly if you place a pen vertically between the palms of your hands and move your left hand toward your body; the pen turns counterclockwise, cyclonically.

Under the right conditions (described later in this chapter), a wavelike kink forms on the front, as shown in Figure 12.1b. The wave that forms is known as a **frontal wave** or an incipient cyclone. Watching the formation of a frontal wave on a weather map is like watching a water wave from its side as it approaches a beach: it first builds, then breaks, and finally dissipates, which is why a middle-latitude cyclonic storm system is known as a wave cyclone. Figure 12.1b shows the newly formed wave with a cold front pushing southward and a warm front moving northward. The region of lowest pressure (called the central pressure) is at the junction of the two fronts. As the cold air displaces the warm air upward along the cold front, and as overrunning occurs ahead of the warm front, a narrow band of precipitation forms (green-shaded area). Steered by the winds aloft, the system typically moves east or northeastward and gradually becomes a fully developed open wave in 12 to 24 hours (see Figure 12.1c). The central pressure is now much lower, and several isobars encircle the wave's apex. These more tightly packed isobars create a stronger cyclonic flow as the winds swirl counterclockwise and inward toward the low's centre. Precipitation forms in a wide band *ahead* of the warm front and *along the narrow band* of the cold front. The region of warm air between the cold and warm fronts is known as the **warm sector**. Here the weather tends to be partly cloudy,

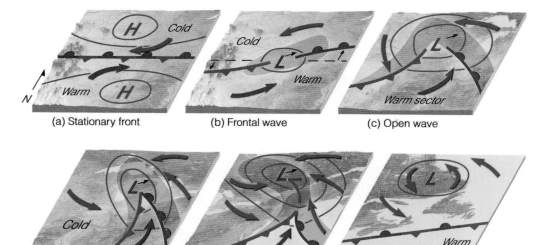

(a) Stationary front   (b) Frontal wave   (c) Open wave

(d) Mature (initial occlusion)   (e) Advanced occlusion   (f) Cutoff cyclone

● FIGURE 12.1 The idealized life cycle of a middle-latitude cyclone (a through f) in the Northern Hemisphere based on the polar-front theory. As the life cycle progresses, the system moves eastward or northeastward in a dynamic fashion. The small arrow next to each L shows the direction of storm movement.

WEATHER WATCH

The great "Blizzard of 1888," which immobilized most of the eastern United States and parts of Canada, taking the lives of over 400 people, was so imprinted on those who survived that many babies born after the storm were named "Storm," "Snowflake," and even "Blizzard."

although scattered showers may develop if the air is conditionally unstable.

Energy for the storm is derived from several sources. As the air masses try to attain equilibrium, warm air rises and cold air sinks, transforming potential energy into kinetic energy—energy of motion. Condensation supplies energy to the system in the form of latent heat. And as the surface air converges toward the low's centre, wind speeds may increase, producing an increase in kinetic energy.

As the open wave moves eastward, its central pressure continues to decrease, and the winds blow more vigorously as the wave quickly develops into a mature cyclone. The faster moving cold front constantly inches closer to the warm front, squeezing the warm sector into a smaller area, as shown in Figure 12.1d. In this model, the cold front eventually overtakes the warm front and the system becomes occluded. At this point, the storm is usually most intense, with clouds and precipitation covering a large area.

The point of occlusion where the cold front, warm front, and occluded front (analysed in Canada as a TROWAL) all come together in Figure 12.1e is referred to as the *triple point.* Notice that in this region, the cold and warm fronts appear similar to the open-wave cyclone in Figure 12.1c. It is here that a new wave (called a **secondary low**) will occasionally form, move eastward, and intensify into a cyclonic storm. The centre of the intense storm system shown in Figure 12.1e gradually dissipates because cold air now lies on both sides of the occluded front. The warm sector is still present but is far removed from

the centre of the storm. Without the supply of energy provided by the rising warm, moist air, the old storm system dies out and gradually disappears (see Figure 12.1f). We can think of the sequence of a developing wave cyclone as a whirling eddy in a stream of water that forms behind an obstacle, moves with the flow, and gradually vanishes downstream. The entire life cycle of a wave cyclone can last from a few days to over a week.

● Figure 12.2 shows a series of wave cyclones in various stages of development along the polar front in winter. Such a succession of storms is known as a *"family" of cyclones.* Observe that to the north of the front are cold anticyclones; to the south over the Atlantic Ocean is the warm, semipermanent Bermuda high. The polar front itself has developed into a series of loops, and at the apex of each loop is a cyclonic storm system. The cyclone over the northern plains (Low 1) is just forming; the one along the East Coast (Low 2) is an open wave; and the occluded system near Iceland (Low 3) is dying out. If the average rate of movement of a wave cyclone from birth to decay is 25 knots, then it is entirely possible for a storm to develop over the central part of North America, intensify into a large storm over the Maritimes, become occluded over the ocean, and reach the coast of England in its dissipating stage less than a week after it formed. ● Figure 12.3 is a visible satellite image of clouds and two middle-latitude cyclones in different stages of development along the polar front. Superimposed on the image are the weather fronts. Look again at Figure 12.1 and determine what stages of development the two cyclones are in.

Although most middle-latitude cyclones move east or northeastward as they develop, some weather systems appear to stall over an area. In these systems, the fronts are nearly stationary (called *quasistationary*) and middle-latitude cyclones develop and move along the fronts without moving them out of an area. This can have serious consequences. For example, if the system is producing freezing rain, then the freezing rain may accumulate in an area—such as the case of the ice storm of 1998 that devastated eastern Ontario and

● FIGURE 12.2 A series of wave cyclones (a "family" of cyclones) forming along the polar front.

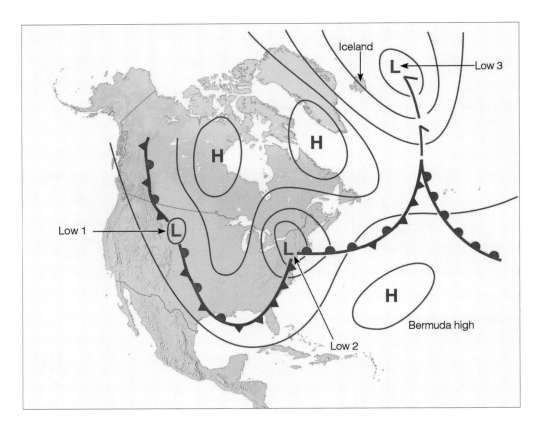

southern Quebec (refer back to Focus on a Special Topic: The Ice Storm of 1998 in Chapter 7 on p. 210). Another example of a quasistationary frontal system with serious consequences is the *pineapple express* system, which can affect the West Coast of North America in winter. For more information on these systems, see Focus on a Special Topic: Pineapple Express, on p. 375 later in this chapter.

It is important to differentiate low-pressure centres that develop in middle-latitude cyclones from thermal low-pressure features that were discussed in Chapter 9 on p. 270.

● FIGURE 12.3 Visible satellite image of the North Pacific with two middle-latitude cyclones in different stages of development during February 2000.

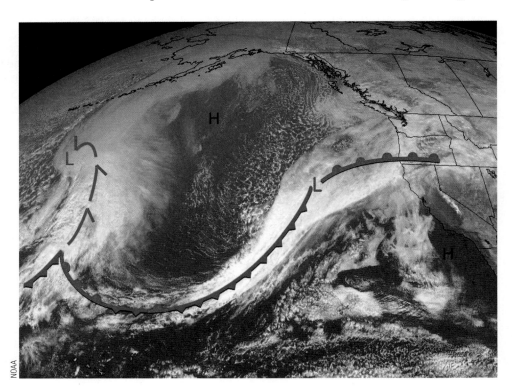

Thermal lows are shallow, warm-core, low-pressure features that form in response to strong surface heating caused by the sun. They usually form in summer over the continents adjacent to cold ocean waters. They do not have organized weather or fronts associated with them and tend to remain stationary, and any movement happens slowly. As we shall see, the low-pressure areas of a middle-latitude cyclone are deep, cold-core structures that are associated with fronts, go through a distinct life cycle (as indicated in Figure 12.1), and are largely responsible for the dramatic weather changes that affect the middle latitudes.

Up to now, we have considered the polar-front model of a developing wave cyclone, which represents a rather simplified version of the stages that an extratropical cyclonic storm system must go through. In fact, few (if any) storms exactly adhere to the model. Nevertheless, the polar-front model serves as a good foundation for understanding the structure of storms. So keep the model in mind as you read the following sections.

Where Do Middle-Latitude Cyclones Tend to Form?

Any development or strengthening of a middle-latitude cyclone is called **cyclogenesis**. There are regions of North America that show a propensity for cyclogenesis, including the Gulf of Mexico, the Atlantic Ocean east of the U.S. states of North and South Carolina, and the eastern slope of high mountain ranges, such as the Rockies and the Sierra Nevada. For example, when a westerly flow of air crosses a north-to-south trending mountain range, the air on the downwind (leeward) side tends to curve cyclonically, as shown in ● Figure 12.4. This curving of air adds to the developing or strengthening of a cyclonic storm. Such storms that form on the leeward side of a mountain are called **lee-side lows**, and their development is called lee cyclogenesis.

Another region of cyclogenesis lies near Cape Hatteras, North Carolina, where warm Gulf Stream water can supply moisture and warmth to the region south of a stationary front, thus increasing the contrast between air masses to a point where storms may suddenly spring up along the front. These cyclones, called **northeasters** or *nor'easters*, normally move northeastward along the Atlantic Coast, bringing high winds and heavy snow or rain to coastal areas. Before the age of modern satellite imagery and weather prediction, such coastal storms would often go undetected during their formative stages, and sometimes an evening weather forecast of "fair and colder" along the eastern seaboard would have to be changed to "heavy snowfall" by morning. Fortunately, with today's weather information-gathering and forecasting techniques, these storms rarely strike by surprise. (Additional information on northeasters is given in Focus on a Special Topic: East Coast Storms on p. 356.)

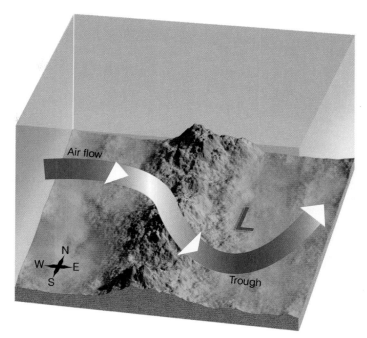

● FIGURE 12.4 As westerly winds blow over a mountain range, the airflow is deflected in such a way that a trough forms on the downwind (leeward) side of the mountain. Troughs and developing cyclonic storms that form in this manner are called *lee-side lows*.

● Figure 12.5 shows the typical paths taken in winter by middle-latitude cyclones and anticyclones. Notice in Figure 12.5a that some of the lows are named after the region where they form, such as the *Hatteras low*, which develops off the coast near Cape Hatteras, North Carolina. The *Alberta clipper* forms (or redevelops) on the eastern side of the Rockies in Alberta and then rapidly skirts eastward near the Canada–U.S. border. Likewise, the *Colorado low* forms (or redevelops) on the eastern side of the Rockies near the state of Colorado. Notice that the lows generally move eastward or northeastward, whereas the highs typically move southeastward and then eastward.

When middle-latitude cyclones deepen rapidly (in excess of 24 hPa in 24 hours), the term *explosive cyclogenesis*, or "*bomb*," is sometimes used to describe them. As an example, explosive cyclogenesis occurred in a storm that developed over the warm Atlantic just east of New Jersey on September 10, 1978. As the central pressure of the storm dropped nearly 60 hPa in 24 hours, hurricane-force winds battered the ocean liner *Queen Elizabeth II* and sank the fishing vessel *Captain Cosmo*.

Some frontal waves form suddenly, grow in size, and develop into huge cyclonic storms. They slowly dissipate, with the entire process taking several days to a week to complete. Other frontal waves remain small and never grow into a giant weather producer. Why is it that some frontal waves develop into huge cyclonic storms, whereas others simply dissipate in a day or so?

This question poses one of the real challenges in weather forecasting. The answer is complex. Indeed, there are many surface conditions that influence the formation of a

FOCUS ON A SPECIAL TOPIC

East Coast Storms

Dr. Chris Fogarty, Halifax, Nova Scotia

East Coast storms (often dubbed "nor'easters" because of the strong northeasterly winds they can bring to the coast) are large storms that usually form over the ocean off the East Coast of the United States and Canada during the fall, winter, and spring. These storms are fueled by the large temperature gradients between the warm ocean and the colder continental landmass. They gather additional energy from the moisture over the ocean and particularly the Gulf Stream—a warm ocean current that flows northward off the U.S. East Coast. These "marine cyclones" are much larger than the **tropical cyclones**, such as hurricanes, discussed in Chapter 15 and can also produce hurricane-force winds (defined as winds greater than 64 knots or 119 km h^{-1}) in their most intense phase. They are a type of middle-latitude cyclone.

Far-reaching warm and cold fronts accompany East Coast storms, which delineate warm and cold air masses (see Chapter 11). Warm air is pushed northward in association with the warm front, whereas cold air flows southward in the wake of the cold front. This migration of air is critical in Earth's energy balance, preventing the high latitudes from becoming excessively cold and the subtropics from becoming excessively warm.

The area of most intense weather associated with these storms is typically to the northwest of the storm centre track. That portion of the storm often spreads over the landmass and can wreak havoc, especially when the precipitation falls as snow. Paralyzing blizzards, damaging winds, flooding, and coastal erosion from surge and large ocean waves are all typical impacts associated with these storms.

A very intense storm in February 2004 crippled Nova Scotia with widespread record-breaking snowfall amounts of 60 to 90 cm and damaging northeasterly winds. A satellite image sequence of that event is shown in • Figure 1 covering a 36-hour period from the storm's formative stage to its mature stage. The storm moved northeastward, passing south of Nova Scotia, which was situated in the most severe (northwestern) part of the storm. The frontal evolution shown in the figure is also representative of many intense East Coast storms.

During the late summer and early fall, tropical cyclones can move northward from the tropics and undergo what is called **extratropical transition**. This represents the change in the storm structure from a non-frontal hurricane (or tropical cyclone) to a frontal middle-latitude cyclone, which can eventually take on a cloud pattern similar to panel (d) in Figure 1.

• FIGURE 1 GOES infrared imagery showing the evolution of a blizzard that struck Nova Scotia between February 18th and 19th in 2004. The centre of the low is labelled by an "L" with an "x," marking the central minimum sea level pressure in hectopascals (shown in yellow). The mid-latitude cyclone that caused this blizzard is shown by the warm and cold fronts in each image. Where the faster moving cold front meets the warm front, an occlusion occurs. The upper occluded front is marked with the hooked TROWAL symbol in panels (c) and (d).

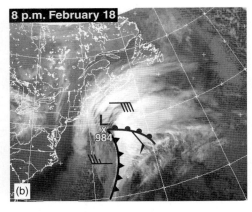

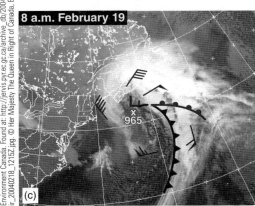

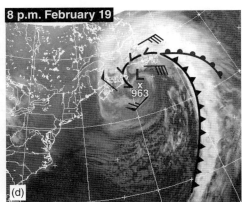

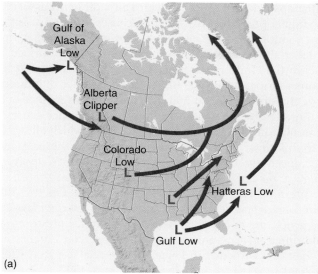

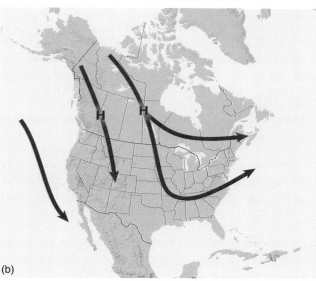

● FIGURE 12.5 (a) Typical paths of winter middle-latitude cyclones. The lows are named after the region where they form. (b) Typical paths of winter anticyclones.

WEATHER WATCH

A powerful middle-latitude cyclone battered the Great Lakes with hurricane-force winds and high seas on November 7 to 13, 1913. Known as the "Black Sunday" storm, it had winds over 76 knots (140 km h⁻¹) and was responsible for sinking 34 ships and killing 270 sailors in lakes Erie and Ontario.

middle-latitude cyclone, including mountain ranges and land–ocean temperature contrasts. However, the real key to the development of a wave cyclone is found in the *upper-wind flow*, in the region of the high-level westerlies. Therefore, before we can arrive at a reasonable answer to our question, we need to see how the winds aloft influence surface pressure systems.

Vertical Structure of Deep Dynamic Lows

In Chapter 9, we learned that thermal pressure systems are shallow systems that weaken with increasing height above the surface. (Look back at Figure 9.13 on p. 270.) On the other hand, developing surface middle-latitude cyclones are deep dynamic lows that usually intensify with height. Hence, they appear on an upper-level chart as either a closed low or a trough.

Suppose that the upper-level low is directly above the surface low as illustrated in ● Figure 12.6. Notice that only at the surface (because of friction) do the winds blow inward toward the low's centre. As these winds converge (flow together), the air "piles up." This piling up of air, called **convergence**, causes air density to increase directly above the surface low. This increase in mass causes surface pressures to rise; gradually, the low fills, and the surface low dissipates. The same reasoning can be applied to surface anticyclones. Winds blow outward, away from the centre of a surface high. If a closed high or ridge lies directly over the surface anticyclone, **divergence** (the spreading out of air) at the surface will remove air from the column directly above the high. The decrease in mass causes the surface pressure to fall and the surface high-pressure area to weaken. Consequently, it appears that if upper-level pressure systems were always located directly above those at the surface (such as shown in Figure 12.6), cyclones and anticyclones would die out soon after they form (if they could form at all). What is it, then, that allows these systems to develop and intensify?

● Figure 12.7 is an idealized model of the vertical structure of a middle-latitude cyclone and anticyclone in the Northern Hemisphere. Note that behind the cold front, there is cold air both at the surface and aloft. This cold, dense air is helping to maintain the surface anticyclone. But remember from Chapter 8 that aloft, in a region of cold air, constant pressure surfaces are squeezed closer together. This squeezing is due to the fact that in the cold, dense air, the atmospheric pressure decreases rapidly with height, causing cold air aloft to be associated with low pressure. Consequently, in the cold air aloft, we find the upper low, which is located behind or to the west of the surface low. Observe also how the surface low

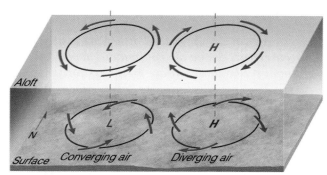

● FIGURE 12.6 If lows and highs aloft were always directly above lows and highs at the surface, the surface systems would quickly dissipate.

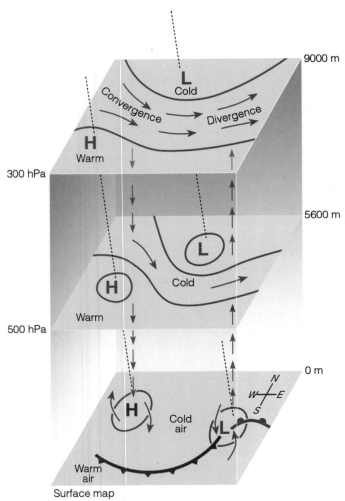

ACTIVE FIGURE 12.7 An idealized vertical structure of a middle-latitude cyclone and anticyclone. Visit the textbook's website to view this and other Active Figures at www.ahrensmeteorology1ce.nelson.com.

tilts toward the northwest as we move up from the surface, showing up as a closed system on the 500 hPa chart and as a trough on the 300 hPa chart. Directly above the surface low, at 300 hPa, the air spreads out and diverges (as indicated by the wind flow). This allows the converging surface air to rise and flow out of the top of the air column just below the tropopause, which acts as a constraint to vertical motions. We now have a mechanism for developing middle-latitude cyclonic storms. *When upper-level divergence is stronger than surface convergence (more air is taken out at the top than is brought in at the bottom), surface pressures drop, and the low intensifies (deepens). By the same token, when upper-level divergence is less than surface convergence (more air flows in at the bottom than is removed at the top), surface pressures rise, and the system weakens.*

We can also use Figure 12.7 to explain the structure of the anticyclone. Notice that at the surface and aloft, warm air lies to the southwest of the surface high. Again, in Chapter 8, we saw that warm air aloft causes the isobaric surfaces to spread

farther apart, which results in warm air aloft being associated with higher pressure. This situation causes the surface anticyclone to tilt toward the southwest—toward the warmer air—at higher altitudes. As we move upward from the surface, we observe that the closed area of high pressure at 500 hPa becomes a ridge at the 300 hPa level. Also notice that directly above the surface high at 300 hPa, there is convergence of air (as indicated by the wind flow lines). Convergence causes an accumulation of air above the surface high, which allows the air to sink slowly and replace the diverging surface air. Hence, *when upper-level convergence of air exceeds low-level divergence (inflow at the top is greater than outflow near the surface), surface pressures rise, and the anticyclone builds.* On the other hand, *when upper-level convergence of air is less than low-level divergence, the anticyclone weakens as surface pressures fall.* (Additional information on the subject of convergence and divergence is provided in Focus on a Special Topic: A Closer Look at Convergence and Divergence on p. 359.)

Look at the wind direction at the 500 hPa level in Figure 12.7. Winds at this altitude tend to steer surface systems in the same direction that the winds are moving. Thus, the surface middle-latitude cyclone will move toward the northeast, whereas the surface anticyclone will move toward the southeast. As we can see in Figure 12.5, these paths indicate the average movement of surface pressure systems in the eastern two-thirds of North America. In general, surface storms travel across North America at about 16 knots (30 km h^{-1}) in summer and about 27 knots (50 km h^{-1}) in winter. The faster winter velocity reflects the stronger upper-level flow* during this time of year.

So far, we have seen that deep pressure systems exist at the surface and aloft throughout much of the troposphere. When the upper-level trough lies to the west of the surface middle-latitude cyclone, the atmosphere is able to redistribute its mass. Regions of low-level converging air are compensated for by regions of upper-level diverging air and vice versa. Cyclones and anticyclones can intensify and, steered by the winds aloft, move away from their region of formation.

Since regions of strong upper-level divergence and convergence typically occur when deep troughs and ridges—waves—exist in the flow aloft, the next section examines these waves and their influence on a developing middle-latitude cyclonic storm.

Upper-Level Waves and Middle-Latitude Cyclones

You may remember from the "dishpan" experiment (found in Focus on an Observation: The "Dishpan" Experiment, in Chapter 10 on p. 302) that, aloft, waves are a fundamental feature of an unevenly heated, rotating sphere, such as Earth.

---

*As a forecasting rule of thumb, surface pressure systems tend to move in the same direction as the wind at the 500 hPa level. The speed at which the surface systems move is about half the speed of the 500 hPa winds.

FOCUS ON A SPECIAL TOPIC

A Closer Look at Convergence and Divergence

We know that convergence is the piling up of air above a region, whereas divergence is the spreading out of air above some region. Convergence and divergence of air may result from changes in wind direction or wind speed. For example, as in ● Figure 2, *speed convergence* occurs when the wind slows down as it moves along, whereas *speed divergence* occurs when the wind speeds up. We can grasp these relationships more clearly if we imagine air molecules to be marching in a band. When the marchers in front slow down, the rest of the band members squeeze together, causing convergence; when the marchers in front start to run, the band members spread apart, or diverge.

Figure 2 shows an upper-air (upper troposphere) chart with a trough, two ridges, and evenly spaced contour lines. Notice that even though the contours are evenly spaced, the winds blow faster in the ridge than they do in the trough.* As the faster flowing air moves away from the ridge and approaches the slower moving air in the trough, the air piles up, producing speed convergence. Where the slower moving air in the trough approaches the faster moving air in the ridge, the air spreads out, producing speed divergence in the airflow.

Convergent and divergent patterns can also be due to changes in wind direction. On an upper-level chart, this type of convergence is called *confluence* (or *directional convergence*). As shown in the upper-level chart in ● Figure 3, confluence occurs when a steady wind flows parallel to contour lines that are becoming closer together. This directional change creates convergence without changes in the wind speed. On the same chart, a type of divergence

*This phenomenon results from the gradient wind concept discussed in Chapter 8 on p. 245.

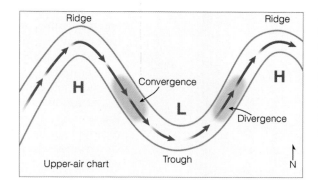

● FIGURE 2 As the faster flowing air in the ridge moves toward the slower flowing air in the trough, the air piles up and converges. This is known as *speed convergence*. As the slower moving air in the trough moves toward the faster flowing air in the ridge, the air spreads apart and diverges. This is known as *speed divergence*.

called *diffluence* (or *directional divergence*) occurs when steady wind flows parallel to contour lines that are becoming further apart. This directional change creates divergence without changes in the wind speed. Notice in Figure 3 that the area of convergence lies above the surface anticyclone (high, H), whereas the area of divergence lies above the surface mid-latitude cyclone (low, L).

However, the assumption of constant wind speed shown in Figure 3 would not normally occur in practice. In reality, as contour lines move closer together, more wind is funneled into a narrower area, and the wind speed also increases. When contour lines spread

farther apart, there is more space for the wind to flow, and the wind speed decreases.* These wind speed changes counteract the convergence or divergence created by the confluence and diffluence. This means that it is difficult to know whether a confluent pattern is causing convergence or a diffluent pattern is causing divergence. Consequently, it is not possible to determine convergence and divergence from the shape of the contour lines on a weather map; you also need to know the wind speed.

*This results from the geostrophic wind concept we discussed previously (in Chapter 8).

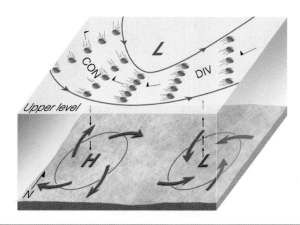

● FIGURE 3 The formation of convergence (CON) and divergence (DIV) of air with a constant wind speed in the upper troposphere. Circles represent air parcels that are moving parallel to the contour lines on a constant pressure chart. The wind-direction shaft lines with flag symbols show the wind speed and direction. Wind speed is indicated by the flag, which remains the same throughout the image, but the direction indicated by the shaft is changing. Below the area of convergence, the air is sinking, and we find the surface high (H). Below the area of divergence, the air is rising, and we find the surface low (L).

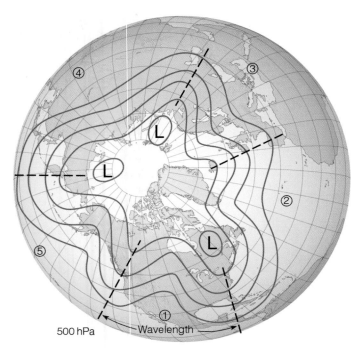

● FIGURE 12.8 A 500 hPa map of the Northern Hemisphere from a polar perspective shows five longwaves encircling the globe. Note that the wavelength of wave number 1 is as great as the width of North America. Solid lines are contours. Dashed lines show the position of longwave troughs.

If we examine an upper-level chart that shows almost the entire Northern (or Southern) Hemisphere, such as ● Figure 12.8, the waves appear as a series of troughs and ridges with significant amplitude that encircle the globe. The distance from trough to trough (or ridge to ridge) is known as the *wavelength*. When

the wavelength is on the order of many thousands of kilometres, the wave is called a **longwave**. At any one time, there are usually between three and six longwaves looping around Earth. The fewer the number of waves, the longer their wavelengths. Since mountain ranges tend to disturb the upper-level wind flow, these waves are often found with a ridge of higher heights over such topographic barriers as the Rockies and the Tibetan Plateau, and a trough of lower heights east of these mountain barriers. Sometimes longwaves exhibit a relatively small amplitude (or north-to-south extent), and the flow is mainly zonal, or west to east. On other occasions, the waves exhibit considerable amplitude and the flow has a strong north-to-south (meridional) component to it.

Longwaves are also known as *planetary waves* and as **Rossby waves**, after C. G. Rossby, a famous meteorologist who carefully studied their motion. Embedded in longwaves are **shortwaves**, which are small disturbances, or ripples, that move with the wind flow (see ● Figure 12.9a). Rossby found that the shorter the wavelength of a particular wave, the faster it moved downstream. Shortwaves tend to move eastward at a speed proportional to the average wind flow near the 700 hPa level, about 3 km above sea level. Longwaves, on the other hand, often remain stationary, move eastward very slowly at less than 4° of longitude per day (about 8 knots), or even move westward (*retrograde**). We can obtain a better idea of this wave movement if we think of longwaves as being huge meanders (loops) in a swiftly flowing stream of water. Water moves through the loops quickly, whereas the loops themselves move eastward very slowly as the fast-flowing water cuts away at one bank and deposits material on the other. Suppose that debris tumbles into the stream, disturbing the flow. The disturbed flow appears

---

*Retrograde wave motion means that the wave is actually moving in the opposite direction of the wind flow.

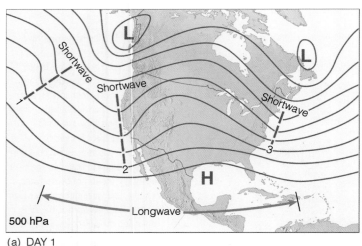

(a) DAY 1

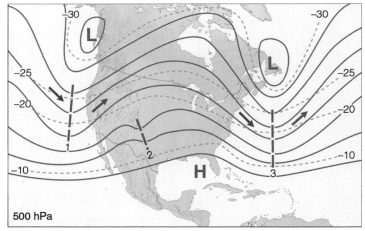

(b) DAY 2 (24 hours later)

ACTIVE FIGURE 12.9 (a) Upper-air chart showing a longwave with three shortwaves (heavy dashed lines) embedded in the flow. (b) Twenty-four hours later, the shortwaves have moved rapidly around the longwave. Notice that the shortwaves labelled 1 and 3 tend to deepen the longwave trough, whereas shortwave 2 has weakened as it moves into a ridge. Notice also that as the longwave deepens in diagram (b), its length actually shortens. Dashed lines are isotherms in °C. Solid lines are contours. Blue arrows indicate cold advection, and red arrows indicate warm advection. Visit the textbook's website to view this and other Active Figures at www.ahrensmeteorology1ce.nelson.com

as a small wrinkle that travels downstream through the loops at a speed near the average stream flow. This wrinkle in the flow is analogous to a shortwave in the atmosphere.

Notice in Figure 12.9b that whereas the longwaves move eastward very slowly, the shortwaves move fairly quickly around the longwaves. Notice also that the shortwaves tend to deepen (i.e., increase in size) when they approach a longwave trough and weaken (become smaller) when they approach a ridge. Moreover, when a shortwave moves into a longwave trough, the trough tends to deepen. (Look at shortwave 3 in Figure 12.9b.)

Where the contour lines in Figure 12.9b are roughly parallel to the isotherms (dashed lines), the atmosphere is said to be **barotropic**. Since the winds at this level more or less parallel the contour lines, in a barotropic atmosphere, the winds blow parallel to the isotherms. By comparison, where the isotherms cross the contour lines, temperature advection occurs, and the atmosphere is said to be **baroclinic**.* Notice in Figure 12.9b that the baroclinic regions tend to be in narrow zones in the vicinity of shortwaves 1 and 3. The shortwaves actually disturb the flow and accentuate the region of baroclinicity.

In the region of baroclinicity, winds cross the isotherms and produce temperature advection. **Cold advection** (or *cold air advection*) is the transport of cold air by the wind from a region of lower (colder) temperatures to a region of higher (warmer) temperatures. In the region of cold advection, since the wind is blowing colder air into the region, the air temperature normally decreases. On the other hand, **warm advection** (or *warm air advection*) is the transport of warm air by the wind from a region of higher (warmer) temperatures to a region of lower (colder) temperatures. In the region of warm advection, the air temperature normally increases because the wind is blowing warmer air into the region. For cold advection

---

*Actually, on a constant pressure surface, baroclinic conditions exist where the air density varies, and barotropic conditions exist where the air density does not vary.

to occur, the wind must blow across the isotherms from colder to warmer regions, whereas for warm advection, the wind must blow across the isotherms from warmer to colder regions.

In the baroclinic regions in Figure 12.9b, observe that strong winds cross the isotherms, producing cold advection (blue arrows) on the trough's west side and warm advection (red arrows) on their east side. Below the baroclinic zone lies the polar front; above it flows the polar-front jet stream. The disturbed flow created by the shortwaves is now capable of aiding in the development or intensification of a surface middle-latitude cyclonic storm. The theory explaining how this phenomenon occurs is known as the *baroclinic wave theory of developing cyclones.*

The Necessary Ingredients for a Developing Middle-Latitude Cyclone

To better understand how a wave cyclone may develop and intensify into a huge middle-latitude cyclonic storm, we need to examine atmospheric conditions at the surface and aloft. Suppose that a portion of a longwave trough at the 500 hPa level lies directly above a surface stationary front, as illustrated in ● Figure 12.10a. On the 500 hPa chart, contour lines (solid lines) and isotherms (dashed lines) parallel each other and are crowded close together. Colder air is located in the northern half of the map, whereas warmer air is located to the south. Winds are blowing at fairly high velocities, which produce a sharp change in wind speed—a strong *wind-speed shear*—from the surface up to this level. Suppose that a shortwave moves through this region, disturbing the flow, as shown in Figure 12.10b. This sets up a kind of instability in the flow (as warmer air rises and colder air sinks) known as **baroclinic instability**.

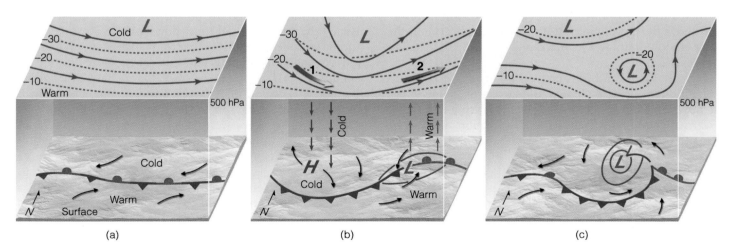

(a)   (b)   (c)

● FIGURE 12.10 An idealized view of the formation of a middle-latitude cyclone during baroclinic instability. (a) A longwave trough at 500 hPa lies parallel to and directly above the surface stationary front. (b) A shortwave by the trough in the 500 hPa height contours enters the area and disturbs the flow aloft, initiating temperature advection (blue arrow, cold advection; red arrow, warm advection). The upper trough intensifies and provides the necessary vertical motions (as shown by vertical arrows) for the development of the surface cyclone. (c) The surface storm occludes, and without upper-level divergence to compensate for surface converging air, the storm system dissipates.

UPPER-AIR SUPPORT With the onset of baroclinic instability, horizontal and vertical air motions begin to enhance the formation of a cyclonic storm. For example, as the flow aloft becomes disturbed, it begins to lend support for the intensification of surface pressure systems as a region of converging air forms above position 1 in Figure 12.10b and a region of diverging* air forms above position 2. The converging air aloft causes the surface air pressure to rise in the region marked *H* in Figure 12.10b. Surface winds begin to blow out away from the region of higher pressure, and the air aloft gradually sinks to replace it. Meanwhile, diverging air aloft causes the surface air pressure to decrease beneath position 2, in the region marked *L* on the surface map. This initiates rising air as the surface winds blow in toward the region of lower pressure. As the converging surface air develops cyclonic spin, cold air flows southward and warm air northward. We can see in Figure 12.10b that the western half of the stationary front is now a cold front and the eastern half is now a warm front. Cold air moves in behind the cold front, whereas warm air slides up along the warm front. These regions of cold and warm advection occur all the way up to the 500 hPa level.

On the 500 hPa chart in Figure 12.10b, cold advection is occurring at position 1 (blue arrow) as the wind crosses the isotherms, bringing cold air into the trough. The cold advection makes the air more dense and lowers the height of the air column from the surface up to the 500 hPa level. (Recall that on a 500 hPa chart, lower heights mean the same as lower pressures.) Consequently, the pressure in the trough lowers and the trough deepens. The deepening of the upper trough causes the contour lines to crowd closer together and the winds aloft to increase. Meanwhile, at position 2 (red arrow), warm advection is taking place, which has the effect of raising the height of a column of air; here the 500 hPa heights increase, and a ridge builds (strengthens). Therefore, *the overall effect of differential temperature advection is to amplify the upper-level wave.* As the trough aloft deepens, its curvature increases, which, in turn, increases the region of divergence above the developing surface storm. At this point, the surface middle-latitude cyclone rapidly develops as surface pressures fall.

Regions of cold and warm advection are associated with vertical motions. Where there is cold advection, some of the cold, heavy air sinks; where there is warm advection, some of the warm, light air rises. Hence, due to advection, air must be sinking in the vicinity of position 1 and rising in the vicinity of position 2.

The sinking of cold air and the rising of warm air provide energy for a developing cyclone as potential energy is transformed into kinetic energy. Further, if clouds form, condensation in the ascending air releases latent heat, which warms the air. The warmer air lowers the surface pressure, which strengthens the surface low even more. So we now have a full-fledged middle-latitude cyclone with all of the necessary ingredients for its development.

Eventually, the warm air curls around the north side of the low, and the storm system occludes* (see Figure 12.10c). Some storms may continue to deepen, but most do not as they move out from under the region of upper-level divergence. Additionally, at the surface, the storm weakens as the supply of warm air is cut off and cold, dry air behind the cold front (called a *dry slot*) is drawn in toward the surface low.

Sometimes an upper-level pool of cold air (which has broken away from the main flow) lies almost directly above the surface low. Occasionally, the upper low will break away entirely from the main flow, producing a **cutoff low**, which often appears as a single contour line on an upper-level chart. When the upper low lies directly above the surface low (as in Figure 12.10c), the storm system is said to be *vertically stacked*. Usually, the isotherms around the upper low parallel the contour lines, which indicate that no significant temperature advection is occurring. Without the necessary energy transformations, the surface system gradually dissipates. As its winds slacken and its central pressure gradually rises, the low is said to be *filling*. The upper-level low, however, may remain stationary for many days. If air is forced to ascend into this cold pocket, for example, through daytime heating of the surface by sunlight creating thermals, convective clouds with showers or even widespread clouds and precipitation may persist for some time, even though the surface storm system itself has moved east out of the picture.

THE ROLE OF THE JET STREAM As we have seen, for middle-latitude cyclones to develop and intensify, there must be upper-level diverging air above the surface storm. A jet stream, such as the polar jet, can provide such areas of divergence. In Chapter 10, we learned that the axis of the polar-front jet stream is linked to the position of the polar front. It is normally above the front at 500 hPa and poleward of the surface front. Although the polar front and the polar jet are usually the strongest and most dominant jet streams in the middle latitudes, there are often fronts such as the arctic front and associated jet streams and mid-latitude cyclones in the colder air north of the polar front that behave in a similar manner. The region of strongest winds in the jet stream is known as a *jet stream core*, or **jet streak**. When the polar jet stream flows in a wavy, west-to-east pattern, a jet streak tends to form, often in the trough of the jet, where pressure gradients are tight. The curving of the jet stream coupled with the changing wind speeds around the jet streak produces regions of strong convergence and divergence of air along the flanks of the jet. (More on this topic is given in Focus on a Special Topic: Jet Streaks and Storms on p. 364.)

Notice that the region of diverging air (marked *D* in ● Figure 12.11a) draws warm surface air upward to the jet, which quickly sweeps the air downstream. Since the air above

---

*Look back at Figure 12.7 (p. 358) and the upper-air chart in Figure 2 and Figure 3 (p. 359) and note the regions of converging air and diverging air on these maps.

---

*If the occluded front should extend west of the low's centre, it is sometimes referred to as a *bentback occlusion*.

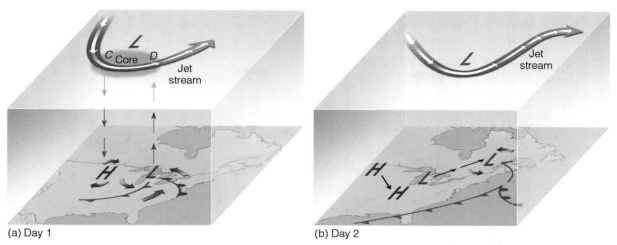

(a) Day 1   (b) Day 2

● FIGURE 12.11   (a) As the polar jet stream and its area of maximum winds (the jet streak, or core) swing over a developing middle-latitude cyclone, an area of divergence (D) draws warm surface air upward, and an area of convergence (C) allows cold air to sink. The jet stream removes air above the surface storm, which causes surface pressures to drop and the storm to intensify. (b) When the surface storm moves northeastward and occludes, it no longer has the upper-level support of diverging air, and the surface storm gradually dies out.

the middle-latitude cyclone is being removed more quickly than converging surface winds can supply air to the storm's centre, the central pressure of the storm drops rapidly. As surface pressure gradients increase, the wind speed increases. Above the high-pressure area, a region of converging air (marked *C* in Figure 12.11a) feeds air downward into the anticyclone. Hence, we find that the *jet streak removes air above the surface cyclone and supplies air to the surface anticyclone.*

As the jet stream steers the middle-latitude cyclonic storm along—toward the northeast in this case—the surface cyclone occludes, and cold air surrounds the surface low (see Figure 12.11b). Since the surface low has moved out from under the pocket of diverging air aloft, the occluded storm gradually fills as surface air flows into the system. In some cases, the jet streak itself can be associated with a shortwave trough rather than a longwave trough, which is illustrated in Figure 12.11. If it is linked to a shortwave trough that moves downwind in concert with the surface low, then the region of upper-level divergence may remain over the surface low for a longer period of time, resulting in prolonged deepening of the surface low.

Since jet streams are strongest and move farther south in winter, we can see why middle-latitude cyclonic storms are better developed and move more quickly during the colder months. During the summer, when the jet streams shift northward, developing middle-latitude storm activity shifts northward and, in Canada, occurs principally over the Yukon, Northwest Territories, and Nunavut.

In general, we now have a fairly good picture of why some surface lows intensify into huge mid-latitude cyclones, whereas others do not. For a surface cyclonic storm to intensify, there must be an upper-level counterpart—a trough of low pressure—that lies to the *west* of the surface low. As shortwaves disturb the flow aloft, they cause regions of differential temperature advection to appear, leading to an intensification of the upper-level trough. At the same time, the polar jet forms into waves and swings slightly south of the developing storm. When these conditions exist, zones of converging and diverging air, along with rising and sinking air, provide energy conversions for the storm's growth. With this atmospheric situation, storms may form even where there are no preexisting fronts.* In regions where the upper-level flow is not disturbed by shortwaves or where no upper trough or jet stream exists, the necessary vertical and horizontal motions are insufficient to enhance cyclonic storm development, and we say that the surface storm does not have the proper upper-air support. The horizontal and vertical motions, cloud patterns, and weather that typically occur with a developing open-wave cyclone are summarized in ● Figure 12.12.

CONVEYOR BELT MODEL OF MIDDLE-LATITUDE CYCLONES
A three-dimensional model of a developing middle-latitude cyclone is illustrated in ● Figure 12.13. The model describes rising and sinking air as travelling along three main "conveyor belts." Just as people ride escalators to higher levels in a department store, so air glides along through a constantly evolving middle-latitude cyclone. According to the **conveyor belt model**, a warm air stream (known as the *warm conveyor belt*—orange arrow in Figure 12.13) originates at the surface in the warm sector, ahead of the cold front. As the warm air stream moves northward, it slowly rises along the sloping warm front, up and over the cold air below. As the rising air cools, water vapour condenses, and clouds form well out

---

*It is interesting to note that the beginning stage of a wave cyclone almost always takes place when an area of upper-level divergence passes over a surface front. However, even if, initially, there are no fronts on the surface map, they may begin to form where air masses with contrasting properties are brought together in the region where the surface air rises and the surrounding air flows inward.

FOCUS ON A SPECIAL TOPIC

Jet Streaks and Storms

Figure 4 shows an area of maximum winds, a *jet streak*, on a 250 hPa chart. Jet streaks have winds of at least 50 knots (93 km h⁻¹) and represent small segments (ranging in length from a few hundred kilometres to over 3000 km) within the meandering jet stream flow.

Jet streaks are important in the development of surface middle-latitude cyclones because areas of convergence and divergence form at specific regions around them. To understand why, consider air moving through a straight jet streak (shaded area) in Figure 5. As the air enters the front of the streak (known as the *entrance region*), it increases in speed; as it leaves the rear of the streak (known as the *exit region*), it decreases in speed. At this elevation in the atmosphere (about 10 km above the surface), the wind flow is nearly in geostrophic balance with the pressure gradient force (directed north) and the Coriolis force (directed south). As the air enters the jet streak, it increases in speed because the contour lines are closer together, causing an increase in the pressure gradient force. The greater force temporarily exceeds the Coriolis

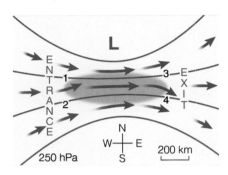

FIGURE 5 Changing air motions within a straight jet streak (shaded area) cause strong convergence of air at point 1 (left entrance region) and strong divergence at point 3 (left exit region).

force, and the air swings slightly to the north across the contour lines, which causes a piling up of air (called a *bottleneck effect*) and strong *convergence at point 1. Weak divergence* occurs at point 2.

Toward the middle of the jet streak, the increase in wind speed causes the Coriolis force to increase and the wind to become nearly geostrophic again. However, as the air exits the jet streak, the pressure gradient force

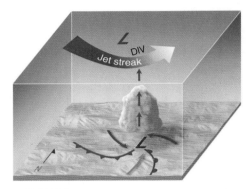

FIGURE 6 An area of strong divergence (DIV) can form with a curving jet streak. Below the area of divergence are rising air, clouds, and the developing middle-latitude cyclonic storm.

is reduced as the contour lines spread farther apart. Hence, the Coriolis force temporarily exceeds the pressure gradient force, causing the air to cross the contour lines and swing slightly to the south. This process produces *strong divergence* at point 3 and *weak convergence* at point 4.

The conditions described so far exist in a straight jet streak that shows no curvature. When the jet stream becomes wavy, and the jet streak exhibits cyclonic curvature (as it does in Figure 4), the areas of weak divergence (at point 2) and weak convergence (at point 4) all but disappear. What we are left with is a curving jet streak that exhibits strong divergence at point 3 (in the left exit region) and strong convergence at point 1 (in the left entrance region). The removal of air in the region of strong divergence causes surface pressures to fall, which results in the development of a surface low-pressure area. Notice in Figure 6 that below the area of strong divergence, the air rises, cools, and, if sufficiently moist, condenses into clouds. This cloudy area is associated with the low-pressure area in a mid-latitude cyclone.

FIGURE 4 A portion of a 250 hPa chart (about 10 km above sea level) that shows the core of the jet—the region of maximum winds (MAX)—called a jet streak.

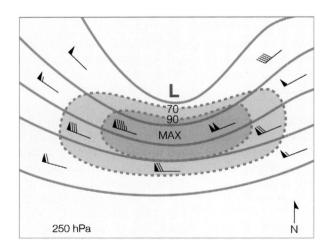

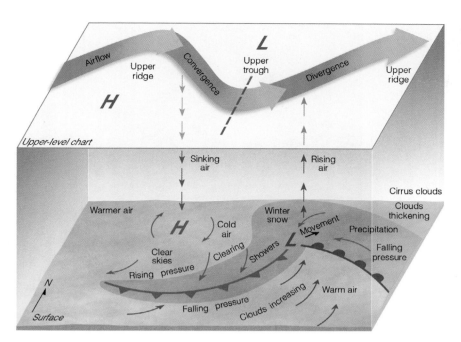

● FIGURE 12.12 Summary of clouds, weather, vertical motions, and upper-air support associated with a developing middle-latitude cyclone.

ahead of the surface low and its surface warm front. From these clouds, steady precipitation usually falls in the form of rain or snow. Aloft, the warm airflow gradually turns toward the northeast, parallel to the upper-level winds.

Directly below the warm conveyor belt, a cold, relatively dry airstream—the *cold conveyor belt*—moves slowly westward (see Figure 12.13). As the air moves west ahead of the

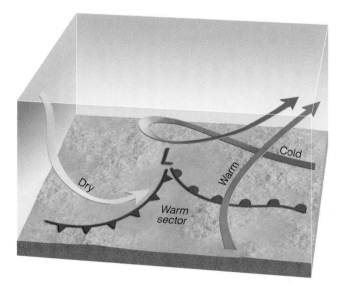

● FIGURE 12.13 The conveyor belt model of a developing middle-latitude cyclone. The warm conveyor belt (in orange) rises along the warm front, causing clouds and precipitation to cover a vast area. The cold conveyor belt (in blue) slowly rises as it carries cold, moist air westward ahead of the warm front but under the rising warm air. The cold conveyor belt lifts rapidly and wraps counterclockwise around the centre of the surface low. The dry conveyor belt (in yellow) brings very dry, cold air downward from the upper troposphere.

warm front, precipitation and surface moisture evaporate into the cold air, making it moist. As the cold, moist airstream moves into the vicinity of the surface low, rising air gradually forces the cold conveyor belt upward. As the cold, moist air sweeps northwest of the surface low, it often brings heavy winter snowfalls to this region of the cyclone. The rising airstream usually turns counterclockwise, around the surface low, first heading south and then northeastward, when it gets caught in the upper airflow. It is the counterclockwise turning of the cold conveyor belt that produces the comma-shaped cloud similar to the one shown in ● Figure 12.14.

The last conveyor belt is a dry one that forms in the cold, very dry region of the upper troposphere. Called the *dry conveyor belt* and shaded yellow in Figure 12.13, this airstream slowly descends from the northwest behind the surface cold front, where it brings generally clear, dry weather and, occasionally, blustery winds. If a branch of the dry air sweeps into the storm, it produces a clear area called a **dry slot**, which appears to pinch off the comma cloud's head from its tail. This phenomenon tends to show up on satellite images as the middle-latitude storm becomes more fully developed (see Figure 12.14).

We are now in a position to tie together many of the concepts we have learned about developing middle-latitude cyclones by examining a monstrous storm that formed during March 1993.

A DEVELOPING MIDDLE-LATITUDE CYCLONE—THE "STORM OF THE CENTURY" A colour-enhanced infrared satellite image of a developing middle-latitude cyclone on the morning of March 13, 1993, is shown in ● Figure 12.15. Notice that its cloud band is in the shape of a comma that covers the entire U.S. eastern seaboard. Such **comma clouds** indicate that the storm is still developing and intensifying. But this storm is not an ordinary wave cyclone—this storm

● FIGURE 12.14 Visible satellite image of a mature mid-latitude cyclone with the three conveyor belts superimposed on the storm. As in Figure 12.13, the warm conveyor belt is in orange, the cold conveyor belt is in blue, and the dry conveyor belt (forming the *dry slot*) is in yellow.

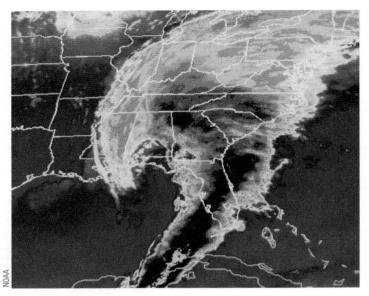

● FIGURE 12.15 A colour-enhanced infrared satellite image that shows a developing middle-latitude cyclone at 2 a.m. EST on March 13, 1993. The darkest shades represent clouds with the coldest and highest tops. Notice that the cloud pattern is in the shape of a comma.

intensified into a superstorm, which some forecasters dubbed "the storm of the century."

The surface weather map for the morning of March 13 (see ● Figure 12.16) shows that the centre of the open wave is over northern Florida. Observe in Figure 12.15 that this position is in the head of the comma cloud. The central pressure of the storm is 975 hPa, which indicates an incredibly deep system, considering that a typical open wave would have a central pressure closer to 996 hPa. A strong cold front stretches from the storm's centre through western Florida. Behind the front, cold arctic air pours into the U.S. Deep South. Ahead of the advancing front, a band of heavy thunderstorms (along a squall line) is pounding Florida with heavy rain, high winds, and tornadoes. Warm, humid air in the warm sector is streaming northward, overrunning cold surface air ahead of the warm front, which is causing precipitation in the form of rain, snow, and sleet to fall over a vast area extending from the U.S. states of Florida to New York.

The 500 hPa chart for the morning of March 13 (see ● Figure 12.17) shows that a deep trough extending southward out of Canada lies to the west of the surface low. A baroclinic atmosphere exists around the trough as isotherms intersect contour lines and strong winds produce differential temperature advection. Notice that warm advection (indicated by red barbs) is occurring on the trough's eastern side, ahead of the surface warm front shown in Figure 12.16. Cold advection (indicated by blue barbs) is occurring on the trough's western side, behind the position of the surface cold front. As

temperature advection deepens the trough, rising and sinking air provide energy for the developing surface storm.

In ● Figure 12.18, we can see that the storm began on March 12 as a frontal wave off the Texas coast. In the upper air, a shortwave, moving rapidly around a longwave, disturbed the flow, setting up the necessary ingredients for the surface storm's development. By the morning of March 13, the storm had intensified into a deep open-wave cyclone centred over Florida. In the upper air, a region of diverging air positioned above the storm caused the storm's surface pressures to drop rapidly. Upper-level southwesterly winds (see Figure 12.17) directed the surface low northeastward, where it became occluded over Virginia during the afternoon of March 13. At this point, the storm's central pressure dropped to an incredibly low 960 hPa, a pressure comparable to a category 3* hurricane. Although the surface winds were quite strong and gusty, they were not as strong as those in a category 3 hurricane because the isobars around the storm were spread farther apart than those in a hurricane and because

---

*As we will see in Chapter 15, a pressure of 960 hPa is equivalent to the pressure in a category 3 hurricane on the Saffir–Simpson scale, which ranges from 1 to 5, with 5 being the strongest.

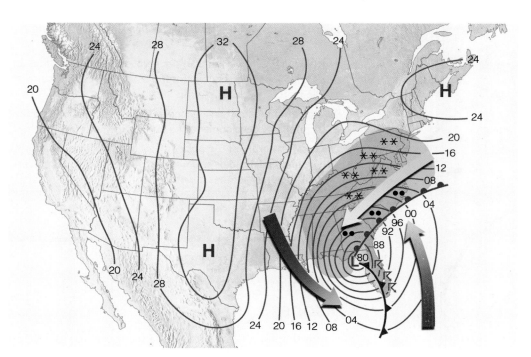

● FIGURE 12.16 Surface weather map for 4 a.m. EST on March 13, 1993. Lines on the map are isobars. A reading of 96 is 996 hPa, and a reading of 00 is 1000 hPa. (To obtain the proper pressure in hectopascals, place a 9 before those readings between 80 and 99 and place a 10 before those readings of 00 to 60.) Green-shaded areas are receiving precipitation. Heavy arrows represent surface winds. The orange arrow represents warm, humid maritime tropical air; the light blue arrow represents cold, moist maritime polar air; and the dark blue arrow represents cold continental arctic air.

surface friction slowed the winds. Higher up, away from the influence of the surface, the winds were much stronger as a wind of 125 knots (232 km h⁻¹) was reported at the top of 1900 m high Mount Washington, New Hampshire, and winds of 114 knots (211 km h⁻¹) were reported at Grand Etang on Cape Breton Island, Nova Scotia.

The upper trough remained to the west of the surface low, and the storm continued its northeastward movement. In Figure 12.18, we can see that by the morning of March 14, the storm (which was now a deep, bentback occluded system) had weakened slightly and was centred along the coast of Maine. Moving out from under its area of upper-level divergence, the

storm weakened even more as it continued its northeastward journey, across Newfoundland and Labrador and out over the North Atlantic. In all, the storm was one of the strongest ever. It blanketed deep snow from the state of Alabama to eastern Canada. Fierce winds piled the snow into huge drifts that closed roads, leaving motorists stranded. The storm shut down every major airport along the East Coast, and more than 3 million people lost electric power at some point during the storm. "The storm of the century" damaged or destroyed hundreds of homes, produced 11 tornadoes in Florida, stranded hikers in North Carolina and Tennessee, caused an estimated $3 billion in damage, and claimed the lives of at least 250 people.

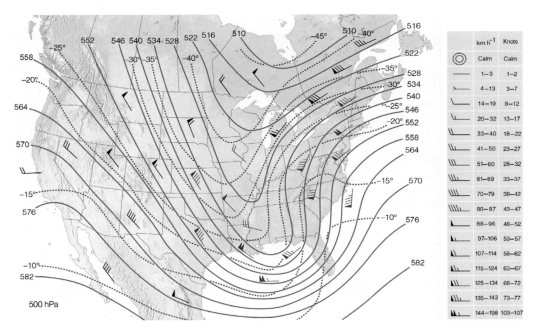

● FIGURE 12.17 The 500 hPa chart for 7 a.m. EST March 13, 1993. Solid lines are contours in decametres, where 564 equals 5640 m. Dashed lines are isotherms in °C. Wind entries in red show warm advection. Those in blue show cold advection. Those in black indicate that no appreciable temperature advection is occurring.

	km h⁻¹	Knots
◎	Calm	Calm
—	1–3	1–2
⌐	4–13	3–7
\	14–19	8–12
\	20–32	13–17
\\	33–40	18–22
\\\	41–50	23–27
\\\	51–60	28–32
\\\\	61–69	33–37
\\\\	70–79	38–42
\\\\\	80–87	43–47
⌐	88–96	48–52
⌐	97–106	53–57
⌐\	107–114	58–62
⌐\	115–124	63–67
⌐\\	125–134	68–72
⌐\\\	135–143	73–77
⌐\\	144–198	103–107

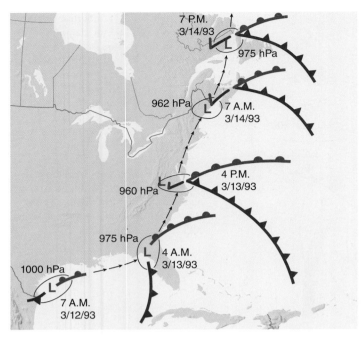

● FIGURE 12.18 The development of the ferocious middle-latitude cyclonic storm of March 1993. A small wave in the western Gulf of Mexico intensifies into a deep open-wave cyclone over Florida. It moves northeastward and becomes occluded over the U.S. state of Virginia, where its central pressure drops to 960 hPa. As the occluded storm continues its northeastward movement, it gradually fills and dissipates. The number next to the storm is its central pressure in hectopascals. Arrows show the direction of movement. Time is Eastern Standard Time.

BRIEF REVIEW

Up to this point, we have looked at the structure and development of middle-latitude cyclones. Before going on, here is a summary of a few of the important ideas presented so far:

- The polar-front (or Norwegian) model of a developing middle-latitude cyclonic storm represents a simplified but useful model of how an ideal storm progresses through the stages of birth, maturity, and dissipation.

- For a surface middle-latitude cyclone to develop or intensify (deepen), the upper-level low must be located to the west of (behind) the surface low.

- For a surface middle-latitude cyclonic storm to form, there must be an area of upper-level diverging air above the surface low. For the surface storm to intensify, the region of upper-level diverging air must be greater than surface converging air (i.e., more air must be removed above the storm than is brought in at the surface).

- When the upper airflow develops into waves, winds often cross the isotherms, producing regions of cold advection and warm advection, which tend to amplify the wave. At the same time, vertical air motions begin to enhance the formation of the surface storm as the rising of warm air and the sinking of cold air provide the proper energy conversion for the storm's growth.

- When a jet stream develops into a looping wave, it provides an area of upper-level diverging air for the development of surface middle-latitude cyclonic storms.

- An area of upper-level divergence in the left-exit region of a jet streak supports development of a surface low and clouds.

- The curving nature of the polar-front jet stream tends to direct surface middle-latitude cyclonic storms northeastward and surface anticyclones southeastward.

Vorticity, Divergence, and Developing Middle-Latitude Cyclones

We know that for a surface middle-latitude cyclone to develop into a deep low-pressure area, there must be an area of strong upper-level divergence situated above the developing storm. Meteorologists, then, are interested in locating regions of divergence on upper-level charts so that developing storms can be accurately predicted. We know from an earlier discussion that divergence and convergence of air are due to changes in either wind speed or wind direction. The problem is that it is a difficult task to measure divergence (or convergence) with any degree of accuracy using upper-level wind information. Therefore, meteorologists must look for something else that can be measured and, at the same time, can be related to regions of diverging (and converging) air. That something is called *vorticity*.

When something spins, it has vorticity. The faster it spins, the greater its vorticity. In meteorology, **vorticity** is a measure of the spin of small air parcels. Although the spin can be in any direction, our concern will be with the spin of horizontally flowing air about a vertical axis, much like an ice skater spins about an imaginary vertical axis. Because our goal is to see how vorticity can be used to identify regions of divergence and convergence, we must give vorticity some quantitative value. When viewed from above, air that spins cyclonically (counterclockwise) has *positive vorticity*, and air that spins anticyclonically (clockwise) has *negative vorticity*.

We can see in ● Figure 12.19 how divergence aloft and the vorticity of surface air are related. Suppose that the air column in Figure 12.19 represents an area of low pressure with weak cyclonic (positive) spin. Further suppose that an invisible barrier separates the column (and the ice skater inside) from the surrounding air. Now suppose that an area of divergence aloft (at the 250 hPa level) moves directly over the air column. Divergence aloft means that within this region, more air is leaving than is entering. This removal of air above the column lowers the atmospheric pressure at the surface. As the surface pressure lowers, the air surrounding the column converges

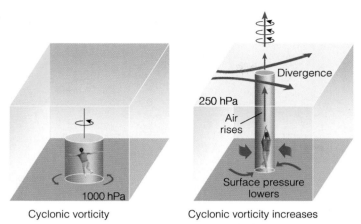

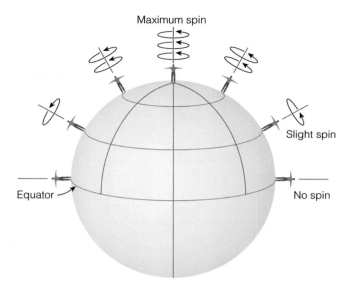

● FIGURE 12.19 When upper-level divergence moves over an area of weak cyclonic circulation, the cyclonic circulation increases (i.e., it becomes more positive), and air is forced upward.

● FIGURE 12.20 Due to the rotation of Earth, the rate of spin of observers about their vertical axes increases from zero at the equator to a maximum at the poles.

on it, pushing inward on its sides. If we assume that the total mass of air in the column does not change, then squeezing the column causes it to shrink horizontally and stretch vertically, forcing the air in the column upward. Meanwhile, as the column stretches, the ice skater's arms are pulled in close, and the skater spins much faster. Hence, the rate at which air flows around the centre of the column must also increase. Consequently, the vorticity of the column increases, becoming more positive. So we can see that divergence aloft causes an increase in the cyclonic (positive) vorticity of surface cyclones, which usually results in cyclogenesis and upward air motions.

Before we consider how vorticity on an upper-level chart ties in with developing middle-latitude cyclones, we need to examine two important types of this phenomenon: Earth's vorticity and relative vorticity.

VORTICITY ON A SPINNING PLANET   Because Earth spins, it has vorticity. In the Northern Hemisphere, **Earth's vorticity** (also called *planetary vorticity*) is always positive because Earth spins counterclockwise about its vertical North Pole axis. The amount of Earth vorticity imparted to any object—even to those that are not moving relative to Earth's surface—depends on the latitude. In ● Figure 12.20, an observer standing on the equator would not spin about his or her own *vertical axis*; farther north, the observer would spin very slowly, whereas at the North Pole, the observer would spin at a maximum rate of one revolution per day. It is now apparent that any object on Earth has vorticity simply because Earth is spinning, and the amount of this Earth vorticity* increases from zero at the equator to a maximum at the poles.

Moving air will generally have additional vorticity relative to Earth's surface. This type of vorticity, called **relative**

vorticity, is the sum of two effects: the curving of the airflow (*curvature*) and the changing of the wind speed over a horizontal distance (*shear*). ● Figure 12.21 illustrates vorticity due to curvature. Air moving through a trough tends to spin cyclonically (counterclockwise), increasing its relative vorticity. In the ridge, the spin tends to be anticyclonic (clockwise), and the relative vorticity of the air increases, but in a negative direction. Whenever the wind blows faster on one side of an air parcel than on the other, a shear force is imparted on the parcel, and it will spin and gain (or lose) relative vorticity. ● Figure 12.22 illustrates relative vorticity due to horizontal wind shear.

The sum of Earth's *vorticity* and the *relative vorticity* is called the **absolute vorticity**. To further illustrate this concept, suppose that you are watching an ice skater spin counterclockwise on a frozen lake. The ice skater possesses positive relative vorticity. If you could suddenly leave Earth and watch the

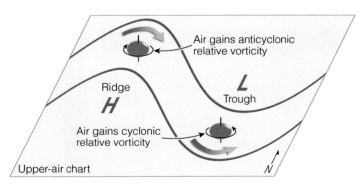

● FIGURE 12.21 In a region where the contour lines curve, air moving through a ridge spins clockwise and gains anticyclonic relative vorticity. In the trough, the air spins counterclockwise and gains cyclonic relative vorticity.

---

*Earth's vorticity at any latitude is equal to the product of twice Earth's angular rate of spin (2) and the sine of the latitude (Φ)—that is, 2 sin Φ. This expression is referred to as the *Coriolis parameter*, and it is usually expressed by the letter *f*. Remember from Chapter 8 that the Coriolis parameter times the wind speed is the Coriolis force that is important for wind.

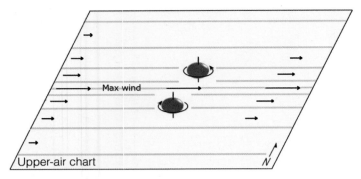

• FIGURE 12.22 Areas of cyclonic (positive) relative vorticity and anticyclonic (negative) relative vorticity can form in a region of strong horizontal wind-speed shear that can occur near a jet stream. Notice that parcels of air on either side of the jet stream have opposite directions of spin.

same spinning skater from space, you would see the ice skater spinning on a rotating platform—Earth. The combination of the skater spinning about her vertical axis (her relative vorticity) plus the small spin imparted to her from Earth (Earth's vorticity) yields the absolute vorticity of the skater. We are now in a position to see how vorticity aloft ties in with divergence and the development of middle-latitude cyclones. (The concept of absolute vorticity can explain why the westerly flow aloft tends to form into waves. This topic is presented in Focus on a Special Topic: Vorticity and Longwaves on p. 371.)

• Figure 12.23 shows an air parcel—a blob of air in this case—moving through a ridge and a trough in the upper troposphere. (For simplicity, we will assume that the parcel has a constant speed and that there is no shear acting on it.) Notice that at every position, the parcel has cyclonic spin. This situation occurs because the parcel's total, or absolute vorticity, is a combination of Earth's vorticity plus its relative vorticity. In middle and high latitudes, Earth's vorticity is great enough to make the parcel's spin cyclonic everywhere on Earth.

At position 1 (in the ridge), the air parcel has only a slight cyclonic spin because the relative vorticity, due to curvature, is anticyclonic and subtracts from Earth's vorticity. At position 2, the relative vorticity due to curvature is zero, which allows Earth's vorticity to spin the parcel faster. At position 3,

WEATHER WATCH

In 1996 between July 19 and 20, a large mid-latitude cyclone, which originated over southern Manitoba, moved across northern Michigan and dramatically evolved into a deep low-pressure system. With a central pressure of 980 hPa, it stalled over the Gaspé region of eastern Quebec, producing large amounts of precipitation. In 50 hours, over 155 mm of rain fell in the Saguenay—Lac-Saint-Jean region, setting the stage for the *Saguenay flood*. Rain from the storm fell on already saturated soil, causing flash flooding of the Saguenay and Ha! Ha! Rivers. This resulted in seven deaths, the evacuation of over 16,000 people, and the destruction of 488 homes. The disaster cost $1.5 billion in damage—Canada's first billion dollar natural disaster.

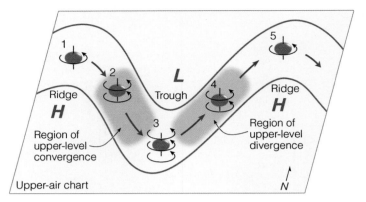

• FIGURE 12.23 The vorticity of an air parcel changes as we follow it through a wave. From position 1 to position 3, the parcel's absolute vorticity increases with time. In this region (shaded blue), we normally experience an area of upper-level converging air. As the air parcel moves from position 3 to position 5, its absolute vorticity decreases with time. In this region (shaded green), we normally experience an area of upper-level diverging air.

the parcel spins even faster as the curvature is cyclonic, which adds to Earth's vorticity. At position 4, the parcel spins more slowly as the curvature is once again zero.

We can see in Figure 12.23 that as the parcel moves from position 1 in the ridge to position 3 in the trough, its absolute vorticity *increases* as it moves along. Within this region is typically found an area of upper-level convergence. To help visualize this, consider that a spinning figure skater increases her rate of spin by pulling in or converging her arms. As the parcel moves from position 3 in the trough to position 5 in the ridge, its absolute vorticity *decreases* as it moves along. Within this region is typically found an area of upper-level divergence. In the figure skater analogy, the rate of spin is decreased when the skater's arms are stretched out, which is analogous to divergence. We can now summarize the information in Figure 12.23 by stating that as a parcel of air moves with the upper-level flow, an *increase in its absolute vorticity over time is related to upper-level converging air, and a decrease in its absolute vorticity with respect to time is related to upper-level diverging air.*[*]

On upper-air charts that show absolute vorticity (such as often on 500 hPa charts), we find that even though the clockwise flow around a high-pressure area produces negative relative vorticity, because of Earth's (positive) vorticity, the absolute vorticity is normally positive everywhere on the map. However, there are regions on the map (about the size of the province of New Brunswick) where the absolute vorticity is considerably greater. An area of high absolute vorticity is referred to as a *vorticity maximum* or *vort max*. A region of low absolute vorticity is known as a *vorticity minimum*.

• Figure 12.24 illustrates how vorticity, divergence, vertical air motions, and surface middle-latitude cyclonic storm development are linked together. The 500 hPa chart (middle chart)

---

[*]Viewed another way, in the upper troposphere, an area of converging air increases the total spin—the absolute vorticity—of air parcels, whereas an area of diverging air decreases the absolute vorticity of air parcels.

FOCUS ON A SPECIAL TOPIC

Vorticity and Longwaves

We know that longwaves develop aloft in the atmosphere and that longwaves are important because they transfer heat and momentum from one latitude to another. But how do these waves form in the first place? The concept of vorticity can help explain longwave formation. For example, let's assume that in the atmosphere there is no divergence or convergence of air, so air columns cannot stretch or contract. Where these conditions prevail, the absolute vorticity of air will be conserved, meaning that the numerical value of the sum of Earth's vorticity and the relative vorticity will not change with time. Thus,

Absolute vorticity = Earth's vorticity + relative vorticity = constant

If ζ_a* is the absolute vorticity, ζ_r the relative vorticity, and f Earth's vorticity, then the expression becomes

$$\zeta_a = \zeta_r + f = \text{constant}$$

Hence, any decrease in Earth's vorticity must be compensated for by an increase in the relative vorticity and vice versa.

Consider, for example, that air in ● Figure 7 is moving horizontally at a constant speed at an altitude near 5.5 km[†] above sea level. At

*The symbol ζ is the Greek letter zeta. Meteorologists often use ζ to represent vorticity.
[†]You may recall that the atmospheric pressure at this level is about 500 hPa.

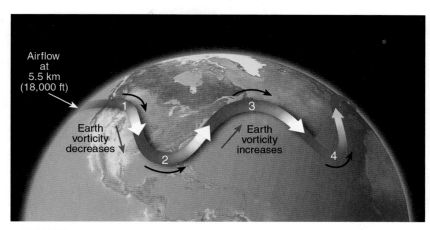

● FIGURE 7 The wavy path of air aloft due to the conservation of absolute vorticity.

this level, divergence is usually near zero, so our initial assumption approaches a real situation. Since there is no wind-speed shear, any change in the relative vorticity will be due to curvature.

Suppose that the airflow is disturbed by a mountain range such that the air at position 1 is flowing southeastward. Heading equatorward, the air moves into a region of decreasing Earth vorticity. To keep the absolute vorticity of the air constant, there must be a corresponding increase in the relative vorticity. Since increasing relative vorticity implies cyclonic curvature, the air turns

counterclockwise at position 2 and heads northeastward. But now the air is moving into a region where Earth's vorticity steadily increases. To offset this increase, the relative vorticity must decrease. The relative vorticity will decrease if the curvature becomes anticyclonic, so at position 3, air turns clockwise and heads toward the equator once again. This again brings the air into a region where Earth's vorticity decreases. To compensate, the air must now turn cyclonically at position 4, and so on. In this manner, a series of upper-level longwaves may develop, encircling the entire Earth.

---

in Figure 12.24 shows a vorticity maximum. If you move from the maximum eastward toward point 1, notice that, as you move along, vorticity decreases with time. Associated with this region, we find upper-level divergence (at 250 hPa) and low-level convergence (at the surface), which implies rising air and the possibility of clouds, precipitation, and cyclonic storm development. It follows that when a vorticity maximum aloft moves toward a stationary front, a wave will form along the front in the region ahead of the vorticity maximum, and a middle-latitude cyclonic storm will have a good chance of developing. Even in the absence of fronts, a zone of organized clouds with precipitation may form in the region ahead of a vorticity maximum. On the other hand, if we move toward the vorticity maximum from position 2, we find

that vorticity increases with time. Here, to the west of the vorticity maximum in Figure 12.24, exist upper-level convergence (at 250 hPa), low-level divergence (at the surface), slowly sinking air, and the generally fair weather we associate with surface high-pressure areas.

VORTICITY ADVECTION To understand the movement and development of mid-latitude cyclones, we need to know how shortwave troughs and ridges move through the middle troposphere and where there are upward and downward vertical motions. These two components of atmospheric motion can be assessed through the concept of **vorticity advection**. (*Recall that vorticity* is the *rate of horizontal spin*, whereas *advection* is the *horizontal transport of a quantity by the wind*.)

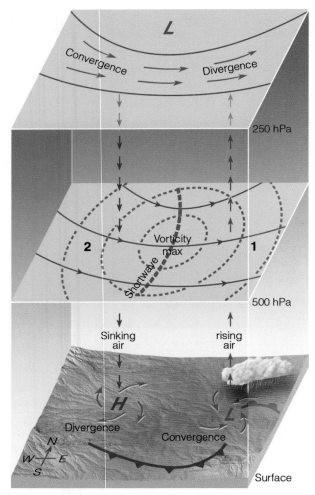

● FIGURE 12.24 A region of high absolute vorticity—a vorticity maximum—on its downwind (eastern) side has diverging air aloft, converging surface air, and ascending air motions. On its upwind (western) side, there is converging air aloft, diverging surface air, and descending air motions.

To assess vorticity advection, we need to examine the winds in conjunction with the vorticity pattern in the middle troposphere. In Figure 12.24, the vorticity pattern is shown by the dashed lines on the middle, 500 hPa map. The pattern varies from a lower vorticity value at *position 2*, to a maximum at the *vorticity maximum*, and back again to a lower value at *position 1*. As shown on the figure, a shortwave trough aligns with the vorticity maximum. The winds blow parallel to the height contours, which are represented by the solid lines, from position 2 toward position 1. Integrating these two components, at position 1, the wind is blowing from high vorticity values toward low vorticity values, so it is moving higher vorticity air into position 1, and, over time, the vorticity will increase there. This is called *positive vorticity advection*. Conversely, at position 2, the wind is blowing from low vorticity toward high vorticity, so it is moving lower vorticity air into position 2, and, over time, the vorticity will decrease there. This is called *negative vorticity advection*. The

result of the vorticity increase at position 1 and decrease at position 2 will be eastward movement of the vorticity maximum and its associated shortwave trough. (Generally, the vorticity maximum moves at half the speed of the wind.)

To assess vertical motion in the atmosphere through vorticity advection, we must consider what happens as positive vorticity advection increases the vorticity at position 1. As the vorticity maximum and shortwave trough approach position 1, the 500 hPa heights will decrease because the 500 hPa trough moves with the shortwave trough. If the surface pressure at the surface low is not changing much, the decreasing heights at 500 hPa mean that thickness of the column of air between 500 hPa and the surface will decrease over time in the region of positive vorticity advection. If there is no change in the number of air molecules in the column, this shrinking column of air is becoming denser and therefore must cool over time. What is it that can cause the column of air below position 1 to cool? The cooling there is generally not caused by cold advection because there is not much thermal advection* right over the surface low. (There is warm advection ahead of the low associated with the warm front and cold advection behind the low associated with the cold front, so there is zero thermal advection over the surface low.) Another way that the air can cool is adiabatic cooling due to upward motion. Remember from Chapter 6 that as air rises, its pressure decreases, and it cools at the dry adiabatic lapse rate of about 10°C per kilometre of lifting. Therefore, upward motion is required to account for the cooling column of air at position 1 due to positive vorticity advection. So we find that, generally, positive vorticity advection at 500 hPa causes upward motion over the surface low, as shown in Figure 12.24. Similarly, negative vorticity advection at 500 hPa aloft induces downward motion over a surface high.

Cloud patterns in visible and infrared satellite images can be very helpful in identifying vorticity maxima in regions where clouds are found. In cloud-free areas, a type of infrared image, called a water vapour image, measures wavelengths of radiation that are emitted by water vapour, so vorticity patterns can be observed in the swirling patterns of moisture even when there are no clouds present. In ● Figure 12.25, observe the cyclonic swirl of water vapour associated with a region of maximum vorticity just off the coast of British Columbia and Washington State.

PUTTING IT ALL TOGETHER—TORONTO'S SNOWSTORM OF THE CENTURY We are now in a position to place much of what we have learned into an actual weather situation. ● Figure 12.26 illustrates the atmospheric conditions that turned an open-wave cyclone into a ferocious storm. The lower map shows a large, mature mid-latitude cyclone stretching between the American East Coast to southern Ontario and Quebec on January 3, 1999. The frontal analysis uses the Meteorological Service of Canada four air mass–three front model, in which the southernmost front is the polar

---

*Recall that warm and cold advection occur when the wind blows warm or cold air into a region.

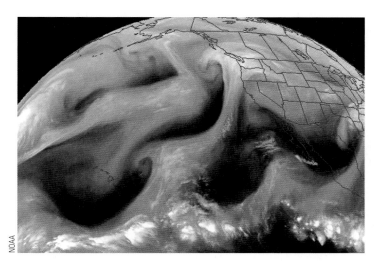

● FIGURE 12.25 This infrared water vapour image shows regions of maximum vorticity as cyclonic swirls of moisture off the coast of British Columbia and Washington State and out over the Pacific. The stretched-out band of clouds toward the bottom of the picture is the intertropical convergence zone.

front that separates maritime tropical from maritime polar air, the middle front is the maritime front separating maritime polar air from maritime arctic air, and the northernmost front is the arctic front separating continental arctic from maritime arctic air. A TROWAL, indicating an occluded system, extends from the triple point where the polar cold and warm fronts meet, connecting the secondary surface low with the low east of Lake Michigan that is associated with the upper low at 500 hPa. (Refer back to Chapter 11 to review fronts and air masses.) Even though three fronts and two low-pressure areas are apparent on the surface map, they are all part of the same middle-latitude cyclone. The counterclockwise circulation around the areas of low pressure is causing warm, moist air from subtropical waters to ride up and over very cold surface air (heavy orange arrow) that is entrenched over the northeastern United States and from Ontario through the Maritime provinces. Rain is falling along the polar cold front and ahead of the polar warm front, which transitions to snow north of the maritime and arctic fronts. In between, freezing rain across western New York State resulted in numerous power outages.

Above the surface, the 500 hPa chart (middle chart of Figure 12.26) shows a broad longwave trough with a shortwave (heavy dashed line) moving through it. Notice that the shortwave is just to the southwest of the two surface lows. This shortwave is the middle-latitude cyclone at 500 hPa. Apparently, the shortwave has disturbed the flow as a strong area of baroclinicity exists to the west of the shortwave. Here isotherms (dashed lines) are intersecting contour lines (solid lines), and cold advection is occurring.

At the 250 hPa level (upper chart in Figure 12.26), the polar-front jet stream (blue arrow) follows the orientation of the surface fronts and passes between the two surface lows. The region in orange represents the zone of strongest winds, the jet streak. The two areas of strong divergence at the right entrance

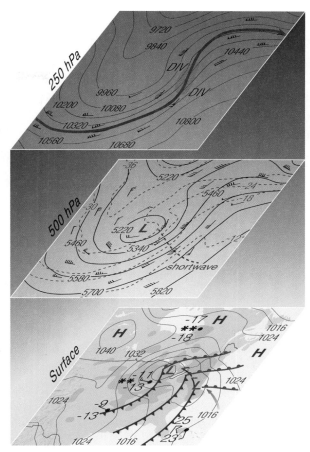

● FIGURE 12.26 The atmospheric conditions for January 3, 1999, at 7 a.m. EST. The bottom chart is the surface weather map. The middle chart is the 500 hPa chart that shows contour lines (solid lines) in metres above sea level, isotherms (dashed lines) in °C, and the position of a shortwave (heavy dashed line). The upper chart is the 250 hPa chart that illustrates contours, winds, and the position of the polar jet stream (dark blue arrow). The letters *DIV* represent an area of strong divergence. The region shaded orange represents the jet stream core—the jet streak.
Source: Environment Canada; NOAA

and left exit of the jet streak (marked by the letters *DIV* on the 250 hPa chart) are almost over the two surface lows.

● Figure 12.27 is a 500 hPa chart that shows absolute vorticity for the same date and time as Figure 12.26. Notice that there are two vorticity maxima. The northern one has a vorticity value of 22 (which is actually $22 \times 10^{-5} \text{ s}^{-1}$) and is associated with the upper low at 500 hPa and the surface low east of Lake Michigan. The southeastern vorticity maximum has a value of 12 and is associated with the surface **secondary low** to the southeast. The shortwave in Figure 12.26 that has been reproduced in Figure 12.27 connects the two vorticity maxima. Notice also that aloft, to the northeast of the two vorticity maxima, lie the regions of strong upper-level divergence associated with the jet stream. Directly below these regions are the southeastern low and the TROWAL. Hence, this is a major storm system that has reached the mature stage and is producing copious amounts of snowfall.

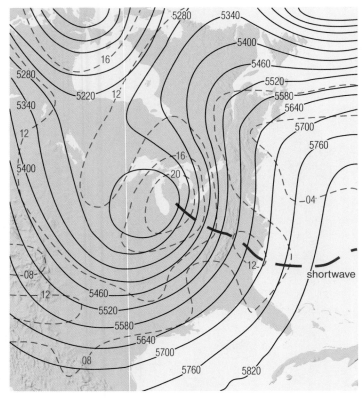

● FIGURE 12.27 The 500 hPa chart for January 3, 1999, at 7 a.m. EST. Solid lines are height contours in metres above sea level. Dashed lines are lines of constant absolute vorticity × 10⁻⁵ s⁻¹.
Data source: NOAA

WEATHER WATCH

On October 12, 1962, the remnants of Typhoon Freda formed an intense middle-latitude cyclone that moved up the West Coast of North America. It spread havoc from California to British Columbia, resulting in at least 46 fatalities—seven of these occurred near Vancouver, in British Columbia's Lower Mainland. The pressure at the centre of the low was recorded at 960 hPa, which is among the lowest pressures recorded for a mid-latitude cyclone. As the storm passed, winds gusting as high as 257 km h⁻¹ were reported from a radar station in southwest Washington State, and in Victoria, winds gusted to 145 km h⁻¹. During the storm, 20 percent of the trees in Vancouver's Stanley Park were blown down.

This particular storm buried much of southern Ontario in 20 to 40 cm of snow, affecting more than five million people. At least 11 people died, and thousands of passengers at the Toronto airports were stranded on some of the busiest days of the year. Chicago was equally debilitated, and this storm is considered the second worst blizzard to hit that city in the 20th century. Southern Quebec and New Brunswick were also hit hard. Four additional storms occurred over the next 12 days, resulting in the snowiest two-week period since 1846 and the greatest January snowfall in downtown Toronto (118.4 cm) ever recorded. The cold air behind the system set many record cold temperatures, such as a record low of −48°C in the state of Maine. This particular storm is well remembered in Canada as it prompted Toronto's Mayor Mel Lastman to declare a state of emergency and call in the Canadian Armed Forces to help dig out the city.

TROPICAL–EXTRATROPICAL LINKAGES Up to this point, we have considered middle-latitude cyclones that develop mainly along the polar front, dividing tropical air from polar air. With its high temperature and humidity, the tropical atmosphere is a potent source of sensible and latent heat for the rest of the atmosphere. Mid-latitude cyclones derive their energy from the temperature and humidity differences across the polar front and transport this heat and moisture toward the poles. Sometimes this interaction between the tropical and extratropical atmosphere is enhanced, resulting in more intense mid-latitude cyclones than would otherwise be expected.

One example of enhancement occurs when the polar front dips close enough to the tropics that very hot, humid, tropical air is entrained by a mid-latitude cyclone. This results in heavy mid-latitude precipitation as an *atmospheric river* of moisture and heat links the subtropics to the middle latitudes. A *pineapple express* (see Focus on a Special Topic: Pineapple Express on p. 375) is an example of this type of system that affects the North American West Coast during winter.

Another way energy and moisture are transferred from the tropics to the middle latitudes occurs when tropical cyclones (also called hurricanes; see Chapter 15) move away from the tropics and begin to weaken. If the remnant hurricane encounters a frontal zone or upper-level trough, its sensible and latent heat can develop into an intense middle-latitude cyclone, which might even be stronger than the original hurricane. Such a transformation from a tropical system to a middle-latitude cyclone is called an **extratropical transition**. Some of the most intense and difficult to forecast East Coast storms start out as hurricanes, but extratropical transitions can play a role in many of the most intense middle-latitude cyclones in other locations, too. Hurricane Hazel in October 1954 was one such storm that caused severe flooding and loss of life in Toronto (see Chapter 15, p. 468).

Polar Lows Up to now, we have concentrated on middle-latitude cyclones, especially those storms that form along the polar front. There are storms, however, that develop over polar water behind (or poleward of) the main polar front. Such storms are called **polar lows**. Although polar lows develop in both hemispheres, our discussion will centre on those storms that form in the Northern Hemisphere in the cold polar air of the North Pacific, North Sea, and North Atlantic, especially in the region south of Iceland. Polar lows that affect Canadian waters have been observed in the Gulf of Alaska, Hudson Bay, Davis Strait, and the Labrador Sea.

With diameters of 500 to 1000 km or less, polar lows are generally smaller in size than their middle-latitude cousin, the wave cyclone that tends to form along the polar front. Some polar lows have a comma-shaped cloud band. Others have a

FOCUS ON A SPECIAL TOPIC

Pineapple Express

Ruping Mo, Pacific Storm Prediction Centre, Meteorological Service of Canada, Vancouver, British Columbia

The "pineapple express" is a meteorological phenomenon that is a warm, moist, southwesterly flow bringing heavy precipitation and sometimes extensive flooding to the North American West Coast. Typically, in the cold season from October to March, an area of low pressure over the Gulf of Alaska maintains a stream of very wet and warm air flowing northeastward from near the Hawaiian Islands (where pineapples are grown) onto the Pacific coast of North America (see ● Figure 8). Such an *atmospheric river* is usually marked by a quasistationary surface front along which mobile cyclones develop and track toward the coast. Periods of heavy precipitation occur when the warm moist air is forced over the coastal mountain ranges. The warm air can also melt the snow pack rapidly in the mountains, further aggravating flooding in some areas.

Figure 8 shows a pineapple express heading toward British Columbia in mid-January 2005. It spread periods of heavy rainfall across the southern portions of the province. At the Vancouver Harbour weather station, 103.4 mm of

rain was recorded within 24 hours on January 17, which was the seventh wettest day in the 80-year period between 1925 and 2005, and another 73.8 mm of rain was recorded on the following day. The heavy and sustained rainfall across the southern coast of British Columbia wiped out the snowpack over the mountains and caused mudslides and flooding, with roads being closed and families evacuated. Further inland and to the north, the precipitation fell as freezing rain or heavy snow, creating severe avalanche hazards along the Trans-Canada Highway and most other mountain passes.

Studies suggest that the pineapple express is usually associated with a distinct weather pattern in the midtroposphere. At 500 hPa, there are higher than normal heights (i.e., high pressure at about 5.5 km above sea level) over the Bering Sea preceding the event, lower than normal 500 hPa heights (and therefore pressure at 5.5 km above sea level) over the Gulf of Alaska throughout the event, and higher than normal 500 hPa heights over the southwestern United States and adjacent eastern Pacific

Ocean during and after the event. This configuration in the midtroposphere provides a persistent storm track from the southwest toward the northeast, whereas the transport of moisture is largely associated with mobile cyclones that track from the subtropical Pacific toward the Pacific coast of North America. There is evidence to suggest that the tropical **Madden–Julian oscillation*** (MJO) exerts a strong impact on the pineapple express. It is thought that tropical temperature patterns associated with the MJO helps create the midtropospheric (500 hPa) pattern needed for the pineapple express to develop. In addition, heavy tropical rainfall of the MJO as it shifts eastward from the eastern Indian Ocean to the western tropical Pacific may feed moisture directly into the pineapple express system; such a system is sometimes called a "tropical punch" by meteorologists.

---

*The Madden–Julian oscillation (MJO) is a pattern of enhanced and suppressed rainfall in the tropics that moves eastward.

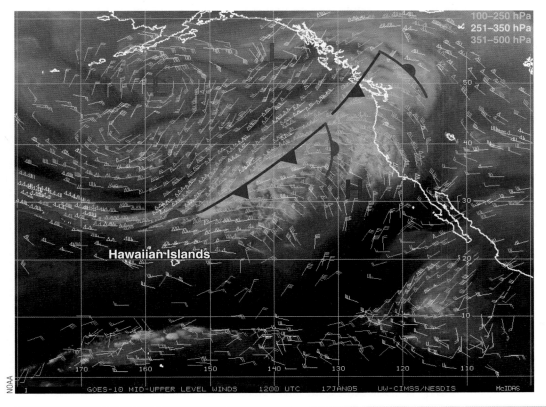

● FIGURE 8 The pineapple express phenomenon, as illustrated by the satellite (GOES-West) water vapour image with derived winds at 4 a.m. PST, January 17, 2005. The water vapour image detects an infrared wavelength that is sensitive to water vapour emissions. Major low and high sea-level pressure centres are marked by the black letters L and H, respectively.

tight spiral of convective clouds that swirls counterclockwise about a clear area, or "eye," which resembles the eye of a tropical hurricane (see ● Figure 12.28). In fact, like hurricanes, these smaller, intense storms often have a warmer central core, strong winds (often gale force or higher), and heavy showery precipitation that, unlike a hurricane,* is in the form of snow.

Polar lows typically form during the winter, from November through March. During this time, the sun is low on the horizon and absent for extended periods. This situation allows the air next to snow- and ice-covered surfaces to cool rapidly and become incredibly cold, forming a **continental arctic** air mass. As this frigid air sweeps off the winter ice that covers much of the Arctic Ocean, it may come in contact with warmer **maritime arctic** air that is resting above a relatively warm ocean current. Where these two masses of contrasting air meet, the boundary separating them is called an **arctic front**.

Along the arctic front, the warmer (less dense) air rises, whereas the much colder (more dense) air slowly sinks beneath it. Recall from our discussion of developing middle-latitude cyclones (p. 361) that the rising of warm air and the sinking of cold air establish a condition known as *baroclinic instability*. As the warm air rises, some of its water vapour condenses, resulting in the formation of clouds and the release of latent heat, which warms the atmosphere. The warmer air has the effect of lowering the surface air pressure. Meanwhile, at the ocean surface, there is a transfer of sensible heat from the relatively warm water to the cold air above. This transfer drives convective updrafts directly from the surface. In addition, it tends to destabilize the atmosphere as heat is gained at

---

*Tropical storms such as hurricanes are covered in Chapter 15. As we will see, the input of heat from the ocean surface into the hurricane increases as the wind speed increases, causing a positive feedback. Moreover, in hurricane environments, the ocean and air temperatures are about the same, whereas in the Arctic, the transfer of sensible heat from the surface is large because the air–sea temperature difference is large.

NASA

● FIGURE 12.28 An enhanced infrared satellite image of an intense polar low situated over the Norwegian Sea, north of the Arctic Circle. Notice that convective clouds swirl counterclockwise about a clear area, or eye. Surprising similarities exist between polar lows and tropical hurricanes described in Chapter 15.

the surface and lost to space at the top of the clouds as they radiate infrared energy upward.

The storm's development is enhanced if an upper trough lies to the west of the surface system and a shortwave disturbs the flow aloft. Similarly, the storm may intensify if a band of maximum winds—a jet streak—moves over the surface storm and a region of upper-level divergence draws the surface air upward. The developing cyclonic storm may attain a central pressure of 980 hPa or lower. Generally, polar lows dissipate rapidly when they move over land.

SUMMARY

In this chapter, we discussed where, why, and how a middle-latitude cyclone forms. We began by examining the early polar-front theory proposed by Norwegian scientists after World War I. We saw that the wave cyclone goes through a series of stages from birth to maturity to finally decay as an occluded storm.

We looked at the important effect that the upper airflow has on the intensification and movement of surface middle-latitude cyclones. We saw that when an upper-level trough lies to the west of a surface low-pressure area and when a shortwave disturbs the flow aloft, horizontal and vertical air motions begin to enhance the formation of the surface storm. Aloft, regions of divergence remove air from above surface middle-latitude cyclones, and regions of convergence supply air to surface anticyclones. The rising of warm air and the sinking of

cold air provide the proper energy conversions for the storm's growth as potential energy is transformed into kinetic energy. A region of maximum winds—a jet streak—associated with jet streams provides additional support as an area of divergence removes air above the surface middle-latitude cyclone, allowing it to develop into a deep low-pressure area. The curving nature of the jet stream tends to direct middle-latitude cyclones northeastward and anticyclones southeastward.

We looked at the concept of vorticity and how it relates to developing middle-latitude cyclones. We found that on the downwind (eastern) side of a vorticity maximum, there is normally diverging air aloft, converging air at the surface, upward vertical motion, and cyclonic storm development. We found that vorticity advection is responsible for the eastward movement of short wave upper troughs, and that positive vorticity advection at 500 hPa results in upward motion. We discussed middle-latitude cyclones that form from the extratropical transition of remnant hurricanes. Finally, we examined polar lows—those storms that form over water in the cold air of the polar regions.

KEY TERMS

The following terms are listed (with page numbers) in the order they appear in the text. Define each. Doing so will aid you in reviewing the material covered in this chapter.

middle-latitude cyclone, 352
polar-front theory, 352
wave cyclones, 352
frontal wave, 352
warm sector, 352
secondary low, 353
cyclogenesis, 355
lee-side lows, 355
northeasters, 355
tropical cyclones, 356
extratropical transition, 356
convergence, 357
divergence, 357
longwave, 360
Rossby waves, 360
shortwaves, 360
barotropic (atmosphere), 361
baroclinic (atmosphere), 361
cold advection, 361
warm advection, 361

baroclinic instability, 361
cutoff low, 362
jet streak, 362
conveyor belt model, 363
dry slot, 365
comma clouds, 365
vorticity, 368
Earth's vorticity, 369
relative vorticity, 369
absolute vorticity, 369
vorticity advection, 371
secondary low, 373
extratropical transition, 374
polar lows, 374
Madden–Julian oscillation, 375
continental arctic (air mass), 376
maritime arctic (air mass), 376
arctic front, 376

QUESTIONS FOR REVIEW

1. On a piece of paper, draw the different stages of a middle-latitude cyclonic storm as it goes through birth to decay according to the polar-front (Norwegian) model.

2. Why do middle-latitude cyclones "die out" after they become occluded?

3. List four regions in North America where middle-latitude cyclones tend to develop or redevelop.

4. Explain this fact: Without upper-level divergence, a surface open wave would probably persist for less than a day.

5. Why do middle-latitude surface low-pressure areas tilt westward with increasing height?

6. If upper-level diverging air above a surface area of low pressure exceeds converging air around the surface low, will the surface low-pressure area weaken or intensify? Explain.

7. What is an Alberta clipper? Where does it form, and how does it move?

8. What are northeasters? Why are they given that name?

9. How are longwaves in the upper-level westerlies different from shortwaves?

10. What are the necessary ingredients for a middle-latitude cyclonic storm to develop into a huge storm system?

11. Why do surface storms tend to dissipate or "fill" when the upper-level low and the surface low become vertically stacked?

12. How does the polar-front jet stream influence the formation of a middle-latitude cyclone?

13. Explain why even though the polar-front jet stream coincides with the polar front, some surface regions are more favourable for the development of middle-latitude cyclones than others.

14. Using a diagram, explain why a surface high-pressure area over Alberta will typically move southeastward, whereas, at the same time, a deep middle-latitude cyclone over the Great Lakes will generally move northeastward.

15. What are the sources of energy for a developing cyclone?

16. How does warm and cold advection aid in the development of a surface middle-latitude cyclone?

17. What are the roles of warm, cold, and dry conveyor belts in the development of a middle-latitude cyclonic storm system?

18. Explain why surface lows tend to deepen when a vorticity maximum aloft moves in from the west.

19. What are polar lows? How and where do they form? What do some of them have in common with tropical hurricanes?

QUESTIONS FOR THOUGHT

1. An English friend of yours says that last night's rain over northern Great Britain was caused by a storm that originally formed east of the Colorado Rockies. Explain how this could happen.

2. Would a middle-latitude cyclone intensify or dissipate if the upper trough were located to the east of the surface disturbance? Explain your answer with the aid of a diagram.

3. Explain why, at 500 hPa, when cold advection is occurring, the air temperature does not drop as fast as it should. (Hint: What type of vertical air motion is also occurring?)

4. Over Earth as a whole, would you expect the atmosphere to be mainly barotropic or baroclinic? Explain.

5. Baroclinic waves seldom form in the tropics. Why not?

6. Suppose that Earth stops rotating. How would this affect Earth's vorticity? What would happen to the absolute vorticity of a moving air parcel? If the parcel were initially moving southwestward, how would its direction change, if at all?

7. Why do Pacific storms often redevelop on the eastern side of the Rocky Mountains?

8. If you only had isotherms on an upper-level chart, how would a cut-off low appear?

9. If polar lows form in cold polar air over water, how is the atmosphere made conditionally unstable so that towering convective cumulus clouds can form?

10. The 500 hPa level (at about 5600 m above the surface) is referred to as the "level of nondivergence." Give an explanation as to what this statement means with reference to what is taking place above and below this level.

PROBLEMS AND EXERCISES

1. On the 250 hPa chart (see ● Figure 12.29), suppose that the winds are blowing parallel to the contour lines.
   (a) On the chart, mark where regions of convergence and divergence are occurring.
   (b) Put an L on the chart where you might expect to observe a developing middle-latitude cyclone at the surface.
   (c) Put an H on the chart where you might expect to observe a surface anticyclone.
   (d) In which directions would the surface cyclonic storm and anticyclone most likely move?
   (e) In terms of convergence and divergence, what are the necessary conditions for the intensification of the

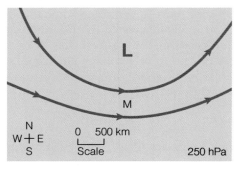

● FIGURE 12.29

surface middle-latitude storm? For the building of the anticyclone?

2. Suppose that there is a region of maximum absolute vorticity at chart position M.
   (a) How would the absolute vorticity be changing with time as an air parcel moves downstream (from position M) with the flow?
   (b) Based on absolute vorticity, circle on the map where you would expect to find upper-level divergence, surface convergence, and a developing middle-latitude cyclone.

View of the results from an early-season snowfall across northern Ontario from the MODIS imager on the Aqua satellite on November 18, 2009. It is not hard to imagine the storm track that gave rise to this pattern.
NASA

Weather Forecasting

13

Weather forecasting is not new—people have been forecasting the weather for thousands of years, long before the advent of weather maps, satellites, and computers. Many early forecasting methods relied on observing the sky and wind or the behaviour of plants and animals. Not all of these methods work, but the ones that do can usually be explained by the science of meteorology. For instance, the following saying has been around at least since biblical times:

> "Red sky at night, sailors delight.
> Red sky in the morning, sailors take warning."

As discussed in Chapter 2 and in Chapter 19, the science behind this proverb is related to the way sunlight is scattered by gases and aerosols as it passes through the atmosphere. Red sky at night means that there is a clear sky to the west where the sun sets, and because weather systems generally move from west to east, this implies that a clear sky and good weather are on their way. Red sky in the morning means that the clear weather has already passed by, so perhaps poor weather is on its way. Another proverb is

> "Halo around the sun or
> moon, rain or snow soon."

As described in Chapter 19, halos are caused by the bending of sunlight through ice crystals—cirrus clouds. We know that cirrus clouds are often seen ahead of a mid-latitude cyclone, so the presence of a halo may indicate an approaching storm.

Other ancient forecasting methods are related to the behaviour of plants or animals. Although living organisms cannot truly forecast the weather, they may be sensitive to environmental conditions that people cannot easily detect. For example, the wings of cicadas cannot vibrate when the humidity is high—so when they are silenced, it may indicate an approaching storm.

Are you aware of other weather proverbs or forecasting methods? If they seem to work, see if you can think of the reasons why.

CONTENTS

To forecast the weather, humans (the *anthrosphere*) use meteorological knowledge to predict the state of the *atmosphere* at some future time. Forecasting is the ultimate test of how well we understand the atmosphere. To reliably make accurate forecasts, we must first learn how the atmosphere functions. Weather forecasts are issued to save lives, property, and crops and to tell us what to expect in our atmospheric environment. In addition, knowing future weather conditions is vital to many human activities.

For example, a summer forecast of extended heavy rain and cool weather would have construction supervisors planning work under protective cover, department stores advertising umbrellas instead of bathing suits, and ice cream vendors vacationing as their business declines. The forecast would alert farmers to harvest their crops before their fields became too soggy to support the heavy machinery needed for the job. And the commuter? Well, the commuter knows that prolonged rain could mean clogged gutters, flooded highways, stalled traffic, blocked railway lines, and late dinners.

On the other hand, a forecast calling for extended high temperatures with low humidity has an entirely different effect. As ice cream vendors prepare for record sales, the dairy farmer anticipates a decrease in milk and egg production. Forest managers prepare warnings of fire danger in parched timber and grasslands. The construction worker is on the job outside once again, but the workday begins in the early morning and ends by early afternoon to avoid the oppressive heat. And the commuter prepares for increased traffic stalls due to overheated car engines.

Put yourself in the shoes of a weather forecaster. It is your responsibility to predict the weather accurately so that thousands (possibly millions) of people will know whether to carry an umbrella, wear an overcoat, or prepare for a winter storm. Since weather forecasting is not an exact science, your predictions will occasionally be incorrect. If your erroneous forecast misleads many people, you may become the target of jokes, insults, and even anger. There are even people who expect you to be able to predict the unpredictable. For example, on Monday, you may be asked whether two Saturdays from now will be a nice day for a picnic or a wedding. And, of course, what about next winter? Will it be bitterly cold?

Unfortunately, accurate answers to such questions are beyond meteorology's present technical capabilities, but "useful" answers may be possible. Will forecasters ever be able to answer such questions confidently? If so, what steps are being taken to improve the forecasting art? How are forecasts made, and why do they sometimes go wrong? These are just a few of the questions we will address in this chapter.

Acquisition of Weather Information

Weather forecasting basically entails predicting how the present state of the atmosphere will change. Consequently, if we wish to make a weather forecast, present weather conditions over a large area must be known and understood. To obtain this information, a network of observing stations is located throughout the world. Over 10,000 land-based stations and hundreds of ships and buoys provide surface weather information four times a day.[*] Most airports and many other stations observe conditions hourly. Additional information, especially upper-air data, is supplied by radiosondes, aircraft, and satellites. Radiosonde data are usually available only at 00:00 and 12:00 UTC, whereas aircraft and satellite observations may occur throughout the day.

A United Nations agency, the *World Meteorological Organization* (WMO), consists of over 175 nations. The WMO is responsible for the international exchange of weather data and certifies that the observation procedures do not vary among nations—an extremely important task because the observations must be comparable. After an observation is taken, it is immediately sent to a communication substation by electronic means, usually over dedicated telephone lines or by satellite relay. From there, the data collected at many observation stations are sent to World Meteorological Centres (located in Melbourne, Australia; Moscow, Russia; and Washington, D.C., U.S.A.). Then worldwide weather information is transmitted electronically to national weather services such as the Canadian Meteorological Centre (CMC), a department of the Meteorological Service of Environment Canada (MSC). CMC is located in Dorval, Quebec, just outside Montreal. Here the massive job of analyzing the data, preparing weather maps and charts, and predicting the weather on a global and national basis begins.[†] From the CMC, this information is transmitted to public and private agencies across the country and worldwide.

The compiled charts, maps, and forecasts are sent electronically to forecasting centres. In Canada, these centres are located in Vancouver, British Columbia (Pacific Storm Prediction Centre); Edmonton, Alberta, and Winnipeg, Manitoba (Arctic and Prairie Storm Prediction Centres); Toronto, Ontario (Ontario Storm Prediction Centre); Montreal, Quebec (Quebec Storm Prediction Centre); and Dartmouth, Nova Scotia, and Gander, Newfoundland (Atlantic Storm Prediction Centre). The Canadian Hurricane Centre is located in the Atlantic Storm Prediction Centre. In addition, there are two aviation forecast centres, the Ice Services Centre in Ottawa, and three aviation defence forecast centres that forecast weather for the Canadian Armed Forces. The Canadian Meteorological Aviation Centres are located in Edmonton and Montreal and forecast weather and hazards at civilian airports and at altitudes for aircraft in flight.

These forecast centres use the data for preparing regional weather forecasts, as well as for warnings of impending severe

---

[*]Observations are usually taken at 00:00, 06:00, 12:00, and 18:00 Coordinated Universal Time (UTC), which is also called *Greenwich Mean Time (GMT)*—local time at the Greenwich Observatory in England. To convert from UTC to your local time, see Appendix F.

[†]By international agreement, data are plotted using the symbols illustrated in Appendix B.

weather. The region serviced by a Storm Predication Centre is a province or region of Canada, whereas the other specialized centres cover a broader geographic area.

The public gets weather forecasts over radio, television, or, increasingly, the Internet. Many stations hire private meteorological companies or professional meteorologists to make their own forecasts aided by MSC material or to modify a forecast produced by a regional Storm Prediction Centre. Other stations hire meteorologically untrained announcers who paraphrase or read the forecasts of the MSC word for word. Canada has two weather-only television stations, more than any other country: *The Weather Network* and its French-language counterpart, *Météo Média*.

Today, forecasters have access to many hundreds of maps and charts, as well as vertical profiles (called *soundings*) of temperature, dew point, and winds. Also available are visible and infrared satellite images, as well as Doppler radar information that can detect and monitor the severity of precipitation and thunderstorms. The output of computer models of weather, called numerical weather prediction (NWP) models, is also a key forecasting tool.

When hazardous weather is likely, the MSC issues advisories in the form of weather watches and warnings. A **watch** indicates that atmospheric conditions favour hazardous weather occurring over a particular region during a specified time period, but the actual location and timing of the occurrence are uncertain. Watches are usually issued well in advance of the event and are replaced by a **warning** when the hazardous weather is either imminent or actually occurring within the specified forecast area. Special Weather Statements are issued to explain a significant weather event to the public that cannot be adequately explained in a normal public forecast. Special Weather Statements are issued when the weather does not warrant a watch or warning. Additional information on watches and warnings is given in ▼ Table 13.1.

▼ Table 13.1   Watches, Warnings, and Special Weather Statements*

SEVERE WEATHER EVENT	WATCH OR WARNING CRITERIA
Tornadoes	**Watch:** when conditions are favourable for the development of severe thunderstorms with one or more tornadoes. **Warning:** when a tornado has been reported or there is evidence based on radar or an observer that one is imminent. Usually issued up to 15 minutes in advance.
Severe thunderstorms	**Watch:** when conditions are favourable for the development of severe thunderstorms with one or more of the following: • wind gusts of 90 km h^{-1} or more • hail of 2 cm in diameter or larger • heavy rainfall (50 mm or more within 1 hour or 75 mm or more within 3 hours) **Warning:** when there is evidence from radar, satellite, or an observer that one or more of the above criteria are occurring or imminent. Usually issued 1 hour in advance.
Blizzards	**Warning:** when winds exceed a locally determined threshold (e.g., 40–50 km h^{-1}) and are expected to reduce visibility below a locally determined threshold (e.g., 400–1000 m) or less over a large area due to blowing snow or snow for at least 4 hours (at least 6 hours north of the treeline). Usually issued 10 hours in advance.
Snow squalls	**Watch:** when conditions are favourable for the development of snow squalls downwind of large water bodies, such as the Great Lakes, with at least one of the following: • localized intense snowfall producing 15 cm or more in 12 hours or less • visibility less than 400 m in snow with or without blowing snow for 3 hours or more **Warning:** when one or more of the above criteria are imminent or occurring. Usually issued 10 hours in advance. (Note that a snowsquall warning may also be issued when similar conditions, but lasting for an hour or less, are associated with passage of a cold front.)
Blowing snow	**Warning:** when blowing snow, caused by winds of at least 30 km h^{-1}, is expected to reduce visibility below 800 m for at least 3 hours. Usually issued 10 hours in advance.
Freezing rain	**Warning:** when freezing rain is expected to pose a hazard to transportation or property or is expected to last for at least 2 hours. Usually issued 10 hours in advance.
Freezing drizzle	**Warning:** when at least 8 hours of freezing drizzle is expected. Usually issued 10 hours in advance.

(Continued)

▼ Table 13.1   (Continued)

SEVERE WEATHER EVENT	WATCH OR WARNING CRITERIA
Strong winds	**Warning:** when sustained winds and/or gusts are expected to exceed regionally determined thresholds ranging from 60 to 90 km h^{-1} for sustained winds to 90 to 110 km h^{-1} for gusts. Usually issued 18 hours in advance.
Wind chill	**Warning:** when wind chill effective temperature is expected to be below a threshold that is locally determined. The threshold ranges from −30 in southern Ontario to −55 in Nunavut and the Northwest Territories. Usually issued 24 hours in advance.
Winter storm	**Watch:** when conditions are favourable for the development of severe and potentially dangerous winter weather, including ● a blizzard ● a major snowfall (25 cm or more within 24 hours) ● a heavy snowfall combined with other winter hazards, such as freezing rain, strong winds, blowing snow and/or extreme wind chill **Warning:** when the conditions above are expected. If blizzard conditions are expected, then a blizzard warning is issued instead. Usually issued 24 hours in advance.
Frost	**Warning:** when widespread frost is expected during the growing season. The criteria vary locally.
Storm surges giving high water levels	**Warning:** when significant wind-forced flooding is to be expected along low-lying coastal areas. Usually associated with a hurricane in Atlantic Canada.
Heavy snowfall	**Warning:** when snowfall amounts are expected to exceed regionally determined thresholds in 6, 12, or 24 hours. Threshold amounts vary widely from 5 cm in 6 hours or 10 cm in 12 hours over southwest British Columbia to 20 cm or more within 24 hours or less in Yukon Territory. Usually issued 24 hours in advance.
Heavy rainfall	**Warning:** when rainfall amounts are expected to exceed regionally determined thresholds in 1, 12, 24, or 48 hours. Examples of thresholds are 25 mm in 1 hour; 50 mm in 12 hours, 50 mm in 24 hours, and 75 mm in 48 hours. (Coastal British Columbia has a 100 mm in 24-hour threshold.) Usually issued 24 hours in advance.
Arctic outflow	**Warning:** when wind chill values of −20 or lower are expected for 6 hours or more in coastal British Columbia due to cold winds from the continent.
Flash freeze	**Warning:** when significant ice is expected to form on roads, sidewalks, or other surfaces over much of a region from freezing of residual water from melted snow or rain due to a sudden temperature drop.
Tsunami	**Warning:** when a tsunami with significant widespread effects is imminent or expected in coastal British Columbia.
Humidex	**Advisory:** when temperatures are expected to exceed 30°C and humidex values exceed 40 in Ontario.
Tropical storm and hurricane	**Watch:** when there is a risk of a tropical storm (winds of 34–63 knots [63–117 km h^{-1}]) or hurricane (winds 64 knots or more [118 km h^{-1}]) within 36 hours. The area affected can be over water or land. **Warning:** tropical storm or hurricane warnings are issued when one of these is expected within 24 hours.
Marine warnings (for marine areas)	**Strong wind warning:** when sustained winds of 20–33 knots (37–62 km h^{-1}) are expected. **Gale warning:** when sustained winds of 34–47 knots (63–87 km h^{-1}) are expected. **Storm warning:** when sustained winds of 48–63 knots (88–117 km h^{-1}) are expected. **Hurricane force wind warning:** when sustained winds of 64 knots (118 km h^{-1}) or more are expected. **Freezing spray warning:** when an ice accretion rate of 0.7 cm h^{-1} or more is occurring or expected.

*It should be noted that watches and warnings for wind chill, heavy rain, or snowfall-related events (e.g., winter storms, blizzards, etc.) may use different criteria in different regions. For example, mountainous areas that experience frequent heavy snow have higher snowfall criteria, whereas areas with infrequent snow have lower snowfall criteria. Similarly, in areas that experience frequent extreme cold, wind chills may have lower (colder) criteria for warnings.

● FIGURE 13.1 The Integrated Forecaster Workstation (IFW) used by weather forecasters at the Meteorological Service of Canada has three displays to provide various weather maps, images, and model outputs, with overlays and animations. The pen tablet monitor in the centre is for graphical input.

© Environment Canada 2010. Photo by Tom Wakefield.

Weather Forecasting Tools

To help forecasters handle all the available charts, imagery, and model outputs and maps, high-speed computer visualization workstations are employed by the Meteorological Service of Canada. The workstation in use today is known as the Integrated Forecaster Workstation (IFW) and is a linux-based computer that supports three large monitors and a digitizing tablet. The IFW system is shown in ● Figure 13.1.

Meteorologists rely on sophisticated computer visualization to manage and make sense of the vast array of information available to make the best possible forecast. Therefore, at the MSC, the IFW computer hardware runs the **NinJo** Workstation meteorological visualization and forecast preparation program. NinJo was developed for use by the government meteorological services in Canada, Germany, Switzerland, and Denmark. It displays meteorological data such as observations, radar, satellite, and model output in layers that can be overlaid, combined, and animated, as illustrated in ● Figure 13.2.

With so much information at the forecaster's disposal, it is essential that the data be easily accessible and in a format that allows several weather variables to be viewed at one time. The **meteogram** is a chart that shows how one or more weather variables has changed at a station over a given period

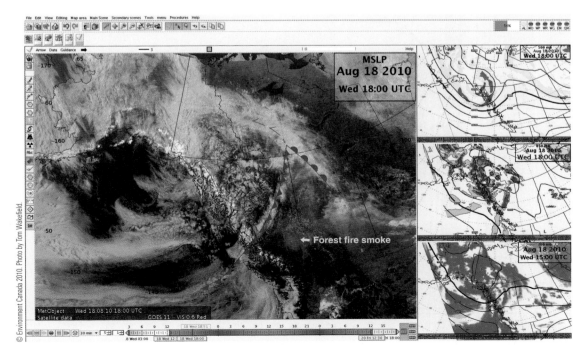

● FIGURE 13.2 Display of visible satellite image with superimposed frontal analysis (left) as well as NWP model output valid at the same time (right) of 500 hPa heights and vorticity, 850 hPa heights and 6-hour precipitation, and 700 hPa heights and total cloud cover.

© Environment Canada 2010. Photo by Tom Wakefield.

of time. As an example, the chart may represent how air temperature, dew point, cloud cover, precipitation, and wind have changed over the past five days, or it may illustrate how these same variables are projected to change over the next five days (see ● Figure 13.3).

Another aid in weather forecasting is the use of soundings—a two-dimensional vertical profile of tempera-

ture, wet-bulb temperature, dew point, and winds (see ● Figure 13.4).* The analysis of a sounding can be especially helpful when making a short-range forecast that covers a

---

*A sounding is obtained from a radiosonde. For additional information on the radiosonde, see the Focus section in Chapter 1 on p. 18. Assessment stability and cloud from soundings are discussed in Chapter 6.

● FIGURE 13.3 Meteogram illustrating predicted weather for 48 hours at Victoria International Airport starting at 4 p.m. PST, January 14, 2010. The forecast is derived from the Global Environmental Mulitscale (GEM) regional model from the Canadian Meteorological Centre of the Meteorological Service of Environment Canada.

Source: Environment Canada. Meteorological Service of Canada. Found at: http://jervis.pyr.ec.gc.ca/archive_db/201001/20100115/GEM-REG/MGRAM/yyi/meteo/GEM-REG_MGRAM_yyi_meteor_2010011500.png. © Published with the permission of Environment Canada, 2010.

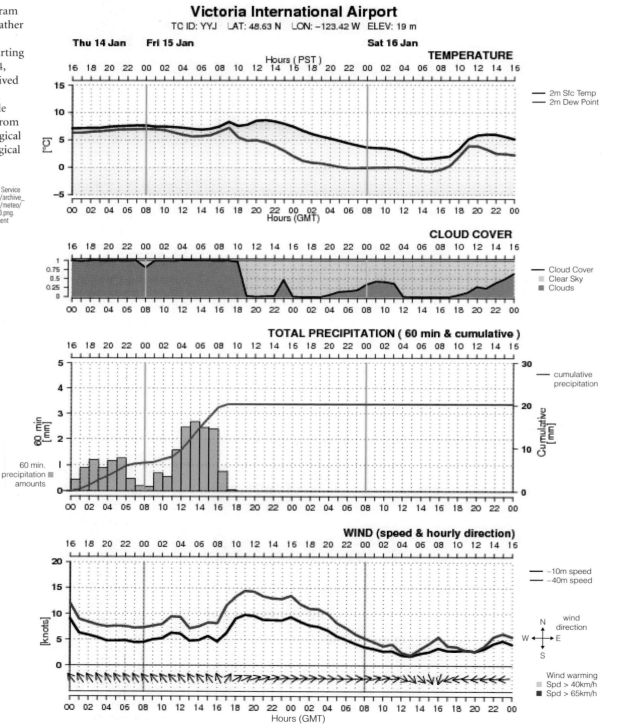

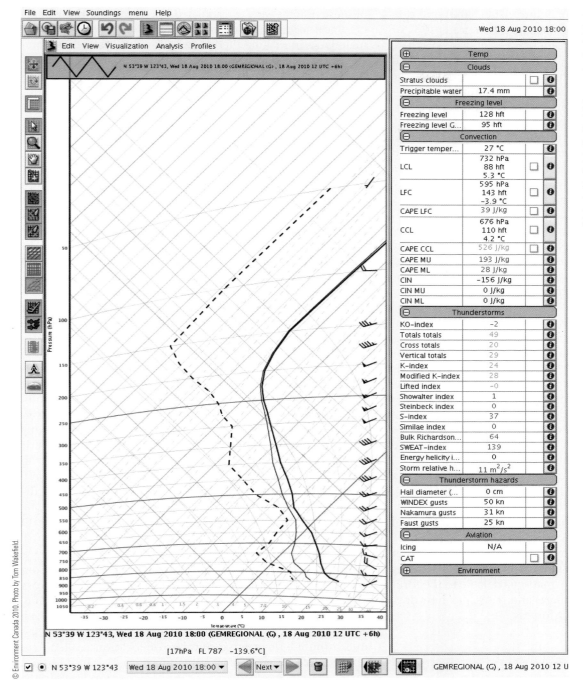

● FIGURE 13.4 A model projected sounding of air temperature, wet-bulb temperature, dew point, and winds on August 18, 2010, at 11:00 a.m. PDT at a location of severe forest fires in central British Columbia just west of the arrow labelled "Forest fire smoke" in Figure 13.2. The sounding is plotted on a tephigram.* On the right side of the sounding are a number of stability indices that tell meteorologists about the potential for convection and thunderstorms.

*Tephigrams are charts for plotting soundings. On these charts, the isotherms (in green) slope from lower left to upper right, and the isobars are the slightly curved horizontal lines that indicate decreasing pressure with altitude. To learn more about them, refer back to Focus on an Advanced Topic: The Tephigram in Chapter 6 on p. 184.

relatively small area, such as the mesoscale. The forecaster examines the closest sounding, as well as the soundings of those sites upwind, to see how the atmosphere might be changing. Computer programs then automatically calculate a number of meteorological indexes from the sounding that can aid the forecaster in determining the likelihood of smaller scale weather phenomena, such as thunderstorms, tornadoes, and hail. Soundings also provide information that can aid in the prediction of fog, air pollution alerts, and the downwind mixing of strong winds. Soundings can be observed from a radiosonde or forecast from an NWP model, as shown in Figure 13.4.

Looking at this sounding in Figure 13.4, a forecaster would see that the profile is quite dry and mostly cloud-free. This can be seen by comparing the temperature (solid line) and dew point (dashed line) and noting that the difference is 10°C or more for most of the profile, except between 600 and 550 hPa, where it is approximately 5°C, indicating possible scattered middle (alto-) clouds. The stability indices calculated on the right side indicate only a slight chance of thunderstorms—there is not enough humidity at low levels to develop significant convective clouds. The winds are southwesterly or westerly everywhere, except they are northwesterly at 750 and 800 hPa, and this is one of the main reasons that forest fire smoke spread eastward, causing air quality advisories in communities in British Columbia and as far as Saskatchewan.

Canada's national Doppler radar network provides very useful information to the forecaster for assessing and monitoring storms—especially convective storms—and their precipitation and wind patterns. NinJo can animate and overlay radar images with other data, as well as perform calculations and analyses of the radar data to alert forecasters of the likelihood of severe convective weather such as thunderstorms, hail, and tornadoes.

Satellite information is also a valuable tool for the forecaster. Visible, enhanced infrared and water vapour images provide a wealth of information, covering inaccessible regions with no other data sources, as well as populated areas. This added information provides a clearer representation of the atmosphere,* allowing better identification of fronts, weather systems, and clouds. Since satellite images are taken every 30 to 60 minutes, they can be animated to produce movies or *loops* of the changing cloud patterns—a very useful tool indeed to allow a meteorologist to monitor the changing weather.

When predicting temperatures, forecasters often look at a chart called the thickness chart. If you are interested in learning how this chart can be used to predict whether falling precipitation will be in the form of rain or snow, read Focus on a Special Topic: The Thickness Chart—A Forecasting Tool on p. 390.

Up to this point, we have examined some of the weather data and tools a forecaster might use in making a weather prediction. With all of this information available to the forecaster, including hundreds of charts and maps, just how does a meteorologist make a weather forecast?

Weather Forecasting Methods

As late as the mid-1960s, all weather maps and charts were plotted by hand and analyzed by individuals. Meteorologists predicted the weather using certain rules that related to the particular weather system in question. For short-range forecasts of six hours or less, surface weather systems were moved along at a steady rate. Upper-air charts were used to predict where surface storms would develop and where pressure systems aloft would intensify or weaken. The predicted positions

---

*Information provided by satellites is located in various sections of this book. For example, see Chapter 5, p. 153, and Chapter 9, p. 287.

of these systems were extrapolated into the future using linear graphical techniques and current maps. Experience played a major role in making the forecast. In many cases, these forecasts turned out to be amazingly accurate. They were good, but with the advent of modern computers, along with our present observing techniques, today's forecasts are even better.

COMPUTERS AND WEATHER FORECASTING: NUMERICAL WEATHER PREDICTION Modern computers can analyze large quantities of data extremely quickly. Each day, the many thousands of observations from around the globe are transmitted to national meteorological centres, such as the Canadian Meteorological Centre, where they are processed by high-speed computers that produce surface and upper-air charts. Meteorologists interpret the weather patterns and then correct any errors that may be present. The final chart, referred to as an **analysis**, represents the state of the atmosphere at the time the data were collected.

Computers not only plot and analyze data, they also predict the weather. The routine daily forecasting of weather by a computer using mathematical equations is called **numerical weather prediction** (NWP).

Because the many weather variables are constantly changing, meteorologists have devised **atmospheric models** that describe the present state of the atmosphere and are able to be stepped forward in time to predict future conditions. These are not physical models that paint a picture of a developing storm; they are, rather, mathematical models consisting of the **governing equations** that describe how atmospheric temperature, pressure, winds, and moisture will change with time. These equations are solved in a computer program, and this is the numerical weather predication model. Actually, the models do not fully represent the real atmosphere but are approximations formulated to retain the most important aspects of the atmosphere's behaviour.

The surface and upper-air observations of temperature, pressure, moisture, winds, and height, as well as data from satellites, aircraft, and radar, are combined with output from an earlier model run to create the model's *initial conditions*. To determine how each of these variables will change, each equation is solved for a small increment of future time—say, one minute—for a large number of locations called grid points, each situated a given distance apart.* In addition, each equation is solved for as many as 50 vertical levels in the atmosphere extending from Earth's surface to the lower stratosphere. The results of these computations are then fed back into the original equations. The computer again solves the equations with the new "data," thus predicting weather over the following minute, and in this way, the model is stepped forward in time. This procedure is done repeatedly until it reaches some desired time in the future, usually 12, 24,

---

*Some models have grid spacing of 0.5 km or less, whereas the spacing in others exceeds 100 km. There are models that describe the atmosphere using a set of mathematical equations with wavelike characteristics rather than at a set of discrete grid points. These are called *spectral models*.

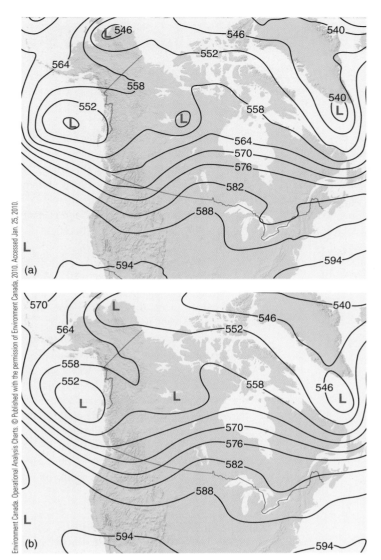

Environment Canada, Operational Analysis Charts. © Published with the permission of Environment Canada, 2010. Accessed Jan. 25, 2010.

• FIGURE 13.5 Two 500 hPa progs valid for 7 p.m. EST, July 12, 2006—from NWP models that were run 48 hours previously. Prog (a) is the MSC GEM regional model, with a resolution (grid spacing) of 15 km, whereas prog (b) is the MSC GEM global model with a resolution of approximately 33 km. Solid lines on each map are height contours in decametres, where 570 dam equals 5700 m. Notice how the two progs (models) agree on the atmosphere's large scale circulation. The main difference between them is in the way the models handle the lows off the West Coast of North America and the southern tip of Greenland. Model (a) has deeper central pressures for both lows and places the western low further to the west and the Greenland low further to the southwest.

36, 48, and out to 120 hours or beyond. The computer then analyzes the model output "data" and produces forecast weather maps showing the projected positions of pressure systems with their isobars or contour lines. The final forecast charts representing the atmosphere at a specified future time are called **prognostic charts** or, simply, **progs**.

Numerical weather prediction requires many hundreds of millions of mathematical calculations just to make a 24-hour forecast. It is for this reason that some of the largest

and most powerful supercomputers in the world are dedicated to the task of predicting the weather. NWP models are in many ways similar to the climate models—also called *general circulation models* or *GCMs*—that are used to make projections about Earth's future climate. For more information about these, read Focus on an Advanced Topic: Climate Models in Chapter 16 on p. 514.

The forecaster uses the progs as a guide to predicting the weather. At present, there are a variety of models (and, hence, progs) from which to choose, each producing a slightly different interpretation of the weather for the same projected time and atmospheric level (see • Figure 13.5). The differences between progs may result from the way the models use the equations, or the distance between grid points, called *resolution*. Some models predict some features better than others: one model may work best in predicting the position of troughs on upper-level charts, whereas another forecasts the position of surface lows quite well—and, frustratingly, this may change from situation to situation. Some models even attempt to forecast the state of the atmosphere 16 or more days into the future. Look at • Figure 13.6 and notice that model (b) in Figure 13.5 with a resolution of 33 km actually did a better job of forecasting the position of the upper low off the West Coast of North America than did model (a) with a finer resolution of 15 km, although model (a) had superior 500 hPa height values near the low.

A good forecaster knows the idiosyncrasies of each model (such as model (a) and model (b) in Figure 13.5) and carefully scrutinizes all the progs. The forecaster then makes a prediction based on the guidance from the computer, a personalized practical interpretation of the weather situation and any local geographic features that influence the weather

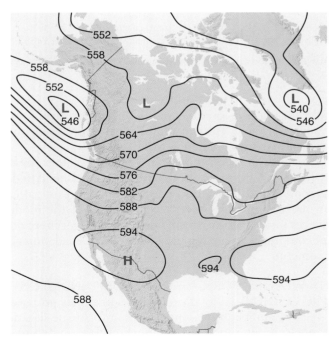

• FIGURE 13.6 The 500 hPa analysis for 7 p.m. EST, July 12, 2006.

FOCUS ON A SPECIAL TOPIC

The Thickness Chart—A Forecasting Tool

The thickness chart can be a valuable forecasting tool for the meteorologist. It can help identify air masses and locate fronts, and a prognostic thickness chart can help predict the daily maximum and minimum temperatures. It can also help predict whether falling precipitation will be in the form of rain or snow. When it is combined with height contours, it can show areas of warm and cold advection. What, then, is a thickness chart?

A thickness chart shows the difference in height between two constant pressure surfaces. The vertical depth or thickness between any two pressure surfaces is related to the average air temperature between the two surfaces. Recall that air pressure decreases more rapidly with height in cold air than in warm air. This fact is illustrated in ● Figure 1, which shows a 1000 hPa pressure surface and a 500 hPa pressure surface. The difference in

height between these two pressure surfaces is called the 1000 hPa to 500 hPa thickness. Notice that the vertical distance (thickness) between these two pressure surfaces is greater in warm air than in cold air. Consequently, warm air produces high thickness, and cold air produces low thickness.* In fact, thickness is directly related to the average layer temperature.

● Figure 2 shows 1000 hPa to 500 hPa thickness lines for a January morning. On the chart, regions of low thickness correspond to cold air in the 1000 to 500 hPa layer, and regions of high thickness correspond to warm air. There are a number of forecasting rules for predicting air temperature and precipitation using this chart. One forecasting rule is that the 540 decametre (540 dam equals 5400 m) thickness line often represents the dividing line between rain and snow, especially for cities receiving precipitation east of the Rocky Mountains. If precipitation is falling, cities with a thickness greater than 540 dam should be

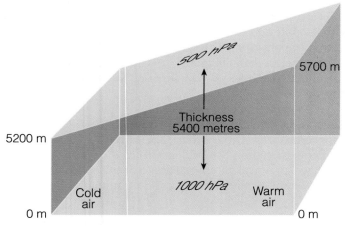

● FIGURE 1 The vertical separation (thickness) between the 1000 hPa pressure surface and the 500 hPa pressure surface is greater in warm air than in cold air.

*The exact relationship between thickness and average layer temperature is given by the hydrostatic law and hypsometric equation, discussed in Focus on an Advanced Topic: The Hydrostatic and Hypsometric Equations in Chapter 8 on p. 250.

within the specific forecast area. ▼ Table 13.2 gives a brief summary of a few "rules of thumb" that a forecaster might use when making a prediction using a prog.

Currently, forecast models predict the weather reasonably well 4 to 6 days into the future. The models tend to do a better job at predicting temperature and jet-stream patterns than precipitation. However, even with all of the modern advances in weather forecasting provided by ever more powerful computers, Meteorological Service of Canada forecasts are sometimes wrong.

WHY WEATHER FORECASTS GO AWRY AND STEPS TO IMPROVE THEM Why do forecasts sometimes go wrong? There are a number of reasons for this unfortunate situation. For one, computer models have inherent flaws that limit the

accuracy of weather forecasts. For example, NWP models idealize the real atmosphere, meaning that each model makes certain assumptions about the atmosphere. These assumptions may be on target for some weather situations and be way off for others. Consequently, the computer may produce a prog that on one day comes quite close to describing the actual state of the atmosphere but is not so close on another. A forecaster who bases a prediction on an "off-day" computer prog may find a forecast of "rain and windy" turning out to be a day of "clear and colder."

Another forecasting problem arises because the majority of models are not global in their coverage, and errors are able to creep in along the model's boundaries. For example, a model that predicts the weather for North America may not accurately represent weather systems that move in along its boundary

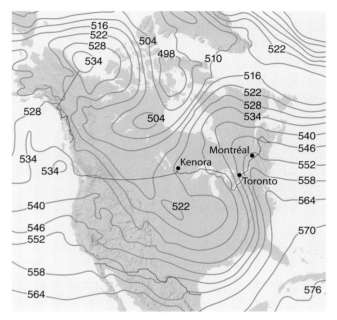

● FIGURE 2 A 1000 hPa to 500 hPa thickness chart for January 25, 2010, at 7:00 a.m. EST. The lines on the chart represent the vertical depth in decametres between the 1000 hPa and 500 hPa pressure surfaces, where 540 equals 5400 m. Low-thickness lines correspond to cold air and high-thickness lines to warm air. For reference, the 5400 m thickness line represents a vertical layer of air 5400 m thick with an average temperature of −7°C. The 520 thickness line is roughly the boundary for arctic air.

Source: Environment Canada. Operational Analysis Charts. Found at: www.weatheroffice.gc.ca/data/analysis/sai_100.gif. Accessed Jan. 25, 2010.

▼ Table 1   Rule of Thumb Relating Snow Level to 1000 to 500 hPa Thickness for Mountainous Areas

1000–500 hPa THICKNESS (decametres)	SNOW LEVEL (metres)
522	Near sea level
528	500
534	1000
540	1500
546	2000

receiving rain, whereas cities with a thickness less than 540 dam should be receiving snow. Look at Figure 2. Kenora, Toronto, and Montreal all received precipitation that day. Would you expect it to be in the form of rain or snow? (Kenora received 6.4 cm of snow, whereas Toronto received 10.8 mm and Montreal received 28.4 mm of rain.) If you live east of the Rockies, find your city on the chart and see whether precipitation would be in the form of rain or snow on this day.

West of the Rockies, the snow level (altitude where the snow turns to rain) decreases by 500 m for every 6 dam decrease in the 1000 to 500 hPa thickness, as shown in ▼ Table 1. This rule works quite well when the atmosphere is well mixed but not if there is a cold, stable layer near the surface.

Changes in the 1000 to 500 hPa thickness can also be related to changes in surface air temperature, even though the thickness actually represents the mean temperature in the layer. The rule of thumb is that, for the same time of day, when the thickness decreases by 2 dam, the surface air temperature decreases by approximately 1°C. Conversely, a 2 dam rise in thickness corresponds to an approximate 1°C rise in surface air temperature.

from the western Pacific. This kind of inaccuracy is probably why model (b) in Figure 13.5—a global model with a lower resolution—more accurately predicted the location of the low off the West Coast than did model (a), which is a nonglobal model with a higher resolution. Obviously, a global model would usually be preferred. But a global model of similar high resolution requires too many computations—the real weather might happen before the model has finished running!

Even though many thousands of weather observations are taken worldwide each day, there are still regions where observations are sparse, particularly over the oceans and at higher latitudes. To help alleviate this problem, the newest GOES satellite, with advanced atmospheric sounders, is providing a more accurate profile of temperature and humidity for the computer models. Wind information now comes from a variety of sources, such as Doppler radar, commercial aircraft, and satellites that translate ocean surface roughness into surface wind speed (see Chapter 9, p. 283).

Earlier, we saw that the computer solves the equations that represent the atmosphere at many locations called grid points, each spaced from 100 km to as low as 0.5 km apart. As a consequence, on computer models with large spacing between grid points (say 60 km), weather systems, such as extensive mid-latitude cyclones and anticyclones, show up on computer progs, whereas much smaller systems, such as thunderstorms, do not. The computer models that forecast for a large area such as North America are, therefore, better at predicting the widespread precipitation associated with a large cyclonic storm than local showers and thunderstorms. In summer, when much of the precipitation falls as local showers,

▼ Table 13.2 A Few Forecasting "Rules of Thumb"*

FORECAST QUESTION	USE OF FORECAST CHART
Cloudy or clear?	On the 700 hPa forecast chart, the 70 percent relative humidity line usually encloses areas that are likely to have clouds.
Will it rain?	(a) On the 700 hPa forecast chart, the 90 percent relative humidity line often encloses areas where precipitation is likely. If upward velocities are present, the chance of measurable precipitation is enhanced. (b) Along the West Coast of North America, precipitation is much more likely north of the 564 dam height contour on the 500 hPa forecast chart.
Will it rain or snow?	On the 850 hPa forecast chart, snow is likely north of the –5°C isotherm, whereas rain is likely south of this line. On the 1000 hPa to 500 hPa thickness chart, the 540 dam thickness line is widely used (east of the Rockies) as the dividing line between rain and snow. Over the mountains, the snow level is generally at sea level for a thickness of 522 dam and increases by 500 m in elevation for every 6 dam increase in 1000 to 500 hPa thickness.
Will the surface low intensify?	For the storm to intensify (deepen), an area of upper-level divergence must be over the surface cyclonic storm. On a 500 hPa forecast chart that shows vorticity, look for a vorticity maximum (vort max) and remember from Chapter 12, p. 372, that to the east of an area of positive vorticity, we usually find upper-level divergence, upward air motions, and cyclonic storm development.

*The forecast charts (progs) found in this table can be obtained from the Internet.

a computer prog may have indicated fair weather, although outside it is pouring rain.

To capture the smaller scale weather features as well as the terrain of the region, the distance between grid points on some models is being reduced. For example, the forecast model known as the Weather Research and Forecasting (WRF) model has a grid spacing as low as 0.5 km or less and is being run daily by government weather agencies and at universities. The MSC uses a high-resolution (small grid spacing) model, called the GEM Limited Area Model (GEM LAM). This model predicts mesoscale atmospheric conditions over a limited region, such as coastal areas and mountains, where terrain might greatly impact the local weather. The GEM LAM is run operationally over western Canada at a horizontal grid spacing of 2.5 km.* During the 2010 Winter Olympics in Vancouver and Whistler, a 1 km resolution GEM LAM model was run to support weather forecasting at Olympic venues.

The problem with models that have a small grid spacing (high resolution) is that as the spacing between grid points decreases, the number of computations increases dramatically. When the distance is halved, there are 16 times as many computations to perform because there are eight times more grid points and twice as many time steps required. If it is not necessary to increase the vertical resolution, then there would still be an eightfold increase in calculations.

Another forecasting problem is that many computer models cannot adequately interpret many of the small-scale processes that influence weather, such as the interactions of

water, ice, surface friction, small convective clouds, and local terrain on weather systems. Although large-scale models now take mountain regions and oceans into account, the processes that occur at scales smaller than the model grid spacing cannot be handled directly by the model but must still be accounted for in some way. This is done by *parameterizing* these factors. This means that the impacts of the small-scale processes are accounted for by simplified mathematical equations that relate the unresolved processes to the variables that are resolved by the model. These mathematical equations are called *parameterizations*.

For example, if the atmosphere has a large middle-latitude cyclone that is a thousand kilometres in size, then a model, such as the GEM Global model with 33 km resolution, can directly simulate the clouds in such a system because the horizontal grid spacing can represent them. Contrast this with the case of a cumulus cloud that is 1 km in diameter. In this situation, the model cannot properly represent that cumulus cloud, its updrafts and downdrafts, or its precipitation. The cumulus cloud would need to be parameterized in the model by examining the atmospheric stability and moisture from the model grid. If the atmosphere was sufficiently unstable, the parameterization would let it overturn and mix. If it was sufficiently moist, the parameterization would let a portion of the grid contain cloud and generate some precipitation. The equations in the model that make these adjustments are the convective parameterizations. Similarly, models have parameterization schemes for other small-scale processes. Since it is difficult to come up with parameterizations that work well in every situation, this is a major source of error in numerical weather prediction models, as well as in climate models.

*This was the horizontal grid spacing as of 2010. As computer power increases, the grid spacing continues to decrease to improve model predictions.

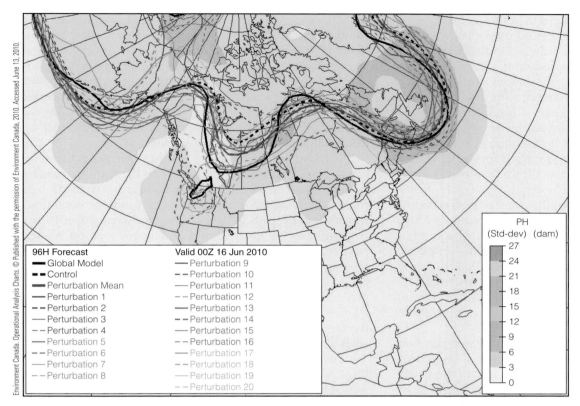

96H Forecast Valid 00Z 16 Jun 2010

— Global Model	— Perturbation 9
▪▪ Control	– – Perturbation 10
— Perturbation Mean	— Perturbation 11
— Perturbation 1	– – Perturbation 12
– ▪ Perturbation 2	— Perturbation 13
— Perturbation 3	– ▪ Perturbation 14
– – Perturbation 4	— Perturbation 15
— Perturbation 5	– – Perturbation 16
– ▪ Perturbation 6	— Perturbation 17
— Perturbation 7	– – Perturbation 18
– – Perturbation 8	— Perturbation 19
	– – Perturbation 20

PH
(Std-dev) (dam)

27
24
21
18
15
12
9
6
3
0

● FIGURE 13.7 Ensemble 500 hPa forecast chart valid at 00:00 UTC on June 16, 2010 (7:00 p.m. EST on June 15, 2010). This is a 96-hour (four-day) forecast that was started six days previous to this time, and it shows only the 558 dam height contour. The chart is constructed by running the model 22 different times, each time beginning with slightly different initial conditions and parameterization schemes. The heavy black line is the GEM Global model, the black dashed line is the ensemble control run, and the other colours are the 20 perturbation runs. The heavy red line represents the average of all the ensembles. The shading indicates the amount of variability (standard deviation) in the 500 hPa heights.

Even with better observing techniques and near-perfect computer models, there are countless small, unpredictable atmospheric fluctuations called **chaos** that limit model accuracy. For example, tiny eddies are much smaller than the grid spacing on the computer model and, therefore, go unaccounted for in the model. These small disturbances, as well as small errors (uncertainties) in the data, generally amplify with time as the computer tries to project the weather farther and farther into the future. After a number of days, these initial imperfections tend to dominate, and the forecast shows little or no accuracy in predicting the behaviour of the real atmosphere. In essence, what happens is that the small uncertainty in the initial atmospheric conditions eventually leads to a huge uncertainty in the model's forecast. This provides a limit on how far in the future we will ever be able to forecast.

Because of the atmosphere's chaotic nature, meteorologists are turning to a technique called **ensemble forecasting** to improve medium- and long-range forecasts. The ensemble approach is based on running several forecast models—or different versions (simulations) of a single model—each beginning with slightly different weather information to reflect the errors inherent in the measurements, as well as errors inherent in parameterizations. Suppose, for example, that a forecast model predicts the state of the atmosphere 24 hours into the future. For the ensemble forecast, the entire model simulation is repeated, but only after the initial conditions are "tweaked" just a little. The "tweaking," of course, represents the degree of uncertainty in the observations.

Repeating this process several times creates an ensemble of forecasts for a range of small initial changes.

● Figure 13.7 shows an ensemble 500 hPa forecast of the 558 dam contour for 7:00 p.m. EST June 15, 2010, as predicted 96 hours previously using the GEM Global model. The chart is constructed by running the model 22 different times, each time starting with slightly different initial conditions and using somewhat different parameterizations. Notice that the contours seem to agree quite well in the position of the trough over Atlantic Canada. A flow like this would bring cool air from northern Quebec to the region, so a forecaster might expect cool and showery conditions. The fact that the ensemble members are in good agreement in this region indicates that we can have high confidence in the forecast. In Charlottetown on Prince Edward Island, the actual daytime high on this day was only 13.2°C with a mixture of sun, clouds, and showers.

Contrast this with the situation over Alberta. Here the contour lines are widely spread, indicating that we should have lower confidence in the features because the atmosphere is less predictable in this region at that time. However, the variability of the 500 hPa heights indicated by the shading in Figure 13.7 is quite low, indicating that the wide spread in the 558 dam contour may be partly due to small variation in the 500 hPa heights in this region. As the forecast goes further and further into the future, the lines look more and more like scrambled spaghetti, which is why an ensemble forecast chart such as this one is often referred to as a *spaghetti plot*.

If, at the end of a specific time, the progs, or model runs, match each other fairly well, as they do over Atlantic Canada in Figure 13.7, the forecast is considered robust. This situation allows the forecaster to issue a prediction with a high degree of confidence. If the progs disagree, as they do over Western Canada in Figure 13.7, the forecaster, with little faith in the computer model prediction, issues a forecast with limited confidence. In essence, *the less agreement among the progs, or model runs, the less predictable the weather.* Consequently, it would not be wise to make outdoor plans for Saturday when on Monday the weekend forecast calls for "sunny and warm" with a low degree of confidence.

The use of ensembles for weather forecasting has become very popular and an active area of weather research. This is because ensemble forecasts can tell us something about how much confidence we can have in the prediction—very useful information indeed. Also, studies have shown that *the forecast made by the average of the ensembles* (the heavy red line in Figure 13.7) *is, on average, more accurate than the forecast made by any one of the ensemble members.* So ensemble forecasting is a way of improving the forecast.

In summary, imperfect numerical weather predictions may result from flaws in the computer models, from errors that creep in along the models' boundaries, from the sparseness of data, and/or from inadequate representation of many pertinent processes, interactions, and inherently chaotic behaviour that occur within the atmosphere.

Up to this point, we have looked primarily at weather forecasts made by high-speed computers running numerical weather prediction models. There are, however, manual forecasting methods, many of which have stood the test of time. Many of these techniques are of value but may give more of a general overview rather than a specific forecast. (Before going on, you may wish to read about how TV weather forecasters present the weather in Focus on an Observation: TV Weathercasters—How Do They Do It? on p. 395.)

OTHER FORECASTING METHODS   Probably the easiest weather forecast to make is a **persistence forecast**, which is simply a prediction that future weather will be the same as present weather. If it is snowing today, a persistence forecast would call for snow through tomorrow. Such forecasts are most accurate for time periods of several hours and become less and less accurate after that.

WEATHER WATCH

The first live person on Canadian television was a weatherperson! On September 8, 1952, the Canadian Broadcasting Corporation's CBLT-TV in Toronto started the country's first English-language broadcasting. Meteorologist Percy Saltzman was the first person to appear on screen. Saltzman continued to present the weather on TV for 30 years. He was famous for his entertaining and informative forecasts, which always ended with him tossing (and catching) his piece of chalk into the air, saying "And that's the weather picture."

Another method of forecasting is the **steady-state**, or **trend**, **forecast**. The principle involved here is that surface weather systems tend to move in the same direction and at approximately the same speed as they have been moving, provided that no evidence exists to indicate otherwise. Suppose, for example, that a cold front is moving eastward at an average speed of 30 km h^{-1} and is 90 km west of your home. Using the steady-state method, we might extrapolate and predict that the front should pass through your area in three hours.

The **analogue method** is yet another form of weather forecasting. Basically, this method relies on the fact that existing features on a weather chart (or a series of charts) may strongly resemble features that produced certain weather conditions sometime in the past. To the forecaster, the weather map "looks familiar," and for this reason, the analogue method is often referred to as **pattern recognition**. A forecaster might look at a prog and say, "I've seen this weather situation before, and this happened." Previous weather events can then be used as a guide to the future. The problem here is that even though weather situations may appear similar, they are never *exactly* the same. There are always sufficient differences in the variables to make applying this method a challenge.

Predicting the weather by **weather types** employs the analogue method. In general, weather patterns are categorized into similar groups or "types," using such criteria as the position of the subtropical highs, the upper-level flow, and the prevailing storm track. As an example, when the Pacific high is weak or depressed southward and the flow aloft is zonal (west to east), surface storms tend to travel rapidly eastward across the Pacific Ocean and into North America without developing into deep systems. But when the Pacific high is to the north of its normal position and the upper airflow is meridional (north–south), looping waves form in the flow, with surface lows usually developing into huge storms. As we saw in Chapter 12, these upper-level longwaves move slowly, usually remaining almost stationary for perhaps a few days to a week or more. Consequently, the particular surface weather at different positions around the wave is likely to persist for some time. ● Figure 13.8 presents an example of weather conditions most likely to prevail with a winter meridional weather type.

The analogue method can be used to predict a number of weather elements, such as maximum temperature. Suppose that in Toronto the average maximum temperature on a particular date for the past 30 years is 10°C. By statistically relating the maximum temperatures on this date to other weather elements—such as the wind, cloud cover, humidity, and 1000 to 500 hPa thickness—a relationship between these variables and maximum temperature can be drawn. By comparing these relationships with current weather information, the forecaster can predict the maximum temperature for the day.

Presently, **statistical forecasts** are made by relating past computer model forecasts with specific weather elements. Known as *Model Output Statistics*, or MOS, these predictions, in effect, are statistically weighted forecast corrections incorporated into the model output. For example, a forecast of tomorrow's maximum temperature for a city might be derived from

TV Weathercasters—How Do They Do It?

Claire Martin, Senior Meteorologist, CBC NEWS: Weather Centre

The job of broadcast meteorology has evolved, much like the actual medium of broadcasting, at breakneck pace over the past two decades. Where once there were monitors and even chalkboards, now there are "green screens" and "virtual sets." TV weathercasting is a slick blend of technology, science, and on-air performance.

First, the technology: most on-air weathercasters use a form of green- (or blue-) screen compositing (commonly referred to as using a "green screen") to appear on air with weather graphics. A green-screen composite image works by shooting a subject in front of an evenly lit, bright, pure green background, as shown in ● Figure 3. The compositing process replaces the entire green area viewed by the camera with another image, known as the background plate. That "other image" is the weather graphic—a satellite or radar map, for example.

Why green? Well, green paint has greater reflectance than blue paint, which can make compositing easier. Also, modern video cameras are usually most sensitive in the green channel and often have the best resolution and detail in that channel. The presenter has to be aware, however, not to dress in the same colour as the background screen (either green or blue) as everything that the camera sees in that colour will be digitally removed and replaced with the background image.

So if the presenter is standing in front of a blank green screen, how does he or she know where to point? Well, TV monitors are positioned at eye level on either side of the edge of the green screen, showing the "true" blend of the background plate and the presenter. The image on the monitor therefore is exactly what an everyday viewer would see if watching television. Note, however, that it can take awhile for a presenter to get used to using this effect— the image on the monitor is not a mirror

● FIGURE 3 Claire Martin of the CBC News Weather Centre in Vancouver, British Columbia, presenting an extended outlook on TV. She is standing in front of the "green screen," but TV viewers see what is shown in the monitor—the green area is electronically replaced by the "background plate"—in this case, a 3-month precipitation outlook for Canada.
Source: Claire Martin

image and hence not the view of himself or herself that a person is used to seeing.

Second, the science: In the early days of weather broadcasting, on-air presenters were simply that—presenters or performers. Nowadays, most on-air weather presenters are degreed meteorologists with a solid understanding of the science. Most do their own forecasting, and nearly all generate their own graphics with sophisticated computer weather graphical packages. The science has become a fundamental part of the broadcast, and the average viewer has come to recognize and even depend on that depth of knowledge.

Finally, the performance: Nearly 90% of Canadians get their local daily weather forecast from the television. It has become a trusted source of weather information for many, and as such, a TV weather presenter needs to

exhibit certain qualities to win and retain an audience. Accuracy is usually quoted as the number one draw for the average viewer, although nearly all understand that weather forecasting is not an exact science and that some predictions may be wrong. A presenter needs to be dependable (on air as frequently as possible) and trustworthy. Many broadcasters tell of times that their confidence in an individual forecast was exceedingly low, but that by passing on that particular information, they gained trust with their audience. Apparently, telling the audience that you have little faith in the forecast, and not hiding that detail, still wins them over. And lastly, the presenter needs to be likeable and "easy to watch." After all, there are 100 channels out there, and the audience can "vote" with just one click.

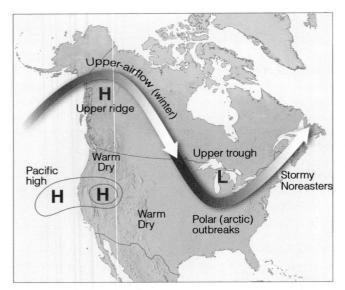

● FIGURE 13.8 Winter weather type showing upper airflow (heavy arrow), the surface position of the Pacific high, and general weather conditions that should prevail.

▼ Table 13.3 Forecast Wording Used by the Meteorological Service of Canada to Describe the Chance of Precipitation (0.2 mm or greater) for Steady Precipitation and for Convective, Showery Precipitation

PERCENT PROBABILITY OF PRECIPITATION	FORECAST WORDING FOR STEADY PRECIPITATION	FORECAST WORDING FOR SHOWERY PRECIPITATION
0 to 20 percent		no mention of precipitation in the forecast
30 to 40 percent	light snow, periods of rain	a few showers, flurries
50 percent	not used	
60 to 70 percent	light snow, periods of rain	showers, flurries
> 80 percent*	rain, snow; rain/ snow at times heavy	showers, flurries; showers/flurries at times heavy

*If the chance of precipitation is 80 percent or more, the percentage is not mentioned, just the precipitation type. For example, a forecast that calls for an 80 percent chance of rain in the afternoon might read like this: "... cloudy today with rain this afternoon...." For an 80 percent chance of rain showers, the forecast might read "... cloudy today with rain showers this afternoon...."

a statistical equation that uses a numerical model's forecast of relative humidity, cloud cover, wind direction, air temperature, and 1000 to 500 hPa thickness. Such statistical equations are derived for many weather elements at cities where forecasts are needed and are usually more accurate than the uncorrected direct numerical weather prediction model output.

When the MSC issues a forecast calling for precipitation, it is often followed by a probability—for example, "60 percent chance of showers." Does this mean (a) that it will shower on 60 percent of the forecast area or (b) that there is a 60 percent chance that it will shower within the forecast area? Neither one! The expression means that there is a 60 percent chance that any random place in the forecast area, such as your home, will receive measurable* precipitation as a shower. Looking at the forecast in another way, if the forecast for 10 days calls for a 60 percent chance of showers, it should shower where you live on six of those days. The verification of the forecast (as to whether or not it actually rained) is usually made at a particular weather station, usually an airport within the forecast region. However, the computer models forecast for a given region, not for an individual location. When the Meteorological Service of Canada issues a forecast calling for a "a few showers," what is the probability (percentage) that it will shower? ▼ Table 13.3 provides this information.

A forecast based on the climate† of a particular region is known as a **climatological forecast**. For example, one way of forecasting the probability of rain in a given city would be to see how often it has rained before during that month. If it rains an average of three days per month, then the climatological forecast would be that there is a 10 percent chance of rain.

---

*Measurable precipitation is defined to be an amount greater than 0.2 mm.

†The climate of a region is represented by the average and variation in daily and seasonal weather elements for a specific interval of time, most often 30 years.

SCRIBE—HOW WORDED PUBLIC FORECASTS ARE PRODUCED IN CANADA We have seen that there are a variety of forecast techniques available to the forecaster. How is all of this information integrated by meteorologists at the Meteorological Service of Environment Canada to come up with the worded public forecast? This is actually done graphically on the Integrated Forecaster Workstation described previously, using a computer program called *Scribe*. Scribe takes output from a numerical weather prediction model that predicts weather elements such as cloud amount; precipitation type, probability, and amount; temperature; wind speed and direction; and any warnings. It places these hourly predictions in a computer window as a set of time series graphs. These graphs can be adjusted interactively by the forecaster using a computer mouse to adjust the computer model prediction as shown in ● Figure 13.9a. When the forecaster is satisfied with the forecast for each weather element, a button is clicked and the worded forecast (in both official languages) is automatically generated (see Figure 13.9b).

TYPES OF FORECASTS Weather forecasts are normally grouped according to how far into the future the forecast extends. For example, a weather forecast for up to a few hours (usually not more than 6 hours) is called a **very short-range forecast**, or *nowcast*. The techniques used in making such a forecast normally involve subjective interpretations of surface observations, satellite imagery, and Doppler radar information. Often weather systems are moved along by the

(a)

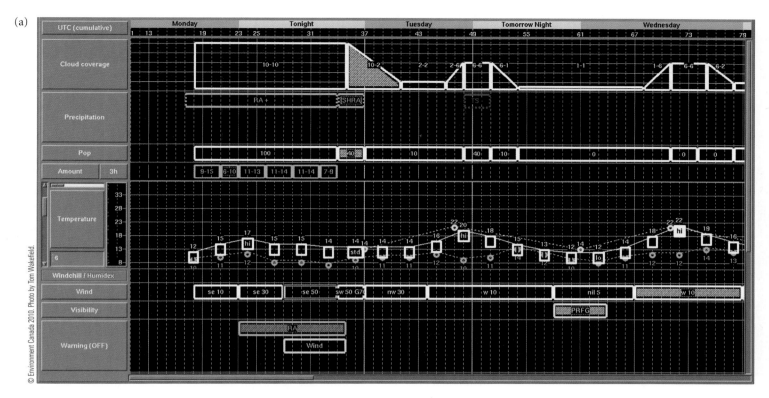

(b)

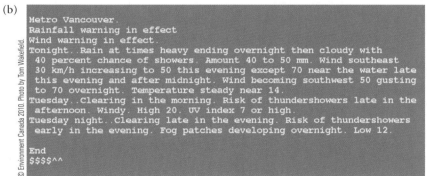

```
Metro Vancouver.
Rainfall warning in effect.
Wind warning in effect.
Tonight..Rain at times heavy ending overnight then cloudy with
  40 percent chance of showers. Amount 40 to 50 mm. Wind southeast
  30 km/h increasing to 50 this evening except 70 near the water late
  this evening and after midnight. Wind becoming southwest 50 gusting
  to 70 overnight. Temperature steady near 14.
Tuesday..Clearing in the morning. Risk of thundershowers late in the
  afternoon. Windy. High 20. UV index 7 or high.
Tuesday night..Clearing late in the evening. Risk of thundershowers
  early in the evening. Fog patches developing overnight. Low 12.


End
$$$$^^
```

● FIGURE 13.9 The Scribe forecast production system.
(a) The graphical window in which a meteorologist can adjust
the forecast weather elements for a city or region. Once the
forecaster is happy with the forecast conditions, a button is
clicked and (b) the worded forecast is automatically produced.

steady-state or trend method of forecasting, with human experience and pattern recognition coming into play.

Weather forecasts that range from about 12 hours to a few days (generally 2 days or 48 hours) are called **short-range forecasts**. The forecaster may incorporate a variety of techniques in making a short-range forecast, such as satellite imagery, Doppler radar, surface weather maps, upper-air winds, and pattern recognition. As the forecast period extends beyond about 12 hours, the forecaster tends to weight the forecast heavily on numerical weather prediction (computer progs) and model output statistics (MOS).

A **medium-range forecast** is one that extends from day 3 to day 7 and beyond to day 14 (168 to 336 hours) into the future. Medium-range forecasts are almost entirely based on numerical weather prediction outputs, such as MOS or ensembles. A forecast that extends beyond 2 days is often called an *extended forecast.*

A forecast beyond about 14 days (336 hours) is called a **long-range forecast**. Although computer progs are available for up to 14 days into the future, they are not accurate in predicting temperature and precipitation and at best show only the broad-scale weather features. Presently, the MSC issues long-range forecasts, called *outlooks*, of whether it is expected to be warmer or colder and wetter or drier than normal for a particular three-month period up to 12 months in the future. The forecast for the first four months is determined using four numerical weather prediction models, each run with 10 ensembles. Beyond four months, a statistical model is used. These are not forecasts in the strict sense but rather an overview of how precipitation and temperature patterns may compare with normal conditions. ● Figure 13.10 gives a typical three-month outlook. Long-range forecasts are made from models that link the atmosphere with the ocean surface temperature. As we saw in Chapter 10, a vast warming (*El Niño*) or

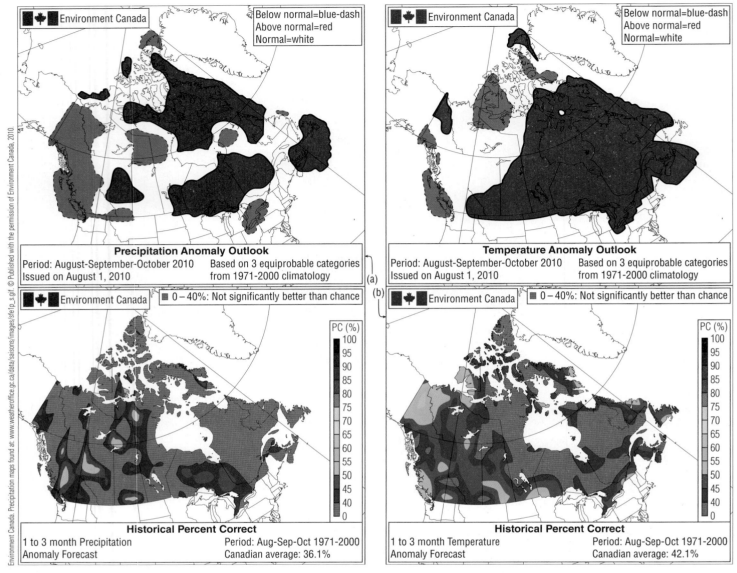

● FIGURE 13.10 The 90-day outlook for (a) precipitation and (b) temperature for August, September, and October 2010 from the Meteorological Service of Environment Canada. In each figure, blue indicates below normal, and red indicates above normal. The historical percent correct is indicated in the bottom panels.

cooling (*La Niña*) of the tropical eastern Pacific can affect the weather in different regions of the world. These interactions, where a warmer tropical Pacific can influence rainfall across North America, are called **teleconnections**.*

These types of interactions between widely separated regions are identified through statistical correlations. For example, look back at Figure 10.23 (p. 314) and observe where seasonally averaged temperature and precipitation patterns over North America tend to depart from normal during El Niño and La Niña events. Using this type of information, the Meteorological Service of Environment Canada

can issue a seasonal outlook of an impending wetter or drier winter, months in advance. Forecasts using teleconnections have shown promise. For example, as the tropical equatorial Pacific became much warmer than normal during the spring and early summer of 1997, forecasters predicted a wet rainfall season over central and southern California. Although the heavy rains did not begin until December, the weather during the winter of 1997–98 was wet and wild: storm after storm pounded the region, producing heavy rains, mudslides, road closures, and millions of dollars in damages.

In most locations throughout North America, the weather is fair more often than rainy. Consequently, there is a forecasting bias toward fair weather, which means that if you made a forecast of "no rain" where you live for each day of the year, your forecast would be correct more than 50 percent of the

*Teleconnections include not only El Niño and La Niña but also other phenomena, such as the Pacific Decadal Oscillation, the North Atlantic Oscillation, and the Arctic Oscillation. (See Chapter 10, p. 310.)

time. But did you show any *skill* in making your correct forecast? What constitutes skill anyway? And how accurate are the forecasts issued by the Meteorological Service of Canada?

Accuracy and Skill in Forecasting

In spite of the complexity and ever-changing nature of the atmosphere, forecasts made for between 12 and 24 hours are usually quite accurate. Those made for between two and five days are fairly good. Beyond about seven days, due to the chaotic nature of the atmosphere, numerical weather prediction accuracy falls off rapidly. Although weather predictions made for up to 3 days are by no means perfect, they are far better than simply flipping a coin. But how accurate are they?

One problem with determining forecast accuracy is deciding what constitutes a right or wrong forecast. Suppose that tomorrow's forecast calls for a minimum temperature of 3°C. If the official minimum turns out to be 5°C, is the forecast incorrect? Is it as incorrect as a forecast that is off by 5°? By the same token, what about a forecast for snow over a large city in a situation when the snow line cuts the city in half, with the southern portion receiving heavy amounts and the northern portion none? Is the forecast right or wrong? At present, there is no clear-cut answer to the question of what an accurate forecast is.

How does forecast accuracy compare with forecast skill? Suppose that you are forecasting the daily summertime weather in Vancouver. It is not raining today, and your forecast for tomorrow calls for "no rain." Suppose that tomorrow it does not rain. You made an accurate forecast, but did you show any skill? For a forecast to show skill, it should be better than one based solely on the current weather (*persistence*) or on the "normal" weather (*climatology*) for a given region.

Meteorological forecasts, then, show skill when they are more accurate than a forecast using only persistence or climatology. Persistence forecasts are usually difficult to improve on for a period of time of several hours or less. Weather forecasts ranging from 12 hours to a few days generally show much more skill than those of persistence. However, as the range of the forecast period increases, because of chaos, the skill drops quickly. The eight- to 14-day mean outlooks both show some skill (which has been increasing over the last several decades) in predicting temperature and precipitation. However, the accuracy of precipitation forecasts is less than that for temperature. Presently, seven-day forecasts now show about as much skill as three-day forecasts did a decade ago. Beyond 15 days, specific forecasts are only slightly better than climatology. However, skill in making forecasts of average monthly temperature and precipitation has approximately doubled from 1995 to 2006.

Forecasting large-scale weather events several days in advance (such as a major blizzard) is far more accurate than forecasting the precise evolution and movement of small-scale, short-lived weather systems, such as tornadoes and severe thunderstorms. ● Figure 13.11 shows the history of improvement of the Canadian Meteorological Centre's numerical weather prediction of the sea-level pressure gradient and

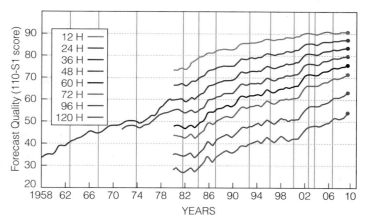

● **FIGURE 13.11** Changes in numerical weather prediction forecast quality at the Canadian Meteorological Centre from 1958 until mid-2009. The "forecast quality" measures the accuracy at predicting sea-level pressure patterns and gradients. The scale is arbitrary, but a forecast quality value of 100 is thought to represent the best realistically possible value. The vertical lines indicate when the computer systems used to make the predictions were upgraded, allowing higher resolution and more realistic predictions. The different coloured lines represent the forecast length, from 12 hours to 5 days (120 hours). Notice that today's five-day forecast is better than a two-day prediction from 30 years ago. Improvements in forecast quality correspond to both improvements in computer power, computer programs, and weather observations.
Source: Environment Canada. 2010.

pattern between 1958 and mid-2009. The figure shows that three-day forecasts of the surface pressure pattern show more skill today than 24-hour forecasts did 24 years ago, 36-hour forecasts did 18 years ago, and 48-hour forecasts did 8 years ago.

Even though the *precise* location where a tornado will form is presently beyond modern forecasting techniques, the general area where the storm is *likely* to form can often be predicted up to three days in advance. With improved observing systems, such as Doppler radar and advanced satellite imagery, the lead time of watches and warnings for severe storms has increased. In fact, the lead time* for tornado warnings has more than doubled over the last decade, with the average lead time today being close to 15 minutes.

Although scientists may never be able to skillfully predict the weather beyond about 15 days using available observations, the prediction of *climatic trends* appears to be more promising. Whereas individual weather systems vary greatly and are difficult to forecast very far in advance, global-scale patterns of winds and pressure frequently show a high degree of persistence and predictable change over periods of a few weeks to a month or more. With the latest generation of supercomputers, general circulation models (GCMs), which now interact with the oceans, are doing a far better job at predicting large-scale atmospheric behaviour than previously. (GCMs are similar to numerical weather prediction

---

Lead time is the interval of time between the issue of the warning and actual observance of the event, in this case, the tornado.

models, but they simulate global patterns of wind, pressure, and temperature and how these phenomena change over longer time periods.) In fact, the new GCMs are able to simulate a number of global patterns quite well, such as *blocking highs*** that can cause precipitation and temperature patterns to deviate considerably from average conditions. As GCMs continue to improve, it is hoped that they will become a reliable tool in the forecasting of seasonal climate, as well as being useful in assessing climate change. (In Chapter 16, we will examine in more detail the climatic predictions based on GCMs, also called *climate models*.)

BRIEF REVIEW

Up to this point, we have looked at the various methods of weather forecasting. Before going on, here is a review of some of the important ideas presented so far:

- A number of tools are available to the forecaster, including surface and upper-air maps, computer progs, meteograms, soundings, Doppler radar, and satellite information.

- The forecasting of weather by computers models is known as *numerical weather prediction*. Mathematical models that describe how atmospheric temperature, pressure, winds, and moisture will change with time are programmed into the computer. The computer takes observed conditions and steps them forward in time to produce a weather forecast. It then draws surface and upper-air forecast charts called *progs*.

- After a number of days, flaws in the computer models—atmospheric chaos and small errors in the initial data—greatly limit the accuracy of weather forecasts.

- Ensemble forecasting is a technique based on running several forecast models (or different versions of a single model), each beginning with slightly different weather information to reflect errors in the measurements. Ensembles give information about the reliability of a forecast, and the ensemble average is generally more accurate than any one of the ensemble members.

- A *persistence forecast* is a prediction that future weather will be the same as the present weather, whereas a *climatological forecast* is based on the climatology of a particular region.

- For a forecast to show skill, it must be better than a persistence forecast or a climatological forecast.

- Weather forecasts for up to a few hours are called *very short-range forecasts*; those that range from about 6 hours to two days are called *short-range forecasts*; *medium-range forecasts* extend from about 2 to 7 days into the future, whereas *long-range forecasts* extend beyond, to about 14 days.

- Seasonal outlooks provide an overview of how temperature and precipitation patterns may compare to normal conditions.

---

*Blocking highs are high-pressure areas that tend to remain nearly stationary for some time, thus "blocking" the west-to-east movement of mid-latitude cyclonic storms.

PREDICTING THE WEATHER FROM LOCAL SIGNS

Because the weather affects every aspect of our daily lives, attempts to predict it accurately have been made for centuries. One of the earliest attempts was undertaken by Theophrastus, a pupil of Aristotle, who in 300 B.C. compiled all sorts of weather indicators in his *Book of Signs*. A dominant influence in the field of weather forecasting for 2000 years, this work consists of ways to foretell the weather by examining natural signs, such as the colour and shape of clouds and the intensity at which a fly bites. Some of these signs have validity and are part of our own weather folklore—"a halo around the moon portends rain" is one of these. Today, we realize that the halo is caused by the bending of light as it passes through ice crystals and that ice crystal–type clouds (cirrostratus) are often the forerunners of an approaching storm (see ● Figure 13.12).

Weather predictions can be made by observing the sky and using a little weather wisdom. If you keep your eyes open and your senses keenly tuned to your environment, you should, with a little practice, be able to make fairly good short-range local weather forecasts by interpreting the messages written in the weather elements. ▼ Table 13.4 is designed to help you with this endeavour.

The movement of clouds at different levels can assist you in predicting changes in the temperature of the air above you. Knowing how the temperature is changing aloft can help you predict the stability of the air (see Chapter 6), as well as whether falling snow will change to rain, or vice versa. (This topic is explored more extensively in Focus on an Observation: Forecasting Temperature Advection by Watching the Clouds on p. 402.)

To further help you forecast the weather, the instant weather forecast chart (see Appendix E) has been prepared by considering the relationship that the pressure and wind have to various weather systems. Although the chart is applicable to much of North America, local influences, such as mountains

© C. Donald Ahrens

● FIGURE 13.12 A halo around the sun (or moon) means that rain is on the way—a weather forecast made by simply observing the sky.

▼ Table 13.4   Forecast at a Glance—Forecasting the Weather from Local Weather Signs*

OBSERVATION	INDICATION	LOCAL WEATHER FORECAST
Surface winds from the S or from the SW; clouds building to the west; warm (hot) and humid (pressure falling)	Possible cool front and thunderstorms approaching from the west	Possible showers; possibly turning cooler; windy
Surface winds from the E or from the SE, cool or cold; high clouds thickening and lowering; halo (ring of light) around the sun or moon (pressure falling)	Possible approach of a warm front	Possibility of precipitation within 12–24 hours; windy (rain with possible thunderstorms during the summer; snow changing to ice pellets or rain in winter)
Strong surface winds from the NW or W; cumulus clouds moving overhead (pressure rising)	A low-pressure area may be moving to the east, away from you, and an area of high pressure may be moving toward you from the west	Continued clear to partly cloudy, cold nights in winter; cool nights with low humidity in summer
Winter night		
(a) If clear, relatively calm with low humidity (low dew point temperature) (b) If clear, relatively calm with low humidity and snow covering the ground (c) If cloudy, relatively calm with low or high humidity	(a) Rapid radiational cooling will occur (b) Rapid radiational cooling will occur (c) Clouds will absorb and radiate infrared (IR) energy to surface	(a) A very cold night (b) A very cold night with minimum temperatures lower than in (a) (c) Minimum temperature will not be as low as in (a) or (b)
Summer night		
(a) Clear, hot, humid (high dew points) (b) Clear and relatively dry	(a) Strong absorption and emission of IR energy to surface by water vapour (b) More rapid radiational cooling	(a) High minimum temperatures (b) Lower minimum temperatures
Summer afternoon		
(a) Scattered cumulus clouds that show extensive vertical growth by mid-morning (b) Afternoon cumulus clouds with limited vertical growth and with tops at just about the same level	(a) Atmosphere is relatively unstable (b) Stable layer above clouds (region dominated by high pressure)	(a) Possible showers or thunderstorms by afternoon with gusty winds (b) Continued partly cloudy with no precipitation; probably clearing by nightfall

*The table shows a few forecasting rules that may be applied when making a short-range local weather forecast.

and large bodies of water, can affect the local weather to such an extent that the large-scale weather patterns on which the chart is based do not always show up clearly. (The chart works best during the fall, winter, and spring, when the weather systems are active. Try it yourself—see if you can use the information in this chart to make your own prediction and then compare this to the MSC forecast.)

Weather Forecasting Using Surface Charts

We are now in a position to forecast the weather, using more sophisticated techniques. Suppose, for example, that we wish to make a short-range weather prediction and the only information available is a surface weather map. Can we make a forecast from such a chart? Most definitely. And our chances of that forecast being correct improve markedly if we have maps available from several days back. We can use these past

FOCUS ON AN OBSERVATION

Forecasting Temperature Advection by Watching the Clouds

We know from Chapter 12 that when cold air is being brought into a region by the wind, we call this cold advection. When warm air is brought into a region, we call this warm advection.

A knowledge of temperature advection aloft is a valuable tool in forecasting the weather. In summer, when the surface is warm, cold advection aloft sets up instability and increases the likelihood of towering cumulus clouds and showers. On the other hand, warm advection aloft usually increases the temperature of the air there, thus making it more stable. During the winter, this often leads to smoke and haze accumulating in the colder air near the surface.

By watching the movement of clouds, we get a good indication of the wind direction at cloud level and the type of advection. Clouds at different levels frequently move in different directions, meaning that the wind direction is changing with height. Wind that changes direction in a clockwise sense (north to northeast to east, etc.) is a *veering* wind. Wind that changes direction in a counterclockwise sense (north to northwest to west, etc.) is a *backing* wind. There are two general rules that will help us determine whether cold or warm advection is occurring in a layer of air above us:

1. winds that back with height (change counterclockwise) indicate *cold advection*
2. winds that veer with height (change clockwise) indicate *warm advection*

As an example, suppose that we observe lower clouds moving from a southerly direction (south wind) and higher clouds moving from a westerly direction (west wind) (see ● Figure 4a). The wind direction is veering with height; warm advection is occurring between the cloud layers, and the air should be getting warmer. If, on another day, we see lower clouds moving from a southerly direction and higher clouds moving from an easterly direction (see Figure 4b), the wind is backing with height and the atmosphere between the cloud layers is probably becoming colder.

An example of the relationship between winds and advection is seen in the vertically shifting winds that accompany weather fronts. ● Figure 5b is a three-dimensional model of a typical open-wave cyclone with its accompanying warm and cold fronts. Behind the cold front, swiftly moving cumulus clouds indicate that a northwesterly wind exists about a kilometre above the surface. Ahead of the advancing warm front, stratiform clouds indicate that southeasterly winds prevail about a kilometre above the ground. We know from Chapter 12 that warm advection takes place *ahead* of the warm front and cold advection *behind* the cold front. In both cases, the advection usually occurs in a layer from the surface up to at least the 500 hPa level.

At the 500 hPa level (see Figure 5a), the position of the upper trough and the region of coldest air is to the west of the surface low. The direction of the wind and the cloud

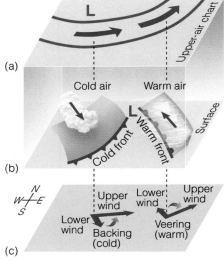

● FIGURE 5 Clouds, winds, and advection associated with a cold and a warm front.

movement are shown by arrows. Because of the upper trough's position, the winds aloft are westerly behind the cold front and southwesterly ahead of the warm front. Figure 5c shows how the wind direction changes from the surface to the 500 hPa level. Behind the cold front, the winds back from northwesterly to westerly as we move upward. Cold advection is taking place as chilling air moves in from the west. Just ahead of the approaching warm front, the wind veers with height from southeasterly at the surface to southwesterly aloft as warm air glides up and over the cool surface air.

We can use this information to improve on a weather forecast. For instance, if you happen to be located ahead of an advancing warm front and the winds above you are veering with height, the chances are that even if precipitation begins as snow, if it is warm enough, it may change to rain as warm air moves in overhead. Behind a cold front where winds are backing with height, the influx of cold air may lower the temperature sufficiently so that rain first becomes mixed with snow and then changes to snow before the storm moves eastward.

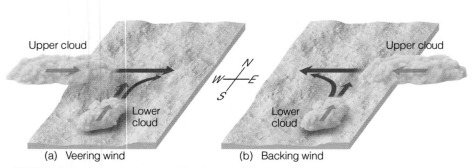

● FIGURE 4 (a) The wind veers with height, suggesting that warm advection is occurring between the cloud layers. (b) The wind is backing with height, and cold advection is occurring between the cloud layers.

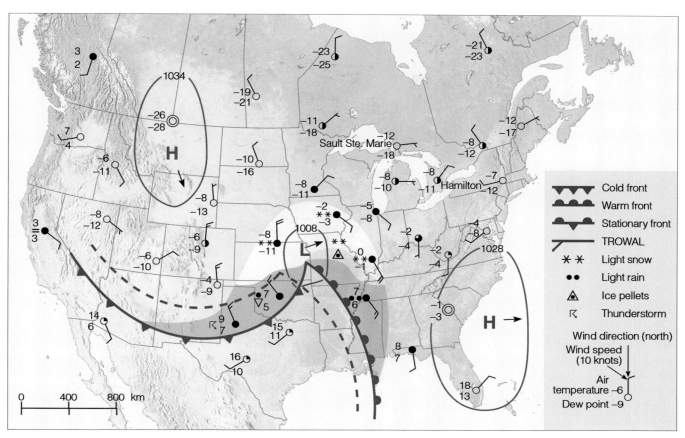

● FIGURE 13.13 An early-winter storm as shown on a surface weather map for 6:00 a.m. Tuesday. Dashed lines indicate positions of weather features six hours ago. Areas shaded green are receiving rain, whereas areas shaded white are receiving snow, and those shaded pink are receiving freezing rain or ice pellets. Note that, for simplicity, only the polar-front location is indicated. In the Canadian frontal model, there would be an additional front—an arctic front north of and parallel to the polar front, which approximately separates rain from snow and freezing rain.

maps to locate the previous position of surface features and use the past trend to predict their movement.

A simplified surface weather map is shown in ● Figure 13.13. The map portrays early winter weather conditions on Tuesday morning at 6:00 a.m. A single isobar is drawn around the pressure centres to show their positions without cluttering the map, and only the location of the main front—the polar front—is shown. Note that an open-wave cyclone is developing over the U.S. Central Plains. The weather conforms to the cyclone model (see Figure 12.12, p. 365), with showers forming along the cold front and light rain, snow, and ice pellets ahead of the warm front. The dashed lines on the map represent the position of the weather systems six hours ago. Our first question is: How will these systems move?

DETERMINING THE MOVEMENT OF WEATHER SYSTEMS

There are several methods we can use in forecasting the movement of surface pressure systems and fronts. The following are a few of these forecasting rules of thumb:

1. For short time intervals, mid-latitude cyclonic storms and fronts tend to move in the same direction and at approximately the same speed as they did during the previous six hours (provided, of course, that there is no evidence to indicate otherwise).

2. Low-pressure areas tend to move in a direction that parallels the isobars in the warm air (the warm air sector) ahead of the cold front.

3. Lows tend to move toward the region of greatest pressure drop, whereas highs tend to move toward the region of greatest rise.

4. If lows are in an elongated trough of low pressure, they tend to move along the trough axis.

5. Lows tend to move with the 500 hPa vorticity maxima and shortwave trough.

6. Lows tend to move along a line of constant 1000 to 500 hPa thickness.

7. Surface pressure systems tend to move in the same direction as the wind at the 500 hPa level. The speed at which surface systems move is about half the speed of the winds at this level.

When the surface map (see Figure 13.13) is examined carefully and when rules of thumb 1 and 2 are applied, it appears that—based on present trends—the low pressure centred over the Central Plains should move northeast.

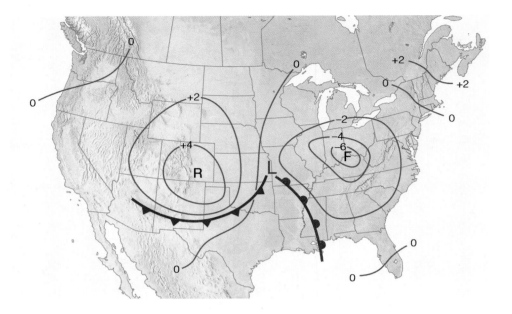

● FIGURE 13.14 Isallobars—lines of equal three-hour pressure change—for 6:00 a.m. Tuesday. The "F" represents the region of greatest pressure fall, whereas the "R" shows the region of greatest pressure rise. A + 2 indicates a rise of 2 hPa over three hours.

If pressure tendencies* are plotted on our map, we can draw lines connecting points of equal pressure change. These lines, called **isallobars**, help us visualize the regions of falling and rising pressure. The distribution of pressure change for our map might look like the one in ● Figure 13.14. Drawn at 2 hPa intervals, the isallobars show a broad region of falling pressure ahead of the warm front, with the largest drop occurring to the northeast of the low-pressure area. This pattern fits with the previous observations and strengthens the prediction that the low centre will move toward the northeast. The area of rising pressure immediately behind the cold front suggests that the high-pressure area over Montana will continue to move southeastward.

Pressure tendencies not only help predict the movement of highs and lows, they also indicate how the pressure systems are changing with time. The rapid fall in pressure in advance of the low and the fact that the low itself is in an area of falling pressure indicate that the storm centre is deepening as it moves. A deepening low means more closely spaced isobars, a greater pressure gradient, and stronger winds—something to take into account when we make our weather forecast. A drop in pressure, on the other hand, in the vicinity of an anticyclone suggests that it is weakening, whereas a rise in pressure means that its central pressure is increasing. Hence, the high-pressure area over Montana is strong (1034 hPa) and will remain so, whereas the high centred off the South Carolina coast is either moving eastward or weakening rapidly, as indicated by the falling pressure in that area.

Before we complete our prediction about the movement of the pressure centres in Figure 13.13, we need to look closely at the high-pressure area off the South Carolina coast. Strong highs, especially slow-moving ones, often retard the eastward

progress of lows, deflecting them either north or south. From all indications—falling pressures and past movement—this anticyclone is weakening and drifting slowly eastward. It should, therefore, pose no immediate problem to the northeastward movement of the storm centre.

Even if we do not have access to pressure tendencies or previous weather maps, we can make an initial approximation of how pressure systems will move by using Figure 12.5 (see p. 357), which shows the average tracks of lows and highs during the winter months. From this diagram, it appears that the cyclones and anticyclones in Figure 13.13 are following rather typical trajectories.

If a 500 hPa chart is available (such as ● Figure 13.15), it would also indicate how the surface pressure systems should move because the winds at this level tend to steer these systems along. From Figure 13.15, it appears that, indeed, the surface low should move northeastward.

A FORECAST FOR HAMILTON AND SAULT STE. MARIE
Our objective now is to make short-range weather forecasts for Hamilton and Sault Ste. Marie, Ontario. To do this, we will project the surface pressure systems, fronts, and current weather into the future by assuming steady-state trends. ● Figure 13.16 gives the 12- and 24-hour projected positions of these features.

WEATHER WATCH

Groundhog Day, February 2, derives from certain religious ceremonies performed before the birth of Christ. Somehow, this date came to be considered the midpoint of winter, and people, in an attempt to forecast what the remaining half would be like, placed the burden of weather prognostication on the backs (or, rather, the shadows) of animals such as the groundhog.

*The *pressure tendency* is the rate at which the pressure is changing during a given time, usually the past three hours.

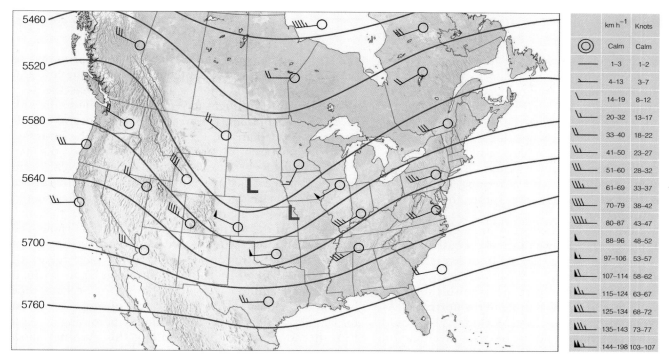

● FIGURE 13.15 A 500 hPa chart for 6:00 a.m. Tuesday showing wind flow. The red "L" represents the position of the surface low. The winds aloft tend to steer surface pressure systems along and, therefore, indicate that the surface low should move northeastward at about half the speed of the winds at this level, or 25 knots (46 km h⁻¹). Solid lines are contours in metres above sea level.

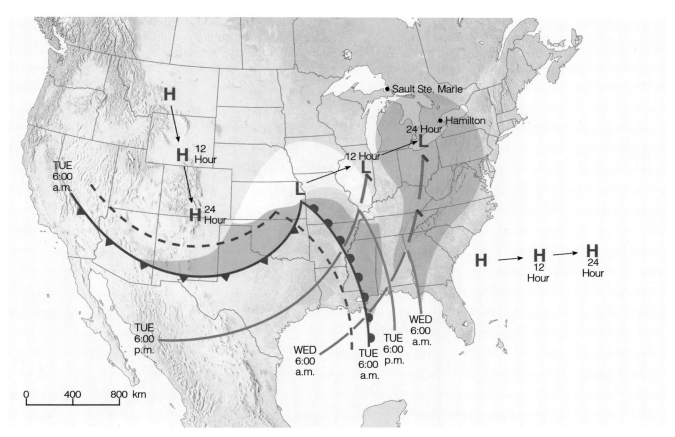

● FIGURE 13.16 Projected 12- and 24-hour movement of fronts, pressure systems, and precipitation from 6:00 a.m. Tuesday until 6:00 a.m. Wednesday. (The dashed lines represent frontal positions six hours ago.) Note that, for clarity, only the polar-front location is indicated. In the Canadian frontal model, there would also be an arctic front, north of and parallel to the polar front, that approximately separates rain from snow and freezing rain.

A word of caution before we make our forecasts. We are assuming that the pressure systems and fronts are moving at a constant rate. This may or may not occur. Low-pressure areas, for example, tend to accelerate until they occlude, after which their rate of movement slows. Furthermore, the direction of moving systems may change due to blocking highs and lows that exist in their path or because of shifting upper-level wind patterns. We will assume a constant rate of movement and forecast accordingly, always keeping in mind that the longer our forecasts extend into the future, the more susceptible they are to error.

If we move the low- and high-pressure areas eastward, as illustrated in Figure 13.16, we can make a basic weather forecast for various regions. As the cold front sweeps southeastward through Texas and the U.S. Gulf states, they might expect showers with a chance of thundershowers followed by clearing, cooling, and winds shifting to the northwest. As the low moves northeastward and occludes, we would expect increasing cloudiness and the onset of precipitation in southwest Ontario and Hamilton. But what about Sault Ste. Marie? Will it see precipitation from this storm?

Snow for Hamilton As the low-pressure centre and TROWAL approach southwest Ontario, we expect to see winds increase from the east, skies become overcast with progressively lower clouds, and then the onset of precipitation. But will it fall as snow, rain, or possibly freezing rain or ice pellets? We can use some clues. First, notice that as the cold front overtakes the warm front and occlusion occurs,

the surface polar cold and warm fronts are projected to remain in the southeast United States, with the low and TROWAL affecting Ontario. This means that the very warmest surface air ought to remain well to the south with the surface warm front. The area of freezing rain and ice pellets (pink-shaded area in Figure 13.13) would be separated from the rain area by the arctic front (not drawn on the map for clarity) that is perhaps 200 km ahead of the polar warm front and may be tied more to the position of the surface low. Therefore, we can perhaps expect that this area should remain south of the Great Lakes. In Figure 13.13, the snow areas ahead of the warm front are receiving light snow. The rate at which snow accumulates depends on its intensity but also on the moisture content of the air—typical values for light snow are about 1.5 cm per hour. From Figure 13.16, it seems that Hamilton will receive snow for perhaps nine hours, so it should receive about 14 cm of snow. A forecast for Hamilton might read: "Mostly cloudy, becoming overcast this afternoon with snow beginning this evening. Snow amounts of 14 cm expected overnight. Winds becoming strong from the east overnight."

Clouds for Sault Ste. Marie Our projected area of precipitation in Figure 13.16 is just to the south of Sault Ste. Marie; however, it is likely that clouds at least from the storm will affect this area. Since our projection may not be accurate, it is also possible that there could be precipitation from the storm affecting the city. If there is any precipitation, it would fall as snow because even the arctic front (not drawn) would

● FIGURE 13.17 Surface weather map for 6:00 a.m. Wednesday. On this map, the arctic front that was omitted in Figures 13.12 and 13.15 has been added.

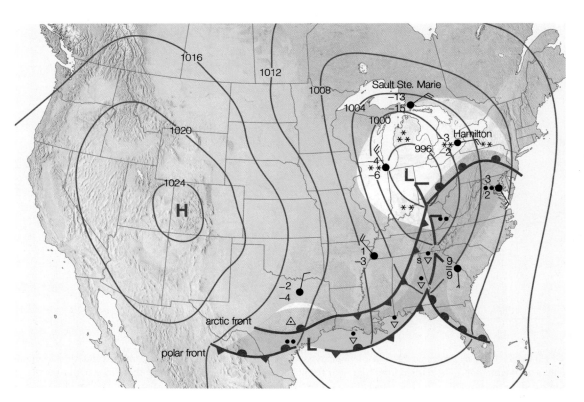

be well to the south. We could indicate uncertainty in the snow forecast like this: "Becoming cloudy this evening with a chance of snow developing overnight. Winds becoming strong northeasterly overnight."

What actually happened? A complete picture of the surface weather systems for 6:00 a.m. Wednesday morning is given in ● Figure 13.17. By comparing this chart to Figure 13.16, we can summarize why our forecasts did not turn out exactly as we had predicted. For one thing, the storm centre south of the Great Lakes moved slower than expected. This reduction in speed often occurs with an occluding system. On the other hand, the precipitation area expanded, which mostly offset the system's slower speed. The forecast for Sault Ste. Marie worked out quite well—the city became cloudy but remained on the edge of precipitation. In Hamilton, they actually received 24 cm of snow instead of the 14 cm forecast. This is because the easterly winds from Lake Ontario enhanced the snowfall amount by picking up moisture from the lake in this early winter situation, producing a lake effect (this is discussed more in Chapter 11 on p. 328).

The subjective forecasting techniques discussed so far are those you can use when making a short-range weather prediction. The following section describes how a meteorologist predicts the weather over a longer time range, for a situation where the weather systems are evolving and not just moving with largely steady-state trends. Here the forecaster must rely heavily on experience as well as more sophisticated tools, which include satellite data, upper-air charts, and computer progs.

A Meteorologist Makes a Prediction

It is early morning on Monday, April 26, 2010, and outside the weather forecast office near Montreal the meteorologist mulls over what is likely to happen with the weather over the next couple of days. It is sunny for now, with normal temperatures, and indications are that this will continue for today at least. But what will tomorrow's weather be like? Will it be similar to today's, or will it change markedly?

A review of the satellite loop and surface weather map analyses over the past couple of days shows that there is a large surface anticyclone or high-pressure area over the Prairies and a series of low-pressure centres with associated fronts tracking across the central and northeastern United States (see ● Figure 13.18). There is also currently a weak low in the Gulf of St. Lawrence with a cold front trailing across New Brunswick. None of these systems seem poised to affect Montreal: the Gulf of St. Lawrence system is already out of the picture, and the U.S. systems should track due east according to the 500 hPa winds and progs. There is an upper trough across Labrador and northern Quebec. Could this system

● FIGURE 13.18 Infrared satellite image from 7 a.m. EST April 26, 2010, with fronts and troughs as well as high- and low-pressure areas superimposed.

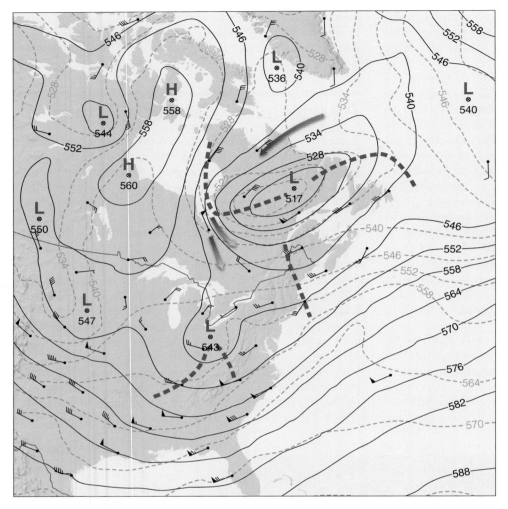

● FIGURE 13.19 The 500 hPa chart for 7 a.m. EST April 26, 2010. Solid black contours are 500 hPa heights in decametres (570 dam = 5700 m); winds blow parallel to these contours. Dashed black contours are 1000 to 500 hPa thickness lines that correspond to the mean temperature in this layer (approximately between sea level and about 5500 m above sea level). Dashed purple lines indicate shortwave troughs. Heavy blue arrows indicate strong cold air advection, and heavy red arrows indicate strong warm air advection. The 500 hPa wind directions are indicated by the wind shafts, with speeds indicated by barbs and flags (each barb is 10 knots; each flag is 50 knots).
Source: Environment Canada. Analyses and Modelling. Found at: http://www.weatheroffice.gc.ca/data/analysis/sai_50.gif

affect the area tomorrow? A look at the 500 hPa chart may help answer this question.

HELP FROM THE 500 HPA CHART ● Figure 13.19 shows the 500 hPa analysis for 7 a.m. Monday, April 26, 2010. While examining the chart, the meteorologist recognizes certain clues that will aid in making the forecast. Overall, the chart shows two streams or regions of flow indicated by the wind observations and height contours. There is a southern stream that is mainly zonal, with westerly winds, across the United States. The 500 hPa low centre south of Lake Erie and its short waves are associated with the surface fronts and lows shown in Figure 13.18 in this region. There is also a northern stream that is meridional, flowing northward from the British Columbia coast, up over a ridge of high pressure over the northern Prairies, Northwest Territories, and Nunavut and then plunging southward again over northern Quebec. On either side of the high are troughs.

Because the shape of this flow around the high resembles the Greek letter omega (Ω), the high and its accompanying ridge are known as an **omega high**. The forecaster recognizes the omega high as a *blocking high*, one that tends to persist in the same geographic location for many days. This blocking pattern also tends to keep the troughs in their respective positions.

However, although the omega high has remained stationary, there have been some changes around the southern and eastern edges of this high-pressure area over the past few days. The upper low, presently south of Lake Erie, has tracked through the bottom of the ridge from Colorado three days previously. Another upper low is right behind it, presently near South Dakota. The northern upper low presently over Labrador split off from an upper low over northern Greenland three days ago and has been moving rapidly southward ever since.

Observe the spacing of the contour lines over northern Quebec around the western side of the Labrador low in Figure 13.19. The close spacing of the contours to the west of the trough and the more widely spaced contours to the east of the trough indicate that stronger winds exist to the west of the trough. The forecaster knows from past experience that this usually means that the trough will deepen. Also note that west of the trough, the winds are blowing across the thickness lines from lower thickness values toward higher thickness values. Recall that low thickness means cold temperature, so this pattern indicates that strong cold advection is occurring as indicated by the blue arrows in Figure 13.19. The heavy dashed purple line on the west side of the trough represents the position of a shortwave trough, which is moving rapidly southward. The injection of cold air and the shortwave into the main trough should cause it to intensify and move southward. At the same time, the trough south of the Great Lakes will continue its eastward movement, to be replaced by the next one in line to the west. It looks like the northern low could link up with one of the lows in the southern stream, crossing the Montreal area at the same time. When this happens, the two streams are said to be *coming into phase*. The forecaster knows that when two streams come into phase like this, it usually indicates conditions favourable for the development and intensification of a mid-latitude cyclonic storm. (You may remember from Chapter 12 that the generation of this type of dynamic instability is called *baroclinic instability*, and it derives its energy from the horizontal differences in temperature. When two streams come into phase, the horizontal temperature gradient is increased, providing more energy to the developing storm.)

One of the main ingredients necessary for the development and intensification of the surface low is divergence of the airflow aloft. The forecaster knows that divergence aloft is associated with a decrease in surface pressure. This decrease, in turn, causes surface air to converge, rise, and condense into widespread cloudiness. But where will regions of divergence, convergence, and rising air be found on tomorrow's map? And how will tomorrow's map be different from today's? Will the northern trough and one of the southern troughs, in fact, come into phase? This is where the computer and the forecaster work together to come up with a prediction.

THE COMPUTER PROVIDES ASSISTANCE The numerical weather prediction model forecasts the atmosphere's future conditions, including the positions of shortwaves and their associated weather systems. It is important to know where the shortwaves will be found because ahead of them, there is usually upper-level divergence, lower-level convergence, rising air, clouds, and precipitation. Hence, predicting the position of a shortwave means predicting regions of inclement weather. (The shortwaves are part of the vorticity maxima discussed in Chapter 12.) Since the numerical weather prediction models simulate all the variables that make up weather, they also project precipitation and clouds.

• Figure 13.20 shows 12-, 24-, and 36-hour predictions of 500 hPa heights with shortwaves, sea-level pressure with 1000 to 500 hPa thickness, and precipitation with clouds from the CMC GEM Regional model. The model is started at 7 a.m. on Monday, April 26, and is therefore a look ahead at the weather through Tuesday evening. Prior to selecting the GEM Regional model, the meteorologist assessed predictions from the CMC GEM Global model, as well as models from the United States

and Europe, and decided, based on recent past performance, that the GEM Regional model has the best handle on the current situation over Quebec.

What does the GEM Regional model indicate? The 500 hPa panels show that the upper low over Labrador does indeed progress southward. As this happens, its shortwave rotates southward and comes into phase with the upper low in the southern stream tracking across the northern United States. The 24-hour prediction for 7 a.m. on Tuesday, April 27, shows that the upper low and shortwave are just west of Montreal, and they remain in this position throughout the day on Tuesday. Since upper-level divergence occurs east of a shortwave trough, this places the Montreal area in a position favourable for upward air motion and cloud development, as well as storm development. The 500 hPa low itself is composed of cold air through a deep column of the atmosphere. When sunlight warms the ground, the atmosphere around the upper low can become quite unstable, generating convective clouds and showers.

The predicted sea-level pressure pattern (see the middle column in Figure 13.20) indicates that the main low-pressure area is offshore, associated with the mid-latitude cyclones in the southern stream. However, a surface trough of low pressure does extend westward across southern Quebec during the day on Tuesday. The GEM Regional prediction of precipitation (see the third column in Figure 13.20) shows an area of heavy precipitation in the Montreal area, starting at around 7 a.m. Tuesday and persisting at least through that day. This precipitation area appears to be caused by the upper low and shortwave trough visible on the 500 hPa prog.

So it seems that precipitation is likely in Montreal, starting sometime on Tuesday morning. Will it fall as rain or snow? Temperatures on Monday morning are in the low teens—so if things are the same on Tuesday, any precipitation will fall as rain. But will things be the same? We have already seen that there is cold advection to the north of Montreal, and the upper low is associated with cold air, so the temperatures should be lower on Tuesday than they are on Monday. But will it be cold enough to snow? One rule of thumb discussed earlier is to look at the 1000 to 500 hPa thickness. The 540 dam line often separates rain from snow. By examining the middle column of progs in Figure 13.20 and the current conditions in Figure 13.19, we can see that the thickness over Montreal is predicted to decline from 543 dam Monday morning to 540 dam by Monday evening, to 534 dam by Tuesday morning, and to 532 dam by Tuesday evening. Therefore, it appears that the precipitation will fall as snow—a late-season snowfall for Montreal. The precipitation rate from the progs seems to be about 2 mm h^{-1}, which would be about 2 cm h^{-1} as snow. The snow looks to last at least 12 hours, between the 24-hour and the 36-hour prog in Figure 13.20, and will likely extend beyond the 36-hour period. Therefore, the precipitation forecast will sound like this: "Increasing cloudiness Monday night with snow beginning Tuesday morning and lasting all day. Accumulations of 24 cm expected during the day."

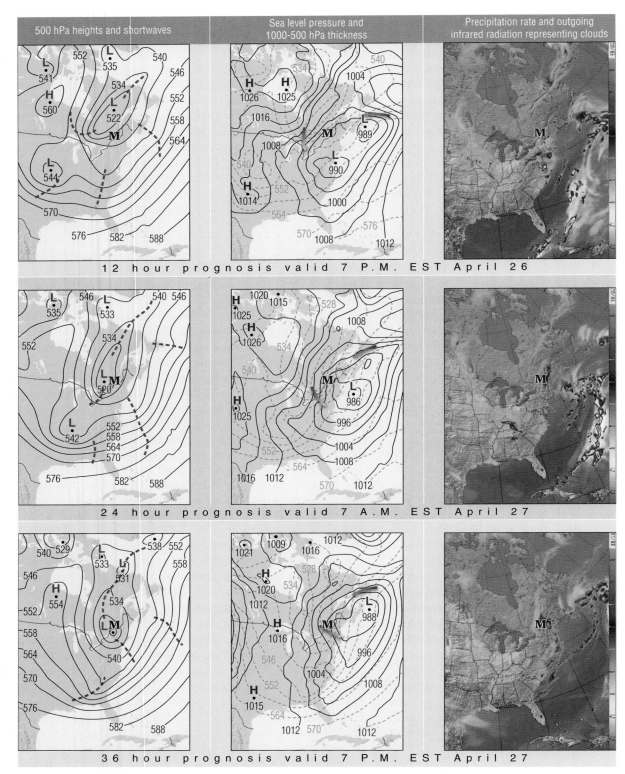

500 hPa heights and shortwaves

Sea level pressure and
1000-500 hPa thickness

Precipitation rate and outgoing
infrared radiation representing clouds

12 hour prognosis valid 7 P.M. EST April 26

24 hour prognosis valid 7 A.M. EST April 27

36 hour prognosis valid 7 P.M. EST April 27

● FIGURE 13.20 Numerical weather prediction progs from the CMC GEM Regional model initialized at 7 a.m. EST April 26, 2010, valid for 12, 24, and 36 hours in the future. The first prog on each row is the 500 hPa chart with shortwaves indicated by purple dashed lines. The next is the forecast mean sea-level pressure, with the 1000 to 500 hPa thickness contours as dashed lines. Thick blue arrows indicate strong cold air advection, and thick red arrows indicate strong warm air advection. The final column shows the forecast precipitation rate (mm h^{-1}) in shades of red and yellow, as well as outgoing infrared radiation coloured to resemble clouds in white and grey. The location of Montreal is indicated by an "M" in each panel.

Data source: Environment Canada. Found at: www.weatheroffice.gc.ca/data/model_forecast/591_100.gif

A VALID FORECAST   The observed conditions in Montreal closely followed the progs. Precipitation started as rain at 4 a.m. on Tuesday, April 27, and became mixed with snow by 6 a.m. and straight snow by 10 a.m. The snow continued until midnight Tuesday, when it became mixed with rain, and largely stopped by 4 a.m. on Wednesday, April 28. Altogether, there was 30 cm of snow and 6.6 mm of rain from this late-season storm.

● Figure 13.21 shows the observed satellite image with fronts and pressure centres superimposed at 7 p.m. on Tuesday, April 27. Note the cloud area over southern Quebec that caused the snow and its frontal system, which developed there in response to the upper low and shortwave. The 500 hPa low at this time is located just south of Montreal, over the State of New York, largely as forecast in Figure 13.20.

So the forecast for Tuesday seemed to work out well. The next challenge will be predicting when the snow ends and what comes next. To make this forecast, it's back to the drawing board—to the computer progs, charts, Doppler radar, and satellite images. The challenge and anticipation of making another forecast are at hand.

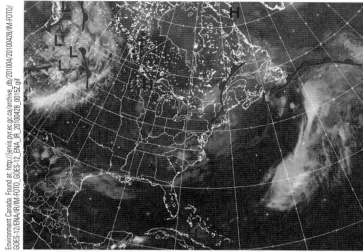

● FIGURE 13.21   Infrared satellite image from 7 p.m. EST April 27, 2010, with fronts and troughs as well as high-and low-pressure areas superimposed.

SUMMARY

Forecasting tomorrow's weather entails a variety of techniques and methods. Persistence, weather maps, satellite imagery, and Doppler radar are all useful when making a very short-range (zero- to six-hour) prediction. For short- and medium-range forecasts, the current analysis, satellite data, pattern recognition, and meteorologist's intuition and experience, along with statistical information and guidance from the many computer progs supplied by the Meteorological Service of Canada and other agencies, all go into making a prediction. For monthly and seasonal long-range forecasts, meteorologists incorporate changes in sea surface temperature in the Pacific and Atlantic oceans into seasonal outlooks of temperature and precipitation in North America.

Different computer progs are based on different atmospheric models that describe the state of the atmosphere and how it will change with time. The atmosphere's chaotic behaviour, along with flaws in the models and tiny errors (uncertainties) in the data, generally amplifies as the computer model tries to project weather further and further into the future. At present, computer progs that predict the weather over a vast region are better at forecasting the position of mid-latitude highs and lows and their development than at forecasting local showers and thunderstorms. To skillfully forecast smaller features, the grid spacing on some models is reduced to as low as 0.5 km.

As new information from atmospheric research programs is used to develop better numerical weather prediction models that are run on the latest generation of computers, it is hoped that the progs will be able to show more skill in predicting the weather even further in the future, perhaps beyond 10 days. More promising at this time is the simulation of large-scale climatic trends by the most recent general circulation models.

The latter part of this chapter does not cover all the methods of weather prediction but, rather, conveys an understanding of the process and problems confronting anyone who attempts to predict the behaviour of this churning mass of air we call our atmosphere.

Most of the forecasting methods in this chapter apply mainly to skill in predicting events associated with large-scale weather systems, such as fronts and mid-latitude cyclones. The next chapter on severe weather deals with the formation and forecasting of smaller scale (mesoscale) systems, such as thunderstorms, squall lines, and tornadoes.

KEY TERMS

The following terms are listed (with page numbers) in the order they appear in the text. Define each. Doing so will aid you in reviewing the material covered in this chapter.

watch, 383
warning, 383
NinJo, 385
meteogram, 385
analysis, 388

numerical weather
   prediction, 388
atmospheric models, 388
governing equations, 388
prognostic charts (progs), 389

QUESTIONS FOR REVIEW

1. What is the function of the Canadian Meteorological Centre?

2. How does a *weather watch* differ from a weather warning?

3. List some of the tools a weather forecaster might use when making a short-range forecast.

4. In what ways has the computer assisted the meteorologist in making weather forecasts?

5. How does a prog differ from an analysis?

6. How are computer-generated weather forecasts prepared?

7. What are some of the problems associated with computer-model forecasts?

8. Make a persistence forecast for your area for this same time tomorrow. Did you use any skill in making this prediction? Explain.

9. Describe four methods of forecasting the weather and give an example for each one.

10. How does pattern recognition aid a forecaster in making a prediction?

11. How can ensemble forecasts improve medium-range weather forecasts?

12. Explain how teleconnections are used in making a long-range seasonal forecast.

13. Do all accurate forecasts show skill? Explain.

14. Would a forecast calling for a 20 percent chance of rain be high enough for you to cancel your plans for a picnic? Explain.

15. Do monthly and seasonal forecasts make specific predictions of rain or snow? Explain.

16. Use the instant weather forecast chart (see Appendix E) as a guide to making a short-range weather forecast when the surface weather elements indicate those shown in ▼ Table 13.5, see below.

17. If low clouds at an elevation of 1000 m above you are moving from the southeast and clouds about 2500 m higher are moving from the southwest, is cold or warm advection taking place between the cloud layers? Explain. (Hint: Read Focus on an Observation: Forecasting Temperature Advection by Watching the Clouds, on p. 402.)

18. List four methods that you could use to predict the movement of a surface mid-latitude cyclone.

19. What is an omega high? What influence does it have on the movement of surface highs and lows?

20. Suppose that where you live, the middle of January is typically several degrees warmer than the rest of the month. If you forecast this "January thaw" for the middle of next January, what type of a weather forecast will you have made?

21. Given a map with isallobars, where will high- and low-pressure systems move?

QUESTIONS FOR THOUGHT

1. Suppose that the chance for a "white Christmas" at your home is 10 percent. Last Christmas was a white one. If for next year you forecast a "nonwhite" Christmas, will you have shown any skill if your forecast turns out to be correct? Explain.

2. Suppose that it is presently warm and raining. A cold front will pass your area in three hours. Behind the front,

▼ Table 13.5   Surface Weather Elements*

TIME	AIR PRESSURE (HPA)	PRESSURE CHANGES	AIR TEMPERATURE (°C)	WIND DIRECTION	CLOUD TYPE	PRECIPITATION
Morning	1017	falling slowly	−2	SE	As	none
Afternoon	1014	rising rapidly	3	NW	Tcu	rain showers
Afternoon	1025	steady	25	NW	Ci	none
Afternoon	990	falling rapidly	−4	NE	Ns	snow
Early morning	1020	rising	−10	NW	Clear	none
Nighttime	1012	falling	−1	SW	Sc	none
Afternoon	1006	falling	11	SE	Ns	rain
Nighttime	1005	rising rapidly	8	W	Cb	rain showers

*To be used in answering Review Question 16.

it is cold and snowing. Make a persistence forecast for your area six hours from now. Would you expect this forecast to be correct? Explain. Now make a forecast for your area using the steady-state, or trend, method.

3. Since computer models have difficulty in adequately considering the effects of small-scale geographic features on a weather map, why don't numerical weather forecasts simply reduce the grid spacing to about 1 km?

4. Explain how the phrase "sensitive dependence on initial conditions" relates to the final outcome of a computer-based weather forecast.

5. You are in Calgary, Alberta, 100 km east of the Rockies in January. The current wind is from the north. Looking at a prog for tomorrow, you see that the wind will be from the west. Will tomorrow's temperature be warmer or cooler than today? Explain. (Hint: You may wish to refer to the section on chinook (foehn) winds on p. 275 in Chapter 9)

PROBLEMS AND EXERCISES

1. From the World Wide Web, the newspaper, or the Meteorological Service of Canada, obtain a copy of next month's 30-day outlook. Based on the temperature and precipitation patterns of this forecast, draw several contour lines representing the upper-air pattern that would be necessary for such a forecast to come true.

2. When a persistent winter pattern at 500 hPa appears similar to that shown in ● Figure 13.22, show on the map where you would forecast the following: (a) a good chance of precipitation; (b) above seasonal temperatures;

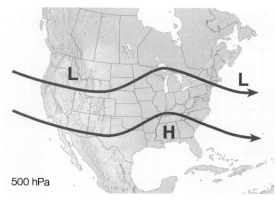

●FIGURE 13.22 Diagram for Exercise 2.

(c) below seasonal temperatures; and (d) generally dry weather. (e) What type of forecasting method did you use to figure out (a) through (d)?

3. In Figure 13.13 (p. 403), mark the position of the following cities: Cleveland, Ohio; Kingston, Ontario; and New Orleans, Louisiana. Based on the projected movement of the surface weather systems in Figure 13.16 (p. 407), make a short-range 24-hour weather forecast for each of these cities. In your forecast, include temperature, pressure, cloud cover, humidity, winds, and precipitation (if any). Compare your forecasts to the actual weather at the end of the period in Figure 13.17 (p. 406).

4. Go outside and observe the weather. Make a weather forecast using the weather signs you observe. Explain the rationale for your forecast.

Thunderstorm over central Alberta. A gust front and associated shelf cloud can be seen ahead of the dark precipitation near the horizon.
Dan Jurak/Getty Images

Thunderstorms and Tornadoes

14

The impenetrable wall of slate-coloured fog swirled and parted, and in front of me was . . . nothing. Where just the day before a subdivision of upscale suburban homes—dubbed Snob Hill by locals—filled the space, now there was nothing but empty basements with their insides sucked clean. No walls, no roofs, no dining room tables, no beds, no knick-knacks or photos. No happy families. Just soldiers patrolling with guns hoisted on their hips, heads swivelling, on the lookout for trouble.

Beirut or Baghdad? Nope. Barrie.

It was Friday, May 31, the day after the Great Tornado of '85 and I was on a military bus filled with shell-shocked reporters, noses pressed to the glass as we silently took in what an angry Mother Nature had brought to town. A series of tornadoes had hopscotched across central Ontario, touching down in Barrie, Fergus, Orangeville, Shelburne, Grand Valley, and Arthur. The whirling clouds delivered death and destruction on an almost unimaginable scale for the region: 12 dead—eight in Barrie alone—and hundreds injured. In Barrie, then a city of 45,000, more than 300 homes were destroyed. Another 300 were damaged, leaving 800 residents homeless.

One of the worst weather systems to ever hit Ontario, the twisters tossed cars like toys, twisted metal factory beams around trees and flattened whole blocks of houses.

The winds in Barrie registered F4 on the tornado scale (F5 is the worst), with winds hitting 331 to 417 km h^{-1}, and mowing a path of utter waste some 600 metres wide.

Gail Swainson, Reporter

The Toronto Star. Ontario twister sparks memories of 1985 tornado. Gail Swainson. Aug. 22, 2009. Found at: http://www.thestar.com/article/684956.

CONTENTS

The devastating tornado described in our opening was actually part of a system consisting of 14 separate tornadoes that affected the U.S. states of Ohio, Pennsylvania, and New York, as well as communities in Ontario. It was so severe that it is still called "Black Friday" by locals and is tied with the Pine Lake, Alberta, tornado of 2000 as the fourth deadliest tornado* in Canadian history. Tornadoes such as these, as well as much smaller ones, are associated with severe thunderstorms. Consequently, we will first examine the different types of thunderstorms. Later, we will focus on tornadoes, examining how and where they form and why they are so destructive.

Thunderstorms

It probably comes as no surprise that a *thunderstorm* is merely a storm containing lightning and thunder. Sometimes a thunderstorm produces gusty surface winds with heavy rain and hail. The storm itself may be a single cumulonimbus cloud, or several thunderstorms may form into a cluster. In some cases, a line of thunderstorms will form that may extend for hundreds of kilometres.

Thunderstorms are *convective storms* that form with rising air. So the birth of a thunderstorm often begins when warm, moist air rises in a conditionally unstable environment.† The rising air may be a parcel of air ranging in size from a large balloon to a city block, or an entire layer, or slab of air, may be lifted. As long as a rising air parcel is warmer (less dense) than the air surrounding it, there is an upward-directed *buoyant force* acting on it. The warmer the parcel is compared to the air surrounding it, the greater the buoyant force and the stronger the convection. The trigger (or "forcing mechanism") needed to start air moving upward may be unequal heating at the surface, the effect of terrain, or the lifting of air along shallow boundaries of converging surface winds. Diverging upper-level winds, coupled with converging surface winds and rising air, also provide a favourable condition for thunderstorm development. Moreover, thunderstorms often form when warm air rises along a frontal zone. Usually, several of these mechanisms work together with vertical wind shear to generate severe thunderstorms.

Summertime convective systems occur in varying degrees of severity, from cumulus clouds to supercell thunderstorms that breed tornadoes. These systems are mechanisms that transfer sensible and latent heat between Earth's surface and the *atmosphere*. On the benevolent side, the precipitation generated by these systems provides moisture to the *hydrosphere* and often much-needed summer rain to the *biosphere*. However, the form and intensity of the precipitation can have disastrous impacts on people and dramatic effects on the landscape when large amounts of water accumulate quickly in an area and run off in a massive stream into low-lying areas or streambeds. This type of event in its severest form can develop into a flash flood, causing death and destruction for people and changing watercourses, sometimes even contaminating wide areas of the landscape. When thunderstorms form hail, there can be significant impacts on the *biosphere* through damage to vegetation, and in severe cases, the hail can become large enough to kill wildlife, livestock, and people. If the vegetation that is damaged happens to be agricultural crops, or if the hail damages structures, then the *anthrosphere* is affected as well. Of course, tornadoes, due to the damage they cause, have an even more significant impact on the anthrosphere in localized areas as wind and pressure changes destroy structures and flying debris causes fatalities.

Although we often see thunderstorms forming where the surface air is quite buoyant (i.e., warm and humid), they may also form when the surface air temperature is no more than 10°C. This latter situation often occurs in winter along the West Coast of North America, when cold air aloft moves over the region. The cold air aloft destabilizes the atmosphere to the point where air parcels, given an initial push upward, are able to continue their upward journey because they remain warmer (less dense) than the colder air surrounding them. The cold air aloft may even produce sufficient instability to generate thunderstorms in wintertime snowstorms.

Most thunderstorms that form over North America are short-lived and produce rain showers, gusty surface winds, thunder and lightning, and sometimes small hail. Many have an appearance similar to the mature thunderstorm shown in ● Figure 14.1. The majority of these storms do not reach severe status. Severe thunderstorms are defined by the Meteorological Service of Canada as having at least one of the following: large hail with a diameter of at least 2 cm, surface wind gusts of 48 knots (90 km h^{-1}) or greater, rainfall of at least 50 mm within one hour or 75 mm within three hours, or production of a tornado.

Scattered thunderstorms (sometimes called "pop-up" storms) that typically form on warm, humid days are often referred to as *ordinary cell thunderstorms** or *air-mass thunderstorms* because they tend to form in warm, humid air masses away from significant weather fronts. Ordinary cell (air mass) thunderstorms can be considered "simple storms" because they rarely become severe, typically are less than a kilometre wide, and go through a rather predictable life cycle from birth to maturity to decay that usually takes less than an hour to complete. However, under the right atmospheric conditions (described later in this chapter), more intense "complex thunderstorms" may form, such as the *multicell thunderstorm* and the *supercell thunderstorm*—a huge, rotating storm that can last

---

*The deadliest Canadian tornado was the 1912 Regina cyclone that killed as many as 40, followed by the Edmonton tornado in 1987 that killed 27. The third deadliest Canadian tornado was the Windsor tornado of 1946 that resulted in 17 fatalities.

†A conditionally unstable atmosphere exists when cold, dry air aloft overlies warm, moist surface air. Additional information on atmospheric instability is given in Chapter 6, beginning on p. 170.

*In convection, the cell may be a single updraft, a single downdraft, or a combination of the two.

● FIGURE 14.1 An ordinary thunderstorm in its mature stage. Note the distinctive anvil top.

C. Donald Ahrens

for hours and produce severe weather such as strong surface winds, large, damaging hail, flash floods, and violent tornadoes.

We will examine the development of ordinary cell (air mass) thunderstorms first before we turn our attention to the more complex multicell and supercell storms.

ORDINARY CELL THUNDERSTORMS **Ordinary cell (air mass) thunderstorms** or, simply, *ordinary thunderstorms* tend to form in a region where there is limited wind shear—that is, where the wind speed and wind direction do not abruptly change with increasing height above the surface. Many ordinary thunderstorms appear to form as parcels of air are lifted from the surface by turbulent overturning in the presence of

wind. Moreover, ordinary storms often form along shallow zones, where surface winds converge. Such zones may be due to any number of things, such as topographic irregularities, sea breeze fronts, or the cold outflow of air from inside a thunderstorm that reaches the ground and spreads horizontally. These converging wind boundaries are normally zones of contrasting air temperature and humidity and, hence, air density.

Extensive studies indicate that ordinary thunderstorms go through a cycle of development from birth to maturity to decay. The first stage is known as the **cumulus stage**, or growth stage. As a parcel of warm, humid air rises, it cools and condenses into a single cumulus cloud or a cluster of clouds (see ● Figure 14.2a). If you have ever watched a thunderstorm

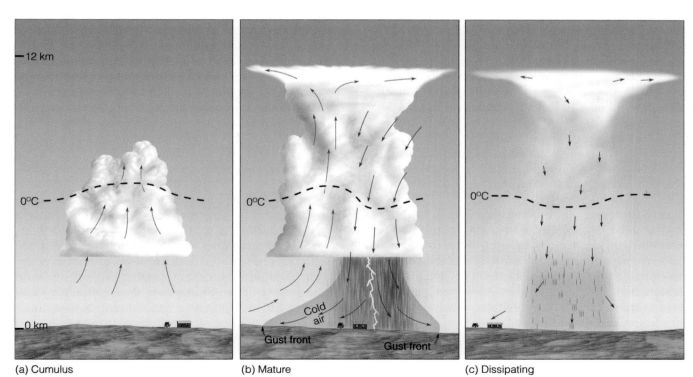

(a) Cumulus (b) Mature (c) Dissipating

ACTIVE FIGURE 14.2 Simplified model depicting the life cycle of an ordinary cell thunderstorm that is nearly stationary as it forms in a region of low wind shear. (Arrows show vertical air currents. The dashed line represents freezing level, 0°C isotherm.) Visit the textbook's website to view this and other Active Figures at www.ahrensmeteorology1ce.nelson.com

develop, you may have noticed that at first the cumulus cloud grows upward only a short distance, and then it dissipates. The top of the cloud dissipates because the cloud droplets evaporate as the drier air surrounding the cloud mixes with it. However, after the water drops evaporate, the air is more moist than before. So the rising air is now able to condense at successively higher levels, and the cumulus cloud grows taller, often appearing as a rising dome or tower.

As the cloud builds, the transformation of water vapour into liquid or solid cloud particles releases large quantities of latent heat, a process that keeps the rising air inside the cloud warmer (less dense) than the air surrounding it. The cloud continues to grow in the unstable atmosphere as long as it is constantly fed by rising air from below. In this manner, a cumulus cloud may show extensive vertical development and grow into a towering cumulus cloud (cumulus congestus) in just a few minutes. During the cumulus stage, there is normally insufficient time for precipitation to form, and the updrafts keep water droplets and ice crystals suspended within the cloud. Also, there is no lightning or thunder during this stage.

As the cloud builds well above the freezing level, the cloud particles grow larger. They also become heavier. Eventually, the rising air is no longer able to keep them suspended, and they begin to fall. While this phenomenon is taking place, drier air from around the cloud is being drawn into it in a process called *entrainment*. The entrainment of drier air causes some of the raindrops to evaporate, which chills the air. The air, now colder and heavier than the air around it, begins to descend as a *downdraft*. The downdraft may be enhanced as falling precipitation drags some of the air along with it.

The appearance of the downdraft marks the beginning of the **mature stage**. The downdraft and updraft within the mature thunderstorm now constitute the cell. In some storms, there are several cells, each of which may last for less than 30 minutes.

During its mature stage, the thunderstorm is most intense. The top of the cloud, having reached a stable region of the atmosphere (which may be the stratosphere), begins to take on the familiar anvil shape as upper-level winds spread the cloud's ice crystals horizontally (see Figure 14.2b). The cloud itself may extend upward to an altitude of over 12 km and be several kilometres in diameter near its base. Updrafts and downdrafts reach their greatest strength in the middle of the cloud, creating severe turbulence. Lightning and thunder are also present in the mature stage. Heavy rain (and occasionally small hail) falls from the cloud. And, at the surface, there is often a downrush of cold air with the onset of precipitation.

Where the cold downdraft reaches the surface, the air spreads out horizontally in all directions. The surface boundary that separates the advancing cooler air from the surrounding warmer air is called a *gust front*. Along the gust front, winds rapidly change both direction and speed. Look at Figure 14.2b and notice that the gust front forces warm, humid air up into the storm, which enhances the cloud's updraft. In the region of the downdraft, rainfall may or may not reach the surface, depending on the relative humidity beneath the storm. In the dry air of the desert in the U.S. Southwest, for example, a mature thunderstorm may look ominous and contain all of the ingredients of any other storm, except that the raindrops evaporate before reaching the ground. However, intense downdrafts from the storm may reach the surface, producing strong, gusty winds and a gust front.

After the storm enters the mature stage, it begins to dissipate in about 15 to 30 minutes. The **dissipating stage** occurs when the updrafts weaken as the gust front moves away from the storm and no longer enhances the updrafts. At this stage, as illustrated in Figure 14.2c, downdrafts tend to dominate throughout much of the cloud. The reason the storm does not normally last very long is that the downdrafts inside the cloud tend to cut off the storm's fuel supply by destroying the humid updrafts. Deprived of the rich supply of warm, humid air, cloud droplets no longer form. Light precipitation now falls from the cloud, accompanied by only weak downdrafts. As the storm dies, the lower level cloud particles evaporate rapidly, sometimes leaving only the cirrus anvil as the reminder of the once mighty presence (see ● Figure 14.3). A single ordinary cell thunderstorm may go through its three stages in one hour or less.

Not only do these thunderstorms produce summer rainfall for a large portion of the Canada, but they also bring with them momentary cooling after an oppressively hot day. The cooling comes during the mature stage, as the downdraft reaches the surface in the form of a blast of welcome relief. Sometimes the air temperature may lower as much as 10°C in just a few minutes. Unfortunately, the cooling effect often is short-lived as the downdraft diminishes or the thunderstorm moves on. In fact, after the storm has ended, the air temperature usually rises, and as the moisture from the rainfall evaporates into the air, the humidity increases, sometimes to a level where it actually feels more oppressive after the storm than it did before.

Up to this point, we have looked at ordinary cell thunderstorms that are short-lived, rarely become severe, and form in a region with weak vertical wind shear. As these storms develop, the updraft eventually gives way to the downdraft, and the storm ultimately collapses on itself. However, in a region where strong vertical wind shear exists, thunderstorms often take on a more complex structure. Strong vertical wind shear can cause the storm to tilt in such a way that it becomes a multicell thunderstorm—a thunderstorm with more than one cell.

MULTICELL THUNDERSTORMS   Thunderstorms that contain a number of cells, each in a different stage of development, are called **multicell thunderstorms** (see ● Figure 14.4). Such storms tend to form in a region of moderate-to-strong vertical wind-speed shear. Look at ● Figure 14.5 and notice

● FIGURE 14.3 A dissipating thunderstorm. Most of the cloud particles in the lower half of the storm have evaporated.

© Howard B. Bluestein

that on the left side of the illustration, the wind speed increases rapidly with height, producing strong wind-speed shear. This type of shearing causes the cell inside the storm to tilt in such a way that the updraft actually rides up and over the downdraft. Note that the rising updraft is capable of generating new cells that go on to become mature thunderstorms. Notice also that precipitation inside the storm does not fall into the updraft (as it does in the ordinary cell thunderstorm), so the storm's fuel supply is not cut off, and the storm complex can survive for a long time. Because the likelihood that a thunderstorm will become severe increases with the length of time the storm exists, long-lasting multicell storms can become intense and produce severe weather.

When convection is strong and the updraft intense (as it is in Figure 14.5), the rising air may actually intrude well into the stable stratosphere, producing an **overshooting top**. As the air spreads laterally into the anvil, sinking air in this region of the storm can produce beautiful mamma clouds. At the surface, below the thunderstorm's cold downdraft, the cold, dense air may cause the surface air pressure to rise—sometimes several hectopascals. The relatively small, shallow area of high pressure is called a *mesohigh* (meaning "mesoscale high").

The Gust Front   When the cold downdraft reaches Earth's surface, it pushes outward in all directions, producing a strong **gust front** that represents the leading edge of the cold, outflowing air. To an observer on the ground, the passage of the gust front resembles that of a cold front. During its passage, the temperature drops sharply and the wind shifts and becomes strong and gusty, with speeds occasionally exceeding

● FIGURE 14.4 This multicell storm complex is composed of a series of cells in successive stages of growth. The thunderstorm in the middle is in its mature stage, with a well-defined anvil. Heavy rain is falling from its base. To the right of this cell, a thunderstorm is in its cumulus stage. To the left, a well-developed cumulus congestus cloud is about ready to become a mature thunderstorm. With new cells constantly forming, the multicell storm complex can exist for hours.

C. Donald Ahrens

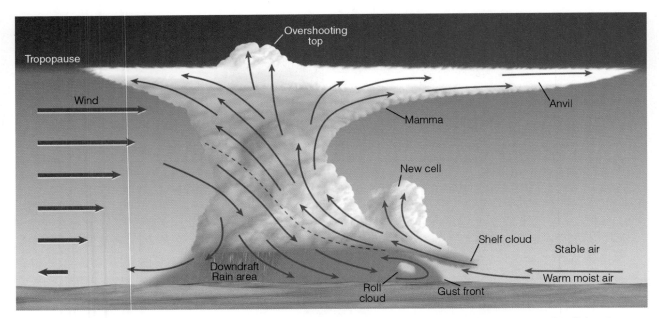

● FIGURE 14.5 A simplified model describing air motions and other features associated with an intense multicell thunderstorm that has a tilted updraft. The severity depends on the intensity of the storm's circulation pattern.

55 knots (102 km h⁻¹). Along the leading edge of the gust front, the air is quite turbulent. Here strong winds can pick up loose dust and soil and lift them into a huge, tumbling cloud (see ● Figure 14.6).* The cold surface air behind the gust front may even linger close to the ground for hours, well after thunderstorm activity has ceased.

As warm, moist air rises along the forward edge of the gust front, a **shelf cloud** (also called an *arcus cloud*) may

---
*In dry, dusty areas or desert regions, the leading edge of the gust front is the haboob described in Chapter 9 (p. 278).

● FIGURE 14.6 A swirling mass of dust forms along the leading edge of a gust front as it moves across western Nebraska.

form, such as the one shown in ● Figure 14.7. These clouds are especially prevalent when the atmosphere is very stable near the base of the thunderstorm. Look again at Figure 14.5 and notice that the shelf cloud is attached to the base of the thunderstorm. Occasionally, an elongated, ominous-looking cloud forms just behind the gust front. These clouds, which appear to slowly spin about a horizontal axis, are called **roll clouds** (see ● Figure 14.8).

When the atmosphere is conditionally unstable, the leading edge of the gust front may force the warm, moist air upward, producing a complex of multicell storms, each with

● FIGURE 14.7 A dramatic example of a shelf cloud (or arcus cloud) associated with an intense thunderstorm. The photograph was taken in the Philippines as the thunderstorm approached from the northwest.

new gust fronts. These gust fronts may then merge into a huge gust front called an **outflow boundary**. Along the outflow boundary, air is forced upward, often generating new thunderstorms (see ● Figure 14.9).

Microbursts   Beneath an intense thunderstorm, the downdraft may become localized so that it hits the ground and spreads horizontally in a radial burst of wind, much like water pouring from a tap and striking the sink below. Such downdrafts are called **downbursts**. A downburst with winds extending only 4 km or less is termed a **microburst**. In spite of its small size, an intense microburst can induce damaging winds as high as 146 knots (270 km h^{-1}). (A larger downburst with winds extending more than 4 km is termed a *macroburst*.) ● Figure 14.10 shows the dust clouds generated from a microburst north of Denver, Colorado. Since a microburst

is an intense downdraft, its leading edge can evolve into a gust front.

Microbursts are capable of blowing down trees and inflicting heavy damage on poorly built structures as well as on sailing vessels that encounter microbursts over open water. In fact, microbursts may be responsible for some damage once attributed to tornadoes. Moreover, microbursts and their accompanying wind shear (i.e., rapid changes in wind speed and wind direction) appear to be responsible for several airline crashes. When an aircraft flies through a microburst at a relatively low altitude, say 300 m above the ground, it first encounters a headwind that generates extra lift. This is position (a) in ● Figure 14.11. At this point, the aircraft tends to climb (it gains lift), and if the pilot noses the aircraft downward, there could be grave consequences for, in a matter of seconds, the aircraft encounters the powerful downdraft

● FIGURE 14.8 A roll cloud forming behind a gust front.

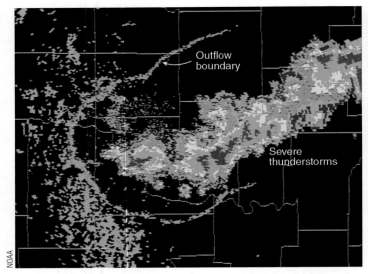

● FIGURE 14.9 Radar image of an outflow boundary. As cool (more dense) air from inside the severe thunderstorms (red and orange colours) spreads outward, away from the storms, it comes in contact with the surrounding warm, humid (less dense) air, forming a density boundary (blue line) called an outflow boundary between cool air and warm air. Along the outflow boundary, new thunderstorms often form.

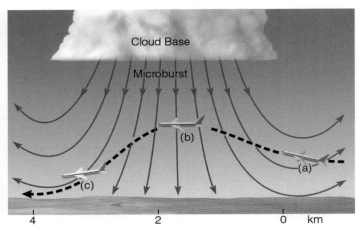

● FIGURE 14.11 Flying into a microburst. At position (a), the pilot encounters a headwind; at position (b), a strong downdraft; and at position (c), a tailwind that reduces lift and causes the aircraft to lose altitude.

(position b), and the headwind is replaced by a tail wind (position c). This situation causes a sudden loss of lift and a subsequent decrease in the performance of the aircraft, which is now accelerating toward the ground.

One accident attributed to a microburst occurred north of Dallas–Fort Worth Regional Airport in August 1985. Just as an aircraft was making its final approach, it encountered severe wind shear beneath a small but intense thunderstorm. The aircraft then dropped to the ground and crashed, killing over 100 passengers. To detect the hazardous wind shear associated with microbursts, many major airports use a high-resolution Doppler radar. The radar uses computer algorithms that are programmed to detect microbursts and low-level wind shear.

The leading edge of a microburst can contain an intense, horizontally rotating vortex that is often filled with dust in a relatively dry region. In dry areas such as eastern Colorado,

● FIGURE 14.10 Dust clouds rising in response to the outburst winds of a microburst north of Denver, Colorado.

many microbursts emanate from virga—rain falling from a cloud but evaporating before reaching the ground. Apparently, in these "dry" microbursts (see Figure 14.10), evaporating rain cools the air. The cooler, heavy air then plunges downward through the warmer, lighter air below. In humid regions, many microbursts are "wet" in that they are accompanied by blinding rain.

Microbursts can be associated with severe thunderstorms, producing strong, damaging winds. But studies show that they can also occur with ordinary cell thunderstorms and with clouds that produce only isolated showers—clouds that may or may not contain thunder and lightning.

Up to this point, you might think that thunderstorm downdrafts are always cool. Most are cool, but, occasionally, they can be extremely hot. For example, during the evening of May 22, 1996, in the town of Chickasha, Oklahoma, a blast of hot, dry air from a dissipating thunderstorm raised the surface air temperature from 31°C to 39°C in just 25 minutes. Such sudden warm downbursts are called **heat bursts**. Apparently, the heat burst originates high up in the thunderstorm and warms by compressional heating as it plunges toward the surface. The heat burst that hit Chickasha was exceptionally strong. Along with the hot air, it was accompanied by high winds that toppled trees, ripped down power lines, and lifted roofs off homes.

Squall-Line Thunderstorms

Multicell thunderstorms may form as a line of thunderstorms, called a **squall line**. The line of storms may form directly along a cold front and extend for hundreds of kilometres, or the storms may form in the warm air 100 to 300 km ahead of the cold front. These *prefrontal squall-line thunderstorms* of the middle latitudes represent the largest and most severe type of squall line, with huge thunderstorms causing severe weather over much of its length (see ● Figure 14.12).*

There is still debate as to exactly how prefrontal squall lines form. Models that simulate their formation suggest that, initially, convection begins along the cold front and then reforms farther away. Moreover, the surging nature of the main cold front itself, or developing cumulus clouds along the front, may cause the air aloft to develop into waves (called gravity waves), much like the waves that form downwind of a mountain chain (see ● Figure 14.13). Ahead of the cold front, the rising motion of the wave may be the trigger that initiates the development of cumulus clouds and a prefrontal squall line. In some instances, low-level converging air is better established ahead of the advancing cold front.

Rising air along the frontal boundary (and along the gust front), coupled with the tilted nature of the updraft, promotes the development of new cells as the storm moves along. Hence, as old cells decay and die out, new ones constantly form, and the squall line can maintain itself for hours

---

*Within a squall line, there may be multicell thunderstorms, as well as supercell storms—violent thunderstorms that contain a single, rapidly rotating updraft. We will look more closely at supercells in the next section.

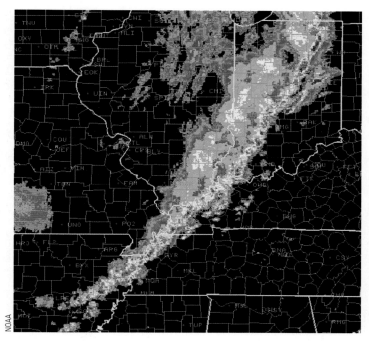

● FIGURE 14.12 A Doppler radar composite showing a prefrontal squall line extending from Indiana southwestward into Arkansas. Severe thunderstorms (red and orange colours) associated with the squall line produced large hail and high winds in October 2001.

on end. Occasionally, a new squall line will actually form ahead of the front as the gust front pushes forward, beyond the main line of storms.

Squall lines that exhibit weaker updrafts and downdrafts tend to be more shallow than prefrontal squall lines and usually have shorter life spans. These storms are referred to as *ordinary squall lines*. Severe weather may occur with them, but, more typically, they form as a line of thunderstorms that exhibit characteristics of ordinary cell thunderstorms. Ordinary squall lines may form along a gust front, with a stationary front, with a weak wave cyclone, or where no large-scale cyclonic storms are present. Many of the ordinary squall lines that form in the middle latitudes exhibit a structure similar to squall lines that form in the tropics.

In some squall lines, the leading area of thunderstorms and heavy precipitation is followed by a region of extensive

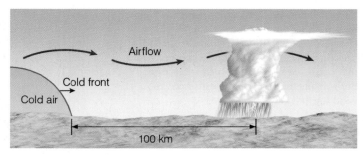

● FIGURE 14.13 Prefrontal squall-line thunderstorms may form ahead of an advancing cold front as the upper airflow develops waves downwind from the cold front.

● FIGURE 14.14 A model describing air motions and precipitation associated with a squall line that has a trailing stratiform cloud layer.

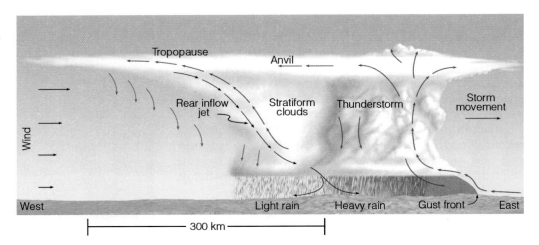

stratified clouds and light precipitation (see ● Figure 14.14). The stratiform clouds represent a region where the anvil cloud trails behind the thunderstorm. Slowly rising air within the region keeps the air saturated. Beneath the rising air, the air slowly descends in association with the falling rain.

The downdraft to the rear of the storm in Figure 14.14 forms as some of the falling precipitation evaporates and chills the air. The heavy cooler air then descends, dragging some of the surrounding air with it. If the cool air rapidly descends, it may concentrate into a rather narrow band of fast-flowing air called the *rear-inflow jet* because it enters the storm from the west (see Figure 14.14). Sometimes the rear-inflow jet will bring with it the stong, upper-level winds from aloft. Should these winds reach the surface, they rush outward producing damaging *straight-line winds** that may exceed 90 knots (167 km h^{-1}). (See ● Figure 14.15.)

As the strong winds rush forward along the ground, they sometimes push the squall line outward so that it appears as a *bow* (or a series of bows) on a radar screen. Such a bow-shaped squall line is called a **bow echo** (see ● Figure 14.16). When the damage associated with the straight-line winds

*Straight-line winds are thunderstorm-generated winds that typically are not associated with rotation.

extends for several hundred kilometres along the squall line's path, the windstorm is called a **derecho** (day-ray-sho), after the Spanish word for "straight ahead." Typically, derechoes form in the early evening and last throughout the night. An especially powerful derecho roared through southern Ontario and New York State during the early morning of July 15, 1995, where it blew down millions of trees in Adirondack State Park.

On July 4, 1999, a derecho originating in Minnesota crossed Lake Superior, reaching Thunder Bay, Ontario, with peak winds near 160 km h^{-1}. There it spawned some small tornadoes before crossing northern Ontario and southern Quebec, passing through Montreal and Sherbrooke. Over 600,000 people lost power in Quebec from windblown trees taking out power lines due to the derecho. In its path from Minnesota through Ontario and Quebec to Vermont, the derecho caused over $100 million in damage, killed four people, and injured 70 others. In North America, most derechoes occur between May and August and are mainly found in the U.S. Ohio Valley, in the Upper Mississippi Valley, and around the Great Lakes, including southwest Ontario. They have been observed as far north as southern Manitoba and northwestern Ontario. In an average year, about 20 derechoes occur in the United States and a smaller number in Canada.

● FIGURE 14.15 A side view of the lower half of a squall-line thunderstorm with the rear-inflow jet carrying strong winds from high altitudes down to the surface. These strong winds push forward along the surface, causing damaging straight-line winds that may reach 100 knots. If the high winds extend horizontally for a considerable distance, the wind storm is called a derecho.

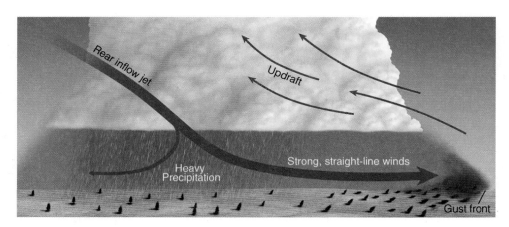

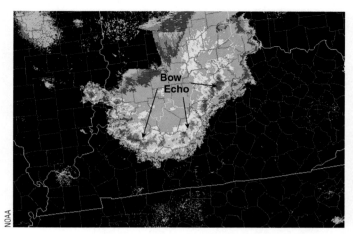

● FIGURE 14.16 The red and orange on this Doppler radar image show an intense squall line moving south-southeastward into Kentucky. The thunderstorms are producing strong straight-line winds called a derecho. Notice that the line of storms is in the shape of a bow. Such bow echos are an indicator of strong, damaging surface winds near the centre of the bow. Sometimes the left (usually northern) side of the bow will develop cyclonic rotation and produce a tornado.

MESOSCALE CONVECTIVE COMPLEXES

Where conditions are favourable for convection, a number of individual multicell thunderstorms may occasionally grow in size and organize into a large, circular convective weather system. These convectively driven systems, called **mesoscale convective complexes (MCCs)**, are quite large—they can be as much as 1000 times larger than an individual ordinary cell thunderstorm, covering an area in excess of 100,000 km² (see ● Figure 14.17).

Within the MCCs, the individual thunderstorms work together to generate a long-lasting weather system that moves slowly (normally less than 20 knots) and often exists for periods exceeding 12 hours. The circulation of the MCCs supports the growth of new thunderstorms as well as a region of widespread precipitation. These systems are beneficial as they provide a significant portion of the growing season rainfall over much of the prairies. However, MCCs can also produce a wide variety of severe weather, including hail, high winds, destructive flash floods, and tornadoes.

Mesoscale convective complexes tend to form during the summer in regions where the upper-level winds are weak, which is often beneath a ridge of high pressure. If a weak cold front should stall beneath the ridge, surface heating and moisture may be sufficient to generate thunderstorms on the cool side of the front. Often moisture from the south is brought into the system by a low-level jet stream found between 1 and 3 km above the surface. In addition, the low-level jet can provide shearing so that multicell storms can form. Most MCCs reach their maximum intensity in the early-morning hours, which is partly due to the fact that the low-level jet reaches its maximum strength late at night or in the early morning. Moreover, at night, the cloud tops cool rapidly by emitting infrared energy to space. Gradually, the

● FIGURE 14.17 An enhanced infrared satellite image showing the cold cloud tops (dark red and orange colours) of a mesoscale convective complex extending from central Kansas across western Missouri. This organized mass of multicell thunderstorms brought hail, heavy rain, and flooding to this area.

atmosphere destabilizes as a vast amount of latent heat is released in the lower and middle parts of the clouds. Within the multicell storm complex, new thunderstorms form as older ones dissipate. With only weak, upper-level winds, most MCCs move southeastward very slowly.

SUPERCELL THUNDERSTORMS

In a region where there is strong vertical wind shear (both speed and direction shear), the thunderstorm may form in such a way that the outflow of cold air from the downdraft never undercuts the updraft. In such a storm, the wind shear may be so strong to create horizontal spin, which, when tilted into the updraft, causes it to rotate. A large, long-lasting thunderstorm with a single, violently rotating updraft is called a **supercell**.* As we will see later in this chapter, it is the rotating aspect of the supercell that can lead to the formation of tornadoes.

● Figure 14.18 shows a supercell with a tornado. The internal structure of a supercell is organized in such a way that the storm may maintain itself as a single entity for hours. Storms of this type are capable of producing an updraft that may exceed 90 knots (167 km h⁻¹), damaging surface winds, and large tornadoes. Violent updrafts keep hailstones suspended in the cloud long enough for them to grow to considerable size—sometimes to the size of grapefruits. Once they are large enough, they may fall out the bottom of the cloud

---

*Smaller thunderstorms that occur with rotating updrafts are referred to as *mini supercells.*

● FIGURE 14.18 A supercell thunderstorm with a tornado sweeps over Texas.

© Warren Faidley/WeatherStock

with the downdraft, or the violent, spinning updraft may whirl them out the side of the cloud or even from the base of the anvil. Aircraft have actually encountered hail in clear air several kilometres from a storm. In some cases, the top of the storm may extend to as high as 18 km above the surface, and the width of the storm may exceed 40 km.

Although no two supercells are exactly alike, for convenience, they are often divided into three types. *Classic (CL) supercells*, for example, are well-balanced storms that produce heavy rain, large hail, high surface winds, and the majority of tornadoes. The classic supercell serves as an excellent model for all supercells and is the one normally shown in diagrams. Supercells that produce heavy precipitation and large hail, which appears to fall in the centre of the storm, are called *HP supercells* (for high precipitation). Such storms often produce extreme downdrafts (downburst) and flash flooding. If tornadoes are present, it is often difficult to see them as they tend to form in the area of heavy precipitation. A supercell characterized by little precipitation, such as the one shown in Figure 14.18, is referred to as an *LP supercell* (for low precipitation). These storms, which are capable of producing tornadoes and large hail, often have a vertical tower that, due to the storm's rotation, resembles a corkscrew.

A model of a classic supercell with many of its features is given in ● Figure 14.19. In the diagram, we are viewing the storm from the southeast, and the storm is moving from southwest to northeast. The rotating air column on the south side of the storm, usually 5 to 10 km across, is called a **mesocyclone** (meaning "mesoscale cyclone"). The rotating updraft associated with the mesocyclone is so strong that precipitation cannot fall through it. This situation produces a rain-free area (called a *rain-free base*) beneath the updraft. Strong, southwesterly winds

aloft usually blow the precipitation northeastward. Notice that large hail, having remained in the cloud for some time, usually falls just north of the updraft, and the heaviest rain occurs just north of the falling hail, with the lighter rain falling in the northeast quadrant of the storm. If low-level humid air is drawn into the updraft, a rotating cloud, called a **wall cloud**, may descend from the base of the storm (see ● Figure 14.20).

We can obtain a better picture of how wind shear plays a role in the development of supercell thunderstorms by observing ● Figure 14.21. The illustration represents atmospheric conditions during the spring over the Central Plains of North America. At the surface, we find an open-wave middle-latitude cyclone with cold, dry air moving in behind a cold front and warm, humid air pushing northward from the Gulf of Mexico behind a warm front. Above the warm surface air, a wedge or "tongue" of warm, moist air is streaming northward. It is in this region that we find a relatively narrow band of strong winds called the *low-level jet*. Winds in the low-level jet may exceed 50 knots. Directly above the moist layer is a wedge of cooler, drier air moving in from the southwest. Higher up, at the 500 hPa level, a trough of low pressure exists to the west of the surface low. At the 250 hPa level, the polar-front jet stream swings over the region, often with an area of maximum wind (a jet streak) above the surface low. At this level, the jet stream provides an area of divergence that enhances surface convergence and rising air. The stage is now set for the development of supercell thunderstorms.

The light yellow area on the surface map (see Figure 14.21) shows where supercells are likely to form. They tend to form in this region because (1) the position of cold air above warm air produces a conditionally unstable atmosphere and because (2) strong vertical wind shear induces rotation.

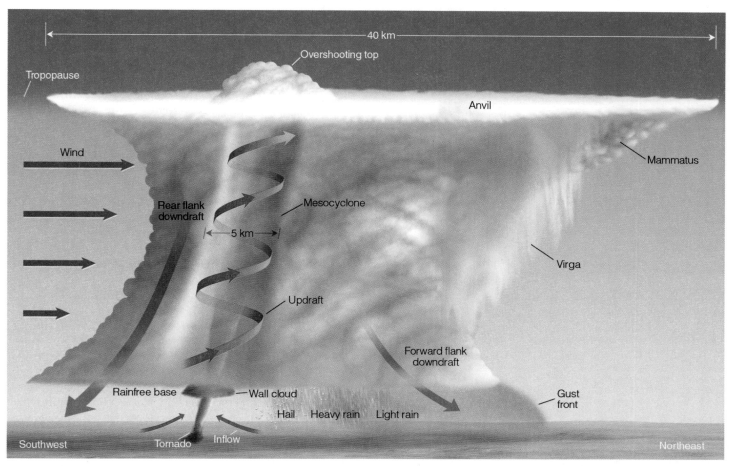

● FIGURE 14.19 Some of the features associated with a classic tornado-breeding supercell thunderstorm as viewed from the southeast. The storm is moving to the northeast.

● FIGURE 14.20 A wall cloud photographed southwest of Norman, Oklahoma.

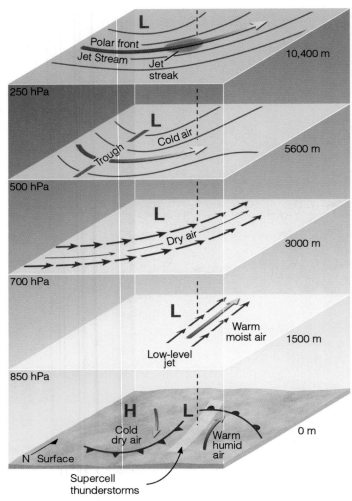

● FIGURE 14.21 Conditions leading to the formation of severe thunderstorms, especially supercells. The area in yellow shows where supercell thunderstorms are likely to form.

Rapidly increasing wind speed from the surface up to the low-level jet provides strong wind-speed shear. Within this region, wind shear causes the air to spin about a horizontal axis. You can obtain a better idea of this spinning by placing a pen (or pencil) in your left hand, parallel to the table. Now take your right hand and push it over the pen away from you. The pen rotates much like the air rotates. If you tilt the spinning pen into the vertical, the pen rotates counterclockwise from the perspective of looking down on it. A similar situation occurs with the rotating air. As the spinning air rotates counterclockwise about a horizontal axis, an updraft from a developing thunderstorm can draw the spinning air into the cloud, causing the updraft to rotate. It is this rotating updraft that is characteristic of all supercells. The increasing wind speed with height up to the 250 hPa level, coupled with the changing wind direction with height from more southerly at low levels to more westerly at high levels, further induces storm rotation.*

---

*As we will see later in this chapter, it is this rotation that sets the stage for tornado development.

Ahead of the advancing cold front, we might expect to observe many supercells forming as warm, conditionally unstable air rises from the surface. Often, however, numerous supercells do not form as the atmospheric conditions that promote the formation of large supercell thunderstorms tend to prevent many smaller ones from forming. To see why, we need to examine the vertical profile of temperature and moisture—a sounding—in the warm air ahead of the advancing cold front.

● Figure 14.22 shows a typical sounding of temperature and dew point in the warm air before supercells form. From the surface up to 800 hPa, the conditionally unstable air is warm and very humid. At 800 hPa, a shallow inversion (or, simply, a very stable layer) acts like a *cap* (or a *lid*) on the moist air below. Above the inversion, the air is cold and much drier. This air is also conditionally unstable as the temperature drops at just about the dry adiabatic lapse rate ($10°C$ km^{-1}). The cooling of this upper layer is due, mainly, to cold air moving in from the west. Cold, dry, unstable air sitting above a warm, humid layer produces *convective instability*, which means that the atmosphere will destabilize even more if a layer of air is somehow forced upward (see Chapter 6, p. 174, for information on this topic).

The lifting of warm surface air can occur at the frontal zones, but the air may also begin to rise anywhere in the region of warm air when the surface air heats up during the day. However, in the morning, the inversion acts as a lid on rising thermals, and only small cumulus clouds form. As the day progresses (and the surface air heats even more), rising air breaks through the inversion at isolated places and clouds build rapidly, sometimes explosively, as the moist air is vented upward through the opening. Thus, we can see that the stable inversion prevents many small thunderstorms from forming. When the surface air is finally able to puncture the inversion, a jet streak associated with the upper-level jet stream (at the 250 hPa level) rapidly draws the moist air up into the cold unstable air, and a large supercell quickly develops to great height.

Most thunderstorms move roughly in the direction of the winds in the middle troposphere. However, most supercell

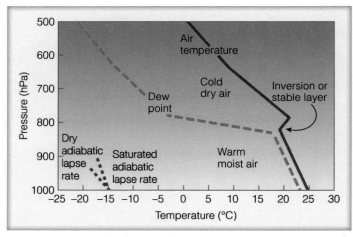

● FIGURE 14.22 A typical sounding of air temperature and dew point that frequently precedes the development of supercell thunderstorms.

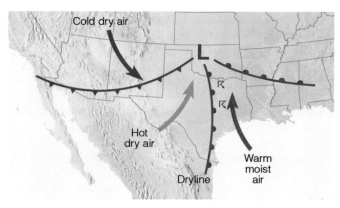

Cold dry air

L

Hot
dry air

Warm
moist
air

Dryline

● FIGURE 14.23 Surface conditions that can produce a dryline with intense thunderstorms.

storms are right-movers; that is, they move to the right of the steering winds aloft. These right-movers tend to move about 30° to the right of the mean wind in the middle troposphere. Apparently, the rapidly rising air of the storm's updraft interacts with increasing horizontal winds that change direction with height (from more southerly to more westerly) in such a way that vertical pressure gradients are able to generate new updrafts on the right side of the storm. Hence, as the storm moves along, new cells form to the right of the winds aloft.*

THUNDERSTORMS AND THE DRYLINE Thunderstorms may form along or just east of a boundary called a *dryline.* Recall from Chapter 11 that the dryline represents a narrow zone where there is a sharp horizontal change in moisture. In the United States, drylines are most frequently observed in the western half of Texas, Oklahoma, and Kansas. In this region, drylines occur most frequently during spring and early summer, where they are observed about 40 percent of the time.

● Figure 14.23 shows springtime weather conditions that can lead to the development of a dryline and intense thunderstorms. The map shows a developing mid-latitude cyclone with a cold front, a warm front, and three distinct air masses. Behind the cold front, modified cool, dry Pacific air (mP) pushes in from the northwest. In the warm air, ahead of the cold front, warm, dry continental tropical (cT) air moves in from the southwest. Farther east, warm but very humid maritime tropical (mT) air sweeps northward from the Gulf of Mexico. The dryline is the north–south oriented boundary that separates the warm, dry air and the warm, humid air.

Along the cold front—where cold, dry air replaces warm, dry air—there is insufficient moisture for thunderstorm development. The moisture boundary lies along the dryline. Because the Central Plains of North America are elevated to the west, some of the hot, dry air from the southwest is able to ride over the slightly cooler, more humid air from the Gulf. This condition sets up a potentially unstable atmosphere just

east of the dryline. Converging surface winds in the vicinity of the dryline, coupled with upper-level outflow, may result in rising air and the development of thunderstorms. As thunderstorms form, the cold downdraft from inside the storm may produce a blast of cool air that moves along the ground as a gust front and initiates the uplift necessary for generating new (possibly more severe) thunderstorms.

BRIEF REVIEW

In the last several sections, we examined different types of thunderstorms. Listed below for your review are important concepts we considered:

● All thunderstorms need three basic ingredients: (1) moist surface air; (2) a conditionally unstable atmosphere; and (3) a mechanism or "trigger" that forces the air to rise.

● Ordinary cell (air mass) thunderstorms tend to form where warm, humid air rises in a conditionally unstable atmosphere and where vertical wind shear is weak. They are usually short-lived and go through their life cycle of growth (cumulus stage), maturity (mature stage), and decay (dissipating stage) in an hour or less. They rarely produce severe weather.

● An ordinary cell thunderstorm dies because its downdraft falls into the updraft, which cuts off the storm's fuel supply.

● As wind shear increases (and the winds aloft become stronger), multicell thunderstorms are more likely to form as the storm's updraft rides up and over the downdraft. The tilted nature of the storm allows new cells to form as old ones die out.

● Multicell storms often form as a complex of storms, such as the squall line (a long line of thunderstorms that form along or out ahead of a frontal boundary) and the mesoscale convective complex (a large circular cluster of thunderstorms).

● The stronger the convection and the longer a multistorm system exists, the greater the chances of the thunderstorm becoming severe.

● Supercell thunderstorms are large, long-lasting violent thunderstorms with a single, rotating updraft that forms in a region of strong vertical wind shear. A rotating supercell is more likely to develop when (a) the winds aloft are strong and change direction from southerly at the surface to more westerly aloft and (b) a low-level jet exists just above Earth's surface.

● Although supercells are likely to produce severe weather, such as strong surface winds, large hail, heavy rain, and tornadoes, not all do.

● A gust front, or outflow boundary, represents the leading edge of cool air that originates inside a thunderstorm, reaches the surface as a downdraft, and moves outward away from the thunderstorm.

● Strong downdrafts of a thunderstorm, called downbursts (or microbursts if the downdrafts are smaller than 4 km), have been responsible for several airline crashes because on striking the surface, these winds produce extreme wind shear—rapid changes in wind speed and wind direction.

*Some thunderstorms move to the left of the steering winds aloft. This movement may happen as a thunderstorm splits into two storms, with the northern half of the storm often being a left-mover and the southern half a right-mover.

● A derecho is a strong straight-line wind produced by strong downbursts from intense thunderstorms that often appear as a bow (bow echo) on a radar screen.

● Intense thunderstorms often form along a dryline, a narrow zone that separates warm, dry air from warm, humid air.

FLOODS AND FLASH FLOODS   Intense thunderstorms are often associated with **flash floods**—floods that rise rapidly with little or no advance warning. Such flooding often results when thunderstorms stall or move very slowly, causing heavy rainfall over a relatively small area. Such flooding occurred over parts of New England and the mid-Atlantic states during June 2006 when a stationary front stalled over the region and tropical moist air, lifted by the front, produced heavy rainfall that caused extensive flooding and damage to thousands of homes. Flooding may also occur when thunderstorms move quickly but keep passing over the same area, a phenomenon called *training* (like railroad cars, one after another, passing over the same tracks). In recent years, flash floods in the United States have claimed an average of more than 100 lives a year and have accounted for untold property and crop damage. (An example of a terrible flash flood that took the lives of more than 135 people is given in Focus on a Special Topic: The Terrifying Flash Flood in the Big Thompson Canyon on p. 431.)

In Canada, flooding is the most costly natural disaster in terms of property damage. The most damaging floods are when river levels rise from spring snowmelt or from prolonged precipitation over a larger area. Manitoba's Red River flood of May 1997 is an example of flooding caused by snowmelt in combination with heavy rain. The Saguenay flood of July 1996 was caused by prolonged precipitation from a stalled middle-latitude cyclone. Large amounts of precipitation associated with tropical cyclones can also trigger major floods, such as with Hurricane Hazel in October 1954.

Flash floods caused by thunderstorms also occur in Canada. On May 30, 1961, a severe thunderstorm deposited up to 254 mm of precipitation in less than one hour near Buffalo Gap, Saskatchewan. The resulting flash flood washed away rail lines and roads and covered hundreds of hectares of farmland in water. On the afternoon of July 1, 2010, Yorkton, Saskatchewan, was hit with severe thunderstorms that deposited at least 75 mm of rain through the evening and overnight. The heavy rain fell on soils that were already saturated from the wet summer weather. A state of local emergency was declared when flooding forced 130 people from their homes.

Flash flooding due to thunderstorms occurred on August 4, 2003, in the Bois Francs region between Trois Rivières and Sherbrooke, Quebec. As an upper-level and surface low stalled over the Great Lakes, a cold front moved over the region during the afternoon. The front organized thunderstorms in the area that resulted in at least 140 mm of rain in the vicinity of Victoriaville, with 139 mm in Tingwick over a two-hour period. The extreme rainfall caused three rivers to overflow, resulting in flooding, riverbank erosion, and damage to roads and buildings, as well as the loss of seven bridges. Five hundred residents had to abandon their homes temporarily and 150 had to be evacuated.

DISTRIBUTION OF THUNDERSTORMS   It is estimated that more than 50,000 thunderstorms occur each day throughout the world. Hence, over 18 million occur annually. The combination of warmth and moisture makes equatorial landmasses especially conducive to thunderstorm formation. Here thunderstorms occur on about one out of every three days. Thunderstorms are also prevalent over water along the intertropical convergence zone, where the low-level convergence of air helps initiate uplift. The heat energy liberated in these storms helps Earth maintain its heat balance by distributing heat poleward (see Chapter 10). Thunderstorms are much less prevalent in dry climates, such as the polar regions and the desert areas dominated by subtropical highs.

● Figure 14.24 shows the average annual number of days having thunderstorms in North America. Notice that they occur most frequently in the southeastern states along the Gulf Coast, with a maximum in Florida. It is here where warm, humid maritime tropical air from the Gulf of Mexico enters North America. A secondary maximum exists just east of the Colorado Rockies. The regions with the fewest thunderstorms are near the Pacific Coast and the Arctic.

In Canada, southwest Ontario has the highest number of thunderstorm days, with a secondary maximum in central Alberta. Why are these two regions prone to thunderstorms? When wind blows from the south, southern Ontario is exposed to humid maritime tropical air from the Gulf of Mexico, which becomes unstable as it is heated from below. Also, in southern Ontario, the configuration of the Great Lakes results in converging lake breezes during summer afternoons. This low-level convergence contributes to upward motion and can trigger thunderstorms in the region. (For more about this, read ahead to Focus on a Special Topic: Canada's Tornado Alley on p. 446.) Although maritime tropical air seldom reaches Alberta, it does experience cold, dry air aloft when cool Pacific air crosses the mountains and loses moisture on the windward slopes—this contributes to instability in the region. Also, the Rocky Mountains and its foothills can cause upward motion, triggering thunderstorms.

In many areas, thunderstorms form primarily in summer during the warmest part of the day when the surface air is most unstable. There are some exceptions, however. During the summer, along the Pacific coast, dry, sinking air produces an inversion that inhibits the development of towering cumulus clouds. In these regions, thunderstorms are most frequent in winter and spring, particularly when cold, moist, conditionally unstable air aloft moves over moist, mild surface air. The surface air remains relatively warm because of its proximity to the ocean. Over the U.S. Central Plains, thunderstorms tend to form more frequently at night. These storms may be caused by a low-level southerly jet stream that forms at night and not only carries humid air northward but also initiates areas of converging surface air, which helps

FOCUS ON A SPECIAL TOPIC

The Terrifying Flash Flood in the Big Thompson Canyon

July 31, 1976, was like any other summer day in the Colorado Rockies as small cumulus clouds with flat bases and dome-shaped tops began to develop over the eastern slopes near the Big Thompson and Cache La Poudre Rivers. At first glance, there was nothing unusual about these clouds as almost every summer afternoon they form along the warm mountain slopes. Normally, strong, upper-level winds push them over the plains, causing rain showers of short duration. But the cumulus clouds on this day were different. For one thing, they were much lower than usual, indicating that the southeasterly surface winds were bringing in a great deal of moisture. Also, their tops were somewhat flattened, suggesting that an inversion aloft was stunting their growth. But these harmless-looking clouds gave no clue that later that evening in the Big Thompson Canyon more than 135 people would lose their lives in a terrible flash flood.

By late afternoon, a few of the cumulus clouds were able to puncture the inversion. Fed by moist, southeasterly winds, these clouds soon developed into gigantic multicell thunderstorms with tops exceeding 18 km. By early evening, these same clouds were producing incredible downpours in the mountains.

In the narrow canyon of the Big Thompson River, some places received as much as 305 mm of rain in the four hours between 6:30 p.m. and 10:30 p.m. local time. This is an incredible amount of precipitation, considering that the area normally receives about 405 mm for an entire year. The heavy downpours turned small creeks into raging torrents, and the Big

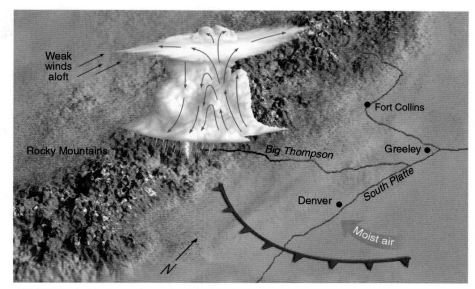

● FIGURE 1 Weather conditions that led to the development of intense thunderstorms that remained nearly stationary over the Big Thompson Canyon in the Colorado Rockies. The arrows within the thunderstorm represent air motions.

Thompson River was quickly filled to capacity. Where the canyon narrowed, the river overflowed its banks and water covered the road. The relentless pounding of water caused the road to give way.

Soon cars, tents, mobile homes, resort homes, and campgrounds were being claimed by the river. Where the debris entered a narrow constriction, it became a dam. Water backed up behind it and then broke through, causing a wall of water to rush downstream.

● Figure 1 shows the weather conditions during the evening of July 31, 1976. A cool front moved through earlier in the day and remained south of Denver. The weak inversion layer associated with the front kept the

cumulus clouds from building to great heights earlier in the afternoon. However, the strong, southeasterly flow behind the cool front pushed unusually moist air upslope along the mountain range. Heated from below, the conditionally unstable air eventually punctured the inversion and developed into a huge multicell thunderstorm complex that remained nearly stationary for several hours due to the weak southerly winds aloft. The deluge may have deposited 190 mm of rain on the main fork of the Big Thompson River in about one hour. Of the approximately 2000 people in the canyon that evening, over 135 lost their lives, and property damage exceeded $35.5 million.

trigger uplift. As the thunderstorms build, their tops cool by radiating infrared energy to space. This cooling process tends to destabilize the atmosphere, making it more suitable for nighttime thunderstorm development.

At this point, it is interesting to compare Figure 14.24 and ● Figure 14.25. Notice that even though the greatest frequency of thunderstorms is near the Gulf of Mexico Coast,

the greatest frequency of hailstorms stretches from the Alberta prairies to the U.S. western Great Plains. One reason for this situation is that conditions over the prairies are more favourable for the development of severe thunderstorms. We also find that in summer along the Gulf Coast, a thick layer of warm, moist air extends upward from the surface. Most hailstones falling into this layer will melt before reaching the

● FIGURE 14.24 The average number of days each year on which thunderstorms are observed throughout North America. (Due to the scarcity of data, the number of thunderstorms is underestimated in the mountainous west.)

Data source: NOAA; The Climates of Canada. D. Phillips. Environment Canada, 1990. Pg. 31 © Her Majesty The Queen in Right of Canada, Environment Canada, 2010.

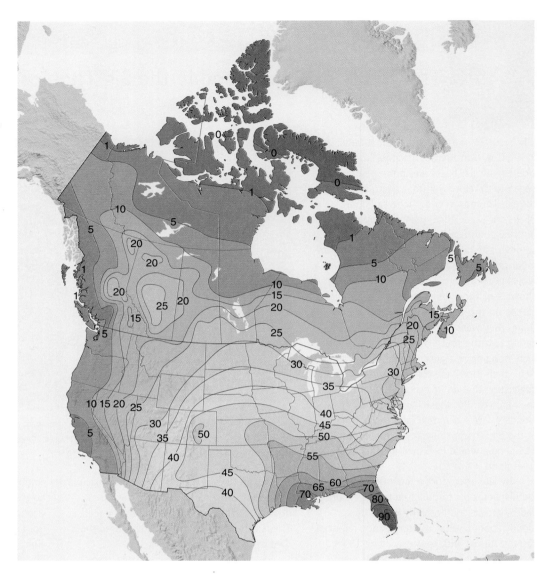

ground. Over the plains, the warm surface layer is much shallower and drier. Falling hailstones do begin to melt, but the water around their periphery quickly evaporates in the dry air. This cools the hailstones* and slows the melting rate so that many survive as ice all the way to the surface.

Now that we have looked at the development and distribution of thunderstorms, we are ready to examine an interesting, although yet not fully understood, aspect of all thunderstorms—lightning.

LIGHTNING AND THUNDER **Lightning** is simply a discharge of electricity, a giant spark, which usually occurs in mature thunderstorms. Lightning may take place within a cloud, from one cloud to another, from a cloud to the surrounding air, or from a cloud to the ground (see ● Figure 14.26). (The majority of lightning strikes occur within the cloud,

*The formation of hail is detailed in Chapter 7 on p. 211.

whereas only about 20 percent or so occur between cloud and ground.) The lightning stroke can heat the air through which it travels to an incredible 30,000 K, which is five times hotter than the surface of the sun. This extreme heating causes the air to expand explosively, thus initiating a shock wave that becomes a booming sound wave—called **thunder**—that travels outward in all directions from the flash.

Light travels so fast that we see light instantly after a lightning flash. But the sound of thunder, travelling at only about 330 m s^{-1}, takes much longer to reach the ear. If we start counting seconds from the moment we see the lightning until we hear the thunder, we can determine how far away the stroke is. Because it takes sound about three seconds to travel 1 km, if we see lightning and hear the thunder 15 seconds later, the lightning stroke occurred 5 km away.

When the lightning stroke is very close—100 m or less—thunder sounds like a clap or a crack followed immediately by a loud bang. When it is farther away, it often rumbles. The

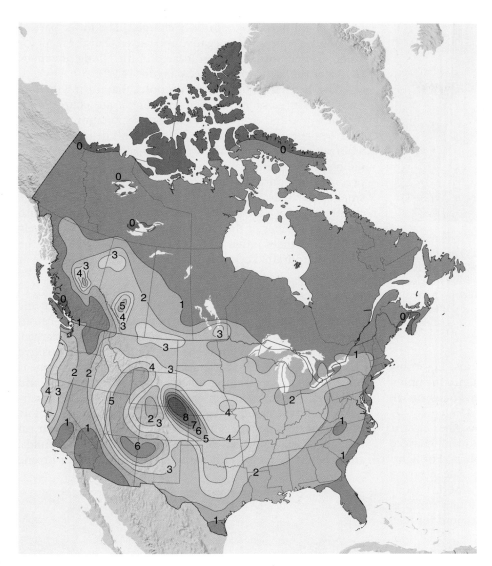

● FIGURE 14.25 The average number of days each year on which hail is observed throughout North America.

Source: NOAA; *The Climates of Canada*. D. Phillips. Environment Canada, 1990. Pg. 31 © Her Majesty The Queen in Right of Canada, Environment Canada, 2010.

rumbling can be due to the sound emanating from different areas of the stroke. Moreover, the rumbling is accentuated when the sound wave reaches an observer after having bounced off obstructions, such as hills and buildings.

In some instances, lightning is seen, but no thunder is heard. Does this mean that thunder was not produced by the lightning? Actually, there is thunder, but the atmosphere refracts (bends) and attenuates the sound waves, making the thunder inaudible. Sound travels* faster in warm air than in cold air. Because thunderstorms form in a conditionally unstable atmosphere, where the temperature normally drops rapidly with height, a sound wave moving outward away from a lightning stroke will often bend upward, away from an observer at the surface. Consequently, an observer closer than about 5 km to a lightning stroke will usually hear thunder, whereas an observer 15 km away will not.

However, even when a viewer is as close as several kilometres to a lightning flash, thunder may not be heard. For one thing, the complex interaction of sound waves and air molecules tends to attenuate the thunder. In addition, turbulent eddies of air less than 50 m in diameter scatter the sound waves. Hence, when thunder from a low-energy lightning flash travels several kilometres through turbulent air, it may become inaudible.

A sound occasionally mistaken for thunder is the **sonic boom**. Sonic booms are produced when an aircraft exceeds the speed of sound at the altitude at which it is flying. The aircraft compresses the air, forming a shock wave that trails out as a cone behind the aircraft. Along the shock wave, the air pressure changes rapidly over a short distance. The rapid pressure change causes the distinct boom. (Exploding fireworks generate a similar shock wave and a loud bang.)

Earlier, we learned that lightning occurs with mature thunderstorms. But lightning may also occur in snowstorms, in dust storms, in the gas cloud of an erupting volcano, and,

*The speed of sound in calm air in units of m s^{-1} is approximately equal to $20\sqrt{T}$, where T is the air temperature in kelvins.

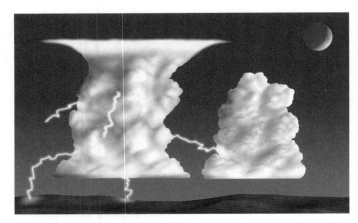

● FIGURE 14.26 The lightning stroke can travel in a number of directions. It can occur within a cloud, from one cloud to another cloud, from a cloud to the air, or from a cloud to the ground. Notice that the cloud-to-ground lightning can travel out away from the cloud and then turn downward, striking the ground many miles from the thunderstorm. When lightning behaves in this manner, it is often described as a *"bolt from the blue."*

on very rare occasions, in nimbostratus clouds. Lightning may also shoot from the top of thunderstorms into the upper atmosphere. More on this topic is given in Focus on an Observation: ELVES in the Atmosphere on p. 435.

What causes lightning? The normal fair-weather electric field of the atmosphere is characterized by a negatively charged surface and a positively charged upper atmosphere. For lightning to occur, separate regions containing opposite electrical charges must exist within a cumulonimbus cloud. Exactly how this charge separation comes about is not totally comprehended; however, there are many theories to account for it.

Electrification of Clouds One theory proposes that clouds become electrified when graupel (small ice particles called *soft hail*) and hailstones fall through a region of supercooled liquid droplets and ice crystals. As liquid droplets collide with a hailstone, they freeze on contact and release latent heat. This process keeps the surface of the hailstone warmer than that of the surrounding ice crystals. When the warmer hailstone comes in contact with a colder ice crystal, an important phenomenon occurs: *there is a net transfer of positive ions (charged molecules) from the warmer object to the colder object.* Hence, the hailstone (larger, warmer particle) becomes negatively charged and the ice crystal (smaller, cooler particle) positively charged as the positive ions are incorporated into the ice crystal (see ● Figure 14.27). The same effect occurs when colder, supercooled liquid droplets freeze on contact with a warmer hailstone and tiny splinters of positively charged ice break off. These lighter, positively charged particles are then carried to the upper part of the cloud by updrafts. The larger hailstones (or graupel), left with a negative charge, either remain suspended in an updraft or fall toward the bottom of the cloud. By this mechanism, the cold upper part of the cloud becomes positively charged, whereas the middle

of the cloud becomes negatively charged. The lower part of the cloud is generally of negative and mixed charge except for an occasional positive region located in the falling precipitation near the melting level (see ● Figure 14.28).

Another school of thought proposes that during the formation of precipitation, regions of separate charge exist within tiny cloud droplets and larger precipitation particles. In the upper part of these particles, we find negative charge, whereas in the lower part, we find positive charge. When falling precipitation collides with smaller particles, the larger precipitation particles become negatively charged and the smaller particles, positively charged. Updrafts within the cloud then sweep the smaller positively charged particles into the upper reaches of the cloud, whereas the larger negatively charged particles either settle toward the lower part of the cloud or updrafts keep them suspended near the middle of the cloud.

The Lightning Stroke Because unlike charges attract one another, the negative charge at the bottom of the cloud causes a region of the ground beneath it to become positively charged. As the thunderstorm moves along, this region of positive charge follows the cloud like a shadow. The positive charge is most dense on protruding objects, such as trees, poles, and buildings. The difference in charges causes an electric potential between the cloud and ground. In dry air, however, a flow of current does not occur because the air is a good electrical insulator. Gradually, the electrical potential gradient builds, and when it becomes sufficiently large (on the order of one

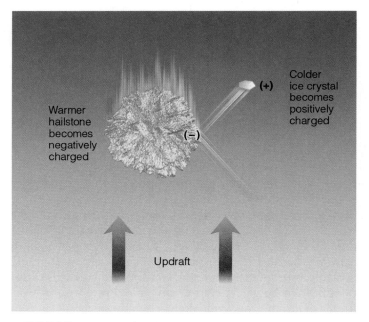

● FIGURE 14.27 When the tiny colder ice crystals come in contact with the much larger and warmer hailstone (or graupel), the ice crystal becomes positively charged and the hailstone negatively charged. Updrafts carry the tiny positively charged ice crystal into the upper reaches of the cloud, while the heavier hailstone falls through the updraft toward the lower region of the cloud.

FOCUS ON AN OBSERVATION

ELVES in the Atmosphere

For many years, airline pilots reported seeing strange bolts of light shooting upward, high above the tops of intense thunderstorms. These faint, mysterious flashes did not receive much attention, however, until they were first photographed in 1989. Photographs from sensitive, low–light-level cameras on board jet aircraft revealed that the mysterious flashes were actually a colourful display called *red sprites* and *blue jets*, which seemed to dance above the clouds.

Sprites are massive but dim light flashes that appear directly above an intense thunderstorm system (see ● Figure 2). Usually red and lasting but a few thousandths of a second, sprites tend to form almost simultaneously with lightning in the cloud below and with severe thunderstorms that have positive cloud-to-ground lightning strokes. (Most cloud-to-ground lightning is negative.) Although it is not entirely clear how they form, the thinking now is that sprites form when positive lightning disrupts the atmosphere's electric field in such a way that charged particles in the upper atmosphere are accelerated downward toward the thunderstorm and upward to higher levels in the atmosphere.

Blue jets usually dart upward in a conical shape from the tops of thunderstorms that are experiencing vigorous lightning activity (see Figure 2). Although faint, blue jets can be seen with the naked eye. They are not well

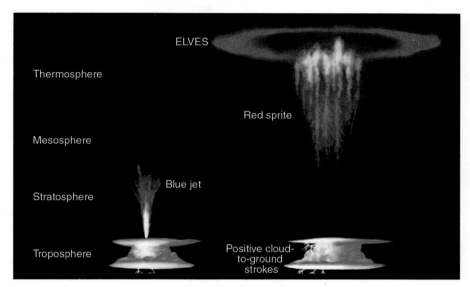

● FIGURE 2 Various electrical phenomena observed in the upper atmosphere.

understood but appear to transfer large amounts of electrical energy into the upper atmosphere.

ELVES,* as illustrated in Figure 2, appear as a faint red halo—too faint to be seen with the naked eye, only with sensitive cameras. They occur in the ionized region of the upper

---

*The acronym ELVES is from **E**missions of **L**ight and **V**ery low frequency perturbations from lightning-induced **E**lectromagnetic pulse **S**ources.

atmosphere. ELVES occur at night and are extremely short-lived. They appear to form when a lightning bolt from an intense thunderstorm gives off a strong electromagnetic pulse (EMP) that causes electrons in the ionosphere to collide with molecules that become excited and give off light.

The roles that red sprites, blue jets, and ELVES play in Earth's global electrical system have yet to be determined.

---

million volts per metre), the insulating properties of the air break down, a current flows, and lightning occurs.

Cloud-to-ground lightning begins within the cloud when the localized electric potential gradient exceeds 3 million volts per metre along a path perhaps 50 m long. This situation causes a discharge of electrons to rush toward the cloud base and then toward the ground in a series of steps (see ● Figure 14.29a). Each discharge covers about 50 to 100 m, then stops for about fifty-millionths of a second, then occurs again over another 50 m or so. This **stepped leader** is very faint and is usually invisible to the human eye. As the tip of the stepped leader approaches the ground, the potential gradient (the voltage per metre) increases, and a current of positive charge starts upward from the ground (usually along elevated objects)

to meet it (see Figure 14.29b). After they meet, large numbers of electrons flow to the ground and a much larger, more luminous **return stroke** several centimetres in diameter surges upward to the cloud along the path followed by the stepped leader (see Figure 14.29c). Hence, the downward flow of electrons establishes the bright channel of upward propagating current. Even though the bright return stroke travels from the ground up to the cloud, it happens so quickly—in one ten-thousandth of a second—that our eyes cannot resolve the motion, and we see what appears to be a continuous bright flash of light (see ● Figure 14.30).

Sometimes there is only one lightning stroke, but more often the leader-and-stroke process is repeated in the same ionized channel at intervals of about four-hundredths of a

● FIGURE 14.28 The generalized charge distribution in a mature thunderstorm.

second. The subsequent leader, called a **dart leader**, proceeds from the cloud along the same channel as the original stepped leader; however, it proceeds downward more quickly because the electrical resistance of the path is now lower. As the leader approaches the ground, normally a less energetic return stroke than the first one travels from the ground to the cloud. Typically, a lightning flash will have three or four leaders, each followed by a return stroke. A lightning flash consisting of many strokes (one photographed flash had 26 strokes) usually lasts less than a second. During this short period of time, our eyes may barely be able to perceive the individual strokes, and the flash appears to flicker.

The lightning described so far (where the base of the cloud is negatively charged and the ground positively charged) is called *negative cloud-to-ground lightning* because the stroke carries negative charges from the cloud to the ground. About 90 percent of all cloud-to-ground lightning is negative. However, when the base of the cloud is positively charged and the ground is negatively charged, a *positive cloud-to-ground lightning* flash may result. Positive lightning, most common with supercell thunderstorms, has the potential to cause more damage because it generates a much higher current level and its flash lasts for a longer duration than negative lightning.

Notice in Figure 14.30 that lightning may take on a variety of shapes and forms. When a dart leader moving toward the

ground deviates from the original path taken by the stepped leader, the lightning appears crooked or forked, and it is called *forked lightning*. An interesting type of lightning is ribbon lightning that forms when the wind moves the ionized channel between each return stroke, causing the lightning to appear as a ribbon hanging from the cloud. If the lightning channel breaks up or appears to break up, the lightning (called *bead lightning*) looks like a series of beads tied to a string. **Ball lightning** looks like a luminous sphere that appears to float in the air or slowly dart about for several seconds. Although many theories have been proposed, the actual cause of ball lightning remains an enigma. **Sheet lightning** forms when either the lightning flash occurs inside a cloud or intervening clouds obscure the flash, such that a portion of the cloud (or clouds) appears as a luminous white sheet. When cloud-to-ground lightning occurs with thunderstorms that do not produce rain, the lightning is often called **dry lightning**. Such lightning often starts forest fires in regions of dry timber.

Distant lightning from thunderstorms that is seen but not heard is commonly called **heat lightning** because it frequently occurs on hot summer nights when the overhead sky is clear. As the light from distant electrical storms is refracted through the atmosphere, air molecules and fine dust scatter the shorter wavelengths of visible light, often causing heat lightning to appear orange to a distant observer.

As the electric potential near the ground increases, a current of positive charge moves up pointed objects, such as antennas and masts of ships. However, instead of a lightning stroke, a luminous greenish or bluish halo may appear above them, as a continuous supply of sparks—a *corona discharge*—is sent into the air. This electric discharge, which can cause the top of a ship's mast to glow, is known as **St. Elmo's fire**, named after the patron saint of sailors. St. Elmo's fire is also seen around power lines and the wings of aircraft. When St. Elmo's fire is visible and a thunderstorm is nearby, a lightning flash may occur in the near future, especially if the electric field of the atmosphere is increasing.

Lightning rods are placed on buildings to protect them from lightning damage. The rod is made of metal and has a pointed tip, which extends well above the structure (see ● Figure 14.31). The positive charge concentration will be maximum on the tip of the rod, thus increasing the probability that the lightning will strike the tip and follow the metal rod harmlessly down into the ground, where the other end is deeply buried.

When lightning enters sandy soil, the extremely high temperature of the stroke may fuse sand particles together,

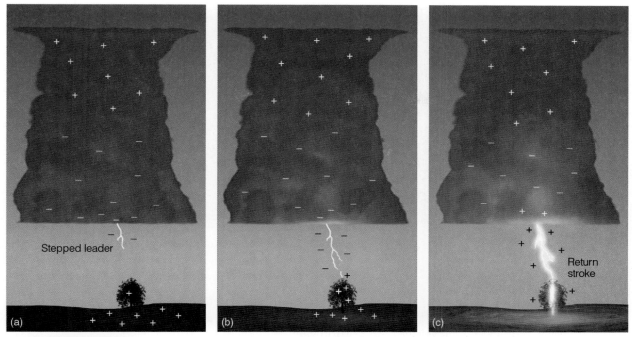

ACTIVE FIGURE 14.29 The development of a lightning stroke. (a) When the negative charge near the bottom of the cloud becomes large enough to overcome the air's resistance, a flow of electrons—the stepped leader—rushes toward Earth. (b) As the electrons approach the ground, a region of positive charge moves up into the air through any conducting object, such as trees, buildings, and even humans. (c) When the downward flow of electrons meets the upward surge of positive charge, a strong electric current—a bright return stroke—carries positive charge upward into the cloud. Visit the textbook's website to view this and other Active Figures at www.ahrensmeteorology1ce.nelson.com

● FIGURE 14.30 Time exposure of an evening thunderstorm with an intense lightning display near Denver, Colorado. The bright flashes are return strokes. The lighter forked flashes are probably stepped leaders that did not make it to the ground.

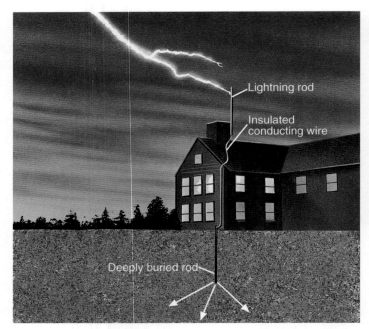

● FIGURE 14.31 The lightning rod extends above the building, increasing the likelihood that lightning will strike the rod rather than some other part of the structure. After lightning strikes the metal rod, it follows an insulated conducting wire harmlessly into the ground.

producing a rootlike system of tubes called a **fulgurite**, after the Latin word for "lightning" (see ● Figure 14.32). When lightning strikes an object such as a car, it normally leaves the passengers unharmed because it usually takes the quickest path to the ground along the outside metal casing of the vehicle. The lightning then jumps to the road through the air or enters the roadway through the tires. If you should be caught in the open in a thunderstorm, what should you do? Of course, seek shelter immediately, but under a tree? If you are not sure, please read Focus on an Observation: Don't Sit Under the Apple Tree on p. 439.

Lightning Detection and Suppression For many years, lightning strokes were detected primarily by visual observation. Today, cloud-to-ground lightning is located by means of an instrument called a *lightning direction-finder*, which works by detecting the radio waves produced by lightning. Such waves are called **sferics**, a contraction from their earlier designation, *atmospherics*. A web of these magnetic devices is a valuable tool in pinpointing lightning strokes throughout Canada and the United States. Lightning detection devices allow scientists to examine in detail the lightning activity inside a storm as it intensifies and moves (see ● Figure 14.33). This gives forecasters a better idea where intense lightning strokes might be expected. In addition, when this information is correlated with satellite images, a more complete and precise structure of a thunderstorm is obtained. Lightning detection on polar orbiting satellites has allowed mapping of global lightning occurrence. In the future, geostationary

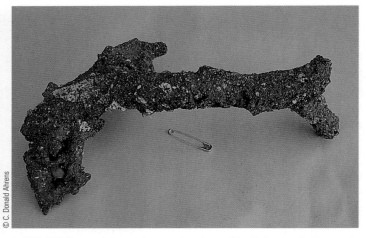

● FIGURE 14.32 A fulgurite that formed by lightning fusing sand particles.

satellites will carry lightning detection sensors as well, which will expand our ability to detect types of lightning worldwide.

Each year, approximately 45 percent of the 8000 forest fires in Canada are caused by lightning. For this reason, tests have been conducted to see whether the number of cloud-to-ground lightning discharges can be reduced. One technique that has shown some success in suppressing lightning involves seeding a cumulonimbus cloud with hair-thin pieces of aluminum about 10 cm long. The idea is that these pieces of metal will produce many tiny sparks, or *corona discharges*, and prevent the electrical potential in the cloud from building to a point where lightning occurs. Although the results of this experiment are inconclusive, many forestry specialists point out that nature itself may use a similar mechanism to prevent excessive lightning damage. The long, pointed needles of pine trees may

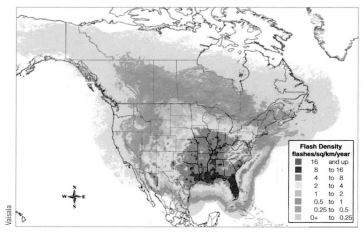

● FIGURE 14.33 The number of lightning flashes per square kilometre per year for cloud-to-cloud and cloud-to-ground lightning combined between 2000 and 2004. Data are from the North American Lightning Detection Network.
Data source: NASA

FOCUS ON AN OBSERVATION

Don't Sit Under the Apple Tree

Because a single lightning stroke may involve a current as great as 100,000 amperes, animals and humans can be electrocuted when struck by lightning. In Canada, there are an average of about 10 fatalities and 92 to 164 injuries due to lightning each year, with most occurring in Ontario. The average yearly death toll in the United States attributed to lightning between 2000 and 2009 is 41.2, with Florida accounting for the most fatalities. Many victims are struck in open places, while riding on farm equipment, playing golf, attending sports events, or sailing in a small boat. Some live to tell about it, as did the retired champion golfer Lee Trevino. Others are less fortunate. About 10 percent of people struck by lightning are killed. Most die from cardiac arrest. Consequently, when you see someone struck by lightning, immediately give CPR (cardiopulmonary resuscitation) as lightning normally leaves its victims unconscious, without a heartbeat and without respiration. Those who do survive often suffer from long-term psychological disorders, such as personality changes, depression, and chronic fatigue.

Many lightning fatalities occur in the vicinity of relatively isolated trees (see ● Figure 3). As a tragic example, in June 2004, three people were killed near Atlanta, Georgia, seeking shelter under

© Johnny Autery

● FIGURE 3 A cloud-to-ground lightning flash hitting a 20 m sycamore tree. It should be apparent why one should not seek shelter under a tree during a thunderstorm.

a tree. Because a positive charge tends to concentrate in upward projecting objects, the upward return stroke that meets the stepped leader is most likely to originate from such objects. Clearly, sitting under a tree during an electrical storm is not wise. What should you do?

When caught in a thunderstorm, the best protection, of course, is to get inside a building. But stay away from electrical appliances and corded phones and avoid taking a shower. Automobiles with metal frames and trucks (but not golf carts) may also provide protection. If no such shelter exists, be sure to avoid elevated places and isolated trees. If you are on level ground, try to keep your head as low as possible but do not lie down. Because lightning channels usually emanate outward through the ground at the point of a lightning strike, a surface current may travel through your body and injure or kill you. Therefore, crouch down as low as possible and minimize the contact area you have with the ground by touching it with only your toes or your heels.

There are some warning signs to alert you to a strike. If your hair begins to stand on end or your skin begins to tingle and you hear clicking sounds, beware—lightning may be about to strike. And if you are standing upright, you may be acting as a lightning rod.

---

act as tiny lightning rods, diffusing the concentration of electric charges and preventing massive lightning strokes.

Now that we have looked at thunderstorms, we are ready to explore a product of a thunderstorm that is one of nature's most awesome phenomena: the tornado, a rapidly spiralling column of air that usually extends down from the base of a cumulonimbus cloud and can strike sporadically and violently.

Tornadoes

A **tornado** is a rapidly rotating column of air that blows around a small area of intense low pressure with a circulation that reaches the ground. A tornado's circulation is present on the ground either as a funnel-shaped cloud or as a swirling cloud of dust and debris. Sometimes called *twisters* or *cyclones*, tornadoes can assume a variety of shapes and forms that range from twisting, ropelike funnels, to cylinder-shaped funnels, to massive black funnels, to funnels that resemble an elephant's trunk hanging from a large cumulonimbus cloud. A **funnel cloud** is a tornado whose circulation has not reached the ground. When viewed from above, the majority of North American tornadoes rotate counterclockwise about their central core of low pressure. A few have been seen rotating clockwise, but those are rare.

The diameter of most tornadoes is between 100 and 600 m, although some are just a few metres wide and others have diameters exceeding 1600 m. In fact, on May 22, 2004, one of the largest tornadoes on record touched down near Hallam, Nebraska, with a diameter of about 4 km. Tornadoes that form ahead of an advancing cold front are often steered by southwesterly winds and, therefore, tend to move from the southwest toward the northeast at speeds usually between 20 and 40 knots (37 to 74 km h⁻¹). However, some have been clocked at speeds greater than 70 knots (130 km h⁻¹). Most tornadoes last only a few minutes and have an average path length of about 7 km.

TORNADO LIFE CYCLE   Major tornadoes usually evolve through a series of stages. The first stage is the dust-whirl stage, where dust swirling upward from the surface marks the tornado's circulation on the ground and a short funnel often extends downward from the thunderstorm's base. Damage during this stage is normally light. The next stage, called the *organizing stage*, finds the tornado increasing in intensity with an overall downward extent of the funnel. During the tornado's mature stage, damage normally is most severe as the funnel reaches its greatest width and is almost vertical (see Figure 14.34). The *shrinking stage* is characterized by an overall decrease in the funnel's width, an increase in the funnel's tilt, and a narrowing of the damage swath at the surface, although the tornado may still be capable of intense and sometimes violent damage. The final stage, called the decay stage, usually finds the tornado stretched into the shape of a rope. Normally, the tornado becomes greatly contorted before it finally dissipates. Although these are the typical stages of a major tornado, minor tornadoes may evolve only through the organizing stage. Some even skip the mature stage and go directly into the decay stage. However, when a tornado reaches its mature stage, its circulation usually stays in contact with the ground until it dissipates.

TORNADO OUTBREAKS   Each year, tornadoes take the lives of many people. The yearly average is less than 100, although over 100 may die in a single day. In recent years, an alarming statistic is that 45 percent of all fatalities occurred in mobile homes. The deadliest tornadoes are those that occur in *families*—that is, different tornadoes spawned by the same thunderstorm. (Some thunderstorms produce a sequence of several tornadoes over two or more hours and over distances of 100 km or more.) Tornado families often are the result of a single, long-lived supercell thunderstorm. When a large number of tornadoes (typically six or more) form over a particular region, this constitutes what is termed a **tornado outbreak**.

A particularly devastating outbreak occurred on May 3, 1999, when 78 tornadoes marched across parts of Texas, Kansas, and Oklahoma. One tornado, whose width at times reached 1.6 km and whose wind speed was measured by Doppler radar at 276 knots (511 km h⁻¹), moved through the southwestern section of Oklahoma City. Within its 64 km path, it damaged or destroyed thousands of homes, injured nearly 600 people, claimed 38 lives, and caused over $1 billion in property damage.

One of Canada's most devastating tornadoes in recent times was the Edmonton tornado of July 31, 1987. To learn more about this devastating storm, see Focus on a Special Topic: The Edmonton Tornado on p. 442. Canada's largest single-day outbreak was the southern Ontario tornado outbreak of August 20, 2009. A total of 18 tornadoes (four were rated F2) were observed in southwestern and central Ontario as well as the Greater Toronto Area.

FIGURE 14.34 Canada's first documented F5 tornado on June 22, 2007, at Elie, Manitoba.

Justin Hobson

One of the most violent outbreaks ever recorded occurred on April 3 and 4, 1974. During a 16-hour period, 148 tornadoes cut through parts of 13 states, killing 307 people, injuring more than 6000, and causing an estimated $600 million in damage. Some of these tornadoes were among the most powerful ever witnessed. The combined path of all the tornadoes during this *superoutbreak* amounted to 4181 km, well over half of the total path for an average year. The greatest loss of life attributed to tornadoes occurred during the tristate outbreak of March 18, 1925, when an estimated 695 people died as at least seven tornadoes travelled a total of 703 km across portions of Missouri, Illinois, and Indiana.

TORNADO OCCURRENCE   Tornadoes occur in many parts of the world, but no country experiences more tornadoes than the United States, which, in recent years, averaged more than 1000 annually and experienced a record 1722 tornadoes during 2004. The greatest number occur in the tornado belt, or *tornado alley*, of the Central Plains, which stretches from central Texas to Nebraska* (see ● Figure 14.35). Canada has the second greatest number of tornadoes, experiencing 60 to 80 each year (see ● Figure 14.36). There is an average of about two fatalities each year in Canada from tornadoes, although this number is highly variable from year to year. In Canada, tornadoes occur most frequently in Alberta through southern Saskatchewan and Manitoba, as well as in southwest Ontario. To learn about Canada's tornado alley stretching between Windsor and Toronto, see Focus on a Special Topic: Canada's Tornado Alley on p. 446.

The Central Plains region of North America is most susceptible to tornadoes because it often provides the proper atmospheric setting for the development of the severe thunderstorms that spawn tornadoes. Here (especially in spring) warm, humid surface air is overlain by cooler, drier air aloft, producing a conditionally unstable atmosphere. When a strong vertical wind shear exists (usually provided by a low-level jet and by the polar jet stream) and the surface air is forced upward, large supercell thunderstorms capable of spawning tornadoes may form. Therefore, tornado frequency is highest during the spring and lowest during the winter, when the warm surface air is normally absent.

The frequency of tornadic activity shows a seasonal shift. For example, during the winter, tornadoes are most likely to form over the southern Gulf states when the polar-front jet is above this region and the contrast between warm and

*Many of the tornadoes that form along the Gulf Coast are generated by thunderstorms embedded within the circulation of hurricanes.

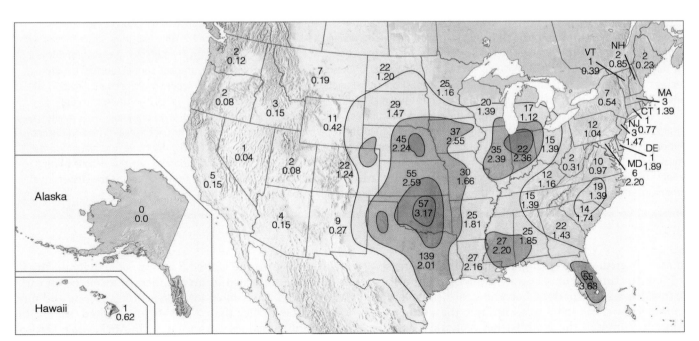

● FIGURE 14.35 Tornado incidence in the United States by state. The upper figure shows the average annual number of tornadoes observed in each state from 1953 to 2004. The lower figure is the average annual number of tornadoes per 10,000 km² in each state during the same period. The darker the shading, the greater the frequency of tornadoes.
Source: NOAA

FOCUS ON A SPECIAL TOPIC

The Edmonton Tornado

Gerhard W. Reuter, Professor, Department of Earth and Atmospheric Science, University of Alberta

July 31, 1987, was a day that changed the life of many Edmontonians and made meteorological history. Dubbed "Black Friday," parts of Edmonton were turned into destruction by severe storms with tennis ball–sized hailstones and a destructive F4 tornado (see ● Figure 4).

The tornado remained on the ground from 2:59 p.m. to 4:25 p.m., cutting a swath of destruction 40 km long and up to 900 m wide in places. The funnel touched down in Leduc, located south of Edmonton. The tornado moved northwards to Mill Woods, levelling many homes. Passing over Sherwood Park, it tossed cars up into the air and damaged oil storage tanks in the refinery areas. Crossing the river, the tornado continued to destroy buildings in Clareview and decimated the Evergreen Mobile Home Park. The tornado killed 27 people, injured more than 300 people, destroyed more than 3000 homes, and caused more than $330 million in property damage. The loss of life, injuries, and property destruction made it the worst natural disaster in Alberta.

What exactly formed this violent tornado that swept through Edmonton? The precise cause of the tornado funnel is not fully understood, but it was a combination of thermal and mechanical effects acting in cumulonimbus clouds. The atmosphere had all the necessary ingredients required for intense storm formation. Edmonton experienced a prolonged heat wave, and the air had extreme humidity values. In fact, July 31, 1987, remains Edmonton's most humid day on

● FIGURE 4
The Edmonton tornado in full force.

The Canadian Press Images/Edmonton Journal/ Steve Simon

record! On July 31, a cold front moved into the area, lifting the boundary layer air to reach cloud condensation. Strong vertical shear in the ambient wind was evident, which is essential for organization of the convective storm and forming a supercell thunderstorm. The storm updraft in the supercell must have reached 80 m s^{-1} (288 km h^{-1}), based on the size of the numerous hailstones, with sizes as large as 12 cm in effective diameter—larger than tennis balls! The hailstones, which were collected in Mill Woods, showed a remarkable variety of shapes, opacities, and sizes. The strong updraft caused extreme convergence near the base of the supercell, and this convergence must have intensified the cyclonic vorticity embedded in the supercell circulation. Conservation of angular momentum as the air spiraled inward must have contributed toward the spin-up of the rotation, causing a tornado.

Video camera recordings of the Edmonton tornado clearly indicate that the tornado consisted of three or four multiple vortices grouping together. Violent tornados with multiple vortices are common when the mother storm exhibits strong convergence.

One of the most interesting and perplexing feature of the Edmonton tornado is its abrupt change of direction of movement. Initially, the tornadic hailstorm storm moved with the midlevel wind blowing eastward. But just when the tornado funnel touched the ground, the storm suddenly changed its direction 90° and pushed directly northward toward Edmonton. We do not know what might have caused this abrupt change of motion direction toward the city. For more information on the Edmonton tornado, see the website of the Edmonton Tornado Atlas at the University of Alberta.

---

cold air masses is greatest. In spring, humid Gulf air surges northward; contrasting air masses and the jet stream also move northward, and tornadoes become more prevalent from the southern Atlantic states westward into the southern Great Plains. In summer, the contrast between air masses lessens, and the jet stream is normally near the Canada–U.S. border; hence, tornado activity tends to be concentrated from the northern plains eastward to New York State. This is illustrated by the earlier peak in the number of tornadoes each month in the United States compared to Canada (see

● Figure 14.37). In the United States, peak activity occurs in May and June, with significant numbers of tornadoes as early as March. In Canada, most tornadoes occur in June and July, with very few earlier than April. In the United States, the most violent tornadoes seem to occur in April, when vertical wind shear tends to be present, as well as when horizontal and vertical temperature and moisture contrasts are greatest. Although tornadoes have occurred at all times of the day and night, they are most frequent in the late afternoon (between 4:00 p.m. and 6:00 p.m.), when the surface air is most

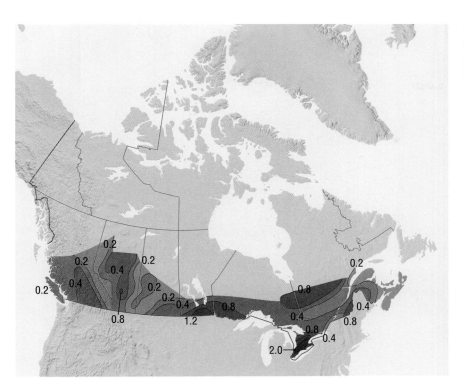

● FIGURE 14.36 Tornado frequency in Canada: number of tornadoes per 10,000 km². Areas north of the shaded area do not have tornado sightings.

The Climates of Canada. D. Phillips. Environment Canada, 1990. Pg. 31. © Her Majesty The Queen in Right of Canada, Environment Canada, 2010.

unstable; they are least frequent in the early morning before sunrise, when the atmosphere is most stable.

TORNADO WINDS   The strong winds of a tornado can destroy buildings, uproot trees, and hurl all sorts of lethal

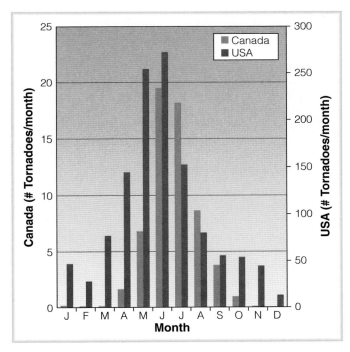

● FIGURE 14.37 Average number of tornadoes during each month in Canada and the United States.

Source: NOAA; Etkin, et al., 2001. "Tornado Climatology of Canada Revisited: Tornado Activity During Different Phases of ENSO." *International Journal of Climatology*. Vol. 21, No. 8. Pg. 915–938.

missiles into the air. People, animals, and home appliances all have been picked up, carried several kilometres, and then deposited. Tornadoes have accomplished some astonishing feats, such as lifting a railroad coach with its 117 passengers and dumping it in a ditch 25 m away. Showers of toads and frogs have poured out of a cloud after tornadic winds sucked them up from a nearby pond. Other oddities include chickens losing all of their feathers, pieces of straw being driven into metal pipes, and frozen hot dogs being driven into concrete walls. Miraculous events have occurred, too. In one instance, a schoolhouse was demolished and the 85 students inside were carried over 100 m without one of them being killed.

Our earlier knowledge of the furious winds of a tornado came mainly from observations of the damage done and the analysis of motion pictures. Today, more accurate wind measurements are made with Doppler radar. Because of the destructive nature of the tornado, it was once thought that it packed winds greater than 500 knots. However, studies conducted after 1973 reveal that even the most powerful twisters seldom have winds exceeding 220 knots, and most tornadoes have winds of less than 125 knots. Nevertheless, being confronted with even a small tornado can be terrifying.

When a tornado is approaching from the southwest, its strongest winds are on its southeast side. We can see why in ● Figure 14.38. The tornado is heading northeast at 50 knots. If its rotational speed is 100 knots, then its forward speed will add 50 knots to its southeastern side (position D) and subtract 50 knots from its northwestern side (position B). Hence, the most destructive and extreme winds will be on the tornado's southeastern side.

Many violent tornadoes (with winds exceeding 180 knots) contain smaller whirls that rotate within them. Such

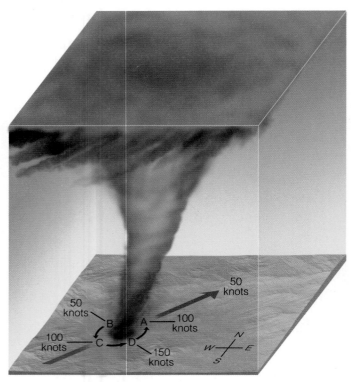

● FIGURE 14.38 The total wind speed of a tornado is greater on one side than on the other. When facing an onrushing tornado, the strongest winds will be on your left side.

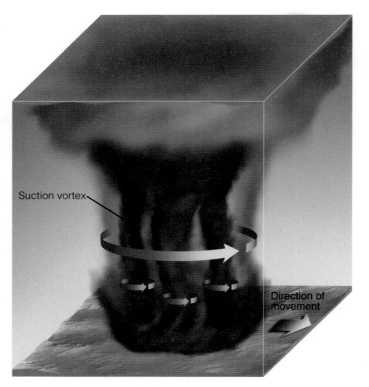

● FIGURE 14.39 A powerful multivortex tornado with three suction vortices.

tornadoes are called *multivortex* tornadoes, and the smaller whirls are called **suction vortices** (see ● Figure 14.39). Suction vortices are only about 10 m in diameter, but they rotate very fast and can do a great deal of damage.

Seeking Shelter   The high winds of the tornado cause the most damage as the walls of buildings buckle and collapse when blasted by the extreme wind force and by debris carried by the wind. Also, as high winds blow over a roof, lower air pressure forms above the roof. The greater air pressure inside the building then lifts the roof just high enough for the strong winds to carry it away. A similar effect occurs when the tornado's intense low-pressure centre passes overhead. Because the pressure in the centre of a tornado may be more than 100 hPa lower than that of its surroundings, there is a momentary drop in outside pressure when the tornado is above the structure. It was once thought that opening windows and allowing inside and outside pressures to equalize would minimize the chances of the building exploding. However, it is now known that opening windows during a tornado actually *increases* the pressure on the opposite wall and increases the chances that the building will collapse. (The windows are usually shattered by flying debris anyway.) So stay away from windows. Damage from tornadoes may also be inflicted on people and structures by flying debris. Hence, the wisest course to take when confronted with an approaching tornado is to *seek shelter* immediately.

At home, take shelter in a basement. In a large building without a basement, the safest place is usually in a small room, such as a bathroom, closet, or interior hallway, preferably on the lowest floor and near the middle of the edifice. Pull a mattress around you as the handles on the side make it easy to hang onto. Wear a bike or football helmet to protect your head from flying debris. At school, move to the hallway and lie flat with your head covered. In a mobile home, leave immediately and seek substantial shelter. If none exists, lie flat on the ground in a depression or ravine. Do not try to outrun an oncoming tornado in a car or truck as tornadoes often cover erratic paths with speeds sometimes exceeding 70 knots (130 km h^{-1}). Stop your car and let the tornado go by or turn around on the road's shoulder and drive in the opposite direction. And do not take shelter under a freeway overpass as the tornado's winds are actually funnelled (strengthened) by the overpass structure. If caught outdoors in an open field, look for a ditch, streambed, or ravine, and lie flat with your head covered.

When tornadoes are likely to form during the next few hours, the Meteorological Service of Canada issues a **tornado watch** to alert the public that tornadoes may develop within a specific area during a certain time period. Many communities have trained volunteer spotters, who look for tornadoes after the watch is issued. (If a tornado is spotted in the watch area, keep abreast of its movement by listening to Weatheradio

▼ Table 14.1   Average Annual Number of Tornadoes and Tornado Deaths by Decade

DECADE	U.S. TORNADOS/YEAR	U.S. DEATHS/YEAR	CANADIAN DEATHS/YEAR
1950–59	480	148	1.3
1960–69	681	94	0
1970–79	858	100	2.1
1980–89	819	52	4.0
1990–99	1220	56	0.3
2000–09	1340*	71	1.7

*More tornadoes are being reported as populations increase and tornado-spotting technology improves.

Canada.) Once a tornado is spotted—either visually or on a radar screen—a **tornado warning** is issued by the responsible Storm Prediction Centre.* In some communities in the United States, sirens are sounded to alert people of the approaching storm. Radio and television stations interrupt regular programming to broadcast the warning. Although not completely effective, this warning system is apparently saving many lives. Despite the large increase in population in areas affected by tornadoes during the past 30 years, tornado-related deaths have generally shown a decrease in the United States (see ▼ Table 14.1).

THE FUJITA SCALE   In the 1960s, the late Dr. T. Theodore Fujita, a noted authority on tornadoes at the University of Chicago, proposed a scale (called the **Fujita scale**) for classifying tornadoes according to their rotational wind speed. The tornado winds are estimated based on the damage caused by the storm. However, classifying a tornado based solely on the damage it causes is rather subjective. But the scale became widely used and is presented in ▼ Table 14.2.

---

*The Meteorological Service of Canada has Storm Predictions Centres in Vancouver, Edmonton, Winnipeg, Toronto, Montreal, and Halifax.

The original Fujita scale, implemented in 1971, was based mainly on tornado damage incurred by a frame house. Because many types of structures are susceptible to tornado damage, a new scale was introduced by the U.S. NOAA and came into effect in February 2007. Called the *enhanced Fujita scale*, or simply the **EF scale**, the new scale attempts to provide a wide range of criteria in estimating a tornado's winds by using a set of 28 damage indicators. These indicators include items such as small barns, mobile homes, schools, and trees. Each item is then examined for the degree of damage it sustained. The combination of the damage indicators along with the degree of damage provides a range of probable wind speeds and an EF rating for the tornado. The wind estimates for the EF scale are given in ▼ Table 14.3.

Statistics reveal that the majority of tornadoes are relatively weak, with wind speeds less than about 177 km h^{-1}. Only a few percent each year are classified as violent, with perhaps one or two EF5 tornadoes reported annually in the United States (although several years may pass without the United States experiencing an EF5). However, it is the violent tornadoes that account for the majority of tornado-related deaths. As an example, both the Barrie tornado of 1985 (see ● Figure 14.40) and the Edmonton tornado of 1987 were powerful F4 tornadoes that claimed 12 and 27 lives, respectively, accounting for nearly all of the tornado fatalities experienced in Canada during the 1980s.

WEATHER WATCH

It may be almost impossible to survive the powerful winds of a violent tornado if you are inside the wrong type of structure, such as a mobile home. During the May 3, 1999, tornado outbreak in Oklahoma, many people who abandoned their unprotected homes in favour of muddy ditches survived largely because the ditches were below ground level and out of the path of wind-blown objects. Many who stayed in the confines of their inadequate homes perished when tornado winds blew their homes away, leaving only the foundations.

▼ Table 14.2   Fujita Scale for Damaging Winds

SCALE	CATEGORY	km h^{-1}	KNOTS	EXPECTED DAMAGE
F0	Weak	64–116	35–62	Light: tree branches broken, sign boards damaged
F1		117–180	63–97	Moderate: trees snapped, windows broken
F2	Strong	181–252	98–136	Considerable: large trees uprooted, weak structures destroyed
F3		253–330	137–179	Severe: trees levelled, cars overturned, walls removed from buildings
F4	Violent	331–417	180–226	Devastating: frame houses destroyed
F5*		418–509	227–276	Incredible: structures the size of automobiles moved over 100 m, steel-reinforced structures highly damaged

*The scale continues up to a theoretical F12. Very few (if any) tornadoes have wind speeds in excess of 509 km h^{-1}.

Canada's Tornado Alley

Patrick King, Environment Canada (retired)

North America's tornado alley of the U.S. Central Plains stretches from central Texas to Nebraska, with peak incidence averaged over all of Oklahoma of 3.2 tornadoes per 10,000 km². Although nowhere in Canada is as high, there is an area of enhanced tornado occurrence in southern Ontario between Windsor and Toronto, which experiences around two tornadoes per 10,000 km². This area is known as Canada's Tornado Alley.

Southern Ontario is a complex peninsula surrounded by three of the Great Lakes, with no point more than about 75 km from one of the lakes. **Lake breezes** develop frequently in the spring and summer as air over the land warms more rapidly than air over the water. A pressure difference develops, and in the absence of a large-scale flow, the lake breeze blows from the lakes to the land (for more information on lake and sea breezes, see p. 270 in Chapter 9). The leading edge of the lake breeze, called the *lake breeze front*, forms parallel to the shoreline and moves inland during the day. Cumulus clouds often develop in upward moving air along the lake breeze front. If the shoreline is not straight, for example, near a peninsula, the lake breezes from different parts of the coastline may

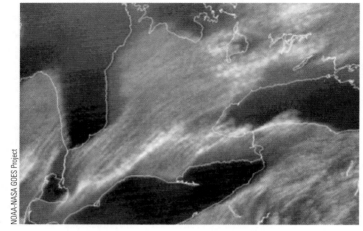

NOAA-NASA GOES Project

● FIGURE 5 Average of six visible satellite images between 10:00 a.m. and 12:30 p.m. EST on July 31, 1994, a warm summer day with moderate southwest flow.

converge, resulting in regions of upward air motion. When this happens, with suitable temperature humidity and stability, thunderstorms may develop.

In the middle latitudes, there is usually a large-scale flow associated with moving high- and low-pressure systems. Therefore, lake breeze circulations typically combine with the large-scale flow to form more complex convergence zones. This is illustrated in ● Figure 5, a composite of six GOES satellite images over a three-hour period on a warm summer's day with moderate southwest flow. The whitest

areas indicate persistent cloudy areas over the three-hour period. Since the cloud lines form parallel to the wind direction, we can also get an idea of the wind flow during this period. In regions where the southwest flow is parallel to a shoreline, a convergence zone at the lake breeze front forms that is indicated by convective cloud formation. In regions where the southwest flow is onshore and crosses the coastline, the lake breeze front may form well inland. In the case shown in ● Figure 6, small thunderstorms developed along the lake breeze fronts.

▼ Table 14.3 Modified (EF) Fujita Scale for Damaging Winds

EF SCALE	km h⁻¹*	KNOTS
EF0	105–137	57–74
EF1	138–177	75–96
EF2	178–218	97–118
EF3	219–266	119–144
EF4	267–322	145–174
EF5	> 322	> 174

*The wind speed is a three-second gust estimated at the point of damage, based on a judgment of damage indicators.

One important reason for the number of deaths and extensive damage being caused by violent tornadoes is that, as the wind speed doubles, the force of the wind exerted on an object increases by a factor of four. Hence, the 300 km h⁻¹ winds of an EF4 tornado exert four times as much force on a building as do the 150 km h⁻¹ winds of an EF1.

Tornado Formation

Although not everything is known about the formation of a tornado, we do know that tornadoes tend to form with intense thunderstorms and that a conditionally unstable

Severe summer weather requires sufficiently warm and moist air and a thunderstorm "initiation" mechanism. Such conditions occur in southern Ontario summer most commonly in a moderate southwesterly flow ahead of an approaching cold front. The southwest flow brings warm, moist maritime tropical air from the Gulf of Mexico, and the cold front provides the lift to initiate convection. Of course, this situation also favours the development of lake breeze fronts, which are very common on potentially severe weather days.

Studies in the United States have shown that tornadoes tend to form where cold fronts interact with preexisting boundaries; could lake breeze boundaries act as triggers to severe weather events in southern Ontario? An indication of this is provided by Figure 6, a plot of afternoon tornado touchdown points in the early part of the season, when lake breezes are most intense owing to the large land/water temperature difference. Coloured dots represent initial points along tornado paths, and lake breeze fronts from Figure 6 are overlaid. Tornadoes appear to be completely suppressed in regions where southwest winds are onshore and enhanced near the lake breeze front locations.

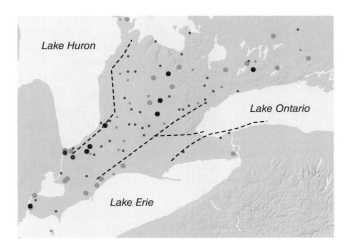

● FIGURE 6 Tornado touchdown points for the years 1917 to 1992 between March 15 and June 15, in the early part of the season, when lake breezes are most intense owing to the large land/water temperature difference. Coloured dots represent tornado intensity from the Fujita scale: purple = F0, orange = F1, green = F2, blue = F3, red = F4. The lake breeze fronts from Figure 6 are overlaid as dashed lines. Source: NOAA

No severe weather occurred in the example in Figure 6. However, with a stronger southwest flow, the Lake Huron and Lake Erie lake breeze fronts can merge near the centre line of the peninsula. Scientists from Environment Canada and Toronto's York University have hypothesized that cold fonts interacting with merged lake breeze fronts may be responsible for the maximum in F2 and stronger tornadoes that are evident in the centre of the peninsula in Figure 6. In fact, all of the touchdown points of F3 and stronger tornadoes in the spring have formed within 20 km of the central axis of the peninsula.

As the lakes warm up, lake breezes become less strong and the relationship between tornado locations and converging lake breeze fronts appears to weaken. In July and August, there are many more weak tornadoes that may occur closer to the lake shores. This is especially evident in Essex County in extreme southwestern Ontario. In spring, almost all of the events are along the Detroit River. In July and August, when the shallow waters of western Lake Erie and Lake St. Clair become very warm, lake breezes will merge in the centre of Essex County, occasionally causing weak tornadoes.

---

atmosphere is essential for their development. Most often they form with supercell thunderstorms in an environment with strong vertical wind shear. The rotating air of the tornado may begin within a thunderstorm and work its way downward, or it may begin at the surface and work its way upward. First, we will examine tornadoes that form with supercells; then we will examine nonsupercell tornadoes.

SUPERCELL TORNADOES Tornadoes that form with supercell thunderstorms are called **supercell tornadoes**. Earlier, we learned that a supercell is a thunderstorm that has a single, rotating updraft that can exist for hours. Recall also that supercells form in a region of strong vertical wind shear that causes the updraft inside the storm to rotate.

In ● Figure 14.41a, notice that there is wind direction shear as the surface winds are southerly and a kilometre or so above the surface they are northerly. There is also wind-speed shear as the wind speed increases rapidly with height. This wind shear causes the air near the surface to rotate about a horizontal axis much like a pencil rotates around its long axis when rolled between your palms. Such horizontal tubes of spinning air are called vortex tubes. (These vortex tubes also form when a strong, southerly, low-level jet exists just above weaker, southerly surface winds.) If the strong updraft of a developing thunderstorm should tilt the rotating tube upward and draw it into the storm, as illustrated in Figure 14.41b, the tilted rotating tube then becomes a rotating air column inside the storm. The rising, spinning air is now part of the storm's

● FIGURE 14.40 The Barrie and central Ontario tornado outbreak of May 31, 1985, was actually 14 separate tornadoes, two of them rated F4. They left a path of destruction resulting in 12 fatalities, 224 injuries, and over 1000 buildings destroyed or damaged.

structure called the mesocyclone—an area of lower pressure (a small cyclone) perhaps 5 to 10 km across. The rotation of the updraft lowers the pressure in the midlevels of the thunderstorm,* which acts to increase the strength of the updraft.

As we learned earlier in the chapter, the updraft is so strong in a supercell (sometimes 90 knots or 167 km h⁻¹) that precipitation cannot fall through it. Southwesterly winds aloft usually blow the precipitation northeastward. If the mesocyclone persists, it can circulate some of the precipitation counterclockwise around the updraft. This swirling precipitation shows up on the radar screen, whereas the area

inside the mesocyclone (nearly void of precipitation at lower levels) does not. The region inside the supercell where radar is unable to detect precipitation is known as the bounded weak echo region (BWER). Meanwhile, as the precipitation is drawn into a cyclonic spiral around the mesocyclone, the rotating precipitation may, on the Doppler radar screen, unveil itself in the shape of a hook, called a **hook echo**, as shown in ● Figure 14.42.

At this point in the storm's development, the updraft, the counterclockwise swirling precipitation, and the surrounding air may all interact to produce the rear-flank downdraft (to the south of the updraft), as shown in ● Figure 14.43. The strength of the downdraft is driven by the amount of precipitation-induced cooling in the upper levels of the storm. The rear-flank downdraft appears to play an important role in producing tornadoes in classic supercells.

When the rear-flank downdraft strikes the ground, it may (under favourable shear conditions) interact with the forward-flank downdraft beneath the mesocyclone to initiate the formation of a tornado. At the surface, the cool air of the rear-flank downdraft (and the forward-flank downdraft) sweeps around the centre of the mesocyclone, effectively cutting off the rising air from the warmer surrounding air. The lower half of the updraft now rises more slowly. The rising updraft, which we can imagine as a column of air, now shrinks horizontally and stretches vertically. This *vertical stretching* of the spinning column of air causes the rising, spinning air to spin faster.* If this stretching process continues, the rapidly rotating air column may shrink into a narrow column of rapidly rotating air—a tornado vortex.

---

*You can obtain an idea of what might be taking place in the supercell by stirring a cup of coffee or tea with a spoon and watching the low-pressure form in the middle of the beverage.

*As the rotating air column stretches vertically into a narrow column, its rotational speed increases. You may recall from Chapter 10 (p. 305) that this situation is called the *conservation of angular momentum*.

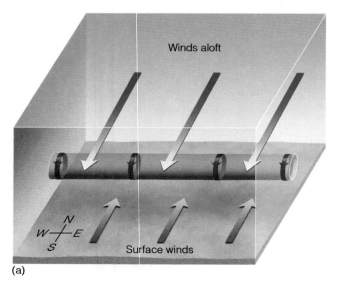

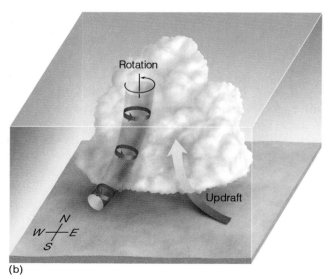

(a)   (b)

● FIGURE 14.41 (a) A spinning vortex tube created by wind shear. (b) The strong updraft in the developing thunderstorm carries the vortex tube into the thunderstorm, producing a rotating air column that is oriented in the vertical plane.

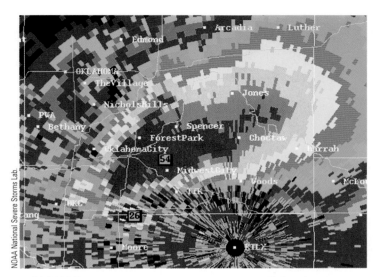

● FIGURE 14.42 A tornado-spawning supercell thunderstorm over Oklahoma City on May 3, 1999, shows a hook echo in its rainfall pattern on a Doppler radar screen. The colours red and orange represent the heaviest precipitation.

As air rushes upward and spins around the low-pressure core of the vortex, the air expands, cools, and, if sufficiently moist, condenses into a visible cloud—the funnel cloud. As the air beneath the funnel cloud is drawn into its core, the air cools rapidly and condenses, and the funnel cloud descends toward the surface. On reaching the ground, the tornado's circulation usually picks up dirt and debris, making it appear both dark and ominous. Although the air along the outside of the funnel is spiralling upward, Doppler radar reveals that, within the core of violent tornadoes, the air is descending toward the extreme low pressure at the ground (which may be 100 hPa lower than

that of the surrounding air). As the air descends, it warms, causing the cloud droplets to evaporate. This process leaves the core free of clouds. Supercell tornadoes usually develop near the right rear sector of the storm, on the southwestern side of a northeastward-moving storm, as shown in Figure 14.43.

Not all supercells produce tornadoes; in fact, perhaps less than 15 percent do. However, recent studies have revealed that supercells are more likely to produce tornadoes when they interact with a preexisting boundary, such as an old gust front (outflow boundary) that supplies the surface air with horizontal spin that can be tilted and lifted into the storm by its updraft. Many atmospheric situations may suppress tornado formation. For example, if the precipitation in the cloud is swept too far away from the updraft, or if too much precipitation wraps around the mesocyclone, the necessary interactions that produce the rear flank downdraft are disrupted, and a tornado is not likely to form. Moreover, tornadoes are not likely to form if the supercell is fed warm, moist air that is elevated above a deep layer of cooler surface air.

As we have seen, the first sign that a supercell is about to give birth to a tornado is the sight of rotating clouds* at the base of the storm. If the area of rotating clouds lowers, it becomes the *wall cloud*. Notice in Figure 14.43 that the tornado extends from within the wall cloud to Earth's surface. Sometimes the air is so dry that the swirling, rotating wind remains invisible until it reaches the ground and begins to pick up dust. Unfortunately, people have mistaken these "invisible tornadoes" for dust devils, only to find out (often

---

*Occasionally, people will call a sky dotted with mamma clouds "a tornado sky." Mamma clouds may appear with both severe and nonsevere thunderstorms as well as with a variety of other cloud types (see Chapter 5). Mamma clouds are not funnel clouds and do not rotate, and their appearance has no relationship to tornadoes.

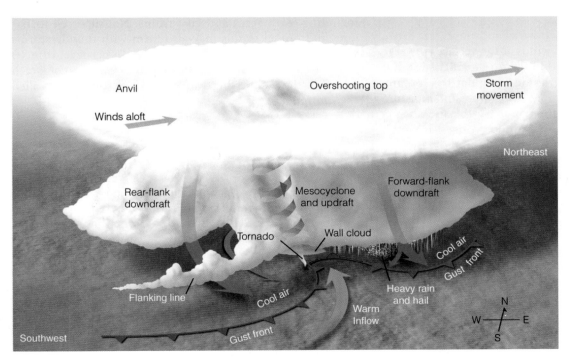

● FIGURE 14.43 A classic tornadic supercell thunderstorm showing updrafts and downdrafts, along with surface air flowing counterclockwise and in toward the tornado. The flanking line is a line of cumulus clouds that form as surface air is lifted into the storm along the gust front.

too late) that they were not. Occasionally, the funnel cannot be seen due to falling rain, clouds of dust, or darkness. Even when not clearly visible, many tornadoes have a distinctive roar that can be heard for several kilometres. This sound, which has been described as "a roar like a thousand freight trains," appears to be loudest when the tornado is touching the surface. However, not all tornadoes make this sound, and when these storms strike, they become silent killers.

Certainly, the likelihood of a thunderstorm producing a tornado increases when the storm becomes a supercell, but not all supercells produce tornadoes. And not all tornadoes come from rotating thunderstorms (supercells).

NONSUPERCELL TORNADOES   Tornadoes that do not occur in association with a preexisting wall cloud (or a mid-level mesocyclone) of a supercell are called **nonsupercell tornadoes**. These tornadoes may occur with intense multicell storms as well as with ordinary cell thunderstorms, even relatively weak ones. Some nonsupercell tornadoes extend from the base of a thunderstorm, whereas others may begin on the ground and build upward in the absence of a condensation funnel.

Nonsupercell tornadoes may form along a gust front where the cool downdraft of the thunderstorm forces warm, humid air upward. Tornadoes that form along a gust front are commonly called **gustnadoes**. These relatively weak tornadoes normally are short-lived and rarely inflict significant damage. Gustnadoes are often seen as a rotating cloud of dust or debris rising above the surface.

Occasionally, rather weak, short-lived tornadoes will occur with rapidly building cumulus congestus clouds. Tornadoes such as these commonly form over east–central Colorado. Because they look similar to waterspouts that form over water, they are sometimes called **landspouts*** (see ● Figure 14.44).

● Figure 14.45 illustrates how a landspout can form. Suppose, for example, that the winds at the surface converge along a boundary, as illustrated in Figure 14.45a. (The wind may converge due to topographic irregularities or any number of other factors, including temperature and moisture variations.) Notice that along the boundary, the air is rising, condensing, and forming into a cumulus congestus cloud. Notice also that along the surface at the boundary, there is horizontal rotation (spin) created by the wind blowing in opposite directions along the boundary. If the developing cloud should move over the region of rotating air, the spinning air may be drawn up into the cloud by the storm's updraft. As the spinning, rising air shrinks in diameter, it produces a tornadolike structure, a *landspout* (see Figure 14.44). Landspouts usually dissipate when rain falls through the cloud and destroys the updraft. Tornadoes may form in this

*Landspouts occasionally form on the backside of a squall line, where southerly winds ahead of a cold front and northwesterly winds behind it create swirling eddies that can be drawn into thunderstorms by their strong updrafts.

● FIGURE 14.44 A well-developed landspout moves over eastern Colorado.

manner along many types of converging wind boundaries, including sea breezes and gust fronts. Nonsupercell tornadoes and funnel clouds may also form with thunderstorms when cold air aloft (associated with an upper-level trough) moves over a region. Common along the West Coast of North America, these short-lived tornadoes are sometimes called *cold-air funnels*.

Severe Weather and Doppler Radar

Most of our knowledge about what goes on inside a tornado-generating thunderstorm has been gathered through the use of Doppler radar. Remember from Chapter 7 that a radar transmitter sends out microwave pulses and that, when this energy strikes an object, a small fraction is scattered back to the antenna. Precipitation particles are large enough to bounce microwaves back to the antenna. Consequently, as we saw earlier, the colourful area on the radar screen in Figure 14.42 represents precipitation intensity inside a supercell thunderstorm.

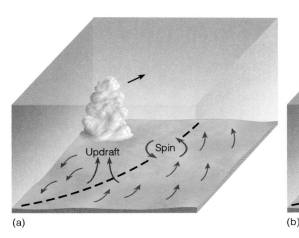
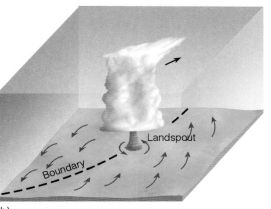

(a)

(b)

● FIGURE 14.45 (a) Along the boundary of converging winds, the air rises and condenses into a cumulus congestus cloud. At the surface, the converging winds along the boundary create a region of counterclockwise spin. (b) As the cloud moves over the area of rotation, the updraft draws the spinning air up into the cloud, producing a nonsupercell tornado, or landspout. (Modified after Wakimoto and Wilson.)

Doppler radar can do more than measure rainfall intensity; it can actually measure the speed at which precipitation is moving horizontally toward or away from the radar antenna. Because precipitation particles are carried by the wind, Doppler radar can peer into a severe storm and reveal its winds.

Doppler radar works on the principle that as precipitation moves toward or away from the antenna, the returning radar pulse will change in frequency. A similar change occurs when the high-pitched sound (high frequency) of an approaching noise source, such as a siren or train whistle, becomes lower in pitch (lower frequency) after it passes by the person hearing it. This change in frequency in sound waves or microwaves is called the *Doppler shift*, and this, of course, is where the Doppler radar gets its name.

A single Doppler radar cannot detect winds that blow parallel to the antenna—at right angles to the radar beam. Consequently, two or more units probing the same thunderstorm are needed to give a complete picture of the winds within the storm. To help distinguish the storm's air motions, wind velocities can be displayed in colour. Colour contouring the wind field gives a good picture of the storm (see ● Figure 14.46).

Even a single Doppler radar can uncover many of the features of a severe thunderstorm. For example, studies conducted in the 1970s revealed, for the first time, the existence of the swirling winds of the mesocyclone inside a supercell storm. Mesocyclones have a distinct image (signature) on the radar display. Tornadoes also have a distinct signature on the radar screen, known as the *tornado vortex signature* (*TVS*), which shows up as a region of rapidly (or abruptly) changing wind directions within the mesocyclone. (Look at Figure 14.46.)

Unfortunately, the resolution of the Doppler radar is not high enough to measure actual wind speeds of most tornadoes, whose diameters are only a few hundred metres or less. However, a new and experimental Doppler system—called *Doppler lidar*—uses a light beam (instead of microwaves) to measure the change in frequency of falling precipitation, cloud particles, and dust. Because it uses a shorter wavelength

of radiation, it has a narrower beam and a higher resolution than does Doppler radar. In an attempt to obtain tornado wind information at fairly close range (less than 10 km), smaller portable Doppler radar units (Doppler on wheels) are peering into tornado-generating storms (see ● Figure 14.47).

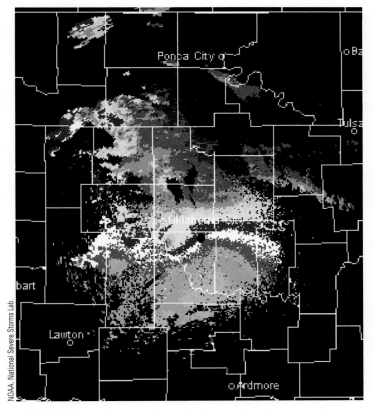

● FIGURE 14.46 Doppler radar display of winds associated with the supercell storm that moved through parts of Oklahoma City during the afternoon of May 3, 1999. The close proximity of the horizontal winds blowing toward the radar (green and blue shades) and those blowing away from the radar (yellow and red shades) indicates strong cyclonic rotation and the presence of a tornado. The radar antenna is situated east of the displayed image.

© Howard B. Bluestein

● FIGURE 14.47 Graduate students from the University of Oklahoma use a portable Doppler radar to probe a tornado near Hodges, Oklahoma.

The Canadian weather radar network consists of 31 Doppler radar units deployed and operated by the Meteorological Service of Canada. They use microwaves with a wavelength of 5 cm and are referred to as WSR-98* radars. In the United States, the network of more than 150 Doppler radar units deployed at selected weather stations within the continental United States is referred to as **NEXRAD** (an acronym for *NEX*t Generation Weather *RAD*ar). The NEXRAD system consists of the WSR-88D Doppler radar. Both the Canadian and U.S. radar networks use computers to control the radars and process their data.

The computers take in data, display them on a monitor, and run computer programs called *algorithms*, which, in conjunction with other meteorological data, detect severe weather phenomena, such as storm cells, hail, mesocyclones, and tornadoes. The output from these algorithms provides a great deal of information to forecasters

that allows them to make better decisions on which thunderstorms are most likely to produce severe weather and possible flash flooding. In addition, the algorithms give advanced and improved warning of an approaching tornado. More reliable warnings, of course, will cut down on the number of false alarms.

Because the Doppler radar shows horizontal air motion within a storm, it can help identify the magnitude of other severe weather phenomena, such as gust fronts, derechoes, microbursts, and wind shears that are dangerous to aircraft. Certainly, as more and more information from Doppler radar becomes available, our understanding of the processes that generate severe thunderstorms and tornadoes will be enhanced, and it is hoped that there will be an even better tornado and severe storm warning system, resulting in fewer deaths and injuries.

In an attempt to unravel some of the mysteries of the tornado, several studies are under way. In one study, scientists using an armada of observational vehicles, aircraft, and state-of-the-art equipment, including Doppler radar, pursued tornado-generating thunderstorms over portions of the U.S. Central Plains during the spring and summer. These observations are providing information on the inner workings of severe thunderstorms. At the same time, laboratory models of tornadoes in chambers (called *vortex chambers*), along with mathematical computer models, are offering new insights into the formation and development of these fascinating storms.

Waterspouts

A **waterspout** is a rotating column of air that is connected to a cumuliform cloud over a large body of water. The waterspout may be a tornado that formed over land and then travelled over water. In such a case, the waterspout is sometimes referred to as a *tornadic waterspout*. Such tornadoes can inflict major damage to ocean-going vessels, especially when the tornadoes are of the supercell variety. Waterspouts not associated with supercells that form over water, especially above warm, tropical coastal waters (such as in the vicinity of the Florida Keys, where almost 100 occur each month during the summer), are often referred to as *fair-weather waterspouts*.* These waterspouts are generally much smaller than an average tornado as they have diameters usually between 3 and 100 m. Fair-weather waterspouts are also less intense as their rotating winds are typically less than 45 knots. In addition, they tend to move more slowly than tornadoes and only last for about 10 to 15 minutes, although some have existed for up to one hour.

Fair-weather waterspouts tend to form in much the same way that landspouts do—when the air is conditionally unstable and cumulus clouds are developing. Some form

---

*The name WSR-98 stands for *Weather Surveillance Radar, 1998*. The Canadian weather radar network was expanded in 1998.

*"Fair-weather" waterspouts may form over any large body of warm water. Hence, they occur frequently over the Great Lakes in summer.

with small thunderstorms, but most form with developing cumulus congestus clouds, whose tops are frequently no higher than 3600 m and do not extend to the freezing level. Apparently, the warm, humid air near the water helps create atmospheric instability, and the updraft beneath the resulting cloud helps initiate uplift of the surface air. Studies even suggest that gust fronts and converging sea breezes may play a role in the formation of some of the waterspouts that form over the Florida Keys.

The waterspout funnel is similar to the tornado funnel in that both are clouds of condensed water vapour with converging winds that rise about a central core. Contrary to popular belief, the waterspout does not draw water up into its core; however, swirling spray may be lifted several metres when the waterspout funnel touches the water. Apparently, the most destructive waterspouts are those that begin as tornadoes over land and then move over water. A photograph of a particularly well-developed and intense waterspout is shown in ● Figure 14.48.

● FIGURE 14.48 A powerful waterspout over the Wadden Sea in northern Netherlands on July 9, 2007.

SUMMARY

In this chapter, we examined thunderstorms and the atmospheric conditions that produce them. Thunderstorms are convective storms that produce lightning and thunder. Lightning is a discharge of electricity that occurs in mature thunderstorms. The lightning stroke momentarily heats the air to an incredibly high temperature. The rapidly expanding air produces a sound called thunder.

The ingredients for the isolated ordinary cell thunderstorm are humid surface air, plenty of sunlight to heat the ground, a conditionally unstable atmosphere, a "trigger" to start the air rising, and weak vertical wind shear. When these conditions prevail, and the air begins to rise, small cumulus clouds may grow into towering clouds and thunderstorms in as little as 30 minutes.

When conditions are ripe for thunderstorm development, and moderate or strong vertical wind shear exists, the updraft in the thunderstorm may tilt and ride up and over the downdraft. As the forward edge of the downdraft (the gust front) pushes outward along the ground, the air is lifted and new cells form, producing a multicell thunderstorm—a storm with cells in various stages of development. Some multicell storms form as a complex of thunderstorms, such as the squall line (which forms as a line of thunderstorms either along or ahead of an advancing cold front), and the mesoscale convective complex (which forms as a cluster of storms). When convection in the multicell storm is strong, it may produce severe weather, such as strong, damaging surface winds, hail, and flooding.

Supercell thunderstorms are large, intense thunderstorms with a single, rotating updraft. The updraft and the downdraft in a supercell are nearly in balance, so the storm may exist for many hours. Supercells are capable of producing severe weather, including strong, damaging tornadoes.

Tornadoes are rapidly rotating columns of air with a circulation that reaches the ground. The rotating air of the tornado may begin within the thunderstorm, or it may begin at the surface and extend upward. Tornadoes can form with supercells, as well as with less intense thunderstorms. The landspout and the gustnado do not form with supercells. Most tornadoes are less than a few hundred metres wide with wind speeds less than 100 knots (185 km h^{-1}), although violent tornadoes may have wind speeds that exceed 250 knots (463 km h^{-1}). A violent tornado may actually have smaller whirls (suction vortices) rotating within it. With the aid of Doppler radar, scientists are probing tornado-spawning thunderstorms, hoping to better predict tornadoes and to better understand where, when, and how they form.

A normally small and less destructive cousin of the tornado is the "fair-weather" waterspout that commonly forms above the warm waters of the Great Lakes and the Florida Keys in summer.

KEY TERMS

The following terms are listed (with page numbers) in the order they appear in the text. Define each. Doing so will aid you in reviewing the material covered in this chapter.

ordinary cell (air mass) thunderstorms, 417
cumulus stage, 417
mature stage, 418
dissipating stage, 418

multicell thunderstorms, 418
overshooting top, 419
gust front, 419
shelf cloud, 420
roll clouds, 420

QUESTIONS FOR REVIEW

1. What is a thunderstorm?
2. What atmospheric conditions are necessary for the development of ordinary cell thunderstorms?
3. Describe the stages of development of an ordinary cell (air mass) thunderstorm.
4. How do downdrafts form in thunderstorms?
5. Why do ordinary cell thunderstorms most frequently form in the afternoon?
6. Explain why ordinary cell thunderstorms tend to dissipate much sooner than multicell storms.
7. How does the Meteorological Service of Canada define a severe thunderstorm?
8. What is necessary for a multicell thunderstorm to form?
9. (a) How do gust fronts form?
   (b) What type of weather does a gust front bring when it passes?
10. (a) Describe how a microburst forms.
    (b) Why is the term wind shear often used in conjunction with a microburst?
11. How do derechoes form?
12. How does a squall line differ from a mesoscale convective complex (MCC)?
13. Give a possible explanation for the generation of a pre-frontal squall-line thunderstorm.
14. How does the cell in an ordinary cell thunderstorm differ from the cell in a supercell thunderstorm?
15. Describe the atmospheric conditions at the surface and aloft that are necessary for the development of most supercell thunderstorms. (Include in your answer the role that the low-level jet plays in the rotating updraft.)

16. What is the difference between an HP supercell and an LP supercell?
17. When thunderstorms are training, what are they doing?
18. In what region in the United States do dryline thunderstorms most frequently form? Why there?
19. Where do the highest frequency of thunderstorms occur in Canada? Why there?
20. Why is large hail more common in Alberta than in Florida?
21. Describe one process by which thunderstorms become electrified.
22. Explain how a cloud-to-ground lightning stroke develops.
23. Why is it unwise to seek shelter under an isolated tree during a thunderstorm? If caught out in the open, what should you do?
24. How does negative cloud-to-ground lightning differ from positive cloud-to-ground lightning?
25. How is thunder produced?
26. What is the primary difference between a tornado and a funnel cloud?
27. Give some average statistics about tornado size, winds, and direction of movement.
28. Why should you not open windows when a tornado is approaching?
29. Why is the central part of the United States more susceptible to tornadoes than any other region of the world?
30. Explain both how and why there is a shift in tornado activity from winter to summer.
31. How does a tornado watch differ from a tornado warning?
32. If you are in a single-story home (without a basement) during a tornado warning, what should you do?
33. Supercell thunderstorms that produce tornadoes form in a region of strong wind shear. Explain how the wind changes in speed and direction to produce this shear.
34. Explain how a nonsupercell tornado, such as a landspout, might form.
35. Describe how Doppler radar measures the winds inside a severe thunderstorm.
36. How has Doppler radar helped in the prediction of severe weather?
37. What atmospheric conditions lead to the formation of "fair-weather" waterspouts?

QUESTIONS FOR THOUGHT

1. Why does the bottom half of a dissipating thunderstorm usually "disappear" before the top?
2. Sinking air warms, yet the downdrafts in a thunderstorm are usually cold. Why?
3. Explain why squall-line thunderstorms often form ahead of advancing cold fronts but seldom behind them.
4. A forecaster may say that he or she looks for "right-moving" thunderstorms when predicting severe weather. What does this mean?

5. Why is the old adage "lightning never strikes twice in the same place" wrong?

6. Suppose that while you are on a high mountain ridge a thundercloud passes overhead. What would be the wisest thing to do—stand upright, lie down, or crouch? Explain.

7. If you are confronted in an open field by a large tornado and there is no way that you could outrun it, probably the only thing that you could do would be to run and lie down in a depression. If given the choice, would you run toward your right or left as the tornado approaches? Explain your reasoning.

8. Tornadoes apparently form in the region of a strong updraft rather than in a downdraft, yet they descend from the base of a cloud. Why?

9. Why are left-moving supercell thunderstorms uncommon in the Northern Hemisphere yet very common in the Southern Hemisphere?

PROBLEMS AND EXERCISES

1. On a map of North America, place the surface weather conditions as well as weather conditions aloft (jet stream and so on) that are necessary for the formation of most supercell thunderstorms.

2. On the surface weather map in ● Figure 14.49, at which number would you most likely observe a line of thunderstorms forming? Explain why you chose that location.

3. A multivortex tornado with a rotational wind speed of 125 knots is moving from southwest to northeast at 30 knots. Assume that the suction vortices within this tornado have rotational winds of 100 knots:

   (a) What is the maximum wind speed of this multi-vortex tornado?

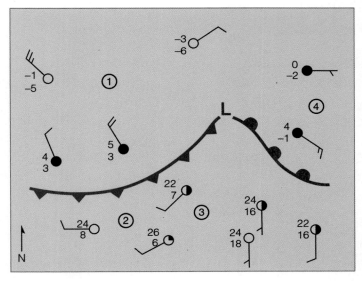

● FIGURE 14.49 Surface weather map for Exercise 2.

   (b) If you are facing the approaching tornado, on which side (northeast, northwest, southwest, or southeast) would the strongest winds be found? The weakest winds? Explain both of your answers.

   (c) According to Table 14.2 and Table 14.3 (pp. 445 and 446), how would this tornado be classified on the old Fujita scale and on the new EF scale?

4. If you see lightning and 10 seconds later you hear thunder, how far away is the lightning stroke?

5. Suppose that several of your friends went on a storm-chasing adventure in the central United States. To help guide their chase, you stay behind with an Internet-connected computer and a cellular phone. Which current weather and forecast maps would you use to guide their storm chase? Explain why you chose those maps.

Hurricane Katrina over the Gulf of Mexico on August 28, 2005. With sustained winds of 152 knots (282 km h^{-1}) and a central pressure near 907 hPa, this large and powerful Category 5 hurricane takes aim at Louisiana and Mississippi.

NOAA

Hurricanes

15

Monday August 29, 2005 "I'm in an airplane high over the Atlantic as hurricane Katrina heads for, then smashes into, the Gulf Coast. Nothing I can do up here. Useless. Last I heard before climbing onto the plane at Heathrow was that Katrina was steaming right for New Orleans as a category 5, the strongest storm there is. Sick with worry. I found out that my wife, Amanda, evacuated to Houston two days ago. She's safe, at least. But what about all my friends? My students? My home? Nothing I can do. And so I sit and wait and try to imagine what's going on in my city so far away.

I've ridden out hurricanes before. The first week I moved to New Orleans back in 1992 as a young graduate student, hurricane Andrew came screaming right for us. . . . Last minute, the hurricane turned away and hit west of the city instead. I wandered the city streets later that night and made a note of the fallen trees, the water-soaked streets, the broken windows. No big deal. A sense of elation, that I was brave and silly enough to stay put rather than run, washed over me. And that remained the pattern in the following years, always a hurricane heading for the city, always the chatter that this was the Big One, always the storm turning away at the last moment and hitting somewhere else, usually Florida or Alabama. Amanda and I stayed back, battened down the hatches, and got invited to hurricane parties. Something exhilarating in staring into the face of mother nature's strength and living to tell. Something addictive in the gamble. . . .

Off the plane and in Toronto I am desperate for information. Yes! Yet again, the hurricane's eye veered at the last minute, took a turn to the east, the worst of it missing my city. Oh, but the destruction I see in Mississippi is bad. I say a prayer for those people, thank the Creator for sparing us once more. . . .

And then Amanda calls me. A levee has been breached on the Industrial Canal. Water is flooding into our bowl. . . . When [Hurricane] Betsy hit decades ago, . . . stories began circulating of people drowning in their attics, pressed up against their roof beams, trapped. That's where the folk wisdom—to always have an axe in the house when a hurricane heads your way—comes from. If your attic fills with water, you've got to chop your way out."

*Award-winning Canadian author (*Three Day Road, Through Black Spruce*) and academic, Joseph Boyden writes about his adopted city, where he is a writer-in-residence at the University of New Orleans.*

Maclean's Magazine. Joseph Boyden. The Drowning of New Orleans. Sept. 7, 2005.

CONTENTS

Born over warm tropical waters and nurtured by a rich supply of water vapour, a hurricane can indeed grow into a ferocious storm that generates enormous waves, heavy rains, and winds that may exceed 150 knots (278 km h^{-1}). What exactly are hurricanes? How do they form? And why do they strike the East Coast of North America more frequently than the West Coast? These are some of the questions we will consider in this chapter.

Tropical Weather

In the tropics, the broad belt around Earth between 23½° north and south of the equator, the weather is much different from that of the middle latitudes. In the tropics, the noon sun is always high in the sky, so diurnal and seasonal changes in temperature are small. The sun rises and sets around the same time all year, and the daily heating of the surface and high humidity favour the development of cumulus clouds and afternoon thunderstorms. Most of these are individual thunderstorms that are not severe. Sometimes, however, they group together into loosely organized systems called *non-squall clusters*. On other occasions, the thunderstorms will align into a row of vigorous convective cells known as a *tropical squall cluster*, or *squall line*. The passage of a squall line is usually noted by a sudden wind gust followed immediately by a heavy downpour that may produce 30 mm of rainfall in about 30 minutes. This deluge is then followed by several hours of relatively steady rainfall. Many of these tropical squall lines are similar to the middle-latitude squall lines described in Chapter 14 on p. 423.

As tropical locations are warm all year long, the weather does not have the four seasons, which are largely determined by temperature variations in the mid- to high latitudes. Rather, most of the tropics are marked by seasonal differences in precipitation. The greatest cloudiness and precipitation occur during the high-sun period, when the intertropical convergence zone moves into the region. Even during the dry season, precipitation can be irregular as periods of heavy rain, lasting for several days, may follow an extreme dry spell.

The winds in the tropics generally blow from the east, northeast, or southeast as the trade winds resulting from the tropical Hadley cell circulation discussed in Chapter 10 (p. 294). Because the variation in sea-level pressure is normally quite small, drawing isobars on a weather map provides little useful information. Instead of isobars, **streamlines** that depict wind flow are drawn. Streamlines are useful because

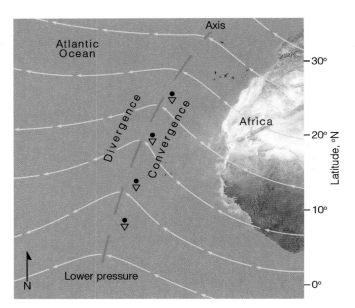

● FIGURE 15.1 A tropical wave (also called an easterly wave) moving off the coast of Africa over the Atlantic. The wave is shown by the bending of streamlines—lines that show wind flow patterns. (The heavy dashed green line is the axis of the trough.) The wave moves slowly westward, bringing fair weather on its western side and rain showers on its eastern side.

they show where surface air converges and diverges. Occasionally, the streamlines will be disturbed by a weak trough of low pressure called a **tropical wave**, or **easterly wave**, because it tends to move from east to west (see ● Figure 15.1).

Tropical waves have wavelengths on the order of 2500 km and travel from east to west at speeds between 10 and 20 knots (18.5 and 37 km h^{-1}). Look at Figure 15.1 and observe that on the western side of the trough (heavy dashed green line), where easterly and northeasterly surface winds diverge, sinking air produces generally fair weather. On its eastern side, southeasterly surface winds converge. The converging air rises, cools, and often condenses into showers and thunderstorms. Consequently, the main area of showers forms behind the trough. Occasionally, an easterly wave will intensify and grow into a hurricane.

Ranked from least to most severe, tropical waves, depressions, storms, and hurricanes transfer latent and sensible energy (moisture and heat) from one Earth system to another and within the atmosphere. These are transferred primarily from the *hydrosphere's* oceans upward into the *atmosphere*. Occasionally, these energy transfers result in tornadoes as hurricanes move inland. As the storms track away from the tropics, this energy is also transferred poleward through the atmosphere, attempting to equalize energy across the globe. Often this sensible and latent heat fuels middle-latitude cyclones, producing some of the larger storms in the mid-latitudes. For humans and our infrastructure (the *anthrosphere*), hurricanes and related storms create some of our most severe weather-related impacts.

Precipitation from these systems can also be intense and severe. It can result in coastal erosion, localized flooding, and other damage as wind-driven fresh and salt water are pushed over the land. But this also has benefits for the *biosphere* as water, soil, and other nutrients that become entrained in these systems move into coastal estuaries. There fresh water mixes with salt water, which creates important nutrient-rich habitat, nursery, and migratory areas for fish, birds, and animals of all types. Further from the coast, fresh water from these storms recharges surface water sources such as rivers and lakes and recharges the groundwater as floodwaters recede. In these ways, other Earth systems are affected by tropical storms.

In this chapter, we will focus primarily on tropical storm phenomena with a main emphasis on tropical cyclones—better known as hurricanes or typhoons. It is worth mentioning that these tropical storm systems are one aspect of *tropical meteorology*. Other aspects are covered elsewhere in this book: monsoon circulation is discussed in Chapter 9. The Hadley cell and several important tropical atmosphere–ocean interactions, such as El Niño–Southern Oscillation, are discussed in Chapter 10.

Anatomy of a Hurricane

A **hurricane** is an intense storm of tropical origin, with sustained winds exceeding 64 knots (119 km h⁻¹),* that forms over the warm northern Atlantic and eastern North Pacific oceans. This same type of storm is given different names in different regions of the world. In the western North Pacific, it is called a **typhoon**; in India, it is known as a cyclone; and in Australia, it is called a *tropical cyclone*. Most North Americans refer to these storms as hurricanes when they originate in the equatorial Atlantic Ocean and as typhoons when they affect the Asian Pacific. By international agreement, **tropical cyclone** is the general term for all hurricane-type storms that originate over tropical waters.

● Figure 15.2 is a photograph of Hurricane Elena situated over the Gulf of Mexico. The storm is approximately 500 km in diameter, which is about average for hurricanes. The area of broken clouds at the centre is its **eye**. Elena's eye is almost 40 km wide. Within the eye, winds are light and clouds are mainly broken. The surface air pressure is very low, nearly 955 hPa.* Notice that the clouds align themselves into spiraling bands (called *spiral rain bands*) that swirl in toward the storm's centre, where they wrap themselves around the eye. Surface winds increase in speed as they blow counterclockwise and inward toward this centre. (In the Southern Hemisphere, the winds blow clockwise around the centre.) Adjacent to the eye is the **eyewall**, a ring of intense thunderstorms that whirl around the storm's centre and may extend upward to almost 18 km above sea level. Within the eyewall, we find the heaviest precipitation and the strongest winds, which, in this storm, are 105 knots (194 km h⁻¹), with peak gusts of 120 knots (222 km h⁻¹).

If we were to venture from west to east (left to right) at the surface through the storm in Figure 15.2, what might we

---

*By definition, any wind exceeding 64 knots is called a *hurricane-force* wind, regardless of whether a hurricane is the cause. Tornadoes and sometimes middle-latitude cyclones can also generate hurricane-force winds.

*An extreme low pressure of 870 hPa was recorded in Typhoon Tip while over the tropical western North Pacific in October 1979, and Hurricane Wilma had a pressure reading of 882 hPa while over the Gulf of Mexico in October 2005.

● FIGURE 15.2 Hurricane Elena over the Gulf of Mexico about 130 km southwest of Apalachicola, Florida, as photographed from the space shuttle *Discovery* in September 1985. Because this storm is situated north of the equator, surface winds are blowing counterclockwise about its centre (eye). The central pressure of the storm is 955 hPa, with sustained winds of 105 knots near its eye.

Eye   Eyewall

Rain-free area

Spiral rain band

NOAA

● FIGURE 15.3 A model that shows a vertical view of air motions and clouds in a typical hurricane in the Northern Hemisphere. The diagram is exaggerated in the vertical.

experience? As we approach the hurricane, the sky becomes overcast with cirrostratus clouds; barometric pressure drops slowly at first and then more rapidly as we move closer to the centre. Winds blow from the north and northwest with ever-increasing speed as we near the eye. The high winds, which generate huge waves over 10 m high, are accompanied by heavy rain showers. As we move into the eye, the winds slacken, rainfall ceases, and the sky brightens as middle and high clouds appear overhead. The atmospheric pressure is now at its lowest point (965 hPa), some 50 hPa lower than the pressure measured on the outskirts of the storm. The brief respite ends as we enter the eastern region of the eyewall. Here we are greeted by heavy rain and strong, southerly winds. As we move away from the eyewall, the pressure rises, the winds diminish, the heavy rain lets up, and, eventually, the sky begins to clear.

This brief, imaginary venture raises many unanswered questions. Why, for example, is the surface pressure lowest at the centre of the storm? And why is the weather clear almost immediately outside the storm area? To help us answer such

questions, we need to look at a vertical view, a profile of the hurricane along a slice that runs through its centre. A model that describes such a profile is given in ● Figure 15.3.

The model shows that the hurricane is composed of an organized mass of thunderstorms* that are an integral part of the storm's circulation. Near the surface, moist tropical air flows in toward the hurricane's centre. Adjacent to the eye, this air rises and condenses into huge cumulonimbus clouds that produce heavy rainfall, as much as 250 mm per hour. Near the top of the clouds, the relatively dry air, having lost much of its moisture, begins to flow outward, away from the centre. This diverging air aloft actually produces an anticyclonic (clockwise in the Northern Hemisphere) flow of air several hundred kilometres from the eye. As this outflow reaches the storm's periphery, it begins to sink and warm, inducing clear skies. In the vigorous convective clouds of the

---

*These huge convective cumulonimbus clouds have surprisingly little lightning (and, hence, thunder) associated with them. Even so, for simplicity, we will refer to these clouds as thunderstorms throughout this chapter.

● FIGURE 15.4 The cloud mass is Hurricane Katrina's eyewall, and the clear area is Katrina's eye photographed inside the eye on August 28, 2005, from a NOAA reconnaissance (hurricane hunter) aircraft. For a satellite image of the storm on this day, look at the opening photograph on p. 456.

eyewall, the air warms due to the release of large quantities of latent heat. This produces slightly higher pressures aloft, which initiate downward air motion within the eye. As the air descends, it warms by compression. This process helps account for the warm air and the absence of convective clouds in the eye of the storm (see ● Figure 15.4).

As surface air rushes in toward the region of much lower surface pressure, it should expand and cool, and we might expect to observe cooler air around the eye, with warmer air farther away. But, apparently, so much heat is added to the air from the warm ocean surface that the surface air temperature remains fairly uniform throughout the hurricane.

● Figure 15.5 is a three-dimensional radar composite of Hurricane Katrina as it passes over the central area of the Gulf of Mexico. Compare Katrina's features with those of typical hurricanes illustrated in Figure 15.2 and Figure 15.3. Notice that the strongest radar echoes (heaviest rain) near the surface are located in the eyewall, adjacent to the eye. ● Figure 15.6 is an example of how surface winds blow around a hurricane in the Northern Hemisphere.

Hurricane Formation and Dissipation

We are now left with an important question: Where and how do hurricanes form? Although there is no widespread agreement on how hurricanes actually form, it is known that certain necessary ingredients are required before a weak tropical disturbance will develop into a full-fledged hurricane.

● FIGURE 15.5 A three-dimensional *TRMM* satellite view of Hurricane Katrina passing over the central Gulf of Mexico on August 28, 2005. The cutaway view shows concentric bands of heavy rain (red areas inside the clouds) encircling the eye. Notice that the heaviest rain (largest red area) occurs in the eyewall. The isolated tall cloud tower (in red) in the northern section of the eyewall indicates a cloud top of 16 km above the ocean surface. Such tall clouds in the eyewall often indicate that the storm is intensifying.

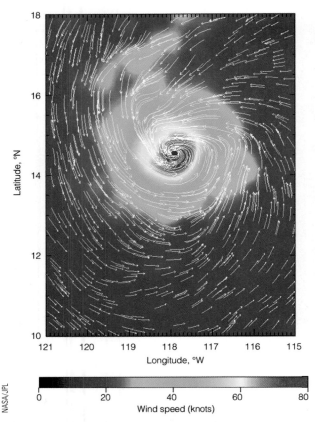

NASA/JPL

● FIGURE 15.6 Arrows show surface winds spinning counterclockwise around Hurricane Dora situated over the eastern tropical Pacific in August 1999. Colours indicate surface wind speeds. Notice that winds of 80 knots (148 km h⁻¹) are encircling the eye (the dark dot in the centre). Wind speed and direction were obtained from the *QuikSCAT* satellite.

NASA

● FIGURE 15.7 Hurricanes form over warm tropical waters. This image shows where sea surface temperatures in the tropical Atlantic exceed 28°C—warm enough for tropical storm development—in May 2002.

THE RIGHT ENVIRONMENT Hurricanes form over tropical waters, where the winds are light, the humidity is high in a deep layer extending up through the troposphere, and the surface water temperature is warm, typically 26.5°C or greater, over a vast area* (see ● Figure 15.7). These conditions usually prevail over the tropical and subtropical North Atlantic and North Pacific oceans during the summer and early fall; hence, the hurricane season normally runs from June through November. ● Figure 15.8 shows the number of tropical storms and hurricanes that formed over the tropical Atlantic during the past 100 years. Notice that hurricane activity picks up in August, peaks in September, and then drops off rapidly.

For a mass of unorganized thunderstorms to develop into a hurricane, the surface winds must converge. In the Northern Hemisphere, converging air spins counterclockwise about an area of surface low pressure. Because this type of rotation will not develop on the equator where the Coriolis force is zero

(see Chapter 8), hurricanes form in tropical regions, usually between 5° and 20° latitude. (In fact, about two-thirds of all tropical cyclones form between 10° and 20° of the equator.)

Hurricanes do not form spontaneously but require some kind of "trigger" to start the air converging. We know, for example, from Chapter 10 that surface winds converge along the intertropical convergence zone (ITCZ). Occasionally, when a wave forms along the ITCZ, an area of low pressure develops, convection becomes organized, and the system grows into a hurricane. Weak convergence also occurs on the eastern side of a tropical wave, where hurricanes have been known to form. In fact, many, if not most, Atlantic hurricanes can be traced to tropical waves that form over Africa. However, only a small fraction of all of the tropical disturbances that form over the course of a year ever grow into hurricanes. Studies suggest that major Atlantic hurricanes are more numerous when the western part of Africa is relatively wet. Apparently, during the wet years, tropical waves are stronger, better organized, and more likely to develop into strong Atlantic hurricanes.

Convergence of surface winds may also occur along a preexisting atmospheric disturbance, such as a front that has moved into the tropics from middle latitudes. Although the temperature contrast between the air on both sides of the front is gone, converging winds may still be present, so thunderstorms are able to organize.

Even when all of the surface conditions are in place for the formation of a hurricane (e.g., warm water, humid air, converging winds, and so forth), the storm may not develop if the weather conditions aloft are not just right. For instance, in the region of the trade winds, especially near latitude 20°, the air is often sinking in association with the subtropical high. The sinking air warms and creates an inversion, known as the

---

*It was once thought that for hurricane formation, the ocean must be sufficiently warm through a depth of about 200 m. It is now known that hurricanes can form in the eastern North Pacific when the warm layer of ocean water is only about 20 m deep.

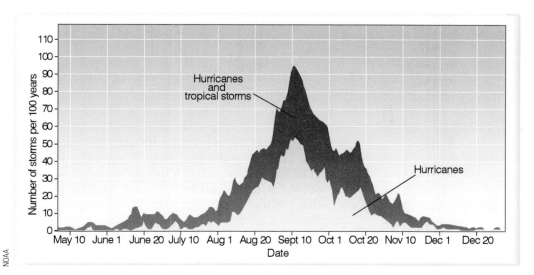

NOAA

● FIGURE 15.8 The total number of hurricanes and tropical storms (red shade) and hurricanes only (yellow shade) that have formed during the past 100 years in the Atlantic Basin—the Atlantic Ocean, the Caribbean Sea, and the Gulf of Mexico.

trade wind inversion. When the inversion is strong, it can inhibit the formation of intense thunderstorms and hurricanes. Also, hurricanes do not form where the upper-level winds are strong, creating strong wind shear. Strong wind shear tends to disrupt the organized pattern of convection and disperses heat and moisture, which are necessary for the growth of the storm.

The situation of strong winds aloft typically occurs over the tropical Atlantic during a major El Niño event. As a consequence, during El Niño, there are usually fewer Atlantic hurricanes than normal. However, the warmer water of El Niño in the northeastern tropical Pacific favours the development of hurricanes in that region. During the cold water episode in the tropical Pacific (known as La Niña),* winds aloft over the tropical Atlantic usually weaken and become easterly—a condition that favours hurricane development.

―――――――――
*El Niño and La Niña are covered in Chapter 10 beginning on p. 310.

THE DEVELOPING STORM The energy for a hurricane comes from the direct transfer of sensible heat and latent heat from the warm ocean surface. For a hurricane to form, a cluster of thunderstorms must become organized around a central area of surface low pressure. But it is not totally clear how this process occurs. One theory proposed that a hurricane forms in the following manner: suppose, for example, that the trade wind inversion is weak and that thunderstorms start to organize along the ITCZ, or along a tropical wave. In the deep, moist, conditionally unstable environment, a huge amount of latent heat is released inside the clouds during condensation. The process warms the air aloft, causing the temperature near the cluster of thunderstorms to be much higher than the air temperature at the same level farther away. This warming of the air aloft causes a region of higher pressure to form in the upper troposphere (see ● Figure 15.9). This situation causes a horizontal pressure gradient aloft that induces the air aloft to

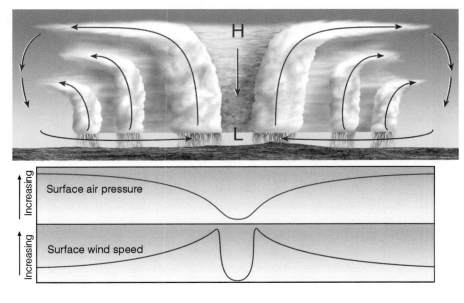

● FIGURE 15.9 The top diagram shows an intensifying tropical cyclone. As latent heat is released inside the clouds, the warming of the air aloft creates an area of high pressure, which induces air to move outward, away from the high. The warming of the air lowers the air density, which, in turn, lowers the surface air pressure. As surface winds rush in toward the surface low, they extract sensible heat, latent heat, and moisture from the warm ocean. As the warm, moist air flows in toward the centre of the storm, it is swept upward into the clouds of the eyewall. As warming continues, surface pressure lowers even more, the storm intensifies, and the winds blow even faster. This situation increases the transfer of heat and moisture from the ocean surface. The middle diagram illustrates how the air pressure drops rapidly toward the eye of the storm. The lower diagram shows how surface winds normally reach maximum strength in the region of the eyewall.

move outward, away from the region of higher pressure in the anvils of the cumulonimbus clouds. This diverging air aloft, coupled with warming of the vertical air column, causes the surface pressure to drop and a small area of surface low pressure to form. The air now begins to spin counterclockwise (Northern Hemisphere) and in toward the region of surface low pressure. As the air moves inward, its speed increases, just as ice skaters spin faster as their arms are brought in close to their bodies (the conservation of angular momentum).

As the air moves over the warm water, small, swirling eddies transfer heat energy from the ocean surface into the overlying air. The warmer the water and the greater the wind speed, the greater the transfer of sensible and latent heat into the air above. As the air sweeps in toward the centre of lower pressure, the rate of heat transfer increases because the wind speed increases. Similarly, the higher wind speed causes greater evaporation rates, and the overlying air becomes nearly saturated. The turbulent eddies then transfer the warm, moist air upward, where the water vapour condenses to fuel new thunderstorms. As the surface air pressure lowers, wind speeds increase, more evaporation occurs at the ocean surface, and thunderstorms become more organized. At the top of the thunderstorms, heat is lost by the clouds radiating infrared energy to space.

The driving force behind a hurricane is similar to that of a heat engine. In a heat engine, heat is taken in at a high temperature, converted into work, and then ejected at a low temperature. In a hurricane, heat is taken in near the warm ocean surface, converted to kinetic energy (energy of motion or wind), and lost at its top through radiational cooling.

In a heat engine, the amount of work done is proportional to the difference in temperature between its input and output regions. The maximum strength a hurricane can achieve is proportional to the difference in air temperature between the tropopause and the surface and to the potential for evaporation from the sea surface. As a consequence, the warmer the ocean surface, the lower the minimum pressure of the storm, and the higher its winds. Because there is a limit to how intense the storm can become, peak wind gusts seldom exceed 200 knots (370 km h^{-1}).

THE STORM DIES OUT   If the hurricane remains over warm water, it may survive for a long time. For example, Hurricane Tina (1992) travelled for thousands of kilometres over deep, warm tropical waters and maintained hurricane-force winds

for 24 days, making it one of the longest lasting North Pacific hurricanes on record. However, most hurricanes last for less than a week.

Hurricanes weaken rapidly when they travel over colder water and lose their heat source. Studies show that if the water beneath the eyewall of the storm (the region of thunderstorms adjacent to the eye) cools by 2.5°C, the storm's energy source is cut off, and the storm will dissipate. Even a small drop in water temperature beneath the eyewall will noticeably weaken the storm. A hurricane can also weaken if the layer of warm water beneath the storm is shallow. In this situation, the strong winds of the storm generate powerful waves that produce turbulence in the ocean water under the storm. Such turbulence creates currents that bring to the surface cooler water from below. If the storm is moving slowly, it is more likely to lose intensity as the eyewall will remain over the cooler water for a longer period.

Hurricanes also dissipate rapidly when they move over a large landmass. Here they not only lose their energy source, but friction with the land surface causes surface winds to decrease and blow more directly into the storm, an effect that causes the hurricane's central pressure to rise. And a hurricane, or any tropical system for that matter, will rapidly dissipate should it move into a region of strong vertical wind shear.

Sometimes remnant hurricanes moving into the middle latitudes may encounter an existing frontal zone or upper-level trough, and the storm can become reinvigorated. They undergo what is known as an **extratropical transition**, where the storm finds new life as an intense middle-latitude cyclone through a transfer of moisture and heat from the decaying hurricane. Sometimes these "new" extratropical storms are more intense than the original hurricane.

Our understanding of hurricane behaviour is far from complete. However, with the aid of computer model simulations and research projects such as RAINEX* (Rainband and Intensity Change Experiment), scientists are gaining new insight into how tropical cyclones form, intensify, and, ultimately, die.

HURRICANE STAGES OF DEVELOPMENT   Hurricanes go through a set of stages from birth to death. Initially, a *tropical disturbance* shows up as a mass of thunderstorms with only slight wind circulation. The tropical disturbance becomes a **tropical depression** when the winds increase to between 20 and 34 knots (37 and 63 km h^{-1}) and several closed isobars appear about its centre on a surface weather map. When the isobars are packed together and the winds are between 35 and 64 knots (64 and 119 km h^{-1}), the tropical depression becomes a **tropical storm**. (At this point, the storm gets a name.) The tropical storm is classified as a hurricane only when its winds exceed 64 knots (119 km h^{-1}).

---

*The *RAINEX* project consisted of reconnaissance aircraft flying into several hurricanes during the hurricane season of 2005. Equipped with sophisticated scientific instruments, including advanced Doppler radar, the mission obtained high-resolution data on each storm's structure, cloud configuration, and winds.

Figure 15.10 shows four tropical systems in various stages of development. Moving from east to west, we see a weak tropical disturbance (a tropical wave) crossing over Panama. Farther west, a tropical depression is organizing around a developing centre with winds less than 25 knots (46 km h^{-1}). In a few days, this system will develop into a hurricane. Farther west is a full-fledged hurricane with peak winds in excess of 110 knots (204 km h^{-1}). The swirling band of clouds to the northwest is Emilia; once a hurricane (but now with winds less than 40 knots), it is rapidly weakening over colder water.

BRIEF REVIEW

Before reading the next several sections, here is a review of some of the important points about hurricanes:

- Hurricanes are tropical cyclones, composed of an organized mass of thunderstorms.

- Hurricanes have peak winds about a central core (eye) that exceed 64 knots (119 km h^{-1}).

- The strongest winds and the heaviest rainfall normally occur in the eyewall—a ring of intense thunderstorms that surround the eye.

- Hurricanes form over warm tropical waters, where light surface winds converge, the humidity is high in a deep layer, and the winds aloft are weak.

- For a mass of thunderstorms to organize into a hurricane, there must be some mechanism that triggers the formation, such as converging surface winds along the ITCZ, a preexisting atmospheric disturbance, such as a weak front from the middle latitudes, or a tropical wave.

- Hurricanes derive their energy from warm tropical oceans and by evaporating water from the ocean's surface. Heat energy is converted to wind energy when the water vapour condenses and latent heat is released inside deep convective clouds.

- When hurricanes lose their source of warm water (either by moving over colder water or over a large landmass), they dissipate rapidly.

- Sometimes the remnants of a hurricane moving into middle latitudes will take on a new life as an intense middle-latitude cyclone through a process called extratropical transition.

Up to this point, it is probably apparent that tropical cyclones called hurricanes are similar to middle-latitude cyclones in that, at the surface, both have central cores of low pressure and winds that spiral counterclockwise (in the Northern Hemisphere) about their respective centres. However, there are many differences between the two systems, which are described in Focus on a Special Topic: How Do Hurricanes Compare to Middle-Latitude Storms? on p. 466.

HURRICANE MOVEMENT   Figure 15.11 shows where most hurricanes are born and the general direction in which they move. Notice that hurricanes that form over the warm, tropical North Pacific and North Atlantic are steered by easterly winds and move west or northwestward at about 10 knots for a week or so. Gradually, they swing poleward around the subtropical high, and when they move far enough north, they become caught in the westerly flow, which curves them to the north or northeast. In the middle latitudes, the hurricane's forward speed normally increases, sometimes to more than 50 knots. The actual path of a hurricane (which appears to be determined by the structure of the storm and the storm's interaction with the environment) may vary considerably. Some take erratic paths and make odd turns that occasionally catch weather forecasters by surprise (see Figure 15.12). There have

FIGURE 15.10 Visible satellite image showing four tropical systems, each in a different stage of its life cycle.

FOCUS ON A SPECIAL TOPIC

How Do Hurricanes Compare to Middle-Latitude Storms?

By now, it should be apparent that a hurricane is much different from the middle-latitude cyclone that we discussed in Chapter 12. A hurricane derives its energy from the warm water and the latent heat of condensation, whereas the mid-latitude storm derives its energy from horizontal temperature contrasts. The vertical structure of a hurricane is such that its central column of air is warm from the surface upward; consequently, hurricanes are called warm-core lows. A hurricane weakens with height, and the area of low pressure at the surface may actually become an area of high pressure above 12 km. Mid-latitude cyclones, on the other hand, are *cold-core lows* that usually intensify with increasing height, with a cold upper-level low or trough often existing above or to the west of the surface low.

A hurricane usually contains an eye where the air is sinking, whereas mid-latitude cyclones are characterized by centres of rising air. Hurricane winds are strongest near the surface, whereas the strongest winds of the mid-latitude cyclone are found aloft in the jet stream.

Further contrasts can be seen on a surface weather map. • Figure 1 shows Hurricane Rita over the Gulf of Mexico and a mid-latitude cyclone over the Maritime provinces. Around the hurricane, the isobars are more circular, the pressure gradient is much steeper, and the winds are stronger. The hurricane has no fronts and is smaller (although Rita is a large Category 5 hurricane). There are similarities between the two systems: both are areas of surface low

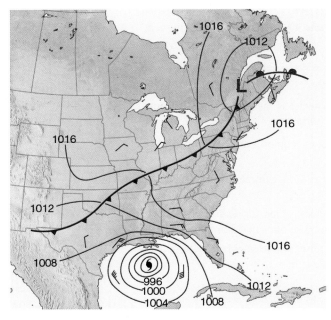

• FIGURE 1 Surface weather map for the morning of September 23, 2005, showing Hurricane Rita over the Gulf of Mexico and a middle-latitude storm system north of New England.

pressure, with winds moving counterclockwise about their respective centres.

It is interesting to note that some northeasters (winter storms that move northeastward along the coastline of North America, bringing with them heavy precipitation, high surf, and strong winds) may actually possess some of the characteristics of a hurricane. For example, a particularly powerful northeaster in January 1989 was observed to have a cloud-free eye, with surface winds in excess of 85 knots (157 km h^{-1}) spinning about a warm inner core. Moreover, some polar lows—lows that develop over polar waters during winter—may exhibit many of the observed characteristics of a hurricane, such as a

symmetrical band of thunderstorms spiraling inward around a cloud-free eye, a warm-core area of low pressure, and strong winds near the storm's centre. In fact, when surface winds within these polar storms reach 58 knots (107 km h^{-1}), they are sometimes referred to as *Arctic hurricanes*.

Even though hurricanes weaken rapidly as they move inland, their circulation may draw in air with contrasting properties. If the hurricane links with an upper-level trough, it may actually make an *extratropical transition* and become a mid-latitude cyclone. Swept eastward by upper-level winds, the remnants of an Atlantic hurricane can become an intense, mid-latitude, autumn storm in Europe.

---

been many instances where a storm heading directly for land suddenly veered away and spared the region from almost certain disaster. As a case in point, Hurricane Elena, with peak winds of 90 knots, moved northwestward into the Gulf of Mexico on August 29, 1985. It then veered eastward toward the west coast of Florida. After stalling offshore, it headed northwest. After weakening, it then moved onshore near Biloxi, Mississippi, on the morning of September 2.

Look at Figure 15.11 and notice that it appears as if hurricanes do not form over the South Atlantic and the eastern South Pacific—directly east and west of South America. Cooler water, vertical wind shear, and the unfavourable position of the ITCZ discourage hurricanes from developing in these regions. Then guess what? For the first time since satellites began observing the South Atlantic, a tropical cyclone formed off the coast of Brazil in March 2004 (see • Figure 15.13). So rare are

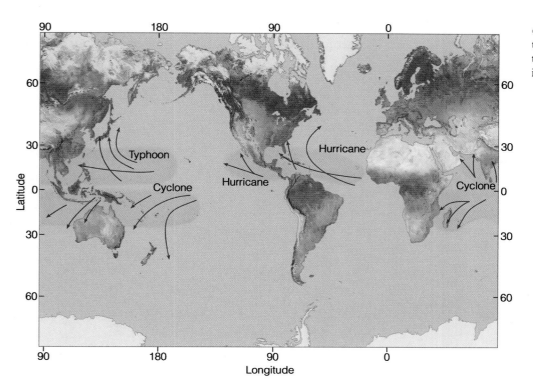

● FIGURE 15.11 Regions where tropical storms form (orange shading), the names given to storms, and the typical paths they take (red arrows).

tropical cyclones in this region that no government agency has an effective warning system for them, which is why the tropical cyclone was not given a name.

As we saw in an earlier section, many hurricanes form off the coast of Mexico over the North Pacific. In fact, this area usually spawns about nine hurricanes each year, which is slightly more than the yearly average of six storms born over

the tropical North Atlantic. We can see in Figure 15.11 that eastern North Pacific hurricanes normally move westward, away from the coast, so little is heard about them. When one does move northwestward, it normally weakens rapidly over the cool water of the North Pacific. Occasionally, however, one will curve northward or even northeastward and slam into Mexico, causing destructive flooding. Hurricane Tico left

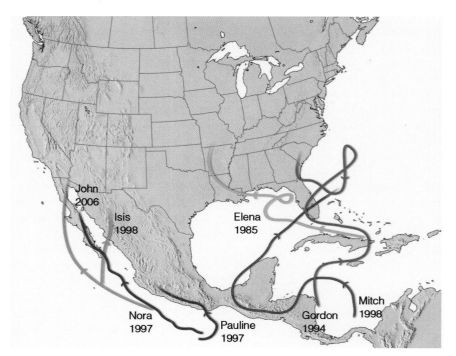

● FIGURE 15.12 Some erratic paths taken by hurricanes.

● FIGURE 15.13 An extremely rare tropical cyclone (with no name) near 28°S latitude spins clockwise over the South Atlantic off the coast of Brazil in March 2004. Due to cool water and vertical wind shear, storms rarely form in this region of the Atlantic Ocean. In fact, this is the only tropical storm ever officially reported there.

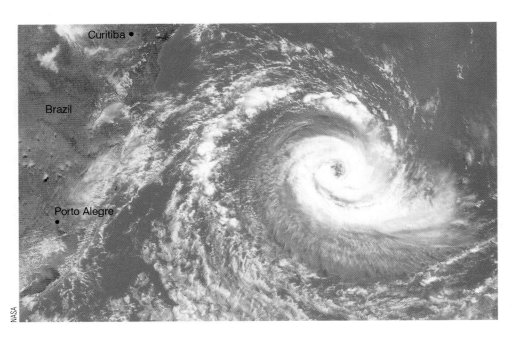

25,000 people homeless and caused an estimated $66 million in property damage after passing over Mazatlán, Mexico, in October 1983. The remains of Tico even produced record rains and flooding in Texas and Oklahoma. Even less frequently, a hurricane will stray far enough north to bring summer rains to southern California and Arizona, as did the remains of Hurricane Nora in September 1997. (Nora's path is shown in Figure 15.12.) The only hurricane on record to reach the West Coast of the United States with sustained hurricane-force winds did so in October 1858 when a hurricane slammed into the extreme southern part of California near San Diego.

The Hawaiian Islands, which are situated in the central North Pacific between about 20° and 23°N, appear to be in the direct path of many eastern Pacific hurricanes and tropical storms. By the time most of these storms reach the islands, however, they have weakened considerably and pass harmlessly to the south or northeast. The exceptions were Hurricane Iwa in November 1982 and Hurricane Iniki in September 1992. Iniki, the worst hurricane to hit Hawaii in the 20th century, battered the island of Kauai with torrential rain, sustained winds of 114 knots that gusted to 140 knots, and 6 m waves that crashed over coastal highways. Major damage was sustained by most of the hotels and about 50 percent of the homes on the island. Iniki (the costliest hurricane in Hawaiian history, with a damage estimate of $1.8 billion) flattened sugarcane fields, destroyed the macadamia nut crop, injured about 100 people, and caused at least seven deaths.

Hurricanes that form over the tropical North Atlantic also move westward or northwestward on a collision course with Central or North America. Most hurricanes, however, swing away from land and move northward, parallel to the coastline of the United States. A few storms, perhaps three per year, move inland, bringing with them high winds, huge waves, and torrential rain that may last for days.

A hurricane moving northward over the Atlantic will normally survive as a hurricane for a much longer time than will its counterpart at the same latitude over the eastern Pacific. The reason for this situation is that an Atlantic hurricane moving northward will usually stay over the warmer water of the Gulf Stream current, whereas an eastern Pacific hurricane heading north will quickly move over much cooler water and, with its energy source cut off, will rapidly weaken.

Atlantic hurricanes can sometimes travel far enough north to impact Canada. They do this either as tropical cyclones, or through an extratropical transition, they become intense middle-latitude cyclones. Indeed, some of Canada's worst weather disasters are related to hurricanes doing just this, such as Hurricane Hazel* in October 1954. Hurricane Hazel spread destruction in its path from the Caribbean, where it killed an estimated 1000 people in Haiti, to North Carolina and onward toward the Great Lakes, causing 95 American deaths. When it reached the state of Pennsylvania, the dying hurricane fed its energy into a middle-latitude cyclone (making an extratropical transistion) and was named "Hazel II." It passed directly over Toronto on October 15. The intense precipitation from the storm caused extensive flooding as river levels rose. Water inundated homes that were built on floodplains in Toronto, claiming 81 Canadian lives.

Hurricane Juan is an example of a hurricane that did reach Canada as a tropical cyclone. As we will read in the Focus on a Special Topic: Hurricane Juan Strikes Halifax (p. 470), Juan had a devastating impact on Nova Scotia.

---

*For more details, readers may find the case study of the Hazel II storm interesting. It is available at Environment Canada's Canadian Hurricane Centre and may be found by searching the centre's web pages.

Naming Hurricanes and Tropical Storms

In an earlier section, we learned that hurricanes are given a name when they reach tropical storm strength. Before hurricanes and tropical storms were assigned names, they were identified according to their latitude and longitude. This method was confusing, especially when two or more storms were present over the same ocean. To reduce the confusion, hurricanes were identified by letters of the alphabet. During World War II, names such as Able and Baker were used. (These names correspond to the radio code words associated with each letter of the alphabet.) This method also seemed cumbersome, so, beginning in 1953, the U.S. National Weather Service began using female names to identify hurricanes. The list of names for each year was in alphabetical order, so the names of the season's first storm began with the letter *A*, the second with *B*, and so on.

From 1953 to 1977, only female names were used. However, beginning in 1978, tropical storms in the eastern Pacific were alternately assigned female and male names, but not just English names as Spanish and French ones were used, too. This practice began for North Atlantic hurricanes in 1979. There are six lists of names that are used in rotation. If a storm causes great damage and becomes infamous as a Category 3 or higher, its name is retired for at least 10 years. The lists of names for each year in all the areas can be found on the U.S. National Hurricane Center website. This list is now updated by a committee of the World Meteorological Organization. In the North Atlantic, only 21 names are used (the letters Q, U, X, Y, and Z are excluded) because until 2005, that was the most hurricanes recorded in a season. If the number of named storms in any year should exceed the names on the list, as occurred in 2005, then tropical storms are assigned names from the Greek alphabet, such as Alpha, Beta, and Gamma. In fact, the last of the 27 named tropical systems in 2005 was Zeta, which actually formed in January 2006.

Devastating Winds, Flooding, and the Storm Surge

When a hurricane is approaching from the south, its highest winds are usually on its eastern (right) side. The reason for this phenomenon is that the winds that push the storm along add to the winds on the east side and subtract from the winds on the west (left) side. The hurricane illustrated in ● Figure 15.14 is moving northward along the East Coast of the United States with winds of 100 knots swirling counterclockwise about its centre. Because the storm is moving northward at about 25 knots, sustained winds on its eastern side are about 125 knots, whereas on its western side, winds are only 75 knots.

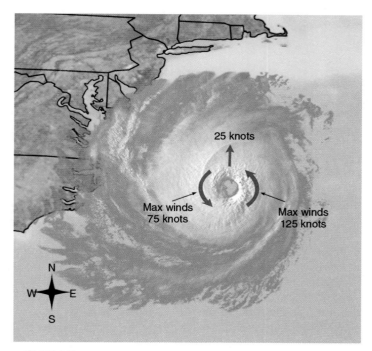

● FIGURE 15.14 A hurricane moving northward will have higher sustained winds on its eastern side than on its western side. If the hurricane moves from east to west, the highest sustained winds will be on its northern side.

Even though the hurricane in Figure 15.14 is moving northward, there is a net transport of water directed westward toward the coast. To understand this behaviour, recall from Chapter 10 that as the wind blows over open water, the water beneath is set in motion. If we imagine the top layer of water to be broken into a series of layers, then we find each layer moving to the *right* of the layer above (Northern Hemisphere). This type of movement (bending) of water with depth (called the *Ekman spiral*) causes a net transport of water (known as *Ekman transport*) to the right of the surface wind in the Northern Hemisphere. Hence, the north wind on the hurricane's left (western) side causes a net transport of water toward the shore. Here the water piles up and rapidly inundates the region.

The high winds of a hurricane also generate large waves, sometimes 10 to 15 m high. These waves move outward, away from the storm, in the form of *swells* that carry the storm's energy to distant beaches. Consequently, the effects of the storm may be felt days before the hurricane arrives.

Although the hurricane's high winds inflict a great deal of damage, the huge waves, high seas, and *flooding* normally cause most of the destruction. The flooding is also responsible for the loss of many lives. In fact, the majority of hurricane-related deaths during the past century have been due to flooding. The flooding is due, in part, to winds pushing water onto the shore and to the heavy rains, which may exceed 630 mm* in 24 hours. Flooding is also aided by the

---

*Hurricanes may sometimes have a beneficial aspect, in the sense that they can provide much-needed rainfall in drought-stricken areas.

Hurricane Juan Strikes Halifax

Dr. Yongsheng Chen, Department of Earth and Space Science and Engineering, York University, Toronto, Ontario

On September 29, 2003, Hurricane Juan made landfall near Halifax, Nova Scotia, as a Category 2 storm on the **Saffir–Simpson scale**. It became the most widely destructive tropical cyclone to hit Atlantic Canada in over a century.

Juan formed from a tropical wave moving off the coast of Africa in mid-September; however, initially strong vertical **wind shear** in the upper levels kept the system disorganized. By September 20, convection around the system greatly increased when the system encountered and interacted with a large upper-level low that was partially associated with the outflow of Hurricane Isabel. While the system moved northwestward around the upper-level low, the mid- and low-level circulations developed. As deep convection became active near the centre on September 24, the system developed a warm core and became a tropical cyclone, developing banding features in its cloud pattern and a distinct outflow. It was named Tropical Depression Fifteen at 7:00 a.m. (EST) on September 24, about 600 km southeast of Bermuda (see Figure 2). The depression steadily strengthened into a tropical storm when its winds exceeded 35 knots (65 km h^{-1}) that evening. Juan attained hurricane status in the morning on September 26 while located 270 km southeast of Bermuda after the deep convection continued to organize, an eye feature developed, and winds exceeded 64 knots (119 km h^{-1}).

Steered by a developing subtropical ridge to its east, Juan moved northward and then northwestward. The warm waters and light

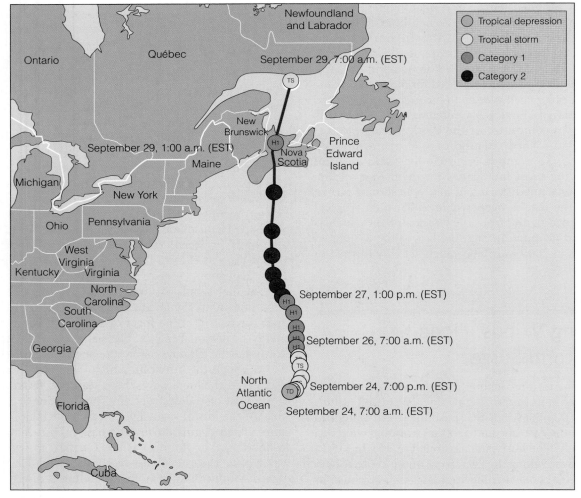

FIGURE 2 The track of Hurricane Juan.

wind shear helped Juan reached its maximum intensity with a one-minute sustained wind of 90 knots (167 km h^{-1}) and a minimum centre sea-level pressure of 969 hPa at 1:00 p.m. (EST) on September 27. Juan then turned again and accelerated to the north. It had little time to weaken over cold waters because of its fast movement (see Figure 2 and ● Figure 3). When Juan made landfall near Halifax, between Prospect and Peggy's Cove at 10:10 p.m. (EST) on September 28, it still retained a maximum wind of 85 knots (157 km h^{-1}) and a minimum pressure of 973 hPa. After quickly sweeping across Nova Scotia, Juan still retained its hurricane strength when it entered the province of Prince Edward Island at 1:25 a.m. (EST) on September 29. It was absorbed by a large extratropical low 12 hours later in the Gulf of St. Lawrence.

The furious wind of Hurricane Juan brought widespread structural and vegetation damage. It was estimated that 100 million trees in Nova Scotia and 1 million trees in Halifax alone were lost. In the landmark Point Pleasant Park, 90 percent of the mature growth was destroyed or irreparably damaged (see ● Figure 4). Thirty-one percent of residential homes in Halifax suffered some degree of damage. The largest significant waves (average of the largest third of waves) recorded were almost 9 m high off Halifax Harbour, with the largest waves recorded at 19.9 m high. Dozens of marinas around Nova Scotia and Prince Edward Island were destroyed, and many small vessels capsized or sank. Coastal flooding was reported as a result of the storm surge, whereas inland flooding was minor because total rainfall was light due to the fast storm motion. Power outages in Nova Scotia and Prince Edward Island left more than 300,000 people without power for up to two weeks. Juan claimed eight lives: four directly and another four in the aftermath. Because of its devastating effects in Canada, the name Juan was retired in the Atlantic basin.

NASA

● FIGURE 3 The MODIS instrument onboard NASA's *Terra* satellite captured this true-colour image of Hurricane Juan approximately 580 km south of Halifax, Nova Scotia. At the time this image was taken on September 28, 2003, Juan was packing sustained winds near 86 knots (160 km h^{-1}) and was moving to the north at 17 knots (32 km h^{-1}).

© Environment Canada, Canadian Hurricane Centre, Chris Fogarty, 2010.

● FIGURE 4 Aerial photograph of tree damage at Point Pleasant Park in Halifax.

▼ Table 15.1   Saffir–Simpson Hurricane (Wind) Scale

SCALE NUMBER	CENTRAL PRESSURE	WINDS		STORM SURGE	
Category	hPa	km h⁻¹	knots	m	Damage
1	≥ 980	119–153	64–82	~ 1.5	Damage mainly to trees, shrubbery, and unanchored mobile homes
2	965–979	154–177	83–95	~ 2.0–2.5	Some trees blown down; major damage to exposed mobile homes; some damage to roofs of buildings
3	945–964	178–209	96–113	~ 2.5–4.0	Foliage removed from trees; large trees blown down; mobile homes destroyed; some structural damage to small buildings
4	920–944	210–249	114–135	~ 4.0–5.5	All signs blown down; extensive damage to roofs, windows, and doors; complete destruction of mobile homes; flooding inland as far as 10 km; major damage to lower floors of structures near shore
5	< 920	> 249	> 135	> 5.5	Severe damage to windows and doors; extensive damage to roofs of homes and industrial buildings; small buildings overturned and blown away; major damage to lower floors of all structures less than 4.5 m above sea level within 500 m of shore

low pressure of the storm. The region of low pressure allows the ocean level to rise (perhaps half a metre), much like a soft drink rises up a straw as air is withdrawn. (A drop of 1 hPa in air pressure produces a rise in ocean levels of 1 cm.) The combined effect of high water (which is usually well above the high-tide level), high winds, and the net Ekman transport toward the coast produces the **storm surge**—an abnormal rise of several metres in the ocean level—which inundates low-lying areas and turns beachfront homes into piles of splinters. The storm surge is particularly damaging when it coincides with normal high tides. Flooding, however, is not just associated with hurricanes as destructive floods can occur with tropical storms that do not reach hurricane strength. An example of such a storm was tropical storm Allison, which inundated parts of Texas in June 2001.

In an effort to estimate the possible damage a hurricane's sustained winds and storm surge could do to a coastal area, the **Saffir–Simpson scale** was developed (see ▼ Table 15.1). The scale numbers (which range from 1 to 5) are based on actual conditions at some time during the life of the storm. Prior to 2009, measures such as central pressure and storm surge, in addition to wind speed, were used in determining the scale. Now, however, only wind speed is used because it is the least influenced by factors other than storm intensity. As the hurricane intensifies or weakens, the category, or scale number, is reassessed accordingly. Major hurricanes are classified as Category 3 and above. In the western Pacific, a typhoon with sustained winds of at least 130 knots (241 km h⁻¹)—at the upper end of the wind speed range in Category 4 on the

Saffir–Simpson scale—is called a **supertyphoon.** ● Figure 15.15 illustrates how the storm surge changes along the coast as hurricanes with increasing intensity make landfall.*

Between 1900 and 2007, a total of 181 hurricanes struck the coastline of the United States. Of these, 73 (40 percent) were major hurricanes—Category 3 or higher. Hence, along the Gulf and Atlantic coasts, on average, about five hurricanes make landfall every three years, two of which are major hurricanes with winds in excess of 95 knots (176 km h⁻¹) and a storm surge exceeding 2.5 m.

● Figure 15.16 shows the tropical storms and hurricanes that approached Canada's coast or coastal waters between 2000 and 2009. In this decade, 62 tropical or post-tropical cyclones† entered the response zone of the Canadian Hurricane Centre and were monitored closely. Of these, a total of 20 made landfall in Canada: one as a Category 2 hurricane (Juan in 2004), three as Category 1 hurricanes, two as tropical storms, and 14 as post-tropical (extratropical) cyclones. This decade is similar to other decades in Canada, as can be seen in ● Figure 15.17.

Although the high winds of a hurricane can devastate a region, considerable damage may also occur from hurricane-spawned tornadoes. About one-fourth of the hurricanes that strike the United States produce tornadoes. In fact, in 2004, six tropical systems produced just over 300 tornadoes in the

---

*Landfall is the position along the coast where the centre of a hurricane passes from ocean to land.

†A post-tropical cyclone is a tropical cyclone that has undergone an extratropical transition.

Normal high tide

Category 1 [1.2-metre rise]

Category 3 [3.6-metre rise]

Category 5 [5.5-metre rise]

● FIGURE 15.15 The changing of the ocean level as different category hurricanes make landfall along the coast. The water typically rises about 1.2 m with a Category 1 hurricane but may rise to 5.5 m (or more) with a Category 5 storm.

southern and eastern United States. The exact mechanism by which these tornadoes form is not totally clear; however, studies suggest that surface topography may play a role by initiating the convergence (and, hence, rising) of surface air. Moreover, tornadoes tend to form in the right front quadrant of an advancing hurricane,* where vertical wind-speed shear is greatest. Studies also suggest that swathlike areas of extreme damage once attributed to tornadoes may actually be due to strong downdrafts (microbursts) associated with the large, intense thunderstorms around the eyewall.

In examining the extensive damage wrought by Hurricane Andrew in August 1992, researchers theorized that the

areas of most severe damage might have been caused by small, whirling eddies perhaps 30 to 100 m in diameter that occur in narrow bands. Many scientists today believe that those rapidly rotating eddies were, in fact, small tornadoes. Lasting for about 10 seconds, the vortices appeared to have formed in a region of strong wind-speed shear in the hurricane's eyewall, where the air was rapidly rising. As intense updrafts stretched the vortices vertically, they shrank horizontally, which induced them to spin faster, perhaps as fast as 70 knots (130 km h^{-1}). When the rotational winds of a vortex are added to the hurricane's steady wind, the total wind speed over a relatively small area may increase substantially. In the case of Hurricane Andrew, isolated wind speeds may have reached 174 knots (322 km h^{-1}) over narrow stretches of south Florida.

*In the northeast quadrant of the hurricane shown in Figure 15.14 on p. 469.

● FIGURE 15.16 Hurricanes and tropical storms that neared Canada between 2000 and 2009.

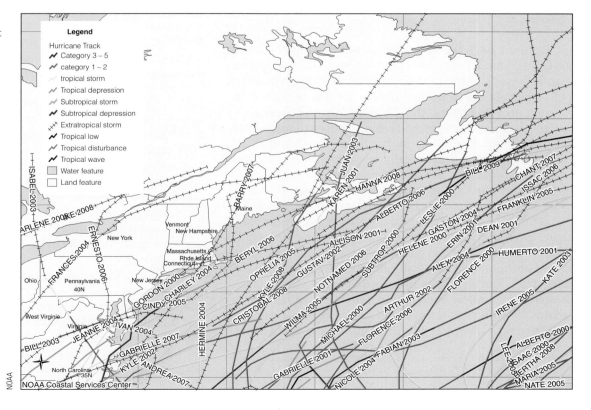

• FIGURE 15.17 The number of Atlantic tropical cyclones making landfall in Canada since 1951.

Source: Environment Canada. Graph: Number of Atlantic Tropical Cyclones Making Landfall in Canada. Found at: http://www.ec.gc.ca/ouragans-hurricanes/default.asp?lang=En&n=E2752604-1

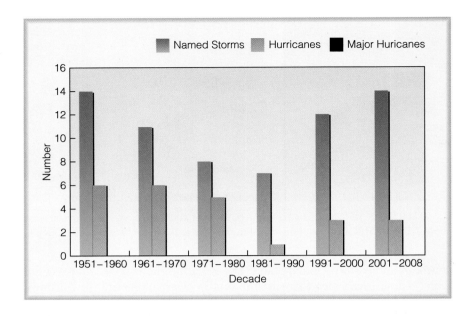

WEATHER WATCH

Are storm surges and tsunamis the same? Although there are similarities between the two, they are actually quite different. A storm surge is an onshore surge of ocean water caused primarily by the winds of a storm (most often a tropical cyclone) pushing sea water onto the coast. Tsunamis are waves generated by disturbances on the ocean floor caused most commonly by earthquakes. As the tsunami wave (or series of waves) moves into shallow water, it builds in height and rushes onto the land (sometimes unexpectedly), swooping up everything in its path, including cars, buildings, and people.

Up until 2005, the annual death toll from hurricanes in the United States, over a span of about 30 years, averaged less than 50 persons.* Most of these fatalities were due to flooding. This relatively low total was due in part to the advanced warning provided by the U.S. National Weather Service and the fact that only a few really intense storms had made landfall during this time. But the hurricane death toll in the United States rose dramatically in 2005 when Hurricane Katrina slammed into Mississippi and Louisiana.

As Hurricane Katrina moved toward the coast, evacuation orders were given to residents living in low-lying areas, including the city of New Orleans. Many thousands of people moved to higher ground, but, unfortunately, many people either refused to leave their homes or had no means of leaving and were forced to ride out the storm. Tragically, more than 1300 people died either from Katrina's huge storm surge and high winds that demolished countless buildings or from the flooding in New Orleans, when several levees broke and parts of the city were inundated with water over 6 m deep. As the population density

continues to increase in vulnerable coastal areas, the potential for another hurricane-caused disaster also increases.

In Canada, there were an estimated 895 deaths between 1900 and 2004 related to hurricanes. Of these, up to 494 were marine deaths at sea. On land, Ontario suffered 99 casualties (81 in 1954 from Hazel, mainly from flooding due to rain), Quebec had 14, Nova Scotia had 24, Newfoundland had 10, Prince Edward Island had two, and New Brunswick had one. Not including Hazel, since 1950, about half of the hurricane-related deaths in Canada have been due to automobile accidents during these storms, about 20 percent due to drowning in coastal waves, and the rest due mainly to rain-caused flooding, post–storm-related deaths, and the deaths of rescue workers. The major hurricanes that have affected Canada, including their death tolls, are listed in ▼ Table 15.2.

SOME NOTABLE HURRICANES **Camille, 1969** Hurricane Camille (1969) stands out as one of the most intense hurricanes to reach the coastline of the United States in the 20th century. With a central pressure of 909 hPa, tempestuous winds reaching 160 knots (296 km h^{-1}), and a storm surge more than 7 m above the normal high-tide level, Camille, as a Category 5 storm, unleashed its fury on Mississippi, destroying thousands of buildings. During its rampage, it caused an estimated $1.5 billion in property damage and took more than 200 lives.

Tip, 1979 Supertyphoon Tip was the largest and most intense tropical cyclone ever recorded. It originated as a tropical depression in the western tropical Pacific near Micronesia and then moved northwestward and intensified. After it passed the island of Guam, Tip's central pressure fell to 870 hPa, a worldwide record low sea-level pressure. At the same time, its winds increased to peak values of 165 knots (305 km h^{-1}). At this time, its diameter was a record-setting 2174 km—the size

---

*In countries outside North America, the annual death toll was considerably higher. Estimates are that more than 3000 people died in Haiti from flooding and mudslides when Hurricane Jeanne moved through the Caribbean in September 2004.

▼ Table 15.2 Notable Canadian Tropical Cyclones

YEAR	HURRICANE	DESCRIPTION	DEATH TOLL
1775	The Newfoundland Hurricane or Independence Hurricane	A large number of mariners, mostly British sailors, lost their lives off the coast of Newfoundland. Some say that this storm was a turning point in the American Revolution.	> 4000 lives of sailors
1869	The Saxby Gale	This post-tropical storm in conjunction with a high tide caused flooding in the Bay of Fundy that nearly covered the isthmus between Nova Scotia and New Brunswick.	Hundreds of lives and farms
1873	The Great Nova Scotia Hurricane	Records indicate that 1200 boats and 900 homes were destroyed in Nova Scotia and 100 more lives were lost in Newfoundland.	600 lives, mostly sailors
1900	The Galveston Hurricane	The Galveston Hurricane caused over 6000 deaths in the United States, making it the deadliest hurricane to affect North America, as well as Canada's deadliest since 1900.	Records are not clear. Estimated at 233; could be up to 250.
1924 1926 1927	The Great August Gales	These were the second (1927), fourth (1926), and sixth (1924) deadliest hurricanes in Canada since 1900 and affected Atlantic Canada.	276 lives, mostly sailors
1935	Newfoundland Hurricane	This hurricane was the first of the season and impacted Newfoundland.	35
1954	Hurricane Hazel	Canada's most noted hurricane and third deadliest since 1900. Rain-induced flooding in Toronto once the hurricane became extratropical accounted for the high number of fatalities.	81
1957	Hurricane Audrey	Audrey was a Category 4 hurricane when it made landfall in Louisiana and Texas, causing 416 fatalities there. After making landfall, it became a powerful extratropical storm affecting Ontario and Quebec. Twelve of the deaths were due to automobile accidents.	15
1959	The Great Escuminac Disaster	This June hurricane became extratropical and intensified over the Gulf of St. Lawrence, where it caused enormous waves, sinking 22 of the 54 fishing boats from the village of Escuminac, New Brunswick.	35 fishermen
1991	The Perfect Storm (also called the Halloween Storm)	The subject of the novel by Sebastian Junger and a movie of the same name, this unusual storm formed from the combination of the remnants of Hurricane Grace and a powerful middle-latitude cyclone.	7 mariners
2003	Hurricane Juan	Halifax, Nova Scotia, was hit by this Category 2 hurricane that downed an estimated 50 to 100 million trees in Nova Scotia and caused over $200 million in damage.	8

Source: Canadian Hurricane Centre.

of a typical middle-latitude cyclone (see ● Figure 15.18). It made landfall in Japan on October 19 and resulted in nearly 100 direct and indirect fatalities.

Andrew, 1992 Another devastating hurricane in the 20th century was Hurricane Andrew. On August 21, 1992, as tropical storm Andrew churned westward across the Atlantic, it began to weaken, prompting some forecasters to surmise that this tropical storm would never grow to hurricane strength.

But Andrew moved into a region favourable for hurricane development. Even though it was outside the tropics near latitude 25°N, warm surface water and weak winds aloft allowed Andrew to intensify rapidly. And in just two days, Andrew's winds increased from 45 to 122 knots, turning an average tropical storm into one of the most intense hurricanes to strike Florida in the past 105 years.

With winds of at least 130 knots (241 km h⁻¹) and a powerful storm surge, Andrew made landfall south of Miami

● FIGURE 15.18 Supertyphoon Tip on October 12, 1979, near the time it reached its record-setting size of 2174 km across and record-setting minimum sea-level pressure of 870 hPa. Typhoon Sarah is located to its west in the South China Sea.

on the morning of August 24 (see ● Figure 15.19). The eye of the storm moved over Homestead, Florida. Andrew's fierce winds completely devastated the area (see ● Figure 15.20) as 50,000 homes were destroyed, trees were leveled, and steel-reinforced tie beams weighing tonnes were torn free of townhouses and hurled as far as several blocks. Swaths of severe damage led scientists to postulate that peak winds may have

approached 174 knots (322 km h^{-1}). Such winds may have occurred with small tornadoes, which added substantially to the storm's wind speed. In an instant, a wind gust of 142 knots (263 km h^{-1}) blew down a radar dome and inactivated several satellite dishes on the roof of the U.S. National Hurricane Center in Coral Gables. Observations reveal that some of Andrew's destruction may have been caused by microbursts in the severe thunderstorms of the eyewall. The hurricane roared westward across southern Florida, weakened slightly, and then regained strength over the warm Gulf of Mexico. Surging northwestward, Andrew slammed into Louisiana as a Category 3 with 120-knot winds on the evening of August 25.

All told, Hurricane Andrew was one of the costliest natural disasters ever to hit the United States. It destroyed or damaged over 200,000 homes and businesses, left more than 160,000 people homeless, caused over $30 billion in damages, and took 53 lives, including 41 in Florida.

Katrina, 2005   Hurricane Katrina was the most costly hurricane ever to hit the United States. Forming over warm tropical water south of Nassau in the Bahamas, Katrina became a tropical storm on August 24, 2005, and a Category 1 hurricane just before making landfall in south Florida on August 25. (Katrina's path is given in Figure 5 on p. 478.) It moved southwestward across Florida and out over the eastern Gulf of Mexico. As Katrina moved westward, it passed over a deep band of warm water called the Loop Current, which allowed Katrina to intensify rapidly. Within 12 hours, the hurricane increased from a Category 3 to a Category 5 storm with winds

● FIGURE 15.19 Colour radar image of Hurricane Andrew as it moves onshore over south Florida on the morning of August 24, 1992. The National Hurricane Center (NHC) is located about 30 km from the centre of the eye.

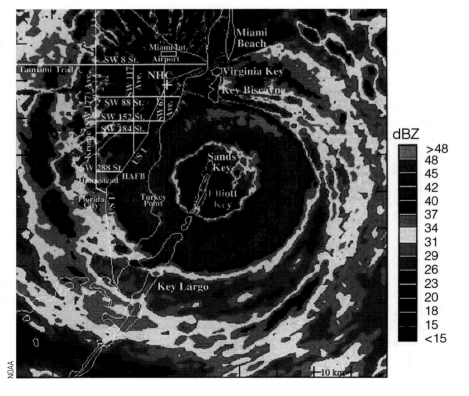

● FIGURE 15.20 A community in Homestead, Florida, devastated by Hurricane Andrew in August 1992.

© Weather Catalog/Media Services

of 282 km h^{-1} and a central pressure of 902 hPa (see the opening photograph of Katrina on p. 456).

Over the Gulf of Mexico, Katrina gradually turned northward toward Mississippi and Louisiana. As the powerful Category 5 hurricane moved slowly toward the coast, its rainbands near the centre of the storm began to converge toward the storm's eye. This process cut off moisture to the eyewall. As the old eyewall dissipated, a new one formed farther away in a phenomenon meteorologists call **eyewall replacement**. The replacement of the eyewall weakened the storm such that Katrina made landfall near Buras, Louisiana, on August 29 (see ● Figure 15.21) as a strong Category 3 hurricane with sustained winds of 110 knots (204 km h^{-1}), a central pressure of 920 hPa, and a storm surge between 6 and 9 m.

Katrina's strong winds and high storm surge on its eastern side devastated southern Mississippi, with Biloxi, Gulfport, and Pass Christian being hit particularly hard (see ● Figure 15.22). The winds demolished all but the strongest structures, and the huge storm surge scoured areas up to 6 km inland.

New Orleans and the surrounding parishes actually escaped the brunt of Katrina's winds as the eye passed just to the east of the city (see Figure 15.21). However, the combination of high winds, large waves, and a huge storm surge caused disastrous breaches in the levee system that protects New Orleans from the Mississippi River, the Gulf of Mexico, and Lake Pontchartrain. When the multiple levees gave way, water up to 6 m deep tragically drowned a large part of the city. This represents the height

of a second-storey commercial building. In neighbourhoods such as the Lower Ninth Ward, one of the lowest and poorest parts of the city, U.S. Coast Guard helicopters pulled residents from their rooftops. Many thousands of people, mostly the poor and disadvantaged, were unable to evacuate the city (see ● Figure 15.23), which was unprepared to deal with this scale of disaster. In shock and devastated, with so many residents unable to return or needing emergency assistance, civil disorder, looting, and violence occurred during the first days after the storm. This required a state of emergency to be declared, with up to 46,000 U.S. National Guard and federal troops providing security and emergency aid to New Orleans. It took 43 days to pump the flood waters out of the city.

Less than a month after Katrina struck, before the city was pumped dry, powerful Hurricane Rita, with sustained winds of 152 knots (282 km h^{-1}), moved over the Gulf of Mexico, south of New Orleans. Strong, tropical storm–force, easterly winds, along with another storm surge, caused some of the repaired levees to break again, flooding parts of the city that just days earlier had been pumped dry. The death toll due to Hurricane Katrina totalled more than 1200, and the devastation wrought by the storm totalled more than $75 billion. Although Katrina may well be the most expensive hurricane on record, tragically, it is not the deadliest.

Other Devastating Hurricanes Before the era of satellites and radar, catastrophic losses of life were more frequent as

FOCUS ON AN OBSERVATION

The Record-Setting Atlantic Hurricane Seasons of 2004 and 2005

Both 2004 and 2005 were active years for hurricane development over the tropical North Atlantic. During 2004, nine storms became full-fledged hurricanes. Of the five hurricanes that made landfall in the United States, three (Charley, Frances, and Jeanne) ploughed through Florida and one (Ivan) came onshore just west of the Florida panhandle (see ● Figure 5), making this the first time since record keeping began in 1861 that four hurricanes had impacted the state of Florida in one year. Total damage in the United States from the five hurricanes exceeded $40 billion.

Then, in 2005, a record 27 named storms developed (the most in a single season), of which 15 (another record) reached hurricane strength. The 2005 Atlantic hurricane season also had four hurricanes (Emily, Katrina, Rita, and Wilma) reach Category 5 intensity for the first time since reliable record keeping began. And Hurricane Wilma had the lowest central pressure ever measured in an Atlantic hurricane—882 hPa. Out of five hurricanes that made landfall in the United States, three (Dennis, Katrina, and Wilma) made landfall in hurricane-wary Florida and one (Ophelia)

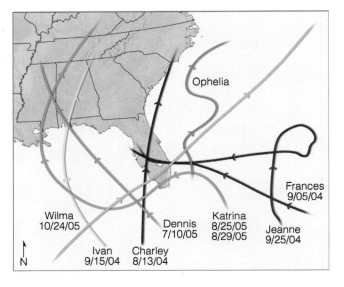

● FIGURE 5 The paths of eight hurricanes that impacted Florida during 2004 and 2005. Notice that in 2004, hurricanes Frances and Jeanne made landfall at just about the same spot along Florida's southeast coast. The date under the hurricane's name indicates the date the hurricane made landfall.

skirted northward along Florida's east coast, giving Florida the dubious distinction of being the only state on record to experience eight hurricanes during the span of 16 months (see Figure 5). Total damage in the United States from the five hurricanes that made landfall exceeded $100 billion.

Apparently, in 2004 and 2005, very warm ocean water and weak vertical wind shear

provided favourable conditions for hurricane development. In previous years, winds associated with a persistent upper-level trough over the eastern United States steered many tropical systems away from the coast before they could make landfall. However, in 2004 and 2005, an area of high pressure replaced the trough, and winds tended to steer tropical cyclones on a more westerly track, toward the coastline of North America.

---

early warning systems were less developed and emergency preparedness and response programs were very localized or nonexistent.

During his fourth voyage in 1502, Christopher Columbus recognized the signs of an approaching hurricane and warned the governor of Santo Domingo (now in the Dominican Republic). His advice was ignored, which resulted in the loss of a Spanish treasure fleet (20 ships and 500 men). Christopher Columbus is also credited with the earliest European documentation of a hurricane as he sheltered his fleet from such a storm on his second voyage in 1494.

In 1900, more than 6000 people lost their lives when a hurricane slammed into Galveston, Texas, with a huge storm surge. Most of the deaths occurred in the low-lying coastal regions as flood waters pushed inland. In October 1893, nearly 2000 people perished on the Gulf Coast of Louisiana as a giant storm surge swept that region. Spectacular losses are not

confined to the Gulf Coast. Nearly 1000 people lost their lives in Charleston, South Carolina, during August of the same year. This hurricane was also the deadliest in Canada since 1900 (look back at Table 15.2). Its remnants crossed through southern Ontario, sinking two ships on Lake Erie, and travelled through the St. Lawrence valley and on to Prince Edward Island and then Newfoundland. On its path, it is estimated to have killed around 250 people in Canada. Most of these were sailors as four ships and 99 smaller fishing vessels were recorded as being lost. Records from the time are unclear as the numbers of deaths from St. Pierre and Miquelon were not well documented.

However, these statistics are small compared to the more than 300,000 lives taken as a killer tropical cyclone and storm surge ravaged the coast of Bangladesh and the low-lying islands of the Ganges Delta with flood waters in 1970. In April 1991, a similar cyclone devastated the area, with

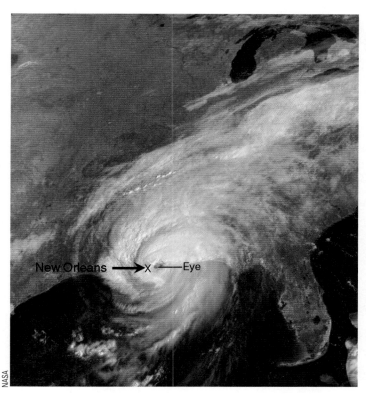

NASA

● FIGURE 15.21 Hurricane Katrina just after making landfall along the Mississippi/Louisiana coast on the morning of August 29, 2005. Shown here, the storm is moving north with its eye due east of New Orleans. At landfall, Katrina had sustained winds of 110 knots, a central pressure of 920 hPa, and a storm surge over 6 m.

reported winds of 127 knots and a storm surge of 7 m. In all, the 1991 storm destroyed 1.4 million houses and killed 140,000 people and 1 million cattle. And again in November 2007, Tropical Cyclone Sidr, a Category 4 storm with winds of 135 knots, moved into the region (see ● Figure 15.24), killing an estimated 3447 people, damaging or destroying over 1 million houses, and flooding more than 2 million acres. Estimates are that Cyclone Sidr adversely affected more than 8.5 million people. Unfortunately, the potential for a repeat of this type of disaster remains high in Bangladesh as many people live along the relatively low, wide flood plain that slopes outward to the bay. And, historically, this region is in a path frequently taken by tropical cyclones.

In late October 1998, Hurricane Mitch became the deadliest hurricane to strike the Western Hemisphere since the Great Hurricane of 1780, which claimed approximately 22,000 lives in the eastern Caribbean. Mitch's high winds, huge waves (estimated maximum height 13.4 m), and torrential rains destroyed vast regions of coastal Central America (for Mitch's path, see Figure 15.12, p. 467). In the mountainous regions of Honduras and Nicaragua, rainfall totals from the storm may have reached 1900 mm. The heavy rains produced floods and deep mudslides that swept away entire villages, including the inhabitants. Mitch caused over $5 billion in damage, destroyed hundreds of thousands of homes, and killed over 11,000 people. More than 3 million people were left homeless or were otherwise severely affected by this deadly storm.

Are major hurricanes on the increase worldwide? Will the intensity of hurricanes increase as the world warms? These questions are addressed in Focus on an Environmental Issue: Hurricanes in a Warmer World on p. 482.

● FIGURE 15.22 High winds and huge waves crash against a boat washed onto Highway 90 in Gulfport, Mississippi, as Hurricane Katrina makes landfall on the morning of August 29, 2005.

© John Bazemore/AP Wide World

● FIGURE 15.23 Flood waters inundate New Orleans, Louisiana, in August 2005 after the winds and storm surge from Hurricane Katrina caused several levee breaks.

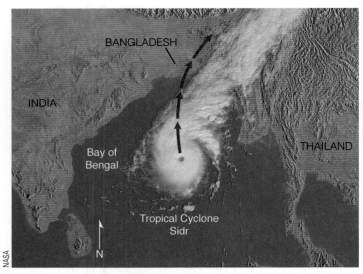

● FIGURE 15.24 Satellite image of Tropical Cyclone Sidr on November 14, 2007, as it moves northward over the Bay of Bengal toward Bangladesh. With winds of 135 knots and a central pressure estimated at 937 hPa, this strong Category 4 storm caused widespread destruction and the loss of many livestock and human lives when it made landfall on November 15, 2007. (The red dashed arrow shows Sidr's path.)

Hurricane Watches, Warnings, and Forecasts

With the aid of ship reports, satellites, radar, buoys, and reconnaissance aircraft, the location and intensity of hurricanes are pinpointed and their movements carefully monitored. When a storm is approaching Canada, the Meteorological Service of Canada's Canadian Hurricane Centre in Dartmouth, Nova Scotia, will issue an Information Statement, a Tropical Storm Watch, a **Hurricane Watch**, a Tropical Storm Warning, or a **Hurricane Warning**. Information Statements are issued when a tropical cyclone is expected to bring gale or storm-force winds (34 to 63 knots or 63 to 117 km h^{-1}) into Canadian territory or territorial waters within three days. Tropical Storm Watches and Hurricane Watches are advisories indicating that particular coastal or inland areas are at risk of experiencing a tropical storm or hurricane within 36 hours. Tropical Storm Warnings and Hurricane Warnings are issued when an area is expected to experience one of these storms within the next day.

The United States has a similar system of watches and warnings that are issued by the U.S. National Hurricane Center in Miami, Florida, or by the Pacific Hurricane Center in Honolulu, Hawaii. Along the U.S. East Coast, the warning is accompanied by a probability. The probability gives the percent chance of the hurricane's centre passing within 105 km of a particular community. The warning is designed to give residents ample time to secure property and, if necessary, to evacuate the area. Because hurricane-force winds can extend a considerable distance on either side of where the storm is expected to make landfall, a hurricane warning is issued for a rather large coastal area, usually about 550 km in length. Since the average swath of hurricane damage is normally about one-third this length, much of the area is "overwarned." As a consequence, many people in a warning area feel that they are needlessly forced to evacuate. The time it takes to complete an evacuation puts special emphasis on the timing and accuracy of the warning.

As a hurricane approaches land, will it intensify, maintain its strength, or weaken? Also, will it continue to move in the same direction and at the same speed? Such questions have challenged forecasters for some time. To forecast the intensity and movement of a hurricane, meteorologists use numerical weather prediction models, similar to those described in Chapter 13 on p. 388.

Information from satellites, buoys, and reconnaissance aircraft (that deploy dropsondes* into the eye of the storm) is fed into the models. The models then forecast the intensity and movement of the storm. There are a variety of forecast models, each one treating some aspect of the atmosphere (such as evaporation of water from the ocean's surface) in a slightly different manner. Often the models do not agree on where the storm will move and on how strong it will be.

The problem of different models forecasting different paths for the same hurricane has been addressed by using the method of *ensemble forecasting*. You may recall from Chapter 13 (p. 393) that an ensemble forecast is based on running several forecast models (or different simulations of the same model), each beginning with slightly different weather information. If the forecast models (or different versions of the same model) all agree that a hurricane will move in a particular direction, the forecaster will have confidence in making a forecast of the storm's movement. If, on the other hand, the models do not agree, then the forecaster will have to decide which model (or models) is most likely correct in forecasting the hurricane's track.

The use of ensemble forecasting along with better forecast models has helped raise the level of skill in forecasting hurricane paths. For example, in the 1970s, the projected position of a hurricane three days into the future was off by an average of 708 km. Today, the average error for the same forecast period has dropped to 278 km. Unfortunately, the forecasting of hurricane intensity has shown little improvement since the early 1990s.

To help predict hurricane intensity, forecasters have been using statistical models that compare the behaviour of the present storm with that of similar tropical storms in the past. The results using these models have not been encouraging. A more recent statistical model uses the depth of warm ocean water in front of the storm's path to predict the storm's intensity. Recall from an earlier discussion (p. 464) that if the reservoir of warm water ahead of the storm is relatively shallow, ocean waves generated by the hurricane's wind will mix deeper, cooler water to the surface. The cooler water will cut off the storm's energy source, and the hurricane will weaken. On the other hand, should a deep layer of warm water exist ahead of the hurricane, cooler water will not be brought to the surface, and the storm will either maintain its strength or intensify as long as other factors remain the same. So knowing the depth of warm surface water* ahead of the storm is important in predicting whether a hurricane will intensify or weaken. As new hurricane-prediction models with greater resolution are implemented, and as our understanding of the nature of hurricanes increases, forecasting hurricane intensification and movement should improve.

Modifying Hurricanes

Because of the potential destruction and loss of lives that hurricanes can inflict, attempts have been made to reduce their winds by seeding them with silver iodide. The idea is to seed the clouds just outside the eyewall with just enough artificial ice nuclei so that the latent heat given off will stimulate cloud growth in this area of the storm. These clouds, which grow at the expense of the eyewall thunderstorms, actually form a new eyewall farther away from the hurricane's centre. As the storm centre widens, its pressure gradient should weaken, which may cause its spiralling winds to decrease in speed.

During project STORMFURY, a joint effort of the U.S. National Oceanic and Atmospheric Administration (NOAA) and the U.S. Navy, several hurricanes were seeded by aircraft. In 1963, shortly after Hurricane Beulah was seeded with silver iodide, surface pressure in the eye began to rise and the region of maximum winds moved away from the storm's centre. Even more encouraging results were obtained from the multiple seeding of Hurricane Debbie in 1969. After one day of seeding,

---

*Dropsondes are instruments (radiosondes) that are dropped from reconnaissance aircraft into a storm. As the instrument descends, it measures and relays data on temperature, pressure, and humidity back to the aircraft. By tracking the dropsonde, data concerning wind speed and wind direction are also obtained.

---

*Sophisticated satellite instruments carefully measure ocean height, which is translated into ocean temperature beneath the sea surface. This information is then fed into the forecast models.

FOCUS ON AN ENVIRONMENTAL ISSUE

Hurricanes in a Warmer World

In Focus on an Observation: The Record-Setting Atlantic Hurricane Seasons of 2004 and 2005 (p. 478), we saw that 2005 was a record year for Atlantic hurricanes, with 27 named storms, 15 hurricanes, and five storms reaching Category 5 status on the Saffir–Simpson scale. Was the record hurricane year 2005 related to global warming?

We know that hurricanes are fuelled by warm tropical water—the warmer the water, the more fuel is available to drive the storm. A mere 0.6°C increase in sea surface temperature will increase the maximum winds of a hurricane by about 5 knots (9 km h^{-1}), everything else being equal. In May 2005, just before the hurricane season got under way, the surface water temperature over the tropical North Atlantic was considerably warmer than normal (see ⬤ Figure 6). Moreover, studies conducted by the National Center for Atmospheric Research (NCAR) in Boulder, Colorado, found that between June and October 2005, sea surface temperatures in the tropical Atlantic were about 0.9°C warmer than the long-time (1901–70) average for that region. The study concluded that about half of the warming (about 0.4°C) was due to global warming caused by increasing concentrations of greenhouse gases in the atmosphere. These findings suggest that global warming (described more completely in Chapter 16) may have had an effect on the intensity of some storms and on the number of tropical storms that formed because the ocean surface remained warmer than normal well past October. Interestingly, during the hurricane season of 2007, only one weak hurricane (Humberto) made landfall in the United States, but two storms (hurricanes Dean and Felix) reached Category 5 status over the warm Gulf of Mexico before weakening and making landfall south of the United States.

Climate models predict that as the world warms, sea surface temperatures in the tropics will rise by about 2°C by the end of this century. Should these projections prove correct, a hurricane forming in today's atmosphere with maximum sustained winds of 125 knots

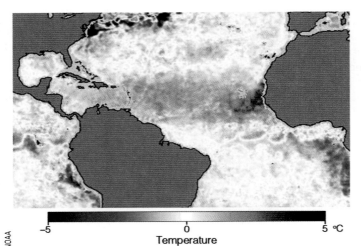

⬤ FIGURE 6 Sea surface temperature departures from the 12-year average (1985–97) on May 30, 2005. Notice that the darker the red, the warmer the surface water.

NOAA

(a Category 4 storm) could, in the warmer world, have maximum sustained winds of 140 knots (a Category 5 storm). As sea surface temperatures rise, will hurricanes become more frequent? Presently, there is no clear answer to this question as some climate models predict more hurricanes, whereas others predict fewer storms, Since Earth's surface is gradually warming, are today's hurricanes more intense than those of the past? Several studies suggest that the frequency of major hurricanes (Category 3 and above) has been increasing (see ⬤ Figure 7). The problem with these studies is that reliable records of tropical cyclones have been available only since the 1970s, when observations from satellites became more extensive. Sophisticated instruments today allow scientists to peer into hurricanes and examine their structure and winds with much greater clarity than in the past. Any trend in hurricane frequency or intensity will likely become clearer when more reliable information on past tropical cyclone activity becomes available.

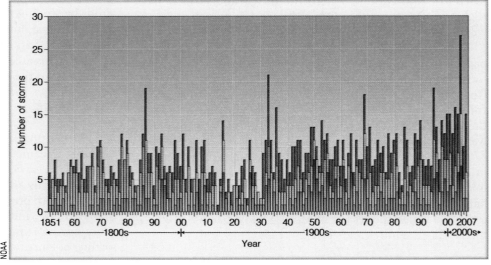

NOAA

⬤ FIGURE 7 The total number of tropical storms and hurricanes (red bars), hurricanes only (yellow bars), and Category 3 hurricanes or greater (green bars) in the Atlantic basin for the period 1851 through 2007.

Debbie showed a 30 percent reduction in maximum winds. But many hurricanes that are not seeded show this type of behaviour. So the question remains: Would the winds have lowered naturally had the storm not been seeded? Several studies even cast doubt on the theoretical basis for this kind of hurricane modification because hurricanes appear to contain too little supercooled water and too much natural ice. Consequently, there are many uncertainties about the effectiveness of seeding hurricanes in an attempt to reduce their winds, and all endeavours to modify hurricanes have been discontinued since the 1970s.

Other ideas have been proposed to weaken the winds of a hurricane. One idea is to place some form of oil (monomo-lecular film) on the water to retard the rate of evaporation and hence cut down on the release of latent heat inside the clouds. Some sailors, even in ancient times, would dump oil into the sea during stormy weather, claiming that it reduced the winds around the ship. At this point, it is interesting to note that a recent mathematical study suggests that ocean spray has an effect on the winds of a hurricane. Apparently, the tiny spray reduces the friction between the wind and the sea surface. Consequently, with the same pressure gradient, the more ocean spray, the higher the winds. If this idea proves correct, limiting ocean spray from entering the air above may reduce the storm's winds. Perhaps ancient sailors knew what they were doing after all.

SUMMARY

Hurricanes are tropical cyclones with winds that exceed 64 knots (119 km h^{-1}) and blow counterclockwise about their centres in the Northern Hemisphere. A hurricane consists of a mass of organized thunderstorms that spiral in toward the extreme low pressure of the storm's eye. The most intense thunderstorms, the heaviest rain, and the highest winds occur outside the eye, in the region known as the eyewall. In the eye itself, the air is warm, winds are light, and skies may be broken or overcast.

Hurricanes (and all tropical cyclones) are born over warm tropical waters where the air is humid, surface winds converge, and thunderstorms become organized in a region of weak upper-level winds. Surface convergence may occur along the ITCZ, on the eastern side of a tropical wave, or along a front that has moved into the tropics from higher latitudes. If the disturbance becomes more organized, it becomes a tropical depression. If central pressures drop and surface winds increase, the depression becomes a tropical storm. At this point, the storm is given a name. Some tropical storms continue to intensify into full-fledged hurricanes as long as they remain over warm water and are not disrupted by strong vertical wind shear.

The energy source that drives the hurricane comes primarily from the warm tropical oceans and from the release of latent heat. A hurricane is like a heat engine in that energy for the storm's growth is taken in at the surface in the form of sensible and latent heat, converted to kinetic energy in the form of winds, and then lost at the cloud tops through radiational cooling.

The easterly winds in the tropics usually steer hurricanes westward. In the Northern Hemisphere, most storms then gradually swing northwestward around the subtropical high. If the storm moves into middle latitudes, the prevailing westerlies steer it northeastward. Because hurricanes derive their energy from the warm surface water and from the latent heat of condensation, they tend to dissipate rapidly when they move over cold water or over a large mass of land, where surface friction causes their winds to decrease and flow into their centres. Sometimes hurricanes moving away from the topics may undergo an extratropical transition and become an intense middle-latitude cyclone or invigorate an existing middle-latitude cyclone. Many of the hurricanes affecting Canada have become extratropical cyclones.

Although the high winds of a hurricane can inflict a great deal of damage, it is usually the huge waves and the flooding associated with precipitation and the storm surge that cause the most destruction and loss of life. The Saffir–Simpson hurricane scale was developed to estimate the potential destruction that a hurricane can cause.

KEY TERMS

The following terms are listed (with page numbers) in the order they appear in the text. Define each. Doing so will aid you in reviewing the material covered in this chapter.

streamlines, 458
tropical wave (easterly wave), 458
hurricane, 459
typhoon, 459
tropical cyclone, 459
eye (of a hurricane), 459
eyewall, 459
trade wind inversion, 463
extratropical transition, 464

tropical depression, 464
tropical storm, 464
Saffir–Simpson scale, 470
wind shear, 470
storm surge, 472
supertyphoon, 472
eyewall replacement, 477
Hurricane Watch, 480
Hurricane Warning, 480

QUESTIONS FOR REVIEW

1. What is a tropical (easterly) wave? How do these waves generally move in the Northern Hemisphere? Are showers found on the eastern or western side of the wave?

2. Why are streamlines, rather than isobars, used on surface weather maps in the tropics?

3. What is the name given to a hurricane-like storm that forms over the western North Pacific Ocean?

4. Describe the horizontal and vertical structure of a hurricane.

5. Why are skies often clear or partly cloudy in a hurricane's eye?

6. What conditions at the surface and aloft are necessary for hurricane development?

7. List three "triggers" that help in the initial stage of hurricane development.

8. (a) Hurricanes are sometimes described as a heat engine. What is the "fuel" that drives the hurricane?

   (b) What determines the maximum strength (the highest winds) that the storm can achieve?

9. Would it be possible for a hurricane to form over land? Explain.

10. If a hurricane is moving westward at 10 knots, will the strongest winds be on its northern or southern side? Explain. If the same hurricane turns northward, will the strongest winds be on its eastern or western side?

11. What factors tend to weaken hurricanes?

12. Distinguish among a tropical depression, a tropical storm, and a hurricane.

13. In what ways is a hurricane different from a middle-latitude cyclone? In what ways are these two systems similar?

14. Why do most hurricanes move westward over tropical waters?

15. If the high winds of a hurricane are not responsible for inflicting the most damage, what is?

16. Most hurricane-related deaths are due to what?

17. Explain how a storm surge forms. How does it inflict damage in hurricane-prone areas?

18. Hurricanes are given names when the storm is in what stage of development?

19. When Hurricane Andrew moved over south Florida during August 1992, what caused the relatively small areas of extreme damage?

20. As Hurricane Katrina moved toward the Louisiana coast, it underwent eyewall replacement. What actually happened to the eyewall during this process?

21. How do meteorologists forecast the intensity and paths of hurricanes?

22. How does a Hurricane Watch differ from a Hurricane Warning?

23. Why have hurricanes been seeded with silver iodide?

24. Give two reasons why hurricanes are more likely to strike Nova Scotia than British Columbia.

QUESTIONS FOR THOUGHT

1. Why are North Atlantic hurricanes more apt to form in October than in May?

2. Would it be possible for a hurricane to form in the tropical North Atlantic or North Pacific during December? Explain.

3. Would the winds of a hurricane decrease more quickly as the storm moves over cooler water or over warmer land? Explain.

4. Explain why the ocean surface water temperature is usually cooler after the passage of a hurricane. (Hint: The answer is not because the hurricane extracts heat from the water.)

5. Suppose, in the North Atlantic, that an eastward-moving ocean vessel is directly in the path of a westward-moving hurricane. What would be the ship's wisest course—to veer to the north of the storm or to the south of the storm? Explain.

6. Suppose that this year five tropical storms develop into full-fledged hurricanes over the North Atlantic Ocean. Would the name of the third hurricane begin with the letter C? Explain.

7. You are in Darwin, Australia (on the north shore), and a hurricane approaches from the north. Where would the highest storm surge be, to the east or west? Explain.

8. Occasionally, when a hurricane moves inland, it will encounter a mountain range. Describe what will happen when this occurs. What will happen to the hurricane's intensity? What will cause the most damage (winds, storm surge, flooding)? Why?

9. Give several reasons how a hurricane that once began to weaken can strengthen.

10. Suppose that a tropical storm in the Gulf of Mexico moves westward across Central America and out over the Pacific. If the storm maintains tropical storm strength the entire time, do you feel it should be given a new name over the Pacific? Explain your reasoning.

PROBLEMS AND EXERCISES

1. A hurricane just off the coast of Maine is moving northeastward, parallel to the eastern seaboard. Suppose that you live in Nova Scotia along the east coast:

   (a) How will the surface winds in your area change direction as the hurricane's centre passes due east of you? Illustrate your answer by making a sketch of the hurricane's movement and the wind flow around it.

   (b) If the hurricane passes east of you, the strongest winds would most likely be blowing from which direction? Explain your answer. (Assume that the storm does not weaken as it moves northeastward.)

   (c) The lowest sea-level pressure would most likely occur with which wind direction? Explain.

2. Use the Saffir–Simpson hurricane scale (see Table 15.1, p. 472) and the text material to determine the category of Hurricane Elena in Figure 15.2 (p. 459).

The warming in the Arctic has caused sea ice to thin by between 15 and 40 percent over the past 30 years. If the warming in this region continues as predicted, sea ice during the summer months may be totally absent by the middle of this century.

Earth's Changing Climate

16

" **B** ut a broad array of scientists, including authors of the report and independent experts, said the latest analysis was the most sobering view yet of a century of transition—after thousands of years of relatively stable climate conditions—to a new norm of continual change."

The New York Times *on the Release of Intergovernmental Panel on Climate Change (IPCC) Fourth Assessment Report, February 2, 2007, Elisabeth Rosenthal and Andrew C. Revkin reporting.*

Elizabeth Rosenthal and Andrew C. Revkin. "Panel Issues Bleak Report on Climate Change." *The New York Times.* Feb. 2, 2007.

CONTENTS

Climate is the combination of long-term meteorological conditions and patterns that characterize an area and time period. It integrates the complex interactions of all Earth systems—the *hydrosphere* (*cryosphere*), *lithosphere*, *biosphere* (*anthrosphere*), and *atmosphere*. Varying any of these components can cause the climate to change.

Evidence shows that climate has changed many times in the past, and this continues today. At global scales, over geologic time, Earth has gone through multiple ice ages and warm periods, radically changing the entire globe. At regional scales, over much shorter times, the growth of urban environments changes the climate in and around our cities, so it differs from that of the surrounding regions. Sometimes these differences are striking—nights in the city are several degrees warmer than nights in the surrounding rural areas. Other times, the differences are more subtle—smoke and haze over a city change the area's radiative properties, affecting local temperatures, radiation values, visibility, and air quality. Microclimates can change at very small scales. Unexpected examples of this occur when the climate of a small area changes enough to affect the biological or seasonal behaviour of plants or animals. ● Figure 16.1 shows how the microclimate around a tree is modified by the light and warmth of a city street lamp.

Climate results from interactions among all of Earth's systems, which involve complex, intricate interrelationships between elements within each system and interrelationships between different Earth systems. Few, if any, elements are isolated from the others. This is why our picture of Earth's changing climate is incomplete. Since climate results from these complex interactions, let's take a closer look at systems and how they work.

Systems are driven by the flows of energy and matter from one location to another. These flows link systems together, allowing a change in one part to cause changes in another or throughout the whole system. Within complex systems, several interactions can function as a unit to continue the operation of the system. Interactions can cause positive or negative feedbacks. When a system is disturbed, negative feedbacks reduce the disturbance and stabilize the system, returning it to its predisturbance state. In contrast, positive feedbacks magnify the disturbance and destabilize the system.

An example of a positive feedback is the **water vapour–greenhouse effect feedback**, where an initial increase in air temperature sets in motion a series of processes that result in a further air temperature increase. In this case, higher temperatures cause more evaporation, which increases atmospheric humidity, causing more outgoing infrared energy to be absorbed by the humid air (the water vapour–greenhouse effect). In turn, this emits more infrared radiation to the surface, further increasing the air temperature (the positive feedback). We will revisit feedbacks and their influence on climate later in this chapter when we consider causes of climate change.

● FIGURE 16.1 Altering the microclimate. Notice in the picture that the leaves are still on the tree near the streetlight. Apparently, this sodium vapour lamp emits enough warmth and light during the night to trick the leaves into behaving as if it were September rather than the middle of November.

Climate change has global implications for all Earth systems and the linkages between them. It can fundamentally alter the flows of energy and moisture on Earth and change how they become distributed across the globe. Changes in the atmosphere's radiative properties affect the global absorption and emission of energy, which causes different patterns of heat and moisture circulation over both oceans and land. These changes have far-reaching impacts, and the complexity of feedbacks within and between systems is just beginning to be appreciated and better understood.

Warming ocean temperatures, salinity changes, and changes to ocean chemistry through acidification are also active areas of research. These changes (to the hydrosphere) have a profound impact on other physical and biological systems. For example, as atmospheric CO_2 levels rise, oceans absorb more CO_2, causing ocean water to become more acidic and reducing the ocean's ability to absorb atmospheric CO_2. More acidic water conditions cause already calcified ocean material, such as corals, to dissolve and inhibit the ability of calcifying marine organisms to produce their shells. Although not fully understood, these impacts already appear to be occurring with detrimental consequences for the marine biosphere.

Additionally, changes in the timing and amount of precipitation, wind, sunshine, or other weather elements affect all Earth's systems. Air temperature affects whether precipitation falls as snow or rain, and this changes how it moves through the hydrosphere. Falling as rain, it is not stored within the cryosphere, and this may limit water availability for later use. Presently, ice covers about 10 percent of Earth's land surface, has a volume of 26 million cubic kilometres, and represents 75 percent of Earth's stored fresh water. Most of it is contained in the ice sheets of Greenland and the Antarctic. If it all melted, sea levels would rise about 65 metres, causing severe impacts for the biosphere and anthrosphere. Much smaller impacts are currently seen around the globe as warmer temperatures are melting Earth's glaciers, ice sheets, and ice caps, and through the thermal expansion of water producing global sea-level rise. Based on the highest warming estimates, by the end of this century, a 4°C rise in global average air temperature would cause a sea-level rise of 43 cm. Although this seems minor, it will directly impact unprotected low-lying coastal areas, increasing the severity of impacts from storm surges. As human settlements tend to concentrate in these areas, this will increase mitigation and impact costs.

Higher sea levels will increase the impacts from extreme weather. Increasingly, extreme weather events are occurring with greater frequency and more severity. This is especially noticeable where the atmosphere affects the lithosphere and hydrosphere through flooding, storm surges, high winds, or more frequent extreme storms. These disrupt the landscape, changing erosion, landslide, and rock disintegration processes and rates.

These global-scale climate changes can affect all living things as they have evolved to suit past conditions and patterns. In the biosphere, the timing of seasonal events is important—coming too early, or too late, can alter biological life cycles, remove key food webs, change feeding patterns, and alter the composition of species in an area. Gradual changes can allow many species to successfully acclimatize to new conditions, but rapid change makes adaptation more difficult. Hence, the pace of climate change is also an important factor in the biosphere's and anthrosphere's ability to successfully adapt.

Climate change is occurring now at an alarming rate, and impacts are being observed globally. Scientific evidence indicates that anthropogenic activities combined with growing populations are the cause. Polar regions are showing the greatest temperature changes, with Arctic sea ice breaking up earlier, permafrost thawing, ice sheets melting, glaciers rapidly retreating, and sea levels rising worldwide. Extreme weather (drought, heat, storms, hurricanes, tornadoes), fires, and epidemic pest infestations such as the mountain pine beetle in western Canada are also more intense and more frequent. In this chapter, we will first examine the evidence for past climate change and then investigate the causes of climate change due to natural variations and human intervention.

Reconstructing Past Climates

Understanding previous climates and how they evolved helps decipher climate change. During the last 150 years, reasonably accurate weather measurements were regularly collected. Now they form an instrumental record that allows us to document weather elements and climate in the recent past. Reconstructing earlier climates involves more ingenuity. Within the period of recorded history, documentation of past droughts, floods, crop yields, rain, snow, dates for lakes freezing, and severe storms can yield important information showing how climate has changed. Where human records do not exist, scientists examine records contained in the environment that substitute, or act as a *proxy*, for instrumental records. Called *climate proxies*, they take many forms, but all contain a natural archive of past climatic events and some way of dating these events. The following list indicates the range and type of techniques that are used:

1. Ice cores, drilled from glaciers or ice sheets, can provide records of temperature; precipitation; lower atmospheric gas composition and chemistry (oxygen isotope* ratios, air trapped in the ice); volcanic eruptions; solar variability; and sea surface productivity. Ice features capture past climatic characteristics by accumulating layers of past precipitation over time. As the snow becomes buried, it compresses, recrystallizing as ice with air pockets from the past. Ice coring is an important technique providing detailed information on many properties.

2. Tree ring size, spacing, and patterns (dendroclimatology) indicate long, annually precise records of regional temperature and precipitation. By accurately matching ring patterns between trees and comparing them to those of other tree species, inferences about seasonal climate can be made.

3. Sediment cores record the amount, type, and deposition pattern of sediment that settles to the bottom of oceans and lakes. These can be used to interpret past ecological events and climate. For example, each layer's thickness indicates seasonality and climate variations. Sometimes the layers can be dated annually. Any pollen, vegetation, volcanic ash, or organisms trapped in layered patterns within the sediment are also indicative of certain climates or time periods. The depth of these features in the sediment also provides date information.

4. Pollen from ice cores, soil deposits, and ocean or lake sediments are indicators of past ecological communities and their climatic conditions.

5. Oxygen–isotope ratios incorporated in corals during their formation record past water temperatures and salinity. Similarly, otoliths, the tiny "ear stones" formed

---

*Isotopes are atoms whose nuclei have the same number of protons but different numbers of neutrons. Isotope ratios become preserved in some natural archives and often act as markers of different climatic and environmental periods.

by calcium carbonate deposition in the inner ears of almost every species of fish, also contain oxygen–isotopes that indicate water temperature.

6. Calcium carbonate layers in stalactites found in caves can be dated, and various isotopes in their chemistry can be related to past precipitation and temperature.

7. Borehole temperature profiles can be related to past surface temperatures because they record temperature signals as they propagate downward into the ground.

8. Glacial deposits and their formations provide clear evidence of past climate. The type of sediment (till, erratic, etc.), its location, and its arrangement on the landscape indicate the glacial history. Incorporated biological material can also provide dates. Combining these types of data allows for detailed climatic reconstruction.

9. Other geologic evidence also yields climate information. Particular rock types, coal beds, wind-blown dust or loess deposits, sand dunes, fossils, the presence of volcanic ash layers, impacts from large fires in the past, and evidence of water level changes, especially in closed basin lakes, can be used to interpret past climates.

Radiocarbon dating and other dating techniques determine reference dates that tie a climate proxy record to a timeline and provide the basis for creating a chronology (an ordered sequence of events). For many proxy records, a chronology is based on the concept that younger material is superimposed on older material. Additionally, there must be at least one independently dated property in the record. This allows us to understand how long ago an event occurred. The depth of any property in the sequence can then be dated by its distance from the reference property.

Often these reference properties are small amounts of organic material that can provide a radiocarbon date. Now let's look at a few of these natural archives more closely.

A wealth of information comes from ice cores. Extracted with hollow-centred drills that penetrate into deep expanses of ice, cores come from ice sheets in Antarctica, Greenland, other high-latitude ice caps, and mountain ice fields such as the Yukon's Mount Logan or British Columbia's Mount Waddington. Where temperatures are low enough, over an entire year, more snow falls than melts, and snow accumulation over successive years compacts and slowly transforms into ice. Trapped bubbles of ancient air can be analyzed to determine the composition of the atmosphere in the past (see ● Figure 16.2). Ice also preserves the ratio of oxygen–isotopes that were present at the time the precipitation fell.

These oxygen–isotope ratios indicate whether conditions were colder or warmer. Here is how this works. Most of the oxygen atoms in water are composed of eight protons and eight neutrons, giving oxygen an atomic weight of 16. However, about one out of every thousand oxygen atoms contains an extra two neutrons, giving this oxygen an atomic weight of 18. Water molecules that have "heavy" oxygen 18 atoms condense more easily and at a higher temperature than water composed of oxygen 16 atoms. During cool periods, more of the water vapour containing oxygen 18 tends to condense and precipitate out of the air. It deposits as rain or snow shortly after it evaporates, before reaching the location of the ice. This depletes oxygen 18 from the water vapour, so the ratio of oxygen 18 to oxygen 16 in the precipitation (and, consequently, in the ice core) is low in cool conditions. Conversely, during warm periods, more water vapour containing

● FIGURE 16.2 Variations in temperature (red line, °C), carbon dioxide (black line, ppmv = parts per million by volume), and methane (blue line, ppbv = parts per billion by volume). Data are derived from Antarctic ice cores: gas concentrations from air bubbles trapped within the ice sheets, temperatures from the analysis of oxygen–isotopes.

Source: From *Climate Change 2001: The Scientific Basis*, 2001, by J. T. Houghton, et al. Copyright © 2001 Cambridge University Press. Reprinted by permission of the Intergovernmental Panel on Climate Change.

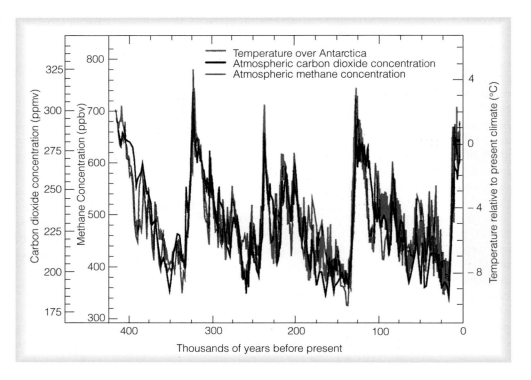

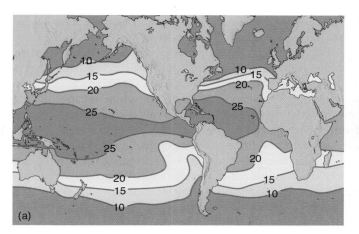
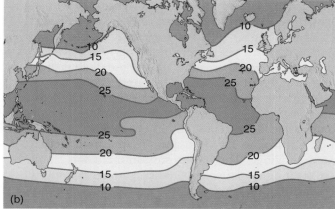

● FIGURE 16.3 (a) August sea surface isotherms (°C) during the last glacial period 18,000 years ago. (b) August sea surface isotherms today. During the last glacial period (a), the Gulf Stream moved more east to west, depriving northern Europe of its warmth and causing a greater north–south ocean surface temperature gradient.

oxygen 18 remains in the air. It can condense and precipitate at the ice core location. Consequently, the ratio of oxygen 18 to oxygen 16 is high in warm conditions. Therefore, examining oxygen–isotope ratios at different depths in an ice core provides an archive of past temperatures.

Ice cores also allow us to decipher climate change causes. For example, layers of sulphuric acid that become trapped in ice can be indicators of volcanic explosions, which cause cooling. Certain eruptions naturally eject huge amounts of sulphur dioxide into the stratosphere. These transform into sulphuric acid—acid snow—and fall to Earth, becoming preserved in the ice sheets. Through their continuous record, the Greenland ice cores show that humans have also caused sulphuric acid to be trapped in ice. The beryllium isotope (^{10}Be), also found in ice cores, indicates fluctuating solar activity as another cause of climate change. Also, various types of dust trapped in the ice indicate arid or wet climatic conditions.

Dendrochronology, or tree ring dating, provides precise regional climate information. It is most useful where trees experience annual growth cycles that are susceptible to temperature stresses or moisture deficiencies during the growing season. As a tree grows, it produces a layer of wood cells under its bark and forms an annual growth ring. The thickness of each ring indicates year-to-year growing conditions. Moisture stress is indicated by the density of the late wood part of the tree ring (wood grown in summer to fall, when generally moisture is less). The presence of frost rings (dark disorganized cells within the ring) indicates short cold periods. In various regions of the world, tree rings from long-dead trees have been correlated to decipher precipitation and temperature patterns extending hundreds, sometimes thousands, of years into the past. The chemistry of the wood itself provides additional information about a changing climate. Additionally, the wood can also be dated using radiocarbon dating techniques, which provide reference dates for tree chronologies.

Ocean sediments contain the remains of calcium carbonate shells belonging to organisms that once lived near the ocean surface. Because certain organisms can live within only a narrow range of temperature, the distribution and types of organisms, as well as the oxygen–isotope ratio incorporated into the calcium carbonate in their shells, indicate the temperature of the water at the time those organisms lived.

Additionally, the oxygen–isotope ratio of these shells indicates the extent of glaciation. When ocean water evaporates, molecules containing the heavy oxygen 18 isotope tend to be left in liquid form. Consequently, during periods of glacial advance, the oceans contained less water but had a higher concentration of oxygen 18. As marine shells are constructed from the oxygen atoms in the ocean water, their ratio of oxygen 18 to oxygen 16 indicates water levels that can be related to the amount of ice. A higher ratio of oxygen 18 to oxygen 16 in the shells suggests less ocean water, more ice, and a colder climate. A lower ratio suggests more ocean water, less ice, and a warmer climate. By viewing the oxygen–isotope ratios in the entire sediment core, a long-term picture of Earth's ocean surface temperature and the extent of glaciation for various times in the past can be reconstructed (see ● Figure 16.3).

All of these climate reconstruction techniques still give an incomplete and sometimes confusing picture of past climates. So let's see what they can reveal.

Prehistoric Climates

Throughout most of Earth's history, the global average temperature was somewhere between 8°C and 15°C warmer than it is today. During much of this time, the polar regions were ice-free. However, these warm conditions were interrupted by several extended periods of glaciation known as **ice ages**. Geologic evidence suggests that Earth's major glaciations occurred roughly 2200 million years ago (mya), 700 mya, 300 mya, and 2.5 mya.

Reconstructing ancient climatic conditions has many challenges, uncertainties, and debates. These issues expand as we attempt to decipher older time periods. Consequently, those involved in this research painstakingly seek evidence from multiple sources in attempts to clarify data interpretations. The most recent glaciation is best understood. It dominated the **Pleistocene** epoch and continues into today. Before we examine it, let's briefly describe the climatic conditions that led up to the Pleistocene.

About 65 mya, as the dinosaurs disappeared, tectonic plates were moving.* The Indian subcontinent was sliding into Asia, creating the Himalayas; other continents were still moving apart but held positions that were close to the locations they occupy today; and Earth was warmer than it is now. Polar ice caps did not exist.

Beginning 55 mya, Earth entered a long cooling trend. It took millions of years for polar ice to appear, but the ice grew thicker as the average global temperature continued to fall. By about 10 mya, a deep blanket of ice covered the Antarctic. In the Northern Hemisphere, at the same time, snow began to accumulate and turn into ice in the high mountain valleys.

At 2.58 mya, continental glaciers appeared in the Northern Hemisphere, marking the beginning of the Pleistocene, which is commonly called the *Ice Age*. But the Pleistocene was not a period of continuous glaciation. Instead, ice advanced (grew) and retreated (melted back) multiple times over large portions of North America and Europe. The warmer intervals between glacial periods lasted 10,000 years or more and are

---

*Plate tectonics and how it affects climate are discussed later in this chapter. See page 498.

called **interglacials**. During interglacials, the average global temperatures were slightly higher than at present.

From a variety of data, Earth scientists estimate that, globally, 13 major and eight minor glacial advances and retreats occurred during the last 2.58 million years. This is an area of active research as certainty when estimating the number and extent of glacial intervals is challenging. Signals from earlier glacial activity are usually obliterated by later glaciations. Also, different types of geologic evidence do not always correlate well, making differentiating the timing and extent of some glacial events difficult. We are still in an ice age, but in a comparatively warmer interglacial, where glaciers have retreated. However, only 18,000 years ago (ya), Earth was at the peak of the last glaciation.

North American glaciers reached their maximum thickness and extent about 18,000 to 22,000 ya. At that time, average temperatures in Greenland were about 10°C lower than they are today, and tropical average temperatures were about 4°C lower. In mountainous areas, alpine glaciers extended their icy fingers down river valleys and coalesced, whereas over flatter terrain, ice caps grew to form larger ice sheets. These all joined and formed huge continental ice sheets. As shown in ● Figure 16.4, ice covered large portions of the land in the Northern Hemisphere. Most of northern Europe, almost all of Canada, the northern United States, and southern mountainous regions were under ice. In North America at that time, ice was up to 2 km thick. It covered 97 percent of Canada and extended beyond the border into the northern United States from coast to coast. At its southern extent, ice reached the Ohio River Valley (south of the Great Lakes, at about 40°N), covering New York State, Connecticut,

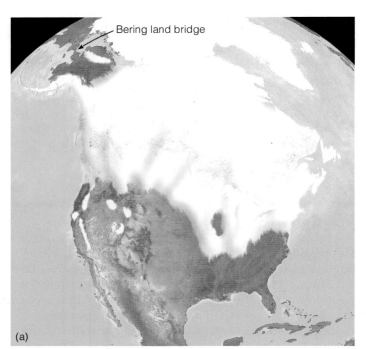

● FIGURE 16.4 Extent of glaciation about 18,000 years ago over (a) North America and (b) western Europe.

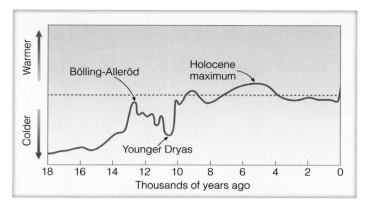

● FIGURE 16.5 Relative air temperature variations (warmer and cooler periods) during the past 18,000 years. Compiled from a variety of proxy sources, these data give only an approximation of temperature changes. In some regions, the timing of these changes may vary somewhat from what is shown in the diagram.

and everything north of this on the eastern coast. Small parts of the Yukon, the nonalpine United States, and most of Alaska remained ice-free. A great deal of water was held in these continental ice sheets, and sea level was estimated to be 120 m lower than it is now. Vast areas of land that are underwater today were exposed. The *Bering land bridge* (shown in Figure 16.4a) is a key example. Its strip of land, connecting Siberia to Alaska, allowed human and animal migration from Asia to North America.

The ice began to retreat about 14,000 ya as surface temperatures slowly rose, producing a warm spell called the *Bölling–Alleröd* period (see ● Figure 16.5). Then, about 12,700 ya, the average temperature suddenly dropped. Northeastern North America and northern Europe reverted back to glacial conditions—a period known as the **Younger Dryas**.* About 1000 years later, the Younger Dryas cold spell ended abruptly as temperatures rose rapidly in many areas. This marks the beginning of the current interglacial called the **Holocene** epoch.

Then, beginning about 8000 ya, a cooling, which was not experienced worldwide, caused the mean temperature to drop by as much as 2°C over Central Europe. Evidence from Europe's mountainous areas indicates that the **timber line** fell about 200 m. As this cold period ended, it was followed by a warming trend that melted North America's continental ice sheets and peaked by about 6000 ya. Figure 16.5 shows how this warm interlude relates to the current interglacial. Because it favoured the development of plants, it is known as the **mid-Holocene maximum**, or *climatic optimum*. About 5000 ya, a slightly cooler than average period started, causing the return of extensive alpine glaciers but no continental ones. This continued almost until today, but as shown in the figure, within the last few hundred years, there has been a return to warmer temperatures. In the next sections, we will look more closely at this period and the last 1000 years.

How fast can climate change? This is also an active area of research. Interestingly, Greenland's ice core data reveal that rapid climatic shifts have occurred. Over central Greenland, around the end of the Younger Dryas, conditions varied from ice age to a much warmer state in as little as three years. Data also show that similar rapid climatic shifts occurred several times toward the end of the Ice Age. What causes such rapid changes in temperature? One possible explanation is given in the Focus on a Special Topic: The Ocean Conveyer Belt and Climate Change on p. 494.

Climate During the Past 1000 Years

Analysis of the available global climate data for the last 1000 years indicates that temperature changes and trends varied in different regions of the world. Both the types of changes and their timing varied, but over the entire period, the range of temperature variation is within a degree of the 1961–90 average global temperature. This makes determining global- or large-scale climate differences or trends difficult.

Climate reconstructions for the last millennium come from a variety of climate proxy records combined with the last 150 years of temperature measurements. They are dominated by proxies for the Northern Hemisphere's mid- to high latitudes, with many of these proxies best suited to deciphering summer conditions. These issues pose some challenges and limitations for distinguishing global changes from larger scale phenomena that affect the proxy record (e.g., El Niño or Pacific Decadal Oscillation [PDO]). Consequently, the most conclusive climate change analyses relate to the Northern Hemisphere and focus on European impacts. For these reasons, research is actively focused on improving proxy records worldwide. It is important to understand these issues when interpreting ● Figure 16.6 and reading this section.

Assembled by different research groups, Figure 16.6 shows 12 different temperature reconstructions for the past 1300 years. In addition to thermometer records, multiple climate proxies (tree rings, pollen, borehole records, sediment cores, corals, ice cores, glacier extents, historical records) are pooled for each reconstruction. As a group, they show how the Northern Hemisphere's average surface air temperature during the last 1300 years compares to the average surface air temperature between 1961 and 1990.* For example, temperatures similar to the 1961–90 average are the same as the average and have a 0°C temperature anomaly. Those that are cooler will have negative temperature anomalies, whereas those that are warmer will be positive. Each temperature reconstruction uses somewhat different data and reconstruction techniques, so a greater degree of overlap implies more confidence that the data represent actual temperature changes.

---

*This exceptionally cold spell is named after the *Dryas*, an arctic flower.

*This is more formally called a *temperature anomaly* with respect to (wrt) the 1961–90 global average surface air temperature.

FOCUS ON A SPECIAL TOPIC

The Ocean Conveyor Belt and Climate Change

During the last glacial period, the climate around Greenland (and probably other areas of the world, such as northern Europe) underwent shifts from ice age temperatures to much warmer conditions in a matter of years. What could bring about such large fluctuations in temperature over such a short period of time? It now appears that a vast circulation of ocean water, known as the conveyor belt, plays a major role in the climate picture.

● Figure 1 illustrates the movement of the ocean conveyor belt, or *thermohaline circulation*.* The conveyor-like circulation begins in the North Atlantic near Greenland and Iceland, where salty surface water is cooled through contact with cold Arctic air masses. The cold, dense water sinks and flows southward through the deep Atlantic Ocean, around Africa, and into the Indian and Pacific oceans. In the North Atlantic, the sinking of cold water draws warm water northward from lower latitudes. As this water flows northward, evaporation increases the water's salinity (dissolved salt content) and density. When this salty, dense water reaches the far regions of the North Atlantic, it gradually sinks to great depths. This warm part of the conveyor delivers an incredible amount of tropical heat to the northern Atlantic. During the winter, this heat is transferred to the overlying atmosphere, and evaporation moistens the air. Strong westerly winds then carry this warmth and moisture into northern and western Europe, where it causes winters to be much warmer and wetter than would normally be expected for this latitude.

Ocean sediment records, along with ice core records from Greenland, suggest that the

*Thermohaline circulations are ocean circulations produced by differences in temperature and/or salinity. Changes in ocean water temperature or salinity create changes in water density.

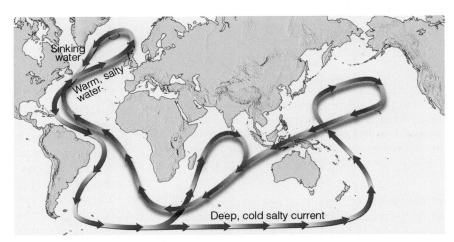

● **FIGURE 1** The ocean conveyor belt. In the North Atlantic, cold, salty water sinks, drawing warm water northward from lower latitudes. The warm water provides warmth and moisture for the air above, which is then swept into northern Europe by westerly winds that keep the climate of that region milder than one would normally expect. When the conveyor belt stops, winters apparently turn much colder over northern Europe.

giant conveyor belt has switched on and off during the last glacial period. Such events have apparently coincided with rapid changes in climate. For example, when the conveyor belt is strong, winters in northern Europe tend to be wet and relatively mild. However, when the conveyor belt is weak or stops altogether, winters in northern Europe appear to turn much colder. This switching from a period of milder winters to one of severe cold shows up many times in the climate record. One such event—the Younger Dryas—illustrates how quickly climate can change and how western and northern Europe's climate can cool within a matter of decades and then quickly return to milder conditions.

One mechanism that is thought to switch the conveyor belt off is a massive influx of fresh water. For example, recent research by Richard Peltier at the University of Toronto has confirmed that about 11,000 years ago during the Younger Dryas event, fresh water from a huge

glacial lake began to flow out the MacKenzie River and into the Arctic and then the North Atlantic ocean. This massive inflow of fresh water reduced the salinity (and, hence, density) of the surface water to the point that it stopped sinking. The conveyor shut down for about 1000 years, during which time severe cold engulfed much of northern Europe. The conveyor belt started up again when fresh water began to drain down the Mississippi rather than into the North Atlantic. It was during this time that milder conditions returned to northern Europe.

Will increasing levels of CO_2 have an effect on the conveyor belt? Some climate models predict that as CO_2 levels increase, more precipitation will fall over the North Atlantic. This situation reduces the density of the sea water and slows down the conveyor belt. In fact, if CO_2 levels double (from its current value), computer models predict that the conveyor belt will slow and that Europe will not warm as much as the rest of the world.

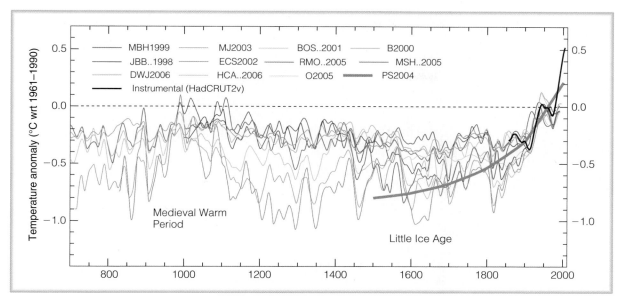

• FIGURE 16.6 Twelve different climate reconstructions showing the variation in the Northern Hemisphere's average temperature for the last 1300 years compared to the 1961–90 average (dashed line). Values below zero indicate that the temperature is cooler than the 1961–90 average. The instrumental (thermometer) record is shown in black. Reconstructions are based on a combination of climate proxies (tree rings, pollen, borehole records, sediment cores, corals, ice cores, glacier extents, and historical records).

Source: *Climate Change 2007: The Physical Science Basis.* Working Group I Contribution to the Fourth Assessment Report of the Intergovernmental Panel on Climate Change, Fig. 6.10 (b). Cambridge University Press.

Roughly 1000 ya, during the *Medieval Warm Period* (or *Medieval Climatic Optimum*), the Northern Hemisphere was slightly warmer than it has been until recently (see Figure 16.6). Even within Europe, some regions were warmer than others. Globally, this period was variable, with some areas warming, others cooling, and some exhibiting moisture changes (floods or droughts) rather than temperature changes. Sparse data from outside North America and Europe make global analysis inconclusive. Depending on the location, the timing and extent of the events varied.

Over western Europe, this warmer, tranquil period saw vineyards flourish and even England produce wine, suggesting warm springs and warm, dry summers. These conditions aided the Vikings in colonizing Iceland and Greenland and in exploring North America for settlement. Then this gave way to a slightly cooler, more variable period around 1150 A.D. Extremely cold winters were followed by relatively warm ones. Vineyards failed. Floods, droughts, and major

storms were documented. Europe experienced several famines in the 1300s. Around 1600 A.D., a deeper cooling began. In many European areas, winters were long and severe; summers were short and wet. Alpine glaciers expanded. England's vineyards vanished, and farming became impossible in the more northern latitudes. The advancing ice pack isolated the Viking colony in Greenland, which collapsed. Although there is no evidence that this cold spell existed worldwide, it has come to be known as the **Little Ice Age**.

Temperature Trends from Measurements

In the 1900s, the average global surface temperature began to rise. As shown in • Figure 16.7, between about 1910 and 1945, the average temperature rose nearly 0.5°C. Temperatures fluctuated for roughly the next 25 years, during which several warmer years alternated with several cooler ones; but the overall result was a very slight cooling. The late 1970s saw the start of a strong warming trend that continues today.

Attempts to explain these trends have considered natural oscillations such as the Pacific Decadal Oscillation (PDO) and El Niño patterns* as these are visible within the temperature trends of the last 100 years. There are periods when these natural climatic cycles seem to match observed temperature

WEATHER WATCH

In 1728, an Inuk in a kayak "discovered" Scotland when he made landfall near Aberdeen. It is thought that a stronger than normal thermal high over Greenland during the Little Ice Age could have resulted in northwesterly winds that carried him from Greenland to Scotland. A larger than normal number of ice floes could have provided him with drinking water as well as seals for food, enabling him to survive the voyage. Three other known Inuit landings in Scotland and England were documented during the Little Ice Age between 1613 and 1728.

*Pacific Decadal Oscillation (PDO) and El Niño are discussed in Chapter 10 (see pp. 310–315).

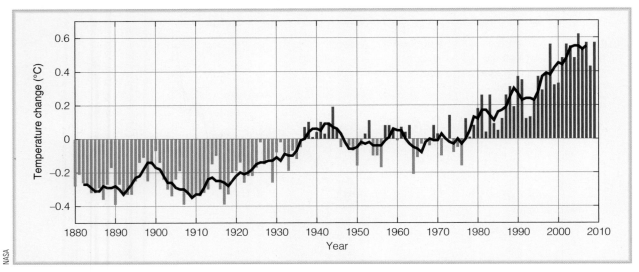

NASA

● **FIGURE 16.7** The red and blue bars represent the variation of the global annual average temperature from the 1951–80 average surface temperature (for the period 1880 through 2009). Red bars are positive values, which indicate that the global annual average temperature is warmer than the 1951–80 average. Blue bars are negative values, which indicate that the global annual average temperature is cooler than the 1951–80 average. The dark solid line shows the temperature trend.

trends, but, overall, these oscillations do not account for the observed temperature increases.

For global annual average temperatures, the years between 2000 and 2009 are the warmest on record (see Figure 16.6). The decade from 1990 to 2000 is the next warmest, with 1998 and 2005 being the warmest in over 1000 years.* Additionally, the 13 years between 1997 and 2009 are among the 14 warmest in the instrumental record. The increase in average temperature experienced over the Northern Hemisphere during the 20th century is greater than that seen during the Medieval Warm Period and is the largest increase in temperature of any century during at least the past 1300 years.

The changes in air temperature shown in Figure 16.7 are derived from three main sources: air temperatures over land, air temperatures over ocean, and sea surface temperatures. Over this 130-year record, some data control issues have developed and need to be considered, such as recording stations being moved and measuring techniques having evolved. Marine observing stations are scarce, making it more difficult to characterize temperatures over oceans. Urbanization tends to expand around the locations of the earlier measuring stations and tends to artificially raise average temperatures, especially in developed nations because of the urban heat island effect. When urban warming is taken into account and improved sea surface temperature information is incorporated into the data, the warming during the century from 1906 to 2005 was 0.74°C. Over the past 50 years, this **global warming** trend has not only continued but has also increased to about 1.3°C per century. A global increase in temperature of 0.74°C may seem small, but reconstructed global temperatures have

not varied by more than 2°C during the past 10,000 years. Consequently, an increase of 0.74°C becomes significant.

Although referred to as a global average warming, it is not uniformly distributed around the world. The greatest warming has occurred in the Arctic and over the mid-latitude continents in winter and spring. Moreover, most of the warming has occurred at night—a situation that has lengthened the frost-free seasons in many mid- and high-latitude regions, although, since the 1980s, the warming has been more equally distributed between day and night. Conversely, a few areas have not warmed in recent decades, such as areas of the oceans in the Southern Hemisphere and parts of Antarctica.

As we will see later in this chapter, most of the recent warming is due to an enhanced greenhouse effect caused by increasing levels of greenhouse gases, such as CO_2.* These are among the questions we will address in the following sections. Now let's consider the possible causes for changing climates and what might be responsible for the most recent changes.

BRIEF REVIEW

Before going on to the next section, here is a brief review of some of the facts and concepts we have covered so far:

● Earth's climate is constantly undergoing change. Evidence suggests that throughout much of Earth's history, the Earth's climate was much warmer than it is today.

---

*The exceptionally warm year of 1998 happened to coincide with a major El Niño warming of the tropical Pacific Ocean, whereas the warm year of 2005 did not.

*Earth's atmospheric greenhouse effect is due mainly to the absorption and emission of infrared radiation by gases, such as water vapour, CO_2, methane, nitrous oxide, and chlorofluorocarbons. Refer back to Chapter 2 for additional information on this topic.

- The most recent glacial period (or Ice Age) was in effect about 2.58 million years ago. During this time, glacial advances were interrupted by warmer periods called interglacial periods. In North America, continental glaciers reached their maximum thickness and extent about 18,000 to 22,000 years ago and disappeared completely from North America by about 6000 years ago.

- The Younger Dryas event represents a time about 12,000 years ago when northeastern North America and northern Europe reverted back to glacier conditions.

- During the century from 1906 to 2005, Earth's surface temperature increased by about 0.74°C. This global warming has not only continued but over the last five decades has increased to about 1.3°C per century.

Possible Causes of Climate Change

Why does Earth's climate change? There are three imposed or *external causes* that control climate from outside the climate system. As expected, natural phenomena can change climate through all of these mechanisms; however, human activities can impact the last two. Furthermore, each *external cause* interacts with the *internal causes* of climate change—altered ocean and air circulation—to redistribute heat and energy within the climate system and link with other Earth systems. We experience internal causes as altered weather patterns and ocean currents. The external causes of climate change are

1. Incoming solar radiation, which controls the energy entering the system and affects the amount of energy available to Earth's systems. Changes here affect the energy flow that drives ocean and atmospheric circulation.

2. Composition of the atmosphere, which controls how shortwave and longwave radiation pass through the atmosphere. This determines the amount and type of energy available at Earth's surface. Changes here affect the form and amount of energy; where it is absorbed, emitted, and transformed; and ocean and atmospheric circulation.

3. Earth's surface, which encompasses the existence, distribution, and location of landmasses; their elevation; and their surface cover. Earth's surface connects to its geothermal subsurface—energy from Earth's interior—a nonradiative energy source. Volcanoes are an astounding reminder of this link. Briefly, the type of surface (water, ice, rock, vegetation) affects the albedo, which determines how much solar energy enters the surface, affecting heat distribution and circulation patterns, even on a global scale. How landmasses are distributed on Earth and their elevation change ocean currents and winds. Landmass latitude determines whether the climate is polar or tropical. Mountains have a similar climatic impact. As we shall see, over geologic time, these factors have had an enormous effect on Earth's climate.

Acting together, the external and internal causes produce a changed climate. As discussed earlier, the processes, intricate interrelationships, and vast number of possible interactions between climate elements make determining the details of climate change impacts extremely complex. No climatic element within the system is isolated from the others; this is why our understanding is incomplete. With this caveat, we will now consider key causes of climate change and what they tell us.

CLIMATE CHANGE: FEEDBACK MECHANISMS   In Chapter 2, we learned that Earth and its atmosphere interact in a delicate balance between incoming and outgoing energy. When this balance is disturbed, the climate can undergo a series of changes called *feedbacks*. As we learned at the start of this chapter, feedback mechanisms can describe interactions that determine how elements of climate behave when the balance is upset. Recall that **positive feedback mechanisms** magnify a disturbance and destabilize the system, whereas **negative feedback mechanisms** reduce a disturbance by returning the system toward its predisturbance condition. At any time, many feedbacks act simultaneously within a system.

The example discussed at the beginning of this chapter, the water vapour–greenhouse effect feedback,[*] is a positive feedback. It magnifies an initial temperature increase, further destabilizing the system. If this feedback was the only operating process, Earth's temperature would increase until the oceans evaporated away. Such a chain reaction is called a *runaway greenhouse effect*. This may have occurred on our neighbouring planet, Venus. Conversely, if temperature is decreasing, this positive feedback amplifies the cooling as well.

A runaway greenhouse effect is unlikely on Earth because strong negative feedbacks, such as the *temperature–infrared radiation feedback*, tend to stabilize it. Here is how this works. As the planet warms, Earth emits more infrared radiation[†] out of the system, to space. Losing more infrared radiation limits the rise in temperature and stabilizes the climate. This is the strongest negative feedback in the climate system. It works in concert with other negative feedbacks and diminishes the possibility of a runaway greenhouse effect.

Another climate-changing positive feedback is the **snow–albedo feedback**, where increasing air temperature melts away polar snow and ice cover more quickly, lowering the surface albedo (reflectivity), so it absorbs more solar energy. This further raises the temperature, causing more previously unaffected areas to melt (see ● Figure 16.8). This feedback is significant in polar latitudes, where increasing global average temperatures are escalating the loss of snow and ice cover. The snow–albedo feedback explains why both the observed

---

[*]Remember, in this positive feedback that increasing air temperature causes more evaporation, which increases the air's humidity. Humid air absorbs more outgoing infrared energy and emits more of it back to the surface, causing an additional air temperature increase.

[†]A lot more! Recall in Chapter 2 (p. 42) that the Stefan–Boltzmann law indicates that the outgoing infrared radiation from a surface increases by the fourth power of the surface's temperature. Consequently, doubling Earth's surface temperature results in 16 times more emitted energy.

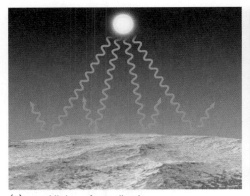

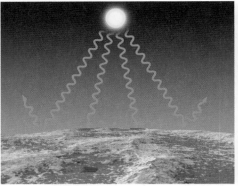

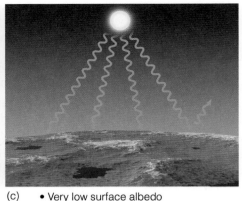

(a)
- High surface albedo
- Low absorption of sunlight
- Gradual surface warming

(b)
- Lower surface albedo
- Higher absorption of sunlight
- Surface warming increases

(c)
- Very low surface albedo
- Much higher absorption of sunlight
- Surface warming enhanced

● FIGURE 16.8 On a warming planet, the snow–albedo positive feedback would enhance the warming. (a) In polar regions, snow reflects much of the sun's energy back to space. (b) If the air temperature were to gradually increase, some of the snow would melt, less sunlight would be reflected, and more would reach the ground, warming it more quickly. (c) The warm surface would enhance the snow melt, which, in turn, would accelerate the rise in temperature.

recent warming and predicted future warming are higher in polar latitudes than elsewhere on Earth. This can also become a runaway feedback that ends only when all the permanent polar ice has melted. This has occurred during Earth's past and a lack of polar ice is the most common climate configuration over geologic time.

However, the snow–albedo feedback also generates a positive feedback on a cooling planet. Suppose that Earth were in a slow cooling trend. Lower temperatures would cause greater snow cover in middle and high latitudes, and these would increase the albedo, so more incoming sunlight would be reflected back to space. The resulting lower temperatures could further increase snow cover, causing even more air temperature reduction. If left unchecked, this could produce a runaway ice age, which many geologists believe likely happened 2.4 billion and 600 million ya.

A negative feedback that acts to stabilize Earth's climate on geologic time scales is the *chemical weathering–CO_2 feedback*. It works as follows. Under high atmospheric CO_2 conditions, the air is warm, with a lot of evaporation and precipitation. Rocks will more rapidly chemically weather because weathering rates increase with both temperature and moisture. Since the chemical weathering of certain rocks removes CO_2 from the air and eventually deposits the carbon as calcium carbonate minerals in ocean sediments, this will reduce the air's CO_2 concentration, which will cool and stabilize Earth's temperature. As the CO_2 and temperature fall, the chemical weathering decreases, and CO_2 removal from the air diminishes.

There are many other feedback mechanisms that link Earth systems and characterize the climate. These can either counteract or amplify climate change tendencies. Although we do not foresee runaway greenhouse effects in the near future, large positive feedbacks may be causing accelerated melting of ice in polar regions, especially in Greenland.

CLIMATE CHANGE: PLATE TECTONICS AND MOUNTAIN BUILDING Earlier in this chapter, we identified Earth's surface as an external cause of climate change. Let's examine this in more detail.

Our planet's surface is always evolving. Slowly shifting landmasses and changing ocean floors shape and redistribute Earth's continents in a process known as **plate tectonics**. According to the theory describing this motion, Earth's crust is composed of two kinds of multiple, huge plates that fit together like pieces of a global jigsaw puzzle. These plates slide over a partially molten zone under the crust and move in relation to each other. As they slide, the lighter continental plates float higher than the denser oceanic plates, forming various land–ocean configurations. They move extremely slowly, only a few centimetres per year, but over geologic time, they change Earth dramatically.

Tectonic forces help explain climate changes that occur over tens of millions to billions of years. Rocks that preserve indicators of past climates, as well as landscape features, fossils, and pockets of ancient tropical soils, all provide evidence that Earth's surface has changed radically in the last 2 billion years. On several occasions, Earth's landmasses have combined into a single supercontinent that later split apart into different land configurations. These periods of global land redistribution correspond to very different climatic, hydrological, and biological regimes. The most recent and best understood supercontinent formed roughly 300 mya and devolved into the continents we have today. Two earlier such events are suggested to have peaked around 600 mya and 900 mya, although precisely defining dates for supercontinent formation has many uncertainties. The seemingly cyclic nature of these events leads some scientists to think that plumes in Earth's molten interior may cause the tectonic motions and associated climate changes.

Despite much effort, little is firmly established about Earth's climate before the break-up of the supercontinent that existed between about 1.1 billion ya and 750 mya. However, evidence indicates that conditions were very different from those today. Land had formed but was concentrated below the equator; oxygen was reaching current levels; Earth's atmospheric ozone was still forming, so ultraviolet radiation restricted life to the oceans; and the first marine multicelled organisms started to appear.* As this supercontinent rifted apart, and the subsequent one reformed, extensive glaciations existed until about 600 mya. Some researchers refer to this period as *Snowball Earth*,[†] a time when ice covered tropical as well as polar regions. Then, very quickly, the ice melted, and invertebrate life in forms that we recognize today originated and proliferated across the globe. The reasons for these dramatic changes are not yet clear, but active research suggests the following combination of causes: solar radiation absorption changes caused by the latitude of the land affecting albedo, volcanic activity releasing greenhouse gases, altered ocean and atmospheric circulation patterns caused by land redistribution, early life forms changing the composition of the atmosphere, and feedbacks between various Earth systems.

Earth's last supercontinent, which formed between 250 and 300 mya, was associated with ice age conditions similar to those we have today, but the continents were concentrated in the Southern Hemisphere. By 150 mya, it was rifting apart again, changing the distribution of land and oceans, as illustrated in ● Figure 16.9. Around 50 mya, landmasses were concentrated in middle and high latitudes—as they are today. In this arrangement, ice sheets are more likely to form because there is a greater likelihood that more sunlight will be reflected

---

*Here is a brief Earth timeline. Numbers are approximate and represent billions of years ago (bya):

Earth's age: 4.53 bya	Single-celled organisms: 2 bya
Oldest rocks: 4.03 bya	Simple multicelled organisms: 1.6 bya
Earliest organic structures: 3 bya	Marine invertebrates: 0.54 bya

[†]Although still controversial as the mechanisms and details are not clear, the theory of *Snowball Earth*, when ice or slush completely covered Earth, is theoretically possible and explains confusing geologic evidence indicating that ancient equatorial zones were once glaciated. Difficulties in associating landmass locations with the timing of climatic events caused the debate.

back into space from the snow that falls over the continent in winter. Less sunlight absorbed by the surface lowers the air temperature, which allows for more snow cover and the formation of continental glaciers* over thousands of years.

As illustrated above, tectonic processes influence climate. Let's explore some of these in more detail. The various arrangements of the continents may influence the path of ocean currents, which can alter the transport of heat from low to high latitudes and change both the global wind system and the climate in middle and high latitudes. As an example, suppose that plate movement "pinches off" a large body of high-latitude ocean water so that warm water transport into the region is cut off. In winter, the surface water would eventually freeze over and allow snow to accumulate on top of it, thereby setting up conditions that could lead to even lower temperatures.

There are other mechanisms by which tectonic processes influence climate. Where oceanic and continental plates meet, ocean plates slide under continental plates (called a subduction zone) and are melted by heat and pressure. The molten rock produces pressure that builds volcanic mountain chains and then may gradually work its way to the surface as volcanic eruptions that spew ash, water vapour, carbon dioxide, sulphur dioxide, and other gases into the atmosphere. The release of these gases (called *degassing*) can also take place where continental plates separate and new crustal rock is forming.

Some scientists speculate that climate change over millions of years might be related to the rate at which the plates move and, hence, to the amount of CO_2 in the air. For example, during times of rapid spreading, an increase in volcanic activity vents large quantities of CO_2 into the atmosphere, which enhances the atmospheric greenhouse effect, causing global temperatures to rise. Millions of years later, when spreading rates decrease, less volcanic activity means that less CO_2 degasses into the atmosphere, weakening the greenhouse effect, which, in turn, causes global temperatures to drop. The accumulation

---

*The amplified cooling that takes place over the snow-covered land is the snow–albedo feedback mentioned earlier.

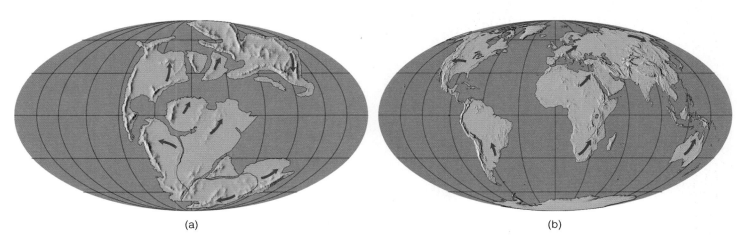

(a)   (b)

● FIGURE 16.9 Geographic distribution of (a) landmasses about 150 million years ago and (b) today. Arrows show the relative direction of continental movement.

of ice and snow over portions of the continents may promote additional cooling by reflecting more sunlight back to space.

A chain of volcanic mountains forming above a subduction zone may disrupt the airflow over them. This is particularly so where dominant wind directions flow over the mountains. Similarly, mountain building that occurs when two continental plates collide, such as the formation of the Himalayan mountains and Tibetan highlands, can have a marked influence on global atmospheric circulation patterns and, hence, on the climate of an entire hemisphere.

Up to now, we have examined how climatic variations can take place over geologic time scales owing to the movement of continents and the associated restructuring of landmasses. These factors were likely important in Earth's earlier climate but do not explain the variations seen during the current Ice Age in which the distribution of landmasses has not changed much. We will now turn our attention to variations in Earth's orbit that may account for climatic fluctuations that take place on a time scale of tens of thousands of years.

CLIMATE CHANGE: VARIATIONS IN EARTH'S ORBIT
Another external cause of climate change involves a change in the amount of solar radiation reaching Earth. The **Milankovitch theory**, named for the astronomer Milutin Milankovitch in the 1930s, states that climatic changes are due to variations in Earth's orbit. The theory states that as Earth travels through space, three separate cyclic movements combine to produce variations in the amount of solar energy that falls on Earth.

The first cycle concerns changes in the shape (**eccentricity**) of Earth's orbit as it revolves about the sun. Notice in ● Figure 16.10 that the orbit changes from being elliptical (dashed line) to being nearly circular (solid line). To go from circular to elliptical and back again takes about 100,000 years.

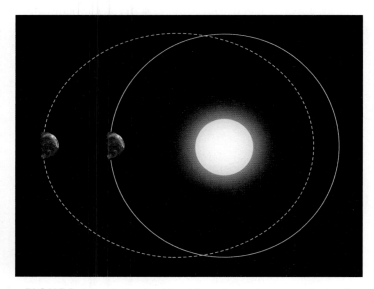

● FIGURE 16.10 For Earth's orbit to change from a nearly circular orbit (solid line) to an elliptical orbit (dashed line) and back again takes almost 100,000 years. This is called a change in eccentricity. (This diagram is highly exaggerated.)

The greater the eccentricity of the orbit (i.e., the more elliptical the orbit), the greater the variation in solar energy received by Earth between its closest and farthest approach to the sun.

Presently, we are in a period of low eccentricity, so our annual orbit around the sun is more circular. Earth is presently closer to the sun in January and farther away in July (see Chapter 3, p. 67). The 3 percent difference in distance is responsible for a nearly 7 percent increase in the solar energy received at the top of the atmosphere in January compared with July. When the difference in distance is 9 percent (a highly elliptical orbit), the difference in solar energy between July and January will be about 20 percent.* A more eccentric orbit will change the length of the seasons by changing the time interval between the spring and autumnal equinoxes.

The second cycle accounts for the way Earth wobbles as it rotates on its axis like a spinning top. This wobble, known as the **precession** of Earth's axis, completes a cycle approximately every 23,000 years. Presently, Earth is closer to the sun in January and farther from it in July. Precession will cause the reverse to be true in about 11,000 years (see ● Figure 16.11). If everything else remains the same, 11,000 years from now, seasonal variations should be greater than they are currently in the Northern Hemisphere and less than at present in the Southern Hemisphere. In about 23,000 years, we will be back to our current position.

The third cycle repeats about every 41,000 years and relates to changes in tilt (**obliquity**) with respect to Earth's orbit. As shown in ● Figure 16.12, during each 41,000-year cycle, it varies from about 22° to 24½°. Presently, Earth's orbital tilt is 23½° and decreasing. Consequently, the tropics (Cancer and Capricorn) are moving equatorward at a rate of about 14 m per year. Smaller tilts create less seasonal variation between summer and winter in middle and high latitudes; thus, Canadian winters will tend to be milder and summers cooler.

Ice sheets over high latitudes of the Northern Hemisphere are more likely to form when less solar radiation reaches the surface in summer. Less sunlight promotes lower summer temperatures. During the cooler summer, snow from the previous winter may not totally melt. The accumulation of snow over many years increases the albedo of the surface. Less sunlight reaches the surface, summer temperatures continue to fall, more snow accumulates, and continental ice sheets gradually form. It is interesting to note that when all of the Milankovich cycles are taken into account, the present trend should be toward *cooler summers* over high latitudes of the Northern Hemisphere.

In summary, Milankovitch cycles combine to cause variations in the solar radiation received at Earth's surface through

1. changes in the shape (eccentricity) of Earth's orbit about the sun

---

*Although large percentage changes in solar energy can occur between summer and winter, the global annual average solar energy received by Earth (due to orbital changes) does not vary. It is the distribution of incoming solar energy between the hemispheres that changes, not the totals.

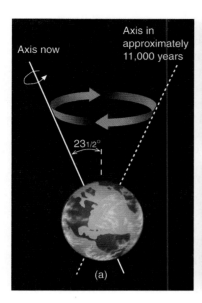

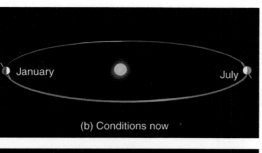

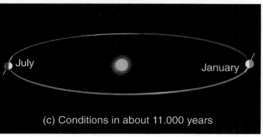

●FIGURE 16.11 (a) Earth's axis of rotation wobbles slowly as it moves through space, tracing out the path of a cone. This is called precession. (b) Presently, Earth is closer to the sun in January (Northern Hemisphere winter). (c) In about 11,000 years, due to precession, Earth will be closer to the sun in July (Northern Hemisphere summer).

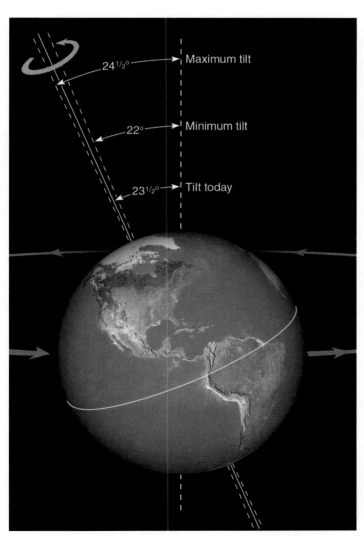

●FIGURE 16.12 Earth currently revolves around the sun while tilted on its axis by an angle of 23½°. During a period of 41,000 years, this angle of tilt ranges from about 22° to 24½°.

2. wobbling (precession) of Earth's axis of rotation
3. changes in the tilt (obliquity) of Earth's axis

In the 1970s, scientists found strong evidence in deep-ocean sediments that variations in climate during the past several hundred thousand years were closely associated with the Milankovitch cycles. More recent studies have strengthened this finding. For example, studies conclude that during the past 800,000 years, ice sheets have peaked about every 100,000 years—a conclusion that corresponds to variations in Earth's eccentricity. Superimposed on this situation are smaller ice advances that show up at intervals of about 41,000 and 23,000 years. These data support orbital variations as a *forcing factor*—external causes for the frequency of glaciation—as they appear to control the severity of the climatic variation.

But orbital changes are not entirely responsible for ice buildup and retreat. Evidence from trapped air bubbles contained in Greenland and Antarctic ice sheets (representing thousands of years of snow accumulation) reveals that CO_2 levels were about 30 percent lower during colder glacial periods when compared to warmer interglacials. Methane follows a similar pattern. These lines of evidence suggest that lower atmospheric CO_2 levels amplify the cooling initiated by orbital changes. Likewise, increasing CO_2 levels at the end of the glacial period may amplify the warming and help account for the rapid ice sheet melting.*

The latest research shows that temperature changes thousands of years ago actually preceded the CO_2 changes by about 200 to 1000 years. This indicates that CO_2 is a positive feedback in the climate system, where higher temperatures, perhaps due to Milankovitch forcing, lead to higher CO_2 levels that further amplify the initial warming, acting in a manner similar to the water vapour–greenhouse effect feedback. Consequently, CO_2 is an internal, natural part of Earth's climate

---

*Past peak CO_2 concentrations were about 325 ppm. These are considerably lower than the current atmospheric CO_2 concentration of 389 ppm.

system. Just why atmospheric CO_2 levels have varied as glaciers expanded and contracted is not clear, but changes in the ocean's biological activity appear to be part of the reason.

Also, changing CO_2 levels may indicate a shift in ocean circulation patterns. Such shifts, brought on by changes in precipitation and evaporation rates, may alter the distribution of heat around the world. This, in turn, could affect the global winds, which may explain why alpine glaciers in the Southern Hemisphere expanded and contracted in tune with Northern Hemisphere glaciers during the last ice age, even though the Southern Hemisphere (according to the Milankovitch cycles) was not in an orbital position for glaciation.

Still other factors may work in conjunction with Earth's orbital changes to explain the temperature variations between glacial and interglacial periods. Some of these are the

1. amount of dust and other aerosols in the atmosphere
2. reflectivity of the ice sheets
3. concentration of other greenhouse gases
4. changing characteristics of clouds
5. rebounding of land, having been depressed by ice

Hence, the Milankovitch cycles, in association with other natural factors, may explain the advance and retreat of ice over periods of 10,000 to 100,000 years. But what caused the Ice Age to begin in the first place? And why have ice ages been so infrequent during geologic time? The Milankovitch theory does not attempt to answer these questions.

CLIMATE CHANGE: ATMOSPHERIC PARTICLES Microscopic liquid and solid particles (aerosols) that enter the atmosphere from both human-induced (anthropogenic) and natural sources can affect climate. Their effect is complex and depends on a number of factors, such as the particles' size, shape, colour, chemical composition, and vertical distribution above the surface. In this section, we will examine aerosols in the lower atmosphere and then look at the effect of volcanic aerosols in the stratosphere.

Aerosols in the Troposphere

Aerosols enter the lower atmosphere in a variety of ways—from factory and automobile emissions, agricultural burning, wildland fires, and dust storms. Many of these sources are due to combustion. Other aerosols are not injected directly into the atmosphere but form when gases convert to particles. Some particles (such as soil dust and sulphate particles) mainly reflect and scatter incoming sunlight, causing daytime cooling of surface air. Other particles (such as smoky soot) readily absorb sunlight, which warms the air around them. Certain aerosols also selectively absorb and emit infrared energy back to the surface, producing a net warming of the surface air at night. However, the overall effect of anthropogenic, or human-produced aerosols is to *cool the surface*.

In recent years, the effect of highly reflective **sulphate aerosols** on climate has been extensively researched. In the lower atmosphere, the majority of these particles come from the combustion of sulphur-containing fossil fuels, but emissions from smouldering volcanoes can also be a significant

source of tropospheric sulphate aerosols. Sulphur pollution has more than doubled globally since preindustrial times. It enters the atmosphere mainly as sulphur dioxide gas and transforms into tiny sulphate droplets or particles. These aerosols usually remain in the atmosphere for just a few days, so they do not have time to spread around the globe. Hence, they are not well mixed, and their effect is felt mostly over the Northern Hemisphere, especially over polluted regions. Over the oceans, a major source of sulphate aerosols comes from tiny drifting aquatic plants called phytoplankton that produce dimethylsulphide (DMS) gas when they die. DMS slowly diffuses into the atmosphere, where it oxidizes to form sulphur dioxide, which then converts to sulphate aerosols.

Sulphate aerosols not only scatter incoming sunlight back to space, they also serve as cloud condensation nuclei. Consequently, they have the potential for altering the physical characteristics and behaviour of clouds. For example, if the number of sulphate aerosols and condensation nuclei inside a cloud increases, the available moisture is divided between more nuclei, a situation that produces many more but smaller cloud droplets. These smaller, more plentiful droplets tend to persist longer, reflect more sunlight, and therefore reduce the amount of sunlight that reaches the surface, lowering surface air temperature.

In summary, sulphate aerosols reflect incoming sunlight, lower surface air temperature during the day, and modify clouds by increasing their reflectivity and their longevity. Sulphate pollution has increased significantly over industrialized areas of eastern Europe, northeastern North America, and China. Consequently, the cooling effect from sulphate particles may explain why the Northern Hemisphere has warmed less than the Southern Hemisphere over the past several decades and some industrialized countries, such as the United States, have experienced less warming than the rest of the world. Also, increased aerosols may account for the observation that until the late 1970s, most of the global warming has occurred at night and not during the day, especially over polluted areas.

Research is ongoing, and the overall effect of tropospheric aerosols on the climate system is not fully understood. Information regarding the possible effect on climate from particles injected into the atmosphere during nuclear war is given in Focus on a Special Topic: Nuclear Winter—Climate Change Induced by Nuclear War, on p. 504.

WEATHER WATCH

Could atmospheric particles have contributed to the demise of the dinosaurs? Evidence supports the theory that about 65 million years ago, a giant meteorite slammed into Earth, sending billions of tonnes of dust and debris into the upper atmosphere. Such particles greatly reduced the amount of sunlight reaching Earth's surface, rapidly causing colder, darker conditions on a global scale. These appear to have persisted, disrupting the food chain and plausibly accounting for the mass extinctions of species, which at that time were dominated by large, plant-eating dinosaurs.

USGS

● FIGURE 16.13 Mount Pinatubo (Philippines) erupts on June 15, 1991. Such large, sulphur-rich volcanic eruptions affect climate as sulphur dioxide gases are blown into the stratosphere and transform into tiny reflective sulphate particles that reflect solar energy before it reaches the surface.

Volcanic Eruptions and Aerosols in the Stratosphere

During volcanic eruptions, fine particles of ash and dust as well as gases can be ejected into the stratosphere (see ● Figure 16.13). Eruptions that are rich in sulphur dioxide gas have the greatest impact on climate because the gas transforms into fine sulphate aerosols over a period of about two months. The aerosols form a dense layer of haze that may reside in the stratosphere for several years, absorbing and reflecting a portion of the sun's radiation back to space. The absorption of solar radiation, along with the absorption of infrared radiation from Earth, warms the stratosphere. The reflection of incoming sunlight by the haze cools the air at Earth's surface, especially in the hemisphere where the eruption occurs.

The largest volcanic eruptions of the 20th century that impacted climate were El Chichón in Mexico (April 1982) and Mount Pinatubo* in the Philippines (June 1991). Pinatubo ejected more than twice the sulphur dioxide ejected by El Chichón, an estimated 18 million tonnes. As shown in ● Figure 16.14, within three months, sulphur dioxide from Mount Pinatubo girdled the world's equatorial regions and then gradually worked its way around the globe. Mathematical models predict that average hemispheric temperatures can drop by about 0.2° to 0.5°C for one to three years after eruptions like this one. In as early as 1992, actual temperature changes associated with the Pinatubo eruption indicated that the mean global surface temperature had decreased by about 0.5°C (see ● Figure 16.15), producing two of the coolest years of the 1990s—1991 and 1992. The cooling would likely have been even greater, but the eruption coincided with a major El Niño (warming) event that began in 1990 and lasted until early 1995 (see Chapter 10, p. 310, for information on El Niño).

As previously noted, sulphur-rich volcanic eruptions warm the lower stratosphere. During winter, the tropical stratosphere can become much warmer than the polar stratosphere, and this situation produces a strong horizontal pressure gradient resulting in strong west-to-east (zonal) stratospheric winds. These winds descend into the upper troposphere, where they direct milder maritime surface air from the ocean onto the continents. This milder ocean air produces warmer winters over Northern Hemisphere continents during the first or second winter after an eruption occurs.

---

*The eruption of Mount Pinatubo in 1991 was estimated to be 10 times greater than that of North America's Mount St. Helens in 1980. Located near the Washington–Oregon border, Mount St. Helens was a lateral explosion that pulverized a portion of the volcano's north slope. The ensuing dust and ash had very little sulphur, so there was virtually no effect on global climate as the volcanic material was dominantly contained in the lower atmosphere. Ash affected airplane and car engines and fell out quite rapidly over a large cone-shaped area that extended eastward from the volcano through the United States and into central British Columbia and Alberta.

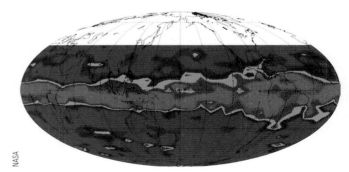

NASA

● FIGURE 16.14 Mount Pinatubo's sulphur dioxide plume (dark red and green areas) as measured by the Upper Atmosphere Research Satellite on September 21, 1991. Only three months after the eruption, in the stratosphere (at an altitude of nearly 25 km), the plume girdles the equator.

Nuclear Winter—Climate Change Induced by Nuclear War

Recent studies reinforce research from the 1980s* indicating that nuclear war would drastically and negatively impact Earth's climate. Simulations using general circulation models[†] show that the consequences of nuclear war would be even more drastic and longer lasting than previously thought. Even a superpower (United States, Russia) nuclear exchange involving only a third of stockpiled weapons would create deadly, dark, below-freezing conditions for a number of years—*nuclear winter*. Such conditions are equal to or colder than those seen during the peak of the last major glaciation—the Ice Age, 18,000 years ago. Even a "minor" nuclear exchange—a regional conflict involving two nations using 100 Hiroshima-sized bombs exploded in the air over large cities (representing only 0.03 percent of the explosive power of the current global arsenal)—would initiate a climatic cooling with

*A four-year study (1983–87) involving more than 300 scientists from more than 30 countries conducted by the Scientific Committee On Problems of the Environment (SCOPE) of the International Council of Scientific Unions has detailed the climatic, environmental, and agricultural effects of nuclear winter.

[†]Model simulations use a state-of-the-art general atmospheric circulation model, ModelE, from the NASA Goddard Institute for Space Studies, with added capability for calculating aerosol transport and removal and realistically simulating the climate response to large volcanic eruptions. It is coupled to a full ocean general circulation model that allows the surface ocean to respond quickly and deeper ocean responses on yearly time scales.

temperatures lower than those seen in the last 1000 years (including those of the Little Ice Age, during the 1400s to 1800s).

Researchers calculate that a nuclear war would raise an enormous pall of thick, sooty smoke from massive fires that would burn for days, even weeks, following an attack. Drifting smoke would be caught in the upper-level westerlies and circle the Northern Hemisphere's middle latitudes. Unlike dust, which mainly scatters and reflects incoming solar radiation, soot absorbs sunlight. Consequently, for months, years, or even a decade after the war, sunlight would have difficulty penetrating the smoke layer (the severity of radiation reduction depending on the size of the nuclear exchange, as shown in ● Figure 2). Initially, this would cause darkness or, at best, twilight at midday. Reduced solar energy would cause land surface air temperatures to drop below freezing, even during the summer, resulting in extensive damage to plants, the loss of most agriculture, and the death of millions to billions of people by starvation. Simulation of even the smallest war indicates that about a decade would pass before temperatures return to preblast conditions—much longer than the impacts seen from the eruption of Mount Pinatubo (see Figure 2).

As the lower troposphere cools, the sooty, upper troposphere warms because smoke particles in this region absorb solar energy. This causes a strong, stable temperature inversion that extends from the surface into the high atmosphere. This would also adversely affect

the weather by suppressing convection, altering precipitation processes and patterns, and causing major changes in the general wind patterns. Heating the upper part of the smoke cloud would cause it to rise into the stratosphere and drift around the world. About one-third of the smoke would remain in the atmosphere for up to a decade; the other two-thirds would be washed out by precipitation in about a month. This smoke lofting, combined with persisting sea ice formed during the initial cooling, would produce climatic change that would remain for more than a decade.

Virtually all research on nuclear winter, including models and analogue studies, confirms this gloomy scenario. Smoke from massive urban fires and large forest fires that rises high into the atmosphere has caused lower surface air temperatures under the smoke, confirming part of the theory. Simulations of past volcanic eruptions using the same models provide further confidence in the results.

Recognition of these impacts in the 1980s improved superpower relations, hastened the end of the Cold War, and may have led to the reduction in the 1980 nuclear arsenal by about one-third. But the danger remains. The current nuclear arsenal is more than sufficient to cause nuclear winter. Also, as other nations develop nuclear capability, the potential for war expands. It will not disappear until the global nuclear weapon totals number in the hundreds, not in the thousands.

An infamous cold spell often linked to volcanic activity occurred during the year 1816, known as the "year without a summer" or "eighteen hundred and froze-to-death." In Europe that year, bad weather contributed to a poor wheat crop, resulting in famine. In North America, unusual blasts of cold arctic air moved through Canada and the northeastern United States between May and September. The cold spells brought heavy snow in June and killing frosts in July and August. In the warmer days that followed each cold snap, farmers replanted, only to have another cold outbreak damage the planting. The unusually cold summer was followed by a bitterly cold winter.

What caused the "year without a summer? Apparently, a rather stable longwave pattern in the atmosphere produced unseasonably cold summer weather over eastern North America and western Europe. The cold weather followed the massive eruption in 1815 of Mount Tambora in Indonesia. In addition, a smaller volcanic eruption occurred in 1809, from which the climate system may not have fully recovered.

In an attempt to correlate sulphur-rich volcanic eruptions with long-term global climate trends, scientists measure the acidity of annual ice layers in Greenland and Antarctica. Generally, higher ice acidities are associated with greater concentrations of sulphuric acid from sulphate aerosols in the atmosphere.

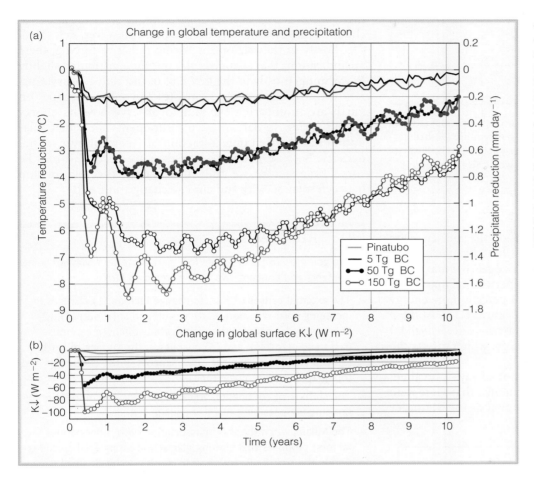

FIGURE 2 The modelled climate response over 10 years for three nuclear wars producing smoke in megatonnes (Tg) of black carbon (BC). (a) The reduction in the global average surface air temperature (red line) and precipitation (black line). Unaffected global average precipitation is 3.0 mm/day, so in year 5 (150 Tg case), the drop of 1.2 mm/day represents a 40 percent decrease in global average precipitation. (b) The reduction in incoming shortwave radiation (K↓) at Earth's surface in watts per square metre for each simulation compared to the reduction caused by Mount Pinatubo's 1991 volcanic eruption—the largest of the 20th century.

Source: Alan Robock, Luke Oman, and Georgiy L. Stenchikov, 2007b: "Nuclear winter revisited with a modern climate model and current nuclear arsenals: Still catastrophic consequences."*Journal of Geophysical Research.* Vol. 112, D13107, doi:2006JD008235. Fig. 2.

Relatively acidic ice has been identified in layers that formed from 1350 to 1700, a time that corresponds to a cooling trend over Europe referred to as the Little Ice Age. Such findings suggest that sulphur-rich volcanic eruptions may have played an important role in triggering this comparatively cool period and possibly triggered others in the geologic past. Moreover, sediment cores drilled in the northern Pacific Ocean reveal the sequence of volcanic events. They show that volcanic eruptions on the nearby continent were at least 10 times larger 2.6 mya, roughly when glaciation began in the Northern Hemisphere.

CLIMATE CHANGE: VARIATIONS IN SOLAR OUTPUT

Measurements made by sophisticated radiometers aboard satellites suggest that the sun's energy output (called brightness) may vary slightly—by a fraction of 1 percent—with sunspot activity.

WEATHER WATCH

The year without a summer (1816) even had its effect on literature. Inspired (or perhaps dismayed) by the cold, gloomy summer weather along the shores of Lake Geneva, Mary Shelley wrote the novel *Frankenstein*.

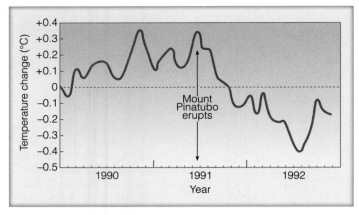

● FIGURE 16.15 The deviation in global average air temperature (red line) from the 1981–90 average global air temperature (dashed line) for the years 1990 to 1992. The cooling trend after Mount Pinatubo erupted in June 1991 is obvious. Average global air temperatures had decreased almost 0.5° C by July 1992.

Source: Data courtesy of John Christy, University of Alabama, Huntsville, and R. Spencer, NASA Marshall Space Flight Center.

Sunspots are huge magnetic storms on the sun. They are cooler regions on the sun's surface that look like dark spots. Occurring in cycles, their number and size reach a maximum approximately every 11 years. During periods of maximum sunspot activity, the sun emits about 0.1 percent more energy than during sunspot minimums (see ● Figure 16.16). At these times, the greater number of bright areas (called *faculae*) around the sunspots radiate more energy, which offsets the effect of the dark spots.

The 11-year sunspot cycle has not always prevailed. Between 1645 and 1715, it is well documented that there were few, if any, sunspots. This period is known as the **Maunder**

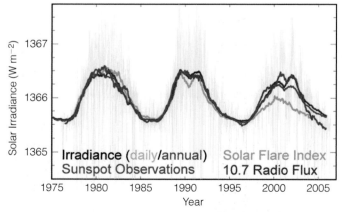

● FIGURE 16.16 Changes in solar radiation reaching Earth outside the atmosphere (yellow, daily values; black, annual average) in watts per square metre as measured by the Earth Radiation Budget Satellite. The blue curve represents the yearly average number of sunspots. As sunspot activity increases from minimum to maximum, the sun's energy output increases by about 0.1 percent. The solar flare index and 10.7 cm radio flux are two other measures of solar output.

Source: V. Ramanathan, B.R. Barstrom and E.F. Harrison, "Climate and earth's radiation budget." *Physics Today.* May 1988. Fig. 5.

minimum.* It is interesting to note that the minimum occurred during the "Little Ice Age," a cool spell in the temperature record experienced mainly over Europe. Some scientists suggest that a reduction in solar energy output was, in part, responsible for this cold spell.

Fluctuations in solar output may account for climatic changes over time scales of decades and centuries. Although many theories have been proposed linking solar variations to climate change, none have been proven. However, as we shall see shortly in the section discussing radiative forcing, there are indications that changed solar output contributed slightly to warming, mostly in the first half of the 20th century. Currently, instruments aboard satellites and solar telescopes on Earth are monitoring the sun to observe how its energy output varies. To date, these measurements show that solar output has changed only a fraction of 1 percent over several decades. Because many years of data are needed, it may be some time before we fully understand the relationship between solar activity and climate change on Earth.

BRIEF REVIEW

Before going on to the next section, here is a brief review of some of the facts and concepts we have covered so far:

● The external causes of climate change include (1) changes in incoming solar radiation (Earth's orbit and solar output); (2) changes in the composition of the atmosphere (greenhouse gases and aerosols); and (3) changes in the surface of Earth (albedo and changes in the arrangements of the continents and topography over geologic time).

● The shifting of continents, volcanic activity, and mountain building are possible causes of climate change.

● The Milankovitch theory (in association with other natural factors) proposes that altering glacial and interglacial episodes during the past 2.58 million years are the result of small variations in the tilt of Earth's axis and in the geometry of Earth's orbit around the sun.

● Trapped air bubbles in the ice sheets of Greenland and Antarctica reveal that CO_2 levels and methane levels were lower during colder glacial periods and higher during warmer interglacial periods. But even when the levels were higher, they still were much lower than they are today.

● Sulphate aerosols in the troposphere reflect incoming sunlight, which tends to lower Earth's surface temperature during the day. Sulphate aerosols may also modify clouds by increasing their reflectivity.

● Volcanic eruptions rich in sulphur may be responsible for cooler periods that span years and decades in the geologic past.

● Fluctuation in solar output (brightness) may account for climatic changes over time scales of decades and centuries.

*Named after E. W. Maunder, a British solar astronomer who first discovered the reduced sunspot period in the late 1880s.

In previous sections, we saw how increasing levels of CO_2 may have contributed to changes in global climate spanning thousands and even millions of years. Today, we are undertaking an uncontrolled global experiment by injecting vast quantities of greenhouse gases into our atmosphere without fully understanding the long-term consequences. The next section describes how CO_2 and other trace gases appear to be enhancing Earth's greenhouse effect, changing and warming the climate.

Current and Future Climate Change

Since the beginning of the 20th century, the average global surface air temperature has risen by more than 0.8°C. Is this warming due to increasing greenhouse gases and an enhanced greenhouse effect? What will happen in the future as greenhouse gases continue to increase? Before we address these questions, let's first examine how greenhouse gas concentrations have changed in the past.

GREENHOUSE GAS TRENDS   We know from Chapter 2 that CO_2 is a greenhouse gas that strongly absorbs infrared radiation and plays a major role in warming the lower atmosphere. As ● Figure 16.17 shows, CO_2 has been increasing steadily in the atmosphere, primarily due to the burning of fossil fuels. Other significant anthropogenic sources of CO_2 include cement production and deforestation. In 2010, the annual average CO_2 was about 389 ppm. If CO_2 levels continue to increase at the same rate that they have been (1.9 ppm per year), atmospheric concentrations will rise to about 560 ppm by the end of this century. Of course, the actual future CO_2 concentrations are quite uncertain because they largely depend on anthropogenic emissions, which are difficult to predict. To complicate the picture, trace greenhouse gases such as methane (CH_4), nitrous oxide (N_2O), and chlorofluorocarbons (CFCs) also absorb infrared radiation.* Collectively, these gases are approaching CO_2 in their ability to enhance the atmospheric greenhouse effect.

In Figure 16.17, notice that the atmospheric concentration of both methane and nitrous oxide have also increased dramatically over the last 250 years. Since the mid-1990s, the atmospheric concentration of chlorofluorocarbons has been decreasing after being phased out due to their impact on stratospheric ozone. However, their substitute compounds, which are also greenhouse gases, have been increasing. Moreover, the total amount of surface ozone probably increased by more than 30 percent since 1750. Because the concentration of this greenhouse gas varies greatly from region to region and depends on the production of photochemical smog, it has probably led to a small increase in the greenhouse effect.

To understand the effect of these increasing greenhouse gas concentrations on past and future temperature, we need to first review a few concepts from Chapter 2.

*Refer back to Chapter 1 and Table 1.1 (p. 6) for additional information on the concentration of these gases, as well as their sources, sinks, and residence times.

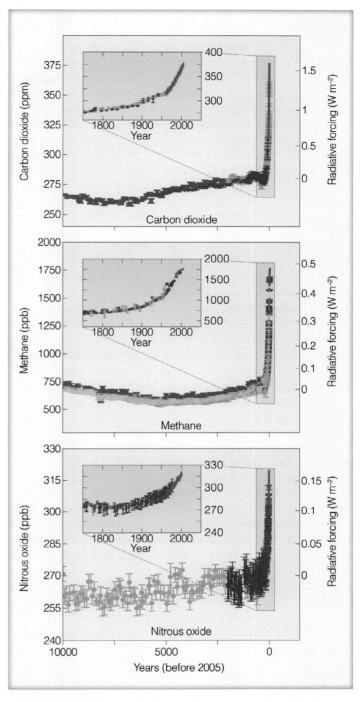

● FIGURE 16.17 Changes in the greenhouse gases carbon dioxide, methane, and nitrous oxide indicated from ice core and modern data.

Source: *Climate Change 2007: The Physical Science Basis.* Working Group I Contribution to the Fourth Assessment Report of the Intergovernmental Panel on Climate Change, Fig. 6.4. Pg. 448. Cambridge University Press.

RADIATIVE FORCING   We know from Chapter 2 that our world without water vapour, CO_2, and other greenhouse gases would be about 33°C colder than at present. With an average surface temperature of about −18°C, much of the planet would be uninhabitable. In Chapter 2, we also learned that when the rate of the incoming solar energy balances the rate of outgoing infrared energy from Earth's surface and atmosphere, the Earth–atmosphere system is in a state of *radiative equilibrium*. Increasing concentrations of greenhouse gases can disturb this equilibrium and are, therefore, referred to as **radiative forcing agents. Radiative forcing*** is a way of comparing the relative impact of different greenhouse gases as well as other external and internal causes of climate change. The total radiative forcing provided by extra CO_2 and other greenhouse gases has increased by about 2.63 W m^{-2} since 1750, with CO_2 contributing about 63 percent of the increase. To put this number into context, it represents over 1 percent of the 240 W m^{-2} that Earth receives on average from the sun. So it is very likely that part of the warming during the last century is due to increasing levels of greenhouse gases. But what role does natural climate variability play in climate change? And since levels of CO_2 have increased by more than 38 percent since the early 1800s, why has the observed increase in global temperature been relatively small?

We know that the climate may change due to natural events. For example, changes in the sun's energy output and volcanic eruptions rich in sulphur are two major natural radiative forcing agents that are external causes of climate change. Studies show that since the middle 1700s, changes in the sun's energy output may have contributed a small positive forcing (about 0.12 W m^{-2}) on the climate system, most of which occurred during the first half of the 20th century.

On the other hand, as discussed previously, volcanic eruptions that inject sulphur dioxide that transforms into sulphate aerosols into the stratosphere produce a negative forcing, which lasts for a few years after the eruption. Because several major eruptions occurred between 1880 and 1920, as well as between 1960 and 1991, the combined change in radiative forcing due to both volcanic activity and solar activity over the past 25 to 45 years was negative and contributed to cooling Earth's surface in that time period. By 2005, however, it appears that the net effect of volcanoes is negligible—at least until the next major eruption occurs.

Sulphate aerosols produced by human activity also have a net cooling effect. This reduction is estimated to have a direct radiative forcing of −0.5 W m^{-2}. Since sulphate aerosols are also cloud condensation nuclei that can change cloud albedo and cloud lifetime, they have an additional indirect radiative forcing of −0.7 W m^{-2}. Thus, aerosols have partially offset the positive radiative forcing from greenhouse gas increases, so the net human-caused radiative forcing since 1750 is 1.6 W m^{-2} (see • Figure 16.18 for an accounting of these and other radiative forcings).

How do these changed radiative forcings affect climatic properties such as temperature and precipitation? The use of

climate models can help answer this question. (Before going on to the next section, you may want to look at Focus on an Advanced Topic: Radiative Forcing—The Ins and Outs on p. 510.)

CLIMATE MODELS AND RECENT TEMPERATURE TRENDS
Climate models are the primary tool used by climate scientists to see the impact of changed radiative forcing agents on Earth systems. The newest, most sophisticated climate models—called **general circulation models** or **global climate models** (GCMs)—take into account a number of important relationships, including the interactions between the oceans and the atmosphere, the processes by which CO_2 is removed from the atmosphere, and the cooling effect produced by sulphate aerosols in the lower atmosphere. The models also include both positive and negative feedbacks affecting the climate system, such as the *water vapour–greenhouse effect feedback* described on p. 488. More information on these models is given in Focus on an Advanced Topic: Climate Models on p. 514.

Earth's average surface temperature increased by about 0.74°C from 1906 to 2005. How does this observed temperature change over the last century compare to temperature changes derived from climate models using different forcing agents? Before we look at what climate models reveal, it is important to realize that the interactions between Earth and its atmosphere are so complex that it is difficult to prove with 100 percent certainty that Earth's present warming trend is due entirely to increasing concentrations of greenhouse gases. This is because any human-induced signal of climate change is superimposed on a background of natural climatic variations (internal causes of climate change), such as the El Niño–Southern Oscillation (ENSO) phenomenon (discussed in Chapter 10). Consequently, in the temperature observations, it is difficult to separate the greenhouse gas radiative forcing from the variation due to natural climate variability and other external forcings. However, today's more sophisticated climate models are much better at representing natural climate variability as well as taking into account forcing agents that are both natural and human induced.

• Figure 16.19 shows the predicted changes in surface air temperature from 1901 to 2005 made by different climate models using natural forcings only and natural plus human-caused (anthropogenic) forcings. The black line presents the actual observed changes in global mean surface air temperature as differences from the 1901–50 mean. Notice that when both positive anthropogenic forcings of greenhouse gases and negative forcings of sulphate aerosols are included in the models (see Figure 16.19a), the projected temperature change and the observed temperature change closely match. However, if the anthropogenic forcings are omitted from the models (see Figure 16.19b), they do a poor job of tracking the observed temperatures after about 1960.

Climate studies using computer models such as these have led scientists to conclude that most of the warming during the latter decades of the 20th century is very likely due to increasing levels of greenhouse gases. In fact, the Intergovernmental Panel on Climate Change (IPCC), a committee of

---

*Radiative forcing is interpreted as an increase (positive) or a decrease (negative) in net radiant energy observed over an area at the tropopause. All factors being equal, an *increase in radiative forcing* will induce surface *warming*, whereas a *decrease* will induce surface *cooling*.

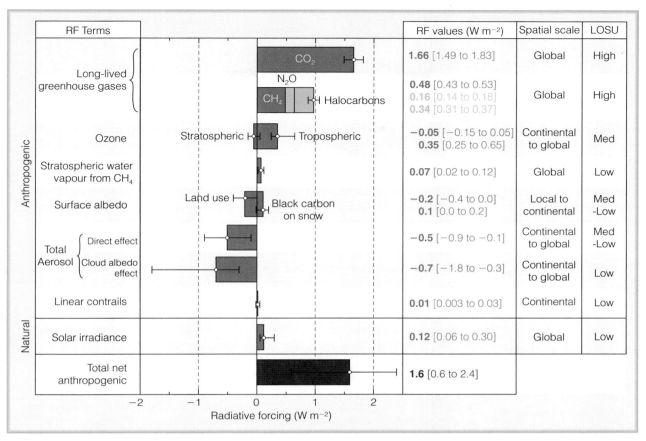

● FIGURE 16.18 Summary of the main factors affecting radiative forcing of climate in 2005 relative to 1750. Human activities caused significant changes in greenhouse gases, ozone, water vapour, surface albedo, aerosols, and condensation trails from aircraft (contrails). The only significant natural forcing increase over this period was due to an increase in solar radiation. The error bar beside each estimate indicates the range of uncertainty. The level of scientific understanding (LOSU) is indicated in the last column.

Source: Time Series of Aerosol Column Optical Depth at the Barrow, Alaska, ARM Climate Research Facility for 2008. Fourth Quarter 2009 ARM and Climate Change Prediction Program Metric Report. Fig. 1. Pg. 1.

over 2000 leading Earth scientists, considered the issues of climate change in a report published in 1990 and updated in 1992, in 1995, in 2001, and again in 2007. The 2007 Fourth Assessment report of the IPCC states that "most of the observed increase in globally averaged temperatures since the mid-20th century is very likely due to the observed increase in anthropogenic greenhouse gas concentrations."[*]

FUTURE CLIMATE CHANGE: PROJECTIONS Today's climate models project that the surface air temperature will increase substantially by the end of this century due to increasing levels of greenhouse gases (see ● Figure 16.20). Notice, however, that the climate models do not all predict the same amount of warming. The climate models are set up to run different future greenhouse gas scenarios describing how greenhouse gas emissions will change with time (see ▽ Table 16.1). Also, each climate model is independent and responds slightly differently to the forcing agents, resulting in the variation around each projection shown in Figure 16.20.

The IPCC in its 2007 report concluded that, depending on the emission scenario, the global average temperature this century would increase by an additional 1.8 to 4.0°C (with a likely range between 1.1 and 6.4°C). If, during this century, the surface temperature should increase by 2°C, the warming would be three times greater than that experienced during the 20th century. An increase of 4°C or more would have potentially devastating effects worldwide. Therefore, it is likely that the warming over the 21st century will be much larger than the warming experienced during the 20th century and probably greater than any warming during the past 10,000 years.

UNCERTAINTIES ABOUT GREENHOUSE GASES There are, however, uncertainties in predicting the future climate because it is not known how quickly greenhouse gases will increase in the future. Besides uncertainties in future anthropogenic emissions of greenhouse gases, it is not completely known how water and land will ultimately affect rising levels of CO_2. Currently, the oceans and the vegetation on land absorb about half of the CO_2 emitted by human sources and therefore play a major role in the climate system. Yet the exact

---

[*]In the report, "very likely" means a greater than 90 percent probability.

Radiative Forcing—The Ins and Outs

To examine radiative forcing from a slightly different perspective, we need to remember a few important concepts from Chapter 2. First, recall that all objects emit radiation and that hotter objects emit more radiant energy. Also recall that the relationship between temperature and radiation (called the Stefan–Boltzmann law) is

$$E = \sigma T^4$$

where E is the energy being emitted ($W\ m^{-2}$), T is the object's temperature (K), and σ is the Stefan–Boltzmann constant ($\sigma = 5.67 \times 10^{-8}\ W\ m^{-2}\ K^{-4}$).

Earth is in a state of radiative equilibrium when incoming energy from the sun that is absorbed by the Earth system equals infrared energy emitted by the Earth system back to space. The radiative equilibrium temperature of Earth is about −18°C.* Remember that due to Earth's greenhouse effect, this temperature (−18°C) is about 33°C lower than Earth's observed average surface air temperature. To better understand radiative forcing, let's see how much energy Earth is emitting while in radiative equilibrium.

If Earth's equilibrium temperature is −18°C, that converts to 255.15 K. Putting this into the Stefan–Boltzmann equation, we obtain

$$E = \sigma T^4$$

$$E = (5.67 \times 10^{-8} \left(\frac{W}{m^2 K^4}\right)(255.15K)^4$$

$$E = 240 \left(\frac{W}{m^2}\right)$$

Thus, the Earth system in radiative equilibrium (and behaving as a blackbody) would be emitting an average of 240 watts of infrared energy over each square metre of surface area, out to space. The equilibrium temperature of −18°C, therefore, is the effective average temperature that would be "seen" by an observer in space who measures the infrared radiation emitted by the Earth system and converts it back into a temperature using the Stefan–Boltzmann equation. This temperature is lower than Earth's surface air temperature because the infrared radiation leaving the Earth system is a combination of radiation emitted by the warmer surface—which is mostly absorbed by the atmosphere due to the greenhouse effect—with radiation emitted by the much colder atmosphere.

Over Earth as a whole, outgoing infrared energy equals incoming solar energy (see ● Figure 3). Consequently, without an atmospheric greenhouse effect, Earth's surface would (on average) emit 240 $W\ m^{-2}$ upward and, at the same time, receive 240 $W\ m^{-2}$ from the sun. But due to the greenhouse effect, this equilibrium of 240 $W\ m^{-2}$ is achieved only at the *top* of the atmosphere. Moreover, as greenhouse gases slowly increase in concentration, they alter this balance by gradually absorbing more and more of Earth's infrared radiation,

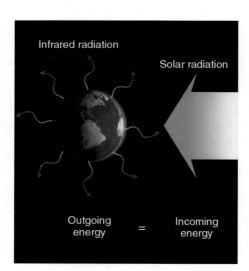

● FIGURE 3 In radiative equilibrium, over Earth as a whole, incoming solar energy equals outgoing infrared energy.

thereby preventing this energy from escaping into space. So without any changes in the climate system, outgoing energy would gradually drop below a value of 240 $W\ m^{-2}$ and incoming solar radiation would exceed outgoing infrared energy, so the Earth system warms. The warmed Earth will emit more infrared energy upward, until outgoing energy again equals 240 $W\ m^{-2}$ and a balance is achieved.

As long as everything else remains the same, a sudden doubling of the current levels of atmospheric CO_2 (about 389 ppm) would result in an outgoing radiation reduction (imbalance) of about 5.5 $W\ m^{-2}$ at the top of the atmosphere. To restore this imbalance, Earth's surface and lower atmosphere must warm by about 1.4°C so that more infrared energy is directed upward.* However, this does not account for feedbacks in the system, such as the water vapour–greenhouse effect positive feedback. Current climate models predict that a sudden doubling in CO_2 would actually result in 2.1 to 4.4°C of global average surface air temperature increase—a much larger rise than would be caused by CO_2 alone. Therefore, as levels of CO_2 and other greenhouse gases increase, they alter the amount of infrared energy lost to space and, in effect, force the atmosphere to respond by increasing the surface air temperature.

Any change in average net radiation that occurs at the top of the atmosphere (actually the top of the troposphere) and is due to some change in the climate system (such as increasing levels of CO_2) is called *radiative forcing*. Therefore, greenhouse gases (which are increasing in concentration) are referred to as *radiative forcing agents*.[†]

---

*This temperature is calculated at an average distance from the sun with no atmospheric greenhouse effect.

*Remember that a small increase in temperature results in a great deal more energy emitted, as $E \propto T^4$.

[†]Additional examples of radiative forcing agents include other external causes of climate change such as changes in solar output and land surface modifications.

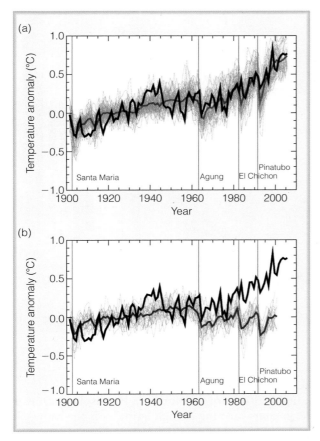

● FIGURE 16.19 Projected global mean surface air temperature changes from different climate models compared to observations (heavy black line). Temperature values are anomalies from the 1901–50 mean value. Positive values are temperatures above the 1901–50 mean, whereas negative values are temperatures below this mean. Simulations are forced with (a) both anthropogenic and natural forcings and (b) natural forcings only. The thin yellow curves in (a) are obtained from 58 simulations by 14 atmosphere–ocean general circulation models. Their mean temperature is in a heavy red line. The thin blue curves in (b) are from 19 simulations produced by five models, and their mean is in a heavy blue line. Vertical lines represent the dates of major volcanic eruptions.

Source: *Climate Change 2007: The Physical Science Basis.* Working Group I Contribution to the Fourth Assessment Report of the Intergovernmental Panel on Climate Change. Cambridge University Press.

effect they will have on rising levels of CO_2 and climate change is not totally clear. For instance, the microscopic plants (phytoplankton) dwelling in the oceans extract CO_2 from the atmosphere during photosynthesis and store it below the ocean surface when they die. Will a warming Earth trigger a large blooming of these microscopic plants, in effect reducing the rate at which atmospheric CO_2 is increasing?

Current models show that warming the Earth tends to *reduce* both ocean and land intake of CO_2. Therefore, more CO_2 should remain in the atmosphere to further enhance global warming. An example of how rising temperatures can play a role in altering the way landmasses absorb and emit CO_2 is found in the arctic tundra. There increased temperatures in recent years are thawing more frozen soil in summer than previously. As a result, deep layers of exposed decaying peat moss release CO_2 into the atmosphere. Until recently, this region absorbed more CO_2 than it released. Now, however, much of the tundra is a *source* of CO_2 instead of a *sink*.

At present, deforestation accounts for about one-fifth of the observed increase in atmospheric CO_2. Hence, changes in land use could influence levels of CO_2 concentrations, especially if the practice of deforestation is replaced by reforestation. Furthermore, it is unknown what future steps countries will take in limiting the emissions of CO_2 from the burning of fossil fuels.

THE QUESTION OF CLOUDS   As the atmosphere warms and more water vapour is added to the air, global cloudiness might increase as well. How would clouds—which come in a variety of shapes and sizes and form at different altitudes—affect the climate system? Clouds reflect incoming sunlight back to space, a process that tends to cool the climate, but clouds also emit infrared radiation to Earth, which tends to warm it. Just how the climate will respond to changes in cloudiness will depend on the type of clouds that form and their physical properties, such as liquid water (or ice) content, depth, and droplet size distribution. For example, high, thin cirriform clouds (composed mostly of ice) tend to promote a net warming effect: they allow a good deal of sunlight to pass through (which warms Earth's surface), yet because they are cold, they warm the atmosphere around them by absorbing more infrared radiation from Earth than they emit upward. Low stratiform clouds, on the other hand, tend to promote a net cooling effect. Since they are composed mostly of water droplets, they reflect much of the sun's incoming energy, which cools Earth's surface, and because their tops are relatively warm, they radiate away much of the infrared energy they receive from Earth. Satellite data confirm that clouds presently have a net cooling effect on our planet, so without clouds, our atmosphere would be warmer.

Additional clouds in a warmer world would not necessarily have a net cooling effect, however. Their influence on the average surface air temperature would depend on their extent and on whether low or high clouds predominate. Consequently, the feedback from clouds could potentially enhance or reduce the warming produced by increasing greenhouse gases. Most models show that as the surface air warms, there will be more convective-type clouds and an increase in cirrus clouds. This situation would tend to provide a small positive feedback on the climate system.*

CONSEQUENCES OF CLIMATE CHANGE: THE POSSIBILITIES   If the world continues to warm as predicted by climate models, where will most of the warming take place? Climate

---

*In addition to the amount and distribution of clouds, the way in which climate models calculate the optical properties of a cloud (such as albedo) can have a large influence on the model's calculations. Also, there is much uncertainty as to how clouds will interact with aerosols and what the net effect will be.

● FIGURE 16.20 Global average projected surface air temperature changes (°C) above the 1980–99 average (zero line) for the years 2000 to 2100. Temperature changes inside the graph and to the right of the graph are based on multiple climate models with different scenarios. Each scenario describes how the average temperature will change based on different concentrations of greenhouse gases and various forcing agents (see Table 16.1). The black line shows global temperature change during the 20th century. The orange line shows projected temperature change where greenhouse gas concentrations are held constant at the year 2000 level. The vertical grey bars on the right side of the figure indicate the likely range of temperature change for each scenario. The thick solid bar within each grey bar gives the best estimate for temperature change for each scenario.

Source: *Climate Change 2007: The Physical Science Basis.* Working Group I Contribution to the Fourth Assessment Report of the Intergovernmental Panel on Climate Change. Cambridge University Press.

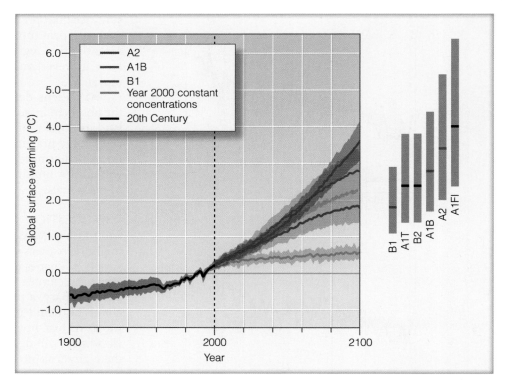

models predict that land areas will warm more rapidly than the global average, particularly in the northern high latitudes in winter (see ● Figure 16.21a). We can see in Figure 16.21b that the greatest surface warming for the period 2001–06 compared to the period 1951–80 occurred over landmasses in the high latitudes of the Northern Hemisphere. One reason for the greater warming at high latitudes is the snow–albedo feedback, discussed previously. These observations of global average temperature change suggest that climate models are on target with their warming projections.

▼ Table 16.1   Projected Average Surface Air Temperature Ranges and Best Temperature Estimates for the Decade 2090–99 Using Six Scenarios*

NAME OF SCENARIO	LIKELY TEMPERATURE RANGE, °C	ESTIMATED TEMPERATURE CHANGE, °C	SCENARIO DESCRIPTION
Constant year 2000 concentrations	0.3–0.9	0.6	
B1	1.1–2.9	1.8	Energy production, technology, and economy all focus on increased efficiency and minimal resources. Growth rate is high. Energy consumption is very low.
A1T	1.4–3.8	2.4	Energy produced using mostly nonfossil sources. Economic and technological growth is rapid. Energy consumption is high.
B2	1.4–3.8	2.4	Energy produced by the most effective means available. Economic and technological development is slow. Energy consumption is moderate.
A1B	1.7–4.4	2.8	Energy produced using a balance of fossil fuels and nonfossil sources. Economic and technological growth is rapid. Energy consumption is high.
A2	2.0–5.4	3.4	Energy is produced by the simplest means available. Global economic and technological growth is slow. Energy consumption is high.
A1FI	2.4–6.4	4.0	Energy produced using mostly fossil fuels. Economic and technological growth is rapid. Energy consumption is high.

*Temperature changes are relative to the average surface air temperature for the period 1980–99.

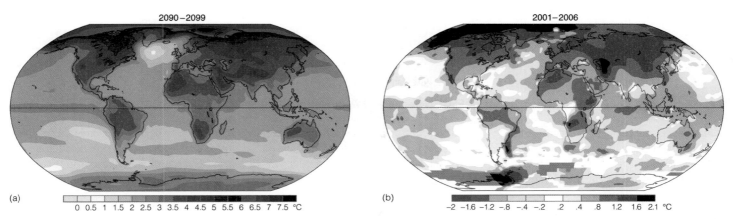

As high-latitude regions of the Northern Hemisphere continue to warm, modification of the land may actually enhance the warming through a positive feedback. For example, the dark green boreal forests of the high latitudes absorb up to three times as much solar energy as does the snow-covered tundra. Consequently, the winter temperatures in subarctic regions are, on average, much higher than they would be without trees. If warming allows the boreal forests to expand into the tundra, the forests may accelerate the warming in that region. As the temperature rises, organic matter in the soil should decompose at a faster rate, adding more CO_2 to the air, which might accelerate the warming even more. Trees that grow in a climate zone defined by temperature may become especially hard hit as rising temperatures place them in an inhospitable environment. In a

weakened state, they may become more susceptible to insects and disease.

As the world warms, total rainfall must increase to balance the increase in evaporation. But precipitation will not be evenly distributed as some areas will get more precipitation and others less (see ● Figure 16.22). Notice in Figure 16.22a that the models project an increase in winter precipitation over high latitudes of the Northern Hemisphere and a decrease in precipitation over areas of the subtropics. A decrease in precipitation in this region could have an adverse effect by placing added stress on agriculture. Some models even suggest that changes in global patterns of precipitation might cause more extreme rainfall events, such as floods and severe drought. In fact, it is interesting to note that during the warming of the 20th century, there was an increase in precipitation by as much

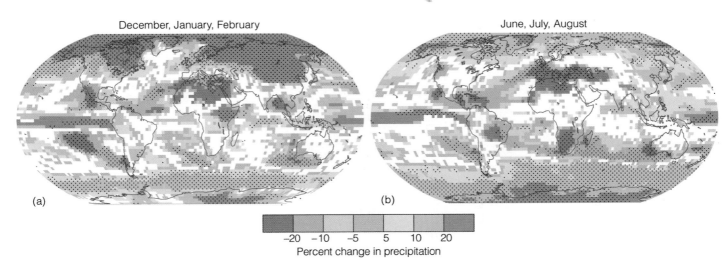

FOCUS ON AN ADVANCED TOPIC

Climate Models

René Laprise. Department of Earth Science and Atmosphere. University of Quebec.

Climate models are computer programs that simulate how the components of the climate system work and interact. The climate system is composed of the *atmosphere*, the *hydrosphere*, and the land surface (part of the *lithosphere* and *biosphere*). These components interact by exchanging *heat* through turbulence at the surface and through radiation transfers; *water* through evaporation and evapotranspiration by vegetation and condensation and precipitation from the atmosphere onto the surface; and *momentum* in winds and currents between the atmosphere, the oceans, and Earth's surface.

Climate models use geophysical properties such as Earth's orbital parameters and rotation rate, the composition of the atmosphere, the distribution of continents, the depth of the oceans and the height of the mountains, land surface types for soils and vegetation, and the intensity of the sun. With these properties and using the known laws of physics, referred to as the **governing equations**, which determine exchanges of momentum (i.e., wind and currents), energy, and mass (i.e., of gases in the air as well as water), climate models calculate the time evolution of many variables. In the atmosphere, these variables include winds, temperature, pressure,

humidity, clouds, and precipitation. In the oceans, variables such as currents, water temperature, salinity, and sea ice are predicted. On land, variables such as ground temperature and water content, river flows, lake temperatures, glaciers, and snow cover are represented. By making simulations with different emissions of greenhouse gases and aerosols, as well as changes in land use, climate models can tell about the impact of anthropogenic influences on climate.

The governing equations are physical laws of nature expressed as a set of complex mathematical equations known as "coupled, nonlinear, partial differential equations." These equations cannot be solved by hand or by using a calculator or spreadsheet, so computations are done in a computer program, called a "climate model." To do this, the atmosphere is divided into a three-dimensional (3D) grid (see ● Figure 4), and the equations are solved by the climate model at each grid point, or node. This is very similar to a **numerical weather prediction** model (see Chapter 13, p. 388) used in forecasting daily weather.

Climate simulations are extremely demanding in computing power and data storage. This is because climate models are

actually simulating the evolution of the weather that makes the climate; there is, unfortunately, no other good way of simulating climate. In climate models, the weather conditions are advanced in time by steps of a few minutes, so all of the equations are solved each grid point every few minutes. This means that millions of time steps are needed to simulate 30 years to represent the climate. To cover the whole Earth and create a global climate model (GCM—also called a general circulation model), with grid points that are 200 km apart, there are hundreds of grid points in the east–west direction as well as in the north–south direction. Combined with 10 or 20 levels in the vertical, this means that there are hundreds of thousands of nodes over which several coupled equations need to be solved at each time step. Such a gargantuan computing load requires the use of today's fastest supercomputers. The amount of data to store is also phenomenal: even with saving the results every six hours, a 30-year simulation takes several gigabytes of computer storage.

The use of a 3D grid in climate models has two effects. First, the governing equations need to be simplified and approximated so that they can be solved. Second, the distance between the grid points (called the *model resolution*) limits the size of phenomena that can be simulated: for example, cumulonimbus clouds in thunderstorms are about 10 km in size, and they cannot be resolved (or represented in the model) if the model resolution is a few hundred kilometres, which is typical in global climate models. Even though these small-scale processes are not resolved in the climate model, their effect on the larger scales still needs to be accounted for. This is done by *parameterizing* these small-scale processes—this means that their effect on the larger scale processes is found by using data or equations that relate the small-scale processes to the large-scale conditions that are resolved in the climate model. These parameterizations introduce substantial uncertainties in climate simulations.

Ideally, one would like to use much finer resolutions, but this would take too much computing power as the number of calculations in a climate model increases very rapidly with

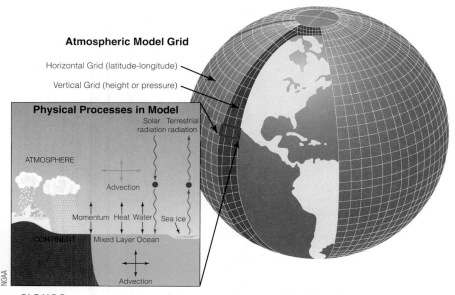

Atmospheric Model Grid

Horizontal Grid (latitude-longitude)

Vertical Grid (height or pressure)

Physical Processes in Model

Solar Terrestrial
radiation radiation

ATMOSPHERE

Advection

Momentum Heat Water ⎱ Sea ice ⎰

CONTINENT Mixed Layer Ocean

Advection

NOAA

● FIGURE 4 Schematic diagram of the physical processes and the grid in a climate model.

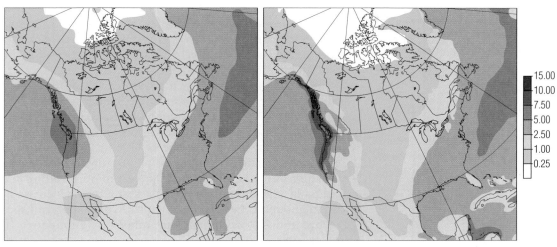

● FIGURE 5 Wintertime average precipitation (in units of millimetres per day) as simulated for 1961–90 from an ensemble of five global and five regional model runs. Left panel simulated by the 400 km Canadian Coupled Global Climate Model (CGCM3). Right panel simulated by the 45 km Canadian Regional Climate Model (CRCM4.2.3).
CRCM. Environment Canada.

finer resolutions. For example, simply going from a 200 km to a 100 km grid resolution implies nearly a 10-fold increase in computations. So if a simulation took one week of computer time with a 200 km grid, it would take two years of computations with a 100 km grid—too long a waiting time for most!

Yet the assessment of climate change impacts from human influences often requires information at finer scales, such as a river basin for hydrological applications. An approach developed in the late 1980s that is increasingly used consists of increasing resolution over only an area of interest, such as North America, Canada, or a particular province. Such simulators are known as regional climate models (RCMs) because their domain does not cover the entire globe. Low-resolution GCMs are used to specify conditions along the perimeters of high-resolution RCMs. RCMs can

use grid meshes of tens of kilometres with much less computing time than a GCM with comparable resolution. An example of precipitation from a GCM compared to that from an RCM for the period 1961–90 is shown in ● Figure 5. Notice the finer scale variation in the RCM, especially over the mountainous west. The difference in precipitation for a future climate scenario for the period 2041–70 is shown in ● Figure 6.

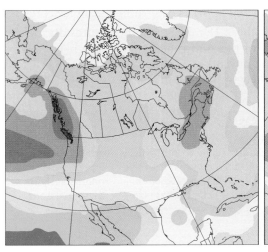
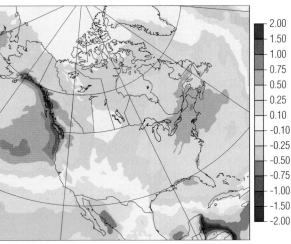

● FIGURE 6 Projection of precipitation changes (in units of millimetres per day) for winter, obtained by an ensemble of five global and five regional model simulations, based on the greenhouse gas and aerosol emissions scenario A2 (see Table 16.1). The changes correspond to the period 2041–70 and are expressed as a difference from the period 1961–90. Left panel simulated by the 400 km Canadian Coupled Global Climate Model (CGCM3). Right panel simulated by the 45 km Canadian Regional Climate Model (CRCM4.2.3).
CRCM. Environment Canada.

The mountain pine beetle kills pine trees by introducing a fungus when it colonizes them to lay its larvae each summer. The beetle larvae are killed by a midwinter cold blast of −40°C. A lack of cold winters in British Columbia's interior during the late 1990s and 2000s has allowed the mountain pine beetle epidemic there to grow to a point where, in 2009, it affected about 16.3 million hectares— an area about three times the size of Nova Scotia. Carried by wind, the beetle epidemic started spreading into Alberta in the mid-2000s. With further climate change, these sorts of insect disturbances are expected to occur more frequently in the future.

as 10 percent over the middle- and high-latitude land areas of the Northern Hemisphere. In contrast, over subtropical land areas, a decrease in precipitation has occurred. It also appears that there has been an increase in the frequency of heavy precipitation events during the last 50 years or so.

In mountainous regions of western North America, where much of the precipitation falls in winter, with warmer temperatures, precipitation might fall more often as rain, causing a decrease in snow-melt runoff to fill reservoirs during the spring.

Other consequences of future climate change will likely be a rise in sea level as glaciers over land (especially Antarctica and Greenland) recede and the oceans continue to expand as they slowly warm. During the 20th century, sea level rose about 17 cm, and today's improved climate models estimate that sea level will rise an additional 18 to 59 cm by the end of this century, depending on the greenhouse gas emission scenario. The rise in sea level will depend on how much the temperature rises and on how quickly the ice in Greenland and Antarctica melts. Rising ocean levels could have a damaging influence on coastal ecosystems. In addition, coastal groundwater supplies might become contaminated with salt water. And as we saw in Chapter 15, as sea surface temperatures increase (other factors being equal), the intensity of hurricanes will likely increase as well.[*]

Another consequence of melting glaciers is an eventual decrease in the supply of meltwater to glacier-fed rivers. This water supply threat is especially critical for the Indus in Tibet and Pakistan as well as the Brahmaputra, a major tributary of the Ganges in Tibet, India, and Bangladesh. These rivers and others in the region sustain hundreds of millions of people. This is also of consequence for many of the rivers in western Canada that receive critical late summer flow from glacial meltwater.

In polar regions, as elsewhere around the globe, rising temperatures produce complex interactions among temperature, precipitation, and wind patterns. Hence, in polar areas, more snow might actually fall in the warmer (but still cold) air, causing snow to build up or, at least, stabilize over the continent

of Antarctica. Over Greenland, which is experiencing rapid melting of ice and snow, any increase in precipitation will likely be offset by rapid melting, so the ice sheets are expected to continue to shrink. Presently, in the Arctic, warming has caused sea ice[*] to shrink and thin continuously since the 1970s. At the end of the summer melt season in September 2007, the extent of Arctic sea ice was at a record minimum (see ● Figure 16.23). If the warming in this region continues at its present rate, polar sea ice in summer may be totally absent by the middle of this century. To learn more about changes in Arctic sea ice and its impacts, read Focus on a Special Topic: Changing Sea Ice in the Arctic and Its Impact on p. 518.

Increasing levels of CO_2 in a warmer world might have additional consequences. For example, higher levels of CO_2 might act as a "fertilizer" for some plants, accelerating their growth. Increased plant growth consumes more CO_2, which might retard the increasing rate of CO_2 in the atmosphere. On the other hand, the increased plant growth might cause some insect populations to grow and eat more, resulting in a net loss in vegetation. It is possible that a major increase in CO_2 might upset the balance of nature, with some plant species becoming so dominant that others are eliminated. In tropical areas, where many developing nations are located, the warming may actually decrease crop yield, whereas in cold climates, where crops are now grown only marginally, the warming effect may actually increase crop yields. In a warmer world, higher latitudes might benefit from a longer growing season, and extremely cold winters might become less numerous, with fewer bitterly cold spells.

Following are some conclusions about climate change and its future impact on our climate system summarized from the 2007 Fourth Assessment Report of the Intergovernmental Panel on Climate Change (IPCC):

● The primary source of the increased atmospheric concentration of carbon dioxide since the preindustrial period is from fossil fuel use, with land use change providing another significant, but smaller, contribution. The atmospheric concentration of carbon dioxide in 2005 exceeds by far the natural range over the last 650,000 years (180 to 300 ppm) as determined from ice cores.

● Average Northern Hemisphere temperatures during the second half of the 20th century were very likely higher than during any other 50-year period in the last 500 years and likely the highest in at least the past 1300 years.

Inuit elders in Canada's Arctic report that the weather today is harder to understand than it was in their youth. Previously, elders could predict local weather accurately from environmental signs, but now that the northern climate has changed, these traditional forecasting methods do not work as well.

---

[*]For more information on hurricanes and climate change, read Focus on an Environmental Issue: Hurricanes in a Warmer World on p. 482.

---

[*]Sea ice is formed by the freezing of sea water.

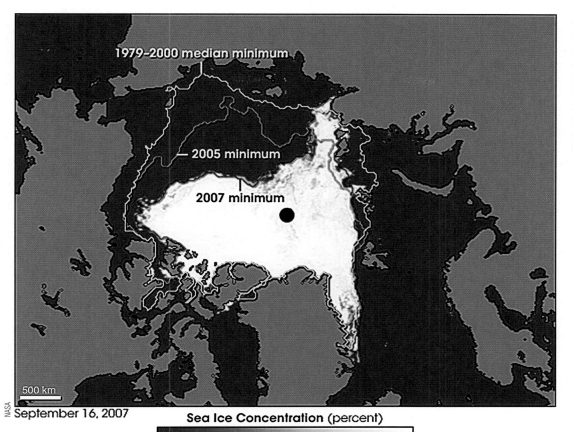

● FIGURE 16.23 The extent of Arctic sea ice on September 16, 2007, when the ice cover was at its record minimum. The green line indicates the previous minimum from 2005, whereas the yellow line indicates the median minimum sea ice extent over the period between 1979 and 2000.

- Temperatures of the most extreme hot nights, cold nights, and cold days are likely to have increased due to anthropogenic forcing. It is likely that heat waves have become more frequent, and it is more likely than not that anthropogenic forcing has increased the risk of heat waves.

- The 100-year linear trend in temperature between 1906 and 2005 was 0.74°C (with a likely range between 0.56 and 0.92°C). The linear warming between 1956 and 2005 was 0.13°C per decade, nearly double the 100-year trend.

- Widespread changes in extreme temperatures have been observed over the last 50 years. Cold days, cold nights, and frost have become less frequent, whereas hot days, hot nights, and heat waves have become more frequent.

- The average atmospheric water vapour content has increased since at least the 1980s over land and ocean as well as in the upper troposphere. The increase is broadly consistent with the extra water vapour that warmer air can hold.

- Observations since 1961 show that the average temperature of the global ocean has increased to depths of at least 3000 m and that the ocean has been absorbing more than 80 percent of the heat added to the climate system. Such warming causes sea water to expand, contributing to most of the observed sea-level rise.

- Average arctic temperatures increased at almost twice the global average rate in the past 100 years. Arctic sea ice has shrunk by 7.4 percent per decade in summers since 1978.

- The maximum area of seasonally frozen ground has decreased by 7 percent in the Northern Hemisphere and by as much as 15 percent during spring.

- Extratropical storm tracks are projected to move poleward, with consequent changes in wind, precipitation, and temperature patterns, continuing the broad pattern of observed trends over the last half-century.

- Based on a range of models, it is likely that future tropical cyclones (typhoons and hurricanes) will become more intense, with larger peak wind speeds and more heavy precipitation associated with ongoing increases in tropical sea surface temperatures.

- New analyses of balloon-borne and satellite measurements of lower and midtropospheric temperature show warming rates that are similar to those of the surface temperature record and are consistent within their respective uncertainties, largely reconciling a discrepancy noted in the Third Assessment report.

- Global average sea level rose at an average rate of 1.8 mm per year (with a likely range between 1.3 and 2.3 mm per

FOCUS ON A SPECIAL TOPIC

Changing Sea Ice in the Arctic and Its Impact

William Gough. Department of Geography and Planning. University of Toronto

Sea ice coverage in the Arctic normally ranges from 8 to 16 million km^2. It has an annual cycle, with ice extent peaking in March and reaching a minimum in September. In recent years, however, there has been a noticeable decline in sea ice extent. In 2007, the sea extent in September was just over 4 million km^2, the lowest in recorded history. The years 2005, 2008, and 2009 were also well below the 1979–2000 average, as shown in ● Figure 7. In other regions of the Arctic and subarctic, which are characterized by seasonal sea ice, such as Hudson Bay, the ice-free season is increasing by a week per decade. These changes are linked to concurrent changes in Arctic air temperature.

These climatic changes are having far-ranging impacts, such as stress on the region's biota and Inuit communities, as well as new transportation opportunities. One example of an impact on the regional biota is the plight of polar bears in southwestern and southern Hudson Bay. These bears feed on seals during the ice-covered season (November to July) and then come to land during the ice-free season (August to October), where females den inland and males lounge along the shore. The animals lose weight during this time. An increase in the ice-free season has added an additional stress on the bears, and the average weight and birth rate are decreasing. Once seasonal sea ice ceases in Hudson Bay, possibly as early as 2050, these animals will be cut off from their main food source. Other polar bears throughout the Arctic will face similar climate-induced stresses.

Inuit communities have been keenly aware of a changing Arctic climate over the past two decades, noting a variety of changes in sea ice

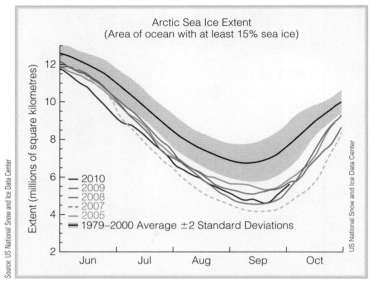

Source: US National Snow and Ice Data Center

● FIGURE 7 Arctic sea ice extent.

conditions, timing of the seasons, and the changing nature of seasonal transitions. On a community level, these changes have caused increased concerns about transportation safety and food security. The transportation issues arise from encountering sea ice conditions that are new in nature. For example, in Igloolik in Northern Foxe Basin, the formation of sea ice in the fall was historically quite rapid. However, in recent years, the ice has been forming over a longer period of time, sometimes forming, melting, and re-forming, adding uncertainty to the identification of safe ice for transportation by snowmobile or dogsled. Food security relates to the availability of traditional or "country" foods and the increased reliance on western foods. Country foods are typically obtained either by hunting or fishing. In a

changing climate, traditional timing and locations for such harvests have either not yielded the same results as in the past or the locations are not accessible at the appropriate times due to delayed sea ice formation.

One potentially positive result of the decreased sea ice extent is the increase in navigable water. In the 17th and 18th centuries, the quest for the Northwest Passage, access to the Pacific from the Atlantic, via the Arctic ocean, was an unrealized dream. However, with the changing sea ice conditions, such a passage is opening up, with the potential for regular commercial navigation through Arctic waters. The increased ice-free season in Hudson Bay will allow the existing port of Churchill, Manitoba, to operate for more than two or three months per year.

year) over 1961 to 2003. The rate was faster over 1993 to 2003: about 3.1 (2.4 to 3.8) mm per year.

- More intense and longer droughts have been observed over wider areas since the 1970s, particularly in the tropics and subtropics. It is more likely than not that human influence has contributed toward increases in area affected by drought and the frequency of heavy precipitation events. Drought-affected areas are projected to increase in extent.

- Mountain glaciers and snow cover have declined on average in both hemispheres. Widespread decreases in glaciers and ice caps have contributed to sea-level rise (ice caps do not include contributions from the Greenland and Antarctic ice sheets).

- Both past and future anthropogenic carbon dioxide emissions will continue to contribute to warming and sea-level rise for more than a millennium due to the time scales required for removal of this gas from the atmosphere.

- The observed widespread warming of the atmosphere and ocean, together with ice mass loss, supports the conclusion that it is extremely unlikely that global climate change of the past 50 years can be explained without external forcing and very likely that it is not due to known natural causes alone.

CLIMATE CHANGE: LAND USE CHANGES   All climate models predict that the climate will change and Earth's surface will warm as humanity continues to emit greenhouse gases into the air. But are humans changing the climate by other activities as well? Modification of Earth's surface taking place right now could be influencing the climate of certain regions. For example, studies show that about half the rainfall in the Amazon River Basin is returned to the atmosphere through evaporation and through transpiration from the leaves of trees. Consequently, clearing large areas of tropical rain forests in South America to create open areas for farms and cattle ranges will most likely cause a decrease in evaporative cooling. This decrease, in turn, could lead to a warming in that area of at least several degrees Celsius. In turn, the reflectivity of the deforested area will change. Similar changes in albedo result from the overgrazing and excessive cultivation of grasslands in semiarid regions, causing an increase in desert conditions (a process known as **desertification**).

Currently, billions of hectares of the world's range and cropland, along with the welfare of millions of people, are affected or threatened by desertification. Annually, millions of hectares are reduced to a state of near- or complete uselessness. The main cause is overgrazing, although overcultivation, poor irrigation practices, and deforestation also play a role. The effect this will have on climate, as surface albedos increase and more dust is swept into the air, is uncertain. (For a look at how a modified surface influences the inhabitants of a region in Africa, read Focus on a Special Topic: The Sahel—An Example of Climatic Variability and Human Existence on p. 520.)

It is interesting to note that some scientists think humans may have been altering climate way before modern civilizations came along. For example, emeritus Professor William Ruddiman of the University of Virginia suggests that humans have been influencing climate for the past 8000 years. Although some climate scientists vehemently oppose his ideas, Ruddiman speculates that without preindustrial farming, which produces methane and some carbon dioxide, we would have

WEATHER WATCH

In our warmer world, many freshwater lakes in northern latitudes are freezing later in the fall and thawing earlier in the spring than they did in years past. A recent study around the Great Lakes region found that the ice-free period on small lakes in the region was increasing at about 1.5 days per decade since the 1920s.

entered a naturally occurring ice age. He even suggests that the Little Ice Age of the 15th through 19th centuries in Europe was human induced because plagues, which killed millions of people, caused a reduction in farming. The reasoning behind this idea goes something like this: as forests are cleared for farming, levels of CO_2 and methane increase, producing a stronger greenhouse effect and a rise in surface air temperature. When catastrophic plagues strike—the bubonic plague, for instance—high mortality rates cause farms to be abandoned. As forests begin to take over the untended land, levels of CO_2 and methane drop, causing a reduction in the greenhouse effect and a corresponding drop in air temperature. When the plague abates, the farms return, forests are cleared, levels of greenhouse gases go up, and surface air temperatures rise.

CLIMATE CHANGE: EFFORTS TO CURB   The most obvious way to curb global warming is to reduce greenhouse gas emissions by reducing the use of fossil fuels, such as oil and coal. Using alternative energy such as solar collectors and wind power—the world's two fastest growing energy sources—could also help with this endeavour.

In an attempt to mitigate the impact humans have on the climate system, representatives from 160 countries met at Kyoto, Japan, in 1997 to work out a formal agreement to limit greenhouse gas emissions in industrialized nations. The international agreement—called the *Kyoto Protocol*—was adopted in 1997 and was put into force in February 2005.

The protocol set mandatory targets for reducing greenhouse gas emissions. Although the percentage by which each country reduces its emissions varies, the overall goal was to reduce greenhouse gas emissions in developed countries by 5.2 percent below 1990 levels during the 5-year period of 2008 through 2012. Although the plan gained worldwide acceptance, the United States never signed the protocol, and most countries did not meet their Kyoto targets.

The Kyoto Protocol was to be updated at the Copenhagen Summit in December 2009, which resulted in the Copenhagen Accord that 138 countries have since signed. The Copenhagen Accord has an aspirational goal of limiting global temperature increases to 2°C, with a long-term goal of 1.5°C. It includes nonbinding emission targets that are specified by each country to be implemented by 2020 in developed countries. Canada has matched the U.S. commitment of lowering its greenhouse gas emissions by 17 percent from 2005 levels—which represents an increase of 2.5 percent from the 1990 levels used as a baseline in the Kyoto Protocol. The Copenhagen Accord also includes a transfer of funds to developing countries for implementation of adaptation and mitigation strategies.

Studies suggest that one way of limiting global warming would be to inject sulphate aerosols into the stratosphere, where they would reflect sunlight. Using climate models, it was found that injecting an amount similar to the emission from the 1991 Mount Pinatubo eruption every one to four years, in conjunction with reducing greenhouse gases, could provide a "grace period" of up to 20 years before major cutbacks in greenhouse gas emissions would be required. Of course,

FOCUS ON A SPECIAL TOPIC

The Sahel—An Example of Climatic Variability and Human Existence

The Sahel is in North Africa, located between about 14° and 18°N latitude (see ● Figure 8). Bounded on the north by the dry Sahara and on the south by the grasslands of the Sudan, the Sahel is a semiarid region of variable rainfall. Precipitation totals may exceed 500 mm in the southern portion, whereas in the north, rainfall is scanty. Yearly rainfall amounts are also variable as a year with adequate rainfall can be followed by a dry one.

During the winter, the Sahel is dry, but as summer approaches, the intertropical convergence zone (ITCZ), with its rain, usually moves into the region. The inhabitants of the Sahel are mostly nomadic people who migrate to find grazing land for their cattle and goats. In the early and middle 1960s, adequate rainfall led to improved pasturelands; herds grew larger, and so did the population. However, in 1968, the annual rains did not reach as far north as usual, marking the beginning of a series of dry years and a severe drought.

The decrease in rainfall, along with overgrazing, turned thousands of square kilometres of pasture into barren wasteland. By 1973, when the severe drought reached its climax, rainfall totals were 50 percent of the long-term average, and perhaps 50 percent of the cattle and goats had died. The Sahara Desert had migrated southward into the northern fringes of the region, and a great famine had taken the lives of more than 100,000 people.

Although low rainfall years have been followed by wetter ones, relatively dry conditions have persisted over the region for the past 40 years or so. The overall dryness of the region has caused many of the larger, shallow lakes (such as Lake Chad) to shrink in size. The wetter years of the 1950s and 1960s appear to be due to the northward displacement of the ITCZ. The drier years, however, appear to be more related to the

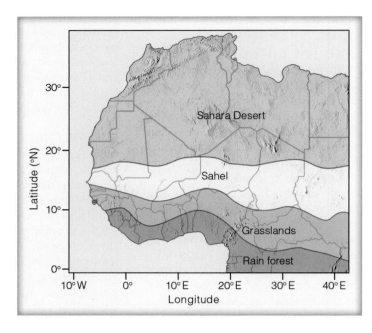

● FIGURE 8 The semiarid Sahel of North Africa is bounded by the Sahara Desert to the north and grasslands to the south.

reduced intensity of rain that falls during the so-called rainy season. Some scientists feel that reduced rain intensity is due to a feedback mechanism whereby less rainfall and reduced vegetation cover modify the surface and promote a positive feedback relationship: the surface changes act to reduce convective activity, which, in turn, reinforces the dry conditions. For example, when vegetation is removed from the surface (perhaps through overgrazing or excessive cultivation), the surface albedo (reflectivity) increases, and the surface temperature drops. Reduced surface temperature would result in less convective activity and drier conditions that would further reduce vegetation. However, some studies show that less vegetation cover does not always result in a higher albedo.

Since the mid-1970s, the Sahara Desert has not progressively migrated southward into the Sahel. During dry years, the desert does migrate southward, but in wet years, it retreats.

Consequently, desertification is not presently overtaking the Sahel, nor is the albedo of the region showing much year-to-year change.

So the question remains: Why did the Sahel experience such devastating drought during the 1970s and 1980s? Recent studies suggest that the dry periods were due to a cooler North Atlantic Ocean. The cooler ocean appears to be the result of sulphate aerosols that enhance the formation of highly reflective clouds above the water. The increase in cloud reflectivity cooled the ocean surface, which, in turn, influenced the circulation of the atmosphere in such a way that the ITCZ did not, on average, move as far north. The sulphate pollution* apparently originated over North America, suggesting that human activities on one continent could potentially cause climate variability on another, with the end result being a disastrous famine.

---

*Recent studies show a correlation between sulphate particles ejected into the stratosphere from volcanic eruptions and past dry spells in the Sahel.

injecting sulphate particles into the stratosphere might have additional consequences, such as changing the temperature of the upper atmosphere and affecting the fragile ozone layer.

Although the idea of injecting the stratosphere with sulphate particles has not been given much credence by climate scientists, the idea of reducing the impact of climate change through global scale technological fixes (called *geoengineering*) is intriguing. The science of geoengineering is fairly new and poses many technological as well as ethical challenges for society.

Cutting down on the emissions of greenhouse gases and pollutants has several potentially positive benefits. A reduction in greenhouse gas emissions could slow down the enhancement of Earth's greenhouse effect and reduce global warming, while at the same time, the reduction in air pollutants might reduce acid rain, diminish haze, and slow the production of photochemical smog. Even if the warming produced by greenhouse gases proves to be less than what modern climate models project, these measures would certainly benefit humanity.

SUMMARY

In this chapter, we considered some of the many ways Earth's climate changes. First, we saw that Earth's climate has undergone considerable change during the geologic past. Some of the evidence for a changing climate comes from tree rings (dendrochronology), chemical analysis of oxygen isotopes in ice cores and fossil shells, and geologic evidence left behind by advancing and retreating glaciers. The evidence from these suggests that throughout much of the geologic past (long before humanity arrived on the scene), Earth was warmer than it is today. There were cooler periods, however, during which glaciers advanced over large sections of North America and Europe.

We examined some of the possible causes of climate change, noting that the problem is extremely complex as a change in one variable in the climate system almost immediately changes other variables. Several complementary theories have been proposed to account for natural climate change over geologic time. One theory suggests that the shifting of the continents, along with volcanic activity and mountain building, may account for variations in climate that take place over millions of years.

The Milankovitch theory proposes that alternating glacial and interglacial episodes during the past 2.5 million years are the result of small variations in the tilt of Earth's axis and in the geometry of Earth's orbit around the sun. Another theory suggests that certain cooler periods in the geologic past may have been caused by volcanic eruptions rich in sulphur. Still another theory postulates that climatic variations on Earth might be due to variations in the sun's energy output.

We looked at temperature trends between 1906 and 2005 and found that, over this span of time, Earth has warmed by about 0.74°C. It is very likely that most of the warming during the last 50 years is due to increasing concentrations of greenhouse gases. Sophisticated climate models project that as levels of CO_2 and other greenhouse gases continue to increase, Earth will warm substantially by the end of this century. The average warming is projected to be an additional 1.8 to 4.0°C (with a likely range between 1.1 and 6.4°C), depending on human greenhouse gas emissions. The models also predict that as Earth warms, there will be a global increase in atmospheric water vapour, an increase in global precipitation, a more rapid melting of sea ice, and a rise in sea level.

KEY TERMS

The following terms are listed (with page numbers) in the order they appear in the text. Define each. Doing so will aid you in reviewing the material covered in this chapter.

water vapour–greenhouse
   effect feedback, 488
dendrochronology, 491
ice ages, 491
Pleistocene, 492
interglacials, 492
Younger Dryas (event), 493
Holocene, 493
timber line, 493
mid-Holocene
   maximum, 493
Little Ice Age, 495
global warming, 496
positive feedback
   mechanisms, 497
negative feedback
   mechanisms, 497

snow–albedo feedback, 497
plate tectonics, 498
Milankovitch theory, 500
eccentricity, 500
precession, 500
obliquity, 500
sulphate aerosols, 502
Maunder minimum, 506
radiative forcing agents, 508
radiative forcing, 508
general circulation
   model (global climate
   model), 508
governing equations, 514
numerical weather
   prediction, 514
desertification, 519

QUESTIONS FOR REVIEW

1. What methods do scientists use to determine climate conditions that have occurred in the past?

2. Explain how the changing climate influenced the formation of the Bering land bridge.

3. How does today's average global temperature compare to the average temperature during most of the past 1000 years?

4. What is the Younger Dryas episode? When did it occur?

5. How does a positive feedback mechanism differ from a negative feedback mechanism? Is the water vapour–greenhouse feedback considered positive or negative? Explain.

6. How does the theory of plate tectonics explain climate change over periods of millions of years?

7. Describe the Milankovitch theory of climatic change by explaining how each of the three cycles alters the amount of solar energy reaching Earth.

8. Given the analysis of air bubbles trapped in polar ice during the past 400,000 years, were CO_2 levels generally higher or lower during colder glacial periods? Were methane levels higher or lower at this time?

9. How do sulphate aerosols in the lower atmosphere affect surface air temperatures during the day?

10. Describe the scenario of nuclear winter.

11. Volcanic eruptions rich in sulphur warm the stratosphere. Do they tend to warm or cool Earth's surface? Explain.

12. Explain how variations in the sun's energy output might influence global climate.

13. Climate models predict that increasing levels of CO_2 will cause the mean global surface temperature to rise significantly by the year 2100. What other greenhouse gases are also causing temperature increases?

14. Describe some of the natural radiative forcing agents and their effect on climate.

15. (a) Describe how clouds influence the climate system.
    (b) Which clouds would tend to promote surface cooling: high clouds or low clouds? Why?

16. In Figure 16.20 (p. 512), explain why the actual rise in surface air temperature (grey line) is much more than the projected rise in air temperature due to natural forcing only.

17. Why do climate scientists now believe that most of the warming experienced during the last 50 years was due to increasing levels of greenhouse gases?

18. List some of the consequences that climate change might have on the atmosphere and its inhabitants.

19. Is CO_2 the only greenhouse gas we should be concerned with for climate change? If not, what are the other gases?

20. Explain how the ocean's conveyor belt circulation works. How does the conveyor belt appear to influence the climate of northern Europe? (Hint: The answer is found in Focus on a Special Topic: The Ocean Conveyor Belt and Climate Change on p. 494.)

21. How is climate projected to change over your region of Canada toward the end of this century?

QUESTIONS FOR THOUGHT

1. Ice cores extracted from Greenland and Antarctica have yielded valuable information on climate changes during the past few hundred thousand years. What do you feel might be some of the limitations in using ice core information to evaluate past climate changes?

2. When glaciation was at a maximum (about 18,000 years ago), was global precipitation greater or less than at present? Explain your reasoning.

3. Consider the following climate change scenario. Warming global temperatures increase saturation vapour pressures over the ocean. As more water evaporates, increasing quantities of water vapour build up in the troposphere. More clouds form as the water vapour condenses. The clouds increase the albedo, resulting in decreased amounts of solar radiation reaching Earth's surface. Is this scenario plausible? What type(s) of feedback(s) is(are) involved?

4. Explain why periods of glacial advance in the higher latitudes tend to occur with warmer winters and cooler summers.

5. Explain two different ways that an increase in sulphate particles might lower surface air temperatures.

6. Are ice ages in the Northern Hemisphere more likely when
    (a) the tilt of Earth is at a maximum or a minimum?
    (b) the sun is closest to Earth during summer in the Northern Hemisphere or during winter?
    Explain your reasoning for both (a) and (b).

7. Most climate models show that the poles will warm faster than the tropics. What effect will this have on winter storms in mid-latitudes?

8. The oceans are a major sink (absorber) of CO_2. One hypothesis states that as warming increases, less CO_2 will be dissolved in the oceans. Would you expect Earth to cool or to warm further? Why?

PROBLEMS AND EXERCISES

1. If the annual precipitation near Hudson Bay (latitude 55°N) is 380 mm per year, calculate how long it would take snow falling on this region to reach a thickness of 3000 m. (Assume that all the precipitation falls as snow, that there is no melting during the summer, and that the annual precipitation remains constant. To account for compaction of the snow, use a snow to water depth ratio of 5 to 1.)

2. On a warming planet, the snow–albedo feedback produces a positive feedback. Make a diagram (or several diagrams) to illustrate this phenomenon. With another diagram, show that the snow–albedo feedback produces a positive feedback on a cooling planet.

In mountainous regions, a variety of climatic types can exist within a relatively short distance. Here, in Alberta, aspen change colour in a continental-type climate, whereas the high peaks with a polar climate experience perpetual snow.

iStockphoto

Global Climate

17

The climate is unbearable. . . . At noon today the highest temperature measured was −33°C. We really feel that it is late in the season. The days are growing shorter, the Sun is low and gives no warmth, katabatic winds blow continuously from the south with gales and drifting snow. The inner walls of the tent are like glazed parchment with several millimetres thick ice-armour. . . . Every night several centimetres of frost accumulate on the walls, and each time you inadvertently touch the tent cloth a shower of ice crystals falls down on your face and melts. In the night huge patches of frost from my breath spread around the opening of my sleeping bag and melt in the morning. The shoulder part of the sleeping bag facing the tent-side is permeated with frost and ice, and crackles when I roll up the bag. . . . For several weeks now my fingers have been permanently tender with numb fingertips and blistering at the nails after repeated frostbites. All food is frozen to ice and it takes ages to thaw out everything before being able to eat. At the depot we could not cut the ham, but had to chop it in pieces with a spade. Then we threw ourselves hungrily at the chunks and chewed with the ice crackling between our teeth. You have to be careful with what you put in your mouth. The other day I put a piece of chocolate from an outer pocket directly in my mouth and promptly got frostbite with blistering of the palate.

Ove Wilson. Ove Wilson was a Swedish medical officer on the Norwegian–British–Swedish Antarctic Expedition of 1949–52.

Ove Wilson quoted in: David M. Gates. *Man and His Environment.* (New York, N.Y.: Harper and Row.) 1972. Pg. 175.

Mushroom: Ian McAllister/Getty Images; Spruce and Mountainside: Shutterstock

CONTENTS

Our opening comes from a report by a team of Norwegian, British, and Swedish scientists on their encounter with one of nature's cruelest climates—that of Antarctica. Their experience illustrates the profound effect that climate can have on even ordinary events, such as eating a piece of chocolate. Although we may not always think about it, climate affects nearly everything in the middle latitudes, too. For instance, it determines the amount of precipitation that is input to the *hydrosphere* and whether or not the precipitation is in the form of snow, providing input to the *cryosphere*. Climate influences the shape of landscapes—the *lithosphere*—as well as the type of vegetation, agriculture, and animals that inhabit an area—the *biosphere*. Climate determines our housing and clothing, how we feel and live, and even where we choose to reside—the *anthrosphere*. Entire civilizations have flourished in favourable climates and have moved away from, or perished in, unfavourable ones. We learned early in this text that climate is the average of the day-to-day weather over a long duration. But the concept of climate is much larger than this because it encompasses, among other things, the variability, including daily and seasonal extremes of weather, within specified areas.

When we speak of climate, then, we must be careful to specify the location we are talking about. For example, residents of a town may boast that their community has mild winters with air temperatures seldom below freezing. This may be true a few metres above the ground in an instrument shelter, but near the ground, the temperature may drop below freezing on many winter nights. This small climatic region near or on the ground is referred to as a **microclimate**. Because a much greater extreme in daily air temperatures exists near the ground than several metres above, the microclimate for small plants is far more harsh than the thermometer in an instrument shelter would indicate.

When we examine the climate of a small area of the Earth's surface, we are looking at the **mesoclimate**. The size of the area may range from a few hectares to several square kilometres. Mesoclimate includes regions such as forests, valleys, beaches, and towns. The climate of a much larger area, such as a province or a country, is called **macroclimate**. The climate extending over the entire Earth is often referred to as **global climate**.

In this chapter, we will concentrate on the larger scales of climate. We will begin with the factors that regulate global climate; then we will discuss how climates are classified. Finally, we will examine the different types of climate.

A World with Many Climates

The world is rich in climatic types. From the teeming tropical jungles to the frigid polar "barrens," there seems to be an almost endless variety of climatic regions. The factors that produce the climate in any given place—the **climatic**

WEATHER WATCH

Even "summers" in Antarctica can be brutal. In 1912, during the Antarctic summer, Robert Scott of Great Britain not only lost the race to the South Pole to Norway's Roald Amundsen but also perished in a blizzard trying to return. Temperature data taken by Scott and his crew showed that the summer of 1912 was unusually cold, with air temperatures remaining below −34°C for nearly a month. These exceptionally low temperatures eroded the men's health and created an increase in frictional drag on the sleds the men were pulling. Just before Scott's death, he wrote in his journal that "no one in the world would have expected the temperatures and surfaces which we encountered at this time of year."

---

controls—are the same that produce our day-to-day weather. Briefly, the controls are the

1. intensity of sunshine and its variation with latitude
2. distribution of land and water
3. ocean currents
4. prevailing winds
5. positions of high- and low-pressure areas
6. mountain barriers
7. topography: altitude, aspect, and slope
8. characteristics of Earth's surface: albedo, vegetation, soil type, and moisture

We can ascertain the effect that these controls have on climate by observing the global patterns of two key weather elements: temperature and precipitation.

GLOBAL TEMPERATURES • Figure 17.1 shows mean annual temperatures for the world. Notice that in both hemispheres, the isotherms are oriented east–west, reflecting the fact that locations at the same latitude receive nearly the same amount of solar energy. In addition, the annual solar energy decreases from low to high latitudes; hence, annual average temperatures tend to decrease from equatorial toward polar regions.* Notice as well that the temperatures tend to be cooler over mountainous regions—this is especially noticeable over the Himalayas and Tibetan plateau, as well as the Andes in South America.

The bending of the isotherms along the coastal margins is due in part to the unequal heating and cooling properties of land and water and to ocean currents and upwelling. For example, along the west coasts of North and South America, ocean currents transport cool water equatorward. In addition to this, the wind in both regions often blows toward the equator, parallel to the coast. This situation favours upwelling of cold water (see Chapter 10), which cools the coastal margins. In the area of the eastern North Atlantic Ocean (north of 40°N), the poleward bending of the isotherms is due to the

---

*Average global temperatures for January and July are given in Figures 3.19 and 3.20.

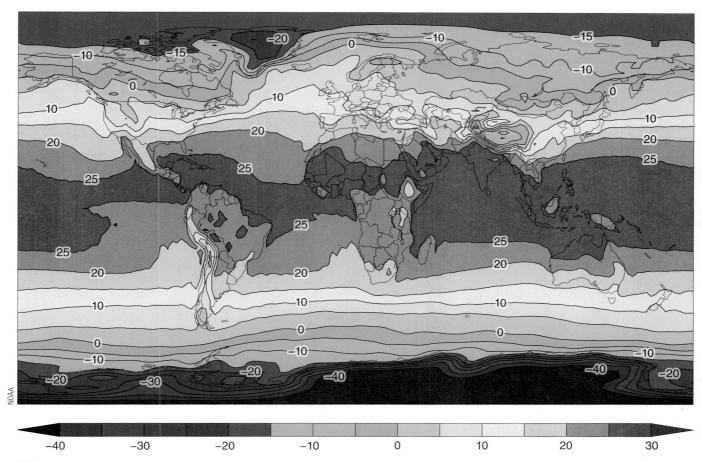

● FIGURE 17.1 Average annual sea-level temperatures throughout the world from 1980–2009 in degrees Celsius.

Gulf Stream and the North Atlantic Drift, which carry warm water northward.

The fact that landmasses heat up and cool off more quickly than large bodies of water means that the variation in temperature between summer and winter will be far greater over continental interiors than along the west coastal margins of continents. Because of this, the climates of interior continental regions will be more extreme as they have, on average, higher summer temperatures and lower winter temperatures than their west coast counterparts. In fact, west coast climates are typically quite mild for their latitude.

The most extreme high temperatures do not occur in the tropics but rather in the subtropical deserts of the Northern Hemisphere. Here the subsiding air associated with the subtropical anticyclones produces generally clear skies and low humidity. In summer, the high sun beating down on a relatively bare landscape produces scorching heat.

The lowest mean temperatures occur over large landmasses at high latitudes. The coldest areas in the Northern Hemisphere are found in the interior of Siberia and Greenland, whereas the coldest area of the world is Antarctica. During part of the year, the sun is below the horizon; when it is

above the horizon, it is low in the sky and its rays do not effectively warm the surface. Consequently, the land remains snow- and ice-covered year-round. The snow and ice reflect perhaps 80 percent of the sunlight that reaches the surface. Much of the absorbed solar energy is used to transform the ice and snow into water vapour. The relatively dry air and Antarctica's high elevation permit rapid radiational cooling during the dark winter months, producing extremely cold surface air. The extremely cold Antarctic helps explain why, overall, the Southern Hemisphere is cooler than the Northern Hemisphere. Another reason the Southern Hemisphere is cooler is that there is less land area in tropical and subtropical areas compared to in the Northern Hemisphere.

GLOBAL PRECIPITATION Appendix G shows the worldwide general pattern of annual precipitation, which varies from place to place. There are, however, certain regions that stand out as being wet or dry. For example, equatorial regions are typically wet, whereas the subtropics and the polar regions are relatively dry. The global distribution of precipitation is closely tied to the general circulation of winds in the

● FIGURE 17.2 A vertical cross section along a line running north to south illustrates the main global regions of rising and sinking air and how each region influences precipitation.

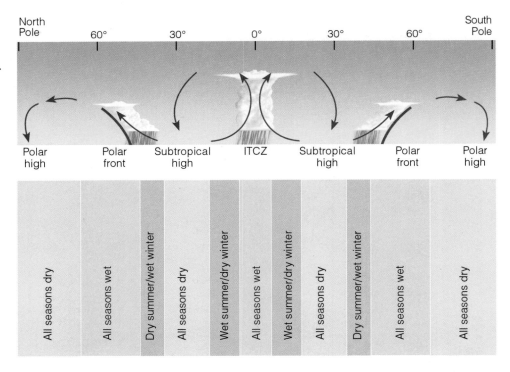

atmosphere (see Chapter 10) and to the distribution of mountain ranges and high plateaus.

● Figure 17.2 shows in simplified form how the general circulation influences the north-to-south distribution of precipitation to be expected on a uniformly water-covered Earth. Precipitation is greatest where the air rises and least where it sinks. Hence, one expects a great deal of precipitation in the tropics and along the polar front and little near subtropical highs and at the poles. Let's look at this in more detail.

In tropical regions, the trade winds converge along the intertropical convergence zone (ITCZ), producing rising air, towering clouds, and heavy precipitation all year long. Poleward of the equator, near latitude 30°, the sinking air of the subtropical highs produces a "dry belt" around the globe. The Sahara Desert of North Africa is in this region. Here annual rainfall is exceedingly light and varies considerably from year to year. Because the major wind belts and pressure systems shift with the season—northward in July and southward in January—the area between the rainy tropics and the dry subtropics is influenced by both the ITCZ and the subtropical highs.

In the cold air of the polar regions, there is little moisture, so there is little precipitation. Winter storms drop light, powdery snow that remains on the ground for a long time because of the low evaporation rates. In summer, a ridge of high pressure tends to block storm systems that would otherwise travel into the area; hence, precipitation in polar regions is meagre in all seasons.

There are exceptions to this idealized pattern. For example, in middle latitudes, the migrating position of the subtropical anticyclones also has an effect on the west-to-east distribution of precipitation. The sinking air associated with these systems is more strongly developed on their eastern side. Hence, the air along the eastern side of an anticyclone tends to be more stable; it is also drier as cooler air moves equatorward because of the circulating winds around these systems. In addition, along coastlines, cold, upwelling water cools the surface air even more, adding to the air's stability. Consequently, in summer, when the Pacific high moves to a position centred off the California coast, a strong, stable subsidence inversion forms above coastal regions. With the strong inversion and the fact that the anticyclone tends to steer storms to the north, the West Coast of North America experiences little rainfall during the summer months.

On the western side of subtropical highs, the air is less stable and more moist as warmer air moves poleward. In summer, over the North Atlantic, the Bermuda high pumps moist tropical air northward from the Gulf of Mexico into the eastern two-thirds of the United States. The humid air is conditionally unstable to begin with, and by the time it moves over the heated ground, it becomes even more unstable. If conditions are right, the moist air will rise and condense into cumulus clouds, which may build into towering thunderstorms.

In winter, the subtropical North Pacific high moves south, allowing storms travelling across the ocean to affect the West Coast of North America. The Bermuda high also moves south in winter. Across much of North America, intense winter storms develop and travel eastward, frequently dumping heavy precipitation as they go. Usually, however, the heaviest precipitation is concentrated along the eastern seaboard as moisture from the Gulf of Mexico moves northward

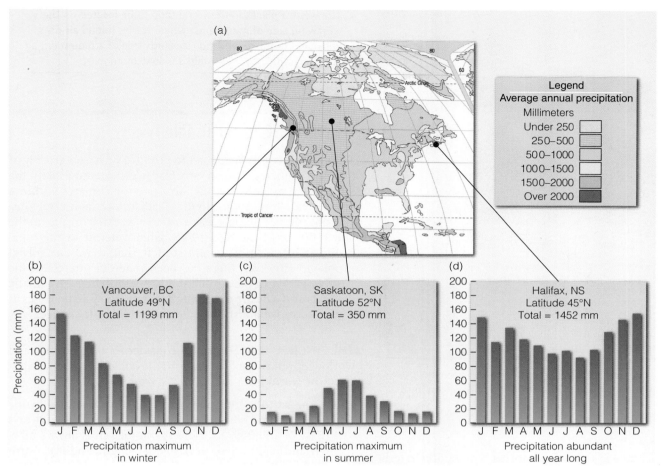

● FIGURE 17.3 Average annual precipitation across North America along with variation in annual precipitation for Vancouver, Saskatoon, and Halifax.

Source: Environment Canada. Climate Normals.

ahead of these systems. Therefore, cities on the prairies typically receive more rainfall in summer and those on the West Coast have maximum precipitation in winter, whereas cities in the east usually have abundant precipitation all year long. ● Figure 17.3 shows the average annual precipitation across North America as well as the contrast in seasonal precipitation among a West Coast city (Vancouver, British Columbia), a prairie city (Saskatoon, Saskatchewan), and an East Coast city (Halifax, Nova Scotia).

Mountain ranges disrupt the idealized pattern of global precipitation (1) by promoting convection (because their

slopes are warmer than the surrounding air) and (2) by forcing air to rise along their windward slopes (*orographic uplift*). Consequently, the windward side of mountains tends to be "wet." As air descends and warms along the leeward side, there is less likelihood of clouds and precipitation. Thus, the leeward (downwind) side of mountains tends to be "dry." As Chapter 6 points out, a region on the leeward side of a mountain where precipitation is noticeably less is called a *rain shadow.*

A good example of the rain shadow effect occurs on Vancouver Island. In this region, winter storms that bring most precipitation generally have winds from the southwest. Situated inland, but on the exposed western side of southern Vancouver Island, the Nitinat River Hatchery station annually receives an average of 3702 mm of precipitation (see ● Figure 17.4). The precipitation amount there is enhanced as southwesterly winds cause uplift on the windward slopes of the Vancouver Island Mountains. In contrast, only 110 km away at the Victoria Gonzales Heights station on the southeastern tip of Vancouver Island, the mean annual precipitation is less than one-sixth as much. The reason for

WEATHER WATCH

Henderson Lake, located on Vancouver Island between Tofino and Port Alberni, is one of the wettest places measurements have been made in North America. It has an annual average precipitation of over 6502.4 mm based on 14 years of measurements. In 1997, it had 9479 mm of precipitation—a Canadian record. This is over 200 times the average of 43 mm for Death Valley, California.

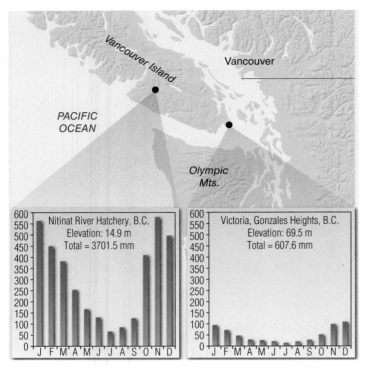

● FIGURE 17.4 The effect of the Vancouver Island Mountains and the Olympic Mountains in Washington State on average annual precipitation.

Data source: Environment Canada.

the large difference is that when storms with southwesterly winds are dumping copious amounts of rain at Nitinat River, southeastern Vancouver Island is downwind and therefore in the rain shadow of the high mountains of the Olympic Peninsula of Washington State.

● Figure 17.5 shows a classic example of how topography produces several rain shadow effects. (Additional information on precipitation extremes is given in Focus on an Observation: Precipitation Extremes on p. 534.)

BRIEF REVIEW

Before going on to the section on climatic classification, here is a brief review of some of the facts we have covered so far:

● Climatic controls are the factors that govern the climate of any given region.

● The hottest places on Earth tend to occur in the subtropical deserts of the Northern Hemisphere, where clear skies and sinking air, coupled with low humidity and a high summer sun beating down on a relatively barren landscape, produce extreme heat.

● The coldest places on Earth tend to occur in the interior of high-latitude landmasses. The coldest areas of the Northern Hemisphere are found in the interior of Siberia and Greenland, whereas the coldest area of the world is Antarctica.

● The wettest places in the world tend to be located on the windward side of mountains, where warm, humid air rises upslope. On the downwind (leeward) side of a mountain, a "dry" region, known as a rain shadow, often exists.

Climatic Classification

The climatic controls interact to produce such a wide array of different climates that no two places experience exactly the same climate. However, the similarity of climates within a given area allows us to divide Earth into climatic regions. Classification of climate was an active research area from the late 1800s until the mid-1900s, when interest in the topic dropped off as most scientists felt that it was a "solved problem." However, there has been renewed interest in the past 20 years with the advent of more comprehensive climatic data sets and a desire to monitor changes in climatic zones as a way of assessing historical and future climatic change.

THE ANCIENT GREEKS   By considering temperature and worldwide sunshine distribution, the ancient Greeks categorized the world into three climatic regions:

1. A low-latitude tropical (or torrid) zone, bounded by the northern and southern limits of the sun's vertical rays (23½°N and 23½°S). Here the noon sun is always high, day and night are of nearly equal length, and it is warm year-round.
2. A high-latitude polar (or frigid) zone, bounded by the Arctic or Antarctic Circle. This zone is cold all year long due to long periods of winter darkness and a low summer sun.
3. A middle-latitude temperate zone, sandwiched between the other two zones. This zone has distinct summer and winter, so it exhibits characteristics of both extremes.

Such a sunlight- or temperature-based climatic scheme is too simplistic. It excludes precipitation, so there is no way to differentiate between wet and dry regions. Better classifications of climate would take into account more of the important meteorological factors.

THE KÖPPEN SYSTEM   A widely used classification of world climates based on the annual and monthly averages of temperature and precipitation was devised by the German scientist Wladimir Köppen (1846–1940). Initially published in 1918, the original **Köppen classification system** has since been modified and refined. Faced with the lack of adequate observing stations throughout the world, Köppen related the distribution and type of native vegetation to the various climates. In this way, climatic boundaries could be approximated where no climatological data were available.

Köppen's scheme employs five major climatic types; each type is designated by a capital letter:

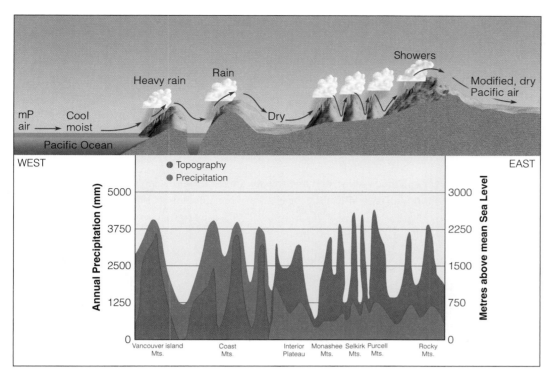

● FIGURE 17.5 The effect of topography on average annual precipitation along a line running from the Pacific Ocean through southern British Columbia to the Rocky Mountains.

Adapted from: D. Phillips. The Climates of Canada. Environment Canada. 1990.

A *Tropical climates (hot and moist)*: all months have an average temperature above 18°C. Since all months are warm, there is no real winter season.

B *Dry climates*: deficient precipitation most of the year. Potential evaporation and transpiration exceed precipitation.

C *Temperate climates (moist mid-latitude climates with mild winters)*: warm-to-hot summers with mild winters. The average temperature of the coldest month is below 18°C and above 0°C.

D *Cold climates (moist mid-latitude climates with severe winters)*: warm summers and cold winters. The average temperature of the warmest month exceeds 10°C, and the coldest monthly average drops to 0°C or below.

E *Polar climates*: extremely cold winters and summers. The average temperature of the warmest month is below 10°C. Since all months are cold, there is no real summer season.

In mountainous country, where rapid changes in elevation bring about sharp changes in climatic type, delineating the climatic regions is more difficult. For this reason, in some climatic classifications, these regions are designated by the letter *H*, for *highland climates*.

● Figure 17.6 gives a simplified overview of the major climatic types throughout the world, according to Köppen's system. Superimposed on the map are some of the climatic controls. These include the average annual positions of the semipermanent high- and low-pressure areas, the average position of the intertropical convergence zone in January

and July, the major mountain ranges and deserts of the world, and some of the major ocean currents. Notice how the climatic controls impact the climate in different regions of the world. As we would expect, due to changes in the intensity and amount of solar energy, polar climates are found at high latitudes and tropical climates at low latitudes. Dry climates tend to be located on the downwind side of major mountain chains and near 30° latitude, where the subtropical highs (with their sinking air) are found. Climates with more moderate winters (C climates) tend to be equatorward of those with severe winters (D climates). Along the west coasts of North America and Europe, warm ocean currents and prevailing westerly winds modify the climate such that coastal regions experience much milder winters than do regions farther inland.

Keep in mind that within the Köppen system, each major climatic group contains subgroups that describe special regional characteristics, such as seasonal changes in temperature and precipitation. The complete Köppen climatic classification system, as revised in 2007, as well as the criteria for the various subgroups, is given in ▼ Table 17.1.

Köppen's system has been criticized primarily because his boundaries (which relate vegetation to monthly temperature and precipitation values) do not always correspond to the natural boundaries of each climatic zone. In addition, the Köppen system implies that there is a sharp boundary between climatic zones, when, in reality, there is a gradual transition.

The Köppen system has been revised several times, most notably by the German climatologist Rudolf Geiger, who

▼ Table 17.1   Köppen's Climatic Classification System

LETTER SYMBOL				
1st	**2nd**	**3rd**	**Climatic Characteristics**	**Criteria**
A			Tropical (hot and moist)	All months have an average temperature of 18°C or higher
	f		Tropical wet (rain forest)	Wet all seasons; all months have at least 60 mm of rainfall
	w		Tropical wet and dry (savanna)	Winter dry season; rainfall in driest month is less than 60 mm and less than $100 - P/25$ (P is mean annual rainfall in mm)
	m		Tropical monsoon	Short dry season; rainfall in driest month is less than 60 mm but equal to or greater than $100 - P/25$
B			Dry	Potential evaporation and transpiration exceed precipitation. The dry/humid boundary is where the mean annual precipitation, P, is less than 10 times $P_{threshold}$. ($P < 10 \times P_{threshold}$) $P_{threshold}$ is defined according to the following rules: $P_{threshold \cdot} = 2T_{av}$ if 70% or more of the annual precipitation occurs in the cooler 6 months (dry summers) $P_{threshold \cdot} = 2\,T_{av} + 28$ if 70% or more of the annual precipitation occurs in the warmer 6 months (dry winters) $P_{threshold} = 2\,T_{av} + 14$ when neither season has more than 70% of the annual precipitation T_{av} is the mean annual temperature in °C
	S		Semiarid (steppe)	$P \geqslant 5 \times P_{threshold}$
	W		Arid (desert)	$P < 5 \times P_{threshold}$
		h	Hot and dry	Mean annual temperature is 18°C or higher
		k	Cool and dry	Mean annual temperature is below 18°C
C			Temperate (moist with mild winters)	Average temperature of the hottest month is more than 10°C and average temperature of the coolest month is between 0°C and 18°C
	w		Dry winters	Average rainfall of wettest summer month at least 10 times as much as in the driest winter month
	s		Dry summers	Average rainfall of driest summer month less than 40 mm and average rainfall of wettest winter month at least 3 times as much as in driest summer month
	f		Wet all seasons	Criteria for Cw and Cs cannot be met
		a	Hot summers	Average temperature of warmest month at least 22°C
		b	Warm summers	Average temperature of all months below 22°C and at least 4 months of the year with average above 10°C
		c	Cool summers	Average temperature of all months below 22°C and 1 to 3 months of the year with average above 10°C
D			Cold (moist with severe winters)	Average temperature of coldest month is 0°C or below; average temperature of warmest month is greater than 10°C
	w		Dry winters	Same as under Cw
	s		Dry summers	Same as under Cs
	f		Wet all seasons	Same as under Cf (criteria for Dw and Ds cannot be met)
		a	Hot summers	Same as under C_a
		b	Warm summers	Same as under C_b
		c	Cold summers	Not (a, b, or d)
		d	Cold summers and very cold winters	Not (a or b) and average temperature of coldest month is less than -38°C
E			Polar	Average temperature of warmest month is below 10°C
	T		Tundra	Average temperature of warmest month is greater than 0°C
	F		Ice cap	Average temperature of warmest month is 0°C or below

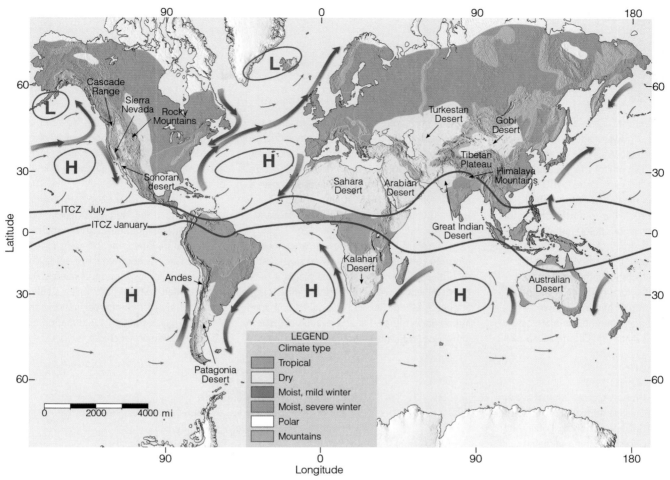

● FIGURE 17.6 A simplified overview of the major climatic types according to Köppen, along with some of the climatic controls. The large Hs and Ls on the map represent the average position of the semipermanent high- and low-pressure areas. The solid red lines show the average position of the intertropical convergence zone (ITCZ) in January and July. The ocean currents in red are warm, whereas those in blue are cold. The major mountain ranges and deserts of the world also are included.

worked with Köppen on amending the climatic boundaries of certain regions. For this reason, the system is sometimes referred to as the Köppen–Geiger classification. Despite criticisms of the Köppen system, it remains the most popular system in use, even 100 years after its introduction. A recent revision to the system was made in 2007 by M. C. Peel and coauthors using modern climatic data sets and is presented in ● Figure 17.7 (p. 536).

THORNTHWAITE'S SYSTEM   To correct some of the deficiencies of the Köppen classification system, the American climatologist C. Warren Thornthwaite (1899–1963) devised a new classification system in the early 1930s. Both systems used temperature and precipitation measurements, and both related natural vegetation to climate. However, to emphasize the importance of precipitation (P) and evaporation (E) on plant growth, Thornthwaite developed a *P/E ratio*, which is essentially monthly precipitation divided by

monthly evaporation. The annual sum of the P/E ratios gives the **P/E index**. Using this index, the Thornthwaite system defines five major humidity provinces and their characteristic vegetations: rain forest, forest, grassland, steppe, and desert.

To better describe the moisture available for plant growth, Thornthwaite proposed a new classification system in 1948 and slightly revised it in 1955. His new scheme emphasized the concept of *potential evapotranspiration** (PE), which is the amount of moisture that would be lost from the soil and vegetation if the moisture were available.

Thornthwaite incorporated potential evapotranspiration into a moisture index that depends essentially on the differences between precipitation and PE. The index is high in moist climates and negative in arid climates. An index of 0 marks the boundary between wet and dry climates.

**Evapotranspiration* refers to the evaporation from soil and transpiration of plants.

FOCUS ON AN OBSERVATION

Precipitation Extremes

Most of the "wettest" places in the world are located on the windward side of mountains. For example, Mount Waialeale on the island of Kauai, Hawaii, has one of the greatest annual average rainfall amounts on record: 11,684 mm. Cherrapunji, on the crest of the southern slopes of the Khasi Hills in northeastern India, receives an average of 10,800 mm of rainfall each year, the majority of which falls during the summer monsoon, between April and October. Cherrapunji, which holds the greatest 12-month rainfall total of 26,470 mm, once received 3800 mm of rain in just five days.

Record rainfall amounts are often associated with tropical storms. On the island of La Réunion (about 650 km east of Madagascar in the Indian Ocean), a tropical cyclone dumped 1350 mm of rain on Belouve in 12 hours. Heavy rains of short duration often occur with severe thunderstorms that move slowly or stall over a region. On July 4, 1956, 30 mm of rain fell from a thunderstorm on Unionville, Maryland, in one minute.

Snowfalls tend to be heavier where cool, moist air rises along the windward slopes of mountains. One of the snowiest places in North America is located at the Paradise Ranger Station in Mount Rainier National Park, Washington State. Situated at an elevation of 1646 m above sea level, this station receives an average of 1758 cm of snow annually. However, a record annual snowfall amount of 2896 cm was recorded at the Mount Baker ski area in northern Washington State during the winter of 1998–99.

As we noted earlier, the driest regions of the world lie in the frigid polar region, on the leeward side of mountains, and in the belt of subtropical high pressure, between 15° and 30° latitude. Arica in northern Chile holds the world record for lowest annual rainfall: 0.8 mm. In the United States, Death Valley, California, averages only 45 mm of precipitation annually. • Figure 1 gives additional information on world precipitation records.

KEY TO MAP

1	World's greatest annual average precipitation	11,872 mm	Mawsynram, India
2	Greatest annual precipitation in Canada	9479 mm	Henderson Lake, British Columba, 1997
3	Greatest 1-month rainfall total	9300 mm	Cherrapunji, India, July 1861
4	Greatest 12-hour rainfall total	1350 mm	Belouve, La Réunion Island, February 28 1964
5	Greatest 24-hour rainfall in Canada	489.2 mm	Ucluelet Brynnor Mines, British Columbia, October 6, 1967
6	Greatest 1-minute rainfall total in the world	38 mm	Barot, Guadeloupe, November 26, 1970
7	Lowest annual average rainfall in the world	0.8 mm	Arica, Chile
8	Lowest annual precipitation in Canada	2.7 mm	Arctic Bay, Nunavut, 1949
9	Greatest snowfall in one season in the world	2896 cm	Mount Baker, Washington State, U.S.A., 1998–99
10	Greatest snowfall in one season in Canada	2446.5 cm	Mount Copeland (Revelstoke), British Columbia, 1971–72
11	Greatest snowfall in 24 hours	193 cm	Silverlake, Boulder, CO, April 14–15, 1921
12	Greatest snowfall in 24 hours in Canada	145 cm	Tahtsa Lake, B.C., February 11, 1999.

CLIMATIC REGIONS IN CANADA Many climatic classification systems have been devised, often for specific applications in particular locations. Environment Canada has devised 11 **Canadian climatic regions**—areas that have broadly similar climatic conditions (see • Figure 17.8, p. 538). British Columbia is the most geographically and climatically diverse region in Canada. Managers of lands and forests there need to understand what species will grow most successfully in various parts of the province. For this reason, the *Biogeoclimatic Ecological Classification (BEC)* system has been developed that subdivides the land into 14 **biogeoclimatic zones**, as well as 97 subzones and 152 variants (see • Figure 17.9, p. 538). The system does not use climatic data directly but is based on site assessments of vegetation, soils, and topography to infer climate and identify areas with uniform climate.

The Global Pattern of Climate

Figure 17.7 gives a more detailed view of how the major climatic regions and subregions of the world are distributed based mainly on the work of Köppen and updated with recent data sets. (The major climatic types and their subdivisions are given in Table 17.1.) We will first examine humid tropical climates in low latitudes, and then we will look at middle-latitude and polar climates. Bear in mind that each climatic region has many subregions of local climatic differences wrought by factors such as topography, elevation, and large bodies of water. Remember, too, that boundaries of climatic regions represent gradual transitions. Thus, the major climatic characteristics of a given region are best observed away from its periphery.

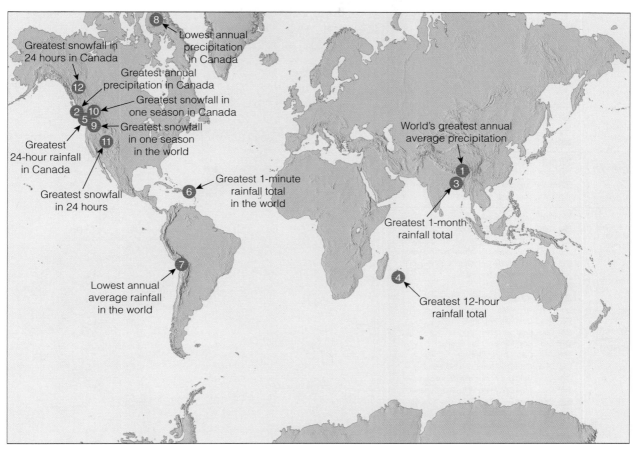

Greatest snowfall in 24 hours in Canada

Lowest annual precipitation in Canada

Greatest annual precipitation in Canada

Greatest snowfall in one season in Canada

Greatest snowfall in one season in the world

Greatest 24-hour rainfall in Canada

Greatest snowfall in 24 hours

Greatest 1-minute rainfall total in the world

World's greatest annual average precipitation

Greatest 1-month rainfall total

Lowest annual average rainfall in the world

Greatest 12-hour rainfall total

● FIGURE 1 Some precipitation records throughout the world.

TROPICAL MOIST CLIMATES (GROUP A) Group A climates have the following key features:

> *General characteristics*: year-round warm temperatures (all months have a mean temperature above 18°C); abundant rainfall (typical annual average exceeds 1500 mm)
>
> *Extent*: northward and southward from the equator to about latitude 15° to 25°
>
> *Major types* (based on seasonal distribution of rainfall): *tropical wet* (Af), *tropical monsoon* (Am), and *tropical wet and dry* (Aw)

At low elevations near the equator, in particular the Amazon lowland of South America, the Congo River Basin of Africa, and the East Indies from Sumatra to New Guinea,

high temperatures and abundant yearly rainfall combine to produce a dense, broadleaf, evergreen forest called a **tropical rain forest**. Here many different plant species, each adapted to differing light intensity, present a crudely layered appearance of diverse vegetation. In the forest, little sunlight is able to penetrate to the ground through the thick crown cover. As a result, little plant growth is found on the forest floor. However, at the edge of the forest, or where a clearing has been made, abundant sunlight allows for the growth of tangled shrubs and vines, producing an almost impenetrable jungle (see ● Figure 17.10).

Within the **tropical wet climate*** (Af), seasonal temperature variations are small (normally less than 3°C) because the

---

*The tropical wet climate is also known as the *tropical rain forest climate*.

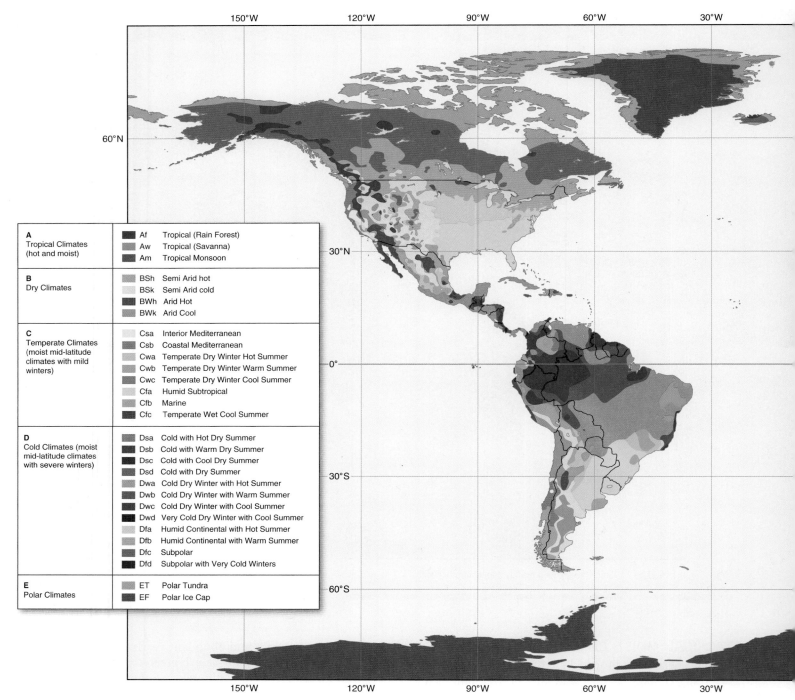

● FIGURE 17.7 Worldwide distribution of climatic regions according to the Köppen–Geiger classification.

Source: M. C. Peel, B. L. Finlayson, and T. A. McMahon 2007. "World map of the Köppen-Geiger climate classification." Hydrology and Earth System Sciences. Fig. 10. Pg. 1642

noon sun is always high and the number of daylight hours is relatively constant. However, there is a greater variation in temperature between day (average high about 32°C) and night (average low about 22°C) than there is between the warmest and coolest months. This is why people remark that winter comes to the tropics at night. The weather here is monotonous and sultry. There is little change in temperature from one day to the next. Furthermore, almost every day,

towering cumulus clouds form and produce heavy, localized showers by early afternoon. As evening approaches, the showers usually end and skies clear. Typical annual rainfall totals are greater than 1500 mm, and in some cases, especially along the windward side of hills and mountains, the total may exceed 4000 mm.

The high humidity and cloud cover tend to keep maximum temperatures from reaching extremely high values. In

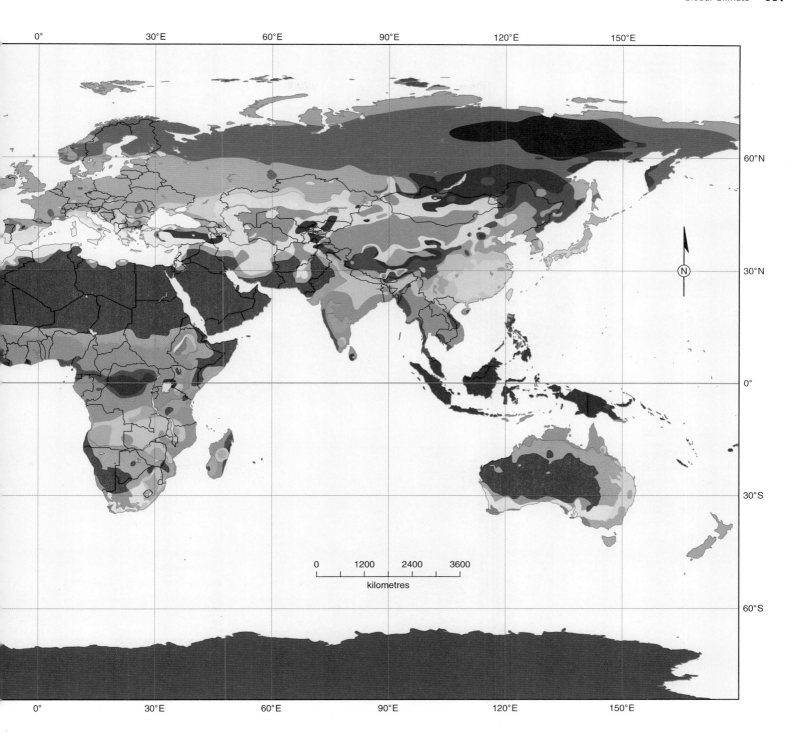

fact, summer afternoon temperatures are normally higher in middle latitudes than here. Nighttime radiational cooling can produce saturation and, hence, a blanket of dew, and, occasionally, fog covers the ground.

An example of a station with a tropical wet climate (Af) is Iquitos, Peru (see • Figure 17.11). Located near the equator (latitude 4°S), in the low basin of the upper Amazon River, Iquitos has an average annual temperature of 25°C with an

annual temperature range of only 2.2°C. Notice also that the monthly rainfall totals vary more than do the monthly temperatures. This is due primarily to the migrating position of the intertropical convergence zone (ITCZ) and its associated wind-flow patterns. Although monthly precipitation totals vary considerably, the average for each month exceeds 60 mm; consequently, no month is considered deficient of rainfall.

● FIGURE 17.8 The climatic regions of Canada.

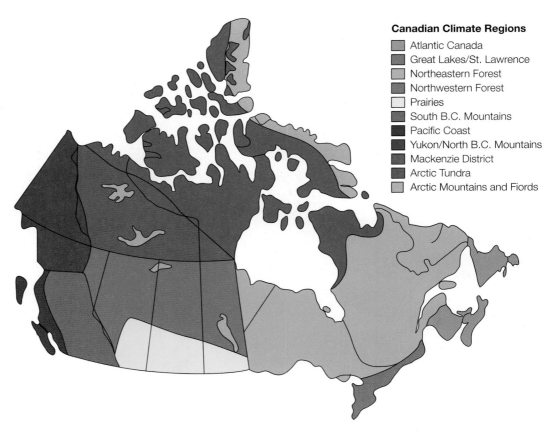

Canadian Climate Regions

- Atlantic Canada
- Great Lakes/St. Lawrence
- Northeastern Forest
- Northwestern Forest
- Prairies
- South B.C. Mountains
- Pacific Coast
- Yukon/North B.C. Mountains
- Mackenzie District
- Arctic Tundra
- Arctic Mountains and Fiords

● FIGURE 17.9 Biogeoclimatic zones of British Columbia. This classification uses site assessments of vegetation, soil, and topography to infer climatic zones. The zones are named after their dominant mature vegetation type.

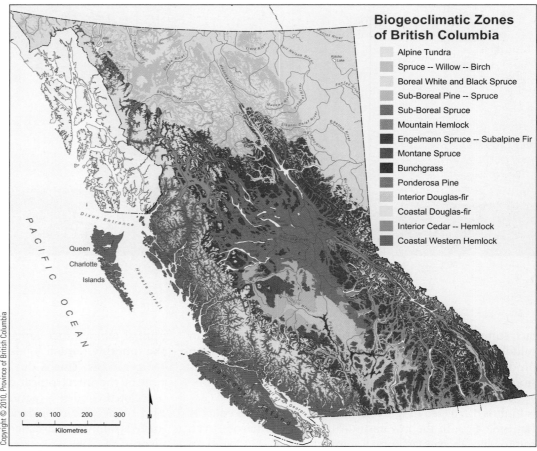

Biogeoclimatic Zones of British Columbia

- Alpine Tundra
- Spruce -- Willow -- Birch
- Boreal White and Black Spruce
- Sub-Boreal Pine -- Spruce
- Sub-Boreal Spruce
- Mountain Hemlock
- Engelmann Spruce -- Subalpine Fir
- Montane Spruce
- Bunchgrass
- Ponderosa Pine
- Interior Douglas-fir
- Coastal Douglas-fir
- Interior Cedar -- Hemlock
- Coastal Western Hemlock

● FIGURE 17.10 Tropical rain forest near Iquitos, Peru. (Climatic information for this region is presented in Figure 17.11.)

Take a minute and look again at Figure 17.10. From the photograph, one might think that the soil beneath the forest's canopy would be excellent for agriculture. Actually, this is not true. As heavy rain falls on the soil, the water works its way downward, removing nutrients in a process called leaching. Strangely enough, many of the nutrients needed to sustain the lush forest actually come from dead trees that decompose. The roots of the living trees absorb this matter before the rains leach it away. When the forests are cleared for agricultural purposes or for the timber, what is left is a thick red soil called laterite. When exposed to the intense sunlight of the tropics, the soil may harden into a bricklike consistency, making cultivation almost impossible.

Köppen classified tropical regions where the monthly precipitation drops below 60 mm for perhaps one or two months as **tropical monsoon climates** (Am). Here yearly rainfall totals are similar to those of the tropical wet climate, usually exceeding 1500 mm a year. Because the dry season is brief and copious rains fall throughout the rest of the year, there is sufficient soil moisture to maintain the tropical rain forest through the short dry period. Tropical monsoon climates can be seen in Figure 17.7 along the coasts of Southeast Asia and India and in northeastern South America.

Poleward of the tropical wet region, total annual rainfall diminishes, and there is a gradual transition from the tropical wet climate to the **tropical wet-and-dry climate** (Aw), where a distinct dry season prevails. Even though the annual precipitation usually exceeds 1000 mm, the dry season, where the

WEATHER WATCH

Hot and humid Belem, Brazil—a city situated near the equator with a tropical wet climate—had an all-time record high temperature of 37°C, exactly the same as the highest temperature ever measured in Fort Simpson, Northwest Territories (nearly 62°N), a city with a Dfc subpolar climate.

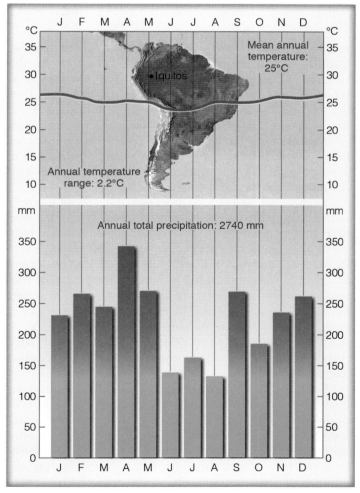

● FIGURE 17.11 Temperature and precipitation data for Iquitos, Peru, latitude 4°S. This station has a tropical wet climate (Af). (This type of diagram is called a climograph. It shows monthly mean temperatures with a solid red line and monthly mean precipitation with blue bar graphs.)

● FIGURE 17.12 Baobob and acacia trees illustrate typical trees of the East African grassland savanna, a region with a tropical wet-and-dry climate (Aw).

© J. L. Medeiros

monthly rainfall is less than 60 mm, lasts for more than two months. Because tropical rain forests cannot survive this "drought," the jungle gradually gives way to tall, coarse **savanna grass** scattered with low, drought-resistant deciduous trees (see ● Figure 17.12). The dry season occurs during the winter (low sun period), when the region is under the influence of the subtropical highs. In summer, the ITCZ moves poleward, bringing with it heavy precipitation, usually in the form of showers. Rainfall is enhanced by slow-moving shallow lows that move through the region.

Tropical wet-and-dry climates not only receive less total rainfall than the tropical wet climates, but the rain that does occur is much less reliable as the total rainfall often fluctuates widely from one year to the next. In the course of a single year, for example, destructive floods may be followed by serious droughts. As with tropical wet regions, the daily range of temperature usually exceeds the annual range, but the climate here is much less monotonous. There is a cool season in winter, when the maximum temperature averages 30°C to 32°C. At night, the low humidity and clear skies allow for rapid radiational cooling, and by early morning, minimum temperatures drop to 20°C or below.

From Figure 17.7 (p. 536), we can see that the principal areas with a tropical wet-and-dry climate (Aw) are those located in western Central America, in the region both north and south of the Amazon Basin (South America), in south-central and eastern Africa, in parts of India and Southeast Asia, and in northern Australia. In many areas (especially within India and Southeast Asia), the marked variation in precipitation is associated with the monsoon—the seasonal reversal of winds.

As we saw in Chapter 9, monsoon circulation is due in part to differential heating between landmasses and oceans. During winter in the Northern Hemisphere, winds blow outward, away from a cold, shallow, high-pressure area centred over continental Siberia. These downslope, relatively dry northeasterly winds from the interior provide India and Southeast Asia with generally fair weather and the dry season. In summer, the wind-flow pattern reverses as air flows into a developing thermal low over the continental interior. The humid air from the water rises and condenses, resulting in heavy rain and the wet season. (A more detailed look at the winter and summer monsoons is shown in Figure 9.30 on p. 282.)

An example of a station with a tropical wet-and-dry climate (Aw) is given in ● Figure 17.13. Located at latitude 11°N in West Africa, Timbo, Guinea, receives an annual average 1630 mm of rainfall. Notice that the rainy season is during the summer, when the ITCZ has migrated to its most northern position. Note also that practically no rain falls during the months of December, January, and February, when the region comes under the domination of the subtropical high-pressure area and its sinking air.

The monthly temperature patterns at Timbo are characteristic of most tropical wet-and-dry climates. As spring approaches, the noon sun is slightly higher, and the more intense sunshine produces greater surface heating and higher afternoon temperatures—usually above 32°C and occasionally above 38°C—creating hot, dry, desertlike conditions. After this brief hot season, a persistent cloud cover and the evaporation of rain tend to lower the temperature during the summer. The warm, muggy weather of summer often resembles that of the tropical wet climate (Af). The rainy summer is followed by a warm, relatively dry period, with afternoon temperatures usually climbing above 30°C.

Poleward of the tropical wet-and-dry climate, the dry season becomes more severe. Clumps of trees are more isolated, and the grasses dominate the landscape. When the potential annual water loss through evaporation and transpiration

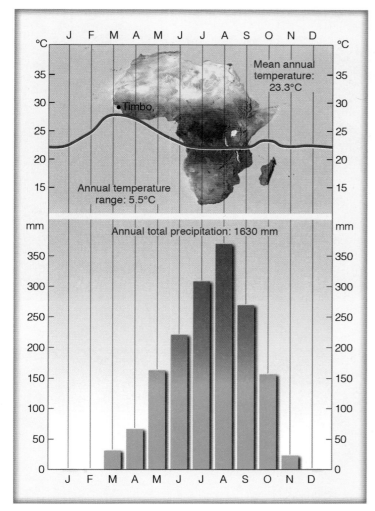

● FIGURE 17.13 Climatic data for Timbo, Guinea, latitude 11°N. This station has a tropical wet-and-dry climate (Aw).

exceeds the annual water gain from precipitation, the climate is described as dry.

DRY CLIMATES (GROUP B)

Group B climates have the following key features:

General characteristics: deficient precipitation most of the year; potential evaporation and transpiration exceed precipitation

Extent: the subtropical deserts extend from roughly 20° to 30° latitude in large continental regions of the middle latitudes, often surrounded by mountains

Major types: arid (BW), the "true desert," and semiarid (BS)

A quick glance at Figure 17.7 (p. 536) reveals that, according to Köppen, the dry regions of the world occupy more land area (about 26 percent) than any other major climatic type. Within these dry regions, a deficiency of water exists. Here the potential annual loss of water through evapo-

ration is greater than the annual water gained through precipitation. Thus, classifying a climate as dry depends not only on precipitation totals but also on temperature, which greatly influences evaporation. For example, 350 mm of precipitation in a hot climate will support only sparse vegetation, whereas the same amount of precipitation in much colder central Canada will support a conifer forest. In addition, a region with a low annual rainfall total is more likely to be classified as dry if the majority of precipitation is concentrated during the warm summer months, when evaporation rates are greater.

Precipitation in a dry climate is both meagre and irregular. Typically, the lower the average annual rainfall is, the greater its variability. For example, a station that reports an annual rainfall of 50 mm may actually measure no rainfall for two years; then, in a single downpour, it may receive 100 mm.

The major dry regions of the world can be divided into two primary categories. The first includes the area of the subtropics (between latitude 15° and 30°), where the sinking air of the subtropical anticyclones produces generally clear skies. The second is found in the continental areas of the middle latitudes. Here, far removed from a source of moisture, areas are deprived of precipitation. Dryness here is often accentuated by mountain ranges that produce a rain shadow effect.

Köppen divided dry climates into two types based on their degree of dryness: the *arid* (BW)* and the *semiarid*, or steppe (BS). These two climatic types can be divided even further. For example, if the climate is hot and dry with a mean annual temperature above 18°C, it is either BWh or BSh (the *h* is for *heiss*, meaning "hot" in German). On the other hand, if the climate is cold (in winter, that is) and dry with a mean annual temperature below 18°C, then it is either BWk or BSk (where the *k* is for *kalt*, meaning "cold" in German).

The **arid climates** (BW) occupy about 12 percent of the world's land area. From Figure 17.7 (p. 536), we can see that this climatic type is found along the west coast of South America and Africa and over much of the interior of Australia. Notice also that a swath of arid climate extends from northwest Africa all the way into central Asia. In North America, the arid climate extends from northern Mexico into the southern interior of the United States and northward along the leeward slopes of the Sierra Nevada. This region includes both the Sonoran and Mojave deserts and the Great Basin.

The southern desert region of North America is dry because it is dominated by the subtropical high most of the year, and winter storm systems tend to weaken before they move into the area. The northern region is in the rain shadow of the Sierra Nevada Mountains. These regions are deficient in precipitation all year long, with many stations receiving less than 130 mm annually. As noted earlier, the rain that does fall is spotty, often in the form of scattered summer afternoon showers. Some of these showers can be downpours that change a gentle gully into a raging torrent of water. More often than not, however, the rain evaporates into the dry air before

---

*The letter *W* is for *Wüste,* the German word for "desert."

● FIGURE 17.14 Rain streamers (virga) are common in dry climates as falling rain evaporates into the drier air before ever reaching the ground.

ever reaching the ground, and the result is rain streamers (virga) dangling beneath the clouds (see ● Figure 17.14).

Contrary to popular belief, few deserts are completely without vegetation. Although meagre, the vegetation that does exist must depend on the infrequent rains. Thus, most of the native plants are **xerophytes**—those capable of surviving prolonged periods of drought (see ● Figure 17.15). Such vegetation includes various forms of cacti and short-lived plants that spring up during the rainy periods.

In low-latitude deserts (BWh), intense sunlight produces scorching heat on the parched landscape. Here air temperatures are as high as anywhere in the world. Maximum daytime readings during the summer can exceed 50°C, although 40°C to 45°C are more common. In the middle of the day, the relative humidity is usually between 5 and 25 percent. At night, the air's relatively low water vapour content allows for rapid radiational cooling. Minimum temperatures often drop below 25°C. Thus, arid climates have large daily temperature ranges, often between 15°C and 25°C and occasionally higher.

During the winter, temperatures are more moderate, and minimums may, on occasion, drop below freezing. The variation in temperature from summer to winter produces large annual temperature ranges. We can see this in the climate record for Phoenix, Arizona (see ● Figure 17.16), a city in the southwestern United States with a BWh climate. Notice that the average annual temperature in Phoenix is 22°C and that the average temperature of the warmest month (July) reaches a sizzling 32°C. As we would expect, rainfall is meagre in all months. There is, however, a slight maximum in July and August. This is due to the summer monsoon, when more humid, southerly winds are likely to sweep over the region and develop into afternoon showers and thunderstorms (see Figure 9.32, p. 283).

● FIGURE 17.15 Creosote bushes and cacti are typical of the vegetation found in the arid southwestern American deserts (BWh).

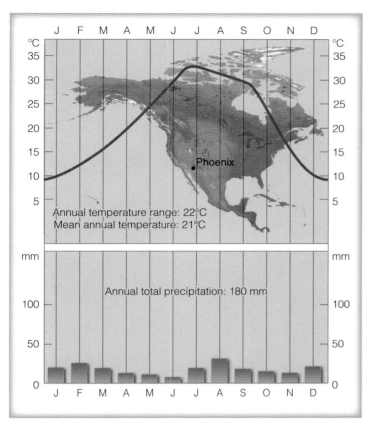

● FIGURE 17.16 Climatic data for Phoenix, Arizona, latitude 33.5°N. This station has an arid climate (BWh).

In middle-latitude deserts (BWk), average annual temperatures are lower. Summers are typically warm to hot, with afternoon temperatures frequently reaching 40°C. Winters are usually extremely cold, with minimum temperatures sometimes dropping below −35°C. Many of these deserts lie in the rain shadow of an extensive mountain chain, such as the Sierra Nevada, Cascade, and Coastal mountains in North America; the Himalayan Mountains in Asia; and the Andes in South America. The meagre precipitation that falls comes from an occasional summer shower or a passing mid-latitude cyclonic storm in winter.

Again, refer to Figure 17.7 and notice that around the margins of the arid regions, where rainfall amounts are greater, the climate gradually changes into **semiarid** (BS). This region is called **steppe** and typically has short bunch grass, scattered low bushes, trees, or sagebrush (see ● Figure 17.17). In North America, this climatic region includes most of the U.S. Great Plains extending into the southern Canadian prairies, the southern coastal sections of California, and the northern valleys of the Great Basin. As in the arid region, northern areas experience lower winter temperatures and more frequent snowfalls. Annual precipitation is generally between 200 and 400 mm. The climatic record for Medicine Hat, Alberta (see ● Figure 17.18), exemplifies the semiarid (BSk) climate.

As average rainfall amounts increase, the climate gradually changes to one that is more humid. Hence, the semiarid (steppe) climate marks the transition between the arid and the humid climatic regions. (Before reading about moist climates, you may wish to read Focus on an Observation: A Desert with Clouds and Drizzle on p. 544 to learn about deserts that experience drizzle but little rainfall.)

TEMPERATE CLIMATES (GROUP C) Group C climates have the following key features:

General characteristics: moist with mild winters (i.e., average temperature of the coldest month is between 0°C and 18°C)

Extent: on the eastern and western regions of most continents, from about 25° to 40° latitude

Major types: humid subtropical (Cfa), marine (Cfb), and dry-summer subtropical, or Mediterranean (Csa, Csb)

● FIGURE 17.17 Cumulus clouds forming over the steppe grasslands of western North America, a region with a semiarid climate (BS).

FOCUS ON AN OBSERVATION

A Desert with Clouds and Drizzle

We already know that not all deserts are hot. By the same token, not all deserts are sunny. In fact, some coastal deserts experience considerable cloudiness, especially low stratus and fog.

Amazingly, these coastal deserts are some of the driest places on Earth. They include the Atacama Desert of Chile and Peru, the coastal Sahara Desert of northwest Africa, the Namib Desert of southwestern Africa, and a portion of the Sonoran Desert in Baja, California (see ● Figure 2). On the Atacama Desert, for example, some regions go without measurable rainfall for decades. And Arica, in northern Chile, has an annual rainfall of only 0.8 mm.

The cause of this aridity is, in part, due to the fact that each region is adjacent to a large body of relatively cool water. Notice in Figure 2 that these deserts are located along the western coastal margins of continents, where a subtropical high-pressure area causes prevailing winds to bring in cool water from higher latitudes along the coast. In addition, these winds help accentuate the water's coldness by initiating upwelling—the rising of cold water from lower levels. The combination of these conditions tends to produce coastal water temperatures between 10°C and 15°C, which is quite cool for such low latitudes. As surface air sweeps across the cold water, it is chilled to its dew point, often producing a blanket of fog and low clouds, from which drizzle falls. The drizzle, however, accounts for very little rainfall. In most regions, it is

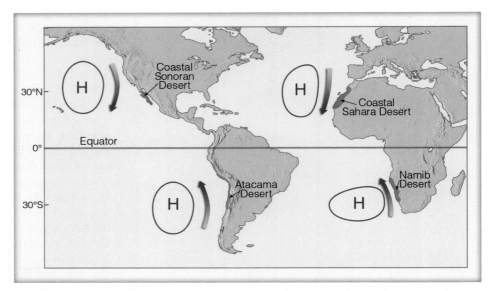

● **FIGURE 2** Location of coastal deserts (dark orange shade) that experience frequent fog, drizzle, and low clouds. (Blue arrows indicate prevailing winds and the movement of cool ocean currents.)

only enough to dampen the streets with a mere trace of precipitation.

As the cool, stable air moves inland, it warms, and the water droplets evaporate. Hence, most of the cloudiness and drizzle is found along the immediate coast. Although the relative humidity of this air is high, the dew point temperature is comparatively low (often near that of the coastal surface water). Inland, further warming causes the air to rise. However, a stable subsidence inversion, associated with the subtropical highs, inhibits vertical motions by

capping the rising air, causing it to drift back toward the ocean, where it sinks, completing a rather strong sea breeze circulation. The position of the subtropical highs, which tend to remain almost stationary, plays an additional role by preventing the intertropical convergence zone with its rising, unstable air from entering the region.

And so we have a desert with clouds and drizzle—a desert that owes its existence, in part, to its proximity to rather cold ocean water and, in part, to the position and air motions of a subtropical high.

The Group C climates of the middle latitudes have distinct summer and winter seasons. Additionally, they have ample precipitation to keep them from being classified as dry. Although winters can be cold and air temperatures can change appreciably from one day to the next, no month has a mean temperature below 0°C. If it did, it would be classified as a D climate—one with severe winters.

The first C climate we will consider is the **humid subtropical climate** (Cfa).* Notice in Figure 17.7 (p. 536) that

Cfa climates are found principally along the east coasts of continents, roughly between 25° and 40° latitude. They dominate the southeastern section of the United States, as well as eastern China and southern Japan. In the Southern Hemisphere, they are found in southeastern South America and along the southeastern coasts of Africa and Australia.

A trademark of the humid subtropical climate is its hot, muggy summers. This sultry summer weather occurs because Cfa climates are located on the western side of subtropical highs, where maritime tropical air from lower latitudes is swept poleward into these regions. Generally, summer dew point temperatures are high (often exceeding 23°C), and so is

---

*In the Cfa climate, the "f" means that all seasons are wet and the "a" means that summers are long and hot. A more detailed explanation is given in Table 17.1 on p. 532.

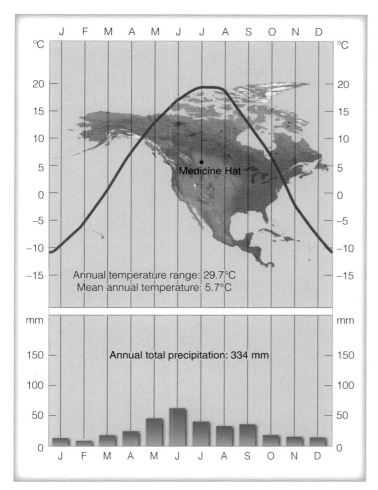

• FIGURE 17.18 Climatic data for Medicine Hat, Alberta, latitude 50°N. Medicine Hat has a semiarid climate (BSk).
Data source: Environment Canada. Canadian Climate Normals.

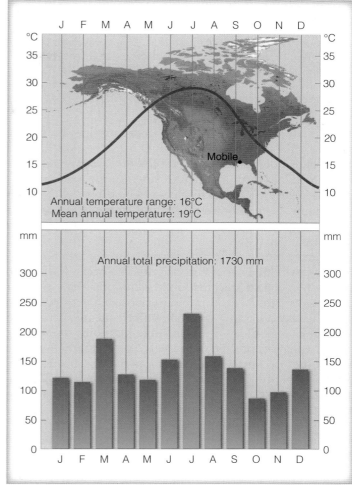

• FIGURE 17.19 Climatic data for Mobile, Alabama, latitude 30°N. This location has a humid subtropical climate (Cfa).

the relative humidity, even during the middle of the day. The high humidity combines with the high air temperature (usually above 32°C) to produce more oppressive conditions than are found in equatorial regions. Summer morning low temperatures often range between 21°C and 27°C. Occasionally, a weak summer cool front will bring temporary relief from the sweltering conditions. However, devastating heat waves, sometimes lasting many weeks, can occur when an upper-level ridge moves over the area.

Winters tend to be relatively mild, especially in the lower latitudes, where air temperatures rarely dip much below freezing. Poleward regions experience winters that are colder and harsher. Here frost, snow, and ice storms are more common, but heavy snowfalls are rare. Winter weather can be quite changeable as almost summerlike conditions can give way to cold rain and wind in a matter of hours when a middle-latitude storm and its accompanying fronts pass through the region.

Humid subtropical climates experience adequate and fairly well-distributed precipitation throughout the year,

with typical annual averages between 800 and 1650 mm. In summer, when thunderstorms are common, much of the precipitation falls as afternoon showers. Tropical storms entering the United States and China can substantially add to the summer and autumn rainfall totals. Winter precipitation most often occurs with eastward-trekking middle-latitude cyclonic storms. In the southeastern United States, the abundant rainfall supports a thick pine forest that becomes mixed with oak at higher latitudes. The climatic data for Mobile, Alabama, a city with a Cfa climate, is given in • Figure 17.19.

Glance back at Figure 17.7 (p. 536) and observe that C climates extend poleward along the western side of most continents from about latitude 40° to 60°. These regions are dominated by prevailing winds from the ocean that moderate the climate, keeping winters considerably milder than stations located at the same latitude farther inland. In addition to this, summers are quite cool. When the summer season is both short and cool, the climate is designated as Cfc. Equatorward, where summers are longer (but still cool), the climate is classified as west coast marine, or simply **marine**, Cfb.*

Where mountains parallel the coastline, such as along the west coasts of North and South America, the marine influence is restricted to narrow belts. Unobstructed by high mountains, prevailing westerly winds pump ocean air over much of western Europe and thus provide this region with a marine climate (Cfb).

During much of the year, marine climates are characterized by low clouds, fog, and drizzle. The ocean's influence produces adequate precipitation in all months, with much of it falling as light or moderate rain associated with maritime polar air masses. Snow does fall, but, frequently, it turns to slush after only a day or so. In some locations, topography greatly enhances precipitation totals. For example, along the West Coast of North America, coastal mountains not only force air upward, enhancing precipitation, they also slow the storm's eastward progress, which enables the storm to drop more precipitation on the area.

Along the northwest coast of North America, rainfall amounts decrease in summer. This phenomenon is caused by the northward migration of the subtropical Pacific high, which is located southwest of this region. The summer decrease in rainfall can be seen by examining the climatic record of Port Hardy (see ● Figure 17.20), a station situated along the northeast coast of Vancouver Island. The data illustrate another important characteristic of marine climates: the low annual temperature range for such a high-latitude station. The ocean's influence keeps daily temperature ranges low as well. In this climatic type, it rains on many days, and when it is not raining, skies are usually overcast. The heavy rains produce a dense forest of western hemlock and red cedar.

Moving equatorward of marine climates, the influence of the subtropical highs becomes greater and the summer dry period is more pronounced. Gradually, the climate changes from marine to one of **dry-summer subtropical** (Cs: Csa and Csb), or **Mediterranean**, because it also borders the coastal areas of the Mediterranean Sea. (Here the lowercase "s" stands for "summer dry.") Along the West Coast of North America, southern Vancouver Island—because it has rather dry summers—marks the transition between the marine climate and the dry-summer subtropical climate to the south.

The extreme summer aridity of the Mediterranean climate is caused by the sinking air of the subtropical highs. In

*In the Cfb climate, the "b" means that summers are cooler than in those regions experiencing a Cfa climate. The temperature criteria for the various subregions are given in Table 17.1 (p. 532).

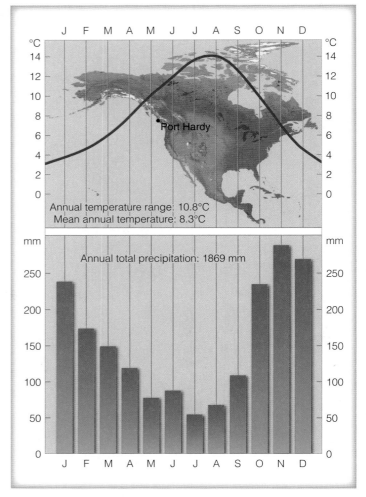

● FIGURE 17.20 Climatic data for Port Hardy, British Columbia, latitude 51°N. This location has a marine climate (Cfb).
Data source: Environment Canada. Canadian Climate Normals.

addition, these anticyclones divert summer storm systems poleward. During the winter, when the subtropical highs recede toward the equator, mid-latitude storms from the ocean frequent the region, bringing with them much needed rainfall. Consequently, Mediterranean climates are characterized by mild, wet winters and mild-to-hot, dry summers.

Where surface winds parallel the coast, upwelling of cold water helps keep the water itself and the air above it cool all summer long. In these coastal areas, which are often shrouded in low clouds and fog, the climate is called *coastal Mediterranean* (Csb). Here summer daytime maximum temperatures usually reach about 21°C, whereas overnight lows often drop below 15°C. Inland, away from the ocean's influence, summers are hot and winters are a little cooler than coastal areas. In this *interior Mediterranean climate* (Csa), summer afternoon temperatures usually climb above 34°C and occasionally above 40°C.

● Figure 17.21 contrasts the coastal Mediterranean climate of San Francisco, California, with the interior Mediterranean climate of Sacramento, California. Although Sacramento is

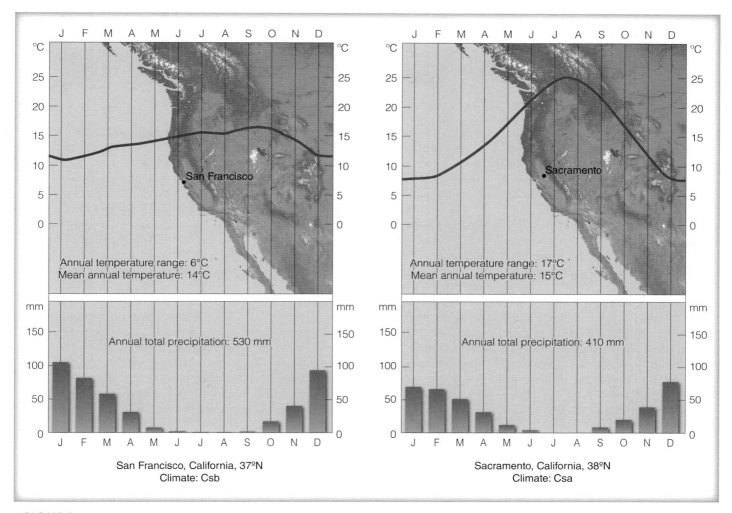

● FIGURE 17.21 Comparison of a coastal Mediterranean climate, Csb (San Francisco, at left), with an interior Mediterranean climate, Csa (Sacramento, at right).

only 130 km inland from San Francisco, Sacramento's average July temperature is 9°C higher. As we would expect, Sacramento's annual temperature range is considerably higher, too. Although Sacramento and San Francisco both experience an occasional frost, snow in these areas is a rarity.

In Mediterranean climates, yearly precipitation amounts range between 300 and 900 mm. However, much more precipitation falls on surrounding hillsides and mountains. Because of the summer dryness, the land supports only a scrubby type of low-growing woody plants and trees called chaparral (see ● Figure 17.22).

At this point, we should note that summers are not as dry along the Mediterranean Sea as they are along the West Coast of North America. Moreover, coastal Mediterranean areas are also warmer due to the lack of upwelling in the Mediterranean Sea.

Before leaving our discussion of C climates, note that when the dry season is in winter, the climate is classified as Cw. Over northern India and portions of China, the relatively dry winters are the result of northerly winds from continental

regions circulating southward around the cold Siberian high. Many lower latitude regions with a Cw climate would be tropical if it were not for the fact that they are too high in elevation and, consequently, too cool to be designated as tropical.

When a moist climate turns dry, drought often results. To learn more about drought, read Focus on a Special Topic: Drought on the Canadian Prairies on p. 548.

COLD CLIMATES (GROUP D)   Group D climates have the following key features:

General characteristics: warm-to-cool summers and cold winters (i.e., average temperature of warmest month exceeds 10°C, and the coldest monthly average drops below 0°C); winters are severe, with snowstorms, blustery winds, and bitter cold; climate is determined by the characteristics of a large continent

Extent: the middle latitudes, poleward of temperate (moist subtropical mid-latitude) climates

Drought on the Canadian Prairies

Ronald Stewart, Department of Environment and Geography, University of Manitoba

When a region's average precipitation is dramatically lower than normal for an extended period of time, drought may result. The word drought refers to a period of abnormally dry weather that produces a number of negative consequences on natural systems and human activities. This general term can be broken down further to meteorological drought (the persistent dryness portion of the above definition) and hydrological drought (characterized by unusual deficiency of groundwater and/or streamflow).

Drought is a relatively common feature of the North American climate system, and all regions of the continent are affected from time to time. However, drought tends to be most common and severe over the central regions of the continent. The Canadian Prairies are therefore prone to drought.

Droughts in the Canadian Prairies are distinctive in North America. The large-scale atmospheric circulations are influenced by blocking from high mountains to the west and long distances from all warm ocean-derived atmospheric water sources; growing season precipitation is generated by a highly complex combination of frontal and convective systems; seasonality is severe and characterized by rela-

tively long, snow-covered, and short growing seasons; local surface runoff is mainly produced by snowmelt water; and there is substantial water storage potential in the poorly drained, postglacial topography.

Despite these common features, there are still large variations in the nature of droughts over the Prairies. For example, the large-scale weather pattern causing drought is not always the same. It is often linked with a high-pressure system over the region, but it can also be associated with sustained **cold air advection**. Similarly, summer temperatures can be hot under the high-pressure conditions or cool under other conditions. Clouds are generally common under all conditions, although their character changes. As well, droughts can be calm or windy with dust, depending on the persistence of the weather pattern. It is not yet clear whether this dust significantly affects the character of drought through the reduction in solar radiation reaching the surface because of the many suspended particles or through the alteration of precipitation production processes with the many more potential cloud condensation nuclei.

One example of prairie drought is the recent one that began in 1999 with cessation

of the meteorological drought in 2004 or 2005, depending on the location across the region, and cessation of the hydrological drought later (see Figure 3). This event produced the worst drought for at least a hundred years in parts of the Canadian Prairies. Even in the dust bowl of the 1930s, no single year over the central Prairies was drier than in 2001. The drought affected agriculture, recreation, tourism, health, hydroelectricity, and forestry in the Prairies. The gross domestic product fell some $5.8 billion, and employment losses exceeded 41,000 jobs for 2001 and 2002. This drought also contributed to a negative or zero net farm income for several provinces for the first time in 25 years, with agricultural production over Canada dropping an estimated $3.6 billion in 2001–02 Previously reliable water supplies, such as streams, wetlands, dugouts, reservoirs, and groundwater, were placed under stress and often failed. The number of natural prairie ponds in May 2002 was the lowest on record. The dry conditions furthermore led to a huge dieback of trees across the northern Prairies.

Although the 1999–2005 drought was most profoundly felt on the Prairies, all regions of Canada were affected. For example, in 2003,

---

Major types: humid continental with hot summers (Dfa), humid continental with warm summers (Dfb), and subpolar (Dfc)

The D climates are controlled by large landmasses. Therefore, they are found only in the Northern Hemisphere. Look at the climate map, Figure 17.7 (p. 536), and notice that D climates extend across North America and Eurasia, from about latitude 40°N to almost 70°N. In general, they are characterized by cold winters and warm-to-cool summers.

As we know, for a station to have a D climate, the average temperature of its coldest month must dip below 0°C. This is not an arbitrary number. Studies suggest that an average monthly temperature of 0°C* or below for the coldest month corresponds to persistent winter snow cover in North America. Hence, D climates experience a great deal of winter

snow that stays on the ground for extended periods. When the temperature drops to a point such that no month has an average temperature of 10°C, the climate is classified as polar (E). Köppen found that the average monthly temperature of 10°C tended to represent the minimum temperature required for tree growth. So no matter how cold it gets in a D climate (and winters can get extremely cold), there is enough summer warmth to support the growth of trees.

There are two basic types of D climates: the **humid continental** (Dfa and Dfb) and the **subpolar** (Dfc). Humid continental climates are observed from about latitude 40°N to 50°N (60°N in Europe). Here precipitation is adequate and fairly evenly distributed throughout the year, although interior stations experience maximum precipitation in summer. Annual precipitation totals usually range from 500 to 1000 mm. Native vegetation in the wetter regions includes forests of spruce, fir, and pine, as well as oak and other deciduous trees. In autumn, nature's pageantry unveils itself as the leaves of deciduous trees turn brilliant shades of red, orange, and yellow (see Figure 17.23).

---

*Köppen found that, in Europe, a coldest month average temperature below −3°C marked the southern limit of persistent snow cover in winter, so some classifications use −3°C as the threshold to separate C from D climates instead of 0°C.

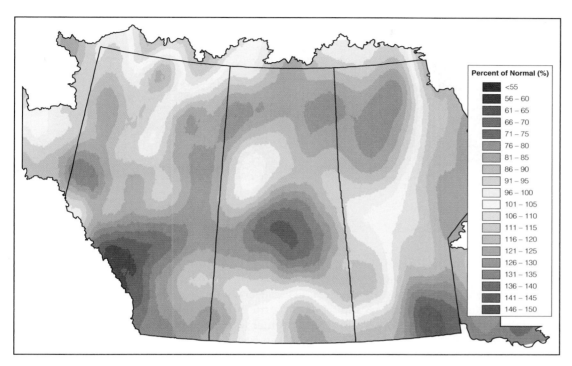

● **FIGURE 3** Precipitation deviation from normal over the Canadian Prairies in the early drought period, the September 1, 2000, to August 31, 2001, agricultural year.

Source: Created by Phillip Harder using data from Agriculture and Agri-Food Canada and Environment Canada.

Percent of Normal (%)

<55
56 – 60
61 – 65
66 – 70
71 – 75
76 – 80
81 – 85
86 – 90
91 – 95
96 – 100
101 – 105
106 – 110
111 – 115
116 – 120
121 – 125
126 – 130
131 – 135
136 – 140
141 – 145
146 – 150

the dry interior of British Columbia experienced massive fires, including one that affected Kelowna. The Great Lakes were very low during this period, with major impacts on shipping, tourism, and other sectors. The Atlantic provinces also suffered from dry conditions and the effects of low water levels.

There is much to learn about drought over Canada that can arise from different mechanisms but with the same devastating results.

Given the importance of drought, a comprehensive research network, referred to as the Drought Research Initiative (DRI), was established in 2005 to address the problem, with an initial focus on the Prairies.

● **FIGURE 17.22** In the Mediterranean-type climates of North America, typical chaparral vegetation includes chamisa, manzanita, and foothill pine.

● FIGURE 17.23 The leaves of deciduous trees burst into brilliant colour during autumn over the countryside of Ontario's Algonquin Park, a region with a humid continental climate.

Raimund Linke/Getty Images

Humid continental climates are subdivided on the basis of summer temperatures. Where summers are long and hot,* the climate is described as *humid continental with hot summers* (Dfa). Here summers are often hot and humid, especially in the southern regions. Midday temperatures often exceed 32°C and occasionally 40°C. Summer nights are usually warm and humid as well. The frost-free season normally lasts from five to six months, long enough to grow a wide variety of crops. Winters tend to be windy, cold, and snowy. Farther north, where summers are shorter and not as hot,† the climate is described as *humid continental with long warm summers* (Dfb). In Dfb climates, summers are not only cooler but also much less humid. Temperatures may exceed 35°C for a time, but extended hot spells lasting many weeks are rare. The frost-free season is shorter than in the Dfa climate and normally lasts between three and five months. Winters are long, cold, and windy. It is not uncommon for temperatures to drop below −30°C and stay below −18°C for days and sometimes weeks. Autumn is short, with winter often arriving right on the heels of summer. Spring, too, is short as late-spring snowstorms are common, especially in the more northern latitudes.

● Figure 17.24 compares the Dfa climate of Windsor, Ontario, with the Dfb climate of Winnipeg, Manitoba. Notice that both cities experience a large annual temperature range. This is characteristic of climates located in the northern interior of continents. In fact, as we move poleward, the annual temperature range increases. In Windsor, it is 27.2°C, whereas 900 km further north in Winnipeg, it is 37.3°C. Windsor's proximity to the Great Lakes also reduces the annual tempera-

ture range there. The summer precipitation maximum expected for these interior continental locations shows up in Figure 17.24, especially for Winnipeg. Most of the summer rain is in the form of convective showers, although an occasional weak frontal system can produce more widespread precipitation, as can a cluster of thunderstorms—the mesoscale convective complex, described in Chapter 14. The weather in both climatic types can be quite changeable, especially in winter, when a brief warm spell is replaced by blustery winds and temperatures plummeting well below −30°C.

When winters are severe and summers are short and cool, with only one to three months having a mean temperature exceeding 10°C, the climate is described as subpolar (Dfc). From Figure 17.7, we can see that in North America, this climate occurs in a broad belt across Canada and Alaska; in Eurasia, it stretches from Norway over much of Siberia. The exceedingly low temperatures of winter account for these areas being the primary source regions for continental polar and arctic air masses. Extremely cold winters coupled with cool summers produce large annual temperature ranges, as exemplified by the climatic data in ● Figure 17.25 for Churchill Falls, Labrador.

Precipitation is comparatively light in the subpolar climates, especially in the interior regions, with most places receiving less than 500 mm annually; however, this is not the case for Churchill Falls. A good percentage of the precipitation falls when weak cyclonic storms move through the region in summer. The total snowfall is usually not large, but the cold air prevents melting, so snow stays on the ground for months at a time. Because of the low temperatures, there is a low annual rate of evaporation that ensures adequate moisture to support the boreal* forests of conifers and birches

---

*"Hot" means that the average temperature of the warmest month is above 22°C and at least four months have a monthly mean temperature above 10°C.
†"Not as hot" means that the average temperature of the warmest month is below 22°C and at least four months have a monthly mean temperature above 10°C.

---

*The word *boreal* comes from the ancient Greek *boreas* meaning "wind from the north."

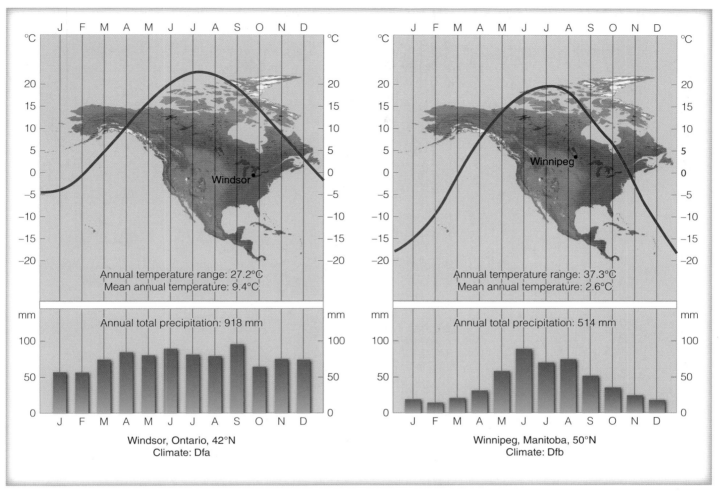

● FIGURE 17.24 Comparison of a humid continental hot summer climate, Dfa (Windsor, at left), with a humid continental cool summer climate, Dfb (Winnipeg, at right).

Data Source: Environment Canada. Canadian Climate Normals.

known as **taiga** (see ● Figure 17.26). Hence, the subpolar climate is known also as a *boreal climate* and as a *taiga climate*.

In the taiga region of northern Siberia and Asia, where the average temperature of the coldest month drops to a frigid −38°C or below, the climate is designated Dfd. Where the winters are considered dry, the climate is designated Dwd.

POLAR CLIMATES (GROUP E)   Group E climates have the following key features:

General characteristics: year-round low temperatures (i.e., average temperature of the warmest month is below 10°C)

Extent: northern coastal areas of North America and Eurasia, Greenland, and Antarctica

Major types: polar tundra (ET) and polar ice caps (EF)

In the **polar tundra** (ET), the average temperature of the warmest month is below 10°C but above freezing. (See ● Figure 17.27, the climatic data for Igloolik, Nunavut.) Here the ground is permanently frozen to depths of hundreds of

metres, a condition known as **permafrost**. Summer weather, however, is just warm enough to thaw out the upper metre or so of soil. Hence, during the summer, the tundra in some locations turns swampy and muddy. Annual precipitation on the tundra is meagre, with most stations receiving less than 200 mm. In lower latitudes, this would constitute a desert, but in the cold polar regions, evaporation rates are very low and moisture remains adequate. Because of the extremely short growing season, *tundra vegetation* consists of mosses, lichens, dwarf trees, and scattered woody vegetation, fully grown and only several centimetres tall (see ● Figure 17.28).

Even though summer days are long, the sun is never very high above the horizon. Additionally, some of the sunlight that reaches the surface is reflected by snow and ice, whereas some is used to melt the frozen soil. Consequently, in spite of the long hours of daylight, summers are quite cool. The cool summers and the extremely cold winters produce large annual temperature ranges.

When the average temperature for every month drops below freezing, plant growth is impossible, and the region is

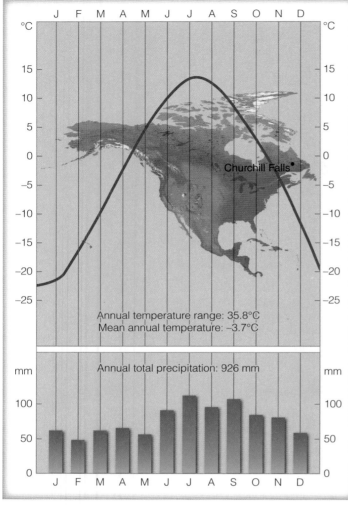

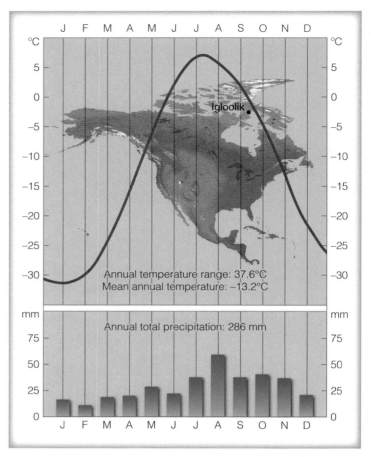

● FIGURE 17.27 Climatic data for Igloolik, Nunavut, latitude 69°N. This is a location with a polar tundra climate (ET).
Data Source: Environment Canada. Canadian Climate Normals.

● FIGURE 17.25 Climatic data for Churchill Falls, Labrador, latitude 53°33′ N. This station has a subpolar climate (Dfc).
Data Source: Environment Canada. Canadian Climate Normals.

● FIGURE 17.26 Coniferous forests (taiga) such as this occur where winter temperatures are low and precipitation is abundant.

● FIGURE 17.28 Tundra vegetation in Nunavut. This type of tundra is composed mostly of sedges and dwarfed wildflowers that bloom during the brief growing season.

perpetually covered with snow and ice. This climatic type is known as **polar ice cap** (EF). It occupies the interior ice sheets of Greenland and Antarctica, where the depth of ice in some places measures thousands of metres. In this region, temperatures are never much above freezing, even during the middle of "summer." The coldest places in the world are located here. Precipitation is extremely meagre, with many places receiving less than 100 mm annually. Most precipitation falls as snow during the "warmer" summer. Strong downslope katabatic winds frequently whip the snow about, adding to the climate's harshness. The data in ● Figure 17.29 for Eismitte, Greenland, illustrate the severity of an EF climate.

HIGHLAND CLIMATES   It is not necessary to visit the polar regions to experience a polar climate. Because temperature decreases with altitude, climatic changes experienced when climbing 300 m in elevation are about equivalent in high latitudes to horizontal changes experienced when travelling 300 km northward. (This distance is equal to about 3° latitude.) Therefore, when ascending a high mountain, one can travel through many climatic regions in a relatively short distance. It is for this reason that some climatic classifications include a sixth group, called *highland climates* (H).

Look back at Figure 17.9 on p. 538 to see how the climate and vegetation change with elevation over mountainous British Columbia. Across the southern part of the province, the Köppen climatic classification changes 10 times and includes all major climatic types except A (tropical). Moving from west to east, the Cfb (Marine) climate on the west coast of Vancouver Island changes to Csb (Mediterranean) on the east coast of southern Vancouver Island and the Vancouver area; it then transitions back to a marine climate and then a Dsb climate (cold with

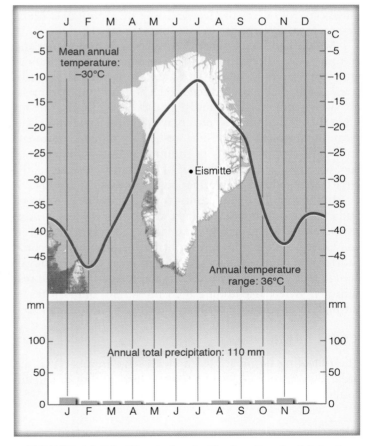

● FIGURE 17.29 Climatic data for Eismitte, Greenland, latitude 71°N. Eismitte is located in the interior of Greenland at an elevation of approximately 3000 m above sea level and has a polar ice cap climate (EF).

warm, dry summers) through the coastal mountains to a BSk (semiarid cold or steppe climate) or BWk (arid, cold desert) climate in the desertlike southern interior of British Columbia in the rain shadow of the Coast Mountains. Moving eastward over the southern interior mountains, the climate changes to a Dfb (cold humid continental with warm summers) to a Dfc (subpolar) climate over the Rocky Mountains. At the higher mountain elevations above the tree line, the climate changes to ET (polar tundra) and to EF over glaciers and ice caps. In southern Alberta, a BSk (semiarid steppe) climate is again found. (See Figure 17.5, p. 531, for the topography and precipitation patterns across the southern part of British Columbia.)

SUMMARY

In this chapter, we examined global temperature and precipitation patterns, as well as the various climatic regions throughout the world. Tropical climates are found in low latitudes, where the noon sun is always high, day and night are of nearly equal length, every month is warm, and no real winter season exists. Some of the rainiest places in the world exist in the tropics, especially where warm, humid air rises upslope along mountain ranges.

Dry climates prevail where potential evaporation and transpiration exceed precipitation. Some deserts, such as the Sahara, are mainly the result of sinking air associated with the subtropical highs, whereas others, due to the rain shadow effect, are found on the leeward side of mountains. Many deserts form in response to both of these effects.

Middle latitudes are characterized by a distinct winter and summer season. Winters tend to be milder in lower latitudes and more severe in higher latitudes. Along the east coast of some continents, summers tend to be hot and humid as moist air sweeps poleward around the subtropical highs. The air often rises and condenses into afternoon thunderstorms in this humid subtropical climate. The west coasts of many continents tend to be drier, especially in summer, as the combination of cool ocean water and sinking air of the subtropical highs, to a large degree, inhibits the formation of cumuliform clouds.

In the middle of large continents, such as North America and Eurasia, summers are usually wetter than winters. Winter temperatures are generally lower than those experienced in coastal regions. As one moves northward, summers become shorter and winters longer and colder. Polar climates prevail at high latitudes, where winters are severe and there is no real summer. When ascending a high mountain, one can travel through many climatic zones in a relatively short distance.

KEY TERMS

The following terms are listed (with page numbers) in the order they appear in the text. Define each. Doing so will aid you in reviewing the material covered in this chapter.

microclimate, 526
mesoclimate, 526
macroclimate, 526
global climate, 526
climatic controls, 526

Köppen classification
    system, 530
P/E index, 533
Canadian climatic
    regions, 534

biogeoclimatic zones, 534
tropical rain forest, 535
tropical wet climate, 535
tropical monsoon
    climates, 539
tropical wet-and-dry
    climate, 539
savanna grass, 540
arid climates, 541
xerophytes, 542
semiarid climate, 543
steppe, 543
humid subtropical
    climate, 544

marine climate, 546
dry-summer subtropical
    (Mediterranean)
    climate, 546
cold air advection, 548
humid continental
    climate, 549
subpolar climate, 549
taiga, 551
polar tundra climate, 551
permafrost, 551
polar ice cap climate, 553

QUESTIONS FOR REVIEW

1. What factors determine the global pattern of precipitation?
2. Explain why, in North America, precipitation typically is a maximum along the West Coast in winter, a maximum on the Prairies in summer, and fairly evenly distributed between summer and winter along the East Coast.
3. Why are the lowest temperatures in polar regions observed in the interior of large landmasses?
4. What climatic information did Köppen use in classifying climates?
5. How did Köppen define tropical climate? How did he define a polar climate?
6. According to Köppen's climatic system (Figure 17.7, p. 536), what major climatic type is most abundant in each of the following areas?
   (a) North America
   (b) South America
   (c) throughout the world
7. What is the primary factor that makes a dry climate "dry"?
8. In which climatic region would each of the following be observed: tropical rain forest, xerophytes, steppe, taiga, tundra, and savanna?
9. What are the controlling factors (the climatic controls) that produce the following climatic regions?
   (a) tropical wet and dry
   (b) Mediterranean

(c) marine

(d) humid subtropical

(e) subpolar

(f) polar ice cap

10. How do C-type climates differ from D-type climates?

11. Why are large annual temperature ranges characteristic of D-type climates?

12. Why are D climates found in the Northern Hemisphere but not in the Southern Hemisphere?

13. Explain why a tropical rain forest climate will support a tropical rain forest, whereas a tropical wet-and-dry climate will not.

14. Why are marine climates (Cfa) usually found on the west coast of continents?

15. What is the primary distinction between a Cfa and a Dfa climate?

16. Explain how arid deserts can be found adjacent to oceans.

17. Why did Köppen use the 10°C average temperature for the warmest month to distinguish between D and E climates?

18. What accounts for the existence of a BWk climate in the western Great Basin of North America?

19. Ulukhaktok, Northwest Territories (71°N latitude), receives a mere 162 mm of precipitation annually. Explain why its climate is not classified as arid or semiarid.

20. Explain why subpolar climates are also known as boreal climates and taiga climates.

QUESTIONS FOR THOUGHT

1. Why do cities directly east of the Rockies (such as Calgary, Alberta—432 mm) receive much more precipitation than cities east of the Coastal Mountains (such as Clinton [168 mm], and Kamloops [279 mm], British Columbia)?

2. What climatic controls affect the climate in your area?

3. Vancouver, Prince Rupert, and Halifax are all coastal cities, yet Halifax has a continental rather than a marine climate. Explain why.

4. Why are many structures in polar regions built on pilings?

5. Why are summer afternoon temperatures in a humid subtropical climate (Cfa) often higher than in a tropical wet climate (Af)?

6. Why are humid subtropical climates (Cfa) found in regions bounded by 20° and 40° (N or S) latitudes and nowhere else?

7. In which of the following climatic types is virga likely to occur most frequently: humid continental, arid desert, or polar tundra? Explain why.

8. As shown in Figure 17.21 (p. 547), San Francisco and Sacramento, California, have similar mean annual temperatures but different annual temperature ranges. What factors control the annual temperature ranges at these two locations?

9. Why is there a contrast in climatic types on either side of the Rocky Mountains but not on either side of the Laurentian Mountains?

10. Over the past 100 years or so, the Earth has warmed by more than 0.7°C. If this warming should continue over the next 100 years, explain how this rise in temperature might influence the boundary between C and D climates. How would the warming influence the boundary between D and E climates?

PROBLEMS AND EXERCISES

1. Suppose that a city has the mean annual precipitation and temperature given in ▼ Table 17.2. Based on Köppen's climatic types, how would this climate be classified? On a map of North America, approximately where would this city be located? What type of vegetation would you expect to see there? Answer these same questions for the data in ▼ Table 17.3.

2. Compare the following climatic classifications for your area:

(a) Ancient Greeks

(b) Köppen system

(c) Thornthwaite's system

Which classification system is best for your area's mesoclimate? Macroclimate?

3. On a blank map of the world, roughly outline where Köppen's major climatic regions are located.

▼ Table 17.2

	JAN	FEB	MAR	APR	MAY	JUN	JUL	AUG	SEP	OCT	NOV	DEC	YEAR
Temperature (°C)	4	6	10	16	20	25	27	26	23	17	9	6	17
Precipitation (mm)	124	106	134	93	96	81	101	83	68	63	86	104	1143

▼ Table 17.3

	JAN	FEB	MAR	APR	MAY	JUN	JUL	AUG	SEP	OCT	NOV	DEC	YEAR
Temperature (°C)	−8	−8	−2	6	13	18	21	20	16	9	2	−5	8
Precipitation (mm)	48	38	56	66	74	91	97	76	79	74	71	48	813

A thick layer of smoke and haze covers Santiago, Chile.
© M. I. G./Baeza/Photo Researchers

Air Pollution

ir pollution makes the earth a less pleasant place to live. It reduces the beauty of nature. This blight is particularly noticed in mountain areas. Views that once made the pulse beat faster because of the spectacular panorama of mountains and valleys are more often becoming shrouded in smoke. When once you almost always could see giant boulders sharply etched in the sky and the tapered arrowheads of spired pines, you now often see a fuzzy picture of brown and green. The polluted air acts like a translucent screen pulled down by an unhappy God.

Louis J. Battan. Dr. Battan (1923–86) was a professor of meteorology and author of 13 professional and popular books on meteorology. The American Meteorological Society has named an award for educational authors after him.

Louis J. Battan. *The Unclean Sky; A Meteorologist Looks at Air Pollution.* (New York, Anchor Books.) 1966.

CONTENTS

Every deep breath fills our lungs mostly with gaseous nitrogen and oxygen. Also inhaled, in minute quantities, may be other gases and particles, some of which could be considered pollutants. These contaminants come from car exhaust, chimneys, forest fires, factories, power plants, and other sources related to human activities.

Virtually every large city has to contend in some way with air pollution, which clouds the sky, injures plants, and damages property. Some pollutants merely have a noxious odour, whereas others can cause severe health problems. The cost is high. In North America, for example, outdoor air pollution takes its toll in health care and lost work productivity at an annual expense that runs into billions of dollars. Estimates are that, worldwide, nearly 1 billion people in urban environments are continuously being exposed to health hazards from air pollutants.

Air pollution is an impact of the *anthrosphere* on all the other Earth systems. Pollutants emitted into air by human activity alter the chemical composition of the *atmosphere*. Some pollutants, such as *greenhouse gases* such as carbon dioxide and methane, alter the atmosphere's ability to absorb and emit radiation. This can have profound impacts on global climate, as discussed in Chapter 16. Other pollutants may be harmful to animals and plants in Earth's *biosphere* or affect the anthrosphere through impacts on human health and property. Air pollutants that are washed out by precipitation join the *hydrosphere* and *cryosphere*, where they may have further effects on the biosphere and *lithosphere* through contamination of soils.

This chapter takes a look at this serious contemporary concern. We will begin by briefly examining the history of problems in this area and then go on to explore the types and sources of air pollution, as well as the weather that can produce an unhealthy accumulation of pollutants. Finally, we investigate how air pollution influences the urban environment and how it brings about unwanted acid precipitation.

A Brief History of Air Pollution

Strictly speaking, air pollution is not a new problem. More than likely, it began when humans invented fire, the smoke of which choked the inhabitants of poorly ventilated caves. In fact, very early accounts of air pollution characterized the phenomenon as "smoke problems," the major cause being people burning wood and coal to keep warm.

To alleviate the smoke problem in old England, King Edward I issued a proclamation in 1273 forbidding the use of sea coal, an impure form of coal that produced a great deal of soot and sulphur dioxide when burned. One person was reputedly executed for violating this decree. In spite of such restrictions, the use of coal as a heating fuel grew during the 15th and 16th centuries.

As industrialization increased, the smoke problem worsened. In 1661, the prominent scientist John Evelyn wrote an essay deploring London's filthy air. And by the 1850s, London had become notorious for its "pea soup" fog, a thick mixture of smoke and fog that hung over the city. These fogs could be dangerous. In 1873, one was responsible for as many as 700 deaths. Another in 1911 claimed the lives of 1150 Londoners. To describe this chronic atmospheric event, a physician, Harold Des Voeux, coined (around 1911) the word *smog*, meaning a combination of smoke and fog.

Little was done to control the burning of coal as time went by, primarily because it was extremely difficult to counter the basic attitude of the powerful industrialists: "Where there's muck, there's money." London's acute smog problem intensified. Then, during the first week of December 1952, a major disaster struck. The winds died down over London and the fog and smoke became so thick that people walking along the street literally could not see where they were going. People wore masks over their mouths and found their way along the sidewalks by feeling the walls of buildings. This particularly disastrous smog lasted 5 days and took nearly 4000 lives, prompting Britain's Parliament to pass the Clean Air Act in 1956. Additional air pollution incidents occurred in England during 1956, 1957, and 1962, but due to the strong legislative measures taken against air pollution, London's air today is much cleaner, and "pea soup" fogs are a thing of the past.

Air pollution episodes were by no means limited to Great Britain. During the winter of 1930, for instance, Belgium's highly industrialized Meuse Valley experienced an air pollution tragedy when smoke and other contaminants accumulated in a narrow, steep-sided valley. The tremendous buildup of pollutants caused about 600 people to become ill, and, ultimately, 63 died. Not only did humans suffer, but cattle, birds, and rats also fell victim to the deplorable conditions.

The industrial revolution brought air pollution to North America as homes and coal-burning industries belched smoke, soot, and other undesirable emissions into the air. Soon large industrial cities, such as St. Louis and Pittsburgh (which became known as the "Smoky City"), began to feel the effects of the ever-increasing use of coal. As early as 1911, studies documented the irritating effect of smoke particles on the human respiratory system and the "depressing and devitalizing" effects of the constant darkness brought on by giant, black clouds of smoke. By 1940, the air over some cities had become so polluted that automobile headlights had to be turned on during the day.

The first major documented air pollution disaster in the United States occurred at Donora, Pennsylvania, during October 1948, when industrial pollution became trapped in the Monongahela River Valley. During the ordeal, which lasted 5 days, more than 20 people died and thousands became ill.* Several times during the 1960s, air pollution levels became dangerously high over New York City. Meanwhile, on the West Coast, in cities such as Los Angeles, the ever-rising automobile population, coupled with the large petroleum processing

---

*Additional information about the Donora air pollution disaster is given in the Focus section on p. 584.

plants, was instrumental in generating a different type of pollutant, photochemical smog—one that forms in sunny weather and irritates the eyes. Toward the end of World War II, Los Angeles had its first (of many) smog alert.

Air pollution episodes in Los Angeles, New York, and other large American cities led to the establishment of much stronger emission standards for industry and automobiles. The U.S. Clean Air Act of 1970, for example, empowered the U.S. federal government to set emission standards that each state was required to enforce. That act was revised in 1977 and again in 1990 to include stricter emission requirements for automobiles and industry. Canada has largely followed the United States, introducing its own Clean Air Act in 1970, which was focused on regulating asbestos, lead, mercury, and vinyl chloride. The Canadian Environmental Protection Act (1999) is the current federal legislation setting national standards for air quality in Canada. Air pollution is a provincial responsibility, so each province and territory also has its own legislation to regulate air pollution. The Canadian Council of Ministers of the Environment, made up of federal and provincial environment departments, has set Canada-wide standards for pollutants such as ozone and fine particulate matter, which came into effect in 2010.

Types and Sources of Air Pollutants

Air pollutants are airborne substances (solids, liquids, or gases) that occur in concentrations high enough to threaten the health of people and animals, to harm vegetation and structures, or to toxify a given environment. Air pollutants come from both natural sources and human activities. Examples of natural sources include wind picking up dust from Earth's surface and carrying it aloft, volcanoes belching tonnes of ash and gases into our atmosphere, and forest fires producing vast quantities of drifting smoke (see ● Figure 18.1).

● FIGURE 18.1 Smoke from forest fires at Okanagan Mountain Park, near Kelowna, British Columbia, on September 2, 2003. The fires destroyed 238 homes and caused the evacuation of 30,000 people at one point. The image is from the *ASTER* satellite with simulated true colour. Red areas are active fires.

Human-induced pollution enters the atmosphere from both *fixed sources* and *mobile sources*. Fixed sources encompass industrial complexes, power plants, homes, office buildings, and so forth; mobile sources include motor vehicles, ships, and jet aircraft. Certain pollutants are called **primary air pollutants** because they enter the atmosphere directly—from smokestacks and tailpipes, for example. Other pollutants, known as **secondary air pollutants**, form only when a chemical reaction occurs between a primary pollutant and some other component of air, such as water vapour or another pollutant. ▼ Table 18.1 summarizes some of the sources of primary air pollutants.

There are two main categories of primary air pollutants: **hazardous air pollutants** and **criteria air contaminants**.

Hazardous air pollutants are toxic in low concentrations and therefore generally emitted only by accident or in small quantities. There are hundreds of these, including compounds such as benzene, toluene, and a host of compounds containing heavy metals such as lead. Also included in this group are persistent organic pollutants such as DDT (a pesticide), PCBs (polychlorinated biphenyls), and many others. These can persist and bioaccumulate in the environment. This means that they do not break down easily, and when they are ingested by organisms, they accumulate and magnify up the food chain, potentially becoming toxic for higher level organisms such as mammals and predators. The transport and deposition of hazardous air pollutants to the Arctic, where they affect the fragile arctic biosphere, is a particularly serious problem.

▼ Table 18.1   Some of the Sources of Primary Air Pollutants

	SOURCES	POLLUTANTS
Natural		
	Volcanic eruptions	Particles (dust, ash), gases (SO_2, CO_2)
	Forest fires	Smoke, unburned hydrocarbons, CO_2, nitrogen oxides, ash
	Dust storms	Suspended particulate matter
	Ocean waves	Salt particles
	Vegetation	Hydrocarbons (VOCs),* pollens
	Hot springs	Sulphurous gases
Human caused		
Industrial	Pulp and paper mills	Particulate matter, sulphur oxides, reduced sulphur
	Coal fired power plant	Ash, sulphur oxides, nitrogen oxides
	Oil burning power plant	Sulphur oxides, nitrogen oxides, CO
	Refineries	Hydrocarbons, sulphur oxides, CO
	Sulphuric acid manufacture	SO_2, SO_3, and H_2SO_4
	Phosphate fertilizer manufacture	Particulate matter, gaseous fluoride
	Iron and steel mills	NO_x, SO_x, VOC, particulate matter, CO, metal oxides, smoke, fumes, dust, organic and inorganic gases
	Base metal smelting	SO_2, particulate matter, heavy metals, dioxins
	Manufacturing plastics	Gaseous resin
	Varnish/paint plants	Acrolein, sulphur compounds
Personal	Automobiles, trucks, lawn and garden equipment, etc	CO, nitrogen oxides, hydrocarbons (VOCs), particulate matter
	Home furnaces/fireplaces	CO, particulate matter
	Open burning of refuse	CO, particulate matter

*VOCs are volatile organic compounds; they represent a class of organic compounds, most of which are hydrocarbons.

Criteria air contaminants are emitted in large quantities as a byproduct of human activity. They have various health impacts, but usually at much higher concentrations than hazardous air pollutants. Air pollutants have far-reaching health effects but mainly affect the body's respiratory system (lungs) and cardiovascular system (heart).

Human activity results in emission of many compounds that are not directly harmful to health or the environment. These are not normally considered air pollutants; however, many have indirect harmful effects. Examples of these include radiatively active gases such as carbon dioxide and methane, which play a role in global warming (see Chapter 16).

CRITERIA AIR CONTAMINANTS   Criteria air contaminants include six pollutants or types of pollutants: particulate matter (PM), sulphur oxides (SO_x), nitrogen oxides (NO_x), carbon monoxide (CO), volatile organic compounds (VOCs), and ammonia (NH_3).

• Figure 18.2 shows that carbon monoxide is the most abundant primary air pollutant in Canada. The primary source for carbon monoxide, nitrogen oxides, and volatile organic compounds is transportation (motor vehicles and so on). Emissions from upstream oil and gas—these are emissions related to the production of oil and gas—are a major source of volatile organic compounds, nitrogen oxides,

Criteria Air Contaminants in Canada

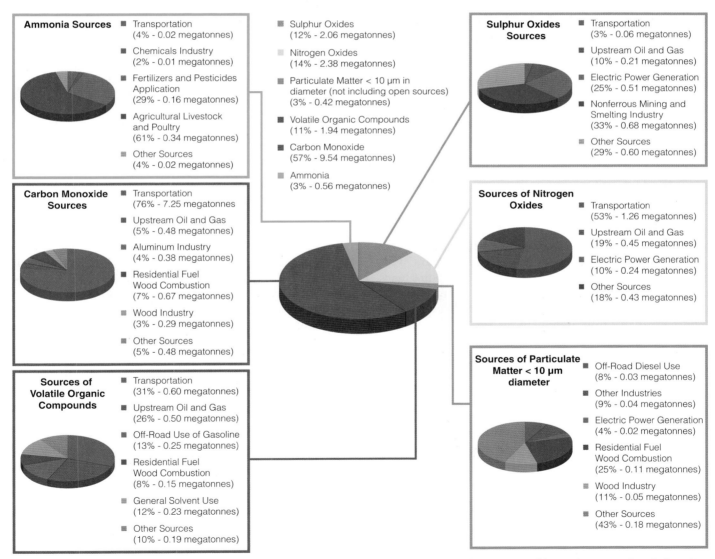

• FIGURE 18.2 Emissions of the main air pollutants—the criteria air contaminants—in Canada, as estimated by Environment Canada for 2005. The central figure shows the total amount of each pollutant emitted, and the satellite figures show the main sources. Note that the particulate matter emissions do not include open sources such as dust from roads, agriculture, or construction. One megatonne is equal to 1 million tonnes.

Data source: Environment Canada. Found at: http://www.ec.gc.ca/air/default.asp?lang=En&n=0D5AD9F6-1

carbon monoxide, and sulphur oxides. Other significant sources include electric power generation mainly from combustion of fossil fuels, burning of wood, industries, and agriculture.

Particulate Matter   The term **particulate matter** represents a group of solid particles and liquid droplets that are small enough to remain suspended in the air. Collectively known as **aerosols**, this grouping includes solid particles that may irritate people but are usually not poisonous, such as soot (tiny solid carbon particles), dust, smoke, and pollen. Particulate matter can be emitted by a variety of processes, many of which either involve some form of combustion, such as wood or diesel fuel burning, or can be categorized as dust sources. Combustion sources generally emit very small particles—usually less than 1 μm* in diameter. Dust sources, also called *open sources*, commonly result from machinery or equipment interacting with Earth's surface. Major open sources include vehicles driving on paved and unpaved roads, equipment operating at construction or industrial sites, and agricultural operations. These dust sources generally create larger particles—larger than 1 or 2 μm in diameter—and are basically composed of suspended soil. Of the estimated 17.5 million tonnes of total particulate matter emitted by human activity over Canada each year, the vast majority (16.8 million tonnes) is from these open sources. Particulate matter can also be formed in the atmosphere as secondary pollutants, mainly from the transformation of nitrogen oxides into ammonium nitrate and sulphur oxides into ammonium sulphate. These secondary particulates are generally small—less than 1 μm in diameter.

One main problem is that particulate pollution may remain in the atmosphere for some time depending on the size and the amount of precipitation that occurs. For example, larger, heavier particles with diameters greater than about 10 μm tend to settle to the ground in about a day or so after being emitted, whereas fine, lighter particles with diameters less than 1 μm can remain suspended in the lower atmosphere for several weeks.

Finer particles with diameters smaller than 10 μm are referred to as *PM10*. These particles pose the greatest health risk as they are small enough to penetrate the lung's natural defence mechanisms. Moreover, winds can carry these fine particles great distances before they finally reach the surface. In fact, suspended particles from sources in Europe and the former Soviet Union are believed to be responsible for the brownish cloud layer called *Arctic haze* that forms over the Arctic each spring. And strong winds over northern China can pick up dust particles and sweep them eastward, where they may settle on North America. This *Asian dust* can reduce visibility, produce spectacular sunrises and sunsets, and coat everything with a thin veneer of particles. For more information on how Asian dust can reach North America, read Focus on an

Environmental Issue: Long-Range Transport of Dust and Air Pollution to Western Canada on p. 564.

In Canada, open sources (dust) generate the majority of the estimated 5.8 million tonnes of PM_{10} each year. The remaining 0.42 million tonnes come from many sources, including residential wood burning; industry; electric power generation, mainly from coal; and off-road diesel, for example, from railway locomotives (see Figure 18.2).

Studies show that particulate matter with diameters less than 2.5 μm, called *$PM_{2.5}$*, is especially dangerous. For one thing, it can penetrate farther into the lungs. Moreover, these tiny particles frequently consist of toxic or carcinogenic (cancer-causing) combustion products. Of recent concern are the $PM_{2.5}$ particles found in diesel soot. Relatively high amounts of these particles have been measured inside school buses, with higher amounts observed downwind of traffic corridors and truck terminals.

Because it often dramatically reduces visibility in urban environments, particulate matter pollution is the most noticeable (see ● Figure 18.3). The ability of particulate matter (also called aerosols) to reduce the transmission of light by scattering through the depth of the atmosphere is measured by the **aerosol optical thickness**, denoted by the Greek letter τ (tau). Perfectly clear air with no scattering would have τ = 0, and as τ increases, the percentage of sunlight that is transmitted goes down and the air takes on a hazy appearance:

$$percent\ transmission = 100 \times e^{-\tau}$$

In this equation, the symbol *e* is a natural constant approximately equal to 2.7183. Figure 18.3 shows what conditions look like for two different values of aerosol optical thickness (τ).

In some cases, particulate matter can have quite diverse chemical compositions. Some of the more dangerous substances include asbestos fibres, arsenic, tiny liquid droplets of sulphuric acid, PCBs, oil, and various pesticides. These types of particulate matter pollution can immediately influence the human respiratory system and can be considered hazardous air pollutants. Particulate matter collected in cities sometimes includes iron, copper, nickel, and lead. Once inside the lungs, it can make breathing difficult, particularly for those suffering from chronic respiratory disorders. Lead particles are especially dangerous as they tend to fall out of the atmosphere and become absorbed into the body through ingestion of contaminated food and water supplies. Lead accumulates in bone and soft tissues and in high concentrations can cause brain damage, convulsions, and death. Even at low doses, lead can be particularly dangerous to fetuses, infants, and children, who, when exposed, may suffer central nervous system damage. Levels of lead in the environment have been dramatically reduced thanks to measures such as the introduction of unleaded gasoline since the 1970s.

Particulate pollution not only adversely affects the lungs, but recent studies suggest that particulate matter can also interfere with the normal rhythm of the human heart. Apparently, as this type of pollution increases, there is a subtle change in a person's heart rate. For a person with an existing

---

*Recall that 1 μm is one-millionth of a metre. (The thickness of this page is about 100 μm.)

(a)

Courtesy Environment Canada

(b)

Courtesy Environment Canada

● FIGURE 18.3 A view looking southeast over British Columbia's Lower Fraser Valley at Chilliwack. (a) On a clear day (25 September 2010 at 4:30 P.M. PDT) with $PM_{2.5}$ values of 1.1 $\mu g\ m^{-3}$ and aerosol optical thickness, $\tau = 0.062$, giving 94 percent light transmission. (b) On a day when particulate matter and other pollutants reduce visibility (5 August 2010 at 1:30 P.M. PDT) with $PM_{2.5}$ values of 24.6 $\mu g\ m^{-3}$ and an aerosol optical thickness, $\tau = 0.964$, giving 38 percent transmission.

cardiac problem, a change in heart rate can produce serious consequences. In fact, one study estimated that, each year, particulate pollution may be responsible for as many as 10,000 heart disease fatalities in the United States.

Rain and snow remove many particles from the air; even the minute particles are removed by ice crystals and cloud droplets. In fact, numerical simulations of air pollution suggest that the predominant removal mechanism occurs when these particles act as nuclei for cloud droplets and ice crystals. Moreover, a long-lasting accumulation of suspended particles (especially those rich in sulphur) is not only aesthetically unappealing but also has the potential for affecting the climate as some particles reflect incoming sunlight, whereas others absorb outgoing infrared energy from Earth's surface.

Many of the suspended particles are hygroscopic, so water vapour readily condenses onto them. As a thin film of water forms on the particles, they grow in size. When they reach a diameter between 0.1 and 1.0 μm, these *wet haze* particles effectively scatter incoming sunlight to give the sky a milky white appearance. The particles are usually ammonium sulphate or ammonium nitrate secondary particulate matter that forms from sulphur oxide and nitrogen oxide

WEATHER WATCH

On any given day, estimates are that about 10 million tonnes of solid particulate matter are suspended in our atmosphere. And in a polluted environment, a volume of air about the size of a sugar cube can contain as many as 200,000 tiny particles.

gases from combustion processes in combination with ammonia. The hazy air mass may become quite thick, and on humid summer days, it often becomes a well defined layer (e.g. look ahead at Figure 18.17).

Sulphur Oxides **Sulphur dioxide (SO_2)** is a colourless gas that comes primarily from the burning of sulphur-containing fossil fuels (such as coal and oil). Its primary source includes power plants, heating devices, smelters, petroleum refineries, and pulp mills. However, it can enter the atmosphere naturally during volcanic eruptions and as sulphate particles from ocean spray.

Sulphur dioxide readily oxidizes to form the secondary pollutants *sulphur trioxide* (SO_3) and, in moist air, highly corrosive *sulphuric acid* (H_2SO_4). Sulphur dioxide and sulphur

FOCUS ON AN ENVIRONMENTAL ISSUE

Long-Range Transport of Dust and Air Pollution to Western Canada

Dr. Ian McKendry, Department of Geography, Atmospheric Science Program, University of British Columbia

Until the late 1990s, it was a widely held view in western Canada that air pollution was of purely local or, at worst, regional origin. This was considered true not only for gaseous pollutants such as ozone, carbon monoxide, or sulphur dioxide but also for the tiny particles suspended and transported in the atmosphere (referred to collectively as particulate matter [PM] or *aerosol*).

In 1998, our understanding of sources of aerosols in Canada changed unexpectedly. A huge springtime dust storm in the arid region of China (Gobi and Taklamakan deserts) alerted scientists to the possibility that aerosols in Canada may originate from much further afield than the continent of North America. The dust storms in China in April 15 to 19, 1998, produced a cloud of dust that travelled across the Pacific Ocean in about five days, producing hazy white skies and spectacular sunsets from California to northern British Columbia. Furthermore, across southern British Columbia, the hourly concentrations of particles < 10 μm in diameter (called PM_{10}) reached ~100 μg m^{-3}, an unusually high value that is five times greater than average concentrations. Dust plumes emanating from another major dust storm in China (dubbed by scientists as "The Perfect Dust Storm"—see ● Figure 1) in April 2001 were widely observed across the whole of North America and confirmed the importance of this trans-Pacific transport pathway (see ● Figure 2).

Since the 1998 and 2001 Gobi dust events, researchers have discovered that not

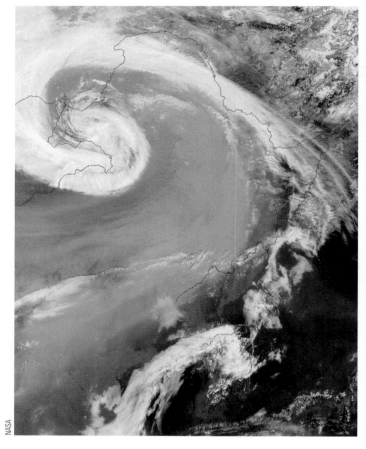

● FIGURE 1 The Perfect Dust Storm, April 7, 2001, from the *MODIS Terra* Satellite. The yellow dust cloud is visible around the spiralling low-pressure centre in the northwest corner of the image, north of the Korean Peninsula. Part of Japan is visible in the lower right corner of the image.

NASA

only does dust travel across the Pacific from the deserts of Asia, but also aerosols and gaseous pollutants produced by the rapidly growing industrial, energy (coal-fired power plants), and transportation sectors of East Asia follow the

same atmospheric pathway. Like desert dust, aerosols and gaseous pollutants produced by human activities (e.g., sulphate aerosols, ozone, and carbon monoxide) also find their way to Canada, particularly during springtime,

---

trioxide together are referred to as SO_x, or *oxides of sulphur*. Winds can carry these aerosols great distances before they reach Earth as undesirable contaminants, making them key ingredients in acid rain (to be discussed at the end of this chapter). When inhaled into the lungs, high concentrations of sulphur dioxide aggravate respiratory problems, such as asthma, bronchitis, and emphysema. Sulphur dioxide in large quantities can cause injury to certain plants, such as lettuce and spinach, sometimes producing bleached marks on their leaves and reducing their yield. Sulphur dioxide is emitted by

a variety of sources, including the metal, mining, and smelting industries, as well as electric power generation from fossil fuels (see Figure 18.2).

Other, nonoxidized forms of sulphur,* such as hydrogen sulphide (H_2S), are not emitted in large quantities but can cause local odour problems. Hydrogen sulphide can also be deadly in very high concentrations. These conditions seldom occur except by accident and in confined spaces, such as mine

---

*These are called *reduced sulphur* compounds.

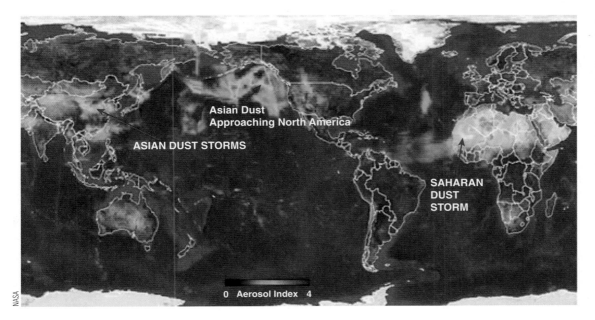

● FIGURE 2 Global imagery from the Total Ozone Mapping Spectrometer Satellite (TOMS) showing dust in the North Pacific approaching the coast of British Columbia on April 13, 2001, from Asian dust storms. Saharan dust storms are also shown streaming aerosol into the mid-Atlantic (dust aerosol is shown by green, yellow, and red hues).

when the meteorology is conducive to such long-range transport. This cocktail of pollutants has the potential to add to and even overwhelm local emissions, is beyond the reach of Canadian regulations, and may have significant implications for human health.

Concern with the environmental impact of long-range transport of dust and pollutants to Canada has prompted several research initiatives designed to observe and ultimately predict the impact of long-range pollutant transport on Canada. These initiatives include the installation by Environment Canada of a high-altitude

chemistry site on Whistler peak in British Columbia and the deployment of a network of lidar systems across Canada to detect atmospheric aerosol layers using powerful laser technology (the Canadian Operational Research Lidar Network or CORALNet). With these tools to assist them, researchers have recently identified episodes of dust transport from the Sahara desert eastward to British Columbia, plumes of volcanic aerosol from eruptions in the Kamchatka Peninsula region, and smoke plumes from forest fires not only in California and Oregon but also from as far afield as Siberia.

This example illustrates that the air we breathe is inextricably linked to the global atmospheric circulation, whereby each gulp of air has a history that includes geographic contact with sources of pollutants across the entire planet. This has important international policy implications and serves to illustrate that we are truly part of a global community whereby air pollution can no longer be considered "local." Furthermore, our own emissions are likely carried far downwind and ultimately contribute to the global pollutant burden, an extreme case of "what goes around comes around."

---

shafts and some industrial settings. Sources of these reduced sulphur compounds include "sour" gas oil wells, pulp mills, and oil refineries.

Nitrogen Oxides   *Nitrogen oxides* are gases that form when some of the nitrogen in the air reacts with oxygen during the high-temperature combustion of fuel. The two primary nitrogen pollutants are **nitrogen dioxide (NO_2)** and **nitric oxide (NO)**, which, together, are commonly referred to as NO_x—or, simply, *oxides of nitrogen*.

Although both nitric oxide and nitrogen dioxide are produced by natural bacterial action, their concentration in urban environments is between 10 and 100 times greater than in nonurban areas. In moist air, nitrogen dioxide reacts with water vapour to form corrosive nitric acid (HNO_3), a substance that adds to the problem of acid rain, which we will address later.

It is estimated that human activity results in nearly 2.4 million tonnes of NO_x emissions in Canada each year. Over half of this is from transportation (motor vehicles), 19 percent is from the production of oil and gas, and about

10 percent is from electric power generation, mainly from burning fossil fuels (see Figure 18.2).

High concentrations are believed to contribute to heart and lung problems, as well as lowering the body's resistance to respiratory infections. Studies on test animals suggest that nitrogen oxides may encourage the spread of cancer. Moreover, nitrogen oxides are highly reactive gases that play a key role in producing ozone and other ingredients of photochemical smog.

Carbon Monoxide   **Carbon monoxide (CO)**, a major pollutant of city air, is a colourless, odourless, poisonous gas that forms during the incomplete combustion of carbon-containing fuels. As we saw earlier, carbon monoxide is the most plentiful of the criteria air contaminants (see Figure 18.2a). Environment Canada estimates that over 9.5 million tonnes of carbon monoxide enter the air annually over Canada alone—three-quarters of it from the burning of fossil fuels in transportation. However, due to stricter air quality standards and the use of emission-control devices, carbon monoxide levels have decreased by about 40 percent since the early 1970s.

Fortunately, carbon monoxide is quickly removed from the atmosphere by microorganisms in the soil, for even in small amounts, this gas is dangerous. Hence, it poses a serious problem in poorly ventilated areas, such as highway tunnels and underground parking garages. Because carbon monoxide cannot be seen or smelled, it can kill without warning. Here is how: normally, your cells obtain oxygen through a blood pigment called *hemoglobin*, which picks up oxygen from the lungs, combines with it, and carries it throughout your body. Unfortunately, human hemoglobin prefers carbon monoxide to oxygen, so if there is too much carbon monoxide in the air you breathe, your brain will soon be starved of oxygen, and headache, fatigue, drowsiness, and even death may result.*

Volatile Organic Compounds   **Volatile organic compounds (VOCs)** represent a class of organic compounds that are mainly **hydrocarbons**—individual organic compounds composed of hydrogen and carbon. At room temperature, they occur as solids, liquids, and gases. Even though thousands of such compounds are known to exist, methane (which occurs naturally and poses no known dangers to health, although it is a potent greenhouse gas) is the most abundant. Other VOCs include benzene, formaldehyde, and some chlorofluorocarbons. Environment Canada estimates that over 1.9 million tonnes of VOCs are emitted by human activity into the air over Canada each year. About 31 percent of this total comes from vehicles used for transportation, about 26 percent from the production of oil and gas, and the rest from off-road gasoline combustion such as lawn mowers, as well as wood burning, solvent use, and other sources (see

Figure 18.2). There are also huge natural sources of VOCs: these are called *biogenic* emissions, and they come from plants and trees such as coniferous forests.

Certain VOCs, such as benzene (an industrial solvent) and benzo[a]pyrene (a product of burning wood, tobacco, and barbecuing), are known to be carcinogens—cancer-causing agents. Since some VOCs are toxic, this category can overlap the hazardous air pollutant category. Although many VOCs are not intrinsically harmful, some will react with nitrogen oxides in the presence of sunlight to produce secondary pollutants, such as ozone, which are harmful to human health.

Ammonia   Ammonia (NH_3) is a colourless gas that has a pungent odour at higher concentrations and is poisonous if inhaled in large amounts. At lower concentrations, it can irritate the nose, throat, and eyes. Ammonia can combine with oxides of sulphur and oxides of nitrogen to form secondary particulates: ammonium sulphate and ammonium nitrate. Nearly all of the 0.56 million tonnes emitted in Canada comes from agriculture: livestock and poultry production as well as fertilizer and pesticides.

OZONE IN THE TROPOSPHERE   As mentioned earlier, the word **smog** originally meant the combining of smoke and fog. Today, however, the word mainly refers to the type of smog that forms in large cities, such as Los Angeles. Because this type of smog forms when chemical reactions take place in the presence of sunlight (called *photochemical reactions*), it is termed **photochemical smog**, or *Los Angeles–type smog*. When the smog is composed of sulphurous smoke and foggy air, it is usually called *London-type smog*.

The most important component of photochemical smog is the pungent-smelling, bluish gas **ozone (O_3)**. Ozone is a powerful oxidant* that reacts easily with many substances and materials. It irritates eyes and the mucous membranes of the respiratory system, aggravating chronic diseases, such as asthma and bronchitis. Even in healthy people, exposure to relatively low concentrations of ozone for six or seven hours during periods of moderate exercise can significantly reduce lung function. This situation is often accompanied by symptoms such as chest pain, nausea, coughing, and pulmonary congestion. Ozone also attacks rubber, retards tree growth, and damages crops. Each year in North America, ozone is responsible for crop yield losses of several billion dollars.

We will see later that ozone forms naturally in the stratosphere through the combining of molecular oxygen and atomic oxygen. There *stratospheric ozone* provides a protective shield against the sun's harmful ultraviolet (UV) rays. However, near the surface, in polluted air, ozone—often referred to as *tropospheric (or ground-level) ozone*—is a secondary pollutant that is not emitted directly into the air. Rather, it forms from a complex series of chemical reactions involving other pollutants, such as nitrogen oxides and volatile organic compounds

---

*Should you become trapped in your car during a snowstorm, and you have your engine and heater running to keep warm, roll down the window just a little. This action will allow the escape of any carbon monoxide that may have entered the car through leaks in the exhaust system.

*An *oxidant* is a substance (such as ozone) whose oxygen readily combines chemically with another substance.

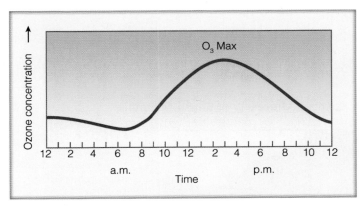

● FIGURE 18.4 Average hourly concentrations of ozone measured at six major cities over a two-year period.

(hydrocarbons). Because sunlight is required to produce ozone, and higher temperature increases its rate of formation, concentrations of tropospheric ozone are normally higher during the afternoons (see ● Figure 18.4) and during the summer months, when sunlight is more intense.

In polluted air, ozone production occurs along the following lines. Sunlight (with wavelengths shorter than about 0.41 μm) dissociates nitrogen dioxide into nitric oxide and atomic oxygen, which may be expressed by

$$NO_2 + \text{solar radiation} \rightarrow NO + O$$

The atomic oxygen then combines with molecular oxygen (in the presence of a third molecule, M) to form ozone, as

$$O_2 + O + M \rightarrow O_3 + M$$

The ozone is then destroyed by combining with nitric oxide; thus,

$$O_3 + NO \rightarrow NO_2 + O_2$$

If sunlight is present, however, the newly formed nitrogen dioxide will break down into nitric oxide and atomic oxygen. The atomic oxygen then combines with molecular oxygen to form ozone again. Since the reactions above both create and then destroy ozone, not much can accumulate if these are the only reactions taking place. Consequently, large concentrations of ozone can form in polluted air only if some of the nitric oxide reacts with other gases *without removing ozone in the process.* For this to happen, certain hydrocarbons (emitted by automobiles and industrial sources) and the hydroxyl radical must come into play.

The formation of the hydroxyl radical (OH) begins when UV radiation (at wavelengths of about 0.31 μm and below) dissociates some of the ozone into molecular oxygen and atomic oxygen:

$$O_3 + \text{UV radiation} \rightarrow O_2 + O$$

The atomic oxygen formed is in an excited state, which means that it can react with a variety of other molecules, including water vapour, to produce two hydroxyl radical molecules; thus,

$$O + H_2O \rightarrow OH + OH$$

The OH is called a "radical" because it contains an unpaired electron. This situation allows the OH molecule to react with many other atoms and molecules, including unburned or partially burned hydrocarbons (RH) released into the air by automobiles and industry, as

$$OH + RH \rightarrow R\cdot + H_2O$$

The product R· represents an organic hydrocarbon that can have a complex molecular structure. The R· is then able to react with molecular oxygen to form $RO_2\cdot$, a reactive molecule that removes nitric oxide by combining with it to form nitrogen dioxide, shown by the expression

$$RO_2\cdot + NO \rightarrow NO_2 + \text{other products}$$

In this manner, nitric oxide can react with hydrocarbons to form nitrogen dioxide *without removing ozone.* Hence, the reactive hydrocarbons in polluted air allow ozone concentrations to increase by preventing nitric oxide from destroying the ozone as rapidly as it is formed.

The hydrocarbons (VOCs) also react with oxygen and nitrogen dioxide to produce other undesirable contaminants, such as *PAN* (peroxyacetyl nitrate)—a pollutant that irritates eyes and is extremely harmful to vegetation—and organic compounds. Ozone, PAN, and small amounts of other oxidating pollutants are the ingredients of photochemical smog. Instead of being specified individually, these pollutants are sometimes grouped under a single heading called *photochemical oxidants.*

In addition to the human-related sources discussed earlier, many hydrocarbons (VOCs) also occur naturally in the atmosphere as they are given off by vegetation. Oxides of nitrogen drifting downwind from urban areas can react with these natural hydrocarbons and produce smog in relatively uninhabited areas. This phenomenon has been observed downwind of cities such as Vancouver, Toronto, Los Angeles, and New York. Some regions have so much natural (background) hydrocarbon that it may be difficult to reduce ozone levels as much as desired.

In spite of vast efforts to control ozone levels in some major metropolitan areas, the results have been generally disappointing because ozone, as we have seen, is a secondary pollutant that forms from chemical reactions involving other pollutants. Ozone production should decrease in most areas when emissions of *both* nitrogen oxides and hydrocarbons (VOCs) are reduced. However, the reduction of only one of these pollutants will not necessarily diminish ozone production because the oxides of nitrogen act as a catalyst for producing ozone in the presence of hydrocarbons (VOCs). Generally, reductions in NO_x in rural areas will lower ozone, and decreases in VOCs are needed to decrease ozone in urban areas. To gain a better understanding of two areas in Canada that have elevated ozone levels, read Focus on an Environmental Issue: Smog in Southern Ontario and British Columbia's Lower Fraser Valley on p. 568.

Up to now, we have concentrated on ozone in the troposphere, primarily in a polluted environment. The next section examines the formation and destruction of ozone in the upper atmosphere—in the stratosphere.

FOCUS ON AN ENVIRONMENTAL ISSUE

Smog in Southern Ontario and British Columbia's Lower Fraser Valley

Dr. Douw Steyn, Department of Earth and Ocean Sciences, University of British Columbia

In Canada, two regions are notably subjected to photochemical smog: southern Ontario, from Windsor to Kingston and up to Parry Sound, and the Lower Fraser Valley (LFV) of British Columbia, from Vancouver to Hope and up the slopes of the North Shore Mountains. As with all other instances of this kind of pollution, the precursor pollutants are emitted from urban, transportation, industrial, and biogenic sources into an atmosphere containing background concentrations of smog constituents (mainly ozone). Once emitted, the precursors and background substances undergo chemical reactions driven by ultraviolet solar radiation to produce full-blown smog. The smog is mixed by turbulence and dispersed by regional winds. Inevitably, smog occurs under synoptic meteorological conditions which suppress vertical mixing and horizontal dispersion. The smog will persist over an episode usually lasting a few days and determined by the continued existence of the synoptic weather conditions.

Oceanic background levels of ozone are typically 20 parts per billion (ppb), whereas continental background ozone levels are around 35 ppb. Smog pollution afflicts most

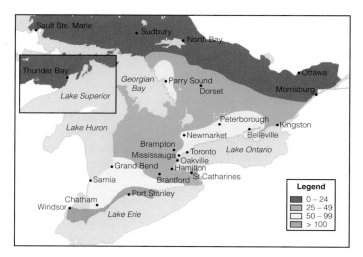

● FIGURE 3 Spatial distribution of one-hour ozone exceedances above 80 ppb in southern Ontario during 2005.

Data source: Ontario Ministry of the Environment. 2006. Air Quality in Ontario 2005 Report.

heavily urbanized regions worldwide, and recent research indicates that Northern Hemisphere background ozone concentrations are increasing due to growing regional emissions of the precursors and relatively long lifetimes of smog constituents when not in contact with earth surface materials.

Smog in southern Ontario arises from local emissions, as well as emissions from

sources as far away as the U.S. Midwest. The sources are traffic, chemical and other industries, and thermal power generation. It is estimated that more than 50% of ozone in southern Ontario arises from sources in the United States., with spatial patterns as indicated in ● Figure 3. Because of the importance of long-range transport of ozone, it is not surprising that smog in southern Ontario is associated

---

OZONE IN THE STRATOSPHERE Recall from Chapter 1 that the stratosphere is a region of the atmosphere that lies above the troposphere between about 10 and 50 km above Earth's surface. The atmosphere is stable in the stratosphere as there exists a strong temperature inversion—the air temperature increases rapidly with height (look back at Figure 1.11, p. 17). The inversion is due, in part, to the gas ozone that absorbs ultraviolet radiation at wavelengths below about 0.3 μm.

In the stratosphere, above middle latitudes, ozone is most dense at an altitude near 25 km (see ● Figure 18.5, p. 570). Even at this altitude, its concentration is quite small as there are only about 12 ozone molecules for every million air molecules (12 ppm*). Although diffuse, this layer of ozone is significant because it shields Earth's inhabitants from harmful

amounts of UV solar radiation. This protection is fortunate because UV radiation at wavelengths below 0.3 μm has enough energy to cause skin cancer in humans. Also, UV radiation at 0.26 μm can destroy acids in DNA (deoxyribonucleic acid), the substance that transmits the hereditary blueprint from one generation to the next.

If the concentration of stratospheric ozone decreases, the following are expected to occur:

● an increase in the number of cases of skin cancer

● a sharp increase in eye cataracts and sun burning

● suppression of the human immune system

● an adverse impact on crops and animals due to an increase in UV radiation

● a reduction in the growth of ocean phytoplankton

● a cooling of the stratosphere that could alter stratospheric wind patterns, possibly affecting the destruction of ozone

---

*With a concentration of ozone of only 12 parts per million in the stratosphere, the composition of air here is about the same as it is near Earth's surface—mainly 78 percent nitrogen and 21 percent oxygen.

with a particular synoptic weather pattern—while the region is under the influence of the "rear" side of a slowly travelling anticyclone. Efforts to control and manage smog in southern Ontario are enormously complicated because emission sources lie outside the jurisdiction of the Ontario Ministry of the Environment, the agency with regulatory responsibility for air quality in this region.

Smog in the LFV arises almost entirely from local emission sources (mostly traffic, biogenic, and marine) located in the LFV, which forms an almost isolated airshed. Episodes of elevated ozone in the LFV occur under anticyclonic conditions at the height of summer and are generally accompanied by sea/land breeze circulations, interacting with valley and slope winds. These local wind systems result in a fairly tight spatial pattern, as shown in ● Figure 4. Air quality in the LFV falls under the jurisdiction of Metro Vancouver. This agency has implemented a sequence of air quality management plans since 1994, with the result that local emissions of NO_x and VOC have reduced by one-third, whereas exceedances of ozone objectives have dropped dramatically.

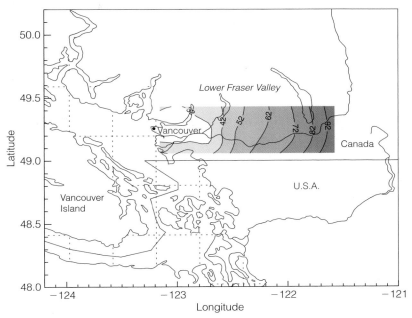

● FIGURE 4 Spatial distribution of ozone in the Lower Fraser Valley of British Columbia during a smog episode (July 29, 2003 at 7 p.m. local time). Contoured ozone values are measured in parts per billion.

Stratospheric Ozone: Production–Destruction Ozone (O_3) forms naturally in the stratosphere by the combining of atomic oxygen (O) with molecular oxygen (O_2)—in the presence of another molecule. First, atomic oxygen (O) is created when high-energy UV radiation with wavelengths less than 0.24 μm splits apart O_2 in the upper atmosphere:

$$O_2 + UV \rightarrow O + O$$

This reaction effectively removes all UV radiation with wavelengths less than 0.24 μm. The O molecules are very reactive and soon bond with an O_2 molecule to form ozone:

$$O + O_2 \rightarrow O_3$$

Although it forms mainly above 25 km, ozone gradually drifts downward by mixing processes, producing a peak concentration in middle latitudes near 25 km. (In polar regions, its maximum concentration is found at lower levels.) Ozone is broken down into molecular and atomic oxygen by absorbing

UV radiation with wavelengths between 0.24 and 0.32 μm (see ● Figure 18.6):

$$O_3 + UV \rightarrow O_2 + O$$

The atomic oxygen (O) then joins a molecular oxygen (O_2) to form ozone again.

If there is sufficient ozone in the stratosphere, this reaction removes nearly all UV radiation with wavelengths between 0.24 and 0.32 μm, effectively protecting Earth's surface from this damaging UV radiation.

Ozone is destroyed by colliding with other atoms and molecules. For example, ozone and atomic oxygen combine to form two oxygen molecules, as

$$O_3 + O \rightarrow 2O_2$$

Likewise, the combination of two ozone molecules destroys ozone, as

$$O_3 + O_3 \rightarrow 3O_2$$

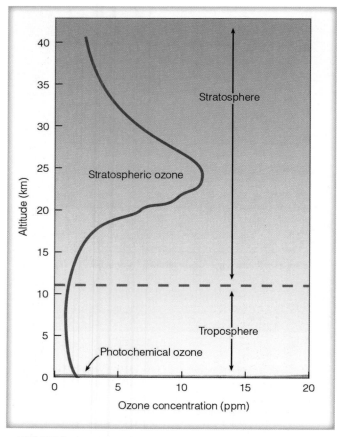

● FIGURE 18.5 The average distribution of ozone above Earth's surface in the middle latitudes.

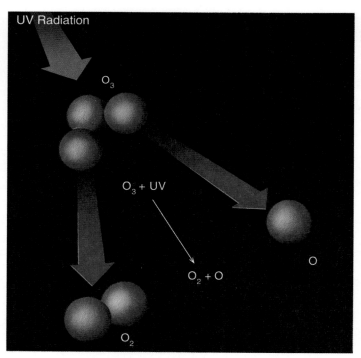

● FIGURE 18.6 An ozone molecule absorbing ultraviolet radiation can become molecular oxygen (O_2) and atomic oxygen (O).

These equations also represent the net result of a number of complex chemical reactions that include trace gases of nitrogen, hydrogen, and chlorine. For example, two natural destructive gases for ozone are nitric oxide (NO) and nitrogen dioxide (NO_2), which, as we have seen, are collectively known as *oxides of nitrogen*. The origin of these gases begins at Earth's surface as soil bacteria produce N_2O (nitrous oxide). This gas gradually finds its way into the stratosphere, where, above about 25 km, solar energy converts some of it into ozone-destroying oxides of nitrogen. In the stratosphere, just a small amount of nitric oxide can destroy a large amount of ozone. The following sequence of chemical reactions will illustrate why. In step 1, the nitric oxide quickly combines with ozone to form nitrogen dioxide and molecular oxygen. Then, in step 2, the nitrogen dioxide combines with atomic oxygen to form nitric oxide and molecular oxygen. Thus,

$$Step\ 1 \quad NO + O_3 \rightarrow NO_2 + O_2$$

$$Step\ 2 \quad NO_2 + O \rightarrow NO + O_2$$

The nitric oxide (NO) released in step 2 is now ready to start destroying ozone again.

Stratospheric ozone is maintained by a delicate natural balance between production and destruction. Could this balance be upset?

Stratospheric Ozone: Upsetting the Balance

The concentration of ozone in the stratosphere may be changed by natural events. In the upper atmosphere, both cosmic rays* and solar particles can produce secondary electrons with sufficient energy to separate molecular nitrogen (N_2) into two nitrogen atoms (N). The nitrogen atoms combine with free atomic oxygen to form nitric oxide, which, in turn, rapidly destroys ozone. Furthermore, large volcanic eruptions, such as the Philippines' Mount Pinatubo in mid-June 1991, can inject ozone-destroying chemicals into the stratosphere. And scientists, using measurements from satellites, discovered that even changes in UV radiation from the sun can cause small variations in the amount of stratospheric ozone.

It is now apparent that human activities are also altering the amount of stratospheric ozone. Concerns involve emissions of chemicals at Earth's surface, such as nitrous oxide emitted from nitrogen fertilizers (which may drift into the stratosphere, where it could destroy ozone) and *chlorofluorocarbons* (CFCs). Until the late 1970s, they were the most widely used propellants in spray cans, such as deodorants and hairsprays. In the troposphere, these gases are quite safe, being nonflammable, nontoxic, and unable to chemically combine with other substances.† Hence, these gases slowly diffused upward in the troposphere without being destroyed.

---

*Cosmic rays are high-energy atomic nuclei and atomic particles that travel through space at extremely high speeds. Most cosmic rays are believed to come from exploding stars (hypernovae), although some are produced in solar flares.

†Recall from Chapter 3 that CFCs do act as strong greenhouse gases in the troposphere.

They apparently enter the stratosphere in two ways: near breaks in the tropopause, especially in the vicinity of jet streams, and in building thunderstorms, especially those that develop in the tropics along the intertropical convergence zone and penetrate the lower stratosphere.

Once CFC molecules reach the middle stratosphere, UV energy that is normally absorbed by ozone breaks them up, releasing atomic *chlorine* in the process. Note in the following sequence of reactions how chlorine destroys ozone rapidly. In step 1, atomic chlorine (Cl) combines with ozone, forming chlorine monoxide (ClO)—a new substance—and molecular oxygen (O_2). Almost immediately, the chlorine monoxide combines with free atomic oxygen (step 2) to produce chlorine atoms and molecular oxygen. The free chlorine atoms are now ready to combine with and destroy more ozone molecules. Chlorine acts as a *catalyst* in the destruction of ozone. This means that although it speeds up the reactions that destroy ozone, it is not consumed in those reactions. Estimates are that a single chlorine atom removes as many as 100,000 ozone molecules before it is taken out of action by combining with other substances:

$$Step\ 1 \quad Cl + O_3 \rightarrow ClO + O_2$$

$$Step\ 2 \quad ClO + O \rightarrow Cl + O_2$$

Fortunately, chlorine atoms do not exist in the stratosphere forever. They are removed as chlorine monoxide combines with nitrogen dioxide to form chlorine nitrate, $ClONO_2$ (step 3). In step 4, free chlorine atoms combine with methane to form hydrogen chloride (HCl) and a new substance, CH_3:

$$Step\ 3 \quad ClO + NO_2 \rightarrow ClONO_2$$

$$Step\ 4 \quad CH_4 + Cl \rightarrow HCl + CH_3$$

Since the average lifetime of a CFC molecule is about 50 to 100 years, any increase in the concentration of CFCs is long-lasting and a genuine threat to the concentration of ozone. Given this fact and the additional knowledge that CFCs contribute to Earth's greenhouse effect, an international agreement called the *Montreal Protocol* was signed in 1987. This agreement established a timetable for diminishing CFC emissions and the use of bromine compounds (halons), which destroy ozone at a rate 50 times greater than do chlorine compounds.* The *Montreal Protocol* is a great success of international environmental diplomacy—a serious environmental problem was identified and its cause pinpointed by scientists. Afterward, a global solution involving all nations was found to solve it, all within about a decade.

During November 1992, representatives of more than half the world's nations met in Copenhagen to update and revise the treaty. The provisions of the meeting called for a quicker phase-out of the previously targeted ozone-destroying chemicals and the establishment of a permanent fund to help

*There are many chemical reactions that involve chlorine and bromine and the destruction of ozone in the stratosphere. The example given so far is just one of them.

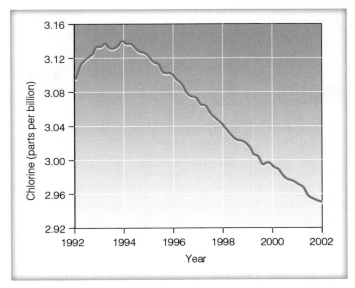

● **FIGURE 18.7** Global average concentration of atmospheric chlorine.
Data source: NOAA

Third World nations find the technology to develop ozone-friendly chemicals. The phase-out appears to be working as global concentrations of atmospheric chlorine and bromine have been decreasing (see ● Figure 18.7).

Although the use of CFCs has decreased 96 percent between 1986 and 2005, there are still millions of kilograms in the troposphere that will continue to slowly diffuse upward. In a 1991 study, an international panel of over 80 scientists concluded that the ozone layer thinned by about 3 percent during the summer from 1979 to 1991 over heavily populated areas of the Northern Hemisphere. Moreover, studies show that stratospheric ozone concentrations above the Canada have declined since the 1980s by about 7 percent in late winter and early spring.

Satellite measurements in 1992 and 1993 revealed that ozone concentrations had dropped to record low levels over much of the globe. The decrease appears to stem from ozone-destroying chemicals and from the 1991 volcanic eruption of Mount Pinatubo that sent tons of sulphur dioxide gas into the stratosphere, where it formed tiny droplets of sulphuric acid. These droplets not only enhance the ozone destructiveness of the chlorine chemicals but also alter the circulation of air in the stratosphere, making it more favourable for ozone depletion. During the mid-1990s, wintertime ozone levels dropped well below normal over much of the Northern Hemisphere. This decrease apparently was due to ozone-destroying pollution along with natural cold stratospheric weather patterns that favoured ozone reduction.

Presently, there are two major substitutes for CFCs, *hydrochlorofluorocarbons* (HCFCs) and *hydrofluorocarbons* (HFCs). The HCFCs contain fewer chlorine atoms per molecule than CFCs and, therefore, pose much less danger to the ozone layer, whereas HFCs contain no chlorine. These gases may have to be phased out, however, as both are greenhouse gases that can enhance global warming.

FOCUS ON AN ENVIRONMENTAL ISSUE

The Ozone Hole

In 1974, two chemists from the University of California at Irvine, F. Sherwood Rowland and Mario J. Molin, warned that increasing levels of CFCs would eventually deplete stratospheric ozone on a global scale. Their studies suggested that ozone depletion would occur gradually and would perhaps not be detectable for many years to come. It was surprising, then, when British researchers identified a year-to-year decline in stratospheric ozone over Antarctica. Their findings, corroborated later by satellites and balloon-borne instruments, showed that since the late 1970s, ozone concentrations have diminished each year during the months of September and October. This decrease in stratospheric ozone over springtime Antarctica is known as the *ozone hole*. In years of severe depletion, such as in 2006, the ozone hole covers almost twice the area of the Antarctic continent (see ● Figure 5).

To understand the causes behind the ozone hole, scientists in 1986 organized the first *National Ozone Expedition*, NOZE-1, which set up a fully instrumented observing station near McMurdo Sound, Antarctica. During 1987, with the aid of instrumented aircraft, NOZE-2 got under way. The findings from

these research programs helped scientists put together the pieces of the ozone puzzle.

The stratosphere above Antarctica normally has among the world's highest ozone concentrations. Most of this ozone forms over the tropics and is brought to the Antarctic by stratospheric winds. During September and October (spring in the Southern Hemisphere), a belt of stratospheric winds called the *polar vortex* encircles the Antarctic region near 66°S latitude, essentially isolating the cold Antarctic stratospheric air from the warmer air of the middle latitudes. During the long dark Antarctic winter, temperatures inside the vortex can drop to −85°C. This frigid air allows for the formation of *polar stratospheric clouds*. These ice clouds are critical in facilitating chemical interactions among nitrogen, hydrogen, and chlorine atoms, the end product of which is the destruction of ozone.

In 1986, the NOZE-1 study detected unusually high levels of chlorine compounds in the stratosphere, and in 1987, the instrumented aircraft of NOZE-2 measured enormous increases in chlorine compounds when it entered the polar vortex. These findings, in conjunction with other chemical discoveries, allowed scientists to pinpoint *chlorine* from CFCs as the main cause of the ozone hole.

Even with a decline in ozone-destroying chemicals, the largest Antarctic ozone hole observed to date occurred in September 2006 (see Figure 5). Apparently, these yearly variations in the size and depth of the ozone hole are mainly due to changes in polar stratospheric temperatures.

In the Northern Hemisphere's polar Arctic, airborne instruments and satellites during the late 1980s and 1990s measured high levels of ozone-destroying chlorine compounds in the stratosphere. By 1997, springtime ozone levels in the Arctic were about 40 percent below average (see ● Figure 6). But observations could not detect an ozone hole like the one that forms over the Antarctic.

Apparently, several factors inhibit massive ozone loss in the Arctic. For one thing, in the stratosphere, the circulation of air over the Arctic differs from that over the Antarctic. Then, too,

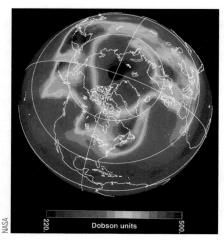

● FIGURE 6 Colour image of total ozone amounts over the Northern Hemisphere for March 24, 1997. Notice that minimum ozone values (purple shades) appear over a region near the North Pole. The colour scale on the bottom of the image shows total ozone values in Dobson units (DU). A Dobson unit is the physical thickness of the ozone layer if it were brought to Earth's surface (500 DU equals 5 mm).

the Arctic stratosphere is normally too warm for the widespread development of the polar stratospheric clouds that help activate chlorine molecules. However, it appears that a very cold Arctic stratosphere and ozone-destroying chemicals were responsible for the low readings in 1997. Moreover, in January 2000, more polar stratospheric clouds formed over the Arctic and lasted longer than during any previous year. This situation contributed to significant ozone loss.

Ozone depletion is not just confined to the stratospheric Arctic and Antarctic. Over southern Canada, the ozone layer has thinned by an average of 7 percent since the 1980s, with the greatest depletion in the late winter and early spring.

We still have much to learn about stratospheric ozone and the processes that both form and destroy it. Presently, atmospheric studies are providing more information so that a more complete assessment of the ozone problem will become available in the future. Studies show that the ozone hole is still there—some years it is stronger, and some years it is weaker.

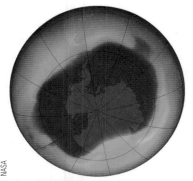

ACTIVE FIGURE 5 Ozone distribution over the South Pole on September 29, 2006, as measured by ozone monitoring equipment on NASA's *Aura* satellite. Notice that the lowest ozone concentration or ozone hole (purple shades) covers most of Antarctica. Visit the textbook's web site to view this and other Active Figures at www.ahrensmeteorology1ce.nelson.com

At present, most scientists believe that ozone levels in the stratosphere throughout the world (except over Antarctica) will return to pre-1980 levels by 2050. Over Antarctica, ozone concentrations will likely remain low until about 2070. In fact, ozone concentrations over springtime Antarctica have been exceptionally low. This sharp drop in ozone is known as the **ozone hole**. (More information on the ozone hole is provided in Focus on an Environmental Issue: The Ozone Hole on p. 572.

Air Pollution: Trends and Patterns

Over the past decades, strides have been made to improve the quality of the air we breathe. ● Figure 18.8 shows the estimated emission trends over Canada for the criteria air contaminants, as well as lead and mercury, based on the National Pollutant Release Inventory collected by Environment Canada. Notice that emissions of most pollutants have decreased since 1985, with lead showing the greatest reduction, primarily due to phasing out of leaded gasoline.

Although the situation has improved, we can see from Figure 18.8 that much more needs to be done as large quantities of pollutants still spew into our air. A large part of the problem of pollution control lies in the fact that even with stricter emission laws, increasing numbers of automobiles (estimates are that more than 250 million vehicles are on the road today in the United States and approximately 19 million in Canada) and other sources can overwhelm control efforts.

In Canada, National Ambient Air Quality Objectives (NAAQOs) are set through negotiation between the provinces and the federal government under the Canadian Environmental Protection Act, and Canada-wide Standards are determined by the Canadian Council of Ministers of the Environment. Since air quality is a provincial responsibility in Canada, individual provinces may set more stringent objectives. In the United States, clean air standards are established by the U.S. Environmental Protection Agency.

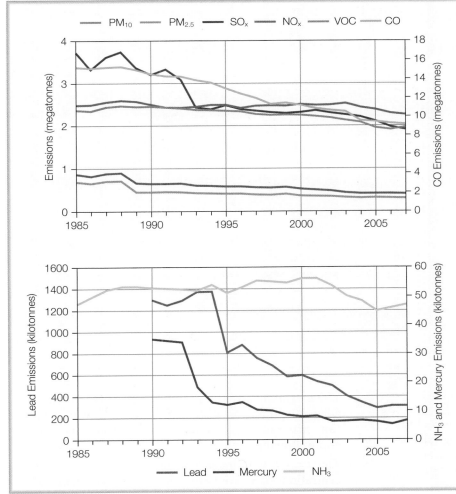

● FIGURE 18.8 Emission estimates of nine pollutants in Canada from 1985 through 2007. Note that one megatonne is equal to one million tonnes, whereas a kilotonne is equal to 1000 tonnes.

Data source: Environment Canada. National Pollutant Release Inventory.

▼ Table 18.2 (a) National Ambient Air Quality Objectives

POLLUTANT	AVERAGING PERIOD	MAXIMUM DESIRABLE	MAXIMUM ACCEPTABLE	MAXIMUM TOLERABLE
Ozone (O_3)	1-hour	100 $\mu g\ m^{-3}$	160 $\mu g\ m^{-3}$	300 $\mu g\ m^{-3}$
	24-hour	30 $\mu g\ m^{-3}$	50 $\mu g\ m^{-3}$	—
	Annual	—	30 $\mu g\ m^{-3}$	—
Carbon monoxide (CO)	1-hour	15,000 $\mu g\ m^{-3}$	35,000 $\mu g\ m^{-3}$	—
	8-hour	6000 $\mu g\ m^{-3}$	15,000 $\mu g\ m^{-3}$	20,000 $\mu g\ m^{-3}$
Sulphur dioxide (SO_2)	1-hour	450 $\mu g\ m^{-3}$	900 $\mu g\ m^{-3}$	—
	24-hour	150 $\mu g\ m^{-3}$	300 $\mu g\ m^{-3}$	800 $\mu g\ m^{-3}$
	Annual	30 $\mu g\ m^{-3}$	60 $\mu g\ m^{-3}$	—
Hydrogen sulphide (H_2S)	1-hour	1 $\mu g\ m^{-3}$	15 $\mu g\ m^{-3}$	—
	24-hour	—	5 $\mu g\ m^{-3}$	—
Nitrogen dioxide (NO_2)	1-hour	—	400 $\mu g\ m^{-3}$	1000 $\mu g\ m^{-3}$
	24-hour	—	200 $\mu g\ m^{-3}$	300 $\mu g\ m^{-3}$
	Annual	60 $\mu g\ m^{-3}$	100 $\mu g\ m^{-3}$	—
Total suspended particulate matter (PM)	24-hour	—	120 $\mu g\ m^{-3}$	400 $\mu g\ m^{-3}$
	Annual (geometric mean)	60 $\mu g\ m^{-3}$	70 $\mu g\ m^{-3}$	—

(b) Reference Levels and Canada-wide Standards

POLLUTANT	AVERAGING PERIOD	CRITERIA
Respirable particulate matter 10 µm or less (PM_{10})	24-hour	A reference level 24-hour average of 25 $\mu g\ m^{-3}$ has been established.
Respirable particulate matter 2.5 µm or less ($PM_{2.5}$)	24-hour	A reference level 24-hour average of 15 $\mu g\ m^{-3}$ has been established.
$PM_{2.5}$ Canada-wide Standard*	24-hour	The annual 98th percentile 24-hour average value of $PM_{2.5}$ averaged over three years must be less than 30 $\mu g\ m^{-3}$. In other words, this standard is not met if the 24-hour average exceeds 30 $\mu g\ m^{-3}$ more than 2 percent of the time each year averaged over a three-year period.
Ozone Canada-wide Standard*	8-hour	The annual 4th highest 8-hour average of ozone averaged over a three-year period must be less than 65 ppb.

*Note that the Canada-wide Standards for ozone and $PM_{2.5}$ came into effect starting in 2010.

In Canada, three ranges of air quality are defined:

- The *maximum desirable level* is the long-term goal for air quality and provides a basis for an antidegradation policy for unpolluted parts of the country and for the continuing development of pollution control technology.

- The *maximum acceptable level* is intended to provide adequate protection against effects on soil, water, vegetation, materials, animals, visibility, and personal comfort and well-being.

- The *maximum tolerable level* denotes time-based concentrations of air contaminants beyond which, owing to a diminishing margin of safety, appropriate action is required without delay to protect the health of the general population.

The National Ambient Air Quality Objectives and Canada-wide Standards currently in effect are given in ▼ Table 18.2. The table shows the concentration of each pollutant needed to exceed the NAAQO or Canada-wide Standard.

To indicate the air quality in a particular region, Health Canada and Environment Canada, working with the provinces, began developing the **Air Quality Health Index (AQHI)*** in 2001, which is now fully operational across Canada. The index is calculated based on the relative risks of a combination of ozone (O_3), particulate matter ($PM_{2.5}$), and nitrogen dioxide (NO_2) and then scaled to give a range between zero and 10. The relative weight given to each of these pollutants is based on health studies. Other pollutants were considered, but it was found that these three gave the best correlation with increases in mortality at the 10 Canadian cities used to develop the index. The AQHI formula is

$$AQHI = \left(\frac{10}{10.4}\right) \times (100 \times (e^{0.000537 \times O_3} - 1 + e^{0.000487 \times PM_{2.5}} - 1 + e^{0.000871 \times NO_2} - 1))$$

In this formula, all pollutants are three-hour average concentrations in units of parts per billion (ppb) for the gases NO_2 and O_3 and in units of micrograms per cubic metre ($\mu g\ m^{-3}$) for $PM_{2.5}$. When the AQHI is between 1 and

*The AQHI replaced the Air Quality Index (AQI). The AQI was based on the maximum value of a set of pollutants, whereas the AQHI is calculated based on all of ozone, nitrogen dioxide, and particulate matter less than 2.5 μm in diameter.

Mexico City lies in a broad basin surrounded by tall mountains. With 20 million inhabitants and 5.5 million vehicles travelling in and around the city daily, Mexico City exceeds the country's ground-level ozone standards about 284 days per year on average.

3, the health risk is low. Values between 4 and 6 indicate a moderate health risk, with a recommendation that people who are at risk should consider reducing strenuous outdoor activity. AQHI values from 7 to 10 indicate a high health risk, with the recommendation that at-risk populations reduce or reschedule strenuous outdoor activities and the general population consider reducing these activities. When the index exceeds 10, individuals who are at risk should avoid strenuous outdoor activities, and the general population should reduce them (see ▼ Table 18.3).

Higher emission standards, along with cleaner fuels (such as natural gas), have made the air over our large cities cleaner today than it was years ago. In fact, total emissions in North America and elsewhere have generally been declining steadily in recent decades. However, there are still issues with some pollutants at some locations in Canada in meeting the Canada-wide Standard for $PM_{2.5}$, which came into effect in 2010 (see ● Figure 18.9). The control of ozone in polluted air is still a problem in many locations in Canada as well (see ● Figure 18.10), especially in southern Ontario and, to a lesser extent, the Lower Fraser Valley of British Columbia. (Refer to Focus on an Environmental Issue: Smog in Southern Ontario and British Columbia's Lower Fraser Valley on p. 568.)

▼ Table 18.3 The Air Quality Health Index (AQHI)

HEALTH RISK	AIR QUALITY HEALTH INDEX (AQHI)	HEALTH MESSAGES	
		AT-RISK POPULATION*	**GENERAL POPULATION**
Low	1–3	**Enjoy** your usual outdoor activities.	**Ideal** air quality for outdoor activities
Moderate	4–6	**Consider reducing** or rescheduling strenuous activities outdoors if you are experiencing symptoms.	**No need to modify** your usual outdoor activities unless you experience symptoms such as coughing and throat irritation
High	7–10	**Reduce** or reschedule strenuous activities outdoors. Children and the elderly should also take it easy.	**Consider reducing** or rescheduling strenuous activities outdoors if you experience symptoms such as coughing and throat irritation
Very high	> 10	**Avoid** strenuous activities outdoors. Children and the elderly should also avoid outdoor physical exertion.	**Reduce** or reschedule strenuous activities outdoors, especially if you experience symptoms such as coughing and throat irritation.

*People with heart or breathing problems may be at greater risk.

Source: Environment Canada. AQHI Categories and Messages. Found at: http://www.ec.gc.ca/cas-aqhi/default.asp?lang=En&n=79A8041B-1. © Her Majesty The Queen in Right of Canada, Environment Canada, 2010.

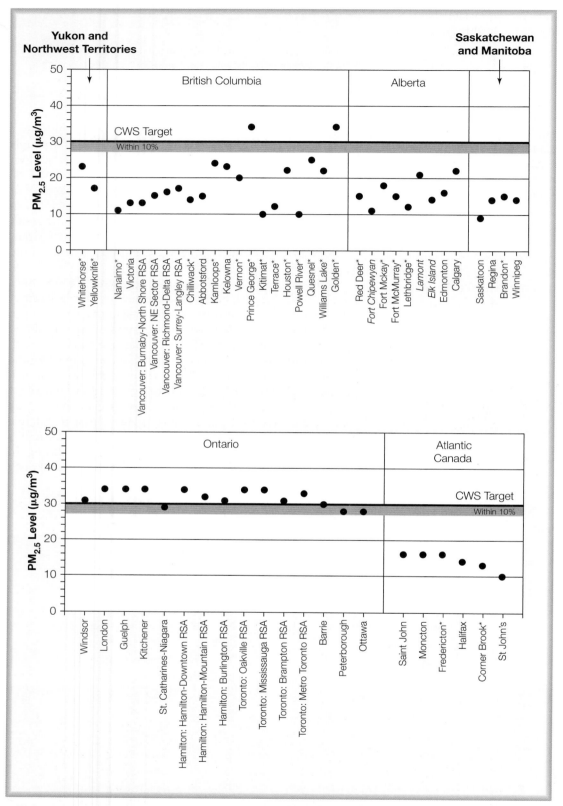

● FIGURE 18.9 Three-year average of the annual 98th percentile of 24-hour $PM_{2.5}$ concentration at stations across Canada between 2003 and 2005. The heavy line at 30 μg m^{-3} indicates the Canada-wide Standard for $PM_{2.5}$. Stations with an asterisk (*) indicate communities with a population less than 100,000 that are not subject to the Canada-wide Standard. Stations in *italics* are in rural locations and also are not subject to the Canada-wide Standard.

Source: Canadian Council of Ministers of the Environment. *Canada-wide Standards for Particulate Matter and Ozone: Five Year Report: 2000–2005.* Fig. 3. Pg. 7. November 2006.

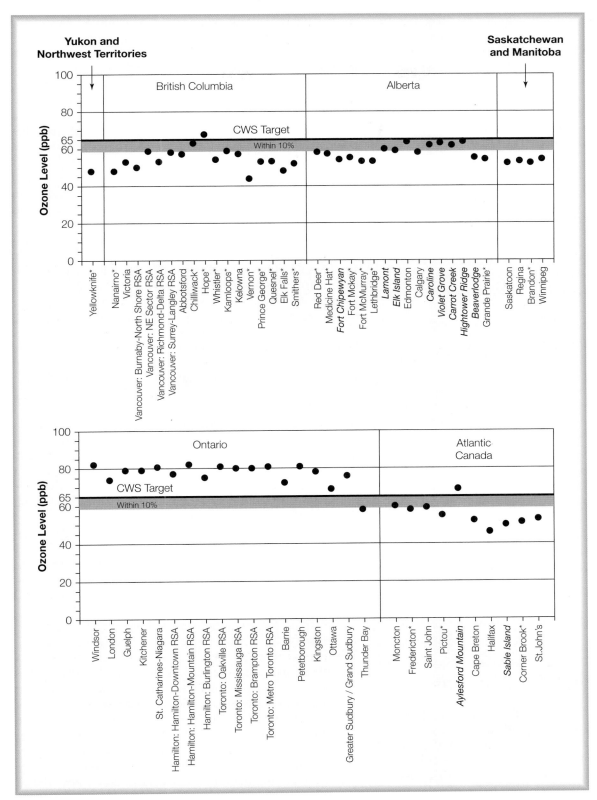

● FIGURE 18.10 Three-year average of the annual fourth highest eight-hour ozone concentration at stations across Canada between 2003 and 2005. The heavy line at 65 ppb indicates the Canada-wide Standard. Note that several stations, mainly in Ontario, were not in compliance with the Canada-wide Standard during these years. Stations with an asterisk (*) have populations less than 100,000 and are not subject to the Canada-wide Standard. Stations in *italics* are in rural locations and also are not subject to the Canada-wide Standard.

Source: Canadian Council of Ministers of the Environment. *Canada-wide Standards for Particulate Matter and Ozone: Five Year Report: 2000–2005.* Fig. 7. Pg. 12. November 2006

Because ozone is a secondary pollutant, its formation is controlled by the concentrations of other pollutants, namely nitrogen oxides and hydrocarbons (VOCs). Moreover, weather conditions play a vital role in ozone formation as ozone reaches its highest concentrations in hot, sunny weather, when surface winds are light and a stagnant, high-pressure area covers the region. As a result of these factors, year-to-year ozone trends are quite variable.

BRIEF REVIEW

Before going on to the next several sections, here is a brief review of some of the important points presented so far:

* Near the surface, primary air pollutants (such as particulate matter, CO, SO_2, NO, NO_2, VOCs, and NH_3) enter the atmosphere directly, whereas secondary air pollutants (such as O_3) form when a chemical reaction takes place between a primary pollutant and some other component of air.

* The word "smog" (coined in London in the early 1900s) originally meant the combining of smoke and fog. Today, the word mainly refers to photochemical smog—pollutants that form in the presence of sunlight.

* Stratospheric ozone forms naturally in the stratosphere and provides a protective shield against the sun's harmful ultraviolet rays. Tropospheric (ground-level) ozone that forms in polluted air is a health hazard and is the primary ingredient of photochemical smog.

* Human-induced chemicals, such as chlorofluorocarbons (CFCs), have been altering the amount of ozone in the stratosphere by releasing chlorine, which rapidly destroys ozone.

* Even though the emissions of most pollutants have declined across the developed world since 1970, millions of people are breathing air that does not meet air quality standards.

Up to now, we have concentrated on outdoor air pollution. Focus on an Environmental Issue: Indoor Air Pollution on p. 580 looks at pollutants inside the home.

Factors That Affect Air Pollution

If you live in a region that periodically experiences photochemical smog, you may have noticed that these episodes often occur with clear skies, light winds, and generally warm, sunny weather. Although this may be "typical" air pollution weather, it by no means represents the only weather conditions necessary to produce high concentrations of pollutants, as we will see in the following sections.

THE ROLE OF THE WIND   The wind speed plays a role in diluting pollution. When vast quantities of pollutants are spewed into the air, the wind speed determines how quickly the pollutants mix with the surrounding air and, of course, how fast they move away from their source. Strong winds tend to lower the concentration of pollutants by spreading them apart as they move downstream. Moreover, the stronger the wind, the more turbulent the air. Turbulent air produces swirling eddies that dilute the pollutants by mixing them with the cleaner surrounding air. Hence, when the wind dies down, pollutants are not readily dispersed and tend to become more concentrated (see ● Figure 18.11 on p. 579).

THE ROLE OF STABILITY AND INVERSIONS   Recall from Chapter 6 that atmospheric stability determines the extent to which air will rise. Remember also that an unstable atmosphere favours vertical air currents, whereas a stable atmosphere strongly resists upward vertical motions. Consequently, smoke emitted into a stable atmosphere tends to have its vertical mixing suppressed and can only spread horizontally.

The stability of the atmosphere is determined by the way the air temperature changes with height (the environmental lapse rate). When the measured air temperature decreases rapidly as we move up into the atmosphere, the atmosphere tends to be more unstable and pollutants tend to be mixed vertically, as illustrated in ● Figure 18.12a on p. 579. If, however, the measured air temperature either decreases quite slowly as we ascend or actually increases with height (remember that this is called an *inversion*), the atmosphere is stable. An inversion represents an extremely stable atmosphere where warm air lies above cool air (see Figure 18.12b). Any air parcel that attempts to rise into the inversion will, at some point, be cooler and heavier (more dense) than the warmer air surrounding it. Hence, the inversion acts like a lid on vertical air motions.

The inversion depicted in Figure 18.12b is called a **radiation (or surface) inversion**. This type of inversion typically forms during the night and early-morning hours when the sky is clear and the winds are light. As we saw earlier in Chapter 3, radiation inversions also tend to be well developed during the long nights of winter.

In Figure 18.12b, notice that within the stable inversion, the smoke from the shorter stacks does not rise very high but spreads out, contaminating the area around it. In the relatively unstable air above the inversion, smoke from the taller stack is able to rise and become dispersed. Since radiation inversions are often rather shallow, it should be apparent why taller chimneys have replaced many of the shorter ones. In fact, taller chimneys disperse pollutants better than shorter ones even in the absence of a surface inversion because the taller chimneys are able to mix pollutants throughout a greater volume of air. Although these taller stacks do improve the air quality in their immediate area, they may also contribute to the acid rain problem by allowing the pollutants to be swept great distances downwind.

As the sun rises and the surface warms, the radiation inversion normally weakens and disappears before noon. By

(a)　　　　　　　　　　　　　　　　　　　　　　　(b)

● FIGURE 18.11 If each chimney emits a puff of smoke every second, then where the wind speed is low (a), the smoke puffs are closer together and more concentrated. Where the wind speed is greater (b), the smoke puffs are farther apart and more diluted as turbulent eddies mix the smoke with the surrounding air.

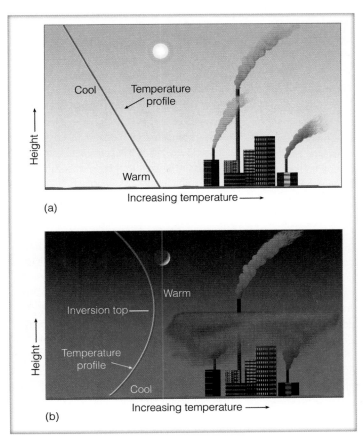

● FIGURE 18.12 (a) During the afternoon, when the atmosphere is most unstable, pollutants rise, mix, and disperse downwind. (b) At night, when a radiation inversion exists, pollutants from the shorter stacks are trapped within the inversion, whereas pollutants from the taller stack, above the inversion, are able to rise and disperse downwind.

afternoon, the atmosphere is sufficiently unstable so that, with adequate winds, pollutants are able to disperse vertically (see Fig. 18.12a). The changing atmospheric stability, from stable in the early morning to conditionally unstable in the afternoon, can have a profound effect on the daily concentrations of pollution in certain regions. For example, on a busy city street corner, carbon monoxide levels can be considerably higher in the early morning than in the early afternoon (with the same flow of traffic). Changes in atmospheric stability can also cause smoke plumes from chimneys to change during the course of a day. (Some of these changes are described in Focus on an Observation: Smokestack Plumes on p. 582.)

Radiation inversions normally last just a few hours, whereas **subsidence inversions** may persist for several days or longer. Subsidence inversions, therefore, are the ones commonly associated with major air pollution episodes. They form as the air above a deep anticyclone slowly sinks (subsides)* and warms.

A typical temperature profile of a subsidence inversion is shown in ● Figure 18.13. Notice that in the relatively unstable air beneath the inversion, the pollutants are able to mix vertically up to the inversion base. The stable air of the inversion, however, inhibits vertical mixing and acts like a lid on the pollution below, preventing it from entering into the inversion.

In Figure 18.13, the region of relatively unstable (well mixed) air that extends from the surface to the base of the inversion is referred to as the **mixing layer**. The vertical extent of the mixing layer is called the **mixing depth**. Observe that

---

*Remember from Chapter 2 that sinking air always warms because it is being compressed by the surrounding air.

FOCUS ON AN ENVIRONMENTAL ISSUE

Indoor Air Pollution

When people think of air pollution, most think of outside air where automobiles, factories, and power plants spew tonnes of contaminants into the air. But, surprisingly, the air we breathe inside our homes and other structures can be much more polluted than the air we breathe outdoors, even in the largest, most industrialized cities! Since we spend so much time indoors, the quality of air we breathe there can have a far greater impact on our health and well-being than the air outdoors (see ● Figure 7).

Many sources of indoor air pollution have been identified, ranging from building materials, pressed wood products, furnishings, and home cleaning products to pesticides, adhesives, and personal care products. In addition, heating sources (such as unvented kerosene heaters, wood stoves, and fireplaces) can release a variety of pollutants into a home. The pollution impact of any heating source depends on several factors, such as its design, age, the level of maintenance it receives, and its location and access to ventilation. For example, a gas cooking stove or heating stove with improper fittings and adjustments can cause a significant emission of carbon monoxide. New carpets and padding (as well as the adhesives used in their installation) can emit volatile organic compounds.

Some pollution sources, such as building materials and foam insulation, produce a constant stream of pollutants, whereas activities such as tobacco smoking emit pollutants into the air on an intermittent basis. A portion of outside air pollution is also brought indoors. Some pollutants enter homes and other structures through cracks and holes in foundations and basements, as is the case with radon.

Radon is a colourless, odourless gas—a natural radioactive compound—that forms as the uranium in soil and rock breaks down. Radon is found everywhere on earth and becomes a problem only when it leaks out of the soil and becomes trapped inside homes and buildings. The radon gas that seeps in through cracks and other openings in a building can accumulate to levels that create a serious health threat. Studies by Health Canada have shown that elevated

● FIGURE 7 With candles burning, a fire in the fireplace, and stain-resistant material on rugs and carpets, there are probably more air pollutants in this living room than there are in a similar-sized volume of air outdoors.

levels of radon in Canadian cities occur in up to 11.3 percent of homes. The level of radon varies greatly between cities and from structure to structure. Inside a home, the radon decays into polonium, a solid substance that attaches itself to dust in the air. The tiny dust particles can be inhaled deep into the lungs, where they attach to lung tissue. As the polonium decays, it damages the lung tissue, sometimes producing mutated cells that may develop into lung cancer.

Another major chemical pollutant found inside our homes is *formaldehyde*. It is a colourless, pungent-smelling gas used widely to manufacture building materials, insulation, and other household products. In most homes, the significant sources of formaldehyde are urea–formaldehyde foam insulation and pressed wood products, such as particle board and plywood paneling. Emissions of formaldehyde from new products can be greatly affected by indoor temperatures and ventilation. Exposure to formaldehyde can cause watery eyes, burning sensations in the nose and throat, breathing difficulties, and nausea. It can also trigger attacks in individuals suffering from asthma.

Another polluter of our indoor air is *asbestos*, a mineral fibre once used in insulation and as a fire retardant. In recent years, manufacturers

have voluntarily reduced the use of asbestos, but much asbestos still remains in furnace and pipe insulation, texturing materials, and floor tiles of older buildings. The most lethal fibres of asbestos are invisible. When inhaled, these tiny particles can accumulate and remain deep in the lungs for extended periods of time, where they damage tissue and potentially cause cancer or a permanent scarring of the lung that can be fatal.

Smoking tobacco indoors can also create an extremely dangerous health situation. Environmental tobacco smoke is a complex mixture of more than 4700 different compounds. These pollutants enter the body as particles and as gases, such as carbon monoxide and hydrogen cyanide. Exposure to tobacco smoke greatly increases the risk of developing lung cancer in both smokers and nonsmokers (studies show that the nonsmoking spouses of smokers experience a 30 percent increase in the occurrence of lung cancer). The small children of smokers are also more likely to fall victim to such illnesses as bronchitis and pneumonia. Heart disease is closely linked to exposure to tobacco smoke, as is premature aging of the skin.

In some regions, a phenomenon known as "*sick building syndrome*" has become the focus of public concern. Several illnesses, such as *Legionnaires' disease* (which is spread through contaminated air conditioning systems), have been attributed to specific building problems, but other, less definite, illnesses have been traced to the indoor office environment as well. These complaints range from dry mucous membranes, sneezing, fatigue, and irritability to forgetfulness and nausea. Unfortunately, the nature of these symptoms and their random occurrence make tracing the source difficult.

Indoor pollutants are responsible for a wide variety of health problems. Irritation of the eyes, nose, and throat; headaches; fatigue; and dizziness are but a few of the maladies attributed to indoor air pollutants. Although these symptoms can be annoying, other life-threatening diseases can occur after prolonged exposure to many of the substances mentioned earlier, such as radon, asbestos, and other toxic compounds.

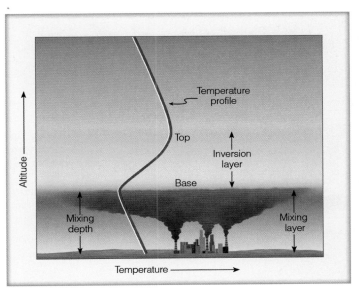

• FIGURE 18.13 The inversion layer acts as a lid on the pollutants below. If the inversion lowers, the mixing depth decreases, and the pollutants are concentrated within a smaller volume.

© J. L. Medeiros

• FIGURE 18.14 A thick layer of polluted air is trapped in the valley. The top of the polluted air marks the base of a subsidence inversion.

if the inversion rises, the mixing depth increases, and the pollutants would be dispersed throughout a greater volume of air; if the inversion lowers, the mixing depth would decrease and the pollutants would become more concentrated, sometimes reaching unhealthy levels. Since the atmosphere tends to be most unstable in the afternoon and most stable in the early morning, we typically find the greatest mixing depth in the afternoon and the most shallow one (if one exists at all) in the early morning. Consequently, during the day, the top of the mixing layer may clearly be visible (see • Figure 18.14). Moreover, during takeoff or landing on daylight flights out of large urban areas, the top of the mixing layer may sometimes be observed.

The position of the semipermanent Pacific high off the West Coast of North America, particularly during summers, contributes greatly to the air pollution in that region. The Pacific high promotes subsiding air, which warms the air aloft. Surface winds around the high promote upwelling of ocean water. Upwelling—the rising of cold water from below—makes the surface water cool, which, in turn, cools the air above. Warm air aloft and the cool, surface (marine) air together produce a strong and persistent subsidence inversion—one that exists 80 to 90 percent of the time over many West Coast cities, such as Vancouver and Los Angeles, between June and October, the smoggy months. The pollutants trapped within the cool marine air are occasionally swept inland by a sea breeze. This action carries smog from the coastal regions into the interior valleys, producing a *smog front* (see • Figure 18.15).

THE ROLE OF TOPOGRAPHY   The shape of the landscape (topography) plays an important part in trapping pollutants. We know from Chapter 3 that, at night, cold air tends to drain downhill, where it settles into low-lying basins

and valleys. The cold air can have several effects: it can strengthen a preexisting surface inversion and it can carry pollutants downhill from the surrounding hillsides (see • Figure 18.16).

Valleys prone to pollution are those completely encased by mountains and hills. The surrounding mountains tend to block the prevailing wind. With light winds and a shallow mixing layer, the poorly ventilated cold valley air can only slosh back and forth like a murky bowl of soup.

Air pollution concentrations in mountain valleys tend to be greatest during the colder months. During the warmer months, daytime heating can warm the sides of the valley to the point that upslope valley winds vent the pollutants upward, like a chimney. Valleys susceptible to stagnant air exist in just about all mountainous regions.

The pollution problem in several large cities is at least partly due to topography. For example, the cities of Los Angeles and Vancouver are surrounded on three sides by hills and mountains. During summer, cool marine air from off the ocean moves inland and pushes against the hills, which tend to block the air's eastward progress. Unable to rise, the cool

WEATHER WATCH

Cubatao, Brazil, just may be the most polluted city in the world. Located south of São Paulo, this heavily industrialized area of 100,000 people lies in a coastal valley—known by local residents as "the valley of death." Temperature inversions and stagnant air combine to trap the many pollutants that spew daily into the environment. Recently, nearly one-third of the downtown residents suffered from respiratory disease, and more babies are born deformed there than anywhere else in South America.

FOCUS ON AN OBSERVATION

Smokestack Plumes

We know that the stability of the air (especially near the surface) changes during the course of a day. These changes can influence the pollution near the ground as well as the behaviour of smoke leaving a chimney. • Figure 8 illustrates different smoke plumes that can develop with adequate wind but different types of stability.

In Figure 8a, it is early morning, the winds are light, and a radiation inversion extends from the surface to well above the height of the smokestack. In this stable environment, there is little up-and-down motion, so the smoke can only spread horizontally rather than vertically. When viewed from above, the smoke plume resembles the shape of a fan. For this reason, it is referred to as a *fanning smoke plume*.

Later in the morning, the surface air warms quickly and destabilizes as the radiation inversion gradually disappears from the surface upward (see Figure 8b). However, the air above the chimney is still stable, as indicated by the presence of the inversion.

Consequently, vertical motions are confined to the region near the surface. Hence, the smoke mixes downwind, increasing the concentration of pollution at the surface—sometimes to dangerously high levels. This effect is called *fumigation*. Here again, we can see why a taller smokestack is preferred. A taller stack extends upward into the stable layer, producing a fanning plume that does not mix downward toward the ground.

If daytime heating of the ground continues, the depth of atmospheric instability increases. Notice in Figure 8c that the inversion has completely disappeared and that the near-surface layer is unstable. Light-to-moderate winds combine with rising and sinking air to cause the smoke to move up and down in a wavy pattern, producing a *looping smoke plume*.

The continued rising of warm air and sinking of cool air, usually with strong winds and mechanical turbulence,* can cause the

---
*Recall from Chapter 9 that mechanical turbulence or *forced convection* is due to turbulent eddies formed by strong winds interacting with a rough surface.

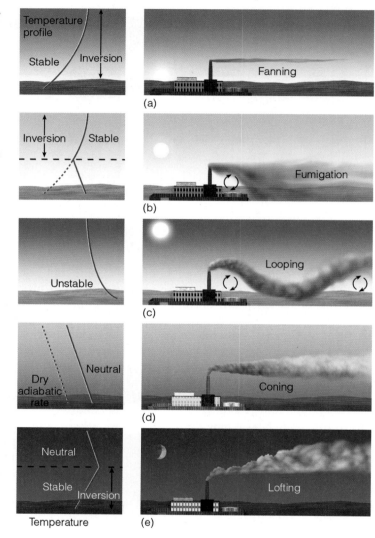

• FIGURE 8 As the vertical temperature profile changes during the course of a day (a through e), the pattern of smoke emitted from the stack changes as well.

temperature profile to equal that of the dry adiabatic rate (see Figure 8d). In this neutral atmosphere, vertical and horizontal motions are about equal, and the smoke from the stack tends to take on the shape of a cone, forming a *coning smoke plume*.

After sunset, the ground cools rapidly and the radiation inversion reappears. When the top of the inversion extends upward to slightly above the stack, stable air is near the ground with neutral air above (see Figure 8e). Because the stable air in the inversion prevents the smoke from mixing downward, the smoke is

carried upward, producing a *lofting smoke plume*. Thus, smoke plumes provide a clue to the stability of the atmosphere, and knowing the stability yields important information about the dispersion of pollutants.

Of course, other factors influence the dispersion of pollutants from a chimney, including the pollutants' temperature and exit velocity, wind speed and direction, and, as we saw in an earlier section, the chimney's height. Overall, taller chimneys, greater wind speeds, a less stable atmosphere, and higher exit velocities result in a lower concentration of pollutants.

● FIGURE 18.15 The leading edge of cool, marine air carries pollutants into Riverside, California, as an advancing smog front.

air settles in the basin, trapping pollutants from industry and automobiles. Baked by sunlight, the pollutants become the infamous photochemical smog. In Vancouver, the sea breeze carries NO_x and VOCs from the city inland up the Lower Fraser Valley, where ozone levels rise. However, photochemical smog can form in non-valley terrain as well under the right circumstances (see ● Figure 18.17).

SEVERE AIR POLLUTION POTENTIAL The greatest potential for an episode of severe air pollution occurs when all of

the factors mentioned in the previous sections come together simultaneously. The following are ingredients for a major buildup of atmospheric pollution:

● many sources of air pollution (preferably clustered close together)

● a deep high-pressure area that becomes stationary over a region

● light surface winds that are unable to disperse the pollutants

● a strong subsidence inversion produced by the sinking of air aloft

● a shallow mixing layer with poor ventilation

● a valley where the pollutants can accumulate

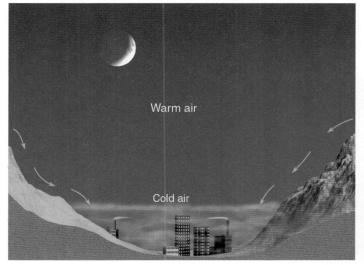

Warm air

Cold air

● FIGURE 18.16 At night, cold air and pollutants drain downhill and settle in low-lying valleys.

● FIGURE 18.17 A thick layer of smog covers the city of Toronto.

Five Days in Donora—An Air Pollution Episode

On Tuesday morning, October 26, 1948, a cold surface anticyclone moved over the eastern half of the United States. There was nothing unusual about this high-pressure area; with a central pressure of only 1025 hPa, it was not exceptionally strong (see ● Figure 9). Aloft, however, a large blocking-type ridge formed over the region, and the jet stream, which moves the surface pressure features along, was far to the west. Consequently, the surface anticyclone became entrenched over Pennsylvania and remained nearly stationary for five days.

The widely spaced isobars around the high-pressure system produced a weak pressure gradient and generally light winds throughout the area. These light winds, coupled with the gradual sinking of air from aloft, set the stage for a disastrous air pollution episode.

On Tuesday morning, radiation fog gradually settled over the moist ground in Donora, a small town nestled in the Monongahela Valley of western Pennsylvania. Because Donora rests on bottom land, surrounded by rolling hills, its residents were accustomed to fog but not to what was to follow.

The strong radiational cooling that formed the fog, along with the sinking air of the anticyclone, combined to produce a strong temperature inversion. Light, downslope winds spread cool air and contaminants over Donora from the community's steel mill, zinc smelter, and sulphuric acid plant.

The fog, with its burden of pollutants, lingered into Wednesday. Cool drainage winds during the night strengthened the inversion and added more effluents to the already filthy air. The dense fog layer blocked sunlight from reaching the ground. With essentially no surface heating, the mixing depth lowered and the pollution became more concentrated. Unable to mix and disperse both horizontally and vertically, the dirty air became confined to a shallow, stagnant layer.

Meanwhile, the factories continued to belch impurities into the air (primarily sulphur dioxide and particulate matter) from stacks no higher than 40 m tall. The fog gradually thickened into a moist clot of smoke and water droplets. By Thursday, the visibility had decreased to the point where one could barely see across the street. At the same time, the air had a penetrating, almost sickening, smell of sulphur dioxide. At this point, a large percentage of the population became ill.

The episode reached a climax on Saturday as 17 deaths were reported. As the death rate mounted, alarm swept through the town. An emergency meeting was called between city officials and factory representatives to see what could be done to cut down on the emission of pollutants.

The light winds and unbreathable air persisted until, on Sunday, an approaching storm generated enough wind to vertically mix the air and disperse the pollutants. A welcome rain then cleaned the air further. All told, the episode had claimed the lives of 22 people. During the five-day period, about half of the area's 14,000 inhabitants experienced some ill effects from the pollution. Most of those affected were older people with a history of cardiac or respiratory disorders.

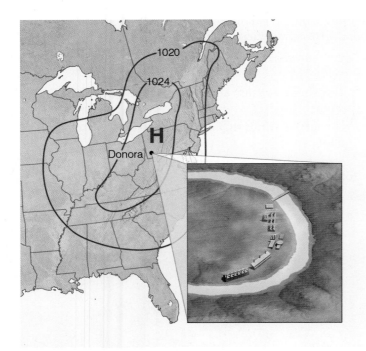

● FIGURE 9 Surface weather map that shows a stagnant anticyclone over the eastern United States on October 26, 1948. The insert map shows the town of Donora on the Monongahela River.

- clear skies so that radiational cooling at night will produce a surface inversion, which can cause an even greater
buildup of pollutants near the ground
- for photochemical smog, adequate sunlight to produce
secondary pollutants, such as ozone

Light winds and poor vertical mixing can produce a condition known as **atmospheric stagnation**. The atmospheric
conditions resulting in stagnation usually occur during particular weather patterns (see • Figure 18.18). Subsidence
inversions, clear skies, and weak winds are associated with
surface anticyclones (areas of high pressure) and ridges. We
also find that frontal inversions ahead of warm fronts can
trap pollutants. If these conditions persist over several days,
then air pollution levels can build up in areas where there are
pollutant sources and can lead to some of the worst air pollution disasters on record, such as the one in the valley city of
Donora, Pennsylvania, where in 1948, 17 people died within
14 hours. (Additional information on this disaster is found
in Focus on an Observation: Five Days in Donora—An Air
Pollution Episode on p. 584.) Conversely the strong, gusty
winds and generally less stable air behind cold fronts result in
good dispersion—the passage of such a cold front may literally "clear the air" after an air pollution episode.

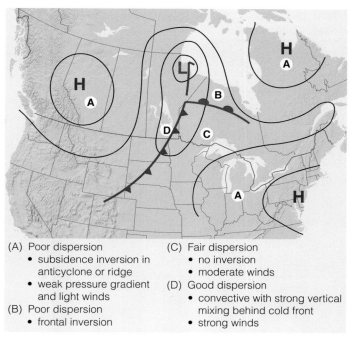

(A) Poor dispersion
- subsidence inversion in anticyclone or ridge
- weak pressure gradient and light winds

(B) Poor dispersion
- frontal inversion

(C) Fair dispersion
- no inversion
- moderate winds

(D) Good dispersion
- convective with strong vertical mixing behind cold front
- strong winds

• FIGURE 18.18 Weather patterns associated with poor, moderate, and good air pollution dispersion.

Air Pollution and the Urban Environment

For more than 100 years, it has been known that cities are generally warmer than surrounding rural areas. This region of city
warmth, known as the **urban heat island**, can influence the
concentration of air pollution. However, before we look at its
influence, let's see how the heat island actually forms.

The urban heat island is due to industrial and urban
development. In rural areas, a large part of the incoming solar
energy is used to evaporate water from vegetation and soil. In
cities, where less vegetation and exposed soil exist, the majority
of the sun's energy is absorbed by urban structures and
asphalt. Hence, during warm daylight hours, less evaporative
cooling in cities allows surface temperatures to rise higher
than in rural areas.*

At night, the solar energy (stored as vast quantities of heat
in city buildings and roads) is slowly released into the city air.
Additional city heat is given off at night (and during the day)
by vehicles and factories, as well as by industrial and domestic
heating and cooling units. The release of heat energy is retarded
by the tall vertical city walls that do not allow infrared (longwave) radiation to escape as readily as do the relatively level

surfaces of the surrounding countryside. This is because the
warm buildings absorb a portion of the infrared radiation
emitted by the surface, and they emit infrared radiation back
to the surface, keeping it warmer. The slow release of heat tends
to keep nighttime city temperatures higher than those of the
faster cooling rural areas. Overall, the heat island is strongest

1. at night, when compensating sunlight is absent
2. during the winter, when nights are longer and there is
more heat generated in the city
3. when the region is dominated by a high-pressure area
with light winds, clear skies, and less humid air

Over time, increasing urban heat islands affect climatological
temperature records, producing artificial warming in climatic
records taken in cities.

The constant outpouring of pollutants into the environment may influence the climate of a city. Certain particles
reflect solar radiation, thereby reducing the sunlight that
reaches the surface. Some particles serve as nuclei on which
water and ice form. Water vapour condenses onto these particles when the relative humidity is as low as 70 percent,
forming haze that greatly reduces visibility. Moreover, the
added nuclei increase the frequency of city fog.*

Studies suggest that precipitation may be greater in cities
than in the surrounding countryside. This phenomenon may
be due in part to the increased roughness of city terrain,
brought on by large structures that cause surface air to slow
and gradually converge. This convergence of air over the city

---

*The cause of the urban heat island is quite involved. Depending on the location,
time of year, and time of day, any or all of the following differences between cities
and their surroundings can be important: albedo (reflectivity of the surface), surface roughness, emissions of heat, emissions of moisture, emissions of particles
that affect net radiation and the growth of cloud droplets, and geometry of the
urban surface that aborbs and radiates infrared radiation to the surface.

---

*The impact that tiny liquid and solid particles, aerosols, may have on a larger scale
is complex and depends on many factors.

▼ Table 18.4   Contrast of the Urban and Rural Environment (Average Conditions)*

CONSTITUENTS	URBAN AREA (CONTRASTED TO RURAL AREA)
Mean pollution level	higher
Mean sunshine reaching the surface	lower
Mean temperature	higher
Mean relative humidity	lower
Mean visibility	lower
Mean wind speed	lower
Mean precipitation	higher
Mean amount of cloudiness	higher
Mean thunderstorm (frequency)	higher

*Values are omitted because they vary greatly depending on city, size, type of industry, and season of the year.

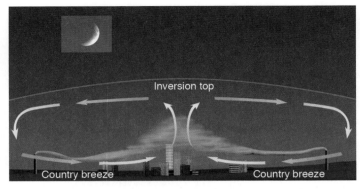

● FIGURE 18.19 On a clear, relatively calm night, a weak country breeze carries pollutants from the outskirts into the city, where they concentrate and rise due to the warmth of the city's urban heat island. This effect may produce a pollution (or dust) dome from the suburbs to the centre of town.

causes upward motion. At the same time, city heat warms the surface air, making it more unstable, which enhances rising air motions, which, in turn, aid in forming clouds and thunderstorms. This process helps explain why both clouds and storms tend to be more frequent over cities. ▼ Table 18.4 summarizes the environmental influence of cities by contrasting the urban environment with the rural.

On clear, still nights, when the heat island is pronounced, a small thermal low-pressure area forms over the city. Sometimes a light breeze—called a **country breeze**—blows from the countryside into the city. If there are major industrial areas along the city's outskirts, pollutants are carried into the heart of town, where they become even more concentrated. Such an event is especially likely if an inversion inhibits vertical mixing and dispersion (see ● Figure 18.19).

Pollutants from urban areas may even affect the weather downwind from them. In a controversial study conducted at La Porte, Indiana—a city located about 50 km downwind of the industries of south Chicago—scientists observed that La Porte had experienced a notable increase in annual precipitation since 1925. Because this rise closely followed the increase in steel production, it was suggested that the phenomenon was due to the additional emission of particles or moisture (or both) by industries to the west of La Porte.

A study conducted in St. Louis, Missouri (the Metropolitan Meteorological Experiment, or *METROMEX*), indicated that the average annual precipitation downwind from this city increased by about 10 percent. These increases closely followed industrial development upwind. This study also demonstrated that precipitation amounts were significantly greater on weekdays (when pollution emissions were higher) than on weekends (when pollution emissions were lower). Corroborative findings have been reported for Paris, France, and for other cities as well. However, in areas with marginal

humidity to support the formation of clouds and precipitation, studies suggest that the rate of precipitation may actually decrease as excess pollutant particles (nuclei) compete for the available moisture, similar to the effect of overseeding a cloud, discussed in Chapter 7. Moreover, studies using satellite data indicate that fine airborne particles, concentrated over an area, can greatly reduce precipitation.

Acid Deposition

Air pollution emitted from industrial areas, especially products of combustion, such as oxides of sulphur and nitrogen, can be carried many kilometres downwind. Either these particles and gases slowly settle to the ground in dry form (*dry deposition*) or they are removed from the air during the formation of cloud particles and then carried to the ground in rain and snow (*wet deposition*). **Acid rain** and *acid precipitation* are common terms used to describe wet deposition of acidic pollutants by both rain and snow, whereas **acid deposition** encompasses both dry and wet acidic substances. How, then, do these substances become acidic?

Emissions of sulphur dioxide (SO_2) and oxides of nitrogen may settle on the local landscape, where they transform into acids as they interact with water, especially during the formation of dew or frost. The remaining airborne particles may transform into tiny dilute drops of sulphuric acid (H_2SO_4) and nitric acid (HNO_3) during a complex series of chemical reactions involving sunlight, water vapour, and other gases. These acid particles may then fall slowly to earth, or they may adhere to cloud droplets or to fog droplets, producing **acid fog**. They may even act as nuclei on which the cloud droplets begin to grow. When precipitation occurs in the cloud, it carries the acids to the ground. Because of this, precipitation is becoming increasingly acidic in many parts of the world, especially downwind of major industrial areas.

Airborne studies conducted during the mid-1980s revealed that high concentrations of pollutants that produce acid rain can be carried great distances from their sources. For example, scientists in one study discovered high concentrations of pollutants about 600 km off the East Coast of North America. It is suspected that they came from industrial East Coast cities. Although most pollutants are washed from the atmosphere during storms, some may be swept over the Atlantic, reaching places such as Bermuda and Ireland. Acid rain knows no national boundaries.

Although studies suggest that acid precipitation may be nearly worldwide in distribution, regions noticeably affected are eastern North America, central Europe, and Scandinavia. Sweden contends that most of the sulphur emissions responsible for its acid precipitation are coming from factories in England. In some places, acid precipitation occurs naturally, such as in northern Canada, where natural fires in exposed coal beds produce tremendous quantities of sulphur dioxide. By the same token, acid fog can form by natural means.

Precipitation is naturally somewhat acidic. The carbon dioxide occurring naturally in the air dissolves in precipitation, making it slightly acidic, with a pH between 5.0 and 5.6. Consequently, precipitation is considered acidic when its pH is below about 5.3 (see ● Figure 18.20). In the northeastern United States and southeastern Canada, where emissions of sulphur dioxide are primarily responsible for the acid precipitation, typical pH values range between 4.2 and 4.7 (see ● Figure 18.21). These emissions come primarily from

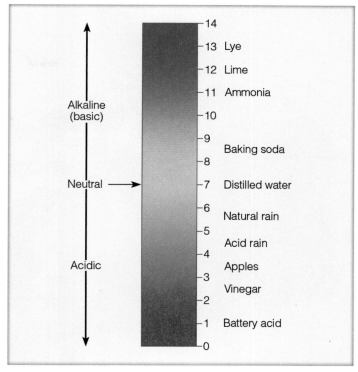

● FIGURE 18.20 The pH scale ranges from 0 to 14, with a value of 7 considered neutral. Values greater than 7 are alkaline and below 7 are acidic. The scale is logarithmic, which means that rain with pH 3 is 10 times more acidic than rain with pH 4 and 100 times more acidic than rain with pH 5.

● FIGURE 18.21 Precipitation average pH values over eastern North America between 1996 and 2000.

Environment Canada. 2004 Canadian Acid Deposition Science Assessment. Fig. 3.2. Pg. 18. © Her Majesty The Queen in Right of Canada, Environment Canada, 2010.

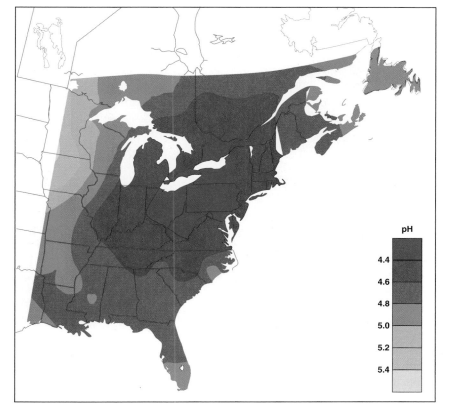

Map legend

Low: Areas primarily comprised of noncarbonate bedrock and coarse textured shallow soils.

Moderate: Areas primarily comprised of noncarbonate bedrock and/or shallow to deep soils.

High: Areas primarily comprised of noncarbonate bedrock and/or deep, fine-textured soils.

Unrated: Dominated by peatlands.

Unrated: Dominated by permanent ice and snow fields.

● FIGURE 18.22 The potential of soils and bedrock to reduce the acidity of acid rain. The red-shaded area circling Hudson Bay corresponds to the Canadian Shield geologic formation.

Source: Environment Canada (1987) Acid Rain: A National Sensitivity Assessment. Inland Waters and Lands Directorate. Ottawa, ON. © Her Majesty The Queen in Right of Canada, Environment Canada, 2010.

combustion of fossil fuels for electric power generation, as well as nonferrous mining and smelting, upstream oil and gas production, and transportation. Major sources are located mostly in the U.S. states of the Midwest, Great Lakes, Ohio River Valley, and East Coast, as well as the provinces of Ontario and Quebec.

High concentrations of acid deposition can damage plants and water resources (freshwater ecosystems seem to be particularly sensitive to changes in acidity). Concern centres chiefly on areas where interactions with alkaline soil are unable to neutralize the acidic inputs. ● Figure 18.22 shows the ability of soils to reduce the acidity of rain across Canada. Note that areas of southeastern Canada associated with the Canadian Shield geologic formation have poor ability to absorb acid rain. Unfortunately, many of these areas have acid precipitation, as indicated in Figure 18.21. Studies indicate that thousands of lakes in the United States and Canada are so acidified that entire fish populations may have been adversely affected. In an attempt to reduce acidity, lime (calcium carbonate, CaCO) is being poured into some lakes. Natural alkaline soil particles can be swept into the air, where they neutralize the acid. If it were not for airborne alkaline dust, China and the western United States would have a more severe acid rain problem than they do.

Many trees in Germany show signs of a blight that is due, in part, to acid deposition. Apparently, acidic particles raining down on the forest floor for decades have caused a chemical imbalance in the soil that, in turn, causes serious deficiencies in certain elements necessary for the trees' growth. The trees

© JK Enright/Alamy

● FIGURE 18.23 The effects on trees from acid rain as well as metal contamination of soils from metal smelting at Sudbury, Ontario. The vegetation and lakes around Sudbury have recovered considerably since emissions were drastically reduced in the 1970s.

are thus weakened and become susceptible to insects and drought. The same type of processes may be affecting North American forests, but at a much slower pace, as many forests at higher elevations from southeastern Canada to South Carolina appear to be in serious decline (see ● Figure 18.23). Moreover, acid precipitation is a problem in the mountainous west, where high mountain lakes and forests seem to be most affected.

Also, acid deposition is eroding the foundations of structures in many cities throughout the world. In Rome, the acidity of rainfall is beginning to disfigure priceless outdoor fountain sculptures and statues. The estimated annual cost of this damage to building surfaces, monuments, and other structures is more than $2 billion.

Control of acid deposition is a difficult political problem because those affected by acid rain can be quite distant from those who cause it. Technology can control sulphur emissions (e.g., stack scrubbers and fluidized bed combustion)

and nitrogen emissions (catalytic converters on cars), but some people argue that the cost is too high. If the United States turns more to coal-fired power plants, which are among the leading sources of sulphur oxide emissions, many scientists believe that the acid deposition problem will become more acute.

In recognition that long-range transport of atmospheric pollutants is a transnational problem, Canada and the United States entered into an air quality agreement in 1991. The agreement was initially to monitor progress at reducing emissions of sulphur dioxide and nitrogen oxides, the precursors to acid rain. It was expanded in 2000 to add ozone, as well as VOCs, an ozone precursor. In 2007, discussions were started to consider adding particulate matter to the agreement. The U.S. Clean Air Act of 1990 imposed a reduction in that country's emissions of sulphur dioxide and nitrogen dioxide. Canada has imposed pollution control standards and set a goal of reducing industrial air pollution by 50 percent.

SUMMARY

In this chapter, we found that air pollution has plagued humanity for centuries. Air pollution problems began when people tried to keep warm by burning wood and coal. These problems worsened during the industrial revolution as coal became the primary fuel for both homes and industry. Even though many cities do not meet all of the air quality objectives, the air over cities in the developed world is cleaner today than it was years ago due to stricter emission standards and cleaner fuels.

We examined the types and sources of air pollution and found that primary air pollutants enter the atmosphere directly, whereas secondary pollutants form by chemical reactions that involve other pollutants. The secondary pollutant ozone is the main ingredient of photochemical smog—a smog that irritates the eyes and forms in the presence of sunlight. In polluted air near the surface, ozone forms during a series of chemical reactions involving nitrogen oxides and hydrocarbons (VOCs). In the stratosphere, ozone is a naturally occurring gas that protects us from the sun's harmful ultraviolet rays. We examined how ozone forms in the stratosphere and how it may be altered by natural means. We also learned that human-released gases, such as chlorofluorocarbons, work their way into the stratosphere, where they release chlorine that rapidly destroys ozone, especially in polar regions.

We looked at the Air Quality Health Index, as well as pollutant levels across Canada. We found that a number of areas across Canada still have days considered unhealthy according to the Canada-wide Standards. We also looked at the main factors affecting air pollution and found that most air pollution episodes occur when the winds are light, skies are clear, the mixing layer is shallow, the atmosphere is stable, and a

strong inversion exists. These conditions usually prevail when a high-pressure area stalls over a region.

We observed that, on average, urban environments tend to be warmer and more polluted than the rural areas that surround them. We saw that pollution from industrial areas can modify environments downwind from them as oxides of sulphur and nitrogen are swept into the air, where they may transform into acids that fall to the surface. Acid deposition, a serious problem in many regions of the world, knows no national boundaries—the pollution of one country becomes the acid rain of another.

KEY TERMS

The following terms are listed (with page numbers) in the order they appear in the text. Define each. Doing so will aid you in reviewing the material covered in this chapter.

air pollutants, 559
primary air pollutants, 560
secondary air pollutants, 560
hazardous air pollutants, 560
criteria air contaminants, 560
particulate matter, 562
aerosols, 562
aerosol optical thickness, 562
sulphur dioxide (SO_2), 563
nitrogen dioxide (NO_2), 565
nitric oxide (NO), 565
carbon monoxide (CO), 566
volatile organic compounds (VOCs), 566

hydrocarbons, 566
smog, 566
photochemical smog, 566
ozone (O_3), 566
ozone hole, 573
Air Quality Health Index (AQHI), 575
radiation (surface) inversion, 578
subsidence inversions, 579
mixing layer, 579
mixing depth, 579
atmospheric stagnation, 585
urban heat island, 585

QUESTIONS FOR REVIEW

1. What are some of the main sources of air pollution?

2. List a few of the substances that fall under the category of particulate matter.

3. How does PM_{10} particulate matter differ from that called $PM_{2.5}$? Which poses the greatest risk to human health?

4. List two ways particulate matter is removed from the atmosphere.

5. Describe the primary sources and some of the health problems associated with each of the following pollutants:
   (a) carbon monoxide (CO)
   (b) sulphur dioxide (SO_2)
   (c) volatile organic compounds (VOCs)
   (d) nitrogen oxides

6. How does London-type smog differ from Los Angeles–type smog?

7. What is the main component of photochemical smog? What are some of the adverse health effects of photochemical smog? Why is the production of photochemical smog more prevalent during the summer and early fall than during the middle of winter?

8. In polluted air, (a) describe the role that NO plays in the production of tropospheric (ground-level) ozone and (b) the role that NO plays in the destruction of tropospheric (ground-level) ozone. What role do hydrocarbons (VOCs) play in the production of ozone and other photochemical oxidants?

9. Describe the main processes that account for stratospheric ozone production and destruction. What natural and human-produced substances could alter the concentration of ozone in the stratosphere?

10. Why is stratospheric ozone beneficial to life on earth, whereas tropospheric (ground-level) ozone is not?

11. (a) How are CFCs related to the destruction of stratospheric ozone?
    (b) If all of the ozone in the stratosphere were destroyed, what possible effects might this have on Earth's inhabitants?

12. Explain how scientists believe that the Antarctic "ozone hole" forms.

13. Describe some of the sources of pollution found inside a home or building.

14. Why are high levels of radon inside a home so dangerous?

15. Why is a light wind, rather than a strong wind, more conducive to high concentrations of air pollution?

16. How does atmospheric stability influence the accumulation of air pollutants?

17. Why is it that polluted air and inversions seem to go hand in hand?

18. Give several reasons why taller smokestacks are better than shorter ones at improving the air quality in their immediate area.

19. How does the mixing depth normally change during the course of a day? As the mixing depth changes, how does it affect the concentration of pollution near the surface?

20. How does topography influence the concentration of pollutants in cities such as Vancouver? In mountainous terrain?

21. List the factors that can lead to a major buildup of atmospheric pollution.

22. What is an urban heat island? Is it more strongly developed at night or during the day? Explain.

23. What causes the "country breeze"? Why is it usually more developed at night than during the day? Would it be more easily developed in summer or winter? Explain.

24. How can pollution play a role in influencing the precipitation downwind of certain large industrial complexes?

25. Why is acid deposition considered a serious problem in many regions of the world? How does precipitation become acidic?

QUESTIONS FOR THOUGHT

1. (a) Suppose that clouds of nitrogen dioxide drift slowly from a major industrial complex over a relatively unpopulated area. If the area is essentially "free" of hydrocarbons, would you expect high levels of tropospheric (ground-level) ozone to form? Explain.
   (b) Now suppose that the clouds of nitrogen dioxide drift slowly over an area that has a high concentration of hydrocarbons (VOCs) from both natural and industrial sources. Would you expect high levels of tropospheric (ground-level) ozone to form under these conditions? Explain your reasoning.

2. For least polluted conditions, what would be the best time of day for a farmer to burn agricultural debris? Explain why you chose that time of day.

3. Why are most severe air pollution episodes associated with subsidence inversions rather than radiation inversions?

4. What surface and upper-air conditions lead to atmospheric stagnation? What synoptic (large-scale) weather patterns?

5. Table 18.4 (p. 586) shows that cloudiness is generally greater in urban areas than in rural areas. Since clouds reflect a great deal of incoming sunlight, they tend to keep daytime temperatures lower. Why, then, during the day, are urban areas generally warmer than surrounding rural areas?

6. Acid snow can be a major problem. Explain why, for a lake in Ontario, acid snow can be a greater problem than acid rain, even when both have the same pH.

7. What atmospheric conditions would produce a fumigation-type plume downwind of a smokestack during the *middle* of the afternoon?

8. Why do we want to reduce high ozone concentrations at Earth's surface while, at the same time, we do not want to reduce ozone concentrations in the stratosphere?

9. Give a few reasons why, in industrial areas, nighttime pollution levels might be higher than daytime levels.

10. A large industrial smokestack located within an urban area emits vast quantities of sulphur dioxide and nitrogen dioxide. Following criticism from local residents that emissions from the stack are contributing to poor air quality in the area, the management raises the height of the stack from 10 m to 100 m. Will this increase in stack height change any of the existing air quality problems? Will it create any new problems? Explain.

11. If the sulphuric acid and nitric acid in rainwater are capable of adversely affecting soil, trees, and fish, why doesn't this same acid adversely affect people when they walk in the rain?

PROBLEMS AND EXERCISES

1. Keep a log of the daily Air Quality Health Index (AQHI) readings in your area and note the pollutants listed in the index. Also, keep a record of the daily weather conditions, such as cloud cover, high temperature for the day, average wind direction and speed, etc. See if there is any relationship between these weather conditions and high AQHI or pollutant readings.

2. Suppose that the AQHI reading is 6 and is mainly due to high concentrations of ozone.
   (a) How would the air be described on this day?
   (b) What would be the general health effects, and what precautions should a person take under these conditions?

3. Suppose that the air temperature at the surface is 30°C. Further suppose that the air temperature decreases at the dry adiabatic rate (10°C/km) up to the base of a strong subsidence inversion, situated about 2000 m above the surface. If the surface air temperature increases to 40°C, and the air temperature continues to decrease at the dry adiabatic rate up to the base of the inversion, will the mixing depth increase or decrease? Explain your answer with a diagram. Will pollutants found within the mixing layer be more concentrated or less concentrated? Explain.

4. Rain with a pH of 2 is how much more acidic than rain with a pH of 5?

5. Keep a log of daily AQHI readings for one urban and one rural location for days on which both areas have similar weather conditions; note the weather conditions. Compare them. How do the weather conditions influence the AQHI in both locations? Do the weather conditions contribute to greater differences in AQHI readings between the two locations? Explain.

Sunlight bending through ice crystals in cirriform clouds produces a 22° halo at sunrise.

Light, Colour, and Atmospheric Optics

19

The sky is clear, the weather is cold, and the year is 1818. Near Baffin Island in Canada, a ship with full sails enters unknown waters. On board are the English brothers James and John Ross, who are hoping to find the elusive "Northwest Passage," the waterway linking the Atlantic and Pacific oceans. On this morning, however, their hopes would be dashed for directly in front of the vessel, blocking their path, is a huge, towering mountain range. Disappointed, they turn back and report that the Northwest Passage does not exist. About 75 years later, Admiral Perry met the same barrier and called it "Crocker Land."

What type of treasures did this mountain conceal— gold, silver, precious gems? The curiosity of explorers from all over the world had been aroused. Speculation was the rule, until in 1913, the American Museum of Natural History commissioned Donald MacMillan to lead an expedition to solve the mystery of Crocker Land. At first, the journey was disappointing. Where Perry had seen mountains, MacMillan saw only vast stretches of open water. Finally, ahead of his ship was Crocker Land, but it was more than 200 miles farther west from where Perry had encountered it. MacMillan sailed on as

far as possible. Then he dropped anchor and set out on foot with a small crew of men.

As the team moved toward the mountains, the mountains seemed to move away from them. If they stood still, the mountains stood still; if they started walking, the mountains receded again. Puzzled, they trekked onward over the glittering snow-fields until huge mountains surrounded them on three sides. At last, the riches of Crocker Land would be theirs. But in the next instant, the sun disappeared below the horizon, and, as if by magic, the mountains dissolved into the cold arctic twilight. Dumbfounded, the men looked around only to see ice in all directions—not a mountain was in sight. There they were, the victims of one of nature's greatest practical jokes, for Crocker Land was a mirage.

CONTENTS

The sky is full of visual events. Optical illusions (mirages) can appear as towering mountains or wet roadways. In clear weather, the sky can appear blue, whereas the horizon appears milky white. Sunrises and sunsets can fill the sky with brilliant shades of pink, red, orange, and purple. At night, the sky is black, except for the light from the stars, planets, and moon. The moon's size and colour seem to vary during the night, and the stars twinkle. To understand what we see in the sky, we will take a closer look at sunlight, examining how it interacts with the *atmosphere* to produce an array of atmospheric visuals.

White and Colours

We know from Chapter 2 (see Figures 2.9 and 2.10, pp. 42 and 43) that nearly half of the solar radiation that reaches the atmosphere is electromagnetic radiation in the form of visible light and that visible light occurs between 0.4 and 0.7 μm wavelengths.* As sunlight enters the atmosphere, it is absorbed, reflected, scattered, or transmitted. How objects at the surface respond to this energy depends on their general nature (their colour, density, composition) and the wavelength of light that strikes them. But what makes it possible for us to see light? Why do we see various colours? What kinds of visual effects do we observe because of the interaction between light and matter? In particular, what can we see when light interacts with our atmosphere? These are the questions we will discuss in this chapter.

We perceive light because radiant energy emitted from the sun travels outward in the form of electromagnetic waves. When these waves reach the human eye, they stimulate antennalike nerve endings in the eye's retina that are either *rods* or *cones*. The rods respond to wavelengths of light in the visible range (0.4 to 0.7 μm) and give us the ability to distinguish light from dark. If we possessed only rod-type receptors, we would see only in black and white. White light occurs when we see all wavelengths in the visible spectrum. Black is the total absence of light.

The cones in our eyes respond to specific wavelengths of visible light and immediately fire an impulse through the nervous system to the brain, which we perceive as the sensation of colour. In the order of decreasing wavelengths, the colours we see are red (longest wavelength, 0.7 μm), orange, yellow, green, blue, indigo, and violet (shortest wavelength, 0.4 μm).† Colour blindness is caused by missing or malfunctioning cones. Radiation in wavelengths shorter than 0.4 μm or longer than 0.7 μm is beyond the visible range and does not stimulate vision in human eyes. White light is perceived when all the visible wavelengths strike the cones of our eyes with nearly equal intensity. This is why the midday sun usually appears white. A star that is cooler than our sun radiates most of its energy at slightly longer wavelengths and appears redder, whereas a star that is much hotter than our sun radiates more energy at shorter wavelengths and appears bluer. A star whose temperature is about the same as the sun's appears white.

Through reflection, objects that are not hot enough to emit radiation in the form of visible wavelengths can still be white or have a colour. The everyday objects we see as red absorb all the visible radiation wavelengths except red; the red light is reflected from the object to our eyes. Blue objects return blue light because they absorb all the visible wavelengths except blue. White objects reflect all wavelengths equally because they do not absorb any visible light. Surfaces that absorb all visible wavelengths and reflect no light at all appear black because no radiation strikes our rods or cones. So when we see white and colours, we know that either reflected or emitted light must be reaching our eyes.

White Clouds and Scattered Light—Nonselective Scattering

An interesting optical feature occurs when we watch the underside of a puffy, growing cumulus cloud change colour from white to dark grey or black. When we see this, we usually associate it with rain. But why is the cloud initially white? Why does it change colour? And why do we associate it with rain? To answer these questions, let's investigate the concept of *scattering* in this and the following sections.

When sunlight bounces off a surface at the same angle at which it strikes the surface, we say that the light is **reflected** and call this phenomenon *reflection* (see ● Figure 19.1). However, the atmosphere has a variety of constituents that tend to deflect solar radiation from its path and send it out in all directions. We know from Chapter 2 that radiation that is reflected in this way is **scattered*** and that scattered light is

---

*The concept of scattered light is illustrated in Figure 2.12, p. 47.

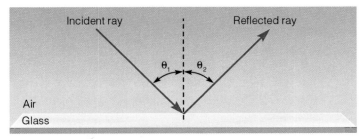

● FIGURE 19.1 For a ray of light striking a flat, smooth surface, the angle at which the incident ray strikes the surface (the angle of incidence, or θ_1) is equal to the angle at which the reflected ray leaves the surface (the angle of reflection, or θ_2). This phenomenon is called Snell's law.

---

*Reported in nanometres (nm), this is 400 to 700 nm. In angstroms (Å), it is 4000 to 7000 Å as 1 Å = 10^{-10} m. Angstroms are commonly used for light measurements and are accepted for use in the SI system but are not formally defined as an SI unit.

†Two common phrases (mnemonics) that are used to remember the order of colours in the visible spectrum are "Roy G. Biv" or "Richard Of York Gave Battle In Vain." These remind us that the colour spectrum is "Red, Orange, Yellow, Green, Blue, Indigo, Violet" if listed from longest to shortest wavelengths.

also called *diffuse light*. During the scattering process, no energy is gained or lost, so no temperature change occurs. In the atmosphere, there is a mixture of air molecules, fine particles of dust, water molecules, organic compounds that are released by vegetation, smoke, and pollutants. In the visible wavelength range, different particle sizes cause different types of scattering. Additionally, as we will see in later sections, scattering causes different phenomena depending on whether we are viewing a glowing object through the scattered light,

such as the sun or moon, or viewing the scattered light directly.

We can categorize three types of scattering processes that operate in the atmosphere: **nonselective scattering**, **Mie scattering**, and **Rayleigh scattering**. As shown in ▼ Table 19.1, they are differentiated by the size of the particle (D) causing the light to scatter relative to the wavelength of light being scattered (λ). In the following sections, we will look at each of these and their effects in more detail.

▼ TABLE 19.1   Types of Scattering of Visible Light*

TYPE OF SCATTERING	TYPE OF PARTICLE	PARTICLE DIAMETER (D) IN MICROMETERS (μm)	NATURE OF SCATTERING (VARIES WITH PARTICLE SIZE (D) AND THE WAVELENGTH OF LIGHT BEING SCATTERED (λ))	RESULTING PHENOMENA
Rayleigh	molecules (mostly air: nitrogen and oxygen)	0.0001 to 0.001	The wavelength of light being scattered is very much larger than the size of the particle doing the scattering (D << λ). Shortest wavelengths of visible light (blues/violets) are scattered better than longer wavelengths (reds.)	Blue sky, blue haze, yellow, orange sunsets
Mie†	Aerosols‡ (dust, salt, pollen, smoke, pollutants)	0.01 to 1.0‡	The wavelength of light being scattered is about the same size as the particle doing the scattering (D ~ λ). Scattering is most effective at wavelengths closest to the particle size. High concentrations of similarly sized particles can preferentially scatter different parts of the visible light spectrum.	**If small particles dominate:** blues scatter, red light remains: blue haze, blue mountains, red sunsets, red sun/moon† **If all particle sizes in this range exist:** white to grey haze and reduced visibility **If large particles dominate:** reds scattered, blue light remains: red skies, blue moon†
Nonselective (also called geometric)	Cloud droplets to rain drops	10 to 100	The wavelength of light being scattered is very much smaller than the size of the particle doing the scattering (D >> λ). All wavelengths scattered equally	White clouds, fog

*The type of scattering behaviour changes depending on the size of the particle doing the scattering (D) relative to the wavelength of light being scattered.(λ).

†See the accompanying text for a more complete explanation.

‡Aerosols can span large size ranges. Smaller aerosols can cause Rayleigh scattering, whereas large aerosols can cause nonselective scattering.

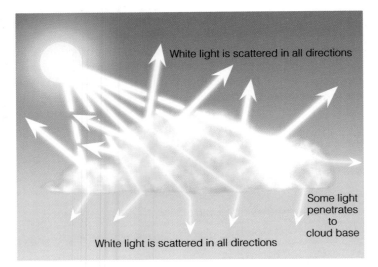

● FIGURE 19.2 Cloud droplets scatter visible light in all directions, and the light from these billions of droplets turns a cloud white.

Particles that are large relative to the wavelength of light being scattered, such as cloud droplets of 10 to 100 μm in diameter, effectively scatter all wavelengths of visible radiation more or less equally and produce white light. We call this phenomenon *nonselective* (also known as *geometric scattering*). Because of their droplet size, clouds are poor absorbers of sunlight. Hence, when we look at a cloud, it appears white because countless cloud droplets scatter all the wavelengths of visible sunlight in all directions (see ● Figure 19.2).

So why do clouds change colour? Even small clouds can be optically thick; a term used to indicate how much the path of light is affected as it passes through. ● Figure 19.3 shows

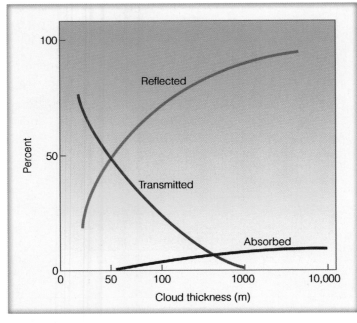

● FIGURE 19.3 The average percentage of radiation that is reflected, absorbed, and transmitted by clouds of various thicknesses.

Ever see a green thunderstorm? Severe thunderstorms that form over the Prairies often appear green. The green colour may be due to reddish sunlight (especially at sunset) penetrating the storm and then being scattered by cloud particles composed of water and ice. With much of the red light removed, the scattered light casts the underside of the cloud as a faint greenish hue.

how light penetration (transmitted light), reflection, and absorption vary with cloud thickness. Notice that almost no light is transmitted as clouds reach 1000 m thick (a typical thickness for single clouds such as stratus or stratocumulus). As clouds grow larger and taller, more sunlight is reflected and less penetrates all the way through. Since little sunlight reaches the underside of the cloud, little light is scattered, and the cloud base appears dark. At the same time, if droplets near the base of the cloud grow larger (beyond 100 μm in diameter), they become less effective scatterers and better absorbers. As a result, the small amount of visible light that reaches the cloud base is absorbed rather than scattered, which makes the cloud appear even darker. Depending on conditions, these cloud droplets may even grow large and heavy enough to fall to Earth as rain. This leads to our association of dark, threatening clouds frequently producing rain and is why clouds appear so dark.

Blue Skies and Hazy Days—Rayleigh and Mie Scattering

Another optical effect is Earth's blue sky. It appears blue because light that stimulates the sensation of blue colour reaches our eye's retina, but what causes this? As with *nonselective scattering*, this is also a consequence of the size of the particle doing the scattering.

Individual air molecules are much smaller than cloud droplets—their diameters are small even when compared to the wavelengths of visible light. Each molecule of oxygen and nitrogen in the air is a *selective scatterer*—each scatters the shorter wavelengths of visible light much more effectively than the longer ones. This selective scattering is also known as *Rayleigh scattering* (see Table 19.1).

As sunlight enters the atmosphere, the shorter visible light wavelengths of violet, blue, and green are scattered by atmospheric gases more than the longer wavelengths of yellow, orange, and especially red.* In fact, violet light is scattered about 16 times more than red light. Consequently, as we view the sky, scattered violet, blue, and green wavelengths strike our eyes from all directions. Because our eyes are more

---

*The reason for this is that the intensity of Rayleigh scattering varies as $\frac{1}{\lambda^4}$ where λ is the wavelength of radiation.

● FIGURE 19.4 The sky appears blue because billions of air molecules selectively scatter the shorter wavelengths of visible light more effectively than the longer ones. This causes us to see scattered blue light coming from all directions.

sensitive to blue than violet or green, these wavelengths produce the sensation of blue coming from all around us, as seen in ● Figure 19.4. Therefore, when we look at the sky, it appears blue (see ● Figure 19.5).

When the particles doing the scattering are about the same size as the wavelengths of light, Mie scattering occurs (see Table 19.1). Mie scattering can seem complex because different optical effects can result depending on the distribution of particle sizes that are present in the air and whether one is observing a glowing object or just the scattered light. Mie scattering is transitional between Rayleigh scattering and nonselective scattering as aerosols, dust, pollens, pollutants, and other fine particles occur between the particle sizes associated with these processes. (Generally, this group of atmospheric constituents is just referred to as aerosols.) So when particles are a bit smaller than the wavelengths of visible

light, the shorter wavelengths (blues) are preferentially scattered in a way that can look like Rayleigh scattering but is caused by a high concentration of similarly sized small particles scattering blue light. If particles are just a bit larger than the wavelengths of visible light, then the longer wavelengths (reds) are scattered preferentially. (In the next section, we will see how these processes create interesting phenomena.) Most of the time, there is a broad range of aerosol particle sizes in the air that span all the wavelengths in the visible spectrum, so they scatter light equally and produce neutral, milky-coloured skies that are white, grey, or beige coloured.

The selective scattering of blue light by small aerosols can make distant mountains appear blue through the process of Mie scattering. This is suggested in the names of landmarks such as the U.S. Blue Ridge Mountains (see ● Figure 19.6) or the Blue Mountains of Australia. Mie scattering occurs when the particles that cause the scattering are larger than the molecular constituents that result in Rayleigh scattering. As shown in Table 19.1, these larger particles are mostly aerosols that are smaller than cloud droplet sizes. The blue haze that creates this phenomenon is the scattered light that is thought to result from light interacting with high concentrations of tiny, naturally occurring compounds. These 0.2 μm diameter particles selectively scatter blue light very effectively. They are produced when terpenes* are released by vegetation and combine chemically with small amounts of ozone. This process can produce blue hazy landscapes in areas that are far removed from air pollution.

When our eyes are bombarded by all wavelengths of visible light, the sky appears milky, the visibility lowers, and we call the day "hazy." If the relative humidity is high enough, soluble particles (nuclei) will "pick up" water vapour and grow into haze particles. Thus, the colour of the sky gives us

---

*Terpenes comprise a large group of organic compounds (mostly hydrocarbons), which are dominantly found in plants as essential oils. Generally, we associate them with naturally occurring aromatic compounds.

● FIGURE 19.5 Blue skies, white clouds, atmospheric particle sizes, and scattering. Rayleigh scattering causes the selective scattering of blue light by air molecules (i.e., very small particles), producing a blue sky. Nonselective scattering causes all wavelengths of visible light to be scattered by liquid cloud droplets (i.e., large particles), producing the white clouds.

© C. Donald Ahrens

● FIGURE 19.6 The Blue Ridge Mountains in Virginia, U.S.A. The blue haze is caused by Mie scattering, where large amounts of similarly sized aerosols, which are smaller than the wavelengths of visible light, selectively scatter blue light. Notice that the scattered blue light causes the most distant mountains to become almost indistinguishable from the sky even though two different processes are responsible for the blue colour.

© C. Donald Ahrens

a hint about how much material is suspended in the air: the more particles, the more scattering, and the whiter the sky becomes. Since most of the suspended particles are near the surface, the horizon often appears white. On top of a high mountain, when we are above many of these haze particles, the sky usually appears a deep blue.

Haze can scatter light from the rising or setting sun, so we see bright light beams, or **crepuscular rays**, radiating across the sky. A similar effect occurs when the sun shines through a break in a layer of clouds (see ● Figure 19.7). Dust, tiny water droplets, or haze in the air beneath the clouds scatter sunlight, making that region of the sky appear bright with rays. Because these rays seem to reach downward from clouds, some people will remark that the "sun is drawing up water." In England, this same phenomenon is referred to as

"Jacob's ladder." No matter what these sunbeams are called, it is the scattering of sunlight by particles in the atmosphere that makes them visible.

Red Suns and Blue Moons— Interesting Effects of Rayleigh and Mie Scattering

Near sunrise or sunset, rays coming directly from the sun strike the atmosphere at a low angle. They must pass through much more atmosphere than at any other time during the day. For example, when the sun is 4° above the

● FIGURE 19.7 The scattering of sunlight by dust and haze produces these white bands of crepuscular rays.

NCAR/UCAR/NSF

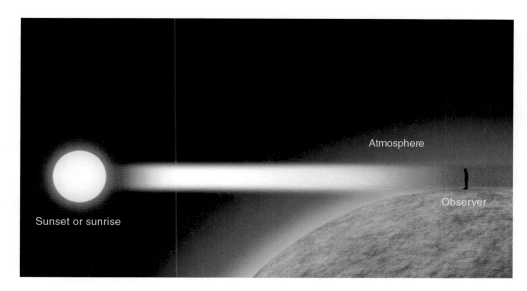

● FIGURE 19.8 Selective scattering by a thick section of atmosphere causes the sun to appear either yellow, orange, or red at sunrise and sunset. The size of the particles in the atmosphere causes the different resulting colours through different scattering processes. Scattering by air molecules results in yellow-orange colours (Rayleigh scattering); scattering by aerosols creates redder sunsets.

horizon, sunlight must pass through an atmosphere that is more than 12 times thicker than when the sun is directly overhead. This increases the number of interactions that light can have on its way through the atmosphere and consequently affects the phenomena we see. By the time sunlight has penetrated this large amount of air, most of the shorter wavelengths of visible light have been scattered away by air molecules (through Rayleigh scattering); only the yellow, orange, and red wavelengths make it through the atmosphere from the glowing sun. On reaching our eyes, these waves produce a bright yellow-orange sunset (see ● Figure 19.8).

Bright, yellow-orange sunsets occur only when the atmosphere is fairly clean, as it would be after a recent rain, when particles are removed and scattering is caused only by air molecules. If the atmosphere contains many fine particles whose diameters are a little larger than air molecules (into the Mie scattering size range), the slightly longer (yellow) wavelengths are also scattered away. Only the orange and red wavelengths penetrate through, and the sun can appear orange to red. When the atmosphere becomes loaded with particles that are even a bit larger, only the longest (red) wavelengths are able to penetrate, and Mie scattering creates a red sun.

Natural events may produce red sunrises and sunsets over the oceans. For example, the scattering characteristics of small, suspended salt particles and water molecules are responsible for the brilliant red suns that can be observed from a beach (see ● Figure 19.9). Volcanic eruptions rich in sulphur create particles in the Mie scattering range that can produce red sunsets, too. Such red sunsets are actually produced by a highly reflective cloud of sulphuric acid droplets, formed from sulphur dioxide gas injected into the stratosphere during powerful eruptions, like that of the Mexican volcano El Chichón in 1982 and the Philippine volcano Mount Pinatubo in 1991. These fine particles,

© C. Donald Ahrens

● FIGURE 19.9 Red sunset near the coast of Iceland. The reflection of sunlight off the slightly rough water is producing a *glitter path*.

moved by the winds aloft, circled the globe, producing beautiful sunrises and sunsets for months and even years after the eruptions. These same volcanic particles in the stratosphere can turn the sky red after sunset as some of the red light from the setting sun bounces off the bottom of the particles back to Earth's surface. Generally, these volcanic red sunsets occur about an hour after the actual sunset (see ● Figure 19.10).

Occasionally, the atmosphere becomes so laden with dust, smoke, and pollutants that even red wavelengths are

© C. Donald Ahrens

● FIGURE 19.10 This bright red sky over California was produced by the sulphur-rich particles that erupted from the Mount Pinatubo volcano in September 1992. The photograph was taken about an hour after sunset.

unable to pierce the filthy air. This causes an eerie effect as no visible light enters the eye, and the sun literally disappears before it reaches the horizon.

So, as we have seen, the scattering of light by the atmospheric particles responsible for Mie scattering can cause some rather unusual sights. If the aerosols (salt, dust, smoke, pollens, pollutants) are roughly uniform in size and great enough in number, they can selectively scatter the sun's rays in certain wavelengths, resulting in interesting colour effects. Recall from the earlier discussion that Mie scattering is most effective when the particle size is about the same as the wavelength of light. If the particles are all similarly sized small aerosols, near the wavelengths of blue light, these particles will preferentially scatter the blues, leaving the longer (red) wavelengths. This creates the blue haze in scattered light that we discussed earlier; but when looking at glowing objects, it causes the sun to be orange to red—even at noon. It can even make the moon appear red.

Alternatively, if there is a higher concentration of larger aerosols (e.g., 0.9 μm), these particles more closely match longer visible wavelengths (reds), so the reds are preferentially scattered and shorter wavelengths (blues) remain. This causes a reddening of the sky and glowing objects to appear blue. It is responsible for blue-coloured suns and the occurrence of fairly rarely observed "blue moons" and explains the origin of the expression "once in a blue moon."

In summary, the scattering of light by small particles in the atmosphere causes many familiar effects: white clouds, blue skies, hazy skies, crepuscular rays, variously coloured suns, and colourful sunsets. In the absence of any scattering, we would simply see a white sun against a black sky—not as attractive or interesting.

Refraction, Twinkling, and Twilight

Light that passes through a substance is said to be transmitted. On entering a denser substance, transmitted light slows in speed. If it enters the substance at an angle, the light's path also bends. This bending is called **refraction**. The amount of refraction depends primarily on two factors: the

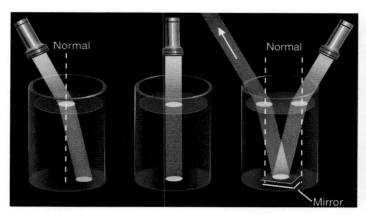

● FIGURE 19.11 The behaviour of light as it enters and leaves a more dense substance, such as water.

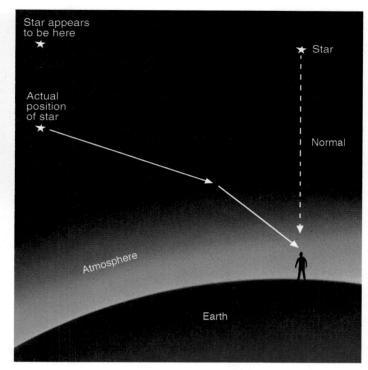

● FIGURE 19.12 The bending of starlight by the atmosphere causes stars that are not directly overhead to appear to be higher in the sky than they really are.

density of the material and the angle at which the light enters the material.

Refraction can be demonstrated in a darkened room by shining a flashlight into a beaker of water (see ● Figure 19.11). If the light is held directly above the water so that the beam strikes the surface of the water straight on, no bending occurs. But if the light enters the water at some angle, it bends toward the *normal*, which runs perpendicular to the air–water boundary and is the dashed line in Figure 19.11. (A *normal* is simply a line that intersects any surface at a right angle. We use the concept of a normal as it shows how much bending occurs as light enters and leaves various substances.) In the figure, a small mirror on the bottom of the beaker reflects the light upward, and this reflected light bends away from the normal as it re-enters the air. We can summarize the light's behaviour as follows: *light that travels from a less dense to a more dense medium loses speed and bends toward the normal, whereas light that travels from a more dense to a less dense medium increases in speed and bends away from the normal.*

Within the atmosphere, light refraction causes a variety of visual effects. For example, at night, the light we see from stars that are directly above us is not bent, but starlight that enters Earth's atmosphere at an angle is bent. In fact, a star whose light enters the atmosphere just above the horizon has more atmosphere to penetrate. Consequently, it is refracted the most. ● Figure 19.12 shows that the bending is toward the normal as the light enters Earth's more dense atmosphere. By the time the "bent" starlight reaches our eyes, the star appears to be higher than it actually is because our eyes cannot detect that the path of light is bent. We see light coming from a particular direction, mentally extend a line to that direction, and interpret the star to be in that position. So the next time you take a midnight stroll, point to any star near the horizon and remember that the star's actual position is not where it appears to be. Its true position is just a bit lower, about half a degree toward the surface according to ▼ Table 19.2.

As starlight enters the atmosphere, it often passes through regions of differing air density. Each region deflects and bends the tiny beam of starlight, constantly changing the star's apparent position and causing it to appear to twinkle or flicker. This condition is known as **scintillation**. Planets, being much closer to us, appear larger and usually do not twinkle because their size is greater than the angle at which their light deviates as it penetrates the atmosphere. However, when planets are near the horizon, they

▼ Table 19.2　Amount of Atmospheric Refraction (Bending) in Minutes (′) as Viewed at Sea Level under Standard Atmospheric Conditions*

OBJECT'S ELEVATION ABOVE HORIZON (DEGREES)	REFRACTION (MINUTES)
0°	35.0′
5°	10.0′
20°	2.6′
40°	1.2′
60°	0.6′
90°	0.0′

*60′ = 1°.

● FIGURE 19.13 The bending of sunlight by the atmosphere causes the sun to appear to rise about two minutes earlier and set about two minutes later than it would otherwise.

sometimes twinkle as this is where the bending of their light is greatest.

The refraction of light by the atmosphere has some other interesting consequences. For example, the atmosphere gradually bends the rays from a rising or setting sun and moon. Because light rays from the lower part of the sun (or moon) are bent more than those from the upper part, the sun appears to flatten out on the horizon, taking on an elliptical shape (Figure 19.17, p. 606, shows this effect). Also, since light is bent more on the horizon, the sun and moon both appear to be higher than they really are. Consequently, they both appear to rise about two minutes earlier and set about two minutes later than they would if there were no atmosphere (see ● Figure 19.13).

Twilight is the extra light that extends both ends of our days. Just after sunset, we call it dusk, whereas just before sunrise, we know it as dawn. It is most apparent under clear skies and occurs as the atmosphere refracts and scatters sunlight from below the horizon, before it reaches our eyes. Different disciplines define twilight in different ways. *Civil twilight* lasts from sunset (or sunrise) until the centre of the sun is 6° below the horizon—this is the definition of twilight we use in our daily lives. Nautical twilight is used by mariners. It occurs when the sun is between 6° and 12° below the horizon but is defined as ending (or beginning) when the light is too dim to see the horizon for navigation. Astronomical twilight is the longest. It ends (or starts) when the sun is between 12° and 18° below the horizon, meaning that it extends from sunset (or sunrise) until observation of the faintest stars is possible.

How long twilight lasts depends on the latitude and season. As discussed in Chapter 3 (p. 68), the longest and shortest day lengths occur on the summer and winter solstices. The length of twilight depends on how quickly the sun disappears below the horizon (as we have seen, this is 6° below the horizon for civil twilight). How quickly the sun disappears depends on the angle the path of the sun makes with the horizon as it sets. When the angle is perpendicular to the horizon, twilight is shortest; when the angle is flattest, twilight is longest. The angle the sun makes when it sets changes with latitude and season. For example, at the equator, on fall equinox (September 22), the sun drops below the horizon at a perpendicular angle, making this the shortest twilight at 21 minutes. On the same date, at higher latitudes, the sun sets at a flatter angle, so it takes longer to reach 6° below the horizon and twilight lasts longer (until you reach the pole where the sun is at the horizon).

The length of twilight also changes with the seasons. At summer solstice, the sun sets at its shallowest angle and twilight is longest until reaching the Arctic Circle, where there is no twilight and the sun is above the horizon for 24 hours (beyond the Arctic Circle, the sun never sets for even longer periods). Equinoxes actually have the shortest twilight because the sun sets closest to a perpendicular angle at this time of year. At winter solstice, the sun sets at an angle that is closer to perpendicular than that for summer solstice, but the angle the sun sets is still not as close to perpendicular as it is during the equinoxes. The reason for this seemingly confusing fact is that during Earth's yearly orbit around the sun, both the sun's setting angle and the sun's north–south position on the horizon change. So even though the sun is lower in the sky at winter solstice, it is also setting in a more southerly location on the horizon, making the sun's setting angle between the flatter angle it has in summer and the steeper angle it has at the equinox. For Canada, seasonally representative twilight and day lengths are as shown in ▼ Table 19.3.

In polar regions, the concept of a *polar day* is used to represent the period during the summer when the sun is continuously in the sky. The length of a polar day varies with the latitude. In the Northern Hemisphere, the Arctic Circle (66.5°N) represents the lowest latitude, where the sun never sets in summer and never rises in winter. At the Arctic Circle on the summer solstice, the sun is in the sky for

▼ Table 19.3   Day and Civil Twilight Length for Some Locations in Canada at Solstices and Equinoxes

| Latitude °N | Place | DAWN AND DUSK CIVIL TWILIGHT LENGTH | | | SUNRISE–SUNSET LENGTH | |
		Summer Solstice (June 21) (minutes)	Winter Solstice (Dec. 21) (minutes)	Equinox (Mar. 20/Sept. 22) (minutes)	Summer Solstice (hours)	Winter Solstice (hours)
42.3	Windsor, Ontario	35	32	28	15.28	9.08
45.5	Montreal, Quebec	38	34	30	15.69	8.70
49.3	Vancouver, British Columbia	43	38	32	16.25	8.19
53.5	Edmonton, Alberta	53	43	35	17.05	7.46
64.8	Dawson, Yukon Territory	sunset to sunrise	85	49	21.08	3.72
68.3	Inuvik, Northwest Territories	all light	11:20 until 16:27	57	24	0
74.7	Resolute (Qausuittuq), Nunavut	all light	all dark	80	24	0

Data source: National Research Council of Canada. Sunrise/Sunset/Sun Angle Calculator.

24 hours, and, conversely, for the winter solstice, the sun never rises for 24 hours on this day. North from the Arctic Circle, the length of a polar day (or night) increases until you reach the geographic North Pole, where it is light (or dark) for six months each year. These periods of light and darkness are important cultural markers for residents of these regions and are identified by various traditional names in different parts of the world—commonly in Canada, from Scandinavian origin, it is known as the midnight sun. Below the Arctic Circle, at 60.5°N (about the latitude of White-horse, Yukon Territory) at midsummer, morning and evening twilight can converge, producing a *white night*—where the twilight lasts all night long. Places between this latitude and the Arctic Circle have longer white nights. Russia and other high-latitude countries have white night festivals to celebrate the long period of daylight.

In general, without the atmosphere, there would be no refraction or scattering, and the sun would rise later and set earlier than it now does. Instead of twilight, darkness would arrive immediately when the sun disappears below the horizon. Imagine how summer evening events would change if we had instant darkness.

BRIEF REVIEW

Up to this point, we have examined how light can interact with our atmosphere. Before going on, here is a review of some of the important concepts and facts we have covered:

- When light is scattered, it is sent in all directions—forward, sideways, and backward.
- The type of scattering that occurs, nonselective or selective (Rayleigh and Mie), depends on the relative difference between the size of the particle doing the scattering and the wavelength of light being scattered.
- White clouds, blue skies, hazy skies, crepuscular rays, and colourful skies are the result of scattered sunlight.
- Glowing objects such as the sun or moon take on the colour of the light that remains after preferential scattering has removed some wavelengths.
- The bending of light as it travels through regions of differing density is called refraction.
- As light travels from a less dense substance (such as outer space) and enters a more dense substance at an angle (such as our atmosphere), the light bends downward, toward the normal. This effect causes stars, the moon, and the sun to appear just a tiny bit higher than they actually are.
- Refraction results in twilight. The length of twilight varies with latitude and season.

The Mirage: Seeing Is Not Believing

In the atmosphere, when an object appears to be displaced from its true position, we call this phenomenon a **mirage**. A mirage is not a figment of the imagination—our minds are not playing tricks on us, but the atmosphere is.

Atmospheric mirages are created by light passing through and being bent by air layers of different densities. Such changes in air density are usually caused by sharp changes in air temperature. The greater the rate of temperature change, the greater the light rays are bent. For example, on a warm, sunny day, black road surfaces absorb a great deal of solar energy and become very hot. Air in contact

C. Donald Ahrens

● FIGURE 19.14 The road in the photograph appears wet because blue skylight is bending up into the camera as the light passes through air of different densities.

with these hot surfaces warms by conduction, and because air is a poor thermal conductor, we find much cooler air only a few metres higher. On hot days, these road surfaces often appear wet (see ● Figure 19.14). Such "puddles" disappear as we approach them, and advancing cars seem to swim in them. Yet we know the road is dry. The apparent wet pavement above a road is the result of blue skylight refracting up into our eyes as it travels through air of different densities. A similar type of mirage occurs in deserts during the hot summer. Many thirsty travellers have been disappointed to find that what appeared to be a water hole was in actuality hot desert sand.

Sometimes these "watery" surfaces appear to shimmer. The shimmering results as rising and sinking air near the ground constantly change the air density. As light moves through these regions, its path also changes, causing the shimmering effect.

When the air near the ground is much warmer than the air above, objects may not only appear to be lower than they really are but are also often inverted. These mirages are called **inferior** (lower) **mirages**. The tree in ● Figure 19.15

certainly does not grow upside down. So why does it look that way? It appears to be inverted because light reflected from the top of the tree moves outward in all directions. Rays that enter the hot, less dense air above the sand are refracted upward, entering the eye from below. The brain is fooled into thinking that these rays came from below the ground, which makes the tree appear upside down. Some light from the top of the tree travels directly toward the eye through air of nearly constant density and, therefore, bends very little. These rays reach the eye "straight on," and the tree appears upright. Hence, off in the distance, we see a tree and its upside-down image beneath it. (Some of the trees in Figure 19.14 show this effect.)

The atmosphere can play optical jokes on us in extremely cold areas, too. In polar regions, air next to a snow surface can be much colder than the air many metres above. Because the air in this cold layer is very dense, light from distant objects entering it bends toward the normal in such a way that the objects can appear to be shifted upward. This phenomenon is called a **superior** (upward) **mirage**. ● Figure 19.16 shows the atmospheric conditions favourable for a superior mirage.

● FIGURE 19.15 Inferior mirage over hot desert sand.

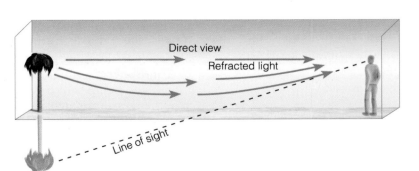

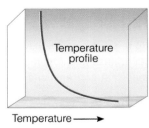

FOCUS ON AN OBSERVATION

The *Fata Morgana*

A special type of superior mirage is the *Fata Morgana*, a mirage that transforms a fairly uniform horizon into one of vertical walls and columns with spires (see ● Figure 1). According to legend, Fata Morgana (Italian for "fairy Morgan") was the half-sister of King Arthur. Morgan, who was said to live in a crystal palace beneath the water, had magical powers that could build fantastic castles out of thin air. Looking across the Straits of Messina (between Italy and Sicily), residents of Reggio, Italy, on occasion would see buildings, castles, and sometimes whole cities appear, only to vanish again in minutes. The *Fata Morgana* is observed where the air temperature increases with height above the surface, slowly at first, then more rapidly, and then slowly again. Consequently, mirages such as the *Fata Morgana* are frequently seen where warm air rests above a cold surface, such as above large bodies of water and in polar regions.

● FIGURE 1 The *Fata Morgana* mirage over water. The mirage is the result of refraction—light from small islands and ships is bent in such a way as to make them appear to rise vertically above the water.

Superior mirages, called *hillingars* by early Norsemen, may have contributed to their discovery of Iceland and Greenland because land normally below the horizon is raised up into view. (A special type of superior mirage, the **Fata Morgana**, is described in Focus on an Observation: The *Fata Morgana* above.)

WEATHER WATCH

In 1939, Captain John Bartlett was able to see the mountains of Iceland from over 500 km away—a record distance. This was due to the hillingar effect (a superior mirage).

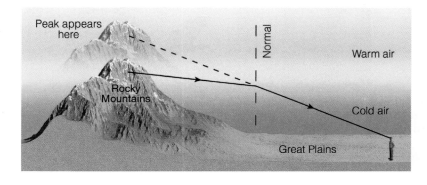

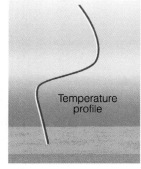

● FIGURE 19.16 The formation of a superior mirage. When cold air lies close to the surface with warm air aloft, light from distant mountains is refracted toward the normal as it enters the cold air. This causes an observer on the ground to see mountains higher and closer than they really are.

Temperature ⟶

© Pekka Parviainen

● FIGURE 19.17 The very light green on the upper rim of the sun is the green flash. Also, observe how the atmosphere makes the sun appear to flatten on the horizon into an elliptical shape.

The Green Flash

Occasionally, a flash of green light—called the **green flash**—may be seen near the upper rim of a rising or setting sun (see ● Figure. 19.17). Remember from our earlier discussion that when the sun is near the horizon, its light must penetrate a thick section of atmosphere. This thick atmosphere refracts sunlight, with purple and blue light bending the most and red light the least. Because of this bending, more blue light should appear along the top of the sun.* But because the atmosphere selectively scatters blue light, very little reaches us, and we see green light instead—this is called a *green rim*.

Usually, the green rim is too faint to see with the human eye. However, under certain atmospheric conditions, such as when the surface air is very hot or when an inversion exists, the setting sun makes a mirage. Most commonly, an inferior mirage of the setting sun forms, and just as the sun sets, the inverted top of the mirage sun and the main sun image superimpose and the green rim is magnified. When this happens, a momentary flash of green light appears, often just before the sun disappears from view. When the observer views the setting sun from above an inversion layer, then a kind of mirage forms that magnifies the green light above the setting sun before it sets, as seen in Figure 19.17.

Halos, Sundogs, and Sun Pillars

A ring of light encircling and extending outward from the sun or moon is called a **halo**. Such a display is produced when sunlight or moonlight is refracted as it passes through ice

*Looking directly at the sun, especially at midday, can cause irreparable damage to the eye. Normally, we get only glimpses or impressions of the sun from the corner of our eye.

crystals. Hence, the presence of a halo indicates that *cirriform clouds* are present.

The most common type of halo is the 22° halo—a ring of light 22° from the sun or moon.* Such a halo forms when tiny, suspended, column-type ice crystals (with diameters less than 20 μm) become randomly oriented as air molecules constantly bump against them. The refraction of light rays through these crystals forms a halo like the one shown in ● Figure 19.18. Less common is the 46° halo, which forms in a fashion similar to that of the 22° halo (see ● Figure 19.19). With the 46° halo, however, the light is refracted through column-type ice crystals that have diameters in a narrow range between about 15 and 25 μm.

*Extend your arm and spread your fingers apart. An angle of 22° is about the distance from the tip of the thumb to the tip of the little finger.

© T. Ansel Toney

● FIGURE 19.18 A 22° halo around the sun, produced by the refraction of sunlight through ice crystals.

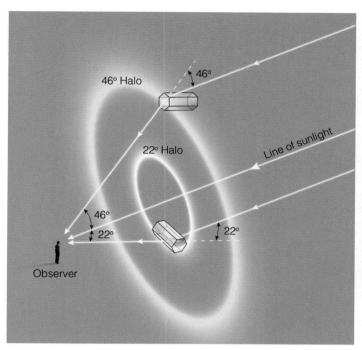

● FIGURE 19.19 The formation of a 22° halo and a 46° halo with column-type ice crystals.

Occasionally, a bright arc of light may be seen at the top of a 22° halo (see ● Figure 19.20). Since the arc is tangent to the halo, it is called a **tangent arc**. Apparently, the arc forms as large six-sided (hexagonal), pencil-shaped ice crystals fall with their long axes horizontal to the ground. Refraction of sunlight through the ice crystals produces the bright arc of light. When the sun is on the horizon, the arc that forms at the top of the halo is called an *upper tangent arc*. When the sun is above the horizon, a *lower tangent arc* may form on the lower part of the halo beneath the sun. The shape of the arcs changes greatly with the position of the sun.

● FIGURE 19.20 A halo with an upper tangent arc.

● FIGURE 19.21 Refraction and dispersion of light through a glass prism.

A halo is usually seen as a bright, white ring, but there are refraction effects that can cause it to have colour. To understand this, we must first examine refraction more closely. When white light passes through a glass prism, it is refracted and split into a spectrum of visible colours (see ● Figure 19.21). Each wavelength of light is slowed by the glass, but each is slowed a little differently. Because longer wavelengths (red) slow the least and shorter wavelengths (violet) slow the most, red light bends the least, and violet light bends the most. The breaking up of white light by "selective" refraction is called **dispersion**. As light passes through ice crystals, dispersion causes red light to be on the inside of the halo and blue light on the outside.

When hexagonal platelike ice crystals with diameters larger than about 30 μm are present in the air, they tend to fall slowly and orient themselves horizontally (see ● Figure 19.22). (The horizontal orientation of these ice crystals prevents a ring halo.) In this position, the ice crystals act as small prisms, refracting and dispersing sunlight that passes through them. If the sun is near the horizon in such a configuration that it, the ice crystals, and the observer are all in the same horizontal plane, the observer will see a pair of brightly coloured spots, one on either side of the sun. These coloured spots are called **sundogs**, *mock suns*, or **parhelia**—meaning "with the sun" (see ● Figure 19.23). The colours usually grade from red (bent least) on the inside closest to the sun to blue (bent more) on the outside.

Whereas sundogs, tangent arcs, and halos are caused by *refraction* of sunlight *through* ice crystals, **sun pillars** are caused by *reflection* of sunlight *off* ice crystals. Sun pillars appear most often at sunrise or sunset as a vertical shaft of light extending upward or downward from the sun (see ● Figure 19.24). Pillars may form as hexagonal platelike ice crystals fall with their flat bases oriented horizontally. As the tiny crystals fall in still air, they tilt from side to side like a falling leaf. This motion allows

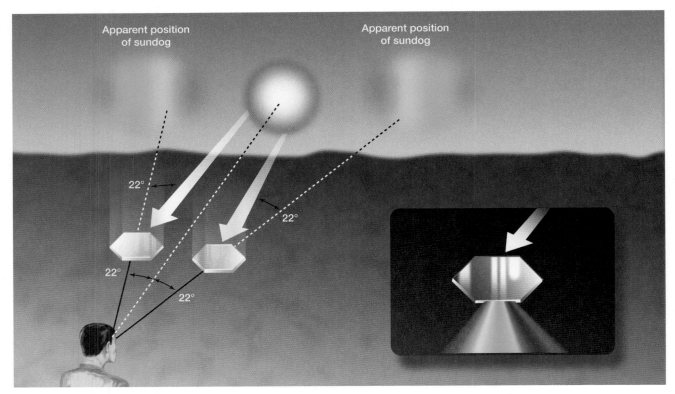

● FIGURE 19.22 Platelike ice crystals falling with their flat surfaces parallel to Earth produce sundogs.

● FIGURE 19.23 The bright areas on each side of the sun are sundogs.

NASA

● FIGURE 19.24 A brilliant red sun pillar extending upward above the sun, produced by the reflection of sunlight off ice crystals.

sunlight to reflect off the tipped surfaces of the crystals, producing a relatively bright area in the sky above or below the sun. Pillars may also form as sunlight reflects off hexagonal pencil-shaped ice crystals that fall with their long axes oriented horizontally. As these crystals fall, they can rotate about their horizontal axes, producing many orientations that reflect sunlight. So look for sun pillars when the sun is low on the horizon and cirriform (ice crystal) clouds are present. ● Figure 19.25 is a summary of some of the optical phenomena that form when cirriform clouds are present.

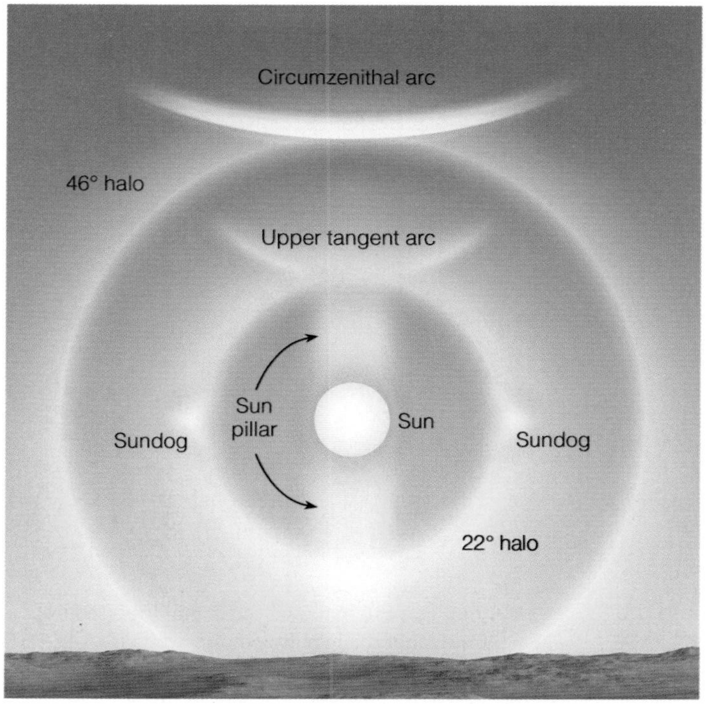

Rainbows

Now we come to one of the most spectacular light shows observed on the Earth—the rainbow. **Rainbows** occur when rain is falling in one part of the sky, and the sun is shining in another. (Rainbows also may form by the sprays from waterfalls and water sprinklers.) To see the rainbow, we must face the falling rain with the sun at our backs. Look at ● Figure 19.26 closely and note that when we see a rainbow in the evening, we are facing east toward the rain shower. Behind us—in the west—it is clear. Because clouds tend to move from west to east in middle latitudes, the clear skies in the west suggest that the showers will give way to clearing. However, when we see a rainbow in the morning, we are facing west, toward the rain shower. It is a good bet that the clouds and showers will move toward us and it will rain soon. These observations explain why the following weather rhyme became popular:

> Rainbow in morning, sailors take warning,
> Rainbow at night, a sailor's delight.*

When we look at a rainbow, we are looking at sunlight that has entered the falling drops and, in effect, has been redirected back toward our eyes. Exactly how this process happens requires some discussion.

As sunlight enters a raindrop, it slows and bends, with violet light refracting the most and red light the least (see

FIGURE 19.25 Optical phenomena that form when cirriform ice crystal clouds are present. (A picture of the circumzenithal arc is in Figure 2 on p. 612.)

---

*This rhyme is often used with the words "red sky" in the place of rainbow. The red sky makes sense when we consider that it is the result of red light from a rising or setting sun being reflected from the underside of clouds above us. In the morning, a red sky indicates that it is clear to the east and cloudy to the west. A red sky in the evening suggests the opposite.

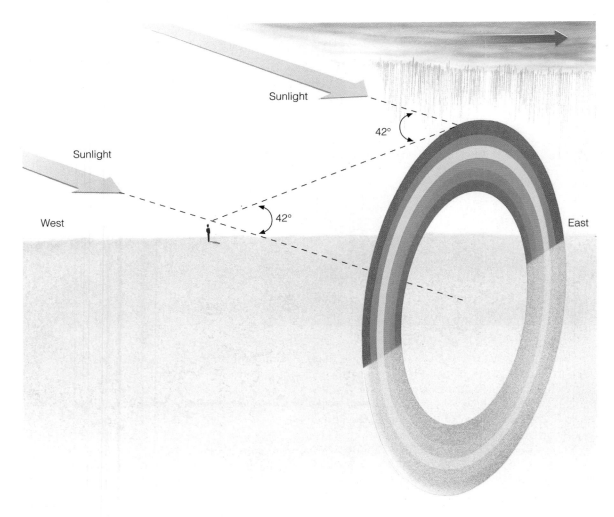

● **FIGURE 19.26** When you observe a rainbow, the sun is always to your back. In middle latitudes, a rainbow in the evening indicates that clearing weather is ahead.

Figure 19.27). Although most of this light passes right on through the drop and is not seen by us, some of it strikes the backside of the drop at such an angle that it is reflected within the drop. The angle at which this occurs is called the *critical angle*. The critical angle is measured between a line that is normal (i.e. at 90°) to the drop surface and the light ray. For water, this angle is 48°. Light that strikes the back of a raindrop at an angle exceeding the critical angle bounces off the back of the drop and is *internally reflected* toward our eyes (see Figure 19.27a). Because each light ray bends differently from the rest, each ray emerges from the drop at a slightly different angle. For red light, the angle is 42° from the beam of sunlight; for violet light, it is 40° (see Figure 19.27b). The light leaving the drop is, therefore, dispersed into a spectrum of colours from red to violet. Since we see only a single colour from each drop, it takes myriads of raindrops (each refracting and reflecting light back to our eyes at slightly different angles) to produce the brilliant colours of a *primary rainbow*.

Figure 19.27b might lead us to believe erroneously that red light should be at the bottom of the bow and violet at the

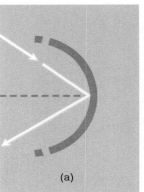
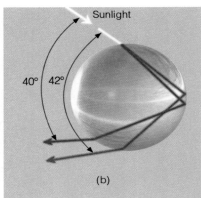

● **FIGURE 19.27** Sunlight internally reflected and dispersed by a raindrop. (a) The light ray is internally reflected only when it strikes the backside of the drop at an angle greater than the critical angle for water. (b) Refraction of the light as it enters the drop causes the point of reflection (on the back of the drop) to be different for each colour. Hence, the colours are separated from each other when the light emerges from the raindrop.

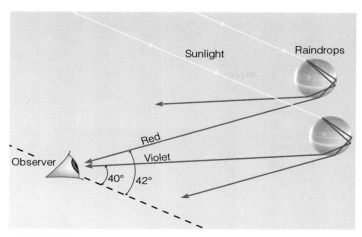

● FIGURE 19.28 The formation of a primary rainbow. The observer sees red light from the upper drop and violet light from the lower drop.

top. A more careful observation of the behaviour of light leaving two drops (see ● Figure 19.28) shows us why the reverse is true. When violet light from the *lower drop* reaches an observer's eye, red light from the same drop travels elsewhere, toward the waist. Notice that red light reaches the observer's eye from the *higher drop*. Because the colour red comes from higher drops and the colour violet from lower drops, the colours of a primary rainbow change from red on the outside (top) to violet on the inside (bottom).

Frequently, a larger second (secondary) rainbow with its colours reversed can be seen above the primary bow (see ● Figure 19.29). Usually, this *secondary bow* is much fainter than the primary one. The secondary bow is caused when sunlight

● FIGURE 19.29 A primary and a secondary rainbow.

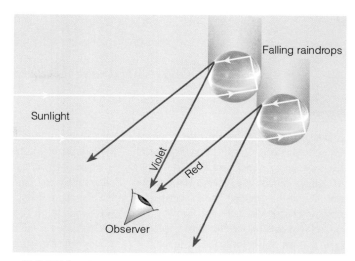

● FIGURE 19.30 Two internal reflections are responsible for the weaker, secondary rainbow. Notice that the eye sees violet light from the upper drop and red light from the lower drop.

enters the raindrops at an angle that allows the light to make two internal reflections in each drop. Each reflection weakens the light intensity and makes the bow dimmer. ● Figure 19.30 shows that the colour reversals—with red now at the bottom and violet on top—are due to the way the light emerges from each drop after going through two internal reflections.

As you look at a rainbow, keep in mind that only one ray of light is able to enter your eye from each drop. Every time you move, whether it be up, down, or sideways, the rainbow moves with you. The reason why this happens is that, with every movement, light from different raindrops enters your eye. The bow you see is not exactly the same rainbow that the person standing next to you sees. In effect, each of us has a personal rainbow to ponder and enjoy! (Can a rainbow actually form on a day when it is not raining? The answer to this question is given in Focus on an Observation: Can It Be a Rainbow If It Is Not Raining? on p. 612.)

Coronas, Glories, and *Heiligenschein*

When the moon is seen through a thin veil of clouds composed of tiny spherical water droplets, a bright ring of light, called a **corona** (meaning crown), may appear to rest on the moon (see ● Figure 19.31). The same effect can occur with the sun, but, due to the sun's brightness, it is usually difficult to see.

The corona is due to **diffraction**—the bending of light as it passes around objects. To understand the corona, imagine water waves moving around a small stone in a pond. As the waves spread around the stone, the trough of one wave may meet the crest of another wave. This situation results in the waves cancelling each other, thus producing calm water. (This is known as *destructive interference*.) Where two crests come together (*constructive interference*), they produce a much larger wave. The same thing happens

FOCUS ON AN OBSERVATION

Can It Be a Rainbow If It Is Not Raining?

Up to this point, we have seen that the sky is full of atmospheric visuals. One that we closely examined was the rainbow. Look back at Figure 19.29 (p. 611) and then look at • Figure 2. Is the colour display in Figure 2 a rainbow? The colours definitely show a rainbowlike brilliance. For this reason, some people will call this phenomenon a rainbow. But, remember, for a rainbow to form, it must be raining, and on this day, it is not.

The colour display in Figure 2 is due to the refraction of light through ice crystals. Earlier in this chapter, we saw that the refraction of light produces a variety of visuals, such as halos,

tangent arcs, and sundogs. (Refer back to Figure 19.25 on p. 609.) The colour display in Figure 2 is a type of refraction phenomenon called a *circumzenithal arc*. The photograph was taken in the afternoon, during late winter, when the sun was about 30° or so above the western horizon and the sky was full of ice crystal (cirrus) clouds. The arc was almost directly overhead, at the zenith (a direction that is normal to Earth's surface)—hence its name, "circumzenithal."

The circumzenithal arc forms about 45° above the sun as platelike ice crystals fall with their flat surfaces parallel to the ground.

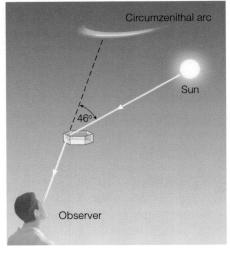

• **FIGURE 3** The formation of a circumzenithal arc.

Remember that this is similar to the formation of a sundog (see Figure 19.22, p. 608), except that in the formation of a circumzenithal arc, sunlight enters the top of the crystal and exits one of its sides (see • Figure 3). This is why the short-lived circumzenithal arc can form only when the sun is lower than 32° above the horizon. When the sun is higher than this angle, the refracted light cannot be seen by the observer.

There is a wide variety of other refraction phenomena (too many to describe here) that may at first glance appear as a rainbow. Keep in mind, however, that rainbows form only when it is raining in one part of the sky and the sun is shining in another.

• **FIGURE 2** Is this a rainbow? The photograph was taken looking almost straight up.

when light passes around tiny cloud droplets. Where light waves constructively interfere, we see bright light; where destructive interference occurs, we see darkness. Sometimes the corona appears white, with alternating bands of light and dark. On other occasions, the rings have colour (see • Figure 19.32).

The colours appear when the cloud droplets (or any kind of small particles, such as volcanic ash) are of uniform size. Because the amount of bending due to diffraction depends on the wavelength of light, the shorter wavelength blue light appears on the inside of a ring, whereas the longer wavelength red light appears on the outside. These colours may repeat over and over,

becoming fainter as each ring is farther from the moon or sun. Also, the smaller the cloud droplets, the larger the ring diameter. Therefore, clouds that have recently formed (such as thin altostratus and altocumulus) are the best corona producers.

WEATHER WATCH

During the summer, rainbows are occasionally seen after a thunderstorm. Because of this fact, the Shoshone Indians viewed the rainbow as a giant serpent that would sometimes rub its back on the icy sky and hurl pieces of ice (hail) to the ground.

• FIGURE 19.31 The corona around the moon results from the diffraction of light by tiny liquid cloud droplets of uniform size.

When different-sized droplets exist within a cloud, the corona becomes distorted and irregular. Sometimes the cloud exhibits patches of colour, often pastel shades of pink, blue, or green. These bright areas produced by diffraction are called **iridescence** (see • Figure 19.33). Cloud iridescence is most often seen within 20° of the sun and is often associated with clouds such as cirrocumulus and altocumulus.

• FIGURE 19.32 Corona around the sun photographed in Colorado. This type of corona, called Bishop's ring, is the result of diffraction of sunlight by tiny volcanic particles emitted from the volcano El Chichón in 1982.

• FIGURE 19.33 Cloud iridescence.

Like the corona, the **glory** is also a diffraction phenomenon. When an aircraft flies above a cloud layer composed of water droplets less than 50 μm in diameter, a set of coloured rings, called the *glory*, may appear around the shadow of the aircraft (see • Figure 19.34). The same effect can happen when you stand with your back to the sun and look into a cloud or fog bank as a bright ring of light may be seen around the shadow of your head. In this case, the glory is called the *Brocken bow*, or

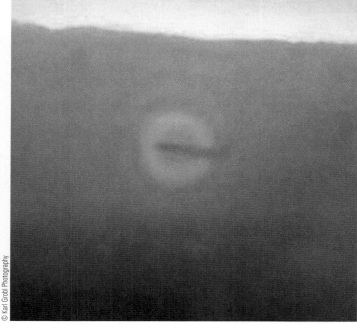

• FIGURE 19.34 The series of rings surrounding the shadow of the aircraft is called the glory.

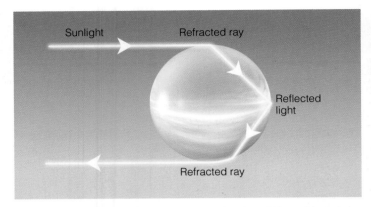

● FIGURE 19.35 Light that produces the glory follows this path in a water droplet.

Brocken spectre, because it is frequently observed when looking down on valley fog from the Brocken Mountains in Germany.

For the glory and the Brocken bow to occur, the sun must be at your back so that sunlight can be returned to your eye from the water droplets. Sunlight that enters the small water droplet along its edge is refracted and then reflected off the backside of the droplet. The light then exits at the other side of the droplet, being refracted once again (see ● Figure 19.35). However, the processes of two refractions and a reflection are not quite enough to bounce the light back to your eyes—by themselves, they do not change the light path by as much as 180°. For the light to be returned to your eyes, the light actually clings ever so slightly to the edge of the droplet—the light travels along the surface of the droplet as a *surface wave* for a short distance, and this additional diffraction process allows it to return to your eyes as a bright glory. Diffraction of light coming from the edges of the droplets produces the ring of light we see as the glory and the Brocken bow. The colourful rings may be due to the various angles at which different colours leave the droplet.

On a clear morning with dew on the grass, stand facing the dew with your back to the sun and observe that around the shadow of your head is a bright area—the **Heiligenschein** (German for halo). The Heiligenschein forms when sunlight, which falls on nearly spherical dew drops, is focused and

● FIGURE 19.36 The *Heiligenschein* is the ring of light around the shadow of the observer's head.

reflected back toward the sun along nearly the same path that it took originally. (Light reflected in this manner is said to be retroreflected.) The light, however, does not travel along the exact path; it actually spreads out just enough to be seen as bright white light around the shadow of your head on a dew-covered lawn (see ● Figure 19.36).

SUMMARY

The scattering of sunlight in the atmosphere can produce a variety of atmospheric visuals, from hazy days and blue skies to crepuscular rays and blue moons. Refraction (bending) of light by the atmosphere causes stars near the horizon to appear higher than they really are. It also causes the sun and moon to rise earlier and set later than they otherwise would. Refraction and scattering result in twilight. Twilight varies with latitude and season.

Mirages form when refraction of light displaces objects from their true positions. Inferior mirages cause objects to

appear lower than they really are, whereas superior mirages displace objects upward. Under certain atmospheric conditions, mirages amplify green light near the upper rim of a rising or setting sun, producing the illusive green flash.

Halos and sundogs form from the refraction of light through ice crystals. Sun pillars are the result of sunlight reflecting off gently falling ice crystals. The refraction, reflection, and dispersion of light in raindrops create a rainbow. To see a rainbow, the sun must be to your back, and rain must be falling in front of you. Diffraction of light produces coronas, glories, and cloud iridescence. We can see the Heiligenschein on a clear morning when sunlight falls on nearly spherical dew drops.

KEY TERMS

The following terms are listed (with page numbers) in the order they appear in the text. Define each. Doing so will aid you in reviewing the material covered in this chapter.

reflected light, 594
scattered light, 594
nonselective scattering, 595
Mie scattering, 595
Rayleigh scattering, 595
crepuscular rays, 598
refraction (of light), 600
scintillation, 601
twilight, 602
mirage, 603
inferior mirage, 604
superior mirage, 604
Fata Morgana, 605

green flash, 606
halo, 606
tangent arc, 607
dispersion (of light), 607
sundogs (parhelia), 607
sun pillars, 607
rainbows, 609
corona, 611
diffraction, 611
iridescence, 613
glory, 613
Heiligenschein, 614

QUESTIONS FOR REVIEW

1. (a) Why are cumulus clouds normally white?
   (b) Why do the undersides of building cumulus clouds frequently change colour from white to dark grey or even black?
2. Explain why the sky is blue during the day and black at night.
3. What can make a setting (or rising) sun appear red?
4. Why do stars "twinkle"?
5. Explain why the sky at the horizon appears white on a hazy day.
6. How does light bend as it enters a more dense substance at an angle? How does it bend on leaving the more dense substance? Make a sketch to illustrate your answer.
7. Since twilight occurs without the sun being visible, how does it tend to lengthen the day?
8. On a clear, dry, warm day, why do dark road surfaces frequently appear wet?
9. What atmospheric conditions are necessary for an inferior mirage? A superior mirage?
10. What process (refraction or scattering) produces crepuscular rays?
11. (a) Describe how a halo forms.
    (b) How is the formation of a halo different from that of a sundog?
12. At what time of day would you expect to observe the green flash?
13. Explain how sun pillars form.

14. What process (refraction, scattering, diffraction) is responsible for lengthening the day?
15. Why can a rainbow be observed only if the sun is toward the observer's back?
16. Why are secondary rainbows higher and much dimmer than primary rainbows? Explain your answer with the aid of a diagram.
17. Explain why this rhyme makes sense: Rainbow in morning, joggers take warning. Rainbow at night (evening), jogger's delight.
18. Suppose you look at the moon and see a bright ring of light that appears to rest on its surface.
    (a) Is this ring of light a halo or a corona?
    (b) What type of clouds (water or ice) must be present for this type of optical phenomenon to occur?
    (c) Is this ring of light produced mainly by refraction or diffraction?
19. Would you expect to see the glory when flying in an aircraft on a perfectly clear day? Explain.
20. Explain how light is able to reach your eyes when you see
    (a) a corona
    (b) a glory
    (c) the *Heiligenschein*
21. How would you distinguish a corona from a halo?
22. What process is primarily responsible for the formation of cloud iridescence—reflection, refraction, or diffraction of light?

QUESTIONS FOR THOUGHT

1. Explain why on a cloudless day the sky will usually appear milky white before it rains and a deep blue after it rains.
2. How long does twilight last on the moon? (Hint: The moon has no atmosphere.)
3. Why is it often difficult to see the road while driving on a foggy night with your high beam lights on?
4. What would be the colour of the sky if air molecules scattered the longest wavelengths of visible light and passed the shorter wavelengths straight through? (Use a diagram to help explain your answer.)
5. Explain why the colours of the planets are not related to the temperatures of the planets, whereas the colours of the stars are related to the temperatures of the stars.
6. If there were no atmosphere surrounding the Earth, what colour would the sky be at sunrise? At sunset? What colour would the sun be at noon? At sunrise? At sunset?
7. Why are rainbows seldom observed at noon?
8. On a cool, clear summer day, a blue haze often appears over the Great Smoky Mountains of Tennessee. Explain why the blue haze usually changes to a white haze as the relative humidity of the air increases.

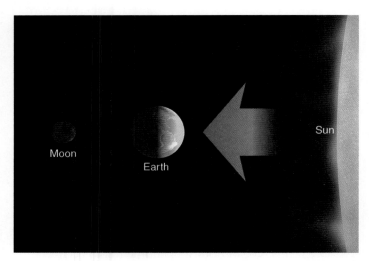

● FIGURE 19.37

9. During a lunar eclipse, Earth, the sun, and the moon are aligned as shown in ● Figure 19.37. Earth blocks sunlight from directly reaching the moon's surface, yet the surface of the moon will often appear a pale red colour during a lunar eclipse. How can you account for this phenomenon?

10. Explain why smoke rising from a cigarette often appears blue yet appears white when blown from the mouth.

11. During Ernest Shackleton's last expedition to Antarctica, on May 8, 1915, seven days after the sun had set for the winter, he saw the sun reappear. Explain how this event—called the Novaya Zemlya effect—can occur.

12. Could a superior mirage form over land on a hot, sunny day? Explain.

13. Explain why it is easier to get sunburned on a high mountain than in the valley below. (The answer is not that you are closer to the sun on top of the mountain.)

14. Why are stars more visible on a clear night when there is no moon than on a clear night with a full moon?

15. During the day, clouds are white, and the sky is blue. Why, then, during a full moon, do cumulus clouds appear faintly white, whereas the sky does not appear blue?

PROBLEMS AND EXERCISES

1. Choose a three-day period in which to observe the sky five times each day. Record in a notebook the number of times you see halos, crepuscular rays, coronas, cloud iridescence, sundogs, rainbows, and other phenomena.

2. Make your own rainbow. On a sunny afternoon or morning, turn on the water sprinkler to create a spray of water drops. Stand as close to the spray as possible (without getting soaked) and observe that as you move up, down, and sideways, the bow moves with you.
   (a) Explain why this happens.
   (b) Also, explain with the use of a diagram why the sun must be at your back to see the bow.

3. Take a large beaker or bottle and fill it with water. Add a small amount of nonfat powdered milk and stir until the water turns a faint milky white. Shine white light into the beaker and, on the opposite side, hold a white piece of paper.
   (a) Explain why the milk has a blue cast to it and why the light shining on the paper appears ruddy.
   (b) What do you know about the size of the milk particles?
   (c) How does this demonstration relate to the colour of the sky and the colour of the sun—at sunrise and sunset?

Units, Conversions, Symbols, and Equations

LENGTH

1 kilometre (km)	= 1000 m
	= 3281 ft
	= 0.62 mi
	= 0.54 nautical miles (nm)
1 statute mile (mi)	= 5280 ft
	= 1609 m
	= 1.61 km
	= 0.87 nm
1 nautical mile (nm)	= 1852 m
(minute of latitude)	= 1.852 km
	= 1.15 mi
	= 6076 ft
	= 1'
1 metre (m)	= 100 cm
	= 3.28 ft
	= 39.37 in
1 foot (ft)	= 12 in
	= 30.48 cm
	= 0.305 m
1 centimetre (cm)	= 0.39 in
	= 0.01 m
	= 10 mm
1 inch (in)	= 2.54 cm
	= 0.08 ft
1 millimetre (mm)	= 0.1 cm
	= 0.001 m
	= 0.039 in
1 micrometre (μm)	= 0.0001 cm
	= 0.000001 m
1 degree latitude	= 111 km
	= 60 nm
	= 69 mi

SPEED

1 knot	= 1 nm h^{-1}
	= 0.51 m s^{-1}
	= 1.852 km h^{-1}
	= 1.15 mi h^{-1}
1 kilometre per hour (km h^{-1})	= 0.54 knot
	= 0.62 mi h^{-1}
	= 0.28 m s^{-1}
1 metre per second (m s^{-1})	= 1.94 knots
	= 2.24 mi h^{-1}
	= 3.60 km h^{-1}
1 mile per hour (mi h^{-1})	= 0.87 knot
	= 0.45 m s^{-1}
	= 1.61 km h^{-1}

AREA

1 square centimetre (cm^2)	= 0.15 in^2
1 square inch (in^2)	= 6.45 cm^2
1 square metre (m^2)	= 10.76 ft^2
1 square foot (ft^2)	= 0.09 m^2

VOLUME

1 cubic centimetre (cm^3)	= 0.06 in^3
1 cubic inch (in^3)	= 16.39 cm^3
1 litre (l)	= 1000 cm^3
	= 0.264 U.S. gallon (gal)
	= 0.220 imperial gallon (imp gal)

MASS

1 gram (g)	= 0.035 ounce
	= 0.002 lb
1 kilogram (kg)	= 1000 g
	= 2.2 lb
1 tonne	= 1000 kg

FORCE

1 newton (N)	= 1 kg m s^{-2}
	= 10^5 dynes
	= 0.2248 lb
1 dyne	= 1 g cm s^{-2}
	= 2.2481 × 10^{-6} pound (lb)

ENERGY

1 joule (J)	= 1 N m
	= 0.239 cal
	= 10^7 erg
1 calorie (cal)	= 4.186 J
	= 4.186 × 10^7 erg
1 erg	= 1 dyne cm^{-1}
	= 2.388 × 10^{-8} cal

POWER

1 watt (W)	= 1 J s^{-1}
	= 14.3353 cal min^{-1}
1 cal min^{-1}	= 0.06973 W
1 horse power (hp)	= 746 W

POWERS OF TEN

POWER	NUMBER	NAME	PREFIX	ABBREVIATION
10^{-12}	0.000000000001	one-trillionth	pico	p
10^{-9}	0.000000001	one-billionth	nano	n
10^{-6}	0.000001	one-millionth	micro	μ
10^{-3}	0.001	one-thousandth	milli	m
10^{-2}	0.01	one-hundredth	centi	c
10^{-1}	0.1	one-tenth	deci	d
10^2	100	one hundred	hecto	h
10^3	1000	one thousand	kilo	k
10^6	1,000,000	one million	mega	M
10^9	1,000,000,000	one billion	giga	G
10^{12}	1,000,000,000,000	one trillion	tera	T

GREEK ALPHABET

LOWER CASE	CAPITAL	NAME	LOWER CASE	CAPITAL	NAME
α	A	alpha	ν	N	nu
β	B	beta	ξ	Ξ	xi
γ	Γ	gamma	o	O	omicron
δ	Δ	delta	π	Π	pi
ε	E	epsilon	ρ	P	rho
ζ	Z	zeta	σ	Σ	sigma
η	H	Eta	τ	T	tau
θ	Θ	theta	υ	Υ	upsilon
ι	I	iota	φ	φ	phi
κ	K	kappa	χ	X	chi
λ	Λ	lambda	ψ	ψ	psi
μ	M	Mu	ω	Ω	omega

TEMPERATURE

fahrenheit to celsius	celsius to fahrenheit
°C = 5/9 (°F —32)	°F = 9/5(°C) + 32

kelvin to celsius	celsius to kelvin
°C = K − 273.15	K = °C + 273.15

TABLE A.1 SI UNITS* AND THEIR SYMBOLS

QUANTITY	NAME	UNITS	SYMBOL
length	metre	m	m
mass	kilogram	kg	kg
time	second	s	s
temperature	kelvin	K	K
density	kilogram per cubic metre	kg m^{-3}	kg m^{-3}
speed	metre per second	m s^{-1}	m s^{-1}
force	newton	kg m s^{-2}	N
pressure	pascal	N m^{-2}	Pa
energy	joule	N m	J
power	watt	J s^{-1}	W

*SI stands for Système International, which is the international system of units and symbols.

TABLE A.2 SOME USEFUL EQUATIONS AND CONSTANTS

NAME	EQUATION	CONSTANTS AND ABBREVIATIONS
Gas law (equation of state)	$P = \rho R_d T$	$R_d = 287$ J kg^{-1} K^{-1} (gas constant for dry air) P = air pressure (Pa) ρ = air density (kg m^{-3}) T = temperature (K)
Gas law for water vapour	$e = \rho_v R_v T$	e = actual vapour pressure (Pa) ρ_v = vapour density (kg m^{-3}) $R_v = 461$ J kg^{-1} K^{-1} (gas constant for water vapour)
Stefan-Boltzmann law	$E = \sigma T^4$	$\sigma = 5.67 \times 10^{-8}$ W m^{-2} K^{-4} E = radiation emitted in W m^{-2}
Wien's law	$\lambda_{max} = \dfrac{W}{T}$	$W = 2897$ μm K λ_{max} = wavelength (μm)
Solar constant		1376 W m^{-2}
Geostrophic wind equation	$V_g = \dfrac{1}{2\Omega \sin \varphi \rho} \dfrac{\Delta P}{\Delta x}$	V_g = geostrophic wind (m s^{-1}) $\Omega = 7.29 \times 10^{-5}$ radian s^{-1} $\star$ φ = latitude Δx = horizontal distance (m) ΔP = pressure difference (Pa)
Coriolis parameter	$f = 2\Omega \sin \varphi$	g = force of gravity (9.8 m s^{-2})
Hydrostatic equation	$\dfrac{\Delta P}{\Delta z} = -\rho g$	Δz = change in height (m)
Circumference of a circle	$C = 2\pi r$	C = circumference $\pi = 3.141593$ r = radius
Area of a circle	$A_c = \pi r^2$	
Area of a sphere	$A_s = 4\pi r^2$	
Volume of a sphere	$V = \dfrac{4}{3}\pi r^3$	
Radius of Earth	Minimum Average Maximum	6353 km 6371 km 6384 km
Earth-Sun Distance	Minimum Average Maximum	1.47×10^8 km 1.496×10^8 km 1.52×10^8 km
Radius of Sun	Average	6.96×10^5 km

* Radians are considered a "unitless" measure of angle, and 2π radians equal 360°.

Weather Map Symbols and the Station Model

SIMPLIFIED SURFACE-STATION MODEL

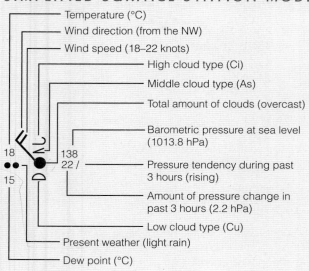

- Temperature (°C)
- Wind direction (from the NW)
- Wind speed (18–22 knots)
- High cloud type (Ci)
- Middle cloud type (As)
- Total amount of clouds (overcast)
- Barometric pressure at sea level (1013.8 hPa)
- Pressure tendency during past 3 hours (rising)
- Amount of pressure change in past 3 hours (2.2 hPa)
- Low cloud type (Cu)
- Present weather (light rain)
- Dew point (°C)

UPPER-AIR MODEL (500 hPa)

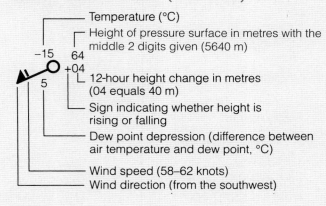

- Temperature (°C)
- Height of pressure surface in metres with the middle 2 digits given (5640 m)
- 12-hour height change in metres (04 equals 40 m)
- Sign indicating whether height is rising or falling
- Dew point depression (difference between air temperature and dew point, °C)
- Wind speed (58–62 knots)
- Wind direction (from the southwest)

PRESSURE TENDENCY

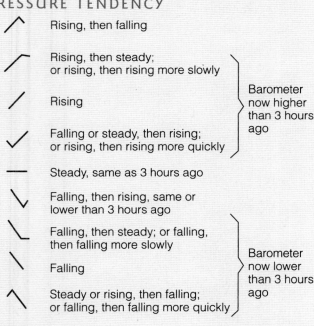

Rising, then falling

Rising, then steady; or rising, then rising more slowly

Rising

Falling or steady, then rising; or rising, then rising more quickly

⎫ Barometer now higher than 3 hours ago

Steady, same as 3 hours ago

Falling, then rising, same or lower than 3 hours ago

Falling, then steady; or falling, then falling more slowly

Falling

Steady or rising, then falling; or falling, then falling more quickly

⎬ Barometer now lower than 3 hours ago

FRONT SYMBOLS

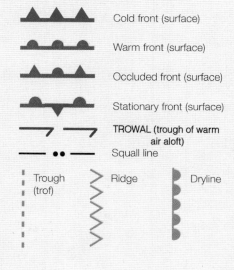

Cold front (surface)

Warm front (surface)

Occluded front (surface)

Stationary front (surface)

TROWAL (trough of warm air aloft)

Squall line

Trough (trof)

Ridge

Dryline

WIND SYMBOLS*

	METRES PER SECOND	KNOTS	KILOMETRES PER HOUR
Calm ⊚	Calm	Calm	Calm
	0.5–1	1–2	1–3
	1.1–3.6	3–7	4–13
	3.7–6.2	8–12	14–19
	6.3–8.8	13–17	20–32
	8.9–11.3	18–22	33–40
	11.4–13.9	23–27	41–50
	14–16.5	28–32	51–60
	16.6–19	33–37	61–69
	19.1–21.6	38–42	70–79
	21.7–24.2	43–47	80–87
	24.3–26.8	48–52	88–96
	26.9–29.3	53–57	97–106
	29.4–31.9	58–62	107–114
	32–34.5	63–67	115–124
	34.6–37	68–72	125–134
	37.1–39.6	73–77	135–143
	53–55	103–107	144–198

© Brooks/Cole, Cengage Learning

* Quick speed estimates can be obtained as follows:
1 wind barb $\cong$ 5 m s^{-1} = 10 knots

CLOUD COVERAGE IN EIGHTHS

Clear
1/8
Scattered
3/8
4/8
5/8
Broken
7/8
Overcast
Obscured
Missing

PRESENT WEATHER SYMBOLS

- World Meteorological Organization code
- weather map present weather symbol
- weather phenomenon name
- (METAR abbreviation for weather reports)

Code	Phenomenon (METAR)
4	Smoke/Volcanic Ash (FU)
5	Haze (HZ)
6	Widespread dust at or nearby stn (DU)
7	Blowing dust, sand, or spray at or nearby stn (BLDU)
8	Dust or sand whirl(s), at or nearby stn, within past hour (PO)
9	Duststorm or sandstorm within sight now or at stn within past hour (DS/SS)
10	Mist (BR)
11	Patches or shallow fog or ice fog at the stn (BCFG)
12	Shallow fog or ice fog at the stn (MIFG)
13	Lightning, no thunder heard (LTNG)
14	VIRGA: Precipitation within sight, not reaching the (VIRGA)
15	Precipitation, distantly visible (VCUP)
16	Precipitation, nearby (VCUP)
17	Thunderstorm, no precipitation (TS)
18	Squalls (SQ)
19	Funnel clouds (FC)
20	Drizzle or snow grains, recently (REDZ)
21	Rain, recently (RERA)
22	Snow, recently (RESN)
23	Rain and snow or ice pellets, recently (RERASN)
24	Freezing drizzle or freezing rain, recently (REFZRA)
25	Rain shower(s), recently (RESHRA)
26	Shower(s) of snow or rain and snow, recently (RESHSN)
27	Shower(s) of hail or of rain and hail, recently (RESHPL)
28	Fog or ice fog, recently (REFG)
29	Thunderstorm, (with or without precip), recently (RETS)
30	Duststorm or sandstorm - decreasing (DS)
31	Duststorm or sandstorm - no change (DS)
32	Duststorm or sandstorm - begun or increasing (DS)
33	Severe duststorm or sandstorm - decreasing (+DS)
34	Severe duststorm or sandstorm - no change (+DS)
35	Severe duststorm of sandstorm - begun or increasing (+DS)
36	Drifting snow (DRSN)
37	Heavy drifting snow (+DRSN)
38	Blowing snow (BLSN)
39	Heavy blowing snow (+BLSN)
40	Fog/ice fog nearby (VCFG)
41	Fog/ice fog, patchy (BCFG)
42	Fog/ice fog, sky visible, thinning (FG)
43	Fog/ice fog, sky invisible, thinning (FG)
44	Fog/ice fog, sky visible, no change (FG)
45	Fog/ice fog, sky invisible, no change (FG)
46	Fog/ice fog, sky visible, begun or thickening (FG)
47	Fog/ice fog, sky invisible, begun or thickening (FG)
48	Fog, depositing rime, sky visible (FZFG)
49	Fog, depositing rime, sky invisible (FZFG)
50	Drizzle, intermittent, slight (-DZ)
51	Drizzle, continuous, slight (-DZ)
52	Drizzle, intermittent, moderate (DZ)
53	Drizzle, continuous, moderate (DZ)
54	Drizzle, intermittent, heavy (+DZ)
55	Drizzle, continuous, heavy (+DZ)
56	Drizzle, freezing, slight (-FZDZ)
57	Drizzle, freezing, moderate/heavy (+FZDZ)
58	Drizzle and rain, slight (-DZRA)
59	Drizzle and rain, moderate/heavy (+DZRA)
60	Rain, intermittent, slight (-RA)
61	Rain, continuous, slight (-RA)
62	Rain, intermittent, moderate (RA)
63	Rain, continuous, moderate (RA)
64	Rain, intermittent, heavy (+RA)
65	Rain, continuous, heavy (+RA)
66	Rain, freezing, slight (-FZRA)
67	Rain, freezing, moderate or heavy (+FZRA)
68	Rain or drizzle and snow, slight (-RASN)
69	Rain or drizzle and snow, moderate/heavy (+RASN)
70	Snow, intermittent, slight (-SN)
71	Snow, continuous, slight (-SN)
72	Snow, intermittent, moderate (SN)
73	Snow, continuous, moderate (SN)
74	Snow, intermittent, heavy (+SN)
75	Snow, continuous, heavy (+SN)
76	Diamond dust (IC)
77	Snow grains (SG)
78	Isolated starlike snow crystals (IC)
79	Ice pellets (PL)
80	Rain shower(s), slight (-SHRA)
81	Rain shower(s), moderate or heavy (+SHRA)
82	Rain shower(s), violent (+SHRA)
83	Shower(s) of rain and snow slight (-SHRASN)
84	Shower(s) of rain and snow, moderate or heavy (+SHRASN)
85	Snow shower(s), slight (-SHSN)
86	Snow shower(s), moderate or heavy (+SHSN)
87	Shower(s) of pellets, w/wo rain or rain and snow mixed - slight (-SHGS)
88	Shower(s) of pellets, w/wo rain or rain and snow mixed - moderate or heavy (+SHGS)
89	Shower(s) of hail, w/wo rain or rain and snow mixed, no thunder - slight (-SHGR)
90	Shower(s) of hail, w/wo rain or rain and snow mixed, no thunder - moderate/heavy (+SHGR)
91	Slight rain - recent thunderstorm (-RARETS)
92	Moderate or heavy rain - recent thunderstorm (+RARETS)
93	Slight snow, or rain and snow mixed, or hail - recent thunderstorm (-SHRETS)
94	Moderate or heavy snow, or rain and snow mixed, or hail - recent thunderstorm (+SHRETS)
95	Thunderstorm with rain and/or snow, slight (TSRA)
96	Thunderstorm with hail (TSGR)
97	Thunderstorm, heavy, with rain and/or snow, no hail (+TSRA)
98	Thunderstorm and duststorm/sandstorm (TSDS)
99	Thunderstorm, heavy, with hail (+TSGR)

MAJOR CLOUD GENERA (TYPE) SYMBOLS

Cloud formation process is indicated by colour:

Stratiform	Cumuliform	Both Stratiform and Cumuliform	Neither specified

High Clouds

0	1	2
Cirrus **Ci**	Cirrocumulus **Cc**	Cirrostratus **Cs**

Middle Clouds

3	4	5
Altocumulus **Ac**	Altostratus **As**	Nimbostratus **Ns**

Low Clouds

6	7	8	9
Stratocumulus **Sc**	Stratus **St**	Cumulus **Cu**	Cumulonimbus **Cb**

DETAILED CLOUD SPECIES AND VARIETIES SYMBOLS

High Cloud Symbols

0	1	2	3	4
No CH clouds	Cirrus fibratus (wispy)	Cirrus spissatus dense in patches	Cirrus spissatus cumulonimbogenitus (formed out of anvil)	Cirrus uncinus or fibratus (progressively invading the sky)

5	6	7	8	9
Cirrus (often in bands) or Cirrostratus (less than 45° above horizon)	Cirrus (often in bands) or Cirrostratus (more than 45° above horizon)	Cirrostratus covering the whole sky	Cirrostratus not covering sky but not invading	Cirrocumulus alone or Cirrocumulus predominant among the CH clouds

Middle Cloud Symbols

0	1	2	3	4
No CM clouds	Altostratus translucidus (mostly transparent)	Altostratus opacus or Nimbostratus	Altocumulus translucidus at a single level	Patches of Altocumulus translucidus, changing, one or more levels

5	6	7	8	9
Altocumulus translucidus in bands	Altocumulus cumulogenitus (or cumulonimbogenitus)	Altocumulus (multilayers)	Altocumulus castellanus or floccus	Altocumulus of a chaotic sky generally at several levels

Low Cloud Symbols

0	1	2	3	4
No CL clouds	Cumulus humilis or Cumulus fractus (no vertical development)	Cumulus mediocris or congestus, (moderate vertical development)	Cumulonimbus calvus, (no outlines nor anvil)	Stratocumulus cumulogenitus (formed by spreading of cumulus)

5	6	7	8	9
Stratocumulus	Stratus nebulosus (continuous sheet)	Stratus fractus or Cumulus fractus (bad weather)	Cumulus and Stratocumulus (multilevel)	Cumulonimbus capillatus (often with an anvil)

Beaufort Wind Scale

▽ TABLE C.1 Wind Speed from Surface Observation

BEAUFORT NUMBER	DESCRIPTION	MARINE FORECAST TERM	OBSERVATIONS OVER LAND	OBSERVATIONS OVER WATER	WIND SPEED		
					m s^{-1}	knots	km h^{-1}
0	Calm		Smoke rises vertically	Sea surface like a mirror, but not necessarily flat	<0.5	< 1	0–2
1	Light air		Direction of wind shown by drifting smoke, but not by wind vanes	Ripples with the appearance of scales are formed, but without foam crests	0.5–1.7	1–3	2–6
2	Light breeze		Wind felt on face; leaves rustle; wind vanes moved by wind; flags stir	Small wavelets, still short but more pronounced; crests do not break; horizon line always very clear if visibility is good	1.8–3.2	4–6	7–11
3	Gentle breeze		Leaves and small twigs move; wind will extend light flag	Large wavelets; crests begin to break; foam appears glassy; scattered whitecaps	3.3–5.3	7–10	12–19
4	Moderate breeze		Wind raises dust and loose paper; small branches move; flags flap	Small waves, becoming longer with frequent whitecaps	5.4–8.4	11–16	20–30
5	Fresh breeze		Small trees with leaves begin to sway; flags ripple	Moderate waves, with a pronounced long form and many whitecaps; some spray	8.5–10.9	17–21	31–39
6	Strong breeze	Strong Wind	Large tree branches in motion; whistling heard in telegraph wires; umbrellas used with difficulty	Large waves begin to form with extensive white foam crests; some spray is likely	11–14	22–27	40–50
7	Near Gale (High wind)	Strong Wind	Whole trees in motion; inconvenience felt walking against wind; flags extend	Sea heaps up and white foam from breaking waves begins to be blown in streaks	14.1–17.1	28–33	51–61

BEAUFORT NUMBER	DESCRIPTION	MARINE FORECAST TERM	OBSERVATIONS OVER LAND	OBSERVATIONS OVER WATER	WIND SPEED		
					m s⁻¹	knots	km h⁻¹
8	Gale	Gale Force Wind	Wind breaks twigs off trees; walking is difficult	Moderately high waves of greater length; edges of crests begin to break into spindrift with foams blown in well-marked streaks along the direction of the wind	17.2–20.7	34–40	62–74
9	Strong gale	Gale Force Wind	Slight structural damage occurs (signs and antennas blown down)	High waves with dense streaks of foam along the direction of the wind; wave crests begin to topple and roll over; spray may affect visibility	20.8–24.3	41–47	75–87
10	Storm, (Whole gale)	Storm Force Wind	Trees uprooted; considerable damage occurs	Very high waves with long overhanging crests and dense white streaks of foam. Surface of the sea appears white; the tumbling of the sea becomes heavy and shock-like; visibility is reduced	24.4–28.5	48–55	88–102
11	Violent Storm	Storm Force Wind	Winds produce wide-spread damage	Exceptionally high waves; sea completely covered with long white patches of foam; visibility is reduced	28.6–32.6	56–63	103–117
12	Hurricane	Hurricane Force Wind	Winds produce extensive damage	Air filled with foam and spray; sea entirely white with foam; visibility is seriously impaired	≥ 32.7	≥ 64	≥ 118

Humidity

▼ TABLE D.I   Equations and Constants for Humidity Calculations

NAME	EQUATION	CONSTANTS
Psychrometric Equation†	$e = e^*_{Tw} - \gamma(T - T_w)$	e = actual vapour pressure (hPa)
		e^*_{Tw} = saturation vapour pressure at the wet bulb temperature (hPa): using Figure D.1, find the e value that corresponds to the wet bulb temperature reading
		γ = psychrometric constant which has different values for unfrozen or frozen conditions. The psychrometric constant is:
		0.66 hPa °C^{-1} over water at sea level
		0.582 hPa °C^{-1} over ice at sea level
		T = air, or dry bulb temperature (°C)
		T_w = wet bulb temperature (°C)
Mixing ratio	$r = \dfrac{0.622 \times e}{P - e}$	r = mixing ratio (g g^{-1}) r is often multiplied 1000 to give units of g kg^{-1}
		P = air pressure (hPa)
Specific humidity	$q = \dfrac{0.622 \times e}{P - (0.378 \times e)}$	q = specific humidity
Absolute humidity (vapour density)	$\rho_v = \dfrac{e}{R_v T}$	ρ_v = absolute humidity or vapour density (kg m^{-3})
		R_v = gas constant for water vapour (461 J kg^{-1} K^{-1})
Relative humidity	$RH = \dfrac{e}{e^*} \times 100\%$	RH = relative humidity (%)
		e^* = saturation vapour pressure (hPa)

† Use this equation to find the actual vapour pressure from psychrometer readings of air temperature (T) and wet bulb temperature (T_w). To determine e^*_{Tw}, use Figure D.1 as follows. Find the T_w value on the x-axis and follow this value vertically to the line, then read across to find the saturation vapour pressure value. This is e^*_{Tw}.

▼ TABLE D.2   Dew-Point Temperature (°C) as determined by psychrometer readings. To obtain the dew point, read the number at the intersection of the temperature and the wet-bulb depression. For example, a temperature of 10°C with a wet-bulb depression of 3°C produces a dew-point temperature of 4°C. (Readings are appropriate for pressures near 1000 hPa.)

	WET-BULB DEPRESSION (DRY-BULB TEMPERATURE MINUS WET-BULB TEMPERATURE) (°C)															
	0.5	1.0	1.5	2.0	2.5	3.0	3.5	4.0	4.5	5.0	7.5	10.0	12.5	15.0	17.5	20.0
−20	−25	−33														
−17.5	−21	−27	−38													
−15	−19	−23	−28													
−12.5	−15	−18	−22	−29												
−10	−12	−14	−18	−21	−27	−36										
−7.5	−9	−11	−14	−17	−20	−26	−34									
−5	−7	−8	−10	−13	−16	−19	−24	−31								
−2.5	−4	−6	−7	−9	−11	−14	−17	−22	−28	−41						
0	−1	−3	−4	−6	−8	−10	−12	−15	−19	−24						
2.5	1	0	−1	−3	−4	−6	−8	−10	−13	−16						
5	4	3	2	0	−1	−3	−4	−6	−8	−10	−48					
7.5	6	6	4	3	2	1	−1	−2	−4	−6	−22					
10	9	8	7	6	5	4	2	1	0	−2	−13					
12.5	12	11	10	9	8	7	6	4	3	2	−7	−28				
15	14	13	12	12	11	10	9	8	7	5	−2	−14				
17.5	17	16	15	14	13	12	12	11	10	8	2	−7	−35			
20	19	18	18	17	16	15	14	14	13	12	6	−1	−15			
22.5	22	21	20	20	19	18	17	16	16	15	10	3	−6	−38		
25	24	24	23	22	21	21	20	19	18	18	13	7	0	−14		
27.5	27	26	26	25	24	23	23	22	21	20	16	11	5	−5	−32	
30	29	29	28	27	27	26	25	25	24	23	19	14	9	2	−11	
32.5	32	31	31	30	29	29	28	27	26	26	22	18	13	7	−2	
35	34	34	33	32	32	31	31	30	29	28	25	21	16	11	4	
37.5	37	36	36	35	34	34	33	32	32	31	28	24	20	15	9	0
40	39	39	38	38	37	36	36	35	34	34	30	27	23	18	13	6
42.5	42	41	41	40	40	39	38	38	37	36	33	30	26	22	17	11
45	44	44	43	43	42	42	41	40	40	39	36	33	29	25	21	15
47.5	47	46	46	45	45	44	44	43	42	42	39	35	32	28	24	19
50	49	49	48	48	47	47	46	45	45	44	41	38	35	31	28	23

AIR (DRY-BULB) TEMPERATURE (°C)

▼ TABLE D.3   Relative Humidity (Percent) as determined by psychrometer readings. To obtain the relative humidity, read the number at the intersection of the temperature and the wet-bulb depression. For example, a temperature of 15°C with a wet-bulb depression of 4°C produces a relative humidity of 66 percent. (Readings are appropriate for pressures near 1000 hPa.)

AIR (DRY-BULB) TEMPERATURE (°C)	WET-BULB DEPRESSION (DRY-BULB TEMPERATURE MINUS WET-BULB TEMPERATURE) (°C)																		
	0.5	1.0	1.5	2.0	2.5	3.0	3.5	4.0	4.5	5.0	7.5	10.0	12.5	15.0	17.5	20.0	22.5	25.0
−20	70	41	11															
−17.5	75	51	26	2														
−15	79	58	38	18														
−12.5	82	65	47	30	13													
−10	85	69	54	39	24	10												
−7.5	87	73	60	48	35	22	10											
−5	88	77	66	54	43	32	21	11	1									
−2.5	90	80	70	60	50	42	37	22	12	3								
0	91	82	73	65	56	47	39	31	23	15								
2.5	92	84	76	68	61	53	46	38	31	24								
5	93	86	78	71	65	58	51	45	38	32	1							
7.5	93	87	80	74	68	62	56	50	44	38	11							
10	94	88	82	76	71	65	60	54	49	44	19							
12.5	94	89	84	78	73	68	63	58	53	48	25	4						
15	95	90	85	80	75	70	66	61	57	52	31	12						
17.5	95	90	86	81	77	72	68	64	60	55	36	18	2					
20	95	91	87	82	78	74	70	66	62	58	40	24	8					
22.5	96	92	87	83	80	76	72	68	64	61	44	28	14	1				
25	96	92	88	84	81	77	73	70	66	63	47	32	19	7				
27.5	96	92	89	85	82	78	75	71	68	65	50	36	23	12	1			
30	96	93	89	86	82	79	76	73	70	67	52	39	27	16	6			
32.5	97	93	90	86	83	80	77	74	71	68	54	42	30	20	11	1		
35	97	93	90	87	84	81	78	75	72	69	56	44	33	23	14	6		
37.5	97	94	91	87	85	82	79	76	73	70	58	46	36	26	18	10	3	
40	97	94	91	88	85	82	79	77	74	72	59	48	38	29	21	13	6	
42.5	97	94	91	88	86	83	80	78	75	72	61	50	40	31	23	16	9	2
45	97	94	91	89	86	83	81	78	76	73	62	51	42	33	26	18	12	6
47.5	97	94	92	89	86	84	81	79	76	74	63	53	44	35	28	21	15	9
50	97	95	92	89	87	84	82	79	77	75	64	54	45	37	30	23	17	11

▼ TABLE D.4 Values of saturation vapour pressure at different temperatures, as well as values of vapour pressure at different dew-point temperatures, over water and ice. To obtain the saturation vapour pressure over water or ice, read the number to the right of the temperature. Similarly to obtain the vapour pressure over water or ice, read the number to the right of the dew point.

AIR TEMPERATURE (°C)	SATURATION VAPOUR PRESSURE OVER WATER (hPa)	SATURATION VAPOUR PRESSURE OVER ICE (hPa)	AIR TEMPERATURE (°C)	SATURATION VAPOUR PRESSURE OVER WATER (hPa)
DEW POINT TEMPERATURE (°C)	VAPOUR PRESSURE OVER WATER (HPA)	VAPOUR PRESSURE OVER ICE (HPA)	DEW POINT TEMPERATURE (°C)	VAPOUR PRESSURE OVER WATER (HPA)
-45	0.11	0.07	0	6.11
-44	0.13	0.08	1	6.57
-43	0.14	0.09	2	7.06
-42	0.15	0.10	3	7.58
-41	0.17	0.11	4	8.13
-40	0.19	0.13	5	8.72
-39	0.21	0.14	6	9.35
-38	0.23	0.16	7	10.02
-37	0.26	0.18	8	10.73
-36	0.29	0.20	9	11.48
-35	0.32	0.22	10	12.28
-34	0.35	0.25	11	13.13
-33	0.38	0.28	12	14.03
-32	0.42	0.31	13	14.98
-31	0.46	0.34	14	15.99
-30	0.51	0.38	15	17.05
-29	0.56	0.42	16	18.18
-28	0.62	0.47	17	19.38
-27	0.67	0.52	18	20.64
-26	0.74	0.57	19	21.98
-25	0.81	0.63	20	23.39
-24	0.88	0.70	21	24.88
-23	0.97	0.77	22	26.45
-22	1.06	0.85	23	28.10
-21	1.15	0.94	24	29.85
-20	1.26	1.03	25	31.69
-19	1.37	1.14	26	33.63
-18	1.49	1.25	27	35.67
-17	1.62	1.37	28	37.82
-16	1.76	1.51	29	40.08
-15	1.91	1.65	30	42.45
-14	2.08	1.81	31	44.95
-13	2.25	1.99	32	47.57
-12	2.44	2.17	33	50.33
-11	2.65	2.38	34	53.22
-10	2.87	2.60	35	56.26
-9	3.10	2.84	36	59.45
-8	3.35	3.10	37	62.79
-7	3.62	3.38	38	66.29
-6	3.91	3.69	39	69.96
-5	4.22	4.02	40	73.81
-4	4.55	4.38	41	77.83
-3	4.90	4.76	42	82.05
-2	5.28	5.18	43	86.46
-1	5.68	5.63	44	91.07

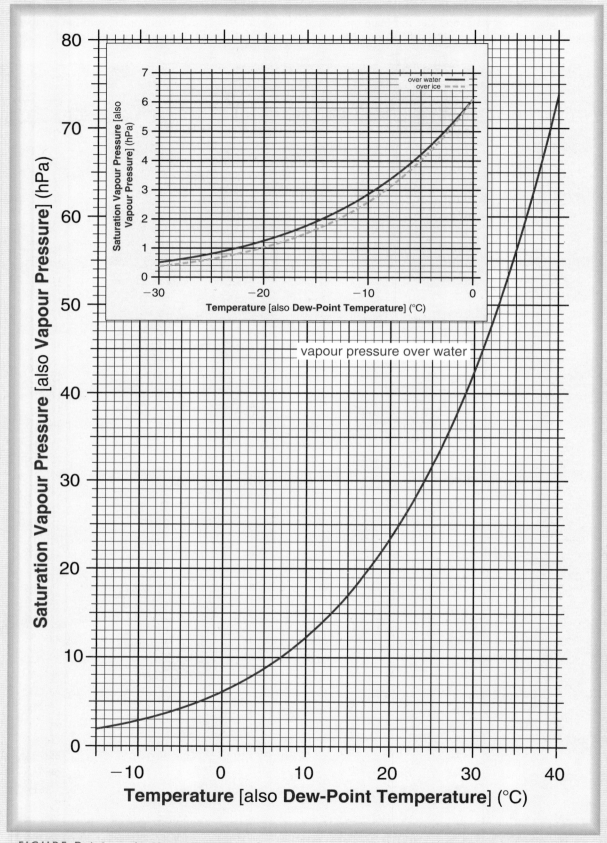

● **FIGURE D.1** Saturation Vapour Pressure as a function of Temperature (as shown by the blue labels). The same values represent Vapour Pressure as a function of Dew Point (as shown by the purple labels).

Instant Weather Forecast Chart

This chart is a guide to forecasting the weather. It is applicable to most of North America, especially the eastern two-thirds. It works best during the fall, winter, and spring, when the weather systems are most active.

▼ TABLE E.1 Instant Weather Forecast Chart

SEA-LEVEL PRESSURE (hPa)	PRESSURE TENDENCY	SURFACE WIND DIRECTION	SKY CONDITION	24-HR WEATHER FORECAST (SEE WEATHER FORECAST CODE)
1023 or higher	rising, steady, or falling	any direction	clear, high clouds, or Cu	1, 18 (in winter, 14)
1022 to 1016	rising or steady	SW, W, NW, N	clear, high clouds or Cu	1, 18
	falling or steady	SW, S, SE	clear, high clouds	1, 3, 17, 5
	falling	SW, S, SE	middle or low clouds	6, 17
	falling	E, NE	middle or low clouds	6, 14
	falling or steady	E, NE	clear or high clouds	3, 5, 14
1015 to 1009	rising	SW, W, NW, N	clear	1, 14
			overcast	2, 16
			precipitation	11, 2, 16
	falling	any direction	clear	3, 17 (dry climate summer: 1, 15)
	falling or steady	SW, S, SE	high clouds	3, 17, 5
	falling	SW, S, SE	middle or low clouds	7
	falling	E, NE	middle or low clouds	7, 12, 14
	falling	SE, E, NE	overcast, precipitation	9
	falling	S, SW	overcast, precipitation	10, 13
1008 or below	rising	SW, W, NW, N	clear	1, 12
	rising	SW, W, NW, N	overcast	2, 12, 16
	rising	SW, W, NW, N	overcast with precipitation	11, 12, 16
	rising	NE	overcast	4, 12, 13, 14
	rising	NE	overcast with precipitation	11, 12, 13, 14
	rising or steady	SW, S, SE	clear	3, 6, 8, 12, 15
	falling	SW, S, SE	overcast	7, 8, 12, 13
	falling	SW, S, SE	overcast with precipitation	8, 10, 12, 13, 16
	falling	N	overcast	4, 14
	falling or steady	E, NE	overcast	7, 12, 14
	falling	E, NE	overcast with precipitation	8, 9, 12, 13

Weather Forecast Codes

CODE	CONDITION	CODE	CONDITION	CODE	CONDITION
1	clear or scattered clouds	7	precipitation possible within 8 hours	13	possible wind shift to W, NW, or N
2	clearing	8	possible period of heavy precipitation	14	continued cool or cold
3	increasing clouds	9	precipitation continuing	15	continued mild or warm
4	continued overcast	10	precipitation ending within 12 hours	16	turning colder
5	precipitation possible within 24 hours	11	precipitation ending within 6 hours	17	slowly rising temperatures
6	precipitation possible within 12 hours	12	windy	18	little temperature change

Changing UTC and GMT to Local Time

The system of time used in meteorology is Coordinated Universal Time (UTC), which is also known as Greenwich Mean Time (GMT), and as Zulu (Z) Time. This is the time measured on the prime meridian (0° longitude) in Greenwich, England.

▼ TABLE F.1   Time Zone Conversions

STANDARD TIME		DAYLIGHT SAVING TIME	
Pacific Standard Time (PST)	UTC-8 h	Pacific Daylight Time (PDT)	UTC-7 h
Mountain Standard Time (MST)	UTC-7 h	Mountain Daylight Time (MDT)	UTC-6 h
Central Standard Time (CST)	UTC-6 h	Central Daylight Time (CDT)	UTC-5 h
Eastern Standard Time (EST)	UTC-5 h	Eastern Daylight Time (EDT)	UTC-4 h
Atlantic Standard Time (AST)	UTC-4 h	Atlantic Daylight Time (ADT)	UTC-3 h
Newfoundland Standard Time (NST)	UTC-3.5 h	Newfoundland Daylight Time (NDT)	UTC-2.5 h

● FIGURE F.1 The Standard Time Zones Of North America.*

Source: National Research Council Canada.

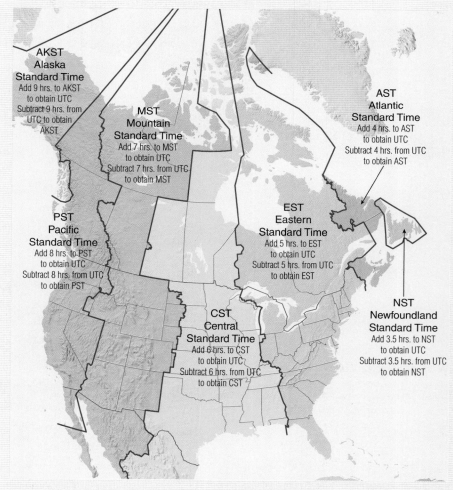

*During daylight saving time add one hour less to local time to obtain UTC. Note that the following areas do not have daylight saving time: the region of northeast British Columbia in the MST zone, the Province of Saskatchewan, and Southampton Island in northern Hudson Bay.

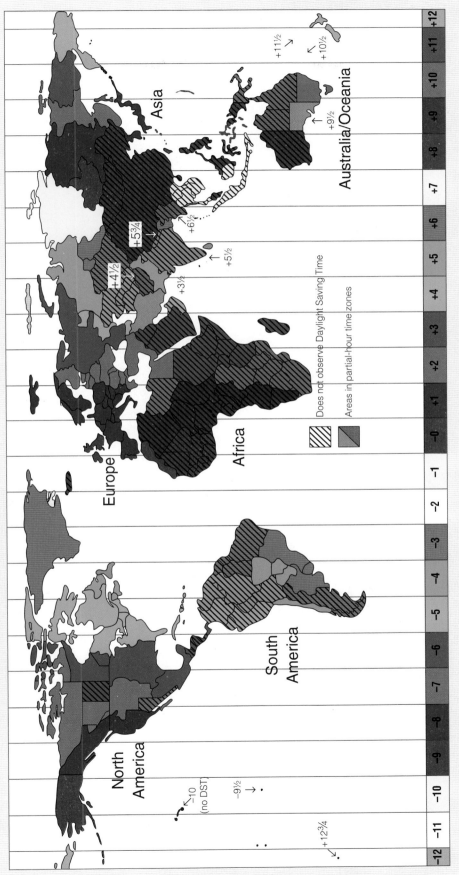

● FIGURE F.2 World Time Zones.

Source: National Institute of Standards and Technology.

Average Annual Global Precipitation

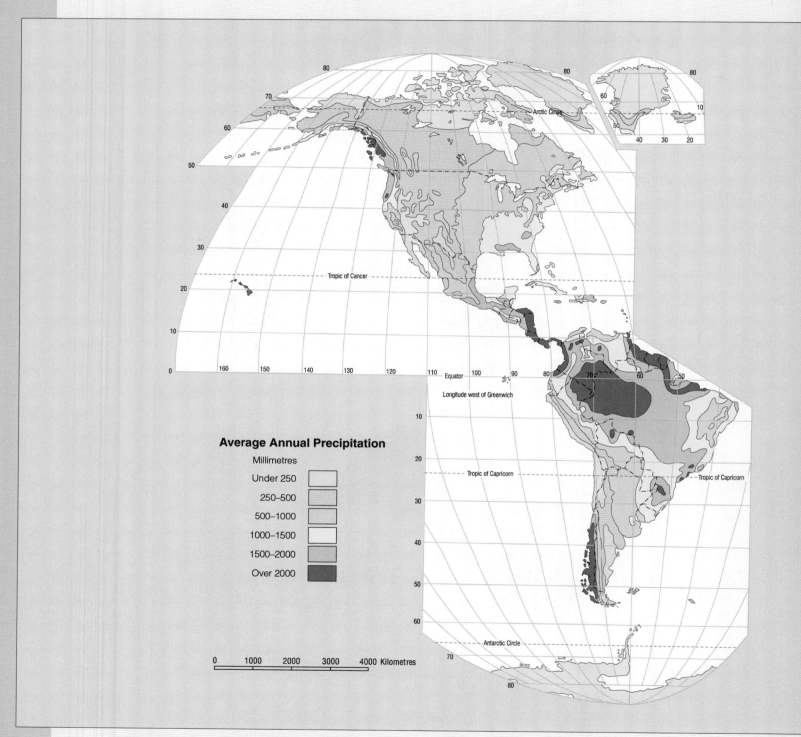

Average Annual Precipitation

Millimetres

Under 250	
250–500	
500–1000	
1000–1500	
1500–2000	
Over 2000	

0 1000 2000 3000 4000 Kilometres

• FIGURE G.1 World map of average annual precipitation.

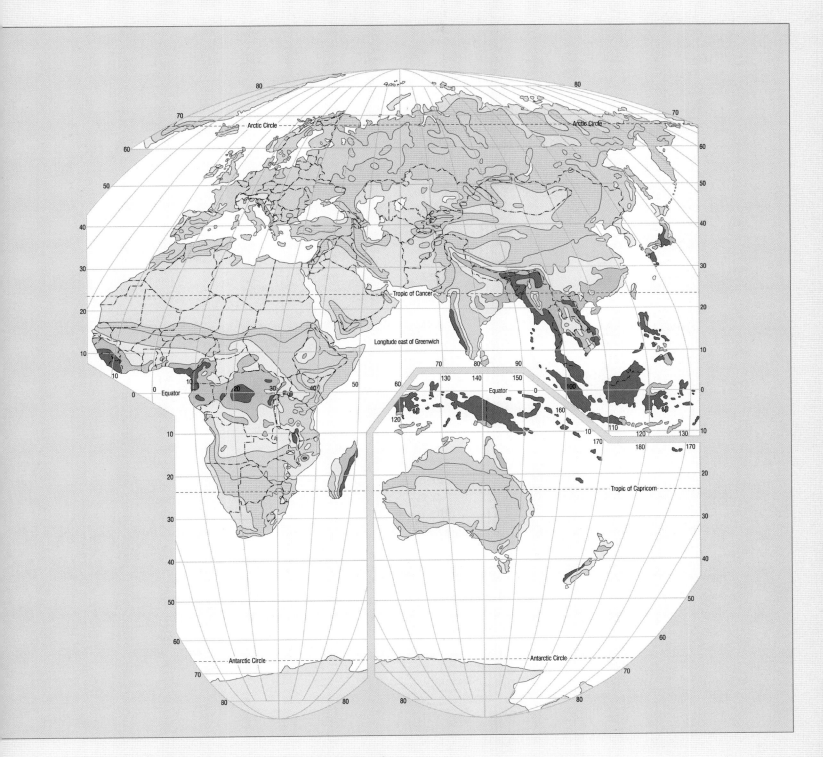

A Western Paragraphic Projection developed at Western Illinois University

Standard Atmosphere

ALTITUDE			PRESSURE (hPa)	TEMPERATURE (°C)	DENSITY (kg m^{-3})
METRES	FEET	KILOMETRES			
0	0	0.0	1013.25	15.0	1.225
500	1,640	0.5	954.61	11.8	1.167
1,000	3,280	1.0	898.76	8.5	1.112
1,500	4,921	1.5	845.59	5.3	1.058
2,000	6,562	2.0	795.01	2.0	1.007
2,500	8,202	2.5	746.91	−1.2	0.957
3,000	9,842	3.0	701.21	−4.5	0.909
3,500	11,483	3.5	657.80	−7.7	0.863
4,000	13,123	4.0	616.60	−11.0	0.819
4,500	14,764	4.5	577.52	−14.2	0.777
5,000	16,404	5.0	540.48	−17.5	0.736
5,500	18,045	5.5	505.39	−20.7	0.697
6,000	19,685	6.0	472.17	−24.0	0.660
6,500	21,325	6.5	440.75	−27.2	0.624
7,000	22,965	7.0	411.05	−30.4	0.590
7,500	24,606	7.5	382.99	−33.7	0.557
8,000	26,247	8.0	356.51	−36.9	0.526
8,500	27,887	8.5	331.54	−40.2	0.496
9,000	29,528	9.0	308.00	−43.4	0.467
9,500	31,168	9.5	285.84	−46.6	0.440
10,000	32,808	10.0	264.99	−49.9	0.413
11,000	36,089	11.0	226.99	−56.4	0.365
12,000	39,370	12.0	193.99	−56.5	0.312
13,000	42,651	13.0	165.79	−56.5	0.267
14,000	45,932	14.0	141.70	−56.5	0.228
15,000	49,213	15.0	121.11	−56.5	0.195
16,000	52,493	16.0	103.52	−56.5	0.166
17,000	55,774	17.0	88.497	−56.5	0.142
18,000	59,055	18.0	75.652	−56.5	0.122
19,000	62,336	19.0	64.674	−56.5	0.104
20,000	65,617	20.0	55.293	−56.5	0.089
25,000	82,021	25.0	25.492	−51.6	0.040
30,000	98,425	30.0	11.970	−46.6	0.018
35,000	114,829	35.0	5.746	−36.6	0.008
40,000	131,234	40.0	2.871	−22.8	0.004
45,000	147,638	45.0	1.491	−9.0	0.002
50,000	164,042	50.0	0.798	−2.5	0.001
60,000	196,850	60.0	0.220	−26.1	0.0003
70,000	229,659	70.0	0.052	−53.6	0.00008
80,000	262,467	80.0	0.010	−74.5	0.00002

Hurricane Tracking Chart

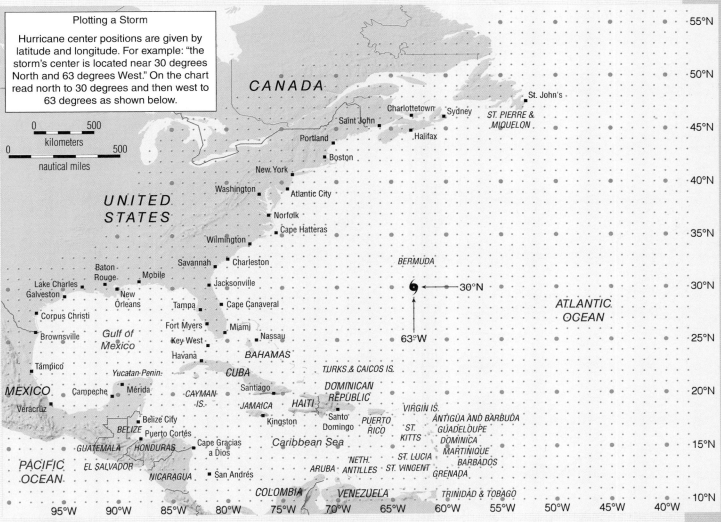

Plotting a Storm

Hurricane center positions are given by latitude and longitude. For example: "the storm's center is located near 30 degrees North and 63 degrees West." On the chart read north to 30 degrees and then west to 63 degrees as shown below.

• FIGURE I.1

Tephigram

● FIGURE J.1 The tephigram, the adiabatic chart used in Commonwealth countries. Lines of **equal pressure (isobars, hPa) are in in black;** equal temperature (isotherms, °C) are in green; dry adiabatic lapse rate (°C) are in red; saturated adiabatic lapse rate (°C) are in blue; and saturation mixing ratio (g kg⁻¹) are in orange.

modified after M. Ambaum.

Absolute humidity   The mass of water vapour in a given volume of air. It represents the density of water vapour in the air.

Absolute vorticity   *See* Vorticity.

Absolute zero   A temperature reading of −273.15°C, or 0 K. Theoretically, there is no molecular motion at this temperature.

Absolutely stable atmosphere   An atmospheric condition that exists when the environmental lapse rate is less than the moist adiabatic rate. This results in a lifted parcel of air being colder than the air around it.

Absolutely unstable atmosphere   An atmospheric condition that exists when the environmental lapse rate is greater than the dry adiabatic rate. This results in a lifted parcel of air being warmer than the air around it.

Accretion   The growth of a precipitation particle by the collision of an ice crystal or snowflake with a supercooled liquid droplet that freezes on impact.

Acid deposition   The depositing of acidic particles (usually sulphuric acid and nitric acid) at Earth's surface. Acid deposition occurs in dry form (*dry deposition*) or wet form (*wet deposition*). Acid rain and acid precipitation often denote wet deposition. (*See* Acid rain.)

Acid fog   *See* Acid rain.

Acid rain   Cloud droplets or raindrops combining with gaseous pollutants, such as oxides of sulphur and nitrogen, to make falling rain (or snow) acidic—pH less than 5.0. If fog droplets combine with such pollutants, *acid fog* is formed.

Actual vapour pressure   *See* Vapour pressure.

Adiabatic process   A process that takes place without a transfer of heat between the system (such as an air parcel) and its surroundings. In an adiabatic process, compression always results in warming, and expansion results in cooling.

Advection   The horizontal transfer of any atmospheric property by the wind.

Advection fog   Occurs when warm, moist air moves over a cold surface and the air cools to below its dew point.

Advection–radiation fog   Fog that forms as relatively warm, moist air moves over a colder surface that cooled mainly by radiational cooling.

Aerosols   Tiny, suspended, solid particles (dust, smoke, etc.) or liquid droplets that enter the atmosphere from either natural or human (anthropogenic) sources, such as the burning of fossil fuels. Sulphur-containing fossil fuels, such as coal, produce *sulphate aerosols*.

Aggregation   The clustering together of ice crystals to form snowflakes.

Air density   *See* Density.

Air glow   A faint glow of light emitted by excited gases in the upper atmosphere. Air glow is much fainter than the aurora.

Air mass   A large body of air that has similar horizontal temperature and moisture characteristics.

Air-mass thunderstorm   *See* Ordinary thunderstorm.

Air-mass weather   A persistent type of weather that may last for several days (up to a week or more). It occurs when an area comes under the influence of a particular air mass.

Air parcel   *See* Parcel of air.

Air pollutants   Solid, liquid, or gaseous airborne substances that occur in concentrations high enough to threaten the health of people and animals, to harm vegetation and structures, or to toxify a given environment.

Air pressure (atmospheric pressure)   The pressure exerted by the mass of air above a given point, usually expressed in hectopascals (hPa), kilopascals (kPa), or millibars (mb).

Air Quality Health Index (AQHI)   An index of air quality that is based on measurements of ambient ozone, particulate matter, and nitrogen dioxide. Intervals on the scale relate to potential health effects.

Aitken nuclei   *See* Condensation nuclei.

Albedo   The fraction of solar radiation returning from a surface compared to that which strikes it. It is the reflectivity of a surface to solar radiation.

Aleutian Low   The subpolar low-pressure area that is often centred near the Aleutian Islands on charts that show mean sea-level pressure.

Altimeter   An instrument that indicates the altitude of an object above a fixed level. Pressure altimeters use an aneroid barometer with a scale graduated in altitude instead of pressure.

Altocumulus   A middle cloud, usually white or grey. Often occurs in layers or patches with wavy, rounded masses or rolls.

Altocumulus castellanus   An altocumulus cloud showing vertical development. Individual cloud elements have towerlike tops, often in the shape of tiny castles.

Altostratus   A middle cloud composed of grey or bluish sheets or layers of uniform appearance. In the thinner regions, the sun or moon usually appears dimly visible.

Anabatic wind   A gentle upslope wind created by daytime heating of the slope and air above it during fair weather.

Analogue forecasting method   A forecast made by comparison of past large-scale synoptic weather patterns that resemble a given (usually current) situation in its essential characteristics.

Analysis   The drawing and interpretation of the patterns of various weather elements on a surface or upper-air chart.

Anemometer   An instrument designed to measure wind speed.

Aneroid barometer   An instrument designed to measure atmospheric pressure. It contains no liquid.

Angular momentum   The product of an object's mass, speed, and radial distance of rotation.

Annual range of temperature   The difference between the warmest and coldest months at any given location.

Anticyclone   An area of high atmospheric pressure around which the wind blows clockwise in the Northern Hemisphere and counterclockwise in the Southern Hemisphere. Also called a "high", and represented with the symbol "H".

Arctic front   In northern latitudes, the semipermanent front that separates deep, cold arctic air from the more shallow, less cold polar air.

Arctic Oscillation (AO)   A reversal of atmospheric pressure over the Arctic that produces changes in the upper-level westerly winds over northern latitudes. These changes in upper-level winds influence winter weather patterns over North America, Greenland, and Europe.

Arcus cloud   *See* Shelf cloud.

Arid climate   An extremely dry climate—drier than the semiarid climate. Often referred to as a "true desert" climate.

Asbestos   A general name for the fibrous variety of silicate minerals that are incombustible and resist chemicals and heat. Once used for fireproofing, electrical insulation, building materials, chemical filters, and brake linings.

Atmosphere   The envelope of gases that surround a planet and are held to it by the planet's gravitational attraction. Earth's atmosphere is mainly nitrogen and oxygen.

Atmospheric boundary layer   The layer of air from Earth's surface usually up to about 1 km where the wind is influenced by friction of Earth's surface and objects on it. Also called the *planetary boundary layer* or the *friction layer*.

Atmospheric greenhouse effect   The warming of the atmosphere by absorbing and emitting infrared radiation while allowing shortwave radiation to pass through. The gases mainly responsible for Earth's atmospheric greenhouse effect are water vapour and carbon dioxide. Also called the *greenhouse effect*.

Atmospheric models   Simulation of the atmosphere's behaviour by mathematical equations or by physical models.

Atmospheric stagnation   A condition of light winds and poor vertical mixing that can lead to a high concentration of pollutants. Air stagnations are most often associated with fair weather, an inversion, and the sinking air of a high-pressure area.

Atmospheric window   The wavelength range between 8 and 11 μm in which little absorption of infrared radiation takes place.

Attenuation   Any process in which the rate of flow of a beam of energy decreases (mainly due to absorption or scattering) with increasing distance from the energy source.

Aurora   Glowing light display in the nighttime sky caused by excited gases in the upper atmosphere giving off light. In the Northern Hemisphere, it is called the *aurora borealis* (northern lights); in the Southern Hemisphere, it is called the *aurora australis* (southern lights).

Autumnal equinox   The equinox at which the sun approaches the Southern Hemisphere and passes directly over the equator. Occurs on September 22 or 23.

AWS   Acronym for Automated Weather Station. A system using a *data logger* and electronic sensors to provide continuous weather observations such as wind, temperature, humidity, pressure, precipitation, cloud height, visibility, etc.

Back-door cold front   A cold front moving south or southwest along the Atlantic coast of North America.

Backing wind   A wind that changes direction in a counterclockwise sense (e.g., north to northwest to west and so on). Opposite of a *veering wind*.

Ball lightning   A rare form of lightning that may consist of a reddish, luminous ball of electricity or charged air.

Banner cloud   A cloud extending downwind from an isolated mountain peak, often on an otherwise cloud-free day.

Baroclinic (atmosphere)   The state of the atmosphere in which surfaces of constant pressure intersect surfaces of constant density. On an isobaric chart, isotherms cross the height contour lines, and temperature advection exists.

Baroclinic instability   A type of instability arising from a meridional (north to south) temperature gradient, a strong vertical wind-speed shear, temperature advection, and divergence in the flow aloft. Many mid-latitude cyclones develop as a result of this instability.

Barograph   A recording barometer.

Barometer   An instrument that measures atmospheric pressure. The two most common barometers are the *mercury barometer* and the *aneroid barometer*.

Barotropic (atmosphere)   A condition in the atmosphere where surfaces of constant density parallel surfaces of constant pressure and there is no thermal advection.

Bead lightning   Lightning that appears as a series of beads tied to a string.

Bergeron process   *See* Ice-crystal process.

Bermuda high   *See* Subtropical high.

Billow clouds   Broad, nearly parallel lines of wavelike clouds oriented at right angles to the wind. Also called Kelvin–Helmholtz wave clouds.

Bimetallic thermometer   A temperature-measuring device usually consisting of two dissimilar metals that expand and contract differentially as the temperature changes.

Blackbody   A hypothetical object that absorbs all of the radiation of a given wavelength that strikes it. It also emits radiation at a maximum rate for its given temperature.

Black frost   *See* Dry freeze.

Black ice   A thin sheet of ice that appears relatively dark when seen on asphalt. It may form as supercooled droplets, drizzle, or light rain come in contact with a road surface that is below freezing. Also, thin, dark-appearing ice that forms on freshwater or saltwater ponds or lakes.

Blizzard   As defined by Environment Canada, a weather condition lasting for four hours or more, characterized by winds of at least 40 km h^{-1} and visibility reduced to less than 1 km in snow and/or blowing snow.

Boulder winds   Fast-flowing, local downslope winds that may attain speeds of 100 knots (185 km h^{-1}) or more. They are especially strong along the eastern foothills of the Rocky Mountains near Boulder, Colorado USA.

Boundary layer   *See* Atmospheric boundary layer.

Bow echo   A line of thunderstorms on a radar screen that appears in the shape of a bow. Bow echoes are often associated with damaging straight-line winds and small tornadoes.

Bowen ratio (β)   The ratio of the sensible heat flux to the latent heat flux ($\beta = Q_H/Q_E$). In hot, dry climates, the Bowen ratio is typically much greater than 1, whereas in moist climates, it is normally less than 1.

Brocken bow   A bright ring of light seen around the shadow of an observer's head as the observer peers into a cloud or fog bank. Formed by the *diffraction* of light.

Broken   A sky condition (abbreviated BKN) in which the sky is six-tenths (five-eights) to nine-tenths (seven-eighths) covered by clouds.

Buoyant force (buoyancy)   The upward force exerted on an object or air parcel by the density difference usually caused by a temperature difference between the parcel and the surrounding air.

Buys-Ballot's law   A law describing the relationship between surface wind direction and pressure distribution. In the Northern Hemisphere, if you stand with your back to the wind and then turn clockwise about 30°, lower pressure will be to your left. In the Southern Hemisphere, stand with your back to the wind and then turn counterclockwise about 30°; lower pressure will be to your right.

California Current   The ocean current that flows southward along the West Coast of the United States from Washington State to Baja, California.

California Norther   A strong, dry, northerly wind that blows in late spring, summer, and early fall in northern and central California. Its warmth and dryness are due to heating caused by downslope compression.

Canadian High   A wintertime semipermanent high-pressure centre located over Canada.

Cap cloud   *See* Pileus cloud.

Carbon dioxide (CO_2)   A colourless, odourless gas whose concentration is about 0.038 percent (385 ppm) in a volume of air near sea level. It is a selective absorber of infrared radiation making it an important atmospheric greenhouse gas. Solid CO_2 is called dry ice.

Carbon monoxide (CO)   A colourless, odourless, toxic gas that forms during the incomplete combustion of carbon-containing fuels.

Ceiling   The height of the lowest layer of clouds when the weather reports describe the sky as broken or overcast.

Ceiling balloon   A small balloon used to determine the height of the cloud base. The height is computed from the balloon's ascent rate and the time required for it to disappear into the clouds.

Ceilometer   An instrument that automatically records cloud height.

Celsius scale   A temperature scale with zero assigned to the temperature at which water freezes and 100 assigned to the temperature at which water at sea level boils.

Centripetal acceleration   The inward-directed acceleration on a particle moving in a curved path.

Centripetal force   The radial force required to keep an object moving in a circular path. It is directed toward the centre of that curved path.

Chaos   The property describing a system that exhibits erratic behaviour in that very small changes in the initial state of the system rapidly lead to large and apparently unpredictable changes sometime in the future.

Chinook wall cloud   A bank of clouds over the Rocky Mountains that signifies an approaching chinook.

Chinook wind   A warm, dry wind on the eastern side of the Rocky Mountains. Similar winds in the Alps are called *foehn*.

Chlorofluorocarbons (CFCs)   Compounds consisting of methane (CH_4) or ethane (C_2H_6) with some or all of the hydrogen replaced by chlorine or fluorine. Used in fire extinguishers, as refrigerants, as solvents for cleaning electronic microcircuits, and as propellants. CFCs destroy ozone in the stratosphere and contribute to the atmospheric greenhouse effect.

Cirrocumulus   A high cloud that appears as a white patch of clouds without shadows. It consists of very small elements in the form of grains or ripples.

Cirrostratus   High, thin, sheetlike clouds composed of ice crystals. They frequently cover the entire sky and often produce a halo.

Cirrus   A high cloud composed of ice crystals in the form of thin, white, featherlike clouds in patches, filaments, or narrow bands.

Clear   A sky condition where there are no clouds; abbreviated CLR.

Clear air turbulence (CAT)   Turbulence encountered by aircraft flying through cloudless skies. Thermals, wind shear, and jet streams can be factors in producing CAT.

Clear ice   A layer of ice that appears transparent because of its homogeneous structure and small number and size of air pockets.

Climate   The accumulation of daily and seasonal weather events over a long period of time.

Climatic controls   The relatively permanent factors that govern the general nature of the climate of a region.

Climatic optimum   *See* Mid-Holocene maximum.

Climatological forecast   A weather forecast, usually a month or more in the future, which is based on the climate of a region rather than on current weather conditions.

Cloud   A visible aggregate of tiny water droplets and/or ice crystals in the atmosphere above Earth's surface.

Cloudburst   Any sudden and heavy rain shower.

Cloud levels   Also called *cloud étages*. Part of the cloud classification system where clouds are divided into high (Ci, Cc, Cs), middle (Ac, As), and low (St, Sc, Cu, Ns, Cb) cloud based on their altitude above ground.

Cloud seeding   The introduction of artificial substances (usually silver iodide or dry ice) into a cloud for the purpose of either modifying its development or increasing its precipitation.

Cloud streets   Lines or rows of cumuliform clouds.

Coalescence   The merging of cloud droplets into a single, larger droplet.

Cold advection (cold air advection)   The transport of cold air by the wind from a region of lower temperatures to a region of higher temperatures.

Cold air damming   A shallow layer of cold air that is trapped by mountains. The classic example is cold air trapped between the Atlantic Coast and the Appalachian Mountains.

Cold fog   *See* Supercooled cloud.

Cold front   A transition zone where a cold air mass advances and replaces a warm air mass.

Cold occlusion   *See* Occluded front.

Cold wave   A rapid fall in temperature that often requires increased protection for agriculture, industry, commerce, and human activities.

Collision–coalescence process   The process of producing precipitation by liquid particles (cloud droplets and raindrops) colliding and joining (coalescing).

Comma cloud   A band of organized cumuliform clouds that looks like a comma on a satellite image.

Computer enhancement   A process whereby the temperatures of radiating surfaces are assigned different shades of grey (or different colours) on an infrared satellite image. This allows specific features to be more clearly delineated.

Condensation   The process by which water vapour becomes a liquid.

Condensation level   The level above the surface marking the base of a cumuliform cloud.

Condensation nuclei   Also called *cloud condensation nuclei*. Tiny particles on whose surfaces condensation of water vapour begins in the atmosphere. Small nuclei less than 0.2 µm in radius are called *Aitken nuclei*; those with radii between 0.2 and 1 µm are *large nuclei*, whereas *giant nuclei* have radii larger than 1 µm.

Conditionally unstable atmosphere   An atmospheric condition that exists when the environmental lapse rate is less than the dry adiabatic lapse rate but greater than the saturated adiabatic lapse rate. Also called *conditional instability*.

Conduction   The transfer of heat by the random motion of molecules. Transfer is always from warmer to colder regions.

Constant height chart (constant level chart)   A chart showing variables, such as pressure, temperature, and wind, at a specific altitude above sea level. Variation in horizontal pressure is depicted by isobars. The most common constant-height chart is the surface chart, which is also called the *sea-level chart* or *surface weather map*.

Constant pressure chart (isobaric chart)   A chart showing variables, such as temperature and wind, on a constant pressure surface. Variations in height are usually shown by lines of equal height (contour lines).

Contact freezing   The process by which contact with a nucleus such as an ice crystal causes supercooled liquid droplets to change into ice.

Continental arctic air mass   An air mass, abbreviated cA, characterized by extremely low temperatures and very dry air.

Continental polar air mass   An air mass, abbreviated cP, characterized by low temperatures and dry air. Not as cold as arctic air masses and seldom analyzed by Canadian meteorologists.

Continental tropical air mass   An air mass, abbreviated cT, characterized by high temperatures and low humidity.

Contour line A line that connects points of equal value. Generally contour lines refer to points of equal elevation above a reference level, most often sea level.

Contrail (condensation trail) A cloudlike streamer frequently seen forming behind aircraft flying in clear, cold, humid air.

Convection Motions in a fluid that result in the transport and mixing of the fluid's properties. In meteorology, convection usually refers to atmospheric motions that are predominantly vertical, such as rising air currents due to surface heating. The rising of heated surface air and the sinking of cooler air aloft are called *free convection*. (Compare to *forced convection*.)

Convective instability Instability arising in the atmosphere when a column of air exhibits warm, moist, nearly saturated air near the surface and cold, dry air aloft. When the lower part of the layer is lifted and saturation occurs, it becomes unstable.

Convergence An atmospheric condition that exists when the winds cause a horizontal net inflow of air into a specified region.

Conveyor belt model (for middle-latitude storms) A three-dimensional picture of a middle-latitude cyclone and the various air streams (called conveyor belts) that interact to produce the weather associated with the storm.

Cooling degree-day A form of degree-day used in estimating the amount of energy necessary to reduce the temperature of warm air to a specified base temperature, usually for the purposes of space cooling. A day that accumulates one cooling degree-day is a day on which the average temperature is 1° above a desired base temperature.

Coriolis force (effect) An apparent force observed on any free-moving object in a rotating system. On Earth, this deflective force results from Earth's rotation and causes moving particles (including the wind) to deflect to the right in the Northern Hemisphere and to the left in the Southern Hemisphere.

Corona (optic) A series of coloured rings concentrically surrounding the disk of the sun or moon. Smaller than the halo, the corona is often caused by the diffraction of light around small water droplets of uniform size.

Cosine law of illumination A law stating that the radiation intensity striking an inclined surface (R) is equal to the intensity if the radiation were striking the surface directly (R_0) times the cosine of the angle between the radiation beam and a line that is perpendicular to the surface called the zenith angle (Z). Mathematically, this is $R = R_0$ cosine (Z).

Country breeze A light breeze that blows into a city from the surrounding countryside. It is best observed on clear nights when the urban heat island is most pronounced.

Crepuscular rays Alternating light and dark bands of light that appear to fan out from the sun's position, usually at twilight.

Criteria Air Contaminants A group of air pollutants that are emitted in large quantities by human activity. They usually include particulate matter, sulphur oxides, nitrogen oxides, carbon monoxide, and volatile organic compounds and sometimes ammonia.

Cumuliform Clouds that are "heap" clouds, or clouds with vertical development.

Cumulonimbus An exceptionally vertically developed cloud, often with a top in the shape of an anvil. The cloud is frequently accompanied by heavy showers, lightning, thunder, and sometimes hail. It is also known as a thunderstorm cloud.

Cumulonimbus incus (anvil) A supplementary feature of a cumulonimbus cloud forming an anvil shape when the upper part of the cloud spreads out as it reaches the stratosphere.

Cumulus A cloud in the form of individual, detached domes or towers that are usually well defined. It has a flat base with a bulging upper part that often resembles cauliflower. Cumulus clouds of fair weather are called *cumulus humilis*. Those that exhibit much vertical growth are called *cumulus congestus* or *towering cumulus*. Those that are fragmented, ragged, and often found under a larger cloud base are called *cumulus fractus*.

Cumulus stage The initial stage in the development of an ordinary cell thunderstorm in which rising, warm, humid air develops into a cumulus cloud.

Curvature effect In cloud physics, as cloud droplets decrease in size, they exhibit a greater surface curvature that causes a more rapid rate of evaporation.

Cutoff low A cold, upper-level low that has become displaced out of the basic westerly flow and lies to the south of this flow.

Cyclogenesis The development or strengthening of middle-latitude (extratropical) cyclones.

Cyclone An area of low pressure around which the winds blow counterclockwise in the Northern Hemisphere and clockwise in the Southern Hemisphere.

Daily range of temperature The difference between the maximum and minimum temperatures for any given day.

Dalton's law of partial pressures The total pressure of a mixture of gases is equal to the sum of the partial pressures of each individual gas in the mixture.

Dart leader The discharge of electrons that proceeds intermittently toward the ground along the same ionized channel taken by the initial lightning stroke.

Data logger A special-purpose, programmable microcomputer that can measure electrical signals, convert them into usable data with recognizable units, process these data (i.e., average them, find the maximum or minimum, identify when the measurements were made, etc.), and store the data in a digital format. It can also often transmit its stored data by communicating over telephone lines, the Internet, a radio, a cellular phone, or a satellite communication system.

Dendrochronology The analysis of the annual growth rings of trees as a means of interpreting past climatic conditions.

Density The ratio of the mass of a substance to the volume occupied by it. Air density is usually expressed as kg m^{-3}.

Deposition A process that occurs in subfreezing air when water vapour changes directly to ice without becoming a liquid first.

Deposition nuclei Tiny particles (ice nuclei) on which an ice crystal may grow by the process of deposition.

Derecho Strong, damaging, straight-line winds associated with a cluster of severe thunderstorms that most often form in the evening or at night.

Desertification A general increase in the desert conditions of a region.

Desert pavement An arrangement of pebbles and large stones that remains behind as finer dust and sand particles are blown away by the wind.

Dew Water that has condensed onto objects near the ground when their temperatures have fallen below the dew point of the surface air.

Dew cell An instrument used to determine the dew-point temperature.

Dew point (dew-point temperature) The temperature to which air must be cooled (at constant pressure and constant water vapour content) for saturation to occur. It is a measure of humidity.

Dew-point hygrometer An instrument that determines the dew-point temperature of the air.

Diffraction The bending of light around objects, such as cloud and fog droplets, producing fringes of light and dark or coloured bands.

Dispersion The separation of white light into its different component wavelengths.

Dissipating stage   The final stage in the development of an ordinary cell thunderstorm when downdrafts exist throughout the cumulonimbus cloud.

Diurnal   Daily—something that varies over the course of a day.

Divergence   An atmospheric condition that exists when the winds cause a horizontal net outflow of air from a specific region.

Doldrums   The region near the equator that is characterized by low pressure and light, shifting winds.

Doppler lidar   The use of light beams to determine the velocity of objects such as particles carried by wind and falling rain by taking into account the Doppler shift.

Doppler radar   A radar that determines the velocity of falling precipitation either toward or away from the radar unit by taking into account the *Doppler shift*.

Doppler shift (effect)   The change in the frequency of waves that occurs when the emitter or the observer is moving toward or away from the other.

Downburst   A severe localized downdraft that can be experienced beneath a severe thunderstorm. (Compare *microburst* and *macroburst*.)

Downslope windstorm   A strong, gusty wind that blows down the lee slope of a mountain range, reaching peak strength usually near the foot of the mountains and diminishing rapidly away from the mountains. Downslope windstorms result from a mountain wave that can form when wind flows over a mountain chain and resemble the flow of water down the lee side of a river rock.

Drizzle   Small water drops between 0.2 and 0.5 mm in diameter that fall slowly and reduce visibility more than light rain.

Drought   A period of abnormally dry weather sufficiently long enough to cause serious effects on agriculture and other activities in the affected area.

Dry adiabatic lapse rate   The rate of decrease of temperature in a rising unsaturated air parcel and the rate of increase in temperature of a descending unsaturated air parcel. The rate of adiabatic cooling or warming is 9.8°C per 1000 m.

Dry adiabats   Lines on an adiabatic chart such as a tephigram, that show the dry adiabatic rate for rising or descending air. They represent lines of constant potential temperature.

Dry-bulb temperature   The air temperature. It is measured by the dry-bulb thermometer of a psychrometer.

Dry climate   A climate deficient in precipitation where annual potential evapotranspiration exceeds precipitation.

Dry freeze   The internal freezing of vegetation without the protective formation of hoarfrost. Also known as *black frost* from the resulting blackening of affected vegetation.

Dry haze   *See* Haze.

Dry lightning   Lightning that occurs with thunderstorms that produce little, if any, appreciable precipitation that reaches the surface.

Dryline   A boundary that separates warm, dry air from warm, moist air. It usually represents a zone of instability along which thunderstorms form.

Dry slot   On a satellite image, the dry slot represents the relatively clear region (or clear wedge) that appears just to the west of the tail of a comma cloud of a mid-latitude cyclonic storm.

Dry-summer subtropical climate   A climate characterized by mild, wet winters and warm to hot, dry summers. Typically located between 30 and 45° latitude on the western side of continents.   Also called *Mediterranean climate.*

Dust devil (whirlwind)   A small but rapidly rotating wind made visible by the dust, sand, and debris it picks up from the surface. It develops best on clear, dry, hot afternoons.

Earth vorticity   The rotation (spin) of an object about its vertical axis brought on by the rotation of Earth on its axis. Earth's vorticity is a maximum at the poles and zero at the equator.

Easterly wave   A migratory wavelike disturbance in the tropical easterlies. Easterly waves occasionally intensify into tropical cyclones. They are also called *tropical waves.*

Eccentricity (of Earth's orbit)   The deviation of Earth's orbit from elliptical to nearly circular.

Eddy   A small volume of air (or any fluid) that moves differently from the larger flow in which it exists.

Eddy covariance   A technique used by micrometeorologists to measure turbulent fluxes of heat, mass (e.g., water vapour or CO_2), and momentum that relies on measurements of all wind components as well as tracer values (i.e., temperature or water vapour or CO_2 concentration) as frequently as 10 to 20 times per second.

Eddy viscosity   The internal friction produced by turbulent flow.

Ekman spiral   An idealized description of the way the wind-driven ocean currents vary with depth. In the atmosphere, it represents the way the winds vary from the surface up through the friction layer or planetary boundary layer.

Ekman transport   Net surface water transport due to the Ekman spiral. In the Northern Hemisphere, the transport is 90° to the right of the surface wind direction.

Electrical hygrometer   *See* Hygrometer.

Electrical thermometers   Thermometers that use temperature-dependent electrical properties of materials to measure temperature. Thermistors and thermocouples are two common types.

Electromagnetic waves   *See* Radiant energy.

El Niño   An extensive ocean warming that begins along the coast of Peru and Ecuador and extends westward over the tropical Pacific. Major El Niño events, occur once every two to seven years and affect global weather patterns. (*See also* ENSO.)

Embryo   In cloud physics, a tiny ice crystal that grows in size and becomes an ice nucleus.

Energy   The property of a system that generally enables it to do work. Some forms of energy are kinetic, radiant, potential, chemical, electric, and magnetic.

Enhanced Fujita (EF) scale   A modification of the original Fujita scale that describes tornado intensity by observing damage caused by the tornado.

Ensemble forecasting   A forecasting technique that entails running several forecast models (or different versions of a single model), each beginning with slightly different weather information. The forecaster's level of confidence is based on how well the models agree (or disagree) at the end of some specified forecast time.

ENSO (El Niño–Southern Oscillation)   A condition in the tropical Pacific whereby the reversal of surface air pressure at opposite ends of the Pacific Ocean induces westerly winds, a strengthening of the equatorial countercurrent, and extensive ocean warming in the eastern Pacific. (*See* El Niño.)

Entrainment   The mixing of environmental air into a preexisting air current or cloud so that the environmental air becomes part of the current or cloud.

Environmental lapse rate   The rate of decrease in air temperature with elevation. It is most often measured with a radiosonde.

Equilibrium   A state where inputs equal outputs and a system is in balance.

Equilibrium vapour pressure   The necessary vapour pressure around liquid water that allows the water to remain in equilibrium with its environment. Also called *saturation vapour pressure.*

Étages   Cloud levels or altitudes used in cloud classification: high, middle, and low.

Evaporation   The process by which a liquid changes into a gas.

Evaporation fog   Fog produced when sufficient water vapour is added to the air by evaporation and the moist air mixes with cooler air and becomes saturated. The two common types are *steam fog*, which forms when cold air moves over warm water, and *frontal fog*, which forms as warm raindrops evaporate in a cool air mass.

Evapotranspiration The sum of evaporation from the surface and transpiration from plants.

Exosphere The outermost portion of the atmosphere.

Exponential rate of change Describes how a quantity changes over distance or time. An exponential change is when the rate of change of something is proportional to its current value. Mathematically, an exponential change is proportional to e^x, where e is 2.718, and x is time or distance. For example, pressure (and air density) decreases exponentially with height above the surface: $P = P_0\, e^{-z/h}$, where P_0 is the sea-level pressure (averaging 1013.25 hPa), z is the elevation above sea level, and h is the scale height of the atmosphere, which is approximately 8 km.

Extratropical cyclone See *Middle-latitude cyclone.*

Extratropical transition When a tropical cyclone moves away from the tropics and undergoes a transition to a middle-latitude cyclone.

Eye A region in the centre of a hurricane (tropical storm) where the winds are light and skies are clear to partly cloudy.

Eyewall A wall of dense thunderstorms that surrounds the eye of a hurricane.

Eyewall replacement A situation within a hurricane (tropical cyclone) where the storm's original eyewall dissipates and a new eyewall forms outward, farther away from the centre of the storm.

Fahrenheit scale A temperature scale where 32 is the freezing temperature of water and 212 is water's boiling temperature (at sea level).

Fall streaks Falling ice crystals that evaporate before reaching the ground.

Fall wind A strong wind that blows downslope after passing through mountain passes or off plateaus.

Fata Morgana A complex mirage that is characterized by objects being distorted in such a way as to appear as castlelike features.

Feedback mechanism A process whereby an initial change in an atmospheric process will tend to either reinforce the process (*positive feedback*) or weaken the process (*negative feedback*).

Ferrel cell The middle-latitude cell in the three-cell model of the general circulation.

Fetch The distance that the wind travels over open water.

Few A sky condition (abbreviated FEW) in which the sky is three-tenths or less (two-eighths or less) covered by clouds.

Flash flood A flood that rises and falls quite rapidly with little or no advance warning, usually as the result of intense rainfall over a relatively small area.

Flurries of snow See Snow flurries.

Foehn See Chinook wind.

Fog A cloud with its base at Earth's surface.

Forced convection On a small scale, a form of mechanical mixing taking place when twisting eddies of air are able to mix surface air with the air above. It is caused by mechanical forces such as the deflection of wind by a rough surface. On a much larger scale, it can be induced by the lifting of warm air along a front (*frontal uplift*) or along a topographic barrier (*orographic uplift*).

Forked lightning Cloud-to-ground lightning that exhibits downward-directed crooked branches.

Formaldehyde A colourless gaseous compound (HCHO) used in the manufacture of resins, fertilizers, and dyes. Also used as an embalming fluid, a preservative, and a disinfectant.

Free convection See Convection.

Freeze A condition occurring over a widespread area when the surface air temperature remains below freezing for a sufficient time to damage certain agricultural crops. A freeze most often occurs as cold air

is advected into a region, causing freezing conditions to exist in a deep layer of surface air. Also called *advection frost*.

Freezing nuclei Particles that promote the freezing of supercooled liquid droplets.

Freezing rain and freezing drizzle Rain or drizzle that falls through air below zero Celsius but remains in liquid form and then freezes on striking a cold object or the ground. Both can produce a coating of ice on objects, which is called *glaze*.

Friction layer The atmospheric layer near the surface usually extending up to about 1 km where the wind is influenced by friction from Earth's surface and objects on it. Also called the *atmospheric boundary layer* or *planetary boundary layer.*

Front The transition zone between two distinct air masses.

Frontal fog See Evaporation fog.

Frontal inversion A temperature inversion encountered on ascending through a sloping front, usually a warm front.

Frontal thunderstorms Thunderstorms that form in response to forced convection (forced lifting) along a front. Most go through a cycle similar to that of ordinary thunderstorms.

Frontal wave A wavelike deformation along a front in the lower levels of the atmosphere. Those that develop into storms are termed *unstable waves*, whereas those that do not are called *stable waves.*

Frontogenesis The formation, strengthening, or regeneration of a front.

Frontolysis The weakening or dissipation of a front.

Frost (also called hoarfrost) A covering of ice produced by deposition on exposed surfaces when the air temperature falls below the frost point.

Frostbite The partial freezing of exposed parts of the body, causing injury to the skin and sometimes to deeper tissues.

Frost-bulb temperature The lowest temperature that can be reached by sublimating frozen water into the air.

Frost point The temperature at which the air becomes saturated with respect to ice when cooled at constant pressure and constant water vapour content.

Frozen dew The transformation of liquid dew into tiny beads of ice when the air temperature drops below freezing.

Fujita scale A scale developed by T. Theodore Fujita for classifying tornadoes according to the damage they cause and their rotational wind speed. (See also Enhanced Fujita scale.)

Fulgurite A rootlike tube (or several tubes) that forms when a lightning stroke fuses sand particles together.

Funnel cloud A funnel-shaped cloud of condensed water, usually extending from the base of a cumuliform cloud. The rapidly rotating air of the funnel is not in contact with the ground; hence, it is not a tornado.

Galaxy A huge assembly of stars (between millions and hundreds of millions) held together by gravity.

Gap winds A strong, low-level wind through horizontal valleys or a gap in a mountain barrier that connects regions of different air masses.

Gas law The thermodynamic law applied to a perfect gas that relates the pressure of the gas to its density and absolute temperature.

Genera The 10 main cloud classification groups in common use today, which are based on appearance and formation process. The genera are cirrus (Ci), cirrostratus (Cs), cirrocumulus (Cc), altostratus (As), altocumulus (Ac), stratus (St), stratocumulus (Sc), cumulus (Cu), nimbostratus (Ns), and cumulonimbus (Cb).

General circulation of the atmosphere Large-scale atmospheric motions over Earth.

Geostationary satellite A satellite that orbits Earth at the same rate that Earth rotates and thus remains over a fixed place above the equator.

Geostrophic wind   A theoretical horizontal wind blowing in a straight path, parallel to the isobars or contours, at a constant speed, and assuming there is no friction. The geostrophic wind results when the Coriolis force exactly balances the horizontal pressure gradient force.

Giant nuclei   *See* Condensation nuclei.

Glaciated cloud   A cloud or portion of a cloud where only ice crystals exist.

Glaze   A coating of ice, often clear and smooth, that forms on exposed surfaces by the freezing of a film of supercooled water deposited by rain, drizzle, or fog. As a type of aircraft icing, glaze is called clear ice.

Global climate   Climate of the entire globe.

Global climate models   Also called general circulation models (GCMs), are mathematical (computer) models of Earth's general circulation based on the laws of physics (*see* Governing equations) that determine weather and climate. They use information such as greenhouse gas concentrations and calculate the resulting climate. Modern GCMs include models of the oceans, ice, and biosphere, as well as the atmosphere, to account for the interactions between these systems. GCMs are the primary tool used by climatologists to make future climate projections and understand past climates.

Global scale   The largest scale of atmospheric motion. Also called the *planetary scale.*

Global warming   Increasing global surface air temperatures. The term global warming is usually attributed to human activities, such as increasing concentrations of greenhouse gases from the combustion of fossil fuels in automobiles and industrial processes.

Glory   Coloured rings that appear around the shadow of an object.

Governing equations   These are the laws of physics that determine the exchanges and conservation of properties in the atmosphere, such as momentum (wind), energy (radiation, internal energy or heat, latent energy, and potential energy), and mass (of air and its constituents as well as of water and humidity). The solution of these equations on a computer is the basis of both *numerical weather prediction* and *global climate models.*

Gradient wind   A theoretical horizontal wind that blows parallel to curved isobars or contours, at constant speed, and assumes there is no friction.

Graupel   Ice particles between 2 and 5 mm in diameter that form in a cloud often by the process of accretion. Snowflakes that become rounded pellets due to riming are called graupel or snow pellets.

Green flash   A small region of green colour that occasionally appears on the upper part of the sun as it rises or sets.

Greenhouse effect   *See* Atmospheric greenhouse effect.

Greenhouse gas   Trace gases in the atmosphere that absorb and emit infrared radiation. The main greenhouse gases are water vapour, carbon dioxide, methane, nitrous oxide, and ozone.

Ground fog   *See* Radiation fog.

Growing degree-day   A form of the degree-day used as a guide for crop planting and for estimating crop maturity dates. A day that accumulates one growing degree-day is a day on which the average temperature is 1° above the desired base temperature for a particular crop.

Gulf stream   A warm, swift, narrow ocean current flowing along the East Coast of North America.

Gust front   A boundary that separates a cold downdraft of a thunderstorm from warm, humid surface air. On the surface, its passage resembles that of a cold front.

Gustnado   A relatively weak tornado associated with a thunderstorm's outflow. It most often forms along the gust front.

Gyre   A large circular, surface ocean current pattern.

Haboob   A dust or sandstorm that forms as cold downdrafts from a thunderstorm turbulently lift dust and sand into the air.

Hadley cell   A thermal circulation proposed by George Hadley to explain the movement of the trade winds. It consists of rising air near the equator and sinking air near 30° latitude.

Hailstones   Transparent or partially opaque particles of ice ranging from pea-sized to golf ball-sized.

Hailstreak   The accumulation of hail at Earth's surface along a relatively long (10 km), narrow (2 km) band.

Hair hygrometer   *See* Hygrometer.

Halons   A group of organic compounds used as fire retardants. In the stratosphere, these compounds release bromine atoms that rapidly destroy ozone. The production of halons is now banned by the Montreal Protocol.

Halos   Rings or arcs that encircle the sun or moon when seen through an ice crystal cloud or a sky filled with falling ice crystals. Halos are produced by refraction of light.

Hazardous air pollutants   A group of pollutants that are toxic in low concentrations.

Haze   Fine dry or wet dust or salt particles dispersed through a portion of the atmosphere. Individually, these are not visible, but cumulatively, they will diminish visibility. *Dry haze* particles are very small, on the order of 0.1 μm. *Wet haze* particles are larger.

Heat   A form of energy transferred between systems by virtue of their temperature differences.

Heat burst   A sudden increase in surface air temperature often accompanied by extreme drying. A heat burst is associated with the downdraft of a thunderstorm or a cluster of thunderstorms.

Heat capacity   The ratio of the heat absorbed (or released) by a system to the corresponding temperature rise (or fall). The heat capacity per unit mass is called the *specific heat.*

Heating degree-day   A form of the degree-day used as an index for space heating requirements. A day that accumulates one heating degree-day is a day on which the average temperature is 1° below the base temperature for space heating.

Heat lightning   Distant lightning that illuminates the sky but is too far away for its thunder to be heard.

Heatstroke   A physical condition induced by a person's overexposure to high air temperatures, especially when accompanied by high humidity.

Hectopascal   Abbreviated hPa. One hectopascal is equal to 100 N m^{-2} (pascals) or 1 millibar. Sea-level pressure is normally close to 1013.25 hPa.

Heiligenschein   A faint white ring surrounding the shadow of an observer's head on a dew-covered lawn.

Heterosphere   The region of the atmosphere above about 85 km where the composition of the air varies with height.

High   *See* Anticyclone.

High inversion fog   A fog that lifts above the surface but does not completely dissipate because of a strong inversion (usually subsidence) that exists above the fog layer.

Holocene   An interglacial geologic epoch in the current Ice Age that began about 11,700 years ago and continues to the present.

Homogeneous (spontaneous) freezing   The freezing of pure water. For tiny cloud droplets, homogeneous freezing does not occur until the air temperature reaches about −40°C.

Homosphere   The region of the atmosphere below about 85 km where the composition of the air remains fairly constant.

Hook echo   The shape of an echo on a Doppler radar screen that indicates the possible presence of a tornado.

Horse latitudes   The belt of latitudes between about 30° and 35° where winds are predominantly light and the weather is hot and dry.

Humid continental climate   A climate characterized by severe winters and mild to warm summers with adequate annual precipitation. Typically located over large continental areas in the Northern Hemisphere between about 40° and 70° latitude.

Humidex An index used by the Meteorological Service of Canada that combines air temperature and vapour pressure to indicate the combined effects of temperature and humidity on human comfort. The index value approximates the temperature in Celsius for a similar comfort level under dry conditions.

Humidity A general term that refers to the air's water vapour content. (*See* Relative humidity.)

Humid subtropical climate A climate characterized by hot, muggy summers, cool to cold winters, and abundant precipitation throughout the year.

Hurricane A tropical cyclone with winds in excess of 64 knots (119 km h^{-1}).

Hurricane warning A warning given when it is likely that a hurricane will strike an area within 24 hours.

Hurricane watch A hurricane watch indicates that a hurricane poses a threat to an area (often within several days) and residents of the watch area should be prepared.

Hydrocarbons Chemical compounds composed of only hydrogen and carbon—they are included under the general term *volatile organic compounds* (VOCs).

Hydrologic cycle A model that illustrates the movement and exchange of water in the hydrosphere among Earth's systems: hydrosphere, atmosphere, lithosphere, and biosphere.

Hydrophobic The ability to resist the condensation of water vapour. Usually used to describe "water-repelling" condensation nuclei.

Hydrostatic equation An equation that describes how quickly air pressure decreases with altitude. It states that the rate at which the air pressure decreases with height is equal to the air density times the acceleration of gravity. It is a statement of hydrostatic equilibrium.

Hydrostatic equilibrium The state of the atmosphere when there is a balance between the upward vertical pressure gradient force and the downward pull of gravity.

Hygrometer An instrument designed to measure the air's water vapour content. The sensing part of the instrument can be hair (*hair hygrometer*), a plate coated with carbon (*electrical hygrometer*), an infrared sensor (*infrared hygrometer*), or a capacitor that is moisture sensitive (*thin-film capacitance hygrometer*).

Hygroscopic The ability of a substance to retain liquid water by reducing its evaporation and thereby accelerating the net condensation of water vapour. Usually used to describe condensation nuclei that have an affinity for water vapour.

Hypothermia The deterioration in one's mental and physical condition brought on by a rapid lowering of human body temperature.

Hypoxia A condition experienced by humans when the brain does not receive sufficient oxygen.

Ice age A major interval of geologic time marked by a long-term reduction in Earth's surface and atmospheric temperature, resulting in extensive continental ice sheets, polar ice sheets, and alpine glaciers. The current Ice Age is known as the *Pleistocene epoch*.

Ice-crystal (Bergeron) process A process that produces precipitation. The process involves tiny ice crystals in a supercooled cloud growing larger at the expense of the surrounding liquid droplets. Also called the *Bergeron process*.

Ice fog A type of fog that forms at very low temperatures, composed of tiny, suspended ice particles.

Icelandic Low The subpolar low-pressure area that, on average, is centred near Iceland on charts that show mean sea-level pressure.

Ice nuclei Particles that act as nuclei for the formation of ice crystals in the atmosphere.

Ice pellets A type of precipitation consisting of transparent pellets of ice 5 mm or less in diameter.

Indian summer An unseasonably warm spell with clear skies near the middle of autumn. Usually follows a substantial period of cool weather.

Inferior mirage *See* Mirage.

Infrared hygrometer *See* Hygrometer.

Infrared radiation Electromagnetic radiation with wavelengths between about 0.7 and 1000 μm. This radiation is longer than visible radiation but shorter than microwave radiation.

Infrared radiometer An instrument designed to measure the intensity of infrared radiation emitted by an object. Also called an *infrared sensor* or a *pyrgeometer*.

Insolation The incoming *solar radiation* that reaches Earth and the atmosphere.

Instrument shelter A boxlike (sometimes wooden) structure designed to protect weather instruments from direct sunshine and precipitation.

Interglacial period (interglacial) A time interval of relatively mild climate during an ice age when continental ice sheets were absent or limited in extent to Greenland and the Antarctic.

Internal energy Sometimes also called heat energy, it is the kinetic energy within a substance due to the random motion of molecules and is equal to its mass times its heat capacity times its temperature.

Intertropical convergence zone (ITCZ) The boundary zone of converging surface air separating the northeast trade winds of the Northern Hemisphere from the southeast trade winds of the Southern Hemisphere. The ITCZ is characterized by convective clouds and precipitation.

Inversion An increase in air temperature with height. Sometimes called a *temperature inversion*.

Ion An electrically charged atom, molecule, or particle.

Ionosphere An electrified region of the upper atmosphere where fairly large concentrations of ions and free electrons exist.

Iridescence Brilliant spots or borders of colours, most often red and green, observed in clouds up to about 30° from the sun.

Isallobar A line of equal change in atmospheric pressure during a specified time interval.

Isobar A line connecting points of equal pressure.

Isobaric chart (map) *See* Constant pressure chart.

Isobaric surface A surface along which the atmospheric pressure is equal everywhere.

Isotach A line connecting points of equal wind speed.

Isotherm A line connecting points of equal temperature.

Isothermal layer A layer where the air temperature is constant with increasing altitude. In an isothermal layer, the air temperature lapse rate is zero.

Jet maximum *See* Jet streak.

Jet streak A region of high wind speed that moves through the axis of a jet stream. Also called *jet maximum*.

Jet stream Relatively strong winds concentrated within a narrow band in the atmosphere.

Katabatic wind Any wind blowing downslope. It is usually cold.

Kelvin A unit of temperature. A kelvin is denoted by K, and 1 K equals 1°C. Zero kelvin is absolute zero, or −273.15°C.

Kelvin scale A temperature scale with zero degrees equal to the theoretical temperature at which all molecular motion ceases. Also called the *absolute scale*. The units are sometimes called "degrees kelvin"; however, this is incorrect usage, and the correct SI terminology is "kelvins," abbreviated K.

Kinetic energy The energy within a body that is a result of its motion.

Kirchhoff's law   A law that states that good absorbers of a given wavelength of radiation are also good emitters of that wavelength. In other words, at a given wavelength, the emissivity equals the absorptivity.

Knot   A unit of speed equal to one nautical mile per hour. One knot equals exactly 1.852 km h^{-1}.

Köppen classification system   A system for classifying climates developed by W. Köppen that is based mainly on annual and monthly averages of temperature and precipitation. It is sometimes called the Köppen–Geiger climate classification system.

Lake breeze   A wind blowing onshore from the surface of a lake.

Lake-effect snows   Localized snowstorms that form on the downwind side of a lake. Such storms are common in late fall and early winter near the Great Lakes as cold, dry air picks up moisture and warmth from the unfrozen bodies of water.

Laminar boundary layer   The layer of air a few millimetres thick that surrounds all surfaces in which there is laminar flow and exchanges of sensible heat or moisture are by conduction only; there is no convection.

Laminar flow   A nonturbulent flow in which the fluid moves smoothly in parallel layers or sheets.

Land breeze   A coastal breeze that blows from land to sea, usually at night.

Landspout   A relatively weak nonsupercell tornado that originates with a cumiliform cloud in its growth stage and is not associated with a mid-level mesocyclone. Its spin originates near the surface. Landspouts often look like waterspouts over land.

La Niña   A condition where the central and eastern tropical Pacific Ocean turns cooler than normal. The counterpart of El Niño.

Lapse rate   The rate at which an atmospheric variable (usually temperature) decreases with height. (*See* Environmental lapse rate.)

Latent heat   The heat that is either released or absorbed by a unit mass of a substance when it undergoes a change of state, such as during evaporation, condensation, sublimation, deposition, freezing, and melting.

Laterite   A soil formed under tropical conditions where heavy rainfall leaches soluble minerals from the soil. This leaching leaves the soil hard and poor for growing crops.

Lee-side low   Storm systems (extratropical cyclones) that form on the downwind (lee) side of a mountain chain. In North America, lee-side lows frequently form on the eastern side of the Rocky Mountains.

Lenticular cloud   A cloud in the shape of a lens.

Level of free convection   The level in the atmosphere at which a lifted air parcel becomes warmer than its surroundings in a conditionally unstable atmosphere.

Lidar   An instrument that uses a laser to generate intense light pulses that are reflected from atmospheric particles. Lidars have been used to determine the amount of particles in the atmosphere as well as particle movement that has been converted into wind speed. Lidar means **LI**ght **D**etection **A**nd **R**anging. *See* doppler lidar.

Lifting condensation level (LCL)   The level at which a parcel of air, when lifted dry adiabatically, would become saturated.

Lightning   A visible electrical discharge produced by thunderstorms.

Liquid-in-glass thermometer   *See* Thermometer.

Little Ice Age   The period from about 1550 to 1850 when average temperatures over Europe were lower and alpine glaciers increased in size and advanced down mountain canyons.

Local winds   Winds that tend to blow over a relatively small area; often due to regional effects, such as mountain barriers, large bodies of water, local pressure differences, and other influences.

Long-range forecast   Generally used to describe a weather forecast that extends beyond about 14 days into the future.

Longwave radiation   A term most often used to describe the infrared energy emitted by Earth and the atmosphere.

Longwaves in the westerlies   A wave in the upper level of westerlies characterized by a long length (thousands of kilometres) and significant amplitude. Also called *Rossby waves*.

Low   *See* Extratropical cyclone.

Low-level jet streams   Jet streams that typically form near Earth's surface below an altitude of about 2 km and usually attain speeds of less than 60 knots.

Macroburst   A strong downdraft (*downburst*) greater than 4 km wide that can occur beneath thunderstorms. A downburst less than 4 km across is called a *microburst*.

Macroclimate   The general climate of a large area, such as a country.

Macroscale   The normal meteorological synoptic scale for obtaining and analyzing weather information to detect and monitor storms and other weather phenomena. It can cover an area ranging from the size of a continent to the entire globe.

Madden–Julian Oscillation   A tropical pattern of enhanced and suppressed rainfall observed over the Indian and Pacific oceans that propagates eastward (abbreviated MJO). The cycle typically lasts 30 to 60 days.

Magnetic storm   A worldwide disturbance of Earth's magnetic field caused by solar disturbances.

Magnetosphere   The region around Earth in which Earth's magnetic field plays a dominant part in controlling the physical processes that take place.

Mamma clouds   Clouds that look like pouches hanging from the underside of a cloud. Also called *mammatus clouds*.

Marine climate   A climate controlled largely by the ocean. The ocean's influence keeps winters relatively mild and summers cool.

Maritime air   Moist air whose characteristics were developed over an extensive body of water.

Maritime arctic air mass   Cold, moist air that forms over an extensive body of water at high latitudes.

Maritime polar air mass   An air mass characterized by cool temperatures and high humidity that forms over middle-latitude oceans.

Maritime tropical air mass   An air mass characterized by high temperatures and high humidity that forms over tropical oceans.

Mature thunderstorm   The second stage in the three-stage cycle of an ordinary thunderstorm. This mature stage is characterized by heavy showers, lightning, thunder, and violent vertical motions inside cumulonimbus clouds.

Maunder minimum   A period from about 1645 to 1715 when few, if any, sunspots were observed.

Maximum thermometer   A thermometer with a small constriction just above the bulb. It is designed to measure the maximum air temperature.

Mean annual temperature   The average temperature at any given location for the entire year.

Mean daily temperature   The average of the highest and the lowest temperature for a 24-hour period.

Mechanical turbulence   Turbulent eddy motions caused by obstructions, such as trees, buildings, mountains, and so on. *See* Forced convection.

Mediterranean climate   *See* Dry-summer subtropical climate.

Medium-range forecast   Generally used to describe a weather forecast that extends from about three to 14 days into the future.

Mercury barometer   A type of barometer that uses mercury to measure atmospheric pressure. The height of the mercury column is a measure of atmospheric pressure.

Meridional flow   A type of atmospheric circulation pattern in which the north–south component of the wind is pronounced.

Mesoclimate   The climate of an area ranging in size from a few acres to several square kilometres.

Mesocyclone   A vertical column of cyclonically rotating air within a supercell thunderstorm.

Mesohigh   A relatively small area of high atmospheric pressure that forms beneath a thunderstorm.

Mesopause   The top of the mesosphere. The boundary between the mesosphere and the thermosphere, usually near 85 km.

Mesoscale   The scale of meteorological phenomena that range in size from a few kilometres to about 100 km. It includes local winds, thunderstorms, and tornadoes.

Mesoscale convective complex (MCC)   A large, organized, convective weather system composed of a number of individual thunderstorms. The size of an MCC can be 1000 times larger than an individual ordinary cell thunderstorm.

Mesoscale convective system (MCS)   A large cloud system that represents an ensemble of thunderstorms that form by convection and produce precipitation over a wide area.

Mesosphere   The atmospheric layer between the stratosphere and the thermosphere. Located at an average elevation between 50 and 85 km above Earth's surface.

Meteogram   A chart that shows how one or more weather variables has changed at a station over a given period of time or how the variables are likely to change with time.

Meteorology   The study of the atmosphere and atmospheric phenomena as well as the atmosphere's interaction with Earth's surface, oceans, and life in general.

Methane   The gas CH_4, which comprises natural gas and is an important *greenhouse gas*.

Microburst   A strong localized downdraft (downburst) less than 4 km wide that occurs beneath thunderstorms. A strong downburst greater than 4 km across is called a *macroburst*.

Microclimate   The climate near Earth's surface, that, due to local conditions may be distinct from the climate in surrounding areas.

Micrometre (μm)   A unit of length equal to one-millionth of a metre.

Microscale   The smallest scale of atmospheric motions.

Middle-latitude cyclone   A cyclonic storm that most often forms along a front in middle and high latitudes. Also called a *mid-latitude cyclonic storm*, a *depression*, a *low*, and an *extratropical cyclone*. It is not a tropical storm or hurricane.

Middle latitudes   The region of the world typically described as being between 30° and 60° latitude.

Mid-Holocene maximum   A warm period in geologic history about 5000 to 6000 years ago that favoured the development of plants.

Mid-latitudes   *See* Middle latitudes.

Mie scattering   The scattering of light by particles that are similar in size to the wavelength of light. In contrast to *Rayleigh scattering*, Mie scattering is not very wavelength dependent, so the scattered light tends to be white. The Mie scattering of light by a polluted atmosphere is responsible for the sky's hazy appearance.

Milankovitch theory   A theory proposed by Milutin Milankovitch in the 1930s suggesting that changes in Earth's orbit were responsible for variations in solar energy reaching Earth's surface, causing climatic changes.

Millibar (mb)   A unit for expressing atmospheric pressure. Sea-level pressure is normally close to 1013.25 mb. One millibar equals one hectopascal (hPa), which is the correct SI unit used in meteorology.

Minimum thermometer   A thermometer designed to measure the minimum air temperature during a desired time period.

Mini-swirls   Small, whirling eddies perhaps 30 to 100 m in diameter that form in a region of strong wind shear of a hurricane's eyewall. They are believed to be small tornadoes.

Mirage   A refraction phenomenon that makes an object appear to be displaced from its true position. When an object appears higher than it actually is, it is called a *superior mirage*. When an object appears lower than it actually is, it is an *inferior mirage*.

Mixed cloud   A cloud containing both water drops and ice crystals.

Mixing depth   The vertical extent of the mixing layer.

Mixing fog   A fog that forms when two moist but unsaturated air masses at different temperatures mix, resulting in saturation at the new mixed temperature and moisture content. Evaporation fog is a type of mixing fog.

Mixing layer   The unstable atmospheric layer that extends from the surface up to the base of an inversion. Within this layer, the air is well stirred.

Mixing ratio   The ratio of the mass of water vapour in a given volume of air to the mass of dry air.

Moist adiabatic lapse rate   *See* Saturated adiabatic lapse rate.

Moist adiabats   *See* Saturated adiabats.

Molecular viscosity   The small-scale internal fluid friction that is due to the random motion of the molecules within a smooth-flowing fluid, such as air.

Molecule   A collection of atoms held together by chemical forces.

Monsoon depressions   Weak low-pressure areas that tend to form in response to divergence in an upper-level jet stream. The circulation around the low strengthens the monsoon wind system and enhances precipitation during the summer.

Monsoon wind system   A wind system that reverses direction between winter and summer. Usually, the wind blows from land to sea in winter and from sea to land in summer.

Mountain and valley breeze   A local wind system of a mountain valley that blows downhill (mountain breeze) at night and uphill (valley breeze) during the day.

Multicell thunderstorm   A convective storm system composed of a cluster of convective cells, each one in a different stage of its life cycle.

Nacreous clouds   Clouds of unknown composition that have a soft, pearly lustre and that form at altitudes about 25 to 30 km above Earth's surface. They are also called *mother-of-pearl clouds*.

Negative feedback mechanism   *See* Feedback mechanism.

Neutral stability (neutrally stable atmosphere)   An atmospheric condition that exists in unsaturated air when the environmental lapse rate equals the dry adiabatic lapse rate. In saturated air, the environmental lapse rate equals the saturated adiabatic lapse rate.

NEXRAD   An acronym for *Next* Generation Weather *Radar*. The main component of NEXRAD is the WSR 88-D Doppler radar that is in wide use in the United States.

Nimbostratus   A dark grey cloud characterized by more or less continuously falling precipitation. It is rarely accompanied by lightning, thunder, or hail.

NinJo   A forecaster workstation computer program for visualizing meteorological data and preparing weather forecasts.

Nitric oxide (NO)   A colourless gas produced by natural bacterial action in soil and by combustion processes at high temperatures. In polluted air, nitric oxide can react with ozone and hydrocarbons to form other substances. In this manner, it acts as an agent in the production of photochemical smog.

Nitrogen (N_2)   A colourless and odourless gas that occupies about 78 percent of dry air in the lower atmosphere.

Nitrogen dioxide (NO_2)   A reddish-brown gas produced by natural bacterial action in soil and by combustion processes at high temperatures. In the presence of sunlight, it breaks down into nitric oxide and atomic oxygen. In polluted air, nitrogen dioxide acts as an agent in the production of photochemical smog.

Nitrogen oxides (NO$_X$)   Gases produced by natural processes and by combustion processes at high temperatures. In polluted air, nitric oxide (NO) and nitrogen dioxide (NO$_2$) are the most abundant oxides of nitrogen, and both act as agents for the production of photochemical smog.

Noctilucent clouds   Wavy, thin, bluish-white clouds that are best seen at twilight in polar latitudes. They form at altitudes about 80 to 90 km above the surface.

Nocturnal inversion   *See* Radiation inversion.

Nonselective scattering (geometric scattering)   The scattering of light by particles much larger than the wavelength of light. Scatters all wavelengths equally well, so the scattered light usually has a white or grey colour.

Nonsupercell tornado   A tornado that occurs with a cloud that is often in its growing stage, and does not contain a midlevel mesocyclone or wall cloud. Landspouts and gustnadoes are examples of nonsupercell tornadoes.

North Atlantic Oscillation (NAO)   A reversal of atmospheric pressure over the Atlantic Ocean that influences the weather over Europe and over eastern North America.

Northeaster   A name given to a strong, steady wind from the northeast that is accompanied by rain and inclement weather. It often develops when a storm system moves northeastward along the coast of North America. Also called a nor'easter.

Northern lights   *See* Aurora.

Nowcasting   Short-term weather forecasts varying from minutes up to a few hours.

Nuclear winter   The dark, cold, and gloomy conditions that presumably would be brought on by nuclear war.

Nucleation   Any process in which the phase change of a substance to a more condensed state (such as condensation, deposition, and freezing) is initiated about a particle (nucleus).

Numerical weather prediction (NWP)   Forecasting the weather based on the solutions of mathematical *governing equations* by high-speed computers.

Obliquity (of Earth's axis)   The tilt of Earth's axis. It represents the angle from the perpendicular to the plane of Earth's orbit.

Obscured   A sky condition (abbreviated X) in which the sky condition is not visible, usually because of a surface-based layer such as fog.

Occluded front (occlusion)   A complex frontal system that forms when a cold front overtakes a warm front. When the air behind the front is colder than the air ahead of it, the surface manifestation of the front is called a *cold occlusion*. When the air behind the front is milder than the air ahead of it, it is called a *warm occlusion*. The upper-level tongue of warm air associated with the occlusion process is called a *TROWAL*.

Ocean conveyor belt   The global circulation of ocean water that is driven by the sinking of cold, dense water near Greenland and Labrador in the North Atlantic.

Ocean-effect snow   Localized bands of snow that occur when relatively cold air flows over a warmer ocean.

Offshore wind   A breeze that blows from the land out over the water. Opposite of an onshore wind.

Omega high   A ridge in the middle or upper troposphere that has the shape of the Greek letter omega (Ω).

Onshore wind   A breeze that blows from the water onto the land. Opposite of an offshore wind.

Open wave   The stage of development of a wave cyclone (*middle-latitude cyclone*) where a cold front and a warm front exist, but no occluded front. The centre of lowest pressure in the wave is located at the junction of the two fronts.

Orchard heaters   Oil heaters placed in orchards that generate heat and promote convective circulations to protect fruit trees from damaging low temperatures. Also called *smudge pots*.

Ordinary cell thunderstorm   A thunderstorm produced by local convection within a conditionally unstable air mass. It often forms in a region of low wind shear and does not reach the intensity of a severe thunderstorm. Also called an *air-mass thunderstorm*.

Orographic clouds   Clouds produced by lifting along rising terrain, usually mountains.

Orographic uplift   The lifting of air over a topographic barrier. Clouds that form in this lifting process are called *orographic clouds*.

Outflow boundary   A surface boundary formed by the horizontal spreading of cool air that originated inside a thunderstorm.

Outgassing   The release of gases dissolved in hot, molten rock.

Overcast   A sky condition (abbreviated OVC) in which the sky is completely covered by clouds.

Overrunning   A condition that occurs when air moves up and over another layer of air.

Overshooting top   A situation in a mature thunderstorm where rising air, associated with strong convection, penetrates into a stable layer (usually the stratosphere), forcing the upper part of the cloud to rise above its relatively flat anvil top.

Oxygen (O$_2$)   A colourless and odourless gas that occupies about 21 percent of dry air in the lower atmosphere.

Ozone (O$_3$)   An almost colourless gaseous form of oxygen with an odour similar to that of weak chlorine. The highest natural concentration is found in the stratosphere, where it is known as stratospheric ozone. It also forms in polluted air near the surface, where it is the main ingredient of photochemical smog. Here it is called *tropospheric ozone*.

Ozone hole   A sharp drop in stratospheric ozone concentration observed over the Antarctic during the austral spring.

Pacific Decadal Oscillation (PDO)   A reversal in ocean surface temperatures that occurs every 20 to 30 years over the northern Pacific Ocean.

Pacific High   *See* Subtropical high.

Parcel of air   An imaginary small body of air a few metres wide that is used to explain the behaviour of air.

Parhelia   *See* Sundog.

Particulate matter   Solid particles or liquid droplets that are small enough to remain suspended in the air. Also called *aerosols*.

Pattern recognition   An analogue method of forecasting where the forecaster uses prior weather events (or similar weather map conditions) to make a forecast.

P/E index (precipitation–evaporation index)   An index that gives the average effectiveness of precipitation in promoting plant growth.

P/E ratio (precipitation–evaporation ratio)   An expression devised for the purpose of classifying climates; based on monthly totals of precipitation and potential evaporation.

Permafrost   A layer of soil beneath Earth's surface that remains frozen throughout the year.

Persistence forecast   A forecast that the future weather condition will be the same as the present condition.

Photochemical smog   *See* Smog.

Photodissociation   The splitting of a molecule by a photon.

Photon   A discrete quantity of energy that can be thought of as a packet of electromagnetic radiation travelling at the speed of light.

Photosphere   The visible surface of the sun from which most of its energy is emitted.

Pileus cloud   A smooth cloud in the form of a cap. Occurs above, or is attached to, the top of a cumuliform cloud. Also called a *cap cloud*.

Pilot balloon   A small balloon that rises at a constant rate and is tracked by a theodolite to obtain wind speed and wind direction at various levels above Earth's surface.

Planck's constant   A physical constant that relates the energy of a photon and the frequency of its associated electromagnetic wave. The frequency is the speed of light divided by the wavelength of the radiation. It has a value of $6.6260689633 \times 10^{-34}$ J s.

Planetary boundary layer   *See* Atmospheric boundary layer.

Planetary scale   The largest scale of atmospheric motion. Sometimes called the *global scale*.

Plasma   *See* Solar wind.

Plate tectonics   The theory that Earth's surface down to about 100 km is divided into a number of plates that move relative to one another across the surface of Earth. Once referred to as continental drift.

Pleistocene epoch (or Ice Age)   The most recent period of extensive continental glaciation that saw large portions of North America and Europe covered with ice. It began about 2.58 million years ago and continues until the present time.

Polar easterlies   A shallow body of easterly winds located at high latitudes poleward of the subpolar low.

Polar front   A semipermanent, semicontinuous front that separates tropical air masses from polar air masses.

Polar-front jet stream (polar jet)   The jet stream that is associated with the polar front in middle and high latitudes. It is usually located at altitudes between 9 and 12 km.

Polar-front theory   A theory developed by a group of Scandinavian meteorologists that explains the formation, development, and overall life history of cyclonic storms that form along the polar front.

Polar ice cap climate   A climate characterized by extreme cold where every month's "average" temperature is below freezing.

Polar low   An area of low pressure that forms over polar water behind (poleward of) the main polar front.

Polar orbiting satellite   A satellite whose orbit closely parallels Earth's longitude lines and thus crosses the polar regions on each orbit.

Polar tundra climate   A climate characterized by extremely cold winters and cool summers as the average temperature of the warmest month climbs above freezing but remains below 10°C.

Pollutants   Any gaseous, chemical, or organic matter that contaminates the atmosphere, soil, or water.

Positive feedback mechanism   *See* Feedback mechanism.

Potential energy   The energy that a body possesses by virtue of its position with respect to other bodies in the field of gravity.

Potential evapotranspiration (PE)   The amount of moisture that, if it were available, would be removed from a given land area by evaporation and transpiration.

Potential temperature   The temperature of a parcel of air that is lowered adiabatically (according to the dry adiabatic lapse rate) from its original position to a pressure of 1000 hPa.

Precession (of Earth's axis of rotation)   The wobble of Earth's axis of rotation that traces out the path of a cone over a period of about 23,000 years.

Precipitation   Any form of water particles—liquid or solid—that falls from the atmosphere and reaches the ground.

Pressure   The force per unit area. *See also* Air pressure.

Pressure gradient   The rate of decrease in pressure per unit of horizontal distance. On the same chart, when the isobars are close together, the pressure gradient is steep. When the isobars are far apart, the pressure gradient is weak.

Pressure gradient force (PGF)   The force due to differences in pressure within the atmosphere that causes air to move and, hence, the wind to blow. It is directly proportional to the pressure gradient.

Pressure tendency   The rate of change of atmospheric pressure within a specified period of time, most often three hours. Same as *barometric tendency*.

Prevailing westerlies   The dominant westerly winds that blow in middle latitudes on the poleward side of the subtropical high-pressure areas. Also called *westerlies*.

Prevailing wind   The wind direction most frequently observed during a given period.

Primary air pollutants   Air pollutants that enter the atmosphere directly.

Probability forecast   A forecast of the probability of occurrence of one or more of a mutually exclusive set of weather conditions.

Prognostic chart (prog)   A chart showing expected or forecasted conditions, such as pressure patterns, frontal positions, contour height patterns, and so on.

Prominence   *See* Solar flare.

Psychrometer   An instrument used to measure the water vapour content of the air. It consists of two thermometers (dry bulb and wet bulb). After ventilating the two thermometers, the wet-bulb and dry-bulb temperature readings are converted to dew point and relative humidity with the aid of tables or formulae.

Radar   An electronic instrument used to detect objects (such as precipitation) by their ability to reflect and scatter microwaves back to a receiver. Radar stands for **RA**dio **D**etection **A**nd **R**anging. (*See also* Doppler radar.)

Radiant energy (radiation)   Energy propagated in the form of electromagnetic waves. These waves do not need molecules to propagate them, and in a vacuum, they travel at nearly 300,000 km s^{-1}.

Radiational cooling   The process by which Earth's surface and adjacent air cool by emitting infrared radiation. Also called *radiative cooling*.

Radiation fog   Fog produced over land when radiational cooling reduces the air temperature to or below its dew point. Also known as *ground fog* and *valley fog*.

Radiation inversion   An increase in temperature with height due to radiational cooling of Earth's surface. Also called a *nocturnal inversion*.

Radiative equilibrium temperature   The temperature achieved when an object, behaving as a blackbody, is absorbing and emitting radiation at equal rates.

Radiative forcing   An increase (positive) or a decrease (negative) in net radiant energy observed over an area at the tropopause. An increase in radiative forcing may induce surface warming, whereas a decrease may induce surface cooling.

Radiative forcing agent   Any factor (such as increasing greenhouse gases and variations in solar output) that can change the balance between incoming energy from the sun and outgoing energy from Earth and the atmosphere.

Radiometer   A device for measuring radiation.

Radiosonde   A balloon-borne instrument that measures and transmits pressure, temperature, and humidity to a ground-based receiving station.

Radon   A colourless, odourless, radioactive gas that forms naturally as uranium in soil and rock breaks down.

Rain   Precipitation in the form of liquid water drops that have diameters greater than that of drizzle.

Rainbow   An arc of concentric coloured bands that spans a section of the sky when rain is present and the sun is positioned at the observer's back.

Rain gauge   An instrument designed to measure the amount of rain that falls during a given time interval.

Rain shadow   The region on the leeside of a mountain where the precipitation is noticeably less than on the windward side.

Rawinsonde observation   A radiosonde observation that includes wind data.

Rayleigh scattering   Light scattering by particles, such as air molecules, that are much smaller than the wavelength of light. The amount of Rayleigh scattering depends on wavelength (λ). It is proportional to λ^{-4}, so the shortest wavelengths are more strongly scattered. The scattering of sunlight's shortest wavelengths (blue) is the main reason why the sky is blue.

Reflected light   *See* Reflection.

Reflection   The process whereby a surface turns back a portion of the radiation that strikes it. When the radiation that is turned back (reflected) from the surface is visible light, the radiation is referred to as *reflected light.*

Refraction   The bending of light as it passes from one medium to another.

Relative humidity   The ratio of the amount of water vapour in the air compared to the amount required for saturation (at a particular temperature and pressure). The ratio of the air's actual vapour pressure to its saturation vapour pressure.

Relative vorticity   *See* Vorticity.

Return stroke   The luminous lightning stroke that propagates upward from Earth to the base of a cloud.

Ribbon lightning   Lightning that appears to spread horizontally into a ribbon of parallel luminous streaks when strong winds are blowing parallel to the observer's line of sight.

Ridge   An elongated area of high atmospheric pressure.

Rime   A white or milky granular deposit of ice formed by the rapid freezing of supercooled water drops as they come in contact with an object in below-freezing air.

Riming   *See* Accretion.

Roll cloud   A dense, roll-shaped, elongated cloud that appears to slowly spin about a horizontal axis behind the leading edge of a thunderstorm's gust front.

Rossby waves   *See* Longwaves in the westerlies.

Rotor cloud   A turbulent cumuliform type of cloud that forms on the leeward side of large mountain ranges. The air in the cloud rotates about an axis parallel to the range.

Rotors   Turbulent eddies that form downwind of a mountain chain, creating hazardous flying conditions.

Saffir–Simpson scale   A scale relating a hurricane's central pressure and winds to the possible damage it is capable of inflicting.

St. Elmo's fire   A bright electric discharge that is projected from (usually pointed) objects when they are in a strong electric field, such as during a thunderstorm.

Sand dune   A hill or ridge of loose sand shaped by the winds.

Santa Ana wind   A warm, dry wind that blows from an elevated desert plateau into southern California from the east. Its warmth is derived from compressional heating.

Saturated adiabatic lapse rate   The rate of change of temperature in a rising or descending saturated air parcel. The rate of cooling for a rising parcel varies with temperature and pressure. It is less than the dry adiabatic lapse rate since latent heat release partly offsets the adiabatic cooling. A typical value is 6°C km^{-1}.

Saturated adiabats   Lines on a tephigram (or other thermodynamic chart) that show the saturated adiabatic lapse rate for rising and descending air.

Saturation (of air)   An atmospheric condition whereby the level of water vapour is the maximum possible at the existing temperature and pressure.

Saturation vapour pressure   The maximum amount of water vapour necessary to keep moist air in equilibrium with a surface of pure water or ice. It represents the maximum amount of water vapour that the air can hold at any given temperature and pressure. (*See* Equilibrium vapour pressure.)

Savanna   A tropical or subtropical region of grassland and drought-resistant vegetation. Typically found in tropical wet-and-dry climates.

Scales of motion   The hierarchy of atmospheric circulations from tiny gusts to giant storms.

Scattered   A sky condition (abbreviated SCT) in which the sky is four-tenths to five-tenths (three-eighths to four-eighths) covered by clouds.

Scattering   The process by which small particles in the atmosphere deflect radiation from its path into different directions.

Scintillation   The apparent twinkling of a star due to its light passing through regions of differing air densities in the atmosphere.

Sea breeze   A coastal local wind that blows from the ocean onto the land. The leading edge of the breeze is termed a *sea-breeze front.*

Sea-breeze convergence zone   A region where sea breezes that started in different regions, flow together and converge.

Sea-breeze front   The horizontal boundary that marks the leading edge of the cooler marine air associated with a sea breeze.

Sea-level pressure   The atmospheric pressure at mean sea level.

Secondary air pollutants   Pollutants that form when a chemical reaction occurs between a primary air pollutant and some other component of air. Tropospheric ozone is a secondary air pollutant.

Secondary low   A low-pressure area (often an open wave) that forms near, or in association with, a main low-pressure area.

Seiches   Standing waves that oscillate back and forth over an open body of water.

Selective absorbers   Substances such as water vapour, carbon dioxide, methane, clouds, and snow that absorb radiation only at particular wavelengths.

Semiarid climate   A dry climate where potential evaporation and transpiration exceed precipitation. Not as dry as the arid climate. Typical vegetation is short grass.

Semipermanent highs and lows   Areas of high pressure (anticyclones) and low pressure (extratropical cyclones) that tend to persist throughout the year at a particular latitude belt. In the Northern Hemisphere, they typically shift slightly northward in summer and slightly southward in winter.

Sensible heat   Heat we can feel and measure with a thermometer.

Sensible temperature   The sensation of temperature that the human body feels in contrast to the actual temperature of the environment as measured with a thermometer.

Severe thunderstorms   Intense thunderstorms capable of producing heavy showers, flash floods, hail, strong and gusty surface winds, and tornadoes. The Meteorological Service of Canada describes a severe thunderstorm as one that produces a tornado, has hail with a diameter of at least 2 cm, or has surface wind gusts of 48 knots (90 km h^{-1}) or greater.

Sferics   Radio waves produced by lightning. A contraction of *atmospherics.*

Shear   *See* Wind shear.

Sheet lightning   Occurs when the lightning flash is not seen, but the flash causes the cloud (or clouds) to appear as a diffuse, luminous white sheet.

Shelf cloud   A dense, arch-shaped, ominous-looking cloud that often forms along the leading edge of a thunderstorm's gust front, especially when stable air rises up and over cooler air at the surface. Also called an *arcus cloud.*

Shelterbelt   A belt of trees or shrubs arranged as a protection against strong winds.

Short-range forecast   Generally used to describe a weather forecast that extends from about six hours to a few days into the future.

Shortwave (in the atmosphere) A small wave that moves around longwaves in the same direction as the air flow in the middle and upper troposphere. Shortwaves are also called *shortwave troughs*.

Shortwave radiation A term most often used to describe the radiant energy emitted from the sun in the visible and near-ultraviolet wavelengths.

Shower Intermittent precipitation from a cumuliform cloud, usually of short duration but often heavy.

Siberian High A strong, shallow area of high pressure that forms over Siberia in winter.

Sleet Mixed rain and snow. In the United States, *sleet,* refers to ice pellets.

Smog Originally, smog meant a mixture of smoke and fog. Today, smog refers to air that has restricted visibility due to pollution or pollution formed in the presence of sunlight—photochemical smog.

Smog front (or smoke front) The leading edge of a sea breeze that is contaminated with smoke or pollutants.

Smudge pots *See* Orchard heaters.

Snow A solid form of precipitation composed of ice crystals in a complex hexagonal form.

Snow–albedo feedback A positive feedback whereby increasing surface air temperatures enhance the melting of snow and ice in polar latitudes. This reduces Earth's albedo and allows more sunlight to be absorbed by Earth's surface, which causes the air temperature to rise even more.

Snowflake An aggregate of ice crystals that falls from a cloud.

Snow flurries Light showers of snow that fall intermittently.

Snow grains Precipitation in the form of very small, opaque grains of ice. The solid equivalent of drizzle.

Snow pellets White, opaque, approximately round ice particles between 2 and 5 mm in diameter that form in a cloud from either ice crystals sticking together or the process of accretion. Also called graupel.

Snow rollers A cylindrical spiral of snow shaped somewhat like a barrel and produced by the wind; like a wind-formed snow ball.

Snow showers *See* Snow flurries.

Snow squalls Heavy snow showers usually accompanied by wind and blowing snow similar to a blizzard but usually with less duration. There are two types: lake-effect snow squalls and snow squalls associated with a cold front.

Snow water equivalent (SWE) *See* Water equivalent.

Solar constant The rate solar energy is received on a surface at the outer edge of the atmosphere which is perpendicular to the sun's rays when Earth is at its mean distance from the sun. The value of the solar constant is about 1376 W m^{-2}.

Solar declination The latitude, between 23½° north and 23½° south, where the sun is directly overhead at solar noon. It depends on the day of the year.

Solar flare A rapid eruption from the sun's surface that emits high-energy radiation and energized charged particles.

Solar wind An outflow of charged particles from the sun that escapes the sun's outer atmosphere at high speed.

Solute effect The dissolving of hygroscopic particles, such as salt, in pure water, thus reducing the relative humidity required for the onset of condensation.

Sonic boom A loud, explosive sound caused by the shock wave emanating from an aircraft or something travelling at or above the speed of sound.

Sounding An upper-air observation, such as a radiosonde observation. A vertical profile of an atmospheric variable such as temperature or winds.

Source regions Regions where air masses originate and acquire their temperature and moisture properties.

Southern Oscillation (SO) The reversal of surface air pressure at opposite ends of the tropical Pacific Ocean that occur during major El Niño events.

Species Cloud classification terms used to describe the structure and appearance of clouds. The 14 species are fibratus, uncinus, spissatus, castellanus, floccus, stratiform, nebulosus, lenticularis, fractus, humilis, mediocris, congestus, calvus, and capillatus.

Specific heat The ratio of heat absorbed (or released) by a unit mass to its corresponding temperature rise (or fall). In the SI system, the units of specific heat are J kg^{-1} K^{-1}.

Specific humidity The ratio of water vapour mass in a given parcel to the total mass of air in the parcel.

Spin-up vortices Small, whirling tornadoes about 30 to 100 m in diameter that form in a region of strong wind shear in a hurricane's eyewall.

Squall line A line of thunderstorms that form along or ahead of a cold front.

Stable air *See* Absolutely stable atmosphere.

Standard atmosphere A hypothetical vertical distribution of atmospheric temperature, pressure, and density in which the air is assumed to obey the gas law and the hydrostatic equation. The lapse rate of temperature in the troposphere for the standard atmosphere is 6.5°C km^{-1}.

Standard atmospheric pressure A pressure of 1013.25 hPa, 760 mm Hg, 1013.25 mb, 29.92 in Hg, or 14.7 lb in^{-2}.

Stationary front A front that is nearly stationary. The wind blows almost parallel and from opposite directions on each side of the front. Also called a *quasistationary front.*

Station pressure The actual air pressure computed at the observing station.

Statistical forecast A forecast based on a mathematical or statistical examination of past observations.

Steady-state forecast A weather prediction based on the past movement of surface weather systems. It assumes that the systems will move in the same direction and at approximately the same speed as they have been moving. Also called *trend forecasting.*

Steam fog *See* Evaporation fog.

Stefan–Boltzmann law A law of radiation that states that the amount of radiant energy emitted from a unit surface area of an object (ideally a blackbody) is proportional to the fourth power of the object's absolute temperature. $E = \sigma T^4$

Steppe An area of grass-covered, treeless plains that has a semiarid climate.

Stepped leader The initial discharge of electrons that proceeds intermittently toward the ground in a series of steps in a cloud-to-ground lightning stroke.

Storm surge An abnormal rise of the sea along a shore, primarily due to the winds of a storm, especially a hurricane.

Stratiform Clouds that are horizontally layered with little vertical development.

Stratocumulus A low cloud, predominantly stratiform, with low, lumpy, rounded masses, often with blue sky between them.

Stratosphere The layer of the atmosphere above the troposphere and below the mesosphere (between about 10 and 50 km), generally characterized by an increase in temperature with height.

Stratospheric polar night jet A jet stream that forms near the top of the stratosphere over polar latitudes during the winter months.

Stratus A low, grey cloud layer with a rather uniform base. Precipitation, if any, from stratus is most commonly drizzle or fine snow. Stratus clouds that are fragmented and ragged and found under a larger cloud base are called *stratus fractus.*

Streamline A line that shows the wind-flow pattern.

Sublimation   The process whereby ice changes into water vapour without melting.

Subpolar climate   The Northern Hemisphere climate that borders the polar climate. It is characterized by severely cold winters and short, cool summers. Also known as *taiga climate* and *boreal climate*.

Subpolar low   A belt of low pressure located between 50° and 70° latitude. In the Northern Hemisphere, this "belt" consists of the *Aleutian low* in the North Pacific and the *Icelandic low* in the North Atlantic. In the Southern Hemisphere, it exists around the periphery of the Antarctic continent.

Subsidence   The slow sinking of air, usually associated with high-pressure areas.

Subsidence inversion   A temperature inversion produced by compressional warming—the adiabatic warming of a layer of sinking air.

Subtropical front   A zone of temperature transition in the upper troposphere over subtropical latitudes, where warm air carried poleward by the Hadley cell meets the cooler air of the middle latitudes.

Subtropical high   A semipermanent high in the subtropical high-pressure belt centred near 30° latitude. The *Bermuda high* is located over the Atlantic Ocean off the East Coast of North America. The *Pacific high* is located off the West Coast of North America.

Subtropical jet stream   The jet stream typically found between 20° and 30° latitude, and at altitudes between 12 and 14 km.

Suction vortices   Small, rapidly rotating whirls perhaps 10 m in diameter that are found within large tornadoes.

Sulphate aerosols   *See* Aerosols.

Sulphur dioxide (SO_2)   A colourless gas that primarily forms during the burning of sulphur-containing fossil fuels.

Summer solstice   The time of year in each hemisphere (north or south) when the sun is highest in the sky and directly overhead at 23½°. It occurs at approximately June 21 in the Northern Hemisphere; when the sun is directly overhead at latitude 23½°N, the Tropic of Cancer.

Sundog   A coloured luminous area produced by refraction of light through ice crystals that appears on either side of the sun. Also called *parhelia*.

Sun pillar   A vertical streak of light extending above (or below) the sun. It is produced by the reflection of sunlight off ice crystals.

Sunspots   Relatively cooler areas on the sun's surface. They represent regions of an extremely high magnetic field.

Supercell storm   A severe thunderstorm that consists primarily of a single rotating updraft. Its organized internal structure allows the storm to maintain itself for several hours. Supercell storms can produce large hail and dangerous tornadoes.

Supercell tornadoes   Tornadoes that occur within supercell thunderstorms that contain well-developed, midlevel mesocyclones.

Supercooled cloud (or cloud droplets)   A cloud composed of liquid droplets at temperatures below 0°C. When the cloud is on the ground, it is called *supercooled fog* or *cold fog*.

Superior mirage   *See* Mirage.

Supersaturation   A condition where the atmosphere contains more water vapour than is needed to produce saturation with respect to a flat surface of pure water or ice and the relative humidity is greater than 100 percent.

Supertyphoon   A tropical cyclone (typhoon) in the western Pacific that has sustained winds of 130 knots or greater.

Supplementary features and accessory clouds   A set of nine features or cloud forms that are part of and depend on the existence of a major cloud type or genera. These are incus, mamma, virga, praecipitatio, arcus, tuba, pileus, velum, and pannus.

Surface inversion   *See* Radiation inversion.

Surface map   A map that shows the distribution of sea-level pressure with isobars and weather phenomena. Also called a *surface chart*.

Synoptic scale   The typical weather map scale that shows features such as high- and low-pressure areas and fronts over a distance spanning a continent. Also called the *cyclonic scale*.

System   A set of interacting interrelated elements forming a complex whole. The atmosphere can be thought of as a system (as can the whole Earth).

Taiga (boreal forest)   The open northern part of the coniferous forest. Taiga also refers to subpolar climate.

Tangent arc   An arc of light tangent to a halo. It forms by refraction of light through ice crystals.

Tcu   *See* Towering cumulus.

Teleconnections   A linkage between weather changes occurring in widely separated regions of the world.

Temperature   The degree of hotness or coldness of a substance as measured by a thermometer. It is also a measure of the average speed or kinetic energy of the atoms and molecules in a substance.

Temperature inversion   An increase in air temperature with height, often simply called an *inversion*.

Terminal velocity   The constant speed obtained by a falling object when the upward drag on the object balances the downward force of gravity.

Texas norther   A strong, cold wind from between the northeast and the northwest associated with a cold outbreak of polar air that brings a sudden drop in temperature. Sometimes called a *blue norther*.

Theodolite   An instrument used to track the movements of a pilot balloon.

Theory of plate tectonics   *See* Plate tectonics.

Thermal   A small, rising parcel of warm air produced when Earth's surface is heated unevenly.

Thermal belts   Horizontal zones of vegetation found along hillsides that are primarily the result of vertical temperature variations.

Thermal circulations   Air flow resulting primarily from the heating and cooling of air.

Thermal lows and thermal highs   Areas of low and high pressure that are shallow in vertical extent and are produced primarily by surface temperatures.

Thermal tides   Atmospheric pressure variations due to the uneven heating of the atmosphere by the sun.

Thermal turbulence   Turbulent vertical motions that result from surface heating and the subsequent rising and sinking of air. Also called *free convection*.

Thermocline   A thin layer in the ocean in which temperature changes rapidly with depth. The thermocline separates the well-mixed upper layer near the surface from the deep water below and is usually located within a few hundred metres of the ocean surface.

Thermograph   An instrument that measures and records air temperature.

Thermometer   An instrument for measuring temperature. The most common is liquid-in-glass, which has a sealed glass tube attached to a glass bulb filled with liquid.

Thermosphere   The atmospheric layer above the mesosphere (above about 85 km) where the temperature increases rapidly with height.

Thunder   The sound due to rapidly expanding gases along the channel of a lightning discharge.

Thunderstorm   A convective storm (cumulonimbus cloud) with lightning and thunder. Thunderstorms can be composed of an ordinary cell, multicells, or a rapidly rotating supercell.

Timber line   A defined region at high altitudes or latitudes beyond which dense tree growth does not occur. The *tree line* is the altitude or latitude beyond which no trees grow.

Tipping bucket rain gauge   A rain gauge that records rainfall by collecting rain in a chamber (bucket) that tips when the chamber fills with 0.2 mm of rain.

Tornado   An intense, rotating column of air that often protrudes from a cumuliform cloud in the shape of a funnel or a rope whose circulation is present on the ground. (*See* Funnel cloud.)

Tornado outbreak   A series of tornadoes that forms within a particular region. Often associated with widespread damage and destruction.

Tornado vortex signature (TVS)   An image of a tornado on the Doppler radar screen that shows up as a small region of rapidly changing wind directions inside a mesocyclone.

Tornado warning   A warning issued when a tornado has actually been observed either visually or on a radar screen. It is also issued when the formation of tornadoes is imminent.

Tornado watch   A forecast issued to alert the public that tornadoes may develop within a specified area.

Towering cumulus   A vertically developed cumulus cloud often associated with showers and abbreviated as Tcu. It is the same as *cumulus congestus* and may grow into a *cumulonimbus*.

Trace (of precipitation)   An amount of precipitation less than 0.2 mm.

Trace gases   Any gas making up less than 1 percent of the atmosphere (all gases except nitrogen and oxygen).

Trade wind inversion   A temperature inversion frequently found in the subtropics over the eastern portions of the tropical oceans.

Trade winds   The winds that occupy most of the tropics and blow from the subtropical highs to the equatorial low.

Transpiration   The process by which water in plants is transferred as water vapour to the atmosphere.

Tropical cyclone   The general term for storms (cyclones) that form over warm tropical oceans.

Tropical depression   A mass of thunderstorms and clouds generally with a cyclonic wind circulation of between 20 and less than 34 knots.

Tropical disturbance   An organized mass of thunderstorms with a slight cyclonic wind circulation of less than 20 knots.

Tropical easterly jet   A jet stream that forms on the equatorward side of the subtropical highs at altitudes near 15 km.

Tropical monsoon climate   A tropical climate with a brief dry period of perhaps one or two months.

Tropical rain forest   A type of forest consisting mainly of lofty trees and a dense undergrowth near the ground.

Tropical storm   Organized thunderstorms with a cyclonic wind circulation between 34 and less than 64 knots.

Tropical wave   A migratory wavelike disturbance in the tropical easterlies. Tropical waves occasionally intensify into tropical cyclones. They are also called *easterly waves*.

Tropical wet-and-dry climate   A tropical climate poleward of the tropical wet climate where a distinct dry season occurs, often lasting for two months or more.

Tropical wet climate   A tropical climate with sufficient rainfall to produce a dense tropical rain forest.

Tropopause   The boundary between the troposphere and the stratosphere.

Tropopause jets   Jet streams found near the tropopause, such as the polar front and subtropical jet streams.

Troposphere   The layer of the atmosphere extending from Earth's surface up to the tropopause (about 10 km above the ground).

Trough   An elongated area of low atmospheric pressure.

TROWAL   An acronym for **TRO**ugh of **W**arm air **AL**oft. It is the surface projection of a tongue of warm air aloft and in Canada is drawn on surface weather maps instead of an *occluded front*.

Turbulence   Any irregular or disturbed flow in the atmosphere that produces gusts and eddies.

Twilight   The time at the beginning of the day immediately before sunrise and at the end of the day after sunset when the sky remains illuminated.

Typhoon   A hurricane (tropical cyclone) that forms in the western Pacific Ocean.

Ultraviolet (UV) radiation   Electromagnetic radiation with wavelengths longer than X-rays but shorter than visible light.

Unstable air   *See* Absolutely unstable atmosphere.

Upper-air front   A front that is present aloft but usually does not extend down to the ground. Also called an *upper front* and an *upper-tropospheric front*.

Upslope fog   Fog formed as moist, stable air flows upward over a topographic barrier.

Upslope precipitation   Precipitation that forms due to moist, stable air gradually rising along an elevated plain. Upslope precipitation is common over the region east of the Rocky Mountains.

Upwelling   In a water body, the rising of water (usually cold) toward the surface from deeper regions.

Urban heat island   The increased air temperatures in urban areas as contrasted to the cooler surrounding rural areas.

Valley breeze   *See* Mountain breeze.

Valley fog   *See* Radiation fog.

Vapour pressure   The pressure exerted by the water vapour molecules in a given volume of air. (*See also* Saturation vapour pressure.)

Varieties   In cloud classification, varieties are special characteristics of arrangement and transparency in clouds. The nine cloud varieties are intortus, vertebratus, undulatus, radiatus, lacunosus, duplicatus, translucidus, perlucidus, and opacus.

Veering wind   The wind that changes direction in a clockwise sense—north to northeast to east, and so on.

Verglas   A coating of ice, usually clear and smooth, formed by freezing of supercooled water onto surfaces. *See* glaze.

Vernal equinox   The time when the sun is directly above the equator as it moves from the southern to the northern hemisphere. Occurs around March 20.

Very short-range forecast   Generally used to describe a weather forecast that is made for up to a few hours (usually less than six hours) into the future. *See* Nowcasting.

Virga   Precipitation that falls from a cloud but evaporates before reaching the ground. (*See* Fall streaks.)

Viscosity   The resistance of fluid flow. (*See* Molecular viscosity and Eddy viscosity.)

Visible radiation (light)   Radiation with a wavelength between 0.4 and 0.7 μm. This region of the electromagnetic spectrum is called the *visible region*.

Visible region   *See* Visible radiation.

Visibility   The greatest distance at which an observer can see and identify prominent objects.

Volatile organic compounds (VOCs)   A class of organic compounds that are released into the atmosphere from sources such as motor vehicles, paints, and solvents. VOCs (which include hydrocarbons) contribute to the production of secondary pollutants, such as ozone.

Vorticity   A measure of the spin of a fluid, usually applied to small air parcels. *Absolute vorticity* is the combined vorticity due to Earth's rotation and the vorticity due to the air's circulation relative to the Earth. *Relative vorticity* is due to curving air flow and horizonal wind shear.

Vorticity advection   The transport of vorticity by the wind. *Positive vorticity advection* (PVA) occurs when the wind blows from high vorticity toward low vorticity, resulting in an increase in vorticity over time at a location. *Negative vorticity advection* (NVA) occurs when the wind blows from low vorticity toward high vorticity, resulting in a decrease in vorticity over time at a location.

Wall cloud   An area of rotating clouds that extends beneath a supercell thunderstorm and from which a funnel cloud may appear. Also called a *collar cloud* and a *pedestal cloud.*

Warm advection (or warm air advection)   The transport of warm air by the wind from a region of higher temperatures to a region of lower temperatures.

Warm-core low   A low-pressure area that is warmer at its centre than at its periphery. Tropical cyclones as well as thermal lows exhibit this temperature pattern.

Warm front   A front that moves in such a way that warm air replaces cold air.

Warm occlusion   See Occluded front.

Warm sector   The region of warm air within a wave cyclone that lies between a retreating warm front and an advancing cold front.

Water equivalent   The depth of water that results from melting a snow sample. Approximately 10 cm of snow will melt to 1 cm of water. Sometimes also called *snow water equivalent* (SWE).

Waterspout   A column of rotating wind over bodies of water.

Water vapour   Water in a vapour (gaseous) form. Also called *moisture* and *humidity.*

Water vapour–greenhouse effect feedback   A positive feedback where increasing surface air temperatures cause increasing water evaporation from the oceans. Increasing concentrations of atmospheric water vapour enhance the greenhouse effect, which cause surface air temperatures to rise even more. Also called the *water vapour–temperature rise feedback.*

Watt (abbreviated as **W**)   A flux or rate of energy per time; the unit of power. In SI units 1 watt equals 1 joule per second.

Wave cyclone   An extratropical cyclone that forms and moves along a front. The circulation of winds about the cyclone tends to produce a wavelike deformation on the front.

Wavelength   The distance between successive crests, troughs, or identical parts of a wave.

Weather   The condition of the atmosphere at any particular time and place.

Weather elements   The elements of *air temperature, air pressure, humidity, clouds, precipitation, visibility,* and *wind* that determine the present state of the atmosphere, the weather.

Weather type forecasting   A forecasting method where weather patterns are categorized into similar groups or types.

Weather types   Certain weather patterns categorized into similar groups. Used as an aid in weather prediction.

Weather warning   A forecast indicating that hazardous weather is either imminent or actually occurring within the specified forecast area.

Weather watch   A forecast indicating that hazardous weather may occur over a particular region during a specified time period.

Weighing rain gauge   A rain gauge that records rainfall by weighing the collected water over a given time and converting the amount of water to rainfall depth.

Westerlies   The dominant westerly winds that blow in the middle latitudes on the poleward side of the subtropical high-pressure areas.

Wet-bulb depression   The temperature difference in degrees between the air temperature (dry-bulb temperature) and the wet-bulb temperature.

Wet-bulb temperature   The lowest temperature that can be obtained by evaporating water into the air.

Wet haze   *See* Haze.

Whirlwinds   *See* Dust devils.

Wien's law   A radiation law that states that the wavelength of maximum emitted radiation by an object (ideally a blackbody) is inversely proportional to the object's absolute temperature.

Wind   Air in motion relative to Earth's surface.

Wind chill   The cooling effect of temperature and wind, expressed as the loss of body heat. Also called *wind-chill index* or *wind chill effect.* It is expressed as an equivalent temperature that relates the amount of cooling due to wind and temperature to the temperature a person would experience under calm conditions.

Wind direction   The direction from which the wind is blowing.

Wind machines   Fans placed in orchards for the purpose of mixing cold surface air with warmer air above.

Wind profiler   A Doppler radar capable of measuring the turbulent eddies that move with the wind. Because of this, it is able to provide a vertical picture of wind speed and wind direction.

Wind rose   A diagram that shows the percentage of time that the wind blows from different directions at a given location over a given time.

Wind-sculptured trees   Trees whose branches are bent, twisted, and broken on one side by strong prevailing winds. Also called *flag trees.*

Wind shear   The rate of change of wind speed or wind direction over a given distance.

Wind sock   A cone-shaped fabric tube that is flown to indicate wind direction and estimates of wind speed. Also called a *wind cone.*

Wind vane   An instrument used to indicate wind direction.

Windward side   The side of an object facing into the wind.

Wind waves   Water waves that form due to the flow of air over the water's surface.

Winter chilling   The amount of time that the air temperature during the winter must remain below a certain value so that fruit and nut trees will grow properly during the spring and summer.

Winter solstice   The time of year in each hemisphere (north or south) when the sun is lowest in the sky and furthest from being directly overhead. In the Northern Hemisphere, this is approximately December 21 and the sun is over 23½°S latitude (the Tropic of Capricorn).

Xerophytes   Drought-resistant vegetation.

Younger-Dryas event   A cold episode that took place about 11,000 years ago, when average temperatures dropped suddenly and portions of the Northern Hemisphere reverted back to glacial conditions.

Zonal wind flow   A wind that has a predominant west-to-east component.

INDEX

Content that is contained in a footnote is indicated by an "n" following the page number.